MW01071589

ENCYCLOPEDIA OF
MATHEMATICS

JAMES TANTON, PH.D.

The Derryfield School Library
2108 River Road
Manchester, N.H. 03104

Facts On File, Inc.

Encyclopedia of Mathematics

Copyright © 2005 by James Tanton, Ph.D.

All rights reserved. No part of this book may be reproduced or utilized in any form or by any means, electronic or mechanical, including photocopying, recording, or by any information storage or retrieval systems, without permission in writing from the publisher. For information contact:

Facts On File, Inc.
132 West 31st Street
New York NY 10001

Library of Congress Cataloging-in-Publication Data

Tanton, James Stuart, 1966–
Encyclopedia of mathematics/James Tanton.
p. cm.
Includes bibliographical references and index.
ISBN 0-8160-5124-0
1. Mathematics—Encyclopedia. I. Title.
QA5.T34 2005
510′.3—dc22   2004016785

Facts On File books are available at special discounts when purchased in bulk quantities for businesses, associations, institutions, or sales promotions. Please call our Special Sales Department in New York at (212) 967-8800 or (800) 322-8755.

You can find Facts On File on the World Wide Web at http://www.factsonfile.com

Text design by Joan M. Toro
Cover design by Cathy Rincon
Illustrations by Richard Garratt

Printed in the United States of America

VB Hermitage 10 9 8 7 6 5 4 3 2 1

This book is printed on acid-free paper.

CONTENTS

Acknowledgments
v

Introduction
vi

A to Z Entries
1

Feature Essays:
"History of Equations and Algebra"
9

"History of Calculus"
57

"History of Functions"
208

"History of Geometry"
226

"History of Probability and Statistics"
414

"History of Trigonometry"
510

Appendixes:

Appendix I
Chronology
539

Appendix II
Bibliography and Web Resources
546

Appendix III
Associations
551

Index
552

ACKNOWLEDGMENTS

My thanks to James Elkins for reading a substantial portion of the manuscript, for his invaluable comments, and for shaping my ideas in writing a number of specific entries. Thanks also go to Frank K. Darmstadt, executive editor at Facts On File, for his patience and encouragement, and to Jodie Rhodes, literary agent, for encouraging me to pursue this project. I also wish to thank Tucker McElroy and John Tabak for taking the time to offer advice on finding archives of historical photographs, and the staff of The Image Works for their work in finding photographs and granting permission to use them. But most of all, thanks to Lindy and Turner for their love and support, always.

INTRODUCTION

Mathematics is often presented as a large collection of disparate facts to be absorbed (memorized!) and used only with very specific applications in mind. Yet the development of mathematics has been a journey that has engaged the human mind and spirit for thousands of years, offering joy, play, and creative invention. The Pythagorean theorem, for instance, although likely first developed for practical needs, provided great intellectual interest to Babylonian scholars of 2000 B.C.E., who hunted for extraordinarily large multidigit numbers satisfying the famous relation $a^2 + b^2 = c^2$. Ancient Chinese scholars took joy in arranging numbers in square grids to create the first "magic squares," and Renaissance scholars in Europe sought to find a formula for the prime numbers, even though no practical application was in mind. Each of these ideas spurred further questions and further developments in mathematics—the general study of Diophantine equations, semi-magic squares and Latin squares, and public-key cryptography, for instance—again, both with and without practical application in mind. Most every concept presented to students today has a historical place and conceptual context that is rich and meaningful. The aim of Facts On File's *Encyclopedia of Mathematics* is to unite disparate ideas and provide a sense of meaning and context.

Thanks to the encyclopedic format, all readers can quickly find straightforward answers to questions that seem to trouble students and teachers alike:

- Why is the product of two negative numbers positive?
- What is π, and why is the value of this number the same for all circles?
- What is the value of π for a shape different than a circle?
- Is every number a fraction?
- Why does the long-division algorithm work?
- Why is dividing by a fraction the same as multiplying by its reciprocal?
- What is the value of i^i?
- What is the fourth dimension?

This text also goes further and presents proofs for many of the results discussed. For instance, the reader can find, under the relevant entries, a proof to the fundamental theorem of algebra, a proof of Descartes's law of signs, a proof that every number has a unique prime factorization, a proof of Bretschneider's formula (generalizing Brahmagupta's famous formula), and a derivation of Heron's formula. Such material is rarely presented in standard mathematical textbooks. In those instances where the method of proof is beyond the scope of the text, a discussion as to the methods behind the proof is at least offered. (For instance, an argument is presented to show how a formula similar to Stirling's formula can be obtained, and the discussion of the Cayley-Hamilton theorem shows that every matrix satisfies at least some polynomial equation.) This encyclopedia aims to be satisfying to those at all levels of interest. Each entry contains cross-references to other items, providing the opportunity to explore further context and related ideas. The reader is encouraged to browse.

As a researcher, author, and educator in mathematics, I have always striven to share with my students the sense of joy and enthusiasm I experience in thinking about and doing mathematics. Collating, organizing, and describing the concepts a high-school student or beginning college-level student is likely to encounter in the typical mathematics curriculum, although a daunting pursuit, has proved to be immensely satisfying. I have enjoyed the opportunity to convey through the writing of this text, hopefully successfully, a continued sense of joy and delight in what mathematics can offer.

Sadly, mathematics suffers from the ingrained perception that primary and secondary education of the subject should consist almost exclusively of an acquisition of a set of skills that will prove to be useful to students in their later careers. With the push for standardized testing in the public school system, this mind-set is only reinforced, and I personally fear that the joy of deep understanding of the subject and the sense of play with the ideas it contains is diminishing. For example, it may seem exciting that we can produce students who can compute 584×379 in a flash, but I am saddened with the idea that such a student is not encouraged to consider why we are sure that 379×584 will produce the same answer. For those students that may be naturally inclined to pause to consider this, I also worry about the response an educator would give upon receiving such a query. Is every teacher able to provide for a student an example of a system of arithmetic for which it is no longer possible to assume that $a \times b$ and $b \times a$ are always the same and lead a student through a path of creative discovery in the study of such a system? (As physicists and mathematicians have discovered, such systems do exist.) By exploring fundamental questions that challenge basic assumptions, one discovers deeper *understanding* of concepts and finds a level of creative play that is far more satisfying than the performance of rote computation. Students encouraged to think this way have learned to be adaptable, not only to understand and apply the principles of a concept to the topic at hand, but also to apply those foundations and habits of mind to new situations that may arise. After all, with the current advances of technology in our society today, we cannot be sure that the rote skill-sets we deem of value today will be relevant to the situations and environments students will face in their future careers. We need to teach our students to be reflective, to be flexible, and to have the confidence to adapt to new contexts and new situations. I hope that this text, in some small ways, offers a sense of the creative aspect to mathematical thinking and does indeed gently encourage the reader to think deeply about concepts, even familiar ones.

Encyclopedia of Mathematics contains more than 800 entries arranged in alphabetical order. The aim of the historical notes, culture-specific articles, and the biographical portraits included as entries, apart from providing historical context, is to bring a sense of the joy that mathematics has brought people in the past. The back matter of this text contains a timeline listing major accomplishments throughout the historical development of mathematics, a list of current mathematics organizations of interest to students and teachers, and a bibliography.

AAA/AAS/ASA/SAS/SSS Many arguments and proofs presented in the study of GEOMETRY rely on identifying similar triangles. The SECANT theorem, for instance, illustrates this. Fortunately, there are a number of geometric tests useful for determining whether or not two different triangles are similar or congruent. The names for these rules are acronyms, with the letter *A* standing for the word *angle,* and the letter *S* for the word *side.* We list the rules here with an indication of their proofs making use of the LAW OF SINES and the LAW OF COSINES:

a. *The AAA rule:* If the three interior angles of one triangle match the three interior angles of a second triangle, then the two triangles are similar.

The law of sines ensures that pairs of corresponding sides of the triangles have lengths in the same ratio. Also note, as the sum of the interior angles of any triangle is 180°, one need only check that *two* corresponding pairs of interior angles from the triangles match.

b. *The AAS and ASA rules:* If two interior angles and one side-length of one triangle match corresponding interior angles and side-length of a second triangle, then the two triangles are congruent.

By the AAA rule the two triangles are similar. Since a pair of corresponding side-lengths match, the two triangles are similar with scale factor one, and are hence congruent. (Note that any two right triangles sharing a common hypotenuse and containing a common acute angle are congruent: all three interior angles match, and

the AAS and ASA rules apply. This is sometimes called the "HA congruence criterion" for right triangles.)

c. *The SAS rule:* If two triangles have two sides of matching lengths with matching included angle, then the two triangles are congruent.

The law of cosines ensures that the third side-lengths of each triangle are the same, and that all remaining angles in the triangles match. By the AAS and ASA rules, the triangles are thus congruent. As an application of this rule, we prove EUCLID's isosceles triangle theorem:

> The base angles of an isosceles triangle are equal.

> Suppose *ABC* is a triangle with sides *AB* and *AC* equal in length. Think of this triangle as representing two triangles: one that reads *ABC* and the other as *CBA*. These two triangles have two matching side-lengths with matching included angles, and so, by the SAS rule, are congruent. In particular, all corresponding angles are equal. Thus the angle at vertex *B* of the first triangle has the same measure as the corresponding angle of the second triangle, namely, the angle at vertex *C*.

This result appears as Proposition 5 of Book I of Euclid's famous work *THE ELEMENTS.*

d. *The SSS rule:* If the three side-lengths of one triangle match the three side-lengths of a second triangle, then the two triangles are congruent.

The law of sines ensures that all three interior angles match, and so the SAS rule applies.

EUCLIDEAN GEOMETRY takes the SAS rule as an AXIOM, that is, a basic assumption that does not require proof. It is then possible to justify the validity of the remaining rules by making use of this rule solely, and to also justify the law of sines. (The fact that the sine of an angle is the same for all right triangles containing that angle relies on SAS being true.)

See also CONGRUENT FIGURES; SIMILAR FIGURES.

abacus Any counting board with beads laid in parallel grooves, or strung on parallel rods. Typically each bead represents a counting unit, and each groove a place value. Such simple devices can be powerful aids in performing arithmetic computations.

The fingers on each hand provide the simplest "set of beads" for manual counting, and the sand at one's feet an obvious place for writing results. It is not surprising then that every known culture from the time of antiquity developed, independently, some form of counting board to assist complex arithmetical computations. Early boards were simple sun-baked clay tablets, coated with a thin layer of fine sand in which symbols and marks were traced. The Greeks used trays made of marble, and the Romans trays of bronze, and both recorded counting units with pebbles or beads. The Romans were the first to provide grooves to represent fixed place-values, an innovation that proved to be extremely useful. Boards of this type remained the stan-

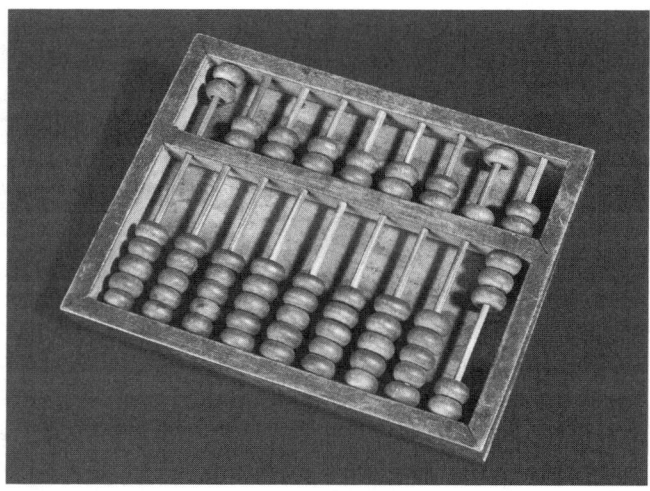

A Chinese abacus from before 1600. Notice that two beads, each representing five units, are placed in each column above the bar. *(Photo courtesy of the Science Museum, London/Topham-HIP/The Image Works)*

dard tool of European merchants and businessmen up through the Renaissance.

The origin of the word *abacus* can be traced back to the Arabic word *abq* for "dust" or "fine sand." The Greeks used the word *abax* for "sand tray," and the Romans adopted the word *abacus*.

The form of the abacus we know today was developed in the 11th century in China and, later, in the 14th century in Japan. (There the device was called a *soroban.*) It has beads strung on wires mounted in a wooden frame, with five beads per wire that can be pushed up or down. Four beads are used to count the units one through four, and the fifth bead, painted a different color or separated by a bar, represents a group of five. This provides the means to represent all digits from zero to nine. Each wire itself represents a different power of ten. The diagram at left depicts the number 35,078.

Addition is performed by sliding beads upward ("carrying digits" as needed when values greater than 10 occur on a single wire), and subtraction by sliding beads downward. Multiplication and division can be computed as repeated addition and subtraction. Historians have discovered that the Chinese and Japanese scholars also devised effective techniques for computing square and cube roots with the aid of the abacus.

The abacus is still the popular tool of choice in many Asian countries—preferred even over electronic

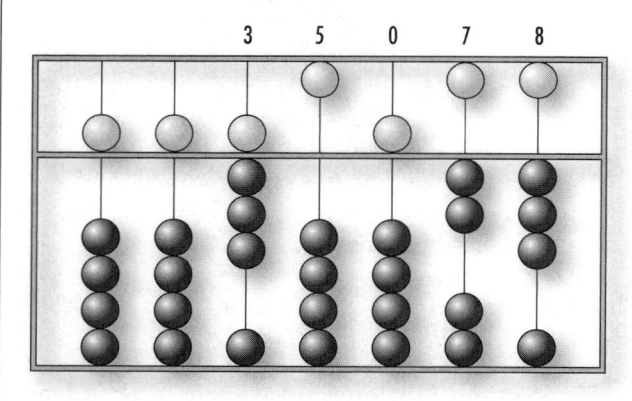

A simple abacus

calculators. It is a useful teaching device to introduce young children to the notion of place-value and to the operations of basic arithmetic.

See also BASE OF A NUMBER SYSTEM; NAPIER'S BONES.

Abel, Niels Henrik (1802–1829) Norwegian *Algebra*

Born on August 5, 1802, Niels Abel might have been one of the great mathematicians of the 19th century had he not died of tuberculosis at age 26. He is remembered, and honored, in mathematics for putting an end to the three-century-long search for a SOLUTION BY RADICALS of the quintic equation. His theoretical work in the topics of GROUP THEORY and ALGEBRA paved the way for continued significant research in these areas.

Abel's short life was dominated by poverty, chiefly due to the severe economic hardships his homeland of Norway endured after the Napoleonic wars, exacerbated by difficult family circumstances. A schoolteacher, thankfully, recognized Abel's talent for mathematics as a young student and introduced him to the works of LEONHARD EULER, JOSEPH-LOUIS LAGRANGE, and other great mathematicians. He also helped raise money to have Abel attend university and continue his studies. Abel entered the University of Christiania in the city of Christiania (present-day Oslo), Norway, in 1821.

During his final year of study, Abel began working on the solution of quintic equations (fifth-degree polynomial equations) by radicals. Although scholars for a long time knew general formulae for solving for QUADRATIC, cubic, and QUARTIC EQUATIONs using nothing more than basic arithmetical operations on the COEFFICIENTS that appear in the equation, no one had yet found a similar formula for solving quintics. In 1822 Abel believed he had produced one. He shared the details of his method with the Danish mathematician Ferdinand Degen in hopes of having the work published by the Royal Society of Copenhagen. Degen had trouble following the logic behind Abel's approach and asked for a numerical illustration of his method. While trying to produce a numerical example, Abel found an error in his paper that eventually led him to understand the reason why general solutions to fifth- and higher-degree equations are impossible. Abel published this phenomenal discovery in 1825 in a self-published pamphlet "Mémoire sur les équations algébriques où on démontre l'impossibilité de la résolution de l'équation générale du cinquième degré" (Memoir on the impossibility of algebraic solutions to the general equations of the fifth degree), which he later presented as a series of seven papers in the newly established *Journal for Pure and Applied Mathematics* (commonly known as *Crelle's Journal* for its German founder August Leopold Crelle). At first, reaction to this work was slow, but as the reputation of the journal grew, more and more scholars took note of the paper, and news of Abel's accomplishment began to spread across Europe. A few years later Abel was honored with a professorship at the University of Berlin. Unfortunately, Abel had contracted tuberculosis by this time, and he died on April 6, 1829, a few days before receiving the letter of notification.

In 1830 the Paris Academy awarded Abel, posthumously, the Grand Prix for his outstanding work. Although Abel did not write in terms of the modern-day concepts of group theory, mathematicians call groups satisfying the COMMUTATIVE PROPERTY "Abelian groups" in his honor. In 2002, on the bicentenary of his birth, the Norwegian Academy of Science and Letters created a new mathematics prize, the Abel Prize, similar to the Nobel Prize, to be awarded annually.

Research in the field of commutative algebra continues today using the approach developed by Abel during his short life. His influence on the development of ABSTRACT ALGEBRA is truly significant.

absolute convergence

A SERIES $\sum_{n=1}^{\infty} a_n$ containing positive and negative terms is said to converge absolutely if the corresponding series with all terms made positive, $\sum_{n=1}^{\infty} |a_n|$, converges. For example, the series $1 - \frac{1}{2} + \frac{1}{4} - \frac{1}{8} + \frac{1}{16} - \frac{1}{32} + \cdots$ converges absolutely because the corresponding series $1 + \frac{1}{2} + \frac{1}{4} + \frac{1}{8} + \frac{1}{16} + \frac{1}{32} + \cdots$ converges. (*See* CONVERGENT SERIES.) The "absolute convergence test" reads:

If $\sum_{n=1}^{\infty} |a_n|$ converges, then the original series $\sum_{n=1}^{\infty} a_n$ also converges.

It can be proved as follows:

Let $p_n = |a_n| - a_n$. Then each value p_n is either zero or equal to $2|a_n|$, depending on whether a_n is positive or negative. In particular we have that $0 \leq p_n \leq 2a_n$.

Consequently, $0 \leq \sum_{n=1}^{\infty} p_n \leq 2\sum_{n=1}^{\infty} |a_n|$ and so, by the

COMPARISON TEST, $\sum_{n=1}^{\infty} p_n$ converges. Consequently

so does $\sum_{n=1}^{\infty} a_n = \sum_{n=1}^{\infty} \left(|a_n| - p_n\right) = \sum_{n=1}^{\infty} |a_n| - \sum_{n=1}^{\infty} p_n$.

This test does not cover all cases, however. It is still possible that a series $\sum_{n=1}^{\infty} a_n$ containing positive and negative terms might converge even though $\sum_{n=1}^{\infty} |a_n|$ does not. For example, the alternating HARMONIC SERIES $1 - \frac{1}{2} + \frac{1}{3} - \frac{1}{4} + \frac{1}{5} - \frac{1}{6} + \cdots$ converges, yet $1 + \frac{1}{2} + \frac{1}{3} + \frac{1}{4} + \frac{1}{5} + \frac{1}{6} + \cdots$ does not. A series that converges "on the condition that the negative signs remain present," that is, one for which $\sum_{n=1}^{\infty} a_n$ converges but $\sum_{n=1}^{\infty} |a_n|$ does not, is called "conditionally convergent." Manipulating conditionally convergent series can lead to all sorts of paradoxes. For example, the following argument "proves" that $1 = 2$:

Start with the observation that:

$$1 - \frac{1}{2} + \frac{1}{3} - \frac{1}{4} + \frac{1}{5} - \frac{1}{6} + \cdots = \ln 2 \approx 0.69$$

(This follows from the study of the harmonic series or from MERCATOR'S EXPANSION.) Consequently:

$$2\ln 2 = 2\left(1 - \frac{1}{2} + \frac{1}{3} - \frac{1}{4} + \frac{1}{5} - \frac{1}{6} + \frac{1}{7} - \frac{1}{8} + \cdots\right)$$

$$= 2 - 1 + \frac{2}{3} - \frac{1}{2} + \frac{2}{5} - \frac{1}{3} + \frac{2}{7} - \frac{1}{4} + \cdots$$

Collecting terms with a common denominator gives:

$$2\ln 2 = (2-1) - \frac{1}{2} + \left(\frac{2}{3} - \frac{1}{3}\right) - \frac{1}{4} + \left(\frac{2}{5} - \frac{1}{5}\right) + \cdots$$

$$\cdots = 1 - \frac{1}{2} + \frac{1}{3} - \frac{1}{4} + \frac{1}{5} - \cdots = \ln 2$$

and so $2 = 1$.

Paradoxes like these show that it is not permissible to rearrange the order of terms of a conditionally convergent series. Mathematicians have shown, however, that rearranging the terms of an absolutely convergent series is valid.

See also ABSOLUTE VALUE.

absolute value (**modulus**) Loosely speaking, the absolute value of a REAL NUMBER is the "positive version of that number." Vertical bars are used to denote the absolute value of a number. For example, the absolute value of negative three is $|-3| = 3$, and the absolute value of four is $|4| = 4$. The absolute value of a real number a is typically envisioned three ways:

1. $|a|$ equals a itself if a is positive or zero, and equals $-a$ if it is negative. (For example, $|-3| = -(-3) = 3$ and $|3| = 3$.)
2. $|a|$ equals the positive square root of a^2. (For example, $|-3| = \sqrt{(-3)^2} = \sqrt{9} = 3$.)
3. $|a|$ is the distance between the points a and 0 on the real number line. (For example, $|-3| = 3 = |3|$ since both -3 and 3 are three units from the origin.) More generally, if a and b are two points on the number line, then the distance between them on the number line is given by $|a - b|$. (For example, the points 4 and -7 are $|4-(-7)| = |4 + 7| = 11$ units apart.)

By examining each of the cases with a and b positive or negative, one can check that the absolute value function satisfies the following properties:

  i. $|a + b| \leq |a| + |b|$
  ii. $|a - b| \leq |a| + |b|$
  iii. $|a \cdot b| = |a| \cdot |b|$

Knowing the absolute value of a quantity determines the value of that quantity up to sign. For example, the equation $|x+2| = 5$ tells us that either $x + 2 = 5$

or $x + 2 = -5$, and so x equals either 3 or –7. Alternatively, one can read the equation as $|x - (-2)| = 5$, interpreting it to mean that x is a point a distance of five units from –2. Five units to the left means x is the point –7; five units to the right means x is 3.

The notion of absolute value was not made explicit until the mid-1800s. KARL WEIERSTRASS, in 1841, was the first to suggest a notation for it—the two vertical bars we use today. Matters are currently a little confusing, however, for mathematicians today also use this notation for the length of a VECTOR and for the MODULUS of a COMPLEX NUMBER.

abstract algebra Research in pure mathematics is motivated by one fundamental question: what makes mathematics work the way it does? For example, to a mathematician, the question, "What is 263×178 (or equivalently, 178×263)?" is of little interest. A far more important question would be, "Why should the answers to 263×178 and 178×263 be the same?"

The topic of abstract algebra attempts to identify the key features that make ALGEBRA and ARITHMETIC work the way they do. For example, mathematicians have shown that the operation of ADDITION satisfies five basic principles, and that all other results about the nature of addition follow from these.

1. *Closure:* The sum of two numbers is again a number.
2. *Associativity:* For all numbers a, b, and c, we have: $(a + b) + c = a + (b + c)$.
3. *Zero element:* There is a number, denoted "0," so that: $a + 0 = a = 0 + a$ for all numbers a.
4. *Inverse:* For each number a there is another number, denoted "–a," so that: $a + (-a) = 0 = (-a) + a$.
5. *Commutativity:* For all numbers a and b we have: $a + b = b + a$.

Having identified these five properties, mathematicians search for other mathematical systems that may satisfy the same five relations. Any fact that is known about addition will consequently hold true in the new system as well. This is a powerful approach to matters. It avoids having to re-prove THEOREMS and facts about a new system if one can recognize it as a familiar one in disguise. For example, MULTIPLICATION essentially satisfies the same five AXIOMS as above, and so for any fact about addition, there is a corresponding fact about multiplication. The set of symmetries of a geometric figure also satisfy these five axioms, and so too all known results about addition immediately transfer to interesting statements about geometry. Any system that satisfies these basic five axioms is called an "Abelian group," or just a GROUP if the fifth axiom fails. GROUP THEORY is the study of all the results that follow from these basic five axioms without reference to a particular mathematical system.

The study of RINGS and FIELDS considers mathematical systems that permit two fundamental operations (typically called addition and multiplication). Allowing for the additional operation of scalar multiplication leads to a study of VECTOR SPACES.

The theory of algebraic structures is highly developed. The study of vector spaces, for example, is so extensive that the topic is regarded as a field of mathematics in its own right and is called LINEAR ALGEBRA.

acceleration *See* VELOCITY.

actuarial science The statistical study of life expectancy, sickness, retirement, and accident matters is called actuarial science. Experts in the field are called actuaries and are employed by insurance companies and pension funds to calculate risks and relate them to the premiums to be charged. British mathematician and astronomer, Edmund Halley (1656–1742) was the first to properly analyze annuities and is today considered the founder of the field.

See also LIFE TABLES.

acute angle An ANGLE between zero and 90° is called an acute angle. An acute-angled triangle is one whose angles are all acute. According to the LAW OF COSINES, a triangle with side-lengths a, b, and c and corresponding angles A, B, C opposite those sides, satisfies:

$$\cos C = \frac{a^2 + b^2 - c^2}{2ab}$$

The angle C is acute only if $\cos C > 0$, that is, only if $a^2 + b^2 > c^2$. Thus a triangle a, b, c is acute if, and only if, the following three inequalities hold:

$$a^2 + b^2 > c^2$$
$$b^2 + c^2 > a^2$$
$$c^2 + a^2 > b^2$$

See also OBTUSE ANGLE; PERIGON; PYTHAGORAS'S THEOREM; TRIANGLE.

addition The process of finding the sum of two numbers is called addition. In the elementary ARITHMETIC of whole numbers, addition can be regarded as the process of accumulating sets of objects. For example, if a set of three apples is combined with a set of five apples, then the result is a set of eight apples. We write: 3 + 5 = 8.

Two numbers that are added together are called addends. For instance, in the equation 17 + 33 = 50, the numbers 17 and 33 are the addends, and the number 50 is their sum. Addition can also be regarded as the process of increasing one number (an addend) by another (called, in this context, an augend). Thus when 17 is augmented by 33 units, the result is 50.

The PLACE-VALUE SYSTEM we use today for writing numbers simplifies the process of adding large integers. For instance, adding together 253 and 589 yields 2 + 5 = 7 units of 100, 5 + 8 = 13 units of 10, and 3 + 9 = 12 units of 1. So, in some sense, it is reasonable to write the answer to this addition problem simply as 7 | 13 | 12 using a vertical bar to separate units of powers of 10. Since 13 units of 10 is equivalent to one unit of 100 and three units of 10, this is equivalent to 8 | 3 | 12. Noting, also, that 12 units of one 12 is equivalent to one unit of 10 and two single units, this can be rewritten as 8 | 4 | 2. Thus we have: 253 + 589 = 842.

The latter process of modifying the figures into single-digit powers of 10 (that is, in our example, the process of rewriting 7 | 13 | 12 as 8 | 4 | 2) is called "carrying digits." Students in schools are usually taught an algorithm that has one carry digits early in the process of completing an addition problem rather than leaving this work as the final step. Either method is valid. (The term "carry a digit" dates back to the time of the ABACUS, where beads on rods represented counts of powers of 10 and the person had to move—"carry"—counters from one rod to another if any count was greater than a single digit.)

The process of addition can be extended to NEGATIVE NUMBERS (yielding an operation called SUBTRAC-

TION), the addition of FRACTIONS (completed with the aid of computing COMMON DENOMINATORS), REAL NUMBERS, COMPLEX NUMBERS, VECTORS, and MATRIX addition. The number ZERO is an additive IDENTITY ELEMENT in the theory of arithmetic. We have that $a + 0 = a = 0 + a$ for any number a.

The sum of two real-valued functions f and g is the function $f + g$ whose value at any input x is the sum of the outputs of f and g at that input value: $(f + g)(x) = f(x) + g(x)$. For example, if $f(x) = x^2 + 2x$ and $g(x) = 5x + 7$, then $(f + g)(x) = x^2 + 2x + 5x + 7 = x^2 + 7x + 7$.

A function with the property that $f(x + y) = f(x) + f(y)$ for all inputs x and y is called "additive." For example, $f(x) = 2x$ is additive.

The addition formulae in TRIGONOMETRY assert:

$$\sin(x + y) = \sin x \cos y + \cos x \sin y$$
$$\cos(x + y) = \cos x \cos y - \sin x \sin y$$
$$\tan(x + y) = \frac{\tan x + \tan y}{1 - \tan x \tan y}$$

The symbol + used to denote addition is believed to have derived from a popular shorthand for the Latin word *et* meaning "and" and was widely used by mathematical scholars in the late 15th century. The symbol first appeared in print in Johannes Widman's 1489 book *Behennde unnd hüpsche Rechnung auf fallen Kauffmannschaften* (Neat and handy calculations for all tradesmen).

See also ASSOCIATIVE; CASTING OUT NINES; COMMUTATIVE PROPERTY; DISTRIBUTIVE PROPERTY; MULTIPLICATION; SUMMATION.

affine geometry The study of those properties of geometric figures that remain unchanged by an AFFINE TRANSFORMATION is called affine geometry. For example, since an affine transformation preserves straight lines and RATIOS of distances between POINTs, the notions of PARALLEL lines, MIDPOINTs of LINE segments, and tangency are valid concepts in affine geometry. The notion of a CIRCLE, however, is not. (A circle can be transformed into an ELLIPSE via an affine transformation. The equidistance of points on the circle from the circle center need not be preserved.)

Affine geometry was first studied by Swiss mathematician LEONHARD EULER (1707–83). Only postulates

1, 2, and 5 of EUCLID'S POSTULATES remain valid in affine geometry.

affine transformation   Any map from the PLANE to itself that transforms straight LINES into straight lines and preserves RATIOS of distances between POINTS (so that the midpoint of a line segment, for instance, remains the midpoint after the transformation) is called an affine transformation. One can prove that any affine transformation must be a LINEAR TRANSFORMATION followed by a translation. Thus an affine transformation T is completely specified by a MATRIX A and a VECTOR **b** so that $T(\mathbf{x}) = A\mathbf{x} + \mathbf{b}$ for any vector **x** representing a point in the plane.

An affine transformation T satisfies the relation:

$$T(s\mathbf{x} + t\mathbf{y}) = sT(\mathbf{x}) + tT(\mathbf{y})$$

for any two vectors **x** and **y** and any two real numbers s and t such that $s + t = 1$. This is sometimes taken as the definition of what it means for a transformation to be affine.

Affine transformations generally do not preserve the lengths of line segments nor the measure of ANGLES between segments. It is possible to transform a CIRCLE into an ELLIPSE, for instance, via an affine transformation.

See also AFFINE GEOMETRY.

Agnesi, Maria Gaëtana (1718–1799) Italian *Calculus*   Born on May 16, 1718, to a wealthy family of silk merchants, Maria Agnesi is best remembered for her influential expository text outlining the methods and techniques of the newly invented CALCULUS. Written with such clarity and precision, *Istituzioni analitiche* (Analytical institutions) garnered her international fame. Agnesi is considered the first major female mathematician of modern times, and she holds the distinction of being the first woman to be awarded a professorship of mathematics on a university faculty.

Agnesi demonstrated remarkable academic talents as a young child. By age 13 she had mastered many languages and had published translations of academic essays. Although little consideration was given to educating women at the beginning of the 18th century, Agnesi's father encouraged her intellectual develop-

ment, provided tutors of the highest quality, and provided forums for her to display her talents to Italian society. In preparation for these events, Agnesi had prepared discourses on a wide variety of topics in science and philosophy, which she published as a collection of 190 essays at age 20.

After the death of her mother, Agnesi undertook the task of instructing her younger brothers in the subject of mathematics. In 1738 she began preparing a textbook for their use, and found the topic so compelling that she devoted her complete intellectual attention to mathematics. Ten years later, her famous two-volume text *Istituzioni analitiche* was published.

The work was the first comprehensive overview of the subject of calculus. Although designed for young students beginning their studies of the subject, Agnesi's work was recognized as providing hitherto unnoticed connections between the different approaches of SIR ISAAC NEWTON (1642–1727) and GOTTFRIED WILHELM LEIBNIZ (1646–1716), independent coinventors of the subject. Her piece also provided, for the first time, clear explanations of previously confusing issues in the topic. Her text collated and explained the work of other contributors to the subject from several different countries, a task no doubt facilitated by her skills in translation. Her talents, not just as an expository writer, but also as a great scholar in mathematics, were apparent. Mathematicians at the time recognized her text as a significant contribution to the further development of the topic of calculus. In 1750 Agnesi was appointed the chair of mathematics at the University of Bologna in recognition of her great accomplishment. Curiously, she never officially accepted or rejected the faculty position. It is known that she never visited the city of Bologna, even though her name appears on university records over a span of 45 years.

After the death of her father in 1752, Agnesi withdrew from mathematics and devoted her life to charitable work, helping the sick and the poor. In 1783 she was made director of a women's poorhouse, where she remained the rest of her life. Having given all her money and possessions away, Agnesi died on January 9, 1799, with no money of her own and was buried in a pauper's grave. The city of Milan, where she had lived all her life, today publicly honors her gravesite.

John Colson, Lucasian professor of mathematics at Cambridge, published an English translation of *Istituzioni analitiche* in 1801. He said that he wanted to give the youth of Britain the same opportunity to benefit from this remarkable text as the young scholars from Italy had been able to enjoy.

Alembert, Jean Le Rond d' (1717–1783)

French *Differential equations, Analysis, Philosophy* Born on November 17, 1717, in Paris, France, scholar Jean le Rond d'Alembert is best remembered for his 1743 treatise *Traité de dynamique* (Treatise on dynamics), in which he attempted to develop a firm mathematical basis for the study of mechanics. D'Alembert pioneered the study of partial DIFFERENTIAL EQUATIONs and their use in physics. He is also noted for his work on vibrating strings.

After briefly pursuing theology and medicine at the Jansenist Collège des Quatre Nations, d'Alembert settled on mathematics as his choice of academic study.

Jean Le Rond d'Alembert, an eminent mathematician of the 18th century, pioneered the study of differential equations and their application to mechanics. *(Photo courtesy of Topham/ The Image Works)*

He graduated from the Collège in 1735 to then pursue interests in fluid mechanics. In 1740 he presented a series of lectures on the topic to members of the Paris Academy of Science, which earned him recognition as a capable mathematician and admittance as a member of the academy. He remained with the institution for his entire career.

D'Alembert came to believe that the topic of mechanics should be based on logical principles, not necessarily physical ones, and that its base is fundamentally mathematical. In his 1743 treatise *Traité de dynamique*, he attempted to refine the work of SIR ISAAC NEWTON (1642–1727) and clarify the underpinnings of the subject. The following year d'Alembert published a second work, *Traité de l'équilibre et du mouvement des fluides* (Treatise on the equilibrium and movement of fluids), that applied his results to the study of fluid motion and introduced some beginning results on the study of partial differential equations. He developed these results further over the following years. In 1747 d'Alembert submitted a paper "Réflexions sur la cause générale des vents" (Reflections on the general cause of air motion) for consideration for the annual scientific prize offered by the Prussian Academy. He did indeed win.

At the same time d'Alembert also began work as a writer and science editor for the famous French *Encyclopédie ou dictionnaire raisonné des sciences, des arts, et des métiers* (Encyclopedia and dictionary of the rationales of the sciences, arts, and professions), taking responsibility for the writing of the majority of mathematical entries. The first volume of the 28-volume work was published in 1751.

D'Alembert published several new mathematical results and ideas in this epic work. For instance, in volume 4, under the entry *differential*, he suggested, for the first time, that the principles of CALCULUS should be based on the notion of a LIMIT. He went so far as to consider defining the derivative of a function as the limit of a RATIO of increments. He also described the new ratio test when discussing CONVERGENT SERIES.

D'Alembert's interests turned toward literature and philosophy, and administrative work, in the latter part of his life. He was elected as perpetual secretary of the Académie Française in 1772. He died 11 years later on October 29, 1783.

His work in mathematics paved the way for proper development of the notion of a limit in calculus, as well as advancement of the field of partial differential equations. In mechanics, he is honored with a principle of motion named after him, a generalization of Sir Isaac Newton's third law of motion.

See also DIFFERENTIAL CALCULUS.

algebra The branch of mathematics concerned with the general properties of numbers, and generalizations arising from those properties, is called algebra. Often symbols are used to represent generic numbers, thereby distinguishing the topic from the study of ARITHMETIC. For instance, the equation $2 \times (5 + 7) = 2 \times 5 + 2 \times 7$ is a (true) arithmetical statement about a specific set of numbers, whereas, the equation $x \times (y + z) = x \times y + x \times z$ is a general statement describing a property satisfied by *any* three numbers. It is a statement in algebra.

Much of elementary algebra consists of methods of manipulating equations to either put them in a more convenient form, or to determine (that is, solve for) permissible values of the variables that appear. For instance, rewriting $x^2 + 6x + 9 = 25$ as $(x + 3)^2 = 25$ allows an easy solution for x: either $x + 3 = 5$, yielding $x = 2$, or $x + 3 = -5$, yielding $x = -8$.

The word *algebra* comes from the Arabic term *al-jabr w'al-muqābala* (meaning "restoration and reduction") used by the great MUHAMMAD IBN MŪSĀ AL-KHWĀRIZMĪ (ca. 780–850) in his writings on the topic.

History of Equations and Algebra

Finding solutions to equations is a pursuit that dates back to the ancient Egyptians and Babylonians and can be traced through the early Greeks' mathematics. The RHIND PAPYRUS, dating from around 1650 B.C.E., for instance, contains a problem reading:

> A quantity; its fourth is added to it. It becomes fifteen. What is the quantity?

Readers are advised to solve problems like these by a method of "false position," where one guesses *(posits)* a solution, likely to be wrong, and adjusts the guess according to the result obtained. In this example, to make the division straightforward, one might guess that the quantity is four. Taking 4 and adding to it its fourth gives, however, only $4 + 1 = 5$, one-third of the desired answer of 15. Multiplying the guess by a factor of three gives the solution to the problem, namely 4×3, which is 12.

Although the method of false position works only for LINEAR EQUATIONS of the form $ax = b$, it can nonetheless be an effective tool. In fact, several of the problems presented in the Rhind papyrus are quite complicated and are solved relatively swiftly via this technique.

Clay tablets dating back to 1700 B.C.E. indicate that Babylonian mathematicians were capable of solving certain QUADRATIC equations by the method of COMPLETING THE SQUARE. They did not, however, have a general method of solution and worked only with a set of specific examples fully worked out. Any other problem that arose was matched with a previously solved example, and its solution was found by adjusting the numbers appropriately.

Much of the knowledge built up by the old civilizations of Egypt and Babylonia was passed on to the Greeks. They took matters in a different direction and began examining all problems geometrically by interpreting numbers as lengths of line segments and the products of two numbers as areas of rectangular regions. Followers of PYTHAGORAS from the period 540 to 250 B.C.E., for instance, gave geometric proofs of the DISTRIBUTIVE PROPERTY and the DIFFERENCE OF TWO SQUARES formula, for example, in much the same geometric way we use today to explain the method of EXPANDING BRACKETS. The Greeks had considerable trouble solving CUBIC EQUATIONS, however, since their practice of treating problems geometrically led to complicated three-dimensional constructions for coping with the product of three quantities.

At this point, no symbols were used in algebraic problems, and all questions and solutions were written out in words (and illustrated in diagrams). However, in the third century, DIOPHANTUS OF ALEXANDRIA introduced the idea of abbreviating the statement of an equation by replacing frequently used quantities and operations with symbols as a kind of shorthand. This new focus on symbols had the subtle effect of turning Greek thinking away from geometry. Unfortunately, the idea of actually using the symbols to solve equations was ignored until the 16th century.

The Babylonian and Greek schools of thought also influenced the development of mathematics in ancient India. The scholar BRAHMAGUPTA (ca. 598–665) gave solutions to

(continues)

History of Equations and Algebra
(continued)

quadratic equations and outlined general methods for solving systems of equations containing several variables. (He also had a clear understanding of negative numbers and was comfortable working with zero as a valid numerical quantity.) The scholar Bhāskara (ca. 1114–85) used letters to represent unknown quantities and, in working with quadratic equations, suggested that all positive numbers have two square roots and that negative numbers have no (meaningful) roots.

A significant step toward the development of modern algebra occurred in Baghdad, Iraq, in the year 825 when the Arab mathematician MUHAMMAD IBN MŪSĀ AL-KHWĀRIZMĪ (ca. 780–850) published his famous piece *Hisab al-jabr w'al-muqābala* (Calculation by restoration and reduction). This work represents the first clear and complete exposition on the art of solving linear equations by a new practice of performing the same operation on both sides of an equation. For example, the expression $x - 3 = 7$ can be "restored" to $x = 10$ by adding three to both sides of the expression, and the equation $5x = 10$ can be "reduced" to $x = 2$ by dividing both sides of the equation by five. Al-Khwārizmī also showed how to solve quadratic equations via similar techniques. His descriptions, however, used no symbols, and like the ancient Greeks, al-Khwārizmī wrote everything out in words. Nonetheless, al-Khwārizmī's treatise was enormously influential, and his new approach to solving equations paved the way for modern algebraic thinking. In fact, it is from the word *al-jabr* in the title of his book that our word *algebra* is derived.

Al-Khwārizmī's work was translated into Latin by the Italian mathematician FIBONACCI (ca. 1175–1250), and his efficient methods for solving equations quickly spread across Europe during the 13th century. Fibonacci translated the word *shai* used by al-Khwārizmī for "the thing unknown" into the Latin term *res.* He also used the Italian word *cosa* for "thing," and the art of algebra became known in Europe as "the cossic art."

In 1545 GIROLAMO CARDANO (1501–76) published *Ars magna* (The great art), which included solutions to the cubic and QUARTIC EQUATIONS, as well as other mathematical discoveries. By the end of the 17th century, mathematicians were comfortable performing the same sort of symbolic manipulations we practice today and were willing to accept negative numbers and irrational quantities as solutions to equations. The French mathematician FRANÇOIS VIÈTE (1540–1603) introduced an efficient system for denoting powers of variables and was the first to use letters as coefficients before variables, as in "$ax^2 + bx + c$," for instance. (Viète also introduced the signs "+" and "−," although he never used a sign for equality.) RENÈ DESCARTES (1596–1650) introduced the convention of denoting unknown quantities by the last letters of the alphabet, x, y, and z, and known quantities by the first, a, b, c. (This convention is now completely ingrained; when we see, for example, an equation of the form $ax + b = 0$, we assume, without question, that it is for "x" we must solve.)

The German mathematician CARL FRIEDRICH GAUSS (1777–1855) proved the FUNDAMENTAL THEOREM OF ALGEBRA in 1797, which states that every POLYNOMIAL equation of degree n has at least one and at most n (possibly complex) roots. His work, however, does not provide actual methods for finding these roots.

Renaissance scholars SCIPIONE DEL FERRO (1465–1526) and NICCOLÒ TARTAGLIA (ca. 1500–57) both knew how to solve cubic equations, and in his 1545 treatise *Ars magna,* Cardano published the solution to the quartic equation discovered by his assistant LUDOVICO FERRARI (1522–65). For the centuries that followed, mathematicians attempted to find a general arithmetic method for solving all quintic (fifth-degree) equations. LEONHARD EULER (1707–83) suspected that the task might be impossible. Between the years 1803 and 1813, Italian mathematician Paolo Ruffini (1765–1822) published a number of algebraic results that strongly suggested the same, and just a few years later Norwegian mathematician NIELS HENRIK ABEL (1802–29) proved that, indeed, there is no general formula that solves all quintic equations in a finite number of arithmetic operations. Of course, some degree-five equations can be solved algebraically. (Equation of the form $x^5 - a = 0$, for instance, have solutions $x = \sqrt[5]{a}$.) In 1831 French mathematician ÉVARISTE GALOIS (1811–32) completely classified those equations that can be so solved, developing work that gave rise to a whole new branch of mathematics today called GROUP THEORY.

In the 19th century mathematicians began using variables to represent quantities other than real numbers. For example, English mathematician GEORGE BOOLE (1815–64) invented an algebra symbolic logic in which variables represented sets, and Irish scholar SIR WILLIAM ROWAN HAMILTON (1805–65) invented algebraic systems in which variables represented VECTORS or QUATERNIONS.

With these new systems, important characteristics of algebra changed. Hamilton, for instance, discovered that multiplication was no longer commutative in his systems: a product $a \times b$ might not necessarily give the same result as $b \times a$. This motivated mathematicians to develop abstract AXIOMS to explain the workings of different algebraic systems. Thus the topic of ABSTRACT ALGEBRA was born. One outstanding contributor in this field was German mathematician AMALIE NOETHER (1883–1935), who made important discoveries about the nature of noncommutative algebras.

See also ASSOCIATIVE; BABYLONIAN MATHEMATICS; CANCELLATION; COMMUTATIVE PROPERTY; EGYPTIAN MATHEMATICS; FIELD; GREEK MATHEMATICS; INDIAN MATHEMATICS; LINEAR ALGEBRA; RING.

In modern times the subject of algebra has been widened to include ABSTRACT ALGEBRA, GROUP THEORY, and the study of alternative number systems such as MODULAR ARITHMETIC. BOOLEAN ALGEBRA looks at the algebra of logical inferences, matrix algebra the arithmetic of MATRIX operations, and vector algebra the mechanics of VECTOR operations and VECTOR SPACES.

An algebraic structure is any set equipped with one or more operations (usually BINARY OPERATIONs) satisfying a list of specified rules. For example, any group, RING, FIELD, or vector space is an algebraic structure. In advanced mathematics, a vector space that is also a field is called an "algebra."

See also BRACKETS; COMMUTATIVE PROPERTY; DISTRIBUTIVE PROPERTY; EXPANDING BRACKETS; FUNDAMENTAL THEOREM OF ALGEBRA; HISTORY OF EQUATIONS AND ALGEBRA (essay); ORDER OF OPERATION.

algebraic number A number is called algebraic if it is the root of a POLYNOMIAL with integer coefficients. For example, $(1/2) (5 + \sqrt{13})$ is algebraic since it is a solution to the equation $x^2 - 5x + 3 = 0$. All RATIONAL NUMBERS are algebraic (since a fraction a/b is the solution to the equation $bx - a = 0$), and all square, cube, and higher roots of integers are algebraic (since $\sqrt[n]{a}$ is a solution to $x^n - a = 0$).

At first thought it seems that all numbers are algebraic, but this is not the case. In 1844 French mathematician JOSEPH LIOUVILLE made the surprising discovery that the following number, today called "Liouville's constant," cannot be a solution to any integer polynomial equation:

$$L = \sum_{n=1}^{\infty} \frac{1}{10^{n!}} = 0.1100010000000000000000100...$$

Numbers that are not algebraic are called "transcendental."

In 1873 French mathematician Charles Hermite (1822–1901) proved that the number e is transcendental, and, nine years later in 1882 German mathematician CARL LOUIS FERDINAND VON LINDEMANN established that π is transcendental. In 1935 Russian mathematician Aleksandr Gelfond (1906–68) proved that any number of the form a^b is transcendental if a and b are both alge-

braic, with a different from 0 or 1, and b irrational. (Thus, for example, $2^{\sqrt{3}}$ is transcendental.)

The German mathematician GEORG CANTOR (1845–1918) showed that the set of algebraic numbers is COUNTABLE. As the set of real numbers is uncountable, this means that most numbers are transcendental. The probability that a real number chosen at random is algebraic is zero. Although it was proven in 1929 that e^π is transcendental, no one to this day knows whether or not π^π is algebraic.

In analogy with algebraic numbers, a FUNCTION $y = f(x)$ is called "algebraic" if it can be defined by a relation of the form

$$p_n(\mathrm{x})y^n + p_{n}-1(x)y^n - 1 + ... + p_1(x)y + p_0(x) = 0$$

where the functions $p_i(x)$ are polynomials in x. For example, the function $y = \sqrt{x}$ is an algebraic function, since it is defined by the equation $y^2 - x = 0$. A transcendental function is a function that is not algebraic. Mathematicians have shown that trigonometric, logarithmic, and exponential functions are transcendental.

See also CARDINALITY.

algorithm An algorithm is a specific set of instructions for carrying out a procedure or solving a mathematical problem. Synonyms include "method," "procedure," and "technique." One example of an algorithm is the common method of LONG DIVISION. Another is the EUCLIDEAN ALGORITHM for finding the GREATEST COMMON DIVISOR of two positive integers. The word *algorithm* is a distortion of "al-Khwārizmī," the name of a Persian mathematician (ca. 820) who wrote an influential text on algebraic methods.

See also BASE OF A NUMBER SYSTEM; MUHAMMAD IBN MŪSĀ AL-KHWĀRIZMĪ.

alternating series A SERIES whose terms are alternately positive and negative is called an alternating series. For example, the GREGORY SERIES $1 - \frac{1}{3} + \frac{1}{5} - \frac{1}{7} + ... = \frac{\pi}{4}$ is an alternating series, as is the (divergent) series: $1 - 1 + 1 - 1 + 1 - 1 + ...$ Alternating series have the form $\sum_{n=1}^{\infty} (-1)^{n-1} a_n = a_1 - a_2 + a_3 - a_4 + ...$, with each a_i positive number.

In 1705, GOTTFRIED WILHELM LEIBNIZ noticed that many convergent alternating series, like the Gregory series, have terms a_i that decrease and approach zero:

i. $a_1 \geq a_2 \geq a_3 \geq \ldots$
ii. $a_n \rightarrow 0$

He managed to prove that *any* alternating series satisfying these two conditions does indeed converge, and today this result is called the "alternating series test." (One can see that the test is valid if one physically paces smaller and smaller steps back and forth: a_1 feet forward, a_2 feet backward, a_3 feet forward, and so on. This motion begins to "hone in" on a single limiting location.) We see, for example, that the series

$$1 - \frac{1}{4} + \frac{1}{9} - \frac{1}{16} + \frac{1}{25} - \frac{1}{36} + \ldots$$ converges. Unfortunately, the alternating series test gives us no indication as to what the value of the sum could be. Generally, finding the limit value is a considerable amount of work, if at all possible. The values of many "simple" alternating series are not known today. (One can show, however, that the above series above converges to $\pi^2/12$. *See* CONVERGENT SERIES.)

See also ZETA FUNCTION.

altitude A line segment indicating the height of a two- or three-dimensional geometric figure such as a POLYGON, POLYHEDRON, CYLINDER, or CONE is called an altitude of the figure. An altitude meets the base of the figure at a RIGHT ANGLE.

Any TRIANGLE has three distinct altitudes. Each is a LINE segment emanating from a vertex of the triangle meeting the opposite edge at a 90° angle. The LAW OF SINES shows that the lengths h_a, h_b, and h_c of the three altitudes of a triangle ABC satisfy:

$$h_a = c \sin \beta = b \sin \gamma$$
$$h_b = a \sin \gamma = c \sin \alpha$$
$$h_c = b \sin \alpha = a \sin \beta$$

where a, b, and c are the side-lengths of the triangle, and α, β, and γ are the angles at vertices A, B, and C, respectively. Here h_a is the altitude meeting the side of length a at 90°. Similarly, h_b and h_c are the altitudes meeting sides of length b and c, respectively. It also follows from this law that the following relation holds:

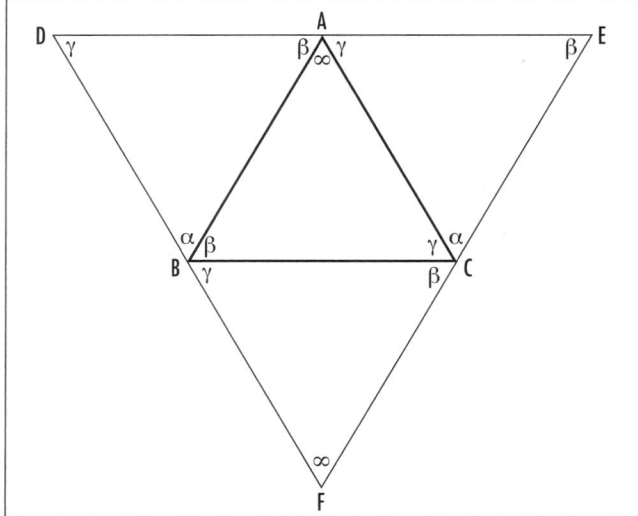

Proving altitudes are concurrent

$$h_a h_b h_c = \frac{(abc)^3}{8r^3}$$

where r is the radius of the circle that contains the points A, B, and C.

The three altitudes of a triangle always meet at a common point called the orthocenter of the triangle. Surprisingly, this fundamental fact was not noticed by the geometer EUCLID (ca. 300 B.C.E.). The claim can be proved as follows:

Given a triangle ABC, draw three lines, one through each vertex and parallel to the side opposite to that vertex. This creates a larger triangle DEF.

By the PARALLEL POSTULATE, alternate angles across parallel lines are equal. This allows us to establish that all the angles in the diagram have the values as shown. Consequently, triangle DAB is similar to triangle ABC and, in fact, is congruent to it, since it shares the common side AB. We have that DA is the same length as BC. In a similar way we can show that AE also has the same length as BC, and so A is the midpoint of side DE of the large triangle. Similarly, B is the midpoint of side DF, and C the midpoint of side EF. The study of EQUIDISTANT points establishes that the perpendicular bisectors of any triangle are CONCURRENT, that is, meet at a point. But the perpendicular bisectors of triangle *DEF* are precisely the altitudes of triangle *ABC*.

One can also show that the three altitudes of a triangle satisfy:

$$\frac{1}{h_a} + \frac{1}{h_b} + \frac{1}{h_c} = \frac{1}{R}$$

where *R* is the radius of the largest circle that sits inside the triangle.

See also EULER LINE.

amicable numbers (friendly numbers) Two whole numbers *a* and *b* are said to be amicable if the sum of the FACTORs of *a*, excluding *a* itself, equals *b*, and the sum of the factors of *b*, excluding *b* itself, equals *a*. For example, the numbers 220 and 284 are amicable:

284 has factors 1, 2, 4, 71, and 142, and their sum is 220
220 has factors 1, 2, 4, 5, 10, 11, 20, 22, 44, 55, and 110, and their sum is 284

The pair (220, 284) is the smallest amicable pair. For many centuries it was believed that this pair was the only pair of amicable numbers. In 1636, however, PIERRE DE FERMAT discovered a second pair, (17296, 18416), and in 1638, RENÉ DESCARTES discovered the pair (9363584, 9437056). Both these pairs were also known to Arab mathematicians, perhaps at an earlier date.

By 1750, LEONHARD EULER had collated 60 more amicable pairs. In 1866, 16-year-old Nicolò Paganini found the small pair (1184, 1210) missed by all the scholars of preceding centuries. Today more than 5,000 different amicable pairs are known. The largest pair known has numbers each 4,829 digits long.

See also PERFECT NUMBER.

analysis Any topic in mathematics that makes use of the notion of a LIMIT in its study is called analysis. CALCULUS comes under this heading, as does the summation of infinite SERIES, and the study of REAL NUMBERS.

Greek mathematician PAPPUS OF ALEXANDRIA (ca. 320 C.E.) called the process of discovering a proof or a solution to a problem "analysis." He wrote about "a method of analysis" somewhat vaguely in his geometry text *Collection*, which left mathematicians centuries later wondering whether there was a secret method hidden behind all of Greek geometry.

The great RENÉ DESCARTES (1596–1650) developed a powerful method of using algebra to solve geometric problems. His approach became known as analytic geometry.

See also ANALYTIC NUMBER THEORY; CARTESIAN COORDINATES.

analytic number theory The branch of NUMBER THEORY that uses the notion of a LIMIT to study the properties of numbers is called analytic number theory. This branch of mathematics typically deals with the "average" behavior of numbers. For example, to answer:

On average, how many square factors does a number possess?

one notes that all numbers have 1 as a factor, one-quarter of all numbers have 4 as a factor, one-ninth have the factor 9, one-sixteenth the factor 16, and so on. Thus, on average, a number possesses $1 + \frac{1}{4} + \frac{1}{9} + \frac{1}{16} + \cdots = \frac{\pi^2}{6} \approx 1.64$ square factors. This particular argument can be made mathematically precise.

See also ANALYSIS; ZETA FUNCTION.

angle Given the configuration of two intersecting LINES, line segments, or RAYS, the amount of ROTATION about the point of intersection required to bring one line coincident with the other is called the angle between the lines. Simply put, an angle is a measure of "an amount of turning." In any diagram representing an angle, the lengths of the lines drawn is irrelevant. For example, an angle corresponding to one-eighth of a full turn can be represented by rays of length 2 in., 20 in., or 200 in.

The image of a lighthouse with a rotating beam of light helps clarify the concept of an angle: each ray or line segment in a diagram represents the starting or ending position of the light beam after a given amount of turning. For instance, angles corresponding to a quarter of a turn, half a turn, and a full turn appear as follows:

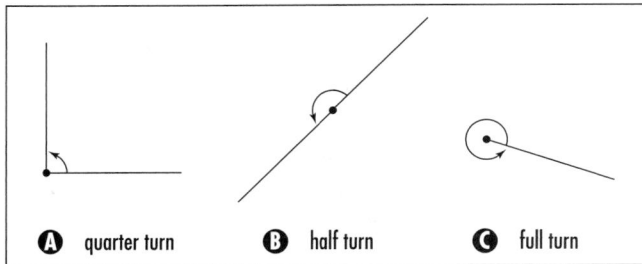

Ⓐ quarter turn **Ⓑ** half turn **Ⓒ** full turn

Some simple angles

Mathematicians sometimes find it convenient to deem an angle measured in a counterclockwise sense as positive, and one measured in a clockwise sense as negative.

Babylonian scholars of ancient times were aware that the year is composed of 365 days but chose to operate with a convenient calendar composed of 12 months of 30 days. Thus the number 360 came to be associated with the notion of a full cycle. Today, harking back to the Babylonians, angles are measured in units of degrees, in which a full rotation corresponds to 360 degrees (written 360°). Thus a half turn corresponds to 180°, and a quarter turn to 90°. A single degree corresponds to 1/360 of a turn.

Each degree is divided into 60 smaller units called minutes, denoted with an apostrophe, and each minute is divided into 60 smaller units called seconds, denoted with two apostrophes. Thus, for instance, 15°46′23″ represents an angle of 15 degrees, 46 minutes, and 23 seconds.

Mathematicians prefer to use a unit of angle measurement independent of the circumstance that we live on the Earth, i.e., one that is natural to mathematics. The chosen unit of measurement is called a radian. Working with the simplest CIRCLE possible, namely, a circle of radius one, mathematicians match the measure of a full turn with the distance around that circle, namely 2π, the circumference of the circle. Thus one full rotation equals 2π radian. A half turn is measured as half of this, namely, π radian, and a quarter turn as $\frac{1}{4} \cdot 2\pi = \frac{\pi}{2}$ radian.

To convert between degree and radian measures, one simply notes that 360 degrees corresponds to 2π radian. (Thus one degree equals $\frac{2\pi}{360} \approx 0.017$ radian, and one radian equals $\frac{360}{2\pi} \approx 57.3°$.)

A unit of measurement called a "gradian" is sometimes used in surveying. In this system, a full turn is considered 400 gradian (and, consequently, a quarter turn is divided into exactly 100 parts). This system is rarely used in mathematics, if at all.

Angles are classified according to their measure:

- An angle of zero degree is called a null angle.
- An angle between 0° and 90° is called acute.
- An angle of 90° is called a right angle. (It is the angle formed when one makes a perfect right turn.)
- An angle between 90° and 180° is called obtuse.
- An angle of 180° is called a straight angle.
- An angle between 180° and 360° is called a reflex angle.
- An angle of 360° is called a PERIGON or a round angle.

Two angles are said to be congruent if they have the same measure. If two angles have measures summing to a right angle, then they are said to be complementary, and two angles are supplementary if their measures sum to a straight angle. Special names are also given to angles that appear in a diagram involving a TRANSVERSAL.

The angle of elevation of a point P above the ground relative to an observer at position Q on the ground is defined to be the angle between the line connecting Q to P, and the line connecting Q to the point on the ground directly below P. If P lies below ground level, then an angle of depression is computed similarly.

The angle between two VECTORs is computed via the DOT PRODUCT. Using TRIGONOMETRY one shows that the angle A between two lines in the plane of slopes m_1 and m_2, respectively, is given by:

$$\tan A = \frac{m_1 - m_2}{1 + m_1 m_2}$$

(This follows by noting that the angle the first line makes with the x-axis is $\tan^{-1}(m_1)$ and the angle of the second line is $\tan^{-1}(m_2)$. Thus the angle we seek is $A = \tan^{-1}(m_1) - \tan^{-1}(m_2)$. The trigonometric identity $\tan(\alpha - \beta) = \tan(\alpha - \beta) = \frac{\tan \alpha - \tan \beta}{1 + \tan \alpha \tan \beta}$ now establishes the result.) Consequently, if $m_1 m_2 = -1$, the lines are PERPENDICULAR. The angle between two intersecting curves in a plane is defined to be the angle between the TANGENT lines to the curves at the point of intersection.

The link between the measure of an angle and the length of arcs of a unit circle to define radian measure can be extended to associate a measure of "angle" with regions on a unit SPHERE. A SOLID ANGLE of a region is simply the measure of surface AREA of that region projected onto a unit sphere. Units of solid angle are called steradians. The full surface area of the sphere is 4π steradian.

See also BABYLONIAN MATHEMATICS; DIHEDRAL; SLOPE; TRIANGLE; TRISECTING AN ANGLE.

angle trisection   *See* TRISECTING AN ANGLE.

annulus (plural, annuli)   An annulus is the region between two CONCENTRIC circles in a plane. The AREA of the annulus is the difference of the areas of the two circles.

If a cyclist rides a perfect circle, the region between the tracks made by the front and rear wheels is an annulus. If the bicycle is r feet long (axel to axel), then, surprisingly, the area of this annulus is πr^2 feet squared, irrespective of the size of the circle the cyclist traces.

antidifferentiation (integration)   The process of finding a function with a given function as its DERIVATIVE is called antidifferentiation. For example, x^2 is an antiderivative of $2x$, since $\frac{d}{dx}(x^2) = 2x$. The MEAN VALUE THEOREM shows that two antiderivatives of the same function differ only by a constant. Thus all the antiderivatives of $2x$, for example, are functions of the form $x^2 + C$.

The antiderivative of a function $f(x)$ is denoted $\int f(x)dx$ and is called the indefinite integral of f. It is defined up to a constant, and so we write, for example, $\int 2x\, dx = x^2 + C$. (The constant C is referred to as a "constant of integration.") The notation is deliberately suggestive of a definite integral of a function, $\int_a^b f(x)dx$, for the area under the curve $y = f(x)$ over the INTERVAL $[a,b]$. The FUNDAMENTAL THEOREM OF CALCULUS shows that the two notions are intimately connected.

See also INTEGRAL CALCULUS.

antilogarithm   *See* LOGARITHM.

antipodal points (antipodes)   Two points on a SPHERE at the opposite ends of a diameter are said to be antipodal. For example, the north and south poles are antipodal points on the EARTH, as are any two points EQUIDISTANT from the equator, with longitudes differing by 180°.

The famous Borsuk-Ulam theorem, first conjectured by Stanislaw Ulam and then proved by Karol Borsuk in 1933, states:

> Let f be a continuous function that assigns two numerical values to each and every point on the surface of a sphere. Then there must exist two antipodal points which are assigned precisely the same pair of values.

An amusing interpretation reads:

> At any instant there exist two antipodal points on the Earth's surface of precisely the same air temperature and air pressure.

Although the proof of this theorem is difficult, a one-dimensional version of the result follows as an easy consequence of the INTERMEDIATE-VALUE THEOREM.

apex (plural, apices)   The point at the top of a POLYGON or a POLYHEDRON, such as the vertex of a triangle opposite its BASE or the vertex of a pyramid, is called the apex of the figure. The distance from the base of the figure to its apex is called the height of the figure.

Apollonius of Perga (ca. 262–190 B.C.E.) Greek *Geometry* Born in Perga, Greek Ionia, now Antalya, Turkey, Apollonius worked during the Golden Age of Greek mathematics and has been referred to throughout history as the Great Geometer. His famous work, *The Conics*, written in eight volumes, greatly influenced the development of mathematics. (The names ELLIPSE, PARABOLA, and HYPERBOLA for the three CONIC SECTIONS, for instance, are said to have been coined by Apollonius.) Copies of the first four volumes of this work, written in the original Greek, survive today. Arabic translations of the first seven volumes also exist.

Little is known of Apollonius's life other than what can be gleaned from incidental comments made in the prefaces of his books. As a young man it is known that he traveled to Alexandria to study with the followers of EUCLID, who then introduced him to the topic of conics.

The first volume of *The Conics* simply reviews elementary material about the topic and chiefly presents results already known to Euclid. Volumes two and three present original results regarding the ASYMPTOTES to hyperbolas and the construction of TANGENT lines to conics. While Euclid demonstrated a means, for instance, of constructing a circle passing through any three given points, Apollonius demonstrated techniques for constructing circles tangent to any three lines, or to any three circles, or to any three objects be they a combination of points, lines, or circles. Volumes four, five, six, and seven of his famous work are highly innovative and contain original results exploring issues of curvature, the construction of normal lines, and the construction of companion curves to conics. Apollonius also applied the theory of conics to solve practical problems. He invented, for instance, a highly accurate sundial, called a hemicyclium, with hour lines drawn on the surface of a conic section.

Apollonius also played a fundamental role in the development of Greek mathematical astronomy. He proposed a complete mathematical analysis of epicyclic motion (that is, the compound motion of circles rolling along circles) as a means to help explain the observed retrograde motion of the planets across the skies that had confused scholars of his time.

Apollonius's work was extraordinarily influential, and his text on the conics was deemed a standard reference piece for European scholars of the Renaissance. JOHANNES KEPLER, RENÉ DESCARTES, and SIR ISSAC NEWTON each made reference to *The Conics* in their studies.

See also CIRCUMCIRCLE; CYCLOID.

Apollonius's circle

Let A and B be two points of the plane and let k be a constant. Then the set of all points P whose distance from A is k times its distance from B is a CIRCLE. Any circle obtained this way is referred to as one of Apollonius's circles. Note that when $k = 1$ the circle is "degenerate," that is, the set of all points EQUIDISTANT from A and B is a straight line. When k becomes large, the Apollonius's circle approaches a circle of radius 1.

To see that the locus of points described this way is indeed a circle, set A to be the origin $(0,0)$, B to be the point $(k + 1, 0)$ on the x-axis, and P to be a general point with coordinates (x,y). The DISTANCE FORMULA then gives an equation of the form $\sqrt{x^2+y^2} = k\sqrt{(x-k-1)^2+y^2}$. This is equivalent to $\left(x - \dfrac{k^2}{k-1}\right)^2 + y^2 = \left(\dfrac{k}{k-1}\right)^2$, which is indeed the equation of a circle, one of radius $\dfrac{k}{k-1}$. APOLLONIUS OF PERGA used purely geometric techniques, however, to establish his claim.

Apollonius's theorem

If a, b, and c are the side-lengths of a triangle and a median of length m divides the third side into two equal lengths $c/2$ and $c/2$, then the following relation holds:

$$a^2 + b^2 = \frac{c^2}{2} + 2m^2$$

This result is known as Apollonius's theorem. It can be proved using two applications of the LAW OF COSINES as follows:

> Let B be the ANGLE between the sides of length a and c. Then $m^2 = a^2 + (c/2)^2 - ac\cos(B)$ and $b^2 = a^2 + c^2 - 2ac\cos(B)$. Solving for $ac\cos(B)$ in the first equation and substituting into the second yields the result.

See also MEDIAN OF A TRIANGLE.

apothem (short radius)

Any line segment from the center of a regular POLYGON to the midpoint of any of its sides is called an apothem. If the regular polygon has n sides, each one unit in length, then an exercise in TRIGONOMETRY shows that each apothem of the figure has length $r = \dfrac{1}{2\tan\left(\dfrac{180}{n}\right)}$.

An analog of PI (π) for a regular polygon is the RATIO of its PERIMETER to twice the length of its apothem. For a regular n-sided polygon, this ratio has value $n\tan(180/n)$. The SQUEEZE RULE shows that this quantity approaches the value π as n becomes large.

See also LONG RADIUS.

applied mathematics The study and use of the mathematical techniques to solve practical problems is called applied mathematics. The field has various branches including STATISTICS, PROBABILITY, mechanics, mathematical physics, and the topics that derive from them, but the distinction from PURE MATHEMATICS might not be sharp. For instance, the general study of VECTORS and VECTOR SPACES can be viewed as either an abstract study or a practical one if one later has in mind to use this theory to analyze force diagrams in mechanics.

Many research universities of today possess two departments of mathematics, one considered pure and the other applied. Students can obtain advanced degrees in either field.

approximation A numerical answer to a problem that is not exact but is sufficient for all practical purposes is called an approximation. For example, noting that 2^{10} is approximately 1,000 allows us to quickly estimate the value of $2^{100} = (2^{10})^{10}$ as 10^{30}. Students are often encouraged to use the fraction 22/7 as an approximate value for π.

Mathematicians use the notation "$\approx$" to denote approximately equal to. Thus, for example, $\pi \approx 22/7$.

Physicists and engineers often approximate functions by their TAYLOR SERIES with the higher-order terms dropped. For example, $\sin x \approx x - \dfrac{x^3}{3!} + \dfrac{x^5}{5!}$, at least for small values of x. The theory of INTEGRAL CALCULUS begins by approximating areas under curves as sums of areas of rectangles.

See also ERROR; FACTORIAL; NUMERICAL DIFFERENTIATION; NUMERICAL INTEGRATION.

Arabic mathematics Mathematical historians of today are grateful to the Arabic scholars of the past for preserving, translating, and honoring the great Indian, Greek, and Islamic mathematical works of the scholars before them, and for their own significant contributions to the development of mathematics. At the end of the eighth century, with the great Library of Alexandria destroyed, Caliph al-Ma'mun set up a House of Wisdom in Baghdad, Iraq, which became the next prominent center of learning and research, as well as the repository of important academic texts. Many scholars were employed by the caliph to translate the mathematical works of the past and develop further the ideas they contained. As the Islamic empire grew over the following seven centuries, the culture of intellectual pursuit also spread. Many scholars of 12th-century Europe, and later, visited the Islamic libraries of Spain to read the texts of the Arabic academics and to learn of the advances that had occurred in the East during the dark ages of the West. A significant amount of mathematical material was transmitted to Europe via these means.

One of the first Greek texts to be translated at the House of Wisdom was EUCLID's famous treatise, *THE ELEMENTS*. This work made a tremendous impact on the Arab scholars of the period, and many of them, when conducting their own research, formulated theorems and proved results precisely in the style of Euclid. Members of the House of Wisdom also translated the works of ARCHIMEDES OF SYRACUSE, DIOPHANTUS OF ALEXANDRIA, MENELAUS OF ALEXANDRIA, and others, and so they were certainly familiar with all the great Greek advances in the topics of GEOMETRY, NUMBER THEORY, mechanics, and analysis. They also translated the works of Indian scholars, ĀRYABHATA and BHĀSKARA, for instance, and were familiar with the theory of TRIGONOMETRY, methods in astronomy, and further topics in geometry and number theory. Any Arab scholar who visited the House of Wisdom had, essentially, the entire bulk of human mathematical knowledge available to him in his own language.

Arab mathematician MUHAMMAD IBN MŪSĀ AL-KHWĀRIZMĪ (ca. 800) wrote a number of original texts that were enormously influential. His first piece simply described the decimal place-value system he had learned from Indian sources. Three hundred years later, when translated into Latin, this work became the primary source for Europeans who wanted to learn the new system for writing and manipulating numbers. But more important was al-Khwārizmī's piece *Hisab al-jabr w'al-muqābala* (Calculation by restoration and reduction), from which the topic of "algebra" ("al-jabr") arose. Al-Khwārizmī was fortunate to have all sources of mathematical knowledge available to him. He began to see that the then-disparate notions of "number" and "geometric magnitude" could be unified as one whole by developing the concept of algebraic objects. This represented a significant departure from Greek thinking, in which mathematics is synonymous with geometry. Al-Khwārizmī's insight provided a means to study both arithmetic and geometry under a single framework, and his methods of algebra paved the way for significant developments in mathematics of much broader scope than ever previously envisioned.

The mathematician al-Mahani (ca. 820) developed refined approaches for reducing geometric problems to algebraic ones. He showed, in particular, that the famous problem of DUPLICATING THE CUBE is essentially an algebraic issue. Other scholars brought rigor to the subject by proving that certain popular, but complicated, algebraic methods were valid. These scholars were comfortable manipulating POLYNOMIALs and developed rules for working with EXPONENTs, They solved linear and QUADRATIC equations, as well as various SYSTEMS OF EQUATIONS. Surprisingly, no one of the time thought to ease matters by using symbols to represent quantities: all equations and all manipulations were described fully in words each and every time they were employed.

With quadratic equations well understood, the scholar OMAR KHAYYAM (ca. 1048–1131) attempted to develop methods of solving degree-three equations. Although he was unable to develop general algebraic methods for this task, he did find ingenious geometric techniques for solving certain types of cubics with the aid of CONIC SECTIONS. He was aware that such equations could have more than one solution.

In number theory, Thabit ibn Qurra (ca. 836–901) found a beautiful method for generating AMICABLE NUMBERS. This technique was later utilized by al-Farisi (ca. 1260–1320) to yield the pair 17,296 and 18,416, which today is usually attributed to LEONHARD EULER (1707–83). In his writing, Omar Khayyam referred to earlier Arab texts, now lost, that discuss the equivalent of PASCAL's TRIANGLE and its connections to the BINOMIAL THEOREM. The mathematician al-Haytham (ca. 965–1040) attempted to classify all even PERFECT NUMBERs.

Taking advantage of the ease of the Indian system of decimal place-value representation, Arabic scholars also made great advances in numeric computations. The great 14th-century scholar JAMSHID AL-KASHI developed effective methods for extracting the nth root of a number, and evaluated π to a significant number of decimal places. Scholars at the time also developed effective methods for computing trigonometric tables and techniques for making highly accurate computations for the purposes of astronomy.

On a theoretical note, scholars also advanced the general understanding of trigonometry and explored problems in spherical geometry. They also investigated the philosophical underpinnings of geometry, focusing, in particular, on the role the famous PARALLEL POSTULATE plays in the theory. Omar Khayyam, for instance, attempted to prove the parallel postulate—failing, of course—but did accidentally prove results about figures in non-Euclidean geometries along the way. The mathematician Ibrahim ibn Sinan (908–946) also introduced a method of "integration" for calculating volumes and areas following an approach more general than that developed by Archimedes of Syracuse (ca. 287–212 B.C.E.). He also applied his approach to the study of CONIC SECTIONS and to optics.

See also BASE OF A NUMBER SYSTEM; HISTORY OF EQUATIONS AND ALGEBRA (essay); HISTORY OF GEOMETRY (essay); HISTORY OF TRIGONOMETRY (essay).

arc Part of a continuous curve between two given points on the curve is called an arc of the curve. In particular, two points on a CIRCLE determine two arcs. If the circumference of the circle is divided by them into two unequal parts, then the smaller portion is usually called the minor arc of the circle and the larger the major arc.

Archimedean spiral *See* SPIRAL OF ARCHIMEDES.

Archimedes of Syracuse (287–212 B.C.E.) Greek *Geometry, Mechanics* Born in the Greek colony of Syracuse in Sicily, Archimedes is considered one of the greatest mathematicians of all time. He made considerable contributions to the fields of planar and solid GEOMETRY, hydrostatics, and mechanics. In his works *Measurement of a Circle* and *Quadrature of the Parabola*, Archimedes solved difficult problems of mensuration in planar geometry by inventing an early technique of INTEGRAL CALCULUS, which he called the "method of exhaustion." This allowed him to compute areas and lengths of certain curved figures. Later, in his works *On the Sphere and Cylinder* and *On Conoids and Spheroids*, he applied this technique to also compute the volume and surface area of the sphere and other solid objects. In his highly original work *On Floating Bodies*, Archimedes developed the mathematics of hydrostatics and equilibrium, along with the LAW OF THE LEVER, the notion of specific gravity, and techniques for computing the CENTER OF GRAVITY of a

A 1547 woodcut depicting Archimedes' realization that a tub of water can be used to compute the volumes and densities of solid figures *(Photo courtesy of ARPL/Topham/The Image Works)*

variety of bodies. In mathematics, Archimedes also developed methods for solving cubic equations, approximating square roots, summing SERIES, and, in *The Sand Reckoning*, developed a notation for representing extremely large numbers.

Except for taking time to study at EUCLID's school in Alexandria, Archimedes spent his entire life at the place of his birth. He was a trusted friend of the monarch of the region, Hiero, and his son Gelon, and soon developed a reputation as a brilliant scientist who could solve the king's most troublesome problems. One famous story asserts that the king once ordered a goldsmith to make him a crown, and supplied the smith the exact amount of metal to make it. Upon receiving the newly forged crown, Hiero suspected the smith of ill doing, substituting some cheaper silver for the gold, even though the crown did have the correct weight. He could not prove his suspicions were correct, however, and so brought the problem to Archimedes. It is said that while bathing and observing the water displaced by his body Archimedes realized, and proved, that the weight of an object suspended in liquid decreases in proportion to the weight of the liquid it displaces. This principle, today known as Archimedes' principle, provided Archimedes the means to indeed prove that the crown was not of solid gold. (It is also said that Archimedes was so excited upon making this discovery that he ran naked through the streets shouting, "Eureka! Eureka!")

Archimedes is also purported to have said, "Give me a place to stand and I shall move the Earth." Astonished by the claim, King Hiero asked him to prove it. Archimedes had, at this time, discovered the principles of the levers and pulleys, and set about constructing a mechanical device that allowed him, single-handedly, to launch a ship from the harbor that was too large and heavy for a large group of men to dislodge.

Dubbed a master of invention, Archimedes also devised a water-pumping device, now known as the Archimedes screw and still used in many parts of the world today, along with many innovative machines of war that were used in the defense of Sicily during the not-infrequent Roman invasions. (These devices included parabolic mirrors to focus the rays of the sun to burn advancing ships from shore, catapult devices, and spring-loaded cannons.) But despite the fame he received for his mechanical inventions, Archimedes believed that pure mathematics was the only worthy pursuit. His accomplishments in mathematics were considerable.

Archimedes of Syracuse, regarded as one of the greatest scientists of all time, pioneered work in planar and solid geometry, mechanics, and hydrostatics. *(Photo courtesy of the Science Museum, London/Topham-HIP/The Image Works)*

By bounding a circle between two regular polygons and calculating the ratio the perimeter to diameter of each, Archimedes found one of the earliest estimates for the value of π, bounding it between the values 3 10/71 and 3 1/7. (This latter estimate, usually written as 22/7, is still widely used today.) Archimedes realized that by using polygons with increasingly higher numbers of sides yielded better and better approximations, and that by "exhausting" all the finite possibilities, the true value of π would be obtained. Archimedes also used this method of exhaustion to demonstrate that the length of any segment of a parabola is 4/3 times the area of the triangle with the same base and same height.

By comparing the cross-sectional areas of parallel slices of a sphere with the slices of a cylinder that encloses the sphere, Archimedes demonstrated that the volume of a sphere is 2/3 that of the cylinder. The volume of the sphere then follows: $V = (2/3)(2r \times \pi r^2) = (4/3)\pi r^3$. (Here r is the radius of the sphere.) Archimedes regarded this his greatest mathematical achievement, and in his honor, the figures of a cylinder and an inscribed sphere were drawn on his tombstone.

Archimedes also computed the surface area of a sphere as four times the area of a circle of the same radius of the sphere. He did this again via a method of exhaustion, by imagining the sphere as well approximated by a covering of flat tiny triangles. By drawing lines connecting each vertex of a triangle to the center of the sphere, the volume of the figure is thus divided into a collection of triangular pyramids. Each pyramid has volume one-third its base times it height (essentially the radius of the sphere), and the sum of all the base areas represents the surface area of the sphere. From the formula for the volume of the sphere, the formula for its surface area follows.

One cannot overstate the influence Archimedes has had on the development of mathematics, mechanics, and science. His computations of the surface areas and volumes of curved figures provided insights for the development of 17th-century calculus. His understanding of Euclidean geometry allowed him to formulate several axioms that further refined the logical underpinnings of the subject, and his work on fluids and mechanics founded the field of hydrostatics. Scholars and noblemen of his time recognized both the theoretical and practical importance of his work. Sadly, Archimedes died unnecessarily in the year 212 B.C.E. During the conquest of Syracuse by the Romans, it is

A 19th-century engraving of Archimedes' water screw, a device for pumping water *(Photo courtesy of AAAC/Topham/The Image Works)*

said that a Roman soldier came across Archimedes concentrating on geometric figures he had drawn in the sand. Not knowing who the scholar was, or what he was doing, the soldier simply killed him.

Archytas of Tarentum (ca. 428–350 B.C.E.) Greek *Geometry, Philosophy* The Greek scholar Archytas of Tarentum was the first to provide a solution to the classic DUPLICATING THE CUBE problem of antiquity. By reducing the challenge to one of constructing certain

ratios and proportions, Archytas developed a geometric construct that involved rotating semicircles through certain angles in three-dimensional space to produce a length essentially equivalent to the construct of the cube root of two. (Creating a segment of this length is the chief stumbling block to solving the problem.) Although his innovative solution is certainly correct, it uses tools beyond what is permissible with straightedge and compass alone. In the development of his solution, Archytas identified a new mean between numbers, which he called the HARMONIC MEAN.

Archytas lived in southern Italy during the time of Greek control. The region, then called Magna Graecia, included the town of Tarentum, which was home to members of the Pythagorean sect. Like the Pythagoreans, Archytas believed that mathematics provided the path to understanding all things. However, much to the disgust of the Pythagoreans, Archytas applied his mathematical skills to solve practical problems. He is sometimes referred to as the Founder of Mechanics and is said to have invented several innovative mechanical devices, including a mechanical bird and an innovative child's rattle.

Only fragments of Archytas's original work survive today, and we learn of his mathematics today chiefly through the writings of later scholars. Many results established by Archytas appear in EUCLID's famous text THE ELEMENTS, for instance.

arc length   To measure the length of a curved path, one could simply lay a length of string along the path, pull it straight, and measure its length. This determines the arc length of the path. In mathematics, if the curve in question is continuous and is given by a formula $y = f(x)$, for $a < x < b$ say, then INTEGRAL CALCULUS can be used to find the arc length of the curve. To establish this, first choose a number of points $(x_1, y_i, ..., (x_n, y_n)$ along the curve and sum the lengths of the straight-line segments between them. Using the DISTANCE FORMULA, this gives an approximate value for the length s of the curve:

$$s \approx \sum_{i=1}^{n} \sqrt{\left(x_i - x_{i-1}\right)^2 + \left(y_i - y_{i-1}\right)^2}$$

Rewriting yields:

$$s \approx \sum_{i=1}^{n} \sqrt{1 + \left(\frac{y_i - y_{i-1}}{x_i - x_{i-1}}\right)^2} \left(x_i - x_{i-1}\right)$$

The MEAN-VALUE THEOREM shows that for each i there is a value c_i between x'_{i-1} and x_i so that $\frac{y_i - y_{i-1}}{x_i - x_{i-1}} = f'(c_i)$, and so the length of the curve is well approximated by the formula:

$$s \approx \sum_{i=1}^{n} \sqrt{1 + \left(f'(c_i)\right)^2} \left(x_i - x_{i-1}\right)$$

Of course, taking more and more points along the curve gives better and better approximations. In the limit, then, the true length of the curve is given by the formula:

$$s = \lim_{n \to \infty} \sum_{i=1}^{n} \sqrt{1 + \left(f'(c_i)\right)^2} \left(x_i - x_{i-1}\right)$$

This is precisely the formula for the integral of the function $\sqrt{1 + \left(f'(x)\right)^2}$ over the domain in question. Thus we have:

The arc length of a continuous curve $y = f(x)$ over the interval $[a,b]$ is given by $s = \int_a^b \sqrt{1 + \left(f'(x)\right)^2} \, dx$

Alternatively, if the continuous curve is given by a set of PARAMETRIC EQUATIONS $x = x(t)$ and $y = y(t)$, for $a < t < b$ say, then choosing a collection of points along the curve, given by $t_1, ..., t_n$ say, making an approximation to the curve's length, and taking a limit yields the formula:

$$s = \lim_{n \to \infty} \sqrt{\left(x(t_i) - x(t_{i-1})\right)^2 + \left(y(t_i) - y(t_{i-1})\right)^2}$$

$$= \lim_{n \to \infty} \sqrt{\left(\frac{x(t_i) - x(t_{i-1})}{t_i - t_{i-1}}\right)^2 + \left(\frac{y(t_i) - y(t_{i-1})}{t_i - t_{i-1}}\right)^2} \left(t_i - t_{i-1}\right)$$

$$= \int_a^b \sqrt{\left(\frac{dx}{dt}\right)^2 + \left(\frac{dy}{dt}\right)^2} \, dt$$

In a similar way one can show that if the continuous curve is given in POLAR COORDINATES by formulae and $x = r(\theta)\cos(\theta)$ and $y = r(\theta)\sin(\theta)$, for $a < \theta < b$, then the arc length of the curve is given by:

$$s = \int_a^b \sqrt{r^2 + \left(\frac{dr}{d\theta}\right)^2} \, d\theta$$

The presence of square-root signs in the integrands often makes these integrals very difficult, if not impossible, to solve. In practice, one must use numerical techniques to approximate integrals such as these.

See also NUMERICAL INTEGRATION.

area   Loosely speaking, the area of a geometric figure is the amount of space it occupies. Such a definition appeals to intuitive understanding. In general, however, it is very difficult to explain precisely just what it is we mean by "space" and the "amount" of it occupied. This is a serious issue. (See Banach-Tarski paradox on following page.)

As a starting point, it seems reasonable to say, however, that a 1 × 1 square should have "area" one. We call this a basic unit of area. As four of these basic units fit snugly into a square with side-length two, without overlap, we say then that a 2 × 2 square has area four. Similarly a 3 × 3 square has area nine, a 4 × 4 square area 16, and so on.

A 3 × 6 rectangle holds 18 basic unit squares and so has area 18. In general, a rectangle that is l units long and w units wide, with both l and w whole numbers, has area $l \times w$:

$$\text{area of a rectangle} = \text{length} \times \text{width}$$

This is a fundamental formula. To put the notion of area on a sound footing, we use this formula as a defining law: the area of *any* rectangle is to be the product of its length and its width.

Although it is impossible to fit a whole number of unit squares into a rectangle that is 5 3/4 units long and $\sqrt{7}$ units wide, for example, we declare, nonetheless, that the area of such a rectangle is the product of these two numbers. (This agrees with our intuitive idea that, with the aid of scissors, about $5\ 3/4 \times \sqrt{7} \approx 15.213$ unit squares will fit in this rectangle.)

From this law, the areas of other geometric shapes follow. For example, the following diagram shows that the area of an acute TRIANGLE is half the area of the rectangle that encloses it. This leads to the formula:

$$\text{area of a triangle} = {}^1\!/_2 \times \text{base} \times \text{height}$$

This formula also holds for obtuse triangles.

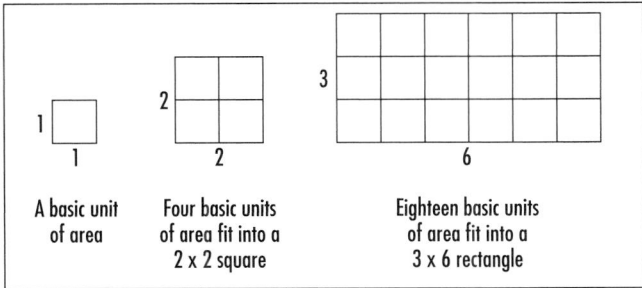

A basic unit of area

Four basic units of area fit into a 2 x 2 square

Eighteen basic units of area fit into a 3 x 6 rectangle

Area

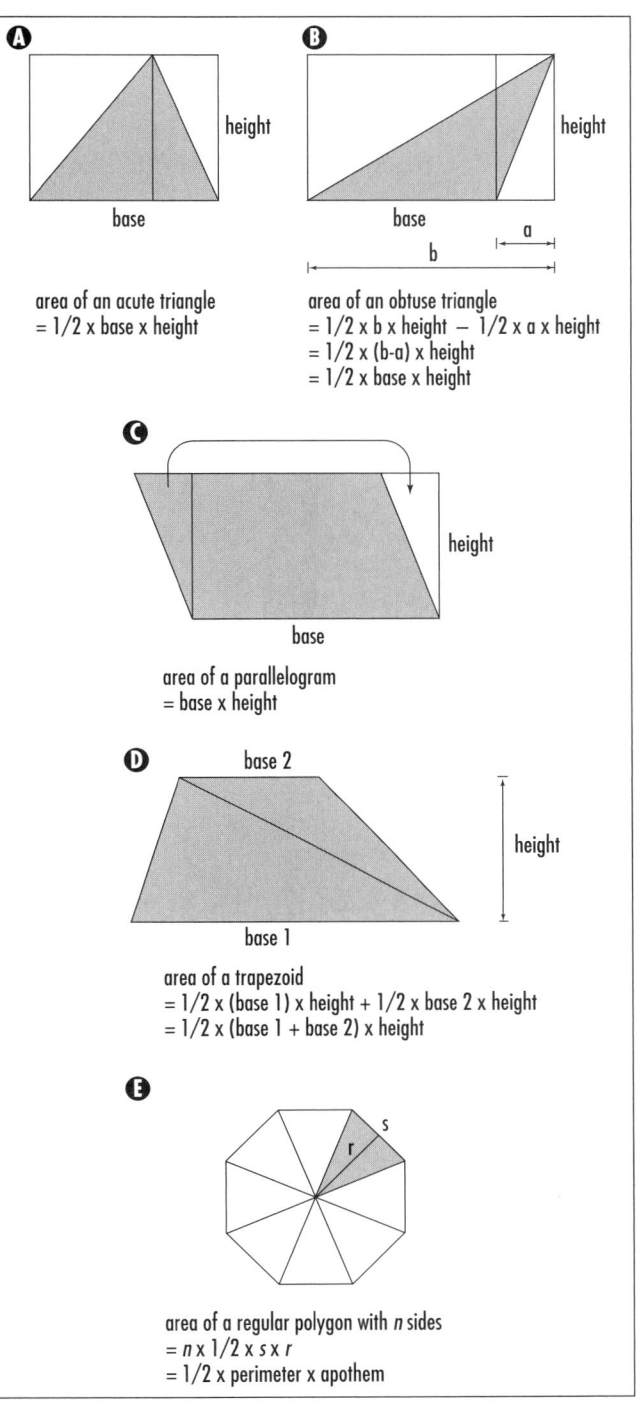

The areas of basic shapes

By rearranging pieces of a PARALLELOGRAM, we see that its area is given by the formula:

$$\text{area of a parallelogram} = \text{base} \times \text{height}$$

In general we can calculate the area of any POLYGON as the sum of the areas of the triangles that subdivide it. For example, the area of a TRAPEZOID is the sum of the areas of two triangles, and the area of a regular POLYGON with *n* sides is the sum of the areas of *n* triangles.

Curved Figures

It is also possible to compute the area of curved figures. For example, slicing a circle into wedged-shape pieces and rearranging these slices, we see that the area of a circle is close to being the area of a rectangle of length half the circumference and of width *r*, the radius of the circle.

If we work with finer and finer wedged-shape pieces, the approximation will better approach that of a true rectangle. We conclude that the area of a circle is indeed that of this ideal rectangle:

$$\text{area of a circle} = \tfrac{1}{2} \times \text{circumference} \times r$$

(Compare this with the formula for the area of a regular polygon.) As PI (π) is defined as the ratio of the circumference of a circle to its diameter, $\pi = \dfrac{\text{circumference}}{2r}$, the area of a circle can thus be written: area $= \tfrac{1}{2} \times 2\pi r \times r$. This leads to the famous formula:

$$\text{area of a circle} = \pi r^2$$

The methods of INTEGRAL CALCULUS allow us to compute areas of other curved shapes. The approach is analogous: approximate the shape as a union of rectangles, sum the areas of the rectangular pieces, and take the LIMIT of the answers obtained as you work with finer and finer approximations.

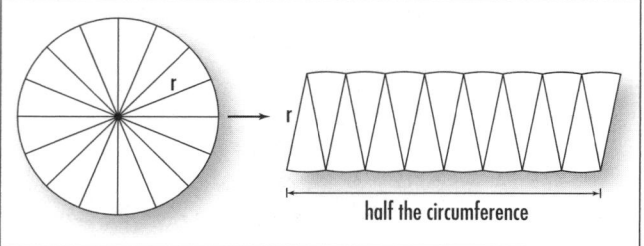

Establishing the area of a circle

Theoretical Difficulties

Starting with the principle that a fundamental shape, in our case a rectangle, is *asserted* to have "area" given by a certain formula, a general theory of area for other geometric shapes follows. One can apply such an approach to develop a measure theory for measuring the size of other sets of objects, such as the notion of the surface area of three-dimensional solids, or a theory of VOLUME. One can also develop a number of exotic applications.

Although our definition for the area of a rectangle is motivated by intuition, the formula we developed is, in some sense, arbitrary. Defining the area of a rectangle as given by a different formula could indeed yield a different, but consistent, theory of area.

In 1924 STEFAN BANACH and Alfred Tarski stunned the mathematical community by presenting a mathematically sound proof of the following assertion:

> It is theoretically possible to cut a solid ball into nine pieces, and by reassembling them, without ever stretching or warping the pieces, form TWO solid balls, each exactly the same size and shape as the original.

This result is known as the Banach-Tarski paradox, and its statement—proven as a mathematical fact—is abhorrent to our understanding of how area and volume should behave: the volume of a finite quantity of material should not double after rearranging its pieces! That our intuitive understanding of area should eventually lead to such a perturbing result was considered very disturbing.

What mathematicians have come to realize is that "area" is not a well-defined concept: not every shape in a plane can be assigned an area (nor can every solid in three-dimensional space be assigned a volume). There exist certain nonmeasurable sets about which speaking of their area is meaningless. The nine pieces used in the Banach-Tarski paradox turn out to be such nonmeasurable sets, and so speaking of their volume is invalid. (They are extremely jagged shapes, FRACTAL in nature, and impossible to physically cut out.) In particular, interpreting the final construct as "two balls of equal *volume*" is not allowed.

Our simple intuitive understanding of area works well in all practical applications. The material presented in a typical high-school and college curriculum, for example, is sound. However, the Banach-Tarski paradox points out that extreme care must be taken when exploring the theoretical subtleties of area and volume in greater detail.

See also SCALE.

Argand, Jean Robert (1768–1822) Swiss *Complex number theory* Born on July 8, 1768, in Geneva, Switzerland, amateur mathematician Jean Argand is remembered today for his famous geometrical interpretation for COMPLEX NUMBERS. An ARGAND DIAGRAM uses two perpendicular axes, one representing a real number line, the second a line of purely complex numbers, to represent complex numbers as points in a plane.

It is not well known that Argand, in fact, was not the first to consider and publish this geometric approach to complex numbers. The surveyor Casper Wessel (1745–1818) submitted the same idea to the Royal Danish Academy in 1797, but his work went unnoticed by the mathematics community. At the turn of the century, Argand independently began to interpret the complex number *i* geometrically as a rotation through 90°. He expounded on the convenience and fruitfulness of this idea in a small book, *Essai sur une manière de représenter les quantités imaginaires dans les constructions géometriques* (Essay on a method for representing imaginary quantities through a geometric construction), which he published privately, at his own expense, in 1806. He never wrote his name in the piece, and so it was impossible to identify the author. By chance, French mathematician Jacques Français came upon the small publication and wrote about the details of the work in an 1813 article, "A Memoir on the Geometric Representation of Imaginary Numbers," published in the *Annales de Mathématiques*. He requested that the unknown originator of the ideas come forward and receive credit for the work. Argand made himself known by submitting his own article to the same journal, presenting a slightly modified and improved approach to his methods. Although historians have since discovered that the mathematicians JOHN WALLIS (1616–1703) and CARL FRIEDRICH GAUSS (1777–1855) each considered their own geometric interpretations of complex numbers, Argand is usually credited as the discoverer of this approach.

Argand was the first to develop the notion of the MODULUS of a complex number. It should also be noted that Argand also presented an essentially complete proof of the FUNDAMENTAL THEOREM OF ALGEBRA in his 1806 piece, but has received little credit for this accomplishment. Argand was the first to state, and prove, the theorem in full generality, allowing all numbers involved, including the coefficients of the polynomial, to be complex numbers.

Argand died on August 13, 1822, in Paris, France. Although not noted as one of the most outstanding mathematicians of his time, Argand's work certainly shaped our understanding of complex number theory. The Argand diagram is a construct familiar to all advanced high-school mathematics students.

Argand diagram (complex plane) *See* COMPLEX NUMBERS.

argument In the fourth century B.C.E., Greek philosopher ARISTOTLE made careful study of the structure of reasoning. He concluded that any argument, i.e., a reasoned line of thought, consists, essentially, of two basic parts: a series of PREMISEs followed by a conclusion. For example:

If today is Tuesday, then I must be in Belgium.
I am not in Belgium.
Therefore today is not Tuesday.

is an argument containing two premises (the first two lines) and a conclusion. An argument is valid if the conclusion is true when the premises are assumed to be true.

Any argument has the general form:

If [premise 1 *AND* premise 2 *AND* premise 3 *AND*...], *then* [Conclusion]

Using the symbolic logic of FORMAL LOGIC and TRUTH TABLES, the above example has the general form:

$$p \rightarrow q$$
$$\neg q$$
$$Therefore \ \neg p$$

The argument can thus be summarized: $((p \rightarrow q) \wedge (\neg q)) \rightarrow (\neg p)$.

One can check with the aid of a truth table that this statement is a tautology, that is, it is a true statement irrespective of the truth-values of the component statements p and q. (In particular, it is true when both premises have truth-value T.) Thus the argument presented above is indeed a valid argument.

An argument that does not lead to a tautology in symbolic logic is invalid. For example,

If a bird is a crow, then it is black.
This bird is black.
Therefore it is a crow.

is an invalid argument: $((p \to q) \land q) \to p$ is not a tautology. (Informally, we can assert that a black bird need not be a crow.)

The following table contains the standard forms of argument commonly used, along with some invalid arguments commonly used in error.

Valid Arguments	Invalid Arguments
Direct Reasoning (modus ponens)	Fallacy of the Converse
$p \to q$	$p \to q$
p	q
therefore q	therefore p
Contrapositive Reasoning (modus tollens)	Fallacy of the Inverse
$p \to q$	$p \to q$
$\neg q$	$\neg p$
therefore $\neg p$	therefore $\neg q$
Disjunctive Reasoning	Misuse of Disjunctive Reasoning
$p \lor q$ $p \lor q$	$p \lor q$ $p \lor q$
$\neg p$ $\neg q$	p q
therefore q therefore p	therefore $\neg q$ therefore $\neg p$
Transitive Reasoning	
$p \to q$	
$q \to r$	
therefore $p \to r$	

In the mid-1700s LEONHARD EULER invented an elegant way to determine the validity of syllogisms, that is, arguments whose premises contain the words *all, some,* or *no.* For example,

All poodles are dogs.
All dogs bark.
Therefore all poodles bark.

is a syllogism, and Euler would depict such an argument as a diagram of three circles, each representing a set mentioned in one of the premises. The validity of the argument is then readily apparent:

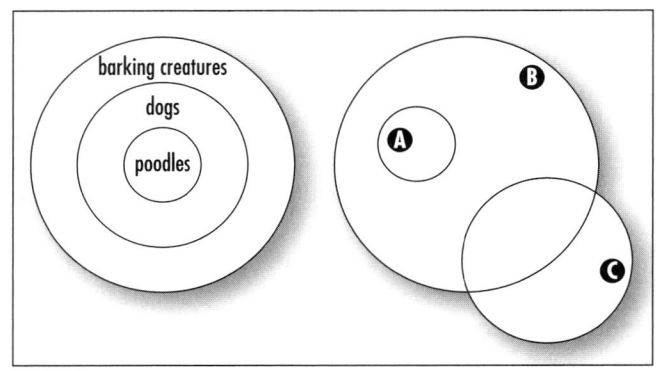

Euler diagrams

An argument of the following structure, for example, can be demonstrated as invalid by arranging circles as shown:

All *A*s are *B*s.
Some *B*s are *C*s.
Therefore, some *A*s are *C*s.

Any diagram used to analyze the validity of an argument is called an Euler diagram.

See also DEDUCTIVE/INDUCTIVE REASONING; QUANTIFIER.

Aristotle (384–322 B.C.E.) Greek *Logic, Philosophy, Physics, Medicine* Born in Stagirus, Macedonia, Aristotle is remembered in mathematics for his systematic study of deductive logic. In laying down the foundations of FORMAL LOGIC, Aristotle identified the fundamental LAWS OF THOUGHT, the laws of reasoning, and the fundamental principles that lie at the heart of any mathematical ARGUMENT. His work in this area so deeply affected the attitudes and approaches of scientific thinking that Western intellectual culture as a whole is often referred to as Aristotelian.

At age 17, Aristotle joined PLATO's Academy in Athens and remained there for 20 years. He worked closely with Plato, and also EUDOXUS, nephew of Plato. The equivalent of a modern-day research university, the Academy brought together scholars from all disciplines and provided a culturally rich environment that encouraged learning and promoted the advancement of knowledge. Due to internal politics, however, Aristotle decided to leave the Academy after Plato's death in 347 B.C.E.

After tutoring the heir of King Phillip II, the future Alexander the Great, for a number of years, Aristotle returned to Athens in 335 B.C.E. to found his own school, the Lyceum. He intended the school to be as broad-based as possible, exploring a wide range of subjects but with prominence given to the study of the natural world. While at the Lyceum, Aristotle wrote 22 texts covering an astonishing range of topics: logic, physics, astronomy, meteorology, theology, metaphysics, ethics, rhetoric, poetics, and more. He founded a theory of kinematics, a study of space, time, and motion, and he established principles of physics that remained unchallenged for two millennia.

With regard to mathematics, Aristotle is remembered for his writings in logic, a subject he identified as the basis of all scientific thought. He invented the syllogism, a form of argument that comes in three parts: a major premise, a minor premise, and a conclusion. Although a straightforward notion for us today, this work represented a first fundamental step toward understanding the structure of reasoning. He presented the following line of thought as an example of a syllogism:

Every Greek is a person.
Every person is mortal.
Therefore every Greek is mortal.

Aristotle recognized that any line of reasoning following this form is logically valid by virtue of its structure, not its content. Thus the argument:

Every planet is made of cheese.
Every automobile is a planet.
Therefore every automobile is made of cheese.

for example, is a valid argument, even though the validity of the premises may be in question. Removing content from structure was a sophisticated accomplishment. Aristotle called his field of logic "analytics" and described his work on the subject in his book *Prior and Posterior Analytics*. He wanted to demonstrate the effectiveness of logical reasoning in understanding science.

Aristotle also discussed topics in the philosophy of mathematics. He argued, for instance, that an unknowable such as "infinity" exists only as a potentiality, and never as a completed form. Although, for example, from any finite set of prime numbers one can always construct one more, speaking of the set of prime numbers as a single concept, he argued, is meaningless.

(Today we say that Aristotle accepted the "potentially infinite" but rejected the "actual infinite.")

It is recorded that Aristotle would often walk through the gardens of the Lyceum while lecturing, forcing his pupils to follow. His students became known as the peripatetics, the word *peripatetic* meaning "given to walking." Copies of Aristotle's lecture notes taken by the peripatetics were regarded as valuable scholarly documents in their own right and have been translated, copied, and distributed across the globe throughout the centuries.

Political unrest forced Aristotle to leave Athens again in 322 B.C.E. He died soon afterward at the age of 62 of an unidentified stomach complaint.

Aristotle's analysis of critical thinking literally shaped and defined the nature of logical thought we exercise today in any academic pursuit. One cannot exaggerate the profundity of Aristotle's influence. By identifying valid modes of thought and clarifying the principles of logical reasoning, Aristotle provided the tools necessary for sensible reasoning and astute systematic thinking. These are skills today deemed fundamental to basic goals of all levels of education.

See also CARDINALITY; DEDUCTIVE/INDUCTIVE REASONING; PARADOX.

arithmetic   The branch of mathematics concerned with computations using numbers is called arithmetic. This can involve a number of specific topics—the study of operations on numbers, such as ADDITION, MULTIPLICATION, SUBTRACTION, DIVISION, and SQUARE ROOTs, needed to solve numerical problems; the methods needed to change numbers from one form to another (such as the conversion of fractions to decimals and vice versa); or the abstract study of the NUMBER SYSTEMS, NUMBER THEORY, and general operations on sets as defined by GROUP THEORY and MODULAR ARITHMETIC, for instance.

The word *arithmetic* comes from the Greek work *arithmetiké*, constructed from *arithmós* meaning "number" and *techné* meaning "science." In the time of ancient Greece, the term *arithmetic* referred only to the theoretical work about numbers, with the word *logistic* used to describe the practical everyday computations used in business. Today the term *arithmetic* is used in both contexts. (The word *logistics* is today a predominantly military term.)

See also BASE OF A NUMBER SYSTEM; FUNDAMENTAL THEOREM OF ARITHMETIC; ORDER OF OPERATION.

arithmetic–geometric-mean inequality *See* MEAN.

arithmetic mean *See* MEAN.

arithmetic sequence (arithmetic progression) A SE-
QUENCE of numbers in which each term, except the first,
differs from the previous one by a constant amount is
called an arithmetic sequence. The constant difference
between terms is called the common difference. For
example, the sequence 4, 7, 10, 13, ... is arithmetic with
common difference 3. An arithmetic sequence can be
thought of as "linear," with the common difference
being the SLOPE of the linear relationship.

An arithmetic sequence with first term a and common difference d has the form:

$$a, a + d, a + 2d, a + 3d, \ldots$$

The nth term a_n of the sequence is given by $a_n = a + (n - 1)d$. (Thus, the 104th term of the arithmetic
sequence 4,7,10,13,..., for example, is $a_{104} = 4 + [103 \times 3] = 313$.)

The sum of the terms of an arithmetic sequence is
called an arithmetic series:

$$a + (a + d) + (a + 2d) + (a + 3d)+\ldots$$

The value of such a sum is always infinite unless the
arithmetic sequence under consideration is the constant
zero sequence: 0,0,0,0,...

The sum of a *finite* arithmetic sequence can be
readily computed by writing the sum both forward and
backward and summing column-wise. Consider, for
example, the sum $4 + 7 + 10 + 13 + 16 + 19 + 22 + 25 + 28 + 31$. Call the answer to this problem S. Then:

$$4 + 7 + 10 + 13 + 16 + 19 + 22 + 25 + 28 + 31 = S$$
$$31 + 28 + 25 + 22 + 19 + 16 + 13 + 10 + 7 + 4 = S$$

and adding columns yields:

$$35 + 35 + 35 + 35 + 35 + 35 + 35 + 35 + 35 + 35 = 2S$$

That is, $2S = 10 \times 35 = 350$, and so $S = 175$. In general,
this method shows that the sum of n equally spaced
numbers in arithmetic progression, $a + b + \ldots + y + z$, is
n times the average of the first and last terms of the sum:

$$S = n \times \frac{a + z}{2}$$

It is said that CARL FRIEDRICH GAUSS (1777–1855), as
a young student, astonished his mathematics instructor
by computing the sum of the numbers 1 though 100 in
a matter of seconds using this method. (We have $1 + 2 + \ldots + 100 = 100 \times \frac{1 + 100}{2} = 5,050$.)

See also GEOMETRIC SEQUENCE; SERIES.

arithmetic series *See* ARITHMETIC SEQUENCE.

array An ordered arrangement of numbers or symbols
is called an array. For example, a VECTOR is a one-
dimensional array: it is an ordered list of numbers. Each
number in the list is called a component of the vector. A
MATRIX is a two-dimensional array: it is a collection of
numbers arranged in a finite grid. (The components of
such an array are identified by their row and column
positions.) Two arrays are considered the same only if
they have the same number of rows, the same number of
columns, and all corresponding entries are equal. One
can also define three- and higher-dimensional arrays.

In computer science, an array is called an identifier,
and the location of an entry is given by a subscript. For
example, for a two-dimensional array labeled A, the
entry in the second row, third column is denoted A_{23}.
An n-dimensional array makes use of n subscripts.

Āryabhata (ca. 476–550 C.E.) *Indian Trigonometry,
Number theory, Astronomy* Born in Kusumapura,
now Patna, India, Āryabhata (sometimes referred to as
Āryabhata I to distinguish him from the mathematician
of the same name who lived 400 years later) was the
first Indian mathematician of note whose name we
know and whose writings we can study. In the section
Ganita (Calculation) of his astronomical treatise
Āryabhatiya, he made fundamental advances in the theory of TRIGONOMETRY by developing sophisticated
techniques for finding and tabulating lengths of half-
chords in circles. This is equivalent to tabulating values
of the sine function. Āryabhata also calculated the
value of π to four decimal places (π ≈ 62,832/20,000 =
3.1416) and developed rules for extracting square and

cube roots, for summing ARITHMETIC SERIES, and finding SUMS OF POWERS.

As an astronomical treatise, *Āryabhatiya* is written as a series of 118 verses summarizing all Hindu mathematics and astronomical practices known at that time. A number of sections are purely mathematical in context and cover the topics of ARITHMETIC, TRIGONOMETRY, and SPHERICAL GEOMETRY, as well as touch on the theories of CONTINUED FRACTIONS, QUADRATIC equations, and SUMMATION. Āryabhata also described methods for finding integer solutions to linear equations of the form $by = ax + c$ using an algorithm essentially equivalent to the EUCLIDEAN ALGORITHM.

Historians do not know how Āryabhata obtained his highly accurate estimate for π. They do know, however, that Āryabhata was aware that it is an IRRATIONAL NUMBER, a fact that mathematicians were not able to prove until 1775, over two millennia later. In practical applications, however, Āryabhata preferred to use $\sqrt{10} \approx 3.1622$ as an approximation for π.

Scholars at the time did not think of sine as a ratio of side-lengths of a triangle, but rather the physical length of a half-chord of a circle. Of course, circles of different radii give different lengths for corresponding half-chords, but one can adjust figures with the use of proportionality. Working with a circle of radius 3,438, Āryabhata constructed a table of sines for each angle from 1° to 90°. (He chose the number 3,438 so that the circumference of the circle would be close to $21,600 = 360 \times 60$, making one unit of length of the circumference matching one minute of an angle.) Thus, in his table, sine of 90° is recorded as 3,438, and the sine of 30°, for example, as 1,719.

With regard to astronomy, *Āryabhatiya* presents a systematic treatment of the position and motions of the planets. Āryabhata calculated the circumference of the Earth as 24,835 miles (which is surprisingly accurate) and described the orbits of the planets as ELLIPSEs. European scholars did not arrive at the same conclusion until the Renaissance.

associative   A BINARY OPERATION is said to be associative if it is independent of the grouping of the terms to which it is applied. More precisely, an operation * is associative if:

$$a * (b * c) = (a * b) * c$$

for all values of *a*, *b*, and *c*. For example, in ordinary arithmetic, the operations of addition and multiplication are associative, but subtraction and division are not. For instance, $6 + (3 + 2)$ and $(6 + 3) + 2$ are equal in value, but $6 - (3 - 2)$ and $(6 - 3) - 2$ are not. (The first equals $6 - 1 = 5$, and the second is $3 - 2 = 1$.) In VECTOR analysis, the addition of vectors is associative, but the operation of taking CROSS PRODUCT is not.

From the basic relation $a * (b * c) = (a * b) * c$, it follows that all possible groupings of a finite number of fixed terms by parentheses are equivalent. (Use an INDUCTION argument on the number of elements present.) For example, that $(a * b) * (c * d)$ equals $(a * (b * c)) * d$ can be established with two applications of the fundamental relation as follows: $(a * b) * (c * d) = ((a * b) * c) * d = (a * (b * c)) * d$. As a consequence, if the associative property holds for a given set, parentheses may be omitted when writing products: one can simply write $a * b * c * d$, for instance, without concern for confusion.

These considerations break down, however, if the expression under consideration contains an infinite number of terms. For instance, we have:

$$0 = 0 + 0 + 0 + \dots$$
$$= (1 - 1) + (1 - 1) + (1 - 1) + \dots$$

If it is permissible to regroup terms, then we could write:

$$0 = 1 + (-1 + 1) + (-1 + 1) + (-1 + 1) + \dots$$
$$= 1 + 0 + 0 + 0 + \dots$$
$$= 1$$

This absurdity shows that extreme care must be taken when applying the associative law to infinite sums.

See also COMMUTATIVE PROPERTY; DISTRIBUTIVE PROPERTY; RING.

asymptote   A straight line toward which the graph of a function approaches, but never reaches, is called an asymptote for the graph. The name comes from the Greek word *asymptotos* for "not falling together" (*a:* "not;" *sym:* "together;" *ptotos:* "falling"). For example, the function $y = 1/x$ has the lines $x = 0$ and $y = 0$ as asymptotes: *y* becomes infinitely small, but never reaches zero, as *x* becomes large, and vice versa. The function $y = (x + 2)/(x - 3)$ has the vertical line $x = 3$

as asymptote: values of the function become infinitely large as x approaches the value 3 from the right, and infinitely large and negative as x approaches the value 3 from the left.

A function $y = f(x)$ has a horizontal asymptote $y = L$ if $\lim_{x \to \infty} f(x) = L$ or $\lim_{x \to -\infty} f(x) = L$. For example, the function $y = \dfrac{3x^2}{x^2 + x + 1}$ has horizontal asymptote $y = 3$, since $\lim_{x \to \infty} \dfrac{3x^2}{x^2 + x + 1} = \lim_{x \to \infty} \dfrac{3}{1 + \dfrac{1}{x} + \dfrac{1}{x^2}}$

$= \dfrac{3}{1 + 0 + 0} = 3$.

An asymptote need not be horizontal or vertical, however. For example, the function $y = \dfrac{x^3 + x^2 + x + 2}{x^2 + 1}$

$= x + 1 + \dfrac{1}{x^2 + 1}$ approaches the line $y = x + 1$ as x becomes large, thus $y = x + 1$ is a "slant asymptote" for the curve.

If a HYPERBOLA is given by the equation $\dfrac{x^2}{a^2} - \dfrac{y^2}{b^2} = 1$,

then manipulating yields the equation $\left(\dfrac{b}{a}\right)^2 - \left(\dfrac{y}{x}\right)^2 = \dfrac{b^2}{x^2}$.

The right hand side tends to zero as x becomes large, showing that the curve has slant asymptotes given by $\left(\dfrac{b}{a}\right)^2 - \left(\dfrac{y}{x}\right)^2 = 0$, that is, by the lines $y = \dfrac{b}{a}x$ and $y = -\dfrac{b}{a}x$.

Extending the definition, we could say that the curve $y = \dfrac{x^3 + 1}{x} = x^2 + \dfrac{1}{x}$ has the parabola $y = x^2$ as an asymptote.

automaton (plural, automata) An abstract machine used to analyze or model mathematical problems is called an automaton. One simple example of an automaton is a "number-base machine," which consists of a row of boxes extending infinitely to the left. One places in this machine a finite number of pennies in the rightmost box. The machine then redistributes the pennies according to a preset rule.

A "$1 \leftarrow 2$" machine, for example, replaces a pair of pennies in one box with a single penny in the box one place to the left. Thus, for instance, six pennies placed into the $1 \leftarrow 2$ machine "fire" four times to yield a final distribution that can be read as "1 1 0." This result is the number six written as a BINARY NUMBER, and this machine converts all numbers to their base-two representations.

A $1 \leftarrow 3$ machine yields base-three representations, and a $1 \leftarrow 10$ machine yields the ordinary base-ten representations. The process of LONG DIVISION can be explained with the aid of this machine.

Variations on this idea can lead to some interesting mathematical studies. Consider, for example, a $2 \leftarrow 3$ machine. This machine replaces three pennies in one box with *two* pennies in the box one place to the left. In some sense, this is a "base one and a half machine." For instance, placing 10 pennies in this machine yields

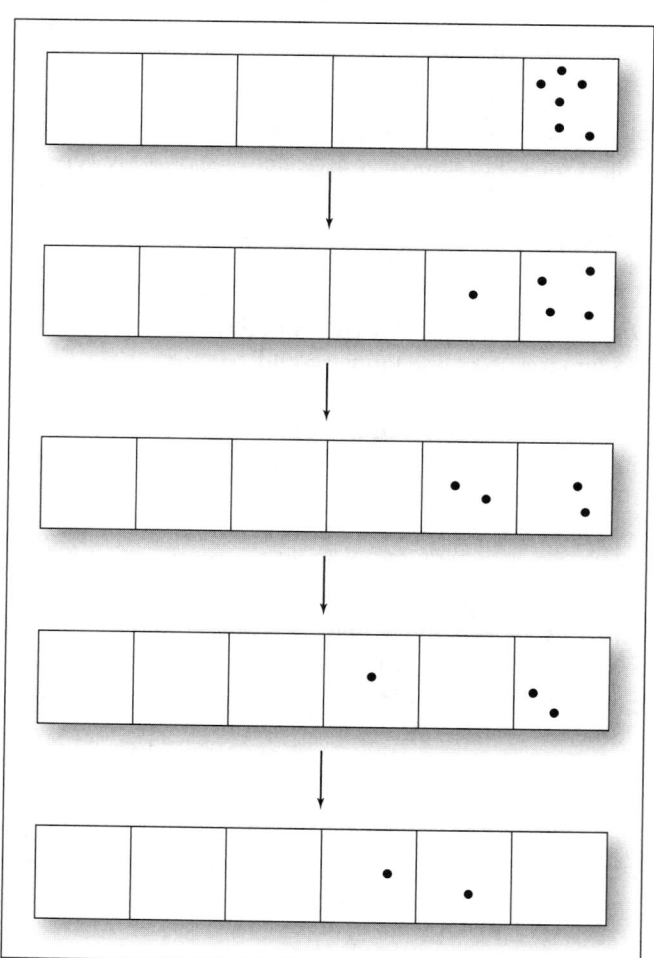

An automaton in action

the representation "2 1 0 1," and it is true that

$$2 \times \left(\frac{3}{2}\right)^3 + 1 \times \left(\frac{3}{2}\right)^2 + 0 \times \left(\frac{3}{2}\right)^1 + 1 \times \left(\frac{3}{2}\right)^0 = 10 .$$

See also BASE OF A NUMBER SYSTEM.

axiom (postulate)   A statement whose truth is deemed self-evident or to be accepted without proof is called an axiom. The name comes from the Greek word *axioma* for "worth" or "quality." The alternative name "postulate" comes from *postulatum*, Latin for "a thing demanded."

One of the great achievements of the great Greek geometer EUCLID and his contemporaries of around 300 B.C.E. was to recognize that not every statement in mathematics can be proved: certain terms remain undefined, and basic rules (postulates) about their relationships must simply be accepted as true. One must develop a mathematical theory with a "big bang," as it were, by simply listing a starting set of assumptions. From there, using the basic laws of reasoning, one then establishes and proves further statements, or THEOREMS, about the system.

For example, in a systematic study of EUCLIDEAN GEOMETRY, the terms *point, line,* and *plane* are undefined, and one begins a systematic study of the subject by studying a list of basic axioms that tells us how these quantities are meant to interrelate. (One axiom of Euclidean geometry, for instance, asserts that between any two points one can draw a line.) All the results presented in a typical high-school text on geometry, for example, are logical consequences of just five principal assumptions.

In SET THEORY, the terms *set* and *element of a set* are undefined. However, rules are given that define the equality of two sets, that guarantee the existence of certain sets, and establish the means of constructing new sets from old ones. In NUMBER THEORY, PEANO'S POSTULATES provide a logical foundation to the theory of numbers and arithmetic.

A statement in a mathematical system that appears true, but has not yet been proved, is called a conjecture.

See also DEDUCTIVE/INDUCTIVE REASONING; ERNST FRIEDRICH FERDINAND ZERMELO.

axiom of choice   First formulated by German mathematician ERNST FRIEDRICH FERDINAND ZERMELO (1871–1953), the axiom of choice is a basic principle of SET THEORY that states that from any given collection C of nonempty sets, it is possible to construct a set S that contains one element from each of the sets in C. The set S is called a "choice set" for C. For example, if C represents the three sets $\{1,2,3\}$, $\{2,4,6,8,...\}$, and $\{5\}$, then $S = \{1,6,5\}$ is a choice set for C. So too is the set $S = \{2,5\}$.

The axiom of choice has been considered counterintuitive when interpreted on a practical level: although it is possible to select one element from each of a finite collection of sets in a finite amount of time, it is physically impossible to accomplish the same feat when presented with an infinite collection of sets. The existence of a choice set is not "constructive," as it were, and use of the axiom is viewed by mathematicians, even today, with suspicion. In 1938 Austrian mathematician KURT GÖDEL proved, however, that no contradiction would ever arise when the axiom of choice is used in conjunction with other standard axioms of set theory.

Zermelo formulated the axiom to prove that every ordered set can be well-ordered. The axiom of choice also proves (and in fact is equivalent to) the trichotomy law, which states that for any pair of REAL NUMBERS a and b, precisely one of the following holds:

   i. $a > b$
   ii. $a < b$
   iii. $a = b$

Although this statement, on one level, appears obvious, its validity is fundamental to the workings of the real numbers and so needs to be properly understood.

See also WELL-ORDERED SET.

B

Babbage, Charles (1791–1871) British *Computation*
Born on December 26, 1791, in London, England,
Charles Babbage is best remembered for his work on
the design and manufacture of a mechanical calculator,
the forerunner of a computer. After first constructing a
"difference machine," Babbage devoted the remainder
of his life to the construction of a superior "analytic
engine" capable of performing all mathematical operations. His work toward this goal laid the foundations
of computer design used today. Partly due to lack of
funding, however, the machine was never completed.

Babbage entered Trinity College, Cambridge, in
1810. While a student, he and a fellow undergraduate
coauthored *Differential and Integral Calculus,* an influential memoir on the history of calculus. After transfer-
ring to Peterhouse College, Babbage received his
bachelor's degree in mathematics in 1814 to then begin
a career in mathematical research. Babbage published a
number of influential papers on the topic of functional
equations and was honored with election to the ROYAL
SOCIETY in 1816. In 1827 he became the Lucasian Professor of Mathematics at Cambridge.

Much of Babbage's theoretical work relied on
consulting tables of logarithms and trigonometric
functions. Aware of the inaccuracy of human calculation, Babbage became interested in the problem of
using a mechanical device to perform complex computations. In 1819 he began work on a small "difference engine," which he completed three years later.
He announced his invention to the scientific community in an 1822 paper, "Note on the Application of

Charles Babbage, an eminent mathematician of the 19th century,
is best known for his design and manufacture of a mechanical
computer. *(Photo courtesy of the Science Museum, London/
Topham-HIP/The Image Works)*

**Charles Babbage completed work on his "difference engine,"
the world's first sophisticated mechanical computer, in 1822.
(Photo courtesy of the Science Museum, London/Topham-
HIP/The Image Works)**

Machinery to the Computation of Astronomical and
Mathematical Tables."

Although the machine was capable of performing
relatively simple, but highly accurate, computations
(using the method of FINITE DIFFERENCES to compute
values of POLYNOMIAL functions), his invention was
well received and was understood to be a first step
toward a new era in computational capabilities. Babbage was awarded a gold medal from the Astronomical
Society and was given a grant from the Chancellor of
the Exchequer to construct a larger, more complex, difference engine.

In 1801 Joseph-Marie Jacquard invented a loom
capable of weaving complex patterns by making use of
a set of instructions set out on cards punched with
holes. Two decades later Babbage decided to follow the
same idea and design a steam-powered engine that
would accept instructions and data from punched
cards. With the assistance of Lord Byron's daughter,
LADY AUGUSTA ADA LOVELACE, Babbage took to work
on creating a sophisticated calculating device. In 1832

he published a book, *On the Economy of Machinery
and Manufactures*, offering a theoretical discussion on
the topic. This could be considered the first published
work in the field of OPERATIONS RESEARCH.

Unfortunately, due to financial and technological
difficulties, the machine was never completed. (The
metalwork technology of the mid-1800s was not capable of the levels of precision Babbage's machine
demanded.) The device in its unfinished state is preserved today in the Science Museum of London.
Although he never realized his dream of building an
operational, mechanical computer, his design concepts
have since been proved correct. It is not an exaggeration to say that the modern computers constructed on
Babbage's theoretical design have revolutionized almost
all aspects of 20th-century life.

Babbage died in London, England, on October
18, 1871.

Babylonian mathematics   The Babylonians of 2000
B.C.E. lived in Mesopotamia, the fertile plain between the
Euphrates and Tigris Rivers in what is now Iraq. We are
fortunate that the peoples of this region kept extensive
records of their society—and their mathematics—on
hardy sun-baked clay tablets. A large number of these
tablets survive today. The Babylonians used a simple sty-
lus to make marks in the clay and developed a form of
writing based on cuneiform (wedge shaped) symbols.

The mathematical activity of the Babylonians seems
to have been motivated, at first, by the practical everyday needs of running their society. Many problems
described in early tablets are concerned with calculating
the number of workers needed for building irrigation
canals and the total expense of wages, for instance. But
many problems described in later texts have no apparent practical application and clearly indicate an interest
in pursuing mathematics for its own sake.

The Babylonians used only two symbols to repre-
sent numbers: the symbol Y to represent a unit and the
symbol $\mathsf{\langle}$ to represent a group of ten. A simple additive system was used to represent the numbers 1
through 59. For example, the cluster $\mathsf{\lll YY}$ represents
"32." A base-60 PLACE-VALUE SYSTEM was then used to
represent numbers greater than 59. For instance, the
number 40,992, which equals $11 \times 60^2 + 23 \times 60 + 12$,
was written: $\mathsf{\langle Y}$ $\mathsf{\ll YYY}$ $\mathsf{\langle YY}$. Spaces were inserted
between clusters of symbols.

Historians are not clear as to why the Babylonians chose to work with a SEXAGESIMAL system. A popular theory suggests that this number system is based on the observation that there are 365 days in the year. When rounded to the more convenient (highly divisible) value of 360, we have a multiple of 60. Vestiges of this number system remain with us today. For example, we use the number 360 for the number of degrees in a circle, and we count 60 seconds in a minute and 60 minutes per hour.

There were two points of possible confusion with the Babylonian numeral system. With no symbol for zero, it is not clear whether the numeral ☖ ☖ represents 61 (as one unit of 60 plus a single unit), 3601 (as one unit of 60^2 plus a single unit), or even 216,060, for instance. Also, the Babylonians were comfortable with

A seventh-century cuneiform tablet from northern Iraq records observations of the planet Venus. *(Photo courtesy of the British Museum/Topham-HIP/The Image Works)*

fractions and used negative powers of 60 to represent them (just as we use negative powers of 10 to write fractions in decimal notation). But with no notation for the equivalent of a decimal point, the symbol ☖ ☖ could also be interpreted to mean 1 + (1/60), or (1/60) + (1/60^2), or even 60 + (1/60^4), for instance. As the Babylonians never developed a method for resolving such ambiguity, we assume then that it was never considered a problem for scholars of the time. (Historians suggest that the context of the text always made the interpretation of the numeral apparent.)

The Babylonians compiled extensive tables of powers of numbers and their reciprocals, which they used in ingenious ways to perform arithmetic computations. (For instance, a tablet dated from 2000 B.C.E. lists all the squares of the numbers from one to 59, and all the cubes of the numbers from one to 32.) To compute the product of two numbers a and b, Babylonian scholars first computed their sum and their difference, read the squares of those numbers from a table, and divided their difference by four. (In modern notation, this corresponds to the computation: $ab = (1/4)[(a + b)^2 - (a - b)^2]$.) To divide a number a by b, scholars computed the product of a and the reciprocal $1/b$ (recorded in a table): $ab = a \times (1/b)$. The same table of reciprocals also provided the means to solve LINEAR EQUATIONS: $bx = a$. (Multiply a by the reciprocal of b.)

Problems in geometry and the computation of area often lead to the need to solve QUADRATIC equations. For instance, a problem from one tablet asks for the width of a rectangle whose area is 60 and whose length is seven units longer than the width. In modern notation, this amounts to solving the equation $x(x + 7) = x^2 + 7x = 60$. The scribe who wrote the tablet then proffers a solution that is equivalent to the famous quadratic formula: $x = \sqrt{(7/2)^2 + 60} - (7/2) = 5$. (Square roots were computed by examining a table of squares.)

Problems about volume lead to cubic equations, and the Babylonians were adept at solving special equations of the form: $ax^3 + bx^2 = c$. (They solved these by setting $n = (ax)/b$, from which the equation can be rewritten as $n^3 + n^2 = ca^2/b^3$. By examining a table of values for $n^3 + n^2$, the solution can be deduced.)

It is clear that Babylonian scholars knew of PYTHAGORAS'S THEOREM, although they wrote no general proof of the result. For example, a tablet now housed in the British museum, provides the following problem and solution:

If the width of a rectangle is four units and the length of its diagonal is five units, what is its breadth?

Four times four is 16, and five times five is 25. Subtract 16 from 25 and there remains nine. What times what equals nine? Three times three is nine. The breadth is three.

The Babylonians used Pythagoras's theorem to compute the diagonal length of a square, and they found an approximation to the square root of two accurate to five decimal places. (It is believed that they used a method analogous to HERON'S METHOD to do this.) Babylonian scholars were also interested in approximating the areas and volumes of various common shapes by using techniques that often invoked Pythagoras's theorem.

Most remarkable is a tablet that lists 15 large PYTHAGOREAN TRIPLES. As there is no apparent practical need to list these triples, this strongly suggests that the Babylonians did indeed enjoy mathematics for its own sake.

See also BASE VALUE OF A NUMBER SYSTEM.

Banach, Stefan (1892–1945) Polish *Analysis, Topology* Born on March 30, 1892, in Kraków, now in Poland, Stefan Banach is noted for his foundational work in ANALYSIS and for generalizing the notion of a VECTOR SPACE to a general theory of a space of functions. This fundamental work allows mathematicians today to develop a theory of FOURIER SERIES, in some sense, in very abstract settings. Banach is also remembered for his work leading to the famous Banach-Tarski paradox that arises in the study of AREA and volume.

Banach began his scholarly career with a university degree in engineering from Lvov Technical University. His academic plans were interrupted, however, with the advent of World War I. During this time Banach was forced to work building roads, although he did manage to find time to also teach at local Kraków schools during this period. Soon after the war Banach joined a mathematics discussion group in Lvov and soon impressed mathematical scholars with his abilities to solve mathematical problems. Within a week of joining the group, Banach had drafted a coauthored research paper on the topic of measure theory, a theory that generalizes the concept of area. Banach continued to produce important results in this field at an extremely rapid rate thereafter.

Banach was offered a lectureship at Lvov Technical University in 1920 and quickly set to work on a doctoral thesis. Despite having no previous official university qualifications in mathematics, Banach was awarded a doctorate in 1922 by the Jan Kazimierz University in Lvov.

Banach's contributions to mathematics were significant. His generalized work on Fourier series founded a branch of mathematics now called functional analysis. It has connections to the fields of measure theory, integration, and SET THEORY. He and his colleague Alfred Tarski presented their famous paradoxical result in 1926 in their paper "Sur la décomposition des ensembles de points en partiens respectivement congruent" (On the decomposition of figures into congruent parts). His 1932 paper, "Théorie des opérations linéaires" (Theory of linear operators), which develops the notion of a normed VECTOR SPACE (that is, a vector space with a notion of length attached to its vectors), is deemed his most influential work. As well as conducting research in mathematics, Banach also wrote arithmetic, algebra, and geometry texts for high-school students.

In 1939 Banach was elected as president of the Polish Mathematical Society. Banach was allowed to maintain his university position during the Soviet occupation later that year, but conditions changed with the 1941 Nazi invasion. Many Polish academics were murdered, but Banach survived, although he was forced to work in a German infectious diseases laboratory, given the task to feed and maintain lice colonies. He remained there until June 1944, but he became seriously ill by the time Soviet troops reclaimed Lvov. Banach died of lung cancer on August 31, 1945.

Banach's name remains attached to the type of vector space he invented, and research in this field of functional analysis continues today. The theory has profound applications to theoretical physics, most notably to quantum mechanics.

bar chart (bar graph) *See* STATISTICS: DESCRIPTIVE.

Barrow, Isaac (1630–1677) British *Calculus, Theology* Born in London, England, (his exact birth date is

Isaac Barrow, a mathematician of the 17th century, is noted chiefly for the inspiration he provided others in the development of the theory of calculus. He may have been the first scholar to recognize and understand the significance of the fundamental theorem of calculus. *(Photo courtesy of ARPL/Topham/The Image Works)*

not known) Isaac Barrow is remembered in mathematics for his collection of lecture notes *Lectiones geometricae* (Geometrical lectures), published in 1670, in which he describes a method for finding tangents to curves similar to that used today in DIFFERENTIAL CALCULUS. Barrow may have also been the first to realize that the problem of finding tangents to curves is the inverse problem to finding areas under curves. (This is THE FUNDAMENTAL THEOREM OF CALCULUS.) The lectures on which his notes were based were extremely influential. They provided SIR ISAAC NEWTON, who attended the lectures and had many private discussions with Barrow, a starting point for his development of CALCULUS.

Barrow graduated from Trinity College, Cambridge, with a master's degree in 1652, but was dissatisfied with the level of mathematics instruction he had received. After leaving the college, Barrow taught himself GEOMETRY and published a simplified edition of

EUCLID's *THE ELEMENTS* in 1655. He became professor of geometry at Gresham College, London, in 1662, and was elected as one of the first 150 fellows of the newly established ROYAL SOCIETY in 1663. He returned to Cambridge that same year to take the position of Lucasian Chair of Mathematics, at Trinity College, and worked hard to improve the standards of mathematics education and interest in mathematical research at Cambridge. With this aim in mind, Barrow gave a series of lectures on the topics of optics, geometry, NUMBER THEORY, and the nature of time and space. His discussions on geometry proved to be highly innovative and fundamentally important for the new perspective they offered. Newton advised Barrow to publish the notes.

In 1669 Barrow resigned from the Lucasian Chair to allow Newton to take over, and he did no further mathematical work. He died in London, England, on May 4, 1677, of a malignant fever. Barrow's influence on modern-day mathematics is oblique. His effect on the development of the subject lies chiefly with the inspiration he provided for others.

base of a logarithm   *See* LOGARITHM.

base of an exponential   *See* EXPONENT; EXPONENTIAL FUNCTION.

base of a number system (radix, scale of a number system)   The number of different symbols used, perhaps in combination, to represent all numbers is called the base of the number system being used. For example, today we use the ten symbols 0, 1, 2, 3, 4, 5, 6, 7, 8, and 9 to denote all numbers, making use of the position of these digits in a given combination to denote large values. Thus we use a base-10 number system (also called a decimal representation system). We also use a place-value system to give meaning to the repeated use of symbols. When we write 8,407, for instance, we mean eight quantities of 1,000 (10^3), four quantities of 100 (10^2), and seven single units (10^0). The placement of each DIGIT is thus important: the number of places from the right in which a digit lies determines the power of 10 being considered. Thus the numbers 8, 80, 800, and 8,000, for instance, all represent different quantities. (The system of ROMAN NUMERALS, for example, is not a place-value system.)

For any positive whole number b, one can create a place-value notational system of that base as follows:

Write a given number n as a sum of powers of b:
$n = a_k b^k + a_{k-1} b^{k-1} + \ldots + a_2 b^2 + a_1 b + a_0$
with each number a_i satisfying $0 \leq a_i < b$. Then the base b representation of n is the k-digit quantity $a_k a_{k-1} \ldots a_2 a_1 a_0$. Such a representation uses only the symbols $0, 1, 2, \ldots, b-1$.

For example, to write the number 18 in base four—using the symbols 0, 1, 2, and 3—observe that $18 = 1 \times 4^2 + 0 \times 4 + 2 \times 1$, yielding the base-4 representation: 102. In the reverse direction, if 5,142 is the base-6 representation of a number n, then n is the number $5 \times 6^3 + 1 \times 6^2 + 4 \times 6 + 2 \times 1 = 1,142$.

One may also make use of negative powers of the base quantity b. For example, using a decimal point to separate positive and negative powers of ten, the number 312.407, for instance, represents the fractional quantity: $3 \times 10^2 + 1 \times 10 + 2 \times 1 + 4 \times \frac{1}{10} + 0 \times \frac{1}{10^2} + 7 \times \frac{1}{10^3}$. In base 4, the number 33.22 is the quantity $3 \times 4 + 3 \times 1 + 2 \times \frac{1}{4} + 2 \times \frac{1}{4^2} = 15 + \frac{2}{4} + \frac{2}{16}$, which is 15.625 in base 10.

The following table gives the names of the place-value number systems that use different base values b. The Babylonians of ancient times used a sexagesimal system, and the Mayas of the first millennium used a system close to being purely vigesimal.

The representation of numbers can be well-represented with the aid of a simple AUTOMATON called a number-base machine. Beginning with a row of boxes extending infinitely to the left, one places in the rightmost box a finite number of pennies. The automaton then redistributes the pennies according to a preset rule. A "1 ← 2" machine, for example, replaces a pair of pennies in one box with a single penny in the box one place to the left. Thus, for instance, six pennies placed into the 1 ← 2 machine "fire" four times to yield a final distribution that can be read as "1 1 0." This result is the number six written as a BINARY NUMBER and this machine converts all numbers to their base-two representations. (The diagram in the entry for automaton illustrates this.) A 1 ← 3 machine yields base-three representations, and a 1 ← 10 machine yields the ordinary base-ten representations.

Long Division

The process of long division in ARITHMETIC can be explained with the aid of a number-base machine. As an example, let us use the 1 ← 10 machine to divide the number 276 by 12. Noting that 276 pennies placed in the 1 ← 10 machine yields a diagram with two pennies in the 100s position, seven pennies in the 10s position, and six pennies in the units position, and that 12 pennies appears as one penny in a box with two pennies in the box to its right, to divide 276 by 12, one must simply look for "groups of 12" within the diagram of 276 pennies and keep count of the number of groups one finds.

base b	number system
2	Binary
3	Ternary
4	Quaternary
5	Quinary
6	Senary
7	Septenary
8	Octal
9	Nonary
10	Decimal
11	Undenary
12	Duodecimal
16	Hexadecimal
20	Vigesimal
60	Sexagesimal

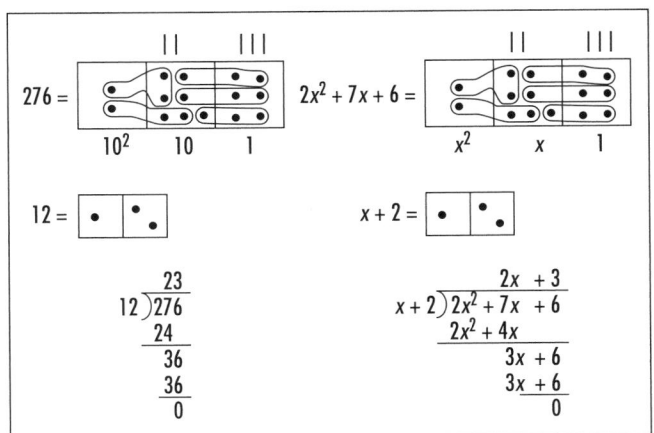

Long division base ten

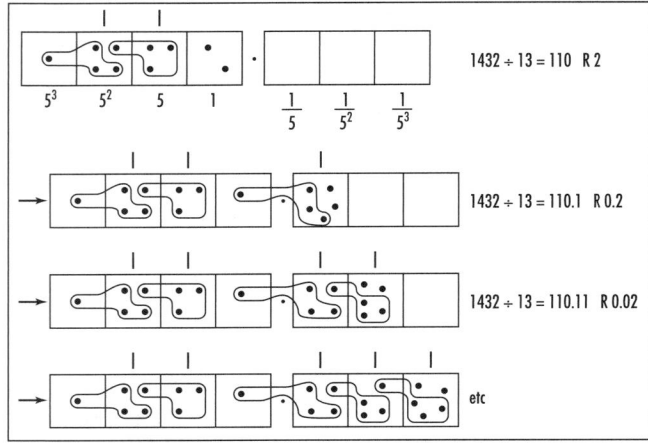

Long division base five

Notice that we find two groups of 12 at the 10s position (that is, two groups of 120), and three groups of 12 at the units position. Thus: $276 \div 12 = 23$. The standard algorithm taught to school children is nothing more than a recording system for this process of finding groups of twelve. Notice too that one does not need to know the type of machine, that is, the base of the number system in which one is working in order to compute a long-division problem. If we simply write the base number as x, and work with a $1 \leftarrow x$ machine, then the same computation provides a method for dividing polynomials. In our example, we see that:

$$(2x^2 + 7x + 6) \div (x + 2) = 2x + 3$$

Thus the division of polynomials can be regarded as a computation of long division. There is a technical difficulty with this: a polynomial may have negative coefficients, and each negative coefficient would correspond to a negative number of pennies in a cell. If one is willing to accept such quantities, then the number-base machine model continues to work. (Note, in this extended model, that one can insert into any cell an equal number of positive and negative pennies without changing the system. Indeed, it might be necessary to do this in order to find the desired groups of pennies.)

The process of long division might produce non zero remainders. For example, in base 5, dividing 1432 by 13 yields the answer 110 with a remainder of 2 units. (In base 10, this reads: $242 \div 8 = 30$ with a remainder of 2.) If one is willing to work with negative powers of five, and "unfire" a group of five pennies, one can continue the long division process to compute, in base 5, that $1432 \div 13 = 110.1111\ldots$

See also BABYLONIAN MATHEMATICS; BINARY NUMBERS; DIGIT; MAYAN MATHEMATICS; NESTED MULTIPLICATION; ZERO.

base of a polygon/polyhedron   The base of a triangle, or of any POLYGON, is the lowest side of the figure, usually drawn as a horizontal edge parallel to the bottom of the page. Of course other edges may be considered the base if one reorients the figure. The base of a POLYHEDRON, such as a cube or a pyramid, is the lowest FACE of the figure. It is the face on which the figure would stand if it were placed on a tabletop.

The highest point of a geometric figure opposite the base is called the APEX of the figure, and the distance from the base to the apex is called the height of the figure.

basis   See LINEARLY DEPENDENT AND INDEPENDENT.

Bayes, Rev. Thomas (1702–1761) English *Probability, Theology*   Born in London, England, (the exact date of his birth is not known), theologian and mathematician Reverend Thomas Bayes is best remembered for his influential article "An Essay Towards Solving a Problem in the Doctrine of Chances," published posthumously in 1763, that outlines fundamental principles of PROBABILITY theory. Bayes developed innovative techniques and approaches in the theory of statistical inference, many of which were deemed controversial at the time. His essay sparked much further research in the field and was profoundly influential. The work also contains the famous theorem that today bears his name.

An ordained minister who served the community of Tunbridge Wells, Kent, England, Bayes also pursued mathematics as an outside interest. As far as historians can determine, he published only two works during his lifetime. One was a theological essay in 1731 entitled "Divine Benevolence, or an Attempt to Prove that the Principal End of the Divine Providence and Government is the Happiness of His Creatures." The other was a mathematical piece that he published anonymously in 1736, "Introduction to the Doctrine of Fluxions, and a

Defense of the Mathematicians Against the Objections of the Author of *The Analyst*," defending the logical foundations of SIR ISAAC NEWTON's newly invented CALCULUS. Despite the apparent lack of published mathematical work, Bayes was nonetheless elected a fellow of the prestigious academic ROYAL SOCIETY in 1742.

Bayes retired from the ministry in 1752 but remained in Tunbridge Wells until his death on April 17, 1761. His friend, Richard Price, discovered the now-famous paper on probability theory among his belongings and submitted it for publication. A second paper, "A Letter on Asymptotic Series from Bayes to John Canton," one on asymptotic series, was also published after Bayes's death. The theoretical approach of inferential statistics Bayes proposed remains an active area of research today.

See also BAYES'S THEOREM; STATISTICS: INFERENTIAL.

Bayes's theorem In his 1763 paper, published posthumously, REV. THOMAS BAYES established a fundamental result, now called Bayes's theorem, that expresses the CONDITIONAL PROBABILITY $P(A|B)$ of an event A occurring given that event B has already occurred in terms of the reverse conditional probability $P(B|A)$. Precisely:

$$P(A \mid B) = P(B \mid A)\frac{P(B)}{P(A)}$$

This formula is easily proved by noting that $P(A \mid B) = \frac{P(A \cap B)}{P(B)}$ and $P(B \mid A) = \frac{P(A \cap B)}{P(A)}$.

More generally, suppose B_1, B_2,..., B_n is a mutually exclusive and exhaustive set of events, that is, a set of nonoverlapping events covering the whole SAMPLE SPACE. Suppose also that we have been told that another event A has occurred. Then the probability that event B_i also occurred is given by:

$$P(B_i \mid A) = \frac{P(A \mid B_i)P(B_i)}{P(A \mid B_1)P(B_1) + \cdots + P(A \mid B_n)P(B_n)}$$

To illustrate: suppose that bag 1 contains five red balls and two white balls, and bag 2 contains seven red balls and four white balls. If a bag is selected at random and a ball chosen from it is found to be red, what is the probability that it came from bag 1?

Here let A be the event "a red ball is chosen" and B_1 and B_2 the events "a ball is selected from bag 1 / bag 2," respectively. Then $P(B_1) = 1/2 = P(B_2)$, $P(A|B_1) = 5/7$, and $P(A|B_2) = 7/11$. Thus the probability we seek, $P(B_1|A)$, is given by:

$$P(B_1 \mid A) = \frac{P(A \mid B_1)P(B_1)}{P(A \mid B_1)P(B_1) + P(A \mid B_2)P(B_2)}$$
$$= \frac{\frac{5}{7} \times \frac{1}{2}}{\frac{5}{7} \times \frac{1}{2} + \frac{7}{11} \times \frac{1}{2}}$$
$$= \frac{55}{104}$$

bearing The ANGLE between the course of a ship and the direction of north is called the ship's bearing. The angle is measured in degrees in a clockwise direction from north and is usually expressed as a three-digit number. For example, a ship heading directly east has a bearing of 090 degrees, and one heading southwest has a bearing of 225 degrees.

The word "bearing" is also used for the measure of angle from north at which an object is sighted. For example, a crewman on board a ship sighting a lighthouse directly west will say that the lighthouse has bearing 270 degrees.

Bernoulli family No family in the history of mathematics has produced as many noted mathematicians as the Bernoulli family from Basel, Switzerland. The family record begins with two brothers, Jacob Bernoulli and Johann (Jean) Bernoulli, respectively, the fifth and 10th children of Nicolaus Bernoulli (1623–1708).

Jacob (December 27, 1654–August 16, 1705) is noted for his work on CALCULUS and PROBABILITY theory, being one of the first mathematicians to properly understand the utility and power of the newly published work of the great WILHELM GOTTFRIED LEIBNIZ (1646–1716). Jacob applied the calculus to the study of curves, in particular to the logarithmic spiral and the BRACHISTOCHRONE, and was the first to use POLAR COORDINATES in 1691. He also wrote the first text concentrating on probability theory *Ars conjectandi* (The art of conjecture), which was published posthumously in 1713. The BERNOULLI NUMBERS also appear, for the first time, in this text. Jacob occupied the chair of mathematics at Basel University from 1687 until his death.

Johann (July 17, 1667–January 1, 1748) is also known for his work on calculus. Being recognized as an expert in the field, Johann was hired by the French nobleman MARQUIS DE GUILLAUME FRANÇOIS ANTOINE L'HÔPITAL (1661–1704) to explain the new theory to him, first through formal tutoring sessions in Paris, and then through correspondence when Johann later returned to Basel. L'Hôpital published the contents of the letters in a 1696 textbook *Analyse des infiniment petits* (Analysis with infinitely small quantities), but gave little acknowledgment to Johann. The famous rule that now bears his name, L'HÔPITAL'S RULE, is due to Johann. Johann succeeded his brother in the chair at Basel University.

The two brothers, Jacob and Johann, worked on similar problems, and each maintained an almost constant exchange of ideas with Leibniz. The relationship between the two siblings, however, was not amicable, and they often publicly criticized each other's work.

Nicolaus (I) Bernoulli (October 21, 1687–November 29, 1759), nephew to Jacob and Johann, also achieved some fame in mathematics. He worked on problems in GEOMETRY, DIFFERENTIAL EQUATIONs, infinite SERIES, and probability. He held the chair of mathematics at Padua University, once filled by GALILEO GALILEI (1564–1642).

Johann Bernoulli had three sons, all of whom themselves became prominent mathematicians:

Nicolaus (II) Bernoulli (February 6, 1695–July 31, 1726) wrote on curves, differential equations, and probability theory. He died—by drowning while swimming—only eight months after accepting a prestigious appointment at the St. Petersberg Academy.

Daniel Bernoulli (February 8, 1700–March 17, 1782), the most famous of the three sons, is noted for his 1738 text *Hydrodynamica* (Hydrodynamics), which laid the foundations for the modern discipline of hydrodynamics. (Daniel's father, Johann, jealous of his son's success, published his own text on hydrodynamics in 1739 but placed on it the publishing date of 1732 and accused his son of plagiarism.) Daniel also worked on the mathematics of vibrating strings, the kinetic theory of gases, probability theory, and

partial differential equations. He was awarded the Grand Prize from the Paris Academy no fewer than 10 times.

Johann (II) Bernoulli (May 28, 1710–July 17, 1790) studied the mathematics of heat flow and light. He was awarded the Grand Prize from the Paris Academy four times and succeeded his father in the chair at Basel University in 1743.

Johann (II) Bernoulli had three sons, two of whom, Johann (III) Bernoulli (November 4, 1744–July 13, 1807) and Jacob (II) Bernoulli (October 17, 1759–August 15, 1789), worked in mathematics. Johann (III) studied astronomy and probability, and wrote on recurring decimals and the theory of equations. He was a professor of mathematics at Berlin University at the young age of 19. Jacob (II) Bernoulli wrote works on the mathematics of elasticity, hydrostatics, and ballistics. He was professor of mathematics at the St. Petersburg Academy, but, like his uncle, drowned at the age of 29 while swimming in the Neva River.

Members of the Bernoulli family had a profound effect on the early development of probability theory, calculus, and the field of continuum mechanics. Many concepts (such as the Bernoulli numbers, a probability distribution, a particular differential equation) are named in their honor.

Bernoulli numbers   *See* SUMS OF POWERS.

Bertrand's paradox   French mathematician Joseph-Louis François Bertrand (1822–1900) posed the following challenge:

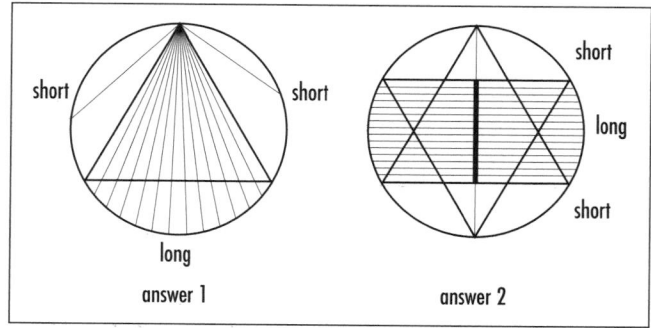

Answering Bertrand's paradox

Imagine an equilateral TRIANGLE drawn inside a CIRCLE. Find the PROBABILITY that a CHORD chosen at random is longer than the side-length of the triangle.

There are two possible answers:

1. Once a chord is drawn we can always rotate the picture of the circle so that one end of the selected chord is placed at the top of the circle. It is clear then that the length of the chord will be greater than the side-length of the triangle if the other end-point lies in the middle third of the perimeter of the circle. The chances of this happening are 1/3, providing the answer to the problem.
2. Rotating the picture of the circle and the selected chord, we can also assume that the chord chosen is horizontal. If the chord crosses the solid line shown, then it will be longer than the side-length of the triangle. One observes that this solid line is half the length of the diameter. Thus the chances of a chord being longer than the side-length of the triangle are 1/2.

Surprisingly, both lines of reasoning are mathematically correct. Therein lies a PARADOX: the answer cannot simultaneously be 1/3 and 1/2.

The problem here lies in defining what we mean by "select a chord at random." There are many different ways to do this: one could spin a bottle in the center of the circle to select points on the perimeter to connect with a chord, or one could roll a broom across a circle drawn on the floor, or perhaps even drop a wire from a height above the circle and see where it lands. Each approach to "randomness" could (and in fact does) lead to its own separate answer. This paradox shows that extreme care must be taken to pose meaningful problems in probability theory. It is very difficult to give a precise definition to "randomness."

Bhāskara II (Bhaskaracharya) (1114–1185) Indian *Algebra, Arithmetic*   Born in Vijayapura, India, Bhāskara (often referred to as Bhāskara II to distinguish him from the seventh-century mathematician of the same name) is considered India's most eminent mathematician of the 12th century. He revised and continued the studies of the great BRAHMAGUPTA, making corrections and filling in gaps in his work, and reached a level of mastery of ARITHMETIC and ALGEBRA that was not matched by a European scholar for several centuries to come. Bhāskara wrote two influential mathematical treatises: *Lilavati* (The beautiful), on the topic of arithmetic, and the *Bijaganita* (Seed arithmetic) on algebra.

Bhāskara was head of the astronomical observatory in Ujjain, the nation's most prominent mathematical research center of the time. Although much of Indian mathematics was motivated by problems and challenges in astronomy, Bhāskara's writings show a keen interest in developing mathematics for its own sake. For example, the text *Lilavati,* consisting of 13 chapters, begins with careful discussions on arithmetic and geometry before moving on to the topics of SEQUENCES and SERIES, fractions, INTEREST, plane and solid geometry, sundials, PERMUTATIONs and COMBINATIONs, and DIOPHANTINE EQUATIONS (as they are called today). For instance, Bhāskara shows that the equation $195x = 221y + 65$ (which he expressed solely in words) has infinitely many positive integer solutions, beginning with $x = 6$, $y = 5$, and $x = 23$, $y = 20$, and then $x = 40$, and $y = 35$. (The x-values increase in steps of 17, and the y-values in steps of 15.)

In his piece *Bijaganita,* Bhāskara develops the arithmetic of NEGATIVE NUMBERS, solves quadratic equations of one, or possibly more, unknowns, and develops methods of extracting SQUARE and CUBE ROOTs of quantities. He continues the discussions of Brahmagupta on the nature and properties of the number ZERO and the use of negative numbers in arithmetic. (He denoted the negative of a number by placing a dot above the numeral.) Bhāskara correctly points out that a quantity divided by zero does not produce zero (as Brahmagupta claimed) and suggested instead that $a/0$ should be deemed infinite in value. Bhāskara solves complicated equations with several unknowns and develops formulae that led him to the brink of discovering the famous QUADRATIC formula.

Bhāskara also wrote a number of important texts in mathematical astronomy and made significant strides in the development of TRIGONOMETRY, taking the subject beyond the level of just a tool of calculation for astronomers. Bhāskara discovered, for example, the famous addition formulae for sine:

$$\sin(A + B) = \sin A \cos B + \cos A \sin B$$
$$\sin(A - B) = \sin A \cos B - \cos A \sin B$$

However, as was the tradition at the time, Bhāskara did not explain how he derived his results. It is conjectured that Indian astronomers and mathematicians felt it necessary to conceal their methods regarding proofs and derivations as "trade secrets" of the art.

Bhāskara's accomplishments were revered for many centuries. In 1817, H. J. Colebrook provided English translations of both *Lilavati* and *Bijaganita* in his text *Algebra with Arithmetic and Mensuration*.

bias   A systematic error in a statistical study is called a bias. If the sample in the study is large, errors produced by chance tend to cancel each other out, but those from a bias do not. For example, a survey on the shopping habits of the general population conducted at a shopping mall is likely to be biased toward people who shop primarily at malls, omitting results from people who shop from home through catalogs and on-line services. This is similar to a loaded die, which is biased to produce a particular outcome with greater than one-sixth probability.

Surprisingly, American pennies are biased. If you delicately balance 30 pennies on edge and bump the surface on which they stand, most will fall over heads up. If, on the other hand, you spin 30 pennies and let them all naturally come to rest, then most will land tails.

See also POPULATION AND SAMPLE.

biconditional   In FORMAL LOGIC, a statement of the form "*p* if, and only if, *q*" is called a biconditional statement. For example, "A triangle is equilateral if, and only if, it is equiangular" is a biconditional statement. A biconditional statement is often abbreviated as *p* iff *q* and is written in symbols as $p \leftrightarrow q$. It is equivalent to the compound statement "*p* implies *q*, and *q* implies *p*" composed of two CONDITIONAL statements. The truth-values of *p* and *q* must match for the biconditional statement as a whole to be true. It therefore has the following TRUTH TABLE:

p	*q*	*p*↔*q*
T	T	T
T	F	F
F	T	F
F	F	T

The two statements *p* and *q* are said to be logically equivalent if the biconditional statement $p \leftrightarrow q$ is true.
See also ARGUMENT.

bijection   *See* FUNCTION.

bimodal   *See* STATISTICS: DESCRIPTIVE.

binary numbers (base-2 numbers)   Any whole number can be written as a sum of distinct numbers from the list of powers of 2: 1, 2, 4, 8, 16, 32, 64, ... (Simply subtract the largest power of 2 less from the given number and repeat the process for the remainder obtained.) For instance, we have:

$$89 = 64 + 25 = 64 + 16 + 9 = 64 + 16 + 8 + 1$$

No power of 2 will appear twice, as two copies of the same power of 2 sum to the next power in the list. Moreover, the sum of powers of 2 produced for a given number is unique. Using the symbol 1 to denote that a particular power of 2 is used and 0 to denote that it is not, one can then encode any given number as a sequence of 0s and 1s according to the powers of 2 that appear in its presentation. For instance, for the number 89, the number 64 is used, but 32 is not. The number 16 appears, as does 8, but not 4 or 2. Finally, the number 1 is also used. We write:

$$89 = 1011001_2$$

(It is customary to work with the large power of 2 to the left.) As other examples, we see that the code 10001011_2 corresponds to the number $128 + \cancel{64} + \cancel{32} + \cancel{16} + 8 + \cancel{4} + 2 + 1 = 139$, and the code 10111_2 to the number $16 + \cancel{8} + 4 + 2 + 1 = 23$. Numbers represented according to this method are called binary numbers. These representations correspond precisely to the representations made by choosing 2 as the BASE OF A NUMBER SYSTEM.

If one introduces a decimal point into the system and interprets positions to the right of the point as negative powers of 2, then fractional quantities can also be represented in binary notation. For instance, 0.101_2 represents the quantity $2^{-1} + 2^{-3} = \frac{1}{2} + \frac{1}{8} = \frac{5}{8}$, and $0.010101..._2$ the quantity $\frac{1}{4} + \frac{1}{16} + \frac{1}{64} + ...$, which, according to the GEOMETRIC SERIES formula, is the fraction $\frac{1}{3}$.

Any DYADIC fraction has a finite binary decimal expansion. As the study of those fractions shows, the binary code of a dyadic can be cleverly interpreted as instructions for folding a strip of paper to produce a crease mark at the location of that dyadic fraction. The process of RUSSIAN MULTIPLICATION also uses binary numbers in an ingenious manner.

Binary numbers are used in computers because the two digits 0 and 1 can be represented by two alternative states of a component (for example, "on" or "off," or the presence or absence of a magnetized region).

It is appropriate to mention that the powers of 2 solve the famous "five stone problem":

> A woman possesses five stones and a simple two-arm balance. She claims that, with a combination of her stones, she can match the weight of any rock you hand her and thereby determine its weight. She does this under the proviso that your rock weighs an integral number of pounds and no more than 31 pounds. What are the weights of her five stones?

As every number from 1 through 31 can be represented as a sum of the numbers 1, 2, 4, 8, and 16, the woman has stones of weights corresponding to these first five powers of 2.

binary operation A rule that assigns to each pair of elements of a set another element of that same set is called a binary operation. For example, the addition of two numbers is a binary operation on the set of real numbers, as is the product of the two numbers and the sum of the two numbers squared. "Union" is a binary operation on sets, as is "intersection," and CROSS PRODUCT is a binary operation on the set of vectors in three-space. However, the operation of DOT PRODUCT is *not* a binary operation on the set of vectors; the results of this operation are numbers, not other vectors.

If the set under consideration is denoted S, then a binary operation on S can be thought of as a FUNCTION f from the set of pairs of elements of the set, denoted $S \times S$, to the set S: $f : S \times S \rightarrow S$.

See also OPERATION; UNARY OPERATION.

binomial Any algebraic expression consisting of two terms, such as $2x + y$ or $a + 1$, is called a binomial.

See also MONOMIAL; POLYNOMIAL; TRINOMIAL.

binomial coefficient *See* BINOMIAL THEOREM.

binomial distribution The distribution that arises when considering the question

> What is the probability of obtaining precisely k successes in n runs of an experiment?

is called the binomial distribution. Here we assume the experiment has only two possible outcomes—"success or failure," or "heads or tails," for example—and that the probability of either occurring does not change as the experiment is repeated. The binomial distribution itself is a table of values providing the answers to this question for various values of k, from $k = 0$ (no successes) to $k = n$ (all successes).

To illustrate: the chance of tossing a "head" on a fair coin is 50 percent. Suppose we choose to toss the coin 10 times. Observe that the probability of attaining any specific sequence of outcomes (three heads, followed by two tails, then one head and four tails, for example) is $\frac{1}{2} \times \frac{1}{2} \times \overset{\text{total of 10 times}}{\cdots} \times \frac{1}{2} = \left(\frac{1}{2}\right)^{10} = \frac{1}{1024}$. In particular, the probability of seeing no heads (all tails) is also 1/1024, as is the chance of seeing 10 heads in a row.

There are 10 places for a single head to appear among 10 tosses, thus the chances of seeing precisely one head out of 10 tosses is $10 \times \left(\frac{1}{2}\right)^{10} = \frac{10}{1024}$, about 1 percent. According to the theory of COMBINATIONS, there are $\binom{10}{2} = \frac{10!}{2!\ 8!} = 45$ ways for two heads to appear among 10 places, and so the probability of seeing precisely two heads among the 10 tosses is $\binom{10}{2} \times \left(\frac{1}{2}\right)^{10} = \frac{45}{1024}$, about 4.4 percent.

Continuing this way, we obtain the binomial distribution for tossing a fair coin 10 times:

Number of Heads	Probability
0	$\binom{10}{0} \times \left(\frac{1}{2}\right)^{10} = \frac{1}{1024} \approx 0.1\%$
1	$\binom{10}{1} \times \left(\frac{1}{2}\right)^{10} = \frac{10}{1024} \approx 1.0\%$
2	$\binom{10}{2} \times \left(\frac{1}{2}\right)^{10} = \frac{45}{1024} \approx 4.4\%$
3	$\binom{10}{3} \times \left(\frac{1}{2}\right)^{10} = \frac{120}{1024} \approx 11.7\%$
4	$\binom{10}{4} \times \left(\frac{1}{2}\right)^{10} = \frac{210}{1024} \approx 20.5\%$
5	$\binom{10}{5} \times \left(\frac{1}{2}\right)^{10} = \frac{252}{1024} \approx 24.6\%$
6	$\binom{10}{6} \times \left(\frac{1}{2}\right)^{10} = \frac{210}{1024} \approx 20.5\%$
7	$\binom{10}{7} \times \left(\frac{1}{2}\right)^{10} = \frac{120}{1024} \approx 11.7\%$
8	$\binom{10}{8} \times \left(\frac{1}{2}\right)^{10} = \frac{45}{1024} \approx 4.4\%$
9	$\binom{10}{9} \times \left(\frac{1}{2}\right)^{10} = \frac{10}{1024} \approx 1.0\%$
10	$\binom{10}{10} \times \left(\frac{1}{2}\right)^{10} = \frac{1}{1024} \approx 0.1\%$

If the coin is biased—say the chances of tossing a head are now only 1/3—then a different binomial distribution would be obtained. For example, the probability of attaining precisely eight heads among 10 tosses is now only $\binom{10}{8}\left(\frac{1}{3}\right)^{8}\left(\frac{2}{3}\right)^{2} \approx 0.003$, about 0.3 percent.

In general, if p denotes the probability of success, and $q = 1 - p$ is the probability of failure, then the binomial distribution is given by the formula $\binom{n}{k}p^k q^{n-k}$, the probability of attaining precisely k successes in n runs of the experiment. This quantity is the kth term of the binomial expansion formula from the BINOMIAL THEOREM:

$$(p+q)^n = \binom{n}{0}p^0 q^n + \binom{n}{1}p^1 q^{n-1} + \binom{n}{2}p^2 q^{n-2} + \cdots + \binom{n}{n}p^n q^0$$

This explains the name of the distribution.

The binomial distribution has mean value (EXPECTED VALUE) $\mu = \sum_{k=0}^{n} k\binom{n}{k}p^k q^{n-k}$, which equals np. (To see this, differentiate the formula $(p+q)^n = \sum_{k=0}^{n}\binom{n}{k}p^k q^{n-k}$ with respect to p.) The standard deviation is $\sigma = \sqrt{npq}$. (See STATISTICS: DESCRIPTIVE.)

The Poisson Distribution

It is difficult to calculate the binomial distribution if n is very large. Mathematicians have shown that the binomial distribution can be well approximated by the NORMAL DISTRIBUTION for large values of n, provided the value p is neither extremely small nor close to one. For these troublesome values of p, SIMÉON-DENIS POISSON showed in 1837 that the values $\frac{\mu^k}{k!}e^{-\mu}$—with $\mu = np$, for $k = 0, 1, 2, \ldots$—provide a sequence of values close to the values one would expect from the binomial distribution. The distribution provided by these approximate values is called a Poisson distribution.

The Geometric Distribution

Alternatively one can ask: *what is the probability that the first success in a series of experiments occurs on the nth trial?* If p is the probability of success and $q = 1 - p$ is the probability of failure, then one obtains a first success on the nth experiment by first obtaining $n - 1$ failures and then a success. The probability of this occurring is: $P(n) = pq^{n-1}$. The distribution given by this sequence of probability values (for $n = 1, 2, 3, \ldots$) is called the geometric distribution. It has mean $\mu = 1/p$ and standard deviation $\sigma = \sqrt{q}/p$. The geometric distribution is a special example of Pascal's distribution, which seeks the probability $P_k(n)$ of the kth success occurring on the nth trial.

See also HISTORY OF PROBABILITY AND STATISTICS (essay).

binomial theorem (**binomial expansion**) The identities $(x + a)^2 = x^2 + 2xa + a^2$ and $(x + a)^3 = x^3 + 3x^2a +$

$3xa^2 + a^3$ are used in elementary ALGEBRA. These are both special cases of the general binomial theorem that asserts, for any positive integer n, we have:

$$(x + a)^n = \sum_{k=0}^{n} \binom{n}{k} x^{n-k} a^k = x^n + \binom{n}{1} x^{n-1} a + \binom{n}{2} x^{n-2} a^2$$

$$+ \cdots + \binom{n}{n-1} xa^{n-1} + a^n$$

Here each number $\binom{n}{k} = \dfrac{n!}{k!(n-k)!}$ is a COMBINATORIAL COEFFICIENT, also called a binomial coefficient.

The binomial theorem is proved by examining the process of EXPANDING BRACKETS, thinking of the quantity $(x + a)^n$ as a product of n factors: $(x + a)(x + a)\ldots$ $(x + a)$. To expand the brackets, one must select an entry from each set of parentheses ("x" or "a"), multiply together all the selected elements, and add together all possible results. For example, there is one way to obtain the term x^n: select x from every set of parentheses. There are n ways to create a term of the form $x^{n-1}a$: select a from just one set of parentheses, and x from the remaining sets. In general there are $\binom{n}{k}$ ways to select k as and $n - k$ xs. Thus, in the expansion, there will be $\binom{n}{k}$ terms of the form $x^{n-k}a^k$.

The combinatorial coefficients are the entries of PASCAL'S TRIANGLE. The binomial theorem applied to $(1 + 1)^n$ explains why the elements of each row of Pascal's triangle sum to a power of two:

$$2^n = (1 + 1)^n = \binom{n}{0} + \binom{n}{1} + \cdots + \binom{n}{n}$$

Applying the theorem to $(1 - 1)^n$ explains why the alternating sum of the entries is zero:

$$0 = (1 - 1)^n = \binom{n}{0} - \binom{n}{1} + \binom{n}{2} - \cdots \pm \binom{n}{n}$$

Applying the theorem to $(10 + 1)^n$ explains why the first few rows of Pascal's triangle resemble the powers of 11:

$11^2 = (10 + 1)^2 = 100 + 2 \times 10 + 1$

$11^3 = (10 + 1)^3 = 1{,}000 + 3 \times 100 + 3 \times 10 + 1$

$11^4 = (10 + 1)^4 = 10{,}000 + 4 \times 1{,}000 + 6 \times 100 + 4 \times 10 + 1$

(The correspondence would remain valid if we did not carry digits when computing higher powers of 11.)

The binomial theorem can be used to approximate high powers of decimals. For example, to estimate $(2.01)^{10}$ we observe:

$2.01^{10} = (2 + 0.01)^{10}$

$= 2^{10} + 10 \times 2^9 \times 0.01 + 2^8 + 45 \times 2^8 \times 0.01^2 + \ldots$

$\approx 1024 + 10 \times 512 \times 0.01 + 45 \times 256 \times 0.00001$

$= 1024 + 51.2 + 1.152$

≈ 1076

In 1665 SIR ISAAC NEWTON, coinventor of CALCULUS, discovered that it is possible to expand quantities of the form $(x + a)^r$ where r is not equal to a whole number. This leads to the generalized binomial theorem:

If r is an arbitrary real number, and $|x| < |a|$, then:

$$(x + a)^r = x^r + rx^{r-1}a + \frac{r(r-1)}{2!} x^{r-2} a^2$$

$$+ \frac{r(r-1)(r-2)}{3!} x^{r-3} a^3 + \cdots$$

The formula is established by computing the TAYLOR SERIES of $f(x) = (x + a)^r$ at $x = 0$. In 1826 Norwegian mathematician NIELS ABEL proved that the series converges for the range indicated. Notice that if r is a positive integer, then the theorem reduces to the ordinary binomial theorem. (In particular, from the $n + 1$'th place onward, all terms in the infinite sum are zero.)

The combinatorial coefficients $\binom{n}{k}$ arising in the binomial theorem are sometimes called binomial coefficients. The generalized combinatorial coefficients appear in expansions of quantities of the type $(x + y + z)^n$ and $(x + y + z + w)^n$, for example.

See also COMBINATION.

bisection method (dichotomous line search, binary line search) Often one is required to find a solution to an equation of the form $f(x) = 0$ even if there are no clear algebraic means for doing so. (For instance, there are no general techniques helpful for solving $\sqrt{x} + \sqrt{x+2} + \sqrt{x+3} - 5 = 0$. The bisection method provides the means to find, at least, approximate solutions to such equations. The method is based on the fact that if two function values $f(a)$ and $f(b)$ of a CONTINUOUS FUNCTION have opposite signs, then, according to the INTERMEDIATE VALUE THEOREM, a ROOT of the equation $f(x) = 0$ lies between a and b. The method proceeds as follows:

1. Find two values a and b ($a < b$) such that $f(a)$ and $f(b)$ have opposite signs.
2. Set $m = (a + b)/2$, the midpoint of the interval, and compute $f(m)$.
3. If $f(m) = 0$, we have found a zero. Otherwise, if $f(a)$ and $f(m)$ have opposite signs, then the zero of f lies between a and m; repeat steps 1 and 2 using these new values. If, on the other hand, $f(m)$ and $f(b)$ have opposite signs, then the zero of f lies between m and b; repeat steps 1 and 2 using these new values. In either case, a new interval containing the zero has been constructed that is half the length of the original interval.
4. Repeated application of this procedure homes in on a zero for the function.

To solve the equation $f(x) = \sqrt{x} + \sqrt{x+2} + \sqrt{x+3} - 5 = 0$, for example, notice that $f(1) = -0.268 < 0$ and $f(2) = 0.650 > 0$. A zero for the function thus lies between 1 and 2. Set $m = 1.5$. Since $f(1.5) = 0.217 > 0$ we deduce that, in fact, the zero lies between 1 and 1.5. Now set $m = 1.25$ to see that the zero lies between 1.25 and 1.5.

One can find the location of a zero to any desired degree of accuracy using this method. For example, repeating this procedure for the example above six more times shows that the location of the zero lies in the interval [1.269,1.273]. This shows that to three significant figures the value of the zero is 1.27.

The bisection method will fail to locate a root if the graph of the function touches the x-axis at that location without crossing it. Alternative methods, such as NEWTON'S METHOD, can be employed to locate such roots.

bisector   Any line, plane, or curve that divides an angle, a line segment, or a geometric object into two equal parts is called a bisector. For example, the equator is a curve that bisects the surface of the EARTH. A straight line that divides an angle in half is called an angle bisector, and any line through the MIDPOINT of a line segment is a segment bisector. If a segment bisector makes a right angle to the segment, then it is called a perpendicular bisector.

Bolyai, János (1802–1860) Hungarian *Geometry* Born on December 15, 1802, in Kolozsvár, Hungary, now Cluj, Romania, János Bolyai is remembered for his 1823 discovery of NON-EUCLIDEAN GEOMETRY, an account of which he published in 1832. His work was independent of the work of NIKOLAI IVANOVICH LOBACHEVSKY (1792–1856), who published an account of HYPERBOLIC GEOMETRY in 1829.

Bolyai was taught mathematics by his father Farkas Bolyai, himself an accomplished mathematician, and had mastered CALCULUS and mechanics by the time he was 13. At age 16 he entered the Royal Engineering College in Vienna and joined the army engineering corps upon graduation four years later.

Like many a scholar throughout the centuries, Farkas Bolyai had worked, unsuccessfully, on the challenge of establishing the PARALLEL POSTULATE as a logical consequence of the remaining four of EUCLID'S POSTULATES. He advised his son to avoid working on this problem. Fortunately, János Bolyai did not take heed and took to serious work on the issue while serving as an army officer. During the years 1820 and 1823 Bolyai prepared a lengthy treatise outlining the details of a new and consistent theory of geometry for which the parallel postulate does not hold, thereby settling once and for all the problem that had troubled scholars since the time of EUCLID: *the parallel postulate cannot be proved a consequence of the remaining postulates of Euclid.*

In Bolyai's system of hyperbolic geometry it is always the case that, for any point P in the plane, there are an infinite number of distinct lines through that point all PARALLEL to any given line not through P. (In ordinary geometry, where the parallel postulate holds, there is only one, and only one, line through a given point P parallel to a given direction. This is PLAYFAIR'S AXIOM.) In his new geometry, angles in triangles sum to less than 180°, and the ratio of the circumference of a circle to its diameter is greater than π.

Just before publishing his work, Bolyai learned that the great CARL FRIEDRICH GAUSS (1777–1855) had already anticipated much of this theory, even though he had not published any material on the matter. Bolyai decided to delay the release of his work. In 1832 he printed the details of his new theory only as a 24-page appendix to an essay his father was preparing. Later, in 1848, Bolyai discovered that Lobachevsky had published a similar piece of work in 1829. Bolyai never published the full version of his original treatise. His short 24-page piece was practically forgotten until Richard Blatzer discussed the work of both Bolyai and Lobachevsky in his 1867 text *Elemente der Mathematik* (Elements of mathematics). At that point, Bolyai's piece was recognized as the first clear account of the mathematics of a new type of geometry. He is today regarded as having independently founded the topic.

Bolyai died on January 27, 1860, in Marosvásárhely, Hungary (now Tirgu-Mures, Romania). In 1945 the University of Cluj honored Bolyai by including his name in its title. It is today known as the Babes-Bolyai University of Cluj.

Bolzano, Bernard Placidus (1781–1848) Czech *Analysis, Philosophy, Theology*

Born on October 5, 1781, in Prague, Bernard Bolzano is remembered as the first mathematician to offer a rigorous description of what is meant by a CONTINUOUS FUNCTION. The related theorem, the INTERMEDIATE-VALUE THEOREM, is sometimes named in his honor.

Bolzano studied philosophy and mathematics at the University of Prague and earned a doctoral degree in mathematics in 1804. He also completed three years of theological study at the same time and was ordained a Roman Catholic priest two days after receiving his doctorate. Choosing to pursue a career in teaching, Bolzano accepted a position as chair of philosophy and religion at the university later that year.

In 1810 Bolzano began work on understanding the foundations of mathematics and, in particular, the logical foundations of the newly discovered CALCULUS. He found the notion of an INFINITESIMAL troublesome and attempted to provide a new basis for the subject free from this concept. In his 1817 paper "Rein Analytischer Beweis" (Pure analytical proof), Bolzano explored the concept of a LIMIT—anticipating the foundational approach of AUGUSTIN-LOUIS CAUCHY offered four

years later—and proved the famous intermediate value theorem. Bolzano was also the first to provide an example of a function that is continuous at every point but differentiable at no point.

Bolzano also anticipated much of GEORG CANTOR'S work on the infinite. In his 1850 article "Paradoxien des Unendlichen" (Paradoxes of the infinite), published by a student two years after his death, Bolzano examined the nature of infinite sets and the paradoxes that arise from them. This piece contains the first use of the word "set" in a mathematical context.

Bolzano died on December 18, 1848, in Prague, Bohemia (now the Czech Republic). His work paved the way for providing rigorous underpinnings to the subject of calculus. In particular, Bolzano identified for the first time in "Rein Analytischer Beweis" the "completeness property" of the real numbers.

Bombelli, Rafael (1526–1572) Italian *Algebra*

Born in Bologna, Italy, in 1526 (the day and month of his birth date are not known), scholar Rafael Bombelli is remembered for his highly influential 1572 book *L'Algebra* (Algebra). In this work, Bombelli published rules for the solution to the QUADRATIC, CUBIC, and QUARTIC EQUATIONS, and was one of the first mathematicians to accept COMPLEX NUMBERS as solutions to equations.

Bombelli began his career as an engineer specializing in hydraulics and worked on a number of projects to turn salt marshes into usable land. Having read the great *Ars magna* (The great art) by GIROLAMO CARDANO (1501–76), Bombelli decided to write an algebra text that would make the methods developed there accessible to a general audience and be of interest and use to surveyors and engineers. He intended to write a five-volume piece but only managed to publish three volumes before his death in 1572.

Bombelli noted that Cardano's method of solving cubic equations often leads to solutions that, at first glance, appear unenlightening. For instance, examination of the equation $x^3 = 15x + 4$ leads to the solution:

$$x = \sqrt[3]{2 + \sqrt{-121}} + \sqrt[3]{2 - \sqrt{-121}}$$

Although scholars at the time rejected such quantities (because of the appearance of the square root of a negative quantity) Bombelli argued that such results should not be ignored and, moreover, that they do lead to real solutions. After developing the algebra of complex numbers, Bombelli could show that the answer presented above, for instance, is just the number $x = 4$ in disguise. Bombelli went on to show that, in fact, every equation of the form $x^3 = ax + b$, with a and b positive, has a real solution, thereby justifying the method of complex numbers.

In addition to developing complex arithmetic, Bombelli developed the basic algebra of NEGATIVE NUMBERS. In particular, he provided a geometric argument to help explain why a negative number times itself must be positive—a notion that still causes many people difficulty today.

Bombelli died in 1572. (The exact date of his death is not known.) GOTTFRIED WILHELM LEIBNIZ (1646–1716), codiscoverer of CALCULUS, taught himself mathematics from Bombelli's *L'Algebra* and described the scholar as "an outstanding master of the analytic art."

Boole, George (1815–1864) British *Logic* Born on November 2, 1815, in Lincolnshire, England, algebraist George Boole is remembered for his highly innovative work in the field of logic. In his pioneering piece, *An Investigation of the Laws of Thought, on Which are Founded the Mathematical Theories of Logic and Probability*, published in 1854, Boole established the effectiveness of symbolic manipulation as a means to represent and perform operations of reasoning. Boole is considered the founder of the field of symbolic logic.

Boole received no formal education in mathematics. As a young man he read the works of JOSEPH-LOUIS LAGRANGE and PIERRE-SIMON LAPLACE, and by age 20 began publishing original results. His early work in the field of DIFFERENTIAL EQUATIONS garnered him national attention as a capable scholar. In 1845 Boole was honored with a gold medal from the ROYAL SOCIETY of London, England's most prestigious academic society. Four years later, in 1849, he was appointed chair of the mathematics department at Queens College, Cork, Ireland, despite having no university degree. Boole stayed at this college for the rest of his life, devoting himself to teaching and research.

Boole began work in mathematical logic before moving to Ireland. At the time, logic was considered to be a topic of interest only to philosophers, but in 1847, in his pamphlet *The Mathematical Analysis of*

George Boole, an eminent mathematician of the 19th century, established the field of mathematical logic. *(Photo courtesy of Topham/The Image Works)*

Logic, Boole successfully argued that the topic has merit in the art of mathematical reasoning. By using symbols to represent statements, Boole developed an "algebra of logic" whose rules and valid manipulations matched the processes of reasoning. Thus mathematical arguments and lines of thought could themselves be reduced to simple algebraic manipulations. For instance, if the symbol x is used to represent "all butterflies," then $1 - x$ represents all that is not a butterfly. If y represents the color blue, then xy is the set of all objects that are both butterflies and blue, that is, all blue butterflies. The expression $(1 - x)(1 - y)$ represents all the nonblue nonbutterflies.

Algebraically the quantity $(1 - x)(1 - y)$ equals $1 - x - y + xy$. One could argue that the INCLUSION-EXCLUSION PRINCIPLE is at play here, saying that the set of nonblue nonbutterflies is the set of all objects that are neither butterflies, nor blue, with an adjustment made to account for the fact that the set of blue butterflies has been excluded twice. These adjustments, however, are awkward. Boole invented a new system of "algebra" that avoids such modifications. The axioms it obeys differ from those of ordinary arithmetic.

The algebra Boole invented proved to be of fundamental importance. It gave 20th-century engineers the means to instruct machines to follow commands and has since been used extensively in all computer design and electrical network theory. In a real sense, Boole was the world's first computer scientist, despite the fact that computers were not invented for another century to come. Boole died unexpectedly in 1864 at the age of 49 from pneumonia. (The exact date of his death is not known.)

Boolean algebra   In the mid-1800s GEORGE BOOLE developed a system of algebraic manipulations suitable for the study of FORMAL LOGIC and SET THEORY, now called Boolean algebra. He assumed that one is given a set of elements, which we will denote x, y, z, ..., on which one can perform two operations, today called Boolean sum, $x + y$, and Boolean product, $x \cdot y$. These operations must satisfy the following rules:

1. The operations are COMMUTATIVE, that is, for all elements x and y we have $x + y = y + x$ and $x \cdot y = y \cdot x$.
2. There exist two special elements, denoted "0" and "1," which, for all elements x, satisfy $x + 0 = x$ and $x \cdot 1 = x$.
3. For each element x there is an inverse element "$-x$" which satisfies $x + (-x) = 1$ and $x \cdot (-x) = 0$.
4. The following DISTRIBUTIVE laws hold for all elements in the set: $x \cdot (y + z) = (x \cdot y) + (x \cdot z)$ and $x + (y \cdot z) = (x + y) \cdot (x + z)$.

One can see that the Boolean operations " + " and " · " are very different from the addition and multiplication of ordinary arithmetic and so cannot be interpreted as such. However, thinking of Boolean addition as the "union of two sets" and Boolean product as "the intersection of two sets," with 0 being the empty set and 1 the universal set, we see that the all four axioms hold, making SET THEORY a Boolean algebra. Similarly, the FORMAL LOGIC of propositional calculus is a Boolean algebra if one interprets addition as the DISJUNCTION of two statements ("or") and product as their CONJUNCTION ("and").

Other rules for Boolean algebra follow from the four axioms presented above. For example, one can show that two ASSOCIATIVE laws hold: $x + (y + z) = (x + y) + z$ and $x \cdot (y \cdot z) = (x \cdot y) \cdot z$.

See also DE MORGAN'S LAWS.

Borromean rings   The term refers to a set of three rings linked together as a set, but with the property that if any single ring is cut, all three rings separate. The design of three such rings appeared on the coat of arms of the noble Italian family, Borromeo-Arese. (Cardinal Carlo Borromeo was canonized in 1610, and Cardinal Federico Borromeo founded the Ambrosian Art Gallery in Milan, Italy.) The curious property of the design attracted the attention of mathematicians.

It is an amusing exercise to arrange four rings such that, as a set, they are inextricably linked together, yet cutting any single ring would set all four free. Surprisingly this feat can be accomplished with *any* number of rings.

See also KNOT THEORY.

bound   A function is bounded if it takes values no higher than some number M and no lower than some second value L. For example, the function $f(x) = \sin x$ is bounded between the values -1 and 1. We call M an upper bound for the function and L a lower bound.

Borromean rings

A function is bounded above if the function possesses an upper bound (but not necessarily a lower bound), and bounded below if it possesses a lower bound (but not necessarily an upper bound). For example, $f(x) = x^2$ is bounded below by the value $L = 0$, since all output values for this function are greater than or equal to zero. The EXTREME-VALUE THEOREM ensures that every CONTINUOUS FUNCTION defined on a closed INTERVAL is bounded.

A set of numbers is bounded above if every number in the set is less than or equal to some value M, bounded below if every number in the set is greater than or equal to some value L, and bounded if it is both bounded above and bounded below. For example, the set $S = \{0.6, 0.66, 0.666, \ldots\}$ is bounded below by 0.6 and bounded above by 1. The smallest possible upper bound for a set is called the least upper bound, and the largest lower bound, the greatest lower bound. The set S has 2/3 as its least upper bound and 0.6 as its greatest lower bound.

The REAL NUMBERS have the property that any subset S of them that is bounded above possesses a least upper bound, and, similarly, any subset that is bounded below possesses a greatest lowest bound. This is a key property that shows that no numbers are "missing" from the real number line. (*See* DEDEKIND CUT.) This is not true of the set of rational numbers, for instance. The set of all rationals whose square is less than 2, for example, is bounded above, by 3/2 for example, but possesses no least upper bound *in the set of rationals:* the square root of two is "missing" from the set of rationals.

A sequence is bounded if, as a set of numbers, it is bounded. A geometric figure in the plane is bounded if it can be enclosed in a rectangle of finite area. For example, a CIRCLE is bounded but a HALF-PLANE is not.

Bourbaki, Nicolas Taking the name of a junior Napoleonic officer, a group of French mathematicians of the 1930s adopted the pseudonym of Nicolas Bourbaki to publish a series of books, all under the title *Éléments de mathématiques* (Elements of mathematics), that attempt to present a complete, definitive, and utterly rigorous account of all modern mathematical knowledge. This project continues today. Contributors to the work remain anonymous and change over the years. To date, over 40 volumes of work have been produced.

The material presented through Bourbaki is austere and abstract. The goal of the founding work was to develop all of mathematics on the axioms of SET THEORY and to maintain the axiomatic approach as new concepts are introduced.

The work, devoid of narrative and motivational context, is difficult to read and not suitable for use as textbooks. During the 1950s and 1960s, however, there were often no graduate-level texts in the developing new fields, and the volumes of Bourbaki were the only sources of reference. It is unlikely that today a graduate student in mathematics would consult the work of Bourbaki.

brachistochrone *See* CYCLOID.

brackets Any pair of symbols, such as parentheses ( ) or braces { }, that are used in an arithmetic or an algebraic expression to indicate that the quantity between them is to be evaluated first, or treated as a single unit in the evaluation of the whole, are called brackets. For example, in the expression $(2 + 3) \times 4$, the parentheses indicate that we are required to first calculate $2 + 3 = 5$ and then multiply this result by 4. In complicated expressions, more than one type of bracket may be used in the same equation. For instance, the expression $3\{2 + 8[2(x + 3) - 5(x - 2)]\}$ is a little easier to read than $3(2 + 8(2(x + 3) - 5(x - 2)))$.

Before the advent of the printing press in the 15th century, the VINCULUM was used to indicate the order of operations. Italian algebraist RAFAEL BOMBELLI (1526–1572) was one of the first scholars to use parentheses in a printed algebraic equation, but it was not until the early 1700s, thanks chiefly to the influence of LEONHARD EULER, GOTTFRIED WILHELM LEIBNIZ, and members of the BERNOULLI FAMILY, that their use in mathematics became standard.

Angle brackets <> are typically only used to list the components of a VECTOR or a finite SEQUENCE. Matters are a little confusing, for in the theory of quantum mechanics, angle brackets are used to indicate the DOT PRODUCT of two vectors (and not the vectors themselves). The left angle bracket "<" is called a "bra" and the right angle bracket ">" a "ket."

In SET THEORY, braces { } are used to list the elements of a set. Sometimes the elements of a sequence are listed inside a set of braces.

If x is a real number, then the bracket symbols $\lfloor x \rfloor$, $\lceil x \rceil$, and $\{x\}$ are used to denote the floor, ceiling, and fractional part values, respectively, of x.

Square brackets [ ] and parentheses ( ) are placed at the end points of an INTERVAL on the real number line to indicate whether or not the end points of that interval are to be included.

See also EXPANDING BRACKETS; FLOOR/CEILING/ FRACTIONAL PART FUNCTIONS; ORDER OF OPERATION.

Brahmagupta (ca. 598–665) Indian *Arithmetic, Geometry, Astronomy* Born in Ujjain, India, scholar Brahmagupta is recognized as one of the important mathematicians of the seventh century. His famous 628 text *Brahmasphutasiddhanta* (The opening of the universe) on the topic of astronomy includes such notable mathematical results as his famous formula for the AREA of a cyclic QUADRILATERAL, the integer solution to certain algebraic equations, and methods of solution to simultaneous equations. This work is also historically significant as the first documented systematic use of ZERO and negative quantities as valid numbers in ARITHMETIC.

Brahmagupta was head of the astronomical observatory at Ujjain, the foremost mathematical center of ancient India, and took an avid interest in the development of astronomical observation and calculation. The first 10 of the 25 chapters of *Brahmasphutasiddhanta* pertain solely to astronomy, discussing the longitude of the planets, lunar and solar eclipses, and the timing of planet alignments. Although rich in mathematical computation and technique, it is the remainder of the work that offers an insight into Brahmagupta's far-reaching understanding of mathematics on an abstract level.

Brahmagupta goes on to describe the decimal PLACE-VALUE SYSTEM used in India at his time for representing numerals and the methods for doing arithmetic in this system. (For instance, he outlines a method of "long multiplication" essentially equivalent to the approach we use today.) Brahmagupta permits zero as a valid number in all of his computations, and in fact gives it the explicit status of a number by defining it as the result of subtracting a quantity from itself. (Until then, zero acted as nothing more than a placeholder to distinguish 203 from 23, for instance.) He also explains the arithmetical properties of zero—that adding zero to a number leaves that number unchanged and multiplying any number by zero

produces zero, for instance. Brahmagupta also detailed the arithmetic of negative numbers (which he called "debt") and suggested, for the first time, that they may indeed be valid solutions to certain problems.

Brahmagupta next explores problems in ALGEBRA. He develops some basic algebraic notations and then presents a series of methods for solving a variety of linear and quadratic equations. For instance, he devised an ingenious technique for finding integer solutions to equations of the form $ax^2 + c = y^2$. (For example, Brahmagupta correctly asserted that $x = 226,153,980$ and $y = 1,766,319,049$ are the smallest positive integer solutions to $61x^2 + 1 = y^2$.) Brahmagupta also presents the famous SUMS OF POWERS formulae:

$$1 + 2 + \cdots + n = \frac{n(n+1)}{2}$$

$$1^2 + 2^2 + \cdots + n^2 = \frac{n(n+1)(2n+1)}{6}$$

$$1^3 + 2^3 + \cdots + n^3 = \frac{n^2(n+1)^2}{4}$$

as well as algorithms for computing square roots.

Unfortunately, as was the practice of writing at the time, Brahmagupta never gave any word of explanation as to how his solutions or formulae were found. No proofs were ever offered.

In the final sections of *Brahmasphutasiddhanta*, Brahmagupta presents his famous formula for the area of a cyclic quadrilateral solely in terms of the lengths of its sides. Curiously, Brahmagupta does not state that the formula is true only for quadrilaterals inscribed in a CIRCLE.

In a second work, *Khandakhadyaka*, written in 665, Brahmagupta discusses further topics in astronomy. Of particular interest to mathematicians, Brahmagupta presents here an ingenious method for computing values of sines.

Brahmagupta's methods and discoveries were extremely influential. Virtually every text that discusses Indian astronomy describes or uses some aspect of his work.

See also BRAHMAGUPTA'S FORMULA.

Brahmagupta's formula Seventh-century Indian mathematician and astronomer BRAHMAGUPTA derived a formula for the AREA of a QUADRILATERAL inscribed in a

CIRCLE solely in terms of the lengths of its four sides. His formula reads:

$$\text{area} = \sqrt{(s-a)(s-b)(s-c)(s-d)}$$

where a, b, c, and d are the four side-lengths and $s = \dfrac{a+b+c+d}{2}$ is the figure's semiperimeter.

If p and q are the lengths of the figure's two diagonals then PTOLEMY'S THEOREM asserts that $pq = ac + bd$. Brahmagupta's formula follows from BRETSCHNEIDER'S FORMULA for the area of a quadrilateral:

$$\text{area} = \frac{1}{4}\sqrt{4(pq)^2 - (b^2 + d^2 - a^2 - c^2)^2}$$

by substituting in this value for pq.

If one of the sides of the quadrilateral has length zero, that is, the figure is a TRIANGLE, then Brahmagupta's formula reduces to HERON'S FORMULA.

See also CYCLIC POLYGON.

braid A number of strings plaited together is called a braid. The theory of braids examines the number of (essentially distinct) ways a fixed number of strings, held initially in parallel, can be braided. One can combine two braids on a fixed number of strings by repeat-

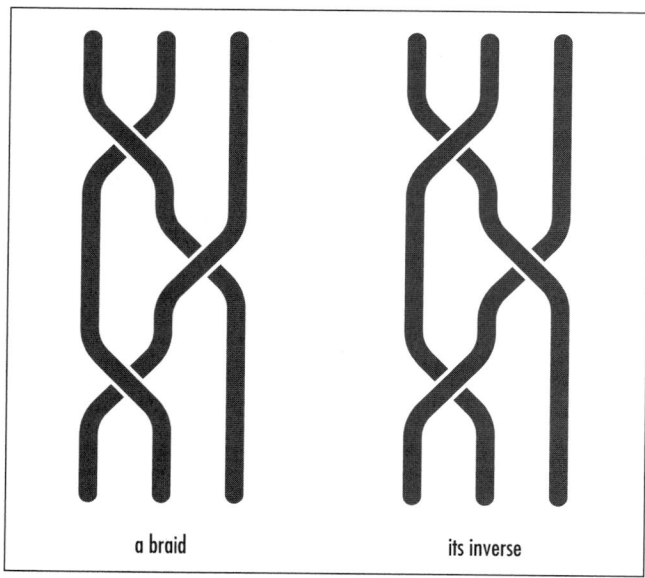

a braid   its inverse

A braid and its inverse

ing the pattern of the second braid at the end of the first braid. If, after completing this maneuver, the act of physically shaking the system of strings settles the strands to the unbraided state, then we say that the two braids are "inverse braids." For example, the two braids shown in the diagram are inverse braids.

If a braid consists of n strings, then the symbol σ_i is used to record the act of switching of the ith string over the $(i + 1)$th string (for $1 \le i \le n - 1$) and σ_i^{-1} for the act of switching of the same two strings but in the opposite sense. A general braid is then described as a string of these symbols (called a "word"). For instance, the two braids shown in the diagram below, at left, are represented by the words $\sigma_1\sigma_2^{-1}\sigma_1$ and $\sigma_1^{-1}\sigma_2\sigma_1^{-1}$, respectively. A braid with no crossings (that is, in which no strings cross) is denoted "1," and the process of combining braids corresponds precisely to the process of concatenating words. Two braids are inverse braids, if, after performing the suggested symbolic manipulations, their resulting concatenated word is 1. For instance, in our example, we have: $\sigma_1\sigma_2^{-1}\sigma_1\sigma_1^{-1}\sigma_2\sigma_1^{-1} = \sigma_1\sigma_2^{-1}\sigma_2\sigma_1^{-1} = \sigma_1\sigma_1^{-1} = 1$. It is possible that two different words can represent the same physical braid. (For instance, on three strings, the braids $\sigma_1\sigma_2\sigma_1$ and $\sigma_2\sigma_1\sigma_2$ are physically equivalent.)

Each set of braids on a fixed number of strings forms a GROUP called a braid group. Austrian mathematician Emil Artin (1898–1962) was the first to study these groups and solve the problem of determining precisely when two different words represent the same braid.

Bretschneider's formula German mathematician Carl Anton Bretschneider (1808–78) wrote down a formula for the AREA of a QUADRILATERAL solely in terms of the lengths of its four sides and the value of its four internal angles. If, reading clockwise around the figure, the side-lengths of the quadrilateral are a, b, c, and d, and the angles between the edges are A, B, C, and D (with the angle between edges d and a being A), then Bretschneider established that the area K of the quadrilateral is given by:

$$K = \sqrt{(s-a)(s-b)(s-c)(s-d) - abcd\,\cos^2(\theta)}$$

Here $s = \dfrac{a+b+c+d}{2}$ is the semiperimeter of the figure and θ is the average of any two opposite angles in the figure: $\theta = \dfrac{B+D}{2}$ or $\theta = \dfrac{A+C}{2}$. (Since the interior angles of a quadrilateral sum to 360 degrees, each quantity yields the same value for the cosine.)

Many texts in mathematics state Bretschneider's result without proof. Although algebraically detailed, the derivation of the result is relatively straightforward. One begins by noting that area K is the sum of the areas of triangles ABC and ADC. We have:

$$K = \frac{1}{2}ab\sin B + \frac{1}{2}cd\sin D$$

Multiplying by four and squaring yields:

$$16K^2 = 4a^2b^2\sin^2 B + 4c^2d^2\sin^2 D + 8abcd\sin B\sin D$$

Call this equation (i). Applying the LAW OF COSINES to each of the two triangles yields the relationship:

$$a^2 + b^2 - 2ab\cos B = c^2 + d^2 - 2cd\cos D$$

which can be rewritten:

$$a^2 + b^2 - c^2 - d^2 = 2ab\cos B - 2cd\cos D$$

Call this equation (ii). Also note that, with the aid of this second equation:

$$
\begin{aligned}
4(s-c)(s-d) &= (a+b-c+d)(a+b+c-d) \\
&= a^2 + b^2 - c^2 - d^2 + 2ab + 2cd \\
&= 2ab\cos B - 2cd\cos D + 2ab + 2cd \\
&= 2ab(1+\cos B) + 2cd(1-\cos D)
\end{aligned}
$$

Similarly:

$$
\begin{aligned}
4(s-a)(s-b) &= (-a+b+c+d)(a-b+c+d) \\
&= 2ab(1-\cos B) + 2cd(1+\cos D)
\end{aligned}
$$

Multiplying these two equations together gives:

$$
\begin{aligned}
16(s-a)&(s-b)(s-c)(s-d) \\
&= 4a^2b^2(1-\cos^2 B) + 4c^2d^2(1-\cos^2 D) \\
&\quad + 4abcd((1+\cos B)(1+\cos D) \\
&\quad + (1-\cos B)(1-\cos D)) \\
&= 4a^2b^2\sin^2 B + 4c^2d^2\sin^2 D + \\
&\quad 8abcd(1+\cos B\cos D)
\end{aligned}
$$

and substituting back into equation (i) produces:

$$
\begin{aligned}
16K^2 &= 16(s-a)(s-b)(s-c)(s-d) - \\
&\quad 8abcd(\cos B\cos D - \sin B\sin D + 1)
\end{aligned}
$$

The following identities from trigonometry:

$$\cos B\cos D - \sin B\sin D = \cos(B+D)$$

and

$$\cos(x) + 1 = 2\cos^2\left(\frac{x}{2}\right)$$

now give:

$$16K^2 = 16(s-a)(s-b)(s-c)(s-d) - 16abcd\cos^2\left(\frac{B+D}{2}\right)$$

which directly yields the famous result.

If, further, the opposite angles of the quadrilateral sum to 180 degrees (in which case the quadrilateral is a CYCLIC POLYGON), then Bretschneider's formula reduces to BRAHMAGUPTA'S FORMULA for the area of a cyclic quadrilateral:

$$K = \sqrt{(s-a)(s-b)(s-c)(s-d)}$$

If one of the sides has length zero, say $d = 0$, then the quadrilateral is a triangle and we have HERON'S FORMULA for the area of a triangle:

$$K = \sqrt{s(s-a)(s-b)(s-c)}$$

This is valid, since every triangle can be inscribed in a circle and so is indeed a cyclic polygon.

Briggs, Henry (1561–1630) British *Logarithms* Born in February 1561 (the exact birth date is not known) in Yorkshire, England, scholar Henry Briggs is remembered for the development of base-10 logarithms, revising the approach first taken by the inventor of logarithms, JOHN NAPIER (1550–1617). Today such common logarithms are sometimes called Briggsian logarithms. In 1617, after consulting with Napier, Briggs published logarithmic values of the first 1,000 numbers and, in 1624, in his famous text *The Arithmetic of Logarithms*, the logarithmic values of another 30,000 numbers, all correct to 14 decimal places.

Briggs graduated from St. John's College of Cambridge University in 1581 with a master's degree and was appointed a lectureship at the same institution 11 years later in 1592 to practice both medicine and mathematics. Four years later Briggs became the first professor of geometry at Gresham College, London, when he also developed an avid interest in astronomy.

Around this time Napier had just developed his theory of logarithms as a mathematical device specifically aimed at assisting astronomers with difficult arithmetical computations. Briggs read Napier's text on the subject in 1614 and, with keen excitement, arranged to visit the Scottish scholar in the summer of 1615. The two men agreed that the theory of logarithms would indeed be greatly simplified under a base-10 system, and two years later Briggs published his first tables of logarithmic values in *Logarithmorum chilias prima* (Logarithms of numbers 1 to 1,000). Later, in his famous 1624 piece, Briggs published logarithmic values for the numbers 1 through 20,000 and from 90,000 to 100,000. (The gap of 70,000 numbers was filled three years later by the two Dutch scholars, Adriaan Vlacq and Ezechiel de Decker.)

Briggs was appointed chair of geometry at Oxford University in 1619. He remained at Oxford pursuing interests in astronomy and classical geometry until his death on January 26, 1630. In his inaugural lecture at Gresham College, ISAAC BARROW (1630–77) expressed gratitude on behalf of all mathematicians for the outstanding work Briggs had accomplished through the study of logarithms.

Brouncker, Lord William (ca. 1620–1684) British Calculus

William Brouncker is best remembered for his work in the early development of CALCULUS and also as one of the first mathematicians in Britain to use CONTINUED FRACTIONS.

Very little is known of Brouncker's early life, including the exact year of his birth, for example, and his nationality. Records do show, however, that he entered Oxford University at the age of 16 to study mathematics, languages, and medicine. He received a doctorate of medicine in 1647, pursuing mathematics and its applications to music, mechanics, and experimental physics as an outside interest throughout his life.

Brouncker held Royalist views and took an active part in the political turmoil of the time in England.

With the restoration of the monarchy in 1660 and the election of King Charles II to the throne, Brouncker was appointed chancellor to Queen Anne and keeper of the Great Seal. Brouncker was also appointed president of the newly created ROYAL SOCIETY of London in 1662.

In mathematics, Brouncker studied infinite SERIES and made a number of remarkable discoveries. He devised a series method for computing LOGARITHMS and a surprising continued-fraction expression for π:

$$\frac{4}{\pi} = 1 + \cfrac{1^2}{2 + \cfrac{3^2}{2 + \cfrac{5^2}{2 + \cdots}}}$$

Brouncker used this formula to correctly calculate the value of π to the 10th decimal place. The great English mathematician JOHN WALLIS later published these results on Brouncker's behalf.

Brouncker also studied DIOPHANTINE EQUATIONS and found a general method for solving equations of the form $nx^2 + 1 = y^2$. LEONHARD EULER later called this equation "Pell's equation," after English mathematician John Pell (1611–84), not realizing that it was Brouncker, not Pell, who had studied it so intensively. Pell's name, unfortunately, remains attached to this equation to this day.

Brouncker died on April 5, 1684, in London, England. He never received fame as a great mathematician for he tended to focus on solving problems posed by others rather than forging original pathways in mathematics research.

brute force The method of establishing the validity of a claim by individually checking each and every instance of the claim is called brute force. Mathematicians much prefer devising general arguments and principles to demonstrate the validity of a claim, rather than resort to this method. Some claims, however, seem amenable only to the technique of brute force. For instance, the following number-naming puzzle can be solved only via brute force:

Think of a number between one and 100. Count the number of letters in its name to obtain a second number. Count the number of letters in the name of the second number to obtain a third number. Continue this way until the chain of numbers obtained this way ends on a number that repeats. This repeating number, in every case, is the number four.

As every number between one and 100 can be written in 12 or fewer letters, the second number obtained will lie between three and 12. One checks, by brute force, that each of these numbers eventually leads to the number four.

At present, the brute-force method is the only known technique guaranteed to yield an optimal solution to the famous TRAVELING-SALESMAN PROBLEM. This problem is of significant practical importance. Unfortunately, even with the fastest computers of today, the brute-force approach cannot be carried out in any feasible amount of time.

The famous FOUR-COLOR THEOREM was solved in the 1970s by reducing the problem to a finite, but extraordinarily large, number of individual cases that were checked on a computer by brute force.

See also GOLDBACH'S CONJECTURE.

Buffon needle problem (Buffon-Laplace problem)
In 1733 French naturalist Georges Buffon proposed the following problem, now known as the Buffon needle problem:

> A needle one inch long is tossed at random onto a floor made of boards one inch wide. What is the probability that the needle lands crossing one of the cracks?

One can answer the puzzle as follows:

Suppose that the needle lands with lowest end a distance x from a crack. Note that $0 \leq x \leq 1$.

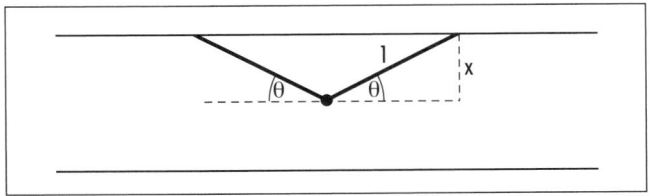

Solving the Buffon needle problem

In the diagram we see that if the needle lands within an angle as indicated by either sector labeled θ, then it will not fall across the upper crack, but it will do so if instead it lands in the sector of angle $\pi - 2\theta$. Thus the probability that the needle will fall across the crack, given that it lands a distance x units below a crack, is

$P_x = \dfrac{\pi - 2\theta}{\pi} = 1 - \dfrac{2}{\pi}\sin^{-1}x$. Summing, that is, integrating, over all possible values of x gives us the total probability P we seek:

$$P = \int_0^1 P_x\,dx = \int_0^1 1 - \frac{2}{\pi}\sin^{-1}x\,dx = 1 - \frac{2}{\pi}\int_0^1 \sin^{-1}x\,dx$$
$$= 1 - \frac{2}{\pi}\Big[x\sin^{-1}x + \sqrt{1-x^2}\Big]_0^1 = 1 - \frac{2}{\pi}\left(\frac{\pi}{2} - 1\right)$$
$$= \frac{2}{\pi}$$

In principle, this problem provides an experimental method for computing PI: simply toss a needle onto the floor a large number of times, say 10,000, and count the proportion that land across a crack. This proportion should be very close to the value $\dfrac{2}{\pi}$. In practice, however, this turns out to be a very tedious approach.

See also MONTE CARLO METHOD.

C

calculus (infinitesimal calculus)   The branch of mathematics that deals with the notion of continuous growth and change is called calculus. It is based on the concept of INFINITESIMALS, exceedingly small quantities, and on the concept of a LIMIT, quantities that can be approached more and more closely but never reached. By treating continuous changes as if they consisted of infinitely small increments, DIFFERENTIAL CALCULUS can be used, for example, to find the VELOCITY of a moving object at any particular instant. INTEGRAL CALCULUS represents the reverse process, finding the aggregate end-result if the continuous change is already known. For example, by integrating the instantaneous velocity of an object over a given time period, one finds the total distance the object moved during this time.

The word *calculus* comes from the Latin word *calx* for "pebble," which in turn is derived from the Greek word *chalis* for "limestone." Small beads or stones arranged in a counting board or on an ABACUS were often used to aid mathematical calculations, and the word *calculus* came to refer to all mathematical activity. Today, however, the word is used almost exclusively to denote the study of continuous change.

See also HISTORY OF CALCULUS (essay).

cancellation   In ARITHMETIC, the process of dividing the numerator and the denominator of a FRACTION by a common factor to produce a simpler fraction is called cancellation. For example, the fraction 18/12, which is $(6 \times 3)/(6 \times 2)$, can be simplified to 3/2 by canceling "6" from the fraction.

An amusing activity in mathematics searches for fractions that remain unchanged by the action of "anomalous cancellation." For example, deleting the digit 6 from each of the numerator and denominator of the fraction 65/26 produces the number 5/2. Curiously, 65/26 does equal five halves. Other fractions with this property include: $\dfrac{64}{16} = \dfrac{4}{1}$, $\dfrac{98}{49} = \dfrac{8}{4}$, and $\dfrac{95}{19} = \dfrac{5}{1}$. Of course deleting digits this way is an invalid mathematical operation, and the equalities obtained here are purely coincidental.

In ALGEBRA the word *cancellation* is used in two settings. Akin to the process of cancellation for fractions, one may simplify a rational expression of the form xy/xz by canceling the factor x from both the numerator and denominator. This gives xy/xz as equal to y/z. Thus, for instance, the expression $\dfrac{x^2+3x+2}{x^2+x}$ $= \dfrac{(x+1)(x+2)}{x(x+1)}$ simplifies to $\dfrac{x+2}{x+1}$.

The process of removing two equal quantities from an equation via subtraction is also called cancellation. For example, in the equation $x + 2y = 5 + 2y$, the term $2y$ can be cancelled to leave $x = 5$.

Both actions fall under the umbrella of a general cancellation law. In an abstract setting, a BINARY OPERATION "*" is said to satisfy such a law if whenever $a*x = b*x$ holds, $a = b$ follows. For the operation of addition, this reads:

$$a + x = b + x \text{ implies } a = b$$

History of Calculus

The study of calculus begins with the study of motion, a topic that has fascinated and befuddled scholars since the time of antiquity. The first recorded work of note in this direction dates back to the Greek scholars PYTHAGORAS (ca. 569–475 B.C.E) and ZENO OF ELEA (ca. 500 B.C.E.), and their followers, who put forward the notion of an INFINITESIMAL as one possible means for explaining the nature of physical change. Motion could thus possibly be understood as the aggregate effect of a collection of infinitely small changes. Zeno, however, was very much aware of fundamental difficulties with this approach and its assumption that space and time are consequently each continuous and thus infinitely divisible. Through a series of ingenious logical arguments, Zeno reasoned that this cannot be the case. At the same time, Zeno presented convincing reasoning to show that the reverse position, that space is composed of fundamental indivisible units, also cannot hold. The contradictory issues proposed by Zeno were not properly resolved for well over two millennia.

The concept of the infinitesimal also arose in the ancient Greek study of area and volume. Scholars of the schools of PLATO (428–348 B.C.E.) and of EUDOXUS OF CNIDUS (ca. 370 B.C.E.) developed a "method of exhaustion," which attempted to compute the area or volume of a curved figure by confining it between two known quantities, both of which can be made to resemble the desired object with any prescribed degree of accuracy. For example, one can sandwich a circle between two n-gons, one inscribed and one circumscribed. As one can readily compute the area of a regular n-gon, the formula for the area of a circle follows by taking larger and larger values of n. (See AREA.) The figure of a regular n-gon as n grows differs from that of a true circle only by an infinitesimal amount. ARCHIMEDES OF SYRACUSE (287–212 B.C.E.) applied this method to compute the area of a section of a PARABOLA, and 600 years later, PAPPUS OF ALEXANDRIA (ca. 300–350 C.E.) computed the volume of a SOLID OF REVOLUTION via this technique. Although successful in computing the areas and volumes of a select collection of geometric objects, scholars had no general techniques that allowed for the development of a general theory of area and volume. Each individual calculation for a single specific example was hailed as a great achievement in its own right.

The resurgence of scientific investigation in the mid-1600s led European scholars to push the method of exhaustion beyond the point where Archimedes and Pappus had left it. JOHANNES KEPLER (1571–1630) extended the use of infinitesimals to solve OPTIMIZATION problems. (He also developed new mathematical methods for computing the volume of wine barrels.) Others worked on the problem of finding tangents to curves, an important practical problem

in the grinding of lenses, and the problem of finding areas of irregular figures. In 1635, Italian mathematician BONAVENTURA CAVALIERI wrote the first textbook on what we would call integration methods. He described a general "method of indivisibles" useful for computing volumes. The principle today is called CAVALIERI'S PRINCIPLE.

French mathematician Gilles Personne de Roberval (1602–75) was the first to link the study of motion to geometry. He realized that the tangent line to a geometric curve could be interpreted as the instantaneous direction of motion of a point traveling along that curve. Philosopher and mathematician RENÉ DESCARTES (1596–1650) developed general techniques for finding the formula for the tangent line to a curve at a given point. This technique was later picked up by PIERRE DE FERMAT (1601–65), who used the study of tangents to solve maxima and minima problems in much the same way we solve such problems today. As a separate area of study, Fermat also developed techniques of integral calculus to find areas between curves and lengths of arcs of curves, which were later developed further by BLAISE PASCAL (1623–62) and English mathematicians JOHN WALLIS (1616–1703) and ISAAC BARROW (1630–77).

At the same time scholars, including Wallis, began studying SERIES and INFINITE PRODUCTS. Scottish mathematician JAMES GREGORY (1638–75) developed techniques for expressing trigonometric functions as infinite sums, thereby discovering TAYLOR SERIES 40 years before BROOK TAYLOR (1685–1731) independently developed the same results.

By the mid-1600s, certainly, all the pieces of calculus were in place. Yet scholars at the time did not realize that all the varied problems being studied belonged to one unified whole, namely, that the techniques used to solve tangent problems could be used to solve area problems, and vice versa. A fundamental breakthrough came in the 1670s when, independently, GOTTFRIED WILHELM LEIBNIZ (1646–1716) of Germany and SIR ISAAC NEWTON (1642–1727) of England discovered an inverse relationship between the "tangent problem" and the "area problem." The discovery of the FUNDAMENTAL THEOREM OF CALCULUS brought together the disparate topics being studied, provided a beautiful and natural perspective on the subject as a whole, and allowed scholars to make significant advances in solving geometric and physical problems with spectacular success. Despite the content of knowledge that had been established up until that time, it is the discovery of the fundamental theorem of calculus that represents the discovery of calculus.

Newton approached calculus through a concept of "flowing entities." He called any quantity being studied a "fluent" and its rate of change a FLUXION. Records show that

(continues)

History of Calculus
(continued)

he had developed these ideas as early as 1665, but he did not publish an account of his theory until 1704. Unfortunately, his writing style and choice of notation also made his version of calculus accessible only to a select audience. Leibniz, on the other hand, made explicit use of an infinitesimal in his development of the theory. He called the infinitesimal change of a quantity x a DIFFERENTIAL, denoted dx. Leibniz invented a beautiful notational system for the subject that made reading and working with his account of the theory immediately accessible to a wide audience. (Many of the symbols we use today in differential and integral calculus are due to Leibniz.) Leibniz formulated his approach in the mid-1670s and published his account of the subject in 1684. Although it is now known that Newton and Leibniz had made their discoveries independently, matters at the time were not clear, and a bitter dispute arose over the priority for the discovery of calculus. In 1712 the ROYAL SOCIETY of England formed a special committee to adjudicate the issue.

Applying the techniques to problems of the real world became the main theme of 18th-century mathematics. Newton's famous 1687 text *Principia* paved the way with its analysis of the laws of motion and the mechanics of the solar system. The Swiss brothers Jakob Bernoulli (1654–1705) and Johann Bernoulli (1667–1748) of the famous BERNOULLI FAMILY, champions of Leibniz in the famous dispute, studied the newly invented calculus and were the first to give public lectures on the topic. Johann Bernoulli was hired to teach differential calculus to the French nobleman GUILLAUME FRANÇOIS DE L' HOPITAL (1661–1704) via written correspondence. In 1696 L'Hopital then published the content of Johann's letters with his own name as author. Italian mathematician MARIE GAETANA AGNESI (1718–99) wrote the first comprehensive textbook dealing with both differential and integral calculus in 1755.

The Swiss mathematician LEONHARD EULER (1707–83) and French mathematicians JOSEPH-LOUIS LAGRANGE (1736–1813) and PIERRE-SIMON LAPLACE (1749–1827) were prominent in developing the theory of DIFFERENTIAL EQUATIONS. Euler also wrote extensively on the subject of calculus, showing how the theory can be applied to a vast range of pure and applied mathematical problems. Yet despite the evident success of calculus, some 18th-century scholars questioned the validity and the soundness of the subject.

The sharpest critic of Newton's and Leibniz's work was the Anglican Bishop of Coyne, George Berkeley (1685–1753). In his scathing essay, "The Analyst," Berkeley demonstrated, convincingly, that both Newton's notion of a fluxion

and Leibniz's concept of an infinitesimal are ill-defined, and that the foundations of the subject are consequently insecure. (Some historians suggest that Berkeley's vehement criticisms were motivated by a personal disdain for the apparent atheism of the type of mathematician who argues that science is certain and that theology is based on speculation.) Mathematicians consequently began looking for ways to put calculus on a sound footing. Significant progress was not made until the 19th century, when French mathematician AUGUSTINE LOUIS CAUCHY (1789–1857) suggested that the notion of an infinitesimal should be replaced by that of a LIMIT. German mathematician KARL WEIERSTRASS (1815–97) developed this idea further and was the first to give absolutely clear and precise definitions to all concepts used in calculus, devoid of any mystery or reliance on geometric intuition. The work of German mathematician RICHARD DEDEKIND (1831–1916) highlighted the role properties of the real number system play in ensuring the validity of the INTERMEDIATE-VALUE THEOREM and EXTREME-VALUE THEOREM and all the essential results that follow from them.

Initially, calculus was deemed a theory pertaining only to continuous change and CONTINUOUS FUNCTIONS. German mathematician BERNHARD RIEMANN (1826–66) was the first to consider, and give careful discussion on, the integration of discontinuous functions. His definition of an integral is the one typically presented in textbooks today. At the end of the 19th century, French mathematician HENRI LÉON LEBESGUE (1875–1941) literally turned Riemann's approach around and developed a concept of integration that can be applied to a much wider class of functions and class of settings. In constructing a Riemann integral, one begins by subdividing the range of inputs, the x-axis, into small intervals and adding areas of rectangles above these intervals of heights given by the function. This is akin to counting the value of a pocketful of coins by taking one coin out at a time, and adding the outcomes as one goes along. Lebesgue suggested, on the other hand, subdividing the range of outputs, the y-axis, into small intervals and measuring the size of the sets on the x-axis for which the function gives the desired output on the y-axis. This is akin to counting coins by first collecting all the pennies and determining their number, all the nickels and ascertaining the size of that collection, and so forth. In order to do this, Lebesgue had to develop a general "measure theory" for determining the size of complicated sets. His new theory proved to be fundamentally important, and it now has profound applications to a wide range of mathematical topics. It proved to be especially important to the sound development of PROBABILITY theory.

See also CALCULUS; DIFFERENTIAL CALCULUS; GRAPH OF A FUNCTION; INTEGRAL CALCULUS; VOLUME.

and for multiplication, assuming that x is not zero:

$$a \times x = b \times x \text{ implies } a = b$$

The cancellation law holds for any mathematical system that satisfies the definition of being a GROUP. With the guaranteed existence of inverse elements, we have $a*x = b*x$ yields $a*x*x^{-1} = b*x*x^{-1}$, which is the statement $a = b$. It holds in MODULAR ARITHMETIC in the following context:

$$ax \equiv bx \pmod{N} \text{ implies } a \equiv b \pmod{N},$$
$$\textit{only if } x \textit{ and } N \textit{ are COPRIME}$$

This follows since the statement $ax \equiv bx \pmod{N}$ holds only if $x(a - b)$ is a multiple of N. If x and N share no prime factors, then it must be the case that the term $a - b$ contains all the prime factors of N and so is a multiple of N. Consequently, $a \equiv b \pmod{N}$. The fact that 4×2 is congruent to 9×2 modulo 10, without 4 and 9 being congruent modulo 10, shows that the cancellation law need not hold if x and N share a common factor.

See also ASSOCIATIVE; COMMUTATIVE PROPERTY; DISTRIBUTIVE PROPERTY.

Cantor, Georg (1845–1918) German *Set theory* Born on March 3, 1845, in St. Petersburg, but raised in Wiesbaden and in Frankfurt, Germany, mathematician Georg Cantor is remembered for his profound work on the theory of sets and CARDINALITY. From the years 1874 to 1895, Cantor developed a clear and comprehensive account of the nature of infinite sets. With his famous DIAGONAL ARGUMENT, for instance, he showed that the set of rational numbers is DENUMERABLE and that the set of real numbers is not, thereby establishing for the first time that there is more than one type of infinite set. Cantor's work was controversial and was viewed with suspicion. Its importance was not properly understood at his time.

Cantor completed a dissertation on NUMBER THEORY in 1867 at the University of Berlin. After working as a school teacher for a short while, Cantor completed a habilitation degree in 1869 to then accept an appointment at the University of Halle in 1869. He worked on the theory of trigonometric SERIES, but his studies soon required a clear understanding of the IRRATIONAL NUMBERS. This need turned Cantor to the study of general sets and numbers.

In 1873 Cantor proved that the set of all rationals and the set of ALGEBRAIC NUMBERS are both COUNTABLE, but that the set of real numbers is not. Twenty years earlier, French mathematician JOSEPH LIOUVILLE (1809–82) established the existence of TRANSCENDENTAL NUMBERS (by exhibiting specific examples of such numbers), but Cantor had managed to show, in one fell swoop, that in fact almost all numbers are transcendental.

Having embarked on a study of the infinite, Cantor pushed forward and began to study the nature of space and dimension. In 1874 he asked whether the points of a unit square could be put into a one-to-one correspondence with the points of the unit interval [0,1]. Three years later Cantor was surprised by his own discovery that this is indeed possible. His 1877 paper detailing the result was met with suspicion and was initially refused publication. Cantor's friend and colleague, the notable JULIUS WILHELM RICHARD DEDEKIND (1831–1916), intervened and urged that the work be printed. Cantor continued work on transfinite sets for a further 18 years. He formulated the famous CONTINUUM HYPOTHESIS and was frustrated that he could not prove it.

Cantor suffered from bouts of depression and mental illness throughout his life. During periods of discomfort, he turned away from mathematics and wrote pieces on philosophy and literature. (He is noted for writing essays arguing that Francis Bacon was the true author of Shakespeare's plays.)

Cantor died of a heart attack on January 6, 1918, while in a mental institution in Halle, Germany. Even though Cantor's work shook the very foundations of established mathematics of his time, his ideas have now been accepted into mainstream thought. Beginning aspects of his work in SET THEORY are taught in elementary schools.

capital *See* INTEREST.

Cardano, Girolamo (Jerome Cardan) (1501–1576) Italian *Algebra* Born on September 24, 1501, in Pavia, Italy, scholar Girolamo Cardano is remembered as the first to publish solutions to both the general CUBIC EQUATION and to the QUARTIC EQUATION in his 1545 treatise *Ars magna* (The great art). Even though these results were due to SCIPIONE DEL FERRO

(1465–1526), Niccolò Tartaglia (1499–1557), and to his assistant Ludovico Ferrari (1522–65), Cardano unified general methods. He was an outstanding mathematician of the time in the fields of algebra, trigonometry, and probability.

At the age of 19, Cardano entered Pavia University to study medicine but quickly transferred to the University of Padua to complete his degree. He excelled at his studies and earned a reputation as a top debater. He graduated with a doctorate in medicine in 1526.

Cardano set up a small medical practice in the village of Sacco, but it was not at all successful. He obtained a post as a lecturer in mathematics at the Piatti Foundation in Milan, where he pursued interests in mathematics while continuing to treat a small clientele of patients.

In 1537 Cardano published two mathematics texts on the topics of arithmetic and mensuration, marking the start of a prolific literary career. He wrote on such diverse topics as theology, philosophy, and medicine in addition to mathematics.

In 1539 Cardano learned that an Italian mathematician by the name of Tartaglia knew how to solve cubic equations, a topic of interest to Cardano since he and Ferrari, his assistant, had discovered methods for solving quartics, if the method for cubics was clear. Tartaglia revealed his methods to Cardano under the strict promise that the details be kept secret. (At the time, Renaissance scholars, such as Tartaglia, were often supported by rich patrons and had to prove their worth in public challenges and debates. Keeping methods secret was thus of key importance.) When Cardano later learned that another scholar by the name of del Ferro had discovered methods identical to those of Tartaglia decades earlier, Cardano no longer felt obliged to keep the solution secret and published full details in his famous 1545 piece *Ars magna* (The great art). Tartaglia was outraged by this act, and a bitter dispute between the two men ensued.

Although Cardano properly credits Ferrari and Tartaglia as the first scholars to solve the cubic equation, it should be noted that Cardano properly identified general approaches that unified previous methods. Cardano also recognized that solutions would often involve complex numbers and was the first scholar to make steps toward understanding these quantities. He died on September 21, 1576, in Rome.

See also Rafael Bombelli.

Cardano's formula (Cardano-Tartaglia formula) *See* cubic equation.

cardinality In common usage, the cardinal numbers are the counting numbers 1, 2, 3, ... These numbers represent the sizes of finite sets of objects. (Unlike the ordinal numbers, however, the cardinal numbers do not take into account the order in which elements appear in a given set.)

In the late 1800s German mathematician Georg Cantor (1845–1918) extended the notion of cardinality to include meaningful examination of the size of infinite sets. He defined two sets to be of the same cardinality if their members can be matched precisely in a one-to-one correspondence. That is, each element of the first set can be matched with one element of the second set, and vice versa. For example, the set of people {Jane, Lashana, Kabeer} is of the same cardinality as the set of dogs {Rover, Fido, Spot}, since one can draw leashes between owners and dogs so that each owner is assigned just one dog, and each dog is leashed to one owner. Each of these sets is said to have cardinality 3. (Both sets are of the same cardinality as the set {1, 2, 3}.) Two sets of the same cardinality are said to be equipotent (equipollent, equinumerable, or, simply, equivalent).

The set of all counting numbers {1, 2, 3, ...} is equipotent with the set of all integers {..., −2, −1, 0, 1, 2, ...}. This can be seen by arranging each set of numbers in a list and matching elements according to their positions in the list:

$$
\begin{array}{cccccc}
1 & 2 & 3 & 4 & 5 & 6 \quad ... \\
\updownarrow & \updownarrow & \updownarrow & \updownarrow & \updownarrow & \updownarrow \\
0 & -1 & 1 & -2 & 2 & -3 \quad ...
\end{array}
$$

This procedure shows that any two sets whose elements can be listed are equipotent. Such a set is said to be denumerable, and Cantor denoted the cardinality of any denumerable set $\aleph_0$, pronounced "aleph null." Cantor's first diagonal argument shows that the set of all counting numbers, the set of integers, and the set of all rational numbers are each denumerable sets and so each have cardinality $\aleph_0$. Mathematicians have shown that the set of all algebraic numbers is also denumerable.

Not all infinite sets, however, are denumerable, as Cantor's second diagonal argument shows. For instance, it is not possible to place the set of all real numbers in a list, and so, in some well-defined sense, the set of real numbers is "more infinite" than the set of counting numbers or the rationals. Cantor denoted the cardinality of the set of real numbers c, for "continuum." We have:

$$\aleph_0 < c$$

In general, the cardinality of one set A is said to be less than the cardinality of another set B if it is possible to match each element of A with a unique element of B, but not vice versa. (Thus A is equipotent with a proper subset of B but not equipotent with B itself.) For example, the cardinality of {Jane, Lashana, Kabeer}, which we denote as 3, is less than the cardinality of the set {Rover, Fido, Spot, Tess, Rue, Tucker, Jet}, which we denote as 7. Although we can match the owners with distinct dogs, we cannot match the dogs with distinct owners.

An infinite set has the property that it is equipotent with a proper subset of itself. (This is usually taken as the definition of what it means for a set to be infinite.) For example, since the set of integers is equipotent with the subset of counting numbers, the set of integers is indeed an infinite set. The graph of the tangent function $y = \tan x$ between $x = -\frac{\pi}{2}$ and $x = \frac{\pi}{2}$ shows that each point on the y-axis is matched with a unique point on the x-axis in the interval $\left(-\frac{\pi}{2}, \frac{\pi}{2}\right)$, and vice versa. Thus the set of all points on the entire number line (the y-axis) has the same cardinality as the set of all points just in a finite interval. The set of all real numbers is indeed an infinite set.

The set of all subsets of a set A is called the "power set" of A, denoted $P(A)$. For example, the power set of $\{a,b,c\}$ is the set of eight elements: $\{\emptyset,\{a\}, \{b\},\{c\},\{a,b\},\{a,c\},\{b,c\},A\}$. (In general, the power set of a set with n elements has 2^n elements.) In some sense, the power set of the set of all counting numbers $A = \{1,2,3,\ldots\}$ matches precisely with the set of all real numbers between zero and one. This can be seen as follows:

> Given a subset B of counting numbers, let x be the real number written as a decimal in base two whose kth digit is 1 if k belongs to B, and 0 otherwise. (Thus, for instance, the subset of odd counting numbers yields the real number x

> = .1010101... .) And conversely, given a real number x between zero and one, create a subset B of counting numbers as dictated by the placement of 1s in its binary expansion. (Thus, for instance, the real number x = .010000... corresponds to the subset {2}.)

(There is one technical difficulty with this correspondence. The numbers 0.01000... and 0.001111..., for instance, represent the same real number, yet correspond to different subsets of the counting numbers. Mathematicians have shown that this difficulty can be obviated.) We have:

$$P(\mathbf{N}) = \mathbf{R}$$

where $\mathbf{N}$ denotes the set of natural numbers and $\mathbf{R}$ the set of real numbers.

The cardinality of the counting numbers is $\aleph_0$, and the above argument shows that the cardinality of its power set is c. This suggests, as for finite sets, that the power set of a set is always of "larger" cardinality than the original set. Cantor used the following argument to prove that this is indeed the case:

> Let A be a set and consider its power set P(A). Since every element a of A gives rise to the element {a} of P(A), we can certainly match the elements A with distinct elements of P(A). We now show, however, that it is not possible to reverse the process and match the elements of P(A) with distinct elements of A.

> Suppose, to the contrary, we have specified a way to associate with each subset of A a distinct element of A. For example, the subset {a,b,c} might be matched with the element a, and the subset {a,c,e,g,...} with b. Notice that the first subset contains the element a with which it is matched, but the second subset does not contain the element b with which it is matched.

> Let U be the set of all elements of A that are used in the above correspondence, but are not elements of the subsets to which they are assigned. (For instance, b above is an element of U, but a is not.) The subset U must be assigned some element of A, call it u. Now ask: Is u a member of U? The set U cannot contain u by the very definition of U.

> But in that case u satisfies the definition of being in U. So by not being in U, u must

be in *U*, and by being in *U*, *u* cannot be in *U*. This absurdity shows that there cannot be a meaningful correspondence that assigns distinct elements of *A* to subsets of *A* after all.

Given an infinite set *A*, Cantor had thus shown that the sets *P*(*A*), *P*(*P*(*A*)), *P*(*P*(*P*(*A*))), … form a never-ending chain of increasingly larger infinite sets. Thus, in a very definite sense, there are infinitely many different types of infinity. At the other end of the spectrum, the study of denumerable sets shows that every infinite set contains a denumerable subset. Thus of all the infinite sets, denumerable sets are the "smallest" type of infinite sets. The CONTINUUM HYPOTHESIS asks whether or not there is an infinite set with cardinality that lies somewhere between that of **N** and *P*(**N**) = **R**.

One might suppose that *P*(**R**), the power set of the set of all points on the real number line, is **R**^{2}, the set of all points in the plane, or, equivalently, that the power set of the set of all points in the unit interval [0,1] is the set of all points inside the unit square [0,1] × [0,1]. Surprisingly, this is not the case: there are just as many points in the unit square as there are in a unit interval. This is seen as follows:

Associate to each point (*x*,*y*) with $0 \le x \le 1$ and $0 \le y \le 1$, each written as an infinite decimal expansion, $x = 0.x_1x_2x_3...$ and $y = 0.y_1y_2y_3...$, the real number $r = 0.x_1y_1x_2y_2x_3y_3...$ in the interval [0,1], and, conversely, match each real number $r = 0.r_1r_2r_3r_4r_5r_6...$ with the point $(0.r_1r_3r_5..., 0.r_2r_4r_6...)$ in the unit square.

(Again there is a technical difficulty caused by those real numbers that have two different decimal representations. For instance, one-half can be written both as 0.5000… and 0.4999… Mathematicians have shown that this difficulty can be obviated.) It turns out that *P*(**R**) corresponds to the set of all possible real-valued functions $y = f(x)$.

See also INFINITY; PEANO'S CURVE.

cardioid The heart-shaped curve traced by a point on the circumference of one circle as it rolls around another circle of equal size is called a cardioid. In POLAR COORDINATES, the cardioid is given by an equation of the form $r = a(1 - \cos\theta)$ where *a* is the common radii of the circles, and in CARTESIAN COORDINATES by $(x^2 + y^2 - ax)^2 = a^2(x^2 + y^2)$. The PARAMETRIC EQUATIONS of the curve are $x = a\cos\theta(1 + \cos\theta)$ and $y = a\sin\theta(1 + \cos\theta)$. The curve has area one-and-a-half times the area of either generating circle, and perimeter eight times the radius.

The cardioid was first studied extensively by Italian mathematician Johann Castillon, who also coined its name in 1741.

See also CYCLOID.

Cartesian coordinates (**orthogonal coordinates, rectangular coordinates**) One of the biggest breakthroughs in the development of mathematics occurred when geometry and algebra were united through the invention of the Cartesian coordinate system. Credited to 17th-century French mathematician and philosopher RENÉ DESCARTES (whose name Latinized reads *Cartesius*), Cartesian coordinates provide a means of representing each point in the plane via a pair of numbers.

One begins by selecting a fixed point *O* in the plane, called the origin, and drawing through it two perpendicular number lines, called axes, one horizontal and one vertical, and both with the point *O* at the zero position on the line. It has become the convention to set the positive side of the horizontal number line to the right of *O*, and the positive side of the vertical number line above *O*, and to call the horizontal axis the *x*-axis, and the vertical one the *y*-axis. The Cartesian coordinates of a point *P* in the plane is a pair of numbers (*x*,*y*) which then describes the location of that point as follows:

The *x*-coordinate, or "abscissa," is the horizontal distance of the point from *O* along the horizontal axis. (A positive distance represents a point to the right of the vertical axis; a negative distance one to the left.) The *y*-coordinate, or "ordinate," is the vertical distance of the point from *O* along the vertical axis. (A positive distance represents a point located above the horizontal axis, and a negative distance one located below.)

For example, if the bottom left corner of this page is the origin of a Cartesian coordinate system, with *x*- and *y*-axes marked in units of inches, then the point with coordinates (4, 1) lies four inches to the right of the left edge of the page, and one inch above the bottom of the page.

Extending this idea to three-dimensions, points in space can be specified by a triple of numbers (x,y,z) representing the distances along three mutually perpendicular number lines. The coordinate axes are usually called the x-, y-, and z-axes. They intersect at a point O, called the origin, which is zero on all three number lines. The axes could be oriented to either form a left-handed or a right-handed system.

Coordinate Geometry

The advent of a coordinate system allowed mathematicians, for the first time, to bring the power of algebra to the study of geometry. For example, straight lines are represented as sets of points (x,y) that satisfy equations of the form $y = mx + b$. Multiplying the SLOPE m of one line with the slope of another quickly ascertains whether or not those two lines are perpendicular, for example. (The product of the slopes of two perpendicular lines is -1.)

French mathematician NICOLE ORESME (1323–82) was the first to describe a way of graphing the relationship between an independent variable and a dependent one, and thus the first to make steps toward uniting geometry and algebra. The explicit construction of what we would call a coordinate system first appeared with the work of French lawyer and amateur mathematician PIERRE DE FERMAT (1601–65). Starting with some horizontal reference line to represent an independent variable x, Fermat would graphically depict the relationship of a second variable y to it as a line segment, held at a fixed angle to the reference line, whose length would vary according to the variable y as it slides along the x-axis. Fermat did not think in terms, however, of identifying a second axis, nor did he require the line segment representing y to be perpendicular to the x-axis.

In his famous 1637 text *La géométrie* (Geometry), René Descartes independently described similar methods for representing algebraic relationships graphically. Because the work of Fermat was not published until after his death, the discovery of coordinate geometry was attributed to Descartes.

Because Fermat and Descartes interpreted the unknown variable y in an algebraic relationship as a physical length, both scholars only ever considered positive coordinates. English mathematician JOHN WALLIS (1616–1703) was the first to introduce the possibility of negative coordinates. The idea of setting a fixed second axis, the y-axis, perpendicular to the x-axis was not popular until the mid 1700s. It was an idea that seemed to evolve gradually. SIR ISAAC NEWTON (1642–1727) is considered the originator of POLAR COORDINATES.

See also COORDINATES; GRAPH OF A FUNCTION.

Cartesian product (cross product, external direct product, product set, set direct product) Given two sets A and B, their Cartesian product, denoted $A \times B$, is the set of all ordered pairs (a,b), where $a \in A$ and $b \in B$. For example, if $A = \{1,2,3\}$ and $B = \{\alpha,\beta\}$, then:

$$A \times B = \{ (1,\alpha), (2,\alpha), (3,\alpha), (1,\beta), (2,\beta), (3,\beta) \}$$

This is different from the set $B \times A$.

If sets A and B are both finite, with n and m elements, respectively, then $A \times B$ is a finite set with nm elements. German mathematician GEORG CANTOR (1845–1918) showed that if A and B are both infinite COUNTABLE sets, then their Cartesian product $A \times B$ is again countable.

The Cartesian product of three sets A, B, and C, denoted $A \times B \times C$, is defined as the set of all ordered triples (a,b,c), with $a \in A$, $b \in B$, and $c \in C$. The Cartesian product of any finite collection of sets is defined similarly. Any SEQUENCE can be thought of as an element of the Cartesian product of a countable number of sets.

If two sets A and B have a particular structure (they might both be GROUPs or VECTOR SPACES, for instance), then it is usually possible to give the Cartesian product $A \times B$ the same structure. For example, if A and B are groups with group operations $*$ and $\bullet$, respectively, then $A \times B$ has the structure of a group with group operation given by:

$$(a_1, b_1) \cdot (a_2, b_2) = (a_1 * a_2, b_1 \bullet b_2)$$

The Klein four-group is the Cartesian product of the two-element group $Z_2 = \{0,1\}$ with itself. (The group operation for Z_2 is addition in mod 2 MODULAR ARITHMETIC.)

See also SET THEORY.

casting out nines The DIVISIBILITY RULES show that the remainder of any number, when divided by 9, is the sum of its digits. For example, 59,432,641 leaves a remainder of 5 + 9 + 4 + 3 + 2 + 6 + 4 + 1 = 34 when divided by 9, which corresponds to a remainder of 3 + 4 = 7. Any sets of digits that sum to 9, such as the 5 and the 4 in the first and third positions of the number above, can be ignored when performing this calculation, for they will not contribute to the remainder.

The method of "casting out nines" is the process of deleting groups of digits summing to 9. The sum of the digits that survive is the remainder that number yields upon division by 9. For example, 59,432,641 → 932,641 → 934 → 34 shows, again, that this number leaves a remainder of 3 + 4 = 7 when divided by 9.

This method is often used to check arithmetical work. For example, we can quickly determine that 563 × 128 cannot equal 72,364. Upon division by 9, 563 leaves a remainder of 5, 128 a remainder of 2, and so their product leaves a remainder of 5 × 2 = 10, which is 1. Yet 72,364 has a remainder of 4.

Long lists of additions, subtractions, and multiplications can be quickly checked this way. Of course, errors may still be present if, by chance, remainders happen to match. For example, casting out nines will not detect that 632 × 723 = 459,636 is incorrect.

Catalan, Eugène Charles (1814–1894) Belgian *Number theory* Born on May 30, 1814, mathematician Eugène Catalan is best remembered for his work in NUMBER THEORY and for the famous series of numbers that bears his name. In 1844 Catalan conjectured that 8 and 9 are the only two consecutive integers that are both nontrivial powers ($8 = 2^3$ and $9 = 3^2$). Establishing this claim, today known as the CATALAN CONJECTURE, stymied mathematicians for over a century. It was only recently resolved.

Catalan's career in academia was turbulent. After entering the École Polytechnique in 1833 he was expelled the following year for engaging in radical political activity. He was later given permission to return to complete his degree and in 1838 was offered a post as lecturer at the institution, which he accepted. His political conduct, however, hampered his ability to advance beyond this entry-level position.

Catalan worked in the field of CONTINUED FRACTIONs and achieved some fame for publishing a simplified solution to LEONHARD EULER's "polygon division problem." This challenge asks for the number of ways to divide a regular polygon into triangles using nonintersecting diagonals. While not being the first to solve the problem (in fact the problem was first stated and solved by 18th-century Hungarian mathematician J. A. Segner and then studied by Euler), Catalan used an approach that was particularly elegant. His 1838 paper on the topic, *"Note sur une équation aux differences finie"* (Note on a finite difference equation), was very influential because of the method it detailed. The sequence of numbers that arise in the study of the problem are today called the CATALAN NUMBERS. They remain his standing legacy.

Catalan died on February 14, 1894, in Liège, Belgium.

Catalan conjecture In his 1844 letter to *Crelle's Journal,* EUGÈNE CHARLES CATALAN conjectured that the integers 8 and 9 are the only two consecutive integers that are both powers ($8 = 2^3$ and $9 = 3^2$). He was not able to prove his claim, and establishing the truth or falsehood of the conjecture became a longstanding open problem. In April 2002, amateur mathematician Preda Mihailescu announced to the mathematics community that he had completed a proof demonstrating Catalan's assertion to be true. Beforehand, Mihailescu had proved a series of related results, all while working at a Swiss fingerprinting company and exploring mathematics as an outside interest. Mathematicians are currently reviewing his final step of the work.

Catalan numbers In 1838 EUGÈNE CHARLES CATALAN studied the problem of finding the number of different ways of arranging *n* pairs of parentheses. For example, there is one way to arrange one set: ( ), two ways to arrange two pairs: ( ) ( ) and (( )), and five ways to arrange three pairs: ((( ))), (( )( )), (( )) ( ), ( ) (( )), and ( ) ( ) ( ). Is there a general formula for the number of ways to arrange *n* pairs of parentheses? This puzzle is today known as "Catalan's problem." As Catalan showed, the solution is given by the formula:

$$C_n = \frac{2 \cdot 6 \cdot 10 \cdots (4n - 2)}{(n + 1)!}$$

yielding the sequence of numbers $C_1 = 1$, $C_2 = 2$, $C_3 = 5$, $C_4 = 14$, $C_5 = 42$, ..., now called the Catalan numbers. It is convenient to set $C_0 = 1$. Some algebraic manipulation shows that the Catalan numbers can also be expressed in terms of BINOMIAL COEFFICIENTS:

$$C_n = \frac{2^n \cdot 1 \cdot 3 \cdot 5 \cdots (2n-1)}{(n+1)!}$$

$$= \frac{2 \cdot 4 \cdot 6 \cdots (2n)}{n!} \cdot \frac{1 \cdot 3 \cdot 5 \cdots (2n-1)}{(n+1)!}$$

$$= \frac{1}{n+1} \cdot \frac{(2n)!}{n!n!}$$

$$= \frac{1}{n+1} \binom{2n}{n}$$

Thus the Catalan numbers can be found in PASCAL'S TRIANGLE as the middle entry of every alternate row divided by one more than the row number, regarding the apex of the triangle as row zero.

One can show that the Catalan numbers satisfy the relationship:

$$C_0 = 1$$
$$C_n = C_0 C_{n-1} + C_1 C_{n-2} + \ldots + C_{n-1} C_0$$

The Catalan numbers appear as the solution to a surprising number of different mathematical problems. We list here just a few examples.

1. *Euler's Polygon Division Problem:* How many ways are there to divide an $(n + 2)$-sided polygon into n triangles using nonintersecting diagonals of the polygon?

A three-sided polygon, that is, a triangle, is already appropriately subdivided. There is one solution to the problem, namely, do nothing. A square can be subdivided into two triangles two different ways. One can check that a pentagon can be so subdivided five different ways. In general the solution to this puzzle is the nth Catalan number.

2. *Laddered Exponents:* How many ways can one interpret a laddered exponent?

For example, 3^2 has only one interpretation: it means $3 \times 3 = 9$. The expression 2^{3^4}, however, can be interpreted two ways: $(2^3)^4 = 4096$ or $(2^{3^4}) = 2417851639229258349412352$. In general, a laddered exponent with $(n + 1)$ terms can be interpreted

C_n different ways. (This problem is equivalent to Catalan's original parentheses puzzle.)

3. *Handshakes across a Table:* In how many different ways can n pairs of people sitting at a circular table shake hands simultaneously? No pair of handshakes may cross.

Two people sitting at a table can shake hands only one way. Four people can accomplish the feat in only two ways. (Diagonal handshakes cross.) In general, n pairs of people can shake hands C_n different ways, for one can interpret two hands shaking as a pair of parentheses.

4. *Stair Climbing:* Starting at the base of a flight of stairs, in how many ways can one take n steps up and n steps down, in any order? (You will necessarily return to the base of the steps on completion of the walk.)

There is one way to take two steps: one step up followed by one step down, and two ways to take four steps: two up, two down, or one up, one down, repeated twice. In general there are C_n ways to accomplish this task. (Thinking of a left parenthesis as an "up step" and a right parenthesis as a "down step," we can see that this puzzle too is equivalent to Catalan's original problem.)

5. *Summation Problem:* Select n numbers from the set $\{1,2,3,\ldots,2n\}$ so that their sum is a multiple of $n + 1$. Can this be done? If so, in how many different ways?

Consider the case $n = 3$, for example. There are five ways to select three numbers from the set $\{1, 2, 3, 4, 5, 6\}$ with sum divisible by four: $1 + 2 + 5 = 8$; $1 + 3 + 4 = 8$; $1 + 5 + 6 = 12$; $2 + 4 + 6 = 12$; $3 + 4 + 5 = 12$. In general, these puzzles can always be solved, and there are C_n ways to do them.

See also CATALAN CONJECTURE.

catenary The shape of the curve formed by a uniform flexible cable hanging freely between two points, such as an electric cable between two telegraph poles, is called a catenary. GALILEO GALILEI (1564–1642) thought this curve to be a PARABOLA, but German scholar Joachim Jungius (1587–1657) later proved that this could not be the case. Jacques Bernoulli (1654–1705) of the famous

BERNOULLI FAMILY was the first to write down the formula for the catenary. Up to constants, it is given by:

$$y = \frac{e^x + e^{-x}}{2} = \cosh(x)$$

This is the hyperbolic cosine function from the set of HYPERBOLIC FUNCTIONs. Engineers, in designing suspension bridges such as the Golden Gate Bridge in San Francisco, make extensive use of this function in their work. The name *catenary* comes from the Latin word *catena* for "chain."

Cauchy, Augustin-Louis (1789–1857) French *Analysis, Calculus, Number theory* Born on August 21, 1789, French scholar Augustin-Louis Cauchy is remembered as one of the most important mathematicians of

Augustin-Louis Cauchy, an eminent mathematician of the 19th century, was the first to use the notion of a "limit," as it is now known, to develop a sound model of continuity and convergence in the theory of calculus. *(Photo courtesy of the Science Museum, London/Topham-HIP/The Image Works)*

his time. With 789 mathematical papers and seven influential textbooks to his credit, Cauchy made significant contributions to the study of NUMBER THEORY, ANALYSIS, GROUP THEORY, DIFFERENTIAL EQUATIONS, and PROBABILITY. In his famous 1821 text *Cours d'analyse* (A course of analysis), Cauchy was the first to provide an exact, rigorous meaning of the terms *derivative* and *integral* as used in CALCULUS through the development of the notion of a LIMIT. Cauchy also properly defined the terms *continuity* and *convergence*. His insistence on the absolute need for rigor and clarity in all of mathematics had a lasting effect and set the standards of rigor required today of all mathematical research.

After graduating from the L'École Polytechnique in 1807 with a degree in mathematics, Cauchy pursued a career in engineering. Noted as a promising practitioner in the field, Cauchy was assigned to the Ourcq Canal project and by age 21 managed to receive a high-ranking commission in Cherbourg as a military engineer.

Despite his busy work life, Cauchy continued to pursue interests in mathematics. In 1811 he proved a result on the geometry of polyhedra, which he submitted for publication. He received considerable praise for this accomplishment and decided to change careers and pursue research in mathematics full time. A year later he returned to Paris and began looking for a faculty position at an academic institution. Cauchy was finally awarded an assistant professorship at the L'École Polytechnique in 1815.

All the while, Cauchy continued to produce and publish mathematical results. The same year as his appointment, Cauchy won the Grand Prize of the L'Académie Royale des Sciences for his outstanding mathematical discoveries on the theory of waves. This recognition garnered him some notice in the scientific community, but real fame came to Cauchy when, another year later, he solved an outstanding problem posed by PIERRE DE FERMAT (1601–1665) on the properties of FIGURATE NUMBERS. Cauchy had now proved himself an expert in a surprisingly large number of disparate fields.

Cauchy published an incredible number of papers during this early period of his life, at a rate of as many as two a week. He was so prolific that the editors of the French journal *Comptes Rendu* imposed a quota on him. In response, Cauchy persuaded a family member, who worked in the publishing field, to create a new journal that contained nothing but papers by him!

The main thrust of much of Cauchy's work was to make mathematics rigorous and precise. He insisted on providing clarity, precision, and rigor in all the courses he taught and in his published work. His famous 1821 text was in fact a course book for students developed with the intention of "doing calculus the correct way." It is fair to say that Cauchy influenced the entire course of mathematical research by pointing out, and demanding, the need for absolute clarity and care in the development of new (and even previously established) ideas. It is said that the great mathematician PIERRE-SIMON LAPLACE (1749–1827), after attending a lecture given by Cauchy on the importance of the convergence of an infinite series, quickly ran home to check the convergence of all the infinite series he had used in his already published popular text *Mécanique céleste* on celestial mechanics.

In 1826 Cauchy published seminal works in the field of number theory, and in 1829 he defined, for the first time, the notion of a complex function of a complex variable.

Cauchy left Paris and the brewing politics of the royal regime in 1830. Upon his return a year later, he refused to swear an oath of allegiance to the new regime and consequently lost his academic position. It was not until the overthrow of Louis Philippe in 1848 that Cauchy regained his university position. Even though Cauchy's publication rate slowed considerably during this trying time, he did accomplish important work on the theory of differential equations and applications to mathematical physics during this period.

Cauchy left a standing mark on the development of calculus with his work on refining the logical basis of the subject and greatly influenced the study of complex functions. A number of fundamental concepts in the field of analysis are named in his honor, including a pair of equations known as the Cauchy-Riemann equations that determine whether or not a complex function is differentiable. Cauchy died in Sceaux, near Paris, France, on May 22, 1857. His collected works, *Oeuvres completes d'Augustin Cauchy* (The complete works of Augustin Cauchy), collated under the auspices of the Académie des Sciences, were published throughout the years 1882 to 1970 in a total of 27 volumes.

Cavalieri, Bonaventura Francesco (1598–1647) Italian *Geometry* Born in Milan, Italy, in 1598 (his exact birth date is not known), mathematician, and disciple of

GALILEO, Bonaventura Cavalieri is best remembered for his 1635 work *Geometria indivisibilibus continuorum* (A new geometry of continuous indivisibles) in which he introduced his famous "method of indivisibles" for determining the areas and volumes of curved figures. This work is considered a forerunner to the entire theory of INTEGRAL CALCULUS.

While still a boy, Cavalieri joined the religious order Jesuati in Milan. In 1616, he transferred to the monastery in Pisa, where he met Galileo and developed an interest in mathematics. Even though Cavalieri taught theology for many years and became a deacon in the order, he actively pursued employment as a mathematician. In 1629 he received a position as a chair of mathematics at Bologna.

By this time Cavalieri had developed his method of indivisibles. Based on ARCHIMEDES' method of exhaustion and JOHANNES KEPLER's theory of the infinitely small, Cavalieri's technique provided a means to rapidly compute the area and volumes of certain geometric figures previously deemed too difficult for analysis. Cavalieri's famous 1635 work describing these methods, however, was not well received and was widely attacked for its lack of rigor. In response, Cavalieri published a revised piece, *Exercitationes geometricae sex,* which successfully settled all concerns. This second piece was acknowledged as a masterpiece and deemed a necessary text of study for all 17th-century scholars.

Cavalieri also studied and wrote extensively on the topics of LOGARITHMS, CONIC SECTIONS, TRIGONOMETRY, optics, and astronomy. He developed a general rule for computing the focal length of lenses, and described the principles and design of a reflecting telescope.

Cavalieri died on November 30, 1647, in Bologna, Italy. His name appears in all high-school geometry textbooks of today for the principle he devised.

See also CAVALIERI'S PRINCIPLE.

Cavalieri's principle Italian mathematician BONAVENTURA CAVALIERI (1598–1647) identified a general principle today known as Cavalieri's principle:

> Solids of equal height have equal volumes if cross-sections made by planes parallel to the bases at the same distances from these bases have equal areas.

It is based on the idea that the volume of a deck of cards, for example, does not change even if the deck is skewed. A close examination of VOLUME explains why Cavalieri's principle is true.

Cayley, Arthur (1821–1895)

Cayley, Arthur (1821–1895) British *Matrix theory, Geometry, Abstract algebra, Analysis* Born on August 16, 1821, in Richmond, England, Arthur Cayley is remembered as a prolific writer, having produced 967 papers in all, covering nearly every aspect of modern mathematics. His most significant work, *Memoir on the Theory of Matrices* (1858) established the new field of matrices and MATRIX algebra. Cayley studied abstract groups and was the first to study geometry in n-dimensional space with n a number greater than three.

Cayley demonstrated a great aptitude for mathematics as a child. A schoolteacher recognized his talent and encouraged Cayley's father to allow him to pursue studies in mathematics rather than leave school and enter the family retail business. Cayley attended Trinity College, Cambridge, and graduated in 1842. Unable to find an academic position in mathematics, Cayley pursued a law degree and practiced law for 14 years. During this time, however, Cayley actively studied mathematics and published over 250 research papers. In 1863 he was finally appointed a professorship in mathematics at Cambridge.

In 1854, while working as a lawyer, Cayley wrote "On the Theory of Groups Depending on the Symbolic Equation $\theta^n = 1$" and other significant papers that defined, for the first time, the notion of an abstract GROUP. At that time, the only known groups were PERMUTATION groups, but Cayley realized that the mathematical principles behind these structures also applied to matrices, number systems, and geometric transformations. This work allowed Cayley to begin analyzing the geometry of higher-dimensional space. This, in turn, coupled with his newly developed matrix algebra, provided the foundation for the theory of quantum mechanics, as developed by Werner Heisenberg in 1925.

Cayley was elected president of the British Association for the Advancement of Science in 1883. From the years 1889 to 1895, Cayley's entire mathematical output was collated into one 13-volume work, *The Collected Mathematical Papers of Arthur Cayley*. This project was supervised by Cayley himself until he died on January 26, 1895, at which point only seven vol-umes had been produced. The remaining six volumes were edited by A. R. Forsyth.

See also CAYLEY-HAMILTON THEOREM.

Cayley-Hamilton theorem

Cayley-Hamilton theorem English mathematician ARTHUR CAYLEY (1821–95) and Irish mathematician SIR WILLIAM ROWAN HAMILTON (1805–65) noted that any square MATRIX satisfies some polynomial equation. To see this, first note that the set of all square $n \times n$ matrices with real entries forms a VECTOR SPACE over the real numbers. For example, the set of 2×2 square matrices is a four-dimensional vector space with basis elements:

$$\begin{pmatrix} 1 & 0 \\ 0 & 0 \end{pmatrix}, \begin{pmatrix} 0 & 1 \\ 0 & 0 \end{pmatrix}, \begin{pmatrix} 0 & 0 \\ 1 & 0 \end{pmatrix}, \text{ and } \begin{pmatrix} 0 & 0 \\ 0 & 1 \end{pmatrix}$$

and any 2×2 matrix $A = \begin{pmatrix} a & b \\ c & d \end{pmatrix}$ is indeed a linear combination of these four linearly independent matrices:

$$A = \begin{pmatrix} a & b \\ c & d \end{pmatrix} = a\begin{pmatrix} 1 & 0 \\ 0 & 0 \end{pmatrix} + b\begin{pmatrix} 0 & 1 \\ 0 & 0 \end{pmatrix} + c\begin{pmatrix} 0 & 0 \\ 1 & 0 \end{pmatrix} + d\begin{pmatrix} 0 & 0 \\ 0 & 1 \end{pmatrix}$$

In general, the set of all $n \times n$ square matrices is an n^2-dimensional vector space. In particular then, for any $n \times n$ matrix A, the $n^2 + 1$ matrices $I, A, A^2, A^3, \ldots, A^{n^2}$ must be linearly dependent, that is, there is a linear combination of these elements that yields the zero matrix:

$$c_0 I + c_1 A + c_2 A^2 + \ldots + c_{n^2} A^{n^2} = 0$$

for some numbers $c_0, \ldots, c_{n^2}$. This shows:

Any $n \times n$ square matrix satisfies a polynomial equation of degree at most n^2.

Cayley and Hamilton went further and proved that any square matrix satisfies its own "characteristic polynomial":

Let A be an $n \times n$ square matrix and set x to be a variable. Subtract x from each diagonal entry of A and compute the DETERMINANT of the resulting matrix. This yields a polynomial in x

of degree n. Then the matrix A satisfies this particular polynomial equation.

For example, consider the 2×2 matrix $A = \begin{pmatrix} 2 & 1 \\ 3 & 4 \end{pmatrix}$.

Then the "characteristic polynomial" of this matrix is:

$$\det \begin{pmatrix} 2-x & 1 \\ 3 & 4-x \end{pmatrix} = (2-x)(4-x) - 3 \cdot 1 = x^2 - 6x + 5$$

One now checks that

$$A^2 - 6A + 5I = \begin{pmatrix} 2 & 1 \\ 3 & 4 \end{pmatrix}^2 - 6\begin{pmatrix} 2 & 1 \\ 3 & 4 \end{pmatrix} + 5\begin{pmatrix} 1 & 0 \\ 0 & 1 \end{pmatrix}$$

$$= \begin{pmatrix} 7 & 6 \\ 18 & 19 \end{pmatrix} - \begin{pmatrix} 12 & 6 \\ 18 & 24 \end{pmatrix} + \begin{pmatrix} 5 & 0 \\ 0 & 5 \end{pmatrix}$$

$$= \begin{pmatrix} 0 & 0 \\ 0 & 0 \end{pmatrix}$$

is indeed the zero matrix.

See also IDENTITY MATRIX; LINEARLY DEPENDENT AND INDEPENDENT.

ceiling function *See* FLOOR/CEILING/FRACTIONAL PART FUNCTIONS.

center of gravity (balance point) The location at which the weight of the object held in space can be considered to act is called the object's center of gravity. For example, a uniform rod balances at its midpoint, and this is considered its center of gravity. A flat rectangular plate made of uniform material held parallel to the ground balances at its center. This point is the figure's center of gravity.

Archimedes' LAW OF THE LEVER finds the balance point P of a system of two masses m_1 and m_2 held in space. The two-mass system can then be regarded as a single mass $m_1 + m_2$ located at P.

The center of gravity of a system of three masses in the space can be found by finding the balance point of just two masses, using ARCHIMEDES' law of the lever, and then applying the law a second time to find the

center of gravity of this balance point and the third mass. This procedure can, of course, be extended to find the center of gravity of any finite collection of masses. (A location computed this way is technically the center of mass of the system. If the force of gravity is assumed to be uniform, then the center of mass coincides with the center of gravity of the system.)

This principle can be extended to locate the center of gravity of arbitrary figures in the plane (viewed as flat, uniformly dense objects held parallel to the ground). If the figure is composed of a finite collection of rectangles glued together, one locates the center of each rectangle, the mass of each rectangle, and then regards the system as a collection of individual masses at different locations. Applying Archimedes' law of the lever as above locates the figure's center of gravity. If a figure can only be approximated as a union of rectangles, one can find the approximate location of the center of gravity via this principle, and then improve the approximation by taking the LIMIT result of using finer and finer rectangles in the approximations. This approach will yield an INTEGRAL formula for the location of the center of gravity.

See also CEVA'S THEOREM; SOLID OF REVOLUTION.

central-limit theorem In the early 1700s scientists from a wide range of fields began to notice the recurring appearance of the NORMAL DISTRIBUTION in their studies and experiments. Any measurement that represents an average value of a sample, or an aggregate value of a series of results, tends to follow this classical bell-shaped distribution. The work of MARQUIS DE PIERRE-SIMON LAPLACE in 1818 and Aleksandr Mikhailovich Lyapunov in 1901, and others, led to the establishment of the central-limit theorem:

> If an experiment involves the repeated computation of the average value of N measurements (a different set of N measurements each time), then the set of average values obtained very closely follows a normal distribution—even if the original experiments do not. The larger the value of N, the better the approximation to a normal curve.

One can go further and say that if the original experiments have mean μ and standard deviation σ, then the collection of average values also has mean μ, but standard deviation $\frac{\sigma}{\sqrt{N}}$. For example, a factory may produce light bulbs packaged in large shipping cartons, 100 per carton. Even though the lifespan of individual light bulbs may vary wildly following no recognizable distribution of values, the central limit theorem asserts that the average lifespan of the bulbs per carton is given by a normal distribution. Since the height of an individual is the aggregate effect of the growth rate of a large number of individual cells, the distribution of heights of men and woman is essentially normal, as is the distribution of heights of most anything that grows—cats, maple trees, or carrots, for example.

Another version of the central-limit theorem is useful when trying to ascertain what proportion p percent of the entire population possesses a certain property (such as "Has blood type AB" or "Will vote Republican next November"). As it is impossible to examine every individual on the globe, or poll every individual in the nation, one can examine a sample of individuals and compute the percentage in this sample with the desired property. The central-limit theorem also asserts:

If many different samples of N individuals are examined, then the distribution of the percentages of those samples possessing a particular property very closely follows a normal distribution (and the larger the value of N, the better is the approximation to a normal curve). This distribution has mean p, the true percentage of the population with this property, and standard deviation $\sigma = \sqrt{\dfrac{p(100-p)}{N}}$.

Both versions of the central-limit theorem allow statisticians to make inferences and predictions based on statistical data.

See also GEORGE PÓLYA; STATISTICS: INFERENTIAL.

Ceva's theorem Let P, Q, and R be, respectively, points on sides BC, CA, and AB of a triangle ABC. (One is permitted to extend one or more sides of the triangle.) Then the lines connecting P to A, Q to B, and R to C are CONCURRENT if, and only if:

$$\frac{BP}{PC} \cdot \frac{CQ}{QA} \cdot \frac{AR}{RB} = 1$$

Here BP, for instance, represents the distance between points B and P, and the ratio $\frac{BP}{PC}$ is considered positive if the direction from B to P is the same as the direction from P to C, and negative if they are in opposite directions.

This result is due to Italian mathematician Giovanni Ceva (1647–1734). It is equivalent to the statement that the operation of finding the CENTER OF GRAVITY of point masses is ASSOCIATIVE. For instance, given three masses at locations A, B, and C in a plane, one could locate the center of mass of the entire system by first computing the center of mass of just two points and then compute the center of mass of that result with the third point. Ceva proved that the same final result is produced no matter which two points are chosen initially.

Ceva's theorem can be proved mathematically by making repeated use of MENELAUS'S THEOREM.

chain rule (function of a function rule) If $y = f(u)$ is a function of a quantity u, which in turn is a function of another quantity x, $u = g(x)$ say, then y itself can be thought of as a function of x as a COMPOSITION of functions: $y = f(g(x))$. The chain rule states that the rate of change of y with respect to the quantity x is given by the formula:

$$\frac{df}{dt} = \frac{\partial f}{\partial x} \cdot \frac{dx}{dt}.$$

This can also be written: $(f(g(x)))' = f'(g(x)) \cdot g'(x)$. For example, to differentiate $y = (x^2 + 2)^{100}$, we can write $y = u^{100}$, where $u = x^2 + 2$ and so:

$$\frac{dy}{dx} = \frac{dy}{du} \cdot \frac{du}{dx} = \frac{d}{du}(u^{100}) \cdot \frac{d}{dx}(x^2+2) = 100u^{99} \cdot 2x$$

$$= 200x(x^2+1)^{99}$$

The chain rule can be proved by making use of the formal definition of a derivative as a LIMIT:

$$\frac{d}{dx}f(g(x)) = \lim_{h \to 0} \frac{f(g(x+h)) - f(g(x))}{h}$$

$$= \lim_{h \to 0} \frac{f(g(x+h)) - f(g(x))}{g(x+h) - g(x)} \cdot \frac{g(x+h) - g(x)}{h}$$

$$= f'(g(x) \cdot g'(x)$$

Intuitively, the concept is easy to grasp if we think of derivatives as rates of change. For example, if y changes a times as fast as u, and u changes b times as fast as x, then we expect y to change ab times faster than x.

The chain rule extends to functions of more than one variable. For example, if $z = f(x, y)$ is a function of two variables with each of x and y a function of t, then one can show that the total derivative of z with respect to t is given by:

$$\frac{df}{dt} = \frac{\partial f}{\partial x} \cdot \frac{dx}{dt} + \frac{\partial f}{\partial y} \cdot \frac{dy}{dt}$$

See also PARTIAL DERIVATIVE.

change of variable   *See* INTEGRATION BY SUBSTITUTION.

chaos   A situation in which a DYNAMICAL SYSTEM can appear to be random and unpredictable is called chaos. More precisely, mathematicians define a dynamical system to be chaotic if the set of all equilibrium points for the system form a FRACTAL.

The term *chaos* was introduced by American mathematician James Yorke and Chinese mathematician Tien-Yien Li in their 1975 seminal paper on iterations of functions on the real number line.

characteristic polynomial   *See* CAYLEY-HAMILTON THEOREM.

Chebyshev, Pafnuty Lvovich (Tchebyshev) (1821–1894) Russian *Number theory, Analysis, Statistics* Born on May 16, 1821, in Okatova, Russia, Pafnuty Chebyshev is remembered for his significant contributions to NUMBER THEORY, ANALYSIS, PROBABILITY theory, and the development of inferential statistics. In 1850 he proved a conjecture posed by French mathematician Joseph-Louis François Bertrand (1822–1900) stating that for any value $n > 3$, there is at least one PRIME between n and $2n - 2$. Chebyshev is also noted for founding an influential school of mathematics in St. Petersburg.

Chebyshev entered Moscow University in 1837 and graduated four years later with an undergraduate degree in mathematics. Driven by an unabashed desire to achieve international recognition, Chebyshev immersed himself in mathematical work. He earned a master's degree in 1846 at the same institution, all the while publishing results on integration theory and methods, the convergence of TAYLOR SERIES, and the development of analysis. Chebyshev also examined the principles of probability theory and developed new insights that prove the main results of the theory in an elementary, but rigorous, way. In particular, Chebyshev was able to offer an elegant proof of SIMÉON-DENIS POISSON's weak LAW OF LARGE NUMBERS.

In 1849 Chebyshev wrote a thesis on the theory of MODULAR ARITHMETIC, earning him a doctorate in mathematics from Moscow University, as well as a prize from the Russian Academy of Science in recognition of its originality and its significance. In his study of prime numbers, Chebyshev not only established Bertrand's conjecture, but also made significant steps toward proving the famous PRIME-NUMBER THEOREM.

Chebyshev was elected as a full professor in mathematics at the University of St. Petersburg in 1850, and, by this time, had indeed achieved international fame. He traveled extensively throughout Europe and collaborated with many scholars on research projects on topics as diverse as mechanics, physics, mechanical inventions, and the construction of calculating machines, as well as continued work in mathematics. He was awarded many honors throughout his life, including membership to the Berlin Academy of Sciences in 1871, the Bologna Academy in 1873, the ROYAL SOCIETY of London in 1877, the Italian Royal Academy in 1880, and the Swedish Academy of Sciences in 1893. Every Russian university elected him to an honorary faculty position, and Chebyshev was even awarded honorary membership to the St. Petersburg Artillery Academy, as well as to the French Légion d'Honneur. He died on December 8, 1894, in St. Petersburg, Russia.

A number of results and concepts are today named in Chebyshev's honor. For example, in analysis, the "Chebyshev polynomials" provide a basis for the VECTOR SPACE of CONTINUOUS FUNCTIONs and have important applications to approximation theory. In statistics and probability theory, CHEBYSHEV's THEOREM provides a "weak law of large numbers."

See also STATISTICS: INFERENTIAL.

Chebyshev's theorem (Chebyshev's inequality) This result, due to the Russian mathematician PAFNUTY LVOVICH CHEBYSHEV (1821–94), can be thought of as an extension of the 68-95-99.7 rule for the NORMAL DISTRIBUTION to one applicable to all distributions. It states that if an arbitrary DISTRIBUTION has mean μ and standard deviation σ, then the probability that a measurement taken at random will have value differing from μ by more than k standard deviations is at most $1/k^2$. This shows that if the value μ is small, then all DATA values taken in an experiment are likely to be tightly clustered around the value μ.

Manufacturers make use of this result. For example, suppose a company produces pipes with mean diameter 9.57 mm, with a standard deviation of 0.02 mm. If manufacturer standards will not tolerate a pipe more than four standard deviations away from the mean (0.08 mm), then Chebyshev's theorem implies that on average about 1/16, that is 6.25 percent, of the pipes produced per day will be unusable.

The LAW OF LARGE NUMBERS follows as a consequence of Chebyshev's theorem.

See also STATISTICS: DESCRIPTIVE.

chicken *See* PRISONER'S DILEMMA.

Ch'in Chiu-shao (Qin Jiushao) (1202–1261) China *Algebra* Born in Szechwan (now Sichuan), China, mathematician and calendar-maker Ch'in Chiu-shao is remembered for his 1243 text *Shushu jiuzhang* (Mathematical treatise in nine sections), which contains, among many methods, an effective technique of iterated multiplication for evaluating polynomial equations of arbitrary degree. (In modern notation, this technique is equivalent to replacing a polynomial such as $4x^3 + 7x^2 - 50x + 9$, for instance, with its equivalent form as a series of nested parentheses: $((4x + 7)x - 50)x + 9$. In this example, only three multiplications are needed to evaluate the nested form of the polynomial compared with the six implied by the first form of the expression. In practice, this technique saves a considerable amount of time.) This approach was discovered 500 years later in the West independently by Italian mathematician Paolo Ruffini (1765–1822) and English scholar William George Horner (1786–1837). Ch'in Chiu-shao also extended this method to find solutions to polynomial equations.

His text is also noted for its development of MODULAR ARITHMETIC. In particular, Ch'in Chiu-shao proved the following famous result, today known as the Chinese remainder theorem:

> If a set of integers m_i are pair-wise COPRIME, then any set of equations of the form $x \equiv a_i \pmod{m_i}$ has a unique solution modulo the product of all the m_i.

For example, this result establishes that there is essentially only one integer x that leaves remainders of 1, 11, and 6, respectively, when divided by 5, 13, and 16 (namely, 726, plus or minus any multiple of $5 \times 13 \times 16 = 1040$).

After serving in the army for 14 years, Ch'in Chiu-shao entered government service in 1233 to eventually become provincial governor of Qiongzhou. His 1243 piece *Shushu jiuzhang* was his only mathematical work.

Chinese mathematics Unfortunately, very little is known about early Chinese mathematics. Before the invention of paper around 1000 C.E., the Chinese wrote on bark or bamboo, materials that were far more perishable than clay tablets or papyrus. To make matters worse, just after the imperial unification of China of around 215 B.C.E., Emperor Shi Huang-ti of the Ch'ih dynasty ordered that all books from earlier periods be burned, along with the burying alive of any scholars who protested. Only documents deemed "useful," such as official records and texts on medicine, divination, and agriculture were exempt. Consequently very little survived beyond this period, although some scholars did try to reconstruct lost materials from memory.

The art of mathematics was defined by ancient Chinese scholars as *suan chu*, the art of calculation. Often the mathematics studied was extremely practical in nature, covering a wide range of applications, including engineering, flood control, and architecture, as well astronomy and divination. Practitioners of the art were capable scientists. Records show, for example, that the Chinese had invented seismographs to measure earthquakes by the year 1000 C.E., and used compasses made with magnetic needles a century later.

Evidence of mathematical activity in China can be dated back to the 14th century B.C.E.. Tortoise shells and cattle bones inscribed with tally marks indicate that the people of the ancient Shang dynasty had developed a base-10 notational system utilizing place value. This establishes that the Chinese were one of the first civilizations to invent a DECIMAL REPRESENTATION system essentially equivalent to the one we use today. Like other civilizations of the time, however, the Chinese had not developed a notation for zero and so wrote, for example, the numbers 43 and 403 the same way, namely as "ǀǀǀǀ ǀǀǀ," relying on context to distinguish the two.

The most important early Chinese mathematical text is *Jiuzhang suanshu* (Nine chapters on the mathematical art) dating from the period of the Han dynasty (206 B.C.E. to 220 C.E.). The author of the work is unknown, but it is believed to be a summary of all mathematical knowledge possessed in China up to the third century C.E., and may well have been the result of several authors contributing to the same work. The text is a presentation of 246 problems replete with solutions and general recipes for solving problems of a similar type. The work is generally very practical in nature, with three chapters devoted to issues of land surveying and engineering, and three to problems in taxation and bureaucratic administration. But the text does describe sophisticated mathematical techniques of an abstract nature, and it offers many problems of a recreational flavor. The document thus also clearly demonstrates that scholars of the time were also interested in the study of mathematics for its own sake.

Jiuzhang suanshu is clearly not written for beginners in the art of mathematics: many basic arithmetic processes are assumed known. Another text of the same period, *Chou pei suan ching* (Arithmetic classic of the gnomon and the circular paths of heaven), describes basic mathematical principles such as working with fractions (and establishing common denominators), methods of extracting square roots, along with basic principles and elements of geometry, and surprisingly, what appears to be a proof of what we today call PYTHAGORAS'S THEOREM. At the very least, the Chinese of this period knew the theorem for right triangles with sides of length 3, 4, and 5 as the diagram of *hsuan-thu*—four copies of a 3–4–5 right triangle arranged in a square—appears in the text. Although this diagram in itself does not constitute a "proof" of Pythagoras's theorem, the idea embodied in the diagram can nonetheless easily be expanded upon to establish a general proof of the result. For this reason, it is believed that the Chinese had independently established the same

famous result. This is certainly verified in the text of *Jiuzhang suanshu*, as many problems posed in the piece rely on the reader making use of the theorem.

Many scholars from the second to the 15th centuries wrote commentaries on the work *Jiuzhang suanshu* and extended many of the results presented there. Perhaps the most famous of these were the commentaries of Liu Hui who, in 263 C.E., offered written proofs of the formulae for the volumes of a square pyramid and a tetrahedron presented in *Jiuzhang suanshu*, as well as developed a more precise value for π than presented in the text. Liu Hui later went on to write *Haidao suanjing* (Sea island mathematical manual), in which he solved problems related to the surveying and mapping of inaccessible objects using a refined method of "double differences" arising from pairs of similar triangles. This extended the work of proportions presented in *Jiuzhang suanshu*. Liu Hui also claimed that the material of *Jiuzhang suanshu* dates back to 1100 B.C.E., but added that the actual text was not written until 100 B.C.E. Historians today differ about how seriously to take his claim.

The text *Jiuzhang suanshu* also contains recipes for extracting square and cube roots, and methods for solving systems of linear equations using techniques very similar to the methods of LINEAR ALGEBRA we use today. Liu Hui's commentary gives justification for many of the rules presented. Although not formal proofs based on axioms, it is fair to describe Liu Hui's justifications as valid informal proofs. It seems that mathematicians for the centuries that followed remained satisfied with simple informal arguments and justifications, and no formal rigor was deemed necessary.

Chinese scholars also made significant contributions to the study of COMBINATORICS, NUMBER THEORY, and ALGEBRA. Mathematician CH'IN CHIU-SHAO (ca. 1200 C.E.) developed inventive methods for evaluating polynomial expressions and solving polynomial equations. He also established the famous "Chinese remainder theorem" of number theory. The work of LI YE of the same period also establishes an algebra of polynomials. Although Chinese mathematicians of the time were familiar with negative numbers, they ignored negative solutions to equations, deeming them absurd.

The famous text *Su-yuan yu-chien* (The precious mirror of the four elements) written by the scholar CHU SHIH-CHIEH (ca. 1300 C.E.) contains a diagram of what has in the West become known as PASCAL'S TRIANGLE.

This text was written 300 years before French mathematician BLAISE PASCAL was born. (Some historians believe that this work in fact dates back 200 years earlier to the writings of mathematician Jia Xian.) Scholars of this time routinely used the triangle to approximate nth roots of numbers using the equivalent of the BINOMIAL THEOREM of today. They preferred their procedural methods of extracting square roots to solve QUADRATIC equations, rather than make use of the general quadratic formula.

Soon after JOHN NAPIER (1550–1617) of the West published an account of his new calculating aid, the NAPIER'S BONES, the Chinese developed an analogous system of graded bamboo rods that could be used to quickly compute long multiplications and divisions. It is not known if the Chinese invented this system independently, or whether the idea was perhaps brought to them by 17th-century Jesuit missionaries. Along with the ABACUS developed in China 500 years earlier, the calculating rods allowed for improved arithmetic computations, especially useful for the precise computations needed in astronomy.

Early scholar ZU CHONGZHI (ca. 500 C.E.) computed the volume of a sphere by a principle identical to that of BONAVENTURA CAVALIERI (1598–1647).

See also MAGIC SQUARE.

chi-squared test   The chi-squared test is a statistical test (*see* STATISTICS: INFERENTIAL) used to determine whether or not two characteristics of a population are independent or associated in some way. For example, imagine a social study looking for a possible correlation between the type of milk people prefer on their cereal and the number of body piercings they possess. Five hundred people were surveyed and the results obtained are displayed in a CONTINGENCY TABLE.

	No piercings	One or two piercings	More than two piercings	
Fat-Free Milk	47	33	22	102
2% Milk	40	80	44	164
Whole Milk	113	37	84	234
	200	150	150	500

Observe, in this study, that 102/500 = 0.204 of the participants are fat-free milk users. If milk choice bears no relationship to body piercings, we would expect then about 0.204 of the 200 folk with no piercings to use fat-free milk. We observed a value of 47 (the observed frequency) but expect a value of 0.204 × 200 = 40.8 (the expected frequency). Similarly, the expected value for fat-free milk users with more than two piercings is 0.204 × 150 = 30.6 and for whole milk users with one or two piercings: (234/500) × 150 = 70.2. In this way we compute all expected frequencies, here shown in parentheses:

	No piercings	One or two piercings	More than two piercings	
Fat-Free Milk	47 (40.8)	33 (30.6)	22 (30.6)	102
2% Milk	40 (65.6)	80 (49.2)	44 (49.2)	164
Whole Milk	113 (52.9)	37 (70.2)	84 (70.2)	234
	200	150	150	500

Denoting the observed frequencies by the letter o and the expected frequencies by e, we compute the chi-squared statistic, χ^2, as:

$$\chi^2 = \sum \frac{(o-e)^2}{e}$$

where the sum is over all entries in the table. (In the 1800s it was customary to convert all differences to a positive value by use of the squaring function rather than the ABSOLUTE VALUE function. This way, techniques of calculus could be readily applied—it is straightforward to differentiate the square function, for example.) A large value for χ^2 indicates that there is considerable discrepancy between observed and expected values, suggesting that the two features of the population are not independent, i.e., that there is a CORRELATION. A small χ^2 value suggests that there is no correlation.

Our particular example yields the value:

$$\chi^2 = \frac{(47-40.8)^2}{40.8} + \frac{(33-30.6)^2}{30.6} + \frac{(22-30.6)^2}{30.6}$$
$$+ \frac{(40-65.6)^2}{65.6} + \frac{(80-49.2)^2}{49.2} + \frac{(44-49.2)^2}{49.2}$$
$$+ \frac{(113-52.9)^2}{52.9} + \frac{(37-70.2)^2}{70.2} + \frac{(84-70.2)^2}{70.2}$$
$$= 120.1$$

The chi-squared distribution is a DISTRIBUTION representing the values one would expect χ^2 to adopt given the assumption that the two features being studied are independent. There is one distribution for each table of given dimensions. Statistics texts usually present lists of values for this statistic. It turns out that the χ^2 value in this example is extraordinarily high, suggesting that this study indicates a strong correlation between milk choice and body piercings. The chi-squared test does not give any information about the nature of the correlation detected, only that it seems to exist. Further examination of the data by alternative methods may provide details of the association.

See also KARL PEARSON.

chord If *A* and *B* are two points on a continuous curve, then a straight-line segment connecting *A* to *B* is called a chord to the curve. This is not to be confused with the ARC of the curve connecting *A* to *B*.

The CIRCLE THEOREMS show that any chord of a circle is bisected by a radius that is perpendicular to it. BERTRAND'S PARADOX shows that the act of selecting chords of a circle at random can lead to philosophical difficulties.

See also CIRCLE.

Chu Shih-Chieh (Zhu Shijie) (ca. 1270–1330) Chinese *Algebra* Often regarded as one of China's greatest mathematicians, Chu Shih-Chieh is remembered for his influential 1303 text *Su-yuan yu-chien* (Precious mirror of the four elements). It contains a diagram of PASCAL'S TRIANGLE, as it has become known in the West, representing one of the earliest appearances of the figure in the history of mathematics. In the work, Chu Shih-Chieh uses the triangle to describe a general method for extracting roots to equations. He also presents a system of notation for polynomials in four unknowns, which he calls the four elements—namely, the celestial, the earthly, the material, and the human—and provides effective techniques for manipulating them to solve problems.

His section on FINITE DIFFERENCES gives formulae for the sums of the first *n* terms of each diagonal of Pascal's triangle (for instance, that $1 + 2 + 3 + \ldots + n$ equals $\frac{n(n + 1)}{2}$, and that $1 + 3 + 6 + 10 + \ldots + \frac{n(n + 1)}{2}$ equals $\frac{n(n + 1)(n + 2)}{6}$), and also provides general techniques for summing arbitrary series. He also provides methods for solving equations via a process of successive approximations.

Many historians claim that Chu Shih-Chieh's impressive work represents the peak of ancient Chinese mathematics, noting that relatively little progress was made for a long time after the publication of this piece. Four years before the release of *Su-yuan yu-chien*, Chu Shih-Chieh wrote a mathematical text intended to help beginners in the subject. Extremely little is known of his personal life.

circle The set of all points in a plane a fixed distance *r* from a given point *O* forms a closed curve in the plane called a circle. The length *r* is called the radius of the circle, and the point *O* its center. If the center point has CARTESIAN COORDINATES $0 = (a,b)$, then the DISTANCE FORMULA shows that any point (x,y) on the circle satisfies the equation:

$$(x - a)^2 + (y - b)^2 = r^2$$

If the point (x,y) on this circle makes an angle θ with a horizontal line through the center of the circle, then we have:

$$x = a + r\cos\theta$$
$$y = b + r\sin\theta$$

These are the PARAMETRIC EQUATIONS of a circle of radius *r* and center (a,b).

The DIAMETER of a circle is the maximal distance between two points on the circle. It equals twice the radius of the circle. A circle is a figure of CONSTANT WIDTH.

The length of the curve closed to form a circle is called the circumference of the circle. Scholars since the time of antiquity have observed that the ratio of the circumference of a circle to its diameter is the same for all circles. This constant value is called PI, denoted π. We have:

$$\pi = 3.14159265\ldots$$

That all circles yield the same value for π is not immediately obvious. This is a property of EUCLIDEAN GEOMETRY of the plane, and a careful study of SCALE explains why it must be true. (The value of π varies from 2 to 3.141592... for different circles drawn on the surface of a SPHERE, for instance, since the diameter of a circle must be measured as the length of a curved line on the surface.)

If C is the circumference of a planar circle and $D = 2r$ is its diameter, then, by definition, $\pi = C/D$. This yields a formula for the circumference of a circle:

$$C = 2\pi r$$

A study of AREA also shows that the area A of a circle is given by:

$$A = \pi r^2$$

It is not immediate that the value π should also appear in this formula.

A study of EQUIDISTANCE shows that it is always possible to draw a circle through any three given points in a plane (as long as the points do not lie in a straight line), or, equivalently, it is always possible to draw a CIRCUMCIRCLE for any given TRIANGLE. APOLLONIUS OF PERGA (ca. 262–190 B.C.E.) developed general methods for constructing a circle TANGENT to any three objects in the plane, be they points, lines, or other circles.

Any line connecting two points on a circle is called a CHORD of the circle. It divides the circle into two regions, each called a segment. A chord of maximal length passes through the center of a circle and is also called a diameter of the circle. (Thus the word *diameter* is used interchangeably for such a line segment and for the numerical value of the length of this line segment.) A *radius* of a circle is any line segment connecting the center of the circle to a point on the circle. Two different radii determine a wedge-shaped region within the circle called a sector. If the angle between the two radii is θ, given in RADIAN MEASURE, then the area of this segment is $\frac{\theta}{2\pi} \cdot \pi r^2 = \frac{1}{2}\theta r^2$. The length of the ARC of the circle between these two radii is $r = \frac{a}{2\sin(A)}$.

Any two points $P = (a_1, b_1)$ and $Q = (a_2, b_2)$ in the plane determine a circle with the line segment connecting P to Q as diameter. The equation of this circle is given by:

$$(x - a_1)(x - a_2) + (y - b_1)(y - b_2) = 0$$

The JORDAN CURVE THEOREM establishes that a circle divides the plane into two regions: an inside and an outside. (This seemingly obvious assertion is not true for circles drawn on a TORUS, for example.) Two intersecting circles divide the plane into four regions; three intersecting circles can be arranged to divide the plane into eight regions; and four mutually intersecting circles can divide the plane into 14 regions. In general, the maximal number of regions into which n intersecting circles divide the plane is given by the formula: $n^2 - n + 2$.

The region formed at the intersection of two intersecting circles of the same radius is called a lens.

There are a number of CIRCLE THEOREMS describing the geometric properties of circles. A circle is a CONIC SECTION. It can be regarded as an ELLIPSE for which the two foci coincide.

If one permits the use of COMPLEX NUMBERS, then any two circles in the plane can be said to intersect. For example, the two circles each of radius one centered about the points $(0,0)$ and $(4,0)$, respectively, given by the equations $x^2 + y^2 = 1$ and $(x - 4)^2 + y^2 = 1$ intersect at the points $(2, i\sqrt{3})$ and $(2, -i\sqrt{3})$.

The three-dimensional analog of a circle is a SPHERE: the locus of all points equidistant from a fixed point O in three-dimensional space. In one-dimension, the analog of a circle is any pair of points on a number line. (Two points on a number line are equidistant from their MIDPOINT.)

The midpoint theorem asserts that all midpoints of line segments connecting a fixed point P in the plane to points on a circle C form a circle of half the radius of C.

See also APOLLONIUS'S CIRCLE; BRAHMAGUPTA'S FORMULA; CYCLIC POLYGON; FAREY SEQUENCE; NINE-POINT CIRCLE; UNIT CIRCLE; VENN DIAGRAM.

circle theorems A CIRCLE is defined as the set of points in a plane that lie a fixed distance r, called the radius, from some fixed point O, called the center. This simple definition has a number of significant geometric consequences:

1. Tangent Theorems

   The point of contact of a TANGENT line with a circle is the point on that line closest to the center point O. As a consequence of PYTHAGORAS'S THEOREM, the line connecting the point of contact to O is at an angle 90° to the tangent line. This proves:

   > The tangent to a circle is PERPENDICULAR to the radius at the point of contact.

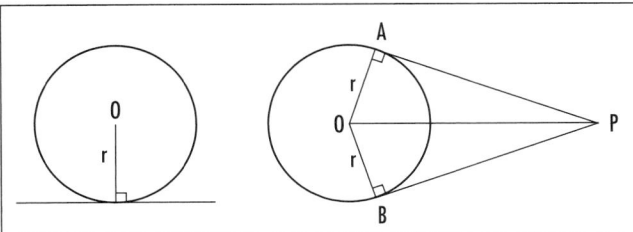

Tangent theorems

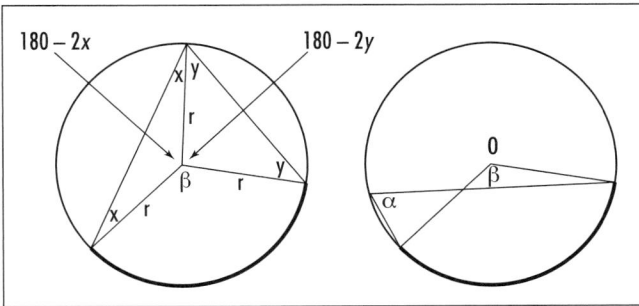

Proving the central-angle/peripheral-angle theorem

This is illustrated in the diagram above, left.

The above diagram to the right shows that two tangents through a common point *P* outside the circle produce line segments *PA* and *PB* of equal length. This follows from the fact that the two triangles produced are both right triangles of the same height with a shared hypotenuse, and are hence congruent. Thus:

> If *PA* and *PB* are tangents to a circle at points *A* and *B*, respectively, then *PA* and *PB* have the same length.

2. Inscribed-Angle Theorems
In the diagram below, left, angles α (the peripheral angle) and β (the central angle) are subtended by the same ARC. Thus we have:

> For angles subtended by the same arc, the central angle is always twice that of the peripheral angle.

This is proved by drawing a radius from the center *O* to the point at which angle α lies to create two

isosceles triangles. Following the left-hand side of the next diagram, and noting that the interior angles of a triangle sum to 180°, we thus have $x + y = \alpha$ and $(180 - 2x) + (180 - 2y) + \beta = 360$, from which it follows that $\beta = 2\alpha$. A modification of this argument shows that the result is still true even if the peripheral angle is located as shown in the right-hand side of the diagram, or if the arc under consideration is more than half the PERIMETER of the circle.

The next three results follow (see diagram below, right):

i. All angles inscribed in a circle subtended by the same arc are equal,
ii. All angles inscribed by a diameter are right angles. (This is known as the theorem of Thales.)

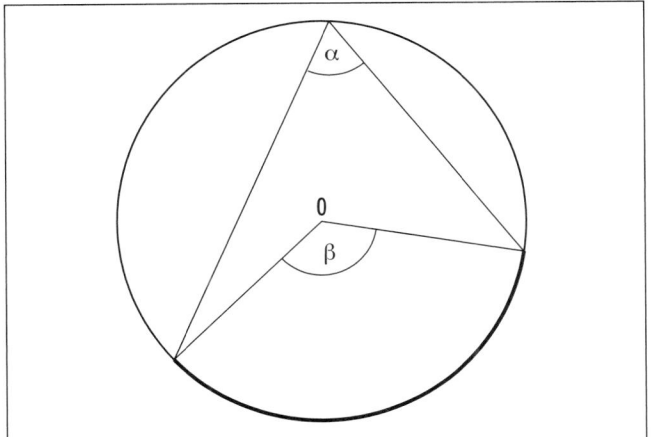

Central and peripheral angles

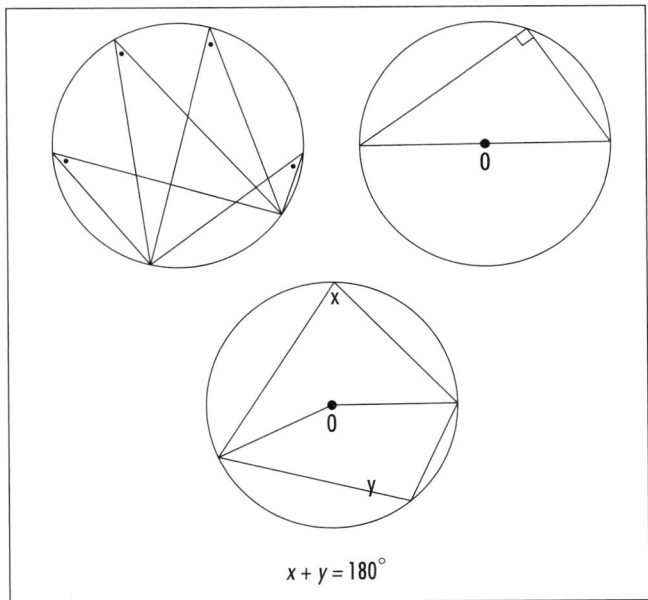

Consequences of the central-angle/peripheral-angle theorem

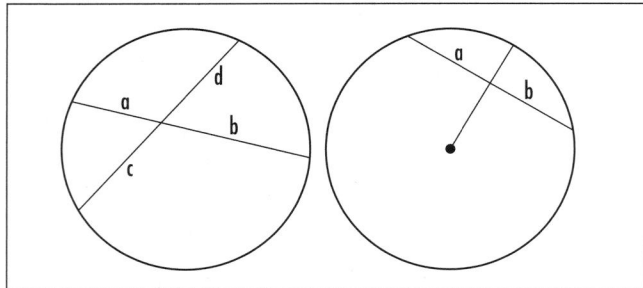

Chord theorems

iii. Opposite angles in a cyclic QUADRILATERAL sum to 180°.

3. Two-Chord Theorem
   In the diagram above, we have: $ab = cd$.

   This is proved by connecting the endpoints of the chords to create two triangles. The inscribed-angle theorems show that these two triangles are similar by the AAA rule. Consequently, $a/c = d/b$.

4. Radius-Chord Theorem
   In the right-hand side of the diagram, we see that:

   The radius of a circle bisects a chord (that is, we have $a = b$) if, and only if, the radius is perpendicular to the chord.

This is proved by drawing two radii to the endpoints of the chord to produce a large isosceles triangle. If the central radius is at 90° to the chord, this produces two congruent right triangles, and so $a = b$. Conversely, if we are told $a = b$, then the LAW OF COSINES, applied to each base angle of the isosceles triangle, shows that the central radius intercepts the chord at an angle of 90°.

See also AAA/AAS/ASA/SAS/SSS; CONGRUENT FIGURES; CYCLIC POLYGON; PTOLEMY'S THEOREM; SECANT; SIMILAR FIGURES.

circumcircle   A circle that passes through all three vertices of the triangle is called a circumcircle for the triangle. That three distinct points in the plane determine a unique circle was first proved by EUCLID in his treatise, *The Elements,* Book III. Euclid also presented a general method for actually constructing the circumcircle of a triangle. (This result was later generalized by APOLLO-NIUS OF PERGA, who showed how to construct a circle tangent to any three points, lines, or circles in the plane.)

The center of the circumcircle of a triangle is called its circumcenter. Clearly, it is a point EQUIDISTANT from the three vertices, and so it must lie on each line of points equidistant from any two vertices. The circumcenter of a triangle can thus be found by drawing the perpendicular bisectors of the sides of the triangle and locating where these three lines meet.

By the LAW OF SINES, the diameter of the circumcircle of a triangle is given by $r = \dfrac{a}{2\sin(A)}$, where a is a side-length of the triangle and A is the angle opposite that chosen side. If the remaining two sides of the triangle have lengths b and c, then the area of the triangle can be written: $\text{area} = \dfrac{1}{2}bc\sin(A)$. By HERON'S FORMULA, this area can also be computed via $\text{area} = \sqrt{s(s-a)(s-b)(s-c)}$ where $s = \dfrac{a+b+c}{2}$. Equating these two equations, solving for $\sin(A)$, and substituting into the formula above produces a formula for the radius of the circumcircle of a triangle solely in terms of its side lengths:

$$r = \frac{abc}{4\sqrt{s(s-a)(s-b)(s-c)}}$$

The circumcircle of a regular POLYGON is that circle that passes through all the vertices of the polygon. The radius of the circumcircle of a square, for example, of side-length x is $r = x/\sqrt{2}$.

See also CONCURRENT; LONG RADIUS; TRIANGLE.

circumscribe/inscribe   If A and B are two geometric figures, with A inside B, drawn so that the two figures have points in common but do not have edges that cross, then we say that figure A is inscribed in B, or, alternatively, that figure B is circumscribed about A. For example, a polygon lying inside a circle with all its vertices on that circle is said to be inscribed in the circle, and a circle inside a polygon touching each side of the polygon is inscribed in the polygon. A circle that passes through all three vertices of a triangle circumscribes that triangle, and the smallest square that surrounds a circle circumscribes that circle.

See also CIRCUMCIRCLE; INCIRCLE.

clock math   *See* MODULAR ARITHMETIC.

closure property   A BINARY OPERATION on a set S is said to be closed if the combination of two elements in that set yields another member of that set. For example, the set of positive whole numbers is closed under addition, since the sum of any two positive integers is itself a positive integer. This set is also closed under multiplication (the product of two positive whole numbers is a positive integer), but not subtraction: if n and m are positive whole numbers, then $n - m$ could be zero or negative and thus no longer in the set of positive whole numbers. (For instance, $3 - 7 = -4$.)

For a more unusual example, consider the set S of all whole numbers that can be expressed as the sum of two SQUARE NUMBERS. We have $S = \{0, 1, 2, 4, 5, 8, 9, 10, 13, 16, 17, 20, 25,...\}$. (For instance, $0 = 0^2 + 0^2$, $5 = 1^2 + 2^2$, $20 = 2^2 + 4^2$, and $25 = 0^2 + 5^2 = 3^2 + 4^2$.) Surprisingly, this set is closed under multiplication. For example, both 5 and 8 belong to S, and so does $5 \times 8 = 40$. (We have $40 = 2^2 + 6^2$.) Also, 10 and 13 belong to S, and so too does 130. (We have $130 = 3^2 + 11^2$.) This general observation follows from the algebraic identity that if $N = a^2 + b^2$ and $M = c^2 + d^2$; then $N \times M = (ac + bd)^2 + (ad - bc)^2$.

This set S is also closed under exponentiation: if N and M are each a sum of two squares, then so is N^M. For example, $5^{13} = (12,625)^2 + (31,250)^2$.

See also SQUARE.

coefficient   A numerical or constant multiplier of the variables in a term of an algebraic expression is called the coefficient of that term. For example, consider the equation $5x^3 - 2x + 7 = 0$, where x is the variable, the coefficient of x^3 is 5, the coefficient of x is -2, and the coefficient of x^2 is zero. In the equation $3 \cos y - 4xy^2 = 7$, the coefficients of $\cos y$ and xy^2 are 3 and -4, respectively.

Sometimes the value of a coefficient is not known and a symbol is used in its stead. For instance, in the expression $ax^2 + bx + c$ with x the variable, the numbers a and b are coefficients (and c is a constant term). Although the values of a, b, and c are not specified, it is understood that their values do not change even as the value of x varies.

In a more general context, the term *coefficient* is used for any number that serves as a measure of some property or characteristic of a set of data or a physical property. For instance, a CORRELATION COEFFICIENT in STATISTICS gives a measure of the extent to which two data sets are interdependent, while the heat coefficient in physics gives a measure as to how well a material conducts heat.

See also BINOMIAL COEFFICIENT; COMBINATORIAL COEFFICIENT; CONSTANT; LEADING COEFFICIENT; POLYNOMIAL.

Collatz's conjecture ("3n + 1" mapping problem)   Consider the following process:

Select a positive integer. If it is odd, triple it and add one; otherwise, divide the number by two. Now perform the same operation again on the result. Repeat this process indefinitely to produce a sequence of numbers.

The number 7, for example, yields the sequence:

7, 22, 11, 34, 17, 52, 26, 13, 40, 20, 10, 5, 16, 8, 4, 2, 1, 4, 2, 1, 4, 2, 1,...

Notice that this sequence finally falls into a 4-2-1 cycle.

In 1937 German mathematician Lothar Collatz conjectured that, no matter the starting integer selected, all sequences lead to the same 4-2-1 cycle. Collatz was unable to prove this claim, but he was also unable to find an example of a starting number that does not behave this way. To this day, no one knows whether or not Collatz's conjecture is true. All integers up to 2.702×10^{16} have been checked.

As a first step toward understanding this problem, mathematicians have proved that 4-2-1 is the only cycle of reasonable size that could possibly appear; it has been established that any other cycle that might appear would be at least 275,000 numbers long.

collinear   Any number of points are said to be collinear if they all lie on the same straight line. Two points are always collinear. Three points in a plane $A = (a_1, a_2)$, $B = (b_1, b_2)$, and $C = (c_1, c_2)$ are collinear only if the lines connecting points A and B and the points connecting A and C have the same SLOPE. This means that the following relationship must hold:

$$\frac{b_2 - a_2}{b_1 - a_1} = \frac{c_2 - a_2}{c_1 - a_1}$$

Three points A, B, and C in three-dimensional space are collinear if the triangle they form has zero area. Equivalently, the three points are collinear if the angle between the VECTORS $\overrightarrow{AB}$ and $\overrightarrow{AC}$ is zero, and consequently the CROSS PRODUCT $\overrightarrow{AB} \times \overrightarrow{AC}$ equals the zero vector.

The collinearity of points in a plane is a topic of interest to geometers. In the mid-1700s, LEONHARD EULER discovered that several interesting points constructed from triangles are collinear, yielding his famous EULER LINE. In 1893 British mathematician James Sylvester (1814–97) posed the question of whether it is possible to arrange three or more points in a plane, not all on a line, so that any line connecting two of the points from the collection passes through a third point as well. Forty years later Tibor Gallai (1912–92) proved that there is no such arrangement.

Two or more distinct PLANES are said to be collinear if they intersect in a common straight line. In this case, the vectors normal to each plane all lie in a plane perpendicular to the common line. Thus one can determine whether or not a collection of planes is collinear by noting whether or not the cross products of pairs of normal vectors are all parallel.

See also GRADIENT; NORMAL TO A PLANE.

combination (selection, unordered arrangement) Any set of items selected from a given set of items without regard to their order is called a combination. Repetition of choices is not permitted. For example, there are six distinct combinations of two letters selected from the sequence A,B,C,D, namely: AB, AC, AD, BC, BD, and CD. (The selection BA, for example, is deemed the same as AB, and the choice AA is not permitted.)

The number of combinations of k items selected from a set of n distinct objects is denoted $\binom{n}{k}$. The number $\binom{4}{2}$, for instance, equals six. The quantity $\binom{n}{k}$ is called a combinatorial coefficient and is read as "n choose k." Given their appearance in the BINOMIAL THEOREM, these numbers are also called binomial coefficients.

One develops a formula for $\binom{n}{k}$ by counting the number of ways to arrange n distinct objects in a row. There are, of course, $n!$ different ways to do this. (*See* FACTORIAL.) Alternatively, we can imagine selecting which k objects are to be arranged in the first k positions along the row (there are $\binom{n}{k}$ ways to do this), ordering those k items (there are $k!$ different ways to do this), and then arranging the remaining $n - k$ objects for the latter part of the row (there are $(n - k)!$ different ways to accomplish this). This yields $\binom{n}{k} k! (n - k)!$ different ways to arrange n objects in a row. Since this quantity must equal $n!$, we have the formula $\binom{n}{k} = \dfrac{n!}{k!(n-k)!}$ for the combinatorial coefficient.

It is appropriate to define $0!$ as equal to one. In this way, the formula just established holds even for $k = n$. (There is just one way to select n objects from a collection of n items, and so $\binom{n}{n} = \dfrac{n!}{n!0!}$ should equal one.) It then follows that $\binom{n}{0} = \dfrac{n!}{0!n!} = 1$. (There is just one way to select no objects.) Mathematicians set $\binom{n}{k}$ to be zero if k is negative or greater than n.

The combinatorial coefficients appear as the entries of PASCAL'S TRIANGLE. They also satisfy a number of identities. We list just four, which we shall phrase in terms of the process of selecting k students to be in a committee from a class of n students.

$$1. \quad \binom{n}{k} = \binom{n}{n-k}$$

(Selecting k students to be in a committee is the same as selecting $n - k$ students not to be in the committee.)

$$2. \quad \binom{n}{k} = \binom{n-1}{k-1} + \binom{n-1}{k}$$

(Any committee formed either includes, or excludes, a particular student John, say. If John is to be on the committee, then one must select $k - 1$ more students from the remaining $n-1$ students. If John is not to be on the committee, then one must select k students from the pool of $n-1$ students that excludes John.)

$$3. \quad \sum_{k=0}^{n}\binom{n}{k}=\binom{n}{0}+\binom{n}{1}+\binom{n}{2}+\cdots+\binom{n}{n}=2^n$$

(There are $\binom{n}{0}+\binom{n}{1}+\cdots+\binom{n}{n}$ possible committees of any size. But this number can also be computed by deciding, student by student, whether or not to put that student in the committee. As there are two possibilities for each student, in or out, there are 2^n possible committees. These counts must be the same.)

$$4. \quad \sum_{k=0}^{n}k\binom{n}{k}=\binom{n}{1}+2\binom{n}{2}+3\binom{n}{3}+\cdots+n\binom{n}{n}=n2^{n-1}$$

(Suppose, in the committee, one student is to be selected as chair. In a committee of size k there are k possible choices for chair. Thus $\sum_{k=0}^{n}k\binom{n}{k}$ counts the total number of committees possible, of any size, with one student selected as chair. But this quantity can also be computed by selecting some student to be chair first—there are n choices for this—and then deciding, student by student, among the remaining $n-1$ students whether that student should be on the committee. This yields $n2^{n-1}$ possibilities.)

Property 1 explains why Pascal's triangle is symmetric. Property 2 shows that each entry in Pascal's triangle is the sum of the two entries above it, and property 3 shows that the sum of all the entries in any row of Pascal's triangle is a power of two.

In 1778 LEONHARD EULER used the notation $\left(\frac{n}{k}\right)$ for the combinatorial coefficients, which, three years later, he modified to $\left[\frac{n}{k}\right]$. In the 19th century, mathematicians started following Euler's original notation, dropping the VINCULUM for the purposes of easing typesetting. Many textbooks today use the notation $_nC_k$, or C^n_k, or even $C(n,k)$, for the combinatorial coefficient $\binom{n}{k}$.

Generalized Coefficients

The generalized combinatorial coefficient $\binom{n}{k_1 \ k_2 \ \ldots \ k_r}$, where $k_1, k_2, \ldots, k_r$ are nonnegative integers summing to n, is defined to be the number of ways one can select, from n items, k_1 objects to go into one container, k_2 objects to go into a second container, and so forth, up to k_r objects to go into an rth container. (Notice that $\binom{n}{k \ n-k}$ is the ordinary combinatorial coefficient.) Mimicking the argument presented above, note that one can arrange n items in a row by first selecting which k_1 items are to go into the first part of the row and ordering them, which k_2 items are to go in the next portion of the row and ordering them, and so on. This shows that $n!=\binom{n}{k_1 \ k_2 \ \ldots \ k_r}k_1!k_2!\ldots k_r!$, yielding the formula:

$$\binom{n}{k_1 \ k_2 \ \ldots \ k_r}=\frac{n!}{k_1!k_2!\ldots k_r!}$$

Generalized combinatorial coefficients show, for example, that there are $\binom{7}{1 \ 1 \ 3 \ 2}=\frac{7!}{1!1!3!2!}=420$ ways to rearrange the letters CHEESES: Of the seven slots for letters, one must choose which slot is assigned for the letter C, which one for the letter H, which three for the letter E, and which two for letter S.

The generalized combinatorial coefficients also appear in generalizations to the BINOMIAL THEOREM. For example, we have the trinomial theorem:

$$(x+y+z)^n=\sum\binom{n}{k_1 \ k_2 \ k_3}x^{k_1}y^{k_2}z^{k_3}$$

where the sum is taken over all triples k_1,k_2,k_3 that sum to n. The proof is analogous to that of the ordinary binomial theorem.

Multi-Choosing

The quantity $\left(\binom{4}{2}\right)$, read as "$n$ multi-choose k," counts the number of ways to select k objects from a collection of n items, where order is not important, but repetition *is* allowed. For example, there are 10 ways to multichoose two objects from the set A, B, C, and D, namely: AB, AC, AD, BC, BD, CD, and AA, BB, CC, and DD. Thus $\left(\!\binom{4}{2}\!\right) = 10$. One can show that a multichoose coefficient equals an ordinary combinatorial coefficient:

$$\left(\!\binom{n}{k}\!\right) = \binom{n+k-1}{k}$$

combinatorial coefficient   *See* COMBINATION.

combinatorics (combinatorial analysis)   The branch of mathematics concerned with the theory and practices of counting elements of sets and the construction of specified arrangements of objects, along with the study of COMBINATIONS and PERMUTATIONS, is called combinatorics. GRAPH THEORY is also regarded as an aspect of combinatorics.

The technique of "double counting," that is, counting the same set of objects in two different ways, is a common practice in combinatorics used to yield interesting results. For example, counting the dots in an $n \times n$ square array along diagonals as opposed to across the rows gives the surprising formula:

$$1 + 2 + 3 + \ldots + (n-1) + n + (n-1) + \ldots + 3 + 2 + 1 = n^2$$

Counting the number of subsets of a set of n elements, either by summing the number of subsets containing, in turn, 0, 1, 2, up to n elements, or by noting that each subset is decided by making n choices between two options—whether or not each element in turn is to be in the subset—yields the formula:

$$\binom{n}{0} + \binom{n}{1} + \binom{n}{2} + \ldots + \binom{n}{n} = 2 \times 2 \times \cdots \times 2 = 2^n$$

EULER'S THEOREM can be considered a result in combinatorial geometry.

See also DISCRETE; FIGURATE NUMBERS.

commensurable   Two quantities having a common measure, meaning that they can be measured in terms of

whole numbers of a common unit, are said to be commensurable. For example, the quantities one month and one week are commensurable because they can both be measured in terms of a whole number of days. In GEOMETRY, two line segments are said to be commensurable if there is another segment whose measure goes evenly, without remainder, into the measures of each segment. For instance, segments of lengths 20 and 12 in. are commensurable for they can each be evenly divided into lengths of 1 (or 2 or 4) in. In general, two segments of lengths a and b units are commensurable if the ratio a/b is a RATIONAL NUMBER. As $\sqrt{2}$ is irrational, segments of length 1 and $\sqrt{2}$ (respectively, the side-length and the diagonal of a unit square) are incommensurable.

A study of the EUCLIDEAN ALGORITHM shows that if given two commensurable line segments of lengths a and b, say, then repeatedly subtracting the shorter length from the longer to produce a new pair of lengths eventually produces two line segments equal in length. This final shared measure is the largest length that divides evenly into the two original segments. (If a and b are whole-number measurements, then the length of the final measure is the GREATEST COMMON DIVISOR of a and b.) If, on the other hand, one can demonstrate that the process of repeatedly erasing the shorter line segment from the longer will continue indefinitely without ever producing two line segments equal in length, then the original two segments cannot be commensurable. Around 425 B.C.E. Greek mathematician THEODORUS OF CYRENE used precisely this observation to prove the irrationality of $\sqrt{2}$.

In NUMBER THEORY, two real numbers a and b are said to be commensurable if their ratio is rational. For instance, the numbers $\sqrt{48}$ and $\sqrt{3}/2$ are commensurable. No one to this day knows whether or not π and e are commensurable. The numbers $\log_5(3)$ and $\log_5(7)$ are incommensurable. (If $\frac{\log_5 7}{\log_5 3} = \frac{p}{q}$ for some whole numbers p and q, then $7^q = 3^p$, which is absurd since every power of 7 is 1 more than a multiple of 3.)

common denominator   Two or more fractions are said to have a common denominator if the denominator of each fraction is the same. For example, the fractions $\frac{5}{12}$ and $\frac{3}{12}$ have a common denominator of 12. It is a straightforward matter to add and subtract fractions with a common denominator. For instance, $\frac{5}{12} + \frac{3}{12} = \frac{8}{12}$ and $\frac{5}{12} - \frac{3}{12} = \frac{2}{12}$.

It is possible to rewrite the terms of an arbitrary collection of fractions so that they all share a common denominator. For instance, the rewriting $\frac{1}{3}$ and $\frac{2}{5}$ as $\frac{5}{15}$ and $\frac{6}{15}$, respectively, shows that the two fractions have a common denominator 15. In fact, any COMMON MULTIPLE of 3 and 5 serves as a common denominator of $\frac{1}{3}$ and $\frac{2}{5}$. For instance, we have $\frac{1}{3} = \frac{10}{30}$ and $\frac{2}{5} = \frac{12}{30}$, and $\frac{1}{3} = \frac{15}{45}$ and $\frac{2}{5} = \frac{18}{45}$.

The LEAST COMMON MULTIPLE of the denominators of a collection of fractions is called the least common denominator of the fractions. For example, the least common denominator of $\frac{1}{3}$ and $\frac{2}{5}$ is 15, and the least common denominator of $\frac{3}{8}$, $\frac{3}{4}$, and $\frac{5}{6}$ is 24. One adds and subtracts arbitrary fractions by rewriting those fractions in terms of a common denominator.

common factor (common divisor)

common factor (common divisor) A number that divides two or more integers exactly is called a common factor of those integers. For example, the numbers 20, 30 and 50 have 2 as a common factor, as well as 1, 5, and 10 as common factors. It is always the case that the largest common factor a set of integers possesses is a multiple of any other common factor. In our example, 10 is a multiple of each of 1, 2, and 5. The value 1 is always a common factor of any set of integers.

The FUNDAMENTAL THEOREM OF ARITHMETIC shows that any number can be uniquely expressed as a product of PRIME factors. Any common factor of two or more integers is a product of primes common to all those integers, and the largest common factor is the product of all the primes in common, with repetition permissible. (This explains why the largest common factor is a multiple of any other common factor.) If the integers have no primes in common, then their largest common factor is one.

See also GREATEST COMMON DIVISOR; RELATIVELY PRIME.

common multiple

common multiple A number that is a multiple of two or more other numbers is called a common multiple of those numbers. For example, 60 is a common multiple of 5, 6, and 10. The lowest number that is a common multiple of a given set of numbers is called their LEAST COMMON MULTIPLE. In our example, 30 is the least common multiple of 5, 6, and 10. Every common multiple is a multiple of the least common multiple.

The FUNDAMENTAL THEOREM OF ARITHMETIC shows that any number can be uniquely expressed as a product of PRIME factors. Any common multiple of two or more integers is the product of, at the very least, all the primes that appear in the factorizations of the given integers, with the necessary repetitions, with perhaps additional factors. With no additional factors present, one obtains the least common multiple.

commutative property

commutative property A BINARY OPERATION is said to be commutative if it is independent of the order of the terms to which it is applied. More precisely, an operation * is commutative if:

$$a*b = b*a$$

for all values of a and b. For example, in ordinary arithmetic, the operations of addition and multiplication are commutative, but subtraction and division are not. For instance, $2 + 3$ and $3 + 2$ are equal in value, but $2 - 3$ and $3 - 2$ are not.

If an operation is both commutative and ASSOCIATIVE, then all products of the same set of elements are equal. For example, the quantity $a*(b*c)$ equals $(a*c)*b$ and $b*(c*a)$. In this case, one is permitted to simply write $a*b*c$, with terms in any order, without concern for confusion.

In SET THEORY, the union and intersection of two sets are commutative operations. In VECTOR analysis, the addition and DOT PRODUCT of two vectors are commutative operations, but the CROSS PRODUCT operation is not. The multiplication of one MATRIX with another is not, in general, commutative.

Geometric operations generally are not commutative. For example, a reflection followed by a rotation does not usually produce the same result as performing the rotation first and then applying the reflection. One could also say that the operations of putting on one's shoes and one's socks are not commutative.

A GROUP is called commutative, or Abelian, if the operation of the group is commutative.

See also NIELS HENRIK ABEL; DISTRIBUTIVE PROPERTY; RING.

comparison test   *See* CONVERGENT SERIES.

completing the square   A QUADRATIC quantity of the form $x^2 + 2bx$ can be regarded, geometrically, as the formula for the area of an incomplete square.

Adding the term b^2 completes the picture of an $(x + b) \times (x + b)$ square. We have:

$$x^2 + 2bx + b^2 = (x + b)^2$$

This process of completing the square provides a useful technique for solving quadratic equations. For example, consider the equation $x^2 + 6x + 5 = 21$. Completing the square of the portion $x^2 + 6x$ requires the addition of the constant term 9. We can achieve this by adding 4 to both sides of the equation. We obtain:

$$x^2 + 6x + 5 + 4 = 21 + 4$$
$$x^2 + 6x + 9 = 25$$
$$(x + 3)^2 = 25$$

from which it follows that $x + 3$ equals either 5 or -5, that is, that x equals 2 or -8.

The process of completing the square generates a general formula for solving all quadratic equations. We have:

The solutions of a quadratic equation $ax^2 + bx + c = 0$, with $a \neq 0$, are given by:

$$x = \frac{-b \pm \sqrt{b^2 - 4ac}}{2a}$$

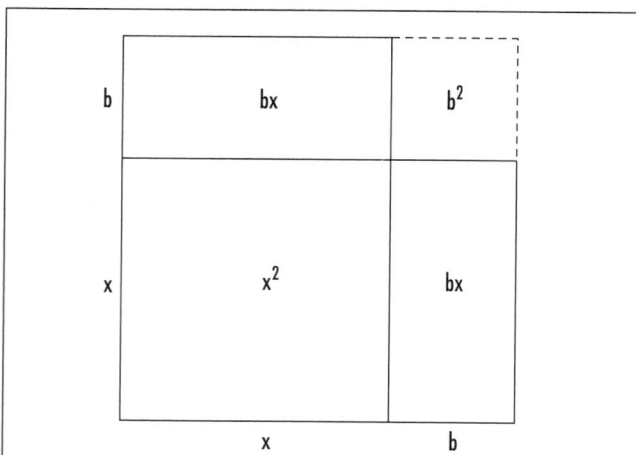

Completing the square

This formula is known as the quadratic formula. To see why it is correct, divide the given equation through by a and add a term to complete the square of resultant portion $x^2 + \frac{b}{a}x$. We have:

$$x^2 + \frac{b}{a}x + \frac{c}{a} = 0$$

$$x^2 + \frac{b}{a}x + \left(\frac{b}{2a}\right)^2 + \frac{c}{a} = \left(\frac{b}{2a}\right)^2$$

$$\left(x + \frac{b}{2a}\right)^2 + \frac{c}{a} = \frac{b^2}{4a^2}$$

$$\left(x + \frac{b}{2a}\right)^2 = \frac{b^2}{4a^2} - \frac{c}{a} = \frac{b^2 - 4ac}{4a^2}$$

$$x + \frac{b}{2a} = \pm\frac{\sqrt{b^2 - 4ac}}{2a}$$

$$x = \frac{-b \pm \sqrt{b^2 - 4ac}}{2a}$$

For example, to solve $x^2 + 6x + 5 = 21$, subtract 21 from both sides of the equation to obtain $x^2 + 6x - 16 = 0$. By the quadratic formula:

$$x = \frac{-6 \pm \sqrt{36 - 4 \cdot (-16)}}{2} = \frac{-6 \pm \sqrt{100}}{2} = \frac{-6 \pm 10}{2} = 2 \text{ or } -8$$

The quadratic formula shows that the two roots r_1 and r_2 of a quadratic equation $ax^2 + bx + c = 0$ (or the single double root if the DISCRIMINANT $b^2 - 4ac$ equals zero) satisfy $r_1 + r_2 = -\frac{b}{a}$ and $r_1 r_2 = \frac{c}{a}$. It also shows that every quadratic equation can be solved if one is willing to permit COMPLEX NUMBERS as solutions. (One may be required to take the square root of a negative quantity.)

There do exist analogous formulae for solving CUBIC EQUATIONS $ax^3 + bx^2 + cx + d = 0$ and QUARTIC EQUATIONS $ax^4 + bx^3 + cx^2 + dx + e = 0$ in terms of the coefficients that appear in the equations. Algebraist NIELS HENRIK ABEL (1802–29) showed that there can be no analogous formulae for solving fifth- and higher-degree equations.

See also FACTORIZATION; FUNDAMENTAL THEOREM OF ARITHMETIC; HISTORY OF EQUATIONS AND ALGEBRA (essay); SOLUTION BY RADICALS.

complex numbers There is no real number x with the property that $x^2 = -1$. By introducing an "imaginary" number i as a solution to this equation we obtain a whole host of new numbers of the form $a + ib$ with a and b real numbers. These new numbers form the system of complex numbers. It is customary to use the variable z to denote an arbitrary complex number: $z = a + ib$. If $b = 0$, then z is a real number. Thus the set of complex numbers includes the set of real numbers. If $a = 0$ so that z is of the form $z = ib$, then z is said to be purely imaginary. In general, if $z = a + ib$, then a is called the real part of z and b the imaginary part of z. We write: $\mathrm{Re}(z) = a$ and $\mathrm{Im}(z) = b$.

The number i is usually regarded as the square root of negative one: $i = \sqrt{-1}$. (One must be careful as there are, in fact, two square roots of this quantity, namely i and $-i$.) The roots of other negative quantities follow: $\sqrt{-9} = \sqrt{-1} \cdot \sqrt{9} = \pm 3i$ and $\sqrt{-30} = \pm i\sqrt{30}$, for instance.

The set of all complex numbers is denoted **C**. Arithmetic can be performed on the complex numbers by following the usual rules of algebra and replacing i^2 by -1 whenever it appears. For example, we have:

$$(2 + 3i) + (4 - i) = 6 + 2i$$
$$(2 + 3i) - (4 - i) = -2 + 4i$$
$$(2 + 3i)(4 - i) = 8 + 12i - 2i - 3i^2$$
$$= 8 + 10i + 3 = 11 + 10i$$

The QUOTIENT of two complex numbers can be computed by the process of RATIONALIZING THE DENOMINATOR:

$$\frac{2+3i}{4-i} = \frac{2+3i}{4-i} \cdot \frac{4+i}{4+i} = \frac{(2+3i)(4+i)}{(4-i)(4+i)} = \frac{5+14i}{4-i^2} = 1 + \frac{14}{5}i$$

One can show that with these arithmetic properties, the set of complex numbers constitutes a mathematical FIELD.

In the early 18th century, French mathematician ABRAHAM DE MOIVRE noticed a striking similarity between complex multiplication and the ADDITION formulae of the sine and cosine functions from TRIGONOMETRY. Given that:

$$(a + ib) + (c + id) = (ac - bd) + i(ad + bc)$$

and:

$$\cos(x + y) = \cos(x)\cos(y) - \sin(x)\sin(y)$$
$$\sin(x + y) = \sin(x)\cos(y) + \cos(x)\sin(y)$$

we obtain the compact formula:

$$(\cos(x) + i\sin(x))(\cos(y) + i\sin(y)) = \cos(x + y) + i\sin(x + y)$$

This observation formed the basis for the famous formula that now bears his name:

$$(\cos(x) + i\sin(x))^n = \cos(nx) + i\sin(nx)$$

A few years later LEONHARD EULER (1707–83) took matters one step further and used the techniques of calculus to establish his extraordinary formula:

$$e^{ix} = \cos(x) + i\sin(x)$$

from which DE MOIVRE'S FORMULA follows easily. (Use $(e^{ix})^n = e^{i(nx)}$.) Moreover, this result shows that de Moivre's formula also holds for noninteger values of n.

That the cosine and sine functions appear as the real and imaginary parts of a simple EXPONENTIAL FUNCTION shows that all of trigonometry can be greatly simplified by rephrasing matters in terms of complex numbers. Although some might argue that complex numbers do not exist in the real world, the mathematics of the complex number system has proved to be very powerful and has offered deep insights into the workings of the physical world. Engineers and physicists phrase a great deal of their work in terms of complex number theory. (Engineers prefer to use the symbol j instead of i.)

It is a surprise to learn that the introduction of a single new number i as a solution to the equation $x^2 + 1 = 0$ provides all that is needed to completely solve *any* POLYNOMIAL equation $a_n x^n + a_{n-1} x^{n-1} + \ldots + a_1 x + a_0 = 0$.

The FUNDAMENTAL THEOREM OF ALGEBRA asserts that a polynomial equation of degree n has precisely n roots (counted with multiplicity) in the complex number system.

It is possible to raise a real number to a complex power to obtain a real result. For example, by EULER'S FORMULA, we have:

$$e^{i\pi} = \cos(\pi) + i\sin(\pi) = -1 + i \cdot 0 = -1$$

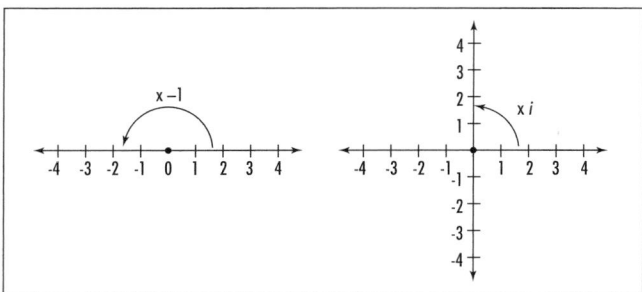

Multiplying by –1 and multiplying by *i*

Also, noting that $e^{i\frac{\pi}{2}} = \cos(\frac{\pi}{2}) + i\sin(\frac{\pi}{2}) = i$, we see that it is also possible to raise a complex number to a complex power to obtain a real result:

$$i^i = \left(e^{i\frac{\pi}{2}}\right)^i = e^{-\frac{\pi}{2}}$$

(Technically, since *i* can also be expressed as $e^{i(\frac{\pi}{2} + 2k\pi)}$ for any whole number *k*, there are infinitely different (real) values for the quantity i^i.)

The Geometry of Complex Numbers

Multiplying the entries of the real NUMBER LINE by –1 has the effect of rotating the line about the point 0 through an angle of 180°. It is natural to ask: multiplication by which number creates a 90° rotation about the point zero?

Call the desired number *x*. Multiplication by *x* twice, that is, multiplication by $x \times x = x^2$, would have the effect of performing two 90° rotations, namely, a

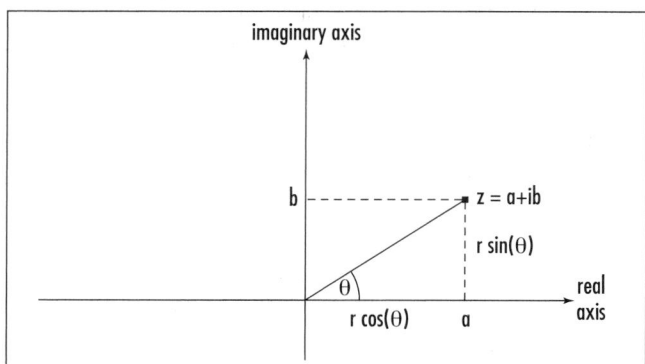

Polar coordinates of a complex number

rotation by 180°. Thus multiplication by x^2 has the same effect as multiplication by –1, and the desired number *x* must therefore satisfy the equation $x^2 = -1$. This shows that $x = i$ and that it is natural to interpret purely imaginary numbers of the form *ib* as members of a vertical number line.

This model provides a natural correspondence between complex numbers and points in the plane. The horizontal axis is called the real axis, the vertical axis the imaginary axis, and an arbitrary complex number $z = a + ib$ appears as the point with coordinates (a,b) on the plane.

This representation of complex numbers as points on a plane is called an Argand diagram in honor of JEAN ROBERT ARGAND (1768–1822) who, along with surveyor Casper Wessel (1745–1818), first conceived of depicting complex numbers in this way. The plane of all complex numbers is also called the complex plane.

The angle θ that a complex number $z = a + ib$ makes with the positive real axis is called the argument of the complex number, and the distance *r* of the complex number from the origin is called its modulus (or, simply, absolute value), denoted $|z|$. PYTHAGORAS'S THEOREM and the DISTANCE FORMULA show that:

$$|z| = \sqrt{a^2 + b^2}$$

Using trigonometry we see that the values *a* and *b* can be expressed in terms of *r* and θ in a manner akin to POLAR COORDINATES. We have:

$$a = r\cos\theta$$
$$b = r\sin\theta$$

With the aid of Euler's formula, this shows that any complex number *z* can be expressed in polar form:

$$z = r\cos\theta + ir\sin\theta = re^{i\theta}$$

From this it follows, for instance, that the product of two complex numbers $z_1 = r_1 e^{i\theta_1}$ and $z_2 = r_2 e^{i\theta_2}$ has modulus $r_1 r_2$ and argument $\theta_1 + \theta_2$:

$$z_1 z_2 = r_1 e^{i\theta_1} r_2 e^{i\theta_2} = r_1 r_2 e^{i(\theta_1 + \theta_2)}$$

It is convenient to define the conjugate of a complex number $z = a + ib$ to be the number $\bar{z} = a - ib$. We have:

$$z \cdot \bar{z} = a^2 + b^2 = |z|^2$$

and, in polar form, if $z = re^{i\theta}$, then $\bar{z} = re^{-i\theta}$. Taking the conjugate of a complex number has the geometric effect of reflecting that number across the real axis.

History of Complex Numbers

The first European to make serious use of the square root of negative quantities was GIROLAMO CARDANO (1501–76) of Italy in the development of his solutions to CUBIC EQUATIONS. He noted that quantities that arose in his work, such as an expression of the form $\sqrt[3]{2 + \sqrt{-121}} + \sqrt[3]{2 - \sqrt{-121}}$ for instance, could be manipulated algebraically to yield a real solution to equations. (We have $\sqrt[3]{2 + \sqrt{-121}} + \sqrt[3]{2 - \sqrt{-121}} = 4$.) Nonetheless, he deemed such a manipulation only as a convenient artifice with no significant practical meaning. French philosopher RENÉ DESCARTES (1596–1650) agreed and coined the term *imaginary* for roots of negative quantities.

During the 18th century, mathematicians continued to work with imaginary roots, despite general skepticism as to their meaning. Euler introduced the symbol i for $\sqrt{-1}$, and Argand and Wessel introduced their geometric model for complex numbers, which was later popularized by CARL FRIEDRICH GAUSS (1777–1855). His proof of the fundamental theorem of algebra convinced mathematicians of the importance and validity of the complex number system.

Irish mathematician SIR WILLIAM ROWAN HAMILTON (1805–65) is credited as taking the final step to demystify the meaning of the complex-number system. He extended the notion of the complex numbers as arising from 90° rotations by showing that *any* rotation in three-dimensional space can naturally and easily be represented in terms of complex numbers. He also noted that the complex numbers are nothing more than ordered pairs of numbers together with a means for adding and multiplying them. (We have $(a,b) + (c,d) = (a + c, b + d)$ and $(a,b)\cdot(c,d) = (ac - bd, ad + bc)$.) In Hamilton's work, the number i became nothing more than the point $(0,1)$.

See also NEGATIVE NUMBERS; STEREOGRAPHIC PROJECTION.

composite Used in any context where it is possible to speak of the multiplication of two quantities, the term *composite* means "having proper factors." For example, the number 12, which equals 3×4, is a COMPOSITE NUMBER, and $y = x^2 + 2x - 3 = (x-1)(x + 3)$ is a composite polynomial (not to be confused with the COMPOSITION of two polynomials).

A quantity that is not composite is called irreducible, or, in the context of number theory, PRIME.

composite number A whole number with more than two positive factors is called a composite number. For example, the number 12 has six positive factors, and so is composite, but 7, with only two positive factors, is not composite. The number 1, with only one positive factor, also is not composite. Numbers larger than one that are not composite are called PRIME.

The sequence 8, 9, 10 is the smallest set of three consecutive composite numbers, and 24, 25, 26, 27, 28 is the smallest set of five consecutive composites. It is always possible to find arbitrarily long strings of composite numbers. For example, making use of the FACTORIAL function we see that the string

$$(n + 1)! + 2, (n + 1)! + 3,\ldots,(n + 1)! + (n + 1)$$

represents n consecutive integers, all of which are composite. (This shows, for example, that there are arbitrarily large gaps in the list of prime numbers.)

See also FACTOR.

composition (function of a function) If the outputs of one function f are valid inputs for a second function g, then the composition of g with f, denoted $g \circ f$, is the function that takes an input x for f and returns the output of feeding $f(x)$ into g:

$$(g \circ f)(x) = g(f(x))$$

For example, if feeding 3 into f returns 5, and feeding 5 into g returns 2, then $(g \circ f)(3) = 2$. If, alternatively, $f(x) = x^2 + 1$ and $g(x) = 2 + \frac{1}{x}$, then

$$\begin{aligned}
(g \circ f)(x) &= g(f(x)) \\
&= g(x^2 + 1) \\
&= 2 + \frac{1}{x^2 + 1}
\end{aligned}$$

Typically $g \circ f$ is not the same as $f \circ g$. In our last example, for instance,

$$(f \circ g)(x) = f\big(g(x)\big)$$

$$= f\left(2 + \frac{1}{x}\right)$$

$$= \left(2 + \frac{1}{x}\right)^2 + 1$$

which is a different function. As another example, if M is the function that assigns to each person of the world his or her biological mother, and F is the analogous biological father function, then $(M \circ F)$(John) represents John's paternal grandmother, whereas $(F \circ M$(John) is John's maternal grandfather.

The notation $g \circ f$ is a little confusing, for it needs to be read backwards. The function f is called the "core function" and needs to be applied first, with the "external function" g applied second. The composition of three functions f, g, and h is written $h \circ f \circ g$ (here h is the external function), and the repeated composition of a function f with itself is written $f^{(n)}$. Thus, for example, $f^{(4)}$ denotes the composition $f \circ f \circ f \circ f$. A set of repeated compositions is called a DYNAMICAL SYSTEM.

Mathematicians have shown that the composition of two CONTINUOUS FUNCTIONs is itself continuous. Precisely, if f is continuous at $x = a$, and g is continuous at $x = f(a)$, then $g \circ f$ is continuous $x = a$.

The composition of two differentiable functions is differentiable. The CHAIN RULE shows that the DERIVATIVE of $g \circ f$ is given by $(g \circ f)'(x) = g'(f(x)) \cdot f'(x)$.

compound interest   *See* INTEREST.

compound statement   *See* TRUTH TABLE.

computer   An electronic device for automatically performing either arithmetic operations on DATA or sequences of manipulations on sets of symbols (as required for ALGEBRA and SET THEORY, for instance), all according to a precise set of predetermined instructions, is called a computer. The most widely used and versatile computer used today is the digital computer in which data are represented as sequences of discrete electronic pulses. As each pulse could either be "on" or "off," it is natural to think of

sequences of 0s and 1s in working in computer theory and, consequently, to work with the system of BINARY NUMBERS to represent data.

A digital computer has a number of separate parts:

1. An input device, such as a keyboard, for entering a set of instructions (program) and data.
2. A central processing unit (CPU) that codes information into binary form and carries out the instructions. (This unit consists of a series of electronic circuit boards on which are embedded a large number of "logic gates," akin to the CONJUNCTION and DISJUNCTION configurations.)
3. Memory units, such as disks and magnetic tape.
4. An output device for displaying results, such as a monitor or a printer.

The study of computer science typically lends itself to the theoretical capabilities of computing machines defined in terms of their programs, not the physical properties of actual computers. The HALTING PROBLEM and the question of being NP COMPLETE, for instance, are issues of concern to scientists in this field.

See also ABACUS; CHARLES BABBAGE; DIGIT.

concave/convex   A curve or surface that curves inward, like the circumference of a circle viewed from the interior, or the hollow of a bowl, for example, is called concave. A curve or surface that curves outward, such as the boundary of a circle viewed from outside the circle, or the surface of a sphere, is called convex.

A geometric shape in the plane or a three-dimensional solid is called convex if the boundary of the shape is a convex curve or surface. For example, triangles, squares, and any regular POLYGON are convex figures. Cubes and spheres are convex solids. Any shape that is not convex is called concave. A deltoid QUADRILATERAL, for example is a concave polygon.

A polygon is convex if each of its interior angles has value less than 180°. Equivalently, a polygon is convex if the figure lies entirely on one side of any line that contains a side of the polygon. A POLYHEDRON is convex if it lies entirely on one side of any plane that contains one of its faces.

A convex figure can also be characterized by the property that, for any two points inside the figure, the line segment connecting them also lies completely within the figure.

concave up/concave down The graph of a function $y = f(x)$ may be described as concave up over an interval if, over that interval, the slope of the tangent line to the curve increases as one moves from left to right. Assuming the function is twice differentiable, this means that the DERIVATIVE $f'(x)$ is a strictly increasing function and, consequently, the double derivative satisfies $f''(x)>0$. (*See* INCREASING/DECREASING.) For example, the double derivative of $f(x) = x^2$ is always positive, $f''(x) = 2>0$, and the parabola $y = x^2$ is concave up. A concave-up graph also has the property that any CHORD joining two points on the graph lies entirely above the graph.

The graph of a function $y = f(x)$ is concave down over an interval if, over that interval, the slope of the tangent line to the curve decreases as one moves from left to right. Assuming the function is twice differentiable, this means that $f'(x)$ is a strictly decreasing function and, consequently, $f''(x)<0$ for all points on the interval. As an example, since the double derivative of $f(x) = x^3$ is negative only for negative values of x ($f''(x) = 6x<0$ for $x<0$), we have that the cubic curve $y = x^3$ is concave down only to the left of the y-axis. A concave-down graph has the property that any chord joining two points on the graph lies entirely below the graph.

A point at which the concavity of the graph changes is called an inflection point or a point of inflection. (Alternative spelling: inflexion.) If $x = a$ is a point of inflection for a twice-differentiable curve $f(x)$, then the double derivative $f''(x)$ is positive to one side of $x = a$ and negative to the other side. It must be the case then that $f''(a) = 0$. The converse need not hold, however. The function $f(x) = x^4$, for example, satisfies $f''(0) = 0$, but the concavity of the curve does not change at $x = 0$.

A study of the concavity of a graph can help one locate and classify local maxima and minima for the curve.

See also GRAPH OF A FUNCTION; MAXIMUM/MINIMUM.

concentric/eccentric Two circles or two spheres are called concentric if they have the same center. Two figures that are not concentric are called eccentric. The region between two concentric circles is called an ANNULUS.

concurrent Any number of lines are said to be concurrent if they all pass through a common point. Many interesting lines constructed from triangles are concurrent. Two lines $a_1x + b_1y = c_1$ and $a_2x + b_2y = c_2$ in the Cartesian plane are concurrent if $a_1b_2 - a_2b_1 \neq 0$.

See also TRIANGLE.

conditional (hypothetical) In FORMAL LOGIC a statement of the form "*If ... then...*" is known as a conditional or an implication. For example, "If a polygon has three sides, then it is a triangle" is a conditional statement.

A conditional statement has two components: *If p, then q*. Statement p is called the antecedent (hypothesis, or premise) and statement q the consequent (or conclusion). A conditional statement can be written a number of different, but equivalent, ways:

If p, then q.
p implies q.
q if p.
p only if q.
p is sufficient for q.
q is necessary for p.

It is denoted in symbols by: $p \rightarrow q$.

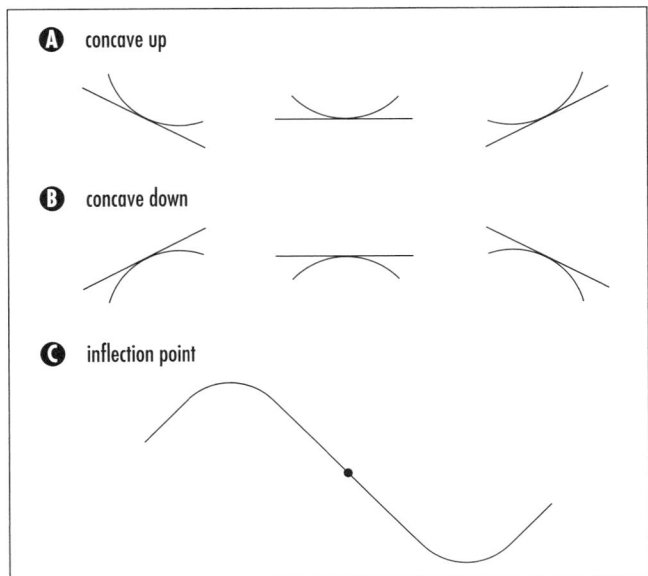

A concave up

B concave down

C inflection point

Concavity

The TRUTH TABLE of the conditional is motivated by intuition. Consider, for example, the statement:

> If Peter watches a horror movie, then he eats popcorn.

The statement is certainly true if we observe Peter watching a horror movie and eating popcorn at the same time (that is, if the antecedent and consequent are both true), but false if antecedent is true but the consequent is false (that is, Peter is watching a horror movie but not eating popcorn). This justifies the first two lines of the truth table below.

The final two lines are a matter of convention. If Peter is not watching a horror movie, that is, if the antecedent is false, then the conditional statement as a whole is moot. FORMAL LOGIC, however, requires us to assign a truth-value to every statement. As watching a romance movie and eating (or not eating) popcorn does not imply that the conditional statement is a lie, we go ahead and assign a truth-value "true" to the final two lines of the table:

p	q	$p{\to}q$
T	T	T
T	F	F
F	T	T
F	F	T

This convention does lead to difficulties, however. Consider, for example, the following statement:

> If this entire sentence is true, then the moon is made of cheese.

Here the antecedent p is the statement: "the entire sentence above is true." The consequent is: "the moon is made of cheese." Notice that p is true or false depending on whether the entire statement $p{\to}q$ is true or false. There is only one line in the truth table for which p and $p{\to}q$ have the same truth-value, namely the first one. It must be the case, then, that p, q, and $p{\to}q$ are each true. In particular, q is true. Logically, then, the moon must indeed be made of cheese.

See also ARGUMENT; BICONDITIONAL; CONDITION— NECESSARY AND SUFFICIENT; SELF-REFERENCE.

conditional convergence *See* ABSOLUTE CONVERGENCE.

conditional probability The probability of an EVENT occurring given the knowledge that another event has already occurred is called conditional probability.

For example, suppose two cards are drawn from a deck and we wish to determine the likelihood that the second card drawn is red. Knowledge of the first card's color will affect our probability calculations. Precisely:

i. If the first card is black, then the probability that the second is red is 26/51 (there are 26 red cards among the remaining 51 cards),
ii. If the first card is red, then the probability that the second is also this color is now only 25/51.

(If we have no knowledge of the color of the first card, then the chances that the second card is red are $1/2$.)

If A and B are two events, then the notation $A|B$ is used to denote the event: "A occurs given that event B has already occurred." The notation $P(A|B)$ denotes the probability of A occurring among just those experiments in which B has already happened. For instance, in the example above:

$$P(\text{the second card is red} \mid \text{the first card is black}) = \frac{26}{51}$$

$$P(\text{the second card is red} \mid \text{the first card is black}) = \frac{25}{51}$$

If, in many runs of an experiment, event B occurs b times, and events A and B occur simultaneously a times, then the proportion of times event A occurred when B happened is a/b. This motivates the mathematical formula for conditional probability:

$$P(A \mid B) = \frac{P(A \cap B)}{P(B)}$$

As an example, suppose we are told that a card drawn from a deck is red. To determine the probability that that card is also an ace we observe:

$$P(\text{ace} \mid \text{red}) = \frac{P(\text{ace and red})}{P(\text{red})} = \frac{P(\text{red ace})}{P(\text{red})} = \frac{\frac{2}{52}}{\frac{1}{2}} = \frac{1}{13}$$

That is, the probability that a red card is an ace is 1/13. Of course, counting the number of times "ace" occurs among the red cards also yields $P(\text{ace} \mid \text{red}) = 2/26 = 1/13$.

Conditional probability is useful in analyzing more complex problems such as the famous TWO-CARD PUZZLE.

If two events A and B are INDEPENDENT EVENTS, then the probabilities satisfy the relation $P(A \cap B) = P(A) \times P(B)$. This shows $P(A \mid B) = P(A)$, that is, the probability of event A occurring indeed is not altered by information of whether or not B has occurred.

See also BAYES'S THEOREM.

condition—necessary and sufficient

In logic, a condition is a proposition or statement p required to be true in order that another proposition q be true. If q cannot be true without p, then we call p a necessary condition. If the validity of p ensures that q is true, then we call p a sufficient condition. For example, for a quadrilateral to be a rectangle it is necessary for it to possess two parallel sides, but this condition is not sufficient. (A trapezoid, for example, has two parallel sides but is not a rectangle.) For a number to be even it is sufficient that the number end with a four, but this condition is not necessary.

If p is a sufficient condition for q, then the CONDITIONAL (implication) $p \rightarrow q$ holds. Mathematicians usually write: q is true if p is true. If p is a necessary condition for q, then the implication $q \rightarrow p$ holds. Mathematicians usually write: q is true *only if* p is true.

If the BICONDITIONAL holds: $p \leftrightarrow q$, then p is necessary and sufficient for q. For example, for a number to be divisible by 10 it is necessary and sufficient that the number end with a zero. Such a statement is usually written: p *if, and only if, q* or, compactly, as "p *iff q.*"

See also FORMAL LOGIC; TRUTH TABLE.

cone

In three-dimensional space, a cone is the surface formed by an infinite collection of straight lines drawn the following way: each line passes through one point of a fixed closed curve inscribed in a plane, called the directrix of the cone, and through a fixed given point above the plane, called the vertex of the cone. The lines drawn are called the generators of the cone.

In elementary work, the directrix is usually taken to be a circle so that the cones produced are circular cones. A circular cone is "right" if its vertex lies directly above the center of the circle, and "oblique" otherwise. Points on the surface of a right circular cone satisfy an equation of the form $x^2 + y^2 = a^2 z^2$, for some constant a.

Technically, the generators of a cone are assumed to extend indefinitely in both directions. Thus an arbitrary cone consists of two identical surfaces meeting at the vertex. Each surface is called *nappe* (French for "sheet") or a half-cone. However, if the context is clear, the word *cone* often refers to just one nappe, or just the part of a nappe between the vertex and the plane of the directrix. The object in this latter case is sometimes called a finite cone. It is bounded and encloses a finite volume.

For a finite cone, the planar region bounded by the directrix is called the base of the cone, and the vertex is called the APEX of the cone. The vertical distance of the apex from the base is called the height of the cone, and the volume V of a finite cone is given by $V = \frac{1}{3}Ah$, where h is the height of the cone and A the area of its base. (*See* VOLUME.) Thus:

> The volume of any cone is one-third of the volume of the CYLINDER that contains it.

ARCHIMEDES OF SYRACUSE (ca. 287–212 B.C.E.) established that the volume of a SPHERE is two-thirds the volume of the cylinder that contains it. (By drawing a cone in this cylinder, Archimedes established that the area of each horizontal slice of a sphere equals the area of the ANNULUS between the cone and the cylinder at the same corresponding height.) The formula for the volume of a sphere readily follows.

See also CONIC SECTIONS.

conformal mapping (equiangular transformation, isogonal transformation)

Any geometrical transformation that does not change the angles of intersection between two lines or curves is called a conformal mapping. For example, in GEOMETRY, reflections, translations, rotations, dilations, and inversions all preserve the angles between lines and curves and so are conformal mappings. MERCATOR'S PROJECTION of the Earth onto a cyclinder preserves every angle on the globe and so is a conformal projection.

See also GEOMETRIC TRANSFORMATION.

congruence Two numbers a and b are said to be congruent modulo N, for some positive integer N, if a and b leave the same remainder when divided by N. We write: $a \equiv b \pmod N$. For example, since 16 and 21 are both 1 more than a multiple of 5, we have $16 \equiv 21 \pmod 5$. Also $14 \equiv 8 \pmod 3$ and $28 \equiv 0 \pmod 7$.

One can equivalently interpret the statement $a \equiv b \pmod N$ to mean: "the difference $a - b$ is divisible by N." Since -2 and 16, for instance, differ by a multiple of 9, we have $-2 \equiv 16 \pmod 9$. Two numbers that are not congruent modulo N are called incongruent modulo N.

One can add, subtract, and multiply two congruences of the same modulus N in the same manner one adds, subtracts, and multiplies ordinary quantities. For example, noting that $14 \equiv 4 \pmod{10}$ and $23 \equiv 3 \pmod{10}$, we do indeed have:

$$14 + 23 \equiv 4 + 3 \pmod{10}$$
$$14 - 23 \equiv 4 - 3 \pmod{10}$$
$$14 \times 23 \equiv 4 \times 3 \pmod{10}$$

Unfortunately the process of division is not preserved under congruence. For example, $14 \div 2$ is not congruent to $4 \div 2$ modulo 10. A careful study of MODULAR ARITHMETIC explains under which circumstances division is permissible.

The arithmetic of congruence naturally occurs in any cyclic phenomenon. For example, finding the day of the week for a given date requires working with congruences modulo 7, and the arithmetic for counting hours as they pass works with congruences modulo 24 or modulo 12. (This leads to the study of CLOCK MATH.)

Certain DIVISIBILITY RULES can be explained via congruences. For example, since $10 \equiv 1 \pmod 9$, any power of 10 is also congruent to 1 modulo 9: $10^n \equiv 1^n = 1 \pmod 9$. Consequently, any number is congruent modulo 9 to the sum of its digits. For example,

$$486 = 4 \times 10^2 + 8 \times 10 + 6 \times 1$$
$$\equiv 4 \times 1 + 8 \times 1 + 6 \times 1 \pmod 9$$
$$= 4 + 8 + 6 \pmod 9$$

Since $4 + 8 + 6$ is a multiple of 9, it follows that 486 is divisible by 9.

See also CASTING OUT NINES; DAYS-OF-THE-WEEK FORMULA.

congruent figures Two geometric figures are congruent if they are the same shape and size. More precisely, two POLYGONS are congruent if, under some correspondence between sides and vertices, corresponding side-lengths are equal and corresponding interior angles are equal. Two different squares with the same side-length, for example, are congruent figures.

Note that two plane figures can be congruent without being identical: one figure may be the mirror image of the other. Two figures are called directly congruent if one can be brought into coincidence with the other by rotating and translating the figure in the plane and oppositely congruent if one must also apply a reflection. Two identical squares, for example, are directly congruent no matter where on the plane they are placed. Two scalene triangles with matching side-lengths might or might not be directly congruent. There are a number of geometric tests to determine whether or not two triangles are congruent as given by the AAA/AAS/ASA/SAS/SSS rules.

In three-dimensional space, two solids are directly congruent if they are identical. If each is the mirror image of the other, they are oppositely congruent.

The term *congruent* is sometimes applied to other geometric constructs to mean "the same." For example, two line segments are congruent if they have equal length, or two ANGLES are congruent if they have equal measure.

See also SIMILAR FIGURES.

conic sections Slicing a right circular CONE with a plane that does not pass through the vertex of the cone produces curves called the conic sections, or simply conics. If the slicing plane is parallel to a straight line that generates the cone, then the resulting conic is a PARABOLA. Otherwise, if the slicing plane passes through just one nappe of the cone, the curve produced is either a CIRCLE or an ELLIPSE, or a HYPERBOLA if the slicing plane cuts both nappes.

If we think of the cone as light rays emanating from a light source held at the vertex, then the shadow cast by a circular ring onto a sheet of card will be a conic section; the particular conic produced depends on the angle at which the card is held. The open ring at the top of a lampshade, for example, casts a hyperbolic shadow on the wall. In the same way, shadows cast by solid balls are conic sections.

The Greek scholars of antiquity were the first to study conic sections. With no practical applications in mind, mathematicians pursued the topic solely for its beauty and its intellectual rewards. Around 225 B.C.E. APOLLONIUS OF PERGA wrote a series of eight books, titled *The Conics,* in which he thoroughly investigated these curves. He introduced the names parabola, ellipse, and hyperbola. ARCHIMEDES OF SYRACUSE (ca. 287–212 B.C.E.) also wrote about these curves. Almost 2,000 years later, scientists began finding applications of conic sections to problems in the real world. In 1604 GALILEO GALILEI discovered that objects thrown in the air follow parabolic paths (if air resistance can be neglected), and in 1609 astronomer JOHANNES KEPLER discovered that the orbit of Mars is an ellipse. He conjectured that all planetary bodies have elliptical orbits, which, 60 years later, ISAAC NEWTON was able to prove using his newly developed law of gravitation. This century, scientists have discovered that the path of an alpha particle in the electrical field of an atomic nucleus is a hyperbola.

The conic sections can be described solely by properties they possess as curves in a plane. We see this by drawing spheres internally tangent to the cone and tangent to the slicing plane defining the curve. For example, given an ellipse, if two internal spheres are tangent to the plane at points F_1 and F_2, then for any point P on the ellipse, its distance from F_1 is the same as its distance from the circle of tangency of the lower sphere, and its distance from F_2 is the same as its distance from the circle of tangency of the upper sphere. Consequently, the sum of its distances from F_1 and F_2 equals the fixed distance between the two spheres as measured along the side of the cone. This property can be used to define an ellipse:

> An ellipse is the set of all points in the plane whose distances from two given points, F_1 and F_2, have a constant sum.

By drawing spheres, one in each nappe of the cone and tangent to the slicing plane of a hyperbola, one can show in an analogous way:

> A hyperbola is the set of all points in the plane whose distances from two given points, F_1 and F_2, have a constant difference.

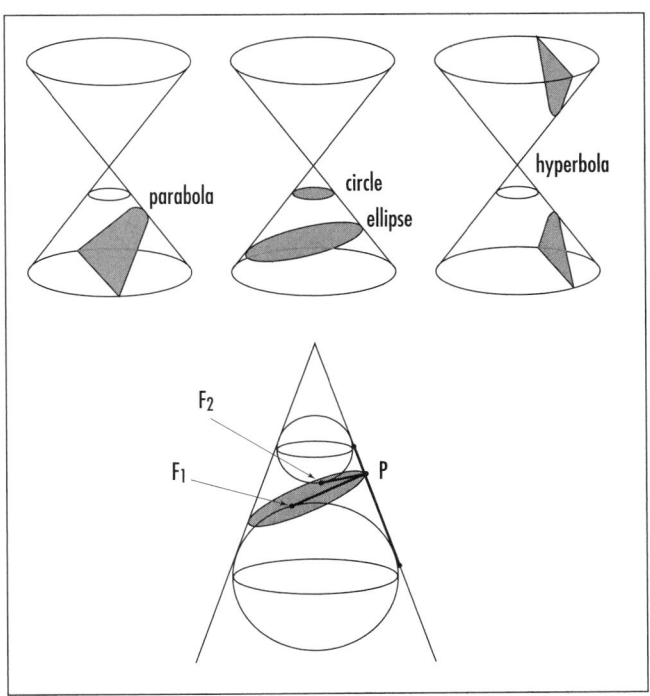

The conic sections

For a parabola, one draws a single sphere tangent to the slicing plane of the parabola and considers the point F at which the sphere touches the plane, and the line L of intersection of the slicing plane with the plane of the circle along which the sphere is tangent to the cone. We have:

> A parabola is the set of all points in the plane the same distance from a point F in the plane and a given line L.

The conic sections have remarkable reflection properties.
See also PROJECTION.

conjunction ("and" statement) In FORMAL LOGIC a compound statement of the form "p and q" is known as a conjunction. For example, "A triangle has three sides and a square has four sides" is a conjunction. A conjunction is denoted in symbols by $p \wedge q$.

For a conjunction as a whole to be considered true, each component (or conjunct) p and q must itself be true. Thus a conjunction has the following TRUTH TABLE:

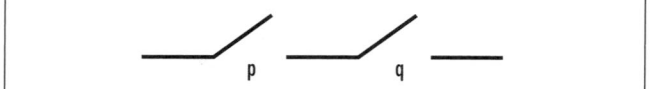

Conjunction circuit

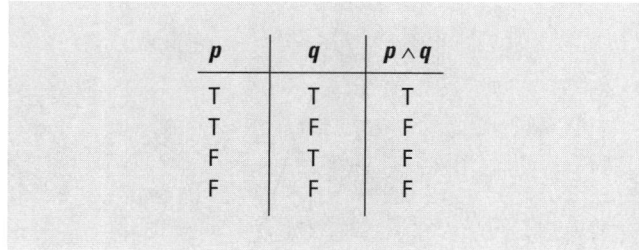

A conjunction can be modeled via a series circuit. If T denotes the flow of current, then current moves through the circuit as a whole if, and only if, both switches p and q allow the flow of current (that is, are closed).

See also DISJUNCTION.

connected   Informally, a geometric object is "connected" if it comes in one piece. For example, a region in the plane is connected if, for any two points in that region, one can draw a continuous line that connects the two points and stays within the region. For example, a HALF-PLANE is connected. The set of all real numbers on the number line different from zero is not connected.

A surface sitting in three-dimensional space, such as a SPHERE or a TORUS, is connected if any two points on the surface can be connected by a continuous path that stays on the surface. A GRAPH is connected if, for any two vertices, there is a contiguous path of edges that connects them.

The connectivity of a geometric object is the number of cuts needed to break the shape into two pieces. For example, a circle (interior and circumference) and a solid sphere each have connectivity one. An ANNULUS and a torus each have connectivity two.

consistent   A set of equations is said to be consistent if there is a set of values that satisfies all the equations. For example, the equations $x + y = 7$ and $x + 2y = 11$ are consistent, since they are satisfied by $x = 3$ and $y = 4$. On the other hand, the equations $x + y = 1$ and $x + y = 2$ are inconsistent.

In FORMAL LOGIC, a mathematical system is said to be consistent if it is impossible to prove a statement to be both true and not true at the same time. That is, a system is consistent if it is free from CONTRADICTION. Mathematicians have proved, for example, that arithmetic is a consistent logical system.

See also ARGUMENT; GÖDEL'S INCOMPLETENESS THEOREMS; LAWS OF THOUGHT; SIMULTANEOUS LINEAR EQUATIONS.

constant   The word *constant* is used in a number of mathematical contexts. In an algebraic expression, any numeric value that appears in it is called a constant. For example, in the equation $y = 2x + 5$ with variables x and y, the numbers 2 and 5 are constants. These numbers may be referred to as absolute constants because their values never change. In general applications, however, constants may be considered to take any one of a number of values. For example, in the general equation of a line $y = mx + b$ the quantities m and b are considered constants even though they may adopt different values for different specific applications.

A specific invariant quantity whose value is determined *a priori*, such as π or e, is also called a constant. In physics, any physical quantity whose value is fixed by the laws of nature, such as the speed of light c, or the universal gravitational constant G, is called a constant.

The constant term in a POLYNOMIAL is the term that does not involve any power of the variable. For example, the polynomials $x^3 - 5x + 7$ and $2z^5 - 3z^2 + z$ have constant terms 7 and 0, respectively.

A constant function is any function f that yields the same output value, a say, no matter which input value is supplied: $f(x) = a$ for all values x. The graph of a constant function is a horizontal line. It is surprising, for instance, to discover that the function given by $f(x) = x^{\frac{1}{\log_{10} x}}$, defined for all positive numbers x, is a constant function. The formula always returns the value 10 no matter which value for the input x is chosen.

constant of integration   The MEAN-VALUE THEOREM shows that any two antiderivatives of a given function differ only by a constant. For example, any antiderivative of the function $f(x) = 2x$ must be of the form $x^2 + C$ for some constant C, called the constant of integration. An antiderivative is usually expressed as an INDEFINITE INTEGRAL. In our example we have $\int 2x\,dx = x^2 + C$.

Care must be taken when working with an arbitrary constant of integration. For example, consider computing the integral $\int \frac{1}{x}\,dx$ via the method of INTEGRATION BY PARTS. Set $u = \frac{1}{x}$ and $v' = 1$ (so that $u' = -\frac{1}{x^2}$ and $v = x$) to obtain:

$$\int \frac{1}{x}\,dx = \frac{1}{x} \cdot x - \int \left(-\frac{1}{x^2}\right) x\,dx$$
$$= 1 + \int \frac{1}{x}\,dx$$

Subtracting the integral under consideration suggests the absurdity: $0 = 1$. Of course, this argument failed to take care of the constants of integration that should appear.

See also ANTIDIFFERENTIATION; INTEGRAL CALCULUS.

constant width A circular wheel has the property that it has constant height as it rolls along the ground. Alternatively, one could say that the width of the curve is the same no matter which way one orients the figure to measure it. Any shape with this property is called a curve of constant width. The so-called Reuleaux triangle, constructed by drawing arcs of circles along each side of an equilateral triangle (with the opposite vertex as center of each circular arc) is another example of such a curve. Wheels of this shape also roll along the ground with constant height.

One can construct wheels of constant height with the aid of a computer. One begins with the PARAMETRIC EQUATIONS of a circle of radius 1 and center (0,0) given by $x(t) = 0 + \cos(t)$ and $y(t) = 0 + \sin(t)$, for $0 \le t < 360°$. Certainly the distance between any two points on this circle that are separated by an angle of 180° is always 2. By changing the location of the center of the circle slightly, we can preserve this distance property, as long as we ensure that the center returns to the same location every 180°. Thus, for example, the equations:

$$x(t) = \frac{1}{10}\sin(2t) + \cos(t)$$
$$y(t) = \frac{1}{8}\cos(2t) + \sin(t)$$

are the parametric equations of another curve with the same constant-width property. (The fractional coefficients were chosen to ensure that the resulting figure is CONVEX.)

French mathematician Joseph Barbier (1839–89) proved that all curves of constant width d have the same perimeter, πd. It is also known that, for a given width, Reuleaux's triangle is the curve of constant width of smallest area.

constructible A geometric figure is said to be constructible if it can be drawn using only the tools of a straightedge (that is, a ruler with no markings) and a compass. The straightedge allows one to draw line segments between points (but not measure the lengths of those segments), and the compass provides the means to draw circles with a given point as center and a given line segment from that point as radius.

The Greek scholars of antiquity were the first to explore the issue of which geometric constructs could be produced with the aid of these primitive tools alone. The geometer EUCLID (ca. 300 B.C.E.) explicitly stated these limitations in his famous text *THE ELEMENTS*. Despite the fact that his exercise has no real practical application (it is much easier to draw figures with rulers to measure lengths and protractors to measure angles), the problem of constructibility captured the fascination of scholars for the two millennia that followed. This illustrates the power of intellectual curiosity alone for the motivation of mathematical investigation. Students in high schools today are still required to study issues of constructibility.

The compass used by the Greeks was different from the one we use today; it would not stay open at a fixed angle when lifted from the page and would collapse. Thus it was not directly possible to draw several circles of the same radius, for instance, simply by taking the compass to different positions on the page. However, in his work *The Elements*, Euclid demonstrated how to accomplish this feat with the Greek collapsible compass. This shows that any construction that can be accomplished with a modern compass can also be accomplished with a collapsible compass. For this reason, it is assumed today that the compass used is a modern one.

A surprising number of constructions can be accomplished with the aid of a straightedge and compass alone. We list here just a few demonstrations.

1. Copy a given line segment *AB* onto a given line *L*

   Place the compass with point at one endpoint of the line segment and pencil tip at the other, thus setting the compass to the length of the line segment. Label one point *A'* on the line *L*, place the point of the compass at *A'* and use it to mark off a point *B'* on the line. The segment *A'B'* is congruent to the segment *AB*.

2. Draw a line perpendicular to a given line *L* through a given point *P* on *L*.

   Set the compass at an arbitrary radius and, with the tip of the compass placed on point *P*, mark off two points *A* and *B* on line *L* equidistant from *P*. Now draw two circles of the same radius, one with center at *A* and the other with center at *B*. These circles intersect at two positions *X* and *Y*. These points *X* and *Y* are the same distance from each of *X* and *Y*, and so, as the study of EQUIDISTANCE proves, the line through *X* and *Y* (and *P*) is the perpendicular bisector of the line segment *AB*. In particular, it is a line through *P* perpendicular to *L*.

3. Construct the perpendicular bisector of a given line segment *AB*.

   The construction described in 2 above accomplishes this feat.

4. Draw a line perpendicular to a given line *L* through a given point *P* not on *L*.

   Set the compass with its point at *P* and draw a large circle that intersects the line *L* at two points *A* and *B*. Now follow the procedure for 2.

5. Draw an equilateral triangle.

   The points *A*, *B*, and *X* described in 2 are the vertices of an equilateral triangle.

6. Copy an arbitrary triangle to a different position on the page.

   Suppose the given triangle has vertices labeled *A*, *B*, and *C*. Set the compass with point at *A* and tip at *B*. Arbitrarily choose a point *A'* elsewhere on the page, and use the compass to draw a circle with center *A'* and radius equal to length of the segment *AB*. Label an arbitrary point on this circle *B*. Use the compass to draw a second circle with center *A'*, but this time with radius equal to the length of *AC*.

We must now select an appropriate point *C'* on this second circle. Draw a third circle with center *B'* of radius equal to the length of *BC*. Label a point of intersection between the final two circles *C'*. Then *A'B'C'* is a congruent copy of the original triangle.

7. Copy a given angle to a different location on the page.

   Simply regard the angle as part of a triangle and follow the instructions of part 6.

8. Construct a line parallel to a given line *L* through a point *P* not on *L*.

   Draw an arbitrary line through *P* that intersects the line *L*. Copy the angle these two lines make at position *P* and draw a third line through *P* at this angle. This produces a diagram of a TRANSVERSAL crossing a pair of lines possessing equal corresponding angles. By the converse of the PARALLEL POSTULATE, the two lines are parallel.

9. Construct a line that divides a given angle precisely in half.

   Draw a circle of arbitrary radius with center at the vertex of the angle. Suppose this circle intersects the rays of the angles at positions *A* and *B*. Now draw two circles of the same radius centered about each of these two points. Let *P* be a point of intersection of the two circles. Then the line connecting the vertex of the angle to *P* is an angle bisector. (The SSS principle of similarity shows that the two triangles produced in the construction are congruent, demonstrating then that the original angle is indeed divided into two equal measures.)

10. Draw a perfect square.

    Draw an arbitrary line segment. This will be the first side of the square. Label its endpoints *A* and *B*. Using part 2, construct a line through *B* perpendicular to the line segment. Use a circle centered about *B* of radius equal in length to *AB* to find a point *C* on this perpendicular line so that *BC* is the same length as *AC*. This provides the second side of the square. Repeat this procedure to construct the remaining two sides of the square.

Not every geometric feat can be accomplished with straightedge and compass alone. For example, although it is possible to also construct a regular pentagon and a regular hexagon with these primitive tools, the construction of a regular heptagon (a seven-sided polygon) is impossible. CARL FRIEDRICH GAUSS (1777–1855) proved that a regular n-gon is constructible if, and only if, n is a number of the form $2^k p_1 p_2 \ldots p_n$, with each p_i a distinct PRIME of the form $2^{2^s} + 1$ (such as 3, 5, 17, 257, and 65,537.) Although one can bisect an angle with straightedge and compass, the problem of TRISECTING AN ANGLE is unsolvable. The two classical problems of SQUARING THE CIRCLE and DUPLICATING THE CUBE also cannot be solved.

Constructible Numbers

A real number r is said to be constructible if, given a line segment on a page deemed to be of unit length, it is possible to construct from it a line segment of length r using only the tools of a straightedge and a compass. For instance, the number 2 is constructible. (Given a line segment AB of length one, use the straightedge to extend the length of the line. Draw a circle of radius equal to the length of AB, centered about B, to intersect the line at a new point C. Then the length of AC is 2.) Any positive whole number is constructible.

Suppose a and b are two constructible numbers with $b > a$. (That is, given a line segment of length 1, we can also produce line segments of lengths a and b.) Then the following is true:

The numbers $a + b$, $b - a$, $a \times b$, $\dfrac{a}{b}$, and $\sqrt{a}$ are constructible.

The diagram at right indicates how to construct these quantities.

(In the third and fourth diagrams, draw lines parallel to the lines connecting the endpoints of the two segments of lengths a and b. Examination of similar triangles shows the segments indicated are indeed of lengths $a \times b$ and a/b, respectively. For the fifth diagram, add lines to produce a large right triangle within the circle with the diameter of length $a + 1$ as hypotenuse. Application of PYTHAGORAS'S THEOREM shows that the segment indicated is indeed of length $\sqrt{a}$.)

It follows now that any rational number is constructible as is any number that can be obtained from the rationals by the application of a finite number of additions, subtractions, multiplications, divisions, and square

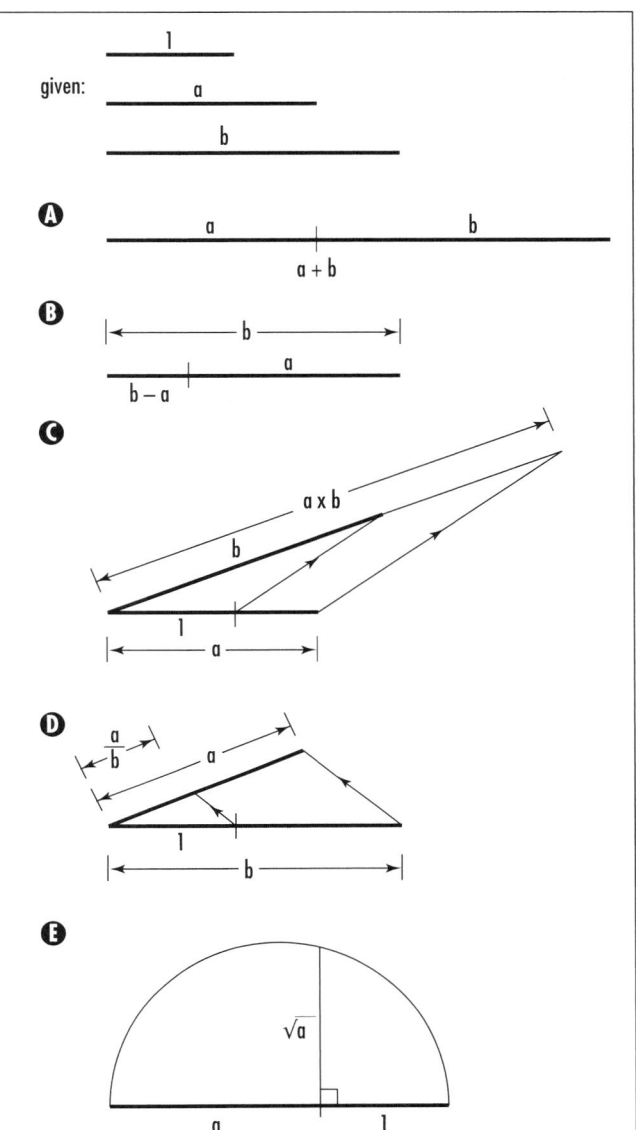

Constructing the sum, difference, product, quotient, and roots of a and b

roots. (For instance, the number

$$\frac{\dfrac{7}{\sqrt{3}} + \sqrt{2 + 13\sqrt{2 + \sqrt{2}}}}{\sqrt{\sqrt{\sqrt{3 + \sqrt{5}}}}}$$

is constructible.) Mathematicians have proved that these are the only types of real numbers that are constructible. Mathematicians have also proved that any number that is constructible is an ALGEBRAIC NUMBER. As π, for instance, is not algebraic, it is not constructible.

See also AAA/AAS/ASA/SAS/SSS.

contingency table A table showing the number of units from a sample having certain combinations of attributes is called a contingency table. For example, a marketing research project records the hair color of participating men and women and presents the results in a contingency table:

Hair Color				
	Black	Brown	Blonde	Red
Male	25	23	8	2
Female	18	16	14	5

The CHI-SQUARED TEST can be used to look for correlations between attributes displayed in such tables.

continued fraction A number that is an integer plus a fraction with denominator that is itself an integer plus a fraction—and continued this way—is called a continued fraction. For example,

$$\frac{1402}{457} = 2 + \cfrac{4}{3 + \cfrac{7}{9 + \cfrac{1}{2 + \frac{3}{5}}}}$$

and

$$\sqrt{2} = 1 + \cfrac{1}{2 + \cfrac{1}{2 + \cfrac{1}{2 + \cfrac{1}{2 + \cdots}}}}$$

are continued-fraction representations of the quantities 1,402/457 and $\sqrt{2}$ (as we shall establish below). The first continued fraction stops after a finite number of steps, and the second continues forever. A continued fraction is said to be in standard form if, like the second example, all the numerators are equal to one, and all the integers involved are positive.

Every positive real number x can be written as a continued fraction in standard form. If $\lfloor x \rfloor$ denotes the largest integer less than or equal to x, and $\{x\}$ the frac-

tional part of x as given by the FRACTIONAL PART FUNCTION, then:

$$x = \lfloor x \rfloor + \{x\} = \lfloor x \rfloor + \cfrac{1}{\cfrac{1}{\{x\}}}$$

As the quantity $1/\{x\}$ itself is a positive real number greater than one, we can, in the same way, write it as an integer plus another fraction with unit numerator. Repeated application of this procedure produces a continued fraction in standard form.

For example, if $x = \dfrac{1,402}{457}$, then we can write

$$x = 3 + \frac{31}{457} = 3 + \cfrac{1}{\cfrac{457}{31}}, \text{ and } \frac{457}{31} = 14 + \frac{23}{31} = 14 + \cfrac{1}{\cfrac{31}{23}},$$

and so on. This produces the standard-form continued fraction:

$$\frac{1,402}{457} = 3 + \cfrac{1}{14 + \cfrac{1}{1 + \cfrac{1}{2 + \cfrac{1}{1 + \frac{1}{7}}}}} = [3,14,1,2,1,7]$$

It is not difficult to show that if $x = a/b$ is a fraction, then $1/\{x\}$ is a new fraction with denominator smaller than b. Repeated application of this procedure must eventually produce a continued fraction with denominator equal to one, so that the procedure terminates. This shows that all numbers that are rational (that is, equal to a fraction) have continued-fraction representations that stop after a finite number of steps. (And, conversely, any such continued fraction "unravels" to produce a quantity that is rational.) Consequently:

All quantities with infinitely long continued-fraction representations are irrational.

For example, one can check that $\sqrt{2} = 1 + \cfrac{1}{1 + \sqrt{2}}$ and substituting this formula into itself gives the continued fraction representation presented above:

$$\sqrt{2} = 1 + \cfrac{1}{1+\sqrt{2}} = 1 + \cfrac{1}{2+\cfrac{1}{1+\sqrt{2}}} = 1 + \cfrac{1}{2+\cfrac{1}{2+\cfrac{1}{1+\sqrt{2}}}}$$

$$= \cdots = 1 + \cfrac{1}{2+\cfrac{1}{2+\cfrac{1}{2+\cfrac{1}{2+\cdots}}}}$$

That this process does not terminate proves that $\sqrt{2}$ is not a fraction. In a similar way, one establishes:

$$\sqrt{3} = 1 + \cfrac{1}{1+\cfrac{1}{2+\cfrac{1}{1+\cfrac{1}{2+\cdots}}}} = [1,\overline{1,2}]$$

$$\sqrt{5} = 2 + \cfrac{1}{4+\cfrac{1}{4+\cfrac{1}{4+\cdots}}} = [2,\overline{4}]$$

and $\sqrt{6} = [2,\overline{2,4}]$, showing that these quantities are also irrational. The number e also has an infinite continued-fraction representation: $e = [2,1,2,1,1,4,1,1,6,1,1,8,1,$ $1,10,...]$, as does the golden mean $\varphi = \frac{1+\sqrt{5}}{2} = [1,1,1,...]$.

Mathematicians have proved that finite continued fractions in standard form have a curious property:

> Reversing the order of the integers that appear in a finite continued fraction produces a new fraction with the same numerator as the original quantity.

For example, as we have seen, $[3,14,1,2,1,7]$ = 1,402/457, and one calculates that:

$$[7,1,2,1,14,3] = 7 + \cfrac{1}{1+\cfrac{1}{2+\cfrac{1}{1+\cfrac{1}{14+\cfrac{1}{3}}}}}$$

equals the fraction 1,402/181.

Continued fractions were systematically studied by LEONHARD EULER (1707–83), and he was the first to formally introduce them in a written text. JOSEPH-LOUIS LAGRANGE (1736–1813) extended much of Euler's work. Continued fractions have proved to be very useful in solving a large selection of DIOPHANTINE EQUATIONS. They also provide excellent rational approximations to irrational numbers. For example, terminating the continued fraction representation for $\sqrt{2}$ after a finite number of steps yields good approximations to the square root of 2:

$$1 = 1, \; 1 + \frac{1}{2} = \frac{3}{2}, \; 1 + \cfrac{1}{2+\cfrac{1}{2}} = \frac{7}{5}, \; 1 + \cfrac{1}{2+\cfrac{1}{2+\cfrac{1}{2}}}$$

$$= \frac{17}{12}, \frac{41}{29}, \frac{99}{70}, \frac{239}{169}, \cdots$$

This particular sequence of fractions, generated by the formula $\frac{a}{b} \rightarrow \frac{a+2b}{a+b}$, was used by Theon of Smyrna as early as the first century C.E. It holds some mysterious properties. For example, every second term of the sequence corresponds to a PYTHAGOREAN TRIPLE:

$$\frac{7}{5} = \frac{3+4}{5} \text{ and } 3^2 + 4^2 = 5^2$$

$$\frac{41}{29} = \frac{20+21}{29} \text{ and } 20^2 + 21^2 = 29^2$$

$$\frac{239}{169} = \frac{119+120}{169} \text{ and } 119^2 + 120^2 = 169^2$$

$$\vdots$$

and every other term, rounding the numerator and denominator each down to the half, yields FIGURATE NUMBERS that are both square and triangular:

$$\frac{3}{2} \rightarrow \frac{1}{1} \text{ and } T_1 = S_1 = 1$$

$$\frac{17}{12} \rightarrow \frac{8}{6} \text{ and } T_8 = S_6 = 36$$

$$\frac{99}{70} \rightarrow \frac{49}{35} \text{ and } T_{49} = S_{35} = 1,225$$

$$\vdots$$

Unraveling the continued-fraction expansions of other irrational numbers can yield analogous discoveries.

continuous function   Informally, a function is said to be continuous if one can draw its graph without ever lifting the pencil from the page. This means that the graph of the function consists of a single curved line with no gaps, jumps, or holes.

More precisely, a function is continuous at a point $x = a$, if the function is defined at the point a, and the LIMIT of $f(x)$ as x approaches a equals the value of the function at a:

$$\lim_{x \to a} f(x) = f(a)$$

(If, for example, a function has values $f(0.9) = 1.9$, $f(0.99) = 1.99$, $f(0.999) = 1.999$, and so on, then one would be very surprised to learn that $f(1)$ equals 18. The function would not be deemed continuous at $x = 1$.) A function that is not continuous at a point is said to be discontinuous, or to have a discontinuity, at that point. A function that is continuous at every point in its domain is called continuous.

Mathematicians have proved that:

i. The sum of two continuous functions is continuous.
ii. The product of two continuous functions is continuous.
iii. The quotient of two continuous functions is continuous at each point where the denominator is not zero.
iv. The COMPOSITION of two continuous functions is continuous.

Since the straight-line graph $f(x) = x$ is continuous, it follows from properties i and ii that any POLYNOMIAL function $p(x) = a_n x^n + \ldots + a_1 x + a_0$ is continuous. By property iii, any RATIONAL FUNCTION is continuous at all points where the denominator is not zero. The functions $\sin x$ and $\cos x$ from TRIGONOMETRY are both continuous. The tangent function, $\tan x = \dfrac{\sin x}{\cos x}$, is continuous at every point other than $\pm\dfrac{\pi}{2}, \pm\dfrac{3\pi}{2}, \pm\dfrac{5\pi}{2}, \ldots$, the locations where cosine is zero.

It is possible to remove a discontinuity of a function at $x = a$ if the limit $\lim_{x \to a} f(x)$ exists. For example, the function $f(x) = \dfrac{x^2 - 1}{x - 1}$ is not defined at $x = 1$, since the quantity $\dfrac{0}{0}$ has no meaning. Nonetheless, algebra shows that the limit of this function as x approaches the value 1 exists:

$$\lim_{x \to 1} \frac{x^2 - 1}{x - 1} = \lim_{x \to 1} \frac{(x-1)(x+1)}{x-1} = \lim_{x \to 1} x + 1 = 2$$

(Dividing through by the quantity $x - 1$ is valid in this calculation since, for values of x close to, but not equal to, 1, the quantity $x - 1$ is not zero.) Consequently, if we declare the value of the function to be 2 at $x = 1$:

$$f(x) = \begin{cases} \dfrac{x^2 - 1}{x - 1} & \text{if x is different from 1} \\ 2 & \text{if x is equal to 1} \end{cases}$$

we now have a continuous function. A discontinuity at $x = a$ for a function f is called removable if $\lim_{x \to a} f(x)$ exists.

The issue of continuity is fundamental to the foundation of CALCULUS. A study of the INTERMEDIATE-VALUE THEOREM and its consequences illustrates this.

continuum hypothesis   A study of DENUMERABLE sets shows that every infinite set contains a denumerable subset. Thus, in a well-defined sense, denumerable sets are the "smallest" types of infinite sets. The DIAGONAL ARGUMENT of the second kind shows that the set of real numbers is not denumerable, that is, in a meaningful sense, the CARDINALITY of the real numbers, denoted c, is "larger" than the cardinality of denumerable sets, which is denoted $\aleph_0$. We have:

$$\aleph_0 < c$$

German mathematician GEORG CANTOR (1845–1918), father of cardinal arithmetic, conjectured that there is no type of infinite set "larger" than an infinite set of denumerable objects, but "smaller" than the continuum of the real numbers. (That is, there is no cardinal number strictly between $\aleph_0$ and c.) This conjecture became known as the continuum hypothesis. It can be stated equivalently as follows:

Any infinite subset of real numbers can either be put in one-to-one correspondence with the set of natural numbers, or can be put in one-to-one correspondence with the entire set of real numbers.

Despite his efforts, Cantor was unable to establish whether or not his continuum hypothesis was true. In 1940 Austrian mathematician KURT GÖDEL proved that the continuum hypothesis cannot be proved false. Unfortunately, as GÖDEL'S INCOMPLETENESS THEOREMS show, this does not mean that the continuum hypothesis is true: there exist statements in mathematics that are undecidable, that is, ones that cannot be proved true and cannot be proved false. It was suspected that the continuum hypothesis might be such an undecidable statement. Twenty-three years later in 1963, American logician Paul Cohen managed to prove that this is indeed the case. Consequently one can either deem the continuum hypothesis as true or as false, an arbitrary choice, and be certain never to run into a mathematical contradiction as a result.

contour integral (curvilinear integral, line integral) If C is a curve in the xy-plane and $z = f(x,y)$ is a function of two variables, then one can attempt to compute the surface AREA (one side) of a "wall" that follows the curve C and has "height" the height of the function above the curve. The integral that computes this, called a contour integral and denoted $\int_c f\,ds$, is constructed by selecting a large number of points p_0, p_1,...,p_n along the curve C and approximating the surface under consideration by a collection of rectangular sections. The ith rectangle can be taken to have base-length the distance between the points p_i and p_{i+1}, which we denote d_i, and height $f(p_i)$. The surface area is thus approximated by the sum $\sum_{i=0}^{n-1} f(p_i)d_i$. Taking the limit as we take finer and finer approximations defines the desired contour integral.

If the curve C is defined by parametric equations: $x = x(t)$ and $y = y(t)$ for some parameter t, $a \le t \le b$, then this procedure gives the contour integral as:

$$\int_C f\,ds = \int_a^b f(x(t),y(t))\sqrt{\left(x'(t)\right)^2 + \left(y'(t)\right)^2}\,dt$$

In physics and in advanced VECTOR calculus, one also considers integrating, for example, the work done in moving a particle along a curve C through a VECTOR FIELD (force field). Such considerations lead to other types of integrals, also called line integrals.

contour line A line on a map that joins points of equal height is called a contour line. Contour lines are usually drawn for equal intervals of height. This gives experienced map readers a clear mental picture of the three-dimensional topography of the land: contour lines close together, for example, indicate that the slope of the land is steep.

In mathematics, contour lines are used to portray the shapes of surfaces sitting in three-dimensional space. For example, all points of the same height $z = c$ on the surface $z = x^2 + y^2$ satisfy the equation $x^2 + y^2 = c$ and so lie on a circle of radius $\sqrt{c}$. This leads to a contour map for the graph of the function $f(x,y) = x^2 + y^2$ consisting of sets of concentric circles about the origin. The surface described is a PARABOLOID with vertex at the origin.

contradiction In FORMAL LOGIC, any statement that yields a TRUTH TABLE with final entries all false is called a contradiction. For example, the statement $(\neg p) \wedge p$ is a contradiction. From any contradiction, it is possible to prove that *any* statement in mathematics is true. For example, one can check that the compound statement $(\neg p) \wedge p \rightarrow q$ is a tautology. Consequently, in mathematics, if one can prove that some statement p and its negation $\neg p$ are both true, then since both $(\neg p) \wedge p$ and $(\neg p) \wedge p \rightarrow q$ are valid, no matter what statement q represents, q is also true by inference. Any contradiction that appears in mathematics would prove, for example, that $1 = 2$, and that every irrational number is a fraction. Mathematicians sincerely hope that mathematics is free from contradiction.

See also CONSISTENT.

contrapositive The contrapositive of a CONDITIONAL statement "*p implies q*" is the statement: "*not q implies not p.*" It is the statement obtained by switching the antecedent with the consequent, and negating each. For example, the contrapositive of the statement, "*If it is a poodle, then it is a dog,*" is "*If it is not a dog, then it is not a poodle.*" The contrapositive of a statement is a logically equivalent form of the statement, and so can be used at any time in its stead. (One shows that $p \rightarrow q$ and $(\neg q) \rightarrow (\neg p)$ have identical TRUTH TABLES.)

In mathematics it is sometimes convenient to prove the contrapositive form of a theorem rather than prove the assertion directly. This approach is called contrapositive reasoning (or *modus tollens*), and the proof presented is a proof by contraposition. For example, the theorem: *if n^2 is odd, then n is odd*, is best proved by assuming that n is even (that is, $n = 2k$ for some integer k) and then showing that n^2 is also even.

The contrapositive of a conditional *"p implies q"* should not be confused with the inverse of the statement: *"not p implies not q."* This variation is *not* a logically equivalent form of the original conditional.

See also ARGUMENT; CONVERSE; PROOF.

convergent improper integral *See* IMPROPER INTEGRAL.

convergent sequence A SEQUENCE of numbers $a_1, a_2, a_3, \ldots$ is said to converge if the terms of the sequence become arbitrarily close to, but do not necessarily ever reach, a particular finite value L. For example, the numbers in the sequence 0.9, 0.99, 0.999,… approach the value 1. We call 1 the LIMIT of this sequence.

Any sequence that converges is called a convergent sequence. If a sequence $\{a_n\}$ converges to limit L, we write $\lim_{n \to \infty} a_n = L$, or, alternatively, $a_n \rightarrow L$ as $n \rightarrow \infty$, which is read as "a_n approaches L as n becomes large." For example, the sequence $\frac{1}{2}, \frac{3}{4}, \frac{7}{8}, \frac{15}{16}, \ldots$ has limit one ($\lim_{n \to \infty} \frac{2^n - 1}{2^n} = 1$), and the sequence $1, -\frac{1}{2}, \frac{1}{3}, -\frac{1}{4}, \frac{1}{5}, \ldots$ has limit zero ($\lim_{n \to \infty} \frac{(-1)^{n+1}}{n} \rightarrow 0$ as $n \rightarrow \infty$). The notions of limit and convergence can be made mathematically precise with an "$\varepsilon - N$ definition" of a limit. (*See* LIMIT.)

A sequence that does not converge is said to diverge. A divergent sequence could have terms that grow in size without bound (1,4,9,16,25,…, for example), terms that oscillate without converging to a limit ($\frac{1}{2}, -\frac{2}{3}, \frac{3}{4}, -\frac{4}{5}, \frac{5}{6}, -\frac{6}{7}, \ldots$, for example), or terms that

oscillate without bound (1,2,1,3,1,4,1,5,1,6,1,7,1,…, for instance).

See also DIVERGENT; INFINITE PRODUCT; SERIES.

convergent series An infinite SERIES $\sum_{n=1}^{\infty} a_n = a_1 + a_2 + a_3 + \ldots$ is said to converge to a value L if the sequence of PARTIAL SUMS, $S_n = a_1 + a_2 + \ldots + a_n$, approaches the value L in the LIMIT as $n \rightarrow \infty$. To illustrate, the series $\sum_{n=1}^{\infty} \frac{1}{2^n} = \frac{1}{2} + \frac{1}{4} + \frac{1}{8} + \frac{1}{16} + \cdots$ has partial sums:

$$S_1 = \frac{1}{2}$$

$$S_2 = \frac{1}{2} + \frac{1}{4} = \frac{3}{4}$$

$$S_3 = \frac{1}{2} + \frac{1}{4} + \frac{1}{8} = \frac{7}{8}$$

$$\vdots$$

$$S_n = 1 - \frac{1}{2^n}$$

$$\vdots$$

which approach the value 1 as n grows. In this sense we say that the series $\sum_{n=1}^{\infty} \frac{1}{2^n}$ converges to 1, and we write:

$$\frac{1}{2} + \frac{1}{4} + \frac{1}{8} + \frac{1}{16} + \cdots = 1$$

If the limit of the partial sums does not exist, then the series is said to diverge. For example, the series $1 - 1 + 1 - 1 + 1 - \ldots$ diverges because the partial sums oscillate between being 1 and 0 and never settle to a particular value. The series $1 + 2 + 3 + 4 + \ldots$ diverges because the partial sums grow arbitrarily large. The series $\sum_{n=1}^{\infty} \frac{1}{\sqrt{n}}$ diverges for the same reason, which can be seen as follows:

$$S_n = \frac{1}{1} + \frac{1}{\sqrt{2}} + \frac{1}{\sqrt{3}} + \cdots + \frac{1}{\sqrt{n}}$$

$$> \frac{1}{\sqrt{n}} + \frac{1}{\sqrt{n}} + \frac{1}{\sqrt{n}} + \cdots + \frac{1}{\sqrt{n}}$$

$$= \frac{n}{\sqrt{n}}$$

$$= \sqrt{n}$$

and so $S_n \to \infty$ as n grows.

There are a number of tests to determine whether or not a given series converges.

The n*th*-Term Test

If a series $\sum_{n=1}^{\infty} a_n$ converges, then it must be the case that $\lim_{n\to\infty} a_n = 0$. Consequently, if the terms a_n of the series do not approach zero, then the series must diverge.

To see why this is true, note that $S_n = a_1 + a_2 + \ldots + a_{n-1} + a_n = S_{n-1} + a_n$. If the partial sums converge to L, then we must have that $\lim_{n\to\infty} a_n = \lim_{n\to\infty}(S_n - S_{n-1}) = L - L = 0$. This test shows, for example, that the series $\sum_{n=1}^{\infty} \frac{n-1}{n} = 0 + \frac{1}{2} + \frac{2}{3} + \frac{3}{4} + \ldots$ diverges because the terms of the series do not become small.

The Comparison Test

This test applies only to series with positive terms.

A series $\sum_{n=1}^{\infty} a_n$ with positive terms converges if each term a_n of the series is less than or equal to the terms of another series with positive terms already known to converge.

A series $\sum_{n=1}^{\infty} a_n$ with positive terms diverges if each term a_n of the series is greater than or equal to the terms of another series with positive terms already known to diverge.

For example, the series $\sum_{n=1}^{\infty} \frac{1}{(\sqrt{n})2^n}$ converges because $\sum_{n=1}^{\infty} \frac{1}{(\sqrt{n})2^n} < \sum_{n=1}^{\infty} \frac{1}{2^n}$, a series which we already know converges. The series $\sum_{n=1}^{\infty} \frac{\sqrt[5]{n+7}}{\sqrt{n}}$ diverges, since

$$\sum_{n=1}^{\infty} \frac{\sqrt[5]{n+7}}{\sqrt{n}} > \sum_{n=1}^{\infty} \frac{1}{\sqrt{n}}, \text{ which we know diverges.}$$

The Ratio Test

This test applies only to series with positive terms.

A series $\sum_{n=1}^{\infty} a_n$ with all terms positive:

    i. converges if $\lim_{n\to\infty} \frac{a_{n+1}}{a_n}$ exists and equals a value smaller than 1

    ii. diverges if $\lim_{n\to\infty} \frac{a_{n+1}}{a_n}$ exists and equals a value greater than 1

If the limit in question actually equals 1, then nothing can be concluded from this test.

This test was first developed by French mathematician AUGUSTIN-LOUIS CAUCHY (1789–1857). It is proved in CALCULUS texts by making clever comparison to a GEOMETRIC SERIES. (Briefly, if for all terms $\frac{a_{n+1}}{a_n} \leq r < 1$, then $a_2 \leq ra_1$, $a_3 \leq ra_2 \leq r^2a_1$, $a_4 \leq ra_3 \leq r^3a_1$, etc., and so $a_1 + a_2 + a_3 + a_4 + \ldots \leq a_1(1 + r + r^2 + r^3 + \ldots)$, which converges.) Consider, for example, the series $\sum_{n=1}^{\infty} \frac{\sqrt{n+1}}{3^n}$.

Here the nth term is given by $a_n = \frac{\sqrt{n+1}}{3^n}$, and we have:

$$\lim_{n\to\infty} \frac{a_{n+1}}{a_n} = \lim_{n\to\infty} \frac{\sqrt{n+2}}{3^{n+1}} \cdot \frac{3^n}{\sqrt{n+1}} = \lim_{n\to\infty} \frac{1}{3} \cdot \sqrt{\frac{n+2}{n+1}}$$

$$= \frac{1}{3} \cdot 1 = \frac{1}{3} < 1. \text{ By the ratio test, this series converges.}$$

The Root Test

This test applies only to series with positive terms.

A series $\sum_{n=1}^{\infty} a_n$ with all terms positive:

    i. converges if $\lim_{n\to\infty} \sqrt[n]{a_n}$ exists and equals a value smaller than 1

    ii. diverges if $\lim_{n\to\infty} \sqrt[n]{a_n}$ exists and equals a value greater than 1

If the limit in question actually equals 1, then nothing can be concluded from this test.

The proof of this test relies on making clever comparison to a geometric series. (If, for all terms, $\sqrt[n]{a_n} \leq r$ for some value r, then $a_n \leq r^n$ and a comparison can be made.) This test is often used if the series contains terms involving exponents. For example, consider the series $\sum_{n=1}^{\infty} \frac{1}{(n^2+n)^n}$. Here the nth term of the series is given by $a_n = \frac{1}{(n^2+n)^n}$, and we have: $\lim_{n\to\infty} \sqrt[n]{a_n} = \lim_{n\to\infty} \frac{1}{n^2+n} = 0 < 1$. By the root test, the series must converge.

The Integral Test

This test applies only to series with positive terms.

Suppose the terms of a series $\sum_{n=1}^{\infty} a_n$ are given by a formula $a_n = f(n)$, where the function $f(x)$ is continuous, positive, and decreasing for $x \geq 1$:

  i. If the IMPROPER INTEGRAL $\int_1^{\infty} f(x)dx$ converges, then the series $\sum_{n=1}^{\infty} a_n$ converges.
  ii. If the improper integral $\int_1^{\infty} f(x)dx$ diverges, then the series $\sum_{n=1}^{\infty} a_n$ diverges.

This can be proved geometrically by drawing rectangles of width 1 just above and just below the graph of $y = f(x)$ for $x \geq 1$, and then comparing the total area of all the rectangles with the area under the curve. To illustrate the test, consider the series $\sum_{n=1}^{\infty} \frac{1}{n^2}$. Since $\int_1^{\infty} \frac{1}{x^2}dx = -\frac{1}{x}\Big]_1^{\infty} = 0 - (-1) = 1$ converges, we have that the series $\sum_{n=1}^{\infty} \frac{1}{n^2}$ converges. In general, one can establish in this way the p-series test.

The p-Series Test

A series of the form $\sum_{n=1}^{\infty} \frac{1}{n^p}$ with p a real number converges if $p > 1$ and diverges if $p \leq 1$.

A series of the form $\sum_{n=1}^{\infty} \frac{1}{n^p}$ is called a p-series.

Absolute Convergence Test

Suppose $\sum_{n=1}^{\infty} a_n$ is a series with both positive and negative terms. If the corresponding series $\sum_{n=1}^{\infty} |a_n|$ with all terms made positive converges, then the original series $\sum_{n=1}^{\infty} a_n$ also converges.

One can use any of the first six tests described above to determine whether or not $\sum_{n=1}^{\infty} |a_n|$ converges. The validity of the absolute convergence test is established in the discussion on ABSOLUTE CONVERGENCE.

Since we have established, for example, that $\sum_{n=1}^{\infty} \frac{1}{n^2} = 1 + \frac{1}{4} + \frac{1}{9} + \frac{1}{16} + \cdots$ converges, the absolute convergence test now assures us that the variant series $1 - \frac{1}{4} + \frac{1}{9} - \frac{1}{16} + \frac{1}{25} - \frac{1}{36} + \frac{1}{49} - \frac{1}{64} + \frac{1}{81} - \frac{1}{100} + \cdots$ also converges (as does any other variation that involves the insertion of negative signs).

The absolute-convergence test does not cover all cases. It is still possible that a series with negative terms, $\sum_{n=1}^{\infty} a_n$, might converge even though $\sum_{n=1}^{\infty} |a_n|$ diverges. This phenomenon is called CONDITIONAL CONVERGENCE.

Alternating-Series Test

If the terms of a series $\sum_{n=1}^{\infty} a_n$ alternate in sign and satisfy

  i. $a_1 \geq a_2 \geq a_3 \geq \ldots$
  ii. $a_n \to 0$

then the series converges.

(See ALTERNATING SERIES.) This test shows, for example, that the alternating HARMONIC SERIES $1 - \frac{1}{2} + \frac{1}{3} - \frac{1}{4} + \frac{1}{5} - \ldots$ converges even though the corresponding series with all terms positive does not.

It is possible to multiply convergent series by constants, and to add and subtract two convergent series. Precisely, if $\sum_{n=1}^{\infty} a_n$ and $\sum_{n=1}^{\infty} b_n$ both converge, and k is a number, then:

$$\sum_{n=1}^{\infty}(ka_n) = k\left(\sum_{n=1}^{\infty} a_n\right)$$

$$\sum_{n=1}^{\infty}(a_n + b_n) = \sum_{n=1}^{\infty} a_n + \sum_{n=1}^{\infty} b_n$$

$$\sum_{n=1}^{\infty}(a_n - b_n) = \sum_{n=1}^{\infty} a_n - \sum_{n=1}^{\infty} b_n$$

These properties can be used to evaluate new infinite sums. For example, in 1740 LEONHARD EULER showed that a particular value of the ZETA FUNCTION is given by $\sum_{n=1}^{\infty}\frac{1}{n^2} = 1+\frac{1}{4}+\frac{1}{9}+\frac{1}{16}+\cdots = \frac{\pi^2}{6}$. It then follows that the alternating form of this series has value:

$$1-\frac{1}{4}+\frac{1}{9}-\frac{1}{16}+\frac{1}{25}-\frac{1}{36}+\cdots = \left(1+\frac{1}{4}+\frac{1}{9}+\frac{1}{16}+\frac{1}{25}+\frac{1}{36}+\cdots\right)$$
$$-2\left(\frac{1}{4}+\frac{1}{16}+\frac{1}{36}+\cdots\right)$$
$$=\frac{\pi^2}{6}-\frac{2}{4}\left(1+\frac{1}{4}+\frac{1}{9}+\cdots\right)$$
$$=\frac{\pi^2}{6}-\frac{1}{2}\cdot\frac{\pi^2}{6}$$
$$=\frac{\pi^2}{12}$$

See also ARITHMETIC SERIES; GEOMETRIC SERIES; POWER SERIES.

converse (reverse implication) The converse of a CONDITIONAL statement *"p implies q"* is the statement: *"q implies p."* It is the statement obtained by reversing the roles of the antecedent and consequent. The converse of a conditional statement might, or might not, be true. For example, the converse of the true statement, *"If a triangle has three equal sides, then it has three equal angles,"* is valid—a triangle with three equal angles does indeed have three equal sides—whereas the converse of the statement, *"If n is divisible by 6, then n is divisible by 2,"* is false—an even number need not be divisible by 6.

See also ARGUMENT; CONTRAPOSITIVE.

convex *See* CONCAVE/CONVEX.

coordinates A set of numbers used to locate a point on a number line, in a plane, or in space are called the coordinates of that point. For example, the coordinates of points on a number line could be given by their distances from a fixed point O (called the origin), with points on one specified side of O being deemed a positive distance from O, and the points on the opposite side of O a negative distance from O.

One way of assigning coordinates to points in the plane is to establish a fixed point O in the plane (again called the origin), and two lines of reference (called axes) that pass through O. Each axis is divided into a positive side and a negative side by O. Given a point P in the plane, one draws lines through P parallel to each of the axes. The distances along which these new lines intersect the axes specify the location of the point P.

When the axes are drawn at right angles, the system is called a Cartesian coordinate system, or a rectangular coordinate system. The axes are usually called the x- and y-axes, and the pair of numbers (x,y) specifying the location of a point P (as x units along one axis, and y units along the second) are called the CARTESIAN COORDINATES of P. In three-dimensional space, the location of points can be specified via three mutually perpendicular (or oblique) axes passing through a common point O.

The idea of assigning sets of numbers to points to specify locations is on old one. By the third century B.C.E., Greek scholars APOLLONIUS OF PERGA and ARCHIMEDES OF SYRACUSE had used longitude, latitude, and altitude to define the position of a point on the Earth's surface. Roman and Greek surveyors labeled maps with grid lines, so as to specify locations via row and column numbers.

See also CYLINDRICAL COORDINATES; DIMENSION; EARTH; POLAR COORDINATES; RIGHT-HANDED/LEFT-HANDED SYSTEM; SPHERICAL COORDINATES.

coplanar   *See* PLANE.

coprime   Another name for RELATIVELY PRIME.

correlation   *See* CORRELATION COEFFICIENT; SCATTER DIAGRAM.

correlation coefficient   Any numerical value used to indicate the extent to which two variables in a study are associated is called a correlation coefficient. For example, a medical study might record the height and shoe size of adult participants suspecting that there might be a relationship between these two features. If two variables are such that when one changes, then the other does so in a related manner (generally the taller an individual, the greater the shoe size on average, say) then the two variables are said to be correlated. A SCATTER DIAGRAM is used to detect possible correlations. If the points of the scatter diagram tend to follow a straight line, then the two variables are linearly correlated.

KARL PEARSON (1857–1936) developed a measure to specifically detect linear relationships. If the DATA values in a study are represented as pairs of values, $(x_1, y_1), \ldots, (x_N, y_N)$, first define:

$$S_{xx} = \frac{1}{N} \sum_{i=1}^{N} (x_i - \overline{x})^2 = \sum_{i=1}^{N} \frac{x_i^2}{N} - \overline{x}^2$$

$$S_{yy} = \frac{1}{N} \sum_{i=1}^{N} (y_i - \overline{y})^2 = \sum_{i=1}^{N} \frac{y_i^2}{N} - \overline{y}^2$$

and

$$S_{xy} = \frac{1}{N} \sum_{i=1}^{N} (x_i - \overline{x})(y_i - \overline{y}) = \sum_{i=1}^{N} \frac{x_i y_i}{N} - \overline{x} \cdot \overline{y}$$

Here $\overline{x}$ is the MEAN of the x-values and $\overline{y}$ the mean of the y-values. The quantities S_{xx} and S_{yy} are called the VARIANCEs and S_{xy} the COVARIANCE of the two variables. Then Pearson's correlation coefficient, denoted R^2, is given by:

$$R^2 = \frac{\left(S_{xy}\right)^2}{S_{xx} S_{yy}}$$

This quantity only adopts values between 0 and 1. A value of $R^2 = 1$ indicates a perfect linear relationship between the two variables, with the points in the associated scatter diagram lying precisely on a straight line. A value $R^2 = 0$ indicates that there is no relationship between the two variables. (In particular, the covariance of the two variables is zero.) All these claims can be proved through a study of the LEAST SQUARES METHOD. An R^2 value close to 1, say 0.9 or higher, indicates that a linear correlation is very likely.

See also RANK CORRELATION; REGRESSION; STATISTICS: DESCRIPTIVE.

countable   Any set, finite or infinite, whose elements can be placed in a list is said to be countable. More precisely, a set S is countable if there is a one-to-one correspondence between the elements of S and a subset of the NATURAL NUMBERS (that is, it is possible to match each element of S with a unique natural number). For example, the set {knife, fork, spoon} is countable because its elements can be matched with the elements of the subset {1, 2, 3} of natural numbers: list knife as first, fork as second, and spoon as third, for instance. The set of all integers is countable, for its elements can be placed in the list:

$$0, 1, -1, 2, -2, 3, -3, \ldots$$

The set of all English words that exist today and might be of use in the future is countable: list all letters of the alphabet (possible one-lettered words), then, in alphabetical order, all combinations of a pair of letters (the two-lettered words), followed by all possible combinations of three letters, and so forth, to produce a well-defined list of all possible strings of letters.

The DIAGONAL ARGUMENT of the first kind shows that the set of all RATIONAL NUMBERS is countable. The diagonal argument of the second kind, however, establishes that the set of REAL NUMBERS is not. In a definite sense then, the set of reals is an infinite set "larger" than the set of rationals.

A set is called DENUMERABLE if it is infinite and countable. Matters are a little confusing, however, for some authors will interchangeably use the terms *countable* and *denumerable* for both finite and infinite sets.

The CARDINALITY of an infinite countable set is denoted $\aleph_0$. A countable set that is not infinite is said to be FINITE.

counterexample An example that demonstrates that a claim made is not true or, at the very least, not always true is called a counterexample. For instance, although the assertion $\sqrt{a+b} = \sqrt{a} + \sqrt{b}$ happens to be true for $a = 1$ and $b = 0$, this identity is false in general, as the counterexample $a = 9$ and $b = 16$ demonstrates. We can thus say that the claim $\sqrt{a+b} = \sqrt{a} + \sqrt{b}$ is an invalid statement.

In the mid-1800s French mathematician Alphonse de Polignac (1817–90) made the assertion:

> Every odd number can be written as the sum of
> a power of two and a PRIME number.

For instance, the number 17 can be written as $17 = 2^2 + 13$, 37 as $37 = 2^3 + 29$, and 1,065 as $1,065 = 2^{10} + 41$. De Polignac claimed to have checked his assertion for all odd numbers up to 3 million, and consequently was convinced that his claim was true. Unfortunately it is not, for the number 127 provides a counterexample to this assertion, as none of the following differences are prime:

$$127 - 2 = 125 = 5 \times 25$$
$$127 - 4 = 123 = 3 \times 41$$
$$127 - 8 = 119 = 7 \times 17$$
$$127 - 16 = 111 = 3 \times 37$$
$$127 - 32 = 95 = 5 \times 19$$
$$127 - 64 = 63 = 3 \times 21$$

(We need not go further, since the next power of two, 128, is larger than 127.) De Polignac overlooked this simple counterexample.

No counterexamples have yet been found to the famous GOLDBACH'S CONJECTURE and COLLATZ'S CONJECTURE.

covariance If a scientific study records numerical information about two features of the individuals or events under examination (such as the height and shoe size of participating adults, or seasonal rainfall and crop yield from year to year), then the DATA obtained from the study is appropriately recorded as pairs of values. If a study has N participants, with data pairs $(x_1, y_1), \ldots, (x_N, y_N)$, then the covariance of the sample is the quantity:

$$S_{xy} = \sum_{i=1}^{N} \frac{(x_i - \overline{x})(y_i - \overline{y})}{N}$$

where $\overline{x}$ is the MEAN x-value and $\overline{y}$ the mean y-value. An exercise in algebra shows that this formula can also be written:

$$S_{xy} = \sum_{i=1}^{N} \frac{x_i y_i}{N} - \overline{x} \cdot \overline{y}$$

The covariance is used to calculate CORRELATION COEFFICIENTS and REGRESSION lines such as for the LEAST SQUARES METHOD, for instance. Pearson's correlation coefficient shows that two variables with covariance zero are independent, that is, the value of one variable has no effect on the value of the other.

See also STATISTICS: DESCRIPTIVE.

Cramer, Gabriel (1704–1752) Swiss *Algebra, Geometry, Probability theory* Born on July 31, 1704, in Geneva, Switzerland, scholar Gabriel Cramer is remembered for his 1750 text *Introduction à l'analyse des lignes courbes algébriques* (Introduction to the analysis of algebraic curves), in which he classifies certain types of algebraic curves and presents an efficient method, today called CRAMER'S RULE, for solving systems of linear equations. Cramer never claimed to have discovered the rule. It was, in fact, established decades earlier by Scottish mathematician COLIN MACLAURIN (1698–1746).

Cramer earned a doctorate degree at the young age of 18 after completing a thesis on the theory of sound, and two years later he was awarded a joint position as chair of mathematics at the Académie de Clavin in Geneva. After sharing the position for 10 years with young mathematician Giovanni Ludovico Calandrini (senior to him by just one year), Cramer was eventually awarded the full chairmanship.

The full position gave Cramer much opportunity to travel and collaborate with other mathematicians across Europe, such as LEONHARD EULER, Johann Bernoulli and Daniel Bernoulli of the famous BERNOULLI FAMILY, and ABRAHAM DE MOIVRE. Cramer wrote many articles in mathematics covering topics as diverse as geometry and algebra, the history of mathematics, and a mathematical analysis of the dates on which Easter falls.

His 1750 text *Introduction à l'analyse des lignes courbes algébriques* was Cramer's most famous work. Beginning chapter one with a discussion on the types of curves he considers, Cramer presents effective techniques for drawing their graphs. In the second chapter he discusses the role of geometric transformations as a means of simplifying the equations to curves (akin to today's approach using the notion of PRINCIPAL AXES). This leads to his famous classification of curves in the third chapter. Here Cramer also discusses the problem of finding the equation of a degree-two curve $ax^2 + bxy + cy^2 + dx + ey = 0$ that passes through five previously specified points in the plane. Substituting in the values of those points leads to five linear equations in the five unknowns a, b, c, d, and e. To solve the problem Cramer then refers the reader to an appendix of the text, and it is here that his famous rule for solving systems of equations appears. Cramer made no claim to the originality of the result and may have been well aware that Colin Maclaurin had first established the famous theorem. (Cramer cited the work of Colin Maclaurin in many footnotes throughout his text, suggesting that he was working closely with the writings of Maclaurin.)

Cramer also served in local government for many years, offering expert opinion on matters of artillery and defense, excavations, and on the reconstruction and preservation of buildings. He died on January 4, 1752, in Bagnols-sur-Cèze, France. Although Cramer did not invent the rule that bears his name, he deserves recognition for developing superior notation for the rule that clarified its use.

Cramer's rule Discovered by Scottish mathematician COLIN MACLAURIN (1698–1746), but first published by Swiss mathematician GABRIEL CRAMER (1704–52), Cramer's rule uses the DETERMINANT function to find a solution to a set of SIMULTANEOUS LINEAR EQUATIONS. An example best illustrates the process.

Consider the set of equations:

$$2x + 3y + z = 3$$
$$x - 2y + 2z = 11$$
$$3x + y - 2z = -6$$

Set A to be the matrix of coefficients:

$$A = \begin{pmatrix} 2 & 3 & 1 \\ 1 & -2 & 2 \\ 3 & 1 & -2 \end{pmatrix}$$

The determinant of this matrix is not zero: $\det(A) = 35$.

A standard property of determinants asserts that if the elements of the first column are multiplied by the value x, then the determinant changes by the factor x:

$$\det \begin{pmatrix} 2x & 3 & 1 \\ x & -2 & 2 \\ 3x & 1 & -2 \end{pmatrix} = x \det(A)$$

Adding a multiple of another column to the first does not change the value of the determinant. We shall add y times the second column, and z times to the third to this first column:

$$\det \begin{pmatrix} 2x+3y+z & 3 & 1 \\ x-2y+2z & -2 & 2 \\ 3x+y-2z & 1 & -2 \end{pmatrix} = x \det(A)$$

But of course this first column equals the column of values of the simultaneous equations:

$$x \det(A) = \det \begin{pmatrix} 2x+3y+z & 3 & 1 \\ x-2y+2z & -2 & 2 \\ 3x+y-2z & 1 & -2 \end{pmatrix}$$
$$= \det \begin{pmatrix} 3 & 3 & 1 \\ 11 & -2 & 2 \\ -6 & 1 & -2 \end{pmatrix} = 35$$

This tells us that the value of x we seek is:

$$x = \frac{35}{\det(A)} = \frac{35}{35} = 1$$

In the same way, the value y is found as the ratio of the determinant of the matrix A with the second column replaced by the column of values of the simultaneous equations and the determinant of A:

$$y = \frac{\det \begin{pmatrix} 2 & 3 & 1 \\ 1 & 11 & 2 \\ 3 & -6 & -2 \end{pmatrix}}{\det(A)} = \frac{-35}{35} = -1$$

and z is a similar ratio of determinants:

$$z = \frac{\det\begin{pmatrix} 2 & 3 & 3 \\ 1 & -2 & 11 \\ 3 & 1 & -6 \end{pmatrix}}{\det(A)} = \frac{140}{35} = 4$$

In a general situation, Cramer's rule states:

If A is the matrix of coefficients of a system of linear equations, then the value of the ith variable x_i in that system of equations is:

$$x_i = \frac{\det(A_i)}{\det(A)}$$

provided the determinant of A is not zero. Here A_i is the matrix A with the ith column replaced with the column of values of the set of equations.

Notice that Cramer's rule shows that there can only be one solution to a system of equations for which the determinant of the coefficient matrix is nonzero.

In the study of determinants, Cramer's rule is used to prove that a matrix A is invertible if, and only if, its determinant is not zero.

critical path   Suppose that we are given a sequence of tasks that need to be accomplished in order to complete a large project, such as building a house or publishing an encyclopedia, and suppose that these tasks have the following properties:

1. There is an order of precedence for certain tasks.
2. Some tasks can be carried out simultaneously.
3. The duration of each task is known.

Then the critical path for the project is the longest (in time) chain of tasks that must be completed in the specified order. The critical path thus puts a bound on the minimum amount of time it takes to complete the entire project.

For example, consider the project of preparing hamburgers and salad for an evening meal. The following table describes the tasks that must be completed, their prerequisite tasks, and their duration.

TASK	Time to Complete (in minutes)	Prerequisite Tasks
W: wash hands	1	None
D: defrost hamburger	10	None
P: shape meat into patties	5	W, D
C: cook hamburgers	10	P
S: wash and slice salad items	8	W
M: mix salad	4	S
T: set table	3	W
E: serve meal	2	C, M, T

The top diagram below provides a useful schematic of the ordering of the tasks. (Their times are written in parentheses.) We see from it that the longest chain, that is, the critical path of the project, is the sequence D-P-C-E requiring 27 min to complete. That all the tasks in this example can indeed be accomplished in exactly 27 min is demonstrated in the second diagram. In general, there is no guarantee that the time dictated by the critical path is actually attainable.

Computers are used to look for critical paths in complex projects.

See also OPERATIONS RESEARCH.

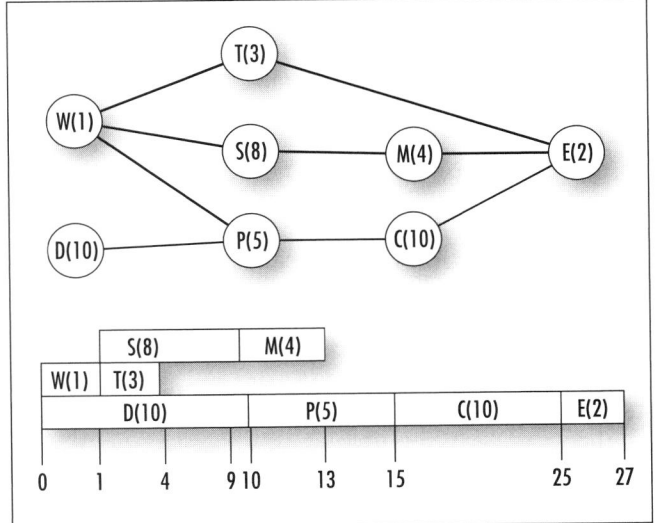

Task diagrams

cross product (**vector product**) Many problems in three-dimensional geometry require physicists and mathematicians to find a VECTOR that is perpendicular to each of two given vectors **a** and **b**. The cross product, denoted **a** × **b**, is designed to be such a vector.

One begins by positioning the two vectors **a** and **b** at the same point in space. These vectors define a plane in space, and one sees that there are two possible directions for a third vector to point so as to be perpendicular to this plane. Mathematicians have settled on the convention of following the "right-hand rule" to determine which direction to choose:

> Take your right hand and point your fingers in the direction indicated by the first vector **a**. Now orient your hand, with your fingers still pointing in this direction, in such a way that your palm faces to the side of the plane containing the vector **b**. (If you curl your fingers, they will consequently turn through the smallest angle that leads from **a** to **b**.) The direction in which your thumb now points is the direction the vector **a** × **b** will take.

Thus **a** × **b** will be a vector that points in one direction while **b** × **a** will point in the opposite direction. (In fact: **b** × **a** = −**a** × **b**.)

Mathematicians have settled on a second convention to define the magnitude of **a** × **b**:

> The magnitude of **a** × **b** is the area of the PARALLELOGRAM defined by the vectors **a** and **b**.

If θ is the smallest angle between **a** and **b**, then the parallelogram defined by the two vectors has side-lengths |**a**| and |**b**|. Taking |**a**| as the base, the height of the parallelogram is then given by |**b**|·sinθ, and consequently the AREA of the parallelogram is |**a**|·|**b**|·sinθ. Thus:

> **a** × **b** is defined to be the vector of magnitude |**a**|·|**b**|·sinθ with direction given by the right-hand rule.

For example, if **i** = <1,0,0> is the unit vector pointing in the direction of the x-axis, and **j** = <0,1,0> the unit vector in the direction of the y-axis, then **i** × **j** is a vector pointing in the direction of the z-axis, with length equal to the area of the unit square defined by **i** and **j**, namely 1. Thus:

$$\mathbf{i} \times \mathbf{j} = <0,0,1> = \mathbf{k}$$

If two vectors **a** and **b** are parallel, then the angle between them is zero and **a** × **b** = **0**.

There is an alternative method for computing cross products. If **a** is given by **a** = <a_1,a_2,a_3> = $a_1\mathbf{i} + a_2\mathbf{j} + a_3\mathbf{k}$ and **b** is given by **b** = <b_1,b_2,b_3> = $b_1\mathbf{i} + b_2\mathbf{j} + b_3\mathbf{k}$, then one can check that the DOT PRODUCT of the vector:

$$(a_2b_3 - a_3b_2)\mathbf{i} + (a_3b_1 - a_1b_3)\mathbf{j} + (a_1b_2 - a_2b_1)\mathbf{k}$$

with each of **a** and **b** is zero. Thus this new vector is perpendicular to both **a** and **b**. Mathematicians have shown that it also has direction given by the right-hand rule and magnitude equal to the parallelogram defined by **a** and **b**. Thus this new vector is indeed the cross product of **a** and **b**:

$$\mathbf{a} \times \mathbf{b} = (a_2b_3 - a_3b_2)\mathbf{i} + (a_3b_1 - a_1b_3)\mathbf{j} + (a_1b_2 - a_2b_1)\mathbf{k}$$
$$= \begin{vmatrix} \mathbf{i} & \mathbf{j} & \mathbf{k} \\ a_1 & a_2 & a_3 \\ b_1 & b_2 & b_3 \end{vmatrix}$$

where, for the final equality, we have written the formula in terms of the DETERMINANT of a 3 × 3 matrix. (To prove that this new vector does indeed match the quantity **a** × **b**, rotate the system of vectors **a**, **b** and **a** × **b** so that **a** points in the direction of the x-axis and **b** lies in the xy-plane. Then, for the rotated system we have: $a_1 = $ |**a**|, $a_2 = 0$, $a_3 = 0$, $b_1 = $ |**b**|cosθ, $b_2 = $ |**b**|sinθ, $b_3 = 0$. One can now readily check that the formula above yields a vector of the required length |**a**|·|**b**|·sinθ pointing in the correct direction. One then argues that the formula continues to hold when the system of three vectors is rotated back to its original position.) Thus, for example, if **a** = <1,4,2> and **b** = <3,0,1>, then **a** × **b** = <4·1 − 2·0, 2·3 − 1·1, 1·0 − 4·3> = <4,5, − 12>. According to the DISTANCE FORMULA, this vector has length $\sqrt{4^2 + 5^2 + (-12)^2} = \sqrt{185}$, which must be the area of the parallelogram formed by **a** and **b**.

In two-dimensions, the determinant $\begin{vmatrix} \mathbf{i} & \mathbf{j} \\ a_1 & a_2 \end{vmatrix} = a_2\mathbf{i} - a_1\mathbf{j} = <a_2, -a_1>$ gives a vector the same length as **a** = <a_1,a_2> and perpendicular to it. In four-dimensional space one can always find a fourth vector perpendicular to each of any given three vectors.

See also ORTHOGONAL; TRIPLE VECTOR PRODUCT; VECTOR EQUATION OF A PLANE.

cryptography The practice of altering the form of a message by codes and ciphers to conceal its meaning to those who intercept it, but not to those who receive it, is called cryptography. If letters of the alphabet and punctuation marks are replaced by numbers, then mathematics can be used to create effective codes.

In 1977 three mathematicians, Ron Rivest, Adi Shamir, and Leonard Adleman, developed a public-key cryptography method in which the method of encoding a message can be public to all without compromising the security of the message. The RSA encryption method, as it is known today, is based on the mathematics of the MODULAR ARITHMETIC and relies on the fact that it is extraordinarily difficult to find the two factors that produce a given large product. It is the primary encryption method used today by financial institutions to transmit sensitive information across the globe.

The RSA encryption method is based on the following result from modular arithmetic:

> Suppose p and q are distinct PRIME numbers. If n is a number with neither p nor q a factor, then $n^{(p-1)(q-1)} \equiv 1 \pmod{pq}$. Moreover, we have $n^{m(p-1)(q-1)+1} \equiv n \pmod{pq}$ for any two natural numbers n and m, even if n is a multiple of p or q.

(We prove this result at the end of this entry.) One proceeds as follows:

1. Encode your message as a string M of numbers.
2. Choose two large prime numbers p and q so that their product $N = pq$ is larger than M. Let $k = (p-1)(q-1)$ and choose a number e with no common factor to k. The numbers N and e can be made public.
3. Raise the number M to the eth power, modulo N. This gives the encoded message M':

$$M' \equiv M^e \pmod{N}$$

4. Since the GREATEST COMMON DIVISOR of k and e is one, the EUCLIDEAN ALGORITHM shows that we can find numbers d and m so that $1 = de - mk$. In particular there is a number d such that $ed \equiv 1 \pmod{k}$. Keep the number d secret.
5. To decode the message, raise M' to the dth power. By the result stated above, this does indeed return the original message M:

$$M'^d = M^{ed} = M^{km+1} = M^{m(p-1)(q-1)+1}$$
$$\equiv M \pmod{N}$$

If one works with very large prime numbers, say, 10,000-digit primes, it is virtually impossible to factor the public number N and find k and d. Thus the RSA system is extremely secure. (This use of large primes also explains the current excitement over the discovery of larger and larger prime numbers.) On the other hand, multiplying and raising large numbers to powers is easy for computers to do, and so the RSA method is also very easy to implement.

Proof of Result

Suppose first that n is not a multiple of p and consider the numbers $1, 2, \ldots, p-1$. Multiply each by n. If, for two numbers x and y in the list, we have $nx \equiv ny \pmod{p}$, then $n(x-y)$ is a multiple of p. This can only happen if x and y are the same number. Thus, up to multiples of p, the products $n \cdot 1, n \cdot 2, \ldots, n \cdot (p-1)$ are distinct and so must represent a rearrangement of the original list, modulo p. Consequently, $n \cdot 1 \cdot n \cdot 2 \cdot \ldots \cdot n(p-1) \equiv 1 \cdot 2 \cdot \ldots \cdot (p-1) \pmod{p}$, and the factor n^{p-1} in the left must be congruent to 1 modulo p: $n^{p-1} \equiv 1 \pmod{p}$. Similarly, we have $n^{q-1} \equiv 1 \pmod{q}$ if n is not a multiple of q. It follows from these two results that if n is not a multiple of p or q, then:

$$n^{(p-1)(q-1)} = (n^{p-1})^{q-1} \equiv 1 \pmod{p}$$
$$n^{(p-1)(q-1)} = (n^{q-1})^{p-1} \equiv 1 \pmod{q}$$

which shows that $n^{(p-1)(q-1)} - 1$ is divisible by both p and q, and hence by pq. This proves the first claim made. The second follows by noting that $n^{m(p-1)(q-1)+1} = n \cdot (n^{p-1})^{m(q-1)} \equiv n \pmod{p}$ is a true statement even if n is a multiple of p, and $n^{m(p-1)(q-1)+1} \equiv n \pmod{q}$ is also true for all values of n. Any quantity $n^{m(p-1)(q-1)+1} - n$ is thus always divisible by both p and q.

cube (hexahedron) The third PLATONIC SOLID, the cube, is the solid figure bounded by six identical square faces that meet at right angles. It has eight vertices and 12 edges. Because the VOLUME of a cube of side-length a is given by a^3, any number raised to the third power is sometimes called a cube. The "cubic numbers" are the cubes of the counting numbers: 0, 1, 8, 27, 64, 125,...

It is possible to subdivide a large cube into 27 smaller cubes with six planar cuts. This number of cuts cannot be improved upon even if one is permitted to "stack" pieces of the cube during the slicing process so as to cut through several pieces at once. (To see this, consider the innermost cube among the 27 little cubes. It has six planar faces, each of which need to be cut. As no single planar cut will slice two of those faces, a minimum of six cuts is indeed required to form the 27 little cubes.) It is possible, however, to dice a large cube into $4 \times 4 \times 4 = 64$ smaller cubes in fewer than nine slices if "stacking" is permitted (six will suffice). Nine slices suffice to dice a cube into $5 \times 5 \times 5 = 125$ smaller cubes. In general, the minimum number of slices required to subdivide a large cube into $n \times n \times n = n^3$ smaller cubes is given by the formula:

$$3 \lceil \log_2 n \rceil$$

making use of the CEILING FUNCTION.

The four-dimensional analog of a cube is a HYPERCUBE.

See also FLOOR/CEILING/FRACTIONAL PART FUNCTIONS; DUPLICATING THE CUBE; PARALLELEPIPED; PRISM.

cube root/*n*th root   The cube root of a number a is a value x such that $x^3 = a$. We write $\sqrt[3]{a}$ for the cube root of a.

Every real number a has exactly one real cube root. For instance, the cube root of 27 is 3 (since $3 \times 3 \times 3$ equals 27), and the cube root of –8 is –2 (since $-2 \times -2 \times -2$ equals –8). If one works within the realm of complex numbers, then the FUNDAMENTAL THEOREM OF ALGEBRA shows that every number has exactly three complex cube roots. For instance, the three cube roots of 27 are $-\frac{3}{2}+i\frac{3\sqrt{3}}{2}$, $-\frac{3}{2}-i\frac{3\sqrt{3}}{2}$, and 3. A study of the *n*th roots of unity shows that the three complex cube roots of a number lie on the vertices of an equilateral triangle in the complex plane.

In general, the *n*th root of a number a is a value x such that $x^n = a$. Again, the fundamental theorem of algebra shows that every number has exactly n complex *n*th roots. If a is a real number, then a has a real *n*th root if a is positive. (For example, a fourth root of 16 is 2.) If a is a negative real number, then a real *n*th root exists only if n is odd. (For example, –243 has a fifth root, namely –3, but no real fourth root.) The real *n*th root of a number a, if it exists, is denoted $\sqrt[n]{a}$.

See also CUBIC EQUATION; ROOT; SQUARE ROOT.

cubic equation   Any degree-three POLYNOMIAL equation of the form $ax^3 + bx^2 + cx + d = 0$ with $a \neq 0$ is called a cubic equation.

During the Renaissance, scholars sought for a general arithmetic formula in terms of the coefficients a, b, c, and d that would solve all cubic equations (one akin to the famous QUADRATIC formula for solving degree-two equations). At the time, however, scholars were not comfortable working with NEGATIVE NUMBERS, or with ZERO as a number, and wrote equations in a form that avoided their appearance. (For instance, the cubic $x^3 - 2x + 5 = 0$ was cast as $x^3 + 5 = 2x$.) Mathematicians consequently thought that there were eight different types of cubic equations to solve.

Italian scholar SCIPIONE DEL FERRO (1465–1526) was the first to make progress in solving certain cubics, but never published his results. Later, the scholar NICCOLÒ TARTAGLIA (ca. 1499–1557) succeeded in solving some additional classes of cubics. GIROLAMO CARDANO (1501–76) published Tartaglia's work (without Tartaglia's consent) in his epic 1545 piece *Ars magna* (The great art) and developed a general approach that solves all cubic equations. The formula he devised is today called "Cardano's formula" or the "Cardano-Tartaglia formula." We describe it here using modern notation:

By dividing the cubic equation through by the leading coefficient a, we can assume that we are working with a cubic of the form:

$$x^3 + Bx^2 + Cx + D = 0$$

for numbers $B = \frac{b}{a}$, $C = \frac{c}{a}$, and $D = \frac{d}{a}$. Substituting $x = y - \frac{B}{3}$ simplifies the equation further to one without a square term:

$$y^3 + py + q = 0$$

(Here $p = C - \frac{B^2}{3}$ and $q = \frac{2B^3}{27} - \frac{BC}{3} + D$.) This form of the cubic is called the reduced cubic, and any solution y to this equation corresponds to a solution $x = y - \frac{b}{3a}$ of the original equation.

Assume that the equation has a solution that can be written as the sum of two quantities: $y = u + v$. Substituting these variables yields:

$$(u^3 + v^3) + (3uv + p)(u + v) + q = 0$$

This equation will be satisfied if we can choose u and v so that $u^3 + v^3 = -q$ and $3uv + p = 0$. This yields a pair of equations for u^3 and v^3:

$$u^3 + v^3 = -q$$

$$u^3 v^3 = \left(-\frac{p}{3}\right)^3$$

Solving for v^3 in the first equation and substituting the result into the second shows that u^3 must satisfy the quadratic equation:

$$\left(u^3\right)^2 + q\left(u^3\right) - \left(\frac{p}{3}\right)^3 = 0$$

and using the quadratic formula, this gives u^3, and consequently $v^3 = -u^3 - q$, to be the two numbers:

$$u^3 \text{ and } v^3 = \frac{-q \pm \sqrt{q^2 + 4\left(\frac{p}{3}\right)^3}}{2} = -\frac{q}{2} \pm \sqrt{\left(\frac{q}{2}\right)^2 + \left(\frac{p}{3}\right)^3}$$

We must now take the cube root of these quantities. Note first that any number M has three cube roots: one real, denoted $\sqrt[3]{M}$, and two imaginary, $w \times \sqrt[3]{M}$, and $w^2 \times \sqrt[3]{M}$, where $w = \frac{-1 + i\sqrt{3}}{2}$. Set:

$$u_1 = \sqrt[3]{-\frac{q}{2} + \sqrt{\left(\frac{q}{2}\right)^2 + \left(\frac{p}{3}\right)^3}}$$

$$u_2 = w \times \sqrt[3]{-\frac{q}{2} + \sqrt{\left(\frac{q}{2}\right)^2 + \left(\frac{p}{3}\right)^3}}$$

$$u_3 = w^2 \times \sqrt[3]{-\frac{q}{2} + \sqrt{\left(\frac{q}{2}\right)^2 + \left(\frac{p}{3}\right)^3}}$$

and

$$v_1 = \sqrt[3]{-\frac{q}{2} - \sqrt{\left(\frac{q}{2}\right)^2 + \left(\frac{p}{3}\right)^3}}$$

$$v_2 = w \times \sqrt[3]{-\frac{q}{2} - \sqrt{\left(\frac{q}{2}\right)^2 + \left(\frac{p}{3}\right)^3}}$$

$$v_3 = w^2 \times \sqrt[3]{-\frac{q}{2} - \sqrt{\left(\frac{q}{2}\right)^2 + \left(\frac{p}{3}\right)^3}}$$

One can now check that the three quantities $u_1 + v_1$, $u_2 + v_2$, and $u_3 + v_3$ represent the three solutions to the reduced cubic $y^3 + py + q = 0$. They constitute Cardano's formula.

The quantity under the square root sign:

$$\Delta = \left(\frac{q}{2}\right)^2 + \left(\frac{p}{3}\right)^3$$

is called the discriminant of the cubic, and it determines the nature of the solutions:

If $\Delta > 0$, then the equation has one real root and two complex roots.
If $\Delta = 0$, then the equation has three real roots, at least two of which are equal.
If $\Delta < 0$, then the equation has three distinct real roots.

In the third case, one is required to take combinations of cube roots of complex numbers to yield, surprisingly, purely real answers. For example, Cardano's method applied to the equation $x^3 = 15x + 4$ yields as one solution the quantity:

$$x = \sqrt[3]{2 + \sqrt{-121}} + \sqrt[3]{2 - \sqrt{-121}}$$

It is not immediate that this number is $x = 4$.

This confusing phenomenon of using complex quantities to produce real results was first explored by Italian mathematician RAFAEL BOMBELLI (1526–72). French mathematician FRANÇOISE VIÈTE (1540–1603) used trigonometric formulae as an alternative approach to identifying the three distinct real roots that appear in this puzzling scenario.

Another Method
French mathematician Viète also developed the following simpler approach to solving cubic equations. This method was published posthumously in 1615.

Take the reduced form of the cubic $y^3 + py + q = 0$ and rewrite it as:

$$y^3 + 3ry = 2t$$

where $r = \dfrac{p}{3}$ and $t = -\dfrac{q}{2}$. Substitute $y = \dfrac{r}{z} - z$ to obtain the simpler equation:

$$(z^3)^2 + 2t(z^3) - r^3 = 0$$

which is a quadratic in z^3. All one need do now is solve for z^3 and extract its three cube roots.

The solution to the cubic equation is needed in the solution to the QUARTIC EQUATION.

See also FUNDAMENTAL THEOREM OF ALGEBRA; HISTORY OF EQUATIONS AND ALGEBRA (essay).

curl   *See* DIV.

curve   A set of points that form a line, either straight or continuously bending, is called a curve. The GRAPH OF A FUNCTION plotted in CARTESIAN COORDINATES, for example, is a curve. A curve can also be considered as the path of a moving particle (and, consequently, PARAMETRIC EQUATIONS can be used to describe it).

As the graph of a function, a curve is called algebraic if it is given by a formula $y = f(x)$ with f an algebraic function, and transcendental if f is transcendental. (*See* ALGEBRAIC NUMBER.) For example, the PARABOLA $y = x^2$ is algebraic, as is any CONIC SECTION, but the sine curve $y = \sin x$ is transcendental.

A curve that lies in a plane is called a planar curve. A curve in three-dimensional space that does not remain in a plane, such as a HELIX, is called skew or twisted. Any curve that lies in three-dimensional space is called a space curve (whether or not it is twisted).

A curve is called closed if one can traverse the curve and return to the same point an indefinite number of times. For example, a CIRCLE is closed. In some settings it is appropriate to allow for a point at infinity, in which case a straight line would also be considered a closed curve. (One can head to infinity in one direction and return from infinity from the other direction and repeat this journey an indefinite number of times.) A curve with ENDPOINTs is called open.

See also ARC; LOCUS.

cusp   *See* TANGENT.

cyclic polygon   A POLYGON is called cyclic if all its vertices (corners) lie on a circle. As every triangle can be inscribed in a circle, all triangles are cyclic. So too are all squares, rectangles, and all regular polygons. However, not every quadrilateral or higher-sided polygon is cyclic

BRAHMAGUPTA (598–665) gave a formula for the area of a cyclic quadrilateral. The CIRCLE THEOREMS show that opposite angles of any cyclic quadrilateral are supplementary. The converse is also true: any quadrilateral with opposite angles summing to 180° is cyclic.

See also BRAHMAGUPTA'S FORMULA; CIRCUMCIRCLE; PTOLEMY'S THEOREM.

cycloid   The shape traced out by a point on the circumference of a circle rolling along a straight line is called a cycloid. (In particular, it is the curve traced out by a piece of gum stuck to the rim of a bicycle wheel.) This curve has many remarkable geometric properties and was studied extensively by mathematicians of the 16th and 17th centuries, and later.

GALILEO GALILEI (1564–1642) was the first to study the curve and gave it its name. In 1644 Galileo's disciple Evangelista Torricelli (who invented the barometer) proved that the area under one arch of the cycloid equals three times the area of the rolling circle. In 1658 English architect Christopher Wren showed that the length of one arch of the cycloid is four times the diameter of the circle. In 1696 Johann Bernoulli of the famous BERNOULLI FAMILY posed and solved the now-famous brachistochrone problem:

> Imagine a small ball starting at a point *A* and rolling down along a curve to a lower point *B* to the right of *A*. The ball is propelled only by the force of gravity. What shape curve connecting *A* to *B* allows the ball to travel between them the fastest?

Surprisingly, a straight line does not give the shortest time, but an upside-down cycloid does.

In 1658 Dutch scientist Christiaan Huygens considered the cycloid in his work on pendulums. He discovered that a simple pendulum in which the bob is forced to follow a cycloid-shaped path always has the same period irrespective of the length of the pendulum. This is called the tautochrone property of the cycloid. (It is also the case that if a ball starts at rest at any point of an inverted cycloid and travels along the curve under the force of gravity, then the time it takes to reach the lowest point of the curve is independent of the starting location of the ball.)

Related curves can be considered by following the path traced by a point on the circumference of a circle as that circle rolls along another circle. If the circle rolls on the *inside* of a fixed circle, then the curve traced is called a hypocycloid. If a circle rolls on the *outside* of a fixed circle, then the curve traced is called an epicycloid. In both cases the fixed circle is called the deferent, and the moving circle is the epicycle.

Some special names are given to the curves created in particular situations. For example, when the two circles have the same radius, the hypocycloid produced is heart-shaped and is called a CARDIOID. When the rolling outer circle has diameter one-fourth that of the fixed circle, the four-pointed curve produced is called an astroid. An epicycloid with five cusps is called a ranunculoid.

The epicycloid was known to APOLLONIUS OF PERGA of the third century B.C.E., who used it in his descriptions of planetary motion.

cylinder

cylinder In three-dimensional space, a cylinder is the surface formed by an infinite collection of parallel straight lines, each passing through one point of a fixed closed curve drawn in a plane. The closed curve is called the directrix of the cylinder, and the lines drawn are called the generators of the cylinder. Often the term *cylinder* is used for the solid figure of finite volume confined between two parallel planes. In this setting, the cylinder has three faces: the two parallel planar regions, each called a base of the figure, and the *lateral surface* given by the straight lines that generate the cylinder. The base of a cylinder need not be a circle. For example, a CUBE satisfies the definition of being a cylinder.

If the lateral surface is at right angles to the base, then the cylinder is called a right cylinder. All other cylinders are called oblique. The height of a cylinder is the perpendicular distance between the two bases.

All horizontal cross-sections of a cylinder are the same size and shape as the base of the cylinder. CAVALIERI'S PRINCIPLE then shows that the volume V of a cylinder is given by $V = Ah$, where h is the height of the cylinder and A is the area of its base.

ARCHIMEDES OF SYRACUSE (ca. 287–212 B.C.E.) showed that the volume of a SPHERE is two-thirds that of the volume of the cylinder that contains it. The formula for the volume of a sphere readily follows.

See also CONE.

cylindrical coordinates

cylindrical coordinates (cylindrical polar coordinates) In three-dimensional space, the location of a point P can be described by three coordinates—r, θ, and z—called the cylindrical coordinates of P, where (r, θ) are the POLAR COORDINATES of the projection of P onto the xy-plane, and z is the height of P above the xy-plane. Cylindrical coordinates are useful for describing surfaces with circular symmetry about the z-axis. For example, the equation of a cylinder of radius 5 with a central axis, the z-axis can be described by the simple equation $r = 5$. (As the angle θ varies between zero and 360°, and the height z varies through all values, points on an infinitely long cylinder are described.) The surface defined by the equation $\theta = c$, for some constant c (allowing r and z to vary), is a vertical HALF-PLANE with one side along the z-axis, and the surface $z = c$ is a horizontal plane.

A point P with cylindrical coordinates (r, θ, z) has corresponding CARTESIAN COORDINATES (x, y, z) given by:

$$x = r \cos \theta$$
$$y = r \sin \theta$$
$$z = z$$

These formulae follow the standard conversion formulae for polar coordinates.

It is usual to present the angle θ in RADIAN MEASURE. In this case, a triple integral of the form $\iiint_V f(x,y,z)dx\, dy\, dz$ over a volume V described in Cartesian coordinates converts to the corresponding integral $\iiint f(r\cos\theta, r\sin\theta, z)\, r\, dr\, d\theta\, dz$ in cylindrical coordinates. The appearance of the term r in the integrand follows for the same reason that r appears in the conversion of a DOUBLE INTEGRAL from planar Cartesian coordinates to polar coordinates.

See also ANGLE; SPHERICAL COORDINATES.

D

data (singular, datum) Information of a numerical nature is called data. For example, records of the daily numbers of visitors to a tourist attraction, or the measurement of growth rates of a yeast culture under different temperature conditions, would be examples of data. Direct counts from observational studies or measurements from experiments like these are called primary or raw data. Numerical information describing the raw data (such as the average value, largest value, range of values, and so on) is sometimes called secondary data. The science of developing methods for collecting, organizing, and summarizing data is called descriptive statistics.

See also STATISTICS; STATISTICS: DESCRIPTIVE.

days-of-the-week formula It is a challenging exercise to derive a mathematical formula that determines the day of the week on which a particular calendar date falls.

As a first step, knowing the day on which January 1 of a year falls, it is reasonably straightforward to determine on which day any other date of that year falls. For example, New Year's Day in the year 2000 was a Saturday. As the days of the week cycle in periods of seven, it follows that January 8, January 15, and January 22 of that year were also Saturdays. Since there are 31 days in January, and because 31 is three more than a multiple of 7, it follows that February 1, 2000, fell on the weekday three days later than Saturday, namely, Tuesday. As there were 29 days in February that year, and 29 is one more than a multiple of 7, March 1 fell on the weekday that directly follows Tues-

day, namely Wednesday; and April 1, 30 days later, fell on a weekday two days after Wednesday, namely Friday. In this way we can determine the weekday of any first day of the month, and from there, the weekday of any particular day of that month.

It is convenient to label Sunday as "day 0," Monday as "day 1," up to Saturday as "day 6." As the weekdays cycle in units of 7, it is also appropriate to ignore all multiples of 7 and work only with the remainders of numbers upon division by 7. (That is, we shall work in a base-7 system of MODULAR ARITHMETIC.) For example, dates 16 and 30 days into a year fall on the same weekday as the day 2 days into the year: all numbers involved here are 2 more than a multiple of 7. We shall call the numbers 16, 30, and 2 "equivalent" and write $16 \equiv 2$ and $30 \equiv 2$, for instance.

January 1, 2000, fell on day 6. As we have noted, February 1 falls $31 \equiv 3$ days later and so lands on day $6 + 3 \equiv 2$, Tuesday. March 1 falls another $29 \equiv 1$ days later, and so lands on day $6 + 3 + 1 \equiv 3$, Wednesday. In general, the following table shows the amount by which a particular date must be adjusted depending on the month in which it lies:

Month	Jan.	Feb.	March	April	May	June
Add	0	3	3 (4)	6 (0)	1 (2)	4 (5)
Month	July	Aug.	Sept.	Oct.	Nov.	Dec.
Add	6 (0)	2 (3)	5 (6)	0 (1)	3 (4)	5 (6)

The numbers in parentheses pertain to leap years. Thus, for example, July 1, 2000, fell on day $6 + 0 = 6$, Saturday; and December 1, 2000, fell on day $6 + 6 \equiv 5$, Friday.

Now it is a matter of determining on which weekday the first day of any given year fell. Assume, for the sake of the mathematical argument, that the Gregorian calendar has been in use for two millennia, and that January 1 in the year 0 was day d. Consider the day the weekday of New Year's Day N years later.

As each ordinary year contains 365 ($\equiv 1$) days, the day on which January 1 falls advances one weekday each year. For each leap year, it advances an additional day. We need to determine the number of leap years over a period of N years.

In general, a leap year occurs every 4 years, yielding $\left\lceil \dfrac{N}{4} \right\rceil$ possible occurrences of a leap year, including year zero. (Here we are making use of the CEILING FUNCTION.) However, no leap year occurs on a year value that is a multiple of 100—and this occurs $\left\lceil \dfrac{N}{100} \right\rceil$ times—except if N is a multiple of 1,000, which occurs $\left\lceil \dfrac{N}{1000} \right\rceil$ times. (The year 1900, for instance, was not a leap year, but the year 2000 was.) The total number of leap years L that occur in a period of N years from year zero is thus given by:

$$ L = \left\lceil \frac{N}{4} \right\rceil - \left\lceil \frac{N}{100} \right\rceil + \left\lceil \frac{N}{1000} \right\rceil $$

For instance, February 29 appeared, in theory,

$$ \left\lceil \frac{2000}{4} \right\rceil - \left\lceil \frac{2000}{100} \right\rceil + \left\lceil \frac{2000}{1000} \right\rceil = 500 - 20 + 2 = 482 \quad \text{times} $$

before the date of January 1, 2000. Thus, the weekday of January 1, year N, is given by:

$$ d + N + L $$

(January 1 year zero, was day d. There is an advance for each of the N years, and an advance of an additional day for each of the L leap years.)

Knowing that New Year's day, 2000, was day 6, we deduce then that the appropriate value of d is given by:

$$ d + 2000 + 482 = 6 $$

That is, working with remainders upon division by 7, $d + 5 + 6 = 6$, yielding $d = 2$. Thus, in our theory, January 1 in the year zero was a Tuesday.

We now have the following ALGORITHM for computing the weekday of any given date. Assume we wish to compute the weekday of the Dth day, of month M, in year N.

1. Consider the year number N and compute its remainder upon division by 7.
2. Compute $L = \left\lceil \dfrac{N}{4} \right\rceil - \left\lceil \dfrac{N}{100} \right\rceil + \left\lceil \dfrac{N}{1000} \right\rceil$ and its remainder upon division by 7.
3. Sum the answers of the previous two steps and add 2. Compute the remainder of this number, if necessary, when divided by 7. This is the weekday number of January 1 of year N.
4. To this weekday number add the day D, subtract 1, and add the appropriate month number from the table above. Look at the remainder upon division by 7, if necessary. This final result is the weekday number of the desired day.

For example, for the date of March 15, 2091, $N = 2091 \equiv 5$, $L = 523 - 21 + 3 = 505 \equiv 1$, yielding January 1 of that year to be day $2 + 5 + 1 \equiv 1$, a Monday. To this we add 14 days (the number of days later is 1 less than the date D) with a month adjustment of value 3 (this is not a leap year). Thus March 15, 2091, will fall on day $1 + 14 + 3 \equiv 4$, a Thursday. (Warning: as the Gregorian calendar was not used before October 15, 1582, this algorithm cannot be applied to dates earlier than this.)

Simplifying the Procedure

This method can be simplified to some extent. Write the year number as *mcyy*, with *m* for millennia, *c* for century, and *yy* as the two-digit year number. More precisely, we mean:

$$ N = 1000m + 100c + yy $$

with $0 \le c \le 9$ and $0 \le yy \le 99$. For example, the year 3261 will be written:

$$ N = 1000 \times 3 + 100 \times 2 + 61 $$

Notice that

$$N = 1000m + 100c + yy \equiv 6m + 2c + yy$$

$$\left\lceil \frac{N}{4} \right\rceil = 250m + 25c + \left\lceil \frac{yy}{4} \right\rceil \equiv 5m + 4c + \left\lceil \frac{yy}{4} \right\rceil$$

$$\left\lceil \frac{N}{100} \right\rceil = 10m + c + \left\lceil \frac{yy}{100} \right\rceil \equiv 3m + c + \left\lceil \frac{yy}{100} \right\rceil$$

$$\left\lceil \frac{N}{1000} \right\rceil = m + \left\lceil \frac{c}{10} + \frac{yy}{1000} \right\rceil$$

yielding:

$$N + L = N + \left\lceil \frac{N}{4} \right\rceil - \left\lceil \frac{N}{100} \right\rceil + \left\lceil \frac{N}{1000} \right\rceil$$

$$= 2m - 2c + yy + \left\lceil \frac{yy}{4} \right\rceil + \left\lceil \frac{c}{10} + \frac{yy}{1000} \right\rceil - \left\lceil \frac{yy}{100} \right\rceil$$

One can check that the quantity $\left\lceil \frac{c}{10} + \frac{yy}{1000} \right\rceil - \left\lceil \frac{yy}{100} \right\rceil$ equals 1 if $yy = 00$ and $c \neq 0$, and zero in every other case. Rewriting steps 1 and 2 from the above algorithm, we have:

January 1 of year *mcyy* falls on day number

$$2 + 2m - 2c + yy + \left\lceil \frac{yy}{4} \right\rceil + \varepsilon$$

where ε is 1 if the year is a turn of the century, but not the turn of a millennium, and zero otherwise.

Thus, for example, consider New Year's Day of the year 3261. The millennium number here is $m = 3$, the century number is $c = 2$, and we have $yy = 61$. Also, $\left\lceil \frac{61}{4} \right\rceil = \lceil 15.25 \rceil = 16$, and, since this is not a turn of the century, $\varepsilon = 0$. Thus New Year's Day of this year will fall on day $2 + (2 \times 3) - (2 \times 2) + 61 + 16 + 0 \equiv 4$, a Thursday. We can now compute the weekday of any other day that year if we wish.

An Alternative Method

Mathematician Chris McManus has observed that whatever day of the week "March 0" (the last day of February) falls, so do 4/4 (April 4), 6/6 (June 6), 8/8 (August 8), 10/10 (October 10), and 12/12 (December 12). The same is true of 9/5 (September 5) and 7/11 (July 11), and their inverses 5/9 and 11/7. (One can remember this with the mnemonic, "I work from 9 to 5 at the 7-to-11 store.") So, for example, if you know that December 20, 2002, was a Friday, and you wish to determine the day on which April 10 of that year fell, quickly compute that December 12 was a Thursday, yielding April 4 also as a Thursday, making April 10 a Wednesday.

decibel Denoted dB, a decibel is a measure of the intensity of sound, with zero decibels representing the lowest intensity at which a sound can be heard.

Decibels follow a base-10 LOGARITHMIC SCALE. This means that each increase of 1 dB represents a 10-fold increase in the intensity of the sound. For example, a note played at 1 dB is 10 times as strong as the softest sound, and one played at 2 dB is 100 times as strong as the softest sound. Normal human speech is at a level of about 60 dB, and a whisper is around 20 dB. The threshold of pain for the human ear is about 90 dB. Rock concerts have been known to reach levels of 120 dB at a distance of 50 m from the sound system.

The decibel level of a sound is computed by the formula:

$$10 \log_{10} \left(\frac{P}{P_0} \right)$$

where P is the intensity of the note being played, and P_0 is the lowest intensity at which that note can be heard.

An interval of 10 dB is called a bel. Sound intensity was originally measured in terms of bel, the name being chosen in honor of the American inventor of the telephone, Alexander Graham Bell. Today, however, the unit of a decibel is considered more useful and is the one most commonly used.

decimal representation *See* BASE OF A NUMBER SYSTEM.

decomposition (factorization) The result of expressing a given object or quantity in terms of simpler components is called a decomposition. For example, the FUNDAMENTAL THEOREM OF ARITHMETIC shows that any natural number decomposes as a product of prime numbers. Also, any rational function decomposes into a sum of PARTIAL FRACTIONS. The process of GAUSSIAN ELIMINATION shows that any square MATRIX A decomposes into the product of a lower triangular matrix L and an upper triangular matrix U. An example of such an LU factorization is:

$$\begin{pmatrix} 2 & 1 \\ 6 & 7 \end{pmatrix} = \begin{pmatrix} 1 & 0 \\ 3 & 1 \end{pmatrix}\begin{pmatrix} 2 & 1 \\ 0 & 4 \end{pmatrix}$$

Any VECTOR decomposes into a sum of basis vectors. For instance:

$$<3, 2, 1> = 3 <1, 0, 0> +2< 0, 1, 0> +< 0, 0, 1>$$

And in GEOMETRY, as any polygon can be divided into triangles, one could say that all polygons "decompose" into a union of triangles.

Studying the simpler pieces in the decomposition of an object can lead to general results about the object. For instance, knowing that the interior angles of a triangle sum to 180° allows us to immediately deduce that the interior angles of any quadrilateral (the union of two triangles) sum to 360°, and that the interior angles of any pentagon (the union of three triangles) sum to 540°.

See also FACTORIZATION.

Dedekind, Julius Wilhem Richard (1831–1916) German *Analysis* Born on October 6, 1831, in Braunschweig, now a part of Germany, Richard Dedekind is remembered for his elegant construction of the REAL NUMBER system, which is based on an idea today known as a DEDEKIND CUT. This work represented an important step in formalizing mathematics. In particular, it offered the means to finally put CALCULUS on a sound mathematical footing.

Dedekind studied NUMBER THEORY and calculus at the University of Göttingen. He earned a doctoral degree in 1852 under the supervision of CARL FRIEDRICH GAUSS (he was Gauss's final pupil), and two years later obtained a habilitation degree granting him the right to be a member of the university faculty.

In 1858 Dedekind accepted a position at the Polytechnikum in Zürich. Dedekind realized that the foundations of calculus, in particular, the properties of the real-number system on which calculus rests, were not properly understood. When faced with the challenge of teaching calculus to students at the Polytechnikum for the first time, Dedekind decided not to sidestep the issue, but rather develop an approach that would properly justify the principles of the subject to himself and to his students. This is when the idea of a Dedekind cut came to him.

Dedekind published the details of this construction several years later in his famous 1872 paper *"Continuity and Irrational Numbers."* This paper was extremely well received and was admired not only for the brilliant ideas it contained, but also for the manner in which those ideas were detailed. Dedekind exhibited a talent for explaining mathematical concepts with exceptional clarity.

In 1862 Dedekind returned to his hometown to accept a position at the Brunswick Polytechnikum. He remained there for the rest of his life. He never married and lived his life with one of his sisters, who also remained unmarried.

Dedekind received many honors for his outstanding work, including election to the Berlin Academy in 1880, the Academy of Rome and the Académie des Sciences, Paris, in 1900, as well as honorary doctorates from the Universities of Zurich, Brunswick, and Oslo. Dedekind died in Brunswick, Germany, on February 12, 1916.

Dedekind made a lasting impact on the modern understanding of the real-number system. Most every college-level course on the topic of the real numbers will discuss in detail the issues Dedekind explored.

Dedekind cut During the 1800s it became clear to mathematicians that in order to prove that CALCULUS is mathematically sound one needs to properly define what is meant by a real number and, moreover, show that the real number system is "complete," in the sense that no points are "missing" from it. This is particularly important for establishing the EXTREME-VALUE THEOREM, the INTERMEDIATE-VALUE THEOREM, and the MEAN-VALUE THEOREM. All the key theorems in calculus rely on these three results.

Although the RATIONAL NUMBERS Q are relatively easy to define, the system of rationals is certainly not complete: the square root of 2, for example, is not a fraction and so is "missing" from the set of rationals. The task of defining exactly what is meant by an irrational number perplexed scholars for a very long time, and the definition of a REAL NUMBER was subject to much debate.

In 1872 JULIUS DEDEKIND had the very simple and elegant idea to simply define the irrationals to be the gaps in the rational number line. He noted that each "gap," like the square root of 2 for example, divides the line of rationals into two pieces—a left piece and a right piece. One can focus one's attention on just the left piece (for those points that are not in it constitute the right piece) and this left piece L satisfies the following three properties:

1. It is not empty, nor is it the whole set of points.
2. If a is a number in L, and b<a, then b also belongs to L.
3. If a is a number in L, it is possible to find another number c also in L but slightly larger than a.

Dedekind simply defined a real number to be any subset L of the rational numbers satisfying these three properties. Such a set is today known as a Dedekind cut.

Every rational number r defines a cut. One can check that the set $r^* = \{a \in Q : a < r\}$ satisfies the three properties. Thus the set of Dedekind cuts "contains" all the rational numbers as sets of this type. It also contains other types of numbers. For example, the square root of 2 is given by the set:

$$L = \{a \in Q : a \text{ is negative, or } a \text{ is positive and } a^2 < 2\}$$

One can check that any union of cuts, in the context of SET THEORY, is again a cut. With this surprisingly simple definition of a real number, Dedekind was able to prove all the properties of the real-number system required for establishing the soundness of calculus. In particular, he was able to show that any collection of real numbers with an upper BOUND necessarily possesses a least upper bound. (This least upper bound is the union of all the cuts listed in the collection.)

deductive/inductive reasoning In the scientific method, there are two general processes for establishing results. The first, called inductive reasoning, arrives at general conclusions by observing specific examples, identifying trends, and generalizing. "The sun has always risen in the past, therefore it will rise tomorrow," for example, illustrates this mode of reasoning.

The inductive process relies on discerning patterns but does not attempt to *prove* that the patterns observed apply to all cases. (Maybe the sun will not rise tomorrow.) For this reason, a conclusion drawn by the inductive process is called a conjecture or an educated guess. If there is just one case for which the conclusion does not hold, then the conjecture is false. Such a case is called a COUNTEREXAMPLE.

To illustrate, in the mid-1700s LEONHARD EULER observed that the product of two consecutive integers plus 41 seems always to yield a PRIME number. For example, $2 \times 3 + 41 = 47$ is prime, as is $23 \times 24 + 41 = 593$ and $37 \times 38 + 41 = 1447$. By inductive reasoning, we would conclude that $n \times (n + 1) + 41$ is always prime. However, this is a false conclusion. The case $n=40$ provides a counterexample: $40 \times 41 + 41 = 41 \times 41 = 1681$ is not prime. (Curiously $n \times (n + 1) + 41$ *is* prime for all values n between −40 and 39.)

Many intelligence tests ask participants to identify "the next number in the sequence." These questions rely on inductive reasoning, but are not mathematically sound. For example, given the challenge:

What number comes next in the sequence: 2 4 6?

any answer is actually acceptable (although the test designers clearly expect the answer "8"). One can check that the POLYNOMIAL

$$-\frac{1}{3}(n-2)(n-3)(n-4) + 2(n-1)(n-3)(n-4)$$
$$-3(n-1)(n-2)(n-4) + \frac{a}{6}(n-1)(n-2)(n-3)$$
$$= \left(\frac{a-8}{6}\right)n^3 + (8-a)n^2 + \left(\frac{11a-76}{6}\right)n + (8-a)$$

for example, has values 2, 4, and 6 when n equals 1, 2, and 3, respectively, and value a when n equals 4. Setting a to be an arbitrary value of your choice gives justification to any answer to this problem. (This particular polynomial was devised using LAGRANGE'S FORMULA.)

On the other hand, deductive reasoning works to *prove* a specific conclusion from one or more general statements using logical reasoning (as given by FORMAL LOGIC) and valid ARGUMENTs. For example, given the statements, "All cows eat grass" and "Daisy is a cow," we can conclude, by deductive reasoning, that Daisy eats grass.

Deductive reasoning does not rely on the premises that are made necessarily being true. For example, "Sydney and Boston are planets, therefore Boston is a planet" is a valid argument, whereas "Either Boston or Venus is a planet, therefore Venus is a planet" is invalid.

Mathematicians are not satisfied with conclusions drawn via inductive reasoning only. They always seek logical proof to conjectures made. But this certainly does not bar mathematicians from making conjectures. For instance, GOLDBACH'S CONJECTURE is an example of an outstanding conjecture still awaiting mathematical proof (or disproof).

deformation In TOPOLOGY, any geometric transformation that stretches, shrinks, or twists a shape, but does not tear or break apart any lines or surfaces that make the shape, is called a (continuous) deformation. For example, it is possible to mold a solid spherical ball made of clay into the shape of a cube without tearing any portions of the clay. In this sense, a cube may be considered a deformation of a sphere. It is not possible, however, to mold a sphere into the shape of a TORUS (donut) without creating a tear. Topologists consequently regard a sphere and a torus as distinct shapes (but a cube and a sphere as the "same" surface).

The notion of a deformation can be made mathematically precise. If, for a fixed set S, one object A is the image of a map f, and a second object B is the image of a second map g:

$$f : S \rightarrow A$$
$$g : S \rightarrow B$$

then B is a deformation of A if there is a continuous function $H(s,t)$ where $s \in S$ and $0 \leq t \leq 1$, so that $H(s,0)$ is the map f, and $H(s,1)$ is the map g. One also says that the map H "deforms A into B."

For example, the function $H(x,t) = t \cos x + (1 - t) \sin x$ continuously transforms a sine curve into a cosine curve.

degree measure *See* ANGLE.

degree of a polynomial The highest power of the variable that appears (with nonzero COEFFICIENT) in a

POLYNOMIAL is called the degree of that polynomial. For instance, the polynomial $4x^3 - 2x + 7$ has degree three, and the polynomial $7w^{57} - 154w^{18} + w^5 - 73w^4 + \pi w^2$ has degree 57. Any nonzero constant can be thought of as a polynomial of degree zero. In some mathematical problems it is convenient to regard the constant 0 as a polynomial of degree "negative infinity." A POWER SERIES, in some sense, is a polynomial of positive infinite degree.

degree of a vertex (valence) In any GRAPH, the number of edges meeting at a particular vertex is called the degree of that vertex. Summing all the degrees of vertices in a graph counts the total number of edges twice. The famous HANDSHAKE LEMMA from GRAPH THEORY is an amusing consequence of this result.

degrees of freedom The number of independent variables needed to specify completely the solution set of a SYSTEM OF EQUATIONS is called the number of degrees of freedom of the system. For example, the mathematical system described by the equations:

$$3x + 2y - z = 7$$
$$x + 4y - 3z = 6$$

has just one degree of freedom: if the value of z is specified, then x and y are given by $x = (8-z)/5$ and $y = (11+8z)/10$.

In physics, the number of degrees of freedom of a mechanical system is the minimum number of coordinates required to describe the state of the system at any instant relative to a fixed frame of reference. For instance, a particle moving in a circle has one degree of freedom: its position is completely specified by the angle between a fixed line of reference and the line connecting the center of the circle to the particle. A particle moving in a PLANE, or on the surface of a SPHERE, has two degrees of freedom.

See also INDETERMINATE EQUATION.

De Moivre, Abraham (1667–1754) French *Geometry, Statistics* Born on May 26, 1667, French scholar Abraham De Moivre is remembered for his pioneering work in the development of analytic geometry and the theory of probability. He was the first to introduce

COMPLEX NUMBERS into the study of TRIGONOMETRY, leading to the famous formula that now bears his name. He was also the first to describe and use the normal frequency curve in statistics.

Immigrating to England in 1685, De Moivre worked as a private tutor in mathematics. He had hoped to receive a faculty position in mathematics but, as a foreigner, was never offered such an appointment. He remained a private tutor all his life, despite the reputation he had garnered as a capable and influential scholar. He was elected a fellow of the ROYAL SOCIETY in 1697 and, in 1710, was appointed to a commission set up by the society to adjudicate on the rival claims of SIR ISAAC NEWTON and GOTTFRIED WILHELM LEIBNIZ as the discoverers of CALCULUS.

De Moivre published two notable texts. The first, *Doctrine of Chance* (1718), carefully examined the underlying principles of PROBABILITY theory and soundly developed fundamental notions such as "statistical independence" and the "probability product law," as well as established foundations for applications to the theory of annuities. The second, *Miscellanea analytica* (1730), successfully identified the principles that later allowed him to write down a formula for the NORMAL DISTRIBUTION, a task that had stymied scholars before this time. This second work also contained the mathematics necessary to establish STIRLING'S FORMULA.

It is said in all seriousness that De Moivre correctly predicted the day of his own death. Noting that he was sleeping 15 minutes longer each day, De Moivre surmised that he would die on the day he would sleep for 24 hours. A simple mathematical calculation quickly yielded the date, November 27, 1754. He did indeed pass away on that day.

See also DE MOIVRE'S FORMULA.

De Moivre's formula (De Moivre's identity) In 1707 French mathematician ABRAHAM DE MOIVRE discovered the following formula, now called De Moivre's formula:

$$(\cos \theta + i \sin \theta)^n = \cos(n\theta) + i \sin(n\theta)$$

For positive integers n the formula can be proved by INDUCTION, making use of the addition formulae for the sine and cosine functions from TRIGONOMETRY. A much simpler approach follows by making use of

EULER'S FORMULA $\cos\theta + i\sin\theta = e^{i\theta}$ and realizing that De Moivre's result is nothing more than a restatement of the exponent rule:

$$\left(e^{i\theta}\right)^n = e^{in\theta}$$

This shows that the De Moivre's formula actually holds for *any* real value for n.

De Morgan, Augustus (1806–1871) British *Algebra, Logic* Born on June 27, 1806, in Madura, India, English citizen Augustus De Morgan is remembered in mathematics for his considerable contributions to FORMAL LOGIC and ALGEBRA. In 1847 he developed a formal system of symbolic manipulations that encapsulated the principles of Aristotelian logic and included the famous laws that now bear his name. He is also remembered for properly defining the process of mathematical INDUCTION and setting this method of proof in a rigorous context.

De Morgan entered Trinity College, Cambridge, at the age of 16, and, at the completion of his bachelor's degree, applied for the chair of mathematics at the newly founded University College, London, at the young age of 21. Despite having no mathematical publications at the time, he was awarded the position in 1827.

De Morgan became a prolific writer in mathematics. His first text, *Elements of Arithmetic*, published in 1830, was extremely popular and saw many improved editions. He later wrote pieces on the topics of CALCULUS and algebra, and his 1849 text *Trigonometry and Double Algebra* was also extremely influential. This latter piece contained a useful geometric interpretation of COMPLEX NUMBERS. De Morgan also wrote literally hundreds of articles for the *Penny Cyclopedia*, a publication put out by the Society for the Diffusion of Useful Knowledge. He presented many original pieces as entries in this work. His precise definition of induction, for instance, appears in an article in the 1838 edition.

Taking an active interest in the general dissemination of mathematical knowledge, De Morgan cofounded in 1866 an academic society, the London Mathematical Society, and became its first president. The society still exists today and works to facilitate and promote mathematical research.

As a collector of odd numerical facts, De Morgan noted that being 43 in the year 1849 was a curious event, given that the number 1,849 is 43 squared. He also observed that all those born in the year 1892 would enjoy a similar coincidence in the year 1936, and those born in 1980 one in the year 2025. (The number 2,025 is 45 squared.)

De Morgan died in London, England, on March 18, 1871.

See also DE MORGAN'S LAWS.

De Morgan's laws If *A* and *B* are two subsets of a universal set, and *A'* and *B'* denote their complements, then the following two identities, known as De Morgan's laws, hold:

$$(A \cap B)' = A' \cup B'$$
$$(A \cup B)' = A' \cap B'$$

These identities can be used to convert any intersection of sets into a union of sets, or vice versa.

These laws can be seen to hold true with the aid of a VENN DIAGRAM (by shading the region outside the intersection or the union of the two sets), or by a formal SET THEORY argument. For instance, to prove the first law, one must establish that any element that belongs to $(A \cap B)'$ also belongs to $A' \cup B'$, and vice versa. This can be done as follows:

If $x \in (A \cap B)'$, then $x \notin A \cap B$, meaning that *x* does not belong to both *A* and *B*. Consequently, *x* belongs to at least one complement *A'* or *B'*, and so $x \in A' \cup B'$.

Conversely, if $y \in A' \cup B'$, then *y* does not belong to one (or both) of *A* and *B*. Consequently, *y* is not an element of $A \cap B$, and so $y \in (A \cap B)'$.

The second law can be proved similarly.

De Morgan's laws can be extended to the intersection or union of more than two sets. We have:

$$(A \cap B \cap C \cap ... \cap Z)' = A \cup B' \cup C' \cup ... \cup Z'$$
$$(A \cup B \cup C \cup ... \cup Z)' = A' \cap B' \cap C' \cap ... \cap Z'$$

When a set *A* is interpreted as "the set of all instances in which a claim *p* is true," and *B* "the set of all instances in which a claim *q* is true," then De Morgan's laws in set theory translate to the following two identities in FORMAL LOGIC:

$$\neg(p \wedge q) = (\neg p) \vee (\neg q)$$
$$\neg(p \vee q) = (\neg p) \wedge (\neg q)$$

A TRUTH TABLE establishes that these corresponding pairs of compound statements are logically equivalent. These equivalences are also called De Morgan's laws.

The formulae presented above were proposed in 1847 by Indian-born British mathematician and logician AUGUSTUS DE MORGAN (1806–71).

denumerable (enumerable, numerable) A COUNTABLE infinite set is said to be denumerable. Thus a denumerable set is any infinite set whose elements can be placed in a list akin to the list of natural numbers 1, 2, 3, ... The first DIAGONAL ARGUMENT shows that the set of RATIONAL NUMBERS is denumerable. The diagonal argument of the second kind establishes that the set of REAL NUMBERS is not. In some definite sense then, the set of real numbers is a "larger" infinite set than the infinite set of rationals. A denumerable set is said to have CARDINALITY $\aleph_0$. Every infinite set contains a denumerable subset. This can be established as follows:

Suppose *X* is an infinite set. Let x_1 be any element of *X*. Since *X* is infinite, this is not the only element of *X*. Let x_2 be another element of *X*. Since *X* is infinite, these are not the only two elements of *X*. Let x_3 be another element of *X*. Thus continuing this way produces a list of elements of *X*: x_1, x_2, x_3, ... This list represents a denumerable subset of *X*.

One can legitimately say, then, that a denumerable set is the "smallest" type of infinite set. That is, $\aleph_0$ is the "smallest" transfinite cardinal number.

See also CONTINUUM HYPOTHESIS.

derivative *See* DIFFERENTIAL CALCULUS.

Desargues, Girard (1591–1661) French *Geometry, Engineering* Born on February 21, 1591, in Lyon, France, mathematician Girard Desargues is considered the founder of PROJECTIVE GEOMETRY, an innovative, non-Greek, approach to geometry. His highly original and famous 1639 text *Brouillon project d'une atteinte aux evenemens des recontres du cone avec un plan*

(Rough draft for an essay on the results of taking plane sections of a cone) outlined the principles of the new theory, contained many new results, and offered, for the first time, a unified theory of the CONIC SECTIONS. His work, however, was largely ignored at the time of its release, and the important impact of its ideas was not properly recognized until the 19th century.

Very little is known of Desargues's personal life. Born into a family of wealth, Desargues certainly had access to an excellent formal education and the freedom to explore scholarly interests. His first works in mathematics, although pertaining to practical themes— the construction of sundials, stone-cutting techniques, and the use of perspective in art—were highly theoretical, densely written, and difficult to read, and consequently were of little use to practitioners in the respective fields. This may explain why the release of his famous 1639 piece, written in equally obscure language, was largely ignored. Only a few copies of the text were printed, only one of which survives today.

Desargues's famous theorem, the "perspective theorem" that bears his name, was published in 1648 by French engraver and painter Abraham Bosse (1602–76) in his treatise *Manière universelle de Mr. Desargues* (General methods of Desargues) on the role of perspective in art. This single result provided a gateway to a whole new approach to geometric thinking. Desargues died in Lyon, France, in September 1661. (The exact date of his death is not known.)

Desargues's theorem Named after its discoverer, French mathematician and engineer GIRARD DESARGUES (1591–1661), this theorem states:

> Suppose two triangles ABC and $A'B'C'$ are positioned in two- or three-dimensional space so that the lines joining the corresponding vertices A and A', B and B', and C and C' pass through a common point (that is, so that the two triangles are in "perspective from a point"). Then, if none of the pairs of sides AB and $A'B'$, AC and $A'C'$, or BC and $B'C'$ is parallel, then the three points of intersection of these pairs of sides lie on a straight line. (That is, the two triangles are in "perspective from a line.")

The case of three dimensions is easiest to prove. Noting that the two triangles cannot lie in parallel planes, one can show that each point of intersection of a pair of sides lies on the line of the intersection of the two planes. The proof of the two-dimensional version of the theorem is delicate. The converse of Desargues's theorem is also true:

> If corresponding sides of two triangles have intersections along the same straight line, then the lines joining corresponding vertices pass through a common point.

Desargues's theorem played a key role in the development of PROJECTIVE GEOMETRY. As noted, the theorem does not hold if some pair of lines under consideration turn out to be parallel. But Desargues observed that this difficulty can be obviated if one were to adjoin to space additional points, "points at infinity," where parallel lines do meet. This inspired Desargues to develop a notion of geometry in which each pair of points determines a unique line and, moreover, each pair of lines determine (intersect at) a unique point. In this system, the notions of "point" and "line" play dual roles, leading to a general principle of duality in this new projective geometry:

> Interchanging the roles of "point" and "line" in any theorem of projective geometry leads to another statement that is also true in projective geometry.

The dual of Desargues's theorem is its converse.
See also PERSPECTIVE.

Descartes, René (1596–1650) French *Geometry, Algebra, Philosophy* Born on March 31, 1596, in La Haye (now Descartes), France, philosopher René Descartes is remembered in mathematics for his 1637 influential work *La géométrie* (Geometry), in which he introduced fundamental principles for incorporating ALGEBRA into the study of GEOMETRY, and vice versa. This work paved the way for developing the notion of a coordinate system and, although not featured in his work, CARTESIAN COORDINATES are today named in his honor. Descartes is also noted for his "rule of signs" for counting the number of solutions to POLYNOMIAL equations, and for promoting the use of symbols in algebraic work. He advocated the use of letters to represent variables, suggesting the convention that letters first in the alphabet should refer to known quantities and letters in the latter part of the alphabet unknown quantities. (For instance, we today, without question, interpret the equation $ax+b=0$ as one containing a variable x, with a and b assumed to be known quantities.) Descartes also developed the index notation for EXPONENTS: x^2, x^3, and the like. As a philosopher, Descartes was interested in exploring the deepest underlying principles of all of scientific knowledge. He felt that mathematics lay at the base of all understanding.

After receiving a law degree from the University of Poitiers in Paris in 1616, Descartes traveled to Holland to enlist in the military school at Breda to study mathematics and mechanics. Duty with the army took him across Europe for a number of years, but in 1628 Descartes returned to Holland and began a comprehensive treatise on physics, optics, and celestial mechanics. He decided, however, not to publish this work after hearing that GALILEO GALILEI (1564–1642) was condemned to house arrest for espousing modern scientific thought. Descartes later modified the piece and published a treatise on science in 1637 in which *La géométrie* appeared as an appendix.

In 1644 Descartes published *Principia philosophiae,* a comprehensive tome exploring all aspects in scientific investigation and knowledge. Divided into four parts—*The Principles of Human Knowledge, The Principles of Material Things, Of the Visible World,* and *The Earth*—Descartes's work argued that mathematics lies at the foundation of all thinking and that all studies of nature and of the universe can be reduced to the mathematical principles of mechanics. This work was extremely influential, but the specific details of some of the theories Descartes outlined in it were problematic. For instance, Descartes believed that forces, such as gravitational forces, could not be transmitted without some kind of ambient medium. Thus, he was forced to conclude that vacuums do not exist, and that the entire universe is filled with matter. He developed a "vortex theory" in which he argued that the planets and the Sun are carried through space by a system of vortices in the ambient medium. This theory of vortices was accepted for nearly 100 years in France until SIR ISAAC NEWTON (1642–1727) demonstrated mathematically that such a dynamical system is impossible.

In 1649 Descartes moved to Stockholm to accept a position to tutor Queen Christina of Sweden in mathe-

René Descartes, an eminent scholar of the 17th century, was the first to apply methods from algebra to solve problems in geometry. The field of analytic geometry was born. *(Photo courtesy of Topham/The Image Works)*

matics. The cold climate did not suit Descartes well, and he died a few months later on February 11, 1650, of pneumonia.

Descartes instigated a philosophical shift as to how mathematics and analytical thought are utilized in the role of scientific investigation. Rather than shape scientific theories around what is observed experimentally, Descartes argued to identify fundamental "self-evident" principles first and to use logical reasoning to understand the causes behind experimental phenomena. His desire to carry the topic of geometry into physics as a part of this process is still felt today: most every branch of modern physics is described in geometric terms.

Descartes's rule of signs Discovered by philosopher and mathematician RENÉ DESCARTES (1596–1650), the "rule of signs" gives a bound on the maximum number of positive roots a POLYNOMIAL equation may possess. This bound is given by the number of sign changes that occur when the terms of the polynomial are written in descending order of degree. As an example, in reading from left to right, the polynomial equation:

$$x^6 + 7x^5 - 2x^4 - 6x^3 - 7x^2 + 8x - 2 = 0$$

has three sign changes (from positive to negative, negative to positive, and back again), and so the equation has at most three positive solutions.

The rule can be extended to count negative roots as well by replacing "x" with "$-x$" (which, in effect, reflects the negative x-axis to the positive side) and applying the same rule to the polynomial that results. In the example above, the resultant polynomial is:

$$(-x)^6 + 7(-x)^5 - 2(-x)^4 - 6(-x)^3 - 7(-x)^2 + 8(-x) - 2 = 0$$

that is,

$$x^6 - 7x^5 - 2x^4 + 6x^3 - 7x^2 - 8x - 2 = 0$$

That there are four sign changes indicates that there are at most four negative solutions to the original equation.

As another example, one can quickly check that the equation $x^6 - 64 = 0$ has at most one positive solution (which must be $x = 2$), and no negative solutions.

Proof of the Rule

Very few mathematics texts present a proof of Descartes's famous result. The argument, unfortunately, is not elementary and relies on techniques of CALCULUS. We present here a proof that also makes use of the principle of mathematical INDUCTION.

Descartes's rule of signs certainly works for polynomial equations of degree one: an equation of the form $ax + b = 0$, with a and b each different from zero, has one positive solution if a and b are of different signs, and no positive solutions if they are the same sign.

Assume Descartes's rule of signs is valid for any polynomial equation of degree n, and consider a polynomial $p(x) = a_{n+1}x^{n+1} + a_nx^n + \ldots + a_1x + a_0$ of degree $n + 1$. Its derivative $p'(x) = (n + 1)a_{n+1}x^n + na_nx^{n-1} + \ldots + a_1$ is a polynomial of degree n and so, by assumption, has at most k positive roots, where k is the number of sign changes that occur. Each root of $p'(x)$ represents a local maximum (hill) or local minimum (valley) of the

original polynomial, and a root of the original polynomial can only occur directly after one such location. Thus the original polynomial has at most k positive roots after the location of the first positive root of $p'(x)$. Whether the graph of $p(x)$ crosses the positive x-axis just before this first local maximum or minimum depends on the signs of $p(0) = a_0$ and $p'(0) = a_1$. If both are positive, then the graph is increasing to a local maximum just to the right of $x = 0$, and there is no additional root. Similarly, there is no additional root if both are negative. Only if a_0 and a_1 have opposite signs could the original equation have $k + 1$ rather than just k positive roots.

As the sign changes of the derivative $p'(x)$ match those of $p(x)$, and with the additional consideration of a possible sign change between a_0, and a_1, we have that the number of sign changes of $p(x)$ does indeed match the number of possible positive roots it could possess. This proves the rule of signs.

We can further note that if a local maximum to a graph occurs below the x-axis, or if a local minimum occurs above the axis, then the polynomial fails to cross the x-axis twice. Thus, the number of positive roots a polynomial possesses could miss the number indicated by the count of sign changes by a multiple of 2. This leads to a more refined version of Descartes's rule of signs:

> Write the terms of a polynomial from highest to lowest powers, and let k be the number of sign changes that occur in reading the coefficients from left to right. Then that polynomial has at most k positive roots. Moreover, the number of positive roots it does possess will be even if k is even, and odd if k is odd.
>
> A bound on the number of negative roots can be found substituting $-x$ for x and applying the same rule to the modified polynomial.

determinant   In the study of SIMULTANEOUS LINEAR EQUATIONS, it is convenient to assign to each square MATRIX (one representing the coefficients of the terms of the simultaneous equations) a number called the determinant of that square matrix. To explain, consider the simple example of a pair of linear equations:

$$ax + by = e$$
$$cx + dy = f$$

Solving for x in the first equation and substituting the result into the second equation yields the solution:

$$x = \frac{de - bf}{ad - bc}$$

$$y = \frac{af - ce}{ad - bc}$$

This, of course, is valid only if the quantity $ad - bc$ is not zero. We call $ad - bc$ the "determinant" of the 2×2 matrix of coefficients:

$$\begin{pmatrix} a & b \\ c & d \end{pmatrix}$$

Notice that this determinant is obtained from the matrix by selecting two elements of the matrix at a time, one located in each row and column of the matrix, and assigning a + or − sign according to whether the order in which the columns are chosen is an even or an odd PERMUTATION:

$$\begin{pmatrix} a & b \\ c & d \end{pmatrix} \quad \text{column 1, column 2} \rightarrow +ad$$

$$\begin{pmatrix} a & b \\ c & d \end{pmatrix} \quad \text{column 2, column 1} \rightarrow -bc$$

In the same way, a system of three simultaneous equations:

$$ax + by + cz = p$$
$$dx + ey + fz = q$$
$$gx + hy + iz = r$$

has a solution provided the quantity $aei - afh + bfg - bdi + cdh - ceg$ is not zero. We call this quantity the determinant of the 3×3 square matrix:

$$\begin{pmatrix} a & b & c \\ d & e & f \\ g & h & i \end{pmatrix}$$

It is obtained from the matrix by selecting three elements of the matrix at a time, one located in each row

and column of the matrix, and assigning a + or − sign according to whether the order in which the columns are chosen is an even or an odd permutation:

$$\begin{pmatrix} a & b & c \\ d & e & f \\ g & h & i \end{pmatrix} \quad \text{Column 1, Column 2, Column 3} \rightarrow +aei$$

$$\begin{pmatrix} a & b & c \\ d & e & f \\ g & h & i \end{pmatrix} \quad \text{Column 1, Column 3, Column 2} \rightarrow -afh$$

$$\begin{pmatrix} a & b & c \\ d & e & f \\ g & h & i \end{pmatrix} \quad \text{Column 2, Column 3, Column 1} \rightarrow +bfg$$

Notice that the signs of the products can equivalently be evaluated in terms of the sign of row permutations.

In general, the determinant of an $n \times n$ matrix is formed by selecting n elements of the matrix, arranged one per row and one per column, forming the product of those entries, assigning the appropriate sign, and adding together all possible results. The determinant of a square matrix A is denoted $\det(A)$ or, sometimes, $|A|$.

The determinant function satisfies a number of key properties:

1. If a column or a row of a matrix A is entirely zero, then $\det(A) = 0$.

(Each product formed in computing the determinant will contain a term that is zero.)

2. If two columns or two rows of the matrix are interchanged, then the sign of $\det(A)$ changes.

(If two columns undergo one more interchange, then the sign of each permutation alters. Since the process of forming the determinant can equivalently be viewed in terms of row permutations, the same is true if two rows are interchanged.)

3. If a matrix A has two identical columns, or two identical rows, then $\det(A) = 0$.

(Interchange those two columns or two rows. The matrix remains unchanged, yet the determinant has opposite sign. It must be the case then that $\det(A) = 0$.)

It is sometimes convenient to think of the determinant of a matrix A as a function of its columns written as vectors $v_1, v_2, ..., v_n$. We write $\det(A) = \det(v_1, v_2, ..., v_n)$. Then, by the above observations, we have:

$$\det(v_1, ..., 0, ..., v_n) = 0$$
$$\det(v_1, ..., v_i, ..., v_j, ..., v_n) = -\det(v_1, ..., v_j, ..., v_i, ..., v_n)$$
$$\det(v_1, ..., v, ..., v, ..., v_n) = 0$$

We also have:

4. $\det(v_1, ..., v + w, ..., v_n) = \det(v_1, ..., v, ..., v_n) + \det(v_1, ..., w, ..., v_n)$

and

5. $\det(v_1, ..., kv, ..., v_n) = k \det(v_1, ..., v, ..., v_n)$

and consequently

6. The value of $\det(A)$ is not altered if a multiple of one column is added to another column: $\det(v_1, ..., v_i + kv_j, ..., v_n) = \det(v_1, ..., v_i, ..., v_n)$

These results follow from the definition of the determinant. The corresponding results about rows are also valid.

CRAMER'S RULE shows that the notion of a determinant is precisely the concept needed to solve simultaneously linear equations. We have:

A system of simultaneous linear equations has a (unique) solution if the determinant of the corresponding matrix of coefficients is not zero.

Cramer's rule goes further and provides a formula for the solution of a system in terms of determinants.

The determinant has another important property. After some algebraic work it is possible to show:

The determinant of the product of two $n \times n$ matrices A and B is the product of their determinants:

$$\det(AB) = \det(A) \times \det(B)$$

The determinant of the IDENTITY MATRIX I is one. If a square matrix A is invertible, then the equation:

$$1 = \det(I) = \det(A \cdot A^{-1}) = \det(A) \cdot \det(A^{-1})$$

shows that $\det(A)$ is not zero and that

$$\det(A^{-1}) = \frac{1}{\det(A)}$$

We have:

If a matrix is invertible, then its determinant is not zero.

The converse is also true:

If the determinant of a matrix is not zero, then the matrix is invertible.

To see why this holds, suppose that A is a matrix with nonzero determinant. Let e_i denote the ith column of the identity matrix. By Cramer's rule, since the determinant is not zero, the system of equations $Ax = e_i$ has a solution $x = s_i$, say. Set B to be the matrix with ith column s_i. Then $AB = I$. This shows at least that A has a "right inverse" B. To complete the proof, let A^T denote the transpose of A, that is, the matrix obtained from A by interchanging its rows and columns. Since the determinant can be viewed equivalently well as a function of the rows of the matrix as its columns, we have that $\det(A^T) = \det(A)$. Since the determinant of A^T is also nonzero, there is a matrix C so that $A^TC = I$. One can check that the transpose of the product of two matrices is the reverse product of their transposes. We thus have: $C^TA = (A^TC)^T = I^T = I$. This shows that the matrix A also has a left inverse C^T. The left and right inverses must be equal, since $C^T = C^TI = C^T AB = IB = B$. Thus the matrix B is indeed the full inverse matrix to A: $AB = BA = I$.

See also INVERSE MATRIX.

diagonal Any line joining two nonadjacent vertices of a POLYGON is called a diagonal of the polygon. For example, a square has two diagonals, each cutting the figure into two congruent right-angled triangles, and a pentagon has five different diagonals. There are no diagonals in a triangle. In general, a regular $n$gon has $\frac{n(n-3)}{2}$ distinct diagonals.

A diagonal for a POLYHEDRON is any line joining two vertices that are not in the same face. A cube, for example, has four distinct diagonals. A diagonal plane for a polyhedron is any plane that passes through two edges that are not adjacent.

diagonal argument The diagram at bottom right shows that it is possible to match the elements of an infinitely long line of objects with the elements of a two-dimensional array of objects in a perfect one-to-one correspondence (meaning that each and every element of the first set is matched with exactly one element of the second set, and vice versa). In some sense, this shows that an infinite two-dimensional array is no "larger" than an infinite one-dimensional array. Similar constructions can be used to match elements of higher-dimensional arrays with the elements of a single infinite line of objects.

The correspondence described is called a diagonal argument of the first kind. It was first introduced by German mathematician GEORG CANTOR (1845–1918) in 1891. This argument shows that the set of positive RATIONAL NUMBERS is of the same CARDINALITY (that is, of exactly the same "size") as the set of NATURAL NUMBERS:

First list an infinite string including all of the rationals with numerator 1: $\frac{1}{1}, \frac{1}{2}, \frac{1}{3}, \dots$. Underneath this string, list all the rationals with numerator 2 that do not reduce to a fraction with numerator 1: $\frac{2}{1}, \frac{2}{3}, \frac{2}{5}, \dots$, and underneath this, write the string of all rationals with numerator 3 that do not reduce to a fraction with numerator 1 or 2: $\frac{3}{1}, \frac{3}{2}, \frac{3}{4}, \frac{3}{5}, \dots$; and so forth. The diagonal argument now shows that the set of all positive rationals can indeed be presented as a single denumerable list:

$$\frac{1}{1}, \frac{2}{1}, \frac{1}{2}, \frac{3}{1}, \frac{2}{3}, \frac{1}{3}, \frac{4}{1}, \dots$$

This shows that the set of rationals is COUNTABLE.

One can also use this argument to show that the union of a countable number of countable sets is itself countable: simply list the elements of each set in an infinite string, one per line of a two-dimensional array, and use the diagonal argument to provide a method of listing all the elements in the array.

Cantor provided a "diagonal argument of the second kind" to show that the set of real numbers is an infinite set that, in some definite sense, is "larger" than the set of counting numbers or the set of rationals. For convenience, work with the set of real numbers in the INTERVAL [0,1]. Each such real number can be written as an infinite decimal, using an infinite string of nines if necessary. For example, 1/3 = 0.3333 ... and 1/2 = 0.5 = 0.49999 ... Cantor's second-diagonal argument proceeds as follows:

Suppose it is possible to produce a complete list of all the real numbers from the interval [0,1], each written as an infinite decimal expansion:

$$0.\mathbf{a_1}\, a_2\, a_3\, a_4\, a_5\, a_6 \dots$$
$$0.b_1\, \mathbf{b_2}\, b_3\, b_4\, b_5\, b_6 \dots$$
$$0.c_1\, c_2\, \mathbf{c_3}\, c_4\, c_5\, c_6 \dots$$
$$0.d_1\, d_2\, d_3\, \mathbf{d_4}\, d_5\, d_6 \dots$$
$$0.e_1\, e_2\, e_3\, e_4\, \mathbf{e_5}\, e_6 \dots$$
$$\vdots$$

Construct another real number $x = 0.\alpha_1\, \alpha_2 \alpha_3 \dots$ as follows:

Set α_1 equal to 1 if a_1 is equal to an 8 or a 9, and equal to 8 if a_1 is any other number; set α_2 equal to 1 if b_2 is

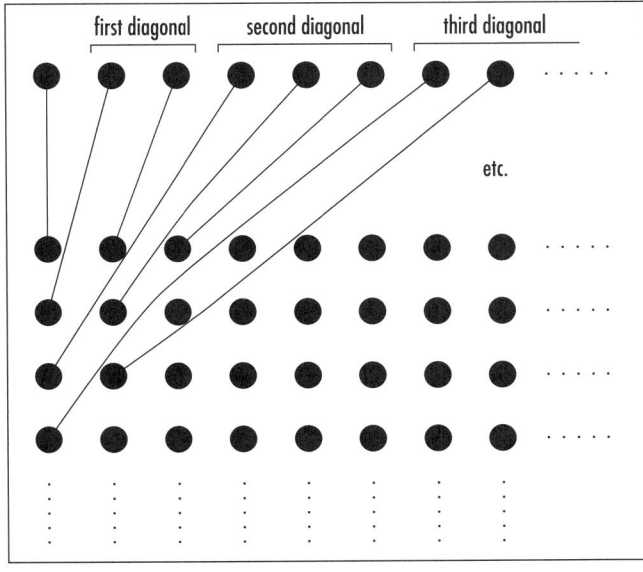

The diagonal correspondence

equal to an 8 or a 9, and equal to 8 if b_2 is any other number; set α_3 equal to 1 if c_3 is equal to an 8 or a 9, and equal to 8 if c_3 is any other number; and so forth.

Then the number x does not appear on the list. It is not the first number on the list, since x and $0.a_1 a_2 a_3 a_4 \ldots$ differ in the first decimal place; nor is it the second number in the list, since x and $0. b_1 b_2 b_3 \ldots$ differ in the second decimal place; nor is it the third, fourth, or 107th number in the list. Thus from any list of real numbers, it is possible to construct another real number that fails to be on the list.

Even if one were to include the number x constructed above on a new list of real numbers, one can repeat the diagonal argument again to produce a new real number y that fails to be on the list. In this way, one can argue that there are always "more" real numbers than can be listed. The set of real numbers is thus of greater cardinality than the set of rational numbers. (It is worth commenting that, at first, it seems easier to simply construct the real number $x = 0.\alpha_1\alpha_2\alpha_3 \ldots$ by selecting α_1 to be any digit different from α_1, α_2 any digit different from b_2, α_3 any digit different from c_3, and so forth. Arbitrary choices, however, could lead to ambiguity and damage the argument. For instance, the number x produced could be 0.50000 … which already appears on the list as 0.49999… The approach taken above carefully obviates this concern.)

Cantor also proved that there is an infinitude of infinite sets all larger than the infinite set of natural numbers.

diameter   The furthest distance between two points on the boundary of a geometric figure is called the diameter of the figure. For example, the diameter of a square of side-length 1 is the distance between two opposite corners of the square. This distance is $\sqrt{2}$. An equilateral triangle of side-length 1 has diameter equal to 1. In this context, the diameter of an object is always a number.

Sometimes the term *diameter* also refers to the line segment itself connecting two boundary points of maximal distance apart. For example, the diameter of a circle is any line segment through the center of the circle connecting two boundary points. A diameter of a SPHERE also passes through its center.

Circles and spheres are figures of CONSTANT WIDTH. *See also* DIAGONAL.

Dido's problem   According to legend, in the year 800 B.C.E., Princess Dido of Tyre fled her Phoenecian homeland to free herself of the tyranny of her murderous king brother. She crossed the Mediterranean and sought to purchase land for a new city upon the shores of northern Africa. Confronted with only prejudice and distrust by the local inhabitants, she was given permission to purchase only as much land as could be surrounded by a bull's hide. The challenge to accept these terms and still enclose enough land to found a city became known as Dido's problem.

The Roman poet Virgil (70–19 B.C.E.), in his epic work *Aeneid*, refers to the legend of Dido and her clever solution to the problem. He claims that Dido cut the hide into very thin strips and pieced them together to form one very long strand, which she then used to enclose a proportion of land of maximal area, as given by the shape of a circle. (More precisely, with coastline as part of the boundary, Dido formed a semicircle with bull-hide strips.) The portion of land she consequently purchased for a minimal price was indeed large enough to build a city. According to Virgil, this story represents the founding of the city of Carthage, which is now a residential suburb of the city of Tunis.

In this story, Princess Dido solved the famous ISOPERIMETRIC PROBLEM:

Of all figures in the plane with a given perimeter, which encloses the largest area?

Mathematical analysis of this problem is difficult. It was not until the late 19th century that mathematicians were finally able to prove that the solution presented in this ancient tale—namely, the circle—is the correct shape.

See also ISOPERIMETRIC PROBLEM.

difference   In ARITHMETIC, the result of subtracting one quantity from another is called the difference of the two quantities. For example, the difference of 105 and 83 is 22. The minus sign is used to denote differences. For instance, we write: 105 − 83 = 22. The difference of two quantities could be negative. We have, for example, $5 - 7 = -2$. The minus sign was first used in a printed text in 1489 by German mathematician Johannes Widman (1462–98).

The absolute difference of two quantities a and b is the ABSOLUTE VALUE of the difference of the two quantities: $|a - b|$. The absolute difference of 13 and 8, for example, is 5, as is the absolute difference of 8 and 13. Some authors use the symbol ~ to denote absolute difference: $8 \sim 13 = 5$.

In SET THEORY, the difference of two sets A and B (also called the relative complement of B in A) is the set of elements that belong to A but not to B. This difference is denoted $A \backslash B$ or $A - B$. For example, $A = \{1,2,3,6,8,\}$ and $B = \{2,4,5,6\}$, the $A \backslash B = \{1,3,8\}$. Also, $B \backslash A = \{4,5\}$.

The symmetric difference of two sets A and B, denoted either $A \triangledown B$, $A + B$, or $A \ominus B$, is the set of all elements that belong to one, but not both, of the two sets A and B. It is the union of the differences $A \backslash B$ and $B \backslash A$. It is also the difference of the union of A and B and their intersection:

$$A \triangledown B = (A \backslash B) \cup (B \backslash A)$$
$$= (A \cup B) - (A \cap B)$$

For the example above, we have: $A \triangledown B = \{1,3,4,5,8\}$.
See also FINITE DIFFERENCES.

difference of two cubes The equation $x^3 - a^3 = (x - a)(x^2 + ax + a^2)$ is called the difference of two cubes formula. One can check that it is valid by EXPANDING BRACKETS. Since the sum of two cubes can also be written as a difference, $x^3 + a^3 = x^3 - (-a)^3$, we have a companion equation $x^3 + a^3 = (x + a)(x^2 - ax + a^2)$.

The DIFFERENCE OF TWO SQUARES and the difference of two cubes formulae generalize for exponents larger than 3. We have:

$$x^n - a^n = (x - a)(x^{n-1} + ax^{n-2} + a^2x^{n-3} + \ldots + a^{n-2}x + a^{n-1})$$

for $n \geq 2$. This shows that the quantity $x - a$ is always a factor of $x^n - a^n$. This observation is useful for factoring numbers. For example, we see that $6^{51} - 1$ is divisible by $6 - 1 = 5$. Since we can also write $6^{51} - 1 = (6^3)^{17} - 1^{17}$, we have that $6^3 - 1 = 215$ is also a factor of $6^{51} - 1$.

If n is odd, then there is a companion formula:

$$x^n + a^n = (x + a)(x^{n-1} - ax^{n-2} + a^2x^{n-3} - \ldots - a^{n-2}x + a^{n-1})$$

This shows, for example, that $2^{12} + 1$ (which equals $(2^4)^3 + 1^3$) is divisible by 17.
See also MERSENNE PRIME.

difference of two squares The equation $x^2 - a^2 = (x - a)(x + a)$ is called the difference-of-two-squares formula. One can check that it is valid by EXPANDING BRACKETS. It can also be verified geometrically: place a small square of side-length a in one corner of a larger square of side-length x. The area between the two squares is $x^2 - a^2$. But this L-shaped region can be divided into two rectangles: one of length x and width $(x - a)$ and a second of length a and width $(x - a)$. These stack together to form a single $(x - a) \times (x + a)$ rectangle. Thus it must be the case that $x^2 - a^2$ equals $(x - a)(x + a)$.

The conjugate of a sum $x + a$ is the corresponding difference $x - a$, and the conjugate of a difference $x - a$ is the corresponding sum $x + a$. Multiplying an algebraic or numeric quantity by its conjugate and invoking the difference-of-two-squares formula can often simplify an expression. For example, if we multiply the quantity $\dfrac{1}{2 - \sqrt{3}}$ by "one," we obtain:

$$\frac{1}{2 - \sqrt{3}} = \frac{1}{2 - \sqrt{3}} \cdot \frac{2 + \sqrt{3}}{2 + \sqrt{3}} = \frac{2 + \sqrt{3}}{2^2 - \left(\sqrt{3}\right)^2}$$

$$= \frac{2 + \sqrt{3}}{1} = 2 + \sqrt{3}$$

(We have "rationalized" the denominator.)

A *sum* of two squares, $x^2 + a^2$, can be regarded as a difference if one is willing to work with COMPLEX NUMBERS. We have: $x^2 + a^2 = x^2 - (ia)^2 = (x - ia)(x + ia)$.
See also DIFFERENCE OF TWO CUBES; RATIONALIZING THE DENOMINATOR.

differential Close to any point x, the graph of a differentiable function $y = f(x)$ is well approximated by a

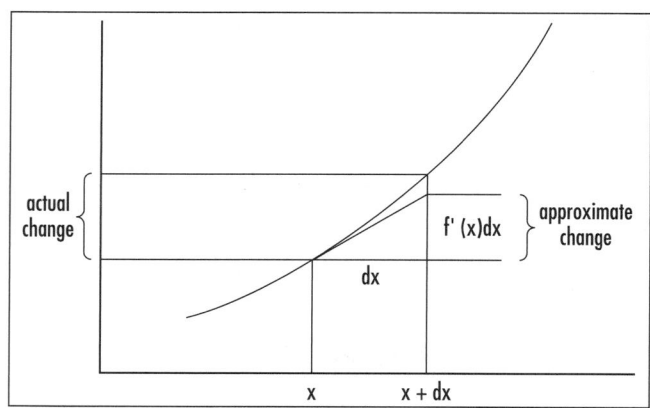

The differential

small straight-line segment tangent to the graph at x. The slope of this tangent line is the DERIVATIVE $f'(x)$. Using the symbol dx to represent a small change in the x-variable, we see that the corresponding change in the y-variable is approximately $dy = f'(x)dx$. The quantities dx and dy are called differentials.

GOTTFRIED WILHELM LEIBNIZ (1646–1716) based his development of the theory of CALCULUS on the idea of a differential. Today we use the notation $\dfrac{dy}{dx}$ for the derivative $f'(x)$, deliberately suggestive of Leibniz's ideas.

See also HISTORY OF CALCULUS (essay); NUMERICAL DIFFERENTIATION.

differential calculus This branch of CALCULUS deals with notions of SLOPE, rates of change and ratios of change. For example, a study of VELOCITY, which can be described as the rate of change of position, falls under the study of differential calculus, as do other concepts that arise in the study of motion.

If a quantity y is a FUNCTION of another quantity x, $y = f(x)$ say, then each change in the x-variable, $x \to x + h$, produces a corresponding change in the y-variable: $f(x) \to f(x + h)$. The ratio of the changes of the two variables is: $\dfrac{f(x + h) - f(x)}{h}$. Graphically, this represents the slope (the "rise" over the "run") of the line segment connecting the two points $(x, f(x))$ and $(x+h, f(x+h))$ on the graph of the curve $y = f(x)$.

The slope of this line segment, for a fixed change h in the x-variable, depends on the shape of the curve and will typically change from point to point. A very

steep curve will give a large rise for a fixed run, for example, whereas a curve that rises slowly will give a low value for slope. In all cases, if the value h is very small, then the slope of the line segment described above approximates the slope of the TANGENT line to the curve at position x. The smaller the value of h, the better is the approximation.

In another setting, if $y = f(t)$ represents the position of a car along a highway at time t, then, over h seconds of travel, the automobile changes position by amount $f(t + h) - f(t)$, and the ratio $\dfrac{f(t + h) - f(t)}{h}$ represents the average rate of change of position, or the average velocity, of the car over h seconds of travel. If the value h is small, then this quantity approximates the actual speed of the car at time t as read by the speedometer. The smaller the value of h, the better is the approximation.

The ratio $\dfrac{f(x + h) - f(x)}{h}$ is called a "Newton quotient" to honor the work of SIR ISAAC NEWTON (1642–1727) in the discovery and development of calculus, and the LIMIT,

$$\lim_{h \to 0} \frac{f(x + h) - f(x)}{h}$$

if it exists, is called the derivative of the function $f(x)$. It represents the slope of the (tangent line to the) graph $y = f(x)$ at position x, or, alternatively, the instantaneous rate of change of the variable $y = f(x)$ at position/time x.

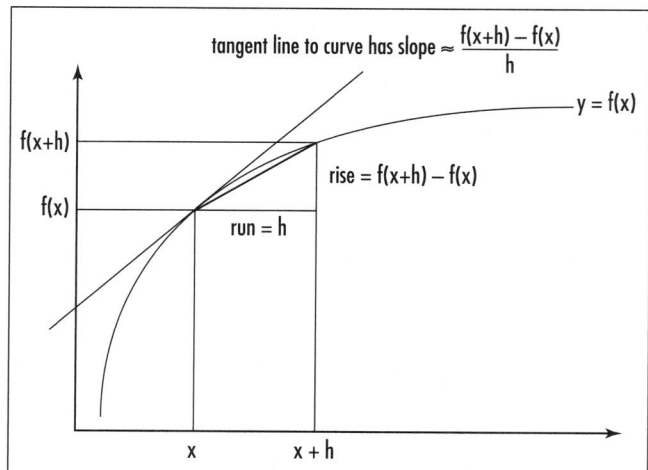

Computing the derivative

The derivative of a function $y = f(x)$ at position x is denoted either $\dfrac{df}{dx}$ or $\dfrac{dy}{dx}$ (or, to make the point at which the derivative is being computed explicit, $\left.\dfrac{df}{dx}\right|_x$ or $\left.\dfrac{dy}{dx}\right|_x$). The derivative is also written $f'(x)$, which is read as "f prime of x."

As an example, the derivative of the function $y = f(x) = x^2$ at position $x = 7$ is given by:

$$\left.\frac{dy}{dx}\right|_{x=7} = \lim_{h \to 0} \frac{f(7+h) - f(7)}{h}$$

$$= \lim_{h \to 0} \frac{(7+h)^2 - 7^2}{h}$$

$$= \lim_{h \to 0} \frac{49 + 14h + h^2 - 49}{h}$$

$$= \lim_{h \to 0} \frac{14h + h^2}{h}$$

$$= \lim_{h \to 0} 14 + h$$

$$= 14$$

That is, the slope of the tangent line to the curve $y = x^2$ at $x=7$ is 14. In general, the derivative of $f(x) = x^2$ at an arbitrary point x is given by: $f'(x) = \lim_{h \to 0} \frac{(x+h)^2 - x^2}{h} = \lim_{h \to 0} 2x + h = 2x$. A considerable amount of algebra is usually needed to compute these limits. The aim is to cancel h in the denominator so as to avoid division by zero.

The process of finding the derivative of a function is called differentiation. A function $y = f(x)$ is called differentiable at a point x if the derivative of the function $f'(x)$ exists at that position. A function is differentiable if its derivative can be computed at every point under consideration. Not every function is differentiable.

For example, the ABSOLUTE VALUE function $y = |x|$ has no well-defined tangent line at its vertex at position $x = 0$, and limit $\lim_{h \to 0} \frac{|0+h| - |0|}{h} = \lim_{h \to 0} \frac{|h|}{h}$ does not exist. (Consider the cases h positive and negative separately.) It can be shown that every differentiable function is continuous, but, as the absolute-value function shows, a continuous function need not be differentiable.

The thrust of differential calculus is thus the computation of the derivatives of functions. The following table shows the derivatives of some standard functions. The PRODUCT RULE, QUOTIENT RULE, and the CHAIN RULE also assist in the computation of derivatives.

$f(x)$	$f'(x)$
k (constant)	0
$mx+b$ (straight line of slope m)	m
x	1
x^r	rx^{r-1}
$\sin x$	$\cos x$
$\cos x$	$-\sin x$
$\tan x$	$\sec^2 x = \dfrac{1}{\cos^2 x}$
$\sec x$	$\sec x \tan x$
$\operatorname{cosec} x$	$-\operatorname{cosec} x \cot x$
$\cot x$	$-\operatorname{cosec}^2 x$
e^x	e^x
e^{kx}	ke^{kx}
a^x	$a^x \ln a$
$\sinh x$	$\cosh x$
$\cosh x$	$\sinh x$
$\ln x$	$\dfrac{1}{x}$

Apart from dealing with issues of rates of change, differential calculus is also used to solve OPTIMIZATION problems, that is, problems of finding the maximum or minimum values for a given function (which are called MAXIMUM/MINIMUM problems).

ANTIDIFFERENTIATION is intimately connected with INTEGRAL CALCULUS, the general problem of computing areas under curves and volumes under surfaces. The FUNDAMENTAL THEOREM OF CALCULUS explains this connection.

The derivative of a function $y = f(x)$ at the point $x = x_1$ can alternatively be defined as the limit:

$$f'(x_1) = \lim_{x_2 \to x_1} \frac{f(x_2) - f(x_1)}{x_2 - x_1}$$

Some authors of mathematics textbooks prefer this definition. Of course, setting $x_1 = x$ and $x_2 = x + h$, it is equivalent to the definition presented above.

See also CONCAVE UP/CONCAVE DOWN; DIFFERENTIAL; DIFFERENTIAL EQUATION; DIRECTIONAL DERIVATIVE; HIGHER DERIVATIVE; HISTORY OF CALCULUS (essay); IMPLICIT DIFFERENTIATION; INCREASING/

DECREASING; MAXIMUM/MINIMUM; MEAN-VALUE THEO-
REM; NUMERICAL DIFFERENTIATION; PARTIAL DERIVA-
TIVE; ROLLE'S THEOREM.

differential equation Any equation that contains
one or more derivatives is called a differential equation.
If such an equation involves just a single independent
variable x and a single dependent variable y and its
derivatives, then it is described as "ordinary." For
example, the equation $\frac{dy}{dx} - 4x + 3 = 0$ is an ordinary
differential equation.

A solution of a differential equation is any function $y = f(x)$ that satisfies the equation. For example,
$y = 2x^2 - 3x$ is a solution to the above equation. So too
is $y = 2x^2 - 3x + 5$. These solutions were found by
writing $\frac{dy}{dx} = 4x - 3$ and integrating: $y = \int 4x - 3\,dx$
$= 2x^2 - 3x + C$. If particular values of x and y are
known, say $y = 0$ when $x = 1$, then a value for C is
determined, in this case $C = 1$, yielding the particular
solution: $y = 2x^2 - 3x + 1$.

The "order" of a differential equation is the order
of the highest derivative that appears in the equation,
and the "degree" of the equation is the power to which
the highest-order derivative is raised. Thus, for example,
$\frac{d^3y}{dx^3} + 2x^4y\frac{dy}{dx} + x^6y^2 = 0$ is a third-order differential

equation of the first degree, and $\left(\frac{d^3y}{dx^3}\right)^2 + \left(\frac{d^2y}{dx^2}\right)^5 + 3 = y$
is a third-order equation of degree 2.

An equation involving more than one independent
variable and PARTIAL DERIVATIVES with respect to those
variables is called a partial-differential equation. For

example, $\left(\frac{\partial^2 z}{\partial x^2}\right)^3 xy + \frac{\partial z}{\partial x} \cdot \frac{\partial z}{\partial y} + xyz = 2$ is a second-order

partial differential equation of degree 3.

Differential equations arise in virtually every
branch of science, economics, and engineering. Any
theory that attempts to describe relationships between
the rates of change of continuously varying quantities
leads to a differential equation. For example, one
model of population growth describes the rate of
change of population size P as proportional to the size
of the population. This leads to the equation $\frac{dP}{dt} = kP$,
where k is some constant. (*See* POPULATION MODELS.)

Finding solutions to differential equations is an essential part of scientific investigation.

There are a number of standard techniques for
solving certain types of differential equations. All
involve rearranging terms or transforming the equation
into a form that can be readily integrated. We outline a
selection of some basic techniques.

Directly Integrable Equations: A first-order equation of
the form:

$$\frac{dy}{dx} = f(x)$$

is directly integrable and has solution given by $y = \int f(x)dx$. Similarly, a second-order equation of the form:

$$\frac{d^2y}{dx^2} = f(y)$$

can be solved by integrating twice.

Separation of Variables: A first-order equation of the
form:

$$f(y)\frac{dy}{dx} = g(x)$$

where f is a function of y only and g is a function of x
only, can be solved by integrating both sides of the
equation with respect to x. This yields:

$$\int f(y)\frac{dy}{dx}dx = \int g(x)dx$$

The method of INTEGRATION BY SUBSTITUTION shows
that the left integral can be interpreted simply as an
integral with respect to the variable y, and so we are
permitted to write:

$$\int f(y)dy = \int g(x)dx$$

We can now evaluate these integrals and solve for y.

For example, to solve $\frac{dy}{dx} = y^2$ write, with an abuse
of notation, $\frac{1}{y^2}dy = dx$ Integration gives $\int\frac{1}{y^2}dy = \int 1\,dx$,

yielding, $-\frac{1}{y} = x - C$, or $y = \frac{1}{C - x}$.

Homogeneous Equations: A first-order equation of the form:

$$\frac{dy}{dx} = f\left(\frac{y}{x}\right)$$

can be solved by substituting $y = vx$. This reduces the equation to one in v and x alone. Separation of variables will now work.

For example, to solve $\frac{dy}{dx} = \frac{x^2 + y^2}{x^2}$, which can be written $\frac{dv}{dx} = 1 + \left(\frac{y}{x}\right)^2$, set $y = vx$ to obtain: $\frac{dv}{dx}x + v = 1 + v^2$. Separation of variables now applies.

Linear Equations: A first-order equation of the form:

$$\frac{dy}{dx} + f(x)y = g(x)$$

can be solved by multiplying through by the integrating factor $e^{\int f(x)dx}$. This yields the equation:

$$e^{\int f(x)dx}\frac{dy}{dx} + e^{\int f(x)dx}f(x)y = e^{\int f(x)dx}g(x)$$

which can be rewritten:

$$\frac{d}{dx}\left(e^{\int f(x)dx}y\right) = e^{\int f(x)dx}g(x)$$

This is directly integrable.

For example, the equation $\frac{dy}{dx} + \frac{y}{x} = x^2$ has integrating factor $e^{\int \frac{1}{x}dx} = e^{\ln x} = x$, and so the equation can be rewritten $\frac{d}{dx}(xy) = x^3$, yielding $xy = \frac{1}{4}x^4 + C$, or $y = \frac{1}{4}x^3 + \frac{C}{x}$.

Basic Second-Order Equations: A second-order equation of the form:

$$\frac{d^2y}{dx^2} = f(y)$$

can be solved by multiplying through by $2\frac{dy}{dx}$ to obtain:

$$\frac{d}{dx}\left(\left(\frac{dy}{dx}\right)^2\right) = 2\frac{dy}{dx}\frac{d^2y}{dx^2} = 2f(y)\frac{dy}{dx}$$

Integrating gives:

$$\left(\frac{dy}{dx}\right)^2 = 2\int f(y)dy + C$$

After taking square roots, one can now separate variables.

It is often the case that no known techniques will solve a particular differential equation that arises in a particular scientific study. In this case, mathematicians will often assume that the solution function $y = f(x)$ can be written as a TAYLOR SERIES: $y = a_0 + a_1x + a_2x^2 + a_3x^3 + \ldots$ By substituting the series into the differential equation, it is usually possible to compute the values of at least the first few coefficients $a_0, a_1, a_2, \ldots$. This approach is called the "method of undetermined coefficients." In this context, the Taylor series used is sometimes called a perturbation function.

digit A symbol that forms part of a number is called a digit. For example, the number 42.768 has five digits. Ten digits are used in decimal notation, namely, 0, 1, 2, 3, 4, 5, 6, 7, 8, and 9. In hexadecimal notation (base 16), the digits are 0, 1, 2, 3, 4, 5, 6, 7, 8, 9, A, B, C, D, E, and F. In the system of binary numbers only two digits are used: 0 and 1. In general, a counting system in base b uses b different digits to represent the numbers 0 through $b - 1$.

In the vernacular, the word *digit* means a finger or a toe. As one learns to count with one's digits it is not surprising that the word has come to be used for specific numbers the fingers represent. The measure of a digit, defined as the width of one finger, about $^3/_4$ of an inch, was a standard Old English unit of LENGTH.

A digital watch displays specific symbols (numbers) as discrete units of time (seconds) pass, and a digital computer processes information supplied to it in the form of numbers. An analog watch, however, uses the sweeping entity of a moving hand as an analog to the passing of time, and an analog computer, now generally considered obsolete, uses a varying voltage to mimic continuously changing input information, such as the rate of flow of oil through a pipeline, for example.

See also BASE OF A NUMBER SYSTEM; ERROR.

dihedral Any geometric construct formed by the intersection of two planes is called dihedral. For example, the line of intersection of two nonparallel planes is called a dihedral line, and the angle between the two planes is called a dihedral angle (or, sometimes, a dihedron). A dihedral angle can be computed by taking the DOT PRODUCT of the two normal vectors to the planes. The dihedral angle of a polyhedron is the angle between two adjacent faces of the solid.

The word *dihedral* comes from the Greek prefix *di-*, meaning "two," and the word *hedra,* meaning "base," "seat," or "surface."

See also NORMAL TO A PLANE.

dilation *See* GEOMETRIC TRANSFORMATION; LINEAR TRANSFORMATION.

dimension The number of coordinates needed to specify the position of a particular point in space is called the dimension of that space. Lines and curves are considered one-dimensional, since the location of any point on a curve can be specified by a single parameter, namely, the distance along the curve at which it lies. Points in two-dimensional space form a surface, and points in three-dimensional space lie within a volume.

A physical theory is said to be multidimensional if it describes the universe with a large number of parameters. Since events in the world occur at specific locations and specific times, four parameters are needed to describe them: three spatial parameters x, y, and z and a fourth parameter t for time. Because of this, "time" is often cited as being the fourth dimension. However, a theory that also considers the electric charge q of objects, say, would be described as "five-dimensional," and one that also considers magnetic strength as "six-dimensional." Physicists have asserted that we live in a universe that is as much as 18-dimensional. There is little meaning here other than that scientists are asserting that all events in the universe can be described fully through 18 different variables. There is no specific reason for the fourth variable to always be interpreted as "time."

To a mathematician, an n-dimensional object is one that is fully described by n parameters. For example, a circle sitting in two-dimensional space is described by an equation with two variables: $x^2+y^2=1$. A SPHERE in three-dimensional space is described by an equation with three variables: $x^2+y^2+z^2=1$. Adding more variables to the equation gives higher-dimensional spheres: the equations $x^2+y^2+z^2+w^2=1$ and $x^2+y^2+z^2+w^2+u^2 = 1$, for instance, might be said to describe "hyperspheres" in four- and five-dimensional space. Although one cannot envision what these objects are, the mathematics of these objects is little different from the mathematics of ordinary circles and spheres.

In a geometric context, "dimension" can be described through the notion of scaling. If we SCALE a geometric object by a factor k, then its size changes accordingly: any line of length a becomes a line of length ka, any planar region of area A becomes a planar region of area k^2A, and any solid of volume V is replaced by a solid of volume k^3V. An object can thus be described as d-dimensional if its "size" scales according to the rule:

$$\text{new size}=k^d \times \text{old size}$$

In this context, it is possible for a geometric object to have fractional dimension. Such an object is called a FRACTAL.

See also HYPERCUBE.

Diocles (ca. 240–180 B.C.E.) *Greek Geometry* Born on Évvoia (Euboea), a Greek island (the exact birth date is not known), Diocles is remembered for his work on CONIC SECTIONS and his innovative solution to the DUPLICATING THE CUBE problem via the invention of a new curve called the cissoid curve.

Almost nothing is known of Diocles' life, and knowledge of his work (until recently) came to us chiefly through references made by scholars after his time. It is known that Diocles was a contemporary of APOLLONIUS OF PERGA (ca. 260–190 B.C.E.) and may have spent considerable time at Arcadia, the intellectual center of Greek culture.

Diocles wrote one significant text, *On Burning Mirrors,* which, although chiefly ignored by Greek scholars, had considerable influence on Arab mathematicians. A translation of this work was only recently discovered in the Shrine Library in Mashhad, Iran, and an English edition of the text was first published in 1976.

The piece is organized as a collection of 16 discussions on original results in geometry, chiefly concerned with the topic of CONIC SECTIONS. One sees that Diocles was the first to prove the reflection property of a PARABOLA, thereby solving an old problem presented by ARCHIMEDES OF SYRACUSE (ca. 287–212 B.C.E.) of finding a mirror surface that produces heat when placed facing the sun. (It is said that Archimedes proposed using curved mirrors to reflect the Sun's rays and burn the sails of enemy ships.) Diocles also describes his "cissoid" curve in this text and a method of constructing, geometrically, the cube root of any given length with its aid. As the construction of the cube root of 2 is the chief stumbling block in the solution of the duplication of the cube problem, the cissoid provides a solution to this classic challenge. Today we describe the cissoid as the plane curve with equation $y^2(2a - x) = x^3$, where a is a constant. The appearance of the cube power makes the construction of cube roots possible.

Some historians suggest Diocles may have used the terms *parabola, hyperbola,* and *ellipse* for the conic sections before Apollonius, the scholar usually credited with the invention of these names. Diocles' work on conics greatly influenced the development of the subject. The exact date of Diocles' death is not known.

Diophantine equation

Any equation, usually in several unknowns, that is studied and required to have only integer-valued solutions is called a Diophantine equation. For example, the JUG-FILLING PROBLEM requires us to find integer solutions to $3x + 5y = 1$, and the classification of PYTHAGOREAN TRIPLES seeks integer solutions to $x^2 + y^2 = z^2$. These are Diophantine problems. FERMAT'S LAST THEOREM addresses the nonexistence of integer solutions to the generalized equation $x^n + y^n = z^n$ for higher-valued exponents. Problems of this type are named after DIOPHANTUS OF ALEXANDRIA, author of the first known book devoted exclusively to NUMBER THEORY.

In 1900 DAVID HILBERT challenged the mathematical community to devise an ALGORITHM that would determine whether or not any given Diophantine equation has solutions. Seventy years later Yuri Matyasevic proved that no such algorithm can exist.

Diophantus of Alexandria

(ca. 200–284 C.E.) Greek *Number theory* Diophantus is remembered as the author *Arithmetica,* the first known text devoted exclusively to the study of NUMBER THEORY. Ten of the original 13 volumes survive today. In considering some 130 problems, Diophantus developed general methods for finding solutions to some surprisingly difficult integer problems, inspiring a field of study that has since become known as DIOPHANTINE EQUATIONS.

Essentially nothing is known about Diophantus's life, not even his place of birth nor the date at which he lived. Author Metrodorus (ca. 500 C.E.), in the *Greek Anthology,* briefly described the life of Diophantus through a puzzle:

> His boyhood lasted one-sixth of his life; his beard grew after one- twelfth more; he married after one-seventh more; and his son was born five years later. The son lived to half his father's age, and the father died four years after the son.

Setting L to be the length of Diophantus's life, we deduce then that the quantity:

$$\frac{L}{6} + \frac{L}{12} + \frac{L}{7} + 5 + \frac{L}{2} + 4$$

equals the total span of his life. Setting this equal to L and solving then yields $L = 84$. Of course the information provided here (that Diophantus married at age 26, lived to age 84, and had a son who survived to age 42) is likely fictitious. The puzzle, however, is fitting for the type of problem Diophantus liked to solve.

In his famous text *Arithmetica* (Arithmetic) Diophantus presents a series of specific numerical problems, with solutions provided, that cleverly lead the reader to an understanding of general methods and general solutions. Diophantus ignored any solution to a problem that was negative or involved an irrational square root. He generally permitted only positive rational solutions. Today, going further, mathematicians call any problem requiring only integer solutions a Diophantine equation.

Some of the problems Diophantus considered are surprisingly difficult. For instance, in Book IV of *Arithmetica* Diophantus asks readers to write the number 10 as a sum of three squares each greater than three. He provides the answer:

$$10 = \left(\frac{1321}{711}\right)^2 + \left(\frac{1285}{711}\right)^2 + \left(\frac{1288}{711}\right)^2$$

Although Diophantus did not use sophisticated algebraic notation, he was the first to use a symbol for an unknown quantity and to introduce a notation for powers of that unknown. He also used an abbreviation for the word *equals*. This represents the first step in history toward moving from verbal algebra to symbolic algebra.

Diophantus's text was profoundly influential and, centuries later, was deemed essential reading for European scholars of the Renaissance. Inspired by an exercise in the text, scholar PIERRE DE FERMAT (1601–65) scrawled the famous comment in the margin of his personal copy of *Arithmetica* that spurred three centuries of intense mathematical research in number theory. This comment became known as FERMAT'S LAST THEOREM.

direction cosines Each point P on the surface of a unit sphere determines a unique direction in three-dimensional space: if O is the center of the sphere, then the ray connecting O to P specifies a direction. Conversely, the direction of any given line in space corresponds to a point P on the unit sphere.

Setting O to be the origin of a CARTESIAN COORDINATE system, the "direction cosines" of any directed line in three-dimensional space are simply the coordinates of the point P on the unit sphere that corresponds to the direction of that line. For example, the direction cosine of the positive x-axis is $(1,0,0)$, and that of the negative z-axis is $(0,0,-1)$.

The use of the word *cosine* in the name of this concept comes from the observation that the direction of a line through O is completely specified by the three angles α, β, and γ it makes with each of positive the x-, y-, and z-axes, respectively. (These angles are assumed to lie between zero and $180°$. They are called the direction angles.) An exercise in geometry then shows that the corresponding point P on the unit sphere has coordinates $(\cos\alpha, \cos\beta, \cos\gamma)$.

The three direction cosines are not independent. Two applications of PYTHAGORAS'S THEOREM show that these numbers satisfy the relation: $\cos^2\alpha + \cos^2\beta + \cos^2\gamma = 1$. Thus any two direction cosines determine the third.

The direction cosines of an arbitrary line are often denoted (l,m,n). The "direction ratios" or "direction numbers" of a line are defined as any set of three numbers in the ratio $l : m : n$. The angle θ between two lines with direction cosines (l_1,m_1,n_1) and (l_2,m_2,n_2) is given by:

$$\cos\theta = l_1 l_2 + m_1 m_2 + n_1 n_2$$

This is simply the DOT PRODUCT of the two VECTORS that describe the directions of the lines.

directional derivative The graph of a function $z = f(x,y)$ is a surface sitting in three-dimensional space. The directional derivative of f at a point $P = (x,y)$ and in the direction given by a VECTOR $\mathbf{v} = < v_1, v_2 >$, denoted $D_{\mathbf{v}}f$, is simply the SLOPE of the surface above the point P in the direction of $\mathbf{v}$. It is assumed that $\mathbf{v}$ is a vector of length 1.

Specifically, if t is a variable, best thought of as "time," then the expression $P + t\mathbf{v}$ represents a straight-line path starting at P pointing in the direction of $\mathbf{v}$, and $f(P + t\mathbf{v})$ is the "slice" of the surface above this line. The directional derivative is then the DERIVATIVE of this quantity with respect to t:

$$D_{\mathbf{v}}f = \frac{d}{dt}f(P+t\mathbf{v})\Big|_{t=0} = \lim_{h\to 0}\frac{f\big((x+hv_1, y+hv_2)\big)}{h}$$

(We require $\mathbf{v}$ to be a vector of unit length so that the "speed" at which we traverse the path $P + t\mathbf{v}$ is 1 unit of length per unit time.)

If we take $\mathbf{v}$ to be the unit vector in the direction of the positive x-axis, $\mathbf{v} = (1,0)$, then $D_{\mathbf{v}}f = \lim_{h\to 0}\frac{f\big((x+h,y)\big)}{h} = \frac{\partial f}{\partial x}$, the PARTIAL DERIVATIVE of the function with respect to x. Similarly, the directional derivative in the direction of the positive y-axis is the partial derivative with respect to y. In general, the CHAIN RULE shows:

$$D_{\mathbf{v}}f = \frac{d}{dt}f(P+t\mathbf{v})\Big|_{t=0} = \frac{\partial f}{\partial x}\cdot\frac{d(x+tv_1)}{dt} + \frac{\partial f}{\partial t}\cdot\frac{d(y+tv_2)}{dt}$$

$$= \frac{\partial f}{\partial x}\cdot v_1 + \frac{\partial f}{\partial y}\cdot v_2$$

which can be rewritten as the DOT PRODUCT of two vectors:

$$D_\mathbf{v}f = \nabla f \cdot \mathbf{v}$$

where $\nabla f = \left\langle \dfrac{\partial f}{\partial x}, \dfrac{\partial f}{\partial y} \right\rangle$ is the GRADIENT of f. This provides the easiest method for computing the directional derivative of a function.

Note that $\nabla f \cdot \mathbf{v} = |\nabla f| \cdot |\mathbf{v}| \, cos(\theta)$, where θ is the angle between the two vectors. Since the cosine function has maximal value for $\theta = 0°$, this shows that the direction $\mathbf{v}$ of steepest slope for a graph at a point P occurs in the direction $\mathbf{v} = \nabla f$. This proves:

The vector ∇f points in the direction in which f increases most rapidly.

Similarly, the cosine function has minimal value for $\theta = 180°$, which shows that the steepest decline occurs in precisely the opposite direction:

The vector $-\nabla f$ points in the direction in which f decreases most rapidly.

These ideas extend to functions of more than just two variables.

direct proof

Most claims made in mathematics are statements of the form:

If the premise A is true, then the conclusion B is true.

A direct proof of such a statement attempts to establish the validity of the claim by assuming that the premise A is true and showing that the conclusion B follows from a series of logical inferences based on A and other previously established known facts. Typically, a direct proof has the form:

1. Assume A is true.
2. Show that A implies B.
3. Conclude that B is true.

The main part of the proof is the demonstration that A implies B.

As a simple example, we prove: if a natural number n is even, then n^2 is a multiple of 4. We will base its proof on the known fact that any even number is a multiple of two (as well as the standard algebraic manipulations).

Proof: Assume that n is even.
Then n can be written in the form $n = 2k$, for some number k.
Consequently, $n^2 = (2k)^2 = 4k^2$, and so is a multiple of four.
This completes the proof.

An INDIRECT PROOF or a PROOF BY CONTRADICTION attempts to establish that the conclusion B must be true by showing that it cannot be false.

See also DEDUCTIVE/INDUCTIVE REASONING; CONTRAPOSITIVE; LAWS OF THOUGHT; PROOF; QED; THEOREM.

Dirichlet, Peter Gustav Lejeune

(1805–1859) German *Analysis, Number theory* Born on February 13, 1805, near Liège, now in Belgium (although he considered himself German), scholar Lejeune Dirichlet is remembered for his significant contributions to the field of ANALYTIC NUMBER THEORY and to the study of FOURIER SERIES. In particular, he is noted for proving that any ARITHMETIC SEQUENCE a, $a+d$, $a+2d$, $a+3d$, … must contain an infinite number of primes, provided the starting number a and the difference d are RELATIVELY PRIME. (This shows, for instance, that there are infinitely many prime numbers that are 7 greater than a multiple of 13.) Dirichlet was the first to provide the modern definition of a FUNCTION we use today and, in the study of trigonometric series, was the first to provide conditions that ensure that a given Fourier series will converge. For this reason, despite the work of JEAN-BAPTISTE JOSEPH FOURIER (1768–1830), Dirichlet is often referred to as the founder of the theory of Fourier series.

Dirichlet graduated from the gymnasium (high school) in Bonn at the age of 16 and went to Paris to study mathematics. He never formally completed an academic program there and consequently never obtained a university degree. In 1825, at the age of 20, Dirichlet received instant fame as a worthy mathematician by publishing a proof that there can be no positive-integer solutions to the fifth-degree equation $x^5 + y^5 = z^5$. This is a special case of FERMAT'S LAST THEOREM, and Dirichlet's work on it represented the first significant step toward solving the general problem since the time of PIERRE DE FERMAT (1601–65), who had established that there are no solutions to the fourth-degree equation, and LEONHARD EULER (1707–83), who had proved that there are no solutions to the third-degree equation. Dirichlet was later able to extend his work to the degree-14 equation, but to no other cases.

In honor of his achievement, Dirichlet was awarded an honorary doctorate from the University of Cologne, and with an advanced degree in hand, Dirichlet then pursued an academic career. He was appointed professor of the University of Berlin in 1828, where he remained for 27 years. In 1855 Dirichlet succeeded CARL FRIEDRICH GAUSS (1777–1855) as chair of mathematics at the University of Göttingen.

Dirichlet developed innovative techniques using the notion of a LIMIT in the study of NUMBER THEORY that allowed him to make significant advances in the field. He presented his famous result on ARITHMETIC SEQUENCES to the Academy of Sciences on July 27, 1837, and published the work in the two-part paper "Recherches sur diverses applications de l'analyse infinitésimale à la théorie des nombres" (Inquiry on various applications of infinitesimal analysis to number theory) during the 2 years that followed. Dirichlet also found applications of this work to mechanics, to the solution of DIFFERENTIAL EQUATIONs, and to the study of Fourier series. He consistently published papers on both number theory and mathematical physics throughout his career. His most notable works include the 1863 book *Vorlesungen über Zahlentheorie* (Lectures on number theory), the 1846 article "Über die Stabilität des Gleichgewichts" (On the stability of the solar system), and the 1857 article "Untersuchungen über ein Problem der Hydrodynamik" (Investigation on a problem in hydrodynamics).

Dirichlet died on May 5, 1859, in Göttingen, Germany. Given the significance of his mathematical work, many mathematicians of today regard Dirichlet as the founder of analytic number theory.

discontinuity *See* CONTINUOUS FUNCTION.

discrete A set of numerical values in which there are no intermediate values is said to be discrete. For example, the set of INTEGERs is discrete, but the set of all REAL NUMBERS is not: between any two real numbers, no matter how close, there is another real number. Any finite set of values is considered discrete.

Since a COUNTABLE set of values can be put in one-to-one correspondence with the integers, a countable set is usually regarded as discrete. This can be confusing, however, since the countable set of RATIONAL NUMBERS, for instance, is discrete in this second sense, but not in the first: between any two rational numbers p and q lies another rational ($\frac{p+q}{2}$, for instance). One must rely on the context of the problem under study to determine whether or not the set of rational numbers should be regarded as discrete.

In STATISTICS and PROBABILITY theory, a set of DATA or set of EVENTs is called discrete if the underlying population is finite or countably infinite. The results of tossing a die, for instance, form a discrete set of events, since the die must land on one of six faces. In contrast, for example the range of heights of Australian women aged 36 is not discrete but continuous.

In GEOMETRY, an ISOMETRY with the property that each point is moved more than some fixed positive distance further away is called a discrete transformation. For example, a translation is discrete, but a rotation or reflection is not.

Discrete geometry is the study of a finite set of points, lines, circles, or other simple figures.

discriminant A QUADRATIC equation of the form $ax^2 + bx + c = 0$ has solutions given by the quadratic formula:

$$x = \frac{-b \pm \sqrt{b^2 - 4ac}}{2a}$$

The quantity under the square root sign, $b^2 - 4ac$, is called the discriminant of the equation. If the discriminant of a quadratic is positive, then the equation has two distinct real roots. For example, the equation $2x^2 - 5x + 2 = 0$ has discriminant 3 and the two real solutions $x = 2$ and $x = 1/2$. If the discriminant of a quadratic is zero, then the equation has just one real root. For instance, $x^2 - 6x + 9 = 0$, with discriminant zero, has only $x = 3$ as a root. (It is a DOUBLE ROOT.) If the discriminant is negative, then the quadratic has no real solutions. It does, however, have (distinct) complex solutions. For example,

$x^2 + x + 1 = 0$, with discriminant -3, has solutions $x = \dfrac{-1 + i\sqrt{3}}{2}$ and $x = \dfrac{-1 - i\sqrt{3}}{2}$.

More generally, the discriminant of any POLYNO-MIAL equation is defined to be the product of the differences squared of all the possible pairs of roots of the equation. For example, if a CUBIC EQUATION has three roots r_1, r_2, and r_3 (possibly repeated), then the discriminant of cubic is the product:

$$(r_1 - r_2)^2 (r_2 - r_3)^2 (r_3 - r_1)^2$$

It is possible to find a formula for the discriminant in terms of the coefficients appearing in the equation. For the case of a quadratic, it turns out to be precisely the quantity $b^2 - 4ac$ described above.

disjunction ("or" statement) A compound statement of the form "p or q" is known as a disjunction. For example, "I visited Sydney or Melbourne" is an example of a disjunction.

Disjunctions can be interpreted in one of two ways. If a disjunction "p or q" is read as

$$p \text{ or } q, \text{ but not both}$$

("I visited just one of the two cities"), then it is said to be "exclusive," and the disjunction is called an "exclusive or" (sometimes denoted XOR). Interpreted as

$$p \text{ or } q, \text{ or possibly both}$$

("I visited at least one of the cities"), then the disjunction is said to be "inclusive" and is called an "inclusive or." In FORMAL LOGIC (and in most of mathematics), disjunctions are always used in the inclusive sense. It is denoted in symbols by $p \vee q$ and has the following TRUTH TABLE:

p	q	$p \vee q$
T	T	T
T	F	T
F	T	T
F	F	F

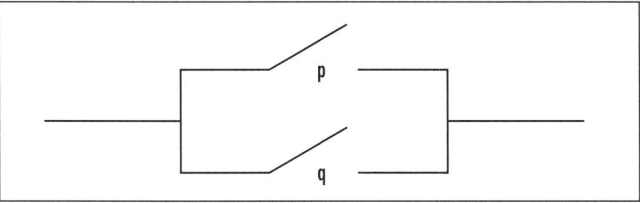

Disjunction circuit

A disjunction can be modeled via a parallel circuit. If T denotes the flow of current, then current moves through the circuit as a whole precisely when one, or both, switches p and q admit current flow.

See also CONJUNCTION.

displacement The distance traveled by a moving object is sometimes called its displacement. Physicists often use the symbol s to denote displacement. The rate of change of displacement is called VELOCITY.

See also DIFFERENTIAL CALCULUS.

distance formula The distance d between two given points $P_1 = (x_1, y_1)$ and $P_2 = (x_2, y_2)$ in the plane is the length of the line segment that connects P_1 to P_2. If one regards this line segment as the hypotenuse of a right triangle with one leg horizontal, that is, parallel to the x-axis, and one leg vertical, parallel to the y-axis, then PYTHAGORAS's THEOREM can be employed to find a formula for d. The length of the horizontal leg is the difference of the x-coordinates $x_2 - x_1$ or $x_1 - x_2$, whichever is positive, and the length of the vertical leg is the difference of the y-coordinates, $y_2 - y_1$ or $y_1 - y_2$. Thus, by Pythagoras's result, we have:

$$d = \sqrt{(x_2 - x_1)^2 + (y_2 - y_1)^2}$$

This is called the two-dimensional distance formula. For example, the distance between the points $(-3, 5)$ and $(2, 1)$ is $\sqrt{(2 - (-3))^2 + (1 - 5)^2} = \sqrt{5^2 + (-4)^2} = \sqrt{41}$. Notice that, as one would expect, the distance formula is symmetric in the sense that the distance between P_1 and P_2 is the same as the distance between P_2 and P_1. The set of all points (x, y) in the plane a fixed distance r from a given point $C = (a, b)$ form a CIRCLE with radius

r and center C. The distance formula gives the equation of such a circle as $r = \sqrt{(x-a)^2 + (y-b)^2}$, or $(x-a)^2 + (y-b)^2 = r^2$.

In three-dimensional space, the distance d between two points $P_1 = (x_1,y_1,z_1)$ and $P_2 = (x_2,y_2,z_2)$ is found via two applications of Pythagoras's theorem. For instance, the distance between the points $O = (0,0,0)$ and $A = (a,b,c)$ is found by first noting that $P = (a,b,0)$ is the point directly below A lying in the xy-plane, and that the triangle OPA is a right triangle. By the two-dimensional distance formula, the distance between O and P is $\sqrt{(a-0)^2 + (b-0)^2} = \sqrt{a^2 + b^2}$. The length of the vertical line connecting P to A is c, and the length of the hypotenuse OA is the distance d we seek. By Pythagoras's theorem we have: $d^2 = \left(\sqrt{a^2 + b^2}\right)^2 + c^2 = a^2 + b^2 + c^2$. Thus: $d = \sqrt{a^2 + b^2 + c^2}$.

A slight modification of this argument shows that the general three-dimensional distance formula for the distance between two points $P_1 = (x_1,y_1,z_1)$ and $P_2 = (x_2,y_2,z_2)$ is:

$$d = \sqrt{(x_2 - x_1)^2 + (y_2 - y_1)^2 + (z_2 - z_1)^2}$$

The set of all points (x,y,z) in space a fixed distance r from a given point $C = (a,b,c)$ form a SPHERE with radius r and center C. The distance formula gives the equation of such a sphere as $(x-a)^2 + (y-b)^2 + (z-c)^2 = r^2$.

The distance formula generalizes to points in n-dimensional space as the square root of the sum of the differences of the n coordinates squared. This works even for one-dimensional space: the distance between two points x_1 and x_2 on the number line is $d = \sqrt{(x_2 - x_1)^2}$. This is precisely the ABSOLUTE VALUE $|x_2 - x_1|$.

The LENGTH of a VECTOR $\mathbf{v} = <a,b,c>$ is given by the distance formula: If we place the vector at location $O = (0,0,0)$ so that its tip lies at $A = (a,b,c)$, then its length is $|\mathbf{v}| = \sqrt{a^2 + b^2 + c^2}$.

Distance of a Point from a Plane in Three-Dimensional Space

The distance of a point P from a plane is defined to be the distance between P and the point N in the plane closest to P. Suppose that the point P has coordinates $P = (x_0,y_0,z_0)$ and the VECTOR EQUATION OF A PLANE is $ax + by + cz + d = 0$ where $\mathbf{n} = <a,b,c>$ is the normal to the plane. Then N is the point (x_1,y_1,z_1) in the plane with vector $\overrightarrow{NP}$ 90° to the plane. This means that the vector $\overrightarrow{NP}$ is parallel to $\mathbf{n}$, and so $\overrightarrow{NP} = k\mathbf{n}$ for some constant k. This

gives the equation $<x_0 - x_1, y_0 - y_1, z_0 - z_1> = k <a,b,c>$, and so $x_1 = x_0 - ka, y_1 = y_0 - kb$, and $z_1 = z_0 - kc$. Since $N = (x_1,y_1,z_1)$ lies in the plane, this point also satisfies the equation of the plane. Algebraic manipulation then shows that $k = \dfrac{ax_0 + by_0 + cz_0 + d}{a^2 + b^2 + c^2}$. By the distance formula, the distance between P and N is:

$$\sqrt{(x_0 - x_1)^2 + (y_0 - y_1)^2 + (z_0 - z_1)^2} = \sqrt{k^2 a^2 + k^2 b^2 + k^2 c^2}$$
$$= |k| \sqrt{a^2 + b^2 + c^2}$$
$$= \frac{|ax_0 + by_0 + cz_0|}{\sqrt{a^2 + b^2 + c^2}}$$

This establishes:

The distance of a point $P = (x_0,y_0,z_0)$ from the plane $ax + by + cz + d = 0$ is given by the formula:

$$\frac{|ax_0 + by_0 + cz_0 + d|}{\sqrt{a^2 + b^2 + c^2}}$$

Distance of a Point from a Line in Two-Dimensional Space

The distance of a point P from a line is defined to be the distance between P and the point N in the line closest to P. The EQUATION OF A LINE is a formula of the form $ax + by + c = 0$. An argument analogous to the one presented above establishes:

The distance of a point $P = (x_0,y_0)$ from the line $ax + by + c = 0$ is given by the formula:

$$\frac{|ax_0 + by_0 + c|}{\sqrt{a^2 + b^2}}$$

See also COMPLEX NUMBERS.

distribution Any table or diagram illustrating the frequency (number) of measurements or counts from an experiment or study that fall within certain preset categories is called a distribution. (*See* STATISTICS: DESCRIPTIVE.) For example, the heights of 1,000 8-year-old children participating in a medical study can be recorded via a histogram. The categories considered are conveniently chosen intervals of height ranges, such as 36.1–40.0 in., 40.1–44.0 in., and so on, for example.

If the DATA gathered is numerical and can adopt a continuous array of values, including fractional values (for example, height can adopt fractional values—48 3/4 in., or 52.837 in. are possible measurements), then one can choose narrower and narrower interval ranges for categories. In the LIMIT, the histogram becomes then the graph of a smooth curve representing the distribution of measurements over a continuous spectrum of values. In some sense, the total area under the curve represents the total number of measurements observed, and the area above an interval [a,b] represents the number of measurements that have value greater than a and less than b. To make this more precise, it is appropriate to scale the distribution so that the total area under the curve is one (that is, one draws histograms with vertical bars of heights representing the *percentage* of measurements recorded within that category—"relative frequencies"—with the total area under such a histogram representing 100 percent). The ideal curve obtained in the limit is called a "probability density function." The area under the curve above an interval [a,b] here represents the PROBABILITY that a measurement taken at random falls within the range [a,b].

A numerical quantity that can adopt a continuous array of values (such as height, weight, or temperature) is usually called a continuous random variable. One ascertains the distribution (probability density function) of a random variable by conducting experiments or studies—for example by recording the heights of 1,000 8-year-old children—or from mathematical reasoning, making use of the CENTRAL-LIMIT THEOREM, the NORMAL DISTRIBUTION, or perhaps the BINOMIAL DISTRIBUTION, for example. Often the distribution of a random, variable is unknown, and "hypothesis testing"

is used to check the validity of an assumption that a SAMPLE follows a particular distribution. This is part of inferential statistics.

If a random variable has probability density function given by a formula $f(x)$, then the area under the curve to the left of a value x gives a new function $F(x)$ called the cumulative distribution function. The quantity $F(x)$ represents the probability that a measurement taken at random has value less than or equal to x.

See also STATISTICS: INFERENTIAL.

distributive property   Given a mathematical system with two operations, such as addition and multiplication, or union and intersection, we say that one operation distributes over the second if applying the first operation to a set of elements combined via the second produces the same result as applying the first operation to the individual members of the combination, and then combining them via the second. For example, in ordinary arithmetic, multiplication distributes over addition. We have, for instance:

$$3 \times (2 + 5 + 4) = 3 \times 2 + 3 \times 5 + 3 \times 4$$

that is, tripling a sum of numbers produces the same result as tripling each individual number and then summing. (In arithmetic, the distributive property corresponds to the operation of EXPANDING BRACKETS.) Notice that addition is not distributive over multiplication, however. For instance: $4 + (6 \times 7) \neq (4 + 6) \times (4 + 7)$. (The first quantity equals $4 + 42 = 46$, whereas the second is $10 \times 11 = 110$.)

In arithmetic, the distributive property is usually expressed as a multiplication applied to the sum of just two terms:

$$a \times (b + c) = a \times b + a \times c$$

That it applies to a sum of three or more terms follows from applying this basic law more than once. For instance:

$$a \times (b + c + d) = a \times ((b + c) + d)$$
$$= a \times (b + c) + a \times d$$
$$= a \times b + a \times c + a \times d$$

Multiplication also distributes over addition "from the right." We have:

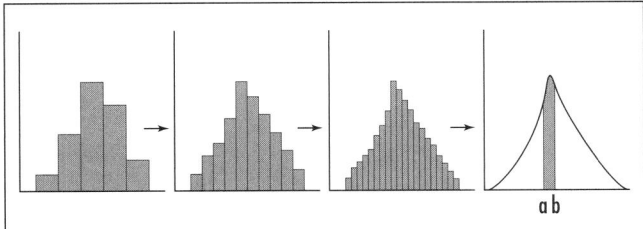

A distribution as a limit

$$(a + b) \times c = c \times (a + b)$$
$$= c \times a + c \times b$$
$$= a \times c + b \times c$$

In SET THEORY, "intersection" distributes over "union" both from the left and from the right:

$$A \cap (B \cup C) = (A \cap B) \cup (A \cap C)$$
$$(A \cup B) \cap C = (A \cap C) \cup (B \cap C)$$

As a mnemonic device, it is helpful to think of the phrase "distributes over" as synonymous with "sprinkles over." We have: multiplication "sprinkles over" additions and intersection "sprinkles over" union. In any RING, the distributive property is the single axiom that combines the two defining operations.

See also ASSOCIATIVE; COMMUTATIVE PROPERTY.

div   A VECTOR FIELD assigns to every point (x,y,z) in space a vector $\mathbf{F} = <f_1(x,y,z), f_2(x,y,z), f_3(x,y,z)>$. The divergence of $\mathbf{F}$, denoted *div* $\mathbf{F}$, is the quantity:

$$div\ F = \frac{\partial f_1}{\partial x} + \frac{\partial f_2}{\partial y} + \frac{\partial f_3}{\partial z}$$

given as a sum of PARTIAL DERIVATIVES. Physicists have shown that this quantity represents the amount of flux leaving an element of volume in space. For example, if $\mathbf{F}$ represents the velocity field of a turbulent fluid, and ρ is the density of the fluid, then $\rho div\mathbf{F}$, calculated at a point P, is the rate at which mass is lost from an (infinitely small) box drawn around P. (At any instant, fluid is flowing into and out of this box.)

The divergence operator is often written as though it is a DOT PRODUCT of two vectors:

$$div\mathbf{F} = \nabla \cdot \mathbf{F}$$

where $\nabla = \left\langle \frac{\partial}{\partial x}, \frac{\partial}{\partial y}, \frac{\partial}{\partial z} \right\rangle$ is the del operator. The CROSS PRODUCT of ∇ with $\mathbf{F}$ is called the curl of $\mathbf{F}$:

$$curl\ \mathbf{F} = \nabla \times \mathbf{F} = \left(\frac{\partial f_3}{\partial y} - \frac{\partial f_2}{\partial z} \right)\mathbf{i} + \left(\frac{\partial f_1}{\partial z} - \frac{\partial f_3}{\partial x} \right)\mathbf{j} + \left(\frac{\partial f_2}{\partial x} - \frac{\partial f_1}{\partial y} \right)\mathbf{k}$$

If $\mathbf{F}$ is the velocity field of a turbulent liquid, then physicists have shown that 1/2 *curl* $\mathbf{F}$, calculated at a point P, is the angular velocity of an element of fluid located at P, that is, it is a measure of the amount of turning it undergoes.

See also GRADIENT.

divergent   This term simply means "does not converge." For example, an infinite SEQUENCE is said to diverge if it has no LIMIT, and an infinite SERIES diverges if the sequence of partial sums diverges.

A divergent sequence is said to be "properly divergent" if it tends to infinity. For example, the sequence 1,2,3, ... is properly divergent (but the sequences 1, –1,1, –1,1, ... and 1, –2,3, –4,5, ... are not). One also describes a series as properly divergent if the corresponding sequence of partial sums has this property. For example, the series 1 + 2 + 4 + 8 + 16 +... is properly divergent.

An INFINITE PRODUCT is divergent if it has value zero or does not converge. An IMPROPER INTEGRAL diverges if the limit defining it does not exist.

See also CONVERGENT SEQUENCE.

divisibility rules   A number is said to be divisible by n if, working solely within the integers, the number leaves a remainder of zero when divided by n. For example, 37 leaves a remainder of 1 when divided by 3, and so is not divisible by 3. On the other hand, 39, leaving a remainder of zero, is divisible by 3.

There are a number of rules to quickly test the divisibility of numbers by small integers. We present divisibility rules for the first 12 integers.

Divisibility by 1
All numbers are divisible by 1.

Divisibility by 2
As all multiples of 10 are divisible by 2, it suffices to check whether or not the final digit of a number is divisible by 2. For example, $576 = 57 \times 10 + 6$. That 6 is a multiple of 2 ensures that 576 is too. We have the rule:

A number is divisible by 2 only if its final digit is 0, 2, 4, 6 or 8.

Divisibility by 3
That 10, 100, 1000,... all leave a remainder of 1 when divided by 3 allows us to quickly determine the remain-

der of any number divided by 3. For example, 3,212 equals $3 \times 1000 + 2 \times 100 + 1 \times 10 + 2 \times 1$ and so leaves a remainder of $3 \times 1 + 2 \times 1 + 1 \times 1 + 2 \times 1 = 8$ when divided by 3. This is the sum of its digits. Of course, a remainder of 8 is equivalent to a remainder of 2. We have:

> The remainder of any number divided by 3 is the sum of its digits. Thus a number is divisible by 3 only if the sum of its digits is a multiple of 3.

One can make repeated use of this rule to check for divisibility. For example, 55,837 leaves a remainder of $5 + 5 + 8 + 3 + 7 = 28$ when divided by 3. This corresponds to a remainder of $2 + 8 = 10$, which, in turn, is a remainder of $1 + 0 = 1$.

This rule shows that scrambling the digits of any multiple of 3 produces a new number that is still a multiple of 3.

Divisibility by 4

As any multiple of 100 is divisible by 4, it suffices to check whether or not the final two digits of a number represent a two-digit multiple of 4. For example, 18,736 equals $187 \times 100 + 36$. As 36 is a multiple of 4 (it can be divided by 2 twice), we are sure then that 18,736 is a multiple of 4. We have:

> A number is divisible by 4 if its final two digits represent a two-digit number that can be divided by 2 twice.

Divisibility by 5

Any number N can be written in the form $N = 10a + b$ where b is the final digit in N. (For example, $739 = 73 \times 10 + 9$.) As 10 is divisible by 5, we need only check whether or not the final digit b is divisible by 5. This gives:

> A number is divisible by 5 only if its final digit is 0 or 5.

Divisibility by 6

For a number to be divisible by 6 it must both be even and a multiple of 3. This gives:

> A number is divisible by 6 only if it is an even number whose digits sum to a multiple of 3.

Divisibility by 7

Every number N is of the form $N = 10a + b$, that is, a multiple of 10 plus a single digit. As $7a$ and $7b$ are clearly divisible by 7, we have that $10a + b$ leaves the same remainder as $10a + b - 7a - 7b = 3(a - 2b)$ does when divided by 7. Thus it suffices to check whether the quantity $3(a - 2b)$ is a multiple of 7. This can only occur if $a - 2b$ is a multiple of 7. Noting that a is the original number N with its final digit removed and b is the final digit of N, we have the rule:

> To test whether or not a number is divisible by 7, remove the last digit and subtract twice that digit from the number remaining. Then the original number is divisible by 7 only if the result of this operation is divisible by 7.

For example, to test whether or not 68,978 is divisible by 7, remove the 8 and subtract twice this, 16, from 6,897, the number remaining. This gives $6,897 - 16 = 6,881$. We can test whether or not 6,881 is a multiple of 7 the same way: $6,881 \rightarrow 688 - 2 = 686$, and once more: $686 \rightarrow 68 - 12 = 56$. That the final result, 56, is divisible by 7 assures us that 68,978 is a multiple of 7.

Divisibility by 8

As any multiple of 1,000 is divisible by 8, it suffices to check whether or not the final three digits of a number represent a three-digit multiple of 8. For example, 648,728 equals $648 \times 1000 + 728$. As 728 can be divided by 2 three times, and hence is a multiple of 8, we have that 648,728 is divisible by 8.

> A number is divisible by 8 if its final three digits represent a three-digit number that can be divided by 2 three times.

Divisibility by 9

Given that 10, 100, 1000,... all leave a remainder of one when divided by 9, the divisibility rule for 9 is identical to that of 3.

> The remainder of any number divided by 9 is the sum of its digits. Thus a number is divisible by 9 only if the sum of its digits is a multiple of 9.

Again, one may make repeated use of this rule. For example, 76,937 leaves a remainder of $7 + 6 + 9 + 3 + 7 = 32$ when divided by 9, which, in turn, corresponds to a remainder of $3 + 2 = 5$. This rule is often used to check arithmetical work via the method of CASTING OUT NINES.

Divisibility by 10

Any number N can be written in the form $N = 10a + b$, where b is the final digit of N. Thus a number is divisible by 10 only if its final digit is a multiple of 10. We have:

> A number is divisible by 10 only if its final digit is a zero.

Divisibility by 11

The numbers 100, 1000,... alternately leave remainders of 1 and –1 when divided by 11. (For example, 100 is 1 more than a multiple of 11, but 1,000 is 1 less.) Thus the remainder of a number when divided by 11 is obtained as the alternate sum of its digits. For example, 69,782, which equals $6 \times 10,000 + 9 \times 1,000 + 7 \times 100 + 8 \times 10 + 2 \times 1$, leaves a remainder $6 \times 1 + 9 \times (-1) + 7 \times 1 + 8 \times (-1) + 2 \times 1 = 6 - 9 + 7 - 8 + 2 = -2$ when divided by 11. (This is equivalent to a remainder of 9.) We have:

> The remainder of any number when divided by 11 is the alternate sum of its digits. Thus a number is divisible by 11 only if the alternate sum of its digits is a multiple of 11.

Divisibility by 12

A number is divisible by 12 only if it is divisible by both 3 and 4. Thus we have:

> A number is divisible by 12 only if its final two digits represent a two-digit multiple of 4, and the sum of all the digits of the number is a multiple of 3.

The divisibility rule for 7 can be extended to other numbers as well. For example, $N = 10a + b$ is divisible by 17 only if $10a + b - 51b = 10(a - 5b)$ is. This, in turn, shows that N is divisible by 17 precisely when quantity $a - 5b$, obtained by deleting and subtracting 5 times the final digit, is divisible by 17. Notice here that 51 is the first multiple of 17 that is 1 more than a multiple of 10.

In the same way we can use that fact that 111 is the first multiple of 37 that is 1 more than a multiple of 10 to obtain a similar divisibility rule for 37, for example. Divisibility rules for all PRIME numbers, except 2 and 5, can be created this way.

division The process of finding the QUOTIENT of two numbers is called division. In elementary arithmetic, the process of division can be viewed as repeated SUBTRACTION. For instance, 60 divided by 12 equals 5 because 12 can be subtracted from this number five times before reaching zero: $60 - 12 - 12 - 12 - 12 - 12 = 0$. We write: $60 \div 12 = 5$. Division can also be described as the process of finding how many subsets or magnitudes are contained within a set or given quantity. For instance, $5 \div 1/2 = 10$ because 10 lengths of one-half are contained in a length of 5 units.

If a number a is divided by a number b to produce a quotient q, $a \div b = q$, then a is called the dividend and b the divisor. The quotient can also be expressed as a FRACTION, a/b, or a RATIO, $a:b$. In general, the quotient q of two numbers a and b satisfies the equation $q \times b = a$. Thus division may also be thought of as the inverse operation to MULTIPLICATION. Thus, since $5 \times 12 = 60$, for instance, 5 is indeed the quotient of 60 and 12. This reasoning also shows that, since $0 \times b = 0$ for any nonzero number b, we have $0 \div b = 0$. Unfortunately, one cannot give meaning to the quantity $0 \div 0$. (Given that $53 \times 0 = 0$, we may be forced to conclude that $0 \div 0 = 53$. At the same time, since $117 \times 0 = 0$, we also have that $0 \div 0 = 117$. We have inconsistency.) It is also not possible to give meaning to the term $a \div 0$ for any nonzero value a. (If $a \div 0 = q$, then $q \times 0 = a$, yielding a CONTRADICTION.)

The LONG DIVISION algorithm provides a means to divide large integers. The process of division can be extended to NEGATIVE NUMBERS, FRACTIONs, REAL NUMBERS, and COMPLEX NUMBERS. In all settings, the number 1 acts as an identity element—provided it operates as a divisor: $a \div 1 = a$ for all numbers a.

The symbol $\div$ is called the "obelus" and first appeared in print in Johann Heinrich Rahn's 1659 text *Teutsche algebra*.

See also DIVISIBILITY RULES; DIVISOR; DIVISOR OF ZERO; EUCLIDEAN ALGORITHM; FACTOR; FACTORIZATION; FACTOR THEOREM; RATIONAL FUNCTION; REMAINDER THEOREM.

divisor Another name for FACTOR.

divisor of zero Two quantities, neither of which are zero, yet multiply together to give zero as their product are called divisors of zero. In ordinary arithmetic, divisors of zero never arise: if $a \times b = 0$, then at least one of a or b must be zero. In MODULAR ARITHMETIC, however, divisors of zero can exist. For example, 3×2 equals zero in arithmetic modulo 6. Two nonzero matrices may multiply to give the zero MATRIX, and the product of two nonzero functions could be the zero function.

The presence of zero divisors in a mathematical system often complicates the arithmetic one can perform within that system. For example, for arithmetic modulo 10, 2×4 equals 2×9, but dividing by 2, a divisor of zero in this system, leads to the erroneous conclusion that 4 and 9 are equal in this system. In general, one can never perform division when a divisor of zero is involved.

dot product (inner product, scalar product) Denoted $\mathbf{a} \cdot \mathbf{b}$, the dot product of two VECTORs $\mathbf{a}$ and $\mathbf{b}$ is the sum of the products of respective components of the vectors. Precisely, if $\mathbf{a} = <a_1, a_2, ..., a_n>$ and $\mathbf{b} = <b_1, b_2, ..., b_n>$, then $\mathbf{a} \cdot \mathbf{b}$ is the number:

$$\mathbf{a} \cdot \mathbf{b} = a_1 b_1 + a_2 b_2 + ... + a_n b_n$$

The result of the dot product operation is always a real number (scalar). For example, the dot product of the two vectors $\mathbf{a} = <1, 4>$ and $\mathbf{b} = <2, 3>$ is $\mathbf{a} \cdot \mathbf{b} = 1 \cdot 2 + 4 \cdot 3 = 14$.

The dot product arises in many situations. For example, a local delicatessen receives the following lunch order form:

How Many?			Cost
2	Ham	@$2.80	———
	Turkey	@$3.15	———
1	Egg Salad	@$1.95	———
3	Tuna	@$2.50	———
	Roast Beef	@$3.60	———
		Total Cost:	———

One could interpret the order as a five-dimensional vector <2,0,1,3,0> to be matched with a five-dimensional "cost vector" <2.80,3.15,1.95,2.50,3.60>. The total

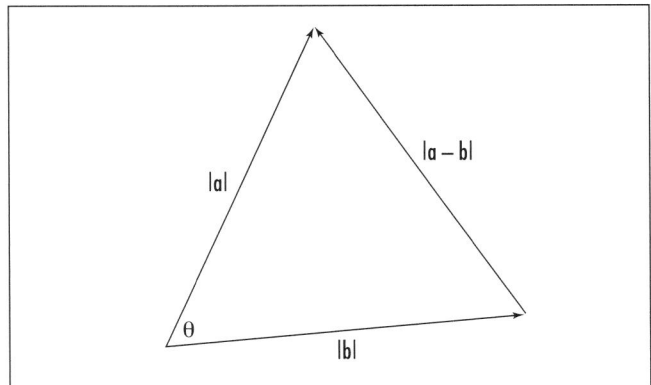

Triangle of vectors

cost of the order is then the dot product of these two vectors:

$$2 \times 2.80 + 0 \times 3.15 + 1 \times 1.95 + 3 \times 2.50 + 0 \times 3.60 = \$15.05$$

Geometrically, the dot product gives a means of computing the angle between two vectors. For example, two two-dimensional vectors $\mathbf{a} = <a_1, a_2>$ and $\mathbf{b} = <b_1, b_2>$, with angle θ between them, form a triangle in the plane with side-lengths given by the DISTANCE FORMULA: $|\mathbf{a}| = \sqrt{a_1^2 + a_2^2}$, $|\mathbf{b}| = \sqrt{b_1^2 + b_2^2}$, and $|\mathbf{a} - \mathbf{b}| = \sqrt{(a_1 - b_1)^2 + (a_2 - b_2)^2}$.

By the LAW OF COSINES we have:

$$|\mathbf{a} - \mathbf{b}|^2 = |\mathbf{a}|^2 + |\mathbf{b}|^2 - 2|\mathbf{a}||\mathbf{b}| \cos(\theta)$$

from which it follows that:

$$\mathbf{a} \cdot \mathbf{b} = a_1 b_1 + a_2 b_2 = |\mathbf{a}||\mathbf{b}| \cos(\theta)$$

Thus the angle between two vectors $\mathbf{a}$ and $\mathbf{b}$ can be computed via the formula:

$$\cos(\theta) = \frac{\mathbf{a} \cdot \mathbf{b}}{|\mathbf{a}||\mathbf{b}|}$$

This formula also holds true for three- and higher-dimensional vectors. For example, consider the unit vector $\mathbf{i} = <1,0,0>$ in three-dimensional space pointing in the direction of the x-axis, and $\mathbf{j} = <0,1,0>$ the corresponding vector pointing in the direction of the y-axis. As the angle between these two vectors is 90°, their dot product **i** · **j** will be zero (**i** · **j** = 1.0 + 0.1 + 0.0 = 0). In general:

Two vectors **a** and **b** are at right angles if, and only if, their dot product **a** · **b** is zero.

The dot product has the following commutative and distributive properties:

$$\mathbf{a} \cdot \mathbf{b} = \mathbf{b} \cdot \mathbf{a}$$
$$\mathbf{a} \cdot (\mathbf{b} + \mathbf{c}) = \mathbf{a} \cdot \mathbf{b} + \mathbf{a} \cdot \mathbf{c}$$

See also CROSS PRODUCT; NORMAL TO A PLANE; ORTHOGONAL; TRIPLE VECTOR PRODUCT; VECTOR EQUATION OF A PLANE.

double integral   The volume under a graph $z = f(x,y)$ of two variables (which is drawn as a surface sitting in three-dimensional space) above a region R in the xy-plane is computed via a double integral, denoted:

$$\iint_R f(x,y)\,dA$$

One approximates this volume by subdividing the region R into small rectangular pieces, drawing a rectangular cuboid above each rectangle with height reaching the surface, and summing the volumes of each of these cuboids. As one takes finer and finer approximations, this process produces better and better approximations to the true volume under the graph. The limit of this process is the double integral:

$$\iint_R f(x,y)\,dA = \lim \sum_{k=1}^{n} f(x_k,y_k)\,dA_k$$

where dA_k denotes the area of the kth rectangular region used to approximate R.

GOTTFRIED WILHELM LEIBNIZ (1646–1716) showed that if the region R is itself a rectangle, say, given by $a \le x \le b$ and $c \le y \le d$, then the double integral can be computed as either of the two iterated integrals:

$$\iint_R f(x,y)\,dA = \int_a^b\left(\int_c^d f(x,y)\,dy\right)dx = \int_c^d\left(\int_a^b f(x,y)\,dy\right)dx$$

(In an iterated integral, one integrates one variable at a time, regarding the second variable as a constant.) This

result holds true for other shaped regions R as well, as long as they are not too complicated.

For example, the volume under the graph $z = xy$ above the rectangle $R = [1,2] \times [2,3]$ is:

$$\iint_R xy\,dA = \int_1^2\int_2^3 xy\,dy\,dx = \int_1^2\left(\frac{1}{2}xy^2\right)_{y=2}^{y=3}dx$$
$$= \int_1^2 \frac{9}{2}x - 2x\,dx = \int_1^2 \frac{5}{4}x\,dx = \frac{15}{8}$$

Notice that the integration is performed from the inside out.

A triple integral $\iiint_V f(x,y,z)\,dV$ of a function of three variables $f(x,y,z)$, computed over a volume V in space, can often be computed as a triple iterated integral, integrating each variable in turn. Again, the order of the integration, typically, does not matter.

See also GEORGE GREEN.

double point   A location on a curve where the curve either crosses itself, or is tangential to itself, is called a double point. In the first case, the point of intersection is called a node, and the curve has two distinct tangents at that point. In the second case, the point of contact is called a tacnode or an osculation. The two tangents to the curve coincide at this point.

See also ISOLATED POINT; SINGULAR POINT; TANGENT.

double root   *See* ROOT.

dummy variable   A variable appearing in a mathematical expression is a dummy variable if it is assigned no specific meaning and if the letter being used for it could equally well be replaced by another letter. An index of SUMMATION, for instance, is a dummy variable: the sum $\sum_{k=1}^{4} k^3$ of denoting $1^3 + 2^3 + 3^3 + 4^3$, for example, could equally well be represented as $\sum_{r=1}^{4} r^3$ or $\sum_{n=1}^{4} n^3$, say. The variable used for the integrand of a DEFINITE INTEGRAL is a dummy variable. The two expressions $\int_0^1 x^2\,dx$ and $\int_0^1 t^2\,dt$, for instance, represent the same definite integral, a number, and so x and t are dummy variables. (The variable x, however, is *not* a dummy variable in the INDEFINITE INTEGRAL $\int x^2\, dx$. This expression is a function of the specific variable x.)

duplicating the cube (Delian altar problem, doubling the cube) One of the problems of antiquity (like SQUARING THE CIRCLE and TRISECTING AN ANGLE) of considerable interest to the classical Greek scholars is the task of constructing a cube whose volume is twice that of a given cube. Legend has it that this problem, known as duplicating the cube, arose during the Greek plague of 428 B.C.E. It is said that the oracle of Delos instructed the people of Athens to double the size of the cubic altar to Apollo as an attempt to appease the god. They were unable to accomplish this feat.

APOLLONIUS OF PERGA (ca. 260–190 B.C.E.) solved the problem with the use of CONIC SECTIONS, but scholars later decided to add the restriction that only the primitive tools of a straightedge (that is, a ruler with no markings) and a compass be used in its solution. The difficulty of the problem increased significantly.

If we assume that the side-length of the original cube is a units long, then one is required to construct a new length b so that $b^3 = 2a^3$. Consequently, $b = \sqrt[3]{2a}$, and so the problem essentially reduces to the challenge of constructing a length $\sqrt[3]{2}$ units long using only a straightedge and compass.

The theory of CONSTRUCTIBLE numbers shows that, in this setting, any quantity of rational length can be constructed, and that if two lengths l_1 and l_2 can be produced, then so too can their sum, difference, product and quotient, along with the square root of each quantity. It seems unlikely that a length of $\sqrt[3]{2}$, being neither rational, nor the square root of a rational number, could be produced. Indeed, in 1837, French mathematician Pierre Laurent Wantzel (1814–48) proved that the number $\sqrt[3]{2}$ is not constructible and, consequently, that the problem of duplicating the cube is unsolvable. (To see that $\sqrt[3]{2}$ is not rational, assume to the contrary that it can be written as a ratio of two integers: $\sqrt[3]{2} = \dfrac{p}{q}$.

Then $2q^3 = p^3$. If the number p has m factors of 2, then the quantity p^3 has $3m$ factors of 2. Consequently, so too must $2q^3$. But this is impossible, as the number of factors of 2 in $2q^3$ must be 1 more than a multiple of 3. This absurdity shows that $\sqrt[3]{2}$ cannot be a ratio of two integers. A similar argument shows that $\sqrt[3]{2}$ does not equal the square root of a rational quantity either.)

Dürer, Albrecht (1471–1528) German *Geometry* Born on May 21, 1471, in Nürnberg, Germany, artist Albrecht Dürer is remembered in mathematics for his significant accomplishments in the development of descriptive GEOMETRY and its applications to the theory of art. In four famous texts, Dürer explained the theory of proportions and described ruler-and-compass techniques for the construction of regular polygons. He explored the art of placing figures in a manner that is pleasing to the eye, thereby beginning a developing theory of PERSPECTIVE, and began a study of the shadows cast by three-dimensional objects. Dürer is noted as the first scholar to publish a mathematics book in German, and also as the first Western scholar to give an example of a MAGIC SQUARE.

Dürer studied painting and woodcut design as a young man. He apprenticed with the leading producer of altarpieces of his time, Michael Wolgemut, until the age of 20 and learned to appreciate the role mathematics could play in the design of artistic works. After reading the works of EUCLID (ca. 300–260 B.C.E.), as well as a number of famous texts on the theory of architecture, Dürer traveled to Italy, the site of the Renaissance revival of mathematics, to study the mathematics of shape, motion, and perspective. Around 1508 Dürer began collating and processing all the material he had studied with the aim of producing one definitive text on the mathematics of the visual arts. This work was never completed, but he did later publish his four volumes on the theory of proportions *Underweysung der Messung mit Zirckel und Richtscheyt in Linien, Ebnen, und gantzen Corporen* (Treatise on mensuration with the compass and ruler in lines, planes, and whole bodies) in 1525.

Dürer is noted for his inclusion of the following array of numbers in the background of his 1514 engraving *Melancholia*:

16	3	2	13
5	10	11	8
9	6	7	12
4	15	14	1

Each row and column, as well as the two main diagonals, sum to 34, and so this array represents a fourth-order magic square. It is the first example of a magic square ever recorded in Western Europe. (It has the added feature of including the year the engraving was completed in the two middle cells of the bottom row.)

Dürer died in Nürnberg, Germany, on April 6, 1528. His theory of proportions allowed artists who succeeded him to easily perform the transformations on figures needed to translate them across a canvas and maintain the correct sense of perspective. (Prior to Dürer, artists accomplished this feat purely by intuition or by trial and error.) In doing so, Dürer had provided a well-thought-out theory of geometric perspective that was also valued by mathematicians.

dyadic Any quantity related to the concept of base 2 is sometimes referred to as dyadic. For example, the dyadic rationals are those fractions whose denominators are powers of 2. Thus $\frac{3}{4} = \frac{3}{2^2}$ and $\frac{173}{1024} = \frac{173}{2^{10}}$, for instance, are dyadic rationals, but $\frac{1}{3}$ is not.

Folding a strip of paper 1 ft long in half produces a crease at the position of the dyadic rational 1/2. If one continues to fold the left or right end of the strip to previously constructed crease marks, then crease marks appear at all the dyadic rationals (and only the dyadic rationals).

This paper-folding activity is intimately connected to the construction of BINARY NUMBERS. For instance, to create a crease mark along a strip of paper at the position of the dyadic rational 13/16, write the numerator 13 in binary:

$$13 = 1101_2$$

(This is equivalent to writing the fraction 13/16 as a binary "decimal:" 13/16 = .1101 in base 2.) Now read the binary expansion backwards, interpreting the digit 1 as the instruction "lift right and fold" and the digit 0 as "lift left and fold." In this example we have:

1 : Lift the right end of the strip and fold to produce a crease at position 1/2.
0 : Lift the left end and fold to the previous crease. This produces a new crease at position 1/4.

1 : Lift the right end and fold to the previous crease. This produces a new crease at position 5/8.
1 : Lift the right end and fold to the previous crease. This produces a new crease at position 13/16, as desired.

In general:

> The binary representation of the numerator of any dyadic rational represents instructions for the construction of that dyadic rational along a strip of paper.

In some limiting sense, this procedure also works for fractions that are not dyadic. For example, the number 1/3 written as decimal in base 2 is .010101... If one were to read this as a set of instructions to "fold right and fold left indefinitely," then the sequence of creases produced do indeed converge to the position 1/3.

dynamical system Any process in which each successive state is a function of the preceding state is called a dynamical system. For instance, the feedback from a microphone as part of a public announcement system is a dynamical system: the amplifier transmits minute erroneous sounds, which the microphone hears and amplifies, which it then hears and amplifies, and so on.

In mathematics, if f is a mapping from a space X to itself, then the ITERATION of f defines a dynamical system—the successive states of the system are the iterates of f arising from a given starting point x:

$$x, f(x), f(f(x)), f(f(f(x))), \ldots$$

For example, the mapping $f(x) = \frac{1}{2}x$ of the real number line to itself, starting with $x = 1$, gives the iterates $1, \frac{1}{2}, \frac{1}{4}, \frac{1}{8}, \ldots$. If X is a circle in the plane and f is the function that rotates that circle 10° clockwise, then the iterates of any point on the circle constitute 36 evenly spaced points on that circle.

The "orbit" of a point x is the sequence of iterates it produces. If the system reaches an equilibrium, that is, tends toward a stable state, or if it cycles between a number of states, then the equilibrium point (or sets of equilibrium points) are called "attractors" of the system. For example, the number zero is an attractor for the system given by $f(x) = \frac{1}{2}x$: no matter the starting value, all iterates converge to the value zero.

The iterates of very simple functions f can exhibit extremely surprising behavior. Take, for instance, the iterates of the function $f(x) = 1 - cx^2$, with initial value $x = 0$. (Here c is a constant. Each value of c determines its own dynamical system.)

Set $c = 0.1$. The first five iterates of the system $x_{n+1} = 1 - 0.1x_n^2$ are:

$$0, 1.000, 0.900, 0.919, 0.916, 0.916$$

The system seems to stabilize to the value 0.916. One checks that the same type of behavior occurs if we repeat this exercise for c set to any value between 0 and 0.75. There is a marked change in behavior for $c = 0.75$, however—the system no longer converges to a single value but rather oscillates between two values: 0.60 and 0.72. We say that the value $c = 0.75$ is a bifurcation point and that the system has undergone "period doubling."

At the value $c = 1.25$, the system bifurcates again to yield systems that oscillate between four separate values. For higher values of c, the system continues to bifurcate, until finally a so-called CHAOS is reached, where the results jump around in a seemingly haphazard manner. This phenomenon is typical of many dynamical systems: the behavior they exhibit is highly dependent on the value of some parameter c. (Such dynamical systems are said to be "sensitive" to the parameter set.)

Researchers have shown that many natural processes that appear chaotic, such as the turbulent flow of gases and the rapid eye movements of humans, can be successfully modeled as dynamical systems, usually with very simple underlying functions defining them. Meteorologists model weather as a dynamical system, which helps them make forecasts. However, extreme sensitivity to parameters can easily lead to erroneous predictions: one small change in the value of just one parameter may produce very different outcomes. The so-called butterfly effect, for instance, claims that the minute changes in air pressures caused by a butterfly flapping its wings might be all that is needed to tip a meteorological dynamical system into chaos.

Iteration of functions with COMPLEX NUMBERS leads to a study of FRACTALs.

_e_ (Euler's number) Swiss mathematician LEONHARD EULER (1707–83) introduced a number, today denoted _e,_ that plays a fundamental role in studies of compound INTEREST, TRIGONOMETRY, LOGARITHMS, and CALCULUS, and that unites these disparate fields. (EULER'S FORMULA, for instance, illustrates this.) The number _e_ has approximate value 2.718281828459045 ... and can be defined in any of the following different ways:

1. The number _e_ is the limit value of the expression $\left(1+\dfrac{1}{n}\right)$ raised to the _n_th power, as _n_ increases indefinitely:

$$e = \lim_{n\to\infty}\left(1+\frac{1}{n}\right)^{n}$$

2. If _L(a)_ denotes the area under the curve _y_ = 1/_x_ above the interval [1,_a_], then _e_ is the location on the _x_-axis for which _L(e)_ = 1.
3. If _f(x)_ is a function that equals its own DERIVATIVE, that is, $\dfrac{d}{dx}f(x) = f(x)$, then _f(x)_ is an EXPONENTIAL FUNCTION with base value _e: f(x) = e^x._
4. _e_ is the value of the infinite sum $1 + \dfrac{1}{1!} + \dfrac{1}{2!} + \dfrac{1}{3!} + ...$

Definition 1 is linked to the problem of computing compound interest. As we show below, definition 2 defines the natural logarithm, and definition 3 arises from studies of natural growth and decay, and consequently the consideration of EXPONENTIAL FUNCTIONS.

The fourth definition arises from the study of TAYLOR SERIES. One proves that all four definitions are equivalent as follows:

First consider the curve _y_ = 1/_x_. It has the remarkable property that rectangles touching the curve and just under it have the same area if the endpoints of the rectangles are in the same ratio _r_. For example, in the first diagram on the opposite page, the rectangles above the intervals [_a,ra_] and [_b,rb_] each have area $\dfrac{r-1}{r}$. By taking narrower and narrower rectangles, all in the same ratio _r_, it then follows that the area under the curve above any two intervals of the form [_a,ra_] and [_b,rb_] are equal.

Following definition 2, let _L(x)_ denote the area under this from position 1 to position _x_. (If _x_ is less than 1, deem the area negative.) Notice that _L(1)_ = 0. Also, set _e_ to be the location on the _x_-axis where the area under the curve is 1: _L(e)_ = 1.

Notice that the area under the curve from 1 to position _ab, L(ab),_ is the sum of the areas under the curve above the intervals [1,_a_] and [_a,ab_]. The first area is _L(a)_ and the second, by the property above, equals _L(b)._ We thus have:

$$L(ab) = L(a) + L(b)$$

This shows that _L_ is a function that converts multiplication into addition, which is enough to prove that it is the LOGARITHMIC FUNCTION base _e_. We have _L(x)_ = log_e(x). This function is called the natural logarithm function and is usually written ln(_x_).

By the FUNDAMENTAL THEOREM OF CALCULUS, the derivative of an area function is the original function and so we have:

$$\frac{d}{dx}\ln(x) = \frac{1}{x}$$

Now consider the corresponding exponential function $y = e^x$. By taking logarithms, we obtain $\ln(y) = x$. Differentiating yields $\frac{1}{y}\frac{dy}{dx} = 1$, and so $\frac{dy}{dx} = y = e^x$. This establishes definition 3 stating that $f(x) = e^x$ is the function that equals its own derivative.

Now that we know the derivative of $y = e^x$, we can compute its Taylor series. We obtain:

$$e^x = 1 + x + \frac{x^2}{2!} + \frac{x^3}{3!} + \cdots$$

Setting $x = 1$ establishes definition 4.

It remains now to establish definition 1. From the graph of the curve $y = 1/x$, it is clear that the region between $x = 1$ and $x = 1 + \frac{1}{n}$ is sandwiched between a rectangle of area $\frac{1}{n} \times \frac{1}{1 + \frac{1}{n}} = \frac{1}{n+1}$ and a rectangle of area $= \frac{1}{n} \times 1 = \frac{1}{n}$. Thus:

$$\frac{1}{n+1} \le \ln\left(1 + \frac{1}{n}\right) \le \frac{1}{n}$$

Multiplying through by n yields:

$$\frac{n}{n+1} \le \ln\left(\left(1 + \frac{1}{n}\right)^n\right) \le 1$$

As n becomes large, the quantity $\frac{n}{n+1}$ approaches the value 1. It must be the case, then, that $\left(1 + \frac{1}{n}\right)^n$ approaches a value for which its logarithm is 1. Consequently $\lim_{n\to\infty}\ln\left(\left(1 + \frac{1}{n}\right)^n\right)$ equals 1. This gives:

$$\lim_{n\to\infty}\left(1 + \frac{1}{n}\right)^n = e$$

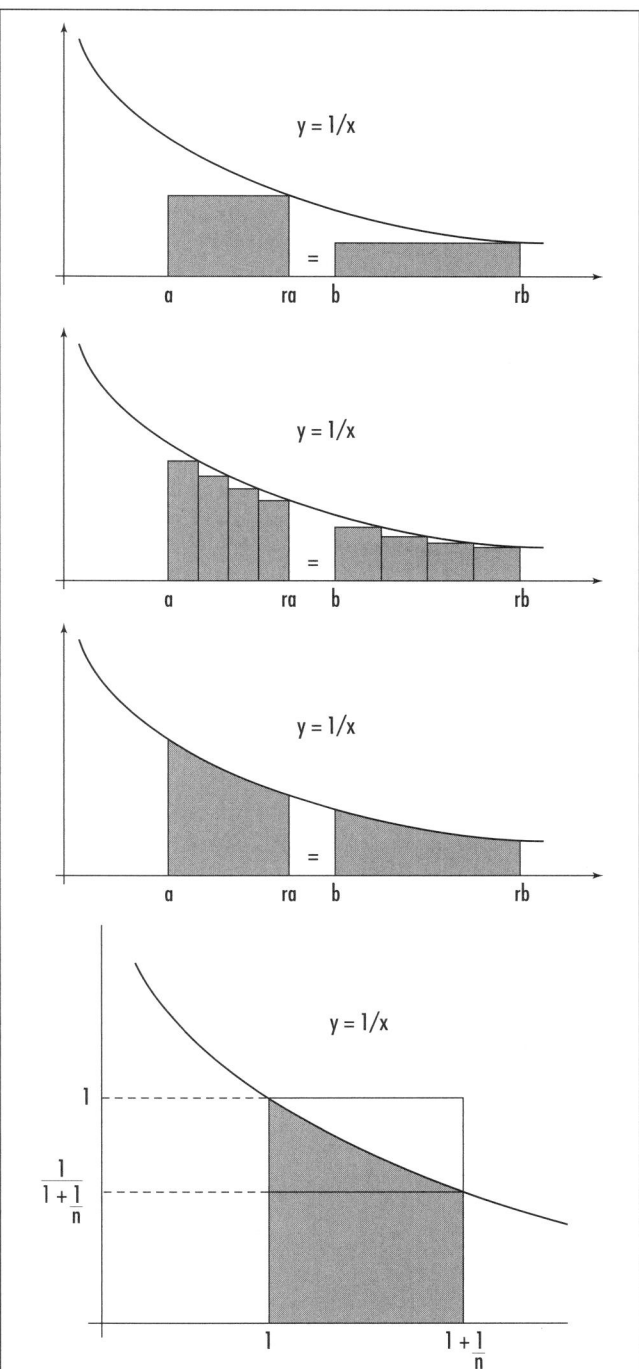

The curve y = 1/x

With regard to the issue of compound interest, it is necessary to compute the limit $\lim_{n\to\infty}\left(1 + \frac{r}{n}\right)^n$.

We have:

$$\lim_{n\to\infty}\left(1+\frac{r}{n}\right)^n = \lim_{n\to\infty}\left(\left(1+\frac{1}{(n/r)}\right)^{(n/r)}\right)^r = e^r$$

This represents the value of an investment of $1 after 1 year accruing continuously compounded interest at an interest rate of r percent per annum.

One can use the fourth definition to prove that e is an IRRATIONAL NUMBER. One begins by assuming, to the contrary, that e is a fraction of the form $\frac{p}{q!}$ and multiplies the formula presented in definition 4 by $q!$. The fractional parts that remain cannot add to a whole number.

In 1873 French mathematician Charles Hermite (1822–1901) proved that e is a TRANSCENDENTAL NUMBER.

Earth   Our planet, the third from the Sun, is often assumed to have the shape of a perfect SPHERE, but detailed measurements show it to be the shape of an oblate spheroid with equatorial radius 3,963 miles (6,378 km) and polar radius 3,950 miles (6,357 km). Its mean orbital distance from the Sun is 9.296×10^7 miles (1.496×10^8 km), one "astronomical unit," and its mass is 1.317×10^{25} lb (5.976×10^{24} kg).

The Earth takes 23 hours, 56 minutes, and 4.1 seconds to complete a sidereal day, that is, one full rotation about its axis as measured relative to the fixed stars. The solar day, the time it takes for a point P on the surface of the Earth initially facing the Sun to return to that position is, by definition, precisely 24 hours. The difference in time measurements is explained by the fact that the Earth advances in its orbit as it rotates: the Earth must turn slightly more than 360° to bring P back to face the Sun as the planet moves forward.

The first known attempt to calculate the circumference of the Earth was made by the Greek scholar ERATOSTHENES OF CYRENE (ca. 275–195 B.C.E.). Eratosthenes observed that at the summer solstice, noon on June 21 of every year, the Sun shone directly to the bottom of a well in the city of Syene (present day Aswan). This meant that the Sun was directly overhead at this time. He also noted that at the same time in Alexandria, a city approximately 500 miles due north of Syene, objects cast shadows, meaning that the Sun was not directly overhead at this location. These observations provided Eratosthenes the means to compute the radius of the Earth. He assumed that the Sun was sufficiently far away from our planet that the rays of light that reach us from it can be regarded as essentially parallel. Eratosthenes measured the angle cast by shadows at Alexandria (by using an object of known height and measuring the length of the shadows cast) and found the angle to be 1/50 of a full turn, about 7.2°. The circumference of the Earth follows: if 1/50th of a full turn corresponds to a distance of 500 miles, then a full turn must correspond to a distance of $50 \times 500 = 25,000$ miles.

Geographical Coordinates
The lines of latitude and longitude form a system of COORDINATES on the Earth. The "parallels of latitude" are circles parallel to the equator, labeled according to a measurement of angle. Stated precisely, if O is the center of the Earth, P is a point on a circle of latitude, P' is the point on the equator directly south of P, and α is the angle POP', then the circle of latitude containing P is labeled α. The latitude of any point can thus be 0° (on the equator) up to 90° north (the North Pole) or 90° south (the South Pole). New York City, for example, has a latitude of 40°45′06″ N.

Conveniently, the North Star lies directly above the North Pole. By measuring the angle of elevation of the North Star (90° at the North Pole, 0° at the equator), one can quickly determine the latitude of any location in the Northern Hemisphere. Using sextants to measure elevation, sailors of the past relied on the North Star to help determine their locations. The meridians, or lines of longitude, run from the North Pole to the South Pole, perpendicular to the circles of latitude. Each meridian is a semicircle. The equator is divided into 360°, with the meridian passing through Greenwich, England (called the prime meridian), deemed angle zero, and the angle of any other meridian ranges from 180° east to 180° west. Stated precisely, the longitude of a point P on the sphere is given by the angle $P'OG'$, where P' is the point on the equator directly south or north of P, and G' is the point on the equator south of Greenwich. New York City, for example, has longitude $73°59'39''$ W.

The longitude of a point P on the Earth's surface can be measured by identifying the time difference between high noon at P and high noon at Greenwich. As the day is divided into 24 1-hour periods, each delay of 1 hour corresponds to 1/24 of a full turn about the Earth's circumference. Thus, at 360/24 = 15° longitude west, for instance, the Sun reaches its highest point in the sky 1 hour later than it does at Greenwich; at 30° longitude west, it occurs 2 hours later, and so forth. After about 1735, sailors were able to carry accurate chronometers to keep track of Greenwich time. By measuring the time of high noon at any given location, sailors could accurately determine the longitude of that location.

See also OBLATE/PROLATE.

eccentricity For any CONIC SECTION, the ratio of the distance of a point on the curve from a fixed point, the focus, to its distance from a fixed line, the directrix, is constant. The value of this ratio, denoted e, is called the eccentricity of the curve, and it gives a measure of the curve's shape. For a PARABOLA, e equals one. If e lies between zero and one, then the curve is an ELLIPSE. If e is greater than one, then the curve is a HYPERBOLA. The eccentricity of a CIRCLE is defined to be zero. (In this context, the eccentricity e is not to be confused with LEONHARD EULER's number e.)

Egyptian fractions Any fraction with unit numerator, such as $\frac{1}{2}, \frac{1}{45}$, and $\frac{1}{598}$, is called an Egyptian fraction. The Egyptians of 4,000 years ago expressed all fractional quantities as sums of *distinct* Egyptian fractions. For example, $\frac{2}{9}$ was written $\frac{1}{6} + \frac{1}{18}$, and $\frac{3}{10}$ as $\frac{1}{5} + \frac{1}{15} + \frac{1}{30}$.

In 1202, FIBONACCI began his own investigation of Egyptian fractions and was the first to prove that every fraction can indeed be expressed as a finite sum of distinct Egyptian fractions. (It is not clear whether the ancient Egyptians ever questioned this.) He showed that subtracting a quantity of the form $1/n$, with n as small as possible, from a given fraction always pro-

duces a new fraction with a smaller numerator. Thus repeated application of this procedure must eventually produce a fraction with unit numerator itself.

As an example, for the fraction 5/17 we have:

$$\frac{5}{17} - \frac{1}{4} = \frac{3}{68} \quad (\frac{1}{3} \text{ is too large a quantity to subtract.})$$

and

$$\frac{3}{68} - \frac{1}{23} = \frac{1}{1564} \quad (\frac{1}{22} \text{ is too large a quantity to subtract.})$$

giving:

$$\frac{5}{17} = \frac{1}{4} + \frac{1}{23} + \frac{1}{1564}$$

Such representations need not be unique. For example, $\frac{3}{10}$ equals both $\frac{1}{5} + \frac{1}{15} + \frac{1}{30}$ and $\frac{1}{4} + \frac{1}{20}$.

See also EGYPTIAN MATHEMATICS.

Egyptian mathematics Our knowledge of ancient Egyptian mathematics from around 2000 B.C.E. comes chiefly from the RHIND PAPYRUS. There we learn, for example, that the Egyptians followed a very natural system for denoting numerals: 1 was a vertical stroke I, 2 was two of them II, 3 was III, and 4 was IIII, and separate symbols were used for 5, 6, 7, 8, and 9, and for 10, 20,..., 100, 200,..., 1000, and so on. All other numbers were represented as groups of these symbols, usually arranged in order from largest to smallest. Like the ROMAN NUMERAL system, the Egyptian system did not use a PLACE-VALUE SYSTEM (the symbol for 5, for example, denoted "5" no matter where it appeared in the number). It is very difficult to do pencil-and-paper calculations without place-value notation, but the Egyptians always used a calculating board, much like an ABACUS, to perform arithmetic calculations, and needed only to record the results. They were therefore not hindered by their cumbersome numerical system. The ancient Egyptians were adept at multiplication, using a method of successive doubling to calculate products. This method is today called EGYPTIAN MULTIPLICATION.

Division problems lead to FRACTIONS. It did not occur to the ancient Egyptians to express fractions with numerators and denominators. In the Rhind papyrus, the mathematician Ahmes simply placed a dot over a number to indicate its reciprocal, except in the case of the fractions $\frac{1}{2}, \frac{1}{3}, \frac{2}{3}$, and $\frac{1}{4}$, each of which had its own symbol. Thus the Egyptians only dealt with fractions of the form $\frac{1}{n}$ (with the exception of two-thirds). Fractions with unit numerators are known today as EGYPTIAN FRACTIONS. All other fractional quantities were expressed as sums of *distinct* Egyptian fractions. For example, $\frac{2}{5}$, which equals $\frac{1}{3} + \frac{1}{15}$, was written $\overset{\bullet}{3} + \overset{\bullet}{15}$, and $\frac{4}{13}$ as $\overset{\bullet}{4} + \overset{\bullet}{18} + \overset{\bullet}{468}$.

The Egyptian's ability to compute such expressions is impressive. The Rhind papyrus provides reference lists of such expressions, and the first 23 problems in the document are exercises in working with such fractional representations.

The ancient Egyptians were adept at solving LINEAR EQUATIONS. They used a method called false position to attain solutions. This involves guessing an answer, observing the outcome from the guess, and adjusting the guess accordingly. As an example, problem 24 of the Rhind papyrus asks:

> Find the quantity so that when 1/7 of itself is added to it, the total is 19.

To demonstrate the solution, the author suggests a guess of 7. That plus its one-seventh is 8, by far too small, but multiplying the outcome by 19/8 produces the answer of 19 that we need. Thus $7 \times (19/8)$ must be the quantity we desire.

The majority of problems in the Rhind papyrus are practical in nature, dealing with issues of area (of rectangles, trapezoids, triangles, circles), volume (of cylinders, for example), slopes and altitudes of pyramids (which were built 1,000 years before the text was written), and number theoretic problems about sharing goods under certain constraints. Some problems, however, indicate a delight in mathematical thinking for its own sake. For example, problem 79 asks:

> If there are seven houses, each house with seven cats, seven mice for each cat, seven ears of grain for each mouse, and each ear of grain would produce seven measures of grain if planted, how many items are there altogether?

This problem appears in FIBONACCI's *Liber abaci*, written 600 years before the Rhind papyrus was discovered. A version of this problem also appears as a familiar nursery-rhyme and riddle, "As I Was Going to St. Ives."

Egyptian multiplication   The RHIND PAPYRUS indicates that the ancient Egyptians of around 2000 B.C.E. used a process of "successive doubling" to multiply numbers. They computed 19×35, for example, by repeatedly doubling 35:

1	35
2	70
4	140
8	280
16	560

Since $19 = 16 + 2 + 1$, summing $560 + 70 + 35 = 665$ gives the product. This method shows that knowledge of the two-times table is all that is needed to compute multiplications. RUSSIAN MULTIPLICATION follows an approach similar to this method.

See also EGYPTIAN MATHEMATICS; ELIZABETHAN MULTIPLICATION; FINGER MULTIPLICATION; MULTIPLICATION; NAPIER'S BONES; RUSSIAN MULTIPLICATION.

eigenvalue (e-value, latent root)   *See* EIGENVECTOR.

eigenvector (e-vector, latent vector, characteristic vector, proper vector)   For a square $n \times n$ MATRIX A, we say a nonzero VECTOR $\mathbf{x}$ is an eigenvector for A if there is a number λ such that $A\mathbf{x} = \lambda\mathbf{x}$. The number λ is called the eigenvalue associated with that eigenvector. If $\mathbf{x}$ is an eigenvector of A, then we have that $(A - \lambda I)\mathbf{x} = 0$, where I is the IDENTITY MATRIX. This shows that the matrix $A - \lambda I$ is not invertible, and so must have zero determinant: $\det(A - \lambda I) = 0$. This is a polynomial equation in λ of degree n, called the "characteristic polynomial" of A. As there can only be at most n solutions to such an equation, we have that an $n \times n$ matrix A has at most n distinct eigenvalues. Mathematicians have proved that associated with each possible eigenvalue there is at least one corresponding eigenvector. Moreover, it has been established that eigenvectors associated with distinct eigenvalues are linearly independent.

The study of eigenvectors and eigenvalues greatly simplifies matrix manipulations. Suppose, for example, a square 3×3 matrix A has three distinct eigenvalues λ_1, λ_2, and λ_3. Set D to be the diagonal matrix

$$D = \begin{pmatrix} \lambda_1 & 0 & 0 \\ 0 & \lambda_2 & 0 \\ 0 & 0 & \lambda_3 \end{pmatrix}$$

and set B to be the matrix whose columns are the corresponding eigenvectors of A. Then it is possible to show that:

$$A = BDB^{-1}$$

This observation allows one to compute high powers of A with very little work. For instance, the quantity A^{100}, for instance, is just a product of three matrices:

$$A^{100} = (BDB^{-1})^{100}$$
$$= BD^{100}B^{-1}$$

noting that D^{100} is simply:

$$D^{100} = \begin{pmatrix} \lambda_1{}^{100} & 0 & 0 \\ 0 & \lambda_2{}^{100} & 0 \\ 0 & 0 & \lambda_3{}^{100} \end{pmatrix}$$

This work also allows mathematicians to define the square root of a matrix:

$$\sqrt{A} = B \begin{pmatrix} \sqrt{\lambda_1} & 0 & 0 \\ 0 & \sqrt{\lambda_2} & 0 \\ 0 & 0 & \sqrt{\lambda_3} \end{pmatrix} B^{-1}$$

(provided the square roots of the eigenvalues are defined), or to define new quantities, such as the logarithm of a matrix:

$$\ln(A) = B \begin{pmatrix} \ln(\lambda_1) & 0 & 0 \\ 0 & \ln(\lambda_2) & 0 \\ 0 & 0 & \ln(\lambda_3) \end{pmatrix} B^{-1}$$

Such actions have proved useful in the study of theoretical physics and engineering.

The prefix *eigen* is German for "characteristic" or "own." The eigenvalues and eigenvectors of a matrix completely characterize the matrix.

See also CAYLEY-HAMILTON THEOREM; INVERSE MATRIX; LINEARLY DEPENDENT AND INDEPENDENT.

Einstein, Albert (1879–1955) German *Relativity, Quantum mechanics* Born on March 14, 1879, in Ulm, Germany, Albert Einstein is recognized as an outstanding mathematical physicist whose work on the special and general theories of relativity completely revolutionized how scientists think about space, matter, and time. Although he regarded himself as a physicist, Einstein's work inspired many significant developments in modern mathematics, including the development of TENSOR analysis as the appropriate means to describe the curvature of space.

The classical school environment did not suit Einstein well. In 1895 he failed the entrance exam for a Swiss technical school, where he hoped to study electrical engineering. After attending a second school at Aarau, he did eventually manage to enter the Zurich

Albert Einstein, an eminent mathematical physicist of the 20th century, revolutionized our understanding of space, matter, and time through his theories of special and general relativity. *(Photo courtesy of Topham/The Image Works)*

school to graduate there in 1900 with a degree in teaching. Unable to find a university position, Einstein accepted a job at the Swiss Patent Office in Bern in 1902 and remained there for 7 years.

During his time at the patent office, Einstein studied theoretical physics in the evenings, without the benefit of close contact with the scientific literature or colleagues, and managed to produce and publish, all in the year 1905, five truly outstanding papers:

- "Über einen die Erzeugung und Verwandlung des Lichtes betreffenden heuristischen Gesichtspunkt" (On a heuristic concerning the production and transformation of light), published in *Annalen der Physik,* March 1905.
- "Die von der molekularkinetischen Theorie der Wärme gefurdete Bewegung von in ruhenden Flüssigkeiten suspendierten Teilchen" (On the movement of small particles suspended in a stationary liquid demanded by the molecular kinetic theory of heat), published in *Annalen der Physik,* May 1905.
- "Zur Elektrodynamik bewegter Körper" (On the electrodynamics of moving bodies), published in *Annalen der Physik,* June 1905.
- "Ist die Trägheit eines Körpes von seinem Energieinhalt abhängig?" (Does the inertia of a body depend upon its energy-content?), published in *Annalen der Physik,* September 1905.
- "Eine neue Bestimmung der Moleküldimensionen" (A new determination of molecular dimension), written in April 1905, published in *Annalen der Physik,* April 1906.

The first paper was concerned with the puzzling photoelectric effect observed by scientists of the time. Heinrich Hertz (1857–94) noticed that the number of electrons released from a section of metal bombarded with a beam of light was determined not by the intensity of the beam, but rather by its wavelength. Max Planck (1858–1947) also observed that electromagnetic energy was emitted from radiating objects according to discrete quantities, again in direct proportion to the wavelength of the radiation. Einstein proposed that light and radiation itself travel in discrete bundles, which he called quanta, and described the mathematics that would consequently explain these phenomena.

In his second paper, Einstein developed mathematical equations that correctly described the motion of atoms and molecules under "Brownian motion." In his third paper, Einstein proposed his theory of special relativity. He noted that because light is able to travel through a vacuum, there is no natural frame of reference for measuring its speed. (The speed of sound, for instance, is measured with respect to the medium of air through which it passes.) Also, since it is impossible to determine whether one is stationary in space or moving through space at a uniform velocity, it follows then that all observers must observe light traveling at the same speed. From this, Einstein developed a series of "thought experiments" that clearly establish that observers traveling at different speeds must hence record different values when measuring quantities such as length and time. (For instance, imagine a light beam bouncing back and forth between two fixed mirrors set at a distance so that the time taken to bounce between the two mirrors is 1 sec. Suppose that a second observer moves past the "clock" with uniform speed. According to this observer, the clock moves past her at uniform speed. A straightforward calculation shows that, since the speed of light is unchanged for this observer, she would see the light beam taking longer than 1 sec to complete a cycle between the two mirrors.) The fourth paper developed the special theory of relativity further, culminating with his famous equation $E = mc^2$, showing that energy and mass are equivalent (with a factor of the speed of light squared incorporated).

Einstein submitted his final paper as a doctoral thesis to the University of Zurich to receive a Ph.D. By 1909 Einstein had been recognized as a leading scientific thinker. He resigned from the patent office and was appointed a full professor at the Karl-Ferdinand University in Prague in 1911. That same year, based on his theory of relativity, Einstein made a prediction about how light rays from distant stars would bend around the Sun, hoping that some day astronomers might be able to observe this effect and verify that his theory of relativity is correct. He also began working on incorporating the role of acceleration (nonuniform motion) into his special theory to develop a general theory of relativity. After a number of false starts, Einstein finally published a coherent general theory in 1915. Four years later, during a solar eclipse, British scientists were able to observe the bending of light rays just as Einstein had predicted. The popular press covered the story, and Einstein immediately received world attention for his achievement.

In 1921 Einstein received the Nobel Prize not for his relativity theory, but, surprisingly, for his work on the photoelectric effect. He also received the Copley

Medal from the ROYAL SOCIETY of London in 1925, and the Gold Medal from the Royal Astronomical Society in 1926.

With the rise of anti-Semitism in Europe, Einstein accepted a position at the Institute of Advanced Study at Princeton, N.J., in 1933. He stayed there until his death on April 18, 1955.

One cannot exaggerate the effect that Einstein's work has had on modern physics. One of his principal goals was to unite the discrete description of particles and matter with the continuous description of electromagnetic radiation and develop a single unified theory of the two. The result is quantum mechanics. Intense work continues today to incorporate other physical forces, such as gravity, into a grand unified theory.

elementary matrix   *See* GAUSSIAN ELIMINATION.

The Elements   No doubt the most influential mathematics text of all time, *The Elements*, written by EUCLID (ca. 300–260 B.C.E.), provided the model for all of mathematical thinking for the two millennia that followed it. Mathematicians agree that this work defines what the pursuit of PURE MATHEMATICS is all about. More than 2,000 editions of *The Elements* have been printed since the production of its first typeset version in 1482.

Written in 13 volumes (called "books") *The Elements* represents a compilation of all the mathematics that was known at the time. Organized in a strict logical structure, Euclid begins the work with a set of basic definitions, "common notions," and axioms (EUCLID'S POSTULATES), and deduces from them, by the process of pure logical reasoning, some 465 propositions (THEOREMS) on the topics of plane geometry, number theory (typically presented in terms of geometry), and solid geometry. The work is revered for its clarity, precision, and rigor.

The work is extremely terse in its presentation. There is no discussion or motivation, and results are simply stated and proved, often referring to a figure accompanying the statement. Each proof ends with a restatement of the proposition studied along with the words, "which was to be proved." The Latin translation of this phrase is *quod erat demonstrandum*, and many mathematicians today still like to end a formal proof with the abbreviation Q.E.D.

Although it is generally believed that no result presented in *The Elements* was first proved by Euclid, the organization of the material and the logical development presented is original. Euclid's choice of beginning postulates shows remarkable insight and a deep wisdom of the subject. His recognition of the need to formulate the controversial PARALLEL POSTULATE, for instance, shows a level of genius beyond all of those who tried to prove his choice irrelevant during the two millennia that followed. (It was not until the 19th century, with the development of NON-EUCLIDEAN GEOMETRY, did mathematicians realize that the parallel postulate was an essential assumption in the development of standard planar geometry.)

The first six books of *The Elements* deal with the topic of plane geometry. In particular, Books I and II establish basic properties of triangles, parallel lines, parallelograms, rectangles, and squares, and Books III and IV examine properties of circles. In Book V, Euclid examines properties of magnitudes and ratios, and applies these results back to plane geometry in Book VI. Euclid presents a proof of PYTHAGORAS'S THEOREM in Book I.

Books VII to X deal with NUMBER THEORY. The famous EUCLIDEAN ALGORITHM appears in book VII, and EUCLID'S PROOF OF THE INFINITUDE OF PRIMES in book IX. Book X deals with the theory of irrational numbers, and Euclid actually proves the existence of these numbers in this work.

The final three volumes of *The Elements* explore three-dimensional geometry. The work culminates with a discussion of the properties of each PLATONIC SOLID and proof that there are precisely five such polyhedra.

elimination method for simultaneous linear equations   Another name for GAUSSIAN ELIMINATION.

Elizabethan multiplication   Also known as the galley method and the lattice method, this multiplication technique was taught to students of mathematics in Elizabethan England. To multiply 253 and 27, for example, draw a 2 × 3 grid of squares. Write the first number along the top, and the second number down the right side. Divide each cell of the grid diagonally. Multiply the digits of the top row, in turn, with each of the digits of the right column, writing the products in

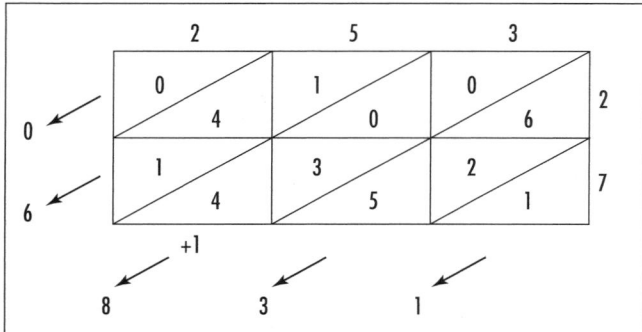

Elizabethan multiplication

the appropriate square cells of the grid as two-digit entries. (Thus compute 3 × 2 as 06, for example.)

Add the entries in each diagonal starting with the bottom right diagonal. Write down the units figure and carry any tens figures that appear to the next diagonal. The answer, 6,831, now appears along the left column and bottom row.

This procedure works for multidigit multiplications of any size. Its success relies on the DISTRIBUTIVE PROPERTY of arithmetic and the process of EXPANDING BRACKETS. In our example,

$$253 \times 27 = (2 \times 10^2 + 5 \times 10 + 3) \times (2 \times 10 + 7)$$
$$= (2 \times 2) \times 10^3 + (2 \times 7) \times 10^2 + (3 \times 2) \times 10$$
$$+ 3 \times 7$$
$$+ (5 \times 2) \times 10^2 + (5 \times 7) \times 10$$

Each diagonal corresponds to a powers-of-10 place, with entries placed in an upper portion of a square cell corresponding to carried figures to the next powers-of-10 position. Try computing a multiplication problem both the Elizabethan way and the usual way, side-by-side, to see that the two methods do not differ.

See also EGYPTIAN MULTIPLICATION; FINGER MULTIPLICATION; MULTIPLICATION; NAPIER'S BONES; RUSSIAN MULTIPLICATION.

ellipse   As one of the CONIC SECTIONS, an ellipse is the plane curve consisting of all points P whose distances from two given points F_1 and F_2 in the plane have a constant sum. The two fixed points F_1 and F_2 are called the foci of the ellipse. An ellipse also arises as the curve produced by the intersection of a plane with a single nappe of a right circular CONE.

Using the notation $|PF_1|$ and $|PF_2|$ for the lengths of the line segments connecting P to F_1 and F_2, respectively, the defining condition of an ellipse can be written:

$$|PF_1| + |PF_2| = d$$

where d denotes the constant sum.

The equation of an ellipse can be found by introducing a coordinate system in which the foci are located at positions $F_1 = (-c,0)$ and $F_2 = (c, 0)$, for some positive number c. It is convenient to write $d = 2a$, for some $a > 0$. If $P = (x,y)$ is an arbitrary point on the ellipse, then the defining condition states:

$$\sqrt{(x + c)^2 + y^2} + \sqrt{(x - c)^2 + y^2} = 2a$$

Moving the second radical to the right-hand side, squaring, and simplifying yields the equation:

$$\sqrt{(x - c)^2 + y^2} = a - \frac{c}{a}x$$

Squaring and simplifying again yields:

$$\frac{x^2}{a^2} + \frac{y^2}{a^2 - c^2} = 1$$

Noting that a is greater than c, we can set the positive quantity $a^2 - c^2$ as equal to b^2, for some positive number b. Thus the equation of the ellipse is:

$$\frac{x^2}{a^2} + \frac{y^2}{b^2} = 1$$

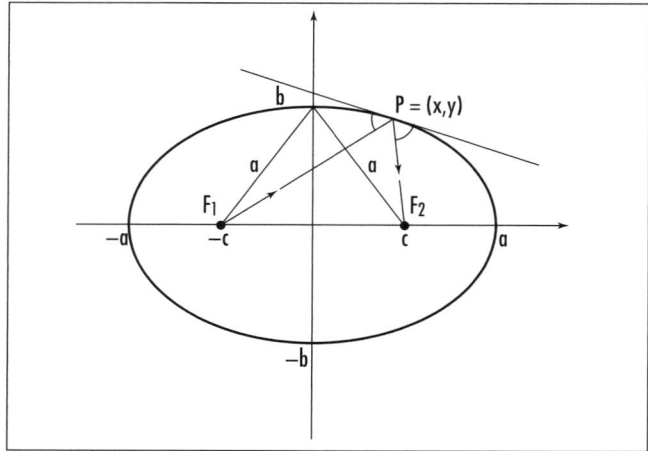

Ellipse

Conversely, one can show that any equation of this form, with $a > b$, does indeed yield an ellipse with foci at positions $(\pm\sqrt{a^2 - b^2}, 0)$, and whose points P have distances from the foci a constant sum $2a$. If, on the other hand, $b > a$, the equation is again an ellipse, but this time with foci along the y-axis at $(0, \pm\sqrt{b^2 - c^2})$. The common sum of distances is $2b$. (If a equals b, the curve is a CIRCLE.)

The equation shows that an ellipse crosses the x-axis at $x = \pm a$ and the y-axis at $y = \pm b$. The numbers a and b are called the semimajor axis and the semiminor axis, respectively.

Ellipses have the following reflection property: any ray of light emanating from one focus is reflected off the side of the ellipse directly toward the other focus. This can be proved by solving an OPTIMIZATION problem. Any room with walls curved in the shape of an ellipse has the property that any whisper uttered at one focus can be heard by anyone located at the second focus: not only do sound waves bounce off the curved wall directed from one focus to the other, but they also travel the same distance and so arrive synchronized at the second focus. The Mormon Tabernacle in Salt Lake City, Utah, and the Whispering Gallery in the U.S. Capitol building in Washington, D.C., are built to have this property.

Elliptical mirrors are used for the treatment of kidney stones. By positioning a mirror so that the kidney stone lies at one focus, medical practitioners can place a high-intensity sound wave generator at the second focus. Waves from the generator pass harmlessly through the patient's body to then concentrate at the stone and destroy it.

An ellipse can be drawn using a pencil, a string and two thumbtacks. Tacking each of the two ends of the string at fixed locations (the foci), one pulls the string taut with the tip of the pencil, and then slowly moves the pencil around, all the while keeping the string taut. The curve traced is an ellipse, with constant sum of the distances from the foci being the length of the string.

In the process of deriving the equation of an ellipse, we presented the equation

$$\sqrt{(x - c)^2 + y^2} = a - \frac{c}{a}x$$

Set $e = \dfrac{c}{a} = \dfrac{\sqrt{a^2 - b^2}}{a} = \sqrt{1 - \dfrac{b^2}{a^2}}$. This is called the ECCENTRICITY of the ellipse and has value between zero and 1. The above equation can be rewritten:

$$\frac{\sqrt{(x - c)^2 + y^2}}{x - \left(-\dfrac{a}{e}\right)} = e$$

The numerator of the quantity on the left side is the distance of a given point P from a focus, and the denominator is the distance of the point P from the vertical line $x = -a/e$, called a directrix of the ellipse. This formulation provides an alternative characterization of the ellipse:

An ellipse is the set of all points P such that the ratio of its distance from a fixed point (the focus) to its distance from a fixed line (the directrix) equals a constant e with value $0 < e < 1$.

The ECCENTRICITY of a circle is defined to be $e = 0$. If $e = 1$, this characterization gives a PARABOLA. For $e > 1$, we have a HYPERBOLA.

See also APOLLONIUS'S CIRCLE; PROJECTION.

ellipsoid Any geometrical surface or solid sitting in three-dimensional space possessing the property that any plane that slices it produces a cross-section that is either an ELLIPSE or a circle is called an ellipsoid. Such a figure has three axes of symmetry.

An ellipsoid, centered about the origin $(0,0,0)$ has equation:

$$\frac{x^2}{a^2} + \frac{y^2}{b^2} + \frac{z^2}{c^2} = 1$$

The points $(\pm a, 0, 0)$, $(0, \pm b, 0)$ and $(0, 0, +c)$ are the locations where the ellipsoid crosses the x-, y-, and z-axes, respectively.

One can create an ellipsoid by rotating an ellipse about one of its axes. This produces a figure with two of the three quantities a, b, c equal in value. An ellipsoid produced in this way is called a spheroid, but not every ellipsoid is a spheroid. If all three quantities a, b, and c have the same value r, the ellipsoid is a SPHERE of radius r.

Mathematicians have shown that the volume of an ellipsoid is given by $(4/3)\pi abc$. (Compare this with the equation for the volume of a sphere.) Since the time of LEONHARD EULER (1707–83), mathematicians have attempted to find a simple formula for the surface area of a general ellipsoid. This has proved to be a very difficult problem, and no closed-form formula exists. (The surface area of an ellipse can only be expressed in terms of a difficult "elliptic integral.")

empty set (null set) Any set that contains no elements is called an empty set. For example, the set of all real numbers greater than three and less than two is empty, as is the set of all people with gills.

A set A is said to be a subset of a set B, written $A \subseteq B$, if all elements of A belong to B. Consequently any empty set is a subset of any other set. In particular, if A and B are both empty, then $A \subseteq B$ and $B \subseteq A$, and the two empty sets are equal. This shows that there is only one empty set. It is usually denoted as $\emptyset$, but it can also be written as { }.

The set with the empty set as its one member is written $\{\emptyset\}$, and the set with the set containing the empty set as its lone member is written $\{\{\emptyset\}\}$. In this way we construct a chain of sets:

$$\emptyset, \{\emptyset\}, \{\{\emptyset\}\}, \{\{\{\emptyset\}\}\}, \ldots$$

which naturally corresponds to the sequence of counting numbers 0, 1, 2, 3, … In this context one could argue that all of mathematics arises from the empty set. It is an interesting exercise then to give a numerical interpretation to a two-member set of the form $\{\emptyset, \{\emptyset\}\}$, for instance.

A set that is not empty is called nonempty.

See also SET THEORY.

endpoint *See* INTERVAL.

epicycle *See* CYCLOID.

epsilon-delta definition *See* LIMIT.

equality Two quantities are said to be equal if, in some meaningful sense, they are equivalent. For example, the quantities 2 + 3 and 5 have the same value and so are equal. The two sets $\{a,b,c\}$ and $\{c,a,b\}$ are equal since they contain the same elements. The symbol = is used to denote the equivalence of two quantities, and so we write 2+3 = 5 and $\{a,b,c\} = \{c,a,b\}$.

Two algebraic expressions are said to be equal if one can be transformed into the other by the standard rules of algebra. For instance, $(x + 1)^2 + 3 = x^2 + 2x + 4$. Two functions are said to be equal if they have the same domains and produce the same output value for each input. For example, the functions $f(x) = 9x$ and $g(x) = x^{\frac{2+\log_3 x}{\log_3 x}}$, defined on positive values of x, are equal.

The symbol = (a pair of parallel line segments to denote equality) was introduced in 1557 by Welsh mathematician ROBERT RECORDE (ca. 1510–58) "because noe 2 thynges can be more equalle."

See also EQUATION.

equating coefficients Two polynomials $f(x) = a_n x^n + a_{n-1} x^{n-1} + \ldots + a_1 x + a_0$ and $g(x) = b_n x^n + b_{n-1} x^{n-1} + \ldots + b_1 x + b_0$ are identical as functions, that is, give the same output values for each input value of x, only if the coefficients of the polynomials match: $a_n = b_n$, $a_{n-1} = b_{n-1}, \ldots, a_0 = b_0$. (The general study of POLYNOMIALS establishes this.) The process of matching coefficients if two polynomials are known to be the same is called "equating coefficients."

For example, if x^2 equals a polynomial of the form $A + B(x - 1) + C(x - 1)(x - 2)$, then, after EXPANDING BRACKETS, we have $x^2 = Cx^2 + (B - 3C)x + (A - B + 2C)$. Equating coefficients yields: $C = 1$, $B - 3C = 0$ (and so $B = 3$), and $A - B + 2C = 0$ (and so $A = 1$). Thus $x^2 = 1 + 3(x - 1) + (x - 1)(x - 2)$. (This technique is often used in the method of PARTIAL FRACTIONS.)

As another example, if α and β are the roots of a quadratic equation of the form $x^2 - mx + n$, then: $x^2 - mx + n = (x - \alpha)(x - \beta) = x^2 - (\alpha + \beta)x + \alpha\beta$. We conclude then that m is the sum of the roots, and n their product.

equating real and imaginary parts Two COMPLEX NUMBERS $a + ib$ and $c + id$ are equal only if $a = c$ and $b = d$. Using this fact is called "equating real and imaginary parts." For example, if $(x + iy)(2 + 3i) = 4 + 5i$, then we must have $2x - 3y = 4$ and $3x + 2y = 5$.

LEONHARD EULER (1707–83) made clever use of this technique to find formulae for PYTHAGOREAN

TRIPLES. Also, using his famous formula $e^{i\theta} = \cos(\theta) + i\sin(\theta)$, today called EULER'S FORMULA, many trigonometric identities can be established quickly by equating real and imaginary parts. As another application, consider the series $\sum_{n=0}^{\infty} \cos(n\theta)$. Since $\cos(n\theta)$ is the real part of $e^{in\theta}$, the series in question is the real part of the GEOMETRIC SERIES $\sum_{n=0}^{\infty} \left(e^{i\theta}\right)^n$. Evaluating, gives

$$\sum_{n=0}^{\infty} \left(e^{i\theta}\right)^n = \frac{1}{1-e^{i\theta}} = \frac{1}{(1-\cos(\theta)) - i\sin(\theta)}$$

$$= \frac{1}{2} + i\frac{\sin(\theta)}{2-2\cos(\theta)}$$

Thus we have $\sum_{n=0}^{\infty} \cos(n\theta) = \frac{1}{2}$.

equation A mathematical statement that asserts that one expression or quantity is equal to another is called an equation. The two expressions or quantities involved are connected by an equals sign, "=." For instance, the statement $(a+b)^2 = a^2 + 2ab + b^2$ is an equation, as are the statements $2x + 3 = 11$ and $10 = 2 \times 5$.

An equation that is true for all possible values of the variables involved is called an IDENTITY. For instance, $y^2 - 1 = (y-1)(y+1)$ is an identity: this equation is true no matter which value is chosen for y. An equation that is true only for certain values of the variables is called a conditional equation. For instance, the equation $2x + 3 = 11$ is conditional, since it is true only if x is four. The equation $2c + d = 6$ is also a conditional, since it holds only for certain values of c and d.

The numbers that make a conditional equation true are called the solutions or roots of the equation. For example, the solution to $2x + 3 = 11$ is $x = 4$. An equation may possess more than one solution, and the set of all possible solutions to a conditional equation is called its solution set. For example, the equation $y^2 = 9$ has solution set $\{3, -3\}$, and the solution set of the equation $2c + d = 6$ is the set of all pairs of numbers of the form $(c, 6 - 2c)$.

The basic principle in solving an equation is to add, subtract, multiply, or divide both sides of the equation by the same number until the desired variable is isolated on one side of the equation. For example, the equation $2x + 7 = 5x + 1$ can be solved by subtract-

ing 1 from both sides to give $2x + 6 = 5x$, then subtracting $2x$ from both sides to obtain $6 = 3x$, and, finally, dividing both sides by 3 to obtain the solution $x = 2$. This approach works well for LINEAR EQUATIONS of one variable. For QUADRATIC equations, and POLYNOMIAL equations of high degree, one may also be required to take square and higher roots in the process of solving the equation. Not all equations, however, can be solved algebraically, in which case one can seek only a GRAPHICAL SOLUTION.

It should be noted that performing the same arbitrary operation on both sides of an equation need not necessarily preserve the validity of the equation. For example, although the statement $9 = 9$ is certainly valid, taking a square root on both sides of this trivial equation could be said to yield the invalid result $-3 = 3$. Although $2(x-1) = 3(x-1)$ is true for the value $x = 1$, dividing through by the quantity $x - 1$ yields the invalid conclusion $2 = 3$. And finally, since $\frac{12}{16} = \frac{3}{4}$, selecting the numerator of each side of the equation yields the absurdity $12 = 3$. Care must be taken to ensure that the operations being used in solving an equation do indeed preserve equality.

Even if the application of the same operation on both sides on an equation is deemed valid, such an act may nonetheless yield a new equation not necessarily exactly equivalent to the first. For instance, starting with $a = b$, squaring both sides yields the equation $a^2 = b^2$, which now means $a = b$ or $a = -b$. Mathematicians use the symbol "$\Rightarrow$" to denote that one equation leads to a second, but that the second need not imply the first. For example, it is appropriate to write: $a = b \Rightarrow a^2 = b^2$. (But $a^2 = b^2 \nRightarrow a = b$.) The symbol "$\Leftrightarrow$" is used to indicate that two equations are equivalent, that is, that the first implies the second, and that the second implies the first. For example, we have: $4x = 12 \Leftrightarrow x = 3$.

See also CUBIC EQUATION; HISTORY OF EQUATIONS AND ALGEBRA (essay).

equation of a line A straight LINE in two-dimensional space has the property that the ratio of the difference in y-coordinates of any two points on the line (rise) to the difference of their x-coordinates (run) is always the same. That is, the SLOPE of a straight line is constant and can be computed from any two given points on the line. Precisely, if (a_1, b_1) and (a_2, b_2) are two points on the line, and (x,y) are the coordinates of an arbitrary point on the line, then we have:

$$\frac{y - b_1}{x - a_1} = \frac{b_2 - b_1}{a_2 - a_1}$$

This provides the "two-point form" equation of the line. For example, the equation of the line passing through the points $(2,3)$ and $(-1,5)$ is $\frac{y - 3}{x - 2} = \frac{5 - 3}{-1 - 2} = -\frac{2}{3}$.

The quantity $\frac{b_2 - b_1}{a_2 - a_1}$ is the slope m of the line. Thus one can rewrite the two-point form of the equation as $\frac{y - b_1}{x - a_1}$. Rearranging yields:

$$y - b_1 = m(x - a_1)$$

This is called the "point-slope form" equation of the line. For example, the equation of a line of slope 4 that passes through the point $(5,7)$ is simply $y - 7 = 4(x - 5)$. Working with this form of equation is useful if the slope of the line is already specified.

Rearranging the point-slope equation yields $y = mx + (ma_1 + b_1)$. Denoting the constant $ma_1 + b_1$ simply as b yields the equation:

$$y = mx + b$$

Noting that if $x = 0$, then we have $y = b$. This shows that the constant b is the y-intercept of the line. For this reason, the above equation is called the "slope-intercept form" equation of the line. Thus, for example, the equation of a line with slope -1 crossing the y-axis at position 3 is $y = -x + 3$.

One disadvantage of the slope-intercept form is that it does not allow one to write down the equation of a vertical line, that is, one that does not intercept the y-axis at all. Returning to the two-point form and cross-multiplying yields the equation $(a_2 - a_1)(y - b_1) = (x - a_1)(b_2 - b_1)$. EXPANDING BRACKETS and rearranging terms again yields the general equation:

$$cx + dy = r$$

for some constants c, d, and r. This is called the general form of the equation of a line. For instance, the equation of the vertical line three units to the right of the y-axis is obtained by selecting $c = 1$, $d = 0$, and $r = 3$, to yield the equation $x = 3$.

Some authors prefer to divide the general form of the equation of a line through by the constant r and change the names of the remaining labels so that the equation reads:

$$\frac{x}{a} + \frac{x}{b} = 1$$

This is called the intercept form of the equation of a line.

In three-dimensional space a line is specified by a point (a,b,c) on the line and a VECTOR by $\mathbf{v} = <v_1,v_2,v_3>$, representing the direction of the line. Thus the coordinates (x,y,z) of any other point on the line are given by:

$$x = a + tv_1$$
$$y = b + tv_2$$
$$z = c + tv_3$$

for some value of the real number t. These are the PARAMETRIC EQUATIONS of the line. (The parametric equations of a line in two-dimensional space are analogous.)

If the vector $\mathbf{v}$ is computed via the difference of coordinates of the point (a,b,c) and a second point (a_1,b_1,c_1) on the line—i.e., $v_1 = a_1 - a$, $v_2 = b_1 - b$, and $v_3 = c_1 - c$—then solving for t in the parametric equations yields:

$$\frac{x - a}{a_1 - a} = \frac{y - b}{b_1 - b} = \frac{z - c}{c_1 - c}$$

These are the "two-point form" equations of a line in three-dimensional space.

See also DIRECTION COSINES; LINEAR EQUATION; SIMULTANEOUS LINEAR EQUATIONS; SKEW LINES; VECTOR EQUATION OF A PLANE.

equation of a plane   *See* VECTOR EQUATION OF A PLANE.

equiangular   A POLYGON is said to be equiangular if all of its interior angles are equal. For example, a rectangle is equiangular (each interior angle equals $90°$), as is an EQUILATERAL triangle (each interior angle equals $60°$). A polygon is called regular if it is both equiangular and equilateral.

A point (x,y) in the Cartesian plane is said to be a lattice point if both x and y are integers, and a polygon drawn in the plane is said to be a lattice polygon if its vertices lie at lattice points. Mathematicians have proved that it is impossible to draw an equiangular lattice polygon with n sides if n is a number different from 4 or 8. (Any four-sided equiangular lattice polygon is a rectangle, and any eight-sided equiangular lattice polygon has eight interior angles, each equal to 135°.) The square and the octagon are the only two regular lattice polygons.

See also CARTESIAN COORDINATES.

equidecomposable Two geometric figures are said to be equidecomposable if it is possible to dissect one figure into a finite number of pieces that can be rearranged, without overlap, to form the second figure. For example, an equilateral triangle of side-length 1 is equidecomposable with a square of the same area.

In the picture, a is of length $\dfrac{3 - \sqrt[4]{3}}{4}$ and b is of length $\dfrac{\sqrt[4]{3} - 1}{4}$. (The challenge to convert an equilateral triangle into a square by dissection was a puzzle first posed by English puzzlist Henry Ernest Dudeney in 1907. The challenge is also known as Haberdasher's puzzle.)

Scottish mathematician William Wallace (1768–1843) proved that any two polygons of the same area are equidecomposable. Mathematicians have since proved that the result remains valid even for figures with curved boundaries. In particular, Hungarian mathematician Miklov Laczovich demonstrated in 1988 that almost 10^{50} pieces are needed to convert a circle into a square.

Surprisingly, the corresponding result in three dimensions does not hold, even for simple polyhedra.

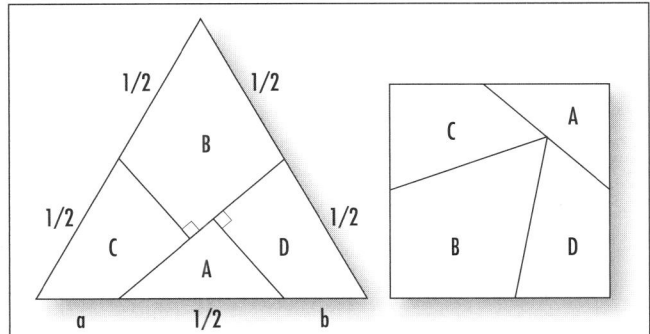

Equidecomposable figures

German mathematician Max Dehn (1878–1952) proved, for instance, that a cube and a regular tetrahedron of the same volume are not equidecomposable.

equidistant Two points P and Q are said to be equidistant from a third point O if they are the same distance from O. We write: $|PO| = |QO|$.

Given a single point O in a plane, the set of all points equidistant from O is a CIRCLE with O as its center. Given two points A and B in a plane, the set of all points equidistant from A and B is the perpendicular bisector of the line segment AB, that is, a straight line perpendicular to AB and passing through the midpoint of AB. (To see this, let M be the midpoint of the line segment AB, and let P be any point in the perpendicular bisector to AB. Suppose that $|PM| = x$ and $|AM| = y = |MB|$. Then, by PYTHAGORAS'S THEOREM, we have $|PA| = \sqrt{x^2 + y^2} = |PB|$, and so P is equidistant from A and B. One can also use Pythagoras's theorem to check that any point not on this line is not equidistant from those two points.)

Given three points A, B, and C in a plane, not in a straight line, there is just one point P equidistant from all three. (To see this, draw the perpendicular bisectors of AB and BC, and let P be the unique point at which they intersect. Then P is equidistant from A and B, and P is also equidistant from B and C. Consequently, P is the same distance from all three points.) Noting that the points A, B, and C can be viewed as the vertices of a TRIANGLE, this proves:

> The three perpendicular bisectors of the sides of any triangle meet at a common point P.

(This observation is used to prove that the three ALTITUDES of any triangle are also CONCURRENT.)

Taking matters further, suppose the common distance of P from each of the three points A, B, and C is r. It then follows that a circle of radius r centered about P passes through each of these points. This proves:

> For any triangle ABC there exists a single circle that passes through each of its vertices A, B, and C.

This circle is called the CIRCUMCIRCLE of the triangle, and the point P, the common point of intersection of the three perpendicular bisectors of the triangle, is called the circumcenter of the triangle.

In three-dimensional space, the set of all points equidistant from a single point O is a sphere with O as its center. Given two points A and B, the set of all points equidistant from them is a plane that passes through the midpoint of the line segment AB and is perpendicular to it. Given three points A, B, and C, not all in a straight line, the set of all points P equidistant from all three is a straight line perpendicular to the plane formed by the three points and passing through the circumcenter of the triangle ABC. There need not be a point equidistant from four given points in three-dimensional space.

The distance of a point from a line is the length of a line segment from the point meeting that line at right angles. Again, using Pythagoras's theorem and similar triangles, one can show that the set of all points equidistant from two intersecting lines in a plane is a pair of perpendicular lines that each pass through the point of intersection of the two lines, and each bisects an angle formed by the lines. Furthermore, arguing as above, one can prove:

In any triangle, the three lines that bisect the interior angles of the triangle meet at a common point.

(Consider first the point of intersection of just two angle bisectors. This point must, in fact, be equidistant from all three sides of the triangle, and so lies on the third angle bisector.)

See also EULER LINE; INCIRCLE.

equilateral   A POLYGON is said to be equilateral if all of its sides have the same length. For example, a square is equilateral, as is a triangle with each interior angle equal to 60°. A polygon is called "regular" if it is both equilateral and EQUIANGULAR.

A point (x,y) in the Cartesian plane is said to be a "lattice point" if both x and y are integers, and a polygon drawn in the plane is said to be a "lattice polygon" if its vertices lie at lattice points. Mathematicians have proved that it is impossible to draw an equilateral lattice polygon with an odd number of sides, although equilateral lattice polygons with any even number of sides do exist. The square and the octagon are the only two regular lattice polygons.

See also CARTESIAN COORDINATES.

equivalence relation   *See* PAIRWISE DISJOINT.

Eratosthenes of Cyrene (ca. 275–195 B.C.E.) Greek *Geometry, Number theory, Astronomy, Geographer* Born in Cyrene, in North Africa, (the exact birth date is not known), Eratosthenes is remembered as the first person to calculate the circumference of the Earth. (*See* EARTH.) Using the known distance between two particular cities, the lengths of shadows cast by the noonday sun at those cities, and simple geometric reasoning, Eratosthenes determined the circumference of the Earth to be 250,000 "stadia." Unfortunately, the exact length of a "stade" is not known today, and so it is not possible to be certain of the accuracy of this result. If we take, as many historians suggest, that the likely length of this unit is 515.6 ft (157.2 m), then Eratosthenes' calculation is extraordinarily accurate.

Eratosthenes traveled to Athens in his youth and spent many years studying there. Around 240 B.C.E. he was appointed librarian of the greatest library of the ancient world, the Library of Alexandria. Early in his scholarly career, Eratosthenes wrote the expository piece *Platonicus* as an attempt to explain the mathematics on which PLATO based his philosophy. Although this work is lost today, scholars of later times referred to it frequently and described it as an invaluable source detailing the mathematics of geometry and arithmetic, as well as the mathematics of music. In this work, Eratosthenes also described the problem of DUPLICATING THE CUBE and provided a solution to it making use of a mechanical device he invented.

Eratosthenes also worked on the theory of PRIME numbers and discovered a famous "sieve" technique for finding primes. This method is still used today and is named in his honor.

Along with measuring the circumference of the Earth, Eratosthenes also devised ingenious techniques for determining the distance of the Earth from the Sun (which he measured as 804 million stadia), the distance between the Earth and the Moon (780,000 stadia), and the tilt of the Earth's axis with respect to the plane in which the Earth circles the Sun (which he measured as 11/83 of 180°, that is, 23° 51′ 15″). Eratosthenes also accurately mapped a significant portion of the Nile River and correctly identified the occurrence of heavy rains near its source as the reason for its erratic flooding near its mouth. He compiled an astronomical catalog listing over 675 specific stars and devised an effective calendar system that included leap years.

As an extraordinarily well-rounded scholar, Eratosthenes wrote literary works on the topics of theater and ethics, and also wrote poetry. His famous poem "Hermes" was inspired by his studies of astronomy.

It is believed that Eratosthenes stayed in Alexandria for the entire latter part of his life. The exact date of his death is not known.

Erdös, Paul (1913–1996) Hungarian *Discrete mathematics, Number theory* Born on March 25, 1913, in Budapest, Hungary, prolific mathematician Paul Erdös (pronounced "air-dish") is remembered as one of the greatest problem-solvers and problem posers of all time. With an uncanny ability to create problems that led to productive new areas of mathematics research, Erdös is credited as founder of the field of "discrete mathematics," the mathematics of computer science. With no permanent home, Erdös traveled the globe multiple times throughout his life, collaborating and writing papers with scholars from all countries. His colleagues invented the term *Erdös number* to describe their close connections to him, assigning an Erdös number of 1 to all those who had coauthored a paper with Erdös, the number 2 to those who had worked with someone who had worked with Erdös, and so on. According to his obituary in the *New York Times*, 458 mathematicians can claim an Erdös number of 1, and over 4,500 scholars an Erdös number of 2.

Erdös entered the University of Budapest in 1930 at the age of 17, and within just a few years, he began making significant contributions to the field of NUMBER THEORY. At age 20 he discovered a new and elementary proof of conjecture of Joseph Bertrand (1822–1900), stating that at least one PRIME lies between any number *n* and its double 2*n*. (Russian mathematician PAFNUTY LIVOVICH CHEBYSHEV established the validity of this claim, by complicated means, in 1850.) Later in life, Erdös also found an elementary proof of the famous PRIME NUMBER THEOREM.

In 1934, at the young age of 21, Erdös was awarded a Ph.D. in mathematics from the University of Budapest. Because of his Jewish heritage, Erdös was forced to leave Hungary, and he accepted the offer of a postdoctoral fellowship in Manchester, England. As the situation in Europe worsened, Erdös decided to move to the United States in 1938.

Erdös never accepted a permanent academic position. He preferred to devote his entire waking hours to the pursuit of mathematics and traveled from one mathematics conference or seminar to another, building up a growing circle of collaborators. In the latter part of his life, Erdös owned nothing more than a suitcase of clothes and traveled from university to university, and from the home of one mathematician to another. He developed a reputation as a brilliant mathematician and an appalling houseguest. Sleeping only 3 to 5 hours a day, Erdös would often wake his mathematical hosts at all hours of the night, eager to get cracking on more mathematical research. By the end of his life, Erdös had worked on over 1,500 mathematical papers. He died in Warsaw, Poland, on September 20, 1996.

Erdös won many prizes during his life, including the 1951 Cole Prize from the American Mathematical Society for his 1949 paper "On a New Method in Elementary Number Theory which Leads to an Elementary Proof of the Prime Number Theorem," and the 1983 Wolf Prize of $50,000 from the Wolf Foundation. He was also awarded a salary from the Hungarian Academy of Sciences. With no need of money, Erdös often gave it away, either to students in need, or as prizes for solving problems he had posed, of which there were many. Mathematicians today are still publishing papers inspired by those challenges.

error The difference between the approximate value of a quantity and the true numerical value of that quantity is called the error of the approximation. There is some confusion in the literature, however, as to how to interpret this definition. If *x* is an approximation of the value *X*, then some texts work with the difference $X - x$ when speaking of the error, whereas other texts use the difference $x - X$. (Thus, the error in using 3.6 as an approximation for 3.59, say, could be deemed as either 0.01 or −0.01.) For this reason, many authors prefer to work with the "absolute error," $|X - x|$, and avoid the issue of sign altogether.

The term *error* is also used for the uncertainty in a measurement. For example, one can typically read temperature only to the nearest degree Fahrenheit. Thus a temperature recording of 75°F should be written, or at least interpreted as, (75 ± 0.5)°F to indicate that there is a possible error of as much as half a degree. When a measurement is written in decimal notation, it is generally understood that the absolute error does not exceed a half unit in the last digit. Thus, for example, a recording of 2.3 indicates that the error does not exceed ±0.05, whereas recording 2.30 indicates that the error does not exceed ±0.005. In this context, the final digit recorded is usually understood to be reliable.

If a number representing a measurement does not have a decimal part, then a dot is sometimes used to indicate up to which point the digits are reliable. For example, in recording a measurement as 2̇300, we are being told that the 3 is the final reliable digit and that the error in this measurement could be as much as ±50 (half the 100s place). A recorded measurement of 230̇0, on the other hand, indicates that the first zero is reliable and that the error in this measurement is at most ±5 (half the 10s place).

The digits up to, and including, the reliable digit are called the significant figures of the measurement. Thus the measurement 2̇300 has two significant figures, for example, whereas the recorded measurement 230̇0 has three significant figures. If a result is expressed in SCIENTIFIC NOTATION, $p \times 10^n$, it is generally understood that all the digits of p are significant. For example, in writing a value 0.0170 as 1.70×10^{-2}, we are indicating that the final 0 is the result of a measurement, and so this digit is reliable. The quantity 0.0170 thus has three significant digits (and the error of this measurement is at most ±0.00005). Similarly a recorded measurement of 0.00030300, for example, has five significant figures. (The initial three zeros of the decimal expansion serve only to place the decimal point correctly. The remaining five digits represent the result of recording a measurement.)

When calculating with approximate values, it is important to make sure that the result does not imply an unrealistic level of precision. For example, if the dimensions of the room are measured as 14.3 ft by 10.5 ft, multiplying length by width gives the area of the room as 150.15 ft². The answer presented this way suggests a level of accuracy up to the nearest 1/100, which is unreasonable given that the initial measurements are made to the nearest 1/10. Generally, the result of a calculation should be presented as no more accurate than the least accurate initial measurement. For example, in adding measurements 230̇0 and 1068, the result should be recorded as 337̇0 (the number

3368 is rounded to the nearest 10). In multiplying 14.3 and 10.5, each with three significant figures, the result should be written 150̇ ft² (again three significant figures).

See also PERCENTAGE ERROR; PRECISION; RELATIVE ERROR; ROUND-OFF ERROR.

Euclid (ca.300–260 B.C.E.) Greek *Geometry* The geometer Euclid is remembered as author of the most famous text in the whole of mathematics, THE ELEMENTS. In 13 books, the work covers all that was known in mathematics at his time, from elementary geometry and number theory, to sophisticated theories of proportion, irrationals, and solid geometry. But Euclid is revered today primarily for his unique approach in organizing the material he presented. Starting with a small set of definitions, "common notions," and AXIOMs (basic statements whose truth seems to be self-evident), Euclid derived by pure logical reasoning some 465 propositions (THEOREMs) in mathematics. This established standards of rigor and powers of deduction that became the model of all further work in mathematics for the two millennia that followed. It can be said that *The Elements* defines what PURE MATHEMATICS is about.

Close to nothing is known of Euclid's life. It is believed that he lived and taught in Alexandria, a Greek city near the mouth of the Nile in what is now Egypt, and may have been chief librarian of the great library at the Alexandria Academy. Many ancient historical texts describing the work of Euclid confuse the mathematician Euclid of Alexandria with philosopher Euclid of Megara, who lived about 100 years earlier. Moreover, Euclid was a very common name at the time, and there were many prominent scholars from a variety fields throughout this period. Because of the subsequent confusion and the lack of specific information about the mathematician Euclid, some historians have put forward the theory that Euclid was not, in fact, a historical figure, but the name adopted by a team of mathematicians at the library of Alexandria who published a complete work under the single name Euclid. (Compare this with the fictitious NICOLAS BOURBAKI of the 20th century). This is not the popular view, however.

The Elements was deemed a standard text of study for Greek and Roman scholars for 1,000 years. It was translated into Arabic around 800 C.E. and studied extensively by Arab scholars. With the revival of scientific interest during the Renaissance, Euclid's work became the model of logical thinking in Europe. More than 2,000 different editions of the text have appeared since the first typeset version produced in the year 1482, and many great scientists of the West, including SIR ISAAC NEWTON (1642–1627) for instance, described their mastering the work of Euclid as a significant part of their development of scientific thinking. Study of *The Elements* was an integral part of the standard U.S. high-school mathematics curriculum up until the 1950s.

Other works attributed to Euclid of Alexandria that have survived today include *Data*, on the properties of figures; *On Divisions*, studying the geometric theory of dividing the areas of figures into certain proportions; and *Optics*, the first Greek work on the theory of PERSPECTIVE. It is also known that Euclid produced at least five other texts in geometry, including a four-book treatise on CONICS, as well as a work on music and another discussing general scientific principles.

See also EUCLID'S POSTULATES.

Euclidean algorithm   In his third book of THE ELEMENTS, EUCLID describes a systematic procedure for finding the GREATEST COMMON DIVISOR of any two positive integers. The method is as follows:

1. Write down the pair of numbers.
2. Subtract the smaller number from the larger.
3. Rewrite the pair of numbers but replace the larger number with the answer from step two.
4. Repeat steps two and three until you have two identical numbers. This repeated value is the greatest common factor of the two original numbers.

As an example, we calculate the greatest common factor of 42 and 60:

$$42{:}60 \to 42{:}18 \to 24{:}18 \to 6{:}18 \to 6{:}12 \to 6{:}6$$

Their greatest common factor is 6. Each step of the procedure produces a pair of numbers with smaller difference. Eventually, a pair with difference zero will result. The Euclidean algorithm is therefore sure to stop after a finite number of calculations.

Why the Algorithm Works
Suppose two numbers a and b eventually produce the value z via this procedure:

$$a:b \to c:d \to \dots \to u:v \to z:z$$

Then it is not too difficult to show that z is indeed the greatest common factor of a and b.

Firstly, if a and b are both multiples of any number n, then so is their difference. This means that all the pairs of numbers produced by this procedure remain multiples of n. In particular, z is a multiple of n. (In the example above, both 42 and 60 are multiples of 3, for example. All the numbers produced via the procedure remain multiples of 3. They are also multiples of 2 and of 6.) This establishes that z is at least as large as any common factor of a and b.

Secondly, working backward through the procedure, we see that the penultimate pair $u:v$ is obtained from $z:z$ via *addition*. (Look at the example above.) Thus both u and v are multiples of z. Working all the way back, we have in fact that all the numbers appearing in the list must be multiples of z, including both a and b. Thus z is a common factor.

These two conclusions show that z is indeed the greatest common factor we seek. As a bonus, the above two paragraphs also show that the greatest common factor of two numbers is a multiple of any other common factor.

Linear Combinations
A surprising consequence of the Euclidean algorithm is that it also gives a constructive method for writing the greatest common factor of two positive integers a and b as a linear combination of the original numbers.

Keeping track of the subtractions performed in the example above, we have:

$$
\begin{aligned}
42:60 &\to 42:18 = (42):(60{-}42)\\
&\to 24:18 = (42)-(60-42):(60-42)\\
&= (2\times42-60):(60-42)\\
&\to 6:18 = (2\times24-60)-(60-42):(60-42)\\
&= (3\times42-2\times60):(60-42)\\
&\to 6:12 = (3\times42-2\times60):(60-42)-\\
&\quad(3\times42-2\times60)\\
&= (3\times42-2\times60):(3\times60-4\times42)\\
&\to 6:6 = (3\times42-2\times60):(3\times60-4\times42)\\
&\quad-(3\times42-2\times60)\\
&= (3\times42-2\times60):(5\times60-7\times42)
\end{aligned}
$$

Thus we can write $6 = 3\times42-2\times60$ (and also $6 = 5\times60-7\times42$.) In general, this shows that it is always possible to write the greatest common factor of two numbers a and b in the form:

$$xa + yb$$

for some integers x and y. This fact is useful for solving the famous JUG-FILLING PROBLEM, for example.

See also FUNDAMENTAL THEOREM OF ARITHMETIC.

Euclidean geometry The GEOMETRY based on the definitions and AXIOMS set out in Euclid's famous work *THE ELEMENTS* is called Euclidean geometry. The salient feature of this geometry is that the fifth postulate, the PARALLEL POSTULATE, holds. It follows from this that through any point in the plane there is precisely one line through that point parallel to any given direction, that all angles in a triangle sum to precisely 180°, and that the ratio of the circumference of any circle to its diameter is always the same value π.

Two-dimensional Euclidean geometry is called plane geometry, and the three-dimensional Euclidean geometry is called solid geometry. In 1899 German mathematician DAVID HILBERT (1862–1943) proved that the theory of Euclidean geometry is free from CONTRADICTION.

See also EUCLID; EUCLID'S POSTULATES; HISTORY OF GEOMETRY (essay); NON-EUCLIDEAN GEOMETRY.

Euclidean space (Cartesian space, n-space) The VECTOR SPACE of all n-TUPLES $(x_1, x_2, \ldots, x_n)$ of real numbers $x_1, x_2, \ldots, x_n$ with the operations of addition and scalar multiplication given by:

$$(x_1, x_2, \ldots, x_n) + (y_1, y_2, \ldots, y_n) = (x_1 + y_1, x_2 + y_2, \ldots, x_n + y_n)$$
$$k(x_1, x_2, \ldots, x_n) = (kx_1, kx_2, \ldots, kx_n)$$

and equipped with the notion of distance between points $\mathbf{x} = (x_1, x_2, \ldots, x_n)$ and $\mathbf{y} = (y_1, y_2, \ldots, y_n)$ as given by the DISTANCE FORMULA:

$$d(\mathbf{x}, \mathbf{y}) = \sqrt{(x_1 - y_1)^2 + (x_2 + y_2)^2 + \ldots + (x_n - y_n)^2}$$

is called a Euclidean space.

Elements of a two-dimensional Euclidean space can be identified with points in a plane relative to a set of CARTESIAN COORDINATE axes. The vector space of all n-tuples of COMPLEX NUMBERS under an analogous set of operations is called a complex Euclidean space.

Euclid's postulates EUCLID of Alexandria (ca. 300–260 B.C.E.) began his famous 13-volume piece *THE ELEMENTS* with 23 definitions (of the ilk, "a point is that which has no part" and "a line is that which has no breadth") followed by 10 AXIOMS divided into two types: five common notions and five postulates. His common notions were:

1. Things that are equal to the same thing are equal to one another.
2. If equal things are added to equals, then the wholes are equal.
3. If equal things are subtracted from equals, then the remainders are equal.
4. Things that coincide with one another are equal to one another.
5. The whole is greater than the part.

Euclid's postulates were:

1. A straight line can be drawn to join any two points.
2. Any straight line segment can be extended to a straight line of any length.
3. Given any straight line segment, it is possible to draw a circle with center one endpoint and with the straight line segment as the radius.
4. All right angles are equal to one another.
5. If two straight lines emanating from the endpoints of a given line segment have interior angles on one given side of the line segment summing to less than two right angles, then the two lines, if extended, meet to form a triangle on that side of the line segment.

(His fourth postulate is a statement about the homogeneity of space, that it is possible to translate figures to different locations without changing their basic structure.)

It is worth noting that Euclid deliberately avoided any direct mention of the notion of infinity. His wording of the second postulate, for instance, avoids the need to state that straight lines can be extended *indefinitely*, and his fifth postulate, also known as the PARALLEL POSTULATE, avoids direct mention of parallel lines, that is, lines that never meet when extended indefinitely.

From these basic assumptions Euclid deduced, by pure logical reasoning, 465 statements of truth (THEO-

REMs) about geometric figures. The systematic approach he followed and the rigor of reasoning he introduced was hailed as a great intellectual achievement. His model of mathematical exploration became the standard for all mathematical research for the next 2,000 years.

Euclid's fifth postulate was always regarded with suspicion. It was never viewed as simple and as self-evident as his remaining four postulates, and Euclid himself did his utmost to avoid using it in his work. (Euclid did not invoke the fifth postulate until his 29th proposition.) Over the centuries scholars came to believe that the fifth postulate could be logically deduced from the remaining four postulates and therefore did not need to be listed as an axiom. Many people proposed proofs for it, including the fifth-century Greek philosopher Proclus, who is noted for his historical account of Greek geometry. Unfortunately his proof was flawed, as were the proofs proposed by Arab scholars of the eighth and ninth centuries, and by Western scholars of the Renaissance.

In 1733 Italian teacher and scholar GIROLAMO SAC-CHERI (1667–1733) believed that because Euclid's axioms model the real world, which he thought to be consistent, they cannot lead to a CONTRADICTION. If the first four postulates do indeed imply that the fifth postulate is also true, then assuming the four postulates together with the negation of the fifth postulate should lead to a logical inconsistency. Unfortunately, in following this tact, Saccheri never came across a contradiction.

In 1795 Scottish mathematician and physicist John Playfair (1748–1819) proposed an alternative formulation of the famous fifth postulate (today known as PLAYFAIR'S AXIOM). This version of the axiom is considerably easier to handle, and its negation is easier to envision. In an attempt to follow Saccheri's approach, Russian mathematician NICOLAI IVANOVICH LOBACHEVSKY (1792–1856) and Hungarian mathematician JÁNOS BOLYAI (1802–1860), independently came to the same surprising conclusion: *the first four of Euclid's postulates together with the negation of Playfair's version of the fifth postulate will not lead to a contradiction.* This established, once and for all, that the fifth postulate is an INDEPENDENT AXIOM and cannot be deduced from the remaining four postulates. More important, by exploring the geometries that result in assuming that the fifth postulate does not hold, scholars were led to the discovery of NON-EUCLIDEAN GEOMETRY.

In the late 1800s the German mathematician DAVID HILBERT (1862–1943) noted that, despite its rigor, Euclid's work contained many hidden assumptions. He also realized, despite Euclid's attempts to describe them, that the notions of "point," "line," and "plane" cannot be properly defined and must remain as undefined terms in any theory of geometry. In his 1899 work *Grundlagen der Geometrie* (Foundations of geometry) Hilbert refined and expanded Euclid's postulates into a list of 28 basic assumptions that define all that is needed in a complete account of Euclid's geometry. His axioms are today referred to as Hilbert's axioms.

See also EUCLIDEAN GEOMETRY; HYPERBOLIC GEOMETRY; SPHERICAL GEOMETRY.

Euclid's proof of the infinitude of primes Around the third century B.C.E., EUCLID proved that there is no such thing as a largest PRIME number, meaning that the list of primes goes on forever. He presented his proof as Proposition IX.20 in his book *THE ELEMENTS*, and he was the first to recognize and prove this fact about prime numbers.

Euclid's proof relies on the observation that any number N is either prime, or factors into primes. His argument proceeds as follows:

Suppose to the contrary that there is a largest prime number p. Then the finite list 2, 3, 5, 7, …, p contains *all* the prime numbers. But consider the quantity:

$$N = 2 \times 3 \times 5 \times 7 \times … \times p + 1$$

It is not divisible by any of prime numbers in our list (it leaves a remainder of one each time), and so it has no prime factor. It must be the case then that N is prime. Thus we have created a new prime number larger than the largest prime p. This absurdity shows that our assumption that there are only finitely many primes must be false.

Euclid's argument is a classic example of a PROOF BY CONTRADICTION. His argument also provides an ALGORITHM for generating new primes from any finite list of primes. For example, from the list of primes 2, 3, 7, Euclid's argument yields $N = 2 \cdot 3 \cdot 7 + 1 = 43$ as a new prime, and from the list 2,3,7,43, we have $N = 2 \cdot 3 \cdot 7 \cdot 43 + 1 = 1807 = 13 \times 139$, yielding 13 as a new prime.

Euclid's argument can be developed further to obtain other interesting facts about prime numbers. For example, we can prove:

There are infinitely many primes that leave a remainder of 3 when divided by 4. (That is, there are infinitely many primes of the form $4n + 3$.)

Again, suppose to the contrary, that the list of such primes is finite: 3, 7, 11,...,p, and this time consider the quantity. $N = 4 \times (3 \times 7 \times 11 \times...\times p) -1$. If this number is prime, then we have found a new prime of the required form. If it is not, then it factors into primes: $N = p_1 \times p_2 \times...\times p_k$. Notice that since N is not divisible by 3, none of its prime factors is equal to 3. It also cannot be the case either that all of the prime factors p_i leave a remainder of 1 when divided by 4 (for then N would also leave a remainder of 1). Thus at least one of these prime factors is a prime of the form $4n + 3$ not already in our list of such primes. It must be the case then that the list of such primes goes on forever.

In a similar way (though it is a little more complicated) one can also prove:

There are infinitely many primes of the form $6n+5$.

(Use $N = 2 \times 3 \times 5 \times...\times p - 1$.) PETER GUSTAV LEJEUNE DIRICHLET (1805–59) generalized these results to prove that if a and b are any two RELATIVELY PRIME whole numbers, then there are infinitely many primes of the form $an + b$.

See also FUNDAMENTAL THEOREM OF ARITHMETIC.

Eudoxus of Cnidus (ca. 408–355 B.C.E.) Greek *Geometry, Number theory, Astronomy* Born in Cnidus, in Asia Minor (now Turkey), Eudoxus is remembered as one of the greatest mathematicians of antiquity. All of his original work is lost, but it is known from later writers that he was responsible for the material presented in Book V of EUCLID's famous treatise *THE ELEMENTS*. In his theory of proportions, Eudoxus developed a coherent theory of REAL NUMBERS using absolute rigor and precision. The full importance of this sophisticated work came to light some two millennia later, when scholars of the 19th century attempted to resolve some

fundamental difficulties with the theory of CALCULUS. They discovered that Eudoxus had already anticipated these fundamental problems and had made significant steps toward resolving them. Eudoxus is also remembered as the first to develop a "method of exhaustion" for computing the AREA of curved figures.

As a young man Eudoxus traveled to Tarentum, now in Italy, to study number theory, geometry, and astronomy with ARCHYTAS OF TARENTUM, a follower of PYTHAGORAS. Both men worked to solve the famous problem of DUPLICATING THE CUBE and, in fact, Eudoxus came up with his own geometric solution to the challenge using special curved lines as an aid. (Although the problem calls for the use of nothing more than a compass and a straight edge, this partial solution was nonetheless a significant achievement.)

Eudoxus studied the theory of proportions. This blend of GEOMETRY and NUMBER THEORY calls two lengths a and b COMMENSURABLE if they are each a whole-number multiple of some smaller length t: $a = mt$ and $b = nt$. In this approach, two ratios $a : b$ and $c : d$ are said to be equal if they are the same multiples of some fundamental lengths t and s: $a = mt$, $b = nt$ and $c = ms$, $d = ns$.

For a long time it was believed that all lengths were commensurable and hence all ratios could be compared. Consequently, the Pythagorean discovery of two incommensurable lengths, namely 1 and $\sqrt{2}$, the side length and the diagonal of a unit square, caused a crisis in the mathematical community. As HIPPASUS OF METAPONTUM (ca. 470 B.C.E.) discovered, there is no small value t such that $1 = mt$ and $\sqrt{2} = nt$. (This is equivalent to the statement that the number $\sqrt{2}$ cannot be written as a fraction n/m.)

Eudoxus came to resolve the crisis of comparing ratios even if they are not commensurable by avoiding all use of a common length t. He defined ratios $a : b$ and $c : d$ to be equal if, for every possible pairs of integers n and m:

i. $ma < nb$ if $mc < nd$
ii. $ma = nb$ if $mc = nd$
iii. $ma > nb$ if $mc > nd$

With this formulation, Eudoxus was able to compare lines of any length, either rational or irrational, and obviate all philosophical difficulties associated with incommensurable quantities. Mathematician JULIUS WILHELM RICHARD DEDEKIND (1831–1916) based his theory of DEDEKIND CUTS on this approach developed by Eudoxus.

In geometry, Eudoxus was the first to establish that the volume of a cone is one-third the volume of the cylinder that surrounds it, and also that the volume of a pyramid is one third the volume of a prism of the same base and height. ARCHIMEDES OF SYRACUSE (ca. 287–212 B.C.E.) made use of these results in his famous treatise *On the Sphere and Cylinder,* citing Eudoxus as the person who first proved them.

Eudoxus maintained an active interest in astronomy throughout his life. He built an observatory in the city of Cnidus and made careful note of the motion of the planets and stars across the night skies. Outside of mathematics, Eudoxus is best known for his ingenious theory of planetary motion based on a system of 27 nested spheres. Using advanced techniques in three-dimensional geometry, Eudoxus was able to use this model to explain the puzzling retrograde motion of the heavenly bodies.

Euler, Leonhard (1707–1783) Swiss *Analysis, Geometry, Number theory, Graph theory, Mechanics, Physics* Born on April 15, 1707, in Basel, Switzerland, genius Leonhard Euler was, beyond comparison, the most prolific mathematician of all time. With over 850 books and papers to his name, Euler made fundamental contributions to virtually every branch of mathematics of his day. He formalized the notion of a FUNCTION (and introduced the notation $f(x)$ for it), and thereby changed the focus of mathematics from a study of fixed curves and lines to a more powerful study of transformation and change. (He was the first, for instance, to regard the special functions from TRIGONOMETRY as functions.) Euler published works on ANALYSIS, DIFFERENTIAL CALCULUS, INTEGRAL CALCULUS, DIFFERENTIAL EQUATIONS, NUMBER THEORY, GEOMETRY, LOGIC, COMBINATORICS, approximations for π, planetary motion and astronomy, navigation, cartography, mechanics, and more. He introduced and made popular many of the standard symbols we use today (such as i for $\sqrt{-1}$, π for PI, E for his famous number, and Σ for SUMMATION). It is simply not possible in a short piece to give proper justice to the phenomenal quantity of contributions Euler made to the study of mathematics.

Euler obtained a master's degree in philosophy from the University of Basel in 1723, following the

Leonhard Euler, the most prolific of famous mathematicians, made significant contributions to almost every field of pure and applied mathematics studied at his time while also establishing new courses of research. Much of the mathematical notation used today was either introduced or made standard by Euler. *(Photo courtesy of Topham/The Image Works)*

path his father set for him to study theology. Euler's interests, however, lay with mathematics, and Euler remained at the university another three years to pursue a course of study in the subject. In 1727 he submitted an entry for the Grand Prize of the Paris Academy of Sciences on the best arrangement of masts on a ship. Euler won second prize, which garnered the attention of the scientific community as an outstanding young graduate. Euler accepted a position at the St. Petersburg Academy of Science that year and was promoted to a full professor of the academy just three years later.

In 1736 Euler published *Mechanica* (Mechanics), a landmark piece that introduced rigorous mathematical techniques as a means for studying the subject. He won the Grand Prize of the Paris Academy in 1738, and again in 1740, and by this time was a highly regarded scholar. At the invitation of Frederick the Great, Euler joined the Berlin Academy of Science in 1741. He remained there for 25 years, assuming the leadership of the academy in 1759.

During his time at Berlin, Euler wrote over 350 articles and several influential books on a wide variety of topics in both pure and applied mathematics. In 1753 he invented the paddle wheel and the screw propeller as means of propelling ships without wind. In 1761 he developed a method of using observational data about the planet Venus to determine the distance of the Earth from the Sun, and also to make measurements of longitude on the Earth's surface.

In 1766 Euler returned to St. Petersburg, but soon afterward fell ill and lost his eyesight. Despite being blind, Euler continued to produce significant pieces of work. (He published approximately 500 books and papers while blind.) Euler also worked to popularize the scientific method and wrote, between the years 1768 and 1772, his famous three-volume piece *Letters to a German Princess* on the topic of popular science. Euler remained at St. Petersburg until his death in 1783.

Euler's name is attached to at least one fundamental concept in nearly every branch of mathematics. For instance, EULER'S THEOREM is a key result in GRAPH THEORY, linking the number of vertices and edges of a graph to the number of regions it produces. The number *e* plays a fundamental role in the theory of differential calculus, differential equations, PROBABILITY theory, STATISTICS, the theory of SUBFACTORIALs and derangements, the study of COMPOUND INTEREST, and the study of COMPLEX NUMBERS through EULER'S FORMULA, for instance. In number theory, EULER'S CONSTANT plays a key role in the study of the HARMONIC SERIES, for instance. Euler found a formula for the "Euler totient function" that provides, for a number *n*, the count of numbers less than *n* that are RELATIVELY PRIME to *n* and showed its importance in the theory of MODULAR ARITHMETIC. Euler's name is intimately associated with the study of the ZETA FUNCTION, with the gamma function in the examination of the FACTORIAL function, with the study of LATIN SQUARES, and with the construction of even PERFECT NUMBERS. (No one to this day knows whether or not examples of odd perfect numbers exist.)

Less well known is Euler's polynomial, $n^2 - n + 41$, which produces a PRIME output for every integer input from −39 through to 40. The 40 × 117 × 240 rectangular block, called "Euler's brick," has the property that

any diagonal drawn on the face of this solid also has integer length. Euler showed that there are infinitely many such blocks with integer side-lengths and integer face diagonals. (No one to this day knows whether or not there exists an Euler brick with internal space diagonals also of integer length.)

Euler died in St. Petersburg, Russia, on September 18, 1783. It is not an exaggeration to say that Euler offered profound insights on practically every branch of mathematics and mathematical physics studied at his time and, moreover, paved the way for many new branches of mathematics research. In 1915 the Euler Committee of the Swiss Academy of Science began collating and publishing his complete works. Divided into four series—mathematics, mechanics and astronomy, optics and sound, and letters and notebooks—76 volumes of work have been released thus far (covering approximately 25,000 pages of written material), and the committee projects another eight volumes of material still to be released.

Eulerian path/circuit   *See* GRAPH THEORY.

Euler line   For any triangle the following are true:

1. The three MEDIANS OF A TRIANGLE meet at a point *G*, the centroid of the triangle.
2. The three ALTITUDES of a triangle meet at a point *H*, the orthocenter of the triangle.
3. The perpendicular bisectors of each side of a triangle meet at a point *O*, the circumcenter of the triangle.

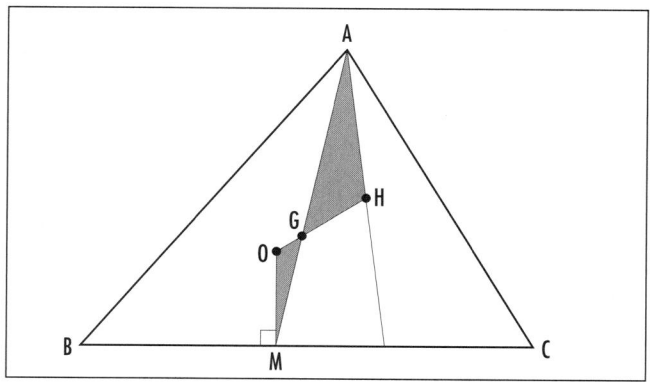

The Euler line

These statements are proved through a study of the median of a triangle, the altitude of a triangle, and the consideration of EQUIDISTANT points, respectively.

In the mid-1700s LEONHARD EULER (1707–83) made the astounding discovery that furthermore, for any triangle, the three points G, H, and O are COLLINEAR, that is, lie on a straight line. This line is called the Euler line of the triangle.

Euler proved this observation as follows: If, by chance, the points O and G coincide, then each median of the triangle is also an altitude. This means that the triangle is symmetric about each median, and so must be equilateral. Consequently, the point H occurs at the same location as O and G, and the three points, trivially, lie on a straight line. If, as is more likely the case, O and G do not coincide, then draw a line through them and consider a point J on this line that is situated so that the length of the segment GJ is twice that of OG. Let M be the midpoint of the base of the triangle.

From a study of the medians of a triangle, we know that length of the segment AG to that of GM in the diagram above is in ratio of 2 to 1. Consequently, the two shaded triangles are similar, and, in particular, angles AJG and GOM match. By the converse of the PARALLEL POSTULATE, lines OM and AJ are parallel. Since OM makes an angle of 90° to the base of the triangle, so too must line AJ, making this line an altitude to the triangle.

Nothing in this argument thus far has relied on vertex A being the object of focus. The same reasoning shows that the altitude from vertex B also passes through the point J, as does the altitude from vertex C. This shows that the point J is in fact the orthocenter H of the triangle. Consequently, O, G, and H do indeed all lie on the same straight line.

Euler's constant In drawing rectangles of width 1 that just cover the curve $y = 1/x$, one sees that the "excess area" above the curve fits inside the first rectangle of height 1, and so sums to a finite value no larger than 1. The amount of excess area, denoted γ, is called Euler's constant. To eight decimal places, it has value 0.57721566. No one knows whether γ is a rational or irrational number.

As the area under the curve from $x = 1$ to $x = n$ is $\int_1^n dx = \ln n$, we have that $1 + \frac{1}{2} + \frac{1}{3} + \ldots + \frac{1}{n-1}$ is approximately equal to $\ln(n)$. More precisely, the sum of the areas of the first n rectangles is given by:

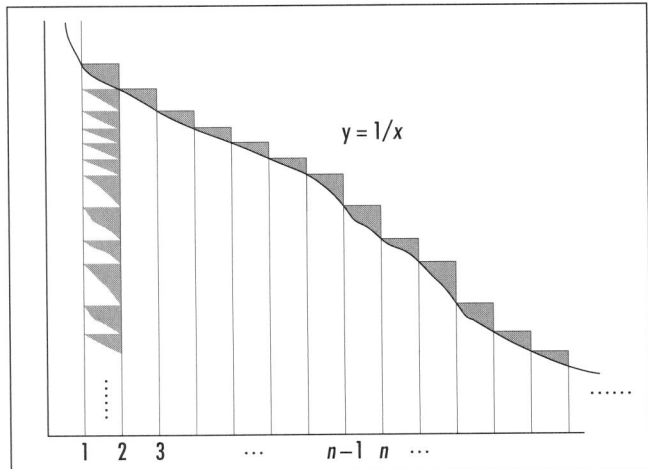

Understanding Euler's constant

$$1 + \frac{1}{2} + \frac{1}{3} + \ldots + \frac{1}{n} = \ln n + \gamma + \text{error}$$

where the "error" is the term $\frac{1}{n}$ minus all the "excess areas" above the curve from position n onward. Notice that these excess areas all fit within the rectangle of height $\frac{1}{n}$, so this error is no bigger than $\frac{1}{n}$. In particular, it is negligibly small if n is large.

See also HARMONIC SERIES.

Euler's formula In 1748 LEONHARD EULER noted that the TAYLOR SERIES for the functions e^x, sin x, and cos x are intimately connected. Since

$$e^x = 1 + \frac{x}{1!} + \frac{x^2}{2!} + \frac{x^3}{3!} + \frac{x^4}{4!} + \frac{x^5}{5!} + \frac{x^6}{6!} +$$

$$\sin x = \frac{x}{1!} - \frac{x^3}{3!} + \frac{x^5}{5!} - \cdots$$

$$\cos x = 1 - \frac{x^2}{2!} + \frac{x^4}{4!} - \cdots$$

setting $x = i\theta$, where i is the square root of –1 and θ is a real number (usually thought of as an angle), yields:

$$e^{i\theta} = 1 + \frac{(i\theta)}{1!} + \frac{(i\theta)^2}{2!} + \frac{(i\theta)^3}{3!} + \frac{(i\theta)^4}{4!} + \frac{(i\theta)^5}{5!} + \cdots$$

$$= 1 + i\frac{\theta}{1!} + i^2\frac{\theta^2}{2!} + i^3\frac{\theta^3}{3!} + i^4\frac{\theta^4}{4!} + i^5\frac{\theta^5}{5!} + \cdots$$

$$= 1 + i\frac{\theta}{1!} - \frac{\theta^2}{2!} - i\frac{\theta^3}{3!} + \frac{\theta^4}{4!} - i\frac{\theta^5}{5!} + \cdots$$
$$= \left(1 - \frac{\theta^2}{2!} + \frac{\theta^4}{4!} - \cdots\right) + i\left(\frac{\theta}{1!} - \frac{\theta^3}{3!} + \frac{\theta^5}{5!} - \cdots\right)$$
$$= \cos(\theta) + i\sin(\theta)$$

The formula $e^{i\theta} = \cos\theta + i\sin\theta$ is today known as Euler's formula. It has some interesting consequences:

Setting $\theta = \pi$, we obtain: $e^{i\pi} = \cos\pi + i\sin\pi = -1 + i \cdot 0 = -1$. Mathematicians often deem this as one of the most beautiful facts of mathematics: it is a remarkably simple equation that connects the mysterious, and pervasive, numbers e, π, i, and -1.

Setting $\theta = \frac{\pi}{2}$ yields $e^{i\frac{\pi}{2}} = i$, which shows that $i^i = \left(e^{i\frac{\pi}{2}}\right)^i = e^{i^2\frac{\pi}{2}} = e^{-\frac{\pi}{2}}$. Thus raising a complex number to a complex power can yield a real answer as a result.

Euler's formula provides a very simple means for deriving (and memorizing) certain identities from TRIGONOMETRY. For example, since

$$e^{iA} \cdot e^{iB} = e^{i(A+B)}$$

we have:

$$(\cos A + i\sin A) \cdot (\cos B + i\sin B) = \cos(A+B) + i\sin(A+B)$$

Expanding the brackets on the left and collecting terms that contain i and those that do not quickly yields:

$$\cos(A+B) = \cos A \cdot \cos B - \sin A \cdot \sin B$$
$$\sin(A+B) = \sin A \cdot \cos B + \cos A \cdot \sin B$$

Similarly, the equations $(e^{iA})^2 = e^{i(2A)}$, $(e^{iA})^3 = e^{i(3A)}$, and so forth yield double-angle and triple-angle formulae, for example.

Euler's formula is also used to represent complex numbers. For example, if z is a point in the complex plane a distance r from the origin, making an angle θ with the x-axis, then its x- and y-coordinates can be written:

$$x = r\cos\theta$$
$$y = r\sin\theta$$

and the complex number is thus:

$$z = x + iy = r\cos\theta + ir\sin\theta = re^{i\theta}$$

This is called the polar form of the complex number. If one multiples two complex numbers, $z = re^{i\theta}$ and $w = se^{i\tau}$, we see that $z \cdot w = rse^{i(\theta+\tau)}$, that is:

The product of two complex numbers is a new complex number whose distance from the origin is the product of the distances from the origin of the two original numbers, and whose angle with the x-axis is the sum of the two angles made by the two original numbers.

Euler's formula makes the derivation of this fact swift and easy.

See also COMPLEX NUMBERS; DE MOIVRE'S FORMULA; E HYPERBOLIC FUNCTIONS.

Euler's theorem (**Euler's formula, Euler-Descartes formula**) A GRAPH is a collection of dots, called vertices, connected in pairs by line segments, called edges, subsequently dividing the plane into a finite number of regions. In 1752 LEONHARD EULER showed that if a graph drawn on the plane has v vertices, e edges, and divides the plane into a total of r regions (this includes the large "outer region"), then:

$$v - e + r = 1 + c$$

where c is the number of "connected components" of the graph, that is, the number of distinct pieces of which it is composed. For example, the graph shown is composed of two "distinct pieces" ($c = 2$) and has nine vertices, 13 edges, and divides the plane into seven distinct regions, and indeed $v - e + r$ equals 3, one more than c.

The formula is easily proved via an INDUCTION argument on the number of edges: if a graph has no edges, then it consists solely of v disconnected points. Thus it has $c = v$ components and divides the plane into just one region. The formula $v - e + r = 1 + c$ holds true. One checks that adding an edge either divides a region into two (thereby increasing the value of r by one), creates an extra region if that edge is a loop (again increasing the value of r by 1), or connects two disconnected components of the graph (thereby decreasing the value of c by 1). In all cases the formula $v - e + r = 1 + c$ remains balanced.

The formula is usually applied to a graph that is connected, that is, has only one component ($c = 1$). In this case the formula reads:

$$v - e + r = 2$$

and this version of the equation is usually called Euler's theorem.

The vertices, edges, and faces of a POLYHEDRON can be thought of as a connected graph. For example, a cube with its top face removed and pushed flat onto a plane yields a graph with eight vertices and 12 edges dividing the plane into six regions. (The large "outer" region corresponds to the top face of the cube that was removed.) We still have: $v - e + r = 2$. In general, for any polyhedron with v vertices, e edges, and f faces we have:

$$v - e + f = 2$$

This was first observed by RENÉ DESCARTES in 1635. Euler had no knowledge of Descartes's work when he developed the formula in the more general setting of graph theory. For this reason, this famous formula is also called the Euler-Descartes formula.

This result holds true only for graphs that lie on the plane (or polyhedra that can be pushed flat onto a plane). One can show that for connected graphs drawn on a TORUS, for example, the formula must be adjusted to read: $v - e + r = 0$. For example, if a polyhedron

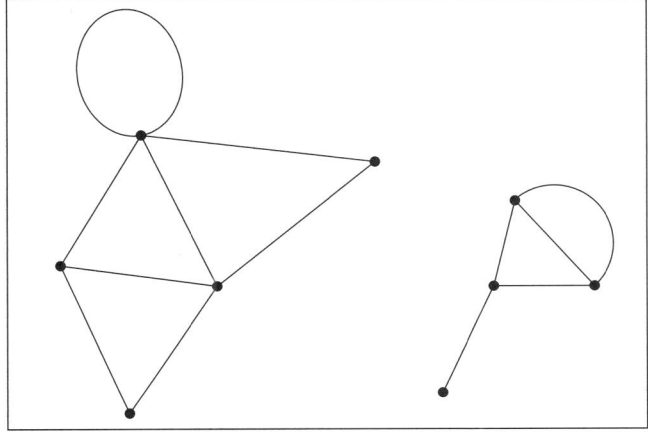

A graph with two components

contains a hole (say, a cube with a square hole drilled through it), one has: $v - e + f = 0$. (There is one technical difficulty here: one needs to be sure that each region or face under consideration is not itself an ANNULUS. One may need to draw in extra edges to break regions into suitable form.)

even and odd functions A function $y = f(x)$ is said to be even if, for each x, the function takes the same value at both x and $-x$, that is, $f(-x) = f(x)$ for all values of x. The graph of an even function is consequently symmetrical about the vertical axis. The functions x^2, $\cos(x)$, and $\dfrac{3x^2 - 5}{1 + x^4}$, for example, are even functions.

A function $y = f(x)$ is said to be odd if, for each x, the function takes opposite values at x and $-x$, that is, $f(-x) = -f(x)$ for all values of x. The graph of an odd function is consequently symmetric with respect to a 180° rotation about the origin. The functions x, x^3, $\sin(x)$, and $\dfrac{x + 5x^3 - 17x^{93}}{3 + 2x^4}$, for example, are odd functions.

Any function $g(x)$ can be expressed as the sum of an even and an odd function. Let:

$$f_{even}(x) = \frac{g(x) + g(-x)}{2}$$

$$f_{odd}(x) = \frac{g(x) - g(-x)}{2}$$

Then $f_{even}(x)$ is even, $f_{odd}(x)$ is odd, and $g(x) = f_{even}(x) + f_{odd}(x)$.

The FOURIER SERIES of any even function contains only cosine terms, and the Fourier series of any odd function only sine terms. The absolute value of any function $f(x)$ is an even function, that is, if $y = |f(x)|$, then y is even.

even and odd numbers Working solely in the realm of the whole numbers, a number is said to be even if it is divisible by 2, and odd if it leaves a remainder of 1 when divided by 2. For example 18 is divisible by 2 and so is even, and 23 leaves a remainder of 1 and so is odd.

There is a physical interpretation to the evenness or oddness of a number: An even number of pebbles, say, represents a pile that can be split into two equal parts—18 pebbles split into two piles of 9, for example. An odd pile of pebbles leaves an extra pebble when one attempts to accomplish this feat: 23 pebbles splits into two piles of 11, with one left over. This interpretation shows that the number zero is even: an empty pile of pebbles can be split into two equal piles of nothing.

Combining together several piles of pebbles, each of which can be evenly split, produces a large pile that still can be so separated. This shows that the sum of any collection of even numbers is even. Combining two odd piles of pebbles produces a result that is even—the two "errant" pebbles combine to produce a result that can be evenly split. The sum of three odd numbers leaves a single errant pebble, however, and so is odd. In summary we have:

> The sum of any number of even numbers is even.
> The sum of an even number of odd numbers is even.
> The sum of an odd number of odd numbers is odd.

These simple ideas are quite powerful. For example, we can quickly ascertain that:

> The sum of the numbers 1 through 100 will be even.

(This sum contains 50 even numbers and 50 odd numbers.)

> It is impossible to make change for a dollar using a combination of 13 pennies, nickels and quarters.

(Thirteen coins of odd denomination will sum to an odd amount. It is impossible then to reach a sum of 100.)

> If 15 arbitrary sheets are torn from a textbook, then the sum of the missing page numbers is necessarily odd.

(Each sheet contains an odd page number on one side and an even page number on the other. The sum of 15 odd numbers and 15 even numbers is necessarily odd.)

> Seventeen cups are placed upside-down on a tabletop. Turning four cups over at a time, it is impossible to reach a state in which every cup is upright.

(To be left upright, each cup must be turned over an odd number of times in the process of the game. Thus an odd number of inversions must occur in all, being a sum of 17 odd numbers. But an odd total will never occur when turning an even number of cups over at a time.)

See also PARITY.

event Any subset of the SAMPLE SPACE of all possible outcomes of an experiment is called an event. For example, the act of casting a die has sample space {1, 2, 3, 4, 5, 6}—all six possible scores—and the event "the score is even" is the subset {2, 4, 6}. An event could be a single outcome ("rolling a two," for example, is the subset {2}), the whole sample space ("rolling a number less then 10"), or the empty set ("rolling a multiple of seven").

The probability of an event E occurring is the ratio of the number of outcomes in that event to the total number of outcomes possible. This ratio is denoted $P(E)$. For example, in casting a die, $P(even) = \frac{3}{6} = \frac{1}{2}$, $P(\{5,6\}) = \frac{2}{6} = \frac{1}{3}$, and $P(multiple\ of\ 7) = \frac{0}{6} = 0$.

SET THEORY is useful for analyzing the chances of combinations of events occurring. If A and B represent two events for an experiment, then:

> The intersection $A \cap B$ represents the event "both A and B occur."
> The union $A \cup B$ represents the event "either A or B occurs."
> The complement A' represents the event "A does not occur."

To illustrate: if, in casting a die, A is the event {2, 4, 6} and B the event {5,6}, then $A \cap B = \{6\}$, $A \cup B = \{2, 4, 5, 6\}$, and $A' = \{1, 3, 5\}$.

The following probability laws hold for two events A and B:

i. $P(A \cap B) = P(A) + P(B) - P(A \cup B)$.
ii. When A and B are disjoint events, that is have no outcomes in common, then $P(A \cup B) = P(A) + P(B)$.
iii. $P(A') = 1 - P(A)$.
iv. When A and B are INDEPENDENT EVENTS, $P(A \cap B) = P(A) \times P(B)$.

The study of PROBABILITY explains these rules.

See also CONDITIONAL; MUTUALLY EXCLUSIVE EVENTS; ODDS.

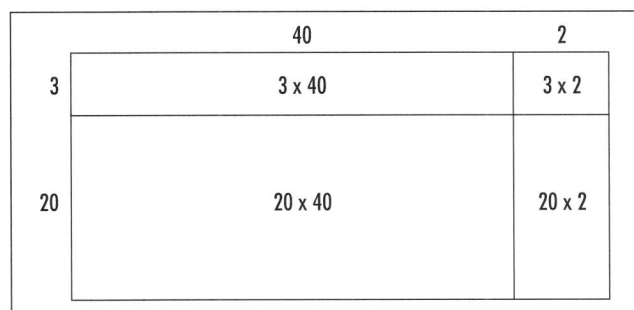

Computing the product of 23 and 42

expanding brackets The geometric figure of a rectangle explains the process of expanding brackets. Take, for example, a 23-by-42 rectangle. Its AREA is given by the product 23 × 42. This product can easily be computed by thinking of 23 as 20 + 3 and 42 as 40 + 2. This corresponds to subdividing the rectangle into four pieces:

We thus have (20 + 3) × (40 + 2) = 20 × 40 + 20 × 2 + 3 × 40 + 3 × 2, which equals 800 + 40 + 120 + 6, or, 966, which is indeed 23 × 42.

Note that each of the four terms in the sum is the product of one number in the first set of parentheses (20 or 3), and one number in the second set (40 and 2), with all possible pairs of numbers appearing. This principle holds in general. For example, the quantity $(x + y)(a + b + c)$ equals the sum of six products: $xa + xb + xc + ya + yb + yc$ (this corresponds to subdividing a rectangle into six pieces), and $(r + s + t + u + v)(k + l + m + n + o + p + q)$ is the sum of 35 individual products.

This principle extends to any number of sets of parentheses. For example:

$$(2 + 3) \times (4 + 5) \times (6 + 7) = 2 \times 4 \times 6 + 2 \times 4 \times 7 + 2 \times 5 \times 6 + 2 \times 5 \times 7 + 3 \times 4 \times 6 + 3 \times 4 \times 7 + 3 \times 5 \times 6 + 3 \times 5 \times 7$$

(Again select one term from each set of parentheses, making sure to include all possible combinations.) This corresponds to subdividing a *cube* into eight pieces.

It also holds for products containing single terms along with sets of parentheses. For example, we have: $(a + b) \times x \times (c + d) = a \times x \times c + a \times x \times d + b \times x \times c + b \times x \times d$.

Many schools teach mnemonic devices for correctly expanding brackets. These can be more compli-

cated than simply understanding the simple process at hand.

See also DISTRIBUTIVE PROPERTY; NESTED MULTIPLICATION.

expected value (expectation, mean) The expected value of a game of chance involving monetary bets is the average or MEAN profit (or loss) per game you would expect if the game were played a large number of times. The expected value illustrates the extent to which a game is set to, or against, your favor. To demonstrate: imagine you have the opportunity to play the following dice game:

You toss a single die and look at the score cast. If a 1 comes up you win $10, and if a 2 appears you win $5. If any other number is cast, you pay a fee of $3 for playing the game.

Is this a game worth playing?

With 600 plays of this game, one would expect close to one-sixth of those rolls (around 100 of them) to yield a 1, and hence a gain of $10, another sixth of the rolls (that is, about 100 rolls) to yield a 2 and a gain of $5, and two-thirds of the rolls (around 400 of them) to result in a loss of $3 (that is, a –$3 profit). The average profit over 600 rolls would thus be:

$$\frac{100 \times 10 + 100 \times 5 + 400 \times (-3)}{600} = \frac{1}{6} \times 10 + \frac{1}{6} \times 5 + \frac{4}{6} \times (-3)$$
$$= 0.50$$

that is, a gain of 50 cents per game. This positive expected value shows that the game is indeed worth playing. Note, however, that one might still lose money while playing the game. What has been demonstrated here is that, for the long run, the game is set to your favor.

Note the appearance of the fractions 1/6, 1/6, and 4/6 in our computation of expected value. These are the probabilities of each identified outcome actually occurring. Such probabilities always appear when computing expected value. In general, if an experiment yields numerical values $x_1, x_2,...,x_n$, with p_i being the probability that outcome x_i occurs, then the expected value of the experiment is given by:

$$x_1 p_1 + x_2 p_2 + ... + x_n p_n$$

For example, in tossing a pair of dice, the expected sum is:

$$2 \times \frac{1}{36} + 3 \times \frac{2}{36} + 4 \times \frac{3}{36} + 5 \times \frac{4}{36} + 6 \times \frac{5}{36} + 7 \times \frac{6}{36}$$
$$+ 8 \times \frac{5}{36} + 9 \times \frac{4}{36} + 10 \times \frac{3}{36} + 11 \times \frac{2}{36} + 12 \times \frac{1}{36} = 7$$

(The possible sums are the numbers 2 through 12, with the probability of casting 2 being 1/36, a 3 being 2/36, and so on.)

Expected value is usually denoted by the letter μ from the GREEK ALPHABET. If an experiment has an infinite number of possible outcomes, then the expected value is given by an infinite sum (SERIES). The BINOMIAL DISTRIBUTION is an example of this. If the random phenomenon can produce a continuous array of values (for example, the height of an individual can be any value, including fractional ones), then the expected value is given by an INTEGRAL:

$$\mu = \int_{-\infty}^{\infty} x \, p(x) \, dx$$

Here $p(x)$ is the probability density function of the random variable under consideration for the given DISTRIBUTION.

The notion of expected value was first developed by Dutch scientist Christiaan Huygens (1629–95) in his treatise *On Reasoning in a Dice Game*.

See also HISTORY OF PROBABILITY AND STATISTICS (essay).

exponent (index)   For a real number b and a positive whole number m, the shorthand b^m is used for the repeated multiplication of b with itself m times: $b^m = b \times b \times ... \times b$ (m times). Thus, for example, $2^5 = 2 \times 2 \times 2 \times 2 \times 2 = 32$, $(-1)^3 = (-1) \times (-1) \times (-1) = -1$, $\left(\frac{1}{3}\right)^2 = \frac{1}{3} \times \frac{1}{3} = \frac{1}{9}$, and $10^1 = 10$. In the expression b^m, m is called the exponent, or the index, and b is called the base of the exponent. We also call b^m a power of b.

The product of two expressions b^m and b^n with the same base b is itself a repeated multiplication of the number b. Precisely:

$$b^m \times b^n = (\overbrace{b \times b \times \cdots \times b}^{m \text{ times}}) \times (\overbrace{b \times b \times \cdots \times b}^{n \text{ times}})$$
$$= \overbrace{b \times b \times \cdots \times b \times b}^{m+n \text{ times}} = b^{m+n}$$

This establishes the multiplication rule for exponents:

To multiply two expressions with the same base, retain the common base and add together the exponents: $b^m \times b^n = b^{m+n}$.

The power rule for exponents, $(b^m)^n = b^{mn}$, follows. (One must add m with itself n times.) The multiplication rule is considered fundamental and allows us to define exponential quantities b^m for values of m other than whole numbers. We follow the principle that the multiplication rule is to always hold.

Consider, for example, the expression 2^0. This quantity has no meaning when interpreted as "the multiplication of two with itself zero times." However, one can assign a meaningful value to this expression by multiplying it with another power of two. For example, according to the multiplication rule, it must be the case that $2^0 \times 2^5 = 2^{0+5} = 2^5$. This says that $2^0 \times 32 = 32$, which tells us that 2^0 must equal one. The multiplication rule thus leads to the rule:

The zero exponent for any nonzero base equals 1: $b^0 = 1$.

To make sense of the quantity 2^{-1}, again invoke the multiplication rule. We have, for example, $2^{-1} \times 2^3 = 2^{-1+3} = 2^2$. This reads: $2^{-1} \times 8 = 4$. It must be the case then that $2^{-1} = 1/2$. Similar calculations show that 2^{-2} must equal 1/4, and that 2^{-3} must equal 1/8. In general, $2^{-m} = \frac{1}{2^m}$. This works for any nonzero base b.

A negative exponent indicates that a reciprocal must be taken: $b^{-n} = \frac{1}{b^n}$.

We can make use of this observation to compute $\frac{5^6}{5^4}$, for example. Rewriting, we have $\frac{5^6}{5^4} = 5^6 \times \frac{1}{5^4} = 5^6 \times 5^{-4} = 5^{6-4} = 5^2 = 25$.

To divide two expressions with the same base, retain the common base and subtract the exponents: $\frac{b^m}{b^n} = b^{n-m}$

The multiplication rule also allows us to make sense of fractional exponents. Consider the quantity $2^{\frac{1}{2}}$. It must be the case that $2^{\frac{1}{2}} \times 2^{\frac{1}{2}} = 2^1 = 2$. Thus $2^{\frac{1}{2}}$ is a value that, when multiplied by itself, equals 2. Consequently $2^{\frac{1}{2}} = \sqrt{2}$. Similarly, $2^{\frac{1}{3}}$ is a value that, when multiplied by itself three times, equals 2, and so $2^{\frac{1}{3}} = \sqrt[3]{2}$, the cube root of 2. In general, $2^{\frac{1}{n}}$ equals the nth root of 2. This works for any nonzero base.

A fractional exponent indicates that a root is to be taken: $b^{\frac{1}{n}} = \sqrt[n]{b}$.

We use the power rule to make sense of other types of fractional exponents. For example, the quantity $27^{\frac{2}{3}}$ can be computed as $27^{\frac{1}{3} \times 2} = \left(27^{\frac{1}{3}}\right)^2 = \left(\sqrt[3]{27}\right)^2 = 3^2 = 9$. In general, we have:

$$b^{\frac{p}{q}} = b^{\frac{1}{q} \times p} = \left(\sqrt[q]{b}\right)^p$$

Finally, to compute a quantity raised to an irrational power, one approximates the exponent by a fraction, computes the corresponding exponential expression, and takes the LIMIT as one uses better and better approximations. For example, writing $\sqrt{2} = 1.414\ldots$, we see that any of the fractions $1, \frac{14}{10}, \frac{141}{100}, \frac{1414}{1000}, \ldots$ can be used to approximate $\sqrt{2}$ with better and better degrees of accuracy. We define $2^{\sqrt{2}}$ to be the limit of the values: $2^1 = 2$, $2^{\frac{14}{10}} = \left(\sqrt[10]{2}\right)^{14} \approx 2.639$, $2^{\frac{141}{100}} = \left(\sqrt[100]{2}\right)^{141} \approx 2.657$, $2^{\frac{1414}{1000}} \approx 2.665, \ldots$.

The multiplication and power rules are valid even for irrational exponents. For example, we have:

$$\left(\sqrt{2}^{\sqrt{2}}\right)^{\sqrt{2}} = \sqrt{2}^{\sqrt{2} \times \sqrt{2}} = \sqrt{2}^2 = 2$$

The Greek mathematician ARCHIMEDES OF SYRACUSE (287–212 B.C.E.) was one of the first scholars to use a special word for the power of a number. He called the quantity 10,000, 10^4, a *myriad,* and he used the phrase "myriad of myriads" for 10,000 squared, $10^4 \times 10^4 = 10^8$. The ancient Greeks, for whom mathematics was synonymous with geometry, called the square of an unspecified quantity a *tetragon number,* meaning a "four-corner number." DIOPHANTUS OF ALEXANDRIA (ca. 200–284 C.E.) used the Greek word *dynamis,* meaning "power," for the square of an unknown, and called a third power a "cube," a fourth power a "power-power," and fifth and sixth powers "power-cube" and "cube-cube," respectively.

It took many centuries for scholars to begin using symbols to denote unknown quantities. German mathematician Michael Stifel (ca. 1487–1567) was the first to develop a notational system for powers of an unspecified quantity x. He denoted the fourth power of x simply as *xxxx.* Other scholars developed alternative notational systems. Scholars eventually settled on the notational system French mathematician and philosopher RENÉ DESCARTES (1596–1650) introduced in 1637, the one we use today. Although Descartes considered only positive integral exponents, later that century the English mathematician SIR ISAAC NEWTON (1642–1727), inspired by the work of JOHN WALLIS (1616–1703), showed that the same notational system can be extended to include negative, fractional, and irrational exponents. LEONHARD EULER (1707–83) later allowed for the possibility of complex exponents.

See also COMPLEX NUMBERS; EXPONENTIAL FUNCTION; LOGARITHM.

exponential function Any function or quantity that varies as the power of another quantity is called exponential. Precisely, if b is a positive number different from one, then the function $f(x) = b^x$ is called the exponential function with base b. The function is defined for all real numbers x. (This would not be the case if b were negative: the value $b^{\frac{1}{2}}$, for example, would not make sense.) The graphs of $y = 2^x$ and $y = \left(\frac{1}{2}\right)^x$

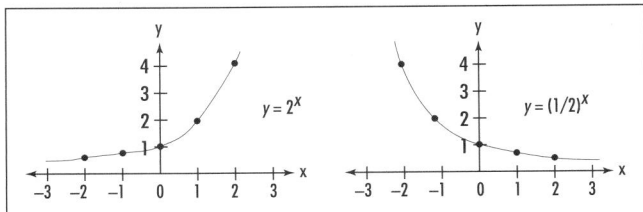

Exponential functions

illustrate the essential differences between the cases $b > 1$ and $0 < b < 1$.

Many types of growth and decay occur at a rate that involves exponential variation. For example, a colony of bacteria might reproduce at a rate that doubles the size of the colony every 24 hours. If initially 500 organisms are present, then after 1 day the culture grows to a size of 1,000 organisms, after 2 days to a size of 2,000 organisms, and by the end of day three there are 4,000 organisms. In general, the population size by the end of day N is given by 500×2^N. Any formula of the form AB^x with $b > 1$ and A constant is said to represent exponential growth. This formula is a simple example of a POPULATION MODEL.

The analogous formula with $0 < b < 1$ represents exponential decay. The decay of radioactive material is a typical example of this. For example, if 50 percent of the radioactivity produced by a nuclear explosion disappears after 5 days, then after 10 days only 25 percent of radioactivity remains, and 12.5 percent remains after 15 days. The percentage of radioactivity present after N days is given by the formula $\left(\frac{1}{2}\right)^{5N}$. The level of radioactivity decreases but will never reach zero.

The DERIVATIVE of an exponential function $f(x) = b^x$ can be computed via IMPLICIT DIFFERENTIATION after first taking a LOGARITHM. Precisely, if $y = b^x$, then $\ln y = x \cdot \ln b$. Differentiation yields $\frac{1}{y} \cdot \frac{dy}{dx} = \ln b$, and so $\frac{dy}{dx} = y \cdot \ln b = b^x \cdot \ln b$. That is:

$$\frac{d}{dx}\left(b^x\right) = b^x \cdot \ln b$$

This formula for the derivative is greatly simplified if one works with base $b = e$, where e is defined to be the number such that $\ln e = 1$. We have:

The derivative of e^x is e^x. Consequently, the graph of the exponential function $f(x) = e^x$ has the property that the slope of the graph at any point is the same as the value of the function at that point.

The function $f(x) = e^x$ is sometimes called *the* exponential function. Because the derivative of this function is particularly simple, it is not surprising that the number e is ubiquitous throughout studies in CALCULUS.

exponential series   The TAYLOR SERIES of the function $f(x) = e^x$, given by $e^x = 1 + x + \frac{x^2}{2!} + \frac{x^3}{3!} + \frac{x^4}{4!} + \ldots$ is called the exponential series. The RATIO TEST from the study of CONVERGENT SERIES shows that this series converges for all values of x. LEONHARD EULER (1707–83) made use of this series to prove his famous formula $e^{ix} = \cos x + i\sin x$, today called EULER'S FORMULA.

expression   Any meaningful combination of symbols that represent numbers, operations on numbers, or other mathematical entities is called an expression. For example, $2 + x$ and a^{b+c} are expressions. One could argue that $x + y = 2$ is an expression, although mathematicians may prefer to call it an equation. Similarly, $\sqrt[3]{8}$ could be called an expression even though it is equivalent to a single number. In FORMAL LOGIC, compound statements are sometimes called expressions. For example, $\neg(p \wedge (q \vee r))$ is an expression.

The word *express* is often used in the sense "to transform." For example, the product $(x - y)(x + y)$ can be expressed equivalently as the DIFFERENCE OF TWO SQUARES: $x^2 - y^2$.

exterior angle   An ANGLE contained between one side of a POLYGON and the extension of an adjacent side is called an exterior angle of the polygon. Since two edges of a polygon meet at a vertex, there are two exterior angles at each vertex. One can easily see, however, that these two angles are equal in value.

The sum of exterior angles in a convex polygon is 360°, since these angles correspond to one full turn. This result is also true for concave polygons if one deems angles that turn to the left as positive and ones that turn to the right as negative.

The EXTERIOR-ANGLE THEOREM, first proved by the geometer EUCLID (ca. 300–260 B.C.E.), states that

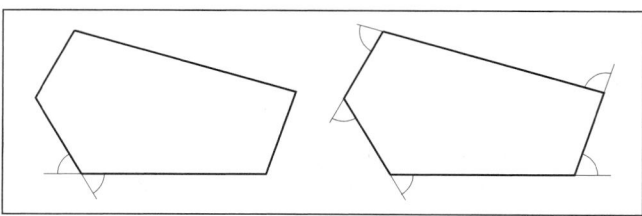

Exterior angles

the exterior angle at one vertex of a triangle is greater in value than that of an interior angle at either of the remaining two vertices.

See also CONCAVE/CONVEX.

exterior-angle theorem In his famous work *THE ELEMENTS*, the geometer EUCLID (ca. 300–260 B.C.E.) established the following result, which he called the exterior-angle theorem:

> For a given triangle *ABC* with interior angles *x* and *y* and exterior angle *z* as shown, we have *z* > *x* and *z* > *y*.

The result is proved as follows:

> Let *M* be the midpoint of side *BC*, and extend a line from point *A* through *M* to a new point *D* so that *AM* and *DM* are the same length. Consider triangles *AMB* and *DMC*. They share two sides of the same lengths and a common angle at *M*. By the SAS principle for similarity, the two triangles are congruent figures. Consequently, angle *MCD* matches angle *MBA*, which is *y*. Since *z* is clearly larger than angle *MCD*, we have that *z* > *y*.
>
> An analogous argument based on a line drawn through the midpoint of side *AC* establishes that *z* is also greater than *x*.

This theorem has one very important consequence.

> If two lines cut by a transversal produce equal alternate interior angles, then the two lines are parallel.

In the diagram above, if the two angles labeled *x* are indeed equal, then the lines cannot meet to the right to form a triangle: the exterior angle *x* cannot be greater

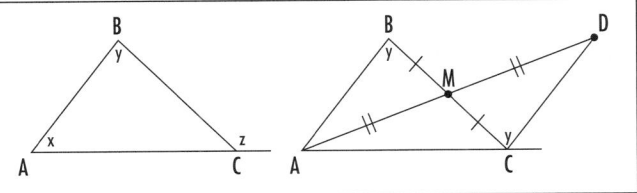

Understanding the exterior angle theorem

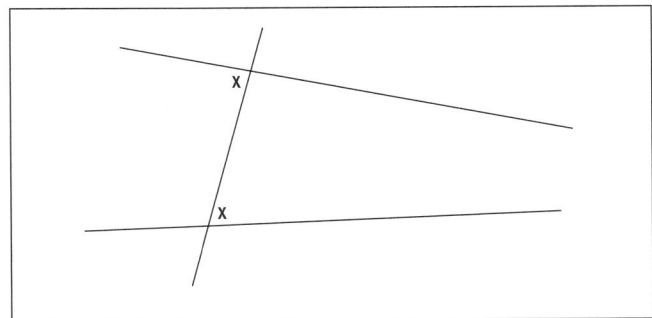

Equal alternate interior angles

than the interior angle *x*. Similarly, the lines cannot meet to the left to form a triangle by the same reason. (Work with the angle 180 – *x*.) It must be the case then that the lines are parallel.

This result is the CONVERSE of Euclid's famous, and controversial, PARALLEL POSTULATE.

It is important to note that Euclid proved the exterior-angle theorem and its consequence without assuming that the parallel postulate holds. If one is willing to assume that the three angles in a triangle always sum to 180° (a statement equivalent to the parallel postulate), then the proof of the exterior angle theorem is TRIVIAL.

extraction The process of finding the ROOT of a number or the solution to an algebraic equation is sometimes called extraction. For example, extracting the square root of 3 is the process of finding its square root. (One might use HERON'S METHOD, for example, to compute $\sqrt{3} = 1.7320508\ldots$)

The term *digit extraction* is often used to describe any method that allows one to compute a specific digit of a number without computing earlier digits. For example, in 1995 mathematician Simon Plouffe discovered the following remarkable formulae for π:

$$\pi = \sum_{n=0}^{\infty} \left(\frac{4}{8n+1} - \frac{2}{8n+4} - \frac{1}{8n+5} - \frac{1}{8n+6} \right) \left(\frac{1}{16} \right)^n$$

It has led mathematicians to a technique that computes the Nth digit of π in base 16 without having to calculate the preceding *N* – 1 digits. (In base 16, the number π appears as 3.243F6A8885A308D … where

A represents 10, B represents 11, and so forth, with F representing 15. Each decimal place is a power of a sixteenth.) Unfortunately no analogous technique is currently known to compute the ordinary base-10 digits of π with ease.

extrapolation The process of estimating the value of a function outside a known range of values is called extrapolation. For example, if the temperature of a cup of tea was initially 200°F and then was measured to be 100°F and 50°F 10 and 20 minutes later, respectively, then one might guess that its temperature after 30 minutes would be 25°F. Methods of extrapolation are normally far less reliable than INTERPOLATION, the process of estimating function values *between* known values. Scientists generally prefer to avoid making predictions based on extrapolation. Meteorologists, however, must use extrapolation techniques to make weather predictions. Long-range forecasts are generally considered unreliable.

See also POPULATION MODELS.

extreme-value theorem This theorem asserts that a CONTINUOUS FUNCTION $f(x)$, defined on a closed INTERVAL $[a,b]$, reaches a maximum value and a minimum value somewhere within that interval. That is, there is a point c in the interval $[a,b]$ such that $f(x) \leq f(c)$, for all x in $[a,b]$, and there is another point d in $[a,b]$ such that $f(x) \geq f(d)$ for all x in $[a,b]$. For example, the extreme-value theorem tells us that, on the interval $[1,5]$ say, the function $f(x) = 3x \cdot \cos(x^2 + \sin x)$ does indeed adopt a largest value. It does not tell us, however, where that maximum value occurs. A point at which a function has a maximum or minimum value is called an extremum.

The theorem is intuitively clear if we think of a continuous function on a closed interval as one whose graph consists of a single continuous piece with no gaps, jumps, or holes: in moving a pencil from the left end point $(a,f(a))$ to the right end point $(b,f(b))$, one would not doubt that there must be a high point on the curve where $f(x)$ reaches its maximum value, and a low point where it attains its minimum value. A rigorous proof of the theorem, however, relies on the notion of the completeness of the real numbers (meaning that no points are missing from the real line). This is a subtle property, one that was not properly understood until the late 1800s with the invention of a DEDEKIND CUT. For example, the function $f(x) = 2x - x^2 = x(2 - x)$ has no maximum value on the interval $[0, 2]$ if the value 1 is excluded from our considerations. Although this seems an artificial example, one still needs to be sure that this type of problem can never occur.

It is vital that the function under consideration be continuous and that the interval under study be closed for the theorem to be true. For example, the (discontinuous) function

$$f(x) = \begin{cases} 2x - x^2 & \text{if } x \neq 1 \\ 0 & \text{if } x = 1 \end{cases}$$

does not reach a maximum value in the closed interval $[0,2]$; nor does the (continuous) function $f(x) = \frac{1}{x}$, defined on the open interval $(0,2)$, since the function is arbitrarily large for values x close to zero.

The INTERMEDIATE-VALUE THEOREM is a companion result to the extreme-value theorem. It asserts that not only does a continuous function on a closed interval actually attain maximum and minimum values, but it also takes on every value between those two extreme values. ROLLE'S THEOREM and the MEAN-VALUE THEOREM follow from the extreme-value theorem if we further assume that the function under consideration is differentiable.

See also DIFFERENTIAL CALCULUS; MAXIMUM/MINIMUM.

face A flat surface on the outside of a solid figure, typically a POLYHEDRON, is called a face of the figure. For example, a cube has six identical faces, and a cylinder has two faces. (The lateral surface of a cylinder is not flat.) In the mid-1700s, Swiss mathematician LEONHARD EULER established that if all the outside surfaces of a figure are flat, then the number of faces f the figure possesses is given by the formula:

$$f = 2 - v + e$$

Here v is the number of vertices and e is the number of edges the figure has.

The angle between two edges of a polyhedron meeting at a common vertex is sometimes called a face angle.

In GRAPH THEORY, any region of plane bounded by edges of a planar graph is sometimes called a face of the graph. EULER'S FORMULA $v - e + f = 2$ also holds for connected planar graphs if one is willing to regard the large unbounded region outside the graph as a face.

See also DIHEDRAL.

factor The term *factor* is used in two mathematical settings: NUMBER THEORY and ALGEBRA. In number theory, if a, b, and n are whole numbers and if a times b equals n, then a and b are called factors of n. For example, 3 and 4 are both factors of 12 (since $12 = 3 \times 4$), as are the numbers 1, 2, 6, and 12 ($2 \times 6 = 12$ and $1 \times 12 = 12$). Any number that divides the given number evenly is a factor. For this reason, factors are sometimes called divisors.

The factors of a given number have a geometric interpretation. For example, one can arrange 12 pebbles into six different rectangular arrays: a 1 by 12 rectangle, a 2 by 6 rectangle, a 3 by 4 rectangle, a 4 by 3 rectangle, a 6 by 2 rectangle, and finally a 12 by 1 rectangle. The dimensions of these rectangles are precisely the factors of 12. This interpretation shows that the factors of a number come in pairs—unless, one of the rectangles formed is a perfect square (in which case, one factor is "paired with itself"). This shows:

> Square numbers have an odd number of factors. All other numbers have an even number of factors.

For example, 36, which equals 6×6, has an odd number of factors: 1 and 36, 2 and 18, 3 and 12, 4 and 9, and 6. This simple observation solves the famous prison warden puzzle:

> A prison warden and 100 inmates, residing in cells numbered 1 through 100, agree to perform the following experiment over 100 days. In the process of the experiment some cell doors will be left unlocked and the prisoners agree not to escape.
>
> On the first day, the prison warden will turn the key of each cell door and leave all the doors unlocked.
>
> On the second day, the warden will turn the key of every second door. This will lock doors numbered 2, 4, ..., 100 and leave the odd-numbered doors open.

On the third day, starting with door number 3, he will turn the key of every third door. This will leave a mixture of doors locked and unlocked.

On the fourth day he will turn the key of every fourth door, on day five the key of every fifth door, and so on, all the way until day 100, when he will turn the key of every 100th door, namely, just the final door. At this time, any prisoner who finds his door open will be allowed to go free.

Which doors will be left unlocked as a result of this experiment?

Observe that the warden will turn the key to door number n on each day d that is a factor of n. For example, the key of door number 39 will be turned on days 1, 3, 13, and 39 only, and the key of door number 25 on days 1, 5, and 25. For a door to be left unlocked at the end of the experiment its key must be turned an odd number of times. As only the square numbers have an odd number of factors we have that prisoners 1, 4, 9, 16, 25, 36, 49, 64, 81, and 100 are set to go free.

Any number n always has 1 and itself as factors. These are called "improper factors." Any other factor of n, if there is one, is called a "proper factor." For example, 1 and 12 are improper factors of 12, and 2, 3, 4, and 6 are proper factors. A number, different from 1, that possesses proper factors is called "composite," or a COMPOSITE NUMBER, and one that does not is called PRIME. For example, the number 25 has a proper factor, namely 5, and so is composite; whereas 7 has only two factors, both of which are improper, and so is prime.

Any proper factor of a composite number, if not itself prime, can be written as a product of two smaller factors. Repeated application of this process shows that any number different from 1, if not already prime, can be written as a product of prime numbers. For example:

$$180 = 30 \times 6 = (6 \times 5) \times (2 \times 3) = 2 \times 3 \times 5 \times 2 \times 3.$$

The FUNDAMENTAL THEOREM OF ARITHMETIC asserts that every integer greater than 1 factors into primes in only one way (up to the order of the primes). A factor-tree is a diagram that describes this factoring process pictorially. Two students drawing different factor trees for the same number will arrive at the same list of primes at the bases of their trees.

In algebra, a factor is a POLYNOMIAL that divides another given polynomial exactly. For example, $2x + 1$ and $x - 1$ are both factors of $2x^2 - x - 1$, since $2x^2 - x - 1 = (2x + 1)(x - 1)$. It is generally agreed that the factors of a polynomial should themselves be nonconstant polynomials with coefficients that are real numbers.

A polynomial that cannot be factored is called irreducible. For example, $x^2 + 1$ cannot be factored over the real numbers and so is irreducible. If one permits COMPLEX NUMBERS as coefficients, then the situation changes: we can write: $x^2 + 1 = (x + i)(x - i)$. The FUNDAMENTAL THEOREM OF ALGEBRA asserts that every polynomial factors in the realm of complex numbers.

See also COMMON FACTOR; DIVISIBILITY RULES; FACTOR THEOREM; GREATEST COMMON DIVISOR; LONG DIVISION; PERFECT NUMBER; PRIME.

factorial If n is a natural number, then the symbol $n!$ (read as "n factorial") denotes the product of all positive integers from 1 to n:

$$n! = 1 \cdot 2 \cdot 3 \cdot \ldots \cdot n$$

For example, $6! = 1 \cdot 2 \cdot 3 \cdot 4 \cdot 5 \cdot 6 = 720$. The order in which the terms of the product are multiplied is immaterial. (We have that $6!$ also equals $6 \cdot 5 \cdot 4 \cdot 3 \cdot 2 \cdot 1$.)

Factorials naturally arise when counting the numbers of ways a collection of objects can be arranged. Given n objects, there are n choices for which object we wish to regard as "first." Once this selection is made, $n - 1$ choices remain for which object to select as second, and after the first and second are chosen, there remain $n - 2$ choices for third. This process continues until two objects remain, yielding two choices for which to regard second-to-last, and just one object to select as last. By the MULTIPLICATION PRINCIPLE, there are thus $n \cdot (n - 1) \cdot (n - 2) \cdot \ldots \cdot 2 \cdot 1 = n!$ different PERMUTATIONS (or arrangements, or orders) of n objects.

If one wishes to arrange just r objects chosen from a collection of n different things, there are n choices for which object to regard as first, $n - 1$ choices for which to regard as second, and so on, to $n - r + 1$ choices for which object to deem rth, yielding a total of $n \cdot (n - 1) \cdot \ldots \cdot (n - r + 1)$ different arrangements of r objects selected from n. This formula can be more compactly written as:

$$\frac{n!}{(n - r)!}$$

When r equals n (arrange all n objects) this formula reads $\frac{n!}{0!}$. To coincide with our previously computed answer of $n!$, it is natural to define $0!$ as equal to 1.

The exclamation-point notation for factorial was first used by Christian Kramp in 1808, in his paper "Élémens d'arithmétique universelle," though other notations for $n!$ popular at the time, and later, included $n_\lrcorner$, n', $\Pi(n)$ and $\Gamma(n+1)$.

LEONHARD EULER (1707–83) attempted to generalize the factorial function to noninteger values. At the age of 22 he discovered the following LIMIT quantity that helps achieve this:

$$\Gamma(x) = \lim_{n \to \infty} \frac{n^x n!}{x(x+1)(x+2)\ldots(x+n)}$$

(He called this expression the gamma function to honor ADRIEN-MARIE LEGENDRE's use of the symbol Γ for factorial.) This expression has the property, as you may check, that $\Gamma(x + 1) = x \cdot \Gamma(x)$ for all positive real values x. Also $\Gamma(1) = \lim_{n \to \infty} \frac{n \cdot n!}{(n+1)!} = \lim_{n \to \infty} \frac{n}{n+1} = 1$. Consequently, $\Gamma(m) = (m - 1)!$ for all positive integers m. Thus, for example, $\Gamma(7) = 6! = 720$. It is now possible to also define quantities such as $\left(\frac{1}{2}\right)!$ and $(\sqrt{2})!$ using this expression.

INTEGRATION BY PARTS shows that the IMPROPER INTEGRAL $\int_0^\infty e^{-t} t^{m-1}\, dt$ also has value $(m - 1)!$. Mathematicians have shown that Euler's gamma function and the corresponding improper integral agree for all positive real values x:

$$\Gamma(x) = \int_0^{+\infty} e^{-t} t^{x-1}\, dt$$

This integral has the unexpected value $\sqrt{\pi}$ when $x = \frac{1}{2}$.

Thus we may conclude, for example, that $\left(-\frac{1}{2}\right)! = \Gamma\left(\frac{1}{2}\right) = \sqrt{\pi}$, and $\left(\frac{1}{2}\right)! = \Gamma\left(\frac{3}{2}\right) = \left(\frac{1}{2}\right) \cdot \Gamma\left(\frac{1}{2}\right) = \frac{\sqrt{\pi}}{2}$.

See also BINOMIAL THEOREM; PERMUTATION; STIRLING'S FORMULA.

factorization The process or the result of writing a number or a POLYNOMIAL as a product of terms is called factorization. For example, the FUNDAMENTAL THEOREM OF ARITHMETIC asserts that every whole number can be written as a product of PRIME numbers and, up to the order of the terms, this factorization is unique. (For instance, $132 = 2 \times 2 \times 3 \times 11$.) Thus every whole number has a unique "prime factorization." The FUNDAMENTAL THEOREM OF ALGEBRA asserts that, in the realm of COMPLEX NUMBERS, every polynomial factors completely into linear terms. (For instance, $2x^3 - x^2 - 13x - 6 = (x - 3)(x + 2)(2x + 1)$ and $x^2 - 4x + 5 = (x - 2 + i)(x - 2 - i)$.) If one wishes to remain in the realm of the REAL NUMBERS, then every polynomial with real coefficients is guaranteed to factor into a product of linear terms and irreducible QUADRATIC terms. (For instance, $x^4 - 1 = (x^2 + 1)(x - 1)(x + 1)$.)

See also DECOMPOSITION; FACTOR THEOREM.

factor theorem The REMAINDER THEOREM shows that if a POLYNOMIAL $p(x)$ is divided by a term of the form $x - a$ for some constant a, then the remainder term is the constant $p(a)$:

$$p(x) = (x - a)Q(x) + p(a)$$

Thus if the value of the polynomial is zero at $x = a$, that is, $p(a) = 0$, then the polynomial factors as $p(x) = (x - a)Q(x)$. This leads to the following factor theorem:

A linear term $x - a$ is a factor of a polynomial $p(x)$ if, and only if, $p(a) = 0$.

For example, for $p(x) = 2x^3 - 4x^2 - 10x + 12$, we have $p(1) = 0$, $p(-2) = 0$, and $p(3) = 0$. Consequently, $x - 1$, $x + 2$, and $x - 3$ are each factors of the polynomial. (In this example we have $p(x) = 2(x - 1)(x + 2)(x - 3)$.) Since 2 and -2 are clearly each zeros of $x^6 - 64$, this polynomial must be divisible by $(x - 2)(x + 2) = x^2 - 4$.

See also FUNDAMENTAL THEOREM OF ALGEBRA.

fair division (**cake cutting**) A classic puzzle asks for a fair way to divide a piece of cake between two greedy brothers. The "you cut, I choose" scheme asks one brother to slice the cake into what he believes to be two equal parts and has the second brother choose one of the two pieces. The first is then guaranteed to receive, in his measure, precisely 50 percent of the cake and the second brother, if he has a different estimation of half, has the advantage of receiving more then 50 percent of the slice according to his own measure. One can avoid this perceived advantage with a different sharing scheme. In the following "knife holding" scheme, each brother is guaranteed to feel that he has the advantage:

> Each brother holds a knife vertically across the cake at the location he believes cuts the cake precisely in half. (If the brothers have different estimation of what constitutes "half," then the knives will be at different locations.) Cutting the cake anywhere between these two positions guarantees each brother a piece, in his estimation, greater than a 50 percent.

This scheme generalizes to sharing among any number of players. For instance, it is possible to share a cake among three players in such a manner that each player honestly believes he is receiving more than one-third of the cake.

> Each player holds a knife vertically across the cake at the location he believes cuts off exactly one-third of the cake from the left. The cake is then cut between the two leftmost positions, giving that piece to the player holding the leftmost knife. This player has received, in his estimation, more than one-third of the cake, and each of the remaining two players believes that more than two-thirds of cake remains. The second two brothers then perform the cake division scheme described above for two players.

Any cake-sharing scheme among n players that guarantees each player, in his estimation, at least $1/n$ of the cake is called a fair-division scheme.

The scheme described above among three players, however, is not "envy free." Although each player believes that he received more than his fair share of the cake, it is not assured that he also believes that he received the largest piece ever cut. (Every fair division scheme between two players is envy free.) Complicated envy-free fair-division schemes do exist for sharing cake among any number of players. There are also fair-division methods for dividing collections of indivisible objects (such as the furniture in an estate) among two or more people using cash payments to even up the final division.

Farey sequence (Farey series) For a given positive whole number n, the sequence of all proper fractions, written in reduced form, with denominators no larger than n and arranged in order of magnitude is called the nth Farey sequence. For example, the fifth Farey sequence is:

$$\frac{0}{1}, \frac{1}{5}, \frac{1}{4}, \frac{1}{3}, \frac{2}{5}, \frac{1}{2}, \frac{3}{5}, \frac{2}{3}, \frac{3}{4}, \frac{4}{5}, \frac{1}{1}$$

These sequences have a number of arithmetic properties. For instance, if $\frac{a}{b}$ and $\frac{c}{d}$ are consecutive terms of a Farey sequence, then the numbers ad and bc, arising from taking their cross product, are always consecutive integers. (For example, the consecutive pair $\frac{3}{5}$ and $\frac{2}{3}$ yield consecutive integers 9 and 10.) Also, if $\frac{a}{b}, \frac{c}{d}, \frac{e}{f}$ are three successive terms of a Farey sequence, then the middle term $\frac{c}{d}$ is the mediant fraction $\frac{a+e}{b+f}$. (For instance, the term between $\frac{3}{5}$ and $\frac{3}{4}$ is $\frac{3+3}{5+4} = \frac{6}{9} = \frac{2}{3}$.) This observation allows one to quickly build up from one Farey sequence to the next: simply compute the mediants between all terms present and retain those whose denominators are not too large.

The Farey sequence was first studied by C. Haros in 1802, but interest in the topic did not stir until British geologist John Farey (1766–1826) published his 1816 piece, "On a Curious Property of Vulgar Fractions" in *Philosophical Magazine*. (Farey was unaware of Haros's work.) In 1938 American mathematician Lester R. Ford presented a remarkable geometric interpretation of the Farey sequence:

> Above each reduced fraction $\frac{a}{b}$ on the number line, draw a circle of radius $\frac{1}{b^2}$ touching the number line at that point. Despite expectation, these circles never overlap, although they do often touch. Moreover, two circles at positions $\frac{a}{b}$ and $\frac{c}{d}$ touch precisely when ad and bc are consecutive integers, and furthermore the largest circle that fits in the space between them above the number line is the circle at the mediant $\frac{a+c}{b+d}$.

(These claims can be proved by making use of the DISTANCE FORMULA to establish that the distance between the centers of two touching circles equals the sum of the radii of the two circles.)

Fermat, Pierre de (1601–1665) French *Number theory, Calculus, Probability theory*   Born on August 17, 1601, in Beaumont-de-Lomagne, France, Pierre de Fermat is remembered as a leading mathematician in the first half of the 17th century, recognized for his founding work in the theory of numbers. Fermat is also responsible for some pioneering work in CALCULUS and the theory of tangents to curves, PROBABILITY theory, and analytic GEOMETRY. His 1679 piece *Isagoge ad locos planos et solidos* (On the plane and solid locus), published posthumously, foreshadowed the work of RENÉ DESCARTES (1596–1650) on the application of algebra to geometry, allowing him to define algebraically important curves such as the HYPERBOLA and the PARABOLA. In optics, he is acknowledged as the first scholar to formulate the "fundamental property of least time," stating that light always follows the shortest paths. Perhaps most notably, Fermat is remembered for the enigmatic comment he scribed in the margin of one of his reading books claiming to have solved a novel problem in number theory. Search for a solution to this problem (if not the one Fermat had in mind) spurred three centuries of important and spirited research in mathematics. FERMAT'S LAST THEOREM was finally resolved in 1994.

Fermat received a bachelor's degree in civil law from the University of Orléans in 1631 and began work as a lawyer for the local parliament of Toulouse that same year. He followed this career path throughout his entire life—accepting a position as a criminal court judge in 1638 and, finally, the high position of king's counselor in 1648. Fermat's work in mathematics was an outside interest.

Fermat first developed a passion for reading and "restoring" classic Greek texts. This meant completing the mathematics of any passages that were missing from the records that survived from ancient times. His work on the text *Plane loci* by APOLLONIUS OF PERGA (ca. 262–190 B.C.E.) garnered the attention of the mathematics community at the time, not only for the restoration work itself, but also for the new geometric methods Fermat had devised for computing tangents to curves and solving maxima/minima problems. Fermat developed a correspondence with French monk MARIN MERSENNE (1588–1648), who served the role of dispersing mathematical information to the notable scholars of the time. Despite the attention Fermat received, he did not seek fame by publishing any of his work. (He published one small piece in his life, which he did anonymously.) Fermat shared his discoveries and results with Mersenne and other scholars, but not his methods for obtaining them. This both inspired and frustrated mathematicians at the time.

In 1654 notable scholar BLAISE PASCAL (1623–62) wrote to Fermat with some mathematical questions about gambling and games of chance. The correspondence that ensued led to the joint development of a new mathematical theory of probability. Fermat is today considered one of the founders of the field. However, Fermat had developed a great interest in the theory of numbers, in particular, the properties of whole numbers. This topic was of little interest to mathematicians at the time—perhaps because of its lack of apparent immediate application—but Fermat attempted to spark interest in the subject by posing challenging questions to his contemporaries. He asked scholars to prove, for instance, that the equation $x^2 + 2 = y^3$ has only one positive integer solution. His colleagues, however, regarded questions such as these as too specific to be of serious concern and often dismissed then. Fermat, on the other hand, realized that understanding the solutions to such specific questions provides a gateway to great insight on the very general and mysterious properties of whole numbers. It was not until Fermat's son Samuel published Fermat's annotated copy of the *Arithmetica* by the classic scholar DIOPHANTUS OF ALEXANDRIA (ca. 200–284 C.E.)—the text containing the famous marginal note—that interest in number theory was revived and Fermat's brilliant work on the topic was fully recognized.

Fermat died in Castres, France on January 12, 1665. The claim posed in the note scrawled in the margin of *Arithmetica* is called Fermat's last theorem. It inspired over three centuries of intense mathematical research in the field of NUMBER THEORY.

See also MAXIMUM/MINIMUM.

Fermat's last theorem   Since ancient times, scholars have been aware of many, in fact infinitely many, different integer solutions to the equation $x^2 + y^2 = z^2$.

(Solutions to this equation are called PYTHAGOREAN TRIPLES.) Around 1637 PIERRE DE FERMAT conjectured that no *positive* integer solutions exist, however, for the equations

$$x^n + y^n = z^n$$

with *n* greater than two. In his copy of the translated works of DIOPHANTUS OF ALEXANDRIA, next to a problem about Pythagorean triples, Fermat wrote his now famous note:

> On the other hand, it is impossible to separate a cube into two cubes, or a fourth power into two fourth powers, or generally any power except a square into two powers with the same exponent. I have discovered a truly admirable proof of this, but the margin is too narrow to contain it.

For over 350 years mathematicians tried to reproduce Fermat's alleged proof. The claim itself became known as Fermat's last theorem, and it was one of the greatest unsolved problems of all time. Although the problem lends itself to no obvious practical applications, attempts to solve it helped motivate the development of a great deal of important mathematics.

It is generally believed that Fermat did not have a proof of the theorem. In his correspondences with colleagues he mentions only the cases *n* equals 3 and 4 and provides no details of proof even for those special cases. Fermat, again as a marginal note in his copy of Diophantus's work, does provide a detailed proof of another challenge posed by Diophantus, one about triangles of rational side length. Although not explicitly mentioned, the proof of the *n* equals 4 case follows readily from mathematical argument he provides. It is thought that Fermat was aware of this.

With the case *n* = 4 taken care of, it is not difficult to see that one need only study the cases where *n* is an odd prime. For example, if it is known that $x^7 + y^7 = z^7$ has no positive integer solutions, then $x^{42} + y^{42} = z^{42}$ can have no positive integer solutions either. (Rewrite the latter equation as $(x^6)^7 + (y^6)^7 = (z^6)^7$.)

In the mid-1700s, LEONHARD EULER proved that the equation with *n* = 3 has no positive integer solutions. The extensive work of MARIE-SOPHIE GERMAIN (1776–1831) during the turn of the century allowed mathematicians to later show that the theorem holds for all values of *n* less than 100. During the 19th and 20th centuries mathematicians developed the fields of algebraic geometry and arithmetic on curves. In 1983, Gerd Faltings proved the so-called Mordell conjecture, an important result with the following immediate consequence: any equation of the form $x^n + y^n = z^n$ with *n* > 3 has, at most, a finite number of positive integer solutions. This led mathematicians a significant step closer to proving Fermat's last theorem for all values of *n*: is it possible to show that that finite number is zero in every case? Finally, in 1995, almost 360 years since Fermat's claim, the English mathematician ANDREW WILES, with the assistance of Richard Taylor, presented a completed proof of Fermat's last theorem. It is, not surprisingly, very long and highly advanced, relying heavily on new mathematics of the century. Needless to say, the proof is certainly beyond Fermat's abilities. Although Wiles's proof is deservedly regarded as a high point of 20th-century mathematics, mathematicians still search for a simplified argument.

Ferrari, Ludovico (1522–1565) Italian *Algebra* Born on February 2, 1522, in Bologna, Italian scholar Ludovico Ferrari is remembered as the first person to solve the QUARTIC EQUATION. He worked as an assistant to GIROLAMO CARDANO (1501–76), who published Ferrari's solution in his famous 1545 work *Ars magna* (The great art).

Assigned to be a servant at the Cardano household at age 14, Ferrari soon impressed his master with his agile mind and with his ability to read and write. Cardano decided to train Ferrari in the art of mathematics. In exchange, Ferrari helped Cardano prepare his manuscripts. Four years after his arrival, and with the blessing of Cardano, Ferrari accepted a post at the Piatti Foundation in Milan as public lecturer in geometry. Ferrari, however, continued to work closely with Cardano.

Ferrari discovered his solution to the quartic equation in 1540, but it relied on the methods of solving the CUBIC EQUATION that had been developed by NICCOLÒ TARTAGLIA (ca. 1499–1557) and revealed to Cardano in secrecy. (Mathematicians at the time were supported by patrons and protected their methods as trade secrets: they were often required to prove their worth by solving challenges no other scholar could solve.) Unable to publish the result without breaking a promise, Ferrari and Cardano felt stymied. However, a few years later, Ferrari learned that another scholar SCIPIONE DEL FERRO (1465–1526) had also developed methods of solving certain types of cubic equations. Although essentially identical to the work of Tartaglia, Ferrari and Cardano decided to publish the solution to the quartic, attributing the work on the cubic needed to del Ferro, with whom no promise of secrecy had been made.

Tartaglia was outraged, and a bitter dispute that lasted for many years ensued between Tartaglia and Ferrari. On August 10, 1548, as was common at the time, Tartaglia challenged Ferrari to an open contest and public debate as an attempt to demonstrate that he was in fact the expert on cubic equations. But it was clear from the contest that Ferrari had a more complete understanding of both cubic and quartic equations. Tartaglia left before the contest was over, and victory was given to Ferrari. He immediately garnered national fame and was given many offers of employment, including a request from the emperor himself to act as royal tutor. Ferrari, however, accepted no position offered at the time, left mathematics, and accepted a lucrative position as tax assessor to the governor of Milan.

Ferrari died in Bolgna, Italy, in October 1565 (the exact date is not known) and is remembered in mathematics solely for his work on the quartic equation.

Ferro, Scipione del (Ferreo, dal Ferro) (1465–1526)

Italian *Algebra* Born on February 6, 1465, in Bologna, Italy, Scipione del Ferro is remembered as the first mathematician to solve the CUBIC EQUATION. Unfortunately none of his writings survive today, and we learn of his work chiefly through the manuscripts of GIROLAMO CARDANO (1501–76) and LUDOVICO FERRARI (1522–65).

Del Ferro was appointed lecturer in arithmetic and geometry at the University of Bologna in 1496, a position he retained for all his life. Little is known of his academic work. Letters to other scholars at the time suggest that del Ferro studied methods for rationalizing rational expressions, ruler-and-compass constructions in geometry, and methods for solving cubic equations.

Mathematicians of del Ferro's time were familiar with the general solution to a QUADRATIC equation of the form $ax^2 + bx + c = 0$. (It should be mentioned, however, that 16th-century scholars did not use zero as a number in an expression, nor permitted the use of negative numbers. Thus the equation $x^2 - 3x + 2 = 0$,

for instance, was written $x^2 + 2 = 3x$.) Mathematicians also knew that, with the appropriate use of substitution, any cubic equation could be reduced to one of two forms: $x^3 + ax = b$ or $x^3 = ax + b$. (Here, again, a and b are positive.) Del Ferro was the first mathematician to solve equations of the first type. Some historians suggest that he was able to solve equations of the second type as well.

Del Ferro recorded all his results in a personal notebook, which he bequeathed to his son-in-law Hannibal Nave, also a mathematician. Nave later shared the contents of the notebook with Cardano and Ferrari. After seeing the method of solving the cubic fully explained, Cardano and Ferrari realized that del Ferro had in fact solved the famous cubic equation some 30 years before NICCOLÒ TARTAGLIA (ca. 1499–1557), their contemporary, had claimed to do the same. In 1545 Cardano published *Ars magna* (The great art), outlining Ferrari's solution to the QUARTIC EQUATION making use of del Ferro's methods for the cubic.

Del Ferro died in Bologna, Italy, some time between October 29 and November 16, 1526. He is remembered in mathematics solely for his work on cubic equations.

Fibonacci (Leonardo Fibonacci, Leonardo of Pisa)

(ca. 1170–1250) Italian *Arithmetic, Number theory* Born in Pisa, Italy (the exact birth date is not known), mathematician Leonardo of Pisa, better known by his nickname Fibonacci, is best remembered for his help in introducing the HINDU-ARABIC NUMERAL system to the merchants and scholars of Europe. He strongly advocated the system in his famous 1202 text *Liber abaci* (The book of counting), a treatise on the techniques and practices of arithmetic and algebra, which proved to be extremely influential. This work also contained a large collection of arithmetical problems, including one that leads to the famous sequence of numbers that bears his name. Considered the most important mathematician of the middle ages, Fibonacci also wrote extensively on the topics of EUCLIDEAN GEOMETRY and DIOPHANTINE EQUATIONS. He is recognized as the first scholar in the West to make advances in the field of NUMBER THEORY since the time of DIOPHANTUS OF ALEXANDRIA.

Although born in northern Italy, Fibonacci was raised and educated in northern Africa, where his father, a merchant and a government representative, held a diplomatic post. As he grew older, Fibonacci also traveled to Greece, Egypt, Syria, France, and Sicily, and took special note of the arithmetic systems used by local merchants of those areas. He became convinced that the number system used by the Arabs with its roots from India—the actual way they wrote numbers and the way they manipulated them to perform calculations—was far superior to any other arithmetic system he had encountered, including the clumsy system of ROMAN NUMERALS in use at the time in Europe. Upon his return to Pisa around the year 1200, Fibonacci began writing his famous piece. Its aim was to simply explain the Hindu-Arabic numerals, the role of a PLACE-VALUE SYSTEM, and illustrate its superior approach. The text begins simply:

> These are the nine figures of the Indians: 9, 8, 7, 6, 5, 4, 3, 2, 1. With these nine figures, and with this sign 0, … any number may be written, as will be shown.

Divided into four sections, the work outlines the methods of addition, multiplication, subtraction, and division. It also discusses fractions (including a discussion on EGYPTIAN FRACTIONS), as well as some geometry and algebra. (Some parts of the text are written from right to left, indicating, perhaps, the extent to which Fibonacci was influenced by Arabic scholars.) Although complete acceptance of the Hindu-Arabic system in the West did not occur until about 300 years later, Fibonacci's work in this area is recognized as the first significant step in this direction.

Fibonacci also wrote extensively in the fields of number theory, trigonometry, and geometry. It is said that the advances Fibonacci presented in his 1225 piece *Liber quadratorum* (The book of square numbers) were of such interest and value that they sparked renewed interest in theoretical mathematics and revived Western mathematics from its slumber during the Middle Ages.

During his life Fibonacci was recognized as a great scholar. Word of his abilities reached the Emperor Frederick II, seated in Palermo, who invited him to compete against other mathematicians of the day in a mathematical tournament. Fibonacci correctly solved all three challenges put before him, garnering him further attention and fame. In 1240 Fibonacci was awarded a salary from the city of Pisa in recognition of his services to the community.

All of Fibonacci's texts, and their reproductions, were written by hand. Copies of *Liber abaci* still survive today.

Fibonacci's name is derived from the shortening of the Latin *filius Bonacci*, meaning the son of Bonaccio, his father's family name. During his life, Fibonacci was also known as Leonardo of Pisa or, in Latin, Leonardo Pisano. Sometimes, Fibonacci also identified himself as Leonardo Bigollo, following the Tuscan word *bigollo* for "a traveler."

By introducing the Hindu-Arabic numeral to Europe, his influence on Western mathematics was profound. He died in the city of Pisa, Italy, likely in the year 1250. (The exact date of death is not known.)

See also FIBONACCI NUMBERS.

Fibonacci numbers Any one of the numbers that appears in the sequence 1, 1, 2, 3, 5, 8, 13, 21, 34, 55, 89, …, where each number, after the second, is the sum of the two preceding numbers, is called a Fibonacci number. If F_n denotes the nth Fibonacci number, then we have

$$F_n = F_{n-1} + F_{n-2}$$

with $F_1 = F_2 = 1$.

These numbers arise from a famous rabbit-breeding problem described in FIBONACCI's text *Liber abaci*:

> How many rabbits would be produced in the nth month if, starting from a single pair, any pair of rabbits of one month produces one pair of rabbits for each month after the next?

(The initial pair of rabbits, for example, do not produce another pair of rabbits until month 3. This same pair produces a new pair for each month thereafter.)

In any month, the totality of rabbits present consists of all pairs of the previous month together with all the new offspring. The number of offspring equals the population size of two months previous. Thus the solution to the problem is the sequence described above.

Any problem whose nth case solution is the sum of the two previous case solutions produces the Fibonacci sequence. For example, there are F_n ways to climb $n - 1$ steps, one or two steps at a time (consider beginning the climb with either a single step or a double step). There are also F_n ways to tile a $1 \times (n - 1)$ row of squares with

1×1 tiles and 1×2 dominoes, and there are F_n sequences of 0s and 1s n-digits long beginning and ending with 1 and containing no two consecutive 0s.

Regarding a 1 as "tails" and 0 and "heads," and ignoring the initial and final 1s, this latter example can be used to show that the PROBABILITY of not getting two heads in a row when tossing a coin n times is $\frac{F_{n+2}}{2^n}$. One can also use it to show that there are F_{n+2} subsets of $\{1,2,\dots,n\}$ lacking two consecutive numbers as members.

Perhaps the most surprising appearances of Fibonacci numbers occur in nature. The seeds in a sunflower's head, for example, appear to form two systems of spirals—often with 55 spirals arcing in a clockwise tilt, and 34 spirals with a counterclockwise tilt. (Large species of sunflowers have 89 and 144 spirals, again consecutive Fibonacci numbers.) This appears to be typical of all natural objects containing spiral floret, petal, or seed patterns: pineapples, pinecones, and even the spacing of branches around the trunk of a tree. The botanical name for leaf arrangement is phyllotaxis.

This pineapple has five diagonal rows of hexagonal scales in one direction, eight rows in the other direction. Note: the sixth (unmarked) row of scales in the left picture is not a new row: it continues the bottommost row of scales.

It is useful to ask whether it is possible to find a value x so that the sequence $1,x,x^2,x^3,\dots$ satisfies the same recursive relationship as the Fibonacci numbers, namely that every term after the second equals the sum of the two preceding terms. This condition therefore requires x to be a number satisfying the equation $1 + x = x^2$. By the QUADRATIC formula there are two solutions:

$$\phi = \frac{1+\sqrt{5}}{2}$$

$$\tau = \frac{1-\sqrt{5}}{2}$$

It follows that any combination of the form $a\phi^n + b\tau^n$ satisfies the same recursive relation as the Fibonacci sequence. Choosing the constants a and b appropriately, so that the first two terms of the sequence produced are both 1, yields the following formula, called Binet's formula, for the nth Fibonacci number:

This pineapple has five diagonal rows of hexagonal scales in one direction and eight rows in the other direction. (Note that the sixth unmarked row of scales in the left picture is not a new row; it is a continuation of the bottommost row of scales.)

$$F_n = \frac{\phi^n - \tau^n}{\sqrt{5}} = \frac{\left(\frac{1+\sqrt{5}}{2}\right)^n - \left(\frac{1-\sqrt{5}}{2}\right)^n}{\sqrt{5}}$$

(It is surprising that this formula yields an integer for every value of n). One can use this result to show that the ratio $\frac{F_n}{F_{n-1}}$ of Fibonacci numbers approaches the value ϕ as n becomes large. This happens to be the GOLDEN RATIO.

The properties of the Fibonacci numbers are so numerous that there is a mathematical periodical, *The Fibonacci Quarterly,* devoted entirely to their continued study.

See also PASCAL'S TRIANGLE; POLYOMINO.

field *See* RING.

Fields medals These are prizes awarded to young researchers for outstanding achievement in mathematics. The awards are regarded as equivalent in stature to Nobel Prizes (which do not exist for mathematics).

"International medals for outstanding discoveries in mathematics" were first proposed at the 1924 International Congress of Mathematicians meeting in Toronto.

Canadian mathematician John Charles Fields (1863–1932) later donated funds to support this idea, and the awards were created and named in his honor. It was agreed that two gold medals would be awarded every four years—at each quadrennial meeting of the International Congress of Mathematicians. The first awards were given in 1936 and, following a wartime hiatus, were resumed in 1950. In 1966, due to the significant expansion of mathematical research, it was agreed that up to four medals could be awarded at any given meeting.

The award itself consists of a cash prize and a medal made of gold. A picture of ARCHIMEDES OF SYRACUSE, along with the quotation, "Transire suum pectus mundoque potiri" (Rise above oneself and take hold of the world), appears on one side of the medal. On the reverse side is the inscription, "Congregati ex toto orbe mathematici ob scripta insignia tribuere" (the mathematicians of the world assembled here pay tribute for your outstanding work).

Following Fields's wish, the awards are presented in recognition of existing work completed by a mathematician, as well as potential for future achievement. For this reason, the awards are usually given to mathematicians under the age of 40.

A board of trustees set up by the University of Toronto administers the awards, and a committee of mathematicians appointed by the International Congress of Mathematicians presents the medals to recipients.

Laurent Lafforgue of the Institut des Hautes Études Scientifiques, Buressre-Yvette, France, and Vladimir Voevodsky of the Institute of Advanced Study, Princeton, New Jersey, were the 2002 recipients of the award. Lafforgue made significant contributions to the so-called Langlands program, a series of far-reaching conjectures proposed by Robert Langlands in 1967 that, if true, would unite disparate branches of mathematics. Voevodsky was awarded the prize for his work in algebraic geometry, a field that unites number theory and geometry.

figurate numbers Arranging dots to create geometric figures leads to a class of numbers called figurate numbers. For example, the triangular numbers are those numbers arising from triangular arrangements of dots, and the square numbers those from square arrays.

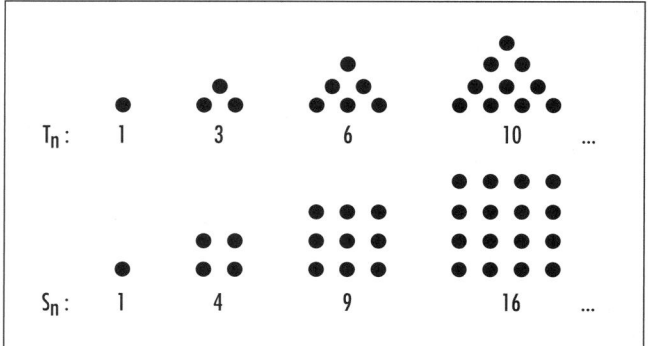

Triangular and square numbers

Other geometric shapes are possible, leading to other sequences of figurate numbers.

Figurate numbers were of special importance to the Pythagoreans of sixth century B.C.E. Believing that everything in the universe could be explained by the "harmony of number," they imparted special importance, even personality, to the figurate numbers. For example, 10, being the sum of the first four counting numbers $1 + 2 + 3 + 4$, in their belief united the four elements—earth, water, fire, and air—and so was to be held in the greatest of reverence. (They named this number *tetraktys*, "the holy four.")

Many arithmetic properties of sums can be readily explained by the figurate numbers. For example, the nth triangular number, T_n, is given by the sum: $1 + 2 + \ldots + n$. As two triangular configurations placed together produce an $n \times (n + 1)$ array of dots, $2T_n$, $= n \times (n + 1)$, we have:

$$T_n = 1 + 2 + \cdots + n = \frac{n(n+1)}{2}$$

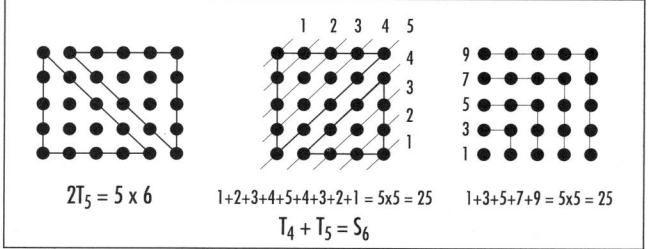

Properties of figurate numbers

The nth square number is given by the formula: $S_n = n \times n = n^2$. Looking at the diagonals of the square reveals the sum:

$$1 + 2 + \ldots + (n - 1) + n + (n - 1) + \ldots + 2 + 1 = n^2$$

Also hidden in a square is the sum of the first n odd numbers. We thus have:

The sum of the first n odd numbers is n^2.

Adding one to each of these summands gives the sum of the first n even numbers. Thus we have:

The sum of the first n even numbers is $n^2 + n$.

There is a nice interplay between the triangular and square numbers. For example, the sum of any two consecutive triangular numbers is always a square number: $1 + 3$ equals 4, $3 + 6$ equals 9, and $6 + 10$ equals 16, for instance. In general:

$$T_{n-1} + T_n = S_n$$

The center diagram in the figure at the bottom of page 194 explains why this is the case.

Similarly, one can arrange eight copies of the one triangle to form a square with its center removed to prove that $8T_n$ is always one less than a square number. In the same way, one can also establish that the following combination of three consecutive triangular numbers is always square: $T_{n-1} + 6T_n + T_{n+1}$.

The numbers $1 = T_1 = S_1$, $36 = T_8 = S_6$ and $1225 = T_{49} = S_{35}$ are both square and triangular, as are 41,616 and 1,413,721. Mathematicians have proved, using CONTINUED FRACTIONS, that there are infinitely numbers with this property.

The alternate triangular numbers 1, 6, 15, 28, … are sometimes called the Bohlen numbers. The nth Bohlen number x is divisible by n and is the unique multiple of n satisfying: $1 + 2 + \ldots + \frac{x}{n} = x$.

In 1796 KARL FRIEDRICH GAUSS proved that every natural number is the sum of at most three triangular numbers, and JOSEPH-LOUIS LAGRANGE in 1770 proved that every natural number is the sum of no more than four square numbers.

One can extend the scope of figurative numbers to three dimensions to produce the cube numbers, tetrahe- dral numbers, and the like. The nth cube number is given by the formula n^3. It is a three-dimensional cubical array of dots, with each layer being a square array of dots. (Thus n layers of n^2 dots gives a total of $n \times n^2 = n^3$ elements.) The nth tetrahedral number is produced by stacking together the first n triangular numbers: $T_1 + T_2 + \ldots + T_n$. This gives the sequence:

$$1, 4 = 1 + 3, 10 = 1 + 3 + 6, 20 = 1 + 3 + 6 + 10, \ldots$$

One can prove that the nth tetrahedral number is given by the formula $\frac{1}{6}n(n + 1)(n + 2)$.

Both the triangular numbers and the tetrahedral numbers appear as diagonals in PASCAL'S TRIANGLE. The successive stacking of tetrahedral numbers produces hypertetrahedral numbers: 1, 5, 15, 35, … These correspond to four-dimensional geometric arrangements of dots. They also appear as a diagonal in Pascal's triangle.

See also NESTED MULTIPLICATION; PERFECT NUMBER; SQUARE; TRIANGLE.

finger multiplication Having memorized the 2-, 3-, 4-, and 5-times tables, there is a popular finger method for computing all values of the 6- through 10-times tables. It is based on the following rule for encoding numbers:

A closed fist represents 5 and any finger raised on that hand adds 1 to that value.

Thus a hand with one finger raised, for example, represents 6. A hand with three fingers raised represents 8. To multiply two numbers between 5 and 10, one then follows these steps:

1. Encode the two numbers, one on each hand.
2. Count 10 for each finger raised.
3. Count the number of *unraised* fingers on each hand and multiply together the two counts.
4. Add the results of steps two and three. This is the desired product.

For example, "6 times 8" is represented as one raised finger on the left hand, three on the right hand. There are four raised fingers in all, yielding the number 40 for step 2. The left hand has four lowered fingers, and the right has two fingers lowered. We compute $4 \times 2 = 8$. Thus the desired product is $40 + 8 = 48$. Similarly, "8 times 8" is computed as $60 + 2 \times 2 = 64$, "9 times 7" as $60 + 1 \times 3 = 63$, and "6 times 7" as $30 + 3 \times 4 = 42$. Notice that one is never required to multiply two numbers greater than five. That the method works is explained by the algebraic identity:

$$(5 + a)(5 + b) = 10(a + b) + (5 - a)(5 - b)$$

The identity $(N + a)(N + b) = 2N(a + b) + (N - a)(N - b)$ shows that we can extend this method to use of a different number of digits on each hand. For example, with $N = 10$, using fingers and toes, one can readily compute 17×18 as "seven raised fingers" and "eight raised toes." Counting each raised digit as 20 ($2N$) we have: $17 \times 18 = 20 \times 15 + 3 \times 2 = 306$.

See also EGYPTIAN MULTIPLICATION; ELIZABETHAN MULTIPLICATION; MULTIPLICATION; NAPIER'S BONES; RUSSIAN MULTIPLICATION.

finite Intuitively, a set is said to be finite if one can recite all the elements of the set in a bounded amount of time. For instance, the set {knife, fork, spoon} is finite, for it takes only a second or two to recite the elements of this set. On the other hand, the set of natural numbers {1, 2, 3, ...} is not finite, for one can never recite each and every element of this set.

Despite our intuitive understanding of the concept, it is difficult to give a precise and direct mathematical definition of a finite set. The easiest approach is to simply define a finite set to be one that is not INFINITE, since the notion of an infinite set can be made clear. Alternatively, since there is a well-defined procedure for mechanically writing down the string of natural numbers 1, 2, 3, ..., one can define a finite set to be any set S whose elements can be put in one-to-one correspondence with a bounded initial segment of the string of natural numbers. For instance, matching "knife" with 1, "fork" with 2, and "spoon" with 3, the set {knife, fork, spoon} is finite because its elements can be matched precisely with the string of natural numbers {1, 2, 3}.

finite differences To analyze the terms of a SEQUENCE, it can be useful to create a table of successive differences (in the sense of "right minus left") between the terms of the sequence, and subsequent differences of the differences. For example, for the sequence 1,2,4,8,15,26,42,64,... we obtain the difference table:

```
1  2  4  8  15  26  42  64 ...
 1  2  4  7  11  16  22 ...
  1  2  3  4   5   6 ...
   1  1  1  1  1 ...
    0  0  0  0 ...
     0  0  0 ...
```

From the pattern that is now apparent, it is clear that the next number in the original sequence will be $64 + 29 = 93$.

The entries in the first row under the original sequence are said to be the "first finite differences"; the second row under the sequence depicts the "second finite differences," and so forth. All the terms that appear in a table of finite differences are completely determined by the values that appear in the leading diagonal. For instance, if the values a, b, c, ... shown below are known, then the remainder of the table must appear as follows:

$$a \quad a+b \quad a+2b+c \quad a+3b+3c+d \quad a+4b+6c+4d+e \ ...$$
$$b \quad b+c \quad b+2c+d \quad b+3c+3d+e \ ...$$
$$c \quad c+d \quad c+2d+e \ ...$$
$$d \quad d+e \ ...$$
$$e \ ...$$

The coefficients that appear in the top row match the entries in each row of PASCAL'S TRIANGLE, which are given by the BINOMIAL COEFFICIENTS. This suggests that it would be enlightening to examine the finite difference tables of the sequences $\binom{n}{0}, \binom{n}{1}, \binom{n}{2}, \cdots$. We obtain:

$$\binom{n}{0}: \quad 1 \quad 1 \quad 1 \quad 1 \quad 1 \ ...$$
$$0 \quad 0 \quad 0 \quad 0 \ ...$$
$$0 \quad 0 \quad 0 \ ...$$
$$0 \quad 0 \ ...$$

$$\binom{n}{1}: \quad 0 \quad 1 \quad 2 \quad 3 \quad 4 \quad \dots$$

$$1 \quad 1 \quad 1 \quad 1 \quad \dots$$

$$0 \quad 0 \quad 0 \quad \dots$$

$$0 \quad 0 \quad \dots$$

$$\binom{n}{2}: \quad 0 \quad 0 \quad 1 \quad 3 \quad 6 \quad \dots$$

$$0 \quad 1 \quad 2 \quad 3 \quad \dots$$

$$1 \quad 1 \quad 1 \quad \dots$$

$$0 \quad 0 \quad \dots$$

Each such sequence produces a table with a straightforward leading diagonal: one that is zero in all places except for the appearance of a single 1. Using this to our advantage, recall that the leading diagonal of the sequence 1, 2, 4, 8, 15, 26, 42, 64, … is 1,1,1,1,0,0,0,…, which is the sum of the leading diagonals for the four sequences given by $\binom{n}{0}, \binom{n}{1}, \binom{n}{2}$, and $\binom{n}{3}$. Thus the nth term in our original sequence must equal the sum of the nth terms of each of the four sequences, namely:

$$\binom{n}{0} + \binom{n}{1} + \binom{n}{2} + \binom{n}{3} = 1 + n + \frac{n(n-1)}{2} + \frac{n(n-1)(n-2)}{3}$$

$$= \frac{n^3 + 5n + 6}{6}$$

Thus we now have a formula for the sequence 1, 2, 4, 8, 15, 26, 42, 64,…

In general, one can use this technique to find a formula for any sequence whose difference table eventually contains a row of constant finite differences. Not all sequences, however, have this property. For example, the difference table for the sequence of FIBONACCI NUMBERS cycles indefinitely:

$$1 \quad 1 \quad 2 \quad 3 \quad 5 \quad 8 \quad 13 \quad 21 \quad 34 \quad \dots$$

$$0 \quad 1 \quad 1 \quad 2 \quad 3 \quad 5 \quad 8 \quad 13 \quad \dots$$

$$0 \quad 0 \quad 1 \quad 1 \quad 2 \quad 3 \quad 5 \quad \dots$$

One must employ alternative techniques to compute formulae for such sequences.

first- and second-derivative tests *See* MAXIMUM/MINIMUM.

Fisher, Sir Ronald Aylmer (1890–1962) British *Statistics, Genetics* Born on February 17, 1890, in London, England, Sir Ronald Fisher is considered the most important statistician of the early 20th century. His landmark 1925 text *Statistical Methods for Research Workers* established methods of designing experiments and analyzing results that have been used extensively by scientists ever since. Fisher was also an able geneticist and made significant contributions to the fields of selection and genetic dominance.

After obtaining a degree in astronomy from Cambridge in 1912, Fisher developed an interest in the theory of errors in astronomical observation. This work led him to a general interest in statistical problems and the analysis of ERROR in all disciplines, including those arising in biology. In 1919 Fisher accepted a position at the Rothamsted Agricultural Experiment Station as a biologist. There he developed his key ideas in the theory of genetics while also founding the theory of experimental design described in his 1925 piece.

Fisher was professor of genetics at University College, London, from 1933 to 1943, and then professor of genetics, University of Cambridge, until 1957. Upon his retirement, Fisher moved to Australia to become a research fellow at the Division of Mathematics and Statistics, CSIRO, Adelaide.

His method of multivariate analysis allowed scientists, for the first time, to properly analyze problems involving more than one variable, and his notion of "likelihood" provided the means to draw general conclusions on the basis of relative probabilities of different events. Fisher also contributed to the science of HYPOTHESIS TESTING by identifying and analyzing new key DISTRIBUTIONS. His work, without doubt, transformed statistics from a general science into a practical and powerful scientific tool. He is considered the founder of modern statistics.

Fisher was elected a fellow of the ROYAL SOCIETY in 1929, and was awarded the Royal Medal of the Society in 1938, the Darwin Medal of the Society in 1948, and the Copley Medal of the Society in 1955. He was knighted in 1952 in recognition of his influential work in statistics and for his development of a statistical theory of natural selection. Fisher died in Adelaide, Australia, on July 29, 1962.

See also HISTORY OF PROBABILITY AND STATISTICS (essay); KARL PEARSON.

fixed point Any point that is mapped to itself by a given TRANSFORMATION is called a fixed point. For example, the points $x = 0$ and $x = 1$ are fixed points for the function $f(x) = x^2$. Any point on the line of REFLECTION for a reflection in a plane is a fixed point for that reflection.

Any continuous map f that maps points in the unit interval [0,1] to points in the same interval must possess a fixed point. (By the INTERMEDIATE-VALUE THEOREM, the graphs of $y = f(x)$ and $y = x$ for $0 \le x \le 1$ must intersect.) This is a special case of a more general result proven by Luitzen Egbertus Jan Brouwer in 1915 stating that, for all values n, any continuous map $f:[0, 1]^n \to [0,1]^n$ must possess at least one fixed point. (For $n = 2$, $[0,1]^2 = [0,1] \times [0,1]$ is the unit square in the plane, and for $n = 3$, $[0,1]^3 = [0,1] \times [0,1] \times [0,1]$ is a unit cube in three-dimensional space.) This theorem has the following amusing consequences:

Consider two square sheets of paper, one lying directly on top of the other. Initially each point of the upper sheet lies directly above its corresponding point on the lower sheet. Now crumple the top sheet and rest the crumpled ball anywhere on the lower sheet. By the Brouwer fixed-point theorem there is still at least one point of the crumpled sheet lying directly above its corresponding point on the bottom sheet.

As a thought experiment, imagine the molecules of the liquid in a cup of coffee as the points in a three-dimensional cube. After the coffee is stirred, the Brouwer fixed point theorem assures that at least one molecule will return to its original location.

See also ITERATION.

floor/ceiling/fractional part functions The floor function, also known as the greatest-integer function, takes a real number x and returns the greatest integer not exceeding x. This quantity is denoted: $\lfloor x \rfloor$. For example, $\lfloor 7.2 \rfloor = 7$, $\lfloor 7.9998 \rfloor = 7$ and $\lfloor 7 \rfloor = 7$. Also, $\lfloor -6.34 \rfloor = -7$.

The ceiling function, also known as the least-integer function, takes a real number x and returns the least integer not smaller than x. This quantity is denoted: $\lceil x \rceil$. For example, $\lceil 7.2 \rceil = 8$, $\lceil 7.998 \rceil = 8$ and $\lceil 7 \rceil = 7$. Also, $\lceil -6.34 \rceil = -6$. The fractional part of a real number x, denoted $\{x\}$, is given by: $\{x\} = x - \lfloor x \rfloor$. For example, $\{7.2\} = 0.2$, $\{7.998\} = 0.998$ and $\{7\} = 0$. Also, $\{-6.34\} = 0.66$.

The names "floor" and "ceiling," as well as the notation for these functions, were introduced by Kenneth Iverson in his 1962 computer science text *A Programming Language*. These functions often arise in applications of counting. For example, there are $\left\lfloor \dfrac{N}{4} \right\rfloor$ multiples of 4 less than, or equal, to N, and, for any two real numbers x and y with $x < y$, the closed interval $[x, y]$ contains $\lfloor y \rfloor - \lceil x \rceil + 1$ integers.

See also DAYS-OF-THE-WEEK FORMULA.

floor function *See* FLOOR/CEILING/FRACTIONAL PART FUNCTIONS.

fluxion In his version of CALCULUS, SIR ISAAC NEWTON thought of variable x as a flowing quantity, or a fluent, and called the rate of change of x the "fluxion" of x. He used the notation $\dot{x}$ for the fluxion of x, $\ddot{x}$ for the fluxion of the fluxion of x, and so forth. Thus if $x = f(t)$, where x is the distance and t the time for a moving object, then $\dot{x}$ is the instantaneous VELOCITY of the object, and $\ddot{x}$ its instantaneous ACCELERATION. Today the term *fluxion* is considered obsolete, and we use the word DERIVATIVE in its stead. The raised-dot notation for derivative, however, is still used by physicists for denoting derivatives with respect to time.

See also CALCULUS; HISTORY OF CALCULUS (essay).

focal chord Any CHORD of a conic curve—a PARABOLA, an ELLIPSE, or a HYPERBOLA—that passes through a FOCUS of the conic is called a focal chord.

See also CONIC SECTIONS; FOCAL RADIUS.

focal radius Any line from the FOCUS of a conic curve—a PARABOLA, an ELLIPSE, or a HYPERBOLA—to a point on the conic is called a focal radius.

See also CONIC SECTIONS; FOCAL CHORD.

focus (plural, foci) Each CONIC SECTION has associated with it one or two special points each called a focus of the conic.

See also ELLIPSE; HYPERBOLA; PARABOLA.

formal logic (symbolic logic) In mathematics, the systematic study of reasoning is called formal logic. It analyzes the structure of ARGUMENTS, as well as the methods and validity of mathematical deduction and proof.

The principles of logic can be attributed to ARISTOTLE (384–322 B.C.E.), who wrote the first systematic treatise on the subject. He sought to identify modes of inference that are valid by virtue of their structure, not their content. For example, "Green and blue are colors; therefore green is a color" and "Cows and pigs are reptiles; therefore cows are reptiles" have the same structure ("A and B, therefore A"), and any argument made via this structure is logically valid. (In particular, the second example *is* logically sound.) This mode of thought allowed EUCLID (ca. 300–260 B.C.E.) to formalize geometry, using deductive proofs to infer geometric truths from a small collection of AXIOMS (self-evident truths).

No significant advance was made in the study of logic for the millennium that followed. This period was mostly a time of consolidation and transmission of the material from antiquity. The Renaissance, however, brought renewed interest in the topic. Mathematical scholars of the time, Pierre Hérigone and Johann Rahn in particular, developed means for representing logical arguments with abbreviations and symbols, rather than words and sentences. GOTTFRIED WILHELM LEIBNIZ (1646–1716) came to regard logic as "universal mathematics." He advocated the development of a "universal language" or a "universal calculus" to quantify the entire process of mathematical reasoning. He hoped to devise new mechanical symbolism that would reduce errors in thinking to the equivalent of arithmetical errors. (He later abandoned work on this project, assessing it too daunting a task for a single man.)

In the mid-1800s GEORGE BOOLE succeeded in creating a purely symbolic approach to propositional logic, that part which deals with inferences involving simple declarative sentences (statements) joined by the connectives:

not, and, or, if … then…, iff

(These are called the NEGATION, CONJUNCTION, DISJUNCTION, CONDITIONAL, and the BICONDITIONAL, respectively.) He successfully applied it to mathematics, thereby making a significant step to achieving Leibniz's goal.

In 1879 the German mathematician and philosopher Gottlob Frege constructed a symbolic system for predicate logic. This generalizes propositional logic by including QUANTIFIERS: statements using words such as *some, all,* and, *no.* (For example, "All men are mortal" as opposed to "This man is mortal.") At the turn of the century DAVID HILBERT sought to devise a complete, consistent formulation of all of mathematics using a small collection of symbols with well-defined meanings. English mathematician and philosopher BERTRAND RUSSELL, in collaboration with his colleague ALFRED NORTH WHITEHEAD, took up Hilbert's challenge. In 1925 they published a monumental work. Beginning with an impressively minimal set of premises ("self-evident" logical principles), they attempted to establish the logical foundations of all of mathematics. This was an impressive accomplishment. (After hundreds of pages of symbolic manipulations, they established the validity of "1 + 1 = 2," for example.) Although they did not completely reach their goal, Russell and Whitehead's work has been important for the development of logic and mathematics.

Six years after the publication of their efforts, however, KURT GÖDEL stunned the mathematical community by proving Hilbert's (and Leibniz's) goal to be futile. He demonstrated once and for all that any formal system of logic sufficiently sophisticated to incorporate basic principles of arithmetic cannot attain all the statements it hopes to prove. His results are today called GÖDEL'S INCOMPLETENESS THEOREMS. The vision to reduce *all* truths of reason to incontestable arithmetic was thereby shattered.

Understanding the philosophical foundations of mathematics is still an area of intense scholarly research.

See also ARGUMENT; DEDUCTIVE/INDUCTIVE REASONING; LAWS OF THOUGHT.

formula Any identity, general rule, or general expression in mathematics that can be applied to different values of one or more quantities is called a formula. For example, the formula for the area A of a circle is:

$$A = \pi r^2$$

where r represents the radius of the circle. The QUADRATIC formula for the roots of a quadratic equation of the form $ax^2 + bx + c = 0$ is:

$$x = \frac{-b \pm \sqrt{b^2 - 4ac}}{2a}$$

foundations of mathematics   The branch of mathematics concerned with the justification of mathematical rules, AXIOMS, and modes of inference is called foundations of mathematics. The paradigm for critical mathematical analysis came from the work of the great geometer EUCLID (ca. 300–260 B.C.E.) who, in his work *THE ELEMENTS*, demonstrated that all geometry known at his time can be logically deduced from a small set of self-evident truths (axioms). LEONHARD EULER (1707–83) produced fundamental results in disparate branches of mathematics and often saw connections between those branches. He too searched for small collections of concepts that were fundamental and, hopefully, common to all fields. In the late 1800s and at the turn of the century with the discovery of RUSSELL'S PARADOX in SET THEORY, mathematicians were led to apparent paradoxes and inconsistencies within the seemingly very basic notions of "set" and "number." This led to the fervent study of the fundamental principles of elementary mathematics and even to the study of the process of mathematical thinking itself (FORMAL LOGIC). In the 1930s Austrian mathematician KURT GÖDEL (1906–78) stunned the mathematical community by proving, essentially, that any formal system of mathematics that incorporates the principles of arithmetic will contain statements that can neither be proved nor disproved, and, in addition, such a system will necessarily be incapable of establishing that it is free from CONTRADICTIONS. Despite these disturbing conclusions, the study of the founding principles of mathematics is still an active area of research today.

See also GEORG CANTOR; GÖDEL'S INCOMPLETENESS THEOREMS; BERTRAND ARTHUR WILLIAM RUSSELL; ALFRED NORTH WHITEHEAD; ERNST FRIEDRICH FERDINAND ZERMELO.

four-color theorem   For centuries, cartographers have known that four colors suffice to color any geographical map (that is, any division of the plane into regions). It is required that regions sharing a common length of boundary be painted different colors (but two regions meeting at a point, such as the states Arizona and Colorado on a map of the United States, may be painted the same tint). Cartographers had also observed that the same is true for any map drawn on a SPHERE (the globe).

The question of whether this observation could be proved true mathematically was first posed by English scholar Francis Guthrie in 1852. Mathematicians AUGUSTUS DE MORGAN (1806–71) and ARTHUR CAYLEY (1821–95) worked to solve the problem and, in 1872, Cayley's student Alfred Bray Kempe (1849–1922) produced the first attempt at a proof of the four-color conjecture. Unfortunately, 11 years later English scholar Percy Heawood (1861–1955) found that Kempe had made an error in his work. In 1890 Headwood later proved that five colors will always suffice to color a planar map, but the proof that just four will actually suffice eluded him. Heawood also looked beyond just planar and spherical maps and made a general conjecture that if a surface contains g holes (such as TORUS with $g = 1$ hole, or a sphere with $g = 0$ holes), then any map drawn on that surface can be colored with

$$\left\lfloor \frac{7 + \sqrt{48g + 1}}{2} \right\rfloor$$

colors, and that there do exist examples of maps on these surfaces that do require precisely this many colors. (The brackets indicate to round down to the nearest integer.)

In 1968 two mathematicians, Gerhard Ringel and J. W. T. Youngs, proved Heawood to be correct for all surfaces with two or more holes and for the torus. Unfortunately, their work did not apply to the case of a sphere and of a KLEIN BOTTLE. It was not until the next decade when, in 1976, mathematicians Kenneth Appel and Wolfgang Haken finally established that four colors do indeed suffice to color any map on a sphere (and hence the plane, since, by placing a small hole in the center of one of the regions to be painted, a punctured sphere can be stretched and flattened onto the plane, and, vice versa, a planar region can be stretched and molded into a punctured sphere).

Appel and Haken's proof was deemed controversial at the time, since it used some 1,200 hours of computer time to check nearly 2,000 complicated spatial arrangements. (It was argued that since no single human being could possibly verify that the computer had completed this work correctly, then the outcome produced is not a valid proof.) Most mathematicians today accept the validity of Appel and Haken's work and consider the famous four-color problem solved. (The search for a simple and elegant proof easily checked by human hand, however, continues.)

It is interesting to note that Heawood's conjecture actually fails for the single case that remains: it is known that six colors suffice to color any map drawn on a Klein bottle ($g = 1$), not seven, as predicted by the formula.

See also FLOOR/CEILING/FRACTIONAL PART FUNCTIONS.

Fourier, Jean Baptiste Joseph (1768–1830) French *Analysis, Engineering* Born on March 21, 1768, French scholar Joseph Fourier is remembered in mathematics for his fundamental contributions to the theory of heat conduction and his study of trigonometric series. His groundbreaking advances in these topics appear in his famous 1822 treatise *Théorie analytique de la chaleur* (Analytic theory of heat). Today, FOURIER SERIES play a fundamental role in the study of physics and engineering, as well as in the development of theoretical mathematics.

Placed in a military school as an orphan at age 10, Fourier exhibited a strong interest in mathematics at this early age. He taught at the same military school for four years before entering the newly established teacher-training school L'École Normale (later renamed L'École Polytechnique) in 1794. There Fourier received instruction in mathematics from JOSEPH-LOUIS LAGRANGE (1736–1813) and excelled in all his studies. In 1797, when Lagrange stepped down as chair of analysis and mechanics, Fourier was appointed the new department chair.

This position was short-lived, however, for the following year Fourier was assigned to Napoleon's army in the invasion of Egypt. He was charged to oversee a variety of archeological and scientific investigations, and it was during this period of his life, while stationed in the Egyptian desert, that Fourier developed a fascination for the mathematics of heat transfer.

Taking a highly original approach, Fourier analyzed conduction by representing complicated oscillating quantities as sums of simpler components. This led him to his theory of trigonometric series. In 1807 Fourier detailed the results of his work in the paper "On the Propagation of Heat in Solid Bodies," but the mathematicians of the day, including Lagrange, were not convinced of the validity of his approach. Fourier rewrote the paper in 1811, and even received a prize for its content, but still received criticism from the mathematics community for its lack of rigor and lack of generality. For another 10 years Fourier worked to establish the mathematical foundations of his work and detail the reasoning behind his methods. Finally, in 1822, after the publication of his famous treatise, all criticisms of his mathematics were settled, and Fourier's approach was accepted as valid and fundamentally important.

Fourier was elected to the powerful position of secretary to the Académie of Sciences that same year, and in 1827 he was honored with membership to the ROYAL SOCIETY of London.

It has been conjectured that Fourier, ironically, may have contracted the exotic illness myxedema while in Egypt, which, among several symptoms, induces an extreme sensitivity to cold. It is said that Fourier kept his living quarters almost unbearably hot and always bundled himself in many layers of clothing. Fourier died of a heart attack on May 16, 1830.

Fourier series have found numerous applications to physics and engineering. Practically every branch of science that contends with the transfer of energy through waves (such as acoustics, seismic studies, wireless communications, spectroscopy) utilizes Fourier's methods. In a more abstract setting, Fourier series are now seen as just one of infinitely many possible approaches to decomposing elements of a VECTOR SPACE of functions in terms of basis elements. This shift in perspective has led to new insights into the study of quantum mechanics.

Fourier series When a musician plays two strings on a violin simultaneously, just one sound wave reaches our ears—the combined effect of the two notes. Nonetheless, the human brain is able to decode the information it receives to "hear" two distinct notes being played. That is, the human brain is able to recognize complicated sound waves as sums of simpler basic sound waves.

In the 18th century, French mathematician and physicist JEAN LE ROND D'ALEMBERT (1717–83), and the Swiss mathematician LEONHARD EULER (1707–83) worked to describe complicated vibrations of strings as sums of simpler functions. The Swiss mathematician Daniel Bernoulli (1700–82) of the famous BERNOULLI FAMILY introduced the use of trigonometric functions in this study, an approach that was later fully developed by French mathematician and physicist JEAN BAPTISTE JOSEPH FOURIER (1768–1830), although his work was motivated by the study of heat conduction. Fourier showed that many functions could be represented as infinite sums of sine and cosine functions.

The result of writing a function as a sum of trigonometric functions is today called a Fourier series. As the trigonometric functions cycle in value every 2π in RADIAN MEASURE, it is assumed in these studies that the functions under consideration are themselves periodic with period 2π.

Assume $f(x)$ is such a function. Then a Fourier series for f is an expression of the form:

$$f(x) = \frac{a_0}{2} + a_1 \cos(x) + a_2 \cos(2x) + \cdots$$
$$+ b_1 \sin(x) + b_2 \sin(2x) + \cdots$$

One finds the values of the constants $a_0, a_1, a_2, \ldots b_1, b_2, \ldots$ by integrating. For example, since $\int_{-\pi}^{\pi} \cos(kx)\, dx = 0 = \int_{-\pi}^{\pi} \sin(kx)\, dx$, we have:

$$\int_{-\pi}^{\pi} f(x)\,dx = \int_{-\pi}^{\pi} \frac{a_0}{2}\,dx + 0 + 0 + \cdots + 0 + 0 + \cdots$$

yielding: $a_0 = \frac{1}{\pi}\int_{-\pi}^{\pi} f(x)\,dx$. Multiplying through by $\sin(x)$ and integrating gives:

$$\int_{-\pi}^{\pi} f(x)\sin(x)dx = \int_{-\pi}^{\pi} \frac{a_0}{2}\sin(x)dx + \int_{-\pi}^{\pi} a_1 \cos(x)\sin(x)dx$$
$$+ \int_{-\pi}^{\pi} a_2 \cos(2x)\sin(x)dx + \cdots$$
$$+ \cdots + \int_{-\pi}^{\pi} b_1 \sin(x)\sin(x)dx$$
$$+ \int_{-\pi}^{\pi} b_2 \sin(2x)\sin(x)dx + \cdots$$
$$= 0 + 0 + 0 + \cdots + b_1\pi + 0 + \cdots b_1$$

showing that $b_1 = \frac{1}{\pi}\int_{-\pi}^{\pi} f(x)\sin(x)dx$.

One can show that the functions $\{1, \cos(x), \cos(2x), \ldots, \sin(x), \sin(2x), \ldots\}$ are ORTHOGONAL in the sense that the integral of the product of any two different functions from this set is zero. This observation allows us to compute all the values $a_0, a_1, a_2, \ldots, b_1, b_2, \ldots$ by this method of multiplying through by a trigonometric function and integrating. We have, in general:

$$a_0 = \frac{1}{\pi}\int_{-\pi}^{\pi} f(x)\,dx$$
$$a_n = \frac{1}{\pi}\int_{-\pi}^{\pi} f(x)\cos(nx)\,dx$$
$$b_n = \frac{1}{\pi}\int_{-\pi}^{\pi} f(x)\sin(nx)\,dx$$

Mathematicians have shown that if $f(x)$ and its DERIVATIVE $f'(x)$ are both CONTINUOUS FUNCTIONs, then the expansion

$$f(x) = \frac{a_0}{2} + a_1 \cos(x) + a_2 \cos(2x) + \cdots$$
$$+ b_1 \sin(x) + b_2 \sin(2x) + \cdots$$

is valid. They have also shown that, if interpreted appropriately, the expansion remains valid even if f or

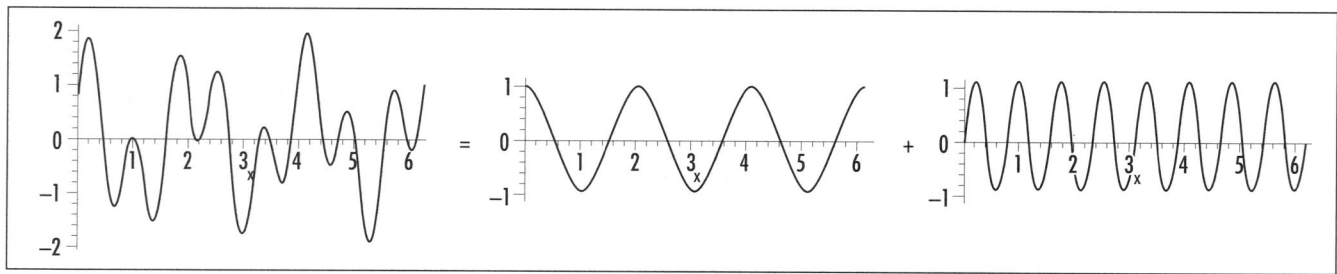

A wave as a sum of a cosine curve and a sine curve

f are not continuous for a finite number of locations in the interval $[-\pi,\pi]$. This shows, for example, that functions that zig-zag like a sawtooth, or jump up and down in value like a staircase, for example, can still be well approximated by a sum of trigonometric functions. For example, take $f(x)$ to be the V-shaped function $f(x) = |x|$ on the interval $[-\pi,\pi]$, with this section of graph repeated over the entire number line to produce the picture of a sawtooth. One checks:

$$a_0 = \frac{1}{\pi}\int_{-\pi}^{\pi} |x|\,dx = \frac{2}{\pi}\int_0^{\pi} x\,dx = \pi$$

$$a_n = \frac{2}{\pi}\int_0^{\pi} x\cos(nx)\,dx = \begin{cases} 0 \text{ for n even} \\ -\dfrac{4}{\pi n^2}\text{for n odd} \end{cases}$$

$$b_n = 0$$

giving:

$$|x| = \frac{\pi}{2} - \frac{4}{\pi}\left(\frac{\cos(x)}{1} + \frac{\cos(3x)}{3^2} + \frac{\cos(5x)}{5^2} + \frac{\cos(7x)}{7^2} + \cdots\right)$$

at least on the interval $[-\pi,\pi]$. Note, as a curiosity, that if we place $x = 0$ into this formula we obtain the remarkable identity:

$$\frac{\pi^2}{8} = 1 + \frac{1}{3^2} + \frac{1}{5^2} + \frac{1}{7^2} + \cdots$$

See also ZETA FUNCTION.

fractal If we SCALE the picture of a geometric object by a factor k, then its size changes accordingly: any line of length a becomes a line of length ka, any planar region of area A becomes a planar region of area k^2A, and any solid of volume V is replaced by a solid of volume k^3V. An object can thus be described as d-dimensional if its "size" scales according to the rule:

$$\text{new size} = k^d \times \text{old size}$$

At the turn of the 20th century, mathematicians discovered geometric objects that are of fractional dimension. These objects are called fractals. One such object is Sierpinski's triangle, devised by Polish mathematician Vaclav Sierpinski (1882–1969). Beginnning with an equilateral triangle, one constructs it by successively removing cen-

tral triangles *ad infinitum*. The final result is an object possessing "self-similarity," meaning that the entire figure is composed of three copies of itself, in this case each at one-half scale. If the dimension of the object is d and the size of the entire object is S, then according to the scaling rule above, the size of each scaled piece is $\left(\dfrac{1}{2}\right)^d \times S$. As the entire figure is composed of three of these smaller figures, we have $S = 3 \times \left(\dfrac{1}{2}\right)^d \times S$. This tells us that $2^d = 3$, yielding $d = \dfrac{\ln 3}{\ln 2} \approx 1.58$. Thus the Sierpinski triangle is a geometric construct that lies somewhere between being a length and an area.

In 1904 Swedish mathematician Nils Fabian Helge von Koch (1870–1924) described a fractal curve constructed in a similar manner. Beginning with a line segment, one draws on its middle third two sides of an equilateral triangle of matching size and repeats this construction *ad infinitum* on all line the segments that appear. The result is called the Koch curve. It too is self-similar: the entire figure is composed of four copies of itself, each at one-third scale. The object has fractal dimension $d = \dfrac{\ln 3}{\ln 2} \approx 1.26$.

The Cantor set, invented by German mathematician GEORG CANTOR (1845–1918), is constructed from a single line segment, by removing its middle third and the middle thirds of all the line segments that subsequently appear. The result is a geometric construct, resembling nothing more than a set of points, but again with the same self-similarity property: the entire construct is composed of two copies of itself, each at one-third scale. The Cantor set is a fractal of dimension $d = \dfrac{\ln 2}{\ln 3} \approx 0.63$. It is not large enough to be considered one-dimensional, but it is certainly "more" than a disconnected set of isolated points.

Fractals also arise in the theory of CHAOS and the study of DYNAMICAL SYSTEMS. French mathematician Gaston Maurice Julia (1892–1978) considered the iterations of functions f that take COMPLEX NUMBERS as inputs and give complex numbers as outputs. If z is a complex number and the set of points $f(z)$, $f(f(z))$, $f(f(f(z)))$, ... are all plotted on a graph, then two possibilities may occur: either the sequence is unbounded, or the points jump about in a bounded region. The set

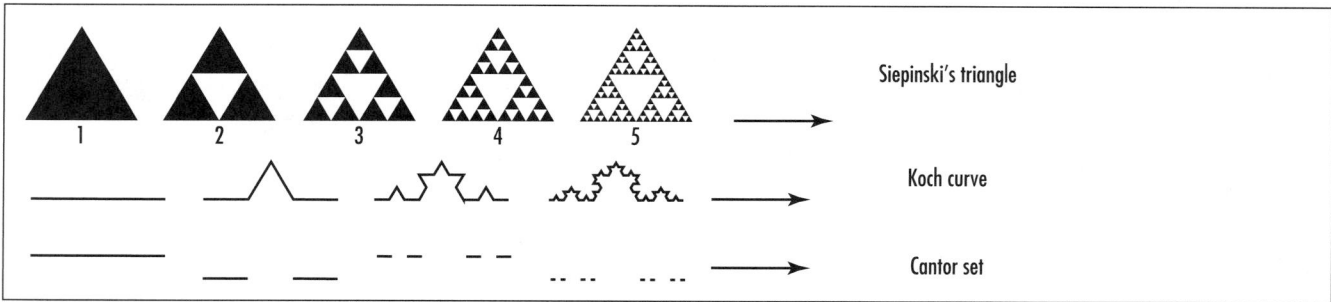

Siepinski's triangle

Koch curve

Cantor set

Fractals

of all values z that lead to a bounded sequence is called the "Julia set" for f. It is a region in the complex plane that is almost always a fractal. Even simple functions such as $f(z) = z^2 + c$, with different values of the constant c, can yield surprisingly varied and beautiful structures.

Polish-born French mathematician Benoit Mandelbrot considered, alternatively, all the complex values c for which the repeated application of the function $f(z) = z^2 + c$, beginning with the point $z = 0$, yields a bounded sequence. Plotting all such points c describes a subset of the complex plane today called the Mandelbrot set. It has an extremely complicated structure. In particular, its boundary is a fractal.

The Sierpinski triangle also arises from an iterative procedure known as the "chaos game:"

> Pick a point at random inside an equilateral triangle (diagram (1) above) and then draw the point halfway between it and one of the vertices of the triangle picked at random. (This point will lie somewhere in the shaded region of diagram (2) above.) Now draw the point halfway between this new point and another vertex picked at random. (This point will lie somewhere in the shaded region of diagram (3) above.) Imagine we continue this process indefinitely. Sierpinski's triangle represents all possible "final" destinations of this point as this game is played an infinite number of times.

Many objects in nature seem to possess the self-similarity properties of fractals. For example, the boundary shape of a cloud looks just as irregular under magnification as it does when looked at directly. Scientists have found it possible to assign fractal dimensions to various objects in nature. The study of fractals has since found applications to the study of crystal formation, fluid mechanics, urban growth, linguistics, economics, and many other diverse areas.

Much of the work in fractal geometry was pioneered by Benoit Mandelbrot, who also coined the term *fractal*.

fractal dimension   *See* FRACTAL.

fraction   Any number written as a QUOTIENT, that is, as one number a divided by another b, is a fraction. We write a/b and call the dividend a the numerator of the fraction and the divisor b the denominator of the fraction. It is assumed that b is not zero.

It is appropriate to regard fractions simply as answers to division problems. For example, if six pies are shared equally among three boys, then each boy receives two pies. We write: $6/3 = 2$. Similarly, if one pie is shared among two boys, then the amount of pie each boy receives is written $1/2$. We, appropriately, call this quantity "half." Clearly if a pies are shared with just one boy, then that boy receives all a pies. This yields the observation:

$$\text{Unit Denominator Rule: } \frac{a}{1} = a$$

Every fractional property can be explained with this simple pie-sharing model. For instance, we have:

$$\text{Cancellation Law: For any nonzero value } x \text{ and}$$
$$\text{fraction } \frac{a}{b} \text{ we have } \frac{xa}{xb} = \frac{a}{b}.$$

For instance, if a pies are shared among b boys, then doubling the number of pies and doubling the number of boys does not alter the amount of pie each boy receives: $\frac{2a}{2b} = \frac{a}{b}$. We have, in effect, "cancelled" a common factor from the numerator and denominator of the fraction. This principle also shows that it is possible to express the same fraction in an infinite number of equivalent forms. For instance, $\frac{3}{5}, \frac{6}{10}, \frac{9}{15}, \frac{30}{50}$, and $\frac{120}{200}$ all represent the same fraction.

In continuing the model of sharing pies, we see that if a_1 pies are shared among b boys, then another a_2 pies are shared among the same boys, then in effect $a_1 + a_2$ pies were shared among those boys. This yields the rule:

Addition of Like Fractions: $\frac{a_1}{b} + \frac{a_2}{b} = \frac{a_1 + a_2}{b}$

thereby providing a method for adding fractions sharing the same denominator. To add fractions with different denominators, convert each fraction to forms that share a COMMON DENOMINATOR and then add. For instance, to sum $\frac{2}{5}$ and $\frac{3}{8}$ write $\frac{2}{5} + \frac{3}{8} = \frac{8 \times 2}{8 \times 5} + \frac{5 \times 3}{5 \times 8} = \frac{16}{40} + \frac{15}{40}$ to obtain the answer $\frac{31}{40}$. This yields the rule:

Addition of Unlike Fractions: $\frac{a}{b} + \frac{c}{d} = \frac{da + bc}{bd}$

Subtraction is performed in a similar manner.

If, in sharing pies, one wished to double the amount of pie each boy receives, one could simply double the number of pies available. This suggests the rule:

Product Rule: $x \times \frac{a}{b} = \frac{xa}{b}$

This shows, along with the cancellation law, multiplying a fraction by its denominator produces its numerator as a result:

Denominator Product Rule: $b \times \frac{a}{b} = \frac{ba}{b} = \frac{a}{1} = a$

The product rule also shows how to multiply for fractions. For instance, to compute $\frac{2}{3} \times \frac{4}{7}$ observe that $\frac{2}{3} \times \frac{4}{7} = \frac{\frac{2}{3} \times 4}{7}$ by the product rule. Multiplying

the numerator and denominator each by three gives $\frac{3 \times \frac{2}{3} \times 4}{3 \times 7}$, which is $\frac{2 \times 4}{3 \times 7}$. This process is summarized:

Multiplication of Fractions: $\frac{a}{b} \times \frac{c}{d} = \frac{ac}{bd}$

The following division rule for fractions is proved in a similar manner:

Division of Fractions: $\dfrac{\frac{a}{b}}{\frac{c}{d}} = \frac{a}{b} \times \frac{d}{c} = \frac{ad}{bc}$

(Multiply the numerator and denominator each by b to obtain $\dfrac{a}{b \times \frac{c}{d}}$. Now multiply the numerator and denominator each by d to yield $\frac{ad}{bc}$.)

Types of Fractions

In writing a generic fraction a/b, it is often assumed that a and b are each a WHOLE NUMBER. If this is indeed the case, then the fraction is called a "common" fraction (or sometimes a "simple" or a "vulgar" fraction). Each common fraction represents a RATIONAL NUMBER. Although there are infinitely many rational numbers, surprisingly, they occupy absolutely no space on the NUMBER LINE.

A common fraction with positive numerator and denominator is called "proper" if its numerator is less than its denominator, and "improper" otherwise. (Thus a proper fraction represents a quantity less than one, and an improper fraction a quantity greater than or equal to one.) A "mixed number" is a number consisting of an integer and a proper fraction. For example, $3\frac{1}{2}$ is a mixed number.

A zero fraction is a fraction with numerator equal to zero. If no pies are shared among b boys, then each boy receives zero pie. Thus every zero fraction is equal to zero: $0/b = 0$.

An "undefined" fraction is a fraction with denominator equal to zero. Such a fraction is invalid, for it cannot have any meaningful value. (If, for instance, $2/0$ had value x, then multiplying through by the denominator yields the absurdity $2 = 0 \times x = 0$. Also, dividing zero by zero could, allegedly, have *any* desired value. For example, one could argue that 0/0 = 17 by noting that $0 \times 17 = 0$ is a true statement. There is no single appropriate value for this quantity.)

A "unit" fraction, also known as an EGYPTIAN FRACTION, is a proper fraction with numerator equal to one. A "complex" fraction is one in which the numerator or denominator, or both, is a fraction. For example, $\frac{3/5}{4/7}$ is a complex fraction. The division-of-fractions rule provides the means to simplify complex fractions.

See also CANCELLATION; CONTINUED FRACTION; DECIMAL REPRESENTATION; PARTIAL FRACTIONS; PERCENTAGE; RATIO; RATIONALIZING THE DENOMINATOR; REDUCED FORM.

fractional part function *See* FLOOR/CEILING/FRACTIONAL PART FUNCTIONS.

frequency In STATISTICS, the absolute frequency of an observed value is the number of times that value appears in a data set. For example, in the sample 4, 6, 3, 4, 4, 1, 7, 3, 7, 9, 8, 4, 4, 5, 2, the absolute frequency of the observation 4 is five, and that of 8 is one. The relative frequency of an observed value is the proportion of times it appears. This is computed by dividing the absolute frequency of the observation by the total number of entries in the data set. For example, the relative frequency of 4 in the above data set is 5/15 = 1/3, and that of 8 is 1/15.

One can compute the relative frequency of entries in an infinite data set by making use of a LIMIT. For example, mathematicians have proved that 5.8% of the powers of two 1, 2, 4, 8, 16, 32, 64, 128,... begin with a 7. (The first power to do so is 2^{46}.) By this they mean that if one were to examine the first N powers of two, approximately 5.8 percent of them begin with a 7. This approximation is made more exact by taking larger and larger values of N.

In physics the term *frequency* is defined as the number of cycles that occur per unit time in a system that oscillates (such as a pendulum, a wave, a vibrating string, or an alternating current). The symbol used for frequency is usually f, although the Greek letter ν is often employed for the frequency of light or other electromagnetic radiation. A unit of frequency is called a hertz (Hz).

frequency distribution *See* STATISTICS: DESCRIPTIVE.

frequency polygon *See* STATISTICS: DESCRIPTIVE.

frieze pattern (band ornament) A design on an infinite strip that consists of repeated copies of a single motif is called a frieze pattern. (In classical architecture, a frieze is a horizontal structure, usually imprinted with decoration, resting along the top of some columns. The modern equivalent is a horizontal strip of wallpaper used to decorate the top portion of a wall just below the ceiling.)

Mathematicians are interested in the SYMMETRY properties of frieze patterns. Each frieze pattern, by definition, is symmetrical under a TRANSLATION T in the direction of the strip. A frieze pattern might also be symmetrical about a horizontal REFLECTION (H), a vertical reflection (V), a ROTATION of 180° about a point in the design (R), a GLIDE REFLECTION (G), or some collection of these five basic transformations. The first frieze pattern shown below possesses all five symmetries.

Not all combinations of T, H, V, R, and G, represent the symmetries of a frieze pattern. For instance, a frieze pattern cannot possess the symmetries T and H alone, for the composition of these two symmetries produces a glide reflection, and so the pattern must possess symmetry G as well. Similarly, any frieze pattern that possesses symmetries H and V must also possess symmetry R, since the combined effect of a vertical and a horizontal reflection is a rotation. Reasoning this way, one can prove that there are only seven combinations of symmetry types a frieze pattern could possess. These are:

T alone	T, H, and G
T and V	T, V, R, and G
T and R	T, H, V, R, and G
T and G	

A frieze pattern

For this reason, mathematicians say that there are only seven possible frieze patterns.

A wallpaper pattern is the two-dimensional analogue of a frieze pattern. Mathematicians have proved that there are exactly 17 different wallpaper patterns.

frustum If a geometric solid is cut by two parallel planes, then the portion of the solid between the two planes is called a frustum. A frustum is also produced by cutting the solid by one plane parallel to the BASE of the solid. For example, a conical frustum, is the shape produced by slicing the top off a cone (with the cut made parallel to the base). A pyramidal frustum is made by slicing off the top of a pyramid, again with the cut made parallel to the base.

The altitude h of a frustum is the distance between the two parallel planes. The volume of a conical or pyramidal frustum is given by the formula:

$$\frac{1}{3}h\left(A_1 + A_2 + \sqrt{A_1 A_2}\right)$$

where A_1 and A_2 are the areas of the upper and lower faces.

See also CONE; SOLID OF REVOLUTION; VOLUME.

F-test *See* STATISTICS: INFERENTIAL.

function (**mapping**) Any procedure or a rule that assigns to each member of one set X one, and only one, element of another set Y is called a function. For example, the relation "is the mother of" is a function from the set X of all the people of the world to the set Y of all women: each person is "assigned" one, and only one, biological mother. The rule that squares numbers takes members of the set $X = \{1,2,3,4\}$ to corresponding elements of the set $Y = \{1,4,9,16\}$.

If f is a symbol used to denote a function from a set X to a set Y, then we write $f: X \rightarrow Y$ and call X the domain of f and Y the codomain of f. We also say that f "maps" elements of X to elements of Y. If a specific element x of the set X is mapped to the element y in Y, then we write $y = f(x)$. (Mathematicians sometimes call x an "input" and the value y the corresponding "output.") For instance, if f is the squaring function described

above, then we have $f(2) = 4$ and $f(3) = 9$. If "M" denotes the "is the mother of" function, and Lawrence's mother is Trenyce, then we can write $M(\text{Lawrence}) = \text{Trenyce}$. Sometimes mathematicians will use arrows with tabs to emphasize the notion of a mapping and write, for example, $M: \text{Lawrence} \mapsto \text{Trenyce}$.

Notice that not all elements of Y need be the result of a mapping. (Not all women are mothers, for example.) Also, it is possible for two or more elements of X to be mapped to the same element of Y. (Two people could have the same mother, for instance.) The set of all possible outputs $f(x)$ for a given function f is called the range (or the IMAGE) of the function. For example, the range of the "is the mother of" function M, is the set of all women who have given birth.

In mathematics one typically studies functions between sets of numbers, usually described by formulae. For example, the squaring function is described by the equation $f(x) = x^2$, or just $y = x^2$. The function $y = 3x + 2$, for example, describes the rule "assign to any number x the number y two more than three times x." In this context, x is called the independent variable and y the dependent variable. (Its value is dependent on the value of x.) A GRAPH OF A FUNCTION is a pictorial representation of a function that operates on numbers.

Equations, such as $y = x^2$ and $y = 3x + 2$, in which the dependent variable is expressed in terms of the independent variable, are said to be in "explicit form." In contrast, expressions for which both the independent and dependent variables appear on one side of the equality sign are said to be in "implicit form." For instance, the equations $2x - y = 5$ and $x^2 + y^2 = 9$ are in implicit form. Implicit equations might, or might not, define y as a function of x. For example, the first equation presented above can be rewritten $y = 2x - 5$, yielding the explicit form of a function. The second relation, however, does not represent a function: there is no unique value of y associated to each value of x. (For example, with $x = 0$, y could be either 3 or –3.)

Any set of ordered pairs (x,y) satisfying some given constraint is called a relation. For example, the set of all ordered pairs (x,y) satisfying the equation $x^2 + y^2 = 9$ is a relation. Graphically, this relation represents a circle of radius three. A relation is a function if it is never the case that two distinct ordered pairs have the same x-coordinate value (that is, no single x-value has two different y-values associated to it). This condition is sometimes called the "vertical line test." For

History of Functions

Since the time of antiquity, scholars were interested in identifying rules or relationships between quantities. For example, the ancient Egyptians were aware that the circumference of a circle is related to its diameter via a fixed ratio that we now call PI, and Chinese scholars, and later the Pythagoreans, knew that the three sides of a right triangle satisfy the simple relationship given by PYTHAGORAS'S THEOREM. Although these results were not expressed in terms of formulae and symbols (the evolution of algebraic symbolism took many centuries), scholars were aware that the value of one quantity could depend on the value of other quantities under consideration. Although not explicit, the notion of a "function" was in mind.

In the mid-1300s French mathematician NICOLE ORESME discovered that a uniformly varying quantity (such as the position of an object moving with uniform velocity, for instance) could be represented pictorially as a "graph," and that the area under the graph represents the total change of the quantity. Oresme was the first to describe a way of graphing the relationship between an independent variable and a dependent one and, moreover, demonstrate the usefulness of the task.

In 1694 German mathematician GOTTFRIED WILHELM LEIBNIZ, codiscoverer of CALCULUS, coined the term *function* (Latin: *functio*) to mean the SLOPE of the curve, a definition that has very little in common with our current use of the word. The great Swiss mathematician LEONHARD EULER (1707–83) recognized the need to make the notion of a relationship between quantities explicit, and he defined the term *function* to mean a variable quantity that is dependent upon another quantity. Euler introduced the notation $f(x)$ for "a function of x," and promoted the idea of a function as a formula. He based all his work in calculus and ANALYSIS on this idea, which paved the way for mathematicians to view trigonometric quantities and logarithms as functions. This notion of function subsequently unified many branches of mathematics and physics.

In 1822 French physicist and mathematician JEAN BAPTISTE JOSEPH FOURIER presented work on heat flow. He represented functions as sums of sine and cosine functions, but commented that such representations may be valid only for a certain range of values. This later led German mathematician PETER GUSTAV LEJEUNE DIRICHLET (1805–1859) to propose a more precise definition:

A function is a correspondence that assigns a unique value of a dependent variable to every *permitted* value of an independent variable.

This, on an elementary level, is the definition generally accepted today.

In the late 19th century, German mathematician GEORG CANTOR (1845–1918) attempted to base all of mathematics on the fundamental concept of a SET. Because the terms *variable* and *relationship* are difficult to specify, Cantor proposed an alternative definition of a function:

A function is a set of ordered pairs in which every first element is different.

This idea is based on the fact that the graph of a function is nothing more than a collection of points (x,y) with no two y-values assigned to the same x-value. Cantor's definition is very general and can be applied not only to numbers but to sets of other things as well. Mathematicians consequently came to think of functions as "mappings" that assign to elements of one set X, called the domain of the function, elements of another set Y, called the codomain. (Each element x of X is assigned just *one* element of Y.) One can thus depict a function as a diagram of arrows in which an arrow is drawn from each member of the domain to its corresponding member of the codomain. The function is then the complete collection of all these correspondences.

Advanced texts in mathematics today typically present all three definitions of a function—as a formula, as a set of ordered pairs, and as a mapping—and mathematicians will typically work with all three approaches.

See also GRAPH OF A FUNCTION.

example, the points $(0,-3)$ and $(0,3)$ on the circle of radius three have the same first-coordinate value, and the vertical line at $x = 0$ intercepts the graph of the circle at two locations. The circle is a relation, but it is not a function.

A function $f:X \to Y$ is said to be "onto" (or "surjective") if every element of the codomain Y is the image of some element of X. (That is, every element y in Y is of the form $y = f(x)$ for some x in X.) For example, the squaring function thought of as a map from the set $\{1,2,3,4\}$ to the set $\{1,4,9,16\}$ is onto. It is not onto, however, when thought of as a map from the set of all real numbers to the set of all real numbers: no negative number results from the squaring operation.

A function $f:X \rightarrow Y$ is said to be "one-to-one" (or "injective") if no two different elements of X yield the same output. For example, the "is the mother of" function is not one-to-one: two different people could have the same mother. The squaring function, thought of as a map from the set $\{1,2,3,4\}$ to the set $\{1,4,9,16\}$, is one-to-one. It is not one-to-one, however, when thought of as a number from the set of all real numbers to the set of all real numbers: the numbers 2 and –2, for example, yield the same output.

A function that is both one-to-one and onto is called a "bijection" (or sometimes a PERMUTATION). A bijection $f:X \rightarrow Y$ has the property that each element of Y "comes from" one, and only one, element of X. Thus it is possible to define the inverse function, denoted f^{-1}: $Y \rightarrow X$, which associates to each element y of Y the element of X from whence it came. Algebraically, if the dependent variable y is given as an explicit formula in terms of x, then the inverse function determines the independent variable x as a formula in terms of y. For example, the inverse of the function $y = 3x + 2$ is given by $x = (y - 2)/3$. (That is, the inverse operation of "tripling a number and adding two" is to "subtract two and then divide by three.") The roles of the variables x and y have switched, and thus the graph of the inverse of a function can be found by switching the x- and y-axes on the graph of the original function.

In the theory of CARDINALITY, bijections play a key role in determining whether or not two sets X and Y have the same "size."

See also ALGEBRAIC NUMBER; HISTORY OF FUNCTIONS (essay).

fundamental theorem of algebra (d'Alembert's theorem)

The following important theorem in mathematics is deemed fundamental to the theory of algebra:

Every polynomial $p(z) = a_n z^n + a_{n-1} z^{n-1} + \ldots + a_1 z + a_0$ with coefficients a_i either real or complex numbers, $a_n \neq 0$, has at least one root. That is, there is at least one complex number α such that $p(\alpha) = 0$.

By the FACTOR THEOREM we must then have $p(z) = (z - \alpha)q(z)$ for some polynomial $q(z)$ of degree $n - 1$. Applying the fundamental theorem of algebra to the polynomial $q(z)$, and again to each polynomial of degree greater than one that appears, shows that the polynomial $p(z)$ factors completely into n (not necessarily distinct) linear factors. We have as a consequence:

Every polynomial $p(z) = a_n z^n + a_{n-1} z^{n-1} + \ldots + a_1 z + a_0$ with coefficients a_i either real or complex numbers, $a_n \neq 0$, factors completely as $p(z) = a_n(z - \alpha_1)(z - \alpha_2)\ldots(z - \alpha_n)$ for some complex numbers $\alpha_1, \alpha_2, \ldots, \alpha_n$.

Consequently, in the FIELD of complex numbers, every polynomial of degree n has precisely n roots (when counted with multiplicity). For instance, the polynomial $z^4 - 2z^3 + 2z^2 - 2z + 1$ factors as $(z - 1)(z - 1)(z - i)(z + i)$ with the root 1 appearing twice. Mathematicians call a field algebraically closed if every degree-n polynomial with coefficients from that field has precisely n roots in that field. The set of complex numbers is thus algebraically closed. (The field of real numbers, however, is not. The polynomial $p(x) = x^2 + 1$, for instance, does not factor within the reals.)

The fundamental theorem of algebra was first conjectured by Dutch mathematician Albert Girard in 1629 in his investigation of imaginary roots. CARL FRIEDRICH GAUSS (1777–1855) was the first to prove the result in his 1799 doctoral thesis. He later re-proved the result several times throughout his life using a variety of different mathematical approaches, and he gave it the name the "fundamental theorem of algebra." In France, the result is known as d'Alembert's theorem to honor the work of JEAN LE ROND D'ALEMBERT (1717–83) and his many (unsuccessful) attempts to prove it.

To prove the theorem, it suffices to consider a complex polynomial with leading coefficient equal to one: $p(z) = z^n + a_{n-1} z^{n-1} + \ldots + a_1 z + a_0$. (Divide through by a_n if necessary.) Notice that if $a_0 = 0$, then the polynomial has one root, namely $z = 0$, and there is nothing more to establish. Suppose then that a_0 is a complex number different from zero.

Using EULER'S FORMULA, regard the variable z as a complex number of the form $z = Re^{i\theta} = R(\cos\theta + i\sin\theta)$, where R is a nonzero real number and θ is an angle. Notice that as θ varies from zero to 360°, $z = R(\cos\theta + i\sin\theta)$ traces a circle of radius R and $z^n = R^n e^{in\theta} = R^n(\cos(n\theta) + i\sin(n\theta))$ wraps around a circle of radius R^n n times. Notice, too, that if R is large, then $p(z) = z^n + a_{n-1} z^{n-1} + \ldots + a_1 z + a_0 = z^n\left(1 + \frac{a_{n-1}}{z} + \cdots + \frac{a_1}{z^{n-1}} + \frac{a_0}{z^n}\right)$

is well approximated as $z^n(1 + 0 + \ldots + 0 + 0) = z^n$. Thus as θ varies from zero to 360°, again for R large, $p(z)$ closely traces a circle of large radius R^n (again winding around n times). If R is sufficiently large, this circle is sure to enclose the point (0,0) in the complex plane. If R shrinks to the value zero, then the trace of $p(z)$ as θ varies is a circle of radius 0 about the point $p(0) = a_0$. That is, the trace of $p(z)$ is the point a_0 in the complex plane. Between these two extremes, there must be some intermediate value of R for which the trace of $p(z)$ passes through the origin (0,0). That is, there is α value on the circle of this radius R for which $p(\alpha) = 0$. This proves the theorem.

fundamental theorem of arithmetic (unique factorization theorem) This fundamental result from arithmetic asserts:

> Every integer greater than one can be expressed as a product of prime factors in one and only one way, up to the order of the factors. (If the number is already prime, then it is a product with one term in it.)

For example, the number 100 can be written as $2 \times 2 \times 5 \times 5$. The fundamental theorem of arithmetic asserts that the number 100 cannot be written as a product of a different set of primes.

Many elementary school children are familiar with the process of factoring with the aid of a factor tree. It is often taken as self-evident that the prime numbers one obtains as factors will always be the same, no matter the choices one makes along the way to construct the tree. However, a proof of this is required.

It is straightforward to see that any number n has, at the very least, some prime factorization: If n is prime, then n is a product of primes with one term in it. If n is not prime, the n can be written as a product of two factors: $n = a \times b$. If both a and b are prime, there is nothing more to do. Otherwise, a and b can themselves be factored. Continue this way. This process stops when all factors considered are prime numbers.

That the prime factorization is unique follows from the following property of prime numbers:

> If a product $a \times b$ equals a multiple of a prime number p, then one of a or b must itself be a multiple of p.

(To see why this is true, suppose that a is not already a multiple of p. Since p is prime, this means that the only factor p and a can have in common is 1. By the EUCLIDEAN ALGORITHM we can thus find numbers x and y so that $1 = px + by$. Multiplying through by b gives: $b = pbx + aby$. The first term in this sum is a multiple of p, and so is the second since ab is. This shows that b must be a multiple of p, if a is not.)

Suppose, for example, we found the following two prime factorizations of the same number:

$$7 \times 13 \times 13 \times 29 \times 29 \times 29 \times 41 = 19 \times 19 \times 23 \times 23 \times 37 \times 61$$

The quantity on the left is certainly a multiple of 7, which means the quantity on the right is too. By the property described above, this means that one of the factors: 19, 23, 37, or 61 is a multiple of seven. Since each of these factors is prime, this is impossible. In general, this line of reasoning shows that the primes appearing in two factorizations of a number must be the same. (It also shows, for example, that no power of 7 could ever equal a power of 13, and that no power of 6 is divisible by 14.)

EUCLID, of around 300 B.C.E., was aware that the prime factorizations of numbers are unique.

Note that, in these considerations, it is vital that 1 not be regarded as prime—otherwise every number would have infinitely many different representations as a product of prime factors. (For example, we could write: $6 = 2 \times 3 = 1 \times 2 \times 3 = 1 \times 1 \times 2 \times 3$, and so on.)

Writing numbers in terms of their prime factorizations helps one quickly identify common factors and common multiples. For example, if $a = p_1^{n_1} p_2^{n_2} \ldots p_k^{n_k}$ and $b = p_1^{m_1} p_2^{m_2} \ldots p_k^{m_k}$, with the numbers n_i and m_i possibly zero (this ensures that each number is expressed via the same list of primes), then the GREATEST COMMON DIVISOR of a and b is the number:

$$\gcd(a,b) = p_1^{\alpha_1} p_2^{\alpha_2} \ldots p_k^{\alpha_k}$$

with each α_i the smaller of n_i and m_i, and the LEAST COMMON MULTIPLE of a and b is:

$$\mathrm{lcm}(a,b) = p_1^{\beta_1} p_2^{\beta_2} \ldots p_k^{\beta_k}$$

with each β_i the larger of n_i and m_i. This proves the relationship:

$$\mathrm{lcm}(a,b) = \frac{a \times b}{\gcd(a,b)}$$

The fundamental theorem of arithmetic is used to prove the irrationality of roots of integers. Precisely, one can show that the square root of a positive integer a is a rational number if, and only if, a is a perfect square (that is, $a = b^2$ for some whole number b). This is established as follows:

Suppose that $\sqrt{a}$ is a fraction. Then $\sqrt{a} = n/m$ for some whole numbers n and m:

Squaring and simplifying gives $am^2 = n^2$. The prime factorization of n gives a prime factorization of n^2 with the property that every prime that appears does so an even number of times. As this is the prime factorization of am^2, and the primes that appear in m^2 already do so an even number of times, it follows that the primes that appear in the factorization of a do so an even number of times as well. Selecting one prime from each pair of primes that appears produces a whole number b with the property $b^2 = a$. Thus the assumption that $\sqrt{a}$ is rational leads us to conclude that a is a perfect square.

Consequently, since the number 2, for instance, is not a perfect square, it cannot be the case that $\sqrt{2}$ is rational. Thus $\sqrt{2}$ must be an irrational number. (A similar argument establishes that $\sqrt[n]{a}$ is rational if, and only if, a is a perfect nth power.)

See also COMMON FACTOR; COMMON MULTIPLE; FACTOR.

fundamental theorem of calculus The theory of CALCULUS develops methods for calculating two important quantities associated with curves, namely, the slopes of tangent lines to curves and the areas of regions bounded by curves. Although isolated problems dealing with these issues have been studied since the time of antiquity (*see* HISTORY OF CALCULUS essay), no unified approach or technique for solving them was developed for a very long time. A breakthrough came in the 1670s when GOTTFRIED WILHELM LEIBNIZ in Germany and SIR ISAAC NEWTON in England independently discovered a fundamental inverse relationship between the tangent problem (differentiation) and the area problem (integration). Their result, today known

as the fundamental theorem of calculus, binds together the two parts of the subject and is no doubt the most important single fact in the whole of mathematics. Leibniz and Newton each recognized the importance of the result, developed the ideas that follow from it, and applied the consequent results to solve problems in science and geometry with what can only be described as spectacular success. Their discovery of the fundamental theorem is, in essence, the discovery of calculus.

The theorem has two parts. Let $y = f(x)$ be a continuous curve. If $A(x)$ represents the area under the curve from position a to position x, then the RATE OF CHANGE of area, $\frac{dA}{dx}$, is given by:

$$\frac{dA}{dx} = \lim_{h \to 0} \frac{A(x+h) - A(x)}{h}$$

Noting that the area under the curve between positions x and $x + h$ can be approximated as a rectangle of width h and height $f(x)$, we have $A(x+h) - A(x) \approx f(x) \cdot h$. Thus $\frac{dA}{dx} \approx \lim_{h \to 0} \frac{f(x) \cdot h}{h} = f(x)$. The details of the argument can be made rigorous to prove:

The rate of change of the area under a curve $y = f(x)$ with respect to x is $f(x)$.

Loosely speaking, "taking the DERIVATIVE of an INTEGRAL returns the original function." In particular, the function $A(x)$ is one of the antiderivatives of $f(x)$. The MEAN-VALUE THEOREM, however, shows that if $F(x)$ is *any* other antiderivative of $f(x)$, then it must differ from $A(x)$ only by a constant: $A(x) = F(x) + c$. Noting that $A(a) = 0$, we obtain: $A(x) = F(x) - F(a)$. In particular, the area under the curve $y = f(x)$ from $x = a$ to $x = b$ is $A(b) = F(b) - F(a)$. As this area is usually denoted $\int_a^b f(x)dx$, this establishes the second part of the fundamental theorem of calculus:

If $f(x)$ is a continuous function, and $F(x)$ is any antiderivative of $f(x)$, then $\int_a^b f(x)dx = F(b) - F(a)$.

This second observation is the key result: it transforms the very difficult problem of evaluating areas (via limits of sums) into the much easier problem of finding antiderivatives. For example, to compute the area under the parabola $y = x^2$ from $x = 3$ to $x = 12$, simply note that $F(x) = \frac{1}{3}x^3$ is an antiderivative of the function in question, $F'(x) = x^2$. The desired area is thus $\frac{1}{3}(12)^3 - \frac{1}{3}(3)^3 = 567$ units squared. The entire thrust of INTEGRAL CALCULUS is consequently transformed to the study of DIFFERENTIAL CALCULUS employed in reverse.

See also ANTIDIFFERENTIATION.

fundamental theorem of isometries   Let *ABC* be a triangle in the plane. Then the location of any point *P* in the plane is completely determined by the three numbers that represent its distances from the vertices of the triangle. Any GEOMETRIC TRANSFORMATION that preserves distances is called an isometry. If an isometry takes points *A, B,* and *C* to locations *A', B',* and *C',* respectively, then it also takes the point P to the unique point P' with matching distances from *A', B',* and *C'.* Thus an isometry is completely determined by its effect on its vertices of any triangle in the plane.

Three REFLECTIONs, at most, are ever needed to map three vertices *A,B,* and *C* of one triangle to three vertices *A', B'* and *C'* of another congruent triangle. (First reflect along the perpendicular bisector of *AA'* to take the point *A* to the location *A'.* Next, reflect along a line through *A'* to take the image of *B* to *B'.* One more reflection may be needed to then send the image of *C* to *C'.*) This observation proves the fundamental theorem of isometries:

> Every isometry of the plane is the composition of at most three reflections.

fuzzy logic   In 1965, Iranian electrical engineer Lofti Zadeh proposed a system of logic in which statements can be assigned *degrees* of truth. For example, whether or not Betty is tall is not simply true or false, but more a matter of degree.

Fuzzy-set theory assigns degrees of membership to elements of fuzzy sets. These degrees range from 1, when the element is in the set, to zero when it is out of the set. Betty's membership in the set of tall people is a matter of degree.

See also LAWS OF THOUGHT.

G

Galilei, Galileo (1564–1642) Italian *Mechanics, Astronomy* Born on February 15, 1564, in Pisa (now in Italy), Galileo Galilei is remembered for his 1638 *Discorsi e dimostrazione matematiche intorno a due nuove scienze* (Dialogues on two new sciences) presenting a new approach to the study of kinematics through a combination of experiment and mathematical theory. He formulated and verified the law of acceleration of falling bodies $\left(s = \frac{1}{2}qt^2 \right)$, established that projectiles follow parabolic paths, and was the first to notice that the period of a pendulum is independent of the weight of the bob. Although not the inventor of the telescope, he was the first to develop a workable design of the device that allowed him to make (outstanding) astronomical observations. In particular, he accrued significant empirical evidence that supported the Copernican theory that the planets travel around the Sun. His belief in this theory brought him in conflict with the Roman Catholic Church and led to his consequent trial and house arrest.

Galileo was sent by his parents to the University of Pisa to study medicine at the age of 17, but this course of study never appealed to him. He left the University without obtaining a degree and began teaching mathematics and mechanics in Sienna. In 1586 he wrote *La balancitta* (The little balance), in which he described the methods of ARCHIMEDES OF SYRACUSE (287–212 B.C.E.) for finding the relative densities of substances with a balance. This piece garnered him significant attention in the scientific community, and in 1589 he was appointed chair of mathematics at the University of Pisa. He remained at this post for 3 years before accepting the prestigious position as professor of mathematics at the University of Padua.

During his 18 years at Padua, Galileo developed his theories of motion. Using inclined planes to slow the rate of descent, Galileo observed that all objects fall at the same rate of acceleration. From this, using mathematics verified by experiment, he deduced his famous law that the distance traveled by a falling object is proportional to the square of the time of its fall. From this it follows that objects tossed in the air move in parabolic arcs.

In 1609 Galileo received word that Dutch scholars had invented a spyglass capable of magnifying images of distant objects. Intrigued, Galileo set to constructing his own version of the spyglass and eventually produced a piece (which he called a "perspicillum") with a magnification power of eight or nine—far superior to the capabilities of the Flemish telescope. Galileo turned his piece to the heavens and commenced a series of remarkable astronomical discoveries.

In his 1610 book *Sidereus nuncius* (Starry messenger), Galileo described the topography of the Moon, established that the Milky Way is composed of stars, and announced his discovery of four moons orbiting Jupiter. For savvy political reasons, he named these moons the "Medicean stars," in honor of Cosimo de Medici, the grand duke of Tuscany. This work garnered Galileo considerable fame and, just one month after the book was released, Galileo was appointed chief mathematician at Pisa and "mathematician and philosopher" to the grand duke of Tuscany.

Later that year Galileo turned his telescope toward Saturn, observed its rings, and then turned toward Venus. He discovered that Venus shows phases like those of the Moon, which could only be possible if this planet were to orbit the Sun, not the Earth. Galileo had gathered convincing evidence to support the Copernican theory of the solar system. Although the religious authority at the time viewed this model as antithetical to the Holy Scripture (in which the Earth is perceived as lying at the center of the universe), church leaders were tolerant of scholars who used alternative theories in the guise of mathematical tools for calculating the orbits of the planets. Galileo, unfortunately, went further and later proclaimed in a private communication that he was convinced that the theory

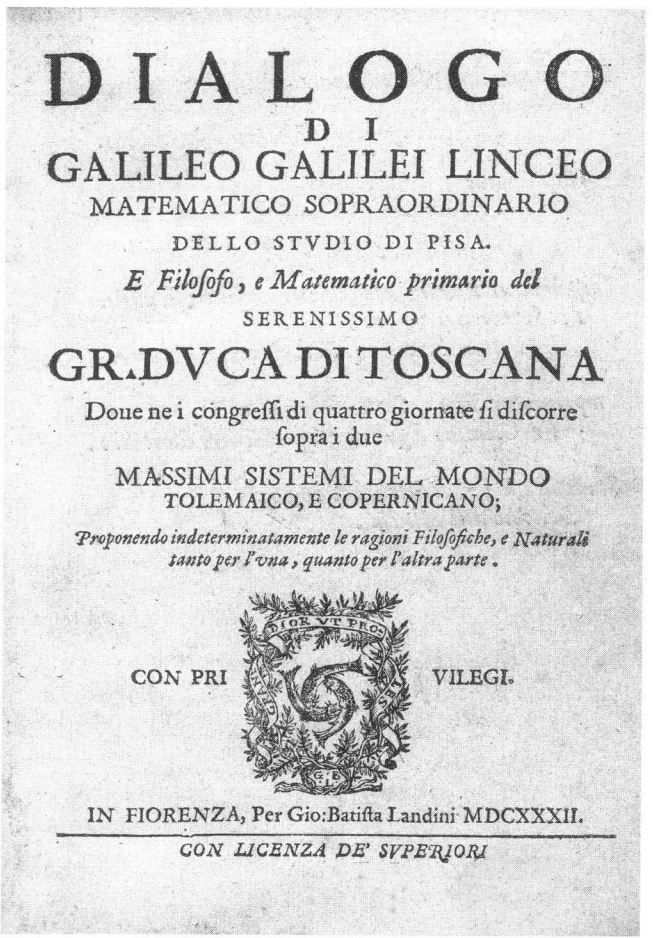

Galileo managed to smuggle his controversial manuscript *Dialogo* out of Italy, where its sale was banned, for publication in Holland. *(Photo courtesy of the Science Museum, London/Topham-HIP/The Image Works)*

Galileo Galilei, an eminent scientist of the 17th century, was the first to combine theory with experiment to understand dynamics. He established that acceleration due to gravity is independent of mass and that projectiles follow parabolic arcs. Galileo also developed the telescope and made significant, and controversial, astronomical observations. *(Photo courtesy of Topham/The Image Works)*

was a statement of physical reality. Unfortunately, the letter was intercepted by the church, and for the 15 years that followed, the Catholic Church reviewed Galileo's work carefully and demanded that he follow church teaching at all times. In 1632, Galileo attempted to publish the text *Dialogo* (Dialogue) on the theory of planetary motion, cleverly written so as to never actually make the claim that the Copernican system is the physical reality), but the church, feeling that the boundary had been crossed, immediately banned its sale, and tried Galileo for heresy. He was found guilty and was placed under house arrest for the remainder of his life. Galileo did, however, manage to publish his famous 1638 piece by having it smuggled out of Italy and printed in Holland.

Galileo died on January 8, 1642. He is remembered for having a profound impact on the very nature of scientific investigation, linking together mathematics, speculative philosophy, and physical experiment in the study of the natural world. On October 31, 1992, on the 350th anniversary of Galileo's death, Pope John Paul II publicly acknowledged that the Catholic Church was in error to convict Galileo of heresy for his theory on the motion of the planets.

Galois, Évariste (1811–1832) French *Abstract algebra* Born on October 25, 1811, in Bourg La Reine, France, scholar Évariste Galois is remembered for his famous results that led not only to the conclusion that there can be no general formula that solves all fifth-degree POLYNOMIAL equations (even though there do exist such formulae for solving QUADRATIC, CUBIC, and QUARTIC EQUATIONs), but more importantly to the classification of which specific equations can be so solved. His seminal 1829 work on the solutions of equations founded the field of GROUP THEORY.

Galois enrolled in his first high-school mathematics course at the age of 17 and just a year later published an original paper, "Démonstration d'un théorème sur les fractions continues périodiques" (Proof of a theorem on periodic continued fractions), on the study of CONTINUED FRACTIONS. His teachers at the time, unfortunately, did not understand his work and did not regard him as a gifted student. (Admittedly, Galois had difficulty articulating his sophisticated ideas). Galois hoped to attend the prestigious École Polytechnique, the leading university of Paris at the time, but failed the entrance exam twice. He enrolled, instead, at the École Normale, but was later expelled for objecting to the university policy prohibiting students from joining the Paris rebellion against King Charles X. He joined the Artillery of the National Guard, a radical wing of the military. In 1831, Galois was imprisoned for wearing its uniform when support for the guard was made illegal.

Previously Galois had read the work of NIELS HENRIK ABEL (1802–29) on the study of algebraic solutions to equations. The SOLUTION BY RADICALS problem was on his mind and, while incarcerated, Galois wrote a manuscript detailing his thoughts on the problem. During this time Galois fell in love with the daughter of the resident physician of the prison, Stephanie-Felice du Motel, but, as letters show, the feelings were not reciprocated. For reasons that are unclear today, Galois fought a duel in her honor soon upon his release. He was wounded in that duel and died the following day, May 31, 1832.

Galois's brother later copied the prison manuscript and sent it to prominent mathematicians of the time. If it were not for his brother's initiative, the bulk of Galois's work would have been lost to us today. French mathematician JOSEPH LIOUVILLE (1809–82) published Galois's seminal piece in 1846.

Galton, Sir Francis (1822–1911) British *Statistics* Born on February 16, 1822, in Sparkbrook, England, Francis Galton is remembered in mathematics for his pioneering work in applying statistical techniques to the analysis of biological problems. Galton's insights and contributions to the nature of statistical analysis paved the way for the development, and consequent widespread use, of statistics throughout the biological and social sciences in the 20th century.

Galton completed a basic bachelor's degree at Cambridge, taking only enough mathematics courses to meet distribution requirements. (He never received serious training in mathematics.) After inheriting a considerable amount of money from his father, Galton journeyed through southwestern Africa and garnered considerable fame as an intrepid explorer. He developed an interest in the study of human hereditary and began considering the issue of selective breeding as a means to improve the human race. (Such pursuits were deemed acceptable at his time.) Galton also developed a simple mathematical model of ancestral hereditary based on the idea that each parent contributes 1/4 of a genetic trait to a child, each grandparent of 1/16 the trait to the child, and so forth, so that the sum of contributions from all ancestors is unity:

$$2 \cdot \frac{1}{4} + 4 \cdot \frac{1}{16} + 8 \cdot \frac{1}{64} + \cdots = \frac{1}{2} + \frac{1}{4} + \frac{1}{8} + \cdots = 1$$

Galton received a knighthood for this work in 1909.

The University of London established the Francis Galton Laboratory for National Eugenics in his honor in the late 1800s. The eminent statistician KARL PEARSON

(1857–1936), much of whose work was greatly influenced by Galton's studies, was once director of the laboratory.

Galton died in Surrey, England, on January 17, 1911.

See also HISTORY OF PROBABILITY AND STATISTICS (essay).

gambler's ruin A class of problems in PROBABILITY theory, all known as problems of gambler's ruin, contains questions of the following ilk:

> A gambler with $5 repeatedly bets $1 on a coin toss. Each time he plays he has a 50 percent chance of winning a dollar and a 50 percent chance of losing a dollar. The gambler will play until he either has $10 or has gone broke. What is the probability that he will succeed in doubling his money?

Problems like these are typically solved as follows: let $P(k)$ be the probability of reaching the $10 goal when the gambler currently has k dollars in hand. We wish to compute $P(5)$. Note three things:

1. $P(0) = 0$. (With zero dollars, the gambler has lost and will not win.)
2. $P(10) = 1$. (With $10 in hand, the gambler has won.)
3. $P(k) = \frac{1}{2}P(k-1) + \frac{1}{2}P(k+1) = \frac{P(k-1) + P(k+1)}{2}$.

   (With k dollars in hand, there is a 50 percent chance the gambler will lose a dollar and will have to try to win with k–1 dollars in hand, and a 50 percent chance that he will win a dollar and will continue play with $k + 1$ dollars in hand.)

Thus we have eleven values, $P(0), P(1),\ldots,P(10)$, with the first value equal to zero, the last equal to 1, and all intermediate values equal to the average of the two values around it. One can check that there is only one sequence of numbers satisfying these conditions, namely: $0, \frac{1}{10}, \frac{2}{10},\ldots,\frac{10}{10} = 1$. This shows that $P(1) = \frac{1}{10}$, $P(2) = \frac{2}{5}$ and, in particular, that $P(5) = \frac{5}{10} = \frac{1}{2}$.

Notice that with one dollar in hand there is only a 1/10 chance that the gambler will win $10 before going broke. One can similarly show that there is only a one in 100 chance that a gambler will win $100, and a one in 1,000 chance that he will win $1000. Casinos are well aware that gamblers are reluctant to stop playing with small profits—especially if a player has had a string of losses and is down to his or her last dollar. The gambler's ruin shows that in all likelihood, gamblers will end up broke before receiving profits they are content with.

See also HARMONIC FUNCTION; RANDOM WALK.

game theory Game theory is the branch of mathematics that attempts to analyze situations involving parties with conflicting interests for which the outcome of the situation depends on the choices made by those parties. Situations of conflict arise in nearly every real-world problem that involves decision making, and game theory has consequently found profound applications to the study of business competition, economics, politics, military operations, property division, and even the study of personal relationships. Although it is difficult to provide a complete analysis of all the types of games that arise, complete solutions exist for simple "matrix games" (which we describe below) involving a small number of players. These simple games can be used to model more-complicated multiplayer situations.

Game theory was first studied in 1921 by French mathematician Félix Edouard Émile Borel (1871–1956), but the importance of the topic was not properly acknowledged until the 1944 release of the monumental work *Theory of Games and Economic Behavior* by JOHN VON NEUMANN (1903–57) and Oskar Morgenstern (1902–77). These two scholars completely analyzed situations of conflict satisfying the following basic conditions:

i. There are a finite number of players.
ii. Each player can select one of a finite number of possible actions. The choice of actions available to each player need not be the same.
iii. All players are aware of the actions, and the consequences, others may choose to take.
iv. Each player is a rational thinker and will select the action that best suits his or her interests.
v. At the play of the game, no participant knows what actions will be taken by the other players.
vi. The outcome of the game can be modeled as a set of payments (positive, zero, or negative) to each of the players.

If the sum of payments to all players is zero, then the game is called a zero-sum game. This means that any player's win must be balanced by a loss for some other players.

The simplest type of game is a two-player zero-sum game. The choice of actions and the possible results of the game can be summarized in a "payoff matrix" A, in which the rows represent the possible actions one player, player R, can take; the columns the possible actions of the second player, player C; and the (i,j)-th entry of the matrix A_{ij} represents the amount player C must pay to player R if R selects row i and C selects column j in the play of the game. (If A_{ij} is negative, then player R pays C a certain amount.) For example, the following table represents a two-person zero-sum game for which each player can take one of three possible actions:

	1	2	3
1	5	−1	1
2	1	0	4
3	−3	−2	2

As a first attempt to analyze this game, consider the actions that player R can take. She hopes that the game will result in the largest positive number possible (in this case 5), for this is the amount player C must pay R if this is the result of the game. Thus player R is tempted to select action 1 (especially since R is aware that player C's optimal outcome of −3 lies in column 1). Of course, player C is aware that player R is likely to think this way, and would be loath to select column 1 as her action for fear of having 5 be the outcome of the game. Player R is aware that player C will think this way, and will suspect then that player C will choose column 2, knowing the she will likely select row 1, and so R, to foil this plan, is tempted to choose an action different from 1 after all. And so on. As one sees, one can quickly enter a never-ending cycle of second guessing.

Instead of aiming to maximize her profit from the game, another approach player R could adopt is to minimize her losses. For example, if player R takes action 1, she could potentially lose 1 point (if C happens to choose column 2). If R chooses row 2, the worst outcome would be no gain or loss, and if she chooses row 3, she could potentially be down three

points. Thus the maximal minimum outcome for player R occurs with the choice of row 2. This line of reasoning is called the "maximin strategy." Similarly, the choice of column 2 results in the minimal maximum outcome in player C's favor. Thus following a "minimax strategy," player C would choose column 2, and the outcome of the game is consequently 0.

The entry 0 in the above payoff matrix is called a saddle point. It is a minimum in its row and a maximum in its column. If a payoff matrix contains a saddle point, then the optimal strategy for each player is to take actions corresponding to the saddle point. The value of the saddle point, for a game that possesses such a point, is called the value of the game.

Not every payoff matrix for a two-person game possesses a saddle point. For instance, the following game corresponding to two possible actions for each player has no saddle point, and following the minimax or the maximin strategies is not optimal for either player.

	1	2
1	1	−2
2	−3	4

(The maximin strategy for player R suggests that she select row 1, and the minimax strategy for player C that she take column 1 yielding the result 1 for the game. Player C, however, can anticipate this and is tempted then to change choice to column 2 to obtain the preferable result of −2. Player R, of course, is aware that player C will likely do this, and so will change her choice to row 2 to obtain the outcome 4, and so forth. The two players again are trapped in an endless cycle of second guessing.)

The appropriate strategy in such a game lacking a saddle point is a mixed strategy, in which each player decides to select an action by random choice, appropriately weighted so as to maximize her expected profit from the game. For instance, suppose player R decides to select row 1 with PROBABILITY p and row 2 with probability $1 − p$. Under this strategy, if player C chooses column 1, then the expected value of player R's profit is $E_1 = 1 \times p + (−3) \times (1 − p) = 4p − 3$. If, on the other hand, player C were to select column 2, then the expected profit for player R is $E_2 = (−1) \times p + 4 \times (1 − p) = 4p − 5p$. These expected outcomes are equal if we choose $p = 7/9$. Thus, by selecting row 1 with this probability, player R can be guaranteed an expected profit of $4\left(\dfrac{7}{9}\right)-3=4-5\left(\dfrac{7}{9}\right)=\dfrac{1}{9}$, regardless of what player C does. Similarly, following a mixed strategy, player C should select column 1 with probability 5/9 so as to minimize her expected loss.

In 1928 von Neumann proved the famous "minimax theorem":

> If each player adopts her best mixed strategy in a zero-sum two-person game, then one player's expected gain will equal the other's expected loss.

The shared expected outcome of such a game is called the "value" of the game.

Games that are not zero-sum games are called variable-sum games. The PRISONER'S DILEMMA and CHICKEN are examples of such partial-conflict games. In them, one searches for dominant strategies, hoping to encounter an equilibrium for the game.

In recent decades the theory of games has successfully been extended to n-person variable-sum games and to games with a continuous range of possible actions and strategies. Game theory is now a standard course offered in university economics departments.

See also FAIR DIVISION.

Gardner, Martin (1914–) American *Recreational mathematics* Born on October 21, 1914, in Tulsa, Oklahoma, freelance writer Martin Gardner is regarded today as solely responsible for cultivating and nurturing interest in 20th-century recreational mathematics. With numerous articles and more than 65 books to his credit, Gardner has achieved worldwide fame as a writer who can make the complex issues of science and mathematics accessible and meaningful to the general audience. Although not a mathematician, Gardner is often credited with having done more to promote and prompt the general pursuit of mathematics than any scholar in the field.

Gardner graduated from the University of Chicago with a bachelor's degree in philosophy in 1936. After serving in the U.S. Navy for four years, Gardner returned to Chicago and began a career as a freelance writer, at first editing and writing short works of fiction. In 1958 Gardner accepted the position of mathematical-games columnist for *Scientific American,* despite never having taken a mathematics course in college. His first piece explaining the mathematics of a Soma cube, a cube dissection puzzle named after a fictitious addictive drug, illustrated his natural ability to discuss and clarify complex issues with ease. This initial piece garnered him national attention, and Gardner remained a columnist for the publication for 25 years. His articles were extremely influential and have since been collected and republished (multiple times) as books.

Well-versed in the practices of illusion and magic, Gardner has also published works on magic, the mathematics of magic, as well as texts that discuss, and debunk, the claims and practices of pseudoscience. He has written two novels, *The Flight of Peter Fromm* (1994) and *Visitors from Oz* (1998), and some works on the topics of philosophy and literature. His best-selling work is *The Annotated Alice—Alice's Adventures in Wonderland and through the Looking Glass* (1965).

Gardner currently lives in Hendersonville, North Carolina, and continues his work as a freelance writer. Despite a lack of formal training in mathematics, Gardner has made some original mathematical discoveries, of a recreational flavor, that have been published in scholarly periodicals.

Gauss, Carl Friedrich (1777–1855) German *Number theory, Geometry, Algebra, Analysis, Statistics, Physics, Astronomy* Born on April 30, 1777, in Brunswick, Duchy of Brunswick (now Germany), Carl Friedrich Gauss is recognized today as the greatest pure mathematician and physicist of his time. His contributions to both fields were enormous. At the age of 18, he invented the LEAST SQUARES METHOD and made the new discovery that a 17-sided regular polygon can be constructed with straightedge and compass alone, signaling that he had accomplished great advances in the theory of CONSTRUCTIBLE numbers. In 1801 he published his masterpiece *Disquisitiones arithmeticae* (Arithmetical investigations) in which he proved the FUNDAMENTAL THEOREM OF ARITHMETIC, the FUNDAMENTAL THEOREM OF ALGEBRA, and introduced the theory of MODULAR ARITHMETIC. In later work, he proposed general solutions to the problem of determining planetary motion, developed theories of statistics that led to the discovery of the Gaussian DISTRIBUTION, and studied potential theory and electricity and magnetism in physics. He also worked out, but did not publish, the principles of HYPERBOLIC GEOMETRY independently of JÁNOS BOLYAI (1802–60) and NIKOLAI IVANOVICH LOBACHEVSKY (1792–1856), and developed the theory of COMPLEX NUMBERS and complex functions.

Gauss demonstrated a talent for mathematics at an early age. (It is said that he astonished his elementary school teachers by summing the integers from 1 to 100 in an instant by observing that the sum amounts to 50 pairs of numbers each summing to 101.) Before leaving high school, Gauss had independently discovered the PRIME NUMBER THEOREM, the BINOMIAL THEOREM, important results in modular arithmetic, and the general arithmetic–geometric-mean inequality. He entered the University of Göttingen in 1795 but, for reasons that are not clear to historians, left before completing a degree to return to his hometown of Brunswick. He submitted a doctoral dissertation to the University of Helmstedt on the topic of the fundamental theorem of algebra, which was accepted *in absentia*. Gauss published his masterpiece *Disquisitiones arithmeticae* soon afterward.

In 1801 Gauss created a sensation in the community of astronomers when he correctly predicted the orbital positions of the asteroid Ceres, discovered January 1, 1801, by Italian astronomer Giuseppe Piazzi. Piazzi had the opportunity to observe an extremely small portion of its orbit before it disappeared behind the sun. Other astronomers published predictions about where it would reappear several months later, as did Gauss, offering a prediction that differed greatly from common opinion. When the comet was observed again December 7, 1801, it was almost exactly where Gauss had predicted. Although Gauss did not reveal his method of prediction at the time, it is known that he used his method of least squares to make the computation. In 1807 Gauss left Brunswick to head the Göttingen observatory, and in 1809 he published his general text on the mathematics of astronomy, *Theoria motus corporum coelestium* (Theory of the motion of heavenly bodies).

While at the observatory, Gauss continued work on mathematics. He began developing a theory of surface curvature and GEODESICs, and studied the convergence of SERIES, integration techniques, STATISTICS, potential theory, and more. In 1818 he was asked to conduct a geodesic survey of the state of Hanover, and 14 years later, he assisted in a project to map the magnetic field of the Earth. His mathematical work on the theory of differential geometry allowed Gauss to prove a number of properties that the Earth's field must possess, which allowed him to correctly predict the value of the field at different locations, as well as the location of the magnetic South Pole. (Moreover, he proved mathematically that there can only be two magnetic poles.)

Gauss remained at Göttingen for the latter part of his career. He was awarded many honors throughout his life, including election as a foreign member of the prestigious ROYAL SOCIETY of London in 1804 and receiving the Copley Medal from the society in 1838. He also won the Copenhagen University Prize in 1822. Gauss published over 300 significant pieces of work, mostly written in Latin, and kept a large number of unpublished notebooks and correspondences that proved to be as mathematically rich as much of his published work.

Gauss died in Göttingen on February 23, 1855. It is impossible to exaggerate the influence Gauss has had on almost every branch of mathematics and mathematical physics. He was a master at solving difficult problems that lay at the heart of complex mathematical ideas, and the very completeness and thoroughness of his work paved the way for significant advances in mathematics. A number of fundamental concepts in number theory, differential geometry, and statistics (such as Gaussian reciprocity, Gaussian curvature, and the Gaussian distribution) are today named in his honor.

Gaussian elimination (pivoting)   Named in honor of CARL FRIEDRICH GAUSS (1777–1855), the process of Gaussian elimination provides the means to find the solution (if one exists) for a system of n SIMULTANEOUS LINEAR EQUATIONS in n unknowns by multiplying selected equations with carefully chosen constants and subtracting equations to eliminate variables. The method is best explained with an example. Consider the following system of three linear equations in three unknowns:

$$y + 3z = 0$$
$$2x + 4y - 2z = 18$$
$$x + 5y + 3z = 14$$

Interchange the first two equations so that the variable x appears in the first row:

$$2x + 4y - 2z = 18$$
$$y + 3z = 0$$
$$x + 5y + 3z = 14$$

Divide the first equation through by 2 so that the coefficient of the leading variable x is 1.

$$x + 2y - z = 9$$
$$y + 3z = 0$$
$$x + 5y + 3z = 14$$

Eliminate the appearance of the variable x from the third equation by subtracting the first equation from it. That is, replace the third equation with the equivalent statement: $(x + 5y + 3z) - (x + 2y - z) = 14 - 9$, that is, $3y + 4z = 5$. We have:

$$x + 2y - z = 9$$
$$y + 3z = 0$$
$$3y + 4z = 5$$

The second equation contains y as the leading variable with a coefficient of 1. Eliminate y from the third equation by subtracting from it three copies of the second equation, that is, replace the third equation with the equivalent statement, $(3y + 4z) - 3(y + 3z) = 5 - 3 \cdot 0$, that is, $-5z = 5$. We now have:

$$x + 2y - z = 9$$
$$y + 3z = 0$$
$$-5z = 5$$

Divide the third equation through by -5 so that the leading variable in it is z with a coefficient of 1:

$$x + 2y - z = 9$$
$$y + 3z = 0$$
$$z = -1$$

The solution to the system of equations is now easy to compute. By a process of back substitution, we see that $z = -1$, from which it follows from the second equation that $y = -3z = 3$, and from the first equation that $x = 9 - 2y + z = 9 - 6 - 1 = 2$. One checks that this is indeed the solution to the original set of equations.

Some Terminology
An elementary row operation is any maneuver on a set of simultaneous linear equations that:

1. Interchanges two equations
2. Multiplies (or divides) an equation through by a nonzero quantity
3. Adds or subtracts a multiple of one equation from another

The process of Gaussian elimination uses elementary row operations to transform a system of linear equations into an equivalent system in "echelon form," that is, one in which each equation leads, in turn, with one of the variables with coefficient 1. Via the process of back substitution, it is then straightforward to determine the solution to the system of equations. The process illustrated above works for any number of equations with the same number of unknowns. It is possible that during the process of Gaussian elimination, a system of equations might yield an absurd statement (such as $0 = 9$, for instance), in which case one would conclude that the system has no solutions, or possibly a vacuous statement (such as, $0 = 0$), in which case one would conclude that the system of equations has infinitely many solutions.

Note that it is possible to take the process of Gaussian elimination further and reduce a system of equations to a system in which each variable appears just once on each line. For instance, in our example, we obtained:

$$x + 2y - z = 9$$
$$y + 3z = 0$$
$$z = -1$$

Subtracting two copies of the second equation from the first yields:

$$x + \quad - 7z = 9$$
$$y + 3z = 0$$
$$z = -1$$

and now adding seven copies of the third equation to the first, and subtracting three copies of the third equation from the second, yields:

$$x \qquad = 2$$
$$y \quad = 3$$
$$z = -1$$

The solution to the system is now apparent.

Consequences for Matrix Theory

Any system of linear equations can be represented via a coefficient MATRIX and a column of constant values. For instance, for our example:

$$y + 3z = 0$$
$$2x + 4y - 2z = 18$$
$$x + 5y + 3z = 14$$

the coefficient matrix, call it A, is given by:

$$A = \begin{pmatrix} 0 & 1 & 3 \\ 2 & 4 & -2 \\ 1 & 5 & 3 \end{pmatrix}$$

the column of constant values, call it **b,** is:

$$\mathbf{b} = \begin{pmatrix} 0 \\ 18 \\ 14 \end{pmatrix}$$

Any elementary row operation performed on the original set of equations corresponds to an operation on the rows of the coefficient matrix A and the column matrix **b.** For instance, in the example above, our first operation was to interchange the first and second rows. This can be accomplished by multiplying A and **b** each by the PERMUTATION matrix

$$\begin{pmatrix} 0 & 1 & 0 \\ 1 & 0 & 0 \\ 0 & 0 & 1 \end{pmatrix}$$

We have:

$$\begin{pmatrix} 0 & 1 & 0 \\ 1 & 0 & 0 \\ 0 & 0 & 1 \end{pmatrix} \begin{pmatrix} 0 & 1 & 3 \\ 2 & 4 & -2 \\ 1 & 5 & 3 \end{pmatrix} = \begin{pmatrix} 2 & 4 & -2 \\ 0 & 1 & 3 \\ 1 & 5 & 3 \end{pmatrix}$$

$$\begin{pmatrix} 0 & 1 & 0 \\ 1 & 0 & 0 \\ 0 & 0 & 1 \end{pmatrix} \begin{pmatrix} 0 \\ 18 \\ 14 \end{pmatrix} = \begin{pmatrix} 18 \\ 0 \\ 14 \end{pmatrix}$$

Similarly, the elementary row operation of dividing the first row through by 2 is accomplished by multiplication with the matrix

$$\begin{pmatrix} 1/2 & 0 & 0 \\ 0 & 1 & 0 \\ 0 & 0 & 1 \end{pmatrix}$$

and the act of subtracting the first equation from the first is accomplished by multiplication with the matrix:

$$\begin{pmatrix} 1 & 0 & 0 \\ 0 & 1 & 0 \\ -1 & 0 & 1 \end{pmatrix}$$

In this way one can see that every elementary row operation corresponds to multiplication by an elementary matrix. This observation has an important consequence.

> The inverse of a square matrix A is simply the product of the elementary matrices that reduce A to the identity matrix.

Our example explains this. We used elementary row operations to reduce the system of equations to the equivalent system:

$$x \qquad\quad = 2$$
$$\quad y \qquad = 3$$
$$\qquad\quad z = -1$$

That is, we found a collection of eight elementary matrices E_1, E_2, ..., E_8 such that application of these eight matrices reduced the matrix of coefficients A to the IDENTITY MATRIX I.

$$E_8 E_7 E_6 E_5 E_4 E_3 E_2 E_1 A = I$$

If we let B be the matrix $E_8 E_7 E_6 E_5 E_4 E_3 E_2 E_1$, then we have $BA = I$, which means that $B = A^{-1}$, the INVERSE MATRIX to A. (As the DETERMINANT of the identity matrix I is 1, the equation $E_8 E_7 E_6 E_5 E_4 E_3 E_2 E_1 A = I$ shows that the determinant of A cannot be zero. Thus, as the study of determinants shows, the matrix A does indeed have an inverse.) Notice that the matrix B is the same elementary row operations applied to the matrix I. This result provides a constructive method for computing the inverse to a matrix A.

> To compute the inverse of a matrix A, write the matrix A and the matrix I side by side. Perform

elementary row operations on the two matrices simultaneously. Reduce A to an identity matrix. The second matrix produced is the inverse matrix to A.

As an example, we compute the inverse matrix to the matrix of coefficients above:

$$
\begin{array}{ccc|ccc}
0 & 1 & 3 & 1 & 0 & 0 \\
2 & 4 & -2 & 0 & 1 & 0 \\
1 & 5 & 3 & 0 & 0 & 1
\end{array}
$$

$$
\begin{array}{ccc|ccc}
2 & 4 & -2 & 0 & 1 & 0 \\
0 & 1 & 3 & 1 & 0 & 0 \\
1 & 5 & 3 & 0 & 0 & 1
\end{array}
$$

$$
\begin{array}{ccc|ccc}
1 & 2 & -1 & 0 & 0.5 & 0 \\
0 & 1 & 3 & 1 & 0 & 0 \\
1 & 5 & 3 & 0 & 0 & 1
\end{array}
$$

$$\vdots$$

$$
\begin{array}{ccc|ccc}
1 & 0 & 0 & 2.2 & 1.2 & -1.4 \\
0 & 1 & 0 & -0.8 & -0.3 & 0.6 \\
0 & 0 & 1 & 0.6 & 0.1 & -0.2
\end{array}
$$

Thus

$$
A^{-1} = \begin{pmatrix}
2.2 & 1.2 & -1.4 \\
-0.8 & -0.3 & 0.6 \\
0.6 & 0.1 & -0.2
\end{pmatrix}
$$

Solving Systems of Linear Equations via Inverse Matrices
In our example:

$$
\begin{aligned}
y + 3z &= 0 \\
2x + 4y - 2z &= 18 \\
x + 5y + 3z &= 14
\end{aligned}
$$

if $\mathbf{x}$ denotes the column vector of the variables:

$$
\mathbf{x} = \begin{pmatrix} x \\ y \\ z \end{pmatrix}
$$

then the system can be compactly written:

$$A\mathbf{x} = \mathbf{b}$$

Multiplying through by the inverse matrix A^{-1} yields:

$$
\frac{1}{N} + \frac{1}{N^2} + \frac{1}{N^3} + \cdots = \frac{1}{N}\left(\frac{1}{1 - \dfrac{1}{N}}\right) = \frac{1}{N-1}
$$

again making the solution to the system apparent. Of course, the work of computing the inverse matrix is equivalent to the original process of Gaussian elimination. This approach, however, has the advantage that it can be readily applied to a different set of constant values $\mathbf{b}$ without repeating the elimination process.

general form of an equation A formula that describes the general relationship between variables without specifying the constants involved is called the general form of the equation. For example, the general form of a QUADRATIC equation in variable x is $ax^2 + bx + c = 0$. (The equation $2x^2 - 3x + 4 = 0$, for instance, is a specific quadratic equation.) The formula for the AREA A of a CIRCLE, $A = \pi r^2$, where r is the radius of the circle, is also an equation in general form.

general linear group (full linear group) The set of all invertible $n \times n$ square matrices with real or complex entries forms a GROUP under the operation of MATRIX multiplication. This group is called the nth-order general linear group and is denoted GL_n. It is straightforward to see that the four group axioms do indeed hold:

Closure. If A and B have inverses, then so does their product AB: $(AB)^{-1} = B^{-1}A^{-1}$.
Identity. The identity matrix I is invertible and so belongs to this set.

Associativity. This rule holds for all matrices, and hence for invertible matrices in particular.

Inverses. The inverse of a matrix A is itself invertible, and so belongs to the set: $(A^{-1})^{-1} = A$.

A study of DETERMINANTs and CRAMER'S RULE shows that a matrix is invertible if, and only if, its determinant is not zero. Thus the nth-order general linear group may be defined as the set of all $n \times n$ square matrices with nonzero determinants. The subset of all square matrices with determinant equal to 1 forms a subgroup of GL_n called the special linear group, denoted SL_n.

geodesic The shortest curve connecting two points on a surface, and lying wholly on that surface, is called a geodesic. For example, PYTHAGORAS'S THEOREM shows that a straight line gives the shortest path between two points on a plane, and thus straight lines are the geodesics of a plane. On a SPHERE, geodesics are sections of great circles. (Planes fly along geodesic arcs across the Earth's surface.) One envisions a geodesic as the path a band of stretched elastic would adopt if held at the two points in question and forced to remain on the surface under study.

On a plane there is always only one geodesic between any two given points. On a sphere there are infinitely many geodesics that connect the two poles of the sphere, for example. (If the two points chosen are not antipodal, however, then the geodesic connecting them is unique.)

A geodesic dome is a domelike structure made of straight-line structural elements held in tension. The straight line segments approximate geodesics of the dome.

geometric distribution *See* BINOMIAL DISTRIBUTION.

geometric mean *See* MEAN.

geometric sequence (geometric progression) A SEQUENCE of numbers in which each term, except the first, is a fixed multiple of the previous one is called a geometric sequence. The constant ratio of terms is

called the common ratio. For example, the sequence $1, 3, 9, 27, \ldots$ is geometric, with common ratio 3. The sequence $1, \frac{1}{2}, \frac{1}{4}, \frac{1}{8}, \ldots$ is also geometric, with common ratio $\frac{1}{2}$.

A geometric sequence with first term a and common ratio r has the form: $a, ar, ar^2, ar^3, \ldots$ The nth term a_n of the sequence is given by $a_n = ar^{n-1}$. (It is common, however, to start the count by calling the first term of the sequence the "zero-th term," so that $a_0 = ar^0 = a$ and the nth term of the sequence is given by the formula: $a_n = ar^n$.) A geometric sequence can also be described as exponential. The common ratio r is the base of the EXPONENTIAL FUNCTION $f(n) = ar^n$.

If the value r lies between -1 and 1, then the terms ar^n of geometric sequence approach the value 0 as n becomes large. If $r = 1$, then the geometric sequence is the constant sequence $a, a, a, \ldots$ For all other values of r, the sequence diverges.

The sum of the terms of a geometric sequence is called a geometric series:

$$a + ar + ar^2 + ar^3 + \ldots$$

In the study of CONVERGENT SERIES, the ratio test shows that this series sums to a finite value if $-1 < r < 1$. The value S of the sum can be computed as follows:

If $S = a + ar + ar^2 + ar^3 + \ldots$, then $rS = ar + ar^2 + ar^3 + ar^4 + \ldots$ Subtracting yields: $S - rS = a$, and so $S = \dfrac{1}{1-r}$

In particular we have $1 + r + r^2 + r^3 + \ldots = \dfrac{1}{1-r}$.

If r is a fraction of the form $\dfrac{1}{N}$ (with $N \geq 2$), then this formula can be rewritten:

$$\frac{1}{N} + \frac{1}{N^2} + \frac{1}{N^3} + \cdots = \frac{1}{N}\left(\frac{1}{1 - \frac{1}{N}}\right) = \frac{1}{N-1}$$

This particular expression can also be justified with a physical demonstration, which we illustrate here for the case $N = 3$:

John takes a piece of paper and tears it into thirds. He hands one piece to Andrea, another to Beatrice, and tears the third piece into thirds again. He hands one piece to Andrea, a second to Beatrice, and tears the third remaining piece into thirds again. He repeats this process indefinitely. Eventually, John will give all the paper away—half to Andrea and half to Beatrice. But the quantity of paper Andrea receives and the quantity Beatrice receives can also be computed as a third, plus a third of a third, plus a third of a third of a third, and so forth. Thus it must be the case that $\frac{1}{3} + \frac{1}{9} + \frac{1}{27} + \cdots$ equals $\frac{1}{2}$.

A repeating decimal can be thought of as a sum of a geometric sequence. For example, the decimal 0.111... equals the series $\frac{1}{10} + \frac{1}{100} + \frac{1}{1000} + \cdots$, which, according to the formula above, is $\frac{1}{9}$. (Consequently the repeating decimal 0.999... equals 9 times this, $9 \times \frac{1}{9}$, which is 1: 0.999... = 1.)

The sum S of just the first n terms of a geometric series, $a + ar + ar^2 + \cdots + ar^{n-1}$ can be computed from the formula $S - rS = a - ar^n$. Provided that $r \neq 1$, this gives:

$$a + ar + ar^2 + \cdots + ar^{n-1} = a\frac{1-r^n}{1-r}$$

(If $-1 < r < 1$, then $r^n \to 0$ as n grows. This again shows that $a + ar + ar^2 + ar^3 + \ldots = a \cdot \frac{1-0}{1-r} = \frac{a}{1-r}$.) For example, the sum $1 + \frac{1}{2} + \frac{1}{4} + \ldots + \frac{1}{512}$ equals $1 \cdot \frac{1 - \left(\frac{1}{2}\right)^{10}}{1 - \frac{1}{2}} = \frac{1023}{512}$. A similar calculation solves a famous "chessboard puzzle":

Legend has it that the game of chess was invented for an Indian maharaja who became so delighted with the game that he wanted to reward the inventor with whatever he desired. The inventor asked for nothing more than one grain of rice on the first square of the chessboard, two grains on the second square, four on the third, and so forth, each square containing double the number of grains than the previous square. Given that there are 64 squares on a chessboard, how many grains of rice did the inventor in fact request?

According to the formula, the inventor asked for $1 + 2 + 4 + 8 + \ldots + 2^{63}$ grains of rice. This equals $1 \cdot \frac{1 - 2^{64}}{1 - 2} = 2^{64} - 1 = 18,446,744,073,709,551,615$ grains, which is the equivalent of about 25 billion cubic miles of rice, an inconceivable quantity. The inventor fooled the maharaja into making a promise he could not possibly honor.

See also ARITHMETIC SEQUENCE; CONVERGENT SEQUENCE; SERIES.

geometric series *See* GEOMETRIC SEQUENCE.

geometric transformation A specified procedure that shifts points in the plane to different positions (and thereby changing the location, and possibly the shapes, of geometric figures) is called a geometric transformation. More precisely, a geometric transformation is a FUNCTION that associates with each point of the plane some other point in the plane. (One may require the function to be one-to-one and onto.)

For example, the function that shifts each point of the plane one unit to the right is a geometric transformation (called a translation). This transformation preserves the shapes of all geometric figures.

Any geometric transformation that preserves distances between points in the plane (and hence the shape and size of geometric figures) is called an isometry or a rigid motion. One that multiplies all distances between points by a constant factor (called the dilation factor) is called a similitude, and a transformation that takes straight lines to straight lines is called a LINEAR TRANSFORMATION. All isometries are linear transformations, for example. These ideas also extend to transformations in three-dimensional space.

While he never made explicit use of the concept in his writings, the idea of a geometric transformation came from the work of the Greek geometer EUCLID (ca. 300–260 B.C.E.).

We list here some classical examples of geometric transformations.

Reflection in a Line

Given a line l in the plane, a reflection in this line takes a point P on one side of l to the corresponding point P'

on the opposite side of *l* such that the segment connecting *P* to *P′* is PERPENDICULAR to *l* and bisected by it. The points on *l* itself are left unmoved. Any reflection is an isometry that transforms geometric figures to their mirror images. The line used in performing the reflection is called the line of reflection.

In a CARTESIAN COORDINATE system, a reflection about the *x*-axis takes a point with coordinates (x,y) to the point $(x,-y)$, and a refection about the *y*-axis changes the sign of the *x*-coordinate: (x,y) becomes $(-x,y)$. The analog of a reflection in a line in three-dimensional space is a reflection in a plane.

Translation

A geometric transformation that moves all points in the plane a fixed distance in a fixed direction is called a translation. No points are left unmoved by a translation. A translation is an isometry, and all geometric figures are transformed to new figures with the same size, shape, and orientation as the originals.

If l_1 and l_2 are two PARALLEL lines in the plane, *d* units apart, a reflection in the first line, followed by a reflection in the second, has the same effect as translating all points in the plane a distance of 2*d* units in a fixed direction perpendicular to the two lines. Thus every translation is equivalent to the COMPOSITION of two reflections.

In a Cartesian coordinate system, a translation takes a point with coordinates (x,y) to the point $(x + a, y + b)$ for some fixed values *a* and *b*.

Rotation

A rotation about a point *O* through an angle θ is the geometric transformation that maps a point *P* in the plane to the point *P′* such that *P* and *P′* are the same distance from *O*, and the angle *POP′* has measure θ. (A counterclockwise turn is applied if θ is positive; a clockwise turn of θ is negative.) Only the location of the point *O* remains unchanged under a rotation, unless the angle θ is a multiple of 360°, in which case all points are fixed.

A rotation is equivalent to two reflections about lines that intersect at *O* making an angle of $\frac{\theta}{2}$ between them. Every rotation is an isometry. The analog of a rotation about a point in three-dimensional space is a rotation about a line.

Glide Reflection

A reflection in a line followed by a translation in a direction parallel to that line is called a glide reflection.

Reflection in a Point

A reflection in a point *O* in the plane is the isometry that takes a point *P* in the plane to the corresponding point *P′* such that *O* lies at the MIDPOINT of the line segment connecting *P* and *P′*. A reflection in a point is equivalent to a rotation of 180° about that point.

Dilation

A dilation with center *O* and dilation factor *k* > 1 is the geometric transformation that leaves *O* fixed, and moves any point *P* further away from *O*, by a factor *k*, along the ray from *O* through *P*. Thus a dilation stretches figures uniformly outward from *O*. It is possible, for example, to convert a square into a rectangle via a dilation. (A dilation with dilation factor *k* between *O* and 1 "shrinks" all points closer to *O*.) A dilation is not an isometry.

Circular Inversion

Also called an "inversion in a circle" or a "reflection in a circle," a circular inversion in a circle, with center *O* and radius *r*, takes a point *P* in the plane a distance *d* from *O*, and maps it to the point *P′* a distance r^2/d from *O* along the same ray from *O* through *P*. Thus points inside the circle are taken outside, and vice versa. Points on the circle itself are left unmoved by the transformation. The image of the center *O* under a circle inversion is undefined.

A circular inversion is not an isometry but proves to be a useful mapping in the study of GEOMETRY. It has the property that circles and straight lines in the plane are converted to new circles and new straight lines.

See also FRIEZE PATTERN; FUNDAMENTAL THEOREM OF ISOMETRIES; SYMMETRY; TRANSFORMATION OF COORDINATES.

geometry The branch of mathematics concerned with the properties of space and of figures, lines, curves, and points drawn in space is called geometry. Plane geometry examines objects drawn in a plane (lines, circles, polygons, and the like), solid geometry deals with figures in three-dimensional space (polyhedra, lines, planes, and surfaces), and SPHERICAL GEOMETRY studies

History of Geometry

The study of GEOMETRY is an ancient one. Records show that Egyptian and Babylonian scholars of around 1900 B.C.E. had developed sound principles of measurement and spatial reasoning in their architecture and in their surveying of land. Both cultures were aware of PYTHAGORAS'S THEOREM and had developed tables of PYTHAGOREAN TRIPLES. (The Egyptians used knotted ropes to construct "3-4-5 triangles" to create RIGHT ANGLES.) Ancient Indian texts on altar construction and temple building demonstrate sophisticated geometry knowledge, and the famous volume *JIUZHANG SUANSHU* (Nine chapters on the mathematical art) from ancient China also includes work on the Pythagorean theorem.

In ancient Greece, mathematical scholars came to realize that many properties of shapes and figures could be deduced logically from other properties. In his epic work *THE ELEMENTS* the Greek geometer EUCLID (ca. 300–260 B.C.E.) collated a large volume of knowledge on the subject and showed that each and every result could be logically deduced from a very small set of basic assumptions (self-evident truths) about how geometry should work. Euclid's work was rigorous and systematic, and the notion of a logical PROOF was born. EUCLID'S POSTULATES and the process of logical reasoning became the model of all further geometric investigation for the two millennia that followed. His method of compiling and organizing all mathematical knowledge known at his time was a significant intellectual achievement. Euclid's rigorous approach was, and still is, modeled in other branches of mathematics. Scholars in SET THEORY, the FOUNDATIONS OF MATHEMATICS, and CALCULUS, for instance, all seek to follow the same process of formal reasoning as the correct approach to achieve proper understanding of these topics.

The next greatest breakthrough in the advancement of geometry occurred in the 17th century with the discovery of CARTESIAN COORDINATES as a means to represent points as pairs of real numbers and lines and curves as algebraic equations. This approach, described by French mathematician and philosopher RENÉ DESCARTES (1596–1650) in his famous 1637 work *La géométrie* (Geometry), united the then-disparate fields of algebra and geometry. Unfortunately, Descartes's interests lay only in advancing methods of geometric construction, not in developing a full algebraic model of geometry. This latter task was pursued by French mathematician PIERRE DE FERMAT (1601–65), who had also outlined the principles of coordinate geometry in an unpublished manuscript that he had circulated among mathematicians before the release of *La géométrie*. Fermat later published the work in 1679 under the title *Isagoge ad locus planos et solidos* (On the plane and solid locus). The application of algebra to the discipline provided scholars a powerful new tool for solving geometric problems, and also provided them with a large number of different types of curves for study.

Fermat's work in geometry inspired work on the theory of DIFFERENTIAL CALCULUS and, later, led to the study of "differential geometry" (the application of calculus to the study of shapes and surfaces). This was developed by the German mathematician and physicist CARL FRIEDRICH GAUSS (1777–1855).

Neither Descartes nor Fermat permitted negative values for distances. Consequently, neither scholar worked with a full set of coordinate axes as we use them today. The

the properties of lines and shapes drawn on the surface of a SPHERE. The word *geometry* comes from the Greek words *ge* meaning "earth" and *metria* meaning "measure." As the origin of the word suggests, the study of geometry evolved from very practical concerns with regard to the accurate measurement of tracts of land, navigation, and architecture.

The Greek mathematician EUCLID (ca. 300–260 B.C.E.) formalized the study of geometry to one of pure logical reasoning and deduction. In his famous work, *THE ELEMENTS*, Euclid collated a considerable volume of Greek knowledge on the subject and showed that all the results known at the time could be deduced from a very small collection of self-evident truths or AXIOMS, which he stated explicitly. Any result that can be deduced from these axioms is today described as Euclidean. Euclid's rigorous approach to geometric investigation remained the standard model of study for two millennia.

See also EUCLIDEAN GEOMETRY; EUCLIDEAN SPACE; EUCLID'S POSTULATES; HISTORY OF GEOMETRY (essay); NON-EUCLIDEAN GEOMETRY; PARALLEL POSTULATE; PROJECTIVE GEOMETRY.

Germain, Marie-Sophie (1776–1831) French *Number theory, Mathematical physics* Born on April 1, 1776, in Paris, France, scholar Marie-Sophie Germain is remembered in mathematics for her significant contributions to the topic of NUMBER THEORY. Most notably, notions of negative distance and negative area were first put forward by Sir Isaac Newton (1642–1727) and Gottfried Wilhelm Leibniz (1646–1716), the coinventors of calculus.

The 19th century saw other major advances in geometry. It had long been noted that Euclid's fifth postulate, the so-called parallel postulate, is not necessary for a great deal of geometry. Many Arab scholars of the first millennium attempted, without success, to show that the fifth postulate could be logically deduced from the remaining four (thereby rendering it unnecessary), as did European scholars of the Renaissance. In 1795 Scottish mathematician John Playfair showed that the fifth postulate is equivalent to the statement that, through any point, one can draw one, and only one, line through that point parallel to a given line. (This is today called Playfair's axiom.) Although not eliminating the need for the fifth postulate, Playfair showed that it could be understood in a more tractable form.

In 1829 Russian mathematician Nikolai Ivanovich Lobachevsky (1792–1856) took a bold step and considered a geometric world in which the fifth postulate is false. He assumed that through a given point *more* than one line could be drawn parallel to a given line. In doing this, Lobachevsky discovered a new, consistent mathematical system free from contradiction, one as logically valid as the geometry of Euclid. (This geometry is today called hyperbolic geometry.) The philosophical impact of Lobachevsky's work was enormous: he had shown that mathematics need not be based on a single set of *physical* truths, and that other equally valid mathematical systems do exist based on alternative, carefully chosen axioms. Lobachevsky had also shown that Euclid's fifth postulate cannot be established as a consequence of the remaining four axioms: he had pre-

sented a valid example of a system in which the first four of Euclid's postulates hold, but the fifth does not.

Surprisingly some of Lobachevsky's ideas were anticipated well before the 19th century. The great Persian mathematician and poet Omar Khayyám (ca. 1048–1122) established a number of results that we recognize today as non-Euclidean. These results were later translated into Latin, and extended upon, by Italian priest Girolamo Saccheri (1667–1733). Unfortunately, neither scholar discovered the validity of non-Euclidean geometry, as each was focused instead on trying to establish Euclid's fifth postulate as a consequence of the remaining four.

The German mathematician Bernhard Riemann (1826–66) discovered an alternative form of non-Euclidean geometry in which Euclid's fifth postulate fails in a different way. In a system of spherical geometry it is *never* possible to draw a line through a given point parallel to a given line.

Riemann's contributions to the advancement of geometry were significant. In his famous 1854 lecture "Über die Hypothesen welche der Geometrie zu Grunde liegen" (On the hypotheses that lie at the foundation of geometry), Riemann put forward the view that geometry can be the study of any kind of space of any number of dimensions, and later developed the mathematics needed to properly describe the shape of space. Albert Einstein (1879–1955) later used this work to develop his theory of relativity.

See also affine geometry; Arabic mathematics; Babylonian mathematics; János Bolyai; dimension; Egyptian mathematics; *The Elements*; geometric; transformation; Greek mathematics; history of equations and algebra (essay); Indian mathematics; perspective; postulate; projective geometry; theorem; topology.

her work paved the way for other scholars in the field to establish the validity of Fermat's last theorem for all values *n* < 100, the only substantial step made toward solving the problem during the 19th century. Germain also worked to develop a mathematical theory of elasticity. She is noted as one of the very few women scholars in mathematics before the 20th century.

At the age of 13, after reading an account of the life and death of Archimedes of Syracuse, Germain said she felt impelled to become a mathematician. She pursued her studies by first teaching herself Latin and Greek and then reading the works of Sir Isaac Newton, Leonhard Euler, and Euclid. As a woman, Germain was not permitted to enter a university, but she did manage to continue her study of mathematics

by reading course notes borrowed from students who attended the École Polytechnique in Paris. Mathematician Joseph-Louis Lagrange (1736–1813) presented a course on analysis at the university, and Germain submitted to him a thesis on the topic under the pseudonym Louis LeBlanc, a male name. Lagrange was greatly impressed by the contents of the paper and sought out its author. He was not at all perturbed to discover that "M. LeBlanc" was a woman. In fact, he was so impressed with her work that he offered her personal counsel in the study of advanced mathematics.

Again under her male pseudonym, Germain later established a correspondence with the German mathematician Carl Friedrich Gauss (1777–1855), with whom she developed and shared much of her work in number theory. Gauss, too, eventually discovered her true identity, but this did not diminish his respect for her as a fine scholar. In 1808 the Institut de France set a competition to find a mathematical theory for the theory of elasticity. Although the necessary mathematics to solve the challenge was not available to scholars at the time, Germain devoted a decade of work toward solving the problem and managed to make significant steps in developing a beginning theory on the subject. Unfortunately, much of her work was ignored by the Institut.

Germain never married nor obtained a professional position; she was supported financially by her father throughout her life. Germain died of breast cancer in Paris, France, on June 27, 1831. Her death certificate listed her occupation merely as "property holder."

glide reflection   A reflection in a line followed by a translation in a direction parallel to that line is called a glide reflection.

See also GEOMETRIC TRANSFORMATION.

gnomon   The L-shaped figure that remains when a PARALLELOGRAM (usually a square) is removed from the corner of a larger similar parallelogram is called a gnomon. For example, in a square 2×2 array of dots, the three dots surrounding any corner dot form the shape of a gnomon. (Notice, incidentally, that $4 = 1 + 3$.) Adding a gnomon of five dots to the 2×2 array produces a square 3×3 array (and we have that $9 = 1 + 3 + 5$), and adding to this a gnomon of seven dots yields a 4×4 array (yielding $16 = 1 + 3 + 5 + 7$). This process shows that, in general, the nth square number equals the sum of the first n odd numbers.

The pointer on a sundial is also called a gnomon. It, and the shadow it casts, together form the shape of an L.

See also FIGURATE NUMBERS.

Gödel, Kurt (1906–1978) Austrian-American *Logic* Born on April 28, 1906, in Brünn, Austria-Hungary (now Brno, Czech Republic), Kurt Gödel is today considered the most important logician of the 20th century. His famous pair of incompleteness theorems stunned the mathematical community, dashing the hopes of all those who had been fervently searching for a set of fundamental axioms from which all mathematics could be logically deduced.

Gödel exhibited a talent for academic work at an early age and had completed the equivalent of a university curriculum before leaving secondary school. In 1923 he entered the University of Vienna to pursue a degree in physics, but changed to mathematics in 1926 when he was introduced to the field of FORMAL LOGIC. It soon became clear to the faculty of the department that Gödel would make considerable contributions to the field. In the summer of 1929 Gödel completed a doctoral dissertation on the topic and was awarded a Ph.D. the following year. His thesis outlined the details of his first famous discovery, but the revolutionary impact of his work was not fully understood by the larger mathematical community until he published the result a year later.

Gödel had proved that any mathematical system sufficiently sophisticated to incorporate the principles of arithmetic will always contain statements that can neither be proved nor disproved. He later also showed that no mathematical system could be proved to be consistent, that is, free from CONTRADICTION, by making use of just the axioms of the system. It had been a 200-year-long dream, since the time of GOTTFRIED WILHELM LEIBNIZ in fact, to place the whole of mathematics on a firm axiomatic base. Gödel had proved, in essence, that such a dream can never be realized.

After joining the faculty of the University of Vienna for three years, Gödel accepted, in 1933, a position at the Institute for Advanced Study in Princeton, New Jersey. After suffering a nervous breakdown, he returned to Vienna after only a year, but later returned to the Institute in 1942 to escape war-torn Europe. Gödel became an American citizen that same year.

Many awards were bestowed upon him throughout his life. In 1950 he was one of two recipients of the first Einstein Award, and he was awarded honorary doctorates from Yale University in New Haven, Connecticut, and Harvard University in Cambridge, Massachusetts, in the two years that followed. He was elected to the National Academy of Sciences in 1955.

It is said that Gödel held fixed opinions about many matters of life, and felt himself to always be right—especially in the disciplines of mathematics and medicine. Gödel was of a nervous disposition, and after enduring severe bleeding from a duodenal ulcer, he decided to maintain an extremely strict diet of his own devising. He developed a general distrust of doctors and, toward the end of his life, became convinced that he was being poisoned. Gödel eventually stopped eating altogether and died of starvation on January 14, 1978.

Gödel's work in mathematical logic strikes at the very core of the subject. His name and his results are familiar to any undergraduate student having taken a first course in formal logic.

See also GÖDEL'S INCOMPLETENESS THEOREMS.

Gödel's incompleteness theorems In 1900, DAVID HILBERT posed 23 problems for future mathematicians. The second of his 23 problems challenged the mathematical community to find a logical base for the discipline of mathematics, that is, to develop a formal system of symbols, with well-defined meaning and well-defined rules of manipulation, on which all of mathematics could be based. Mathematicians, such as BERTRAND ARTHUR WILLIAM RUSSEL and his colleague ALFRED NORTH WHITEHEAD, took on the challenge and made some progress in this direction. But in 1931 KURT GÖDEL stunned the mathematical community by proving once and for all that such an aim could never be achieved.

Gödel's first incompleteness theorem states that any mathematical system sufficiently sophisticated to incorporate the principles of arithmetic will necessarily contain statements (theorems) that can neither be proved nor disproved. Informally, this means that it will never be possible to program a computer to answer all mathematical questions.

His second incompleteness theorem (which is actually a consequence of the first) asserts that no system of mathematical logic incorporating the principles of arithmetic will ever be capable of establishing its own consistency. That is, the proof that a system of mathematics is free from CONTRADICTION will require axioms and principles not contained in the system.

Gödel proved his theorems by assigning numbers to the symbols used in a mathematical system (commas, digits, left and right parentheses, etc.), and then devising an ingenious method of assigning numbers to all mathematical statements (combinations of symbols) in the system. It was then possible to show that the mathematical system contains a statement of the form: "This sentence cannot be proved." Of course, it is impossible to prove or disprove such a statement.

Gödel's work dashed centuries of hope of finding a small set of basic axioms on which to base all of mathematics.

See also CONSISTENT; FORMAL LOGIC; TRUTH TABLE.

Goldbach, Christian (1690–1764) Prussian *Number theory* Born on March 18, 1690 in Königsberg, Prussia (now Kaliningrad, Russia), Christian Goldbach is best remembered in mathematics for the famous unsolved conjecture that bears his name.

Goldbach studied in St. Petersburg and became a professor of mathematics and historian at the university in 1725. He traveled extensively throughout Europe and established personal contact with a number of prominent mathematicians of the time. Although Goldbach studied infinite sums, the theory of equations, and the theory of curves, his most important work was in the field of NUMBER THEORY, much of which was conducted through correspondence with the Swiss mathematician LEONHARD EULER (1707–83). One particular letter later garnered international attention. In it, Goldbach mentioned that every even number seems to be the sum of two PRIME numbers. For instance, he noted that 4 equals 2 + 2, 6 equals 3 + 3, 10 equals 3 + 7, and 1,000 equals 113 + 887. Unable to find an instance where this was not the case, Goldbach asked Euler whether or not this was indeed a property of all even numbers. Euler attempted to resolve the issue but found that he was unable to establish the claim, nor find a COUNTEREXAMPLE to it. Later, English mathematician Edward Waring published the problem in his popular 1770 text *Meditationes algebraicae* (Meditations on algebra), properly attributing the problem to Goldbach. The challenge garnered considerable notoriety. Today called GOLDBACH'S CONJECTURE, the problem remains one of the most famous unsolved challenges in mathematics.

Goldbach died in Moscow, Russia, on November 20, 1764. Apart from the conjecture that bears his name, Goldbach is also remembered for his correspondence with PIERRE DE FERMAT (1601–1655) and for being one of the few mathematicians at the time to understand the implications of Fermat's approach to number theory.

Goldbach's conjecture In his 1742 letter to LEONHARD EULER, Prussian mathematician CHRISTIAN

GOLDBACH (1690–1764) conjectured that every even number greater than 2 is the sum of two PRIME numbers. For example, 4 = 2 + 2, 6 = 3 + 3,..., 20 = 3 + 17,..., 50 = 13 + 37,..., 100 = 3 + 97, and so forth. No one to this day has been able to establish whether this claim holds true for each and every even number. The problem has since become known as Goldbach's conjecture, and it remains one of the most famous unsolved mathematical problems of today. Between March 20, 2000, and March 20, 2002, a $1-million prize was offered to anyone who could solve Goldbach's conjecture. The prize went unclaimed.

Goldbach's conjecture is equivalent to the challenge of proving that every integer greater than 5 is the sum of *three* primes. (If the number is odd, subtract 3 and write the resultant number as a sum of two primes. If it is even, subtract 2.)

The source of the difficulty with this problem is that primes are defined in terms of *multiplication,* while the problem involves *addition.* It is often very difficult to establish connections between these two separate operations on the integers.

In 1931, Russian mathematician L. Schnirelmann (1905–38) proved that every positive integer is the sum of at most 300,000 primes. Although this result seems ludicrous in comparison to the original problem, it is a significant first step to solving the conjecture: it shows at least that it is possible to put a bound on the number of primes representing the integers.

In 1937 another Russian mathematician, Ivan Vinogradoff (1891–1983), proved that every sufficiently large number can be written as a sum of four primes. What constitutes "sufficiently large" is not known (Vinogradoff's work only shows that there cannot be infinitely many integers that require more than four prime summands)—but at least a small bound has been placed on "most" integers. In 1973 Jing-Run Chen proved that every sufficiently large even number is the sum of a prime and a number that is either prime or has two prime factors.

Work on solving the Goldbach conjecture continues. As of the year 2003, it has been confirmed by computer that the conjecture holds true for all numbers up to 6×10^{16}.

See also BRUTE FORCE.

golden ratio (divine proportion, extreme and mean ratio, golden mean, golden section)
A line segment connecting two points A and B is said to be divided by a third point P in the golden ratio if the ratio of the whole length AB to the length AP is the same as the RATIO of the length AP to the length PB:

$$\frac{AB}{AP} = \frac{AP}{PB}$$

The value of this ratio is denoted φ (the Greek letter *phi*) and can be computed as follows:

Set the length PB to be one unit. Then, since $\frac{AP}{PB} = \varphi$, the length AP is φ units, and the length of the entire segment AB is $\varphi + 1$. We have:

$$\frac{\varphi + 1}{\varphi} = \frac{\varphi}{1}$$

yielding the QUADRATIC equation $\varphi^2 = \varphi + 1$. This has two solutions. Selecting the solution that is larger than 1 yields:

$$\varphi = \frac{1 + \sqrt{5}}{2} = 1.618033988...$$

(The second solution to this quadratic equation is $\frac{1 + \sqrt{5}}{2} = 1 - \varphi = -0.618033988....$)

The golden ratio was studied and made famous by the Pythagoreans, the followers of the Greek mathematician PYTHAGORAS (ca. 500 B.C.E.). They discovered it in their study of the PENTAGRAM, the figure that appears when one draws in the diagonals of a regular pentagon. The sides of a pentagram divide each other in the golden ratio.

The following method shows how to construct a point P that divides a given line segment AB into the golden ratio:

Given a line segment AB, draw a perpendicular line through A. Find the MIDPOINT M of AB and locate on the perpendicular line a point C whose distance from A is the same as the length AM. (Use a circle with center A and radius AM as an aid.) Now draw a circle with center C and radius BC to locate a second point D on the perpendicular line such that length BD equals length BC. Let P be the point on the line segment AB such that length AP equals length AD. (Use a circle with center A

and radius AD to find this point.) Then point P divides the segment AB in the golden ratio.

(To see why this works, suppose that the line segment AB is of unit length, and the length of AP is x. Then triangle ABC is a right triangle with hypotenuse $\frac{1}{2} + x$. PYTHAGORAS'S THEOREM now shows that $x = \varphi - 1 = \frac{1}{\varphi}$, giving $\frac{AB}{AP} = \varphi$.)

Noting that the construction of midpoints, circles, and perpendicular lines can all be accomplished with a straight edge (that is, a ruler with no markings) and a compass alone, the above procedure shows that the golden ratio is a constructible number.

Dividing the relation $\varphi^2 = \varphi + 1$ through by φ yields $\varphi = 1 + \frac{1}{\varphi}$. Substituting this formula into itself multiple times establishes:

$$\varphi = 1 + \frac{1}{\varphi} = 1 + \frac{1}{1 + \frac{1}{\varphi}} = 1 + \frac{1}{1 + \frac{1}{1 + \frac{1}{\varphi}}} = \ldots$$

Repeating this process indefinitely shows that the golden ratio has the following simple CONTINUED FRACTION expansion:

$$\varphi = 1 + \cfrac{1}{1 + \cfrac{1}{1 + \cfrac{1}{1 + \cfrac{1}{1 + \ldots}}}}$$

If one terminates this expansion after a finite number of steps, then ratios of the FIBONACCI NUMBERS appear:

$$1 = \frac{1}{1} = \frac{F_2}{F_1}, 1 + \frac{1}{1} = \frac{2}{1} = \frac{F_3}{F_2}, 1 + \frac{1}{1 + \frac{1}{1}} = \frac{3}{2} = \frac{F_4}{F_2}, 1 + \frac{1}{1 + \frac{1}{1+1}}$$

$$= \frac{5}{3} = \frac{F_5}{F_4}, \text{ and so forth. (That this pattern persists can}$$ be proved by INDUCTION.) We have:

$$\lim_{n \to \infty} \frac{F_{n+1}}{F_n} = \varphi$$

An induction argument also proves that:

$$\varphi^n = F_n \varphi + F_{n-1}$$

Substituting the formula $\varphi = \sqrt{1+\varphi}$ into itself multiple times gives an expression for φ as a sequence of nested radicals akin to VIÈTE'S FORMULA for π. We have:

$$\varphi = \sqrt{1 + \sqrt{1 + \sqrt{1 + \sqrt{1 + \ldots}}}}$$

The golden ratio also occurs in TRIGONOMETRY as $\cos\left(\frac{\pi}{5}\right) = \frac{\varphi}{2}$, $\cos\left(\frac{2\pi}{5}\right) = \frac{1}{2\varphi}$, and $\sin\left(\frac{\pi}{10}\right) = \frac{1}{2\varphi}$, for instance.

The number φ also appears in a number of unexpected places in nature and throughout mankind's artistic pursuits. Since the golden ratio is well approximated by the fraction 16/10, some scholars suggest that the ancient Egyptians of 3000 B.C.E. used the golden ratio repeatedly in the construction of their tombs. The "golden chamber" of the tomb of Rameses IV measures 16 ells by 16 ells by 10 ells, that is, approximately the ratio $\varphi : \varphi : 1$; other tombs are found in the approximate ratio $\varphi^2 : \varphi : 1$; and Egyptian furniture found in those tombs often had overall shape based on the ratio $\varphi : 1 : 1$. German artist ALBRECHT DÜRER (1471–1528) wrote a four-volume text, *Treatise on Human Proportions*, detailing occurrences of the ratio φ in the human body. (He claimed, for instance, that ratio of the length of the human face to its width is approximately φ, and also that the elbow divides the human arm, shoulder to fingertip, in the golden ratio.) Artists of that time came to view the golden ratio as a "divine PROPORTION" and used it in all forms of artistic work. The GOLDEN RECTANGLE was deemed the rectangular shape most pleasing to the eye.

golden rectangle Any rectangle whose sides are in the ratio 1 to φ, where $\varphi = \frac{1 + \sqrt{5}}{2}$ is the GOLDEN RATIO, is called a golden rectangle. Such a rectangle has the property that excising the largest square possible from one end of the figure leaves another rectangle in the same proportion. (The remaining rectangle has proportions $\varphi - 1$ to 1. Since the golden ratio satisfies the equation $\varphi^2 = \varphi + 1$, we have: $\frac{1}{\varphi} = \frac{\varphi - 1}{1}$.) By this method, new golden rectangles can be constructed from

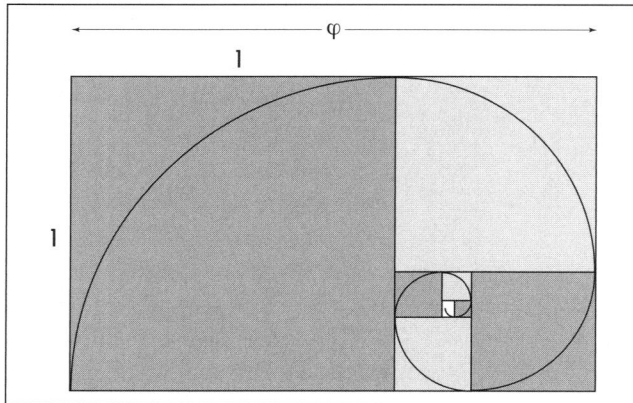

The golden rectangle

a given golden rectangle *ad infinitum*. Drawing arcs between the nonadjacent corners of the squares produces a spiral with the remarkable property that it cuts any radius from the center of the spiral at the same ANGLE (approximately 73°).

Throughout history the golden rectangle has been considered to have a particularly pleasing shape. The pillars of the Greek Parthenon of the 5th century B.C.E. are spaced to produce golden rectangles in its facade (although no one can be certain that the designers intended this to be the case). Many European scholars of the 1500s based much of their work on the golden rectangle, and German engraver ALBRECHT DÜRER (1471–1528) used the golden ratio in his analysis of PROPORTIONS found in the human body. Many rectangles used today in magazine advertisements, for instance, are surprisingly close to golden rectangles.

The golden rectangle also appears in the fifth PLATONIC SOLID, the icosahedron. If all the edges of the figure are assumed to be one unit long, then the distance between any given edge and the (parallel) edge opposite to it on the other side of the figure is φ. Thus any two opposite edges of an icosahedron form two sides of a golden rectangle.

A "golden triangle" is an isosceles triangle whose sides to its base are in the ratio φ to 1. Such a triangle has internal angles 72°, 36°, and 72°. If one of the base angles is bisected, then the figure is divided into two smaller triangles, one of which is a new golden triangle.

See also SPIRAL OF ARCHIMEDES.

googol/googolplex   In 1938 Milton Sirotta, nine-year-old nephew of American mathematician Edward Kasner, coined the term *googol* for the number 1 followed by 100 zeros (10^{100}). At the same time he coined the term *googolplex* for the number 1 followed by a googol zeros ($10^{googol} = 10^{(10^{100})}$). Physicists believe that there are only about 10^{80} particles in the entire universe, considerably less than a googol.

Gosset, William Sealy (1876–1937) British *Statistics*   Born on June 13, 1876, in Canterbury, England, William Gosset is remembered for his important work in statistics, most notably his invention of the t-test, details of which were published in 1908 under the title STUDENT'S T-TEST.

Gosset studied chemistry and mathematics at New College, Oxford, before accepting a position as a chemist at the Guinness brewery in Dublin in 1899. His interests in statistics were motivated by the practical problems of measuring and maintaining production quality. Rather than test each and every sample of product from the brewery, Gosset looked for mathematical techniques that would allow him to deduce reliable information based on just a small number of samples. His famous t-test is the result of his efforts. In an effort to protect trade secrets, the brewery forbade its employees to publish, but Gosset printed his method nonetheless under the pseudonym of "Student."

Gosset's work was deemed of great value to the brewery. In 1922 he was given an assistant in statistics, and he was allowed to build up a small department of statistics within the company. The Student t-test is today considered a fundamental technique in the repertoire of tools used by industry and science when dealing with concerns of quality control and general statistical inference.

In 1935 Gosset left Ireland to manage a new Guinness brewery in London. He died two years later on October 16, 1937. His statistical analysis of variance had a profound effect on the practices of 20th-century industry.

See also HISTORY OF PROBABILITY AND STATISTICS (essay).

grad   *See* GRADIENT.

graph 233

gradient The SLOPE of a line is sometimes called its gradient or grade. For example, the gradient of the line connecting the two points $A = (a_1, a_2)$ and $B = (b_1, b_2)$, denoted m_{AB}, is given by:

$$m_{AB} = \frac{\text{"rise"}}{\text{"run"}} = \frac{b_2 - a_2}{b_1 - a_1}$$

Three points A, B, and C in the plane are COLLINEAR if the gradients m_{AB} and m_{AC} are equal.

In the setting of multivariable calculus—the study of CALCULUS applied to FUNCTIONS of more than one variable—the gradient of a function $f(x,y,z)$ of three VARIABLES, also called the "grad of f," is defined to be the VECTOR of its PARTIAL DERIVATIVES. It is denoted grad(f) or ∇f and is given by:

$$\nabla f = \; < \frac{\partial f}{\partial x}, \frac{\partial f}{\partial y}, \frac{\partial f}{\partial z} >$$

Such a quantity proves to be useful in computing the DIRECTIONAL DERIVATIVE of a function. In physics, ∇f is also used to describe the spatial variation in the magnitude of a force, such as a gravitational force or a magnetic force. The study of directional derivatives shows that the quantity ∇f calculated at a point represents the direction from that point in which the rate of change of the force f is a maximum.

See also DIV.

graph (network) Any diagram of points and line segments connecting pairs of points is called a graph. The points are usually called vertices or nodes, and the line segments are called edges. More than one edge is allowed to connect the same pair of vertices to yield a set of multiple edges. One can also permit an edge connecting a vertex to itself via a loop. Edges can intersect, but the places where they cross are not considered vertices. For example, the following picture is a graph. It has just six vertices and comes in two disconnected pieces.

A graph that comes in just one piece is called connected. This means that it is always possible to travel from any one vertex to any other by traversing a sequence of edges.

Graphs can be used to codify information. For example, a graph might represent the network of possible flight routes between cities, the flow of information between departments in a large organization, or even the set of acquaintances among people attending a party. (Each vertex represents a person in the room, and an edge is drawn between two vertices if the corresponding two people know each other.) The general study of graphs can translate into interesting facts about travel possibilities, streamlining data flow, acquaintanceships, and the like.

The degree (or valence) of a vertex is the number of edges that meet at that vertex. Loops are counted twice.

A graph is called complete if every vertex is connected to each and every other vertex by a single edge. For example, the complete graph on four vertices looks like a square with the two diagonals drawn in. Each vertex has degree three.

A graph is called planar if it can be drawn on a plane without two edges crossing. The complete graph on four vertices is planar if one draws one of the diagonals "outside the square." The THREE-UTILITIES PROBLEM is an example of a graph that is not planar.

A cycle in a graph is a sequence of edges that starts and ends at the same vertex and does not travel over the same edge twice. Finding an "Euler circuit," i.e., is, a cycle that traverses each and every edge in a graph precisely once, is an old problem. (*See* GRAPH THEORY.)

A connected graph containing no loops or cycles is called a tree. These graphs look like a series of forking branches, and hence the name *tree*. Any tree diagram, such as a PROBABILITY, tree is an example of a graph that is a tree. It follows from EULER'S THEOREM that any connected graph with n vertices and $n-1$ edges must be a tree.

See also CRITICAL PATH; EULERIAN PATH/CIRCUIT; HAMILTONIAN PATH/CIRCUIT; TOURNAMENT.

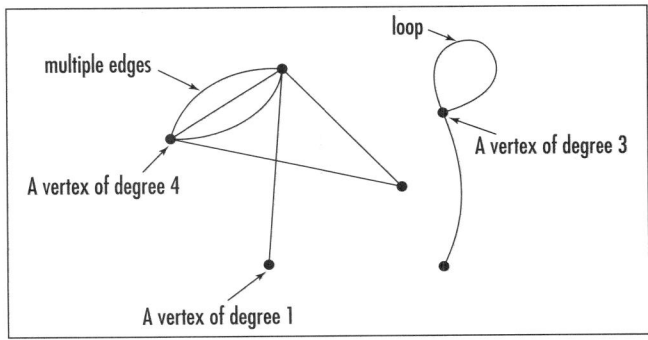

A typical graph

graphical solution Any solution to a pair of simultaneous equations found by plotting the graph of each equation and visually inspecting the location of a point of intersection is called a graphical solution. For example, in plotting the graphs of $y = 2x + 3$ and $y = 9 - 4x$, one sees that the lines intersect at the point $(1,5)$ and so $x = 1$, $y = 5$ is a solution to the pair of equations. A graphical approach is often the only feasible method of solving a complicated pair of equations. Graphing calculators are equipped with a "trace" button that allows one to quickly find the coordinates of points of intersection up to some degree of accuracy.

A single equation $f(x) = 0$ can be solved graphically by locating the x-intercepts of the function. For example, to find the square root of 2, one can plot a graph of the function $f(x) = x^2 - 2$ and attempt to identify the location of the positive x-intercept.

This graphical method makes it clear that a pair of SIMULTANEOUS LINEAR EQUATIONS:

$$ax + by = p$$
$$cx + dy = q$$

has either one solution (the lines intersect at a unique point), no solutions (the lines are parallel), or infinitely many solutions (the lines coincide).

A DIFFERENTIAL EQUATION of the form $\frac{dy}{dx} = f(x,y)$ can be solved graphically. We seek a curve $y = g(x)$ in the plane whose slope $\frac{dy}{dx}$ at any point (x, y) is given by $f(x,y)$. Thus if, for a large selection of points (x,y) across the plane, we draw short line segments of slope $f(x,y)$, the shapes of curves following these slopes may be visually apparent. If we are also told the value of the function $y = g(x)$ at one particular point (x, y), then following the slope of the line segments from that point onward describes a particular solution to the equation.

See also BISECTION METHOD; NEWTON'S METHOD.

graph of a function A drawing or a visual representation that shows the relationship between two or more variables is called a graph. It is usual to draw the graph of a FUNCTION of a single variable $y = f(x)$ on a CARTESIAN COORDINATE system with an x-axis and a y-axis at right angles. The graph of the function is then the set of all points (x,y) that satisfy the equation $y = f(x)$ drawn as a curve in the plane. For example, the set of all points (x, y) that satisfy $y = x^2$ forms a PARABOLA, while the graph of the function $y = 3x$ is a straight line through the origin with slope 3.

A general approach to graphing a function is to make a table of (x, y) pairs that satisfy the equation under consideration and then to locate these points on a coordinate system. If sufficiently many points are drawn, then a smooth curve connecting the dots is likely to be a good representation of the function. Graphing calculators employ this technique when displaying the graph of a function.

As one develops familiarity with basic equations, graphing simple formulae becomes a matter of routine. For example, one can establish that a LINEAR EQUATION of the form $y = mx + b$ yields a straight-line graph of slope m crossing the y-axis at position b, and that an equation of the form $x^2 + y^2 = r^2$ represents a CIRCLE of radius r. To determine the graphs of more-complicated functions, scholars—for many decades—could only resort to the tedious task of making tables and plotting individual points until a general picture emerged. The advent of CALCULUS, however, at the turn of the 18th century brought with it the power to quickly identify and examine the basic shape and structure of complicated graphs. By examining the first DERIVATIVE of a function, for instance, one can determine where the graph increases and decreases, as well as the location of any local maxima and minima. The second derivative provides information about the shape of the curve—whether it is concave up or concave down. This information, together with knowledge of the x- and y-intercepts of the curve and any ASYMPTOTES it might possess, is enough to draw a reasonably accurate picture of the graph without having to plot individual points. Application of these newly discovered techniques from calculus was literally an eye-opening experience for scholars of the time.

French mathematician NICOLE ORESME (ca. 1323–82) was the first to draw the graph of a function and to find an interesting interpretation for the area of the region under it. French lawyer and amateur mathematician PIERRE DE FERMAT (1601–65) developed the idea further, defining a general procedure for associating curves to formulae. Given a relationship between two variables A and B, say, Fermat drew a horizontal reference line for the independent variable A and imagined a second line sliding along this reference line at a fixed angle whose length B varied according to the relationship defined. He did not require that the line representing B be at right angles to the reference line. French mathematician and philosopher RENÉ DESCARTES (1596–1650) independently developed the same technique, but he had the further insight to introduce algebraic symbolism to describe relationships via formulae and to label graphs appropriately. Descartes too, however, envisioned functions as sliding lengths along a single reference line. It was not until decades later that mathematicians began drawing an explicit second coordinate axis, the y-axis, at a fixed 90° angle to the horizontal reference x-axis.

Quantities other than functions can also be graphed. For example, an inequality of the form $y < x + 3$ defines the HALF-PLANE below the line $y = x + 3$. The graph of a function of two variables $z = f(x, y)$ is a surface sitting in three-dimensional space. For example, the graph of the function $z = x^2 + y^2$ is a PARABOLOID.

A logarithmic graph is one in which both axes of the coordinate system are marked in LOGARITHMIC SCALE. Such plots are particularly useful for examining equations of the form $y = ax^n$. A polar-coordinate graph is a plot of an equation of the form $r = f(\theta)$ given in POLAR COORDINATES. Polar-coordinate graph paper assists in plotting such graphs.

Graphs of other types of numerical relationships are studied in descriptive statistics.

See also CONCAVE UP/CONCAVE DOWN; COORDINATES; GRAPHICAL SOLUTION; HISTORY OF FUNCTIONS (essay); INCREASING/DECREASING; MAXIMUM/MINIMUM; STATISTICS: DESCRIPTIVE.

graph theory   The mathematical study of graphs is called graph theory. The field was founded in 1736 by the Swiss mathematician LEONHARD EULER (1707–83) with his solution to the famous seven bridges of Königsberg problem:

> The old German city of Königsberg (now the Russian city of Kaliningrad) was built on the two banks of the river Pregel and on two islands in the river. The different parts of the city were connected by seven bridges. A debate ensued among the residents as to whether it was possible to walk a complete tour of the city crossing each and every bridge precisely once. (It wasn't deemed necessary to return to one's starting location.) Most people felt that

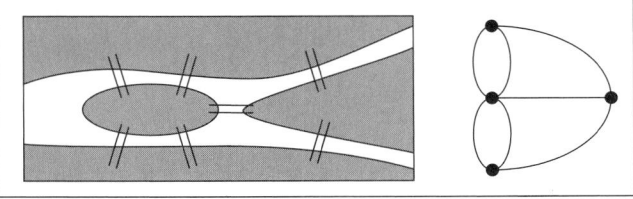

The seven bridges of Königsberg problem

this was impossible, but could this be *proved* to be the case?

Euler's insight into solving the problem came by reducing the city plan to a GRAPH, that is, a diagram of points (vertices) and edges, with each point representing a land mass, and each edge a bridge. Any stroll through the city thus corresponds to a journey along the edges of the graph. The Königsberg bridge problem is therefore equivalent to asking:

> Is it possible to draw the above graph without lifting pencil from page and without tracing over the same edge twice?

Euler was able to solve the more general problem of classifying all those graphs that can be so traced. He observed that, in tracing such a path, any edge drawn entering a vertex must be matched by an edge exiting that vertex. Thus all the edges meeting at each vertex are matched in pairs *unless* one starts or ends a journey at a particular vertex, in which case one edge remains unmatched. Thus a graph that can be so traced must contain just two vertices of odd degree—these will be the start and end of the journey; or no vertices of odd degree—the path starts and ends at the same location. Any graph containing more than two vertices of odd degree cannot be drawn without lifting pencil from page. In particular, the seven bridges of Königsberg problem cannot be solved (all four vertices have odd degree).

Any path that traces through a graph following each edge precisely once is known today as an Eulerian path. If the path starts and ends at the same vertex it is called an Eulerian circuit. Euler went further to show that all connected graphs possessing precisely two vertices of odd degree do indeed have Eulerian paths, and that graphs with all vertices of even degree do possess

Eulerian circuits. If a graph represents transportation routes in a city or across a nation, for example, then seeking Eulerian circuits for the graph can have important practical use.

In 1857 WILLIAM ROWAN HAMILTON explored the issue of finding paths through graphs that do not necessarily trace each edge, but instead visit each vertex in the graph once, and only once. Today such a path is called a Hamiltonian path, or a Hamiltonian circuit if it is a path that returns to the starting vertex. Although mathematicians have developed some theorems that give conditions under which Hamiltonian paths will exist, no one to this day knows a simple algorithm that enables us to find them. Each graph must still be examined individually, and finding a Hamiltonian path—if one exists—is usually a matter of inspired guessing.

Simple counting can lead to important results in graph theory. For example, summing the degrees of all the vertices in a graph counts the total numbers of ends of edges. As every edge has two ends, we have:

> In any graph, the sum of the degrees of all the vertices equals twice the number of edges in the graph.

In particular, the sum of all the degrees of vertices must be an even number. Consequently, there cannot be an odd number of odd numbers in this sum.

> In any graph, the number of vertices of odd degree is even.

This result has an amusing interpretation. Thinking of edges in a graph as handshakes between people, we have established the so-called HANDSHAKE LEMMA:

> At any instant, the number of people on this planet, living or deceased, who have participated in an odd number of handshakes is even.

A POLYHEDRON can be thought of as a graph drawn on a SPHERE—each corner of the polyhedron is a vertex, and each edge of the figure is an edge of the graph. If a polyhedron consists of only t triangular faces, then $3t$ counts the number of edges of the figure twice (each triangle has three edges, and each edge borders two triangles). Thus $3t = 2e$, where e is the number of edges. This shows that t must be an even number. We have:

It is impossible to cover a sphere with an odd number of triangles.

Euler also went on to show that for any graph drawn on a plane or a sphere (with no edges of the graph intersecting):

$$v - e + r = 2$$

where v is the number of vertices the graph possesses, e its number of edges, and r the number of regions defined by the graph—including the large "outer" region if the graph is drawn on a plane. This is called EULER'S FORMULA. If the graph is drawn instead on a TORUS, this formula is modified to read: $v - e + r = 0$.

Graph theory is a discipline under intense continued study. Its many applications vary from the purely theoretical to the very concrete and practical. Routing problems, information-flow problems, and electronic circuit design, for example, can all be effectively analyzed and refined through the study of this field.

See also EULERIAN PATH/CIRCUIT; HAMILTONIAN PATH/CIRCUIT; THREE-UTILITIES PROBLEM; TOPOLOGY; TOURNAMENT; TRAVELING-SALESMAN PROBLEM.

greatest common divisor (greatest common factor, highest common factor)   The largest FACTOR common to a given set of integers is called the greatest common divisor of those integers. For example, the numbers 72, 120, and 180 have factors 1, 2, 3, 4, 6, and 12 in common, with 12 being the greatest common divisor. We write gcd(72, 120, 180) = 12.

One can find the greatest common divisor of two or more integers simply by listing the factors of each integer and identifying the largest factor they have in common. Alternatively, if the prime decompositions of the integers are known, then their greatest common divisor can be determined as the product of the primes they have in common. For example, noting that:

$$72 = 2 \times 2 \times 2 \times 3 \times 3$$
$$120 = 2 \times 2 \times 2 \times 3 \times 5$$
$$180 = 2 \times 2 \times 3 \times 3 \times 5$$

we see that these three numbers share, as prime factors, two 2s and one 3. Their greatest common divisor is thus gcd (72,120,180) = $2 \times 2 \times 3 = 12$. (This second approach shows that any common factor of a given set of integers is a divisor of the greatest common factor.)

The EUCLIDEAN ALGORITHM can be used to find the greatest common divisor of two integers if either of these methods is infeasible. Repeated use of the Euclidean algorithm will find the greatest common divisor of more than two integers.

The Euclidean algorithm also shows that it is always possible to write the greatest common divisor of two integers a and b as a linear combination of a and b, that is, it is always possible to find integers x and y so that:

$$\gcd(a,b) = ax + by$$

Similarly, the greatest common factor of any finite set of integers $a_1, a_2, \ldots, a_n$ can be expressed as a linear combination of the form $\gcd(a_1, a_2, \ldots, a_n) = a_1 x_1 + a_2 x_2 + \ldots + a_n x_n$. In our example:

$$12 = \gcd(72, 120, 180)$$
$$= 72 \times (1) + 120 \times (1) + 180 \times (-1)$$

See also FUNDAMENTAL THEOREM OF ARITHMETIC; JUG-FILLING PROBLEM; RELATIVELY PRIME.

Greek alphabet To honor the mathematical scholars of Greek antiquity, mathematicians today often use letters of the Greek alphabet to represent variables and symbols in equations. Typically, lowercase letters are used to represent variables (such as angles, COMPLEX NUMBERS, and quantities studied in STATISTICS), and uppercase letters are used for standard arithmetical and statistical operations. The uses can vary from author to author, however.

The following table lists the letters of the Greek alphabet along with the common uses of some characters.

See also GREEK MATHEMATICS.

Greek mathematics The ancient Greeks of ca. 600 B.C.E. to ca. 480 C.E. set the current standards of logical rigor in mathematics. Although many ancient cultures practiced and developed mathematics, it was the Greeks who developed the explicit art of "proof" and explored the power of pure deductive reasoning to its fullest.

Upper Case	Lower Case	Name	Pronunciation	Use
A	α	alpha	AL-fuh	α: often denotes an angle
B	β	beta	BAY-tuh	β: often denotes an angle
Γ	γ	gamma	GAM-uh	γ: often denotes an angle Γ: the GAMMA FUNCTION
Δ	δ	delta	DEL-tuh	δ: a small quantity (EPSILON-DELTA DEFINITION) Δ: denotes change
E	ε	epsilon	EP-sil-on	ε: a small quantity (EPSILON-DELTA DEFINITION)
Z	ζ	zeta	ZAY-tuh	ζ: the ZETA FUNCTION
H	η	eta	AY-tuh	
Θ	θ	theta	THAY-tuh	θ: often denotes an angle (POLAR COORDINATES)
I	ι	iota	eye-OH-tuh	ι: a small quantity
K	κ	kappa	KAP-uh	
Λ	λ	lambda	LAM-duh	λ: wavelength (in physics)
M	μ	mu	MYOO	μ: denotes "microns"; Möbius function
N	ν	nu	NYOO	ν: frequency (physics)
Ξ	ξ	xi	kuh-SEYE	
O	o	omicron	OM-ee-KRON	
Π	π	pi	PIE	π: the ratio of a circle to its diameter Π: (infinite) product
P	ρ	rho	ROH	ρ: radius of a sphere (SPHERICAL COORDINATES)
Σ	σ	sigma	SIG-ma	σ: STANDARD DEVIATION Σ: SUMMATION
T	τ	tau	TAU	
Υ	υ	upsilon	OOP-si-LON	
Φ	φ	phi	FEE	φ: often denotes an angle (SPHERICAL COORDINATES)
X	χ	chi	K-EYE	χ: CHI-SQUARED TEST
Ψ	ψ	psi	SIGH	ψ: wave function (physics)
Ω	ω	omega	oh-MAY-guh	ω: a complex number; the first transfinite ORDINAL NUMBER.

We should mention that when speaking of "Greek mathematics," historians include any mathematician who wrote in the Greek language and followed the Greek tradition of mathematical thought. Greek was the common language of the Mediterranean world during ancient times, and many intellectuals from different parts of that region are today considered Greek scholars. For instance, the great Archimedes was from Syracuse, now a part of Italy, and EUCLID (ca. 300–260 B.C.E.) is believed to have lived in Alexandria, Egypt.

There are very few original records of Greek work. Initially, knowledge was transmitted only orally from teacher to student. Around 450 B.C.E. the Greeks adopted the ancient Egyptian practice of writing on papyrus scrolls. Unfortunately, papyrus—a grasslike plant grown in the Nile Delta region—decays rapidly away from the exceptionally dry climate of Egypt. The Greeks combated this problem by repeatedly making copies of their works but, because of the effort involved, copied only those pieces they deemed of utmost importance. The first mathematical work preserved and honored this way was Euclid's masterpiece THE ELEMENTS of ca. 300 B.C.E. Historians have had to rely on commentary made by later scholars to deduce what was accomplished mathematically before the time of Euclid.

Greek scholars approached all of mathematics through the study of GEOMETRY. Even their work on the properties of whole numbers, ratios, and proportions, as well as mechanics and astronomy was done in a geometric style. A "number," for instance, was literally a line segment, and a "ratio" was understood in terms of COMMENSURABLE segments. It is interesting to note that Greek scholars took careful steps to avoid speaking directly of the infinite. (The fifth-century B.C.E. paradoxes on the nature of motion and the infinitely small developed by ZENO OF ELEA deeply affected Greek thinking.) For instance, Euclid stated that any line segment could be extended to any arbitrary length, but never spoke of lines that were infinitely long. In EUCLID'S PROOF OF THE INFINITUDE OF PRIMES, Euclid stated that from any finite list of PRIME numbers one can always construct one more, but never spoke of the set of primes as infinite.

Many historians regard THALES OF MILETUS (ca. 625–547 B.C.E.) as the first Greek mathematician of note. Commentaries suggest that Thales identified, and proved, seven key geometric propositions, including

that the base angles of an ISOSCELES TRIANGLE are always equal and that the inscribed angle from the diameter of a CIRCLE is always a right angle, for instance. The great scholar and mystic PYTHAGORAS lived a century later, and he and his followers are credited with the discovery of the famous result about right triangles (today called PYTHAGORAS'S THEOREM) and the discovery of IRRATIONAL NUMBERS. A great deal of mystery surrounds the life and legend of Pythagoras. He founded a semireligious sect called the Pythagorean Brotherhood (women were equal members) based on certain mystic significances ascribed to whole numbers and their ratios.

The great philosopher PLATO (428–348 B.C.E.) wrote a great deal about mathematics in his famous dialogues, demonstrating a deep personal respect for the subject. The five regular polyhedra—the PLATONIC SOLIDS—are named in his honor. In his philosophical treatises, Plato used the example of mathematics as something that cannot be discovered by the senses, but can nonetheless be discovered by the power of logical reasoning. He also believed mathematics to be an essential part of a cultured person's education. Philosopher ARISTOTLE (384–322 B.C.E.) adopted the same view and used mathematics as examples in his development of FORMAL LOGIC and his analysis of ARGUMENTs.

Today, the Greek scholar Euclid is considered to be the most influential mathematics scholar of all time. In his famous work *The Elements,* Euclid collated all mathematical knowledge known at his time into a single tome. Although an impressive feat, it was the organization of the text that had the greatest impact. Beginning with a small collection of "self-evident truths," Euclid showed that all mathematical knowledge of his time could be deduced by pure logical reasoning alone. This work demonstrated the power of the mind and set the model for all mathematical research in the future. Mathematicians today still work to the standards of rigor as set by Euclid. Next to the Bible, Euclid's *The Elements* is the most widely published book of all time.

After producing *The Elements,* Euclid continued work on the CONIC SECTIONS, on optics, and on general problems in geometry. He continued interest in CONSTRUCTIBLE numbers and no doubt contemplated the classic Greek problem of SQUARING THE CIRCLE. (In *The Elements* Euclid had demonstrated general procedures for squaring arbitrary polygonal figures.) This challenge, as well as the problems of TRISECTING AN ANGLE and DUPLICATING THE CUBE, spurred a great deal of significant further research in mathematics for centuries to come.

ARCHIMEDES OF SYRACUSE (ca. 287–212 B.C.E.) solved the problem of squaring the parabola, as well as made significant advances in computing the areas and volumes of curved figures and solids. (He also "solved" the problem of squaring the circle by making use of his ARCHIMEDEAN SPIRAL. Unfortunately, his method went beyond the use of a straightedge and compass alone, and so is not a permissible solution to the original problem.)

APOLLONIUS OF PERGA (ca. 262–190 B.C.E.) continued work on conic sections and is credited for properly defining an ELLIPSE, a HYPERBOLA, and a PARABOLA. Around the same time, Greek astronomer Hipparchus wrote a table of "chord values" (the equivalent to a modern table of sine values), which he used to solve astronomical problems. This represented the beginning development of TRIGONOMETRY in Greek mathematics, but also marked an end of fervent mathematical development in the Greek tradition. For the five centuries that followed, new developments were limited to straightforward advances in astronomy, trigonometry, and algebra, with just a few notable exceptions.

In the second century C.E., Greek astronomer CLAUDIUS PTOLEMY corrected and extended Hipparchus's table and clarified the mathematics that is used to produce such a table. He is also known as one of the first scholars to make a serious attempt at proving Euclid's PARALLEL POSTULATE. In the third century, DIOPHANTUS OF ALEXANDRIA produced his famous text *Arithmetica* (Arithmetic), from which the study of DIOPHANTINE EQUATIONS was born. In the mid-fourth century, the enthusiastic PAPPUS OF ALEXANDRIA attempted to revive interest in ardent mathematical research of the Greek style. He produced his treatise *Synagoge* (Collections) to act as a commentary and guide to all the geometric works of his time and included in it a significant number of original results, extensions of ideas, and innovative shifts of perspective. Unfortunately, he did not succeed in his general goal. After Pappus, of note is HYPATIA OF ALEXANDRIA (370–415), the first woman to be named in the history of mathematics, credited for writing insightful commentaries on the works of

Apollonius and Diophantus, and PROCLUS (ca. 410–485), who is noted for his detailed commentary on the work of Euclid and his own attempt to prove the parallel postulate.

The beginning of the fifth century marks a clear end to the tradition of Greek mathematics.

See also ARCHYTAS OF TARENTUM; DEDUCTIVE/INDUCTIVE REASONING; ERATOSTHENES OF CYRENE; EUCLID'S POSTULATES; EUCLIDEAN ALGORITHM EUCLIDEAN GEOMETRY; EUDOXUS OF CNIDUS; HERON OF ALEXANDRIA; HIPPASUS OF METAPONTUM; HIPPOCRATES OF CHIOS; HISTORY OF EQUATONS AND ALGEBRA; MENELAUS OF ALEXANDRIA; PAPPUS'S THEOREMS; PTOLEMY'S THEOREM; PYTHAGOREAN TRIPLES; THEODORUS OF CYRENE; ZENO'S PARADOXES.

Green, George (1793–1841)

Green, George (1793–1841) British *Calculus* Born in July 1793 (his exact birth date is not known) in Nottingham, England, George Green is remembered today for his influential 1828 paper "Essay on the Application of Mathematical Analysis to the Theory of Electricity and Magnetism," in which he developed the notion of "potential" and proved a fundamental mathematical result today known as Green's theorem.

Green left school at the age of nine and worked in his father's bakery for the next 30 years. Historians do not know how Green developed an understanding of mathematics, nor how he had access to current work in the field, but in 1828 he published one of the most important scientific papers of the time. Apart from advancing the mathematical understanding of electromagnetism, important for physicists, Green also established a fundamental mathematical technique for computing CONTOUR INTEGRALS and DOUBLE INTEGRALS. The famous theorem that bears his name states that if a region R in the xy-plane is bounded by a curve C, and if functions $P(x,y)$ and $Q(x,y)$ have continuous PARTIAL DERIVATIVES, then:

$$\int_C P(x,y)\,dx + Q(x,y)\,dy = \iint_R \left(\frac{\partial Q}{\partial x} - \frac{\partial P}{\partial y} \right) dA$$

This result appears in every multivariable calculus textbook of today.

After reading the famous 1828 piece, mathematician Sir Edward Bromhead invited Green to continue work in the field of mathematical physics, and over the next six years they together produced three papers on the topics of electricity and hydrodynamics. At the advice of Bromhead, Green entered Cambridge University in 1833, at age 40, to start an undergraduate degree in mathematics. Green published six more papers after completing the program. He died on May 31, 1841, in Nottingham.

Gregory, David (1659–1708) British *Calculus, Geometry* Born on June 3, 1659, in Aberdeen, Scotland, David Gregory is remembered for his expository writings. In 1684 he published *Exercitatio geometrica* (Exercises in geometry), outlining many of the results of JAMES GREGORY, his uncle, on infinite SERIES. He also published some of SIR ISAAC NEWTON's results on mathematics and astronomy, and he was the first to teach the "modern" Newtonian theories at Cambridge. In 1703 he issued the first-ever complete collection of all of EUCLID's works.

Gregory studied at Marischal College, part of the University of Aberdeen, and accepted a professorship of mathematics at the University of Edinburgh in 1683. He lectured on the topics of mathematics, mechanics, and hydrostatics. In 1691 he was elected Savilian Professor at Oxford, the same year he was awarded membership to the ROYAL SOCIETY of London.

In 1702 Gregory published *Astronomiae physicae et geometricae elementa* (Astronomical physics and elements of geometry), which was a popular account of Newton's work, and a year later, his edition of the collected works of Euclid. Gregory strongly supported Newton in the debate over whether it was he, or German mathematician GOTTFRIED WILHELM LEIBNIZ, who had first discovered CALCULUS.

Gregory also completed his own work on the study of series, and published work on the topic of optics. He died on October 10, 1708, in Berkshire, England.

Not noted as an outstanding mathematician, Gregory is remembered primarily for the role he played in preserving the papers and recording the verbal communications passed to him by his uncle James Gregory and by Sir Isaac Newton.

Gregory, James (1638–1675) British *Calculus* Born in November 1638 (the exact date is not known), Scottish mathematician James Gregory independently discov-

ered and explained important concepts in CALCULUS before the subject was later fully developed by founders SIR ISAAC NEWTON and GOTTFREID WILHELM LEIBNIZ. Gregory was aware of the FUNDAMENTAL THEOREM OF CALCULUS, was the first to distinguish between convergent and divergent SERIES, and knew of TAYLOR SERIES 40 years before BROOK TAYLOR published his results. In mathematics he is remembered, in particular, for the series that bears his name. Outside of mathematics, Gregory is best remembered for his theoretical description of a revolutionary type of reflecting telescope.

Gregory was home-schooled in mathematics before undertaking studies at Marischal College in Aberdeen. He studied the mathematics of optics and in 1661 published a text on the subject, *Optica promota* (The advancement of optics), written in a purely mathematical style. Beginning with five postulates and 37 definitions, Gregory developed the theory of reflection and refraction of light in a systematic and rigorous manner, culminating with the description of a new principle for the construction of a telescope.

In 1664 Gregory traveled to Italy to stay at the University of Padua to pursue interests in mathematics. During this time Gregory developed the foundations of "infinitesimal geometry," the details of which he published as *Geometriae pars universalis* (The universal part of geometry). Today we would describe this book as a systematic treatment of topics in calculus.

Gregory was the first to discuss the convergence and divergence of series, work which led him in 1671 to the discovery of series expansions of functions. He refrained from publishing his discoveries, however, having heard a rumor that Newton may have already developed similar results. (Actually Newton was not aware of the key theorem needed to put this theory on a sound footing and so was unable to develop the theory.) Brook Taylor later discovered the same key result and published his description of the topic in 1715.

Throughout his short life, Gregory maintained an interest in astronomy. He died at the age of 36 in late October 1675, shortly after suffering a stroke he incurred while observing the moons of Jupiter with some students. Gregory did not achieve fame as a mathematician during his life. It is only in retrospect that historians today realize the important influence he had in helping Newton develop his ideas.

See also GREGORY SERIES.

Gregory series (Leibniz's series) Named after the Scottish astronomer and algebraist JAMES GREGORY (1638–75), the MACLAURIN SERIES for arctan(x) is sometimes called the Gregory series:

$$\arctan(x) = x - \frac{x^3}{3} + \frac{x^5}{5} - \frac{x^7}{7} + \cdots$$

This expression is valid for $-1 \le x \le 1$. Placing $x = 1$ into the formula yields the following remarkable formula for PI:

$$\frac{\pi}{4} = 1 - \frac{1}{3} + \frac{1}{5} - \frac{1}{7} + \cdots$$

Often the name "Gregory series" is used to mean this particular expression.

See also INVERSE TRIGONOMETRIC FUNCTIONS; MADHAVA OF SANGAMAGRAMMA; TAYLOR SERIES.

group Research in pure mathematics is motivated by one question: what makes mathematics work the way it does? By identifying the key principles that underpin one type of mathematical system—be it geometry and symmetry, or numbers and arithmetic—mathematicians establish connections between disparate fields: known facts about any one system satisfying a set of basic principles translate immediately to analogous facts about a second system satisfying analogous principles. For example, a study of arithmetic shows that the operation of addition satisfies the same four basic principles as multiplication. Thus any known fact about addition is accompanied by a known fact about multiplication (and this corresponding result about multiplication consequently requires no new proof.)

Motivated by the workings of arithmetic, mathematicians define a group to be any set, often denoted G, whose elements can be combined in some way that mimics the addition or the multiplication of the integers. Specifically, if we denote the result of combining the elements a and b of G by the symbol $a*b$, then G is a group if the following four axioms hold:

1. *Closure:* For all a and b in G, the element $a*b$ is also a member of G.

2. *Associativity:* For all a,b,c in G we have: $a*(b*c) = (a*b)*c$.

3. *Existence of an identity:* There is an element e in G so that $a*e = a = e*a$ for all a in G.

4. *Existence of inverses:* For any a in G there is an element b in G such that $a*b = e$ and $b*a = e$.

To honor the work of Norwegian scholar NIELS HENRIK ABEL (1802–29), mathematicians call G "Abelian" if a fifth axiom also holds:

5. *Commutativity:* For all a and b in G we have: $a*b = b*a$.

These axioms do indeed capture the working principles behind both addition and multiplication. For example, interpreting "*" as addition with "e" as the number zero, the set of all integers satisfies the axioms of an Abelian group. (In this setting the inverse of an integer a is usually denoted $-a$.) The set of all real numbers with zero removed forms an Abelian group under multiplication. In this context, $*$ is interpreted as the product operation, "e" is the number 1, and the inverse of an element a is the number $1/a$.

A group could be abstract. For example, the set of four elements $G = \{e,a,b,c\}$ is an Abelian group under an operation $*$ given as follows:

*	e	a	b	c
e	e	a	b	c
a	a	e	c	b
b	b	c	e	a
c	c	b	a	e

The set of elements $G = \{1, -1, i, -i, j, -j, k, -k\}$ with group operation $*$, given by multiplication as QUATERNIONs, is an example of a non-Abelian group. The set of all 2×2 invertible matrices with real entries is also a non-Abelian group under the operation of MATRIX multiplication.

GROUP THEORY is the study of the general structure of groups and all the results that follow from the four (or five) basic axioms. The subject is incredibly rich, and many mathematicians today devote their entire research careers to the further development of this topic.

Group theory has profound applications to physics and science. Any physical system that possesses symmetry of some kind, for instance, can be analyzed through the tools and techniques of this topic. For example, the six symmetries of an equilateral triangle form a (non-Abelian) group: there are two rotations, a clockwise and a counterclockwise rotation of 60°; three reflections, one about each altitude of the triangle; and an identity operation corresponding to conducting no action at all. By declaring the operation "*" to be the effect of performing one action followed by the other, one can check that all four group axioms hold for this system. (The fifth axiom, however, does not hold, since the performance of rotation followed by a reflection, for instance, gives a result different from the performance of that same reflection followed by the rotation.) The analysis of symmetry plays an important role in crystallography and quantum mechanics. Researchers in these fields deem group theory an essential tool in their work.

The set of symmetries of any regular n-sided POLYGON forms a group called the nth dihedral group. It has $2n$ members consisting of n rotations (including the identity element) and n reflections. The set of rotations in and unto itself forms a group of just n elements. Notice that n is a factor of $2n$. In general, any subset H of a group G that itself satisfies the four axioms of a group is called a subgroup of G. French mathematician JOSEPH-LOUIS LAGRANGE (1736–1813) proved that if H is a subgroup of a group G, then the number of elements in H evenly divides the count of the number of elements in G. This result has interesting consequences when applied to systems of symmetry, or to groups arising from the study of NUMBER THEORY and MODULAR ARITHMETIC.

See also ABSTRACT ALGEBRA; FIELD; GENERAL LINEAR GROUP; HOMOMORPHISM; ISOMORPHISM; RING.

group theory   The general study of GROUPs and the results that follow from the basic axioms that define them is called group theory. Many of the key principles behind group theory were first identified by the German mathematician CARL FRIEDRICH GAUSS (1777–1855) in his studies of NUMBER THEORY and MODULAR ARITHMETIC. The development of a group theory as a subject in its own right, however, is usually attributed to the young French mathematician ÉVARISTE GALOIS (1811–32); who devised the innovative tools necessary to study solutions to algebraic equations in depth and from an abstract perspective.

By identifying the abstract principles that make algebra and arithmetic work the way they do, group theory provides a powerful tool for analyzing any mathematical system that satisfies the same basic axioms. Applying group theory to the symmetries of a physical system, for example, can often lead to important consequences in physics.

See also ABSTRACT ALGEBRA; FIELD; GROUP; RING.

H

Hadamard, Jacques (1865–1963) French *Number theory, Analysis, Mathematical physics* Born on December 8, 1865, in Versailles, France, Jacques Hadamard is remembered as one of the two mathematicians who independently proved the famous PRIME NUMBER THEOREM first conjectured by the German mathematician CARL FRIEDRICH GAUSS. Belgian scholar CHARLES-JEAN DE LA VALLÉE-POUSSIN was the second mathematician to prove the result.

Hadamard received a doctoral degree in 1892 from the École Normale Supérieure after completing a dissertation on complex functions and TAYLOR SERIES. The same year he also completed work on the Riemann ZETA FUNCTION and the Riemann hypothesis, earning him the Grand Prix des Sciences Mathématique from the institution. Hadamard soon came to also realize that, with the recent developments in the field of complex functions, all the necessary pieces were now in place to develop a proof of the famous outstanding prime number conjecture made by Gauss. (Vallée-Poussin independently made the same realization.) Hadamard presented his proof of the famous result in 1896. Later that same year Hadamard was appointed as professor of astronomy and rational mechanics at the University of Bordeaux.

Throughout his career Hadamard also made significant contributions to the field of matrix theory by identifying a class of matrices that can be used in COMBINATORICS to create "block designs" and have applications to PROBABILITY theory. He also studied VECTOR SPACE theory (defining the term *functional* for a linear function on a vector space) and contributed to the study of mathematics education. Hadamard wrote over 300 scientific papers. In 1906 he was elected president of the French Mathematical Society, and 3 years later was appointed chair of the Collège de France. He stayed in that position for only 3 years, before accepting the position as professor of analysis at the École Polytechnique. He was elected to the Paris Academy of Sciences in 1912.

Hadamard died in Paris, France, on October 17, 1963. His proof of the prime number theorem is considered his most outstanding achievement.

half-plane The plane that lies on one side of a given line is called a half-plane. If the points on the line itself are considered part of the region, then we say that the half-plane is "closed." An "open" half-plane excludes the points on the line.

If the equation of the given line is $ax + by = c$, then the set of points (x,y) that satisfy $ax + by > c$ form an open half-plane on one side of the line, and those points (x,y) satisfying $ax + by < c$ form an open half-plane on the other side. If c is positive, then this second inequality represents the half-plane that contains the origin (we have $a \cdot 0 + b \cdot 0 = 0 < c$), and if c is negative, then the first inequality contains the origin. If c equals zero, one must substitute in different values for x and y to determine which inequality represents which half-plane. The inequalities $ax + by \geq c$ and $ax + by \leq c$ represent closed half-planes.

In three-dimensional space, a half-space is the region of space that lies on one side of a plane. The half-space can be either closed or open according to whether or not points of the plane should be considered as part of this region.

Hall's matching theorem (Hall's marriage theorem)

In 1935 English mathematician Philip Hall established the following result, now known as Hall's matching theorem:

> Given n sets $A_1, A_2, \ldots, A_n$ with the property that the union of any k of them ($1 \le k \le n$) contains at least k distinct elements, it is always possible to select n distinct objects, one from each set.

The condition placed on these sets is not trivial—some sets could be empty, or the same element could appear in more than one set, for example.

The theorem is certainly true for the case $n = 1$: a single set satisfying the condition of the theorem contains at least one element. Two sets satisfying the conditions of the theorem together contain at least two distinct elements. As neither set is allowed to be empty, one set contains one element, and the other another distinct element. Thus the theorem also holds true for the case $n = 2$. One can build up a general proof of the theorem using a proof by INDUCTION.

The validity of the theorem can be demonstrated with an amusing game of solitaire: divide a shuffled deck of cards into 13 piles of four cards. The challenge is to select an ace from one pile, a two from another, a three from a third, all the way down to king from a 13th pile. Hall's matching theorem ensures that this game can always be won, no matter how the cards are shuffled. (Think of each pile as a set containing one, two, three, or four elements—the distinct denominations that appear in that pile. Among any k piles it must be the case that at least k distinct denominations appear.)

Hall gave another interpretation to his theorem (explaining the alternative name to the result):

> Suppose n women each list the names of the men they would like to marry. As long as any k women mention at least k distinct names among them, $1 \le k \le n$, then it is possible to make satisfactory matches for all.

See also SEMI-MAGIC SQUARE.

halting problem

In 1936 computer theorist Alan Turing contemplated whether or not it would ever be possible to write a computer program that could read any other program and determine whether that program will come to a stop or will run forever (by falling into an infinite loop, for example). This question has since become known as the halting problem. Turing concluded that such a program could not possibly exist. He reasoned via a clever ARGUMENT of SELF-REFERENCE:

> Suppose a program HALT(P) exists that can read a computer program P and print *yes* or *no* according to whether P will or will not halt. Consider then another program, which we will call TROUBLE, that takes a program P and does the following:

TROUBLE (P):
If HALT(P) = "yes" then perform an infinite loop.
If HALT(P) = "no" then halt.

> That is, TROUBLE takes a program P and goes into an infinite loop if P is a program that halts, and it halts if P is a program that does not. Now ask: what does TROUBLE (TROUBLE) do? We have that TROUBLE halts if TROUBLE does not halt, and does not halt if it does! This absurdity shows that no such program HALT(P) could exist.

There are technical difficulties with this argument (one must be careful to properly distinguish between the roles of a program and the *input* of a program, for example), but Turing was able to overcome these concerns and show that this argument is fundamentally sound.

Hamilton, Sir William Rowan (1805–1865)

Irish *Algebra, Graph theory, Number theory, Mathematical physics* Born on August 4, 1805, in Dublin, Ireland, William Rowan is generally considered Ireland's greatest mathematician of all time. He is remembered for his development of an entirely new algebraic system, the QUATERNIONS, which, seven decades later, proved to be crucial for the development of quantum mechanics and the mathematical physics of the internal structure of an atom.

Hamilton was a child prodigy, mastering 12 different languages by the age of 13. During the teen years, he read the works of EUCLID in the original Greek, the works of SIR ISAAC NEWTON in Latin, and the works of PIERRE-SIMON DE LAPLACE in French. He found a subtle error in Laplace's classic text *Mécanique céleste,* and wrote a letter to the astronomer royal of Ireland, John Brinkley, explaining the error and how it should be corrected. Brinkley immediately recognized Hamilton's genius as a rising mathematician and publicly dubbed him the "first mathematician of his age."

Hamilton entered Trinity College at the age of 18 to study optics and mathematics. His original work in these fields as an undergraduate, which included two papers "Systems of Right Lines in a Plane" and "Theory of Systems of Rays," was regarded as so significant and innovative as to warrant his immediate appointment as a professor of astronomy at the college before the completion of his basic degree.

Along with his work in optics, Hamilton made significant contributions to the field of GRAPH THEORY and to the algebra of COMPLEX NUMBERS, publishing results on the latter topic in his 1837 paper "Preliminary and Elementary Essay on Algebra as the Science of Pure Time." In 1842 he took on the difficult challenge of trying to create an algebraic system for three-dimensional space that had the same algebraic properties as the complex numbers in two-dimensional space. [A point (x,y) in the plane can be matched with the complex number $x + iy$. This thus provides a means to "multiply" to two points in space: $(x_1, y_1) \cdot (x_2, y_2) = (x_1x_2 - y_1y_2, x_1y_2 + x_2y_1)$.] Although he was never able to find a solution to this "multiplication of triples" problem, his efforts did lead him to the remarkable discovery of a different type of number system suitable for four-dimensional space. He called this system the quaternions, and found some surprising connections to mathematical physics. In particular he observed that each quaternion corresponds naturally to a physical transformation in three-dimensional space and that the multiplication of two quaternions matches perfectly with the composition of the two physical transformations they represent. Thus the geometry of three-dimensional physical space can be reduced to the algebraic study of the algebra of quaternions. Hamilton was convinced his work would revolutionize mathematical physics. Although it does have applications to the field today, sadly, his work did not have the impact he hoped.

Hamilton received many awards throughout his life, most notably a knighthood in 1835 and election to the National Academy of Sciences in the United States as its first foreign member. He wrote poetry for solace throughout his life, and argued publicly that the language of mathematics is just as artistic as the language expressed through poetry. His close friend poet William Wordsworth (1772–1834) did not agree. Hamilton died near Dublin on September 2, 1865.

Hamiltonian path/circuit   *See* GRAPH THEORY.

ham-sandwich theorem   As a generalization of the INTERMEDIATE-VALUE THEOREM and the two-pancake theorem that follows from it, mathematicians have proved the following result, called the ham-sandwich theorem:

> Given any three objects sitting in three-dimensional space, there exists a single plane that simultaneously slices the volume of each object exactly in half.

For example, there is a single plane that simultaneously slices the Eiffel tower, the planet Neptune, and this book each precisely in half by volume. The theorem gains its name from the following interpretation:

> It is possible, in a single planar cut, to divide each of two pieces of bread and a slab of ham into two pieces of equal volume. This is possible no matter the shape of the food pieces and no matter where in space the three items are placed.

The result generalizes to higher dimensions:

> Given any N objects sitting in N-dimensional space, it is always possible to find an $(N - 1)$-dimensional "plane" that simultaneously slices the "volume" of each object in half.

(With $N = 2$, this is the two-pancake theorem.)

handshake lemma   This amusing result states that, at any instant, the number of people on this planet, living or deceased, who have taken part in an odd total number of handshakes is necessarily even. This lemma can be proved with the aid of GRAPH THEORY.

Hardy, Godfrey Harold (1877–1947) British *Analysis, Number theory* Born on February 7, 1877, in Cranleigh, England, eminent mathematician Godfrey Hardy is remembered for his significant contributions to the fields of NUMBER THEORY, inequalities, and to the study of the Riemann ZETA FUNCTION and the Riemann hypothesis. Hardy also encouraged the Indian mathematician SRINIVASA AIYANGAR RAMANUJAN to come to England, and collaborated with him for five years to produce a number of significant results, most notably on the theory of PARTITIONs. In 1940 Hardy published *A Mathematician's Apology,* which remains, to this day, one of the most vivid and eloquent descriptions of how a mathematician thinks.

Hardy entered Trinity College, Cambridge, in 1896 to pursue an advanced degree in mathematics. After publishing a number of papers on the topics of INTEGRALs, SERIES, and general topics of ANALYSIS, Hardy published, in 1908, the undergraduate textbook *A Course of Pure Mathematics,* which explained, in a rigorous manner, the concepts of function, LIMIT, and the elements of analysis. This work was very influential and is said to have transformed the entire nature of university teaching.

In 1911 Hardy began a 35-year-long collaboration with English mathematician John Littlewood (1885–1977), leading to a change of focus in the field of number theory. Together they produced over 100 joint publications. This work also tied in nicely with the research Hardy was conducting in 1914 with Ramanujan, also on topics in number theory.

Hardy was recognized as an important figure in mathematics. In 1910 he was elected a fellow of the ROYAL SOCIETY of London, received the Royal Medal of the Society in 1920, the De Morgan Medal of the Society in 1929, and the Sylvester Medal of the Society in 1940 for his work in pure mathematics. Seven years later he was awarded the Copley Medal of the Society for his contributions to the field of analysis.

Hardy died in Cambridge, England, on December 1, 1947. He is generally recognized as the leading English pure mathematician of his time. His 1932 book *The Theory of Numbers,* cowritten with E. M. Wright, is considered a classic text and is still used by graduate students and researchers in the field of number theory today.

harmonic function A function defined at a discrete number of locations is said to be harmonic if the value of the function at any one location equals the average of the function values at its neighboring locations. For example, if 10 students sit in a circle, then their "age function" would be harmonic if the age of any one student equals the average age of his or her two neighbors. (A little thought shows that this is only possible if the age of each student is the same. A harmonic function for points arranged in a circle must be constant.)

Harmonic functions play an important role in the study of RANDOM WALKs and calculating odds in gambling. Imagine, for example, a gambler playing a simple game of tossing a coin to either win a dollar or to lose a dollar. We ask: with \$3 in hand, what are her chances of reaching the \$10 mark before going broke? To compute this, let $P(N)$ denote the PROBABILITY of achieving the goal starting with N dollars in hand. Clearly $P(0) = 0$ and $P(10) = 1$. We wish to compute $P(3)$.

The key is to note that $P(N)$ is a harmonic function on its 11 values zero through 10. This follows because there are equal chances for the gambler to lose or win a dollar and thus to next play with either $N - 1$ or $N + 1$ dollars in hand. Consequently:

$$P(N) = \frac{1}{2}P(N-1) + \frac{1}{2}P(N+1) = \frac{P(N-1)+P(N+1)}{2}$$

and so each quantity $P(N)$ is indeed the average of the values just preceding and succeeding it. Some thought shows that the values $P(0) = 0$ up to $P(10) = 1$ must be strictly increasing by equal intervals of one-tenth. The values $P(N)$ are thus $P(N) = \frac{N}{10}$. In particular, $P(3) = \frac{3}{10}$, showing that there is only a 30 percent chance that the gambler will achieve her goal before losing all her cash.

harmonic mean *See* MEAN.

harmonic sequence (harmonic progression) A SEQUENCE of numbers of the form $\frac{1}{a_1}, \frac{1}{a_2}, \frac{1}{a_3}, \ldots$ with integers $a_1, a_2, a_3, \ldots$ forming an ARITHMETIC SEQUENCE is called a harmonic sequence. The numbers $a_1, a_2, a_3, \ldots$ have a constant difference between them. For example, $1, \frac{1}{2}, \frac{1}{3}, \frac{1}{4}, \ldots$ is a harmonic sequence, as is $\frac{1}{4}, \frac{1}{7}, \frac{1}{10}, \frac{1}{13}, \ldots$ and $\frac{1}{25}, \frac{1}{50}, \frac{1}{75}, \frac{1}{100}, \ldots$. The word *harmonic* is used because the nth harmonic produced by a violin string is the tone produced by the string that is $1/n$ times as long.

The corresponding SERIES $\sum_{n=1}^{\infty} \frac{1}{a_n}$ for any harmonic series necessarily diverges. In the study of CONVERGENT SERIES, the comparison test shows that this must be the case. (Compare a series of the form $\sum_{n=0}^{\infty} \frac{1}{a+dn}$ with $\sum_{n=1}^{\infty} \frac{1}{n}$, which we know diverges.)

See also HARMONIC SERIES.

harmonic series The particular infinite sum $1 + \frac{1}{2} + \frac{1}{3} + \frac{1}{4} + \frac{1}{5} + \ldots$ is called the harmonic SERIES. The word *harmonic* is used because the nth harmonic produced by a violin string is the tone produced by the string that is $1/n$ times as long.

Even though the terms of this series approach zero, the series does not sum to a finite value. This can be seen by grouping the terms of the series into sections of length two, four, eight, 16, and so on, and making a simple comparison:

$$1+\frac{1}{2}+\frac{1}{3}+\frac{1}{4}+\frac{1}{5}+\cdots+\frac{1}{16}+\frac{1}{17}+\cdots+\frac{1}{32}+\frac{1}{33}+\cdots$$

$$= 1+(\frac{1}{2})+(\frac{1}{3}+\frac{1}{4})+(\frac{1}{5}+\cdots+\frac{1}{16})+(\frac{1}{17}+\cdots+\frac{1}{32})$$

$$+(\frac{1}{33}+\cdots$$

$$> 1+(\frac{1}{2})+(\frac{1}{4}+\frac{1}{4})+(\frac{1}{16}+\cdots+\frac{1}{16})+(\frac{1}{32}+\cdots+\frac{1}{32})$$

$$+(\frac{1}{64}+\cdots$$

$$= 1+(\frac{1}{2})+(\frac{1}{2})+(\frac{1}{2})+(\frac{1}{2})+(\frac{1}{2})+\cdots$$

$$= \infty$$

That the series diverges means that summing sufficiently many initial terms of the series will produce answers arbitrarily large (although it may take a large number of terms to do this). For example, summing the first four terms produces an answer larger than 2, the first 11 terms an answer larger than 3, the first 13,671 terms an answer larger than 10, and the first

1.53×10^{43} terms an answer larger than 100. Clearly the series diverges to infinity very slowly.

In 1734 LEONHARD EULER showed that, for large values of n, the nth PARTIAL SUM of the harmonic series can be well approximated by a LOGARITHM:

$$1+\frac{1}{2}+\frac{1}{3}+\cdots+\frac{1}{n} \approx \ln(n)+\gamma$$

where $\gamma \approx 0.577$ is a constant (called EULER'S CONSTANT) and the error in this approximation is no larger than $1/n$. (Notice that $\ln(n) + \gamma \to \infty$ as n grows. This again shows that the series diverges.)

The partial sums of the harmonic series are called the harmonic numbers, and are denoted H_n. The first 10 harmonic numbers are:

$$H_1 = 1;$$
$$H_2 = \frac{3}{2}; \quad H_3 = \frac{11}{6} = \frac{11}{2 \cdot 3};$$
$$H_4 = \frac{25}{4 \cdot 3}; \quad H_5 = \frac{137}{4 \cdot 15}; \quad H_6 = \frac{49}{4 \cdot 5}; \quad H_7 = \frac{363}{4 \cdot 35};$$
$$H_8 = \frac{761}{8 \cdot 35}; \quad H_9 = \frac{7129}{8 \cdot 315}; \quad H_{10} = \frac{7381}{8 \cdot 315}.$$

One can use an induction argument to show that if $2^k \leq n < 2^{k+1}$, then the denominator of H_n (written in reduced form) is a multiple of 2^k. Consequently, no denominator (except for the first) can be 1. This proves:

No harmonic number, other than the first, is an integer.

The divergence of the harmonic series solves the amusing rubber-band problem:

An infinitely tiny ant starts at one end of a rubber band, 1 ft long, and crawls a distance of 1 in. toward the other end. It then pauses, and the band is stretched 1 ft longer (to a total length of 2 ft), carrying the ant along with it to the 2-in. position. The ant then crawls for another inch, to the 3-in. position, and pauses while the band is stretched another foot longer. This process of walking an inch and pausing while the band stretches a foot continues indefinitely. (We assume the band is infinitely elastic.) Will the ant ever make it to the end of the rubber band?

Note that in the first leg of the journey, the ant covers 1/12th of the length of the band, and that this proportion remains the same as the band is stretched. During the second leg of the journey, the ant covers now only 1/24th of the length of the band (1 in. of 24 in.), 1/36th of the length during the third length, and so on. Thus after n legs of the journey, the ant covers the fraction $\frac{1}{12}\left(\frac{1}{1}+\frac{1}{2}+\frac{1}{3}+\cdots+\frac{1}{n}\right)=\frac{H_n}{12}$ of the band. The ant reaches the finish only if H_n ever surpasses the value 12. As the harmonic series diverges, this must indeed be the case.

Although the harmonic series diverges, the ALTERNATING SERIES test from the study of CONVERGENT SERIES shows that the alternating harmonic series, given by

$$1-\frac{1}{2}+\frac{1}{3}-\frac{1}{4}+\frac{1}{5}-\frac{1}{6}+\cdots$$

converges. We can use Euler's approximation formula to find its value. First note:

$$
\begin{aligned}
S_{2n} &= 1-\frac{1}{2}+\frac{1}{3}-\frac{1}{4}+\cdots+\frac{1}{2n-1}-\frac{1}{2n} \\
&= \left(1+\frac{1}{2}+\frac{1}{3}+\cdots+\frac{1}{2n-1}+\frac{1}{2n}\right)-2\left(\frac{1}{2}+\frac{1}{4}+\cdots+\frac{1}{2n}\right) \\
&= \left(1+\frac{1}{2}+\frac{1}{3}+\cdots+\frac{1}{2n-1}+\frac{1}{2n}\right)-\left(\frac{1}{1}+\frac{1}{2}+\cdots+\frac{1}{n}\right) \\
&= \ln(2n)+\gamma-\ln(n)-\gamma+\text{error} \\
&= \ln 2+\text{error}
\end{aligned}
$$

where the error is no larger than $2/n$ and so converges to zero as n grows. Consequently, the even partial sums of the series approach the value In(2). The sum of an odd number of terms equals $S_{2n+1}=S_{2n}+\frac{1}{2n+1}=$ $S_{2n+1}=S_{2n}+\frac{1}{2n+1}=\ln 2+\text{error}+\frac{1}{2n+1}$ and so too approaches ln2 as n becomes large. Thus we have:

$$1-\frac{1}{2}+\frac{1}{3}-\frac{1}{4}+\frac{1}{5}-\frac{1}{6}+\cdots=\ln(2)$$

This example is often used to illustrate the difference between conditional and ABSOLUTE CONVERGENCE of a series. In that setting, it can also be used to provide an amusing "proof" that 1 equals 2.

See also MERCATOR'S EXPANSION.

helix A spiral-shaped curve sitting in three-dimensional space is called a helix. The name is the Greek word for a "spiral" or a "twist."

A cylindrical helix lies on a cylinder and cuts across straight lines drawn along the length of the cylinder at a constant angle α. A spiral staircase and the thread on a straight screw are examples of cylindrical helices. A conical helix is a spiral curve on a CONE, and a spherical helix is a spiral on a SPHERE that cuts lines of longitude at a constant angle.

A cylindrical helix has PARAMETRIC EQUATIONS: $x = a\cos t$, $y = a\sin t$, and $z = bt$, where a and b are constants, and t is the parameter. A conical helix is given by: $x = ae^t\cos t$, $y = ae^t\sin t$, and $z = e^t$.

Heron of Alexandria (ca.10–75 C.E.) Greek *Geometry, Number theory, Physics, Engineering* Sometimes called Hero, Heron of Alexandria is remembered in mathematics for his three-volume text *Metrica*, rediscovered in 1896. The work discusses and develops in great detail the principles of geometry, number, and numerical approximation. It also contains the earliest known proof of the famous formula that bears Heron's name. Outside of mathematics, Heron is best known for his contributions to mechanics and fluid mechanics.

Demonstrating a wide range of scientific interests, Heron wrote studies in optics, pneumatics (the study and use of gas and fluid pressures), astronomy, surveying techniques, and planar and solid geometry, but it is the work presented in *Metrica* that proves his genius as a mathematical intellect. Book I of this famous piece computes the areas of triangles, quadrilaterals, and regular polygons, as well as the surface areas of cones, prisms, spheres, and other three-dimensional shapes. His famous formula for the area of a triangle solely in terms of its side-lengths is presented in this section, along with a general method for computing the square root of a number to any prescribed degree of accuracy. (This procedure is today called HERON'S METHOD.

Book II of *Metrica* derives formulae for the volume of each PLATONIC SOLID, as well as the volumes of cones and spherical segments, and Book III extends EUCLID's study of geometric division. Heron also presents a method for determining the cube root of a number in this third volume.

Heron also wrote a number of important treatises on mechanics, many of which survive today. His text *Pneumatica* represents a careful (but, in places, inaccurate) study of pressure in fluids, along with a description of a collection of trick gadgets and toys illustrating specific scientific principles. He also describes designs for over 100 practical machines, including pneumatic pulleys and lifts, wind organs, coin-operated machines, fire engines, and steam-powered engines that operate in a way similar to today's jet engine.

Some of Heron's texts have the appearance of draft lecture notes, leading some historians to suspect that he may have taught at the famous Museum of Alexandria. Little is actually known of Heron's life.

Heron's formula (Hero's formula)   In his work *Metrica*, HERON OF ALEXANDRIA (ca. 100 C.E.) presented a formula for the AREA of a TRIANGLE solely in terms of the side-lengths of the triangle. Today known as Heron's formula, it reads:

$$area = \sqrt{s(s-a)(s-b)(s-c)}$$

where a, b, and c are the sides of the triangle, and $s = \frac{1}{2}(a + b + c)$ is its "semiperimeter." The formula is a special case of BRAHMAGUPTA'S FORMULA, discovered 500 years later.

Heron's formula can be proved as follows: if θ is the angle between the sides of length a and b, then the area of the triangle is given by $area = \frac{1}{2} ab \sin(\theta)$. The LAW OF COSINES asserts that $c^2 = a^2 + b^2 - 2ab \cos(\theta)$. Solving for $\sin(\theta)$ and $\cos(\theta)$, and substituting into the standard identity from TRIGONOMETRY, $\cos^2(\theta) + \sin^2(\theta) = 1$, yields, after some algebraic work, Heron's result.

See also BRETSCHNEIDER'S FORMULA; MEDIAN OF A TRIANGLE; QUADRILATERAL; TRIANGLE.

Heron's method (Hero's method)   In Book I of his volume *Metrica*, HERON OF ALEXANDRIA gives a method for approximating the SQUARE ROOT of a number. It works as follows:

> Estimate the value of the square root. Divide this guess into the number under consideration, and take the average of the result and the initial estimate. This will produce a better approximation to the square root.
>
> Repeated application of this method produces an estimate to any desired degree of accuracy.

In symbols, Heron claims that if x approximates the square root of a number N, then $\dfrac{x + \dfrac{N}{x}}{2}$ is a better approximation. For example, taking 3 as an approximation to the square root of 10, we obtain $\dfrac{3 + \dfrac{10}{3}}{2} \approx 3.1667$ as an improved estimate. Repeating the procedure yields $\dfrac{3.1667 + \dfrac{10}{3.1667}}{2} \approx 3.1623$ as an even better approximation. (In fact, $\sqrt{10} \approx 3.16227766$.)

To show why this method works, let $y = \dfrac{x + \dfrac{N}{x}}{2}$. Then $(y - \sqrt{N}) = \dfrac{1}{2x}(x - \sqrt{N})^2$, which shows that if the error $x - \sqrt{N}$ is small, then the error $y - \sqrt{N}$ will be even smaller. (We are assuming here that x is a value greater than 1.)

This method was known to the Babylonians of 2000 B.C.E. It is also equivalent to NEWTON'S METHOD when applied to the function $f(x) = x^2 - N$.

See also BABYLONIAN MATHEMATICS.

higher derivative   Taking the DERIVATIVE of the same function more than once, if permissible, produces the higher derivatives of that function. The first, second and third derivatives of a function $f(x)$ are denoted, respectively, $f'(x)$, $f''(x)$ and $f'''(x)$, and for $n \geq 4$, the nth derivative as $f^{(n)}(x)$. For example, the third derivative of $f(x) = x^4 + \sin x$ is $f'''(x) = 24x - \cos x$. This notation for the repeated derivative is due to JOSEPH-LOUIS LAGRANGE (1736–1813). GOTTFRIED WILHELM LEIBNIZ (1646–1716), coinventor of CALCULUS, used the notation $\dfrac{d^n f(x)}{dx^n}$ for the higher derivatives, and French mathematician Louis Arbogast (1759–1803) wrote $D^n f(x)$. All three notational systems are used today.

Hilbert, David (1862–1943) German *Formal logic, Geometry, Mathematical physics, Algebraic number theory*   Born on January 23, 1862, in Königsberg, Prussia (now Kaliningrad, Russia), mathematician David Hilbert is remembered as one of the founding fathers of 20th-century mathematics. In 1899 Hilbert published his famous *Grundlagen der Geometrie* (Foundations of geometry), in which he provided a completely rigorous axiomatic foundation of the subject clarifying the hidden assumptions that EUCLID had made in his development of the subject two millennia earlier. Moreover, Hilbert advanced the topic of FORMAL LOGIC and used his results to prove that his approach to geometry is CONSISTENT given that the arithmetic of the real numbers is free of contradictions. In 1900 Hilbert posed 23 problems to the mathematicians of the 20th century that he felt lay at the heart of vital mathematical research. Two of his problems were solved almost immediately, but the remaining 21 challenges did indeed stimulate important mathematical thinking. Many of his challenges are still unsolved today. Hilbert also greatly influenced the development of quantum theory in theoretical physics: his notion of a "Hilbert space" provided the right conceptual framework for the subject. Hilbert also made important contributions to the fields of special relativity and general relativity.

Hilbert received a doctorate of mathematics from the University of Königsberg in 1885 after writing a thesis in ABSTRACT ALGEBRA. He was appointed to a faculty position at the university, where he remained for 10 years before accepting the position as chair of mathematics at the University of Göttingen in 1895. Hilbert taught and worked at Göttingen for the remainder of his career.

His 1897 text *Zahlbericht* (Number theory) was hailed as a brilliant synthesis of current thinking in algebraic number theory, and the original results it contained were acknowledged as outstanding. Hilbert's abilities to grasp the subtleties of a sophisticated mathematical theory, develop penetrating insights, and provide new innovative and stimulating perspectives on a subject were apparent. Throughout his career Hilbert worked on a wide variety of disparate subjects, making groundbreaking contributions to each before moving on to the next. He published his famous work on EUCLIDEAN GEOMETRY in 1899.

In 1900 Hilbert was invited to address the Paris meeting of the International Congress of Mathematicians. During his speech he detailed 10 mathematical problems that he felt were of great importance. (He expanded the list to 23 problems when he published his address.) These problems include the CONTINUUM HYPOTHESIS, GOLDBACH'S CONJECTURE, a search for

the axiomatization of physics, and a search for a general algorithm for solving DIOPHANTINE EQUATIONS. Some important progress, and in many cases, complete solution, has been made on all the challenges posed except for one, the so-called Riemann hypothesis, which asks for the locations of the roots of the ZETA FUNCTION. This remains, perhaps, the most famous unsolved problem of today.

Later in life Hilbert worked on formal logic and on the foundations of theoretical physics. Between 1934 and 1939 he published two volumes of *Grundlagen der Mathematik* (Foundations of mathematics), cowritten with Paul Bernays (1888–1977), which were intended to develop a proof of the consistency of mathematics. (GÖDEL'S INCOMPLETENESS THEOREMS showed, however, that such a goal is unattainable.) His development of functional analysis provided the correct mathematical framework for the theory of quantum mechanics.

Hilbert received many honors throughout his career, including a special citation from the Hungarian Academy of Sciences in 1905. Upon his retirement in 1929, the city of Göttingen named a street after him, and the city of Königsberg, his birthplace, declared him an honorary citizen. He died in Göttingen on February 14, 1943. He is remembered for shaping the very nature of 20th-century research in the pure mathematics.

Hilbert's infinite hotel (Hilbert's paradox)   German mathematician DAVID HILBERT (1862–1943) observed that studies of the infinite can often lead to nonintuitive and surprising conclusions. His famous infinite-hotel paradox illustrates some of his ideas:

> Imagine a hotel with an infinite number of rooms. Suppose every room is occupied. Is it possible for the hotel to accommodate one more guest?

If the rooms are numbered 1, 2, 3, and so on, Hilbert pointed out that, despite the inconvenience, each existing guest can be moved from room n to room $n + 1$, thereby leaving room 1 free for a latecomer. There is indeed room for another single guest. One can go further: suppose a tourist bus, with an infinite number of tourists on board, arrives at the hotel. Moving each hotel guest from room n to room $2n$, instead, leaves all the odd-numbered rooms vacant for the infinite number of new arrivals.

If an infinite number of tourist buses, each containing an infinite number of tourists, arrive at the same time, one can still find room to accommodate the multitude of new guests:

Move all the existing hotel guests to the even-numbered rooms.

Place the tourists from the first bus into the rooms given by the powers of 3: 3, 9, 27,...

Place the tourists from the second bus into the rooms given by the powers of 5: 5, 25, 125,...

Next use all rooms numbered the powers of 7, of 11, of 13, and so on, down along the list of all the PRIME numbers. As, according to EUCLID'S PROOF OF THE INFINITUDE OF PRIMES, there are infinitely many primes, and no two powers of different primes are the same, this allocation scheme does the trick.

See also CARDINALITY; FUNDAMENTAL THEOREM OF ARITHMETIC; INFINITY; PARADOX; TRISTRAM SHANDY PARADOX.

Hindu-Arabic numerals   The numeral system we use today is called the Hindu-Arabic numeral system. Using a base-10 PLACE-VALUE SYSTEM, numbers are expressed via combinations of the symbols 0, 1, 2, 3, 4, 5, 6, 7, 8, and 9, organized so as to represent groupings of powers of 10. For instance, the number 574 represents the five groups of 100, seven groups of 10, and four single units.

This numerical system originated from India around 600 C.E., almost in the exact same form as we use it today. The system was transmitted to the Arabs two centuries later as they worked to translate the Sanskrit works on astronomy into Arabic. The Arab mathematician MUHAMMAD IBN MŪSĀ AL-KHWĀRIZMĪ (ca. 800) wrote an influential treatise describing the Hindu numeral system, and used it in his famous book *Hisab al-jabr w'al-muqābala* (Calculation by restoration and reduction), from whose title the modern word *algebra* is derived. As Western scholars began translating the Arabic texts into Latin, word of the efficient numeration system spread across Western Europe. Italian scholar FIBONACCI (ca. 1170–1250) avidly promoted their use. By the end of the 17th century, the Hindu-Arabic numeral system completely replaced the cumbersome system of ROMAN NUMERALS that were the standard in Europe for over 1,500 years.

See also BASE OF A NUMBER SYSTEM; DECIMAL REPRESENTATION; NUMBER; ZERO.

Hippasus of Metapontum   *See* PYTHAGORAS.

Hippocrates of Chios (ca. 470–410 B.C.E.) Greek *Geometry* Born in Khíos (Chios), Greece, Hippocrates is remembered as the first mathematician to have found the area of a curved figure, namely, that of a LUNE. He also made first steps toward properly analyzing the problem of SQUARING THE CIRCLE.

Very little is known of Hippocrates' life. It is thought that Hippocrates may have first worked as a merchant before developing an interest and talent for geometry later in his life. Although he wrote only one text on the topic (unfortunately lost to us today), many scholars throughout history referred to his work in their own studies.

Hippocrates was interested in the famous problem of squaring the circle, that is, finding a method of constructing a square of the same area as a given circle. Although he was unable to solve this problem, he did manage to show that, in many instances, it is possible to "square a lune." For example, in the diagram below, figure *ABCD* is a square, and the shaded region is the lune formed by one circular arc with center *D* and a second circular arc with center *O*, the center of the square. Hippocrates, using geometric reasoning, showed that the area of the lune matches the area of the smaller square shown.

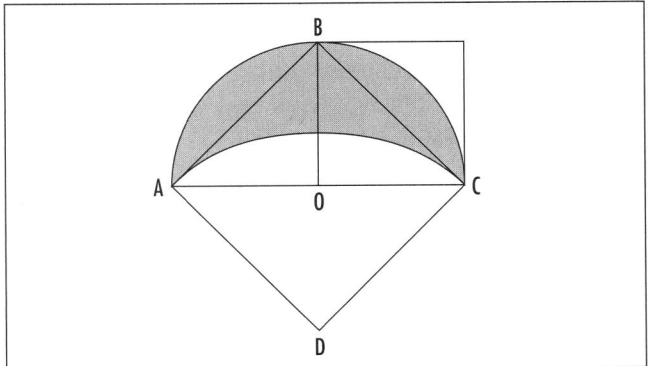

Hippocrates lune

(One can check this algebraically by labeling the radius of the larger circular arc r and computing the side-length of the large square, the area of the sector *DAC*, the area of the semicircle *ABC*, and consequently the area of the lune.)

In applying this type of analysis to a number of different lunes, Hippocrates managed to prove that the ratio of the areas of two different circles is the same as the ratio of their radii squared. This new result was a significant achievement. Some historians believe that EUCLID may have drawn on the work of Hippocrates when he described and proved this result in his famous text *THE ELEMENTS*.

histogram   *See* DISTRIBUTION; STATISTICS: DESCRIPTIVE.

homogeneous   A POLYNOMIAL in several variables $p(x,y,z,...)$ is called homogeneous if all terms in the polynomial have the same total degree. For example, each term of the polynomial $x^3 + 5x^2y - xy^2$, has total degree three, and so represents a homogeneous polynomial.

More generally, a function of several variables $f(x,y,z,...)$ is homogeneous if there is a natural number n such that the following relation holds for any nonzero constant k:

$$f(kx,ky,kz,...) = k^n f(x,y,z,...)$$

For example, the function $x^2 \sin(\frac{x+y}{y}) + xy\cos(\frac{y^2}{x^2})$ is homogeneous.

Identifying homogeneous functions can be helpful in solving DIFFERENTIAL EQUATIONS. Any formula that represents the MEAN of a set of numbers is required to be homogeneous.

In physics, the term *homogeneous* describes a substance or an object whose properties do not vary with position. For example, an object of uniform density is sometimes described as homogeneous. In cooking, when creaming butter and sugar, for instance, one aims to produce a homogeneous mixture.

homomorphism   A map f between two sets S and T, each equipped with some kind of algebraic structure, is called a homomorphism (Greek, *homos*, "same"; *mor-*

phé, "form" or "shape") if it preserves the algebraic relations within the sets. For example, if the algebraic operation is "multiplication" in the set S and "addition" in the set T, and if z is the product of two elements in S, $z = x \times y$, then $f(z)$ should correspond to the sum of the two corresponding elements in T:

$$f(x \times y) = f(x) + f(y)$$

The LOGARITHM function, for example, is a homomorphism from the set of positive real numbers under multiplication to the set of all real numbers under addition.

If f is the doubling function, $f(x) = 2x$, on real numbers, then f is a homomorphism under addition, but not under multiplication:

$$f(x + y) = 2(x + y) = 2x + 2y = f(x) + f(y)$$
$$f(x \times y) = 2 \times x \times y \neq f(x) \times f(y)$$

The squaring map $f(x) = x^2$ preserves multiplication but not addition.

A map between two GROUPS is called a homomorphism if it preserves the algebraic operations of the groups. A map $f : S \to T$ between two RINGS is called a homomorphism if it preserves both the additions and multiplications of the system:

$$f(x + y) = f(x) + f(y)$$
$$f(x \times y) = f(x) \times f(y)$$

for all elements x and y of the ring S.
See also ISOMORPHISM.

Hypatia of Alexandria   (ca. 370–415 C.E.) Greek *Philosophy, Commentary*   Born about 370 in Alexandria, Egypt, Hypatia is remembered as the first woman in the history of mathematics to have made significant contributions to the field, both as a scholar and as a teacher. She was the daughter of mathematician and astronomer Theon of Alexandria. Hypatia wrote influential commentaries on her father's work and also on the work of APOLLONIUS OF PERGA (ca. 263–190 B.C.E.) and DIOPHANTUS OF ALEXANDRIA (ca. 200–284 C.E.).

Little is known of Hypatia's life. She likely received instruction in mathematics from her father, and certainly maintained a close working relationship with him throughout her life. It is known that she assisted him in writing his 11-part commentary on the mathematical work of CLAUDIUS PTOLEMY (85–165 C.E.) and on his production of a revised version of EUCLID's *The Elements.* Hypatia produced her own commentaries on classical pieces, including Diophantus's famous *Arithmetica,* Apollonius's *Conics,* and astronomical works by Ptolemy. All of her work, however, is today lost, and we know of them only through references made by later scholars.

Around 400 C.E. Hypatia headed the Platonist school at Alexandria, where she consulted on scientific matters and lectured on philosophy and mathematics. During this time, Christianity surfaced as the dominant religion of the region, and fanatics felt threatened by her intellect and scholarship. Around 415 C.E. Hypatia was brutally murdered by a group of religious followers who deemed her philosophical views pagan. Many historians suggest that the death of Hypatia marks the beginning of Alexandria's decline as the great center of scholarship and learning of antiquity.

It is worth mentioning that at least one other woman is known to have played an active role in mathematics during the Greek times. In his work *Collection,* PAPPUS OF ALEXANDRIA (300–350 C.E.) gives acknowledgment to a female scholar by the name of Pandrosion. Essentially nothing is known about her.

hyperbola   As one of the CONIC SECTIONS, the hyperbola is the plane curve consisting of all points P whose distances from two given points F_1 and F_2 in the plane have a constant difference. The two fixed points F_1 and F_2 are called the foci of the hyperbola. The hyperbola also arises as the curve produced by the intersection of a plane with the two nappes of a right circular CONE.

Using the notation $|PF_1|$ and $|PF_2|$ for the lengths of the line segments connecting P to F_1 and F_2, respectively, the defining condition of a hyperbola can be written as one of two equations:

$$|PF_1| - |PF_2| = d \text{ or } |PF_2| - |PF_1| = d$$

where d denotes the constant difference. Each equation defines its own curve, or branch, of the same hyperbola.

The equation of a hyperbola can be found by introducing a coordinate system in which the foci are located at positions $F_1 = (-c, 0)$ and $F_2 = (c, 0)$, for some

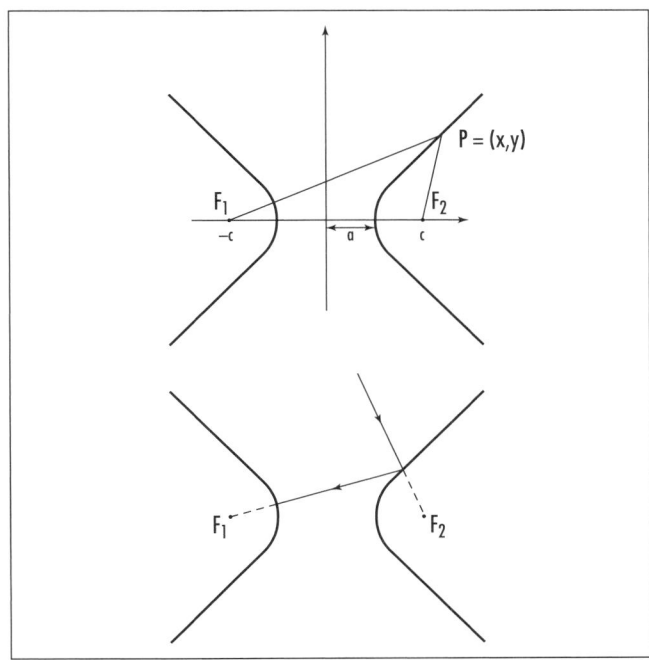

Hyperbola

positive number c. It is convenient to write $d = 2a$, for some $a > 0$. If $P = (x, y)$ is an arbitrary point on the hyperbola, then, according to the DISTANCE FORMULA, the defining conditions read:

$$\sqrt{(x+c)^2 + y^2} - \sqrt{(x-c)^2 + y^2} = \pm 2a$$

Moving the second radical to the right-hand side, squaring, and simplifying yields the equation:

$$\sqrt{(x-c)^2 + y^2} = \pm\left(\frac{c}{a}x - a\right)$$

Squaring and simplifying again yields $\dfrac{x^2}{a^2} - \dfrac{y^2}{c^2 - a^2} = 1$. Noting that c is greater than a, we can set the positive quantity $c^2 - a^2$ as equal to b^2, for some $b > 0$. Thus the equation of the hyperbola is:

$$\frac{x^2}{a^2} - \frac{y^2}{b^2} = 1$$

Conversely, one can show that any equation of this form does indeed yield a hyperbola with foci at positions

$(\pm\sqrt{a^2 + b^2},0)$, and whose points P have distances with common difference $2a$ from the foci. This equation also reveals that the hyperbola has slant ASYMPTOTES $y = \pm\frac{b}{a}x$.

Hyperbolas have the following reflection property: any ray of light that approaches the convex side of one branch along a line pointing toward one focus is reflected directly toward the other focus. (This property can be proved in a similar way that the reflection property of a PARABOLA is proved.)

Hyperbolas appear in the folding of a thin piece of paper. Draw a circle and a dot outside the circle on a sheet of paper. Fold the dot onto the circle and crease the paper. Open up the fold and do this again, this time folding the dot to a different point on the circle. As you do this many times, one branch of a hyperbola will emerge along the side of all the creases. The marked dot is one focus of the hyperbola, and the center of the circle is the other. The radius of the circle is the constant difference of distances of a point on the hyperbola from the two foci.

In the process of deriving the equation of a hyperbola, we presented the equation:

$$\sqrt{(x-c)^2 + y^2} = \left(\frac{c}{a}x - a\right)$$

valid for one branch of the figure at least. Set $e = \frac{c}{a} = \frac{\sqrt{a^2 + b^2}}{a} = \sqrt{1 + \frac{b^2}{a^2}}$. This is called the ECCENTRICITY of the hyperbola and has a value greater than 1. The above equation can be rewritten:

$$\frac{\sqrt{(x-c)^2 + y^2}}{x - \frac{a}{e}} = e$$

The numerator of the quantity on the left side is the distance of a given point P from a focus, and the denominator is the distance of the point P from the vertical line $x = \frac{a}{e}$, called a directrix of the hyperbola. This formulation provides an alternative characterization of the hyperbola:

A hyperbola is the set of all points P such that the ratio of its distance from a fixed point (the focus) to its distance from a fixed line (the directrix) equals a constant $e > 1$.

See also APOLLONIUS'S CIRCLE; ELLIPSE.

hyperbolic functions In analogy to the fact that the sine and cosine functions of TRIGONOMETRY represent the coordinates of a point on the unit circle, $x^2 + y^2 = 1$, the hyperbolic functions represent the coordinates of a point on the right branch of a HYPERBOLA, $x^2 - y^2 = 1$.

Specifically, the x-coordinate of a point on the curve, at an "angle" t, is called the hyperbolic cosine function, and is denoted $\cosh(t)$, and the y-coordinate is called the hyperbolic sine function, denoted $\sinh(t)$. We have:

$$\cosh^2(t) - \sinh^2(t) = 1$$

One can also see that $\cosh(-t) = \cosh(t)$ and $\sinh(-t) = -\sinh(t)$.

Mathematicians sometimes define the hyperbolic sine and cosine functions via the formulae:

$$\cosh(t) = \frac{e^t + e^{-t}}{2}$$

$$\sinh(t) = \frac{e^t - e^{-t}}{2}$$

One can check that these formulae do indeed satisfy the equation of a hyperbola $x^2 - y^2 = 1$ with $x > 0$, and so must indeed be the coordinates of its points. (These equations are reminiscent of the formula $\cos(t) = \frac{e^{it} + e^{-it}}{2}$ and $\sin(t) = \frac{e^{it} + e^{-it}}{2i}$, which follow from EULER'S FORMULA for ordinary trigonometric functions.)

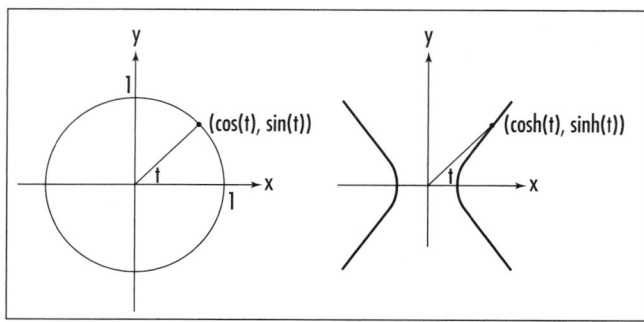

The trigonometric and the hyperbolic functions

Taking the DERIVATIVE we see:

$$\frac{d}{dt}\cosh(t) = \frac{d}{dt}\frac{e^t + e^{-t}}{2} = \frac{e^t - e^{-t}}{2} = \sinh(t)$$

and

$$\frac{d}{dt}\sinh(t) = \cosh(t)$$

In analogy to the ordinary trigonometric functions, mathematicians define four additional hyperbolic functions:

$$\tanh(t) = \frac{\sinh(t)}{\cosh(t)}, \coth(t) = \frac{1}{\tanh(t)},$$

$$\operatorname{sech}(t) = \frac{1}{\cosh(t)}, \operatorname{csch}(t) = \frac{1}{\sinh(t)}$$

The hyperbolic cosine curve arises in nature as the shape of the curve formed by a chain hanging freely between two points. This curve is called a CATENARY.

See also OSBORNE'S RULE.

hyperbolic geometry (Lobachevskian geometry) Independently discovered in 1823 by Hungarian mathematician JÁNOS BOLYAI (1802–60) and in 1829 by Russian mathematician NIKOLAI IVANOVICH LOBACHEVSKY (1792–1856), hyperbolic geometry is a NON-EUCLIDEAN GEOMETRY in which the famous PARALLEL POSTULATE fails in the following manner:

> Through a given point not on a given line, there is more than one line parallel to that given line.

French mathematician JULES HENRI POINCARÉ (1854–1912) later provided a simple model for this geometry and the means to easily visualize geometric results in this theory. The "Poincaré disk" consists of all the points in the interior of the UNIT CIRCLE. A "point" in geometry is any point inside this circle, and a "line" is to be interpreted as a circular arc within the circle with endpoints perpendicular to the boundary of the circle. Any diameter of the boundary circle is also considered a line. Distances are not measured with a traditional ruler: points on the boundary circle are considered to be infinitely far from the center of the circle.

Bolyai and Lobachevsky showed that all but the fifth of EUCLID'S POSTULATES hold in the hyperbolic geometry and, moreover, that this model of geometry is consistent (that is, free of CONTRADICTIONS). This establishes that the parallel postulate cannot be logically deduced as a consequence of the remaining axioms proposed by Euclid.

In hyperbolic geometry, all angles in triangles sum to less than 180°, and the ratio of the circumference of any circle to its diameter is less than π. (Moreover, the value of this ratio is not the same for all circles.) Also, it is possible for two perpendicular lines to be parallel to the same line.

Physicists, following the work of ALBERT EINSTEIN, suggest that the geometry of our universe is hyperbolic: that it appears to us as Euclidean is a result of the fact that we occupy such a small portion of it. (This is analogous to the fact that it is difficult to recognize the Earth as round when living on it.)

See also EUCLIDEAN GEOMETRY; PLAYFAIR'S AXIOM; SPHERICAL GEOMETRY.

hyperboloid The SOLID OF REVOLUTION obtained by rotating a HYPERBOLA about one of its axes is called a hyperboloid. If the rotation is performed about the axis that lies between the two branches of the curve, then the resulting surface is called a hyperboloid of one sheet. It resembles a cylinder "pinched" at the center so as to curve inward. Points on this surface satisfy an equation of the form $\frac{x^2}{a^2} + \frac{y^2}{b^2} = \frac{z^2}{c^2} + 1$ for some constants a, b, and c. If, instead, the rotation is performed about the axis that connects the two foci of the hyperbola, then the resulting surface has two distinct parts, each resembling a bowl. This surface is called a hyperboloid of two sheets. Points on this surface satisfy the equation $\frac{x^2}{a^2} + \frac{y^2}{b^2} = \frac{z^2}{c^2} - 1$.

One can construct a model of a hyperboloid of one sheet by holding two metal rings, one directly above the other, and tying vertical strings from points on the lower ring to their corresponding points on the top ring. If one then rotates the top ring so that the vertical strings begin to tilt, the model begins to "constrict." The resulting surface is a hyperboloid of one sheet. One can also see this surface by tying a string around a handful of uncooked spaghetti strands.

By rotating the top disc of this 1872 string model, one can transform a cylinder of chords into the shape of a hyperboloid. *(Photo courtesy of the Science Museum, London/Topham-HIP/The Image Works)*

When the strands tilt, the surface of a hyperboloid of one sheet appears.

See also ELLIPSOID; PARABOLOID.

hypercube   The higher-dimensional analog of SQUARE in two dimensions and a CUBE in three dimensions is called a hypercube.

A unit square in the plane has four vertices at locations $(0,0),(1,0),(0,1)$ and $(1,1)$. A unit cube in three-dimensional space has eight vertices at locations $(0,0,0)$, $(1,0,0)$, $(0,1,0)$, $(1,1,0)$, $(0,0,1)$, $(1,0,1)$, $(0,1,1)$ and $(1,1,1)$. By analogy, a hypercube of side-length 1 sitting in four-dimensional space has 16 vertices given by: $(0,0,0,0)$, $(1,0,0,0)$, $(0,1,0,0),...,(1,1,1,1)$.

Some people find it helpful to interpret the fourth coordinate as time. In this context, each particular value of the fourth coordinate represents a given instant in time, and the remaining three coordinates describe an ordinary object in three-dimensional space. A four-dimensional hypercube can thus be thought of as an ordinary cube moving through time. (One can also apply this interpretation to ordinary squares and cubes of lower dimension. For example, a line segment is the result of a point sliding horizontally over time, a square is the result of a horizontal line segment sliding in the direction of the positive y-axis over time, and a cube is the result of a two-dimensional square sliding in the direction of the positive z-axis over time.) A four-dimensional hypercube is also called a tesseract.

A square in two-dimensional space has four vertices and four edges. A cube in three-dimensional space has eight vertices and six square faces. One can show that a tesseract has eight vertices and six cubic "faces," and that, in general, an n-dimensional hypercube has 2^n vertices and $2n$ faces, each itself a hypercube of one dimension less.

Four line segments hinged end-to-end fold to make a square. Six squares arranged in a "cross" fold to make a cube. Eight cubes arranged in an analogous way—four in a row with the four remaining cubes attached to the four exposed faces of the second cube—fold in four-dimensional space to make a hypercube. All in all, there are 261 different ways to arrange eight ordinary cubes that fold in four-dimensional space to make a hypercube. Each arrangement is called a NET.

hypotenuse   The longest side of a triangle that contains a 90° angle, that is, a right triangle, is called the hypotenuse of the triangle. PYTHAGORAS'S THEOREM shows that this longest side lies opposite the right angle. (If the three side-lengths of a right triangle are a, b, and c, with c opposite the right angle, then $c^2 = a^2 + b^2$. This establishes that c is indeed larger than both a and b.) The two remaining sides of a right triangle are called the legs of the triangle.

The word *hypotenuse* is derived from the Greek term *hypoteinousa* meaning "under tension" (the prefix *hypo*-means "under" and *teinein* means "to stretch").

Mathematicians have shown that an integer c can be the length of the hypotenuse of a right triangle with integer side-lengths if, and only if, in the prime FACTORIZATION of c, no PRIME that appears an odd number of times is 3 more than a multiple of 4.

See also PYTHAGOREAN TRIPLES.

hypothesis testing   *See* STATISTICS: INFERENTIAL.

I

identity An equation that states that two mathematical expressions are equal for all possible values of any variables that occur in them is called an identity. For example, the DIFFERENCE OF TWO SQUARES formula $x^2 - y^2 = (x - y)(x + y)$ and the trigonometric equation $\sin^2\theta + \cos^2\theta = 1$ are identities. Sometimes the symbol $\equiv$ is used instead of $=$ to indicate that the statement is an identity. It is read as "identically equal to."

A numerical statement, one involving no variables at all, might still be called an identity if it illuminates something interesting about the numbers involved. For example, the following relationship could be called an identity because of the unexpected appearance of the number π:

$$3\arctan\left(\frac{1}{4}\right) + \arctan\left(\frac{1}{20}\right) + \arctan\left(\frac{1}{1985}\right) = \frac{\pi}{4}$$

See also EQUATION; IDENTITY ELEMENT.

identity element (**identity, neutral element**) An element of a set that, combined with any other element of the set, leaves that element unchanged is called an identity element. More precisely, given a BINARY OPERATION "*" on a set S, we say that e in S is an identity element if:

$$a*e = a$$
$$e*a = a$$

for all elements a of the set.

For example, the number zero is an identity element for the set of real numbers under the operation of addition: we have that $a + 0 = a = 0 + a$ for all numbers a. With respect to multiplication, 1 is an identity element, since $a \times 1 = a = 1 \times a$ for all numbers a. (The number zero is sometimes called an "additive identity" and the number 1 a "multiplicative identity.")

In matrix theory the IDENTITY MATRIX is an identity element under multiplication for the set of all square matrices of a particular size. In SET THEORY, the EMPTY SET is an identity under the operation of taking union. Any GROUP is required to have an identity element.

It is impossible for a mathematical system to possess two different identity elements with respect to the same binary operation. If, for instance, e and f are both identity elements for a set S, then $e*f$ equals e (since f is an identity element), and it also equals f (since e is an identity element). Thus e and f are the same.

In some mathematical theories it is desirable to distinguish between a "left identity," that is an element e_L such that $e_L*a = a$ for all elements a, and a "right identity" e_R that satisfies $a*e_R = a$ for all a. Of course, if the theory satisfies the COMMUTATIVE PROPERTY, then e_L and e_R are equal and represent an identity for the system.

See also INVERSE ELEMENT.

identity matrix (**unit matrix**) A square MATRIX with each main diagonal element equal to 1 and all other entries equal to zero is called an identity matrix. Such a matrix is usually denoted I or Id, or even Id_n if the

ORDER OF A MATRIX is n. For example, the 3×3 identity matrix is:

$$I_3 = \begin{pmatrix} 1 & 0 & 0 \\ 0 & 1 & 0 \\ 0 & 0 & 1 \end{pmatrix}$$

If A is a matrix with m rows and n columns, then matrix multiplication shows we have:

$$AI_n = A$$
$$I_m A = A$$

that is, multiplication with the appropriate identity matrix leaves any other matrix unchanged. If all the matrices under consideration have the same number of rows as columns, say n of each, then I_n acts as an IDENTITY ELEMENT for that set under matrix multiplication.

Any 2×2 matrix A of the form:

$$\begin{pmatrix} d & \dfrac{1-d^2}{c} \\ c & -d \end{pmatrix}$$

or

$$\begin{pmatrix} d & c \\ \dfrac{1-d^2}{c} & -d \end{pmatrix}$$

satisfies $A^2 = I_2$. Thus any such matrix can be thought of as a SQUARE ROOT of the 2×2 identity matrix. If one is willing to permit the use of COMPLEX NUMBERS, then the following matrix is an example of a cube root of the 3×3 identity matrix:

$$A = \begin{pmatrix} 0 & 0 & -i \\ i & 0 & 0 \\ 0 & 1 & 0 \end{pmatrix}$$

This matrix satisfies the relation $A^3 = I_3$.

image (**range**) The set of all values that a FUNCTION could adopt is called the image of the function. For example, the image of the function $y = x^2$ defined for

all REAL NUMBERS is the set of all numbers greater than or equal to zero. The term is also used for the output of a specific input for the function. For instance, in the example above, the image of the number 3 is 9.

In dealing with functions of real numbers, the term *range* is preferred over *image*. Often mathematicians will use the word *image* only when thinking of a problem geometrically. For example, if the function is a GEOMETRIC TRANSFORMATION such as a reflection in a line, then one would speak of the image of geometric figures under this transformation. In this example, the image of any circle is another circle, and the image of a straight line is another straight line.

implication *See* CONDITIONAL.

implicit differentiation When two variables x and y satisfy a single equation $F(x,y) = 0$, it may be possible to regard the equation nonetheless as defining y as a function of x, even though no explicit formula of this type may be apparent. (*See* IMPLICIT FUNCTION.) In such a case, one can go further and differentiate the equation as a whole, regarding y as a function of x and using the CHAIN RULE in the process to find a formula for the derivative $\dfrac{dy}{dx}$. This process is known as implicit differentiation. For example, if $xy^3 + 7x^2y = 1$, then differentiating yields:

$$y^3 + x3y^2\frac{dy}{dx} + 14xy + 7x^2\frac{dy}{dx} = 0$$

and so:

$$\frac{dy}{dx} = -\frac{y^3 + 14xy}{3xy^2 + 7x^2}$$

(assuming the denominator is not zero).

Implicit differentiation is useful for finding the derivatives of inverse functions, for example. For instance, if $y = \sin^{-1}(x)$, then $\sin(y) = x$. Differentiating yields $\cos(y)\dfrac{dy}{dx} = 1$, and so, $\dfrac{dy}{dx} = \dfrac{1}{\cos(y)} = \dfrac{1}{\sqrt{1 - \sin^2(y)}}$

$$= \frac{1}{\sqrt{1 - x^2}}.$$

Sometimes a function is more easily differentiated if one first applies a logarithm and then differentiates implicitly. This process is called logarithmic differentiation. For example, to differentiate $y = x^x$, write $\ln(y) = \ln(x)^x = x\ln(x)$. Then $\frac{1}{y} \cdot \frac{dy}{dx} = \ln(x) + x \cdot \frac{1}{x}$, and so,

$$\frac{dy}{dx} = y(\ln(x)+1) = x^x(\ln(x)+1).$$

implicit function A formula of the form $F(x,y) = 0$ is said to define the variable y implicitly as a function of x. For example, the equation $xy = 1$ implicitly defines the dependent variable y as $y = \frac{1}{x}$. However, matters can be confusing, for a single equation may define y as several possible functions of x (for example, the equation $x^2 + y^2 = 1$ suggests that either $y = \sqrt{1-x^2}$ or $y = -\sqrt{1-x^2}$), and in some instances, it might not even be possible to solve for an explicit formula for y (for example, it is not at all clear that $x^2y^2 + y\cos(xy) = 1$ does indeed give a formula for y).

Mathematicians have proved that as long as the function $F(x,y)$ has continuous PARTIAL DERIVATIVES, and that there is some point (a,b) for which $F(a,b) = 0$ and $\frac{\partial F}{\partial y} \neq 0$ at (a,b), then the equation $F(x,y) = 0$ does indeed define y as a function of x, at least for values of x within a small neighborhood of $x = a$. Moreover, the process of IMPLICIT DIFFERENTIATION is valid for these values of x. This result is known as the implicit function theorem.

improper integral (infinite integral, unrestricted integral) In the theory of INTEGRAL CALCULUS it is permissible to extend the notion of a definite integral $\int_a^b f(x)dx$ to include functions $f(x)$ that become infinite in the range under consideration, $[a,b]$, or, alternatively, to consider integration over an infinite range: $[a,\infty)$, $(-\infty,b]$, or even $(-\infty,\infty)$. Such integrals are called improper integrals. One deals with such integrals by restricting the range of integration and taking the limit as the interval expands to the required size.

Consider, for example, the integral $\int_0^1 \frac{1}{\sqrt{x}}dx$. The integrand $\frac{1}{\sqrt{x}}$ becomes unbounded as x approaches the value zero, and so, in the normal way, the integral $\int_0^1 \frac{1}{\sqrt{x}}dx$ is not well defined. However, the function is bounded on the interval $[L,1]$ for any $0 < L < 1$, and the integral $\int_L^1 \frac{1}{\sqrt{x}}dx$ is valid. We have: $\int_L^1 \frac{1}{\sqrt{x}}dx = 2x^{\frac{1}{2}}\Big|_L^1$ = $2 - 2\sqrt{L}$. The improper integral under consideration is then defined to be the LIMIT as $L \to 0$, from above, of these values: $\int_0^1 \frac{1}{\sqrt{x}}dx = \lim_{L\to 0^+}\int_L^1 \frac{1}{\sqrt{x}}dx = \lim_{L\to 0^+} 2 - 2\sqrt{L} = 2$.

In general, if an integrand $f(x)$ is infinite at an end point a, then the improper integral $\int_a^b f(x)dx$ is defined as the limit:

$$\int_a^b f(x)\,dx = \lim_{L\to a^+}\int_L^b f(x)dx$$

or, if infinite at the end point b, then the improper integral is defined as the limit:

$$\int_a^b f(x)dx = \lim_{L\to b^-}\int_a^L f(x)dx$$

If the limit exists, then the improper integral is said to converge. (Thus, for example, $\int_0^1 \frac{1}{\sqrt{x}}dx$ is a convergent improper integral.) If the limit does not exist, then the improper integral is said to diverge. (One can check, for example, that $\int_0^1 \frac{1}{x}dx$ diverges.)

To illustrate the second type of improper integral, consider, for example, the quantity $\int_1^\infty \frac{1}{x^2}dx$. This is to be interpreted as the limit:

$$\int_1^\infty \frac{1}{x^2}dx = \lim_{L\to\infty}\int_1^L \frac{1}{x^2}dx$$

which can be readily computed: $\int_1^\infty \frac{1}{x^2}dx = \lim_{L\to\infty}\int_1^L \frac{1}{x^2}dx = \lim_{L\to\infty}-\frac{1}{x}\Big|_1^L = \lim_{L\to\infty}\left(1-\frac{1}{L}\right)=1$. (We interpret this as saying that the total area under the curve $y = \frac{1}{x^2}$ to the right of $x = 1$ is 1.)

In general, improper integrals of this type are computed as follows:

$$\int_a^\infty f(x)dx = \lim_{L\to\infty}\int_a^L f(x)dx$$
$$\int_{-\infty}^b f(x)dx = \lim_{L\to-\infty}\int_L^b f(x)dx$$
$$\int_{-\infty}^\infty f(x)dx = \lim_{L\to\infty}\int_{-L}^L f(x)dx$$

If the limits exist, then the improper integrals are said to converge. (Thus, for example, $\int_1^\infty \frac{1}{x^2}\,dx$ is a convergent improper integral.) If the limit does not exist, then the improper integral is said to diverge. (One can check, for example, that $\int_1^\infty \frac{1}{\sqrt{x}}\,dx$ diverges.) The integral test used in the study of CONVERGENT SERIES makes use of improper integrals of this type.

It is possible to define even more-general improper integrals by considering integrands that take the value infinity at several places within the range of integration, possibly over an infinite range of integration. One interprets such improper integrals as the sum of several limits.

Incan mathematics   The Incan Empire ruled the region of Peru and its surroundings before the Spanish conquest of the continent in 1532. It was an extraordinarily sophisticated civilization. The Inca people had highly developed systems of agriculture, transportation, road construction, textile production, and administration, yet, surprisingly, had no form of writing. Consequently, with absolutely no cultural records or documents available, very little is known of the mathematics of these people.

It is known that the Inca recorded numerical information with a device known as a *quipu,* which consisted of a collection of strings all tied to a common central object, usually another rope. Administrators simply tied knots in the strings to record numbers, with each string representing a different count. Knots were grouped in clusters to represent quantities of powers of 10. Thus, for example, the number 283 was represented (backward) by a cluster of three knots placed near the free end of a string (three units), a space, a cluster of eight knots in the center of the string (eight 10s), another space, and finally two knots at the top of the string (two 100s). The Incas, in effect, were following the DECIMAL REPRESENTATION system we use today. A large space indicated a digit of zero (to distinguish 208 from 28, for instance).

Strings of different colors were used to signify what the knots on that string were counting. For example, white strings may have represented the counts of cattle, while green strings may have represented the levels of punishment to be executed for committing a certain crime.

Each town in the Incan Empire had its own *quipca-mayoc,* a "keeper of the knots," who recorded census information about the population, the produce, and weapons of the town. *Quipues* recording this information were sent every year, by relay runners, to the capital of the empire, Cuzco.

The extent to which the Incan people manipulated numbers and performed calculations is unclear. A Spanish priest by the name of José de Acosta, who lived among the Incas from 1571 to 1586, describes in his book, *Historia natural y moral de las Indias,* residents shuffling kernels of corn on the ground as though manipulating the beads of an ABACUS. Obviously some kind of numerical computation was being conducted, but Acosta did not understand the method and was unable to explain what exactly was being done.

The Inca thought of numbers in very concrete terms. The concept of "two-ness," for instance, did not exist for the Inca, and the count of 2 applied to very specific things. This is evidenced by the fact that the Inca had several different words for *two,* which were used to distinguish different contexts—the two objects that make a pair, the 2 that came from one object divided in half, and the one object and its specific partner that make a pair, for instance. There was no single all-encompassing word for the state of having two parts.

It is likely that the Inca people kept astronomical records, a topic of interest to most ancient cultures, but there is no direct evidence of this.

incircle   A circle that lies inside a triangle and touches all three sides is called the incircle of that triangle. The center of that circle is called the incenter of the triangle, and its radius is called the inradius of the triangle.

It is relatively straightforward to see that for any triangle there is precisely one circle in the triangle that touches all three sides. (Erase one side of the triangle and note that there are infinitely many circles that fit into the wedge formed by the two remaining sides. In redrawing the third side, observe that precisely one of those circles lies tangent to that additional edge.)

The incenter of a triangle is a point that lies the same distance from each of the three sides of the triangle. Note that any point that is EQUIDISTANT from just two intersecting lines lies on the angle bisector of those lines (that is, on a line that cuts the angle between the two given lines in half). Consequently, we have that the incenter of any triangle lies on each of the angle bisectors of the vertices of the triangle. This establishes:

The three angle bisectors of a triangle are CON-CURRENT.

An excircle for a triangle is a circle that lies outside the triangle and is tangent to one side of the triangle and tangent to extensions of the remaining two sides. A triangle has three distinct excircles.

In a more general context, if it is possible to draw a circle inside a POLYGON tangent to each side of the polygon, then that circle is called an incircle of the polygon. Every regular polygon has an incircle. There is no incircle for a nonsquare rectangle.

See also CIRCUMCIRCLE.

inclination/declination The angle θ between a ray emanating from the origin of a CARTESIAN COORDINATE system and the positive x-axis as measured in an anticlockwise direction is called the inclination of the ray. An angle measured in the clockwise direction is called its declination. These terms are rarely used in mathematics today.

A PLANE that is not horizontal is called an inclined plane, and the angle that the line of greatest slope within the plane makes with the horizontal is called the angle of inclination of the plane.

inclusion-exclusion principle If $n(A)$ denotes the number of elements in a finite set A, then the number of elements in the union $A \cup B$ of two sets is given by:

$$n(A \cup B) = n(A) + n(B) - n(A \cap B)$$

(Counting the number of elements in each of A and B counts the elements that belong to both twice. One must

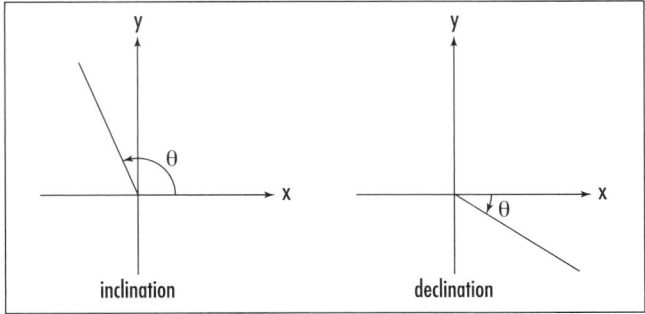

Angles of inclination and declination

compensate for this double count.) Similarly, the number of elements in the union of three sets is given by:

$$\begin{aligned} n(A \cup B \cup C) = {} & n(A) + n(B) + n(C) \\ & -n(A \cap B) - n(B \cap C) \\ & - n(A \cap C + n(A \cap B \cap C) \end{aligned}$$

(One can establish this either by reasoning through which elements are counted multiple times, or by noting that $A \cup B \cup C = (A \cup B) \cup C$ and applying the previous observation:

$$\begin{aligned} n((A \cup B) \cup C) &= n(A \cup B) + n(C) - n((A \cup B) \cap C) \\ &= n(A \cup B) + n(C) - n((A \cap C) \cup (B \cap C)) \end{aligned}$$

Two more applications of the formula for the union of two give the result.)

In general, an INDUCTION argument shows that the number of elements in the union of n sets is given by:

$$\begin{aligned} n(A_1 \cup A_2 \cup \ldots \cup A_k) = {} & n(A_1) + n(A_2 + \ldots + n(A_k) \\ & -n(A_1 \cap A_2) - n(A_1 \cap A_3) - \ldots \\ & - n(A_{k-1} \cap A_k) \\ & + n(A_1 \cap A_2 \cap A_3) + \ldots \\ & + n(A_{k-2} \cap A_{k-1} \cap A_k \\ & \vdots \\ & + (-1)^k n(A_1 \cap A_2 \cap \ldots \cap A_k) \end{aligned}$$

This formula is called the general inclusion-exclusion principle. It can be interpreted as follows:

> The number of elements of a finite set that possess at least one of k possible properties is equal to the number possessing exactly one property, minus the number possessing exactly two properties, plus the number possessing precisely three properties, and so on, up to the count of those elements possessing all k properties.

This powerful counting principle has important applications in PROBABILITY theory.

increasing/decreasing A SEQUENCE of numbers $\{a_n\}$ is said to be increasing if $a_1 \leq a_2 \leq a_3 \leq \ldots$, that is, each term in the sequence is greater than or equal to the one that precedes it. It is called strictly increasing if $a_1 < a_2 < a_3 < \ldots$ The constant sequence $1,1,1, \ldots$, for example, is considered increasing.

Increasing sequences have the pleasing property that their convergence (or divergence) is easy to identify:

An increasing sequence converges if, and only if, it is bounded above.

(*See* BOUND and CONVERGENT SEQUENCE.) This seems clear geometrically. If on the real number line the sequence of numbers $\{a_n\}$ moves steadily to the right yet does not penetrate a barrier at position B, then the numbers must "pile up" at some limit value $L \leq B$. Conversely, if an increasing sequence converges to a value L, then all the terms of the sequence lie to the left of L on the number line and hence are bounded above by L. (These statements can be proved rigorously by making use of the "completeness" of the real number line via the notion of a DEDEKIND CUT.)

A sequence $\{a_n\}$ is said to be decreasing if $a_1 \geq a_2 \geq a_3 \geq \dots$, that is, each term in the sequence is less than or equal to the one that precedes it. It is called strictly decreasing if $a_1 > a_2 > a_3 > \dots$. Similarly one can show that a decreasing sequence converges if, and only if, it is bounded below. A sequence that is either increasing or decreasing is called monotonic.

A real-valued function f is said to be increasing over an interval if, over that interval, greater input values produce greater (or possibly equal) output values, that is, if a and b are two values in the interval with $a < b$, then $f(a) \leq f(b)$. If it is always the case that $f(a) < f(b)$, then the function is said to be strictly increasing on the interval. The MEAN-VALUE THEOREM shows that a differentiable function f is increasing precisely on those intervals where the DERIVATIVE is nonnegative, $f'(x) \geq 0$, and strictly increasing if $f'(x) > 0$.

A real-valued function f is said to be decreasing over an interval if greater input values produces smaller (or possibly equal) output values of the function, that is, if $a < b$, then $f(a) \geq f(b)$. If it is always the case that $f(a) > f(b)$, then the function is said to be strictly decreasing on the interval. The MEAN-VALUE THEOREM shows that a differentiable function f is decreasing precisely on those intervals for which $f'(x) \leq 0$, and strictly decreasing if $f'(x) < 0$.

A function that is either increasing or decreasing on an interval is called monotonic on that interval.

See also DIFFERENTIAL CALCULUS; GRAPH OF A FUNCTION; MAXIMUM/MINIMUM.

increment   A small finite change in the value of a variable or in the value of a function is called an increment. If the variable is x, then an increment of x is usually denoted Δx or δx. If f is a function of x, then the corresponding increment of f, denoted Δf, equals $\Delta f = f(x + \Delta x) - f(x)$. The LIMIT of the ratio $\Delta f / \Delta x$, as the increment Δx decreases to zero, is called the DERIVATIVE of f. A negative increment is sometimes called a decrement.

See also DIFFERENTIAL.

indefinite integral   *See* ANTIDIFFERENTIATION; INTEGRAL CALCULUS.

independent axiom   An AXIOM of a mathematical theory is said to be independent if it cannot be derived from the remaining axioms of a theory as a logical consequence. For instance, the existence of NON-EUCLIDEAN GEOMETRIES establishes that Euclid's famous PARALLEL POSTULATE is independent of his four remaining postulates. In general, this is the approach mathematicians take to show that an axiom A in a mathematical system is independent of the remaining axioms: present an example of another consistent mathematical theory proved to be free of contradictions in which all axioms except A hold and for which A is false. (Consequently, it cannot be the case that these axioms do imply that A is true, for then one has a system in which both A and not A hold.)

See also EUCLID'S POSTULATES.

independent events   Two experiments run in succession are deemed independent if the outcomes obtained from one experiment do not affect the outcomes obtained in the second. For example, the results from casting a die do not influence the results obtained in later tossing a coin. These two actions are independent.

In PROBABILITY theory we say two EVENTS A and B are independent if knowledge of A having occurred has no influence on the likelihood of B next occurring. For example, in two rolls of die casting, an even number on the first roll and casting a 6 on the second are independent events—the chances of rolling a six are 1/6, no matter the result of the first roll.

Two events that are not independent are called dependent. For example, the event "Sally is wearing a raincoat" is likely to be dependent on the event "it is raining." CONDITIONAL PROBABILITY is used to analyze dependent events.

Experiments that involve drawing objects from a finite source, such as balls from a bag, or cards from a deck, *without* replacing the objects withdrawn, often yield dependent events. For example, the events "the first card drawn from a deck of cards is red" and "the second card drawn is red" are dependent—the probability that the second card selected is the desired color could either be 26/51 or 25/51 , depending on whether the first card drawn is black or red.

A careful study of PROBABILITY shows that two independent events A and B satisfy the probability relation:

$$P(A \text{ and } B) = P(A) \times P(B)$$

Mathematicians take this relation as the definition of two events A and B being independent.

indeterminate equation An equation, or a SYSTEM OF EQUATIONS, with an infinite number of solutions is called indeterminate. For instance, the equation:

$$x + 2y = 5$$

is indeterminate (for each real number t there is the solution $x = 5 - 2t$ and $y = t$), as is the system of two equations:

$$a + b + c = 1$$
$$2a + 3b - c = 0$$

in variables a, b, and c. (All triples of the form $(a,b,c) = (t, \frac{1 - 4t}{3}, \frac{2 - t}{3})$ are solutions.)

The single equation $x^2 + y^2 + z^2 + w^2 = 0$ is not indeterminate: it has the unique (real) solution $x = y = z = w = 0$.

See also DEGREES OF FREEDOM.

index (plural, indices) A number that indicates a place or describes the characteristic or the nature of a function is called an index. For example, a SEQUENCE of numbers might be denoted $\{a_n\}$, with the index n denoting the place in the sequence. (Thus a_7, for instance, would represent the seventh number in the list.) An EXPONENT is sometimes called an index. For example, one might say that the function $f(x) = x^5$ has index 5 to indicate that quantities are being raised to the fifth power. The radical function $g(x) = \sqrt[3]{x}$, for example, has index 3, and the logarithmic function $h(x) = \log_{10} x$ has index 10.

In statistics and business, an index is a figure used to show the variation in some quantity over a period of time, usually standardized relative to some base value. For example, a retail price index is used to measure changes in the cost of household items. For the base year, the index is usually set at 100. If, for instance, the cost of dish detergent in the base year is $1.60, and the price rises to $1.76 and to $2.00 in the two subsequent years, then the price index for this item would be 100, 110, and 125 in each of those years, respectively.

It is possible to design an index that gives a general measure of the value of more than one commodity simultaneously. For example, the Dow Jones index of the New York Stock Exchange is an indicator of the worth of a representative selection of industrial, transportation, and utility stocks.

Indian mathematics The entire course of Western mathematics was profoundly affected by a single Indian invention, that of the place-value decimal system. That every possible number can be expressed via a set of just 10 symbols, 0, 1, 2, 3, 4, 5, 6, 7, 8, and 9, by making careful use of the place of each symbol, seems such a simple notion nowadays that it is hard to appreciate its profound importance. This elegant notation system provided the means for Indian scholars to perform complicated arithmetical computations with relative ease, which in turn led to significant developments in numerical techniques, approximation methods, and the theory of arithmetic. Only when other cultures adopted the place-value decimal system from India could they accomplish the same mathematical feats that this culture had already developed.

The earliest dated evidence of mathematical activity in the Indian subcontinent goes back to the Indus civilization of 2500 B.C.E. Bronze weights and graded rods (rulers) from the period show that these people were already working with a decimal system. The Indus people worked with a basic unit of length 1.32 in. long (today called the "Indus inch"), 10 of which make their version of a "foot." Excavations show that the weights and graded rods were used extensively in construction.

The earliest written records of Indian culture are the religious texts the Vedas, dating between 1500 B.C.E. and 800 B.C.E. Although not mathematical in content, appendices to the texts give specific rules for constructing altars, exhibiting a thorough understanding of the basic principles of geometry. Early versions of the digits 0 through 9 were used at this time.

By 600 C.E., the Vedic religion had gone into decline, and Jainism came to the fore. During this period mathematics was driven by the needs of the religion and its demands for careful astronomical observations to pinpoint the exact times of religious observances and the development of an accurate calendar. The decimal representation system was now fully developed, and scholars were able to make precise and surprisingly accurate calculations. The mathematician ĀRYABHATA (ca. 500 C.E.), for instance, had developed a theory of TRIGONOMETRY to aid astronomical calculations, had developed methods for extracting square roots, evaluated π to a high degree of accuracy, and was able to find integer solutions to a large class of equations that arose from astronomical theories.

One written text from this period was discovered in 1881 in the town of Bakhshali, now in Pakistan. Written on birch bark, the Bakhshali manuscript shows that mathematicians were also comfortable with fractions, basic algebraic manipulations (they used a dot to represent an unknown quantity), and sophisticated approximation formulae. For example, the manuscript describes, in words, the following approximation for the square root of a number N:

$$\sqrt{N} = \sqrt{a^2+b} \approx a + \frac{b}{2a} - \frac{\left(\frac{b}{2a}\right)^2}{2\left(a+\frac{b}{2a}\right)}$$

Here a^2 is the largest perfect square smaller than N and $b = N - a^2$. For example, we have:

$$\sqrt{50} = \sqrt{7^2+1} \approx 7 + \frac{1}{14} - \frac{\left(\frac{1}{14}\right)^2}{2(7+\frac{1}{14})} \approx 7.071067821$$

which is correct to seven decimal places.

Two mathematical research centers were formed in India during the era of Jainism, both astronomical observatories. Āryabhata headed the first center at Kusumapura in the northeast of the Indian subcontinent, and mathematician Varahamihira, who also made contributions to astronomy and trigonometry, headed the second center at Ujjain, also in the north.

Varahamihira was succeeded by the seventh-century mathematician BRAHMAGUPTA, who, in his famous work *Brahmasphutasiddhanta* (The opening of the universe), introduced and explained the arithmetic of nonpositive numbers. He was the first mathematician in history to give zero the status of a number, defining it to be the result of subtracting a quantity from itself. Brahmagupta's work also includes a formula for the area of a cyclic quadrilateral in terms of its sides (today called BRAHMAGUPTA'S FORMULA), and presents methods for solving linear and quadratic equations, as well as systems of equations. Brahmagupta also developed sophisticated interpolation techniques for computing sine values in trigonometry.

For the next 200 years, Indian scholars worked to refine further methods of trigonometry and techniques of astronomical calculation. The mathematician BHĀSKARA II of the 12th century made advances in number theory, algebra, combinatorics, and astronomy, and wrote a comprehensive text summarizing the state of mathematics and astronomy in India at his time. Soon afterward, other Indian scholars developed these ideas further. Jaina mathematicians also clarified the standard EXPONENT rules and manipulated exponents in a manner that suggests today that they were also familiar with the basic principles of LOGARITHMs.

The 14th-century scholar MADHAVA OF SANGAMA-GRAMMA made significant advances in ANALYSIS. He produced the infinite series expansions of trigonometric and inverse trigonometric functions (today called TAYLOR SERIES), discovered the BINOMIAL THEOREM, and even produced GREGORY'S SERIES for π, which he used to approximate its value to a considerably accurate degree.

During the first millennium India had very little contact with the cultures of the West. News of the decimal representation system, however, did manage to spread to other countries relatively quickly. A manuscript written in Syria in 662 discusses the new method of calculation, and there is evidence that the decimal system was being used in Cambodia and other Asian countries soon afterward. By the ninth century, the decimal system was in common use in the Islamic world, and from there it was quickly transmitted to Europe. Arab scholars maintained a keen interest in the work of Indian mathematicians for the centuries that followed and took an active role in preserving and translating many Indian texts.

See also ARABIC MATHEMATICS; BASE OF A NUMBER SYSTEM.

indirect proof (proof by contradiction, reductio ad absurdum) Most claims made in mathematics are statements of the form:

If the premise *A* is true, then the conclusion *B* is true.

An indirect proof of such a statement attempts to establish the validity of the claim by assuming that the premise *A* is true and showing, consequently, that the conclusion *B* cannot be false. One does this by exploring the logical consequences of assuming *A* and "not *B*" until, at some point, a contradiction (such as, 1 + 1 = 3, for instance) is reached. Based on the belief that mathematics should be free from absurdities, mathematicians generally accept this approach as sufficient for establishing the validity of *B*.

As an example, we prove: for a natural number *n*, if n^2 is even, then n is even. We base the proof on known facts about even and odd numbers, and the standard algebraic manipulations.

Proof: assume that n^2 is even, and assume, to the contrary, that *n* is not.
Then it must be the case that *n* is odd. Consequently, *n* is one more than a multiple of 2 and can be written in the form 2*k* + 1, for some number *k*.
This means that n^2 is given by $n^2 = (2k + 1)^2 = 4k^2 + 4k + 1 = 2(2k^2 + 2k) + 1$ and so is also one more than a multiple of 2.
We conclude that n^2 is both even and odd—clearly a contradiction.
It cannot be the case, then, that *n* is odd. It must therefore be even.

Notice in this proof that we assumed premise *A* to be true, and arrived at the contradiction that "not *A*" also holds. An indirect proof that arrives at a contradiction of this type is usually called a "proof using the contrapositive": we established the validity of "*A* implies *B*" by demonstrating that the CONTRAPOSITIVE "not *B* implies not *A*" holds. EUCLID'S PROOF OF THE INFINITUDE OF PRIMES is an example of an indirect proof that does not rely on the contrapositive. EUCLID (300–260 B.C.E.) was the first mathematician to extensively employ the technique of indirect proof.

A DIRECT PROOF attempts to establish the validity of a proposition "if *A*, then *B*" by assuming that the premise *A* is true and following its logical consequences until statement *B* is established. Not all propositions, however, are amenable to a direct approach. For example, given that a squared number n^2 is even (that is, $n^2 = 2m$, say) it is not immediate how one should proceed to establish directly that *n* is consequently even.

Any indirect proof relies on the assumption that a statement that cannot be false must be true. Some philosophers and mathematicians who study the LAWS OF THOUGHT seriously question this assumption.

See also CONTRADICTION; DEDUCTIVE/INDUCTIVE REASONING; PROOF; QED; THEOREM.

induction (complete induction, finite induction, mathematical induction) The method of proof known as mathematical induction has been used by scholars since the earliest times, but it was not until 1838, thanks to the work of English logician and mathematician AUGUSTUS DE MORGAN, that the principle was properly identified and described. In a formal context, the principle of mathematical induction asserts:

If a set *S* of numbers satisfies the following two properties:

    i. The number 1 belongs to *S*.
    ii. If a number *k* belongs to *S*, then so does its successor *k* + 1.

then it must be the case that *S* contains all the natural numbers 1, 2, 3,...

(This principle appears as an AXIOM in GIUSEPPE PEANO's set of postulates for the construction of the natural numbers.) Often the set *S* is taken to be a set of natural numbers *n* for which some property or formula *P*(*n*) is true: $S = \{n: P(n) \text{ holds}\}$.

To illustrate the principle of induction, we shall prove that, for every natural number *n*, we have that $1 + 2 + 3 + ... + n$ equals $\frac{1}{2}n(n + 1)$. Let *P*(*n*) represent this assertion and set S to be the set of all natural numbers for which this assertion is true:

$$P(n): 1 + 2 + 3 + \ldots + n = \frac{1}{2}n(n + 1)$$

$$S = \{n: \text{the formula } P(n) \text{ is true}\}$$

Our aim is to show that all natural numbers belong to S (that is, that $P(n)$ is true for all natural numbers n). We must establish two things:

1. The number 1 belongs to S.

Entering $n = 1$ into the formula yields the patently true statement $1 = \frac{1}{2} \times 1 \times (1 + 1)$. Thus $P(1)$ is true and $1 \in S$.

2. Assume that the number k belongs to S. That is, assume that the formula

$$1 + 2 + \ldots + k = \frac{1}{2}k(k + 1)$$

is true. Does it necessarily follow that the corresponding formula for $k + 1$ is valid? That is, can we then deduce, under this assumption, that

$$1 + 2 + \ldots + k + (k + 1) = \frac{1}{2}(k + 1)((k + 1) + 1)$$

will hold?

To achieve this, start with the equation $P(k)$, assumed to be valid, and add the quantity $k + 1$ to both sides. This gives:

$$1 + 2 + \ldots + k + (k + 1) = \frac{1}{2}k(k + 1) + (k + 1)$$

After rearranging and regrouping terms on the right, this reads:

$$1 + 2 + \ldots + k + (k + 1) = \frac{1}{2}(k + 1)(k + 2)$$

which is the statement $P(k + 1)$. By the principle of mathematical induction, it now follows that the set S does indeed contain all the natural numbers, that is, $P(n)$ is a valid formula for all values of n.

The principle of mathematical induction is a powerful technique that can establish the validity of an infinite number of statements in one fell swoop. As illustrated above, it is particularly useful for establish-

ing formulae and equations involving the natural numbers. One disadvantage, however, is that this method of proof gives no indication as to what the appropriate formula to be proved should be: the inductive method of proof offers no insight, for instance, as to why the formula for the sum of the first n counting numbers is $\frac{1}{2}n(n + 1)$. One must arrive at candidates for formulae via other means.

The inductive process is often likened to knocking down a chain of dominoes. To successfully topple a row of standing dominoes one must:

1. Knock down the first domino
2. Be certain that the dominoes are appropriately spaced so that when any one domino falls, it is sure to knock down the next

If one can establish that these two properties hold, then all dominoes in a chain (even an infinitely long one) will fall.

Despite the elegance and simplicity of the principle, one must still apply care when utilizing the principle of mathematical induction. The following amusing argument, for instance, illustrates what can go wrong:

Claim: all horses are the same color.

We "prove" this as follows: Let S be the set of all natural numbers for which the following statement is true.

$P(n)$: if n horses stand in a field, then all horses in that field are the same color.

Clearly $P(1)$ is true: if only one horse is standing in a field, then all horses in that field are the same color. Now make the assumption, despite its absurdity, that $P(k)$ is true, that is, any k horses in a field must be the same color. Now consider $k+1$ horses in a field. Remove one horse, Chester, say. This leaves k horses in the field, which, by our assumption, must all be the same color. Return Chester to the field and remove a different horse. This again leaves k horses in the field, which, by assumption, must all be the same color. This shows that Chester is the same color as the first k horses, and in fact that all $k+1$ horses are the same color. We have, from $P(k)$, established that $P(k+1)$ follows. By the principle of mathematical induction, it now follows that $P(n)$ is true, no matter the value of n. In particular, all the horses in the world are the same color.

To examine this argument, note that it is indeed the case that $P(1)$ is valid, and it is true that $P(k)$ implies $P(k+1)$—for almost all values of k. The fault in the argument above is that although $P(2)$ establishes $P(3)$, and $P(3)$ establishes $P(4)$, and so forth, it is not the case that $P(1)$ establishes $P(2)$: removing one horse from a field containing just two horses does not, alas, establish that both horses are the same color. In our chain of dominoes, all but the first two dominoes are properly spaced: the first domino does not topple the second domino during its fall, and the chain remains standing. This illuminates that one must always be careful that any argument presented in a proof by induction is indeed valid for all values of k.

See also PROOF; SUMS OF POWERS.

inductive reasoning *See* DEDUCTIVE/INDUCTIVE REASONING.

inequality A mathematical statement that one quantity or expression is greater than or less than another is called an inequality. The following symbols are used:

$a > b$, a is greater than b
$a < b$, a is less than b
$a \geq b$, a is greater than or equal to b
$a \leq b$, a is less than or equal to b

Inequalities satisfy a number of ORDER PROPERTIES.

An inequality is called closed or unconditional if it holds for all values of the variables, if any, that appear in the equation. For instance, the inequalities $3 \leq 5$, $x^2 + 1 > x^2$, and $5 + y^2 > 4y$ are closed inequalities for they are always true. Inequalities that are not closed are called open or conditional. The set of values of the variables that appear in the inequality that make the statement true is called the solution set of the inequality. For example, the open inequality $2x + 1 > 7$ has as a solution set the set of all real numbers x for which $x > 3$. The solution set of the open inequality $a^2 + b^2 < 0$ is the EMPTY SET.

Open inequalities can be solved in much the same manner as equations. As the order properties show, one can add or subtract the same quantity to both sides of the inequality and preserve the inequality, or can multiply the inequality through by a positive quantity. Multiplying through by a negative quantity changes the sense of the inequality. (For example, if $a < b$, then subtracting the quantity $a + b$ from both sides yields $-b < -a$. This shows that the effect of multiplying through by -1 is to reverse the sense of the inequality.) For example, one can solve the inequality $2x + 1 > 7$, as follows:

$$2x + 1 > 7$$
$$2x > 6$$
$$\frac{1}{2} \times 2x > \frac{1}{2} \times 6$$
$$x > 3$$

indeed yielding the solution set $\{x : x > 3\}$.

It is worth noting that if $a \cdot b > 0$, then we can be sure that a and b are either both positive or both negative. If, on the other hand, $a \cdot b < 0$, then we can be sure that a and b have opposite signs. These observations are essential for solving inequalities involving a single variable raised to the second power. For example, to solve the open inequality $x^2 + x - 2 > 0$, factor the QUADRATIC to obtain $(x + 1)(x - 2) > 0$ and examine the two possible scenarios. Either $x + 1$ and $x - 2$ are both greater than zero (yielding that x must be greater than 2), or $x + 1$ and $x - 2$ are both less than zero (yielding that x must be less than -1). Thus the solution set to the inequality is the set of all real numbers x with $x > 2$ or $x < -1$.

A single inequality in two variables defines a region in the plane. For example, the inequality $2x + y \geq 3$ is satisfied by the points $(0,3)$, $(5,0)$, and $(2,2)$, for instance. In this example, the complete solution set is the closed HALF-PLANE sitting above the line $y = -2x + 3$ in the plane.

There are a number of standard inequalities in mathematics:

Triangle inequality: For any triangle with side-lengths a,b, and c we have: $a + b > c$.
Arithmetic-geometric mean inequality: For nonnegative numbers $a_1, a_2, \ldots, a_n$ we have:

$$\sqrt[n]{a_1 a_2 \cdots a_n} \leq \frac{a_1 + a_2 + \cdots + a_n}{n}$$

Bernoulli's inequality: For any real number x greater than 1 and positive integer n:

$$(1 + x)^n > 1 + nx$$

(This can be proved by truncating the MACLAURIN SERIES:

$$(1 + x)^n = 1 + nx + \frac{1}{2}n(n-1)x^2$$
$$+ \frac{1}{6}n(n-1)(n-2)x^3 + \ldots,$$

One can also show that the inequality is also true for $-1 < x < 0$).

Cauchy-Schwarz inequality: For any two VECTORS **a** and **b**, their DOT PRODUCT satisfies:

$$|\mathbf{a} \cdot \mathbf{b}| \leq |\mathbf{a}| \times |\mathbf{b}|$$

Thus, for any two lists of numbers $a_1, a_2, \ldots, a_n$ and $b_1, b_2, \ldots, b_n$ we have:

$$\left(\sum_{k=1}^{n} a_k b_k\right)^2 \leq \left(\sum_{k=1}^{n} a_k^2\right)\left(\sum_{k=1}^{n} b_k^2\right)$$

Weierstrass's product inequality: For any list of numbers $a_1, a_2, \ldots, a_n$ with $0 \leq a_k \leq 1$ for all k, we have:

$$(1 - a_1)(1 - a_2)\ldots(1 - a_n) \geq 1 + a_1 + a_2 + \ldots + a_n$$

(This can be proved by INDUCTION on the number of elements in the list.)

Napier's inequality: For any two positive real numbers a and b we have:

$$\frac{1}{b} < \frac{\ln b - \ln a}{b - a} < \frac{1}{a}$$

Exponential inequalities: For a positive real number x and a real number c, we have:

$$x^c < 1 + c(x - 1) \text{ if } 0 < c < 1$$
$$x^c > 1 + c(x - 1) \text{ if } c > 1$$

(This is a generalization of Bernoulli's inequality.)

Isoperimetric inequality: For any closed geometric figure in the plane with perimeter P, its area A is less than the area of a circle of the same perimeter:

$$A \leq \pi\left(\frac{P}{2\pi}\right)^2 = \frac{P^2}{4\pi}$$

Equality holds if, and only if, that figure is a circle.

A mathematical statement that two quantities or expressions are never equal is called an inequation. For instance, the statement $3 \neq 5$ is an inequation.

Mathematicians and physicists often write $a >> b$ if a is significantly larger than b and $a << b$ is a is significantly smaller. For example, $\sqrt{n^2 + 100n} \approx n$ if $n >> 0$. There is a joke among mathematicians to use the symbol $\leq\geq$ to mean "less than, greater than, or possibly equal to" when one is not sure of the numerical answer to a problem.

See also ARITHMETIC-GEOMETRIC MEAN INEQUALITY; ISOPERIMETRIC PROBLEM; TRIANGLE INEQUALITY.

inference   In logic, the general process of developing an ARGUMENT to draw a conclusion from a set of premises is called inference. The process could be deductive or inductive.

In statistics, inference is the process of coming to a conclusion about a population based on a study of a sample. Sometimes the conclusion itself is called an inference. Inferential statistics is the science of making inferences and predictions about a population based on numerical information gathered from a sample.

See also DEDUCTIVE/INDUCTIVE REASONING; POPULATION AND SAMPLE; STATISTICS: INFERENTIAL.

infinite product   The product of an infinite number of factors, $a_1 \times a_2 \times a_3 \times \ldots$, is called an infinite product. The nth number in the product is called the nth term of the product, and the product of the first n terms, $P_n = a_1 \times a_2 \times \ldots \times a_n$ is called the nth partial product. In 1812 CARL FRIEDRICH GAUSS introduced the notation $\prod_{i=1}^{\infty} a_i$ for an infinite product (and $\prod_{i=1}^{n} a_i$ for the nth partial product).

An infinite product might have a value of zero ($1 \times \frac{1}{2} \times \frac{1}{3} \times \frac{1}{4} \times \ldots$, for example), might be infinite in value ($1 \times 2 \times 3 \times 4 \times \ldots$, for example), or could oscillate in value ($1 \times (-1) \times 1 \times (-1) \times \ldots$, for instance). Only if the partial products P_n approach a finite *nonzero* value L as $n \to \infty$ is the infinite product said to converge (to the value L). Otherwise the infinite product diverges.

For example, the infinite product $\prod_{i=2}^{\infty}\left(1 - \frac{1}{i^2}\right)$ has nth partial product:

$$P_n = \left(1-\frac{1}{2^2}\right)\left(1-\frac{1}{3^2}\right)\left(1-\frac{1}{4^2}\right)\cdots\left(1-\frac{1}{n^2}\right)$$

$$= \left(1-\frac{1}{2}\right)\left(1+\frac{1}{2}\right)\left(1-\frac{1}{3}\right)\left(1+\frac{1}{3}\right)\left(1-\frac{1}{4}\right)\left(1+\frac{1}{4}\right)$$

$$\cdots\left(1-\frac{1}{n}\right)\left(1+\frac{1}{n}\right)$$

$$= \frac{1}{2}\cdot\frac{3}{2}\cdot\frac{2}{3}\cdot\frac{4}{3}\cdot\frac{3}{4}\cdot\frac{5}{4}\cdots\cdots\frac{n-1}{n}\cdot\frac{n+1}{n}$$

$$= \frac{n+1}{2n}$$

which approaches the value $\frac{1}{2}$ as $n \to \infty$. Thus we can write $\prod_{i=2}^{\infty}\left(1-\frac{1}{i^2}\right)=\frac{1}{2}$.

By taking the LOGARITHM, we see that an infinite product with positive terms $\prod_{i=1}^{\infty}a_i$ converges if, and only if, the SERIES $\sum_{i=1}^{\infty}\ln a_i$ converges. (This is why mathematicians do not deem convergence to zero as a valid form of convergence for an infinite product: the value of the sum $\sum_{i=1}^{\infty}\ln a_i$ would then be $-\infty$) Mathematicians have shown that a series $\sum_{i=1}^{\infty}a_i$ converges absolutely exactly if $\prod_{i=1}^{\infty}(1+|a_i|)$ converges.

In 1656 English mathematician JOHN WALLIS discovered the following remarkable representation for $\frac{\pi}{2}$ as an infinite product:

$$\frac{\pi}{2}=\frac{2}{1}\cdot\frac{2}{3}\cdot\frac{4}{3}\cdot\frac{4}{5}\cdot\frac{6}{5}\cdot\frac{6}{7}\cdots.$$

This expression is today called WALLIS'S PRODUCT.

See also ABSOLUTE CONVERGENCE; ZETA FUNCTION.

infinitesimal A positive quantity, supposedly infinitely small yet not itself zero, is called an infinitesimal. The abstract existence of such quantities played an important, yet troublesome, role in the early development of CALCULUS. The theoretical difficulties incurred by them were later made moot when AUGUSTIN-LOUIS

CAUCHY (1789–1857) introduced the notion of a LIMIT. Cauchy's work put calculus on a sound theoretical setting and removed the need to ever speak in terms of infinitesimals. The notion is now considered obsolete. (Although mathematicians have recently developed a theory of nonstandard analysis that does, in some sense, include a valid concept of an infinitesimal, after all.)

See also FLUXION; HISTORY OF CALCULUS (essay).

infinity In common usage, the word *infinite* is used to denote something that is unbounded, limitless, and endless. For example, the set of counting numbers 1, 2, 3, … is unbounded (after any given counting number there is always another), and the set of these numbers is said to be infinite. In geometry, a straight line is usually perceived as without end, and so it is infinite in extent. In the physical world, however, there is no clear example of infinite quantity (or, if there were, we as humans would never be able to fully perceive it). Physicists have come to believe, for instance, that there are only a finite number of atoms in the universe (on the order of 10^{87} atoms) and that the universe is closed in shape and not of infinite extent. If this is indeed the case, then it would never be possible to draw an infinite line (there are not enough atoms for the ink), and if the universe is indeed bounded, the line may eventually loop back on itself. Nonetheless, the infinite is an abstract concept we feel we can, on some level, comprehend. It does, however, bring with it many paradoxical difficulties, as demonstrated by ZENO'S PARADOXES, HILBERT'S INFINITE HOTEL, and the TRISTRAM SHANDY PARADOX, for example.

The notion of something that is unlimited arises in mathematics in a number of varied settings. For instance:

1. *Limits:* In 1655 English mathematician John Wallis introduced the symbol ∞ as shorthand for the phrase "becoming large and more positive." This notation is used today, in particular, in the study of LIMITs. For example, the statement $\lim_{n\to\infty}\frac{10n^2-3n+2}{2n^2+n+1}=5$ is to be read: "the quantity $\frac{10n^2-3n+2}{2n^2+n+1}$ approaches the value 5 as n becomes large and positive." The equation $\lim_{n\to\infty}\frac{2-x^3}{x}=-\infty$ reads: "the function $\frac{2-x^3}{x}$ becomes large and negative as x becomes large and positive." This means that for any negative number $-M$ one cares

to choose, one can always find a larger value of x for which $\dfrac{2-x^3}{x}$ is, and remains, less than $-M$ for ever larger values of x.

2. *Geometry:* In DESARGUES'S THEOREM, French mathematician GIRARD DESARGUES (1591–1661) found it convenient to regard parallel lines as intersecting at some point of infinity. Thus in the theory of PROJECTIVE GEOMETRY, the notion of points "infinitely far away" is given meaning and context.

3. *Set theory:* Italian astronomer and physicist GALILEO GALILEI (1564–1642) observed that every counting number can be matched with its square, showing in some sense that the infinite set of counting numbers is no more infinite than the subset of square numbers:

$$1 \quad 2 \quad 3 \quad 4 \quad 5 \quad 6 \quad \ldots$$
$$\updownarrow \quad \updownarrow \quad \updownarrow \quad \updownarrow \quad \updownarrow \quad \updownarrow$$
$$1 \quad 4 \quad 9 \quad 16 \quad 25 \quad 36 \ldots$$

In the 19th century, British algebraist AUGUSTUS DE MORGAN (1806–71) and German mathematicians JULIUS WILHELM RICHARD DEDEKIND (1831–1916) and GEORG CANTOR (1845–1918) realized that this is a common property of infinite sets, and that it is appropriate to use this property as the definition of what it means for a set to be infinite:

A set is infinite if its elements can be matched, without repetition, with the elements of a proper subset of itself.

This definition has the advantage that it makes clear what it means for a set to be finite.

A set is finite if it is not infinite.

(A comment should be made on this point. Although we all have a clear intuitive understanding of what it means for a set to be FINITE, it is not at all easy to provide a direct mathematical description of this concept. However, it is possible to show that no set of the form $\{1,2,3,\ldots,n\}$ is infinite, that is, there is no means to match the elements of this set with the elements of a proper subset of itself without producing repetition. This is done with an INDUCTION argument on n. One can use this as a link between this indirect approach and our intuitive under-

standing.) Cantor went further to develop an astounding theory of CARDINALITY that shows, among other things, that there are many different types of infinite sets, some deserving of being called "more infinite" than others.

The notion of the infinite has been studied and used since antiquity. ARCHIMEDES OF SYRACUSE (ca.287–212 B.C.E.) used the notion of an infinitely small quantity to develop formulae for the areas and volumes of curved figures and solids. (In some vague sense, one can view a circle, for instance, as a regular polygon with infinitely many sides, all infinitely short in length. It is better, however, to view the circle as the LIMIT figure of a sequence of regular polygons.) The geometer EUCLID (ca.300–260 B.C.E.) proved that the set of PRIME numbers is infinite, and ZENO OF ELEA (ca. 490–425 B.C.E.) contemplated the infinite in his studies of time, space, and motion.

Although the scholars of Greek antiquity utilized the infinite, they were wary of it. Euclid, for example, went to great pains to phrase matters in a way that never made mention of a quantity that was actually infinite. For instance, he proved that from any given finite collection of primes, one can always construct one more (rather than state that the set of prime numbers *is* infinite), and in his famous work *THE ELEMENTS*, he never made mention of lines that continue indefinitely; he only spoke of extending line segments further if needed. (This subtle turn of phrase proved to be important to GEORG FRIEDRICH BERNHARD RIEMANN with his 19th-century invention of SPHERICAL GEOMETRY.) ARISTOTLE (384–322 B.C.E.) argued that the "actual infinite" did not exist, and that one can only argue in terms of potentiality: given a finite part, one can always provide more. This point of view held fast for almost two millennia. Cantor's work on the actual infinite, inspired by the beginning ideas of Dedekind and De Morgan, was deemed revolutionary at its time.

inflection (inflexion) *See* CONCAVE UP/CONCAVE DOWN.

inflection point (point of inflection) A point on a curve at which the tangent line to the curve changes from rotating in one sense (clockwise or counterclockwise) to rotating in the opposite sense. The concavity of the curve changes at such a point.

See also CONCAVE UP/CONCAVE DOWN.

information theory The branch of PROBABILITY theory that deals with the transmission, processing, and checking of messages sent electronically is called information theory. The field was established in 1948 by American mathematician CLAUDE EDWARD SHANNON, who first showed that it is possible to encode all types of information (words, sounds, pictures) into simple sequences of 0 and 1 bits that can then be transmitted along a wire as pulses. (Up to then, scientists thought it would be necessary to transmit electromagnetic *waves* along wires.) To adjust for erroneous noise that is often transmitted along with the message, mathematicians have since developed probability techniques that determine the likelihood that the message received is free from errors. Over the decades, mathematicians have also developed efficient redundancy checks to help detect and correct errors. These techniques are far more efficient than simply repeating the message for comparison.

Information theory is also used to measure the amount of information a signal of any kind might contain. Transmitting and correctly receiving a single letter of the alphabet—that is, one of any 26 letters—contains more information, for instance, than the receipt of a single binary digit, 0 or 1. Mathematicians use logarithms to measure information content and say that receipt of a letter of the alphabet contains $\frac{\log_2 26}{\log_2 2} \approx 4.7$ as much information as receipt of a binary digit. (We have $\frac{\log_2 2}{\log_2 2} = 1$.) This assumes that each letter in the alphabet is equally likely to occur. In general, if the probability of the letter *a* being sent is p_1, the letter *b* is p_2, and so forth, then the measure of informational content is given by:

$$- p_1 \log_2 p_1 - p_2 \log_2 p_2 - \ldots - p_{26} \log_2 p_{26}$$

(This agrees with the previous value of 4.7 computed when each probability p_i has value 1/26.) This quantity is a measure of the entropy of the initial data set.

The field of information theory has obvious applications to work in telegraphy, radio transmission, and the like, but it has also recently been used to analyze human speech, in the study of languages, and in cybernetics.

instantaneous value The value of a varying quantity, such as VELOCITY, at a particular instant in time is called its instantaneous value. An instantaneous value is a DERIVATIVE.

integer (directed number, signed number) Any of the positive or negative WHOLE NUMBERS, or ZERO, is called an integer: ...–3,–2,–1,0,1,2,3,... More precisely, once the NATURAL NUMBERS have been defined via PEANO'S POSTULATES, say, one can define an integer to be any quantity that can be expressed as the sum or difference of two natural numbers. (For instance, the number –5 can be regarded, formally, as the pair of natural numbers 3 and 8 written in the form 3–8. It can be defined equally well by the pair 1–6 or the pair 12–17, for example.) The difference of any two integers is always another integer, which constitutes a mathematical RING. The set of integers is denoted **Z** (from the German word *Zahlen* for "numbers"). German mathematician GEORG CANTOR (1845–1918) showed that the set of integers is COUNTABLE and so has CARDINALITY $\aleph_0$.

The TAYLOR SERIES of $f(x) = \dfrac{x}{(1-x)^2}$ has the integers appearing as coefficients:

$$\frac{x}{(1-x)^2} = x + 2x^2 + 3x^3 + 4x^4 + \ldots$$

(This series is valid for –1 < x < 1.) This shows, for instance, that in setting $x = \frac{1}{100}$, the fraction $\frac{10}{81}$ has the integers appearing in turn in each decimal place: $\frac{10}{81} = 0.12345\ldots$ (Unfortunately our practice of "carrying digits" disguises the fact that the pattern we see continues.)

See also FLOOR/CEILING/FRACTIONAL PART FUNCTIONS.

integral *See* ANTIDIFFERENTIATION; INTEGRAL CALCULUS.

integral calculus The calculation of sums of infinitely small quantities, INFINITESIMALS, is called integral calculus. For example, consider the problem of finding the length of a curved path using only a straight yardstick. One could *approximate* the distance along the curve by marking a number of points along the curve, measuring the straight-line segments between them, and summing the lengths of these segments.

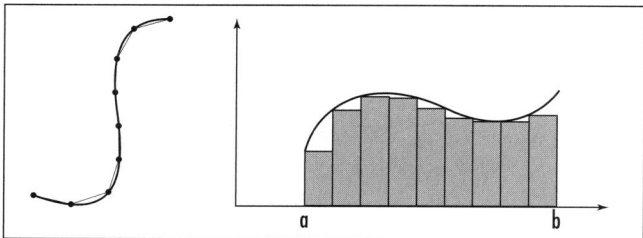

Approximating length and area

An even better approximation could be made using more points and consequently shorter line segments. The *actual* length of the curve would be the LIMIT value of these improved approximations as we use shorter and shorter line segments connecting more and more points on the curve (the sum of infinitely short line segments). Similarly, the area under a curve drawn in the plane can be approximated as the sum of the areas of a finite number of rectangles drawn under the curve. Using narrower and narrower rectangles will give better and better approximations. The *actual* area under the curve is the limit value of these approximations (the sum of infinitely narrow rectangles).

Any process that involves segmenting a quantity into manageable pieces, summing, and taking the limit of these sums as the process is refined falls under the category of integral calculus. Traditionally, integral calculus is first taught as the process of finding the area under a curve $y = f(x)$ over an interval $[a,b]$. The area denoted

$$\int_a^b f(x)\, dx$$

is called a definite integral, and is defined to be the limit, as h tends to zero, of the sums of the areas of rectangles of width at most h, used to approximate the area of the curve as described above. Such an approximation with rectangles is called a Riemann sum, in honor of the German mathematician BERNHARD RIEMANN (1826–66), whose work led mathematicians to show that this approach is indeed mathematically sound, in particular that, under reasonable conditions, all ways of approximating the area under the curve lead to the same limit value.

GOTTFRIED WILHELM LEIBNIZ (1646–1716), one of the inventors of CALCULUS, introduced the symbol $\int$ to represent an integral. He thought of it as an elongated

S denoting sum, and he called the theory of integration *calculus summatorius*. Swiss mathematician Johann Bernoulli (1667–1748) of the famous BERNOULLI FAMILY worked with Leibniz in developing the theory, but he preferred the name *calculus integralis* and the use of a capital letter I as the sign of integration. (In Latin, the word *integralis* means "making up a whole.") The two gentlemen settled on a happy compromise of using Bernoulli's name for the theory and Leibniz's symbol for the integral.

The idea of using a limit to calculate the areas of curved figures, or the lengths of curved paths, has been used by scholars from the time of ARCHIMEDES OF SYRACUSE in the third century B.C.E. to the time of PIERRE DE FERMAT in the middle of the 17th-century. In practice, however, the techniques employed to perform these calculations have always been extremely difficult and complicated. The great achievement of Leibniz and ISAAC NEWTON (1642–1727), independent discoverers of calculus, was to recognize that integration is simply a process of reverse differentiation, today called ANTIDIFFERENTIATION. This result, known as the FUNDAMENTAL THEOREM OF CALCULUS, essentially states that to find the area under a curve $y = f(x)$ over an interval $[a,b]$, look for a function $F(x)$ whose DERIVATIVE is $f(x)$. Then:

$$\int_a^b f(x)dx = F(b) - F(a)$$

The function $F(x)$ is called an antiderivative of f. The right-hand side of this equation is often abbreviated as $F(x)\big|_a^b$. Thus, for example, the area under the parabola $y = x^2$ from $x = 0$ to $x = 2$, is given by $\int_0^2 x^2\, dx = \frac{1}{3}x^3\Big|_0^2 = \left(\frac{1}{3}2^3\right) - \left(\frac{1}{3}0^3\right) = \frac{8}{3}$. This remarkable result obviates all need to work with complicated limits.

To highlight the interplay between integration and reverse differentiation, the antiderivative of a function $f(x)$ is usually denoted $\int f(x)dx$ and is called the indefinite integral of f. It is defined up to a CONSTANT OF INTEGRATION. Thus $\int f(x)dx$ is a *function* whose derivative is $f(x)$ (whereas the definite integral $\int_a^b f(x)dx$ is a *number* equal to the area under the curve $y = f(x)$ over the interval $[a,b]$). The thrust of integral calculus is the development of methods for finding the antiderivatives of functions.

Since the derivative of the sum of two functions is the sum of the derivatives, and the derivative of a function multiplied by a number k is just k times its derivative, we have:

$$\int f(x) + g(x)dx = \int f(x)dx + \int g(x)dx$$
$$\int k \cdot f(x)dx = k \int f(x)dx$$

The corresponding results for definite integrals also hold true. This follows from the fundamental theorem of calculus (since these results are true for the antiderivatives), but they can also be seen to be valid by geometric arguments. (For example, if the values of a graph are multiplied by a number k, that is, if the function $y = f(x)$ is replaced by $y = k \cdot f(x)$, then the area under the graph increases by a factor of k.)

The following table shows some common integrals:

$f(x)$	$\int f(x)dx$
x^r	$\dfrac{x^{r+1}}{r+1} + C$ provided $r \neq -1$
$\dfrac{1}{x}$	$\ln|x| + C$
e^x	$e^x + C$
e^{ax}	$\dfrac{1}{a}e^{ax} + C$
a^x	$\dfrac{a^x}{\ln a} + C$
$\ln x$	$x\ln x - x + C$
$\sin x$	$-\cos x + C$
$\cos x$	$\sin x + C$
$\tan x$	$-\ln|\cos x| + C$
$\csc x$	$\ln|\csc x - \cot x| + C$
$\sec x$	$\ln|\sec x + \tan x| + C$
$\cot x$	$\ln|\sin x| + C$
$\sin^2 x$	$\dfrac{1}{2}x - \dfrac{1}{4}\sin 2x + C$
$\cos^2 x$	$\dfrac{1}{2}x + \dfrac{1}{4}\sin 2x + C$
$\sinh x$	$\cosh x + C$
$\cosh x$	$\sinh x + C$
$\dfrac{f'(x)}{f(x)}$	$\ln|f(x)| + C$

There are a number of techniques for finding the indefinite integrals of more-complicated functions such as INTEGRATION BY PARTS and INTEGRATION BY SUBSTITUTION.

See also ANTIDIFFERENTIATION; CALCULUS; DIFFERENTIAL CALCULUS; DOUBLE INTEGRAL; HISTORY OF CALCULUS (essay); IMPROPER INTEGRAL; NUMERICAL INTEGRATION.

integral test See CONVERGENT SERIES.

integrand The function that is to be integrated is called the integrand. For example, the expression $f(x)$ in either of the integrals $\int f(x)dx$ or $\int_a^b f(x)dx$ is the integrand.

See also ANTIDIFFERENTIATION; INTEGRAL CALCULUS.

integration by parts This technique of integration is useful for finding the integral of the product of two functions. It makes use of the PRODUCT RULE for differentiation in reverse. Specifically, the product rule reads:

$$\frac{d}{dx}(u(x) \cdot v(x)) = u'(x)v(x) - u(x)v'(x)$$

Integrating both sides and rearranging thus yields:

$$\int u(x)v'(x)dx = u(x)v(x) - \int u'(x)v(x)dx$$

(A CONSTANT OF INTEGRATION will appear when all integrals are finally computed.) Thus one can make effective use of this formula if the integral $\int u'(x)v(x)dx$ turns out to be much easier to compute. For example, to evaluate $\int x \cos x \, dx$, write: $u(x) = x$ and $v'(x) = \cos(x)$, yielding $u'(x) = 1$ and $v(x) = \sin(x)$ (again ignoring a constant of integration for the moment) so that:

$$\int x\cos(x)dx = x\sin(x) - \int 1 \cdot \sin(x)dx = x\sin(x) + \cos(x) + C$$

One typically chooses the factor that is easy to differentiate to be $u(x)$ and the factor that is straightforward to integrate for $v'(x)$.

The integration-by-parts formula can be used even if the integrand is composed of a single factor. One can "insert a 1" into the integrand to imagine that there is a second factor equal to the constant function 1. To illustrate, to compute $\int \ln(x)dx$ write: $\int \ln(x)dx = \int \ln(x) \cdot 1dx$ and set:

$$u(x) = \ln(x) \quad v'(x) = 1$$

$$u'(x) = \frac{1}{x} \quad v(x) = x$$

Then:

$$\int \ln(x)\,dx = x\ln(x) - \int \frac{1}{x} \cdot x \, dx = x\ln(x) - x + C$$

The integration-by-parts formula can be applied more than once to complete an integration problem. For example, to evaluate $\int e^x \cos(x)dx$, one application of the technique yields:

$$\int e^x \cos(x)dx = e^x \sin(x) - \int e^x \sin(x)dx$$

Applying the integration-by-parts technique to the second integral yields:

$$\begin{aligned}\int e^x \cos(x)dx &= e^x \sin(x) - \int e^x \sin(x)dx \\ &= e^x \sin(x) - [-e^x \cos(x) - \int e^x (-\cos(x))dx] \\ &= e^x (\sin(x) + \cos(x)) - \int e^x \cos(x)\,dx\end{aligned}$$

Algebra now shows that the integral we seek is given by:

$$\int e^x \cos(x)dx = \frac{1}{2}e^x\left(\sin(x) + \cos(x)\right) + C$$

The method of integration by parts was discovered by English mathematician BROOK TAYLOR (1685–1731).

See also ANTIDIFFERENTIATION; INTEGRAL CALCULUS; INTEGRATION BY SUBSTITUTION; RECURRENCE RELATION.

integration by substitution (change of variable, substitution rule for integration) This technique of integration is used to find the integral of a function easily recognized as a COMPOSITION of two simpler functions. It is essentially the CHAIN RULE for differentiation employed in reverse. Specifically, the chain rule states:

$$\frac{d}{dx}f(u(x)) = f'(u(x))\frac{du}{dx}$$

Integrating both sides yields:

$$\int f'(u(x))\frac{du}{dx}dx = f(u(x)) + C$$

Noting that, since f is the antiderivative of f', the right-hand side of this formula can be thought of as the indefinite integral of f' evaluated with u as the variable. Thus it is valid to write:

$$\int f'(u(x))\frac{du}{dx}dx = \int f'(u)\,du$$

This equation is the change-of-variables equation for integration. For example, we can use it to evaluate $\int 2x(1 + x^2)^4\,dx$. Setting $u(x) = 1 + x^2$, giving $\frac{du}{dx}$, the integral reads:

$$\begin{aligned}\int 2x\left(1+x^2\right)^4 dx &= \int u^4 \frac{du}{dx}\,dx \\ &= \int u^4 \, du \\ &= \frac{1}{5}u^5 + C \\ &= \frac{1}{5}\left(1+x^2\right)^4 + C\end{aligned}$$

Notice that the notation we used here mimics the properties of fractions: we are permitted to replace a term $\frac{du}{dx}\,dx$ under an integral sign by the term dx.

When using this technique, one typically chooses $u(x)$ to be a function that simplifies a complicated part of the INTEGRAND, and then adjusts matters so that the factor $\frac{du}{dx}$ appears explicitly. For example, to evaluate $\int x^2 \sqrt{x^3 - 1}\,dx$, it is natural to set $u(x) = x^3 - 1$. Then $\frac{du}{dx} = 3x^3$, and we have:

$$\begin{aligned}\int x^2 \sqrt{x^3 - 1}\,dx &= \int x^2 \sqrt{u}\,dx \\ &= \int \frac{1}{3}3x^2 \sqrt{u}\,dx \\ &= \frac{1}{3}\int \sqrt{u}\frac{du}{dx}\,dx \\ &= \frac{1}{3}\int \sqrt{u}\,du \\ &= \frac{1}{3}\cdot\frac{2}{3}u^{\frac{3}{2}} + C \\ &= \frac{2}{9}\left(x^3 - 1\right)^{\frac{3}{2}} + C\end{aligned}$$

When computing a definite integral via this technique, it is advisable to compute the indefinite integral first and then, at the end of the process, work with the limits of integration. For example, the definite integral $\int_1^5 x^2 \sqrt{x^3-1}\, dx$ does not equal $\frac{1}{3}\int_1^5 \sqrt{u}\, du$, since the values $x = 1$ to $x = 5$ refer to the x-variable and not to the new u-variable. We compute:

$$\int_1^5 x^2 \sqrt{x^3-1}\, dx = \frac{2}{9}\left(x^3-1\right)^{\frac{3}{2}}\Big|_1^5 = \frac{2}{9}124^{\frac{3}{2}}$$

See also ANTIDIFFERENTIATION; INTEGRAL CALCULUS; INTEGRATION BY PARTS.

intercept A point at which two figures intersect is called an intercept. The term is usually reserved for the locations at which the GRAPH OF A FUNCTION crosses the x-axis and the y-axis in a system of CARTESIAN COORDINATES, to form what are called the x- and y-intercepts. If a straight line has x-intercept $(a, 0)$ and y-intercept $(0, b)$, then the equation of the line is given by $\frac{x}{a} + \frac{y}{b} = 1$. This is called the "intercept form" of its equation.

The y-intercept of an arbitrary function $y = f(x)$ is the point $(0, f(0))$.

interest A fee paid for the use of money is called interest. For instance, mortgage companies charge home buyers a fee for borrowing money, and credit card companies charge customers a fee for the privilege of using company money to make purchases. In reverse, banks pay customers money for maintaining a balance in a savings account. (The customers have, in effect, lent money to the institution.) In any such arrangement, the amount borrowed or invested is called the principal or capital, and the fee, expressed as a percentage rate, that is, as a number of dollars per hundred to be paid each year of the loan, is called the interest rate.

There are two types of interest. The first, simple interest, is computed only on the principal. For example, if a customer borrows $3,000 for 3 years under simple interest computed at 15 percent per annum, then the customer has agreed to pay $0.15 \times 3{,}000 = \$450$ for each year of the loan, plus return the original $3,000 at the end of the 3-year period. The customer thus pays a total of $4,350 at the end of the 3 years.

In general, the total interest I paid on a principal amount P at an annual interest rate R (expressed as a decimal) for T years is given by the formula:

$$I = P \times R \times T$$

(In our example, $I = 3{,}000 \times 0.15 \times 3 = 1{,}350$.) At the end of the loan, the amount A owed is:

$$A = P + I = P + P \times R \times T = P(1 + RT)$$

It is rare today, however, that a lending institution will provide loans with fees computed by simple interest. Typically, one is also expected to pay interest on any interest incurred as the loan progresses.

Compound interest is calculated by adding the interest to the principal and recalculating the interest at the end of agreed "conversion periods." For example, a credit card company may charge 18 percent interest per annum, but will compound the interest monthly (at a rate of $18/12 = 1.5\%$ per month). Thus, after borrowing $1,000, say, a customer is expected to return the principal plus $0.015 \times 1{,}000 = \$15$ in interest, that is, a total of $1,015, at the end of the first month. In failing to do so, the customer will then be expected to return this amount, plus an additional 1.5% on this amount, that is, a total of $1{,}015 \times (1 + 0.015) = \$1{,}030.23$ at the end of the second month. At the end of each month passed, the amount owed increases by a factor of $\left(1+\dfrac{0.18}{12}\right)$

$= (1 + 0.015)$. After 1 year, that is 12 months, the balance on the credit card will thus be $1{,}000 \times \left(1+\dfrac{0.18}{12}\right)^{12} = \$1{,}195.62$.

In general, if a principal amount of P dollars is borrowed at an interest rate of R percent per year (expressed as a decimal), and the interest is compounded n times a year for T years, then the amount A owed at the end of that time period is given by the formula:

$$A = P\left(1+\frac{R}{n}\right)^{nT}$$

Consider now a bank that offers its savings account customers an annual interest rate of 10 percent compounded weekly. A $1,000 investment after 1 year would thus yield $1,000 \times \left(1 + \dfrac{0.10}{52}\right)^{52} = \$1,105.06$.

If the interest is instead computed daily, customers receive a slightly better return: $1,000 \times \left(1 + \dfrac{0.10}{365}\right)^{365}$ = $1,105.16, and an even better return if the interest is compounded hourly, or even every minute. The maximum possible return is achieved when interest is computed every instant.

In the mid-1700s Swiss mathematician LEONHARD EULER tackled the problem of computing continuously compounded interest and showed that as n becomes large, the quantity $A = P(1 + \dfrac{R}{n})^{nT}$ approaches the value Pe^{RT}. (A study of the number e establishes this claim.) The formula:

$$A = Pe^{RT}$$

thus represents the balance of an investment, or a loan, after T years, under the ideal state of interest compounded continuously. Thus our $1,000 investment, after 1 year, compounded continuously at a rate of 10 percent per annum, yields a return of $1,000 \times e^{0.10 \times 1}$ = $1,105.17. Banks today use this formula to compute interest on savings accounts.

The practice of charging a fee for the use of money is an ancient one. Records show that many civilizations, the Hebrews, the Greeks, the Romans, and even the Babylonians of 2000 B.C.E., for instance, charged simple interest on loans. The Romans called the practice *usury* and often fees were as high as 60 percent. Some religious orders, including Christianity, Judaism, and Islam, questioned the ethics of the practice, arguing that one should only be charged a fee for the use of something that could be worn out or lose value due to wear and tear. (At the time, money was not seen to lose any value during the course of a loan.) For many centuries the practice of charging a fee for the use of money was forbidden by these orders.

During the growth of industry and trade during the Middle Ages and the Renaissance, attitudes changed. More and more people requested cash loans to take part in new opportunities, and lenders felt it appropriate to be compensated for not taking part in those opportunities themselves. Lenders again started charging fees. The church relaxed its attitude toward usury, and new establishments, called banks, were formed to handle, store, and loan money. The term *interest* from the Latin phrase *id quod interest* meaning "that which is between" soon replaced the term usury. Today the word *usury* is used only in a negative context of charging illegally high interest rates.

See also E.

interior angle   The ANGLE formed by two sides of a POLYGON lying inside the polygon is called an interior angle. For example, all four interior angles of a RECTANGLE equal 90°.

If an interior angle is greater than 180°, then that angle is called a "re-entrant angle" and the polygon is concave. Any interior angle less than 180° is called "salient."

Each interior angle of an n-sided regular polygon equals $\dfrac{n-2}{n} \times 180°$.

See also CONCAVE/CONVEX; TRANSVERSAL.

intermediate-value theorem (Bolzano's theorem)   Named after the Czech mathematician BERNHARD PLACIDUS BOLZANO (1781–1848), the intermediate-value theorem asserts that if $f(x)$ is a CONTINUOUS function defined on a closed INTERVAL, then this function assumes every value between $f(a)$ and $f(b)$; that is, if N is any number between $f(a)$ and $f(b)$, then there is at least one point c between a and b such that $f(c) = N$. For example, the function $f(x) = x^2$ is continuous on the interval [3,4] and has $f(3) = 9$ and $f(4) = 16$. Since 11 is between 9 and 16, the intermediate value theorem ensures us of the existence of a number c, between 3 and 4, such that $f(c) = c^2 = 11$, that is, it proves that the square root of 11 exists.

The theorem is intuitively clear if we think of a continuous function on a closed interval as one whose graph consists of a single continuous piece with no gaps, jumps, or holes: in moving a pencil from the left endpoint $(a, f(a))$ to the right endpoint $(b, f(b))$, it seems obvious that the pencil tip adopts all "heights" between initial height $f(a)$ and final height $f(b)$. In climbing the face of a mountain, say, one must indeed pass through every intermediate height from the base to the apex. A rigorous proof of the theorem, however, relies on the notion of the "completeness" of the real numbers (meaning that no points, like the square root of 11, are missing from the real line). This is a subtle property, one that was not properly understood until the late 1800s with the construction of a DEDEKIND CUT.

The intermediate-value theorem is useful for locating roots of equations. For example, consider the function $f(x) = x^3 + 2x - 5$. It is continuous on the interval $[1,2]$ and satisfies $f(1) < 0$ and $f(2) > 0$. It follows then that the equation $x^3 + 2x - 5 = 0$ has a root somewhere between 1 and 2. By working with smaller and smaller intervals, one can often use this method to determine the location of a root with a good degree of precision. The BISECTION METHOD, for example, employs this technique.

The intermediate-value theorem has a number of amusing consequences:

In theory, it is always possible to slice a pancake, no matter how irregular its shape, exactly in half with a single straight-line cut.

Hold a knife to the left of the pancake so that 100 percent of the cake lies to its right. Now slide the knife, in parallel, across the pancake until it lies on the other side of the cake. At this location, zero percent of the pancake lies to the right of the knife. By the intermediate-value theorem, there must be some intermediate location for the knife that yields the value of 50 percent lying to its right. That is, there is, in theory, a knife position that cuts the pancake exactly in half.

Note that this result does not depend on the angle we initially hold the knife—vertically, horizontally, or diagonally. We have in fact shown that it is always possible to slice a pancake in half with a knife held at any previously set angle.

In theory, it is always possible to simultaneously slice *two* pancakes each exactly in half with a *single* straight-line cut, no matter the shapes of the pancakes nor their location on the table.

This result is known as the two-pancake theorem. The previous result assures us we are always able to slice the first pancake exactly in half, pointing the knife at any angle we care to choose. The concern is that the knife might or might not cut the second pancake. Sup-

pose we find a direction, deem this angle zero degrees, that slices the first pancake in half, but misses the second pancake entirely, with 100 percent of the second pancake lying to the right of the knife, say. Turn the knife 180°. We now have a knife cut (same line, opposite direction) that slices the first pancake exactly in half, with the second pancake lying entirely to its left, that is, zero percent to the right. By the intermediate-value theorem, there must be an intermediate angle between zero degrees and 180° that slices the first pancake exactly in half, and has 50 percent of the second pancake lying to its right—that is, one that simultaneously slices the second pancake in half as well.

At any instant, there are two points on the Earth's equator directly opposite each other with exactly the same air temperature.

For each position θ degrees longitude on the equator, let $f(\theta)$ be the air temperature at this position minus the air temperature at the opposite point of the equator, at $\theta + 180°$:

$$f(\theta) = temp(\theta) - temp(\theta + 180°)$$

Notice that $f(\theta + 180°)$ equals the same value, but opposite in sign, to $f(\theta)$:

$$f(\theta + 180°) = temp(\theta + 180°) - temp(\theta) = -f(\theta)$$

Thus the function $f(\theta)$ moves from positive to negative values. By the intermediate-value theorem, there must be an intermediate location where the function is zero. This is the desired position on the Earth's equator.

See also EXTREME-VALUE THEOREM; FIXED POINT; HAM-SANDWICH THEOREM; SPHERE.

interpolation The process of estimating the value of a function between two values already known is called interpolation. For example, if the temperature of a cup of tea was initially 200°F, and two minutes later it was 180°F, one might guess that its temperature at the one-minute mark was 190°F, based in the assumption that the temperature decreases steadily over time. This represents that simplest method of interpolation, called linear interpolation: we presuppose that the variation of the function can be described as a straight line passing through the known values.

If a function f has values $f(a)$ and $f(b)$ at locations a and b, the linear interpolation estimates the value of the function at x, between a and b, as:

$$f(x) \approx f(a) + \frac{x-a}{b-a}\big(f(b) - f(a)\big)$$

If the graph of f is smooth and the interval $[a,b]$ is small, then linear interpolation generally gives a good approximation of the true value. If, however, the interval is large, it is less likely that this will remain the case. For example, given that $3^2 = 9$ and $3^4 = 81$, linear interpolation suggests that 3^3 (which equals 27) is approximately $9 + \frac{3-2}{4-2}(81-9) = 45$.

Improved methods of interpolation take into account other data values. For example, JOSEPH-LOUIS LAGRANGE's interpolation method, called LAGRANGE'S FORMULA, estimates function values by fitting a POLYNOMIAL to all the known data values. Alternatively, if it is suspected, for example, that the function is exponential, that is, of the form $y = b^x$, then linear interpolation works well for the logarithm of the function: the relationship $\log y = b \log x$ is linear.

EXTRAPOLATION is the process of estimating function values outside of the range of values observed.

See also REGRESSION.

interval   Any single segment of the number line is called an interval. More precisely, an interval is a set of numbers containing all real numbers between two given numbers. The given numbers are called endpoints, and they might, or might not, be regarded as part of the interval. For example, the set of all real numbers between, and including, 2 and 5 is an interval; as is the set of all real numbers strictly greater than 9 but less than or equal to 23.

An interval is called closed if both endpoints are included. Square brackets are used to indicate that this is the case. For example,

$$[a,b]$$

denotes the set of all real numbers x that satisfy $a \le x \le b$. On a number line, the endpoints are marked by blacked-in circles to indicate that they are included.

An interval is open if neither endpoint is included. Parentheses are used to indicate that this is the intent. For example:

$$(a,b)$$

denotes the set of all real numbers x that satisfy $a < x < b$. On a number line, the endpoints are marked by open circles to indicate that they are not included.

If a variable x satisfies either $a \le x < b$ or $a < x \le b$ then it is said to be located in a half-open interval denoted, respectively:

$$[a,b) \text{ or } (a,b]$$

On a number line, a blacked-in circle marks the endpoint that is included in the interval and an open circle the endpoint that is excluded.

With this notational system, the two intervals described in the opening paragraph are thus denoted $[2,5]$ and $(9,23]$. Some texts use reverse square brackets instead of parentheses to denote the exclusion of endpoints—thus an open interval is denoted $]a,b[$, and a half-open interval as $[a,b[$, for example.

The length of an interval is the distance between its endpoints. For example, the interval $(9, 23]$ has length 14, as do the intervals $[9, 23]$, $[9, 23)$, and $(9, 23)$.

Incorporating the symbol ∞ for infinity allows one to denote unbounded intervals. For example, $[a, \infty)$ denotes the set of all real numbers x larger than or equal to a; $(-\infty, a)$ denotes the set of all real numbers x strictly less than a; and so on. Sometimes mathematicians write $(-\infty, \infty)$ for the set of *all* real numbers. (One typically does not place a square bracket next to the infinity symbol, although it is possible to define a number system that allows for inclusion of ∞ as a valid number for consideration.)

An interval can consist of a single point (the interval $[1,1]$ for example) or can be the empty set (the interval $[4,1]$ for example). The intersection of any two intervals is an interval, but the union of two intervals might or might not be an interval.

See also CARDINALITY; INTERMEDIATE-VALUE THEOREM.

invariant   A property, quantity, or relationship that is not changed by collection of specific operations or transformations, or by how that property is observed, is called an invariant. For example, the distance between two points in a plane does not change under a ROTATION. Thus distance is invariant of rotations. (It is not, however, invariant of DILATIONs, for instance.) An application of EULER'S THEOREM shows that any soccer ball design that uses hexagons and pentagons, no matter how irregular those shapes may be, with three edges meeting at each vertex, must include precisely 12 pentagons. (The number of hexagons can vary.) The count of pentagons is thus an invariant in soccer ball design.

Some invariants can be quite surprising. For example, consider the following theoretical exercise:

> Suppose 10 ft of length is added to a rope that was just long enough to wrap snugly around the equator of the Earth. Imagine that the rope is again wrapped around the equator, but this time—due to its extra length—it hovers just above the ground. How high is the gap between the ground and the suspended rope?

To answer this question, let R denote the radius of the Earth. The length of the extended rope is $2\pi R + 10 = 2\pi\left(R + \dfrac{5}{\pi}\right)$ feet, which corresponds to the circumference of a circle of radius $R + \dfrac{5}{\pi}$ feet. This shows that the rope hovers $\dfrac{5}{\pi} \approx 1.6$ ft off the ground. Notice that the answer, apart from being surprisingly large in value, does not depend on the value of R. This means that the size of the planet is immaterial. Thus if 10 ft of length is added to a rope that fits snugly around the equator of any planet—Mars, Jupiter, or a planet the size of a pea—the extended rope will always hover $\dfrac{5}{\pi}$ ft off the ground. This value is an invariant for the problem.

As another example, consider the following famous "splitting game":

> Write the number 12 at the top of a page and below it write a pair of positive whole numbers that sum to 12, say, 7 and 5. On the side of the paper record the product $7 \times 5 = 35$. Now write below 7 a pair of numbers that sum to 7, say, 3 and 4, and record the product $3 \times 4 = 12$. Continue in this manner, "splitting" each number that appears in the diagram into two and recording the product of the pair of numbers chosen. Do this until the number 1 appears 12 times. The following represents one possible such splitting diagram:

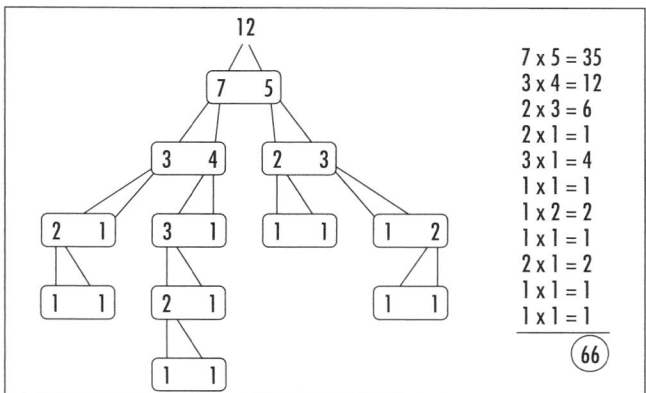

The splitting game

> Now sum all the products recorded. What value is obtained?

Surprisingly, no matter which splitting diagram one constructs, the sum of products is an invariant of the game and will always have value 66. (The number 66 happens to be the 11th TRIANGULAR NUMBER. In general, if one begins this game with a number N, then the invariant that arises in the game is the $(N-1)th$ triangular number.)

inverse element   An element of a set that, when combined with another element produces the IDENTITY ELEMENT of the set, is called an inverse element. More precisely, if a set S comes equipped with a BINARY OPERATION "$*$" and an identity element e, then an inverse for an element a of the set is another element b such that $a*b = b*a = e$.

For example, for the set of numbers under the operation of addition, the inverse of any number a is its negative $-a$. In this context, the identity element is zero and we do indeed have:

$$a + (-a) = (-a) + a = 0$$

Under the operation of multiplication, the identity element is 1, and the inverse of any (nonzero) number a is its reciprocal $\dfrac{1}{a}$:

$$a \times \dfrac{1}{a} = \dfrac{1}{a} \times a = 1$$

Each and every element of a GROUP is required to have an inverse. The inverse of the identity element is itself.

It is not possible for a single element a to have two different inverses b_1 and b_2. This is established by noting that $e = a*b_2$ so that $b_1 = b_1*e = b_1*(a*b_2) = (b_1*a)*b_2 = e*b_2 = b_2$.

From the symmetry of the definition, we have that if b is the inverse of a, then a is also the inverse of b. Consequently, the inverse of an inverse is the original element. Phrased in terms of addition, this reads $-(-a) = a$ and in terms of multiplication as:

$$\frac{1}{\frac{1}{a}} = a$$

The inverse of an element a is often denoted a^{-1}, especially if the binary operation under consideration can be interpreted as a type of multiplication. For example, the inverse of a square MATRIX A, if it exists, is denoted A^{-1}.

inverse function (inverse mapping, reverse function)
A FUNCTION f with domain D and range R, $f : D \to R$, is said to be invertible or to have an inverse function if, for each possible output y of the function, $y \in R$, there is one, and only one, input $x \in D$, that produces that output. We write $x = f^{-1}(y)$ for the input x that produces the given output y. (Thus $x = f^{-1}(y)$ if, and only if, $f(x) = y$.) This then defines a function $f^{-1}: R \to D$, called the inverse function to f. In some sense, the inverse function "undoes" the original function.

For example, consider the function on real numbers that doubles an input and adds 3: $f(x) = 2x + 3$. The output 11 is produced from the input of 4, and so we have $f^{-1}(11) = 4$. In general, an output of y is produced from an input $x = \dfrac{y-3}{2}$, and so $f^{-1}(y) = \dfrac{y-3}{2}$. (This formula is obtained by solving for x in the equation: $y = 2x + 3$ to yield $x = \dfrac{y-3}{2}$.)

Since $f^{-1}(y)$ is the input that produces the output y, and x is the input that produces the output $f(x)$, the following relations hold:

$$f(f^{-1}(y)) = y \text{ for all values } y \text{ in the range of } f$$

and

$$f^{-1}(f(x)) = x \text{ for all } x \text{ in the domain of } f$$

This explains the awkward notation for the inverse function: In the study of the COMPOSITION of functions,

f^m denotes the composite $f_0 \ f_0 ... _0 f$ (m times), and we have $f^m {}_0 f^n = f^{m+n}$. To give meaning to the quantity f^0, this rule states that $f_0 f^0 = f^1 {}_0 f^0 = f^{1+0} = f$, suggesting that we should set $f^0(x) = x$ for all values x. Consequently, the statement $f^{-1} {}_0 f^1 = f^0$ suggests that $f^{-1}(f(x)) = x$ for all x, indicating that f^{-1} is the appropriate notation for the inverse function. The superscript of -1 should not be confused with the operation of inversion. (We write $(f(x))^{-1}$ to denote $\dfrac{1}{f(x)}$, and leave $f^{-1}(x)$ to mean the inverse function of f.)

It is customary to denote the input of a real function as the variable x and the output as the variable y. This can lead to some confusion. For instance, to compute the inverse function of $y = f(x) = x^3 + 2$ we solve for the input x in the equation in terms of the output y to yield, $x = \sqrt[3]{y-2}$, but we interchange the x and y variables so that x denotes the new input and y the new output: $y = \sqrt[3]{x-2}$. This yields the formula $f^{-1}(x) = \sqrt[3]{x-2}$ for the inverse function.

As the formulae $y = f(x)$ and $x = f^{-1}(y)$ represent exactly the same equation, the two formulae yield exactly the same curves when plotted against a pair of x- y-coordinate axes. Following the convention to interchange the x- and y-variables for the second equation to write $y = f^{-1}(x)$ is tantamount to interchanging the x- and y-axes in the graph of the curve. Flipping the graph across the diagonal line $y = x$ returns the y-axis to the vertical position and the x-axis to the horizontal position, but also flips the curve drawn across the diagonal line. Thus the graphs of $y = f(x)$ and $y = f^{-1}(x)$ are mirror images of each other across a diagonal line.

Not every function possesses an inverse function. For example, there is no inverse function to the squaring function $y = x^2$: some outputs arise from more than one possible input. (The output of 4, for instance, arises from the two inputs 2 and -2.) However, it is often possible to restrict a function to a certain portion of its domain and define an inverse function for that restricted domain. For instance, for the squaring function, if we require that only nonnegative inputs are to be considered, then an inverse function does exist: we have $y = \sqrt{x}$ (the positive square root) as inverse function. One defines the INVERSE TRIGONOMETRIC FUNCTIONS, for example, by restricting to a suitable portion of the domain.

If the function $y = f(x)$, then the derivative of the inverse function $y = f^{-1}(x)$ is given by:

$$\frac{d}{dx}f^{-1}(x) = \frac{1}{f'\left(f^{-1}(x)\right)}$$

provided that the quantity in the denominator is not zero. This formula can be established by making use of the CHAIN RULE in the statement $f(y) = x$. (Differentiating gives: $f'(y) \cdot y' = 1$ and so $y' = \frac{1}{f'(y)}$, which is the above formula.)

A simple version of a general inverse-function theorem states that if the derivative of a function $y = f(x)$ is nonzero at a point $x = a$, then an inverse function exists, at least when the domain is restricted to a small interval about a. (Since the derivative of $f(x) = x^2$ is zero at $x = 0$, it is not possible to define an inverse function to the squaring function about the point $x = 0$.)

inverse hyperbolic functions (area hyperbolic functions) Defined in an analogous way to the INVERSE TRIGONOMETRIC FUNCTIONS, the inverse hyperbolic functions are the inverse functions of the HYPERBOLIC FUNCTIONS. For instance, the inverse hyperbolic sine of a number x, written arc sinh x or $\sinh^{-1}x$, is a value a whose hyperbolic sine is x: $\sinh a = x$. Similarly, the inverse hyperbolic cosine of x is a value a with $\cosh a = x$, and the inverse hyperbolic tangent of x is a value a such that $\tanh a = x$. (Technically, for a given value x there are two different values a for which $\cosh a = x$, one positive and one negative. By convention, the positive value is always chosen.)

Since the hyperbolic sine function is defined on all real values and yields all real values as possible outputs, the function $\sinh^{-1} x$ is defined for all real values of x. On the other hand, the hyperbolic cosine function only yields output values greater than or equal to 1, and consequently the inverse hyperbolic cosine function $\cosh^{-1} x$ is defined only for values of $x \geq 1$. Similarly, the inverse hyperbolic tangent function $\tanh^{-1} x$ is defined only for $-1 < x < 1$.

The inverse hyperbolic functions have the following DERIVATIVES:

$$\frac{d}{dx}\sinh^{-1}x = \frac{1}{\sqrt{1+x^2}}$$
$$\frac{d}{dx}\cosh^{-1}x = \frac{1}{\sqrt{x^2-1}}$$
$$\frac{d}{dx}\tanh^{-1}x = \frac{1}{1-x^2}$$

These can be established by making use of the relation $\cosh^2 y - \sinh^2 y = 1$. For instance, to compute the derivative of $y = \sinh^{-1} x$, write $\sinh y = x$ and then differentiate this equation making use of the CHAIN RULE. This yields $\frac{dy}{dx} = 1$, thereby establishing:

$$\frac{dy}{dx} = \frac{1}{\cosh y} = \frac{1}{\sqrt{1+(\sinh y)^2}} = \frac{1}{\sqrt{1+x^2}}$$

as claimed.

It is possible to give alternative formulations of the inverse hyperbolic functions. Noting that $\cosh y = \frac{e^y + e^{-y}}{2}$ and $\sinh y = \frac{e^y - e^{-y}}{2}$, we have $\cosh y + \sinh y = e^y$ or $y = \ln(\cosh y + \sinh y)$. Set $y = \sinh^{-1} x$. Then $\sinh y = x$ and $\cosh y = \sqrt{1 + \sinh^2 y} = \sqrt{1 + x^2}$, yielding:

$$\sinh^{-1} x = \ln(x + \sqrt{x^2 + 1})$$

which is valid for all values of x. Similarly,

$$\cosh^{-1} x = \ln(x + \sqrt{x^2 - 1})$$

valid for $x \geq 1$, and

$$\tanh^{-1} x = \ln\sqrt{\frac{1+x}{1-x}}$$

valid for $-1 < x < 1$.

inverse matrix (matrix inverse) A square MATRIX A is said to be invertible (or nonsingular) if there is a matrix B such that $AB = BA = I$, where I is the IDENTITY MATRIX. The matrix B is called the inverse matrix to A. There is at most one inverse matrix to given matrix A. (If B_1 and B_2 are both inverse matrices, then $B_1 = IB_1 = B_2AB_1 = B_2I = B_2$.) If an inverse matrix for a matrix A exists, then it is denoted A^{-1}. A study of DETERMINANTs shows that a matrix is invertible if, and only if, its determinant is not zero.

The matrix inverse of a 2×2 matrix

$$A = \begin{pmatrix} a & b \\ c & d \end{pmatrix}$$

with determinant $\det(A) = ad - bc$ is the matrix:

$$A^{-1} = \frac{1}{ad-bc}\begin{pmatrix} d & -b \\ -c & a \end{pmatrix}$$

The process of GAUSSIAN ELIMINATION provides a relatively straightforward method for computing the matrix inverse of any square matrix of a larger size.

If, in a system of n SIMULTANEOUS LINEAR EQUATIONS $A\mathbf{x} = \mathbf{b}$, the matrix A of coefficients is invertible, then the system has solution given by $\mathbf{x} = A^{-1}\mathbf{b}$. In particular, if e_j denotes the column vector whose only nonzero entry is a 1 in the jth position, then $x_j = A^{-1}e_j$ is the jth column of A^{-1}. That is, the jth column of this inverse matrix is a solution to the system of equations $Ax_j = e_j$. By CRAMER'S RULE, the ith entry of this column, that is the (i,j)th entry of the inverse matrix, is the ratio of determinants:

$$(A^{-1})_{ij} = \frac{\det(A\,|_{ij})}{\det(A)}$$

where $A|_{ij}$ is the matrix A with the ith column replaced by e_j. Computing $\det(A|_{ij})$ is equivalent, up to a plus or minus sign, to computing the determinant of the matrix obtained from A by deleting its ith column and jth row. This value is sometimes called the (i,j)th cofactor of A.

If A is invertible, then it is impossible to find a nonzero column vector $\mathbf{x}$ such that $A\mathbf{x} = 0$. (Otherwise, $\mathbf{x} = A^{-1}\,0 = 0$.) This observation is important for the study of EIGENVECTORs and EIGENVALUEs.

See also GENERAL LINEAR GROUP.

inverse of a statement   *See* CONTRAPOSITIVE.

inverse square law   Any relationship between two physical variables for which one is proportional to the RECIPROCAL of the square of the other is referred to as an inverse square law. For example, the law of gravitation as developed by SIR ISAAC NEWTON (1642–1727) asserts that the magnitude F of the gravitational force between two bodies of masses m and M is given by:

$$F = G\frac{mM}{r^2}$$

Here G is the gravitational constant (equal to 6.67×10^{-11} m^{3}kg^{-1}sec^{-2}) and r is the distance between the two masses. This is an inverse square law. The illumination provided by a source of light decreases by the inverse of the square of the distance from the source and so too is an inverse square relationship.

inverse trigonometric functions   An INVERSE FUNCTION to any trigonometric function is called an inverse trigonometric function. For instance, the inverse sine of a number x, written arcsin$x$ or $\sin^{-1}x$, is an ANGLE a for whose sine is x: $\sin a = x$. Since the sine curve adopts values only between –1 and 1, the inverse sine function is defined only for values $-1 \le x \le 1$. One should also note that for any value x there are infinitely many angles a with $\sin a = x$. It is usually assumed then that the angle a is chosen so that $-\frac{\pi}{2} \le a \le \frac{\pi}{2}$. (This is called the range of principal values for sine.) Similarly the inverse cosine of a number x with $-1 \le x \le 1$, written arccos x or $\cos^{-1} x$, is an angle a, usually chosen in the principal range for cosine, $0 \le a \le \pi$, with $\cos a = x$. Since the tangent function adopts all real values, the inverse tangent function is defined for any real number x, and arctan x, or $\tan^{-1} x$, is defined to be that angle a in the principal range for tangent, $-\frac{\pi}{2} \le a \le \frac{\pi}{2}$, such that $\tan a = x$.

The inverse trigonometric functions have the following DERIVATIVES:

$$\frac{d}{dx}\sin^{-1} x = \frac{1}{\sqrt{1-x^2}}, \text{ for } x \ne \pm 1$$
$$\frac{d}{dx}\cos^{-1} x = -\frac{1}{\sqrt{1-x^2}}, \text{ for } x \ne \pm 1$$
$$\frac{d}{dx}\tan^{-1} x = \frac{1}{1+x^2}$$

These can be established by making use of the relation $\sin^2 y + \cos^2 y = 1$. For instance, to compute the derivative of $y = \sin^{-1} x$, write $\sin y = x$ and then differentiate this equation making use of the CHAIN RULE. This yields $\cos y \frac{dy}{dx} = 1$, thereby establishing:

$$\frac{dy}{dx} = \frac{1}{\cos y} = \frac{1}{\sqrt{1-(\sin y)^2}} = \frac{1}{\sqrt{1-x^2}}$$

as claimed. The TAYLOR SERIES of the arctan function gives GREGORY'S SERIES.

One can establish a number of forward-inverse identities for the trigonometric functions. As examples we have:

$$\sin(\cos^{-1} x) = \sqrt{1 - x^2}$$

$$\sin(\tan^{-1} x) = \frac{x}{\sqrt{1 + x^2}}$$

$$\cos(\sin^{-1} x) = \sqrt{1 - x^2}$$

$$\cos(\tan^{-1} x) = \frac{1}{\sqrt{1 + x^2}}$$

$$\tan(\sin^{-1} x) = \frac{x}{\sqrt{1 - x^2}}$$

$$\tan(\cos^{-1} x) = \frac{\sqrt{1 - x^2}}{x}$$

For instance, if $a = \cos^{-1} x$, then $\cos a = x = \frac{x}{1}$. Thus angle a appears in a right triangle with hypotenuse 1 and adjacent leg of length x. By PYTHAGORAS'S THEOREM, the length of the opposite leg is $\sqrt{1 - x^2}$ and so $\sin a = \frac{\sqrt{1 - x^2}}{1} = \sqrt{1 - x^2}$, establishing the first relation. The remaining identities are proved similarly.

See also INVERSE HYPERBOLIC FUNCTIONS.

irrational number Any number that cannot be expressed as a RATIO of two integers is called an irrational number. As the study of RATIONAL NUMBERS shows, the irrational numbers are precisely those numbers whose decimal expansions do not terminate or fall into a repeating cycle of values. For example, the number with the decimal expansion 0.113133133313333133333133... is irrational. The study of rational numbers also shows that, in a very real sense, "most" numbers are irrational.

A famous proof, often attributed to Hippasus of Metapontum (ca. 470 B.C.E.), shows that $\sqrt{2}$ is irrational. THEODORUS OF CYRENE (ca. 465–398 B.C.E.) established the same result geometrically, and also showed that the numbers $\sqrt{3}$ through to $\sqrt{17}$ (excluding $\sqrt{4}$, $\sqrt{9}$, and $\sqrt{16}$) are irrational. The FUNDAMENTAL THEOREM OF ARITHMETIC can be used to prove that the mth root of a positive integer n is rational if, and only if, n is already the mth power of an integer. (If $\sqrt[m]{n} = \frac{a}{b}$ for some integers a and b, then $a^m = nb^m$.

Writing each of a, b, and n as a product of primes, and noting that the primes that consequently appear on the left side of this equation must match those that appear on the right, we conclude that each prime factor of n appears in n a multiple of m times. This establishes that $n = c^m$ for some integer c.) The same reasoning shows that a number such as $\log_2 5$ is irrational. (If $\log_2 5 = \frac{a}{b}$, then $2^a = 5^b$, contradicting the fundamental theorem of arithmetic.)

Truncating the decimal expansion of an irrational number produces a rational arbitrarily close to that irrational number. For example, 1.4, 1.41, 1.414, ... is a sequence of rational numbers converging to $\sqrt{2} = 1.41421356...$

In 1737 LEONHARD EULER established that the number e is irrational, and in 1761 JOHANN HEINRICH LAMBERT (1728–77) proved the irrationality of π. No one to this day knows whether or not the numbers 2^e, π^e, and $\pi^{\sqrt{2}}$ are irrational. (It is known that e^π and $e \cdot \pi$ are irrational.) Surprisingly, the rationality or irrationality of EULER'S CONSTANT γ is still not known.

It is possible for an irrational number raised to an irrational power to be rational. For example, if $x = (\sqrt{2})^{\sqrt{2}}$ turns out to be rational, then we have an example of such a phenomenon. If x, on the other hand, is not rational, then it is irrational and $x^{\sqrt{2}} = ((\sqrt{2})^{\sqrt{2}})^{\sqrt{2}} = (\sqrt{2})^2 = 2$ is an example of what we seek. (Unfortunately this indirect line of reasoning does not indicate which of the two possibilities actually occurs.)

See also ALGEBRAIC NUMBER; CONTINUED FRACTION; E; NUMBER; REAL NUMBERS; SURD; TRANSCENDENTAL NUMBER.

isolated point (acnode) A point that satisfies the equation of a curve but is not on the main arc of the curve is called an isolated point. For example, the curve has $y^2 = x^3 - x^2$ has $x = 0$, $y = 0$ as a solution, with no other solution near this position. The point $(0,0)$ is an isolated point for the equation.

See also DOUBLE POINT.

isometry (congruence transformation) A GEOMETRIC TRANSFORMATION, such as a translation, rotation, or a reflection, that preserves the distances between points in space is called an isometry. Isometries thus have the property of preserving the shape and size of geometric figures. (If, for instance, three points *A*, *B*, and *C* represent the vertices of an equilateral triangle of side-length 1, then the images *A′*, *B′*, and *C′* of the three points under an isometry again form an equilateral triangle of side-length 1.)

The FUNDAMENTAL THEOREM OF ISOMETRIES shows that any isometry in the plane is the result of at most three reflections. Any isometry that preserves the location of three planar points leaves all points in the plane unchanged. No isometry can preserve the location of only two points. If an isometry preserves the location of just one point, then that isometry must be a rotation about that point.

isomorphism   A one-to-one and onto correspondence between the elements of two sets equipped with BINARY OPERATIONs that preserved the operations on those sets is called an isomorphism. For example, the map *f:n* → 2*n* matches each natural number with an even number (and, in reverse, each even number is paired with a unique natural number) and, moreover, preserves the operation of addition between the two sets: We have *f* (*n* + *m*) = *f*(*n*) + *f*(*m*) for all natural numbers *n* and *m*. Thus, with respect to addition, *f* is an isomorphism between the two sets. (Note, however, that *f* does not preserve products and so is not an isomorphism with respect to multiplication.)

The LOGARITHM function ln *x* provides a map between the set of all positive real numbers and the set of all real numbers. The logarithm converts multiplications in the first set into additions in the second set: ln(*xy*) = ln *x* + ln *y*. The set of positive reals under multiplication is thus isomorphic to the set of all reals under addition.

The word *isomorphism* is derived from the Greek words *isos* meaning "same" or "equal" and *morphos* meaning "shape" or "structure."

See also HOMOMORPHISM.

isoperimetric problem   A classical problem in mathematics, called the isoperimetric problem, asks:

Of all figures drawn in the plane with a given perimeter, which encloses the largest area?

The origin of this problem dates back to antiquity with the famous legend of Princess Dido and DIDO'S PROBLEM.

The topic of isoperimetrics (prefix *iso:* the same) was systematically studied by the 17th- and 18th-century Swiss family of mathematicians, the BERNOULLI FAMILY. They discovered and classified many curves satisfying certain maximum and minimum properties, but did not solve the isoperimetric problem. LEONHARD EULER (1707–83) also contributed to this field, using the techniques of CALCULUS.

JAKOB STEINER (1796–1863) used purely geometric methods to establish the following partial answer: *if there were a shape that offered maximal area for a given perimeter, then that shape must be a circle.* Steiner thought that he had completely solved the isoperimetric problem, that the answer must be a circle, but several years later German mathematician KARL THEODOR WILHELM WEIERSTRASS (1815–97) pointed out that the assumption that problems always have solutions can lead to absurdities. The following amusing example illustrates his concern:

> Consider the question: of all positive integers, which is the largest? If there is an answer to this problem, then it cannot be a number *n* different from 1, for then *n*² is an integer larger than *n*. This leaves only the integer 1 as the answer to the question.

Weierstrass realized that it could very well be the case that the isoperimetric problem has no solution, despite Steiner's work, and so perhaps even the circle can be replaced with a shape of greater area for the same perimeter. To completely solve the isoperimetric problem, Weierstrass realized one must also prove that a solution to the problem exists. Finally, in 1870, Weierstrass was able to do this by developing a new mathematical theory today called the "calculus of variations." The solution to the problem, known by ancient scholars, was, at long last, established as mathematically correct.

Simple variations of the isoperimetric problem are easier to handle. Consider, for example, the question: of all rectangles of perimeter 40 ft, which has the largest area? This can be answered via an exercise in algebra: Any rectangle has two long sides and two short sides. If the perimeter is 40 ft, then two sides are longer than 10 ft, and two are shorter than 10 ft (unless the figure is a square). Writing the dimensions of the rectangle as 10 + *x* ft long, and 10 – *x* ft wide, for some number *x*, its area is then given by the formula

$(10 - x)(10 + x) = 100 - x^2$. We see now that the area is maximal precisely when x is zero, that is, the figure is a square.

An exercise in geometry shows that of all triangles of a given perimeter, the equilateral triangle has largest area.

See also SOAP BUBBLES; STEINER POINT.

isosceles trapezoid (isosceles trapezium)   A trapezoid with the two nonparallel sides of equal length is called an isosceles trapezoid. The figure has a line of symmetry perpendicular to the parallel sides. Any isosceles trapezoid can be considered a truncated ISOSCELES TRIANGLE.

See also TRAPEZOID/TRAPEZIUM.

isosceles triangle   A triangle possessing two sides of equal length is called isosceles. The name is derived from the Greek terms *iso* meaning "same" and *skelos* meaning "leg." Some scholars insist that a triangle be called isosceles only if two, but not all three, of its sides are equal in length, but this is really only a matter of taste. (A triangle with all three sides equal in length is called equilateral.) The base angles of an isosceles triangle are equal. This can be established as follows:

Suppose triangle *ABC* is isosceles with sides *AB* and *BC* equal in length. Regard a diagram of this triangle as the picture of two triangles, *ABC* and *CBA* superimposed on top of one another. These two triangles share pairs of sides that match in length, and share a common angle (at *B*) between those two sides. Thus, by the SAS principle from the study of similarity, the two triangles are congruent. Consequently, all corresponding angles between the triangles match. In particular, the angle at vertex *A* must match the angle at vertex *B*.

If one slides the vertex *B* in a direction parallel to the base *AC,* then the perimeter of the triangle changes (although the AREA of the figure, $A = \frac{1}{2} \times$ base $\times$ height does not). The study of OPTIMIZATION shows that the sum of distances $|AB| + |BC|$ is at a minimum when the triangle is isosceles. We have:

Of all triangles with a given base and a fixed height, the isosceles triangle has the least perimeter.

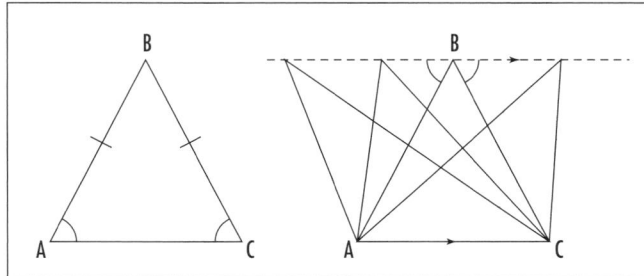

Isosceles triangle

To minimize the entire perimeter of a triangle without changing its area, one can adjust each vertex of the triangle (by sliding the vertex along a line parallel to the opposite side) so that it is an apex of an isosceles triangle. This produces an equilateral triangle and thereby establishes:

Of all triangles of a fixed area, the equilateral triangle has the least perimeter.

It is possible to arrange five points in a plane so that any three points chosen at random among them form the vertices of an isosceles triangle. (The points lie at the vertices of a regular pentagon.) It is impossible to accomplish the same feat with six or more points.

See also AAA/AAS/ASA/SAS/SSS; CONGRUENT FIGURES; TRIANGLE.

iteration   The repeated application of a mathematical procedure in which each step is applied to the output of the previous step is called iteration. For example, HERON'S METHOD for computing the square roots of numbers is an iterative procedure.

The iterates of a function f, beginning with the value x_0, are defined by the sequence of values:

$$x_1 = f(x_0),\ x_2 = f(x_1) = f(f(x_0)),\ x_3 = f(x_2) = f(f(f(x_0))),\ldots$$

If this sequence converges to a value a, say, then:

$$a = \lim_{n\to\infty} x_n = \lim_{n\to\infty} f(x_{n-1}) = f(a)$$

(We are assuming here that f is a CONTINUOUS FUNCTION.) Thus iteration provides a means for solving equations of the form $x = f(x)$. For example, repeatedly pushing the "cosine" button on a calculator will produce values that approach a solution to the equation $x = \cos x$. (As the calculator can only display eight or 10 decimal places, this means that the values shown on the screen will eventually stabilize.) NEWTON'S METHOD is often cited as the most widely used iterative procedure.

See also CONVERGENT SEQUENCE; DYNAMICAL SYSTEM; RECURSIVE DEFINITION.

J

Jacobi, Carl Gustav Jacob (1804–1851) German *Analysis* Born on December 10, 1804, in Potsdam, Prussia (now Germany), Carl Jacobi is remembered for his important work on the theory of elliptic functions and for applying his work in astonishing ways to the theory of numbers. He proved a famous conjecture of PIERRE DE FERMAT (1601–65) stating that every number can be written as the sum of four perfect squares. (Previously, CARL FRIEDRICH GAUSS had proved that every number is the sum of the three triangular FIGURATE NUMBERS.) He also made important contributions to the theory of dynamics in physics and made a careful study of the theory of DETERMINANTs. The generalized CHANGE OF VARIABLE formula for DOUBLE INTEGRALs (and higher-multiple integrals) contains a determinant that today is named in his honor.

Jacobi completed his entire secondary education within 1 year but was forced to wait several years before reaching the minimum of age of 16 to enter the University of Berlin. During this time Jacobi read advanced works in mathematics and conducted research work on polynomial equations. Jacobi received his doctorate from Berlin in 1825 and took a teaching position at the University of Königsberg a year later. By this time, Jacobi had already made some fundamental discoveries in the field of NUMBER THEORY. He commenced his work on elliptic functions soon afterward.

The change-of-variables formula in INTEGRAL CALCULUS states that if f is a function of a variable u, which in turn is a variable of x, then the integral of f with respect to u can be computed as:

$$\int_a^b f(u)\,du = \int_c^d f(u(x))\frac{du}{dx}\,dx$$

Here the limits of integration change to reflect the change of variable. In his study of determinants, Jacobi showed, in two dimensions, that if $f(u,v)$ is a function of two variables, with u and v each functions of x and y, then the appropriate change-of-variable formula becomes:

$$\iint_R f(u,v)\,du\,dv = \iint_S f(u(x,y),v(x,y))\begin{vmatrix} \dfrac{\partial u}{\partial x} & \dfrac{\partial u}{\partial y} \\ \dfrac{\partial v}{\partial x} & \dfrac{\partial v}{\partial y} \end{vmatrix} dx\,dy$$

where R is a region in the uv-plane, S is the corresponding region in the xy-plane, and $\begin{vmatrix} \dfrac{\partial u}{\partial x} & \dfrac{\partial u}{\partial y} \\ \dfrac{\partial v}{\partial x} & \dfrac{\partial v}{\partial y} \end{vmatrix}$ represents the determinant of the 2×2 MATRIX of PARTIAL DERIVATIVES. An analogous result applies for triple- and higher-multiple integrals. Such determinants are today called Jacobians. Although French mathematician AUGUSTIN-LOUIS CAUCHY (1789–1857) had discovered these transformation formulae earlier, it was Jacobi who first developed the theory of functional determinants fully in his comprehensive 1841 publication *De determinantibus functionalibus* (Functional determinants).

Jacobi was passionate about theoretical research and once uttered the following defense for pursuing science for its own sake: "The sole aim of science is the honor of the human mind, and from this point of view a question about numbers is as important as a question about the system of the world."

Jacobi died of smallpox on February 18, 1851.

Jiuzhang suanshu (*Chiu-chang Suan-shu*) *See* CHINESE MATHEMATICS.

Jordan, Marie Ennemond Camille (1838–1922) French *Topology, Abstract algebra, Engineering* Born on January 5, 1838, in Lyon, France, Camille Jordan is remembered for his work on PERMUTATION groups and the theory of equations. He revived interest in the important work of ÉVARISTE GALOIS and extended many of his ideas and offered substantial new contributions. Jordan's later work on ANALYSIS includes the famous theorem that now bears his name.

Jordan trained and worked throughout his life as an engineer. His interests in GROUP THEORY were motivated by the mathematical study of crystal structures. In many respects he defined this field of study with the publication of his 1870 text *Traité des substitutions et des equations algebraique* (Treatise on permutations and algebraic equations), the first book ever published on the topic. This work stimulated significant further mathematical research on the subject for the entire century that followed. Jordan was awarded the Poncelet Prize from the Académie des Science for the piece.

After the release of his famous work Jordan continued work in algebra, establishing important results in the fields of LINEAR ALGEBRA, NUMBER THEORY, DIFFERENTIAL EQUATIONS, mechanics, and FOURIER SERIES. Work in these latter fields led him to the topics of TOPOLOGY and analysis, and the study of continuous groups, which occupied the final part of his career.

Jordan died in Paris, France, on January 22, 1922. From 1961 to 1964 René Garnier and Jean Dieudonné collated all of Jordan's work and released the four-volume series *Oeuvres de Camille Jordan* (The work of Camille Jordan).

One point of confusion should be settled. Although Jordan did work in the field of matrix algebra, the method of GAUSSIAN ELIMINATION, often called Gauss-

Jordan elimination, is not named after Camille Jordan, but rather surveyor Wilhelm Jordan (1842–99). W. Jordan used this elimination method to properly analyze measurement errors. (The "Jordan canonical form" of a matrix, however, is due to Camille Jordan.)

See also JORDAN CURVE THEOREM.

Jordan curve theorem In 1893 French mathematician MARIE ENNEMOND CAMILLE JORDAN stated a fundamental result:

> Any loop drawn in the plane that does not intersect itself divides the plane into two distinct parts: an inside and an outside.

This utterly obvious theorem, now called the Jordan curve theorem, is surprisingly difficult to prove. Its validity says something significant about the geometry and structure of the surface on which the curve is drawn. For example, circles drawn on a plane or a sphere certainly divide those surfaces into two parts, but this need not be the case for a circle drawn on a TORUS, for example: a "loop" that circumnavigates the donut does not separate the surface into two distinct pieces. One must first develop a clear understanding of the geometry of space before results such as the Jordan curve theorem can be proved. This is a subtle and difficult issue. American mathematician Oswald Veblen (1880–1960) was the first to prove the theorem valid.

Josephus problem This famous mathematical puzzle is based on the difficult story of Josephus Flavius, a Jewish military leader fighting against the Romans in the town of Jotapata during the first century C.E. It is said that the battle, which took place in the spring of 67, was lost. Rather than submit their fates to the hands of the Romans, Josephus's men took brief refuge in a cave, where they voted to commit group suicide. Although Josephus did not agree with this plan, he offered no open opposition to it. Instead, he suggested that they proceed in as orderly a fashion as possible. He proposed that they sit in a circle and humanely kill every third man until only one person was left. That person would then commit suicide. Josephus, it is said, cleverly positioned himself so that he would be the final survivor. Rather than commit suicide, he surrendered to the Romans.

This story leads to the following general mathematical problem:

> Suppose N participants, numbered 1, 2, ..., N, sit in a circle and begin counting off every nth person. Each person so selected is called "out" and leaves the game. Counting continues until the last participant is declared the winner of the game. Given the size of the group, is it possible to predict beforehand who the winner will be? Assume the count begins with player 1.

The following table shows the position of the winner, counting every third person, for games involving $N = 1$ through $N = 15$ people:

N	1	2	3	4	5	6	7	8	9	10	11	12	13	14	15
$W(N)$	1	2	2	1	4	1	4	7	1	4	7	10	13	2	5

The numbers arising can be analyzed as follows: A game with N players becomes a game with $N-1$ players as soon as the first person is called out, except that the start of the $N-1$ game is moved three places along. Thus, if $W(N)$ denotes the winner of an N-player game, we have $W(N) = W(N-1) + 3$. One may need to adjust this formula, however, to take into account possible "counting around the full circle." For example:

> Clearly $W(1) = 1$.
>
> Adding 3 gives $W(2) = 4$. But, counting around the circle, we see that position four in a two-player game is really player 2. Thus $W(2) = 2$.
>
> Adding three gives $W(3) = 5$. But, counting around the circle, we see that position five in a three-player game is really player 2. Thus $W(3) = 2$,
>
> and so on.

In this way we can compute all the entries of the table without having to perform the game. Other versions of the game can be analyzed similarly.

Jourdain's paradox

In 1913 French mathematician Philip Jourdain proposed a variation of the famous LIAR'S PARADOX, now sometimes called the card paradox. On one side of a card is printed: "The statement on the other side of this card is true," and on the other

reverse side: "The statement on the other side of this card is false." Neither sentence can be true or false, for the statement on the reverse side implies the opposite.

It is worth pointing out that if, instead, both sentences read: "The statement on the other side of this card is false," then no paradox occurs; it is simply the case that just one of the sentences is actually false, and the other is true. If, on the other hand, both sentences read: "The statement on the other side of this card is true," then both statements could be true, or both could be false.

See also SELF-REFERENCE.

jug-filling problem

A famous category of decanting problems contains puzzlers of the following ilk:

> Given a 3-gallon jug and a 5-gallon jug (without any markings), is it possible to draw exactly 1 gallon of water from a well?

As one is not given the means to measure the exact contents of a partially filled jug, we are left with only three allowable maneuvers:

1. Completely fill an empty jug from the well
2. Completely empty a full jug into the well
3. Pour water from one jug into another, completely filling or emptying one jug in the process

This particular problem has an easy solution: completely fill the 3-gallon jug and pour its contents into the 5-gallon jug. Refill the 3-gallon jug and pour part of its contents to fill up the 5-gallon jug. This leaves (and we are certain of this) precisely 1 gallon of water in the 3-gallon jug and 5 gallons of water in the second. Now empty the 5-gallon jug.

In this solution the 3-gallon jug was filled twice and the 5-gallon jug emptied once. If we count +1 each time a jug is filled and −1 for each time it is emptied, the solution described above can thus be represented by the equation:

$$2 \times 3 + (-1) \times 5 = 1$$

Surprisingly *any* solution to the DIOPHANTINE EQUATION $x \times 3 + y \times 5 = 1$ corresponds to a solution to the jug-filling problem. (For example, $(-3) \times 3 + 2 \times 5 = 1$ represents a solution in which the 5-gallon jug is completely filled twice and the 3-gallon jug completely emptied three times.) Since 1 is the GREATEST COMMON FACTOR of 3 and 5, the EUCLIDEAN ALGORITHM provides solutions to the equation $x \times 3 + y \times 5 = 1$, and hence to the jug-filling problem.

As a variation we ask: is it possible to obtain exactly 1 gallon of water using a 9-gallon jug and a 15-gallon jug? This would require finding a solution to the equation $x \times 9 + y \times 15 = 1$. If there were a solution to this problem, then 1 would be a combination of two multiples of 3, and so itself a multiple of 3. This, of course, is absurd, and there is no solution to this problem. This type of argument can be used to show that d gallons of water can be obtained from an a-gallon jug and a b-gallon jug if, and only if, d is a multiple of the greatest common factor of a and b.

See also BICONDITIONAL.

Julia set   *See* FRACTAL.

al-Kashi, Jamshid Mas'ud (ca. 1380–1429) Iranian *Arithmetic, Astronomy* Born in Kashan, Iran, al-Kashi is best remembered in mathematics for his early use of decimals to represent fractions and to approximate real numbers, two centuries before European scholars developed the same technique. His famous text, *The Key to Arithmetic,* completed March 2, 1427, describes all the mathematics necessary for those studying astronomy, surveying, architecture, and accounting. Al-Kashi developed iterative techniques for extracting nth roots and for solving CUBIC and QUARTIC EQUATIONS, and gave an estimate for the value of π correct to 16 decimal places. (This far surpasses the achievement of the ancient Greek and Chinese scholars who approximated π as a fraction correct, in decimal notation, to six decimal places.)

That al-Kashi and his contemporaries chose to represent quantities as decimals indicates an important and sophisticated shift in the historical understanding of "number." For instance, writing π explicitly as a decimal, just as any number can be so written, changed the status of π from a physical geometric construct (the ratio of the circumference of a circle to its diameter) to an ordinary member of the class of REAL NUMBERS. Al-Kashi adopted a perspective where all of mathematics was thus unified by the study of the real-number system.

Al-Kashi was very interested in astronomy and viewed much of his mathematical work as a tool for assisting astronomical calculations. Techniques of TRIGONOMETRY played a prominent role. In his text *The Treatise on the Chord and Sine,* al-Kashi presented tables of sine values and listed a value of sin(1°) accurate to 16 decimal places. This allowed for extremely accurate astronomical calculations. On a purely mathematical note, al-Kashi also worked on the TRISECTING AN ANGLE problem, noting that the obstacle behind it reduces to one of solving a cubic equation. Applying iterative techniques, al-Kashi was able to provide an innovative approach to the problem, one that would find approximate solutions to any prescribed degree of accuracy. European scholars working on the same problem throughout the Middle Ages and the Renaissance were unaware of al-Kashi's work.

Kendall's method *See* RANK CORRELATION.

Kepler, Johannes (1571–1630) German *Astronomy, Calculus* Born on December 27, 1571, in Württemberg, Germany, Johannes Kepler is best remembered for his three laws of planetary motion: that planets travel in elliptical orbits with the Sun at one focus; that they do so at a rate so that the line joining the planet to the Sun sweeps out equal areas in equal units of time; and that the square of the period of a planetary orbit is proportional to the cube of the width of the ELLIPSE it traverses. These laws were all based on detailed observation of the orbit of Mars. As a mathematician, Kepler is noted for giving the first clear proof of how LOGARITHMS work, for his study of the geometry of the PLATONIC SOLIDS, for his discovery of two nonconvex regular polyhedra, and for his mathematical treatment of the close-packing properties of SPHERES. (He also explained why the honeycomb shape is the most efficient design for dividing a planar region into separate cells.) More importantly, Kepler devised a method for computing the volumes of many SOLIDS OF REVOLUTION with the aid of INFINITESIMALS. Today this is seen as a significant contribution to the development of CALCULUS.

Kepler studied astronomy and theology at the University of Tübingen, Germany. At the time, only six planets were known to astronomers, and all were assumed to be in circular orbit about the Sun. In 1596 Kepler published *Mysterium cosmographicum* (Mys-

Johannes Kepler, an eminent astronomer of the 17th century, is noted for his introduction of three laws of planetary motion. These laws provided Sir Isaac Newton the inspiration to develop his inverse-square law of gravitation. *(Photo courtesy of the Science Museum, London/Topham-HIP/The Image Works)*

tery of the cosmos), in which he presented a mathematical theory explaining the relative sizes of the planets' orbits. Convinced that God had created the universe according to a mathematical plan, Kepler posed that if a sphere were drawn about the path of Saturn and a CUBE inscribed in this sphere, then the orbit of Jupiter lies on a sphere inscribed in this cube. Moreover, inscribing a TETRAHEDRON in this second sphere and a sphere within the tetrahedron captures the orbital path of Mars. Continuing in this way, with a dodecahedron between Mars and Earth, an icosahedron between Earth and Venus, and an octahedron between Venus and Mercury, Kepler produced a model for orbit sizes that is accurate to within 10 percent of observed values, well within experimental error. As there are only five Platonic solids, this model also explained why, supposedly, there were only six planets. Of course Kepler's Platonic model of the solar system is not correct—three more planets were later discovered and, as Kepler himself later established, no orbit of a planet is circular.

Kepler moved to Prague near the turn of the century to work with one of the foremost astronomers of the time, Tycho Brahe (1546–1601). When Brahe died, Kepler succeeded him as imperial mathematician.

Brahe had kept extensive records on the orbit of Mars, and from them, Kepler was forced to conclude that its orbit was an ellipse. He also noted, from the data, that the velocity of the planet altered in such a way that the line connecting the Sun to the planet swept out equal areas in equal times. These two laws, when extended to all planets, are today called Kepler's first two laws. He published them in his 1609 piece *Astronomia nova* (New astronomy). Ten years later he added a third law: the squares of the times taken by the planets to complete an orbit are proportional to the cubes of the lengths of the major axes of their elliptical orbits.

Kepler had no explanation as to why these laws were true other than the compelling evidence of the data. It was not until SIR ISAAC NEWTON (1642–1727) formulated his famous law of gravitation that Kepler's laws could be mathematically deduced.

During his marriage ceremony in 1613 to his second wife Susanna (his first wife Barbara died in 1611), Kepler noticed that the servants would measure the volume of a wine barrel by slipping a rod diagonally through the bunghole and measuring the length that fit. He began to wonder why this method worked. This led him to his study of the solids of revolutions and the publication of his famous 1615 piece *Nova stereometria doliorum* (New stereometry of wine barrels). This work was later developed further by BONAVENTURA CAVALIERI (1598–1647).

In 1614 JOHN NAPIER published his account of logarithms, a mathematical device he invented with the aim of helping astronomers with their large numerical calculations. Kepler was delighted to learn of Napier's work and helped provide a mathematical explanation of Napier's method. Kepler also published tables of logarithmic values accurate to eight decimal places.

Kepler died on November 15, 1630, in Regensburg, Germany. His second law of planetary motion played a crucial role in shaping Sir Isaac Newton's thinking in the development of his theory of mechanics. Although Newton made no direct mention of Kepler's work in his early writings, he credited Kepler as the source of inspiration for his ideas in a lecture he gave to the members of the ROYAL SOCIETY before the release of his famous 1687 work *Principia*.

al-Khwārizmī, Muhammad ibn Mūsā (ca. 780–850)

Arab *Algebra* Born in Khwārizm, (now Khiva), Uzbekistan, Arab scholar Muhammad ibn Mūsā al-Khwārizmī is remembered for writing two extraordinarily influential texts. The first, *Al-jam' w'al-tafriq ib hisab al-hind* (Addition and subtraction in Indian arithmetic), introduced the Indian system of numerals to the West, and a second work that sparked the development of algebra. The title of his second work, *Hisab al-jabr w'al-muqābala* (Calculation by restoration and compensation), when translated into Latin, produced the very word *algebra* we use today.

Al-Khwārizmī was a scholar at the House of Wisdom in Baghdad, a learning academy and library founded by the Caliph al-Mamun in 813. (This institution was the first library constructed anywhere in the world since the destruction of the famous Library of Alexandria.) Housing a large collection of Greek philosophical and scientific texts, the library hired Arab scholars to translate manuscripts into Arabic and to conduct further research in the topics covered. Al-Khwārizmī was one such scholar who took particular interest in the works on mathematics and astronomy.

In his first famous pieces, *Al-jam' w'al-tafriq ib hisab al-hind*, al-Khwārizmī described and explained the advantages of the decimal-place system as used in India for writing numbers and doing arithmetic. This work was translated into Latin some 300 years later and became the primary source for Europeans who wanted to learn the new numeration system. It is interesting to note that although this work uses a symbol for zero (in Arabic called *sifr*) al-Khwārizmī did not regard it as a number and used it only as a placeholder (to distinguish 203 from 23, for instance).

Al-Khwārizmī intended that his second famous piece, *Hisab al-jabr w'al-muqābala*, also be used as a teaching guide. It aimed to offer an array of techniques and methods for solving very practical problems in matters of trade, inheritance, law, surveying, and architecture. Beginning at an extraordinarily elementary level, al-Khwārizmī first defined the natural numbers and the act of counting to then move on to the main topic of the first section of the piece, namely, solving elementary linear and quadratic equations. It is important to note that all of al-Khwārizmī's mathematics was done entirely in words, and no symbols were ever used. He called a number a "unit," an unknown a "root," and a quantity squared a "square."

To solve equations al-Khwārizmī used two operations which he called *al-jabr* and *al-muqābala*. The first, "completion," is the process of removing any negative terms from an equation. In modern notation, this converts $x^2 - 5x = 3$ to $x^2 = 3 + 5x$, for instance. The second, "balancing," is the process of subtracting positive terms of the same power when they occur on each side of an equation. In modern notation, this means rewriting $5x^2 + 7 = 3x^2 + x$, for instance, as $2x^2 + 7 = x$. With these two maneuvers, al-Khwārizmī reduces all linear and quadratic equations into six basic types, which he lists. The remainder of this part of the text is devoted to solving these equations. Today's standard practice of COMPLETING THE SQUARE is described in this section.

The next part of al-Khwārizmī's text consists of applications and worked examples. Because many of the problems were geometric in nature, al-Khwārizmī would dismiss negative solutions to problems and allow only positive answers. He recognized that quadratic equations may yield two solutions to a problem. Al-Khwārizmī also computed areas of simple geometric figures, as well as the volume of a sphere, the cone, and the pyramid, in this section. The final section of the text focuses on the complicated Islamic rules for inheritance and does not develop any new mathematical content.

By the 10th century, Spain was under Islamic control, and European scholars interested in the intellectual culture of the Islamic world traveled to Spain to study the Arabic texts. It was there that Al-Khwārizmī's works were discovered and translated into Latin.

Al-Khwārizmī also wrote texts on astronomy, the sundial, the Jewish calendar, and on geography. He computed the latitudes and longitudes of over 2,400 specific localities in preparation for the construction of an accurate world map.

The word *algorithm* is believed to be derived from al-Khwārizmī's name. Medieval European scholars, attempting to translate the Arab scholar's name into Latin, called the practice of using Hindu-Arabic numerals "algorism," from which, in turn, any general practice or procedure became known as an "algorithm." That a version of his name became part of our Western vocabulary illustrates the extent to which al-Khwārizmī's work influenced the development of arithmetic and algebra in Europe.

Klein, Felix Christian (1849–1925) German *Abstract algebra, Geometry, Topology* Born on April 25, 1849, scholar Felix Klein is remembered for uniting the disparate fields of GEOMETRY and ALGEBRA through the study of GEOMETRIC TRANSFORMATIONS. He showed that the abstract analysis of the algebra of transformations through GROUP THEORY leads to a clear understanding of both EUCLIDEAN GEOMETRY and NON-EUCLIDEAN GEOMETRY. In particular, he proved that ordinary Euclidean geometry can be proved CONSISTENT, that is, free of contradiction, if, and only if, non-Euclidean geometry is consistent. This demonstrated, for the first time, that the controversial non-Euclidean theories of geometry were of equal importance to the theory of ordinary Euclidean geometry. In topology, the KLEIN BOTTLE is named in his honor.

Klein received a doctorate in mathematics from the University of Bonn in 1868 after completing a thesis exploring applications of geometry to mechanics. After a short period of military service, Klein accepted a position as professor of mathematics at Erlangen, in Bavaria, in 1872. It was here that he began his work on the analysis of geometric transformations and the unification of Euclidean and non-Euclidean geometries. His approach to the subject with a focus on "group invariants" was extremely influential, and

work in this field continues today. (It is called Klein's "Erlangen program.")

In 1886 Klein accepted the position as department chair at the University of Göttingen, where he remained until his retirement in 1913. He tutored students from his home in the years that followed.

In addition to conducting high-level research in mathematics, Klein also wrote mathematical pieces intended for the general public and founded a mathematical encyclopedia that he supervised until his death. Klein was honored with election to the ROYAL SOCIETY in 1885 and received the Copley Medal from the Society in 1912. He died on June 22, 1925, in Göttingen, Germany.

See also KLEIN-FOUR GROUP.

Klein bottle The MÖBIUS BAND is a three-dimensional object possessing just one surface and just one edge. A SPHERE and a TORUS (donut shape), on the other hand, are objects with two surfaces—an outside surface and an inside surface—possessing no edges. The Klein bottle is an alternative three-dimensional object with only one surface, akin to that of a Möbius band, but possessing no edges.

One can easily model these surfaces mentioned with a pair of trousers. For example, sewing together the two leg openings produces a circular tube with a hole for the waist. This hole can be patched with a piece of material to produce a complete model of a torus. A Klein bottle is produced by bringing one trouser leg up, over, and through the waist of the trousers and pushing it down through the tube of the second leg before sewing the two leg openings together. Again, the hole for the waistband represents a hole in the surface, but the object produced is nonetheless a (punctured) Klein bottle. Unfortunately it is no longer physically possible to patch the hole with a piece of material. This shows that the Klein bottle does not properly exist in a three-dimensional universe. (One can imagine, however, that the material that would make a patch for the hole of the waist has been "plucked" up into the fourth-dimension. In this sense, the Klein bottle is a valid mathematical object in four-dimensional space.)

One can check with this model that the ideal surface of a Klein bottle would indeed be one-sided: an ant crawling on one side of trouser material could reach any other part of the trousers, on either side of the material, without ever cheating by crossing over the edge of the hole given by the waist band.

If one cuts a pair of trousers sewn to make a (punctured) Klein bottle in half along the line that follows the inner and outer seams of the trousers, then the model falls into two pieces, each of which is easily seen to have arisen from a Möbius band. This shows that a Klein bottle can also be thought of as the union of two Möbius bands adjoined along their edges. (If one physically attempts to sew together two Möbius bands, one soon finds it is necessary to puncture a band to produce a hole akin to the hole for the waist of a pair of trousers.)

The two knots with fewer than five crossings

Klein-four group (viergruppe)   There are essentially only two GROUPS with just four elements. The first is the "cyclic group" C_4 given by the rotational symmetries of a square, $C_4 = \{1, R, R^2, R^3\}$, where R represents a 90° rotation of a square in a clockwise direction and 1 is the IDENTITY ELEMENT, a rotation of zero degrees. As a rotation of 360° is equivalent to no action, we have that $R^4 = 1$. Multiplication in this group is given by the product rule of exponents. We have, for example, that $R^2 \times R^3 = R^5 = R$.

The second group with just four elements is called the Klein-four group. Denoting its elements as 1, a, b, and c, with 1 being the identity element, it has a multiplication table given by:

×	1	a	b	c
1	1	a	b	c
a	a	1	c	b
b	b	c	1	a
c	c	b	a	1

This group can also be represented by a set of symmetries of a square. Set a to mean a reflection about a vertical axis, b a reflection about a horizontal axis, and c a rotation of 180°. We have, for instance, that a vertical reflection followed by a rotation of 180° is the equivalent of a horizontal reflection ($c \times a = b$), and that two reflections about the same axis lead to the identity ($a \times a = b \times b = 1$).

knot theory   The branch of TOPOLOGY that studies the properties of closed loops embedded in three-dimensional space is called knot theory. Each loop studied, called a knot, represents the path traced by a piece of string interlaced in space (without self-intersection) whose free ends have been joined together. If one tangled piece of string can be physically transformed into an exact copy of a second tangled string (or at least a mirror image of that second tangle), then we say that the two knots the strings represent are equivalent. At present, no one knows a general procedure guaranteed to determine with relative ease whether or not two given knots are equivalent.

Knots can be depicted on a two-dimensional page as a picture of a loop that crosses over and under itself. It is conventional to draw the picture of a given knot with as few crossings as possible. A picture with just one or two crossings is equivalent to an untangled loop (the "unknot"). There is only one knot (up to equivalence) with three crossings and only one with four crossings. Each are shown above. There are two distinct knots requiring a minimum of five crossings, three requiring a minimum of six crossings, and seven requiring a minimum of seven crossings. After this, the number of distinct knots with a given minimum number of crossings when drawn on a page grows rapidly.

Knots were first studied by CARL FRIEDRICH GAUSS (1777–1855) and his student Johann Listing (1808–82). In 1877 Scottish physicist Peter Tait classified all knots with up to seven crossings. He also conjectured that no "alternating knot" (that is, one whose path alternately crosses over and under itself) is equivalent to the unknot. A century later, New Zealand mathematician Vaughn Jones proved him to be correct.

In 1928 American mathematician James Waddell Alexander discovered a theoretical means to associate to each knot a POLYNOMIAL in such a manner that if two knots are equivalent, then the polynomials associated to them turn out to be the same. This thus provided an algebraic means to distinguish knots: if two knots have different Alexander polynomials, then they are not equivalent. In the 1960s British mathematician John Horton Conway discovered an explicit method for implementing Alexander's polynomials, and in 1984 Vaughn Jones developed an alternative (simpler) choice of polynomial to associate with each knot. (It was with the "Jones polynomial" that Jones was able to establish Tait's claim.) The same year, eight mathematicians independently discovered another polynomial that generalizes Jones's approach. (This polynomial is called the HOMFLYPT polynomial—an acronym of their names.) Unfortunately, it is not yet known whether Jones's technique provides a perfect correspondence between distinct knots and distinct polynomials—it is feasible that two nonequivalent knots could produce the same Jones's polynomial. (No one has yet seen an example of this.)

Knot theory is an active area of current research. It is used in the practical study of complicated molecules such as DNA, in the analysis of electrical circuits, and in the planning of street and highway networks.

Although the mirror image of a knot is deemed equivalent to the original knot, mathematicians have proved that no knot (except the unknot) can be physically deformed into its mirror image. It is also interesting to note that no knots exist in two-dimensional space or four-dimensional space (meaning that any knot in these spaces can be deformed into the unknot).

See also BRAID.

Koch curve   *See* FRACTAL.

Kolmogorov, Andrey Nikolaevich (1903–1987) Russian *Probability theory, Analysis, Dynamical systems* Born on April 25, 1903, in Tambov, Russia, scholar Andrey Kolmogorov is remembered for his significant contributions to the field of PROBABILITY theory and his groundbreaking work in the study of DYNAMICAL SYSTEMS. He also contributed to the study of FOURIER SERIES, ANALYSIS, and TOPOLOGY. Applying his work to planetary studies, Kolmogorov was also able to establish the mathematical stability of our solar system.

Kolmogorov began his working career as a railway conductor. In his spare time, however, he studied math-

ematics and physics, and managed to write a treatise on SIR ISAAC NEWTON's laws of mechanics. In 1920 he entered Moscow State University. By the time he completed his undergraduate degree he had published eight influential papers on the topics of SET THEORY and analysis. Kolmogorov published another 10 papers before receiving his doctoral degree in 1929.

Two years later Kolmogorov was appointed a professor of mathematics at Moscow State University. At this time, he began work on writing *Grundbegriffe der Wahrscheinlichkeitsrechnung* (Foundations on the theory of probability), in which he attempted, successfully, to build up the entire theory of probability from a finite set of axioms using nothing but logical rigor. Kolmogorov received national fame when the work was published in 1933. The piece was translated into the English language by Nathan Morrison in 1950.

In 1938 Kolmogorov was appointed department head of probability and statistics at the newly established Steklov Mathematical Institute in Russia. Soon afterward Kolmogorov's attention turned to the study of turbulence, and in 1941 he wrote two important papers on the nature of turbulent airflow from jet engines, laying down the founding principles of the theory of dynamical systems. He later applied this work to the study of planetary motion.

Kolmogorov received many awards for his outstanding work. In 1939 he was elected to the U.S.S.R. Academy of Science, and over the following years, he received eight prizes from the state and the academy. He was also elected to a number of foreign academies, including the Royal Statistical Society of London in 1956, the American Academy of Arts and Sciences in 1959, the Netherlands Academy of Sciences in 1963, the ROYAL SOCIETY of London in 1964, and the French Academy of Sciences in 1968. Kolmogorov was also awarded the Balzan International Prize in 1962.

Kolmogorov died in Moscow, Russia, on October 20, 1987. His work paved the way for continued research in the fields of Fourier analysis, "Markov chains" in probability theory, and topological analysis.

Kronecker, Leopold (1823–1891) German *Number theory* Born on December 7, 1823, in Liegitz, Prussia (now Legnica, Poland), mathematician and philosopher Leopold Kronecker is noted as a scholar for his considerable contributions to the field of NUMBER THEORY. As a philosopher, he is also remembered as the first to cast doubts on the existence of TRANSCENDENTAL NUMBERS and on hierarchies of infinite sets, notions first put forward by KARL THEODOR WILHELM WEIERSTRASS (1815–97) and GEORG CANTOR (1845–1918). At one point, Kronecker even went as far as to deny the existence of IRRATIONAL NUMBERS.

Kronecker entered Berlin University in 1841 and received a doctoral degree in mathematics after completing a thesis on the topic of algebraic number theory 4 years later. Rather than pursue an academic career, Kronecker decided to return to Liegitz to help with the family banking business. Over the 10 years that followed, he became quite wealthy and then had the luxury to pursue mathematics on his own without the need to ever accept a university position. In 1855 he returned to Berlin and published a series of influential papers in quick succession, which garnered him national attention. In 1861 Kronecker was elected as a member of the prestigious Berlin Academy, which earned him the right to teach at Berlin University, even though he was not a faculty member. (In 1883, some 20 years later, Kronecker was awarded an official position with the university.) Kronecker was also elected to the Paris Academy in 1868, and to the ROYAL SOCIETY of London in 1884.

During the 1870s Kronecker took a keen interest in the new, indirect approaches mathematicians were using to prove the existence of certain types of numbers. But he soon came to feel that such nonconstructive practices were philosophically flawed. For instance, Cantor had demonstrated that the set of real numbers is "more infinite" than the set of ALGEBRAIC NUMBERs, and hence, numbers that are not algebraic (so-called transcendental numbers) must exist.

In Kronecker's view, mathematics could only ever be based on finite quantities with a finite number of operations applied to them. An argument such as Cantor's, therefore, in Kronecker's thinking, had no meaning. Even when CARL LOUIS FERDINAND VON LINDEMANN (1852–1939) proved in 1882 that the specific number π is transcendental, Kronecker complimented Lindemann on his beautiful proof but added that, in fact, he had accomplished nothing, since transcendental numbers do not exist.

Kronecker was adamant in his views and, in his 1887 piece *Über den Zahlbergriff* (On number theory), he attempted to persuade the mathematical community of the absolute necessity of only using direct and finite

techniques. He is famous for once having remarked that "God created the integers. All else is the work of man" as an attempt to bring mathematical thinking back to concrete principles. Both Weierstrass and Cantor felt under personal attack by Knonecker and thought that he was deliberately undermining their own research programs.

Although many mathematicians today would not agree with Kronecker's views, they were taken seriously at the time, and the ideas that Kronecker put forward were later expanded upon in the century that followed. The practice of using only direct and finite approaches to prove results in mathematics is today called constructivism.

Kronecker died on December 29, 1891, in Berlin, Germany.

Kruskal's count   In the early 1980s, Princeton physicist Martin Kruskal discovered a remarkable mathematical property that all passages of written text seem to possess. This phenomenon is now referred to as Kruskal's count.

To illustrate the principle, review, for example, the familiar nursery rhyme:

> *Twinkle twinkle little star,*
> *How I wonder what you are,*
> *Up above the world so high,*
> *Like a diamond in the sky.*
> *Twinkle twinkle little star,*
> *How I wonder what you are.*

Perform the following steps:

1. Select any word from the first or second line and count the number of letters it contains.
2. Count that many words forward through the passage to land on a new word. (For example, choosing the word *star,* with four letters, will transport you to the word *what.*)
3. Count the number of letters in the new word, and move forward again that many places.
4. Repeat this procedure until you can go no further (that is, counting forward will take you off the nursery rhyme).
5. Observe the final word on which you have landed.

Surprisingly, no matter on which word you start this counting task, the procedure always takes you to the same word in the final line, namely, the word *you.*

Kruskal observed that this same phenomenon seems to occur with *any* sufficiently large piece of text—counting forward in this way from any choice of beginning word lands you at the same place at the end of the page. This provides an amusing activity for several people to perform simultaneously, all working with the same text, but starting with different choices of initial word.

The phenomenon can be explained as follows: imagine that on a first run through the text, all the words encountered are circled. Landing on any one of these circled words in a subsequent run of the experiment will take the player to the same final word. What then is the likelihood that a second run through the experiment will not "hit" any of the previously circled words? If the passage is sufficiently large, say involving 20 steps, the chances of "missing" a circled word every time is likely to be very small, especially since the typical English passage contains reasonably short words. Estimates computing this PROBABILITY show that it likely has a value smaller than 0.1 percent. Thus, with about 99.9 percent certainty, two paths through a passage of text will coincide at some point.

L

Lagrange, Joseph-Louis (1736–1813) Italian-French *Analysis, Mechanics, Abstract algebra, Number theory* Born on January 25, 1736, in Turin, Italy, Joseph-Louis Lagrange is remembered for, among other things, his definitive 1788 text *Mécanique analytique* (Analytical mechanics), in which he lays out a purely formal and completely rigorous exposition of how and why things move. In effect, the work summarizes the entire state of the post-Newton mechanics of the 18th century. The publication is also famous for featuring no figures or diagrams. (Lagrange apparently felt that the formulae he developed captured physical reality sufficiently well that pictures were not necessary.) He also published two important memoirs in 1769 and 1770, which he combined into a single 1798 treatise, *Traité de la résolution des équations numériques de tous les degrés* (Treatise on the resolution of numeric equations of all degree), on the theory of equations, in which he developed general principles for solving POLYNOMIAL equations up to degree four via principles that we today would label GROUP THEORY. The famous "Lagrange's theorem" from the study of groups appears in this work, as does the famous INTERPOLATION formula that bears his name. He also made significant contributions in the field of NUMBER THEORY, proving in 1770, for instance, that every positive integer is the sum of four square numbers. Also, in 1771, he was the first to prove what is today called "Wilson's theorem," that a number n is PRIME if, and only if, $(n - 1)! + 1$ is divisible by n.

Although Lagrange was born in Turin and spent the early part of his life there, he eventually settled in Paris and identified himself as French. He was introduced to the topics of physics and mathematics at the College of Turin. He first studied the principles of the TAUTOCHRONE and, by 1754, had published some important discoveries about the topic. He shared the results with LEONHARD EULER (1707–83), who was suitably impressed. At the young age of 19, Lagrange was appointed professor of mathematics at the Royal Artillery School in Turin. There he made significant progress on the general study of mechanics and wrote a number of influential papers. In 1756 he was elected to the Berlin Academy under Euler's recommendation.

During the 20 years that Lagrange stayed at the Berlin Academy, he worked on a wide variety of topics. He lay down the foundations of a subject to become known as "dynamics" and made significant contributions to the studies of fluid mechanics, to the integration of differential equations, and to the solution of systems of differential equations, as well as to the study of vibrating strings and the propagation of sound. In 1766 he succeeded Euler as director of mathematics at the Berlin Academy, and in 1772 he shared one of the biennial prizes from the Paris Académie des Sciences with Euler for his work on the "three-body problem" from physics. He won a prize again from the Paris Académie in 1774 for his work on the motion of the moon, and again in 1780 for his mathematical study of planetary motion.

In 1787 Lagrange left Berlin to take a position at the Académie des Sciences in Paris. He published his work *Mécanique analytique* the following year. He also worked with the Paris Académie to standardize the weights and measures of the day. The committee formulated the metric system and advocated the general use of a decimal base.

Lagrange's significant achievements were recognized by the emperor Napoleon in 1808, when he was named to the Legion of Honour and a count of the empire. Five years later he was also named Grand Croix of the Ordre Impérial de la Réunion. He died in Paris, France, on April 10, 1813.

Lagrange's impact in mathematics, especially in mathematical physics, is still felt today. Many fundamental concepts in mechanics and multivariable calculus—such as the Lagrangian (the difference between kinetic energy and potential energy of a set of particles), the Lagrangian description (a measure of deformation of a physical body), and Lagrange multipliers in calculus—play a vital role in the current study of these subjects.

See also LAGRANGE'S FORMULA.

Lagrange's formula (Lagrange's interpolation formula) Given a collection of points on the plane, it is sometimes desired to find the formula for a function that passes through each of those points. For example, a scientist may seek a formula for a function that fits all the data values obtained from an experiment. In the late 1700s, Italian-French mathematician JOSEPH-LOUIS LAGRANGE suggested the following INTERPOLATION formula:

If (a_1,b_1), (a_2,b_2), ..., (a_n,b_n) are a collection of points in the plane, with the values a_i distinct, then

$$f(x) = b_1 \frac{(x-a_2)(x-a_3)\cdots(x-a_n)}{(a_1-a_2)(a_1-a_3)\cdots(a_1-a_n)}$$
$$+ b_2 \frac{(x-a_1)(x-a_3)\cdots(x-a_n)}{(a_2-a_1)(a_2-a_3)\cdots(a_2-a_n)}$$
$$+ \cdots + b_n \frac{(x-a_1)(x-a_2)\cdots(x-a_{n-1})}{(a_n-a_1)(a_n-a_2)\cdots(a_n-a_{n-1})}$$

is a POLYNOMIAL, of degree $n-1$, that passes through each of the points.

This formula is today known as Lagrange's formula. One can check that it works by substituting $x = a_1$ to see that $f(a_1) = b_1$, and so on.

As an example, consider the points (1,2), (2,5), and (3,1) in the plane. Lagrange's formula shows that the quadratic

$$f(x) = 2\frac{(x-2)(x-3)}{(1-2)(1-3)} + 5\frac{(x-1)(x-3)}{(2-1)(2-3)} + 1\frac{(x-1)(x-2)}{(3-1)(3-2)}$$
$$= -\frac{7}{2}x^2 + \frac{27}{2}x - 8$$

passes through each of them.

Many intelligence tests ask participants to identify "the next number in the sequence." Lagrange's formula provides a means for justifying absolutely any answer to such a question. For example, the next number in the sequence 2,4,6,... could well be 103. We can argue that the sequence follows the formula: $a_n = \frac{95}{6}n^3 - 95n^2 + \frac{1057}{6}n - 95$. (Apply Lagrange's formula to the points (1,2), (2,4), (3,6), and (4,103).)

Lambert, Johann Heinrich (1728–1777) Swiss-German *Geometry, Analysis, Number theory, Physics* Born on August 26, 1728, scholar Johann Lambert is best remembered as the first to prove, in 1761, that π is an IRRATIONAL NUMBER. He also worked on Euclid's PARALLEL POSTULATE and came close to the discovery of NON-EUCLIDEAN GEOMETRY. Lambert also developed the notation and the theory of HYPERBOLIC FUNCTIONS.

In 1766 Lambert wrote *Theorie der Parellellinien* (On the theory of parallel lines), in which he postulated the existence of surfaces on which triangles have angular sums less than 180°, thereby yielding an example of a geometry in which the parallel postulate would be false. (Such a surface was later discovered. It is called a pseudosphere.) Lambert proved that in this geometry, the sum of the angles of a triangle would not be constant, and in fact would increase (but never to equal 180°) as its area decreases.

In 1737 LEONHARD EULER had proved that e and e^2 are both irrational. In the paper "Mémoire sur quelques propriétés remarquables des quantités transcendantes circulaires et logarithmiques" (Memoir on some remarkable properties of transcendental quantities circular and logarithmic) presented to the Berlin Academy of Sciences in 1761, Lambert provided a proof that if x is a rational number different from zero, then neither e^x nor tan x can be rational. Thus Lambert had succeeded in extending Euler's result in establishing that e raised to a fractional power, $e^{\frac{p}{q}}$, is never rational, and, moreover, noting that $\tan \frac{\pi}{4} = 1$, that $\frac{\pi}{4}$ also cannot be rational. Lambert also conjectured that both e and π are TRANSCENDENTAL NUMBERS. (It was not for another 100 years that, in 1873, French mathematician Charles Hermite proved the transcendence of e, and German mathematician CARL LOUIS FERDINAND VON LINDEMANN, in 1883, the transcendence of π.)

Lambert also worked on the theory of PROBABILITY and, in physics, made important contributions to the study of heat and light. He died in Berlin, Germany, on September 25, 1777. Much of Lambert's work, although significant at his time, can be seen today as having paved the way for others to achieve greater advances.

Laplace, Pierre-Simon, marquis de (1749–1827)

French *Mechanics, Analysis, Differential equations, Probability theory* Born on March 23, 1749, in Normandy, France, scholar Pierre-Simon Laplace is best remembered for his influential five-volume treatise *Traité de mécanique céleste* (Celestial mechanics), published between the years 1799 and 1825. In this work Laplace tried to develop a rigorous mathematical understanding of the motion of the heavenly bodies, including the various anomalies and inequalities that were observed in their orbits. In doing so, Laplace made significant strides in the development of DIFFERENTIAL EQUATIONS, DIFFERENCE equations, PROBABILITY, and STATISTICS. He was the first to extend SIR ISAAC NEWTON's theory of gravitation to the study of the whole solar system. In 1812 Laplace also published his *Théorie analytique des probabilités* (Analytic theory of probability), which advanced the topics of probability and statistics considerably.

At the age of 16 Laplace entered Caen University and soon discovered a love for mathematics. Three years later, without completing his degree, Laplace moved to Paris to work with the mathematician JEAN LE ROND D'ALEMBERT (1717–83), supporting himself as a professor of mathematics at the École Militaire. By the time he was 24, Laplace had produced 13 high-quality papers on the topics of INTEGRAL CALCULUS, mechanics, and physical astronomy, earning him national attention. He continued to produce fundamentally important results, and was soon regarded as one of the most influential scientists of his time.

In 1773 Laplace was elected to the Paris Académie des Sciences and was awarded a senior position there 12 years later. He was assigned to a special committtee of the Paris Académie in 1790 with the charge of standardizing all European weights and measures.

Laplace's work on celestial mechanics was revolutionary. In 1786 he had proved that the small perturbations observed in the orbital motion of the planets will forever remain small, constant, and self-correcting, and in 1796 he was the first to propose the idea that the solar system originated from the contraction and cooling of a large rotating, and consequently flattened, nebula of incandescent gas. In 1799 he published the first two volumes of his famous *Mécanique céleste*, in which he described the general laws of motion of solids and fluids, applied the universal law of gravity to studies of the solar system, and developed methods for analyzing the difference and differential

Pierre-Simon Laplace, an eminent mathematical physicist of the 19th century, made fundamental contributions to the study of planetary motion by applying Sir Issac Newton's theory of gravitation to the entire solar system. He also pioneered work in the theory of probability and statistics. *(Photo courtesy of Topham/The Image Works)*

equations that arise in these studies. Later volumes studied the local forces acting between molecules and the large-scale effect of these local interactions. He applied this work to the study of pressure and density, the nature of gravity, capillary action, sound, refraction, and the theory of heat.

In 1812 Laplace published his famous piece on the topic of probability. Motivated by the analysis of errors in observation, Laplace developed the theory of probability from beginning principles and made clear the mathematical underpinnings of many aspects of the topic. He provided his own definition of probability and used it to justify the basic mathematical manipulations. The work discusses his famous LEAST SQUARES METHOD, the BUFFON NEEDLE PROBLEM, and BAYES'S THEOREM. In analyzing the distribution of errors in scientific observations, Laplace also applied this work to the accurate determination of the masses of Jupiter, Saturn, and Uranus, as well as to improving triangulation methods in surveying, and correct determination of longitude and latitude in geodesy.

Under Napoleon, Laplace became a count of the empire in 1806 and was awarded the title of marquis in 1817. Laplace wrote that he believed that the nature of the universe is completely deterministic, meaning that the motions all objects in the solar system are predetermined by the initial conditions at the start of the universe. When Napoleon asked him where God fit into this view, Laplace is said to have replied: "I have no need of that hypothesis."

Laplace died in Paris, France, on March 5, 1827. It is not possible to exaggerate the influence Laplace had on the development of the mathematical theory of mechanics. A number of fundamental concepts, such as the Laplace operator in potential theory and the Laplace transform in the study of differential equations, are named in his honor.

Latin square An $n \times n$ Latin square is an arrangement of the first n Latin letters A, B, C, D, … in a square array in such a manner that no row or column of that array contains two identical letters. For example, the arrangement below is a 3×3 Latin square:

$$
\begin{array}{ccc}
A & B & C \\
C & A & B \\
B & C & A
\end{array}
$$

A Latin square is called diagonal if, in addition, no letter is repeated in either diagonal of the square. The following, for example, is a diagonal 4×4 Latin square:

$$
\begin{array}{cccc}
A & B & C & D \\
C & D & A & B \\
D & C & B & A \\
B & A & D & C
\end{array}
$$

No 3×3 Latin square is diagonal.

There is just one 1×1 Latin square, two 2×2 Latin squares (neither of which are diagonal), 12 3×3 Latin squares, and 576 4×4 Latin squares. The number of Latin squares of a given order grows extremely rapidly, as the following table shows:

n	Number of $n \times n$ Latin Squares
1	1
2	2
3	12
4	576
5	161,280
6	812,851,200
7	61,479,419,904,000
8	108,776,032,459,082,956,800
9	5,524,751,496,156,892,842,531,225,600
10	9,982,437,658,213,039,871,725,064,756,920,320,000

There are more 15×15 Latin squares than there are atoms in the universe (which physicists calculate to be about 10^{81}).

Two Latin squares of the same order are called mutually orthogonal if all the pairs of letters that appear when the two squares are superimposed are different. For example, the two squares shown below are mutually orthogonal. Here we have used letters of the Greek alphabet for the second square. The resultant array of pairs of elements that appears is called a Graeco-Latin square.

$$
\begin{array}{cccc}
A & B & C & D \\
B & A & D & C \\
C & D & A & B \\
D & C & B & A
\end{array}
+
\begin{array}{cccc}
\alpha & \beta & \gamma & \delta \\
\gamma & \delta & \alpha & \beta \\
\delta & \gamma & \beta & \alpha \\
\beta & \alpha & \delta & \gamma
\end{array}
=
\begin{array}{cccc}
A\alpha & B\beta & C\gamma & D\delta \\
B\gamma & A\delta & D\alpha & C\beta \\
C\delta & D\gamma & A\beta & B\alpha \\
D\beta & C\alpha & B\delta & A\gamma
\end{array}
$$

This square solves the famous "officer problem" first posed by LEONHARD EULER (1707–83):

Arrange 16 officers, each from one of four possible regiments and of one of four possible ranks, in a *4 × 4* array so that no two officers in any given row or column come from the same regiment, nor have the same rank.

Graeco-Latin squares are also sometimes called Euler squares. Mathematicians have proved that there is no solution to the officer problem for the case of 36 officers of six different ranks from six different regiments (nor for the case of four officers from two regiments of two ranks), but that all other versions of the officer problem do have solutions. In other words, $n \times n$ Graeco-Latin squares exist for all values of n except $n = 2$ and $n = 6$.

Graeco-Latin squares are important in the design of experiments in scientific studies. For example, if four species of tomato A, B, C, and D are to be tested with four different fertilizers α, β, γ, and δ, the plots can be laid out according to a Graeco-Latin square to be sure that each species of tomato and each fertilizer appears in each row and column.

law of averages This "law" refers to the incorrect belief that previous outcomes of independent runs of a random trial influence the outcomes of runs yet to occur. For example, after tossing seven "heads" in a row, the supposed law of averages would dictate that "tails" is now a more likely outcome. This of course is not the case. The chance of tossing tails on the eighth toss is still 50 percent. The law of averages is a common misinterpretation of the mathematically correct LAW OF LARGE NUMBERS. Gamblers often feel that after a long string of losses, the chances of winning a next hand must be considerably greater.

law of cosines/law of sines (cosine rule, sine rule) Let a, b, and c be the three side-lengths of a triangle with interior angles as shown.

The law of cosines asserts:

$$c^2 = a^2 + b^2 - 2ab\cos(C)$$

(with analogous statements for angles A and B). When C is a right angle, this result reduces to a statement of PYTHAGORAS'S THEOREM. Hence the law of cosines can be regarded as a generalization of this famous result.

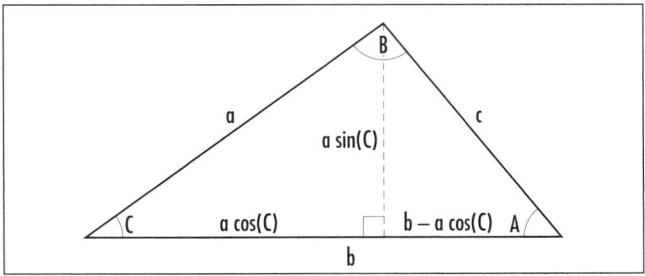

Establishing the law of cosines

The law can be proved by drawing an altitude from the apex of the triangle and applying Pythagoras's theorem to the right-angled triangle containing the altitude and side c in the picture above.

The law of sines asserts:

$$\frac{a}{\sin(A)} = \frac{b}{\sin(B)} = \frac{c}{\sin(C)}$$

This can be proved by drawing the altitudes of the triangle. For example, the altitude above is simultaneously of length $a\sin(C)$ and $c\sin(A)$, thereby establishing part of the law of sines.

Drawing the CIRCUMCIRCLE to the triangle, and calculating the sine of angle A in the shaded triangle shown (with the same peripheral angle A) establishes:

$$\frac{a}{\sin(A)} = D$$

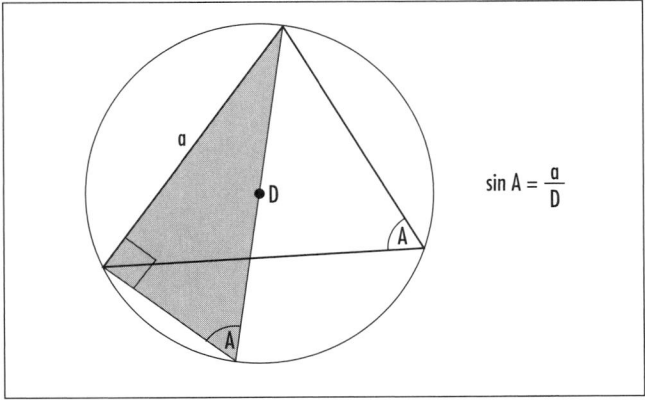

Establishing the law of sines

where D is the diameter of the circumcircle. (Here we make use of the fact that any angle subtended from a diameter is 90°.) Thus the three quantities expressed in the law of sines each equal the diameter of the circumcircle of the triangle, thereby offering an alternative proof of the law.

See also CIRCLE THEOREMS.

law of large numbers   If one tosses a fair coin 10 times, one would expect, on average, five of those tosses to be "heads." Of course, in any single run of 10 tosses, any number of heads is possible, even a string of 10 heads in a row, but this is extremely unlikely. The number of heads actually observed in an experiment is likely to be four, five, or six, close to 50 percent. (The probability of any particular count of heads appearing is described by the BINOMIAL DISTRIBUTION.)

In a run of 100 tosses we would expect the effects of excessive runs of heads, or tails, to "average out" and the proportion of heads obtained to be even closer, on average, to the "true value" of 50 percent—and closer still if we run an experiment of 1,000 or 10,000 tosses, or more. This is the law of large numbers in action. Precisely, this law states the following:

> The more times a random phenomenon is performed, the closer the proportion of trials in which a particular outcome occurs approximates the true probability of that outcome occurring.

If, for example, a 1 never occurred when rolling a die 10 times, we can be assured, however, that the proportion of 1s appearing in another 100, 1,000, 10,000, ... tosses will approach the value one-sixth.

Many gamblers incorrectly interpret the law of large numbers as a method for predicting outcomes of random events. (*See* LAW OF AVERAGES.)

The law of large numbers is a mathematical consequence of CHEBYSHEV'S THEOREM. It can be interpreted as saying that if a random phenomenon produces numerical outcomes with mean value μ, then the mean of N observed values of the phenomenon approaches the value μ as N increases. Chebyshev's theorem is related, for it gives measures of how values are distributed about the mean.

See also MONTE CARLO METHOD; STATISTICS: DESCRIPTIVE.

law of sines (sine rule)   *See* LAW OF COSINES/LAW OF SINES.

law of the lever   ARCHIMEDES OF SYRACUSE (ca. 287–212 B.C.E.) recognized that two weights w_1 and w_2 placed at distances x_1 and x_2, respectively, from the fulcrum (pivot point) of a simple lever will balance when $x_1 w_1 = x_2 w_2$. This principle is called the law of the lever. For example, an adult weighting twice as much as a child will balance on a seesaw if she sits half the distance from the pivot point as the child.

See also CENTER OF GRAVITY.

laws of thought   The Greek philosopher ARISTOTLE (384–322 B.C.E.) identified three laws of logic, all tautologies (meaning that each has a TRUTH TABLE with constant value T) that have since been deemed fundamental descriptions of the way we think. His three laws of thought are:

1. *Law of Noncontradiction:* It is not the case that something can be both true and not true.
   Symbolically: $\neg[p \wedge (\neg p)]$

2. *Law of Excluded Middle:* Each must either be true or not true.
   Symbolically: $p \vee (\neg p)$

3. *Law of Identity:* If something is true, then it is true.
   Symbolically: $p \rightarrow p$

Mathematicians often rely on the law of excluded middle to establish the validity of mathematical results: an INDIRECT PROOF or a PROOF BY CONTRADICTION proves that a statement p is true by showing that it cannot be false. However, not all philosophers (and mathematicians) agree with this approach and question the validity of this second law. For example, as the 20th-century Austrian mathematician KURT GÖDEL showed, there are some statements in mathematics that can neither be proved nor disproved, and are consequently neither true nor false. The constructivist movement accepts results established by DIRECT PROOF only.

To move beyond the law of the excluded middle, logicians have attempted to generalize FORMAL LOGIC to include *three* possible values of truthhood: true, false, and undecided. A further generalization, called FUZZY LOGIC, treats truth as a continuous quantity with values ranging from 0 (utterly false) to 1 (utterly true). As an example, in this theory the sentence:

> This sentence is false.

is assigned a truth value of of $^1/_2$, and is deemed half true and half false. (One arrives at this value as follows: first note that if a statement p is assigned a truth-value v, which is either 0 or 1, then the statement $\neg p$ has opposite value $1 - v$, 1 or 0. We assume that this remains true even if v is of fractional value. If p represents the sentence: "This sentence is false," then its truth establishes its falsehood, $p \rightarrow \neg p$, and its falsehood its truth, $\neg p \rightarrow p$. The truth-value v of the statement oscillates between the values v and $1 - v$. The only stable value for v occurs when $v = 1 - v$, yielding the appropriate truth-value $v = 1/2$.)

See also ARGUMENT; SELF-REFERENCE.

leading coefficient For a POLYNOMIAL $p(x) = a_n x^n + a_{n-1} x^{n-1} + \dots + a_1 x + a_0$, the coefficient a_n of the highest power of the variable is called the leading coefficient of the polynomial. For example, the leading coefficient of the polynomial $2x^3 - 3x + 6$ is 2, and that of $x + 3$ is 1. A polynomial is called monic if its leading coefficient is 1.

In solving a polynomial equation $a_n x^n + a_{n-1} x^{n-1} + \dots + a_1 x + a_0 = 0$, it is often convenient to assume that the polynomial in question is monic. One achieves this by dividing the equation through by a_n. Solving the equation $2x^4 - 8x^3 + 2x - 6 = 0$, for instance, is equivalent to solving $x^4 - 4x^3 + x - 3 = 0$.

If all the coefficients of a monic polynomial are integers, then any rational ROOT to the polynomial must itself be an integer. For example, if the fraction $x = \dfrac{p}{q}$ (written in reduced form) were a solution to the polynomial equation $x^4 - 4x^3 + x - 3 = 0$, then, substituting in this value for x and multiplying through by q^4 yields:

$$p^4 - 4p^3 q + pq^3 - 3q^4 = 0$$

This shows that p^4 is a multiple of q. Since p and q share no common factors, this is only possible if q equals 1. Thus the root $x = \dfrac{p}{q} = p$ is an integer.

A similar argument shows that if the CONSTANT term a_0 of a polynomial is 1, then any rational root of the polynomial (if it has one) must be a fraction of the form $x = \dfrac{1}{q}$. Consequently, any monic polynomial with a constant term of 1 can possess at most one rational (or integer) root, namely, $x = 1$. This proves, for example, that $x = 1$ is the only rational root of the equation:

$$x^7 - x^5 + 2x^2 - 3x + 1 = 0$$

(One checks that $x = 1$ is indeed a solution to this equation.)

least common multiple A number that is a multiple of two or more other numbers is called a COMMON MULTIPLE of those numbers. The smallest common multiple is called their least common multiple, written as "lcm." For example, the least common multiple of 10, 12, and 15 is 60. There is no smaller number that is evenly divisible by each of these numbers. We have lcm(10, 12, 15) = 60.

The least common multiple of a set of integers can be found by splitting each number into prime factors. For example, to find the lcm of 180 and 378, write:

$$180 = 2 \times 2 \times 3 \times 3 \times 5$$
$$378 = 2 \times 3 \times 3 \times 3 \times 7$$

The lcm is then found by multiplying the prime factors together, taking each the maximum number of times it appears in any of the numbers. In our case: lcm(180, 378) = $2 \times 2 \times 3 \times 3 \times 3 \times 5 \times 7$ = 3,780. This method shows that any common multiple of a collection of integers is a multiple of the least common multiple. It also shows that, for two positive integers a and b:

$$\text{lcm}(a,b) = \frac{ab}{\gcd(a,b)}$$

where "gcd" denotes the GREATEST COMMON FACTOR (divisor) of the two numbers. The analogous relationship for three or more integers, however, does not hold in general.

See also FUNDAMENTAL THEOREM OF ARITHMETIC.

least squares method If a SCATTER DIAGRAM indicates a linear correlation between the two variables of interest in a scientific study (for example, the average shoe size of adults might be linearly correlated to height), then one can seek the equation of a line that best fits the data. Such a line is called a regression line. Specifically, if a study produces N pairs of data values, $(x_1, y_1), \ldots, (x_N, y_N)$, then one seeks a linear equation $y = ax + b$ that minimizes the total deviation of data points from that line. This total deviation could be measured as a sum of absolute values:

$$|y_1 - (ax_1 + b)| + |y_2 - (ax_2 + b)| + \ldots + |y_N - (ax_N + b)|$$

(yielding what is called the Chebyshev approximation criterion), but this quantity is difficult to analyze using the techniques of CALCULUS. (The ABSOLUTE VALUE function is not differentiable.)

Another measure of total deviation is the sum of all the individual deviations squared, which, again, is a sum of positive quantities:

$$D = (y_1 - (ax_1 + b))^2 + (y_2 - (ax_2 + b))^2 + \ldots + (y_N - (ax_N + b))^2$$

The task is to choose values for a and b that minimize this sum. This is called the least squares criterion.

A necessary condition for D to adopt a minimal value is that the two partial derivatives $\dfrac{\partial D}{\partial a}$ and $\dfrac{\partial D}{\partial b}$ equal zero, yielding the two normal equations:

$$\frac{\partial D}{\partial a} = -2\sum_{i=1}^{N}(y_i - ax_i - b)x_i = 0 \implies a\sum_{i=1}^{N}x_i^2 + b\sum_{i=1}^{N}x_i = \sum_{i=1}^{N}x_iy_i$$

$$\frac{\partial D}{\partial b} = -2\sum_{i=1}^{N}(y_i - ax_i - b) = 0 \implies a\sum_{i=1}^{N}x_i + Nb = \sum_{i=1}^{N}y_i$$

Dividing through by N and solving for a (the slope) and b (the intercept), we obtain:

$$a = \frac{\left(\displaystyle\sum_{i=1}^{N}x_iy_i\right) - \bar{x}\cdot\bar{y}}{\left(\displaystyle\sum_{i=1}^{N}x_i^2\right) - \bar{x}^2}$$

and

$$b = \bar{y} - a\cdot\bar{x}$$

where $\bar{x}$ is the mean x-value, and $\bar{y}$ is the mean y-value. Setting:

$$S_{xx} = \frac{1}{N}\sum_{i=1}^{N}(x_i - \bar{x})^2 = \frac{1}{N}\left(\sum_{i=1}^{N}x_i^2\right) - \bar{x}^2$$

(this is the VARIANCE of the x-values) and

$$S_{xy} = \frac{1}{N}\sum_{i=1}^{N}(x_i - \bar{x})(y_i - \bar{y}) = \frac{1}{N}\left(\sum_{i=1}^{N}x_iy_i\right) - \bar{x}\cdot\bar{y}$$

(the COVARIANCE of the two variables), these formulae can be more compactly written: $a = \dfrac{S_{xy}}{S_{xx}}$ and $b = \bar{y} - \dfrac{S_{xy}}{S_{xx}}\bar{x}$. Thus the least squares method gives the equation for the line of best fit as:

$$y - \bar{y} = \left(\frac{S_{xy}}{S_{xx}}\right)(x - \bar{x})$$

Measuring the Degree of Fit

The quantity D that was minimized (above) is called the "error sum of squares":

$$D = \sum_{1=i}^{N}\left(y_i - (ax_i + b)\right)^2$$

It reflects the amount of variation of the data points about the regression line. The total corrected sum of squares (SST) of y:

$$SST = \sum_{i=1}^{N}\left(y_i - \bar{y}\right)^2$$

gives a measure of the scattering of the y-values in general. Necessarily, $D \leq SST$. The difference, $SST - D$, called the regression sum of squares, reflects the amount of variation in the y-values explained by the linear regression line $y = ax + b$ when compared with their general distribution. That the quantity $SST - D$ is positive prompts the definition of the CORRELATION COEFFICIENT, R^2, given by $R^2 = \dfrac{SST - D}{SST}$. An exercise in algebra shows:

$$R^2 = \frac{\left(S_{xy}\right)^2}{S_{xx}S_{yy}}$$

This numerical value represents the proportion in total variation in the y variable that can be accounted for by the line of best fit. This proportion has values between 0 and 1, with a value of 1 indicating that all variation is due to a linear fit, that is, the data values lie perfectly on the regression line, and an R^2 value of zero indicates that none of the variation in the y-values is due to a linear correlation. If $R^2 = 0.84$, for example, then we can say that 84 percent of the variation in the y-values is accounted for by a linear relationship with the values of x.

See also REGRESSION.

Lebesgue, Henri-Léon (1875–1941) French *Analysis* Born on June 28, 1875, French mathematician Henri Lebesgue is remembered for his revolutionary ideas in CALCULUS and in the theory of integration. By generalizing the notion of AREA to one of an abstract "measure theory," Lebesgue transformed the object of an integral into a tool applicable to an extraordinarily large class of settings. He published work on this topic at the young age of 27.

Lebesgue studied at the École Normale Supérieure, France, and taught at the University of Nancy for 3 years. He presented his famous work to his university colleagues during his final year there in 1902. The idea behind his approach is relatively simple. One typically computes an integral by subdividing the range of inputs, the x-axis, into small intervals and then adding the areas of rectangles above these intervals of heights given by the function. This is akin to counting the value of a pocket full of coins by taking one coin out at a time and adding the outcomes as one goes along. Lebesgue's approach, however, is to subdivide the range of outputs, the y-axis, into small intervals and to measure the size of the sets on the x-axis for which the function gives the desired output on the y-axis. This is akin to counting coins by first collecting all the pennies and determining their number, then all the nickels and ascertaining the size of that collection, and so forth. Of course the shape of the sets one encounters along the x-axis can be complicated and difficult to measure in size. The work of French mathematicians Émile Borel (1871–1956) and MARIE ENNEMOND CAMILLE JORDAN (1838–1922) in developing so-called measure theory provided Lebesgue the means to do this.

Lebesgue wrote over 50 papers and two books, including his 1902 paper "Intégrale, longeur, aire" (Integrals, lengths, area), his 1910 article "Sur l'intégration des functions discontinues" (On the integration of discontinuous functions), and his 1906 monograph *Leçons sur les séries trigonométriques* (Lectures on trigonometric series). He also made important contributions to the fields of TOPOLOGY and FOURIER SERIES, and was appointed professor at the Sorbonne, University of Paris, in 1910.

At one point in his life, Lebesgue expressed serious unease about continuing the work in integration theory he himself had founded. He feared that by making mathematics abstract, topics in the subject would begin to lose meaningful context. He died in Paris, France, on July 26, 1941.

Lebesgue's revolutionary approach to integration theory is taught to all upper-level college students and graduate students in mathematics today. It is considered a core component of any serious study of analysis.

left derivative/right derivative The DERIVATIVE of a function $f(x)$ at position x is defined as the LIMIT $f'(x) = \lim_{h \to 0} \frac{f(x+h)-f(x)}{h}$. The value of the limit, if it exists, represents the slope of the tangent line to the graph of the function at position x. If the quantity h is restricted to run only through negative values as it approaches the value zero, that is, if the limit above is replaced by a limit from the left, we obtain the left derivative of the function at x:

$$\lim_{h \to 0^-} \frac{f(x+h)-f(x)}{h}$$

Restricting h to run only through positive values produces the right derivative of the function at x:

$$\lim_{h \to 0^+} \frac{f(x+h)-f(x)}{h}$$

(See LIMIT.) The general derivative $f'(x)$ exists if, and only if, the left and right derivatives both exist and agree in value. This need not always be the case. Consider the function $f(x) = x + |x|$, for example, at the position $x = 0$. left derivative (with h negative) is $\lim_{h\to0^-} \dfrac{(h+|h|) - (0+|0|)}{h} = \lim_{h\to0^-} \dfrac{h+(-h)}{h} = 2$ where as the right derivative (with h positive) is $\lim_{h\to0^+} \dfrac{(h+|h|) - (0+|0|)}{h} = \lim_{h\to0^+} \dfrac{h+h}{h} = 2$. The tangent line to the curve has slope zero just to the left of $x = 0$ and slope 2 just to its right. This inconsistency shows that there is no well-defined tangent line to the graph $f(x) = x + |x|$ at $x = 0$.

Legendre, Adrien-Marie (1752–1833) French *Geometry, Number theory* Born on September 18, 1752, in Paris, France, Adrien-Marie Legendre is remembered as a capable French mathematician who devoted his life to the subject but had the unfortunate luck to see the bulk of his work rendered obsolete by the discoveries and abilities of younger, brighter mathematicians.

Legendre published in the fields of NUMBER THEORY, elliptic functions, EUCLIDEAN GEOMETRY, and celestial mechanics. His name is associated with a number of mathematical concepts, including Legendre polynomials, an important class of functions useful for solving certain types of DIFFERENTIAL EQUATIONS; the LEAST SQUARES METHOD; and the Legendre symbol in number theory. He published an elementary geometry text, *Eléments de géométrie* (Elements of geometry), which dominated the teaching of geometry in America and Europe throughout the 19th century. In 1791 the Académie des Sciences undertook the task to standardize weights and measures and directed Legendre to make the necessary astronomical observations to compute the length of a meter.

Little is known of Legendre's early life. Born into a family of wealth, Legendre had access to an excellent formal education and the means to pursue scholarly interests. In 1770 Legendre defended a thesis in mathematics and physics at the Collège Mazarin. His 1782 mathematical research on the path of projectiles earned him an award from the Berlin Academy and garnered him some attention as a scholar of mathematics. A year later Legendre was appointed an adjoint position at the Académie des Sciences in Paris, later renamed the Institut National des Sciences et des Arts. He remained there until 1824, rising appropriately in rank and position throughout the decades. However, due to a political disagreement with the running of the Institut,

Legendre was denied a pension in his retirement and died in poverty 9 years later on January 9, 1833.

Despite his work being seen as obsolete, Legendre did raise a number of fundamental questions in the fields of number theory and elliptical function theory that spurred a great deal of mathematical investigation during the century that followed him.

Leibniz, Gottfried Wilhelm (1646–1716) German *Calculus, Logic* Born on July 1, 1646, in Leipzig, Saxony (now Germany), scholar Gottfried Wilhelm Leibniz is remembered for discovering, and being the first to publish in 1684, the theory of DIFFERENTIAL CALCULUS. In subsequent years he also developed the theory of INTEGRAL CALCULUS and formulated the FUNDAMENTAL THEOREM OF CALCULUS that unites the two fields. This work was accomplished independently of the progress made by

Gottfried Wilhelm Leibniz, an eminent mathematician of the 18th century, was the first to found and publish an account of differential calculus. He later developed the theory of integral calculus and the fundamental theorem of calculus that connects the two subjects. *(Photo courtesy of Feltz/Topham/The Image Works)*

SIR ISAAC NEWTON (1642–1727) on the same topics. Leibniz also devoted a considerable amount of effort into developing a *characteristica generalis,* a universal language, as an attempt to generalize the logical formalism created by ARISTOTLE (384–322 B.C.E.). (Logician GEORGE BOOLE later followed this goal in the 1800s.)

Leibniz entered the University of Leipzig at age 14, as was customary at the time, to commence a 2-year general degree course. Having read the works of Aristotle, Leibniz was already beginning initial work on formalizing and systematizing the process of reasoning. He received degrees in law and in philosophy over the following 6 years, and then began the ambitious project of collating all human knowledge. He began with the study of motion and kinematics, and published, in 1671, his book *Hypothesis physica nova* (New physical hypothesis).

Leibniz traveled to Paris in 1672 and began a study of physics and mathematics with leading scientists in the city at that time. Two years later, he had developed his theory of differential calculus, but was struggling to find a good system of mathematical notation for the theory. In an unpublished 1675 manuscript Leibniz had described the PRODUCT RULE for differentiation and established the rules for differentiating POLYNOMIALs.

Word had reached Newton of the results Leibniz had developed, and Newton immediately wrote to him explaining that he had already discovered the theory a decade earlier. Newton, however, did not provide details of his work. Leibniz courteously replied to Newton but, not realizing that correspondences were delayed by months, Newton suspected Leibniz of dwelling over his letter, reconstructing the missing details, and stealing his ideas. Although it is understood today that Leibniz had accomplished his work completely independently of Newton, a bitter dispute between the two gentlemen ensued, one that lasted for decades.

In 1684 Leibniz published his details of differential calculus in *Nova methodus pro maximis et minimis* (A new method for determining maxima and minima) after finally establishing an effective system of notation for his work—the *d* notation we use today. By this time Leibniz had also developed his theory of integral calculus (along with the familiar ∫ *dx* notation), and began publishing details of the work in 1686. (Newton wrote of his method of "fluxions" in 1671 but failed to get it published.)

Throughout his life, Leibniz also made significant contributions to the study of DIFFERENTIAL EQUATIONs, the theory of equations and the use of a DETERMINANT to solve systems of equations, and generalizing the BINOMIAL THEOREM to more than two variables. Also, in his quest to collate all human knowledge, Leibniz wrote significant treatises on metaphysics and philosophy. He also developed a general "law of continuity" for the universe, suggesting that all that occurs in nature does so in matters of degree, and argued that "mass times velocity squared" is a fundamental quantity that is conserved in physical systems. (This is today called the "law of conservation of energy.")

Leibniz died in Hanover, Germany, on November 14, 1716. It is not possible to exaggerate the effect Leibniz had on the development of analytical theory in the centuries that followed him. His choice of notational system for calculus, for instance, facilitated clear understanding of the subject and easy use of its techniques. Mathematicians today typically use the notation developed by Leibniz rather than that developed by Newton.

Leibniz's theorem   *See* PRODUCT RULE.

lemma   *See* THEOREM.

length   The distance along a line, or the distance in which a figure or solid extends in a certain direction, is called its length. One can measure two lengths for a rectangle to give an indication of its size. (The greater of the two dimensions is usually called its length, and the smaller its breadth.)

Early units for length were given by parts of the body. For example, a "cubit" was defined to be the length of the forearm, measured from the elbow to the tip of the middle finger (about 19 in.); an "ell," still sometimes used for measuring cloth, is the length from the tip of one's nose to the end of an outstretched arm (about 35 in.); a "hand," used for measuring the heights of horses, is the width of a man's hand (about 4 in.); and a "foot" was defined as the distance paced by one step. The ancient Romans considered a foot to be the equivalent of 12 thumb-widths, yielding the word *inch* from the Latin word *unicia* meaning one-twelfth. The Romans also identified 1,000 paces as a *milia passuum,* leading to our concept of one "mile." One-thousandth of a foot is called a "gry." Lengths defined by body measurements are subject to great variation, and it was found that different communities throughout the world were using different standards of length even though the units used were given the same name. During the 19th and 20th centuries an international standard of units was developed, and unambiguous units of length were defined.

In mathematics, the notion of length appears in several settings and can be given precise definitions. For example, on the number line, the length of a line segment connecting a number a to number b is given by the ABSOLUTE VALUE $|b - a|$; the length of a line segment connecting points $A = (a_1, a_2)$ and $B = (b_1, b_2)$ is given by the DISTANCE FORMULA $\sqrt{(b_1 - a_1)^2 + (b_2 - a_2)^2}$; and the length of a VECTOR $\mathbf{a} = <a_1, a_2>$ is given by $\|\mathbf{a}\| = \sqrt{a_1^2 + a_2^2}$. (These latter formulae generalize to three and higher dimensions.)

CALCULUS can be used to find the ARC LENGTH of curved lines in two- and three-dimensional space.

See also SI UNITS.

L'Hôpital, Guillaume François Antoine, marquis de

(1661–1704) French *Calculus* Born in Paris, France, in 1661 (his exact birth date is not known), Guillaume l'Hôpital is remembered as the famed author of the first textbook on the topic of DIFFERENTIAL CALCULUS, *Analyse des infiniment petits* (Analysis with infinitely small quantities), written in 1696. Apart from explaining the methods and details of the newly discovered theory, this work also contains the first formulation of the rule that now bears his name.

L'Hôpital's talent for mathematics was recognized as a boy. At the age of 15 he solved a problem on the CYCLOID put forward by BLAISE PASCAL (1623–1662) and later contributed to the solution of the famous BRACHISTOCHRONE problem. Before pursuing mathematics in earnest, l'Hôpital served as a cavalry officer but soon had to resign due to nearsightedness.

In 1691 l'Hôpital hired Swiss mathematician Johann Bernoulli of the BERNOULLI FAMILY to teach him the newly discovered theory of calculus. This was conducted chiefly by correspondence, and the agreement was made that all content of the letters sent between them would belong to the marquis. This material formed the basis of his 1696 text.

Chapter one of *Analyse des infiniment petits* defines the notion of a DIFFERENTIAL (or difference, as l'Hôpital called it) and provides rules as to how they are to be manipulated. It also outlines the basic principles of differential calculus. The second chapter gives the method for determining the tangent line to a curve, and chapter three deals with MAXIMUM/MINIMUM problems using problems from mechanics and geography as examples. Later chapters deal with cusps, points of inflection, higher-order derivatives, evolutes, and caustics. L'HÔPITAL'S RULE appears in chapter nine.

After l'Hôpital's death Johann Bernoulli complained publicly that not enough credit was given to him for the work contained in the text. (L'Hôpital did write a note of gratitude in the book to Bernoulli, and to GOTTFRIED WILHELM LEIBNIZ, for contributing their ideas.) It is known today, for example, that Bernoulli, not l'Hôpital, discovered l'Hôpital's rule.

L'Hôpital wrote a complete manuscript for a second book *Traité analytique des sections coniques* (Analytical treatise on conic sections), which was published posthumously in 1720. He had also planned to write a third text, one on the topic of integral calculus, but discontinued work on the project when he heard that Leibniz was working on his own book on the topic.

L'Hôpital died in Paris, France, on February 2, 1704.

L'Hôpital's rule (L'Hospital's rule)

Named after GUILLAUME FRANÇOIS ANTOINE L'HÔPITAL (1661–1704), a student of the mathematician Johann Bernoulli of the BERNOULLI FAMILY, l'Hôpital's rule is a method for finding the LIMIT of a ratio of two functions, each of which separately tends to zero. Precisely:

Suppose $f(x)$ and $g(x)$ are two differentiable functions with $f(a) = 0$ and $g(a) = 0$ at some point a. Then the limit of the ratio $f(x)/g(x)$ as $x \to a$ is equal to the limit of the ratio of the derivatives $f'(x)/g'(x)$ as $x \to a$ (provided the derivative of $g(x)$ is never zero, except possibly at $x = a$).

As an example, to compute the limit $\lim_{x \to 1} \dfrac{x^2 - 1}{2x - 2}$ (which looks to be of the form 0/0), one simply takes the derivative of numerator and denominator separately:

$$\lim_{x\to 1}\frac{x^2-1}{2x-2}=\lim_{x\to 1}\frac{2x}{2}=\lim_{x\to 1}x=1$$

As another example we have: $\lim_{x\to 0}\dfrac{\sin x}{x}=\lim_{x\to 0}\dfrac{\cos x}{1}=$ cos (0) = 1.

If we make the assumption that the derivative $g(x)$ is not zero at $x = a$, then the proof of l'Hôpital's rule is relatively straightforward:

Write the limit $\lim_{x\to a}\dfrac{f(x)}{g(x)}$ as $\lim_{h\to 0}\dfrac{f(a+h)}{g(a+h)}$. Since $f(a) = g(a) = 0$, we have

$$\lim_{h\to 0}\frac{f(a+h)}{g(a+h)}=\lim_{h\to 0}\frac{f(a+h)-f(a)}{g(a+h)-g(a)}$$

$$=\lim_{h\to 0}\frac{\dfrac{f(a+h)-f(a)}{h}}{\dfrac{g(a+h)-g(a)}{h}}$$

$$=\frac{f'(a)}{g'(a)}$$

The result is still valid if $g'(a) = 0$, but one must make clever use of the MEAN-VALUE THEOREM to establish it. Johann Bernoulli is the one who discovered and proved l'Hôpital's rule.

It is permissible to apply l'Hôpital's rule to the same limit more than once, that is, to differentiate the numerator and denominators each a number of times. To illustrate, we have:

$$\lim_{x\to 0}\frac{1-\cos x}{x^2}=\lim_{x\to 0}\frac{\sin x}{2x}=\lim_{x\to 0}\frac{\cos x}{2}=\frac{1}{2}$$

The theorem also holds for limits as $x \to \infty$.

The rule also works for a ratio of functions $f(x)/g(x)$ if each function separately tends to infinity as $x \to a$, (that is, we have an indeterminate ratio ∞/∞). One notes that the functions $1/f(x)$ and $1/g(x)$ each tend to zero as $x \to a$, and so:

$$\lim_{x\to a}\frac{f(x)}{g(x)}=\lim_{x\to a}\frac{\left(1/g(x)\right)}{\left(1/f(x)\right)}=\lim_{x\to a}\frac{\left(1/g(x)\right)'}{\left(1/f(x)\right)'}$$

by l'Hôpital's rule. Consequently:

$$\lim_{x\to a}\frac{f(x)}{g(x)}=\lim_{x\to a}\frac{\left(1/g(x)\right)'}{\left(1/f(x)\right)'}$$

$$=\lim_{x\to a}\frac{g'(x)}{\left(g(x)\right)^2}\cdot\frac{\left(f(x)\right)^2}{f'(x)}$$

$$=\left(\lim_{x\to a}\frac{f(x)}{g(x)}\right)^2\cdot\left(\lim_{x\to a}\frac{f'(x)}{g'(x)}\right)^{-1}$$

from which it follows that $\lim_{x\to a}\dfrac{f(x)}{g(x)}=\lim_{x\to a}\dfrac{f'(x)}{g'(x)}$. As examples, we have:

$$\lim_{x\to\infty}\frac{7x^2+2}{x^2+x-3}=\lim_{x\to\infty}\frac{14x}{2x+1}=\lim_{x\to\infty}\frac{14}{2}=7$$

and

$$\lim_{x\to\infty}\frac{\ln x}{x}=\lim_{x\to\infty}\frac{(1/x)}{1}=0$$

liar's paradox The sixth century B.C.E. Cretan prophet Epimenides is purported to have said, "All Cretans are liars," a statement now referred to as the liar's paradox. It is difficult to determine whether Epimenides, himself a Cretan, is telling the truth here. The statement, however, is not a true PARADOX. It may be the case that Epimenides is aware of at least one honest countryman and is making here a false statement via the use of the word *all*. A sharper version of the intended paradox lies in the statement, "This sentence is false."

One may also consider the logical consequences of variations of this statement. For example, what if Epimenides had stated instead: "All Cretans are truth-tellers"?

See also SELF-REFERENCE.

life tables (**mortality tables**) Based on census results and results from medical and social studies, life tables are tables of values indicating the proportion of people of a certain age expected to live to successively higher ages. Different tables are provided for different specific populations and ages (for example, Australian males age 40, Canadian females age 25) and are used extensively by insurance companies to analyze risk in issuing insurance policies, as well as in scientific studies. For example, such tables can be used to compare mortality rates for illnesses in different age groups, or populations with different habits. Life tables are regularly updated to take account of new factors that may alter life expectancy.

In his 1662 pamphlet, *Natural and Political Observations Made upon the Bills of Mortality,* English shopkeeper John Graunt was the first to collate and publish tables of mortality for a specific population. In summarizing government burial records for the years 1604–61, Graunt was able to estimate the number of deaths, decade by decade, to expect among a group of typical 100 Londoners born at the same time. He gave the name "life table" to his display of results. Graunt was also able to make general observations about the population—that women live longer than men, that the death rate is typically constant, and the like—and he was the first to comment on the regularity of social phenomena in this way.

In 1693, relying on records collated by Casper Heumann of Breslau, English astronomer Edmund Halley refined the mathematical techniques used by Graunt to compile a revised and more detailed set of mortality tables, ones suitable for properly analyzing annuities.

ACTUARIAL SCIENCE is the mathematical study life expectancies and other demographic trends. Halley's work is said to be the founding work in this field.

See also HISTORY OF PROBABILITY AND STATISTICS (essay); STATISTICS.

limit   Intuitively, a limit is a quantity that can be approached more and more closely but not necessarily ever reached. For example, the numbers in the SEQUENCE 0.9, 0.99, 0.999,... approach, but never reach, the value 1. We say that the limit of this sequence is one. The function $f(x) = \frac{1}{x}$ takes values closer and closer to zero as x becomes large. We say the limit of this function as x becomes large is zero.

The notion of a limit was first properly identified by the French mathematician AUGUSTIN-LOUIS CAUCHY (1789–1857), arising as a necessary tool for placing the theory of calculus on sound theoretical footing. German mathematician KARL THEODOR WILHELM WEIERSTRASS (1815–97) later developed the idea further and gave the concept of a limit the precise, rigorous definitions we follow today.

The Limit of a Sequence

A sequence of numbers $a_1, a_2, a_3, \ldots$ has limit L if one can demonstrate that for any positive number ε (no matter how small), eventually all the numbers in the sequence will be this close to the value L. That is, one can find a value N so that a_n lies between $L - \varepsilon$ and $L + \varepsilon$ if $n > N$.

This says that no matter which level of precision you care to choose (ε), eventually all the numbers in the list (from a_N onward) will be within a distance ε from L. For example, in the sequence 0.9, 0.99, 0.999,..., all the numbers in the list from the third place onward are within a distance 1/1000 from the value 1, and all numbers from the sixth place onward are within one-millionth of the value 1. In fact, for any small value ε, we can locate a position in the sequence so that from that position onward, all values in the list are within a distance ε from one.

If a sequence $\{a_n\}$ has a limit value L, we write: $\lim_{n \to \infty} a_n = L$. For example, one can show that $\lim_{n \to \infty} \frac{1}{n} = 0$ and $\lim_{n \to \infty} \frac{n^2 - 1}{n^2} = 1$. A careful study of CONVERGENT SEQUENCES shows that not all sequences, however, have a limit.

Any infinite sum (SERIES) can be thought of as a limit of a sequence of PARTIAL SUMs, and any INFINITE PRODUCT the limit of a sequence of partial products.

The Limit of a Function

A function $f(x)$ has limit L as x becomes large if one can demonstrate that for any positive number ε (no matter how small), all the outputs of the function will eventually be this close to the value L. That is, one can find a number N so that the value $f(x)$ lies between $L - \varepsilon$ and $L + \varepsilon$ if $x > N$.

This says that no matter which level of precision you care to choose (ε), eventually all outputs of the function $f(x)$ (from $x = N$ onward) will be within a distance ε from L. For example, all outputs of the function $f(x) = \frac{1}{x}$ are smaller than 0.001 from the point $x = 1,000$ onward. Similarly, all outputs are smaller than 0.000001 from $x = 1,000,000$ onwards. In fact, for any small value ε it is possible to locate a value $x = N$ so that all outputs of the function from that position onward are within a distance ε from zero. (For example, $N = \frac{1}{\varepsilon}$ will do.)

If a function $f(x)$ has a limit value L as x becomes large, we write: $\lim_{x \to \infty} f(x) = L$. One can similarly define the notion of a limit as x becomes large and negative: $\lim_{x \to -\infty} f(x) = L$.

One can also consider the possibility of the outputs of a function $f(x)$ approaching a value L as x approaches a finite value a. Loosely speaking, we say the "limit of $f(x)$ as x tends to a is L" if, as x gets closer and closer to a, the outputs $f(x)$ get closer and closer to L. If this is indeed the case, we write: $\lim_{x \to a} f(x) = L$.

To make this notion precise, we need to assume one is given a desired degree of precision ε, and show that it is indeed possible to specify a "degree of closeness to a" that ensures all the outputs $f(x)$ are within a distance ε of L. This leads to Weierstrass's famous epsilon-delta definition of a limit:

> A function $f(x)$ has "limit L as x tends to a" if one can demonstrate that for any positive number ε (no matter how small), the outputs of the function can be made this close to L by restricting x to values very close, but not equal, to a. That is, one can produce a number δ so that if x, different from a, lies between $a - \delta$ and $a + \delta$, then we can be sure that $f(x)$ lies between $L - \varepsilon$ and $L + \varepsilon$.

This says that no matter which level of precision you care to choose (ε), all outputs of the function $f(x)$ for values x close to a (namely, within a distance δ of a) will be within a distance ε from L. Consider, for example, the function $f(x) = 5x$ for values close to $x = 2$. Notice that all outputs of the function are within a distance 0.1 from 10 if x is within a distance 0.02 from 2. All outputs of the function are within a distance 0.001 from 10 if x is within a distance 0.0002 from 2. In fact, for any small value ε, it is possible to describe a number δ so that if x is within a distance δ from 2, then $f(x) = 5x$ is within a distance ε from 10. (In fact, $\delta = \frac{\varepsilon}{5}$ will do.) This shows $\lim_{x \to 2} 5x = 10$.

If a function $f(x)$ is continuous at $x = a$, then the limit $\lim_{x \to a} f(x)$ exists and equals $f(a)$. This, however, need not always be the case (in which case we say that f is discontinuous at a).

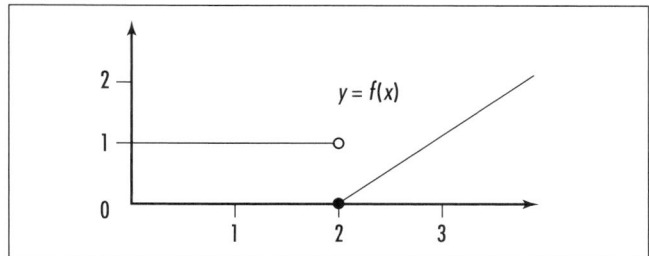

Left and right limits that do not match

It is sometimes convenient to describe limit "just from the left" or "just from the right." Written as $\lim_{x \to a^-} f(x)$, a limit from the left is defined as a value L so that outputs of the function $f(x)$ can be made as close to L as we please by restricting x to values close to and to the left of a (that is, for values of x between $a - \delta$ and a, for some number δ). A limit from the right, written $\lim_{x \to a^+} f(x)$, is a value L so that outputs of the function $f(x)$ can be made as close to L as we please by restricting x to values close to and to the right of a (that is, for values of x between a and $a + \delta$, for some number δ). For example, in the graph above we have $\lim_{x \to 2^-} f(x) = 1$ and $\lim_{x \to 2^+} f(x) = 0$. That the left and right limits do not agree shows that the function is discontinuous at $x = 2$.

The word *limit* is also used in INTEGRAL CALCULUS in terms of a limit of integration. Given a definite integral $\int_a^b f(x)\, dx$, the number a is called the lower limit of integration, and b the upper limit of integration.

See also ASYMPTOTE; CONTINUOUS FUNCTION; DERIVATIVE; DIVERGENT; HISTORY OF CALCULUS (essay); LEFT DERIVATIVE/RIGHT DERIVATIVE; REMOVABLE DISCONTINUITY; ZENO'S PARADOXES.

limit from the left/right *See* LIMIT.

Lindemann, Carl Louis Ferdinand von (1852–1939) German *Number theory* Born on April 12, 1852, scholar Ferdinand von Lindemann is best remembered for his 1882 proof that π is a TRANSCENDENTAL NUMBER. This accomplishment finally settled the age-old problem of SQUARING THE CIRCLE: by proving that π is not a solution to a polynomial equation with integer coefficients, Lindemann had demonstrated the impossibility of constructing a square of the same area of a given circle using the classical tools of a straight-edge and compass alone.

Lindemann wrote a thesis on the topic of NON-EUCLIDEAN GEOMETRY under the direction of CHRISTIAN FELIX KLEIN (1849–1925), and was awarded a doctoral degree from Erlangen in 1873. He completed an advanced habilitation degree in 1877 at the University of Würzburg and was appointed a faculty position at the University of Freiburg that same year. He later transferred to the University of Königsberg, and then eventually accepted a chair at the University of Munich in 1893, where he remained for the rest of his career.

In 1873, the year Lindemann was awarded his doctorate, French mathematician Charles Hermite published his proof that the number e is transcendental. Lindemann traveled to Paris to meet Hermite and to discuss the methods of his proof. Using the famous formula $e^{i\pi} = -1$ of LEONHARD EULER (1707–83), Lindemann realized that Hermite's methods could be extended to also establish the transcendence of π. Lindemann published his proof in his 1882 paper *Über die Zahl* (On numbers).

Lindemann was also interested in physics and contributed to the studies of electrons. He also worked to translate and revise the work of the mathematician JULES HENRI POINCARÉ (1854–1912).

In 1894 Lindemann was elected to the Bavarian Academy of Sciences. He was also praised with an honorary degree from the University of St. Andrews in 1912. He died in Munich, Germany, on March 6, 1939, and will always be remembered in history for bringing a close to the classic problem of squaring the circle.

line A CURVE is sometimes called a line. In GEOMETRY, a line is usually understood to be straight, but it is difficult to properly define what is meant by this. The geometer EUCLID (ca. 300–260 B.C.E.) provided the intuitive definition of a line as a "length with no breadth," but he never attempted to define what is meant by a length or what it means to say that a construct has no breadth. Euclid, however, did state that between any two points A and B in the plane, there is such a thing as a straight line that connects them. Today mathematicians take this as the starting point of geometry, leaving the terms *line* and *point* (and *plane*)

as undefined terms, but taking the properties we expect them to possess (such as "between every two points there is a line that connects them") as AXIOMS for the theory of geometry.

If one is working with a theory of geometry (or of shape and space) in which there is a clear notion of a distance between two points, then one could define a straight line between two points to be the shortest path between those points. For instance, PYTHAGORAS'S THEOREM, in some sense, establishes that straight paths, as we intuitively think of them, are indeed the shortest routes between two points. On the surface of a SPHERE, the shortest paths between points are arcs of great circles, and it is therefore appropriate to deem these as the "straight" paths in SPHERICAL GEOMETRY.

See also COLLINEAR; CONCURRENT; EQUATION OF A LINE; LINEAR EQUATION; SLOPE.

linear algebra The study of matrices and their applications is called linear algebra. As matrices are used to analyze and solve systems of SIMULTANEOUS LINEAR EQUATIONS and to describe LINEAR TRANSFORMATIONS between VECTOR SPACES, this topic of study unites geometric thinking with numerical analysis. As the set of all invertible matrices of a given size form a group, called the GENERAL LINEAR GROUP, techniques of ABSTRACT ALGEBRA can also be incorporated into this work.

See also MATRIX.

linear equation An equation is called linear if no variable appearing in the equation is raised to a power different from 1, and no two (or more) variables appearing in the equation are multiplied together. For example, the equation $2x - 3y + z = 6$ is linear, but the equations $2x^3 - 5y + z^{-1} = 0$ and $4xy + 5xz = 7$ are not.

A function of one variable is said to be linear if it is of the form $f(x) = ax + b$, for some constants a and b. More generally, a function of several variables of the form

$$f(x_1, x_2, \ldots, x_n) = a_0 + a_1 x_1 + a_2 x_2 + \ldots + a_n x_n$$

for some constants $a_0, a_1, a_2, \ldots, a_n$ is called linear.

Any equation of the form $ax + by = c$ represents a LINE in two-dimensional space. (Solving for y, assuming that b is not zero, yields the linear function $y = -\frac{a}{b}x + \frac{c}{b}$.) An equation of the form $ax + by + cz = d$ represents a PLANE in three-dimensional space.

A linear combination of variables x_1, x_2, x_3, … is a sum of the form

$$a_1x_1 + a_2 x_2 + a_3x_3 + \ldots$$

for some constants a_1, a_2, a_3, … In VECTOR SPACE theory, a set of vectors is said to be linearly dependent if some linear combination of those vectors is zero.

In LINEAR ALGEBRA, a MATRIX equation of the form $A\mathbf{x} = \mathbf{b}$ is called a linear equation. It represents a system of SIMULTANEOUS LINEAR EQUATIONS.

A linear differential equation is a DIFFERENTIAL EQUATION of the form:

$$a_0y + a_1 \frac{dy}{dx} + a_2 \frac{d^2y}{dx^2} + \cdots + a_n \frac{d^ny}{dx^n} = f(x)$$

for some constants a_0, a_1, a_2, …, a_n and some fixed function $f(x)$.

In some settings it is appropriate to apply the term *linear* to specific variables appearing in a complicated expression. For instance, the term $5x^2yz$ is linear with respect to y and with respect to z.

See also EQUATION OF A LINE; EQUATION OF A PLANE; LINEAR TRANSFORMATION; LINEARLY DEPENDENT AND INDEPENDENT.

linearly dependent and independent

A collection of functions is said to be linearly dependent if one of them can be expressed as a sum of constant multiples of the other; if this is not possible, then the collection is said to be linearly independent. For example, the functions $f_1(x) = x$, $f_2(x) = x^2 - 2x$, $f_3(x) = x^2$ are linearly dependent, since $f_3(x) = 2f_1(x) + f_2(x)$. The functions $\{x, 7x\}$ are also linearly dependent, since the second function is a constant multiple of the first. On the other hand, the functions $\{x, x^2, x^3\}$ are linearly independent, as are the functions $\{\sin x, \cos x\}$.

A set of VECTORS is said to be linearly dependent if it is possible to write one vector as a combination of the remaining vectors. Equivalently, vectors $\mathbf{v}_1$, $\mathbf{v}_2$, …, $\mathbf{v}_n$ are linearly dependent if it is possible to choose scalars $c_1, c_2, \ldots, c_n$, not all zero, so that

$$c_1\mathbf{v}_1 + c_2\mathbf{v}_2 + \ldots + c_n\mathbf{v}_n = 0$$

(If c_i, say, is not zero, then dividing through by this scalar shows that $\mathbf{v}_i$ is a sum of multiples of the remaining vectors.) If this is not possible, then the vectors are said to be linearly independent. For example, in three-dimensional space, the vectors $\mathbf{i} = <1,0,0>$, $\mathbf{j} = <0,1,0>$ and $\mathbf{k} = <0,0,1>$ are linearly independent—it is not possible to write any one as a sum of multiples of the other two.

A basis for a VECTOR SPACE is a collection of linearly independent vectors with the property that any other vector in the vector space can be written as a sum of multiples of these vectors. For example, the vectors $\mathbf{i}$, $\mathbf{j}$, and $\mathbf{k}$ form a basis for the vector space of three-dimensional vectors for any other vector $\mathbf{a} = <a_1,a_2,a_3>$ that can be expressed as the combination $\mathbf{a} = a_1\mathbf{i} + a_2\mathbf{j} + a_3\mathbf{k}$. It is impossible to express a vector as a combination of basis vectors in two different ways. (To explain: Suppose $\mathbf{v}_1$, $\mathbf{v}_2$, $\mathbf{v}_3$ is a basis for a vector space, and that some vector $\mathbf{a}$ can be expressed as a combination of these vectors in two different ways: $\mathbf{a} = a_1\mathbf{v}_1 + a_2\mathbf{v}_2 + a_3\mathbf{v}_3 = b_1\mathbf{v}_1 + b_2\mathbf{v}_2 + b_3\mathbf{v}_3$. Subtracting gives the equation $(a_1 - b_1)\mathbf{v}_1 + (a_2 - b_2)\mathbf{v}_2 + (a_3 - b_3)\mathbf{v}_3 = 0$. Since the vectors $\mathbf{v}_1, \mathbf{v}_2, \mathbf{v}_3$ are linearly independent, it must be the case that $a_1 = b_1$, $= a_2 = b_2$, and $a_3 = b_3$.)

Mathematicians have proved that every vector space must have a basis, and that the number of vectors in any basis for a particular vector space is always the same. This number is called the dimension of the vector space. In particular, the set of all functions is a vector space and so must have a basis. One candidate for such a basis is the infinite collection of functions $\{1, x, x^2, x^3, x^4, \ldots\}$. This set is certainly linearly independent, and the work of constructing TAYLOR SERIES shows that all "appropriately nice" functions can be expressed as infinite sums of these basic functions. Functions like $\sin(x)$, $\cos(x)$, and $\sin(7x)$ repeat values every 2π and are called periodic. Mathematicians have shown that the collection $\{1, \sin(x), \cos(x), \sin(2x), \cos(2x), \sin(3x), \cos(3x), \ldots\}$ forms a basis for the vector space of all periodic functions. This leads to the study of FOURIER SERIES.

See also ORTHOGONAL.

linear programming The branch of mathematics concerned with finding the maximum or minimum values of linear functions, that is, functions involving variables raised only to the first power, subject to a number of inequalities that must remain true, is called linear programming. This field has profound applications to economics and industry and is an area of active research. While, in principle, OPTIMIZATION problems of this type are straightforward to solve, practical problems may involve well over 100 variables and be difficult to analyze. The challenge is to find efficient techniques for finding solutions.

The principle of linear programming is best illustrated with an example. Suppose, for instance, we wish to find the maximum value of the function $U = 4x - 3y$ subject to the constraints $x \geq 0$, $y \geq 0$, $x \leq 1$, and $y \leq 1$. In this example, the constraints define a unit square in the plane, and we wish to find the point (x, y) in this "feasible region" that provides the largest value for the "objective function" $U = 4x - 3y$.

Reasoning backward, note that each possible value c of the objective function defines a line $4x - 3y = c$ of slope 4/3. As the value of c varies, this line sweeps across the plane. Starting with a large value of c and decreasing its value, we thus seek the first value, of c that produces a line that touches the feasibility region. Clearly, this will occur at one of the vertices of the square. Checking all four vertices, $(0,0)$, $(1,0)$, $(0,1)$, and $(1,1)$, we see that $x = 1$, $y = 0$ gives the largest possible value 4 for U.

In general, the constraint conditions define a polygonal region in space, and the maximal and minimal values of U can only occur at vertices of the region. Linear programming then seeks to find efficient methods for checking which vertices yield the largest and smallest values for U.

See also OPERATIONS RESEARCH.

linear transformation A map $T : V \rightarrow W$ between two VECTOR SPACES V and W is called a linear transformation if the following two conditions hold:

i. $T(\mathbf{a} + \mathbf{b}) = T(\mathbf{a}) + T(\mathbf{b})$ for any two vectors $\mathbf{a}$ and $\mathbf{b}$
ii. $T(k\mathbf{a}) = kT(\mathbf{a})$ for any number k and any vector $\mathbf{a}$

If the vector spaces V and W represent the set of all points in the plane or in three-dimensional space, then a linear transformation is an example of a GEOMETRIC TRANSFORMATION that takes straight lines to straight lines. For instance, rotations and reflections are linear transformations. However, not every geometric transformation is a linear transformation. Although a translation, for example, preserves straight lines in the plane, it does not satisfy the first condition described above and so is not a linear transformation.

If $e_1 e_2,\ldots,e_n$ is a basis for the vector space V, then any vector $\mathbf{a}$ in V can be written as a linear combination of these basis vectors:

$$\mathbf{a} = c_1 e_1 + c_2 e_2 + \ldots + c_n e_n$$

for some numbers $c_1, c_2,\ldots, c_n$. Thus the value of the linear transformation T is completely determined by its values on the basis vectors:

$$T(\mathbf{a}) = T(c_1 e_1 + c_2 e_2 + \ldots + c_n e_n)$$
$$= c_1 T(e_1) + c_2 T(e_2) + \ldots + c_n T(e_n)$$

If $f_1, f_2,\ldots,f_m$ is a basis for the second vector space W, then each vector $T(e_j)$ is a linear combination of these basis vectors:

$$T(e_j) = a_{1j}f_1 + a_{2j}f_2 + \ldots + a_{mj}f_m$$

Thus the numbers a_{ij} completely specify how the linear transformation works. Let A be the MATRIX with (i,j)th entry equal to a_{ij}. This shows that every linear transformation is represented by a matrix. Moreover, if we represent the basis vectors $e_1,e_2,\ldots,e_n$ as the column vectors:

$$e_i = \begin{pmatrix} 1 \\ 0 \\ \vdots \\ 0 \end{pmatrix}, e_2 = \begin{pmatrix} 0 \\ 1 \\ \vdots \\ 0 \end{pmatrix},\ldots,e_n = \begin{pmatrix} 0 \\ 0 \\ \vdots \\ 1 \end{pmatrix}$$

then the matrix A, whose jth column is the sequence of values that result when T is applied to the jth basis vector e_j, satisfies $Ae_j = T(e_j)$, and, in general, for any vector $\mathbf{a}$ we have:

$$T(\mathbf{a}) = A\mathbf{a}$$

That is, we have:

Any linear transformation T from a vector space V to a vector space W is given by multiplication with a matrix A whose jth column is the effect of T on the jth basis vector of V.

For example, consider a rotation R in the plane about the origin through an angle θ. Such a rotation takes a unit vector in the direction of the x-axis, $e_1 = \begin{pmatrix} 1 \\ 0 \end{pmatrix}$, to the vector $\begin{pmatrix} \cos\theta \\ \sin\theta \end{pmatrix}$, and the unit vector in the direction of the y-axis, $e_2 = \begin{pmatrix} 0 \\ 1 \end{pmatrix}$, to the vector $\begin{pmatrix} -\sin\theta \\ \cos\theta \end{pmatrix}$. Thus the matrix representing a rotation through an angle θ is given by:

$$R = \begin{pmatrix} \cos\theta & -\sin\theta \\ \sin\theta & \cos\theta \end{pmatrix}$$

In the same way, a reflection about the x-axis, for instance, is given by the matrix:

$$\begin{pmatrix} 1 & 0 \\ 0 & -1 \end{pmatrix}$$

and a dilation by a factor k as:

$$\begin{pmatrix} k & 0 \\ 0 & k \end{pmatrix}$$

Of course, if one were to work with a different set of basis vectors, the matrix representing the linear transformation would be different.

One can show that if matrix A represents a linear transformation $T: V \rightarrow W$ and matrix B represents a linear transformation $S: W \rightarrow R$, then the matrix product BA represents the composite linear transformation: $S_{\circ}T: V \rightarrow R$. If the matrix A is invertible, then the INVERSE MATRIX A^{-1} represents the inverse linear transformation $T^{-1}: W \rightarrow V$. (This inverse map exists if A is indeed invertible.)

See also AFFINE TRANSFORMATION; MATRIX OPERATIONS.

Liouville, Joseph (1809–1882) French *Number theory, Analysis* Born on March 24, 1809, in Saint-Omer, France, scholar Joseph Liouville is best remembered for his 1844 proof of the existence of TRANSCENDENTAL NUMBERS. Liouville also managed to provide, for the first time, specific examples of numbers that cannot be algebraic. (These numbers are today called Liouville numbers.) He is also noted for his contributions to DIFFERENTIAL EQUATIONs, differential geometry (the study of CALCULUS on three-dimensional shapes and surfaces), complex analysis (calculus applied to complex numbers), and NUMBER THEORY. In 1864 he also edited and published manuscripts left by ÉVARISTE GALOIS (1811–32) on POLYNOMIAL equations. Liouville wrote over 400 mathematical papers during his career, around 200 of which were on the topic of number theory.

Liouville graduated from the École Polytechnique in 1827 with a basic degree in mathematics and mechanics. After taking on a number of different teaching positions, Liouville was eventually appointed professor of analysis and mechanics at that same institution in 1838. Meanwhile, Liouville had already garnered an international reputation for his work on electrodynamics, partial differential equations, and the study of heat, as well as for his establishment of a new mathematics journal, *Journal de Mathématiques Pures et Appliqués* (Journal of pure and applied mathematics), today commonly referred to as *Liouville's Journal*.

Correspondence with mathematicians CHRISTIAN GOLDBACH and Daniel Bernoulli of the BERNOULLI FAMILY sparked Liouville's interest in transcendental numbers. He attempted to prove that the number e was transcendental, but did not succeed. (This feat was later accomplished by French mathematician Charles Hermite in 1873.) However, using the theory of continued fractions, Liouville managed to construct a class of real numbers x with the property that for each natural number n there is a fraction $\frac{p}{q}$ satisfying the inequality:

$$|x - \frac{p}{q}| < \frac{1}{q^n}$$

This, Liouville showed, was enough to establish that x is transcendental. In particular, Liouville showed that the specific number (sometimes now called Liouville's number):

$$\sum_{n=1}^{\infty} \frac{1}{n!} = 0.11000100000000000000000010000...$$

is transcendental.

Liouville died in Paris, France, on September 8, 1882. He established his place in the history of mathematics for pioneering work on the study of REAL NUMBERS.

Liu Hui   *See* CHINESE MATHEMATICS.

Li Ye (Li Chi, Li Zhi) (1192–1279) Chinese *Algebra* Li Ye is remembered for his 1248 text *Ceyuan Haijing* (Sea mirror of circle measurements), in which he introduced the "method of the celestial element"—a system of notation for polynomials in one variable (the celestial element)—and a set of techniques for solving such polynomial equations. Using the single diagram of a circular city wall inscribed inside a large right-angled triangle, *Ceyuan Haijing* leads the reader through 170 geometric problems, cleverly designed to illustrate the techniques of translating geometry into algebra, and then solving the consequent algebraic equations. Over 650 different formulae for triangular areas and segment lengths are presented in this text.

Extremely little is known of Li Ye's life, except that his work apparently earned him some regard. In 1266 Li Ye was appointed a position at the elite Hanlin Academy by the emperor Kublai Khan, grandson of the great Genghis Khan. He remained there only a few years to then retire and live the rest of his life as a hermit.

Lobachevsky, Nikolai Ivanovich (1792–1856) Russian *Geometry*   Born on December 1, 1792, in Novgorod, Russia, Nikolai Lobachevsky is remembered for his 1826 discovery of HYPERBOLIC GEOMETRY and for detailing many of its properties. (This work was conducted independently of the discoveries made by JÁNOS BOLYAI (1802–60) 3 years earlier.) He was the first to publish a description of a NON-EUCLIDEAN GEOMETRY in his 1826 article "A Concise Outline of the Foundations of Geometry."

Lobachevsky entered Kazan State University in 1807 with the intent to study medicine, but soon changed interest to pursue courses in mathematics and physics.

He graduated with a master's degree in 1811 and, three years later, was appointed a lectureship position at the university. In 1822 he was appointed full professor, and in 1827 was named rector of the university. He remained at Kazan State University until his retirement in 1846.

Lobachevsky was introduced to the topic of geometry as a student. Ever since the time of EUCLID (ca. 300–260 B.C.E.), scholars questioned Euclid's choice of axioms as the basis for all of geometry. His fifth postulate, the famous PARALLEL POSTULATE, was deemed of a different nature than the remaining four, and scholars suspected that it could be deduced from them as a logical consequence. This became an outstanding challenge in mathematics. For two millennia scholars attempted to establish the fifth postulate as a THEOREM, but failed.

Upon learning of this problem, Lobachevsky began analyzing the situation for himself. Rather than attempt to prove the fifth postulate, he considered the possibility that it need not follow from the remaining four axioms, and, moreover, allowed for the possibility of a geometry in which the first four axioms do hold but one in which the fifth postulate is blatantly false. With this expanded thinking, Lobachevsky discovered a consistent theory of geometry—hyperbolic geometry—different from Euclidean geometry, but nonetheless valid in its own right. That such a geometry exists showed, once and for all, that the fifth postulate is in fact independent of the remaining four axioms. This was a remarkable achievement.

Lobachevsky presented the results of his discovery to his colleagues at Kazan State University in 1826 and published his article "A Concise Outline of the Foundations of Geometry" in the *Kazan Messenger*. Unfortunately, the St. Petersburg Academy of Sciences decided not to publish his piece as a peer-reviewed article, and Lobachevsky's work did not receive widespread recognition.

Although Lobachevsky managed to publish papers on the topic at later dates, including his 1840 German paper "Geometrische Untersuchungen zur Theorie der Parallellinien" (Geometric investigation on the theory of parallel lines), it was not until after his death on February 24, 1856, that the importance of his work was understood and published in a mainstream forum. Today, Lobachevskian geometry plays a central role in the modern description of space and motion in relativistic quantum mechanics and the general theory of relativity.

local maximum/local minimum *See* MAXIMUM/ MINIMUM.

locus (plural, loci) A set of points satisfying some specified condition is called a locus of points. For example, the locus of all points in the plane at equal distance from a given point is a circle, and the locus of all points EQUIDISTANT from two given points A and B in the plane is a straight line perpendicular to the line segment connecting A and B through its MIDPOINT. Further, if the point A is replaced by a circle, then the locus of all points equidistant from A and B is a HYPERBOLA if B lies outside the circle A, and an ELLIPSE if B lies inside A. The locus of all points equidistant from a line A and a point B is a PARABOLA.

If, in a system of CARTESIAN COORDINATES, a locus of points can be expressed in the form

$$\{(x,y) : f(x,y) = 0\}$$

then the equation $f(x,y) = 0$ is called the equation of the locus. For example, $x^2 + y^2 - 25 = 0$ is the equation of a circle of radius 5 centered about the origin.

logarithm The power (EXPONENT) to which a number b must be raised to obtain a given number N is called the base-b logarithm of N. That is, if $b^x = N$, then we write $\log_b N = x$, which we read as "the power of b that gives N is x." It is assumed that the number b is positive. For example, one must raise the number 10 to a power of 2 to obtain the number 100, and so the base-10 logarithm of 100 is 2:

$$\log_{10} 100 = 2$$

We also have:

$\log_{10} 1,000 = 3$ (the power of 10 that gives 1,000 is 3)

$\log_{\frac{1}{2}}\left(\frac{1}{8}\right) = 3$ (the power of $\frac{1}{2}$ that gives $\frac{1}{8}$ is 3)

$\log_3\left(\frac{1}{9}\right) = -2$ (the power of 3 that gives $\frac{1}{9}$ is −2)

$\log_{43} 1 = 0$ (the power of 43 that gives 1 is zero)

and

$\log_6 \sqrt{6} = \frac{1}{2}$ (the power of 6 that gives $\sqrt{6}$ is 1/2)

Because a quantity b^x is never negative, or zero, there is no logarithm of a negative number or of zero.

Because logarithms are exponents, they obey the same rules as exponents. For example, the multiplication rule for exponents reads $b^x b^y = b^{x+y}$ indicating that, upon multiplication, exponents add. This leads to the rule of logarithms:

1. $\log_b (N \times M) = \log_b N + \log_b M$

(Precisely, if $x = \log_b N$ and $y = \log_b M$, then we have: $b^x = N$ and $b^y = M$. Consequently, $N \times M = b^x \times b^y = b^{x+y}$, which states that $x + y$ is the power of b that gives $N \times M$. That is, $\log_b N \times M = x + y = \log_b N + \log_b M$.)

Similarly, the exponent rule $(b^x)^y = b^{xy}$ leads to the rule:

2. $\log_b(N^y) = y\log_b N$

We also have the rules:

3. $\log_b(b^x) = x$ (The power of b that gives b^x is indeed x.)
4. $b^{\log_b x} = x$ (Indeed, $\log_b x$ is the power of b that gives x.)
5. $\log_b 1 = 0$ (The power of b that gives 1 is zero.)

Logarithms were invented by Scottish mathematician JOHN NAPIER (1550–1617) as a means to simplify arithmetic calculations. For example, rule 1 shows that any multiplication problem can be converted to the much simpler operation of addition using logarithms. This discovery was of great interest to scholars of the Renaissance, in particular astronomers, who were struggling with problems requiring the manipulation of very large numbers. Such computations were extremely tedious and prone to many errors. Inspired by problems dealing with the size of the Earth, Napier felt that working with the number $10^7 = 10,000,000$ would be most helpful to scientists, and he chose the number $b = 1 - \frac{1}{10^7}$ as the base of his logarithms. Napier multiplied all the quantities he worked with by 10^7 to help avoid the appearance of decimals. Today his logarithm of a number N would be written $10^7 \log_{1-\frac{1}{10^7}}\left(\frac{N}{10^7}\right)$.

Although logarithms are now understood as exponents, Napier himself did not think of them in this way. He developed his theory of logarithms geometrically, thinking of them as a ratio of distances traveled by two moving objects, one moving along a straight line at a speed of 1 unit per second, and the other moving along a line segment 1 unit long and with speed changing according to its distance from the endpoint it is approaching. Napier chose the name *logarithm* from the Greek words *logos* for "ratio" and *arithmos* for "number." It was not until the end of the 17th century that mathematicians recognized that logarithms were, in fact, exponents.

After Napier published his work in 1614, English mathematician HENRY BRIGGS (1561–1630) suggested to Napier that, like our number system, logarithms should be based on the number 10. Napier agreed that this would indeed simplify matters, and $b = 10$ was then deemed the preferred base for logarithms. Base-10 logarithms are today called common logarithms or Briggs's logarithms. The common logarithm of N is simply denoted log N or lg N.

In 1624 Briggs published tables listing values of common logarithms for the numbers 1 to 20,000 and 90,000 to 100,000, inclusive. The values for the numbers 20,000–90,000 were completed after Briggs's death by Dutch mathematician Adriaan Vlacq.

Swiss watchmaker Jobst Bürgi, maker of astronomical instruments, also conceived of logarithms to facilitate the multiplication of large numbers. However, since Napier published his work first, the credit for their discovery was not given to Bürgi.

A number with a given value for its logarithm is called the antilogarithm, or antilog, of that value. The base-10 antilog of a value x is 10^x.

The common logarithms of the numbers 3.7, 370, 370,000, for example, differ by whole numbers: log 3.7 ≈ 0.5682, log 370 = log(3.7 × 100) ≈ 0.5682 + 2 = 2.5682, and log 370,000 = log(3.7 × 10⁵) ≈ 5.5682. The decimal part of a logarithmic value is called the mantissa, and the integer part is called the characteristic of the logarithm. (For example, the three logarithms above each have mantissa 0.5682 and characteristics 0, 2, and 5, respectively.) Logarithmic tables from the past listed only the mantissas of numbers from 1 to 10. The logarithm of any other number can then be computed by adding the appropriate integer to represent the power of 10 needed.

In CALCULUS it is convenient to work with logarithms of base e. The number e is an irrational number with value approximately 2.718281828… Logarithms of base e are called natural logarithms, and a logarithm of base e is denoted ln. Thus, $\ln(e^3) = 3$, for example.

If one is willing to work with complex numbers, then it is possible to give meaning to the logarithm of a negative number. For example, EULER'S FORMULA tells us that $e^{i\pi} = -1$. Consequently $\ln(-1) = i\pi$. Going further, extending to logarithms of complex numbers, we have, for instance, $\ln i = \ln\left(\sqrt{-1}\right) = \frac{1}{2}\ln(-1) = \frac{i\pi}{2}$. In this setting, all nonzero numbers—real and complex—have logarithms.

See also E; EXPONENTIAL FUNCTION; LOGARITHMIC FUNCTION; LOGARITHMIC SCALE; SLIDE RULE.

logarithmic function   Any CONTINUOUS FUNCTION $f(x)$, not identically zero, defined for positive values of x with the property that $f(a \cdot x) = f(a) + f(x)$ for all positive values a and x, is called a logarithmic function. Such functions are said to "convert multiplication into addition." The series of observations below shows that every logarithmic function is given by a LOGARITHM: $f(x) = \log_b x$ for some positive base b.

1. All logarithmic functions $f(x)$ satisfy $f(1) = 0$.

(This follows from the observation: $f(1) = f(1.1) = f(1) + f(1)$.)

2. All logarithmic functions satisfy $f(\frac{1}{a}) = -f(a)$. Consequently, the logarithmic functions give both positive and negative outputs.

(This follows from the observation: $0 = f(1) = f(a \cdot \frac{1}{a}) = f(a) + f(\frac{1}{a})$.)

3. For any logarithmic function there is a number b for which $f(b) = 1$.

(We have $f(1) = 0$ and that it is possible to choose a value a such that $f(a)$ is positive. One of the values $f(a)$, $f(a^2) = f(a) + f(a)$, $f(a^3) = f(a) + f(a) + f(a)$, … will be greater than one. Thus it is possible to find a number c with $f(c) > 1$. By the INTERMEDIATE-VALUE THEOREM, there must be a value b between 1 and c, so that $f(b) = 1$.)

4. If $f(x)$ is a logarithmic function with $f(b) = 1$, then $f(x) = \log_b x$.

(For any positive integer n, we have $f(x^n) = f(x) + f(x) + \ldots + f(x) = nf(x)$. Also, $f(x^{-n}) = f(\frac{1}{x^n}) = -f(x^n) = -nf(x)$. For a fraction $\frac{p}{q}$, we have $qf(x^{\frac{p}{q}}) = f\left((x^{\frac{p}{q}})^q\right) = f(x^p) = pf(x)$, which shows $f(x^r) = rf(x)$ for any rational number r. Since every irrational number is a LIMIT of rationals, this property also holds for irrational numbers. In particular, for any value x, $f(b^x) = xf(b) = x$. That is, f is indeed the function that extracts the exponent of a given power of b.)

If a logarithmic function satisfies $f(e) = 1$, then it is called *the* logarithmic function. It is the function given by the natural logarithm: $f(x) = \ln x$. Its derivative is given by $\frac{d}{dx}(\ln x) = \frac{1}{x}$, and consequently $\ln x = \int^x \frac{1}{t} dt$. Thus the logarithmic function can alternatively be defined as the area under the $y = \frac{1}{x}$ curve. (*See* E.)

The derivative of a logarithmic function to an arbitrary base b can be computed via implicit differentiation after first unraveling and then reapplying a logarithm. Specifically, if $y = \log_b x$, then $b^y = x$. Applying the natural logarithm yields $y\ln b = \ln x$. Differentiating, we see that $\frac{dy}{dx}\ln b = \frac{1}{x}$, and so, $\frac{dy}{dx} = \frac{1}{x\ln b}$. Thus we have:

$$\frac{d}{dx}\left(\log_b x\right) = \frac{1}{\ln b} \cdot \frac{1}{x}$$

See also HOMOMORPHISM; LOGARITHMIC SCALE; SLIDE RULE.

logarithmic scale A line marked with distances whose LOGARITHMs are the actual distances along the line is said to be in logarithmic scale. For example, in a base-10 logarithmic scale, if 10 is the label 1 in. along the line, each successive inch will be marked 100, 1,000, etc. (The position labeled 1,000, for example, is physically $\log_{10} 1,000 = 3$ in. along the line.) If a set of coordinate axes are both in logarithmic scale, then the plot of a curve $y = x^n$, for example, will appear as the plot of $\log y = \log(x^n) = n \log x$, that is, as though the variables in question are $\log y$ and $\log x$. In this setting, the graph appears as a straight line of SLOPE n. Scientists often plot graphs on log-log graph paper and measure slope to find the value of n that best fits the data.

The term *logarithmic scale* is also used to describe any quantity that is typically measured in terms of logarithms. In 1935 American seismologist Charles Richter set up the Richter scale to describe the intensity of earthquakes. The scale he devised is logarithmic, base 10, meaning that an earthquake measuring 6 on his scale, for example, is 10^6, or 1 million, times stronger than the small quakes that he set at value zero. The pH measurement of acidity or alkalinity used in chemistry is also logarithmic, as is the unit DECIBEL used to measure sound intensity.

See also SLIDE RULE.

logistic growth *See* POPULATION MODELS.

long division *See* BASE OF A NUMBER SYSTEM.

long radius The distance from the center of a regular POLYGON to one of its vertices is called the long radius of the polygon. This quantity is also the radius of CIRCUMCIRCLE of the polygon. If the polygon has n sides, each 1 unit in length, then an exercise in TRIGONOMETRY shows that the long radius has value

$$r = \frac{1}{2\sin\left(\dfrac{180}{n}\right)}$$

An analog of PI for a regular polygon is the ratio of its PERIMETER to twice the length of it long radius (which equals the diameter of the polygon if n is even). This quantity approaches the value as π as n becomes large. This follows from the fact that, according to the SQUEEZE RULE, $\frac{\sin(x)}{x}$ approaches the value 1 as x, measured in radians, becomes small.

See also APOTHEM; RADIAN MEASURE.

Lovelace, Augusta Ada Byron (1815–1852) *British Computation* Born on December 10, 1815, in Piccadilly, England, Augusta Ada Byron Lovelace is remembered as one of the first people to write a set of instructions for a computing machine. As an assistant to

Augusta Ada Byron Lovelace, assistant to 19th-century scholar Charles Babbage, was the first to develop a theory of mechanical computation and the first to write the equivalent of a computer program. *(Photo courtesy of the Science Museum, London/Topham-HIP/The Image Works)*

the notable CHARLES BABBAGE (1791–1871) in his design and planned construction of an "analytic engine," Lovelace outlined the principles needed in performing mathematics on a machine and, in effect, was the first person to ever write a computer program.

Daughter of the famous poet George Gordon, Lord Byron, Lovelace was encouraged to study mathematics by her mother, herself an adept mathematician. On June 5, 1833, while attending a party, Lovelace met Babbage and learned of his work on mechanical computation. Fascinated by the topic, Lovelace visited Babbage's London studio 2 weeks later to see his first machine, the "difference engine." From that date on, Lovelace worked as Babbage's assistant on the develop-

ment of a superior device capable of receiving instruction and data from punch-cards, and able to perform all possible types of mathematical operations.

European scholars at the time were writing commentaries of Babbage's work, and Lovelace took it upon herself to translate French material into English. In 1843 she published an annotated translation of a work by Luigi Menabrea (1809–96), *Notions sur la machine analytique de Charles Babbage* (Notes on Charles Babbage's analytic engine), in which, through her extensive annotations, she effectively determined the entire theoretical workings of such a machine. (This material was not properly developed by Menabrea.) Lovelace had, in fact, explained how to program a machine to perform abstract mathematics.

On a philosophical note, Lovelace addressed the pressing question of whether calculating machines (computers) would ever be able to think. She argued that machines will only ever be able to do what we know how to order them to perform, and as such will never be able to anticipate results or truths. The latter point, she felt, is what constitutes "thinking," whereas the former does not. This argument has since become known as "Lady Lovelace's objection."

Due to lack of funding, Babbage and Lovelace were never able to complete construction of the new engine. Lovelace's work, nonetheless, is recognized today as having correctly anticipated the theoretical issues of computer science. She died in London, England, on November 27, 1852.

lune   A crescent-shaped figure in the plane formed by two circular arcs is called a lune. Lunes were studied extensively by HIPPOCRATES OF CHIOS in the fifth century B.C.E. As a first attempt to solving the infamous problem of SQUARING THE CIRCLE, he showed that it is possible to square lunes of the type as shown in the illustration below. (That is, Hippocrates outlined a

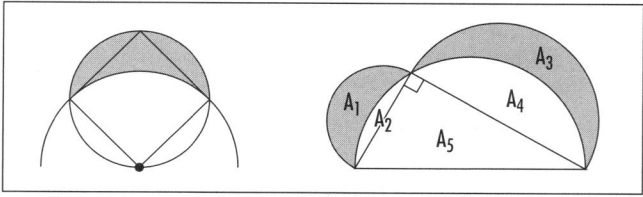

Lunes

specific series of steps that allowed one to construct a square of the same area as this lune using a straight-edge and compass alone.) Mathematicians have since proved that there are only four other types of lunes that can also be so squared.

In the figure on the preceding page, the areas of the two shaded lunes sum to the area of the triangle. Here, each circular arc is a semicircle drawn on the side of a right triangle. This claim follows from the generalized version of PYTHAGORAS's THEOREM, which states that the areas of two semicircles constructed on the two shorter sides of a right triangle sum to the area of the semicircle constructed on the hypotenuse. (Consequently, in our diagram, areas $A_1 + A_2$ and $A_3 + A_4$ sum to $A_2 + A_4 + A_5$, yielding $A_1 + A_3 = A_5$.)

In the study of spherical geometry, the part of the surface of a sphere bounded by two great circles is also called a lune. If the angle between the two great circles is θ degrees, then the surface area of this slice of the sphere is $4\pi r^2 \cdot \dfrac{\theta}{360}$ units squared, where r is the radius of the sphere.

M

Maclaurin, Colin (1698–1746) British *Calculus* Born in February 1698 (his exact birth date is not known), in Kilmodan, Scotland, scholar Colin Maclaurin is considered the foremost British mathematician of the generation that followed Sir Isaac Newton. In his two famous texts *Geometrica organica* (Organic geometry) of 1720, and *Treatise of Fluxions* of 1742, Maclaurin developed and extended the subject of CALCULUS and offered many original results. The famous series that bears his name, however, is a special case of the work of Brook Taylor (1685–1731), as Maclaurin appropriately acknowledged. He also wrote an influential elementary textbook, *A Treatise on Algebra,* on the application of algebra to geometry.

Orphaned at age 10, Maclaurin entered the University of Glasgow in 1709, which, at the time, was deemed an acceptable alternative to a secondary school education. There he studied the works of Euclid, which sparked in him a passion for mathematics. At age 14 he completed the basic degree of master and gave a public lecture on Newton's theory of gravitation, exhibiting a level of scientific knowledge comparable with that of scholars of the day.

In 1717 Maclaurin was appointed professor of mathematics at Marischal College in the University of Aberdeen. This position provided him the opportunity to travel to London in 1719, where he met Sir Isaac Newton (1642–1727) and continued his studies in mathematics and physics. Maclaurin's work was well received, and he was elected a fellow of the Royal Society that same year. Two years later he was also awarded the grand prize from the French Académie des Sciences for his work on the impact of bodies. At the same time, he had written and published his famous treatise *Geometrica organica,* which garnered him considerable regard as a fine scholar in geometry.

In an attempt to settle the criticisms of Newton's newly developed calculus, Maclaurin published his own account of the theory. His lengthy *Treatise of Fluxions* appealed to geometry to bring rigor to Newton's use of FLUXIONS and fluents. He also developed the theory of infinite SERIES, produced new tests of convergence, and discussed POWER SERIES expansions of functions (TAYLOR SERIES). This work was very influential, and in honor of his achievement, his name remains attached to the series he considered—the Maclaurin SERIES. Maclaurin also provided many new applications of calculus in this work.

Maclaurin received a second prize from the Académie des Sciences in 1740, this time for his study of tides. That the prize was also awarded to Leonhard Euler (1707–83) and Daniel Bernoulli (1707–82) of the Bernoulli family that same year shows that Maclaurin was regarded as an equal with the top two mathematicians of his day. He died in Edinburgh, Scotland, on January 14, 1746.

Maclaurin series *See* Taylor series.

Madhava of Sangamagramma (ca. 1350–1425) Indian *Trigonometry, Astronomy* Born near Cochin in southwestern India, Madhava is remembered for his brilliant discoveries in ANALYSIS. He computed, for example, the equivalent of the TAYLOR SERIES of the sine, cosine, and arctangent functions of TRIGONOMETRY, discovered the GREGORY SERIES for π (which he used to compute an approximation of this value correct to 11 decimal places), and, moreover, provided accurate estimates for the error term in truncating the series after a finite number of steps. Madhava also produced the most accurate table of sine values of his time. The methods Madhava used to accomplish these feats are believed to be essentially the same as those developed in CALCULUS by GOTTFRIED WILHELM LEIBNIZ, SIR ISAAC NEWTON, and BROOK TAYLOR. Of course, Madhava had discovered these techniques 300 years prior to their invention of this subject.

Very little is known of Madhava's life, and all of his mathematical writings are lost. Historians have learned of Madhava's mathematical work through the few astronomical texts of his that have survived, and from the commentaries scholars following Madhava made of his work.

It is worth mentioning that from his series expansion for arctangent:

$$\tan^{-1} x = x - \frac{x^3}{3} + \frac{x^5}{5} - \cdots$$

Madhava set $x = 1$ to obtain the familiar Gregory series for π. It is not well known that Madhava also set $x = \frac{1}{\sqrt{3}}$ to obtain the following alternative formula for π:

$$\pi = \sqrt{12}\left(1 - \frac{1}{3 \times 3} + \frac{1}{5 \times 3^2} - \frac{1}{7 \times 3^3} + \cdots\right)$$

magic square A square array of numbers for which the sum of the numbers in any row, column, or main diagonal is the same is called a magic square. The constant sum obtained is called the magic constant of the square. Usually the numbers in a magic square are required to be distinct, and often it is assumed that for an $n \times n$ square, the specific numbers 1, 2, 3, ..., n^2 are used. (It is convenient to designate such a magic square as a standard type.)

The earliest known example is the "Lho shu square" that appears in an ancient Chinese manuscript from the time of Emperor Yu of around 2200 B.C.E. Here the numbers 1 through 9 are arranged in a 3×3 array to produce a magic square of magic constant 15. Up to rotations and reflections, this is the only arrange-

ment of these nine integers that produces a magic square. (Thus we say that there is only *one* 3×3 magic square of standard type.)

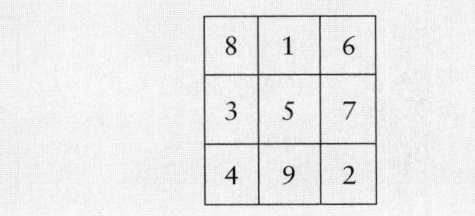

Ancient Chinese scholars, and later Arab scholars, computed examples of standard 4×4, 5×5, and higher-order magic squares. (The 5×5 magic square shown below is attributed to Yang Hui of the 13th century, and the 6×6 magic square to Chêng Ta-wei of the 16th century.) German artist ALBRECHT DÜRER (1471–1528) depicted the 4×4 magic square below in the background of his engraving *Melancholia,* and it is the believed that this is the first introduction of a magic square to the Western world. Famous scientist and statesman Benjamin Franklin (1707–90) was masterful at inventing high-order magic squares and is said to have toyed with new squares whenever political debates became tedious.

16	3	2	13
5	10	11	8
9	6	7	12
4	15	14	1

1	23	16	4	21
15	14	7	18	11
24	17	13	9	2
20	8	19	12	6
5	3	10	22	25

Taken from an 18th-century blockbook version of the Book of Changes, or the I Ching, the left diagram depicts the Lho shu magic square. *(Photo courtesy of C. Walker/Topham/The Image Works)*

27	29	2	4	13	36
9	11	20	22	31	18
32	25	7	3	21	23
14	16	34	30	12	5
28	6	15	17	26	19
1	24	33	35	8	10

It is straightforward to compute the magic constant M of a standard $n \times n$ magic square. Adding the entries in each row produces the magic constant M and, as there are n rows, the quantity nM must equal the sum of all the entries in the table. Thus, as the SUMS OF POWERS formulae show:

$$nM = 1 + 2 + 3 + \cdots + n^2 = \frac{n^2(n^2+1)}{2}$$

yielding:

$$M = \frac{1}{2}n(n^2+1)$$

For $n = 1, 2, 3, 4, 5, \ldots$ this gives the sequence of values $1, 5, 15, 34, 65, \ldots$

There are no 2 × 2 magic squares of standard type, just one standard 3 × 3 magic square, 880 standard 4 × 4 magic squares, and 275,305,224 standard 5 × 5 magic squares. To this day, no one knows the count of 6 × 6 standard magic squares.

General methods are known for constructing standard $n \times n$ magic squares of any size n larger than 2. For example, for any odd value of n, begin by placing the number 1 at any location inside the array and incrementally placing subsequent numbers in the square diagonally above and to the right. Follow a "wrap-around effect" so that paths leading off the top of the array return to the bottom, and those leading off to the right return to the left. When one encounters a square that is already filled, place the subsequent integer in the cell directly below the current cell to then continue on moving diagonally upward to the right. For example, starting with 1 in the center, this method produces the following 5 × 5 magic square. (Notice, as one reads through the sequence of entries, the numbers 6, 11, 16, and 21 were each "bumped" down to a lower diagonal.)

10	12	19	21	3
11	18	25	2	9
17	24	1	8	15
23	5	7	14	16
4	6	13	20	22

This method of construction is known as the Siamese method and is attributed to the French ambassador to Siam (now Thailand) Simon de la Loubere (ca. 1670). Methods for constructing standard magic squares of even order do exist, but are considerably more complicated.

There are a plethora of alternative requirements one could place on a square arrangement of numbers to produce magic squares with alternative remarkable properties. We list here just a few examples:

1. *Semi-Magic Squares:* A square array of numbers that fails to be a magic square only because its main diagonals do not add to the magic constant is called a SEMI-MAGIC SQUARE. These squares have the remarkable property that, when regarded as matrices, the MATRIX sum, product, and inverse of any collection of semi-magic squares is again semi-magic.

2. *Magic Multiplication Squares:* A square array of numbers in which every row, column, and diagonal has the same product is said to be a magic multiplication square. The following array, for instance, is such a magic square.

12	9	2
1	6	36
18	4	3

3. *Magic Division Squares:* A square array of numbers for which for each triple of numbers a, b, and c in a row, column, or diagonal, the quotient $a \div (b \div c)$ is the same as a magic division square. For example, the following array is such a square:

18	9	3
36	6	1
12	4	2

4. *Addition-Multiplication Magic Squares:* A square that is simultaneously a magic square under addition and under multiplication is called an addition-multiplication magic square. The following 8 × 8 array is an example of such a square:

39	34	138	243	100	29	105	152
116	25	133	120	51	26	162	207
119	104	108	23	174	225	57	30
150	261	45	38	91	136	92	27
135	114	50	87	184	189	13	68
216	161	17	52	171	90	58	75
19	60	232	175	54	69	153	78
46	81	117	102	15	76	200	203

5. *Border Squares:* A magic square (under addition) that remains magic when its border is removed is called a border magic square. If one can continue to remove borders from subsequent squares and still preserve the magic property, then we say we have a nested magic square. The following is an example of a nested magic square:

31	30	33	16	15
13	26	27	22	37
14	21	25	29	36
32	28	23	24	18
35	20	17	34	19

26	27	22
21	25	29
28	23	24

6. *Alphamagic Squares:* If replacing each entry in an (additive) magic square with the number of letters in the (English) name of that number produces a new magic square, then we say that we have an alphamagic square. For example, the square below, written in words, is alphamagic (The number of letters in each name is written in parentheses, forming its own magic square.)

five(4)	twenty-two (9)	eighteen (8)
Twenty-eight (11)	fifteen (7)	two (3)
twelve (6)	eight (5)	twenty-five (10)

7. *Magic Rectangles:* A rectangular array of numbers in which the sum of entries in any row or column is the same is called a magic rectangle. If there are n rows and m columns, and the magic constant is M, then summing all the rows shows that nM equals the sum of all entries in the array. Summing all columns gives nM as the sum of all entries in the array, and so we must have $nM = mM$. If n and m are required to be different, this forces the magic constant to be zero. Thus any magic rectangle, such as the one below, must include negative entries.

3	−4	1
−3	4	−1

One can also consider alternative arrangements of numbers and look for magic pentagrams and hexagrams, or magic cubical arrays of number, for instance. This is a popular topic of exploration for general enthusiasts of mathematics.

magnitude   A positive measure of the size of a quantity is called its magnitude. For example, the magnitude of a real number is its ABSOLUTE VALUE. (Thus, for instance, 27 and −27, although of opposite polarity, have the same magnitude.) The magnitude of a VECTOR is its length.

Physicists and astronomers use the phrase ORDER OF MAGNITUDE to refer to the smallest power of 10 needed to represent a quantity. For example, the numbers 4×10^{23} and 9.78×10^{23} are of the same order of magnitude.

See also SCIENTIFIC NOTATION.

Mandelbrot set *See* FRACTAL.

matrix (plural, matrices) A rectangular array of numbers displayed in rows and columns and enclosed in parentheses is called a matrix. (In science, the word *matrix* is used to describe the background material, soil or rock, that holds an object such as a fossil or a crystal in place. In mathematics, the word is used to describe an array that "holds" numbers in place.) An $m \times n$ matrix has m rows and n columns. For example, the object below is a 2×3 matrix:

$$\begin{pmatrix} 5 & -1 & 3 \\ -2 & 0 & 2 \end{pmatrix}$$

A matrix is called square if the number of rows equals the number of columns. A matrix with just one row (a row matrix) or just one column (a column matrix) can be regarded as a VECTOR.

Typically a matrix is denoted with a capital letter. For instance, the matrix above might be called A,

$$\tan^{-1} x = x - \frac{x^3}{3} +$$

and the entry in the ith row and jth column as A_{ij}. For instance, in this example, $A_{11} = 5$ and $A_{23} = 2$.

Matrices arise in the study of SIMULTANEOUS LINEAR EQUATIONS and the study of LINEAR ALGEBRA. Any LINEAR TRANSFORMATION can be represented via a matrix. One can combine matrices according to a set of standard MATRIX OPERATIONS.

See also EIGENVECTOR; GENERAL LINEAR GROUP; IDENTITY MATRIX; INVERSE MATRIX.

matrix operations There are three basic arithmetic operations one can perform on a MATRIX or on a pair of matrices of the same dimension (that is, two matrices with the same number of rows and the same number of columns).

1. *Scalar Multiplication:* To multiply a matrix by a real number k, multiply each entry of the matrix by that number. For example, if

$$A = \begin{pmatrix} 2 & -2 & 0 \\ 0 & -1 & 7 \\ 3 & 4 & 1 \end{pmatrix}$$

then

$$4A = \begin{pmatrix} 8 & -8 & 0 \\ 0 & -4 & 28 \\ 12 & 16 & 4 \end{pmatrix}$$

In general, the formula for the (i,j)th element of the scalar product kA is:

$$(kA)_{ij} = kA_{ij}$$

2. *Matrix Addition:* To sum two matrices of the same dimension, add corresponding entries and enter each sum in the corresponding place in the matrix sum. For example, if

$$A = \begin{pmatrix} -1 & 3 \\ 0 & 5 \end{pmatrix}$$

and

$$B = \begin{pmatrix} 2 & -2 \\ 3 & 4 \end{pmatrix}$$

then

$$A + B = \begin{pmatrix} -1+2 & 3+(-2) \\ 0+3 & 5+4 \end{pmatrix} = \begin{pmatrix} 1 & 1 \\ 3 & 9 \end{pmatrix}$$

In general, the formula for the (i,j)th element of the sum $A + B$ is:

$$(A + B)_{ij} = A_{ij} + B_{ij}$$

The scalar product and the matrix sum satisfy the relation:

$$k(A + B) = kA + kB$$

3. *Matrix Multiplication:* The DOT PRODUCT of two vectors provides a natural way to obtain a single numerical value from two separate lists of numbers: the product of $(x_1, x_2, ..., x_n)$ and $(y_1, y_2, ..., y_n)$ is given by the sum $x_1 y_1 + x_2 y_2 + ... + x_n y_n$. Each row and column of a matrix provides a list of numbers, and so one can create from two matrices A and B a new array of numerical values whose entries are the dot products of the rows or columns of A with the rows or the columns of B. It has proved to be convenient to just use the rows of A and the columns of B, provided that the number of entries in each row of A matches the number of entries in each column of B. In summary:

If A is an $n \times m$ matrix and B an $m \times r$ matrix, then the matrix product AB is the $n \times r$ matrix whose (i,j)th entry is the dot product of the ith row of A with the jth column of B. We have:

$$(AB)_{ij} = A_i 1 B_{1j} + A_{i2} B_{2j} + ... + A_{im} B_{mj}$$

For example, if $A = \begin{pmatrix} 2 & 0 & -1 \\ 1 & 3 & 5 \end{pmatrix}$ and $B = \begin{pmatrix} 1 & 3 \\ 3 & 5 \\ 2 & -1 \end{pmatrix}$,

then AB is the 2×2 matrix:

$$AB = \begin{pmatrix} 2\cdot1+0\cdot3+(-1)\cdot2 & 2\cdot3+0\cdot5+(-1)\cdot(-1) \\ 1\cdot1+3\cdot3+5\cdot2 & 1\cdot3+3\cdot5+5\cdot(-1) \end{pmatrix}$$

$$= \begin{pmatrix} 0 & 7 \\ 20 & 13 \end{pmatrix}$$

In this example, the product BA is not defined.

In many applications it is assumed that all matrices are square matrices, that is, they have equal numbers of rows and columns. In this setting, even though the products AB and BA of two square matrices A and B of the same size may each be defined, they are likely to be unequal. Thus the matrix product does not satisfy the COMMUTATIVE PROPERTY. The IDENTITY MATRIX is a matrix I with the property that $AI=IA=A$ for any square matrix A of a fixed size.

The transpose of a matrix A, denoted A^T, is the matrix obtained from A by interchanging its rows with its columns. The product $A^T B$ is the matrix whose (i,j)th entry is the dot product of the ith column of A with the jth column of B. Similarly, the product AB^T has (i,j)th entry the dot product of the ith row of A with the jth row of B, and $A^T B^T$ has (i,j)th entry the dot product of the ith column of A with the jth row of B. This latter example equals the transpose of the original product BA. We thus have: $(BA)^T = A^T B^T$.

If one LINEAR TRANSFORMATION is represented by a matrix A and a second by the matrix B, then the COMPOSITION of these two transformations is represented by a matrix equal to the product BA. (This is read backwards: the transformation represented by A is applied first and is followed by the second transformation B.) It is precisely the desire to make this observation hold true that first led mathematicians to define the matrix product in the manner described above.

See also DETERMINANT; GENERAL LINEAR GROUP; INVERSE MATRIX.

maximum/minimum   The highest point on the graph of a function is called the maximum point of the graph, and the value of the function at that point is called the maximum value of the function (or its global maximum or absolute maximum). Similarly, the minimum point of the graph is the point at which the graph has its lowest value, and the minimum value of the graph is the value of the function at that point (also called the global minimum or absolute minimum). It is possible for a function to have no maximum value or no minimum value. For example, the function $y = x$, defined over all real numbers, has no maximum or minimum value, and the function $y = \dfrac{1}{1+x^2}$, defined over all real numbers, has no minimum value. The EXTREME-VALUE THEOREM shows, on the other hand, that every continuous function defined on a closed interval necessarily adopts both a maximum and a minimum value on that interval.

A local maximum (also called a relative maximum) for a function is a point on the graph of a function that is higher than all its nearby points on the graph. Clearly, a local maximum need not be the highest point on the graph, although the highest point certainly qualifies as a local maximum. Similarly, a local minimum (or relative minimum) for a graph is the location of a function value smaller than all nearby function values. Again, a local minimum need not be an absolute minimum, although it could be.

If the function in question is differentiable, then the tools and techniques of CALCULUS allow one to readily locate local maxima and minima. It is clear geometrically, for example, that the slope of the tangent line to a graph is zero at a local maximum or local minimum. We have:

If the function $f(x)$ has a local maximum or local minimum at $x = c$, then $f'(c) = 0$

The limit definition of the derivative provides a precise proof of this. If the value $f(c)$ is a local maximum, for example, then, for small h, the value $f(c + h)$ is less than the value $f(c)$. Consequently, if h approaches the value zero by running through positive values just above zero, then the quotient $\dfrac{f(c+h) - f(c)}{h}$ is negative. This

shows that the derivative $f'(c) = \lim_{h \to 0} \dfrac{f(c+h) - f(c)}{h}$ must be ≤ 0. On the other hand, if h approaches zero through negative values, then the quotient is positive, and $f'(c) \geq 0$. It must be the case then that $f'(c) = 0$.

Any value $x = c$ for which $f'(c) = 0$ is called a critical point (or a stationary point) for the function. A study of INCREASING/DECREASING functions establishes:

A critical point $x = c$ is a local maximum for the function f if, and only if, $f(x)$ is increasing just to the left of c and decreasing just to the right of c. Consequently, $x = c$ is a local maximum if, and only if, $f'(c) = 0$ and $f'(x) > 0$ just to the left of c, and $f'(x) < 0$ just to its right.

A critical point $x = c$ is a local minimum for the function f if, and only if, $f(x)$ is decreasing just to the left of c and increasing just to the right of c. Consequently, $x = c$ is a local minimum if, and only if, $f'(c) = 0$ and $f'(x) < 0$ just to the left of c, and $f'(x) > 0$ just to its right.

This observation, called the first-derivative test, allows one to determine whether or not a critical point is a local maximum or a local minimum. (Any critical point at which the derivative does indeed change sign is called a turning point.) As an example, consider the function $f(x) = \dfrac{x}{x^2 + 1}$. To find its critical points we need to solve the equation $f'(x) = 0$. This gives:

$$f'(x) = \frac{\left(x^2+1\right)\cdot 1 - x\cdot(2x)}{\left(x^2+1\right)^2} = \frac{1-x^2}{\left(1+x^2\right)^2} = 0 \text{, yielding } x = -1$$

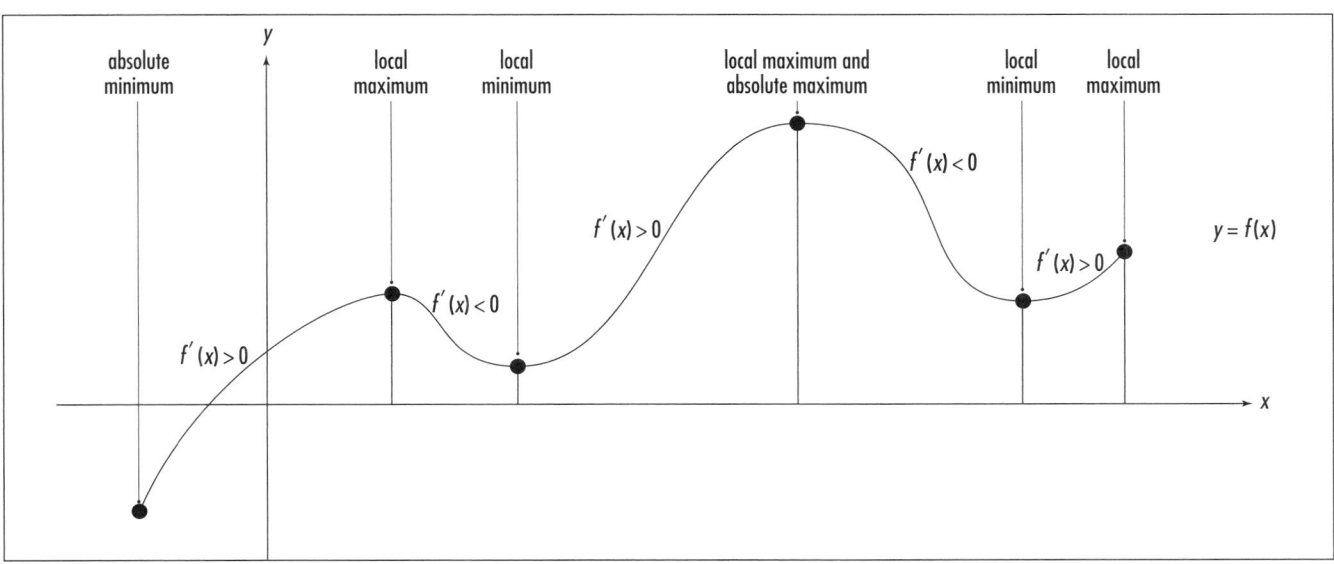

Maxima and minima

and $x = 1$ as critical points. For a value just to the left of $x = -1$ (say, x = −1.1), we see that $f'(x)$ is negative. For a value just to the right of $x = -1$ (say, $x = -0.9$), it is positive. Thus the function has a local minimum at $x = -1$. Similar analysis shows that $x = 1$ is a local maximum.

A study of CONCAVE UP/CONCAVE DOWN functions establishes:

At a local maximum, a function is concave down. Consequently, a local maximum for a function $f(x)$ can be identified as a point $x = c$ with $f'(c) = 0$ and $f''(c) < 0$.

At a local minimum, a function is concave up. Consequently, a local minimum for a function $f(x)$ can be identified as a point $x = c$ with $f'(c) = 0$ and $f''(c) > 0$.

This observation is called the second-derivative test.

For example, consider the challenge of finding two positive numbers whose sum is 100 and whose product is as large as possible. If we let x and y represent two variable positive numbers, then we are being asked to "maximize" the product function $p = xy$, given that $x + y = 100$. Solving for y and substituting yields: $P = x(100 - x) = 100x - x^2$. This expresses P as a function of x alone. A maximum for this function can only occur at a critical point. Solving $\frac{dP}{dx} = 0$ yields $100 - 2x = 0$ or $x = 50$. To classify this critical point, note that $\frac{d^2P}{dx^2} = -2 < 0$. This shows that the curve is concave down. Consequently, the point $x = 50$, the only critical point, is a (global) maximum. The corresponding value for y is $y = 100 - 50 = 50$. The maximum value of the product is thus $50 \times 50 = 2,500$.

See also OPTIMIZATION.

Mayan mathematics   The Mayan culture, based in the Yucatán Peninsula, Central America, spanned the period of 250 C.E. to 900 C.E., but was based on a civilization established in the region as early as 2000 B.C.E. The people of this time had built large cities, replete with construction of all types, including large reservoirs for storing rainwater, sophisticated irrigation systems, and raised fields for farming. The rulers were priests, and religious practices were synchronized with astronomical observations

In 1511, with a force of 11 ships and 508 soldiers, Spanish explorer Hernán Cortés conquered the peoples of the Yucatán. He met next to no resistance. Thirty years later, Bishop Diego de Landa of the Franciscan Order was sent to the New World as a missionary, and he tried his best to help the Maya peoples and protect them from their new Spanish masters. At first he was horrified by the religious practices of the people and the unusual icons that appeared in their written texts, believing the writing to be the lies of the devil. He ordered that all Mayan idols be destroyed and all the Mayan books be burned, which grieved the local people considerably. Perhaps out of remorse, Landa felt the need to write in 1566 a text, *Relación de las cosas de Yucatán* (About things of the Yucatán), recording the writing, customs, religious practices, and history of the Maya people, along with some attempt at justifying his actions. It is from this work, and four remnants of actual Mayan text saved from the fire, that we learn of Mayan mathematics. Much of the mathematics of the people was motivated by computations of the planet cycles and a need to keep an accurate calendar system.

Coming from a tropical climate, the Maya were fully aware that the human body comes with 20 digits, and it is not surprising that these people developed a base-20 system of counting (today called a vigesimal system). Thus, instead of counting in groups of units (1), tens (10), one hundreds (100), and one thousands (1,000), people of this culture counted in terms of units (1) and twenties (20). Motivated by the number of days in the year, the Maya next worked in groups of 360, and then groups 20×360. Thus, for example, a number depicted as 2-12-5-17 in the Mayan system represents the value:

$$17 \times 1 + 5 \times 20 + 12 \times 360 + 2 \times 20 \times 360 = 18,837$$

(The system is thus not strictly vigesimal.)

Numbers were represented as dots and bars, with a bar denoting five units. Thus a diagram depicting a group of two bars and three dots represents the number 13. (Some historians suggest that each dot represents a finger tip, and the horizontal bar an outstretched hand.) The Maya used positional notation to represent large numbers, including a symbol for zero, the picture of a closed fist, or perhaps a conch shell, to represent no digits of a certain power of 20 or 360.

The Maya had two calendar systems. The Tzolkin, the ritual calendar, contained 13 months, one for each god of the upper world, with 20 days per month. Thus the Tzolkin followed a 260-day year (which happens to be the number of days between the two days of the year in which the Sun is directly overhead in the Yucatán Peninsula). The second calendar, the Haab, the agricultural calendar, was the usual 365-day year, divided into 18 months of 20 days, and a short 5-day month called the Wayeb. The two calendars coincided every 18,980 days (the lowest common multiple of 260 and 365) that is, every 52 years, yielding a period that constituted a Mayan "century." The Maya were also fully aware that the most visible planet in the heavens, Venus, returned to its exact same position every 584 days, a number that happens to divide $2 \times 18,980$. That Venus returns to its same position precisely at the passing of exactly two "centuries," that is every 104 years, was of profound religious significance to the Maya people.

It seems that the Maya had no standard methods for multiplying or dividing large numbers, and they seemed never to have developed the concept of a fraction. Yet despite the cumbersome nature of their notational system, the Maya performed some exceptionally accurate astronomical computations. For instance, they correctly calculated the exact length of a year to be 365.242 days, and the length of the lunar month to be 29.5302 days. (These results were not presented in terms of decimals, of course. Records show, for example, that the Maya computed that 149 lunar months span exactly 4,400 days.) The Maya made their astronomical observations using tools no more sophisticated than a pair of sticks tied together at a right angle through which to observe the planets.

mean A mean of two numbers a and b is a number m between a and b that, in some sense, represents the "middle" of the two numbers. The most common measure of "mean" is the average or arithmetic mean given by $m = \dfrac{a + b}{2}$. It represents the location on the number line half way between positions a and b. Alternatively, one could consider the geometric mean given by $m = \sqrt{ab}$. This represents the side-length of a square whose area is the same as that of an $a \times b$ rectangle.

In the fourth century B.C.E., members of the later Pythagorean school identified 10 means, now called the neo-Pythagorean means. For instance, the first two are the arithmetic and geometric means. The third mean,

given by $m = \dfrac{2}{\dfrac{1}{a} + \dfrac{1}{b}}$, is called the harmonic mean, and the fourth, $m = \dfrac{a^2 + b^2}{a + b}$, is the counterharmonic mean. All "means" have the property that if the numbers a and b are each multiplied by k, then m is also multiplied by k.

The notion of mean can be extended to that of more than two numbers. Given n numbers $a_1, a_2, \ldots, a_n$ we set:

$$\text{arithmetic mean: } m = \frac{a_1 + a_2 + \cdots + a_n}{n}$$

$$\text{geometric mean: } m = \sqrt[n]{a_1 \cdot a_2 \cdots a_n}$$

$$\text{harmonic mean: } m = \frac{n}{\dfrac{1}{a_1} + \dfrac{1}{a_2} + \cdots + \dfrac{1}{a_n}}$$

$$\text{counter-harmonic mean: } m = \frac{a_1{}^2 + a_2{}^2 + \cdots + a_n{}^2}{a_1 + a_2 + \cdots + a_n}$$

For example, the arithmetic mean of 3, 8, and 9 is $20/3 = 6\,2/3$, and their geometric mean is $\sqrt[3]{3 \cdot 8 \cdot 9} = 6$. It is a theorem of algebra that, for any set of positive numbers $a_1, a_2, \ldots, a_n$, the arithmetic mean is always greater than or equal to the geometric mean:

$$\frac{1}{n}\left(a_1 + a_2 + \cdots + a_n\right) \geq \sqrt[n]{a_1 \cdot a_2 \cdots a_n}$$

This is called the arithmetic–geometric-mean inequality. (For the case with just two numbers a and b, the statement $\dfrac{1}{2}(a + b) \geq \sqrt{ab}$ is equivalent to the patently true statement $(a - b)^2 \geq 0$.) By applying this inequality to the numbers $\dfrac{1}{a_1}, \dfrac{1}{a_2}, \ldots, \dfrac{1}{a_n}$, the harmonic-geometric inequality follows:

$$\frac{n}{\dfrac{1}{a_1} + \dfrac{1}{a_2} + \cdots + \dfrac{1}{a_n}} \leq \sqrt[n]{a_1 \cdot a_2 \cdots a_n}$$

In statistics, the arithmetic mean μ of a set of observations $a_1, a_2, \ldots, a_n$ is called a sample mean. The

EXPECTED VALUE of a random variable is also called its mean.

See also MEAN VALUE; STATISTICS: DESCRIPTIVE.

mean value   Let f be a CONTINUOUS FUNCTION on a closed interval $[a,b]$. The height of a rectangle whose width is $(b - a)$ and whose area is equal to the area under the curve above the interval $[a,b]$ is called the mean value of the function. The mean value of f is denoted $\bar{f}$ and is given by:

$$\bar{f} = \frac{1}{b-a}\int_a^b f(x)dx$$

Loosely speaking, if one "smoothes out" the rises and falls of the graph of the function, without changing the area under the graph, then the height of the resulting level curve is $\bar{f}$.

If the function f represents, for example, the air temperature at the general post office in Adelaide, Australia, recorded over a 24-hour period, then $\bar{f}$ represents the average temperature at downtown Adelaide that day.

mean-value theorem (**Lagrange's mean-value theorem**)   French mathematician JOSEPH-LOUIS LAGRANGE (1736–1813) was the first to state the following important theorem in CALCULUS, today called the mean-value theorem:

> If a curve is continuous over a closed interval $[a,b]$, and has a tangent at every point between a and b, then there is at least one point in this interval at which the tangent is parallel to the line segment that connects the endpoints $(a,f(a))$ and $(b,f(b))$.

In more stringent mathematical language, this theorem reads:

> If a function $f(x)$ is continuous in the closed interval $[a,b]$, and differentiable in the open interval (a,b), then there exists at least one value c between a and b such that

$$f'(c) = \frac{f(b) - f(a)}{b - a}$$

Note that the quantity $\frac{f(b) - f(a)}{b - a}$ is the SLOPE (rise over run) of the line segment connecting the two endpoints. It also equals the average slope of the curve over the entire interval $[a,b]$. (To see this, note that at any point x, the quantity $f'(x)$ is the slope of the tangent line at that point. Summing, that is integrating, over all values and dividing by the length of the interval under consideration gives the average or mean slope of the curve: $\frac{\int_a^b f'(x)dx}{b - a} = \frac{f(b) - f(a)}{b - a}$.) Thus the mean-value theorem states that for any differentiable function defined on an interval $[a,b]$, there exists a location where the actual slope of the curve equals the average slope of the graph.

The mean-value theorem has four important consequences:

1. If the derivative of a function is always positive, then the function is increasing.

This means that if a and b are two numbers with $a < b$, then we have $f(a) < f(b)$. Since, for some number c, we have $f'(c) = \frac{f(b) - f(a)}{b - a}$ and the quantities $b - a$ and $f'(c)$ are both positive, we must have that $f(b) - f(a)$ is also positive.

2. If the derivative of a function is always negative, then the function is decreasing.

This is established in a manner similar to the above.

3. If the derivative of a function is always zero, then the function is constant in value.

We need to show that for any two values a and b we have that $f(a)$ equals $f(b)$. This follows from the mean-value theorem, since for some value c we have:

$$f(b) - f(a) = f'(c) \cdot (b - a) = 0 \cdot (b - a) = 0$$

4. If two functions $f(x)$ and $g(x)$ have the same derivative, then the two functions differ by a constant, that is, $f(x) = g(x)+C$ for some number C.

Let $h(x) = f(x) - g(x)$. Then the derivative of $h(x)$ is always zero, and so by the third result $h(x) = C$ for some constant value C.

The mean-value theorem itself can be established as a consequence of ROLLE'S THEOREM. One does this by writing down the equation of the line that connects the two endpoints $(a, f(a))$ and $(b, f(b))$ of the function f. It is given by:

$$g(x) = f(a) + \frac{f(b) - f(a)}{b - a}(x - a)$$

(Put in $x = a$ and $x = b$ to see that this is correct.) Now consider the function:

$$h(k) = f(x) - g(x)$$

It is a differentiable function with $h(a) = h(b) = 0$, and so by Rolle's theorem, there is at least one value c between a and b for which $h'(c) = f'(c) - g'(c) = 0$. This yields the equation asserted in the statement of the mean-value theorem.

The mean-value theorem can be thought of as a statement about the nature of a differentiable curve intersecting a straight line. French mathematician AUGUSTIN-LOUIS CAUCHY (1789–1857) later generalized the theorem to one about *any* two differentiable curves intersecting at two points. The result is known as the extended mean-value theorem or Cauchy's mean-value theorem.

> If two functions f and g have the same values at $x = a$ and $x = b$, are continuous in the closed interval $[a,b]$, differentiable in the open interval (a,b), and further if $g(a) \neq g(b)$ and $g'(x)$ is never zero in (a,b), then there is at least one value c between a and b for which:
>
> $$\frac{f(b) - f(a)}{g(b) - g(a)} = \frac{f'(c)}{g'(c)}$$

The theorem is proved in a similar way by making use of the support function:

$$h(x) = f(x) - f(a) - \frac{f(b) - f(a)}{g(b) - g(a)}\big(g(x) - g(a)\big)$$

See also DERIVATIVE; INCREASING/DECREASING.

median *See* STATISTICS: DESCRIPTIVE.

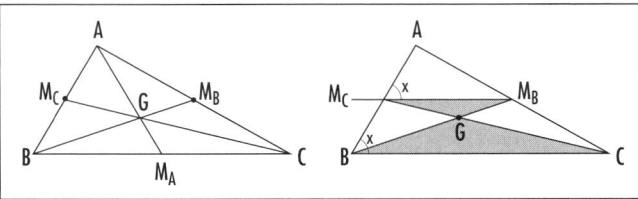

Medians are concurrent

median of a triangle A line segment connecting the MIDPOINT of one side of a triangle to the vertex opposite to that side is called a median of the triangle. Any triangle has three medians. It is considered a fundamental result that the three medians of a triangle always meet at a common point (called the centroid of the triangle and usually denoted G.) To see this, consider a triangle with vertices A, B, and C and midpoints as shown.

First note:

> Any line connecting the midpoints of two sides of a triangle is parallel to the third side of the triangle.

In the diagram above, by the SAS principle, triangles BAC and $M_C A M_B$ are similar, with a scale factor of two, since they share a common angle at A and the two sides of each triangle match in a 2-to-1 ratio. Consequently the angles labeled x are equal, yielding two equal alternate angles, from which it follows from the converse of the PARALLEL POSTULATE that the lines $M_C M_B$ and BC are parallel.

We now have that angles $M_B M_C C$ and $M_C C B$ are equal, as are angles $M_C M_B B$ and $M_B B C$. Consequently, by the AAA principle, the two shaded triangles are similar, again in a 2-to-1 ratio. In particular the line segments BG and $G M_B$ are in this ratio, as are the line segments CG and $G M_C$. This establishes:

> The point of intersection of any two medians of a triangle lies two-thirds of the way along each median.

Consequently, the median AM_A will also intercept median BM_B two-thirds the distance along the length of BM_B, namely, at the same point G. Thus all three medians are indeed CONCURRENT at G.

As a variation of HERON'S FORMULA, the area of a triangle can be expressed solely in terms of the lengths of the medians of the triangle. We have:

$$\text{area} = \frac{4}{3}\sqrt{s(s - m_A)(s - m_B)(s - m_C)}$$

where m_A, m_B, and m_C are the lengths of the three medians and $s = \dfrac{m_A + m_B + m_C}{2}$.

See also AAA/AAS/ASA/SAS/SSS; APOLLONIUS'S THEOREM; EULER LINE.

Menelaus of Alexandria

Menelaus of Alexandria (ca. 70–130) *Greek Geometry, Trigonometry*   Born in Alexandria, Egypt, mathematician Menelaus is noted for his only surviving work *Sphaerica* (Spheres), which contains the earliest known results on SPHERICAL GEOMETRY and spherical trigonometry. By converting spherical results into planar ones, Menelaus also established a number of significant theorems about planar geometry, including the famous result that now bears his name.

Extremely little is known of Menelaus's life. Despite being cited throughout history as a native of Alexandria, it is known that Menelaus spent some portion of his life in Rome. For instance, records from the year 98 list a number of astronomical observations made by Menelaus from that city at that time.

Menelaus's work in spherical geometry was likely inspired by his work in astronomy. The first of the three volumes of *Sphaerica* defines the basic principles of the subject and includes the very precise definition of a spherical triangle as one made by three arcs of great circles, each less than a semicircle. Following the same level of rigor as established by the geometer EUCLID, Menelaus developed the theory of this geometry in considerable depth with precise logical reasoning. (Curiously, Menelaus eschewed any use of PROOF BY CONTRADICTION, an approach that Euclid freely used.) In volume II of *Sphaerica*, Menelaus developed applications to astronomy, and in volume III explored spherical trigonometry and applications to plane geometry.

Arab scholars of the period 850–1500 C.E. translated Menelaus's work and wrote many commentaries on the piece. The same scholars also made reference to other texts by Menelaus, including pieces called *Chords in a Circle* and *Elements of Geometry,* as well as a comprehensive text on the topic of mechanics. Sadly, no copies of these works survive today.

See also ARAB MATHEMATICS; MENELAUS'S THEOREM.

Menelaus's theorem   Suppose a TRANSVERSAL cuts the three sides of triangle $A_1A_2A_3$ at points P_1, P_2, and P_3 as shown:

Then, if A_iP_j represents the length of the line segment connecting point A_i to point P_j, we have:

$$\frac{A_1P_1}{A_2P_1} \cdot \frac{A_2P_2}{A_3P_2} \cdot \frac{A_3P_3}{A_1P_3} = 1$$

This result was first observed by the Greek mathematician MENELAUS OF ALEXANDRIA (ca. 70–130 C.E.). He proved it by drawing lines from each vertex A_i to the transversal to yield five right-angled triangles. Examining the angles within these triangles shows that a number of these triangles are similar. Chasing through all the pairs of sides that consequently are in the same ratio eventually establishes the result.

The converse of Menelaus's theorem is also true:

If P_1, P_2, and P_3 are three points on the sides of a triangle $A_1A_2A_3$, with P_1 on side A_1A_2 (possibly extended), P_2 on side A_2A_3 (possibly extended), and P_3 on side A_3A_1 (possibly extended), satisfying

$$\frac{A_1P_1}{A_2P_1} \cdot \frac{A_2P_2}{A_3P_2} \cdot \frac{A_3P_3}{A_1P_3} = 1$$

then the three points are COLLINEAR.

See also AAA/AAS/ASA/SAS/SSS; CEVA'S THEOREM.

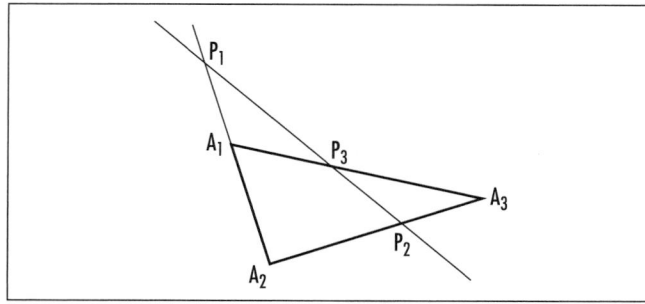

Menelaus's theorem

Mercator's expansion (Mercator's series) The TAY-
LOR SERIES expansion of the natural LOGARITHMIC
FUNCTION is given as follows:

$$\ln(1+x) = x - \frac{x^2}{2} + \frac{x^3}{3} - \frac{x^4}{4} + \dots$$

It is valid for $-1 < x \le 1$. This series is sometimes called
Mercator's expansion in honor of Danish mathematician Nicolaus Mercator (ca. 1619–87) who, in 1668,
was the first to publish this expansion.

The Mercator expansion has the curious of property of apparently proving the absurd statement that 1
equals 2. Setting $x = 1$ yields:

$$\ln 2 = 1 - \frac{1}{2} + \frac{1}{3} - \frac{1}{4} + \frac{1}{5} - \dots$$

and so

$$2\ln 2 = 2 - 1 + \frac{2}{3} - \frac{1}{2} + \frac{2}{5} - \frac{1}{3} + \frac{2}{7} - \frac{1}{4} + \frac{2}{9} - \frac{1}{5} + \dots$$

$$= (2-1) - \frac{1}{2} + \left(\frac{2}{3} - \frac{1}{3}\right) - \frac{1}{4} + \left(\frac{2}{5} - \frac{1}{5}\right) - \dots$$

$$= 1 - \frac{1}{2} + \frac{1}{3} - \frac{1}{4} + \frac{1}{5} - \dots$$

$$= \ln 2$$

This paradox alerted mathematicians to the fact
that it is not always permissible to rearrange the terms
of a series. In 1837 PETER GUSTAV LEJEUNE DIRICHLET
(1815–59) proved that such an operation is valid if the
series is absolutely convergent. Unfortunately, the alternating HARMONIC SERIES expressed above is not.

See also ABSOLUTE CONVERGENCE.

Mercator's projection It is not possible to make a flat
map of the world without incorporating some kind of
distortion. In the mid-1600s, Flemish cartographer Gerhard Kremer (1512–94), known as Mercator, devised a
method for mapping points from the surface of the Earth
onto a planar surface in such a way that all compass
directions, at least, are preserved. Although the distances
and areas are distorted under this PROJECTION, the general shapes of small regions, such as small countries and
small bodies of water, are reasonably well preserved.

The mathematical construct of Mercator's projection is obtained by imagining a cylinder placed around
the sphere of the Earth tangent to the equator and parallel to the axis of the Earth. A point on the surface of
the Earth is mapped to a point on this cylinder by
drawing a line from the center of the Earth and
through this point until it cuts the cylinder. (The North
and South Poles are not mapped.) The cylinder is then
cut and unrolled to form a flat surface.

In Mercator's projection, lines of longitude are the
same distance apart, but lines of latitude get farther
apart from the equator. Mercator adjusted the vertical
spacing of the lines of latitude on his flat map to compensate for this distortion and to make the shapes of
countries resemble more closely their true shape as they
appear on the globe.

Mercator's projection can be given by mathematical
formulae. Under his mapping, a point on the Earth's surface at an angle α latitude and angle β longitude has
CARTESIAN COORDINATES x and y on the plane given by:

$$x = k\alpha$$

$$y = k\log\tan\left(\frac{\beta}{2}\right)$$

for some constant k. The angles between lines on the
surface of the sphere (away from the poles) are preserved under Mercator's projection and so this mapping is an example of a CONFORMAL MAPPING.

See also STEREOGRAPHIC PROJECTION.

Mersenne, Marin (1588–1648) French *Number the-
ory, Theology* Born on September 8, 1588, in Oize,
France, Marin Mersenne is remembered for the list of
PRIME numbers that bear his name. These primes are
intimately connected with the formulation of even PER-
FECT NUMBERs.

Mersenne studied theology as a teenager and at age
23 joined the Minims, a religious order devoted to
prayer and scholarship. Throughout his life Mersenne
pursued interests in NUMBER THEORY, mechanics, and
acoustics. He defended the work of GALILEO GALILEI
(1564–1642) and RENÉ DESCARTES (1596–1650) against
theological criticism, and took on the task of translating
many of Galileo's texts into French. Historians believe

that it is because of Mersenne's efforts that Galileo's work became known outside of Italy.

Mersenne played an important role in 17th-century science, not only for his contribution to number theory and mechanics, but also for his service as a channel of communication between mathematicians: scholars would write to Mersenne for the sole purpose of having their ideas disseminated. Letters from over 78 different correspondents, including PIERRE DE FERMAT (1601–65), Galileo, and Christiaan Huygens were discovered in his monastery cell after his death.

In 1644 Mersenne published *Cogitata physico-mathematica* (Physico-mathematical thoughts), his famous text on number theory. Mersenne was particularly interested in finding a formula that would generate all the prime numbers. Although he failed in this effort, his work did lead him to consider those prime numbers p for which $2^p - 1$ is also prime, now called the MERSENNE PRIMEs. These numbers have proved to be of significant importance in several different branches of number theory.

Mersenne died in Paris, France, on September 1, 1648.

Mersenne prime   A PRIME number of the form $2^n - 1$ is called a Mersenne prime. For example, $2^3 - 1 = 7$ and $2^7 - 1 = 127$ and are Mersenne primes. These numbers were studied by French philosopher and mathematician MARIN MERSENNE (1588–1648) in his attempts to find a formula that would generate all prime numbers. Although he failed in this pursuit, primes of this form are today named in his honor.

Note that if n factors as $n = ab$, then the quantity $2^n - 1$ also factors: $2^{ab} - 1 = (2^a - 1)(2^{a(b-1)} + 2^{a(b-2)} + ... + 2^a + 1)$. Thus in order for $2^n - 1$ to be prime, it must be the case that n is prime. However, not every prime number n leads to a Mersenne prime. For example, although $n = 11$ is prime, $2^{11} - 1 = 2047 = 23 \times 89$ is not. The first few Mersenne primes are 3, 7, 31, 127, 8191, 131071, 524287, 2147483647, ... corresponding to the prime values n equal to 2, 3, 5, 7, 13, 17, 19, 31,...

Only 40 Mersenne primes are currently known, yet despite their scarcity, they still remain a fruitful source of large prime numbers. Almost certainly, when a newspaper proclaims that a new "largest" prime has been found, it turns out to be of the form $2^n - 1$. For example, the largest known prime as of the year 2004 is the Mersenne prime with $n = 20,996,011$. It is a prime number over 6 million digits long. Mersenne primes are intimately connected with PERFECT NUMBERs.

See also DIFFERENCE OF TWO CUBES.

midpoint   A point on a line segment dividing the length of that segment into two equal parts is called the midpoint of the segment. If two points in a plane have CARTESIAN COORDINATES $P = (x_1, y_1)$ and $Q = (x_2, y_2)$, then the midpoint M of the segment connecting P to Q has coordinates $M = (\frac{x_1 + x_2}{2}, \frac{y_1 + y_2}{2})$. Similarly, for two points $P = (x_1, y_1, z_1)$ and $Q = (x_2, y_2, z_2)$ in three-dimensional space, the coordinates of the midpoint M of the line segment connecting them are given by: $M = (\frac{x_1 + x_2}{2}, \frac{y_1 + y_2}{2}, \frac{z_1 + z_2}{2})$.

A line through the midpoint of a line segment and PERPENDICULAR to that segment is called a perpendicular bisector. A study of EQUIDISTANT points shows that the three perpendicular bisectors of the three sides of any triangle meet at a single point (called the circumcenter of the triangle). The three MEDIANs OF A TRIANGLE are also CONCURRENT.

The circle-midpoint theorem asserts that if one draws a circle C in the plane and selects a point P anywhere in the plane, then all the midpoints of line segments connecting P to points on the circle form a circle of half the original radius. This can be seen valid as follows:

> Assume the circle has radius r and is positioned about the origin of a Cartesian coordinate system. Then any point Q on the circle has coordinates $Q = (r \cos\theta, r \sin\theta)$, for some value θ. If the coordinates of P are given by $P = (a, b)$, then the coordinate of the midpoint M is $M = (\frac{a}{2} + \frac{r}{2}\cos\theta, \frac{b}{2} + \frac{r}{2}\sin\theta)$. As θ varies, this describes a circle of radius $\frac{r}{2}$ with center $(\frac{a}{2}, \frac{b}{2})$.

See also BISECTOR; CIRCLE THEOREMS.

midrange   *See* STATISTICS: DESCRIPTIVE.

Möbius, August Ferdinand (1790–1868) German *Topology, Astronomy*   Born on November 17, 1790, in Schulpforta, Saxony (now Germany), scholar August Ferdinand Möbius is remembered in mathematics for his work in GEOMETRY and TOPOLOGY, and his conception of the one-sided surface today called a MÖBIUS BAND.

Möbius entered the University of Leipzig in 1809 to study law but soon changed interests and took up a study of mathematics, astronomy, and physics. In 1813 he traveled to Göttingen to work with CARL FRIEDRICH GAUSS (1777–1855), the finest mathematician of the day, and also director of the astronomical observatory in Göttingen. Möbius completed a doctoral thesis in 1815 in the topic of astronomy and soon afterward completed a habilitation thesis on the topic of trigonometric equations. In 1816 he was appointed chair of astronomy and mechanics at the University of Leipzig. He stayed at Leipzig the remainder of his life.

Möbius published an influential work, *Der barycentrische Calcül* (The calculus of barycenters), on analytic geometry in 1827. In it he outlined a number of significant, and original, advances in the fields of PROJECTIVE GEOMETRY and AFFINE GEOMETRY. He also studied the geometry of the plane of COMPLEX NUMBERS, and showed that any complex function of the form:

$$f(z) = \frac{az + b}{cz + d}$$

with z a complex number, and a, b, c, and d real numbers satisfying $ad - bc \neq 0$, transforms straight lines and circles in the complex plane into new straight lines and circles. Complex functions of the form above are today called Möbius functions.

In 1831 Möbius published important results in NUMBER THEORY. He is best known in this field for his discovery of the Möbius inversion formula, which can be described as follows. Consider a function μ defined on the set of positive integers as follows: $\mu(1) = 1$; $\mu(p) = -1$ if p is PRIME; $\mu(n) = (-1)^k$ if n has k primes factors, all distinct; and $\mu(n) = 0$ if n has a prime factorization that includes repeated primes. (Thus, for instance, $\mu(70) = \mu(2 \cdot 5 \cdot 7) = -1$ and $\mu(20) = \mu(2 \cdot 2 \cdot 5) = 0$.) If f is any function on the positive integers, and, from it, a new function F is defined as $F(n) = \Sigma_{d|n} f(d)$, a sum over all the factors d of n, then Möbius's inversion formula states that one can recover from F the original function f via the rule:

$$f(n) = \sum_{d|n} F(d)\mu\left(\frac{n}{d}\right)$$

Möbius wrote about the one-sided surface that bears his name in an 1858 piece discovered only after his death. His interest in the object was motivated by a general question on the geometry of polyhedra that had caught his interest. Although Möbius was not the first to describe the one-sided surface (German mathematician Johann Listing had considered the object just a few years earlier), his mathematical analysis of the surface was deep and significant.

Möbius also published important works in the fields of astronomy, celestial mechanics, and statics. He died in Leipzig, Germany, on September 26, 1868. His work on number theory has had a profound effect on the nature of research in the subject today. For instance, the concept of the Möbius function has found natural applications to generalized abstract settings in algebraic geometry, combinatorics, and partially ordered sets, thereby providing new insights into the study of numbers.

Möbius band (**Möbius strip**)   The one-sided surface obtained by gluing together the two ends of a long rectangular strip twisted 180° so as to produce a half-twist in the resulting object is called a Möbius band. This surface has the property that if one paints the surface all the way round, one finds that both sides of the original strip are colored. This shows that the surface is indeed one-sided. Also, tracing one's finger along the edge of a Möbius band covers every possible point on the edge of the object, thus showing, in addition, that the surface has only one edge.

In some sense it is impossible to cut a Möbius band in half. For instance, cutting the object along a central line parallel to the edge produces a single connected two-sided object. The reason for this can be seen in the diagram on the following page. (Note that the top half of the original strip is connected to the bottom portion via the segments labeled a that are glued together, and again for the segments labeled b.) In general, if a band of paper with n half-twists is cut in half along the central line, one piece will result if n is odd, and two pieces are produced if n is even. Also, if one attempts to cut a Möbius band into thirds (along two parallel lines that divide the strip in thirds), then two interlocking rings result, one twice the length of the other. The shorter of the two is another Möbius band.

Gluing together two Möbius bands along their edges produces a surface called a KLEIN BOTTLE. One

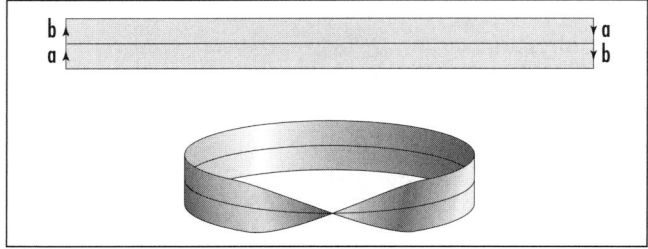

Möbius band

can also obtain a Klein bottle by "folding" a Möbius band in half along the central line parallel to its edge and gluing together the points of the edge that meet. (If one does this for an ordinary band of paper containing no half twists, the resulting surface is a TORUS.) Unfortunately these constructions cannot be fully completed in three-dimensional space, and one must make use of the fourth dimension to obtain sufficient maneuverability.

In a CARTESIAN COORDINATE system, if one rotates a line segment in the yz-plane about the z-axis, the resulting surface of revolution is a band with no half-twists. If that line segment were to rotate about its MIDPOINT 180° during the course of being rotated about the z-axis, then the resulting surface is a Möbius band. For convenience, suppose the line segment is of length 2 and is initially positioned in the yz-plane parallel to the z-axis with midpoint at distance position 2 along the y-axis. Each point on the line segment is determined by a parameter v, between -1 and 1, with $v = 0$ corresponding to the midpoint of the segment. Suppose that when the segment has turned an angle θ about the z-axis, the segment has turned an angle $\frac{\theta}{2}$ about its midpoint. Then a careful analysis of the positions of points along this segment as they are rotated about the z-axis shows that the PARAMETRIC EQUATIONS of the resulting Möbius band are given by:

$$x = \left(2 - v\sin\left(\frac{\theta}{2}\right)\right)\sin(\theta)$$

$$y = \left(2 - v\sin\left(\frac{\theta}{2}\right)\right)\cos(\theta)$$

$$z = v\cos\left(\frac{\theta}{2}\right)$$

Mathematicians have shown that, for any map of regions drawn on a Möbius band, at most six colors

would ever be needed to paint the design so that no two regions sharing a boundary are painted the same color.

The Möbius band was independently discovered by German mathematician Johann Listing (1808–88) and German scholar AUGUST FERDINAND MÖBIUS (1790–1868). The one-sided nature of the band was later exploited by the B.F. Goodrich Company in their design of Möbius-like conveyor belts. By spreading the "wear and tear" on both sides of a strip, these belts lasted twice as long as conventional belts.

See also DIMENSION; FOUR-COLOR THEOREM; SOLID OF REVOLUTION.

mode *See* STATISTICS: DESCRIPTIVE.

modular arithmetic   The numerals on the face of a clock provide a model for an unusual mathematical system called "clock math" or "arithmetic mod 12." One thinks as follows: if it is currently 3:00, then 8 hours later it will be 11:00. We write 3 + 8 = 11, noting nothing unusual here. However, waiting 6 hours from 10:00, say, gives the equation 10 + 6 = 4, for the time at the end of that wait will be 4:00. Following this new interpretation for addition, clock math gives, for example, 4 + 11 = 3, 8 + 2 = 10, and 7 + 7 = 2.

It is convenient to call the number 12 "zero." (After all, in clock math, adding 12 hours to any time does not change the time registered on the clock and so has no effect in this system.) The number 13 is regarded the same as 1, (the 13th hour on a clock lies at the same position as the first hour), the number 14 is 2, and so forth. In general, clock math replaces any number with its excess over a multiple of 12. For example, 26 is two more than a multiple of 12, and so 26 is equivalent to 2. We write $26 \equiv 2 \pmod{12}$. Similarly, $29 \equiv 5 \pmod{12}$, $43 \equiv 7 \pmod{12}$, and $72 \equiv 0 \pmod{12}$. The number -2 is 10 more than a multiple of 12 and so $-2 \equiv 10 \pmod{12}$. The symbol $\equiv$ is called CONGRUENCE.

One can perform multiplication in clock math. We have, for instance, that 3×7, normally 21, equals 9 in clock math: $3 \times 7 = 9$. (This can also be realized in terms of repeated addition: $3 \times 7 = 7 + 7 + 7 = 2 + 7 = 9$.) In the same way, $2 \times 4 = 8$, $4 \times 5 = 8$, and $6 \times 6 = 0$ in clock math. The following table shows all products in this system.

x	0	1	2	3	4	5	6	7	8	9	10	11
0	0	0	0	0	0	0	0	0	0	0	0	0
1	0	1	2	3	4	5	6	7	8	9	10	11
2	0	2	4	6	8	10	0	2	4	6	8	10
3	0	3	6	9	0	3	6	9	0	3	6	9
4	0	4	8	0	4	8	0	4	8	0	4	8
5	0	5	10	3	8	1	6	11	4	9	2	7
6	0	6	0	6	0	6	0	6	0	6	0	6
7	0	7	2	9	4	11	6	1	8	3	10	5
8	0	8	4	0	8	4	0	8	4	0	8	4
9	0	9	6	3	0	9	6	3	0	9	6	3
10	0	10	8	6	4	2	0	10	8	6	4	2
11	0	11	10	9	8	7	6	5	4	3	2	1

Notice that $2 \times 2 = 4$, and so it is appropriate to write $\sqrt{4} = 2$. We also have that $\sqrt{4}$ equals 4, 8, and 10, since $4 \times 4 = 4$, $8 \times 8 = 4$, and $10 \times 10 = 4$ in clock math. In the same way, $\sqrt{1}$ equals 1, 5, 7, or 11, and $\sqrt{9}$ equals 3 or 9. There is no number equivalent to $\sqrt{2}$, for instance, in this system.

It is possible to give an interpretation of some fractions in clock math. Consider 4/5 for instance. In ordinary arithmetic, this fraction is a number x such that $5 \times x = 4$. Looking at the fifth row of the table above, we see that $5 \times 8 = 4$, and so it is appropriate to interpret the fraction 4/5 as the number 8 in clock math. In the same way, 1/5 = 5, 3/7 = 9, and 2/7 = 2. Notice, however, that it is not possible to give an interpretation to the fraction 1/6, for instance, since the number 1 does not appear anywhere in the sixth row of the table. (There is no number x such that $6 \times x = 1$.) That not every fraction is represented in clock math is deemed a deficiency of the system.

Generalized Clock Math: Modular Arithmetic

One can envision a clock with a different number of hours represented on its face. For example, in 5 o'clock math, just five hours are depicted, 0, 1, 2, 3, and 4 (again it is appropriate to deem the fifth hour as the same as the zeroth hour), and all other numbers are replaced by their excess over a multiple of 5. Thus, for example, we have $6 \equiv 1 \pmod 5$ and $32 \equiv 2 \pmod 5$. The product table for mod-5 arithmetic appears as follows. Notice that every digit appears in every (nonzero) row of the table.

x	0	1	2	3	4
0	0	0	0	0	0
1	0	1	2	3	4
2	0	2	4	1	3
3	0	3	1	4	2
4	0	4	3	2	1

In general, in base N modular arithmetic, each number is replaced by its excess over a multiple of N.

If N is a COMPOSITE NUMBER, N equals $a \times b$ say, then the entry in the ath row and bth column of the product table is zero (for example, 3×4 is zero in 12-clock math). Consequently the number zero appears in the ath row more than once, giving insufficient space for all the other digits to appear in that row. On the other hand, if N is PRIME, then each and every nonzero row of the product table does indeed contain every digit, as demonstrated by the product table for 5-clock math. (To see why this is the case, note that for any number a less than the prime N is RELATIVELY PRIME to N, and so, by the EUCLIDEAN ALGORITHM, there exist integers x and y such that $ax + yN = 1$. This shows that ax is 1 more than a multiple of N, and so $ax \equiv 1 \pmod N$. Consequently the number 1 appears in the ath row, xth column, of the product table. So too does the number 2, since $a(2x) = 2ax \equiv 2 \pmod N$, the number 3, $a(3x) = 3ax \equiv 3 \pmod N$, and so forth.) We have:

> If N is a prime number, then all digits appear in each nonzero row of the product table for arithmetic modulo N.

In particular, for any nonzero number a in mod-N arithmetic, the number 1 appears in the ath row of the product table. Thus there is a number x such that $a \times x = 1$. The fraction $\frac{1}{a}$ thus has a valid interpretation in this system. This argument applies to all fractions one may wish to consider. We have:

> If N is a prime number, then all fractions exist in mod-N arithmetic.

This completely classifies all modular arithmetic systems that possess fractions.

Fermat's Little Theorem

Let p be a prime number. Then for every nonzero number a less than p, all the digits zero through $p − 1$ appear in the ath row of the product table for mod-p arithmetic. Ignoring the zeroth multiple of a, this asserts that all the multiples of a, namely, $a \times 1, a \times 2, \ldots, a \times (p − 1)$, correspond, in some order, to the list of digits $1, 2, \ldots, p − 1$. Consequently, the product of the numbers in each list must be the same:

$$a \times 1 \times a \times 2 \times \ldots \times a \times (p − 1) = 1 \times 2 \times \ldots \times (p − 1)$$

Rewriting yields:

$$a^{p−1} \times 1 \times 2 \times \ldots \times (p − 1) = 1 \times 2 \times \ldots \times (p − 1)$$

Multiplying through by the fractions $\frac{1}{2}, \frac{1}{3}$, up to $\frac{1}{p−1}$ (which exist in prime-clock math) gives the famous result first discovered by French lawyer and amateur mathematician PIERRE DE FERMAT (1601–65):

For any prime number p, $a^{p−1} \equiv 1 \pmod{p}$ for all nonzero values $a < p$.

Applying this observation to the specific value $a = 2$ provides a useful method for testing the primality of numbers: *if p is prime, then $2^{p−1} − 1$ is divisible by p. If not, then the number p is not prime.* Unfortunately, some numbers pass the test without being prime. For instance, $2^{340} − 1$ is divisible by 341, even though 341 is composite. (We have $341 = 11 \times 31$.) Composite numbers of this type are called pseudo-primes.

See also CONGRUENCE.

modulus (plural, moduli) The ABSOLUTE VALUE of a quantity, without consideration of its sign or direction, is sometimes called the modulus of the quantity. For example, the numbers −3 and 3, although of opposite parity, have the same modulus of 3. The modulus of a VECTOR is its length, and the modulus of a COMPLEX NUMBER is the length of the vector that represents that complex number. Specifically, if $z = a + ib$ is a complex number, then its modulus, written $|z|$, is the nonnegative real number $\sqrt{a^2 + b^2}$. If the complex number is written in polar form, $z = r\cos\theta + ir\sin\theta = re^{i\theta}$, then its modulus is r.

In the study of LOGARITHMs, the number by which logarithms of one base are multiplied to give logarithms of a different base is called the modulus. For example, the equation:

$$\log_a x \times \log_b a = \log_b x$$

shows that multiplication by the modulus $\log_b a$ converts logarithms of base a to ones of base b. (To see why this works, note that if $y = \log_a x$, then $a^y = x$. Consequently, $\log_b(a^y) = \log_b x$, yielding $y \times \log_b a = \log_b x$.) In particular, multiplication by the number $\log_{10} e \approx 0.434294$ converts natural logarithms into common logarithms.

In MODULAR ARITHMETIC, the number by which quantities are divided is called the modulus of the system. For example, in "clock math," the modulus of the system is 12.

monomial Any algebraic expression consisting of a single term, such as $5x^3 y^2$, is called a monomial.

See also BINOMIAL; POLYNOMIAL; TRINOMIAL.

Monte Carlo method Pioneered by JOHN VON NEUMANN (1903–57) and the Polish mathematician Stanislav Ulam, the Monto Carlo method is a simple probabilistic method that is sometimes employed by applied mathematicians to analyze processes that are too complicated to analyze otherwise. Named after the famous gambling casino, the Monte Carlo method simply uses the LAW OF LARGE NUMBERS to estimate the probability of a desired event occurring. For example, to estimate the probability that five letters chosen at random from the alphabet spell a word in the English language, one could simply perform the experiment a large number of times (that is, have a computer select five letters at random 1,000 times, say) and count the proportion of times an English word is obtained. This proportion gives an estimate of the probability one seeks. Many casinos employ this technique to determine the payout ODDS for many of their complicated games.

The Monte Carlo method is also used to estimate the area of a plane figure with an irregular outline. For example, to estimate the area of an oil spill over the ocean, scientists take an aerial photograph of the entire spill, taking note of the dimensions covered by the photograph, say a 4-by-5-km rectangle. The photograph is then digitized and fed into a computer, which is programmed to select, at random, a large number of points in the photograph, say, 10,000. The computer then counts the number r of these points that hit the oil *spill* shown in the photograph, yielding a proportion $\frac{r}{10,000}$ that well approximates the fraction of the total area covered by oil. For instance, if $r = 7,568$ points land in locations corresponding to the spill, then the area of the entire spill is very close to $\frac{7,568}{10,000} \times 20 = 15.1$ km^2. The greater the number of points selected, the more accurate is the estimate.

In the same way, one can estimate the values of complicated definite INTEGRALs, such as $\int_3^7 \cos^2(x^2 + \sin(x))dx$. For example, select a large number of points at random from a box that contains the curve in question and count the proportion of them that fall below the given curve.

See also BUFFON NEEDLE PROBLEM; PROBABILITY.

Monty Hall problem Named after the host of a popular American TV game show *Let's Make a Deal!*, the Monty Hall problem is a classic puzzler often used to test initiates in the field of PROBABILITY theory. It goes as follows:

> On a game show, three closed doors stand before you. The host informs you that a cash prize lies behind one of the doors, with nothing behind the other two. You select a door, but before you open it, the host quickly opens one of the remaining two doors to show you that the prize is not there. He now gives you the chance to change your mind and open instead the third remaining door. The question is: what should you do? Should you stay with your original choice of door, or switch to the other option? Is there any advantage to switching?

One's typical first reaction to this puzzle is that there is no advantage at all to switching. Since two doors remain with only one containing a prize, the chance of selecting the correct door, either by staying with the chosen door or switching, is always 50 percent. Surprisingly, this reasoning is not correct, for it makes no use of the subtle information the host presents to you, which you can actually use to your advantage.

Suppose, before you play the game, you decide that you will stay with your choice. Then a win for you relies on choosing the correct door initially, and there is a 1-in-3 chance of this happening. If, on the other hand, you play the game with the decision to switch, then winning relies on choosing an *incorrect* door initially (this is where the host's action comes to the fore), and there is a 2-in-3 chance of this being the case. All in all, we see in fact that switching doubles your chances of winning! (This line of reasoning is made all the more convincing if you imagine a game played with 100 closed doors, only one of which conceals a prize. Choosing the correct door initially is very unlikely. However, if the host reveals that 98 of the remaining doors are empty, you will certainly decide that odds are in your favor to switch.)

See also KRUSKAL'S COUNT; TWO-CARD PUZZLE.

Morley's theorem In 1899 mathematician Frank Morley (1860–1937) discovered that the intersections of adjacent pairs of angle trisectors in any triangle always form an equilateral triangle. Precisely stated, given a triangle *ABC*, draw for each vertex a pair of line segments dividing the angle at that vertex into thirds. If the two line segments closest to side *AB* meet at point *D*, the two segments closest to side *BC* at point *E*, and the two segments closest to side *AC* at point *F*, then *DEF* is guaranteed to be an equilateral triangle.

It is remarkable that this elegant fact of EUCLIDEAN GEOMETRY was not discovered until so long after Euclid's time. The proof of this result, although a little long and detailed, requires only very elementary geometric techniques.

multiplication The process of finding the product of two numbers is called multiplication. In elementary arithmetic, multiplication can be defined as the process of finding the total number of elements in a collection of sets where each set in that collection has the same number of elements. Thus, for example, the accumulation of four sets each possessing three objects gives a total of 12 objects. We write: $4 \times 3 = 12$. In this context, multiplication can thus be regarded as a process of repeated addition: $4 \times 3 = 3 + 3 + 3 + 3 = 12$. If one arranges the 12 objects in a rectangular array of four rows of three, then reading the arrangement as three columns of four shows that 3×4 provides the same answer as 4×3. In general, this reasoning shows that products of counting numbers satisfy the

COMMUTATIVE PROPERTY: $a \times b = b \times a$ for all counting numbers a and b.

If a number a is multiplied by a number b to form a product $a \times b$, then the first number a is called a multiplicand and the second number b a multiplier. (Of course, the commutative property of multiplication obviates the need to distinguish the multiplicand from the multiplier.)

The symbol $\times$ for multiplication was used in WILLIAM OUGHTRED'S (1574–1660) 1631 work *Clavis mathematicae* (The key to mathematics), but historians suspect that the symbol was in use up to 100 years earlier. Mathematicians today also use a raised dot to indicate multiplication ($4 \cdot 3 = 12$, for instance), or they simply write symbols side by side if variables are being used ($x \times y = xy$ or $2 \times w = 2w$, for example).

There are a number of methods for computing the product of two large numbers, such as ELIZABETHAN MULTIPLICATION, EGYPTIAN MULTIPLICATION, and RUSSIAN MULTIPLICATION.

The process of multiplication can be extended to NEGATIVE NUMBERS (yielding the necessary consequence that the product of two negative quantities is positive), FRACTIONS, REAL NUMBERS, COMPLEX NUMBERS, and MATRIXes. Two VECTORs can be multiplied by a DOT PRODUCT or a CROSS PRODUCT. The product of two sets is called a CARTESIAN PRODUCT.

The number 1 is a multiplicative IDENTITY ELEMENT in the theory of arithmetic. We have that $a \times 1 = a = 1 \times a$ for any number a.

The product of two real-valued functions f and g is the function $f \cdot g$, whose value at any input x is the product of the outputs of f and g at that input value: $(f \cdot g)(x) = f(x) \cdot g(x)$. For example, if $f(x) = x^2 + 2x$ and $g(x) = 5x + 7$, then $(f \cdot g)(x) = (x^2 + 2x)(5x + 7) = 5x^3 + 17x^2 + 14x$.

The product formulae in TRIGONOMETRY assert:

$$\sin x \cos y = \frac{\sin(x + y) + \sin(x - y)}{2}$$

$$\sin x \sin y = \frac{\cos(x - y) - \cos(x + y)}{2}$$

$$\cos x \cos y = \frac{\cos(x + y) + \cos(x - y)}{2}$$

$$\cos x \sin y = \frac{\sin(x + y) - \sin(x - y)}{2}$$

See also ASSOCIATIVE; DISTRIBUTIVE PROPERTY; INFINITE PRODUCT.

multiplication principle (fundamental principle of counting) Suppose that a task can be broken up into two steps. If the first step can be done in one of a ways, and the second in b different ways (regardless of the result of the first step), then the multiplication principle says that the original task can be done in $a \times b$ ways.

As an example, imagine that five roads connect town A to town B, and seven roads connect town B to town C. Then one has $5 \times 7 = 35$ alternatives for driving from A to C. When rolling a die and tossing a coin, $6 \times 2 = 12$ different outcomes are possible.

The multiplication principle extends to tasks that are composed of more than two steps. For example, with three different sets of shoes, four different trousers, and three different shirts, one has $3 \times 4 \times 3 = 36$ outfits to wear. There are $10 \times 10 \times 10 \times 26 \times 26 \times 26 = 17,256,000$ different license plate numbers composed of three single-digit numbers followed by three letters.

See also FACTORIAL; PERMUTATION.

mutually exclusive events (disjoint events) Two EVENTs are mutually exclusive if they cannot both occur in a single run of an experiment. For example, in tossing a die, the events "rolling a 3" and "rolling an even number" are mutually exclusive, whereas, the events "rolling a multiple of 3" and "rolling an even number" are not.

If A and B are two mutually exclusive events for an experiment, then the probability that either one event or the other occurs is given by the addition law:

$$P(A \cup B) = P(A) + P(B)$$

See also PROBABILITY.

Napier, John (**Jhone Neper**) (1550–1617) British *Logarithms* Born in Edinburgh, Scotland, in 1550 (his exact birth date is not known), John Napier is remembered in mathematics for his work on TRIGONOMETRY and methods of computation, most notably for his invention of LOGARITHMS.

As was usual at the time, Napier entered the St. Andrew's University in 1563 at the age of 13. It is unlikely that Napier studied mathematics at this institution, and historians today are unsure by what means he acquired a working knowledge of the subject. Napier took an active interest in theology and published a number of important works on the role of religion in society. His study of mathematics was only ever a hobby.

Napier was aware that astronomers at his time were hampered by the difficulty of doing large computations by hand, in particular, computing the products of very large numbers. He sought for a method that would simplify the process. Using the curious geometrical model of an object moving along a straight segment of unit length with speed varying according to its distance from the endpoint, Napier found a formula that converted problems of multiplication into simpler computations of addition. Today we would say that Napier worked with the formula:

$$N = 10^7 \left(1 - \frac{1}{10^7}\right)^L$$

John Napier, an eminent mathematician of the 17th century, discovered logarithms. *(Photo courtesy of the Science Museum, London/Topham-HIP/The Image Works)*

where N is a number and L is, what he called the "logarithm" of that number. (Napier inserted factors of 10^7 so as to help avoid the appearance of decimals in computations.) Napier published his results in 1614 in his piece *Mirifici logarithmorum canonis descriptio* (Description of the marvelous rule of logarithms).

If we divide Napier's formula through by 10^7 we obtain:

$$\frac{N}{10^7} = \left(\left(1 - \frac{1}{10^7}\right)^{10^7}\right)^{\frac{L}{10^7}} \approx \left(\frac{1}{e}\right)^{\frac{L}{10^7}}$$

using the fact that $\lim_{n \to \infty} \left(1 + \frac{r}{n}\right)^n = e^r$, where e is Euler's number. Thus, up to a factor of 10^7, we see that Napier's logarithms are close to the logarithms of today, but to the base $1/e$. Under the advice of English scholar HENRY BRIGGS (1561–1630), Napier later modified his method of logarithms to base 10.

Napier invented mechanical devices to assist in the computation, multiplication, division, and the extraction of square and cube roots. He also made significant contributions to the study of spherical trigonometry, finding formulae for the ratio of sides of triangles drawn on spheres. These formulae are today named in his honor.

Napier died in Edinburgh, Scotland, on April 4, 1617.

See also E; NAPIER'S BONES.

Napier's bones (Napier's rods) In 1614, Scottish mathematician JOHN NAPIER designed a set of graduated rods, now called Napier's bones, that can be used to convert all long multiplication problems to easier, and more swiftly solved, problems of addition. Each rod, made of bone or ivory, was engraved with a column of numbers consisting of the multiples of the digit inscribed at the head of the rod. Diagonal lines were drawn to separate the tens and the units of each multiple. A blank rod represented the multiples of zero.

To compute a long multiplication, such as the product 3717×25, one would line up four rods, one for each digit 3, 7, 1, and 7, as shown:

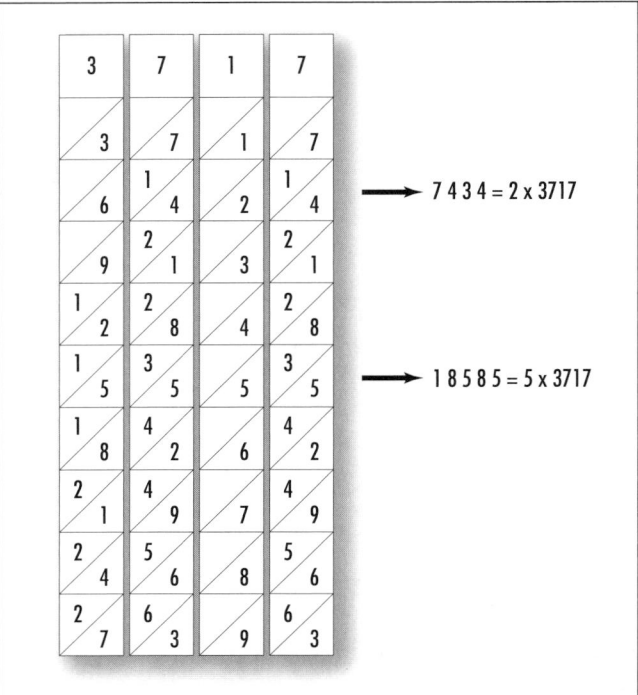

Multiplying with Napier's bones

Looking at the second and fifth rows (corresponding to the digits 2 and 5), and adding along the diagonals in the rows (this corresponds to keeping track of carried digits), we see that $2 \times 3{,}717 = 7{,}434$ and $5 \times 3{,}717 = 18{,}585$. Adding a zero to the tail of the first product gives $20 \times 3{,}717 = 74{,}340$, and so the product we seek is the sum of the two numbers 74,340 and 18,585. This can be swiftly computed with pen and paper.

There is one complication: in looking at any particular row, two numbers in a diagonal may sum to a total with two digits. One must carry the first digit of the sum to the next diagonal to the left. For example, looking at the eighth row we see that $8 \times 3{,}717$ equals 29,736. (The 1 from the sum of 8 and 5 carries to the 6 in the next diagonal to the left.)

See also EGYPTIAN MULTIPLICATION; ELIZABETHAN MULTIPLICATION; FINGER MULTIPLICATION; MULTIPLICATION; RUSSIAN MULTIPLICATION.

Nash, John (1928–) American *Game theory, Topology* Born on June 13, 1928, in Bluefield, West Virginia, John Nash is remembered for his seminal work in nonzero-sum GAME THEORY. Nash's 1949 doctoral dissertation "Non-Cooperative Games" (only 45 pages long) proved to have such a profound impact on the development and study of economic theory that it simply redefined the nature of economic thinking during the four decades that followed. For this, he was honored as a corecipient of the 1994 Nobel Prize in economics. Nash also made extraordinary contributions to the field of TOPOLOGY and solved a number of breakthrough problems in the study of Riemannian geometry.

Nash entered the Carnegie Institute of Technology (now Carnegie-Mellon University) in 1945 with the intent to study chemical engineering, but soon discovered a passion for mathematics. He graduated 3 years later with both bachelor and master degrees in mathematics and entered Princeton University in 1948 to commence work on a doctorate. He graduated just 2 years later.

By 1954 Nash had published five extraordinary papers on the topic of game theory, including "Equilibrium Points in N-person Games" in 1950, "The Bargaining Problem" in 1950, and "Non-Cooperative Games" in 1951, and two truly significant works on the topic of analytic geometry, "Real Algebraic Manifolds" in 1952, and "C^1 Isometric Imbeddings" in 1954. He had clearly established himself as a mathematical genius and was recognized as such by the mathematical community. He was a tenured professor at the Massachusetts Institute of Technology at the age of 29.

In the early 1960s, Nash began developing paranoid schizophrenia and became virtually incapacitated by the disease for the two decades that followed. He lost his position at M.I.T, spent a number of years roaming Europe and America, and returned to Princeton, New Jersey, before receiving institutional help.

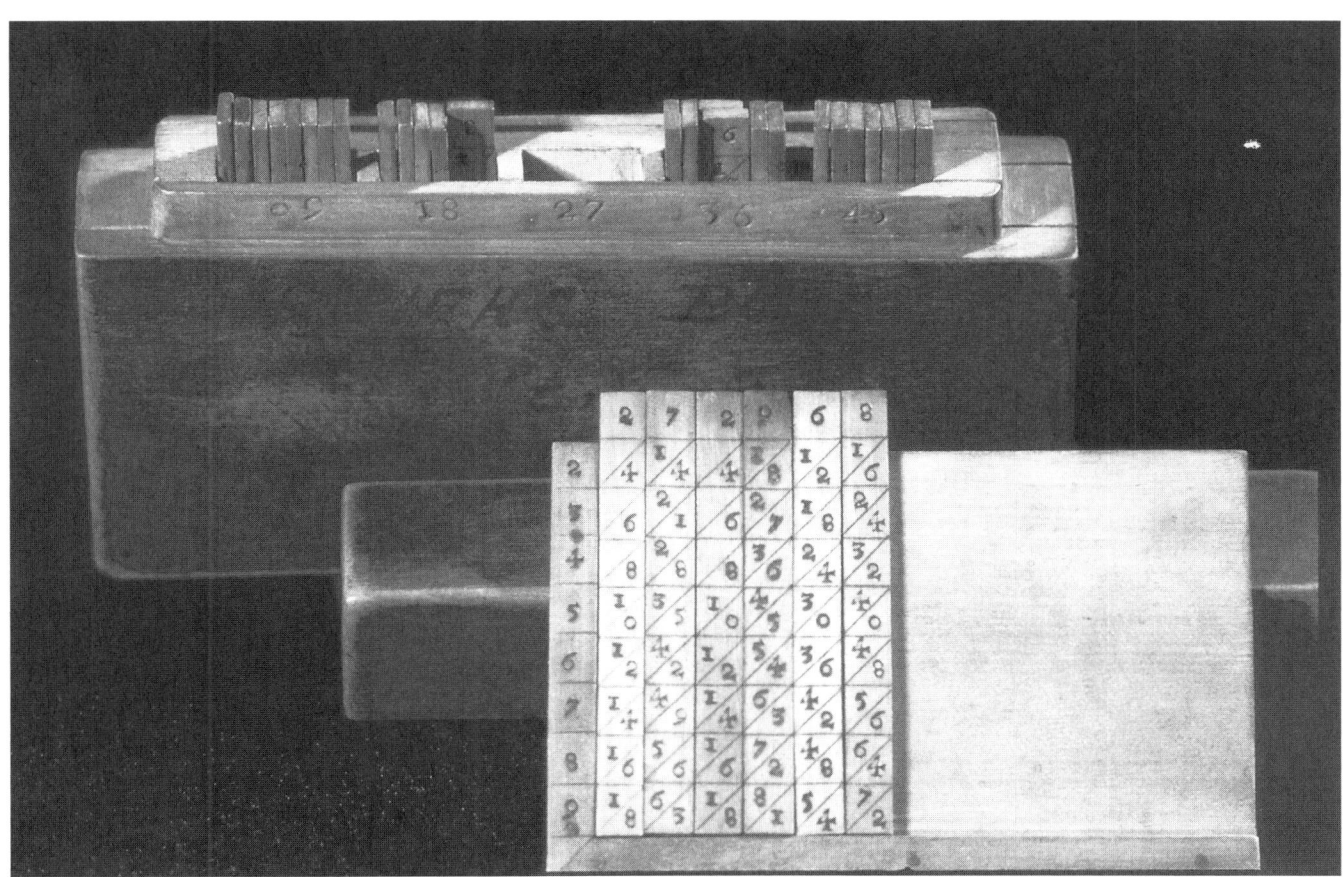

Napier's bones, such as this set of rods dating from 1690, provided the means to simplify the process of long multiplication. *(Photo courtesy of the Science Museum, London/Topham-HIP/The Image Works)*

Slowly, over many years, Nash began to recover and eventually returned to teaching at Princeton and contemplating mathematical work. He delivered a paper at the 10th World Congress of Psychiatry in 1996 describing his illness. Along with the 1994 Noble Prize, Nash was awarded the 1999 Leroy P. Steele Prize by the American Mathematical Association. Nash describes his 45-page dissertation as his "most trivial work" of his career.

natural number (counting number, whole number) Any of the positive whole numbers 1, 2, 3, … is called a natural number or a counting number. The natural numbers are used to count separate objects. The collection of all counting numbers {1,2,3,…}, called the set of natural numbers, is denoted **N**, or sometimes **IN** or **Z⁺**. Some mathematicians choose to include the number ZERO in this set. (There is no standard convention on the matter.)

The set of natural numbers is closed under ADDITION and MULTIPLICATION, that is, the sum or product of any two natural numbers is again a natural number. The set, however, is not closed under SUBTRACTION or DIVISION. For example, the result of subtracting 5 from 3 is no longer a natural number, nor is the result of dividing 7 by 2.

An infinite set of objects whose elements can be arranged in a list akin to the list of natural numbers is said to be COUNTABLE. Not all infinite sets are countable.

See also CLOSURE PROPERTY; DISCRETE; FIGURATE NUMBERS; NUMBER; NUMBER THEORY; ORDINAL NUMBERS; WHOLE NUMBER.

necessary condition *See* CONDITION—NECESSARY AND SUFFICIENT.

negation (not statement) In FORMAL LOGIC a statement of the form "not *p*" is called the negation of the statement *p*. For example, "Shakespeare did not write *Hamlet*" is the negation of the statement that Shakespeare did. In practice it is not always necessary for the term *not* to occur. For example, the negation of $x > y$ is $x \le y$.

The negation of a statement *p* is denoted in symbols as $\neg p$ (or sometimes as ~*p*, −*p*, or even $\bar{p}$). The negation of statement has truth-value opposite to that of the original statement, and so has a TRUTH TABLE:

p	$\neg p$
T	F
F	T

negative numbers Any REAL NUMBER less than ZERO is called a negative number. In practical applications, negative numbers are used to denote quantities that are below some reference point. For example, in the centigrade temperature scale, a temperature of −10° is 10° below the freezing point of water.

The product of two negative numbers is a positive quantity. For instance, $(-1) \times (-1) = 1$. This can be justified by making use of the DISTRIBUTIVE PROPERTY and the fact that the product of any quantity with zero is zero. Specifically, we have:

$$(-1) \times 0 = 0$$
$$(-1) \times (1 + (-1)) = 0$$
$$(-1) \times 1 + (-1) \times (-1) = 0$$
$$-1 + (-1) \times (-1) = 0$$
$$(-1) \times (-1) = 1$$

Negative numbers were generally viewed with suspicion throughout history. Ancient Egyptian and Babylonian scholars ignored negative solutions to their mathematical equations, as did ancient Greek scholars (to whom "number" was directly associated with the notion of physical length). Chinese scholars were comfortable working with negative quantities in intermediate steps toward solving a problem, but they never permitted them as final solutions.

Seventh-century Indian scholar BRAHMAGUPTA is noted as the first scholar to properly determine the arithmetic of negative quantities. He deemed their existence as valid by equating positive quantities with possessions and negative quantities with debt. This work was later expanded upon by scholar BHĀSKARA (ca. 1114–85). Although Arab scholars of this time read and translated the Indian texts, they chose not to work with negative quantities. As 12th-century European scholars garnered much of their mathematical knowledge from the Islamic world, familiarity with negative quantities did not readily come to the West.

In the 16th century, prominent mathematicians such as GIROLAMO CARDANO, Michael Stifel, and FRANÇOIS VIÈTE adamantly rejected the notion of negative numbers, deeming these quantities as "meaningless" and "absurd." This attitude generally persisted for the century that followed, even though scholars found it necessary to work with them algebraically as they solved more sophisticated mathematical equations. By the turn of the 18th century, however, it was generally admitted that negative numbers are a necessary construct in mathematics. LEONHARD EULER (1707–83) was comfortable working with negative quantities.

With the development of ABSTRACT ALGEBRA in the 19th century, the need to assign "meaning" to numbers became less important. Even though some 19th-century scholars such as AUGUSTUS DE MORGAN continued to publish commentary against the validity of negative quantities, their usefulness, and necessity, was generally accepted.

See also BABYLONIAN MATHEMATICS; CHINESE MATHEMATICS; EGYPTIAN MATHEMATICS; GREEK MATHEMATICS; NUMBER; POSITIVE.

nested multiplication To evaluate a POLYNOMIAL such as $p(x) = 2x^3 + 7x^2 - 4x + 3$ for a particular value $x = 5$, say, one simply substitutes 5 for x and performs the required number of multiplications. In this example, $3 + 2 + 1 = 6$ multiplications are needed:

$$p(5) = 2 \times 5 \times 5 \times 5 + 7 \times 5 \times 5 - 4 \times 5 + 3 = 408$$

(It is generally the case that a polynomial contains a TRIANGULAR NUMBER of products.) The number of multiplications required can be significantly reduced if one first rewrites the polynomial in a form known as nested multiplication. In this example we write:

$$p(x) = 2x^3 + 7x^2 - 4x + 3$$
$$= (2x^2 + 7x - 4)x + 3$$
$$= ((2x + 7)x - 4)x + 3$$

Thus $p(5)$ can be computed with just three operations of multiplication:

$$p(5) = ((2 \times 5 + 7) \times 5 - 4) \times 5 + 3$$
$$= ((10 + 7) \times 5 - 4) \times 5 + 3$$
$$= (85 - 4) \times 5 + 3$$
$$= 408$$

Notice that this process simply multiplied the first coefficient by 5, added the second coefficient, multi-

plied the result by 5, added the third coefficient, multiplied by 5, and then added the final coefficient. The process is compactly recorded in a table as follows:

5	2	7	−4	3
	0	10	85	405
	2	17	81	**408**

The first row lists the coefficients of the polynomial and a zero is placed under the first coefficient. One works from left to right adding the entries in the two rows, multiplying the result by 5, and recording that result in the next column. The entry in the bottom right corner is the value $p(5)$.

The remaining numbers on the bottom row have a surprising interpretation. According to the FACTOR THEOREM, if the polynomial $p(x)$ is divided by the term $x - 5$, then the remainder will be $p(5) = 408$. In this example one can check that

$$\frac{p(x)}{x - 5} = 2x^2 + 17x + 81 + \frac{408}{x - 5}$$

The numbers on the bottom row of the table above are precisely the coefficients of the quotient. This same phenomenon occurs for any polynomial of any degree. Examining an abstract example illustrates why this works. (For simplicity we will again work with a cubic equation.)

Consider the polynomial $p(x) = ax^3 + bx^2 + cx + d$ divided by the linear term $x - h$. The process of LONG DIVISION yields the following:

$$
\require{enclose}
\begin{array}{r}
ax^2 \quad + (ah+b)x \quad + ((ah+b)h+c) \\
x-h \enclose{longdiv}{ax^3 + bx^2 \quad\quad + cx \quad\quad\quad + d} \\
\underline{ax^3 - ax^2 h} \\
(ah+b)x^2 + cx \\
\underline{(ah+b)x^2 - (ah+b)xh} \\
((ah+b)h+c)x + d \\
\underline{((ah+b)h+c)x - ((ah+b)h+c)h} \\
((ah+b)h+c)h + d
\end{array}
$$

On the other hand, the method of evaluating $p(h)$ via the process of nested multiplication yields the table:

h	a	b	c		d	
	0	ah	$(ah + b)h$		$((ah + b)h + c)h$	
	a	$ah + b$	$(ah + b)h + c$		$((ah + b)h + c)h + d = p(h)$	

Comparison of the two tables side by side show that the two processes are identical.

The process of drawing a table for evaluating a polynomial $p(x)$ at a value $x = h$, or, equivalently, for dividing $p(x)$ by $x - h$, is called synthetic division. It is a very simple and efficient process. As an example, the following table shows that the polynomial $p(x) = 2x^4 + 3x^3 - 4x^2 + 5x + 4$ has value 6 when evaluated at $x = \frac{1}{2}$, and quotient $2x^3 + 4x^2 - 2x + 4$ when divided by $x - \frac{1}{2}$.

$\frac{1}{2}$	2	3	−4	5	4
	0	1	2	−1	2
	2	4	−2	4	6

Returning to the theme of triangular numbers: Any nested product of the form:

$$(9(\ldots(9(9(9 + 1)+1)+1)\ldots+1)+1)$$

is triangular. For instance, $9 + 1 = 10 = T_4$, $9(9 + 1) + 1 = 91 = T_{13}$, and $9(9(9 + 1) +1) + 1 = 820 = T_{40}$. This follows from the relation $9T_n + 1 = T_{3n+1}$ beginning with $T_1 = 1$. This relation also shows that if the number of nines present in the nested product is n, then the $\frac{3^n - 1}{2}$-th triangular number is produced.

net In geometry, a net is a diagram drawn on a page which, when cut out and folded along the lines indicated, can be used to construct a POLYHEDRON. For example, the following diagram shows nets of a cube and a TETRAHEDRON.

Not every arrangement of six squares on a page produces a net for a cube. (For example, a row of six squares does not fold to form a cube.) One can show that there are 11 essentially different nets for a cube, each representing a fundamentally different way of

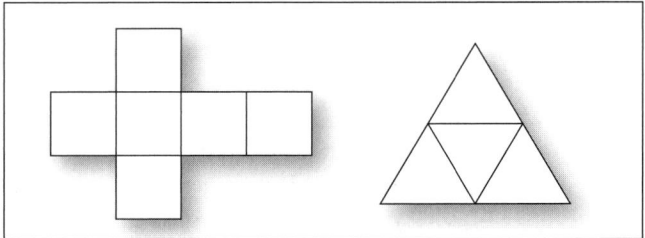

Nets for a cube and a tetrahedron

unfolding the figure. There are just two different ways to unfold a tetrahedron. Mathematicians have shown that there are 261 different ways to unfold a four-dimensional HYPERCUBE into a three-dimensional net of eight connected cubes.

In terms of accounting, the word *net* refers to the profit calculated after deducting all operating expenses. For example, if a small private school receives $100,000 in tuition payments per year as income but incurs an annual operating expense of $97,000 (for salaries, insurance fees, school supplies, and the like), then the school is operating with an annual net profit of $3,000.

In commerce, the term *net* denotes the weight of goods excluding the weight of any wrapper or container holding the goods. For example, many grocery stores subtract the weight of the plastic containers used to hold fresh food items from the total weight of the item being purchased. Customers are charged only for the net weight of the item.

Neumann, John von (1903–1957) Hungarian-American *Game theory, Logic, Analysis, Abstract algebra* Born on December 28, 1903, in Budapest, Hungary, pure and applied mathematician John von Neumann is remembered for his important contributions to a tremendously wide range of topics. In pure mathematics, he worked on problems in SET THEORY and devised a set of axioms for the subject alternative to those proposed by ERNST FRIEDRICH FERDINAND ZERMELO (1871–1953). He also made advances in functional analysis, OPTIMIZATION theory, and GROUP THEORY, and succeeded in providing an axiomatic system for the theory of quantum mechanics. In applied mathematics, von Neumann is best remembered for his 1944 text

The Theory of Games and Economic Behavior, written with Oskar Morgenstern (1902–76), which essentially founded the field of GAME THEORY. He also made major contributions to the development of the modern computer, both as a practical design problem and as a theoretical investigation into the nature and capabilities of AUTOMATA. He was the first to devise a way of storing programs inside a computer.

Von Neumann demonstrated an incredible aptitude for mathematics at a very early age. He received private tutoring in mathematics from faculty members at the University of Budapest while he was a high-school student, and by age 18 he had written and published a coauthored mathematics paper. Although von Neumann was admitted into the mathematics program at the University in 1921, he followed his father's wishes and pursued chemistry at the University of Berlin and at the Technische Hochschule in Zürich. Remarkably, without ever attending a class at the University of Budapest, von Neumann continued his work in mathematics privately, passed the University mathematics examination, wrote an influential thesis on the topic of set theory, and was awarded a degree there in 1926, all the while pursuing a degree in chemical engineering in Zürich. In 1930 von Neumann accepted a professorship in mathematics at Princeton University in the United States.

In 1932 he published his groundbreaking work *Mathematische Grundlagen der Quantenmechanik* (Mathematical foundations of quantum mechanics) on theoretical quantum mechanics, which led him to the study of operator algebras and functional analysis for the years that followed. In 1938 he was awarded the Bôcher Prize from the American Mathematical Society for his work in this field. In the early 1940s von Neumann's interests turned to applied mathematics, and in 1944 he published his famous piece on the topic of game theory. His subsequent work in the logical design of computers and the theory of automata seemed, to von Neumann, a natural extension of all his previous studies.

During and after World War II von Neumann served as a consultant to the Defense Department and to the Los Alamos Scientific Laboratory. In 1955 President Eisenhower appointed him to the Atomic Energy Commission, and in 1956 he was awarded the Enrico Fermi Award for all his contributions in this regard. The same year he was honored in academia with the Albert Einstein Commemorative Award.

Von Neumann was elected to a large number of academic societies throughout his life, including the National Academy of Sciences (the United States), The American Academy of Arts and Sciences, Instituto Lomdardo do Scienze e Lettere (Italy), and the Royal Netherlands Academy of Sciences and Letters. He died in Washington, D.C., on February 8, 1957.

Newton, Sir Isaac (1642–1727) British *Calculus, Mechanics, Dynamics, Optics, Astronomy, Natural philosophy*   Born on January 4, 1642, in Woolsthorpe, England, Sir Isaac Newton is remembered as one of the greatest scientific scholars of all time. He dominated and revolutionized the mathematics and the physics of the 17th century. Newton is responsible for the development of differential and integral CALCULUS and the discovery of the FUNDAMENTAL THEOREM OF CALCULUS that unites the two fields. He also generalized the BINOMIAL THEOREM to incorporate noninteger exponents; developed numerical methods for solving DIFFERENTIAL EQUATIONS, approximating LIMITS, and computing integrals; and discovered many important results in the theory of equations. In physics, he is remembered for his formulation of a system of mechanics capable of precise and accurate descriptions of the motions of all objects and for his universal law of gravitation through an INVERSE SQUARE LAW. He outlined this work in his 1687 work *Philosophiae naturalis principia mathematica* (The mathematical principles of natural philosophy), often referred to simply as *Principia*. This is today regarded as one of the most important scientific works of all time.

Newton entered the Trinity College, Cambridge, in 1661 to study law and philosophy, but he soon discovered an interest in mathematics, optics, and mechanics. A personal notebook from the time reveals that he read, and worked through in detail, the works of EUCLID (ca. 300–260 B.C.E.) and RENÉ DESCARTES (1596–1650) on the topic of geometry, of GALILEO GALILEI (1564–1642) on the mechanics of the universe, and of FRANÇOIS VIÈTE (1540–1603) and JOHN WALLIS (1616–1703) on algebra. Newton extended some of Wallis's techniques for finding the area of curved figures and worked on new approaches to infinite SERIES. He graduated from Cambridge with a bachelor's degree in 1665, but he did not attract the attention of his professors as a particularly gifted scholar at the time.

Sir Isaac Newton, an eminent scientist of the 18th century, dominated and revolutionized the mathematics and physics of the late 1600s and early 1700s. *(Photo courtesy of the Science Museum, London/Topham-HIP/The Image Works)*

The college closed for 2 years due to the plague, and Newton returned to his hometown of Woolsthorpe. There he began work on developing new approaches to mathematics, physics, and optics. He made the fundamental discovery that integration is the reverse process of differentiation, and he used this unifying concept to lay down the foundations for a general theory of calculus. (As such, he completed this work two decades before German scholar GOTTFRIED WILHELM LEIBNIZ (1646–1716) independently made the same discovery.) Newton used the notion of a FLUXION (derivative) as the basis of his approach. He summarized his work in his 1671 piece *De methodis serierum et fluxionum* (The method of infinite series and fluxions) but failed to have the work published. (An English translation of the piece was printed posthumously in 1736.) This seems to have

been a frequent problem for Newton. Historians gather from examination of his letters and personal writings that he had a morbid dislike and fear of criticism and would invariably hold back from publishing his ideas. This led to a bitter dispute, for instance, 20 years later when Leibniz published his discoveries and claimed to have invented calculus.

The University of Cambridge reopened in 1667, and Newton returned to obtain a master's degree 1 year later. By this time, through what he had shared with friends and reported in letters, he had garnered a reputation as a gifted scholar, and 2 years later, at the

PHILOSOPHIÆ
NATURALIS
PRINCIPIA
MATHEMATICA

Autore *JS. NEWTON*, Trin. Coll. Cantab. Soc. Matheseos Professore *Lucasiano*, & Societatis Regalis Sodali.

IMPRIMATUR·
S. PEPYS, *Reg. Soc.* PRÆSES.
Julii 5. 1686.

LONDINI,
Jussu *Societatis Regiæ* ac Typis *Josephi Streater*. Prostant Venales apud *Sam. Smith* ad insignia Principis *Walliæ* in Cœmiterio D. *Pauli*, aliosq; nonnullos Bibliopolas. *Anno* MDCLXXXVII.

Considered the most important scientific work of modern times, Sir Isaac Newton's *Principia* outlines the inverse-square law of gravitation and its application to the mathematical derivation of Johannes Kepler's three laws of planetary motion. *(Photo courtesy of the Science Museum, London/Topham-HIP/The Image Works)* young age of 27, Newton was appointed Lucasian professor at Cambridge.

At this time Newton began serious work on the study of optics. With the aid of a prism he discovered that white light was composed of a spectrum of colors, each refracting through a lens at a different angle. He concluded then that all refracting telescopes were subject to chromatic aberrations and set to work on building a reflecting model. In 1672 he was elected as a fellow of the prestigious ROYAL SOCIETY of London in honor of this work. He published his text *Optiks* (Optics) in 1704.

Newton also began formulating his famous laws of motion and his laws of gravitation during the mid-1660s. In 1684 eminent astronomer Edmond Halley (1656–1742) suggested to Newton that he should investigate the mathematics of gravitational attraction and attempt to derive JOHANNES KEPLER'S famous three laws. Newton apparently replied simply that he already had many years earlier. Only after Halley's persistent urgings did Newton agree to publish the results in what was to become his most famous piece of work, the 1687 piece *Principia*.

Newton also published in his lifetime, somewhat delayed after completing the work, *Enumeratio linearum tertii ordnis* (Enumeration of lines of the third order) in 1704, *Arithmetica universalis* (Universal arithmetic), his collected works in algebra in 1707, and *Analysis per quantitatum series* (Analysis by means of various series) in 1711.

Newton resigned from his position at the University of Cambridge in 1701 to take a prestigious government position in London and participated in very little mathematical research thereafter. In 1703 he was elected president of the Royal Society and was reelected to that position each year until his death on March 20, 1727. In 1705 he was knighted by Queen Anne, becoming the first scientist to be so honored for scholarly achievement.

It is impossible to understate the influence that Newton's work has had on the development of all scientific research. Scholars today agree that the publication of *Principia* marked the beginning of the modern scientific era.

Newton's method In many practical situations, one is required to find a numerical solution to an equation of the form $f(x) = 0$, even if there are no clear algebraic means for solving such an equation. (There are no general methods suitable for solving $\sqrt{x} + \sqrt{x + 2} + \sqrt{x + 3} - 5 = 0$, for instance.) If the function in question is differentiable (*see* DIFFERENTIAL CALCULUS), then one can employ Newton's method to find the approximate location of a ROOT.

One begins by making an initial guess x_0, hopefully missing the correct value of the root by just a small amount. Suppose it turns out that the root is h units away from x_0 (thus $f(x_0 + h) = 0$). Using this value of h, we can approximate the derivative of the function at x_0 as:

$$f'(x_0) \approx \frac{f(x_0 + h) - f(x_0)}{h} = -\frac{f(x_0)}{h}$$

Turning this around, we see that h is approximately $-f(x_0)/f'(x_0)$. Substituting this value for h shows that if x_0 is our initial guess for the root, then the point:

$$x_1 = x_0 - \frac{f(x_0)}{f'(x_0)}$$

is likely to be a much better approximation to the root. (Geometrically, x_1 is the location at which the tangent line to the graph $y = f(x)$ at position x_0 crosses the x-axis.) Repeating this procedure, each time using the outcome just obtained as the next initial guess, produces a sequence of values $x_0, x_1, x_2, \ldots$, with $x_{n+1} = x_n - \frac{f(x_n)}{f'(x_n)}$ yielding successively better approximations of the root. One can perform this procedure until a desired degree of accuracy is obtained.

To illustrate Newton's method, we compute $\sqrt{2}$ to four decimal places. Computing this root is equivalent to solving the equation $x^2 - 2 = 0$. Setting $f(x) = x^2 - 2$, we have $f'(x) = 2x$. Thus the sequence of approximations is given by the formula:

$$x_{n+1} = x_n - \frac{x_n^2 - 2}{2x_n} = \frac{x_n + \frac{2}{x_n}}{2}$$

(Compare with HERON'S METHOD.) With $x_0 = 1$ as our initial guess, we obtain the approximations:

$$x_1 = \frac{3}{2} = 1.5$$

$$x_2 = \frac{17}{12} = 1.416667$$

$$x_3 = \frac{99}{70} = 1.414216$$

$$x_4 = \frac{577}{408} = 1.414213$$

and so on. Hence, to four decimal places, we see $\sqrt{2} = 1.4142$.

If one initially makes *two* guesses x_0 and x_1 to the root of an equation $f(x) = 0$, then one can use the formula $f'(x_1) \approx \dfrac{f(x_1) - f(x_0)}{x_1 - x_0}$ to approximate the slope of the curve at position $x = x_1$. An improved approximation for the root is thus given by $x_2 = x_1 - \dfrac{f(x_1)}{f'(x_1)} \approx x_1 - f(x_1)\dfrac{x_1 - x_0}{f(x_1) - f(x_0)}$. Repeated application leads to the secant method for approximating roots: given two initial guesses x_0 and x_1 set:

$$x_{n+1} = x_n - f(x_n)\frac{x_n - x_{n-1}}{f(x_n) - f(x_{n-1})}$$

Neyman, Jerzy (1894–1981) *Moldavian Statistics* Born on April 16, 1894, in Bendery, Moldavia, statistician Jerzy Neyman is remembered for his fundamental work in the theory of inferential statistics. He made significant contributions to the practice of hypothesis testing and estimation, and was the first to introduce the notion of a confidence interval.

Neyman studied mathematics at Kharkov University and received a doctorate in mathematics and statistics in 1924 from the University of Warsaw. Collaborating with KARL PEARSON's son, E. S. Pearson, Neyman wrote a number of influential papers. The two most prominent pieces "On the Problem of the Most Efficient Tests of Statistical Hypotheses" and "The Testing of Statistical Hypotheses in Relation to Probabilities A Priori" were both published in 1933.

In 1938 Neyman emigrated to the United States, where he worked at the University of California, Berkeley, for the remainder of his life. He died in Oakland, California, on August 5, 1981.

His work in statistics, with the introduction of the confidence interval, revolutionized sampling techniques and practices in agriculture, biology, medicine, and the physical sciences, and his study of stratified populations led to a new field of statistical theory. The structure of the Gallup poll used today, for instance, is based on this work.

See also STATISTICS: INFERENTIAL.

nine-point circle Consider a triangle with vertices A, B, and C. Let M_A, M_B, and M_C be the MIDPOINTs of each of its three sides. Consider too the three altitudes of the triangle, lines that each pass through one vertex of the triangle to meet the opposite side at a right angle. Let H_A, H_B, and H_C be the feet of the three altitudes, one on each side of the triangle. A study of EQUIDISTANT points shows that the three altitudes pass through a common point H called the orthocenter of the triangle. (See ALTITUDE.) Let P_A, P_B, and P_C be the midpoints of the line segments connecting H to each vertex A, B, and C, respectively. We now have nine points associated with the triangle: M_A, M_B, M_C, H_A, H_B, H_C, P_A, P_B, and P_C. Surprisingly, these nine points will always lie on a circle. This circle is called the nine-point circle of the triangle.

This result was discovered by German mathematician Karl Wilhelm Feuerbach (1800–34). The center of the circle turns out to lie on the EULER LINE of the triangle. In fact, if O is the circumcenter of the triangle, that is, the center of the CIRCUMCIRCLE of the triangle, then the center of the nine-point circle Q is the midpoint of the line segment connecting O to H. (With this known, one can then prove that Q is indeed the same distance from each of the nine points listed, thereby establishing the claim that all nine points lie on a circle.)

Feuerbach also proved that the nine-point circle and the INCIRCLE of the triangle touch at just one point and, moreover, that the nine-point circle is TANGENT to each of the three excircles of the triangle.

Given four arbitrary points in the plane, one can construct a nine-point circle for any triangle formed by a set of three of those points. As there are four sets of three points, this yields four nine-point circles in all. Surprisingly, these four nine-point circles all pass through a single common point.

Noether, Amalie (Emmy) (1882–1935) *German Abstract algebra* Born on March 23, 1882, in Erlangen, Bavaria, mathematician Emmy Noether is remembered for her highly creative work in the theory of RINGS and in ABSTRACT ALGEBRA. She was responsible for directing algebra away from detailed arithmetical calculations to a general axiomatic study of structure. She is also noted for her determination to succeed as a woman scholar in the male-dominated realm of early 20th-century academia.

Noether studied languages and music during her school years, and trained to be a teacher of French and English. In 1900 she passed her state certification exams but never worked as an educator. Instead she decided to pursue a career in mathematics, attending lectures at the University, of Erlangen. At the time women were not permitted to enroll as students at the university, and her studies at the university were never deemed official. In 1904 she passed the state matriculation exam and then went to the University of Göttingen. Three years later she was awarded a doctorate from that institution. However, she was not permitted to pursue the habilitation degree that would earn her the appropriate qualification to become a university faculty member. Noether returned to Erlangen.

During the following years Noether published a number of influential papers on the topic of algebra. Her work caught the attention of prominent mathematicians DAVID HILBERT (1862–1943) and CHRISTIAN FELIX KLEIN (1849–1925), who invited her back to the University of Göttingen and fought the university administration to give her the right to pursue an advanced degree. She was finally awarded an official position as a faculty member of the university in 1922.

Mathematicians today consider her 1921 paper "Idealtheorie in Ringbereichen" (The theory of ideals in ring structures) to be of fundamental influence in the development of modern abstract algebra.

Noether was granted many honors and awards for her work. In 1928 she was invited to address the International Mathematical Conference at Bologna, and spoke again at Zürich in 1932, the same year she was a joint recipient of the Alfred Ackermann-Teubner Memorial Prize for the advancement of mathematical knowledge.

With the uprising of the Nazis in 1933 Noether was dismissed from her position at Göttingen because of her Jewish faith. She then accepted a professorship at Bryn Mawr College, Pennsylvania. At this time she also lectured at the Institute of Advanced Study at Princeton, New Jersey.

Noether died at Bryn Mawr on April 14, 1935. She wrote a total of 45 research papers throughout her career and shaped the entire course of study in abstract algebra. Mathematician Bartel Leendert van der Waerden (1903–96) was particularly influenced by her work and continued to promote her ideas after her death.

non-Euclidean geometry After numerous unsuccessful attempts throughout history to establish the PARALLEL POSTULATE as a consequence of the remaining four of EUCLID'S POSTULATES, mathematicians began to contemplate theories of geometry in which the fifth postulate does not hold. Any such theory of GEOMETRY is called a non-Euclidean geometry.

In 1795 Scottish mathematician and physicist John Playfair (1748–1819) presented an alternative, but equivalent, formulation of the parallel postulate: *through any point in the plane, there is precisely one line through that point parallel to any prescribed direction.* Recasting the postulate this way makes it apparent that negation of the famous fifth postulate has two parts. Either:

1. There are no lines through a given point parallel to a given direction.
2. There is more than one line through a given point parallel to a given direction.

From 1826 to 1829 Russian mathematician NICOLAI IVANOVICH LOBACHEVSKY (1792–1856) developed a consistent theory of geometry in which Euclid's fifth postulate fails in the manner described in point 2. Hungarian mathematician JÁNOS BOLYAI (1802–60) independently came to the same surprising conclusion by assuming that through any point there are infinitely many distinct lines parallel to a given direction. (CARL FRIEDRICH GAUSS (1777–18) had also come to similar conclusions, but he did not publish his results.) Such a theory of geometry is today called HYPERBOLIC GEOMETRY.

In the 1850s German mathematician GEORG FRIEDRICH BERNHARD RIEMANN (1826–66) presented an alternative form of geometry in which Euclid's fifth postulate fails in the manner describe in point 1 above. In this theory, today called Riemannian geometry or SPHERICAL GEOMETRY, "lines" are great circles drawn on the surface of a sphere. Consequently, it is impossible to draw a pair of lines that never intersect.

Despite the simplicity of the Riemannian model, it was not immediately obvious to scholars of the 19th century that it was permissible. Euclid's second postulate seems to state that straight lines should be of infinite length. Riemann had the insight to note that "extended indefinitely" does not imply "infinitely long," thus allowing him to consider a geometry in which straight lines loop back on themselves.

In hyperbolic geometry, all angles in a triangle sum to less than 180°, and the ratio of the circumference of a circle to its diameter is greater than π. In spherical geometry, all angles in a triangle sum to more than 180°, and the ratio of the circumference of a circle to its diameter is less than π. EUCLIDEAN GEOMETRY, in which angles in a triangle always sum exactly to 180° and the ratio of the circumference to diameter of any circle is precisely π, can thus be regarded as an intermediate between the two.

See also EUCLID; PLAYFAIR'S AXIOM.

normal distribution In the early 1700s scientists noticed that errors from measurements in scientific experiments repeated many times tended to follow the same form of DISTRIBUTION, even though the studies were conducted in unrelated fields (physics versus biology or sociology, for example). The particular pattern of errors observed is today known as the normal distribution (or Gaussian distribution). The French mathematician ABRAHAM DE MOIVRE, in 1733, was the first to write a mathematical formula for the distribution. Later, mathematicians PIERRE-SIMON LAPLACE (1749–1827), SIMÉON-DENIS POISSON (1781–1840), and CARL FRIEDRICH GAUSS (1777–1855) specified and proved many of its mathematical properties. At the turn of the 20th century Aleksandr Mikhailovich Lyapunov, and later others, refined and developed Laplace's work to establish the CENTRAL-LIMIT THEOREM to explain the frequent occurrence of the normal distribution in all scientific studies.

The normal distribution turns out to be a symmetrical bell-shaped curve. It is scaled so that the total area under the curve is exactly equal to 1. The location of the peak of the curve is called its mean, denoted μ, and the width of the curve is measured by a value σ, called its standard deviation. (See STATISTICS: DESCRIPTIVE and DISTRIBUTION.) On both sides of the peak, the curve uniformly falls steeply downward. The curve then falls

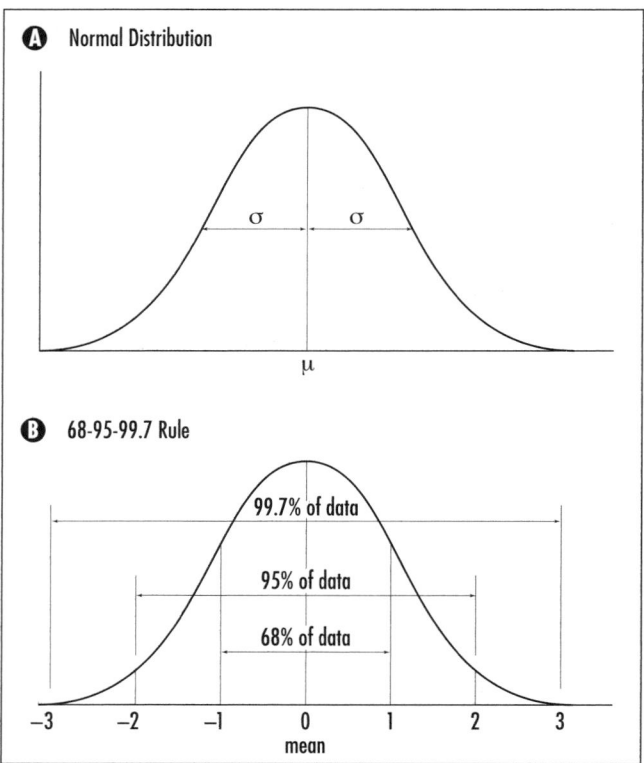

The normal distribution

less steeply, with curvature facing upward. The location at which the curvature changes direction from inward to upward curvature is a distance of 1 standard deviation from the mean, the peak of the curve.

The "68-95-99.7 rule" asserts that 68 percent of the area under the curve lies within a distance of 1 standard deviation on either side of the mean; 95 percent of the area lies within 2 standard deviations of the mean; and 99.7 percent within 3 standard deviations. For example, it might be observed in a medical study that the mean height of women between ages 18 and 24 is normally distributed, with mean μ = 64.5 in. and standard deviation σ = 2.5 inches. We can then deduce that 68 percent of young women are between 64.5 – 2.5 = 62 and 64.5 + 2.5 = 67 in. tall.

If a measurement in an experiment is known to follow a normal distribution, then the probability that a measurement taken at random lies within the range [*a,b*] is found by computing the area under the curve above the interval [*a,b*]. Reference texts in statistics provide tables of area computations for a normal distribution of mean zero and standard deviation 1 (the standard normal distribution). One can convert an arbitrary normal distribution into a standard normal form by use of Z-SCOREs, which then allows one to calculate probabilities for that distribution.

The formula for the curve describing a normal distribution of mean μ and standard deviation σ is:

$$f(x) = \frac{1}{\sigma\sqrt{2\pi}} e^{-\frac{1}{2}\left(\frac{x-\mu}{\sigma}\right)^2} \; ; -\infty < x < \infty$$

The cumulative-distribution function of the standard normal distribution ($\mu = 0$, $\sigma = 1$) is denoted $\phi(z)$ and is given by $\phi(z) = \frac{1}{\sqrt{2\pi}} \int_{-\infty}^{z} e^{-\frac{z^2}{2}} dz$. Statistics references usually list values of this function from $z = 0.0$ to $z = 4.0$. (We have $\phi(0.0) = 0$ and $\phi(4.0) = 0.99997$.) Values of the function for negative values of z can be deduced using the fact that the normal distribution is symmetric in shape.

See also CHEBYSHEV'S THEOREM; STATISTICS: INFERENTIAL.

normal to a curve
In two-dimensional space, a normal to a curve at a point P on the curve is the line through P that lies at right angles to the TANGENT to the curve. For example, a radius of a circle is normal to the circumference of the circle.

See also NORMAL TO A PLANE; NORMAL TO A SURFACE; ORTHOGONAL; PERPENDICULAR.

normal to a plane (normal vector to a plane)
In three-dimensional space, a VECTOR whose direction is PERPENDICULAR to a plane is said to be a normal to the plane. A normal vector is consequently perpendicular to *any* vector that lies in the given plane.

If $ax + by + cz = d$ is the equation of the plane, then $\mathbf{n} = \langle a, b, c \rangle$ is a normal to the plane. This follows from the derivation of the VECTOR EQUATION OF A PLANE. Any (nonzero) scalar multiple of this vector is also a normal to the plane.

See also NORMAL TO A CURVE; NORMAL TO A SURFACE; ORTHOGONAL.

normal to a surface
In three-dimensional space, a normal to a surface at a point P on the surface is a line

through P that is PERPENDICULAR to the TANGENT plane of the surface at P. For example, a line passing through the center of a sphere is normal to the surface at each of the two points it intersects the sphere.

It is assumed that the surfaces under discussion are "smooth," so that at each point there is a well-defined tangent plane. For example, a SPHERE is smooth, as is a TORUS, but the surface of a cube is not. (There is no well-defined tangent plane at one of its corners, for example.)

See also NORMAL TO A CURVE; NORMAL TO A PLANE; ORTHOGONAL.

NP complete
In 1971 computer scientist Steven Cook specified a certain class of computational problems as "equivalently difficult" in the sense that if any one of the problems in this class can be solved "in a reasonable amount of time" on a computer, then all the problems in this class can be so solved. This category of problems is called NP complete.

The famous TRAVELING-SALESMAN PROBLEM is one such problem. It seeks to find the shortest route that visits a number of cities. It is known that if n represents the number of cities in the problem, then, as n increases, the number of possible routes to check grows as a FACTORIAL function in n. These numbers grow extraordinarily fast, faster than any POLYNOMIAL function in n. (A polynomial is a formula of the form $a_r n^r + a_{r-1} n^{r-1} + \ldots + a_1 n + a_0$.) Problems that grow in complexity as a polynomial are considered "solvable in a reasonable amount of time." It is not known whether there is a way to solve the traveling-salesman problem in polynomial time. If it can be so solved, then, as Steven Cook showed, each and every NP problem can also be solved in polynomial time. There are many different problems in the NP class. The fact that no one to this day has found a "fast" algorithm for solving any one of them suggests that the traveling salesman problem, in particular, has no computationally feasible means of solution. The letters NP stand for "nondeterministic polynomial time."

A different issue asks whether a proposed solution to a problem can be checked to be valid within a polynomial amount of time. This leads to the class of "P" problems. It is not known whether the class of P problems (those that take a polynomial amount of time to check solutions) is the same as the class of NP problems (those that are hoped to take a polynomial amount of time to find solutions). This question is usually written: P = NP?

See also POLYNOMIAL TIME.

nth root of unity Any COMPLEX NUMBER z such that $z^n = 1$ is called an *n*th root of unity. When n is 2 or 3, the relevant roots are usually called square roots and cube roots, respectively. There are two square roots of unity, namely, $z = 1$ and $z = -1$, and there are three cube roots of unity: $z = 1$, $z = \frac{-1+i\sqrt{3}}{2}$, and $z = \frac{-1-i\sqrt{3}}{2}$. In general, as the FUNDAMENTAL THEOREM OF ALGEBRA shows, there are n *n*th roots of unity. The number $z = 1$ is an *n*th root of unity for all values of n.

EULER'S THEOREM shows that the *n*th roots of unity are given by the formula:

$$z = e^{i\frac{2\pi k}{n}} = \cos\left(\frac{2\pi}{n}k\right) + i\sin\left(\frac{2\pi}{n}k\right)$$

for $k = 0,1,\ldots,n$ (since $z^n = \left(e^{i\frac{2\pi k}{n}}\right)^n = e^{i2\pi k} = \cos(2\pi k) + i\sin(2\pi k) = 1 + 0 = 1$). This shows that the n *n*th roots of unity all lie on a circle of radius one and are equidistant along that circle at angles that are multiples of $\frac{2\pi}{n}$. That is, when plotted on the complex plane, the n *n*th roots of unity lie at the vertices of a regular *n*gon with the point $z = 1$ on the real axis as one vertex of the polygon.

An *n*th root of unity z is called a primitive if $z^n = 1$, but $z^k \neq 1$ for any positive integer smaller than n. For example, although $z = -1$ is a fourth root of unity, it is not a primitive root, since we also have $(-1)^2 = 1$. The numbers $z = i$ and $z = -i$ are both primitive fourth roots of unity.

An *n*th root of unity z satisfies the equation $z^n - 1 = 0$. Factoring yields:

$$(z - 1)(1 + z + z^2 + z^3 + \ldots + z^{n-1}) = 0$$

If z is different from 1, then it must be the case that the second term of the left side of this expression is zero. This proves:

Each *n*th root of unity different from 1 satisfies $1 + z + z^2 + z^3 + \ldots + z^{n-1} = 0$.

Taking $n = 5$, for example, and EQUATING REAL AND IMAGINARY PARTS, this shows, for instance, that:

$$\cos(0) + \cos\left(\frac{2\pi}{5}\right) + \cos\left(\frac{4\pi}{5}\right) + \cos\left(\frac{6\pi}{5}\right) + \cos\left(\frac{8\pi}{5}\right) = 0$$

and

$$\sin(0) + \sin\left(\frac{2\pi}{5}\right) + \sin\left(\frac{4\pi}{5}\right) + \sin\left(\frac{6\pi}{5}\right) + \sin\left(\frac{8\pi}{5}\right) = 0$$

n-tuple (list) A set of n objects taken in a particular order is called an *n*-tuple. An *n*-tuple of numbers a_1, a_2, ..., a_n, in that order, is usually denoted $(a_1,a_2,\ldots,a_n)$ or $\langle a_1,a_2,\ldots,a_n \rangle$. Two *n*-tuples $(a_1,a_2,\ldots,a_n)$ and $(b_1,b_2,\ldots,b_n)$ are equal if, and only if, corresponding entries match: $a_1 = b_1$, $a_2 = b_2$,..., $a_n = b_n$. Thus, for instance, (2,1.3) and (1,2,3) are distinct 3-tuples.

A 3-tuple is called a "triple," and a 4-tuple a "quadruple." When $n = 2$, an *n*-tuple is called an ordered pair.

An *n*-tuple of numbers can represent a point in *n*-dimensional space, a finite SEQUENCE, or, if angled brackets are used, an *n*-dimensional VECTOR. An ordered PARTITION of a number N is an *n*-tuple of numbers whose entries sum to N.

See also ORDERED SET.

number The development of different types of numbers can be seen as motivated by the need for solving different types of equations. For example, the counting numbers (that is, the NATURAL NUMBERS **N**) suffice for solving any equation of the type $x + 2 = 5$, for instance, but not an equation of the type $x + 5 = 2$. (There is no solution to this equation within the set of counting numbers.) This motivates the introduction of NEGATIVE NUMBERS and the construction of the INTEGERS **Z**. But this set is not always sufficient for solving equations of the type $5x = 3$, for instance. Desiring solutions to equations of this type leads to the construction of FRACTIONs and the set of all RATIONAL NUMBERS **Q**. Unfortunately, again, not all equations can be solved within this system. For example, the equation $x^2 - 2 = 0$ has no rational solution. Extending the set of rational numbers to include solutions to equations of this type introduces IRRATIONAL NUMBERS and the construction of the REAL NUMBER system **R**. Yet this new system also does not suffice for solving all equations. With the introduction of a single additional number, denoted i, to represent an ("imaginary") solution to the equation $x^2 + 1 = 0$, the COMPLEX NUMBERS **C** are born. Surprisingly, as shown by the FUNDAMENTAL THEOREM OF ALGEBRA, the introduction of this single number is all that is needed to solve *any* POLYNOMIAL equation $a_n x^n + \ldots + a_1 x + a_0 = 0$. Thus the complex numbers represent a system of numbers that is algebraically closed in the sense that the construction of no new type of number is needed to solve arithmetic equations.

On a conceptual level, the notion of "number" is intimately connected with the act of counting. Simple counting systems of ancient times used tally marks to record numbers, and over the millennia this basic numeration scheme evolved to the sophisticated PLACE-VALUE SYSTEM we use today. (The ancient Egyptians of around 3000 B.C.E. were perhaps the first to move from the use of tally marks alone.) It was a great intellectual achievement for mankind when the notion of "number" was removed from the specific objects being counted, recognizing, for instance, that two cows, two houses, and two days all share a common property of "two-ness." (Even today we sometimes use different words to count different types of "two." For instance, the words *twins, couple,* and *pair* cannot be used interchangeably to represent two people.) This simple recognition of an abstract commonality between sets of objects was exploited by German mathematician GEORG CANTOR 1845–1918 who, in the late 1800s, developed a general notion of CARDINALITY. With it, Cantor extended the notion of "number" to include counts of sets of infinite size. He established, for instance, that there are an infinite number of different types of infinity and managed to develop a meaningful system of arithmetic for his transfinite numbers.

Irish mathematician SIR WILLIAM ROWAN HAMILTON (1805–65) followed a different route and worked to extend the notion of "number" to represent operations on n-dimensional space. AN ARGAND DIAGRAM shows that the complex numbers have a natural representation as points on a plane. Hamilton sought to give meaning to an arithmetic for points in three- and higher-dimensional space. Although he did not succeed in accomplishing this goal for three-dimensional space, his invention of the QUATERNIONS shows this feat can be done in four-dimensional space. (The octonions provide an arithmetic for eight-dimensional space.)

The following diagram illustrates the relationship between the number systems described:

$$\mathbf{N} \subset \mathbf{Z} \subset \mathbf{Q} \subset \mathbf{R} \subset \mathbf{C} \subset \text{quaternions}$$

See also ARABIC MATHEMATICS; BABYLONIAN MATHEMATICS; BASE OF A NUMBER SYSTEM; CHINESE MATHEMATICS; DECIMAL REPRESENTATION; EGYPTIAN MATHEMATICS; GREEK MATHEMATICS; HINDU-ARABIC NUMERALS; INDIAN MATHEMATICS; MAYAN MATHEMATICS; ROMAN NUMERALS; ZERO.

number line (real line) A straight line, usually horizontal, for which each point on the line represents a REAL NUMBER is called a number line. One assumes that the line extends indefinitely both to the left and to the right. A single point O on the line, called the origin, corresponds to the number ZERO in the real number system, and it is conventional to assume that a point a distance a units to the right of O represents the positive real number a and a point b units to the left of O the negative real number $-b$. The integers are thus represented as evenly spaced points, one unit apart, along the line. A number line is a one-dimensional CARTESIAN COORDINATE system.

The theory of CARDINALITY shows that there are just as many points on the number line as there are points in a two-dimensional plane. The DIAGONAL ARGUMENT shows that the set of RATIONAL NUMBERS (fractions) take up absolutely no space on the number line.

See also DIMENSION.

number systems *See* BASE OF A NUMBER SYSTEM.

number theory (higher arithmetic) The study of the arithmetic properties of numbers is called number theory. The fact that many simple statements about numbers can be extraordinarily difficult to prove, if at all possible, makes this topic an alluring and stimulating subject for mathematicians. (GOLDBACH'S CONJECTURE, for instance, remains unsolved.) CARL FRIEDRICH GAUSS (1777–1855), charmed by the subject and its "inexhaustible wealth," called number theory the "queen of mathematics."

Elementary number theory is the study of those topics in number theory that utilize only the basic techniques of ARITHMETIC and high-school mathematics in their solutions. For example, the classification of the PYTHAGOREAN TRIPLES would be considered a problem in elementary number theory, as would the solution of many DIOPHANTINE EQUATIONS. (The use of the word *elementary* here by no means implies that the level of mathematical sophistication used is elementary.) ANALYTIC NUMBER THEORY incorporates the notion of LIMIT in the study of numbers, and algebraic number theory extends the study of number theory to a general study of ALGEBRAIC NUMBERS and new number systems that include solutions to otherwise unsolvable algebraic equations.

See also ABSTRACT ALGEBRA; CATALAN CONJECTURE; COLLATZ'S CONJECTURE; EUCLID'S PROOF OF THE INFINITUDE OF PRIMES; EUCLIDEAN ALGORITHM; FUNDAMENTAL THEOREM OF ARITHMETIC; PEANO'S POSTULATES; PRIME; PRIME-NUMBER THEOREM.

numerical differentiation The DERIVATIVE of a function $f(x)$ can be well approximated as a "Newton quotient":

$$f'(x) \approx \frac{f(x+h) - f(x)}{h}$$

at least for small values of h. Any use of this formula to approximate the value of a derivative is called numerical differentiation. For example, we can approximate the derivative of $f(x) = x^2$ at $x = 7$ simply as $f'(7) \approx \frac{(7 + 0.1)^2 - 7^2}{0.1} = 14.1$.

Rewriting the formula for the Newton quotient gives:

$$f(x + h) \approx f(x) + hf'(x)$$

If the derivative of the function is known, then this formula can be used to approximate values of f. For example, to estimate square roots, set $f(x) = \sqrt{x}$ to obtain:

$$\sqrt{x + h} \approx f(x) + f'(x)h = \sqrt{x} + \frac{h}{2\sqrt{x}}$$

Thus $\sqrt{38}$, for example, is approximately $\sqrt{36} + \frac{2}{2\sqrt{36}} = 6 + \frac{1}{6} \approx 6.167$.

The second derivative of a function is well approximated by the quotient:

$$f''(x) \approx \frac{f(x+h) - 2f(x) + f(x-h)}{h^2}$$

This follows using the approximation $f''(x) \approx \frac{f'(x+h) - f'(x)}{h}$, with $f'(x+h) \approx \frac{f(x+h) - f(x)}{h}$ and $f'(x) \approx \frac{f(x) - f(x-h)}{h}$.

See also NEWTON'S METHOD.

numerical integration According to the theory of INTEGRAL CALCULUS, the numerical value of a definite integral $\int_a^b f(x)dx$ is determined by finding an antiderivative $F(x)$ to the integrand $f(x)$ and then computing the quantity $F(b) - F(a)$. Although theoretically sound, it is rare in real-world applications that such a procedure can ever be completed. There are two possible complications:

1. An antiderivative to the integrand cannot be found. (Consider the integral $\int_1^2 \frac{e^x}{x} dx$, for instance.)

2. The function $f(x)$ might not be completely specified. (In performing an experiment, one can only ever record a finite number of data values, in which case the values of a function $f(x)$ are known only at a finite number of points.)

Nonetheless, despite these limitations, scientists and engineers often still require a numerical value for the area under the curve $y = f(x)$, at least to some specified degree of accuracy. Numerical integration is any technique that allows one to find an approximate value for a definite integral $\int_a^b f(x)dx$. There are two elementary methods currently in use:

1. *Trapezoidal Rule (also known as the trapezium rule):* Divide the interval $[a,b]$ into $n + 1$ equally spaced points $a = x_0, x_1, \ldots x_{n-1}, x_n = b$. For convenience denote $f(x_i)$ by f_i and let P_i denote the point (x_i, f_i) on the curve above $x = x_i$. The straight-line segment connecting P_i to P_{i+1} can be used as an approximation for the curve $y = f(x)$ between and x_i and x_{i+1}. The area under this part of the curve is thus approximately the area of a trapezoid of width $h = \frac{b-a}{n}$, left edge of height f_i and right edge of height f_{i+1}. This area is given by: $\frac{1}{2}h(f_i + f_{i+1})$.

Adding up the areas of all such trapezoids between a and b gives the trapezoid rule:

$$\int_a^b f(x)\,dx \approx \frac{1}{2}h\big(f_0 + 2f_1 + 2f_2 + \cdots + 2f_{n-1} + f_n\big)$$

Mathematicians have shown that the error in this approximation is close to $\frac{C}{n^2}$, for some constant C. Thus doubling the number of points used to make the approximation increases the accuracy of the result by a factor of 4.

To illustrate the trapezoid rule, we estimate $\int_0^1 \sqrt{1-x}\,dx$ using $n = 4$. Here $f(x) = \sqrt{1-x}$ and $x_0 = 0$, $x_1 = 0.25$, $x_2 = 0.5$, $x_3 = 0.75$, $x_4 = 1$ with $h = 0.25$. We have:

$$f_0 = 1$$
$$f_1 = \sqrt{0.75} = 0.866$$
$$f_2 = \sqrt{0.5} = 0.707$$
$$f_3 = \sqrt{0.25} = 0.500$$
$$f_4 = 0$$

Consequently:

$$\int_0^1 \sqrt{1-x}\,dx \approx \frac{1}{2} \times 0.25 \times (1 + 2 \times 0.866 + 2 \times 0.707$$
$$+ 2 \times 0.500 + 0$$
$$= 0.643$$

(The true value of the integral is 2/3.)

2. *Simpson's Rule:* While the trapezoidal rule uses straight-line segments to approximate the curve, Simpson's rule uses the arcs of parabolas through three points at a time: one through the points P_0, P_1, and P_2, the next through P_2, P_3, and P_4, and so on. (It is assumed that n is even for this method.) Writ-

ing the equations for each of these parabolic arcs and summing the areas under each gives, after some work, the rule:

$$\int_a^b f(x)\,dx \approx \frac{1}{3}h(f_0 + 4f_1 + 2f_2 + 4f_3 + 2f_4 + \cdots$$
$$+ 2f_{n-2} + 4f_{n-1} + f_n)$$

Mathematicians have shown that the error in this approximation is close to $\frac{C}{n^4}$ for some constant C. Thus, doubling the number of points used to make the approximation increases the accuracy of the result by a factor of 16.

To illustrate Simpson's rule, we again estimate $\int_0^1 \sqrt{1-x}\,dx$ using $n = 4$. We have:

$$\int_0^1 \sqrt{1-x}\,dx \approx \frac{1}{3} \times 0.25 \times (1 + 4 \times 0.866 + 2 \times 0.707$$
$$+ 4 \times 0.500 + 0)$$
$$= 0.657$$

If n is a multiple of 3, one can use arcs of cubic curves to establish the rule:

$$\int_a^b f(x)\,dx \approx \frac{3}{8}h(f_0 + 3f_1 + 3f_2 + 2f_3 + 3f_4$$
$$+ 3f_5 + 2f_6 + \cdots + f_n)$$

This is sometimes called Simpson's 3/8-rule.

These methods of approximation are incorrectly attributed to English mathematics teacher THOMAS SIMPSON (1710–61).

See also MONTE CARLO METHOD; NUMERICAL DIFFERENTIATION.

O

oblate/prolate A curved surface similar to a sphere but lengthened or shortened in one direction is called a spheroid. If the length of the diameter from pole to pole is greater than the length of the diameter connecting two opposite points on the equator, then the spheroid is said to be oblate. If, on the other hand, the polar diameter is less than the equatorial diameter, then the spheroid is called prolate. The Earth, for example, is not a perfect sphere but is an oblate spheroid.

See also EARTH; ELLIPSOID; SPHERE.

oblique This term is used in a number of geometric settings to mean either "not at right angles" or "does not contain a right angle." For example, two intersecting lines drawn in the plane are oblique if they meet at an ANGLE different from 90°, or three lines meeting at a point in three-dimensional space are oblique if they are not mutually perpendicular. A single line drawn in the plane is called an oblique line if it is neither horizontal nor vertical, and an oblique coordinate system has axes that are not at right angles.

Any angle that is not a multiple of 90° is called an oblique angle. An oblique triangle is one that does not contain a right angle.

An oblique cone is a cone with its vertex not directly above the center of its base (and so the line connecting the vertex to the center of the base is not at right angles to the base). An oblique prism has lateral edges that are not perpendicular to the base.

See also CARTESIAN COORDINATES; RIGHT ANGLE; SKEW LINES.

obtuse angle An ANGLE between 90° and 180° is called an obtuse angle. A TRIANGLE in which one of its angles is obtuse is called an obtuse triangle. According to the LAW OF COSINES, a triangle with side-lengths a, b, and c, and corresponding angles A, B, C opposite those sides, satisfies:

$$\cos C = \frac{a^2 + b^2 - c^2}{2ab}$$

For the angle C to be obtuse, it must be the case that $\cos C < 0$, that is, $a^2 + b^2 < c^2$. Thus a triangle a, b, c is obtuse if one of the following inequalities holds: $a^2 + b^2 < c^2$, $c^2 + a^2 < b^2$ or $b^2 + c^2 < a^2$.

An angle between 180° and 360° is called a reflex angle. An angle of 180° degrees is a straight angle, and one of 360° is called a full turn or a PERIGON.

See also ACUTE ANGLE; PYTHAGORAS'S THEOREM; TRIANGLE.

obverse Changing the predicate B of a statement of the form "all A are B" from positive to negative, or vice verse, and negating the statement as a whole produces the obverse of the statement: "no A is not B." For example, the obverse of the statement "all men are mortal" is "no man is immortal." Euler diagrams show that any statement of the this type is logically equivalent to its obverse.

See also ARGUMENT.

362

octant   The *xy-*, *yz-*, and *xz*-planes in a three-dimensional CARTESIAN COORDINATE system divide space into eight regions called "octants." The set of points (x,y,z) satisfying $x > 0$, $y > 0$, and $z > 0$ constitute the first or "positive" octant. The remaining octants are labeled the second, third, and fourth octants counterclockwise around the positive *z*-axis, with the fifth, sixth, seventh, and eighth octants underneath these four, again in a counterclockwise sense, with the fifth octant directly under the first.

The two-dimensional analog of an octant is a QUADRANT. For *n* greater than three, the *n*-dimensional analog is called an orthant. Points in the positive orthant have all coordinates positive. Popular science fiction novels often describe space as divided into quadrants. This is the incorrect term for three-dimensional space.

An ANGLE of 45° is also sometimes called an octant.

odds   If in a game of chance there are *a* outcomes that are deemed favorable (thereby constituting a desired EVENT E), and *b* outcomes that are unfavorable (not in E), then the odds in favor of E is the ratio $\frac{a}{b}$, usually written *a:b* or "*a* to *b*", and the odds against E is the ratio $\frac{b}{a}$, written *b:a* or "*b* to *a*."

For example, in casting a die, the odds in favor of rolling a 2 are 1:5. There is just one favorable outcome, namely a 2, and five unfavorable outcomes: {1,3,4,5,6}. Odds are usually written in reduced form. For example, if the odds in favor of an event are 12 to 4, one typically writes 3:1 rather than 12:4.

Odds are closely connected to PROBABILITY computations. If the odds in favor of an event E are *a* to *b*, then one is informed that *a* of a total of *a* + *b* possible outcomes belong to the set E. Thus the probability of event E occurring is given by $P(E) = \frac{a}{a+b}$, and that of it not occurring by $P(not\ E) = \frac{b}{a+b}$. Conversely, if the probability of an event E is known, then the odds in favor of E can be computed as a ratio of probabilities:
$$\frac{P(E)}{P(not\ E)} = \frac{\frac{a}{a+b}}{\frac{b}{a+b}} = \frac{a}{b}$$, and the odds against as the inverse ratio of probabilities.

Odds are typically used at horse races—usually "odds against"—to tell gamblers the payoffs on various bets. For example, advertised odds of 5 to 3 on a particular horse indicates that a $3 bet on that horse will win $5 (plus the return of the original three dollars) if that horse wins the race. It also tells the gambler that this horse is believed to have only a 3/8 chance of winning.

Omar Khayyám (**Umar al-Khāyammī**) (1048–1122) Arab *Algebra, Astronomy*   Born on May 18, 1048, Persian scholar Omar Khayyám is remembered in mathematics for his significant contributions to the advancement of ALGEBRA. In his famous text, *Treatise on Demonstration of Problems of Algebra,* Khayyám developed both geometric and algebraic rules for solving QUADRATIC and CUBIC EQUATIONS, making clever use of CONIC SECTIONS. He was the first mathematician to ever conceive of a general theory for solving cubics and was the first to recognize that ruler-and-compass constructions alone would never suffice as a geometric approach for this goal. Outside of mathematics, Omar Khayyám is best known for his poems, which were freely translated by Edward Fitzgerald in 1859 in the text *The Rubáiyát of Omar Khayyám.*

Khayyám studied philosophy and mathematics at his birthplace of Nishapur, Persia (now Iran), and quickly became an expert scholar. By age 25 he had written three significant texts: one on music, one on arithmetic, and one a first text on the topic of algebra. In 1073 Khayyám moved to the city of Esfahān to help with the construction of an observatory and to head a team of scientists studying astronomy. Over the course of his 18 years at the observatory Khayyám completed extraordinarily accurate tables of astronomical data and tables of trigonometric values. He also computed the length of the year to be 365.24219858156 days, which, at the time, was correct to the sixth decimal place. (The length of a year is changing over time. Today its value is 365.242190 days. The value of the sixth decimal place changes over the course of a century.)

Later in life Khayyám moved to the cultural center of Merv (now Mary in Turkmenistan) and returned to his interests in algebra. He wrote his famous treatise on the topic while there. Khayyám noted that the great Greek scholars of antiquity made no serious study of the theory of cubic equations and decided to take it upon himself to develop this work appropriately. Khayyám created ingenious geometric methods for finding cubic equations, but never, as he hoped he would, found workable algebraic techniques. Nonetheless, Khayyám did manage to classify all cubic equations and was the first to recognize that such equations might possess two different solutions. (It is not clear whether or not Khayyám understood that cubics might, in fact, possess three distinct solutions.)

Much of Khayyám's work on NUMBER THEORY and methods of numerical computation has been lost. His development of PASCAL'S TRIANGLE, as it is called today, and its use in the BINOMIAL THEOREM, for instance, are only mentioned in passing in his famous algebraic work. Khayyám also wrote about the founding principles of geometry. He tried to establish that the famous PARALLEL POSTULATE is a consequence of other postulates (failing, of course), but did discover along the way a number of significant results about non-Euclidean figures. He is best remembered for his contributions to algebra.

Khayyám died in the city of Nishapur, Persia (now Iran), ca. December 4, 1122.

operation Any mechanistic procedure on the elements of a set that produces a unique result for those elements is called an operation on that set. For example, addition is an operation on the set of integers: for any collection of integers there is a unique value called their sum. The act of finding the union of two sets is an operation on the collection of all sets.

A UNARY OPERATION is a rule that associates a result with each element of a set S. The act of squaring a number, for instance, is a unary operation on the set of real numbers. A BINARY OPERATION provides a result for every two elements of a set S. The addition of two integers is a binary operation on the set of all integers.

In ARITHMETIC, addition, subtraction, multiplication, division, and the extraction of square roots are called elementary operations. On the other hand, the rule that associates with each natural number the sum of its digits, for instance, is not elementary.

A symbol used to denote an operation is sometimes called an operator. For instance, $+, -, \times, \div, \sqrt{}$, and $\log_{10}$ are operators.

operations research (**operational research, OR**) The study of the role of mathematics and STATISTICS in solving problems that arise in business, commerce, and the production of goods and services is called operations research. Often an OPTIMIZATION problem will arise in

any attempt to minimize costs or to maximize profits, and techniques of LINEAR PROGRAMMING may be required. The problem of scheduling interrelated tasks can be analyzed via GRAPH THEORY, and the search for a CRITICAL PATH in a schematic diagram can lead to improved business practices. The TRAVELING-SALESMAN PROBLEM also illustrates the role of graph theory in determining the efficient delivery of goods. Problems about production quality and reliability rely heavily on techniques of statistical inference, and the mathematics of GAME THEORY has helped businesses devise market strategies and mutually beneficial trade practices.

opposite A side of a TRIANGLE is said to be opposite a given angle in the triangle if that side is not one of the arms forming the angle. For example, the side opposite the largest angle in a triangle is the longest side of the triangle. (This follows from the LAW OF COSINES.)

Two lines intersecting at a point (often called the vertex) form four angles. Any pair of angles sharing this vertex, but having no arm in common, is called a pair of opposite or vertical angles. They have the same measure.

If a geometric figure has a center of symmetry, then two sides or two angles are said to be opposite if they are joined by a line through this center.

optimization The process of finding the best possible solution to a problem is called optimization. In

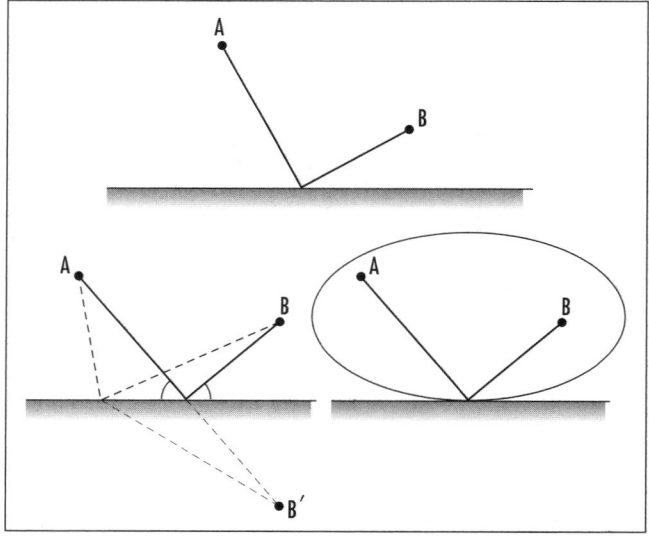

Solving the "shortest path" problem

mathematics this usually requires finding the maximum or minimum value of a function, perhaps subject to some constraints. The techniques of CALCULUS are immensely successful in achieving such goals. (See MAXIMUM/MINIMUM.)

Some optimization problems can be solved geometrically. Consider, for instance, the challenge of finding the shortest path from a point A to a point B that "visits" a straight wall. Noticing that it suffices to consider only paths composed of two straight-line segments, we ask which such pair of segments yields the shortest journey.

Notice that each path from A to B via the wall is matched by a path of the same length from A to the same point of contact on the wall, and then to the mirror image B' of B on the other side of the wall. As the straight path connecting A to B' (making equal angles on either side of the wall) is the shortest route from A to B', it follows that the shortest route from A to B via the wall is the one that bounces off the wall at equal angles. Alternatively, one can solve this problem by drawing ELLIPSES about the points A and B with A and B as foci. The first ellipse that touches the wall gives the location on the wall that yields the shortest path from A to B via the wall.

That these two solutions solve the same problem establishes the well-known reflection property of an ellipse: as any path from one focus A of an ellipse to a point on the curve of the ellipse and back to the second focus B is a solution to the path-walking problem (with the straight wall being the tangent line to the ellipse), it follows that this path bounces off the side of the ellipse at equal angles. Consequently, a beam of light emitted from one focus A of an ellipse follows a path that reflects off the side of the ellipse so as to arrive at the second focus B.

SNELL'S LAW of refraction can also be viewed as the solution to an optimization problem.

See also ISOSCELES TRIANGLE; LINEAR PROGRAMMING; OPERATIONS RESEARCH; PEDAL TRIANGLE; STEINER POINT.

orbit   *See* DYNAMICAL SYSTEM.

ordered set   A set S is said to be partially ordered if it comes equipped with a relation, usually denoted ≤, that allows one to compare the size or the relative positions of elements in the set. The relation ≤ must satisfy the following three conditions:

1. *Reflexivity:* $a \leq a$ for all $a \in S$.
2. *Antisymmetry:* If $a \leq b$ and $b \leq a$, then $a = b$.
3. *Weak Transitivity:* If $a \leq b$ and $b \leq c$, then $a \leq c$.

For example, interpreting ≤ as "less than or equal to" provides a partial order on the set of all real numbers. For instance, $4.6 \leq \sqrt{30}$ and $-1 \leq 0$. The set of all subsets of the set {A,B,C,D,E}, for example, is partially ordered if one interprets ≤ to mean "is a subset of." In this case, for instance, {B,D} ≤ {A,B,D,} and Ø ≤ {C}, but {A,B,C} ≰ {B,C,D,E}. The set of natural numbers may also be partially ordered by interpreting ≤ as "is a factor of." For example, in this setting we have $3 \leq 6$ and $4 \leq 12$, but $5 \nleq 12$. Notice in these last two examples that not all elements in the set can be compared.

A set is called totally ordered if a fourth condition holds:

4. *Trichotomy:* For all $a,b \in S$ either $a \leq b$ or $b \leq a$ holds. (If both hold, then $a = b$.)

The set of all real numbers, or the set of all natural numbers, for example, are both totally ordered under the "less than or equal to" relation. The natural numbers, however, are not totally ordered under the "is a factor of" relation. Neither is the set of all subsets of a given set.

Sometimes a totally ordered set is called a chain or a sequence, since all elements can be arranged on a line, with each element on the line in relation ≤ to each element to its right.

Often it is convenient to write $b \geq a$ to mean $a \leq b$ and $a < b$ to mean $a \leq b$ and $a \neq b$. The statement $b > a$ is interpreted similarly. For instance, in the example of subsets, we have {A} < {A,B} and {C,D} ≥ {D,C}. (In fact, {C,D} = {D,C}.)

Given two elements a and b of a set S with partial order relation ≤, we say that an element u of S is an upper bound for a and b if $a \leq u$ and $b \leq u$. We call u a least upper bound if u is smaller than any other upper bound for a and b; that is, if $a \leq v$ and $b \leq v$ for some other element v, then $u \leq v$. For example, in the example of subsets, the least upper bound of {A,B} and {B,C} exists and equals their union {A,B,C}. Similarly, an element l is a lower bound for a and b if $l \leq a$ and $l \leq b$, and it is a greatest lower bound if l is larger than any other lower bound for a and b. In our example, the greatest lower bound of $\{A,B\}$ and $\{B,C\}$ is their intersection $\{B\}$. A partially ordered set is called a lattice if every pair of elements has a least upper bound and a greatest lower bound. The existence of unions and intersections shows that the set of all subsets of any given set is a lattice. The set of natural numbers under the relation "is a factor of" is also a lattice: the least upper bound of any two natural numbers is their lowest COMMON MULTIPLE, and the greatest lower bound is their GREATEST COMMON FACTOR. Many results from NUMBER THEORY can be interpreted as statements about this order relation of the natural numbers.

Any totally ordered set is a lattice. For example, the least upper bound of any two real numbers a and b is simply the one that is the larger of the two, and their greatest lower bound is the smaller number.

See also BOUND; ORDER PROPERTIES.

order of a matrix (**dimension of a matrix**)   An $m \times n$ MATRIX, that is, a matrix with m rows and n columns, is said to be of order $m \times n$ (read as "m by n"). An $n \times n$ matrix is sometimes called a square matrix of order n.

In GROUP THEORY an element g of a GROUP is said to be of order n if n is the first positive integer such that $g^n = e$, assuming there is such an integer. (Here e is the IDENTITY ELEMENT of the group.) If, in some mathematical work, one is thinking of matrices as elements of a group, one usually reserves the word *order* for a group theoretic meaning, and uses the word *dimension* to describe the size of the matrix. For example, the matrix

$$A = \begin{pmatrix} 0 & 1 & 0 \\ 1 & 0 & 0 \\ 0 & 0 & 1 \end{pmatrix}$$

is a square matrix of dimension 3 and order 2, since $A^2 = I$.

See also IDENTITY MATRIX.

order of magnitude   *See* SCIENTIFIC NOTATION.

order of operation (**operational precedence**)   In evaluating arithmetic computations involving more than one type of operation, mathematicians have assigned an order of precedence as to which operations are exercised first.

It is agreed that any computation involving ADDITION alone is computed in the order as read from left to right. For instance, $8 + 5 + 2$ is computed as $13 + 2$, which is 15. (Although the ASSOCIATIVE property of addition shows that the order of computation in this case does not matter.) As SUBTRACTION can be viewed as the addition of negative quantities, any computation involving both addition and subtraction is thus computed in the same manner, as read from left to right. For instance, $2 - 5 + 7$ is computed as $(-3) + 7$, which is 4. (Again, the associative property shows that following this convention is not vital.)

MULTIPLICATION can be viewed as "repeated addition" and so, in some sense, is a more potent operation than addition and subtraction. It is given precedence over these operations. DIVISION, which can be viewed as multiplication by RECIPROCAL quantities, is given the same status. Thus given any computation involving all four operations, one is expected to compute all multiplications and divisions that appear first (read in a left-to-right manner) and all additions and subtractions second. For instance, one computes:

$$2 + 4 \times 3 \div 6 - 3 \times 3 + 5$$

as $2 + 12 \div 6 - 9 + 5$, which equals $2 + 2 - 9 + 5$, which is zero. (Reading strictly from left to right produces the incorrect answer of 5.)

As EXPONENTs can be viewed as an act of performing repeated multiplications, all powers that appear in a computation are given greater precedence over multiplications and divisions, and so must be computed first. For instance, $2 + 6^2 \div 9$ is computed as $2 + 36 \div 9 = 2 + 4 = 6$.

Often parentheses or BRACKETS are introduced to change the order of operations in a computation. Mathematicians follow the convention that if parentheses are present, one must compute the quantities inside the parentheses first (using the above rules). If multiple sets of parentheses are present, this requires evaluating the innermost parentheses first. For instance, we compute:

$$2 \times (3 + (3^2 + 6) \times 2) + 1$$

as

$$2 \times (3 + (3^2 + 6) \times 2) + 1 = 2 \times (3 + (9 + 6) \times 2) + 1$$
$$= 2 \times (3 + 15 \times 2) + 1$$
$$= 2 \times (3 + 30) + 1$$
$$= 2 \times 33 + 1$$
$$= 66 + 1$$
$$= 67$$

Often schoolchildren are taught a mnemonic device such as "Please Excuse My Dear Aunt Sally" to recall the order of operations: parentheses, exponents, multiplication, division, addition, subtraction.

See also EXPANDING BRACKETS.

order properties   The REAL NUMBERS satisfy a number of basic properties with respect to the order relation <, meaning strictly less than. We have:

1. *Trichotomy Law:* For any two real numbers a and b, exactly one of the following holds: $a < b$, $b < a$, or $a = b$
2. *Transitive Law:* If $a < b$ and $b < c$, then $a < c$.
3. *Addition Law:* If $a < b$, then $a + c < b + c$ for real number c.
4. *Multiplication Law:* If $a < b$ and $c > 0$, then $ac < bc$.
5. *Completeness Law:* Any set of real numbers that is bounded above has a least upper BOUND.

The first two properties are standard features of an ORDERED SET, and the next two properties explain the extent to which addition and multiplication respect the order relation. The final statement is a key property of the real number system. In their attempts to make the theory of CALCULUS mathematically rigorous and precise, mathematicians came to realize that one had to be sure that no numbers are missing from the real number line. (The INTERMEDIATE-VALUE THEOREM, for instance, relies on this.) The fifth law above is designed to ensure this. With his construction of a DEDEKIND CUT, German mathematician JULIUS WILHELM RICHARD DEDEKIND (1831–1916) was able to prove that the real-number system does indeed satisfy this fifth property. (The set of rational numbers, on the other hand, satisfies the first four properties but not the fifth, and so is not complete. Although, for example, every rational number x that satisfies $x^2 \leq 2$ is smaller than 3, there is no smallest rational number that provides a bound for all rationals of this type.)

From the five basic properties listed above, other familiar properties of the real numbers follow. For instance, one can show that if $a < b$, then $-a > -b$. (Use property 3 adding $-a$ and then $-b$ to both sides.) More generally, we have that if $a < b$ and $c < 0$, then $ac > bc$. From this and property 4, it follows that the square of any nonzero number is positive: $a^2 > 0$ for all a. This final observation also shows that there is no analogous order relation for the COMPLEX NUMBERS:

If the complex numbers were ordered, then we would have $i^2 > 0$, yielding the absurd statement $-1 > 0$.

It is often convenient to write $a \leq b$ to mean "less than or possibly equal to." For instance, $3 \leq 5$ and $12 \leq 12$. Often the order properties of the real numbers are phrased in terms of this version of the relation.

See also AXIOM OF CHOICE; WELL-ORDERED SET.

ordinal numbers   In common usage, the adjectives that denote the position of individual objects in a sequence, such as a "first," "second," or "107th," are called ordinal numbers. In the late 1800s German mathematician GEORG CANTOR (1845–1918) noted that the natural numbers 0, 1, 2,... satisfy the following simple order property:

If one writes down any finite number of natural numbers, then there is always a unique earliest natural number that was omitted from the list.

For example, in the list {0,1,4,7} the number 2 is the first natural number missing from this list. Let us use the notation $a,b,c,...|$ for the first natural number missing from the collection $a,b,c, ...$ For example, we have:

$$0,1,4,7 \mid = 2$$
$$0,1,2,3,4 \mid = 5$$
$$0 \mid = 1$$

Cantor considered the possibility of allowing the lists of natural numbers used to be infinite. He defined a quantity ω (omega) as:

$$\omega = 0,1,2,3,... \mid$$

that is, ω is the earliest number greater than any finite natural number. It follows then that $0,1,2,3,...,\omega| = \omega + 1$, the first number after ω, and $0,1,2,3,...,\omega, \omega + 1| = \omega + 2$, and so forth. One can continue and establish that

$$0,1,2,3,\ldots,\omega,\omega+1, \omega+2,\ldots| = \omega + \omega = \omega \times 2$$

and

$$0,1,2,3,\ldots,\omega,\omega+1,\ldots,\omega+\omega,\omega+\omega+1,\ldots| = \omega + \omega + \omega$$
$$= \omega \times 3$$

and, in continuing this process:

$$0,1,2,3,\ldots,\omega,\omega+1,\ldots$$
$$\omega+\omega,\omega+\omega+1,\ldots$$
$$\omega+\omega+\omega,\omega+\omega+\omega+1,\ldots$$
$$\vdots \qquad | = \omega + \omega + \omega + \omega + \cdots$$
$$= \omega \times \omega$$
$$= \omega^2$$

In this way, Cantor managed to develop an extraordinary new type of arithmetic that extends the set of natural numbers to a system that includes transfinite numbers. Since these quantities were derived from the order property of natural numbers, he called these new numbers ordinal numbers. Cantor developed clear and precise rules for doing arithmetic with these numbers, including adding, multiplying, and raising ordinal numbers to powers of each other. Cantor noted that the ordinal number $\omega^{\omega^{\omega^{\cdots}}}$ is the first ordinal number that cannot be obtained from earlier ordinal numbers by a finite number of additions, multiplications, and exponentiations. He called this number ε_0.

Following the approach taken in the study of CARDINALITY, Cantor also showed that the ordinal numbers can alternatively be constructed as follows. First deem two sets, each of whose elements are ordered, to be of the same ordinal number if there is a one-to-one correspondence between the elements of the sets that preserves the order of those elements. (For example, the set of negative whole numbers and the set of positive whole numbers have the same ordinal number via the correspondence $n \leftrightarrow -n$.) Then ω is defined to be the ordinal number of the set of natural numbers.

Oresme, Nicole (ca. 1323–1382) French *Coordinate geometry* Born ca. 1323 in Allemagne, France, (the exact birth date is not known) medieval scholar Nicole Oresme is best remembered in mathematics for studies of motion, and as the first scholar to depict a relationship between two variables as a GRAPH OF A FUNCTION.

Oresme studied ARISTOTLE's theory of motion, kinematics, at the University of Paris under the guidance of philosopher and logician Jean Buridan. He received an arts degree in the early 1340s and later went on to obtain a master's degree in theology at the same institution in 1355. Although Oresme followed a career path dedicated to work in a Catholic order (he was later appointed bishop of Lisieux), Oresme continued to pursue an active interest in the work of Aristotle throughout his life and published important works that influenced scholarly thinking on this subject. Oresme found philosophical difficulties with Aristotle's proposed definitions of time and space, which themselves were dependent on the notion of movement, and proposed alternative definitions independent of this concept.

Perhaps Oresme's most important work is *De configurationibus qualitatum et motuum* (The geometry of qualities and motion), in which he describes, for the first time, a general procedure for representing relationships between variables pictorially. Moreover, he realized that the area under the graph of a uniformly varying quantity represents the total change of the quantity. Oresme had consequently invented a type of coordinate geometry and made first steps toward work in integral CALCULUS. (In this same work, Oresme also proved that the distance traveled over a fixed time by an object moving with constant acceleration is the same as for an object moving at uniform velocity equal to the speed of the first object at the midpoint of the time period. This is a remarkable achievement given that the tools and techniques of calculus were not available to him at the time.)

Oresme also studied infinite SERIES, often using ingenious graphical tricks to establish results. The standard proof of the divergence of the HARMONIC SERIES (by grouping the terms of the series and comparing with sums of one-half) is due to him.

His studies on the motions of the planets also led him to study proportions and RATIONAL NUMBERS. Oresme was the first scholar in the history of mathematics to consider, and work with, fractional exponents. Like the scholars of antiquity Oresme sought for harmony in the universe. He proposed, for instance, that the ratio of the periods of any two heavenly bodies will always have a rational value. (This is not the case.)

Oresme died in Lisieux, France, in 1382. The exact death date is not known.

orientation In mathematics, the term *orientation* is used to refer to a sense of direction or rotation. For example, a line that is labeled with a direction is said to be oriented, and a closed loop drawn in the plane can be assigned either a clockwise or counterclockwise orientation. In two- and three-dimensional space, coordinate axes are said to be either positively or negatively oriented (and are called RIGHT-HANDED/LEFT-HANDED SYSTEMS), and a surface, such as a SPHERE for example, is said to be orientable, since at every location on the surface there is, loosely speaking, a well-defined notion of "up." (For example, as inhabitants of the surface of the Earth we define "up" to the direction pointing away from the center of the Earth.) Not all surfaces are orientable. The MÖBIUS BAND, for example, is a surface that cannot be oriented.

orthogonal (perpendicular) This term is used in any setting to describe two geometric constructs that meet at right angles. For example, two curves, or straight lines, are orthogonal if they intersect at right angles, that is, the angle between the two TANGENTs to the curves at the point of intersection is 90°. Two VECTORS are orthogonal if the angle between them is 90° (and consequently their DOT PRODUCT is zero). Two surfaces can also be said to be orthogonal. For example, a plane passing through the center of a sphere intersects the surface of the sphere orthogonally—the tangent plane to the sphere at any point of intersection is perpendicular to the given plane.

The term *orthogonal* is also used in some generalized settings. For example, two functions f and g are said to be orthogonal over the interval [a,b] if the integral $\int_a^b f(x)g(x)dx$ is zero. (The dot product $\mathbf{a} \cdot \mathbf{b}$ of two vectors $\mathbf{a} = < a_1, a_2,..., a_n >$ and $\mathbf{b} = < b_1, b_2,..., b_n >$ is the sum of the vector components multiplied together: $\mathbf{a} \cdot \mathbf{b} = a_1b_1 + a_2b_2 +...+ a_nb_n$. The above integral is a generalized sum of the components of the functions multiplied together.) The functions $\sin(x)$ and $\cos(x)$, for example, are orthogonal over the interval $[0, 2\pi]$. This is an important observation for the development of FOURIER SERIES.

A MATRIX is said to be orthogonal if its rows represent vectors that, taken any two at a time, have dot product equal to zero.

See also NORMAL TO A CURVE; NORMAL TO A PLANE; NORMAL TO A SURFACE.

Osborne's rule Mathematicians have observed that each trigonometric identity yields an identity for HYPERBOLIC FUNCTIONS if we simply:

1. Replace each trigonometric function with its hyperbolic analog.
2. Change the sign of any term involving the product of two hyperbolic sines (sinh).

This principle is called Osborne's rule. For example, from the trigonometric identity $\cos(x + y) = \cos x \cos y - \sin x \sin y$, we obtain the hyperbolic identity $\cosh(x + y) = \cosh x \cosh y + \sinh x \sinh y$. From $1 + \tan^2 x = \sec^2 x$, which is $1 + \frac{\sin^2 x}{\cos^2 x} = \frac{1}{\cos^2 x}$, follows $1 - \frac{\sinh^2 x}{\cosh^2 x} = \frac{1}{\cosh^2 x}$, or $1 - \tanh^2 x = \text{sech}^2 x$. This principle works because, by EULER'S FORMULA and the definition of the hyperbolic functions, $\cos x = \frac{e^{ix} + e^{-ix}}{2} = \cosh(ix)$ and $\sin x = \frac{e^{ix} - e^{-ix}}{2i} = -i\sinh(ix)$. Thus any identity that holds for sines and cosines will also hold for cosh and sinh, except a factor of $i^2 = -1$ will alter the sign of a product of two hyperbolic sines.

Oughtred, William (1574–1660) British *Logarithms* Born March 5, 1574, English mathematician William Oughtred is best remembered for his work in developing and designing the calculating device known as a SLIDE RULE.

At the turn of the 17th century, scientists were excited by the recent discovery of LOGARITHMs as an aid for converting tedious computations of multiplication and division into simpler operations of addition and subtraction. In 1620 English mathematician Edmund Gunter plotted a LOGARITHMIC SCALE along a 2-ft-long ruler and showed how a pair of calipers could be used to physically add and subtract lengths, and therefore provide, for the first time, a purely mechanical means of computing products and quotients. Inspired by the work of Gunter, Oughtred devised a simpler device consisting of two sliding rulers that accomplished the same feat. In 1632 he published *Circles of Proportion and the Horizontal Instrument*, a short book describing slide rules (and sundials). Oughtred's slide rule became the modern-day equivalent of today's

William Oughtred, a scholar of the 17th century, invented the modern slide rule. *(Photo courtesy of the Science Museum, London/Topham-HIP/The Image Works)*

× for multiplication and :: for proportion. In 1657 Oughtred also wrote one of the first comprehensive texts on TRIGONOMETRY, introducing concise notation for the topic for the first time. He also studied SPHERICAL GEOMETRY and astronomy.

Surprisingly, Oughtred received very little formal education in the topic of mathematics. He received a master's degree from King's College, Cambridge, in 1600, and was ordained an Episcopal minister in 1603. In 1604 he became the vicar of Shalford and, later, the rector of Albury, Surrey.

Oughtred offered mathematics instruction to a number of private students who came to his house and would live there free of charge. The famous mathematician JOHN WALLIS (1616–1703), for instance, was one of his pupils. Oughtred died in Albury, England, on June 30, 1660.

outlier An observation that is deemed unusual, and possibly erroneous because it is widely separated from the main cluster of points in the sample, is called an outlier. For example, in the set {4, 8, 3, 6, 5, 6, 903}, the observation 903 is an outlier. Outliers may be correct observations reflecting some abnormality in the system being studied, or they might be the result of an error in measurement or recording. For example, 903 could be a mistyping of 9, 3.

See also DATA; STATISTICS: DESCRIPTIVE.

pocket calculator for scientists and scholars for centuries to come.

Oughtred was also an accomplished scholar in several branches of mathematics. In 1631 he published an influential text, *Clavis mathematicae* (The key to mathematics), in which he described the HINDU-ARABIC NUMERALS and DECIMAL REPRESENTATION, and also outlined concepts in algebra. (At that time, Roman numerals were still in use.) He introduced the symbols

oval Derived from the Latin word *ovus* for "egg," the term *oval* is a generic name for any curved figure that resembles an elongated circle. There is no precise mathematical definition for this term.

An ELLIPSE is considered an oval. It is defined to be the set of points P whose sum of distances from two fixed points A and B in the plane is constant: $|AP| + |BP| = k$. The set of all points P that have constant product from two fixed points A and B, $|AP| \times |BP| = k$, also describe curves that can look like ovals. These curves were first studied by 17th-century Italian astronomer Giovanni Domenico Cassini (1625–1712) and are today called the ovals of Cassini.

P

pairwise disjoint (independent, mutually exclusive) A collection of sets *A, B, C,* … is said to be pairwise disjoint if the intersection of any two sets from the collection is empty. For example, the sets {1,2}, {3,4}, {5,6}, … are pairwise disjoint, as are the three sets:

A = the set of all even natural numbers
B = the set of all natural numbers 1 greater than a
     power of 100
C = the set of all prime numbers 1 less than a multiple
     of 4

A collection of pairwise disjoint subsets *A, B, C,* … of a larger set *S* is said to be exhaustive if every element of *S* is listed as an element of (just one) specified subset. In this case we also say that the sets *A, B, C,* … partition the set *S*. The first example presented above, for instance, is an exhaustive subset of the natural numbers and so partitions the positive whole numbers. The second example, however, does not. (The number 13, for instance, is not a member of any of the specified subsets.)

Given a partition *A, B, C,* … of a set *S*, two elements *a* and *b* of *S* are said to be equivalent, with respect to that partition, denoted *a* ~ *b*, if they belong to the same subset specified by the partition. For example, the days of the year are partitioned by seven disjoint sets given by the weekday names of the days. For instance, August 1, 1966, and June 30, 2003, are equivalent in this context since they both belong to the subset called "Monday."

The notion of equivalence satisfies three key properties:

i. *Reflexivity:* any element *a* is equivalent to itself: *a* ~ *a*
ii. *Symmetry:* if *a* is equivalent to *b*, then *b* is equivalent to *a*: $a \sim b \Rightarrow b \sim a$
iii. *Transitivity:* if *a* is equivalent to *b* and *b* is equivalent to *c*, then *a* is equivalent to *c*: $a \sim b, b \sim c \Rightarrow a \sim c$

In general, any relationship "~" defined on elements of a set *S* satisfying the three properties listed above is called an equivalence relation on *S*. For example, deeming two words of the English language to be equivalent if they each possess the same number of vowels is an equivalence relation on the set of all words. It turns out that any equivalent relation, no matter how it is defined, arises from a partition of the set on which it is based. For example, the set of all words is partitioned by the sets $W_0, W_1, W_2,$ … where W_k is the set of all words with precisely *k* vowels.

See also DAYS-OF-THE-WEEK FORMULA; SET THEORY.

Pappus of Alexandria (ca. 300–350 C.E.) Greek *Geometry* Born in Alexandria, Egypt, Pappus is considered today to be the last great geometer of antiquity to work in the Greek way of thought and scholarly tradition. He wrote commentaries on the works of EUCLID and CLAUDIUS PTOLEMY, but is most notably remembered for his treatise *Synagoge* (Collections), of which volumes III–VII of the original eight have survived intact today.

These texts contain detailed accounts of much of Greek mathematics, some of which would have otherwise been lost to us.

Extraordinarily little is known of Pappus's life. Many historians suspect that he lived in Alexandria all his life and that he may have headed his own school there.

Pappus's famous piece, *Synagoge,* was written as a general guide to all of Greek geometry, more specifically, a companion guide to read alongside the Greek original works. As such, books I and II cover basic arithmetic and methods of expressing large numbers. Book III examines the ARITHMETIC MEAN, GEOMETRIC MEAN, and HARMONIC MEAN of proportions, properties of basic geometry, and the representation of these means in geometry, as well as some consideration of the PLATONIC SOLIDS. Book IV examines special curves, including the SPIRAL OF ARCHIMEDES, and the problem of TRISECTING AN ANGLE.

In book V Pappus presents a mathematical analysis of the structure of honeycombs, showing that the hexagonal structure uses the least amount of perimeter to partition the plane. He also examines POLYHEDRA. Book VI explores the topic of astronomy, chiefly as a review of the works of Theodosius, Autolycus, Aristarchus, and Euclid on the topic as a means to correct common misinterpretations.

In book VII the following problem, today known as Pappus's problem, appears:

> Given four lines in the plane, find the locus of points P such that the proportion of the product of the distances of P from any two lines to the product of distances to the remaining two lines is constant.

Pappus shows that the resulting curve is always a CONIC SECTION. He also generalizes the problem to consider the case of greater than four lines in the plane. Much of the geometric work of the French mathematician RENÉ DESCARTES (1596–1650) was inspired by this problem.

Book VIII of *Synagoge* deals with mechanics. It begins with a written definition of the CENTER OF GRAVITY of a figure (Pappus is the only Greek scholar to have ever provided a definition of this concept) and discusses a number of geometric results that employ the notion. He also describes the principles of levers, pulleys, wedges, axels, and of the screw.

Pappus wrote *Synagoge* partly as an attempt to revive fervent interest in classical Greek mathematics and the style of mathematical research they conducted. Unfortunately, this did not occur. Pappus's work, nonetheless, greatly influenced the course of thinking in the development of projective geometry one millennium later. Many of his geometric results can be best stated and understood within this setting.

Pappus's theorems Greek mathematician PAPPUS OF ALEXANDRIA (ca. 320 C.E.) established two fundamental results about the surface area and the volume of a SOLID OF REVOLUTION.

> Consider the solid of revolution formed by revolving a curve in the plane about a line that does not intersect the curve. Then the surface area of that solid is equal to the product of the length of the curve and the distance the centroid (CENTER OF GRAVITY) of the curve traveled about the line of revolution. The volume of the solid is equal to the product of the area bounded by the curve (or the area between the curve and the line of revolution) and the distance traveled by the centroid of the curve.

These results can be proved by approximating the curve as a series of straight line segments, establishing the claim for this polygonal path, and then verifying that the result remains valid as one works with finer and finer approximations of the curve. This is a process that involves a LIMIT. It is remarkable that Pappus was able to establish these results before the advent of CALCULUS some 1,350 years later.

In the study of PROJECTIVE GEOMETRY, the following result is also attributed to Pappus:

> Consider a hexagon *ABCDEF* in the plane. This hexagon may be of any shape and may even self-intersect. Suppose the points *A, C,* and *E* lie on one straight line, and the points *B, D,* and *F* lie on another. For each of the three pairs of opposite sides of the hexagon, find the location at which these sides (if extended) intersect. Then these three points of intersection themselves lie on a straight line.

BLAISE PASCAL (1623–1662) later extended this result to prove that if a hexagon has vertices lying on any CONIC SECTION, then the three points of intersection of opposite pairs of sides lie on a straight line. (A pair of two straight lines can be thought of as a degenerate HYPERBOLA.)

parabola   As one of the three CONIC SECTIONS, a parabola is the plane curve consisting of all points P that are equally distant from a given fixed point F, and a given fixed line L. The fixed point is called the focus of the parabola, and the fixed line its directrix. A parabola also arises as the curve produced by the intersection of a plane through a right circular CONE held parallel to the slant side of the cone.

The equation of a parabola can be found by introducing a coordinate system in which the focus is the point $F = (0,a)$, for some positive number a, and the directrix is the horizontal line $y = -a$. If $P = (x,y)$ is an arbitrary point on the parabola, then the DISTANCE FORMULA describes the defining condition as $\sqrt{x^2 + (y-a)^2} = y + a$. Squaring and simplifying yields the equation:

$$y = \frac{1}{4a}x^2$$

Conversely, reversing these steps shows that any equation of the form $y = A(x - p)^2 + q$ is the equation of a parabola with focus $F = (p, q + \frac{1}{4A})$ and directrix $y = q - \frac{1}{4A}$. Thus the graph of any quadratic equation is a parabola.

The reflection property of a parabola states that any incoming ray of light perpendicular to the directrix will be reflected directly to the focus F. On the diagram above right, this means that the angles to the tangent line to the curve A and B are equal. (This can be proved with CALCULUS by noting that, at the point $P = (x,y)$, the slope of the tangent line to the parabola $y = \frac{1}{4a}x^2$ is $m_1 = \frac{x}{2a}$, whereas the slope of the line connecting the point Q to F, is $m_2 = -\frac{2a}{x}$. Since $m_1 m_2 = -1$, these lines are perpendicular. This shows that the tangent line bisects the isosceles triangle FPQ, yielding that angles A and B are equal.) Satellite dishes and reflecting telescopes use dishes with parabolic cross-sections so as to

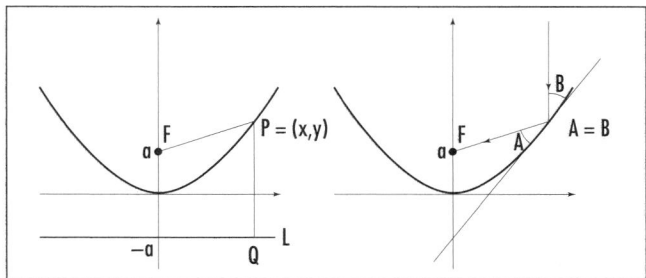

Parabola

focus parallel rays of light to a fixed point, and conversely, search-light reflectors and automobile headlight reflectors, for example, are parabolic: all rays from a bulb positioned at the focus are reflected parallel to the axis of the parabola. (*See* PARABOLOID.)

Parabolas appear in the folding of a thin sheet of paper. Draw a dark straight line on the sheet—this will be the directrix of the parabola—and a dot not on the line, the focus. Fold the dot onto the line and crease the paper. Open up the fold and do this again, this time folding the dot to a different point on the line. As you do this many times, the shape of a parabola emerges along the side of all the creases.

A parabola is said to have ECCENTRICITY e equal to 1. The ratio of the distance of a point P on the curve from a fixed point (the focus) to its distance from a fixed line (the directrix) is always 1.

See also APOLLONIUS'S CIRCLE; ELLIPSE; HYPERBOLA.

paraboloid   The SOLID OF REVOLUTION obtained by rotating a PARABOLA about its axis is called a paraboloid. The points on its surface satisfy an equation of the form $z = b(x^2 + y^2)$, where b is a constant, and each horizontal cross-section, or each CONTOUR LINE, of the solid is a circle. The shape of the figure resembles a bowl.

Techniques of INTEGRAL CALCULUS show that the volume of a section of the solid, up to a height h, is given by $V = \frac{1}{2}\pi a^2 h$, where a is the radius of the circular cross-section at height h. It's surface area is

$$A = \frac{\pi a}{6h^2}\left((a^2 + 4h^2)^{\frac{3}{2}} - a^3\right).$$

Each vertical cross-section of the paraboloid is, of course, a parabola. The common focus of these parabolas is called the focus of the paraboloid. The reflection property of a parabola shows that light rays and beams of sound emitted from the focus of a parabola are reflected out of the paraboloid as a parallel beam. Reflectors in automobile headlights, for example, are parabolic. Conversely, a parallel beam received by a paraboloid reflector is converged into its focus. Satellite dishes and reflecting telescopes use parabolic reflectors.

An elliptic paraboloid has horizontal cross-sections that are ELLIPSES and is given by the equation $z = ax^2 + by^2$, where a and b are constants. A hyperbolic paraboloid (also called a saddle surface) satisfies $z = ax^2 - by^2$. It has horizontal cross-sections that are HYPERBOLAS, and indeed resembles the shape of a horse saddle.

See also ELLIPSOID; HYPERBOLOID.

paradox A statement that seems contradictory or counter to common sense is called a paradox. Some paradoxes in mathematics arise as logical consequences of seemingly plausible premises, such as the Banach-Tarski paradox arising from the study of AREA, BERTRAND'S PARADOX, and RUSSELL'S PARADOX. (The appearance of these particular paradoxes caused mathematicians to reassess the basic foundations of mathematics.) Other paradoxes can be simple self-contradictory statements. For example, it is said that Socrates once reported, "One thing I know is that I know nothing," and George Orwell once wrote, "Freedom is slavery." Both assertions have a flavor similar to the LIAR'S PARADOX.

Other famous paradoxes include:

Grelling's Paradox
Some words in the English language aptly describe themselves. For example, *pentasyllabic,* meaning "having five syllables", is indeed pentasyllabic. Other words are not self-descriptive: for example, *monosyllabic, edible,* and *tangible.* Define a new word *heterological* to mean "a word that does not describe itself." Now ask, *"Is heterological heterological?"*

Berry's Paradox
The number *one million, one hundred thousand, one hundred and twenty one* can be named as "the first number not nameable in under 10 words." (All num-

bers less than 1,100,121 can be named in nine words or less.) But this latter description uses only nine words to indicate this number and so it *can* be named with less than 10 words, despite its definition.

The Barber Paradox
A barber places a sign on the store window. It reads: "This barber shaves all men who do not shave themselves, and only these men." Who shaves the barber?

The Lawyer Paradox
This paradox is traditionally ascribed to the Greek philosopher Protagoras (490–421 B.C.E.)

A law instructor accepts a penniless student under his wing for tuition under the agreement that the student pay the tutor his fees if and only if he wins his first case in court. However, after qualifying as a lawyer, the student takes up a different career and never undertakes a first case. The tutor later sues him for his fees.

The student cleverly decides to represent his own case. This way, he reasons, he need never pay the tutor his fees. If he wins the case, the ruling shall be that he need not pay, whereas if he loses the case, he would be exempted from paying as per his previous agreement with the tutor. Surprisingly, the tutor reasons too that he cannot lose. If the student wins this case, then he must pay the fees according to their previous agreement, whereas, if the student loses, the ruling shall be that he must pay!

The Unexpected Quiz
A school-teacher informs his class that there shall be a surprise quiz some day the following week, and that no one shall be able to deduce the night before that the quiz shall come the next day. Students then reason as follows:

> There can be no quiz on Friday, for on Thursday night, having had no quiz the four previous days, we would know the quiz will come next day. The quiz must therefore be some day other than Friday.
>
> The quiz cannot occur on Thursday either, for on Wednesday night, having had no quiz the three previous days (and with Friday ruled out as a possibility) we would know that the quiz will come next day.
>
> By the same reasoning, the quiz cannot occur on Wednesday, Tuesday, or even Monday.

The students thus logically deduce that there can be no quiz any day the following week. They were legitimately surprised then by a pop quiz that came Wednesday. (Sometimes this paradox is phrased in terms of a prisoner awaiting his execution. In this setting it is called the "unexpected hanging paradox.")

The Wallet Paradox

Two men decide to play the following game:

> Both men open their wallets. Whoever has the least amount of money wins the contents of the other's wallet.

Each man can legitimately argue that he stands to gain more than he can lose, and therefore that the game is biased in his own favor. However, such a simple win/lose game cannot simultaneously be advantageous to both players.

Aristotle's Wheel Paradox

Two wheels of different sizes are glued together so that their centers are aligned. The entire system then rolls along a double track as shown:

In one revolution, both wheels roll the same distance and so, in the diagram, $x = y$. But x is the circumference of the small wheel and y the circumference of the large wheel. Thus we are forced to conclude that two wheels of different sizes have the same circumference!

These paradoxes rely on clever SELF-REFERENTIAL statements (Grelling's paradox, Berry's paradox, the barber paradox, and even the lawyer paradox), hidden false assumptions (students assumed the teacher was telling the truth—they do not know this; winning and losing the wallet game are not equally likely; and could the barber be a woman?), and disguised compound

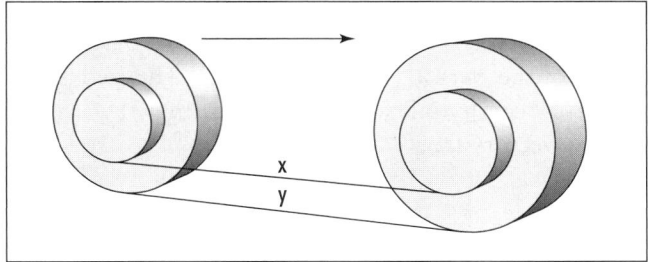

Aristotle's wheels

motions (the small wheel is carried forward by the large wheel as it rotates and so slides along the upper track).

See also CONDITIONAL; HILBERT'S INFINITE HOTEL; JOURDAIN'S PARADOX; TRISTRAM SHANDY PARADOX; ZENO'S PARADOXES.

parallel Two lines in a plane are said to be parallel if they never meet no matter how far they are extended. If two lines labeled L and M are parallel, then we write $L \parallel M$.

The notion of parallel lines is an abstract concept: one cannot physically draw a line infinite in length, nor can one check the entire extent of two infinite lines to determine whether or not they eventually intersect. (We, as human beings, can only conceive of finite quantities and *local* phenomena.) In order to make working with parallel lines feasible, the geometer EUCLID (ca. 300 B.C.E.) introduced an AXIOM, called the PARALLEL POSTULATE, to describe the local behavior of parallel lines. He asserted that, for any TRANSVERSAL crossing a pair of parallel lines, alternate interior angles are equal. (Thus, for instance, the pair of angles labeled x in the diagram on page 376 are indeed equal in measure, as are each pair of right angles about the lines labeled a and b.) From this Euclid was able to establish several properties of parallel lines that we intuitively expect to be true. For instance, Euclid proved:

> Two parallel lines always remain a fixed distance apart. That is, in the diagram on page 376, it must be the case that length a equals length b.

(Observe that the two triangles in the diagram share the same angles and a common diagonal length. The ASA principle now assures that they are CONGRUENT FIGURES. Consequently, the distances a and b must be equal.)

Euclid also observed that the converse of his parallel postulate is true and requires no special assumptions about geometry for its proof. (It follows immediately from the EXTERIOR ANGLE THEOREM.)

> If two lines cut by a transversal produce equal alternate interior angles, then the two lines must be parallel.

This result allows one to readily construct parallel lines. To produce a line through a given point P parallel

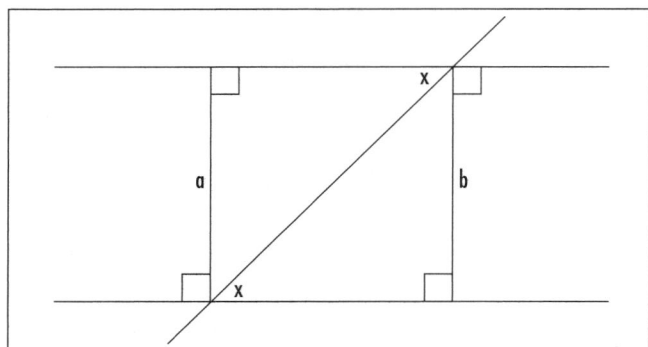

Parallel lines

to a given line *L*, draw an arbitrary line *L′* through *P* that intersects *L*. Measure the angle between *L* and *L′*, and draw another line *L″* through *P*, making this angle with *L′* so as to create a diagram with equal alternate interior angles. By the result above, *L″* is a line through *P* parallel to *L*.

Scottish mathematician John Playfair (1748–1819) proved that for any given point *P* and line *L* in the plane, there is one, and only one, line through *P* parallel to *L*. This result is today known as PLAYFAIR'S AXIOM. Playfair proved that one could take this axiom as a starting point for geometry and establish the parallel postulate from it.

See also AAA/AAS/ASA/SAS/SSS; HISTORY OF GEOMETRY (essay); PARALLELOGRAM; SIMILAR FIGURES.

parallelepiped (parallelopiped) A solid figure with six faces, each of which is a PARALLELOGRAM, is called a parallelepiped. A cube with all six faces square is a special example of a parallelepiped, as is a "rectangular box" with six rectangular faces (also called a cuboid). In these two examples, edges meet at the corners of the solids at 90° angles, but this need not be the case for a general parallelepiped.

The volume of a parallelepiped is the area of the BASE of the figure multiplied by the figure's height. If the three edges meeting at a corner are regarded as vectors **a**, **b**, and **c**, then the volume of the parallelepiped is also given by the absolute value of the TRIPLE VECTOR PRODUCT **a** · **b** × **c**.

A parallelotope is a parallelepiped whose three side-lengths are in the ratio 4:2:1.

See also PRISM.

parallelogram A QUADRILATERAL with opposite sides parallel, and hence equal in length, is called a parallelogram. The area of a parallelogram is "base times height." That is to say, if one pair of parallel sides, of length *b*, are *h* units apart, then the area of the figure is *bh*. (*See* AREA.)

An examination of alternate angles across parallel lines quickly shows that opposite angles in a parallelogram are equal and that adjacent angles are supplementary, that is, sum to 180°. The parallelogram is the only quadrilateral with this property.

The DIAGONALS of a parallelogram bisect one another, and again the parallelogram is the only quadrilateral that has this property. If the two different side-lengths of a parallelogram are *a* and *b*, and the length of its two diagonals are *p* and *q*, then

$$p^2 + q^2 = 2a^2 + 2b^2$$

This follows from applying the LAW OF COSINES to the four triangles defined by the diagonals (two triangles on either side of each diagonal) and adding all four equations. Interpreted as an equation of VECTOR lengths, this establishes the PARALLELOGRAM LAW in vector-space theory.

A parallelogram with all four sides equal in length is called a rhombus (or sometimes a rhomb, diamond, or a lozenge). A rhomboid is a parallelogram with adjacent sides unequal in length.

If all four angles of a parallelogram are equal, then each angle equals 90°, and the figure is a RECTANGLE. A parallelogram that is both equilateral and equiangular is a SQUARE.

See also TRAPEZOID/TRAPEZIUM.

parallelogram law (parallelogram rule) The fact that VECTOR addition is commutative is sometimes called the parallelogram law. This is appropriate, since the sum of the two vectors **a** and **b** in the plane, **a** + **b**, is given as the DIAGONAL of the parallelogram defined by the two vectors. Note that the second diagonal is given by the vector difference **a** − **b**. The geometric properties of a parallelogram establish that the lengths of these vectors satisfy the relation:

$$\|\mathbf{a} + \mathbf{b}\|^2 + \|\mathbf{a} - \mathbf{b}\|^2 = 2\|\mathbf{a}\|^2 + 2\|\mathbf{b}\|^2$$

This equation is also called the parallelogram law.

See also COMMUTATIVE PROPERTY.

parallel postulate (Euclid's fifth postulate) In his famous work *THE ELEMENTS*, the geometer EUCLID (ca. 300 B.C.E.) proposed five basic postulates that describe the principles of geometry on a flat PLANE. His fifth postulate states:

> If two lines L and M subtended from a common line segment have interior angles summing to less than 180°, then those lines, if extended, will meet to form a TRIANGLE.

One can consequently conclude that if those angles sum to more than 180° then the lines, if extended on the other side of the given line segment, will form a triangle on that side. (The sum of the supplementary interior angles will be less than 180°.) Furthermore, if the lines never meet, that is, if they are PARALLEL, then the sum of the two angles x and y shown cannot be less than nor greater than 180°, and so must equal 180°. We have:

> For a pair of parallel lines with a TRANSVERSAL, as shown in the diagram below right, the angles labeled x and y must sum exactly to 180°. Consequently, the alternate interior angles of the transversal are equal.

This is often taken as the statement of Euclid's fifth postulate. For this reason, it is often called the parallel postulate.

As the parallel postulate is intimately connected with the formation of triangles, it is not surprising that one can use it to prove:

> The sum of the interior angles of any triangle equals 180°.

As we show below, it is possible to alternatively take this statement as a basic postulate and use it to establish the parallel postulate. (Thus the parallel postulate and this claim about triangles are logically equivalent starting assumptions.)

In his 1795 text *Elements of Geometry*, Scottish mathematician John Playfair offered a third, logically equivalent version of the parallel postulate:

> Given a line L in the plane and a point P not on that line, then there exists one, and only one, line through P parallel to L.

This alternative version of the postulate is today called PLAYFAIR'S AXIOM. In 1829 it provided Russian mathematician NIKOLAI IVANOVICH LOBACHEVSKY (1792–1856) the inspiration to consider, and discover, a consistent theory of a NON-EUCLIDEAN GEOMETRY.

We now have three assertions:

1. Euclid's parallel postulate
2. Angles in a triangle sum to 180°
3. Playfair's axiom

These three statements are logically equivalent, that is, any one implies the other two. To see this, it suffices to show that (3) implies (2), (2) implies (1), and that (1) implies (3). We will need to make use of the converse of Euclid's parallel postulate:

> If a transversal across a pair of lines yields equal alternate interior angles, then the lines are parallel.

As the EXTERIOR-ANGLE THEOREM shows, this statement can be proved without making use of any of the three statements above.

Consider now the three implications:

(3) implies (2):

Given a triangle with angles a, b, and c as shown in the diagram on page 378, draw two lines at the apex that copy angles a and b. Given that both yield equal alternate interior angles across a transversal, both lines are parallel to the base of the triangle. By Playfair's axiom there is only one such line. Thus angles a, b, and c all lie on the same straight line and so sum to 180°.

(2) implies (1):

In the middle diagram on page 378, suppose that lines L and M are parallel, and so never meet. Could it

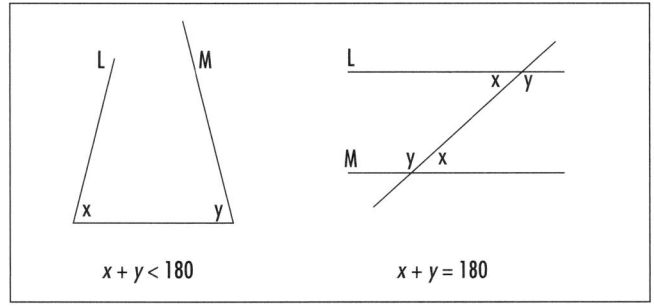

Understanding the parallel postulate

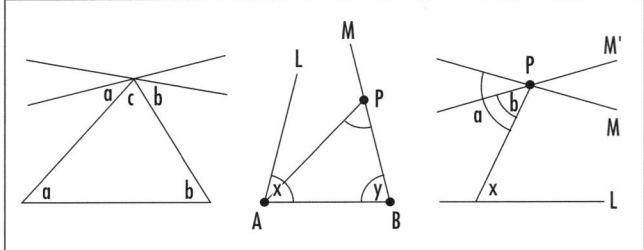

Equivalent versions of the parallel postulate

be that angles x and y sum to less that 180°? Draw a point P on line M, and slide it along M sufficiently far so that the angle of the triangle APB at P is less than $180 - x - y$. Clearly the angle in this triangle at A is less than x. We now have a triangle with three angles that sum to something less than $x + (180 - x - y) + y = 180°$. This is impossible by our assumption that (2) is true. It cannot be the case that L and M are parallel after all.

(1) implies (3):

Discussion on the exterior-angle theorem shows that there certainly exists at least one line through a point P parallel to a given line L. Suppose there are two such lines. Call them M and M' as shown in the diagram above. Draw any line from P to L. Since, by (1), we are assuming that alternate interior angles are equal for parallel lines, both angles a and b shown are equal to x. Consequently M and M' must be the same line.

See also HISTORY OF GEOMETRY (essay).

parameter   *See* PARAMETRIC EQUATIONS.

parametric equations (freedom equations) When we think of a curve as the path traced by a moving point, it is convenient to represent the curve as two equations, one for each coordinate x and y, in terms of a third variable usually denoted t: $x = f(t)$ and $y = g(t)$. The variable t is called a parameter (from the Greek *para*, meaning "together," and *meter*, meaning "measure"). Equations with parameters are called parametric equations. In physical problems, t is usually thought of as time.

Any curve $y = f(x)$ can be expressed in terms of parametric equations: simply set $x = t$ and consequently $y = f(t)$.

Parametric representations, however, are not unique. The curve $y = 4x^2$, for example, can be represented as $x = t$, $y = 4t^2$ and also $x = \frac{t^3}{2}$, $y = t^6$. Infinitely many other representations are possible.

A CIRCLE of radius r and center (m,n) has a parametric representation: $x = m + r \cos t$ and $y = n + r \sin t$. (Notice that $(x - m)^2 + (y - n)^2 = (r \cos t)^2 + (r \sin t)^2 = r^2(\cos^2 t + \sin^2 t) = r^2$, the equation of a circle.) Similarly, an ELLIPSE $\frac{x^2}{a^2} + \frac{y^2}{b^2} = 1$ has parameterization: $x = a \cos t$ and $y = b \sin t$.

There is no general method for transforming an equation of the form $F(x,y) = 0$ into parametric equations. Each individual example needs to be examined carefully, and much ingenuity is often required. (For example, it is not immediately clear what the parametric equations for the curve given by $x^5 + xy + y^5 = 0$ could be.) For the reverse direction, one can attempt to convert a pair of parametric equations into a single equation of the form $F(x,y) = 0$ by solving for t. To illustrate, consider the equations: $x = 3t + 1$ and $y = t - 1$. The first equation yields $t = (x - 1)/3$, and the second, $t = y + 1$. Consequently: $y + 1 = (x - 1)/3$, or $y = x/3 - 4/3$. This shows that the parametric equations under study represent a straight-line path.

The SLOPE of a curve given via parametric equations is $\frac{dy}{dx} = \frac{y'(t)}{x'(t)}$ provided the DERIVATIVE $x'(t)$ is not zero. This follows from the CHAIN RULE for differentiation: if y is a function of x, which itself is a function of t, then $\frac{dy}{dt} = \frac{dy}{dx} \cdot \frac{dy}{dt}$.

Parametric equations are also used to describe the motion of particles in three-dimensional space. One is given three equations in a parameter t, one for each of the three coordinates x, y, and z.

See also ARC LENGTH; CONSTANT WIDTH.

parity Two integers that are either both even or both odd are said to have the same parity. For instance, 17 and 53 have the same parity (both are odd), and 9 and 14 have opposite parity. The study of EVEN AND ODD NUMBERS often makes use of parity to establish a number of sophisticated results.

Sometimes the term *parity* is used in a more general setting as to mean "being in one of two possible states." For example, three hockey pucks labeled A, B, and C lying on a playing field, not in a straight line, can be in one of two possible states: moving from *A* to *B* to *C* could either have a clockwise or counterclockwise sense. Every time a hockey player hits one puck between the other two, the parity of the system changes. This shows, for instance, that a hockey player will never be able to return the three pucks to their original positions if she performs hits of this type 33 times.

As another example, it is impossible for 25 people standing in a 5 × 5 grid of squares, one person per cell, to shift places if each person is asked to move one place over to a neighboring vertical or horizontal cell. A parity argument explains why. If we imagine the cells colored black and white according to a checkerboard pattern, then the 13 people standing in black cells are required to occupy just 12 white cells, rendering the challenge unsolvable. (If, however, players are permitted to take diagonal steps, then the puzzle can be solved.)

partial derivative Given a function $f(x,y)$ of two variables, the DERIVATIVE of *f* with respect to just one of the variables, treating the other variable as a constant, is called a partial derivative of that function. Specifically, the partial derivative of the function with respect to *x*, denoted $\frac{\partial f}{\partial x}$ or sometimes f_x, is the LIMIT:

$$\frac{\partial f}{\partial x} = \lim_{h \to 0} \frac{f(x+h,y) - f(x,y)}{h}$$

and the partial derivative with respect to *y*, denoted $\frac{\partial f}{\partial y}$ or f_y, is:

$$\frac{\partial f}{\partial y} = \lim_{h \to 0} \frac{f(x,y+h) - f(x,y)}{h}$$

Geometrically, the graph of the function $z = f(x,y)$ is a surface sitting in three-dimensional space, and the quantity $\frac{\partial f}{\partial x}$ measures the slope of the graph surface above the point (x,y) in the direction parallel to the *x*-axis, and $\frac{\partial f}{\partial y}$ is the slope of the surface in the direction parallel to the *y*-axis. Since one of the variables is being treated as a constant, the partial derivative of a function can be found by using the normal rules of differentiation. For example, for the function $f(x,y) = x^2y^3 + 2x + y$, we have $\frac{\partial f}{\partial x} = 2xy^3 + 2$ and $\frac{\partial f}{\partial y} = 3x^2y^2 + 1$.

The symbol ∂ for "partial" was introduced in 1788 by the French mathematician JOSEPH-LOUIS LAGRANGE. It resembles the *d* used for ordinary differentiation.

The *n*th partial derivative of a function with respect to the same variable is denoted $\frac{\partial^n f}{\partial x^n}$ and differentiating with respect to two different variables leads to mixed partial derivatives: $\frac{\partial^2 f}{\partial x \partial y}, \frac{\partial^2 f}{\partial y \partial x}, \frac{\partial^3 f}{\partial^2 x \partial y}$ and the like. The quantity $\frac{\partial^2 f}{\partial x \partial y}$, for example, is to be interpreted as "differentiate first with respect to *y*, and then with respect to *x*." (That is, one performs the operations in reverse order as indicated by the denominator.) For instance, if $f(x,y) = x^2y^3 + 2x + y$ then:

$$\frac{\partial^2 f}{\partial x \partial y} = \frac{\partial}{\partial x}\left(\frac{\partial f}{\partial y}\right) = \frac{\partial}{\partial x}\left(3x^2y^2 + 1\right) = 6xy^2$$

and

$$\frac{\partial^2 f}{\partial y \partial x} = \frac{\partial}{\partial y}\left(\frac{\partial f}{\partial x}\right) = \frac{\partial}{\partial y}\left(2xy^3 + 2\right) = 6xy^2$$

The equality

$$\frac{\partial^2 f}{\partial x \partial y} = \frac{\partial^2 f}{\partial y \partial x}$$

is no coincidence. Mathematicians have proved that the order of partial differentiation does not matter as long as all the partial derivatives involved are continuous functions. The notion of a partial derivative extends to functions of more than two variables.

A partial-differential equation is an equation that contains partial derivatives of a function. For example, the function $f(x,y) = e^{xy}$ is a solution to the partial-differential equation:

$$\frac{\partial f}{\partial x} + \frac{\partial f}{\partial y} = (x+y)f$$

See also CHAIN RULE; DIFFERENTIAL EQUATION; DIRECTIONAL DERIVATIVE.

partial fractions Algebraic expressions containing the ratio of two POLYNOMIALS $\frac{p(x)}{q(x)}$, with the degree of

$p(x)$ less than the degree of $q(x)$, can be split into partial fractions for easier handling. This first requires splitting the denominator $q(x)$ into linear and irreducible quadratic factors, each to some index, and then writing $\frac{p(x)}{q(x)}$ as a sum of terms of each of the following type:

1. Corresponding to each (repeated) linear term $(x - k)^n$ in $q(x)$, there is a sum of terms $\frac{A_1}{(x-k)} + \frac{A_2}{(x-k)^2} + \cdots + \frac{A_n}{(x-k)^n}$.

2. Corresponding to each (repeated) irreducible quadratic terms $(ax^2 + bx + c)^m$ in $q(x)$, there is a sum of terms $\frac{B_1 x + C_1}{(ax^2 + bx + c)} + \frac{B_2 x + C_2}{(ax^2 + bx + c)^2} + \cdots + \frac{B_m x + C_m}{(ax^2 + bx + c)^m}$.

Here the numbers A_i, B_i, and C_i are constants. Some examples illustrate the process:

$$\frac{4}{(x+2)(x-3)} = \frac{A}{(x+2)} + \frac{B}{(x-3)}$$

$$\frac{x^2 + 2x + 5}{(x-1)^3} = \frac{A}{(x-1)} + \frac{B}{(x-1)^2} + \frac{C}{(x-1)^3}$$

$$\frac{3x^2 + 5x + 7}{x(x-5)(x^2 + x + 1)^2} = \frac{A}{x} + \frac{B}{(x-5)} + \frac{Cx + D}{(x^2 + x + 1)}$$

$$+ \frac{Ex + F}{(x^2 + x + 1)^2}$$

The values of the constants A, B, C, ... are found by multiplying through by the denominator and then EQUATING COEFFICIENTS. Alternatively, one can substitute appropriate values for the variable x to determine the values of some of these unknowns more quickly. For example, in the first example, after multiplying through, we have: $4 = A(x - 3) + B(x + 2)$. Setting $x = 3$ yields $4 = 0 + 5B$, establishing that B is 4/5, and setting $x = -2$ gives $A = -4/5$. We thus have:

$$\frac{4}{(x+2)(x-3)} = \frac{-\frac{4}{5}}{(x+2)} + \frac{\frac{4}{5}}{(x-3)}$$

$$= \frac{4}{5}\left(\frac{1}{x-3} - \frac{1}{x+2}\right)$$

Mathematicians have proved that every rational function $\frac{p(x)}{q(x)}$, with the degree of $p(x)$ less than the degree of $q(x)$, can indeed be written as a sum of partial fractions, and that the constant terms appearing, A_i, B_i, and C_i, are unique for that rational function. (That is, no rational function can be expressed as a sum of partial fractions in two different ways.) Partial fractions are generally used for solving INTEGRALS and in solving DIFFERENTIAL EQUATIONS. As an example, we have:

$$\int \frac{4}{(x+2)(x-3)} dx = \frac{4}{5} \int \frac{1}{(x-3)} - \frac{1}{(x+2)} dx$$

$$= \frac{4}{5}(\ln|x - 3| - \ln|x + 2|) + C$$

$$= \frac{4}{5} \ln\left|\frac{x-3}{x+2}\right| + C$$

partial order *See* ORDERED SET.

partial sum The nth partial sum S_n of an infinite series $a_1 + a_2 + a_3 + \ldots$ is the sum of just the first n terms of the series: $S_n = a_1 + a_2 + \ldots + a_n$. For example, the first four partial sums of the series $\sum_{n=1}^{\infty} \frac{1}{2^n} = \frac{1}{2} + \frac{1}{4} + \frac{1}{8} + \frac{1}{16} + \frac{1}{32} + \cdots$ are $S_1 = \frac{1}{2}$, $S_2 = \frac{1}{2} + \frac{1}{4} = \frac{3}{4}$, $S_3 = \frac{1}{2} + \frac{1}{4} + \frac{1}{8} = \frac{7}{8}$, and $S_4 = \frac{1}{2} + \frac{1}{4} + \frac{1}{8} + \frac{1}{16} = \frac{15}{16}$.

A series is said to converge to a value L if the partial sums S_n tend to L in the LIMIT as $n \to \infty$. In the above example, the sequence of partial sums $\frac{1}{2}, \frac{3}{4}, \frac{7}{8}, \frac{15}{16}, \ldots$ approaches the value 1. Thus we write: $\frac{1}{2} + \frac{1}{4} + \frac{1}{8} + \frac{1}{16} + \frac{1}{32} + \frac{1}{64} + \cdots = 1$.

See also CONVERGENT SERIES.

partition In NUMBER THEORY a partition of a natural number n is a representation of n as a sum of positive integers. For example, 20 + 15 + 5 and 10 + 10 + 10 + 10 are partitions of the number 40, as is the representation 40 itself. A partition is considered ordered if the order of the terms in the sum is considered important.

For instance, 20 + 15 + 5 and 5 + 20 + 15 are considered to be two different ordered partitions of the number 40.

There is just one ordered partition of the number 1, two ordered partitions of 2, four of 3, and eight ordered partitions of the number 4:

1:1
2:1 + 1 = 2
3:1 + 1 + 1 = 2 + 1 = 1 + 2 = 3
4:1 + 1 + 1 + 1 = 2 + 1 + 1 = 1 + 2 + 1 = 1 + 1 + 2
  = 2 + 2 = 1 + 3 = 3 + 1 = 4

This pattern persists:

A number n has 2^{n-1} ordered partitions.

(One can create an ordered partition of a number n by first writing n as a sum of n 1s, and then deleting some, all, or none of the + signs between the 1s. As there are $n - 1$ plus signs to consider, each providing a choice of whether to leave or to erase, this provides 2^{n-1} possibilities in all.)

If the order of the terms in a partition is not considered important, then the partition is called unordered. (For example, 20 + 15 + 5 and 5 + 20 + 15 are now considered the same partition of the number 40.) The symbol $P(n)$ is usually used to denote the number of unordered partitions a number n possesses. For instance, $P(4) = 5$, since there are just five unordered partitions of the number 4 (namely, 1 + 1 + 1 + 1, 2 + 1 + 1, 2 + 2, 3 + 1 and 4). The following table shows the first 10 partition function values:

n	1	2	3	4	5	6	7	8	9	10
$P(n)$	1	2	3	5	7	11	15	22	30	42

To this day, no one knows an exact formula for this function, and generally very little is known about its behavior.

In the mid-1700s LEONHARD EULER studied this function and found one truly remarkable relationship. Consider those triangular FIGURATE NUMBERS that are divisible by three: 3, 6, 15, 21, 36, 45, … Taking one-third of those numbers yields the sequence: 1, 2, 5, 7, 12, 15, … Declaring $P(0) = 1$ and $P(n)$ to be zero if n is negative, Euler proved that the following formula holds:

$$P(n) - P(n-1) - P(n-2) + P(n-5) + P(n-7)$$
$$- P(n-12) - P(n-15) + \ldots = 0$$

Here the signs between terms alternate in pairs. Notice too that although the left side of this equation appears to be an infinite sum, eventually all the terms in the sum equal zero. Consequently one is required to add and subtract only a finite list of numbers.

This formula allows us to compute higher values of $P(n)$. For instance, to find the value of $P(11)$ use:

$$P(11) - P(10) - P(9) + P(6) + P(4) - P(-1) - P(-4) + \ldots = 0$$

That is,

$$P(11) - 42 - 30 + 11 + 5 - 0 - 0 + \ldots = 0$$

yielding

$$P(11) = 56$$

In the 1930s, mathematicians GODFREY HAROLD HARDY and SRINIVASA AIYANGAR RAMANUJAN proved that if n is large, then the function $P(n)$ is well approximated by the formula:

$$P(n) \approx \frac{1}{4n\sqrt{3}} e^{\sqrt{\frac{2n\pi^2}{3}}}$$

Thus, for instance, the value of $P(1,000)$ is approximately $\dfrac{1}{4000\sqrt{3}} e^{\sqrt{\frac{2000\pi^2}{3}}} \approx 2.4 \times 10^{31}$.

Understanding the properties of the unordered partition function is still an active area of research.

Pascal, Blaise (1623–1662) French *Probability theory, Calculus, Geometry, Philosophy* Born on June 19, 1623, in Clermont-Ferrand, France, philosopher and scientist Blaise Pascal is remembered for his work in GEOMETRY, hydrostatics, and PROBABILITY theory. Although he did not invent the famous arithmetical triangle that bears his name, he did make extensive use of its properties in his development of probability and the study of statistical distributions. He also worked on problems of finding arc lengths and areas of curved figures. He is noted, in particular, for his work on the

CYCLOID and the development of ideas from this work that led to the invention of CALCULUS.

Pascal was fascinated with mathematics as a young teenager and by age 16 had already discovered a number of noteworthy results in the study of PROJECTIVE GEOMETRY. He published his first work in geometry, *Essay on Conic Sections,* in 1640.

In order to assist his father with his work in recording and collecting taxes, between the years 1642 and 1645 Pascal invented and built a mechanical calculator that could perform sophisticated arithmetic computations. Over the next 7 years Pascal oversaw the construction of as many as 50 of these machines but, sadly, few were sold. Production of the "Pascaline" ceased, but Pascal is nonetheless noted as the second person in history to have built such a device. (German scholar Wilhelm Schickard built the first in 1624.)

In the late 1640s Pascal developed an interest in hydrostatics and the properties of atmospheric pressure. He published *New Experiments concerning Vacuums* in 1647—a controversial work as scientists at the time doubted that vacuums exist—and 4 years later his famous piece *Treatise on the Equilibrium of Liquids.*

Blaise Pascal, an eminent scholar of the 17th century, is noted for his pioneering work in the field of probability theory. *(Photo courtesy of Monique Salaber/The Image Works)*

His interest in pure mathematics never wavered during this time, and he continued to develop work on the theory of CONIC SECTIONS. (Sadly, his manuscripts from this period are lost to us today.)

In correspondence with PIERRE DE FERMAT (1601–65), Pascal worked on solving two challenging problems about games of chance: what is the expected number of tosses required to roll a double 6 with a pair of dice? And how should one divvy up the bets laid down for a game of dice, if the game must be halted part way through its play? Pascal and Fermat solved the first problem and the second only for the case of a two-player game. This work marked the beginning development of modern probability theory.

Pascal was a deeply religious man and is also remembered for his famous line of rational thought arguing in favor of believing in the existence of God. Now called "Pascal's wager," Pascal's argument simply stated that if God does not exist, then nothing is lost by believing in Him, whereas if God does exist, all will be lost by not acknowledging His existence.

Pascal computed the arc length and area of one branch of the cycloid curve, and also the volume and surface area of a SOLID OF REVOLUTION obtained by rotating it about the x-axis. This was his final mathematical accomplishment before devoting his life to religious service. He died of cancer on August 19, 1662, at age 39.

Although not remembered for having produced a large body of profoundly creative work, Pascal did nonetheless play seminal roles in the development of several new fields of study (infinitesimal calculus, probability theory, projective geometry, and fluid statics, for instance). He made significant contributions to each field and, perhaps more important, played a significant role in clarifying the foundations of these studies and systematizing the methods they utilize.

Pascal's distribution   *See* BINOMIAL DISTRIBUTION.

Pascal's triangle   The triangular array of numbers with 1 at the apex, with a 1 at the beginning and end of each row, and with the property that each interior number is the sum of the two numbers above it in the preceding row is called Pascal's triangle. The first few rows of the triangle are shown below:

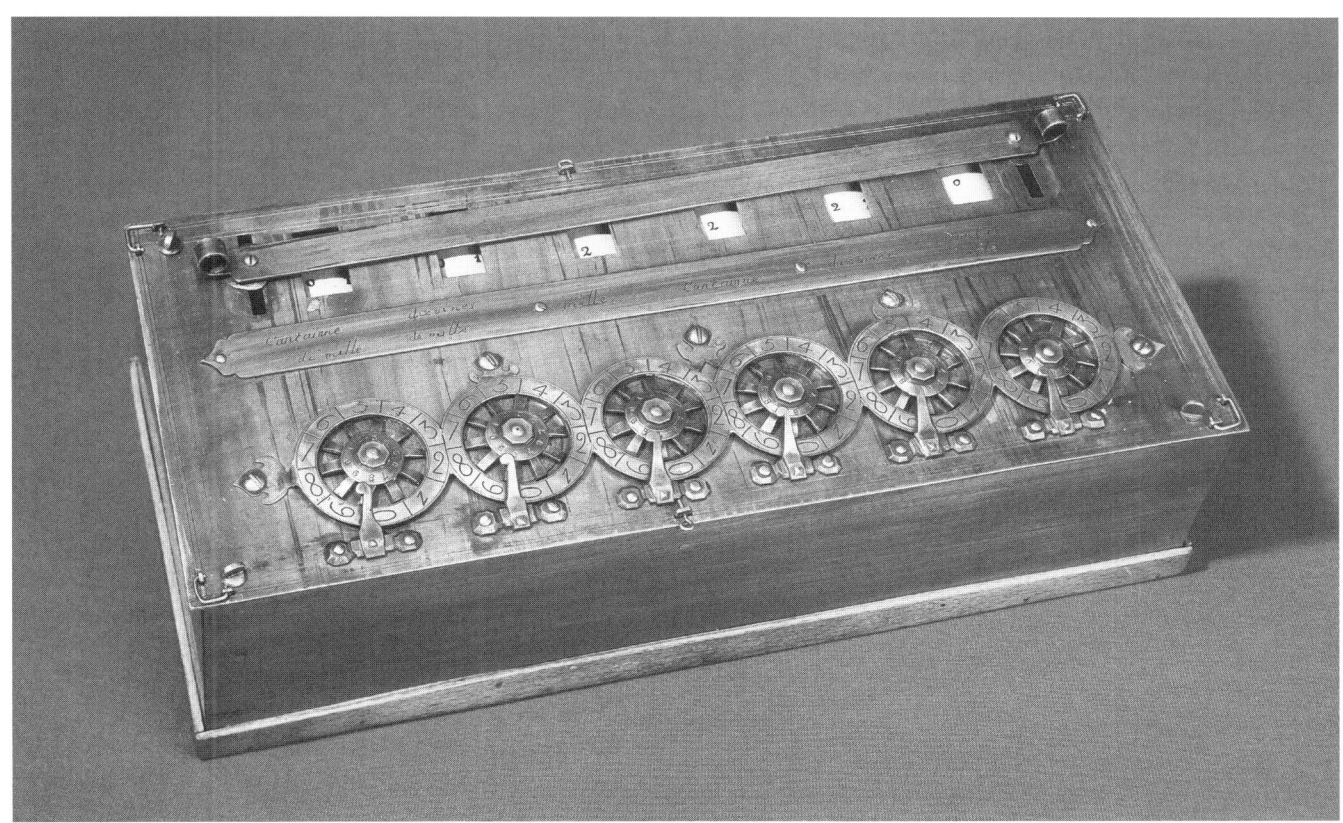

In 1642 Blaise Pascal invented and built the world's first calculating machine. *(Photo courtesy of the Science Museum, London/Topham-HIP/The Image Works)*

```
            1
          1   1
        1   2   1
      1   3   3   1
    1   4   6   4   1
  1   5  10  10   5   1
1   6  15  20  15   6   1
```

The triangle is named after French mathematician BLAISE PASCAL (1623–62), who used it extensively in his development of PROBABILITY theory. However, Pascal never claimed to have invented the triangle. It appears, for example, in CHU SHI-CHIEH's 1303 text *Siyuan Yujian* (The precious mirror of the four elements), and tables of the first few rows of the triangle appear in ancient Arab texts.

To motivate the construction of the triangle, one can ask the following question:

If, in a rectangular grid, one is permitted to move from cell to cell only by taking eastward or southward steps, how many distinct routes are there from the top left cell to any other given cell?

For example, to move to the circled cell shown, one could take three steps east followed by two steps south (which we shall record as EEESS), or, alternatively, one step south, two steps east, one step south, followed by a final step east (SEESE). In fact, *any* arrangement of three Es and two Ss corresponds to a possible path, and one can check that there are 10 such combinations of five letters in all. This shows that there are 10 routes to the circled cell shown. It is convenient to assign the number one to the top left cell. (Starting at the top left cell there is only one course of action needed to reach that cell, namely, no action.)

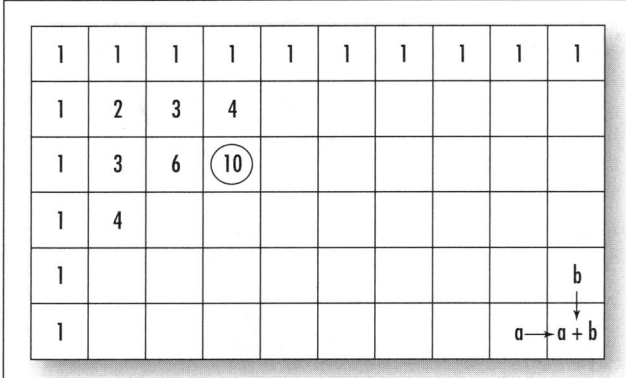

Discovering Pascal's triangle

Also notice, in order to reach any particular cell, one has two options: either head to the cell directly above it and take a final step southward, or head to the cell just to its left and take a final step eastward. Thus the number of paths to a particular cell is the sum of the two counts for the cells above it and to its left. This is the defining feature of Pascal's triangle.

We can say more: the numbers in each cell count the number of ways to arrange a fixed number of Es and Ss. For instance, the number below the circled cell in the picture will be 20. This shows that there must be 20 distinct ways to arrange three Es and three Ss. Alternatively, we could determine this count by asking: How many ways are there to select three places (the positions that the letter E will lie) among a string of six positions? The answer to this question is in the COMBINATORIAL COEFFICIENT $\binom{6}{3} = \frac{6!}{3!3!} = 20$. Counting the top row of Pascal's triangle as row zero and the first entry of each row the zeroth entry of that row, we now see that the kth entry in the nth row of Pascal's triangle corresponds to combinatorial coefficient $\binom{n}{k}$, which equals the number of ways to arrange k Es and $(n - k)$ Ss as a string of n letters. We have:

Pascal's triangle is the triangular array of all combinatorial coefficients.

The BINOMIAL THEOREM shows that the combinatorial coefficients also appear in expansions of expressions of the form $(x + y)^n$. We have:

The entries in the nth row of Pascal's triangle are precisely the coefficients that appear in the expansion of $(x + y)^n$.

For instance, we have $(x + y)^2 = x^2 + 2xy + y^2$, $(x + y)^3 = x^3 + 3x^2y + 3xy^2 + y^3$, and $(x + y)^4 = x^4 + 4x^3y + 6x^2y^2 + 4xy^3 + y^4$.

Pascal's triangle possesses a number of remarkable properties. We list just a sample.

1. The entries in the nth row of Pascal's triangle sum to 2^n.

   For instance, $1 = 1$, $1 + 1 = 2$, $1 + 2 + 1 = 4$, $1 + 3 + 3 + 1 = 8$, and $1 + 4 + 6 + 4 + 1 = 16$. This follows from expanding the quantity $(1 + 1)^n$.

2. The alternating sum of the entries of each row of Pascal's triangle is zero.

   For instance, $1 - 1 = 0$, $1 - 2 + 1 = 0$, $1 - 3 + 3 - 1 = 0$, and $1 - 4 + 6 - 4 + 1 = 0$. This follows from expanding the quantity $(1 + (-1))^n$.

   It is worth observing that both of these results can also be obtained by noting that each entry in Pascal's triangle is the sum of two entries in the preceding row. Thus the direct or the alternating sum of the entries in one row corresponds to a sum that considers each entry in the previous row twice.

3. The FIBONACCI NUMBERS appear as sums of entries in certain diagonals of Pascal's triangle.

   To explain: the apex 1 constitutes the first diagonal. The first 1 in row one constitutes the second diagonal. The first 1 in row two and the final 1 in row three constitute the third diagonal. (Note that $1 + 1 = 2$.) Continuing in this way, we see that $1 + 3 + 1 = 5$, $1 + 4 + 3 = 8$, and $1 + 5 + 6 + 1 = 13$. By writing each term in the diagonal as the sum of the two entries above it, we see that the sum of entries in one diagonal matches the sums of entries of the previous two diagonals.

4. Each row of Pascal's triangle corresponds to the digits of a power of 11.

   For instance, $11^0 = 1$, $11^1 = 11$, $11^2 = 121$, $11^3 = 1,331$, and $11^4 = 14,461$. This follows from expanding the quantity $(10 + 1)^n$. If we were not required to carry digits when performing multiplications, this correspondence would remain exact.

5. Select any entry N in Pascal's triangle and circle all the entries that lie in the parallelogram that has N and the apex as opposite vertices. Then the sum of all entries in that parallelogram is 1 less than the number M directly below N two rows down.

For example, in the triangle shown on previous page, the number 20 is one greater than the sum of numbers bounded in the parallelogram with 6 as the lowest corner. (We have 6 + 3 + 3 + 1 + 2 + 1 + 1 + 1 + 1 = 19.) This property can be proved by noting that the number M is the sum of the two numbers above it, which, in turn, are each the sum of two numbers in the previous row, and so forth. One can match this backward tabulation with the entries of the described parallelogram (except for a single 1).

See also CATALAN NUMBERS.

Peano, Giuseppe (1858–1932) Italian *Foundations of mathematics* Born on August 27, 1858, in Cuneo, Italy, logician Giuseppe Peano is remembered for his influential work in mathematical logic and in the FOUNDATIONS OF MATHEMATICS. In 1889 he published a famous set of axioms (revised in 1899), today called PEANO'S POSTULATES, which defined the NATURAL NUMBERS in terms of sets. A year later he invented space-filling curves, such as PEANO'S CURVE, which were thought impossible at the time. These curves show, in some sense, that there are just as many points on a line as there are in a plane.

Peano graduated from the University of Turin in 1880 with a doctorate in mathematics and his habilitation degree in 1884. He remained at the university as a professor of mathematics throughout his career.

Peano's early work was in the field of DIFFERENTIAL EQUATIONs, where he studied and established significant results on the problem of classifying those equations for which solutions are guaranteed to exist. But his interests changed toward mathematical logic in 1888 with the publication of his text *Geometric Calculus*, the opening chapter of which was devoted to the topic. Peano published his famous axioms defining the natural numbers 1 year later in the small pamphlet *Arithmetices principia, nova methodo exposita*, (Arithmetic principles, exposition of a new method), which he wrote entirely in Latin to the surprise of his colleagues. Soon after this, he presented his work on his famous curve.

Mathematical historians feel that Peano's contributions to mathematics dwindled after this. In 1892 Peano began work on an enormous undertaking to collate all known results in mathematics, essentially as a giant list. This project, which he called *Formulario mathematico* (Mathematical forms), consumed his working hours for a full 16 years. He completed the project in 1908, but

received little attention for it. Although the text contained a great deal of valuable information, it was difficult to read, not only because of its dry nature, but also because Peano chose to write it in a new "universal language" he invented based on a simplified version of Latin. He hoped to develop a universal culture of mathematical exploration united by a common language. His dream was never realized.

Peano died in Turin, Italy, on April 20, 1932. Much of his mathematical work, although significant at the time, is chiefly of historic interest today.

Peano's curve In 1890 Italian mathematician GIUSEPPE PEANO (1858–1932) described a curve that could pass through every point of a square. The curve is constructed by an iterative process. One begins by drawing the diagonal of a square and then breaking the square into nine subsquares and drawing certain diagonals of those subsquares. At the next stage, each subsquare is divided into nine sub-subsquares and the pattern is repeated. The Peano curve is the curve that results when this procedure is repeated indefinitely. One can intuitively see that Peano's construct is an object that passes through every point of the square. (It passes through some points more than once.) For this reason, Peano's curve is described as a space-filling curve. His construct shows, in some sense, that the set of points inside a square is no more infinite than the set of points on a curve.

To make the definition of Peano's curve mathematically precise, for each number t in the interval $[0,1]$ let $P_n(t)$ denote the point in the square that is t units along the length of the curve that is produced in the nth step of the above procedure. (Notice, for example, that $P_n\left(\frac{1}{2}\right) = (0.5, 0.5)$ for all n.) Then define $P(t)$ to be the LIMIT of these points:

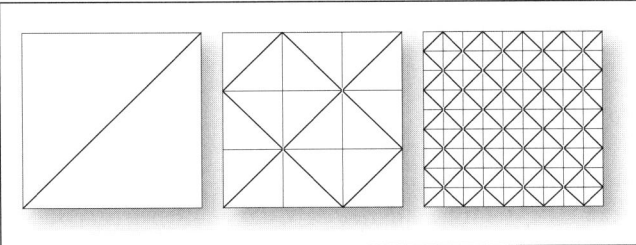

Peano's space-filling curve

$$P(t) = \lim_{n \to \infty} P_n(t)$$

One can show that this formula does indeed represent a CONTINUOUS FUNCTION from the unit interval [0,1] to all points in the unit square. Since any continuous function defined on an interval can be thought of as the PARAMETRIC EQUATION of a curve, it is indeed appropriate to think of Peano's function then as a curve in two-dimensional space.

Notice that if we take the side-length of the square to be 1 unit, then the length of the diagonal is $\sqrt{2}$ units long. Each iteration of the procedure produces a curve three times as long as the curve in the previous step.

See also CARDINALITY; INFINITY.

Peano's postulates In 1889 Italian mathematician GIUSEPPE PEANO (1858–1932) presented a first set of basic postulates that he hoped would characterize precisely the properties of the natural numbers. He revised his system 10 years later to state:

There is a set N whose elements are called "numbers" with the following properties:

1. To every number a one can assign another number a' called its successor. (We normally think of a' as $a + 1$.)
2. No two different numbers have the same successor.
3. There is at least one number which is not the successor of any other number.
4. *Induction axiom:* If a subset M of numbers contains at least one number that is not a successor, but has the property that for any number a in M, its successor a' is in M, then $M = N$. (That is, every number is in M.)

It is worth noting that it follows from these axioms that there is just one number that is not the successor of any other number. (Let e be any number that is not a successor and let M be the set of all numbers that are successors along with the number e. By the induction axiom, $M = N$. It follows then that e is the only number that fails to be a successor.) This special number is usually called ZERO.

From these very basic postulates Peano was able to define an operation of addition on numbers and from there derive all the properties of arithmetic we use today.

See also INDUCTION; NATURAL NUMBERS.

Pearson, Karl (1857–1936) British *Statistics* Born on March 27, 1857, in London, England, mathematician Karl Pearson is remembered for his influence in the development of statistics as applied to biology and the social sciences. Pearson introduced, for the first time, such basic concepts as STANDARD DEVIATION and the notion of a CORRELATION COEFFICIENT. He also developed the invaluable CHI-SQUARED TEST.

Educated at Cambridge University, Pearson held a faculty position at University College, London, for the most part of his career, studying the analysis of heredity and evolution in biology. Beginning in 1893 Pearson published a series of 18 papers all titled "Mathematical

Karl Pearson, an eminent statistician of the 20th century, was the first to develop fundamental concepts such as "standard deviation" and "correlation coefficient." *(Photo courtesy of Topham/Fotomas/The Image Works)*

Contribution to the Theory of Evolution," that, over their 21-year span, outlined the principles of statistical analysis. His work, however, was not without controversy. While other statisticians at the time, including SIR RONALD AYLMER FISHER (1890–1962) and WILLIAM SEALY GOSSET (1876–1937), emphasized the role of causes rather than correlation, and worked with small sample sizes in deducing reliable information, Pearson analyzed large sample sizes and explored overall trends. This difference in philosophy led to a bitter dispute between Pearson and Fisher, sufficiently vehement to cause Fisher to turn down a prestigious position at the college so as to simply avoid working with Pearson. Nonetheless, Pearson's work was duly recognized as significant, and Pearson was awarded the position as the first Galton Professor of Eugenics at the Galton National Laboratory of Eugenics. He was chair there from 1911 to 1933.

Apart from making significant contributions to the field of statistics, Pearson also practiced law, was active in politics, and wrote literary works. In 1892 he published *The Grammar of Science*, a philosophical text that attempted to extend the approach and methods of science to a wide range of general pursuits. This piece even anticipated some of the ideas of relativity theory.

Pearson founded the journal *Biometrics* and was the editor of *Annals of Eugenics*. Because of his fundamental work in the development of modern statistics, many scholars today regard Pearson as the founder of 20th-century statistics. He died in Coldharbour, England, on April 27, 1936.

See also SIR FRANCIS GALTON, HISTORY OF PROBABILITY AND STATISTICS (essay); STATISTICS: INFERENTIAL.

pedal triangle Given a point *P* in a triangle *ABC,* the pedal triangle with respect to *P* is formed by dropping perpendicular lines from *P* to each of the three sides of the triangle. If *X, Y,* and *Z* are the locations at which these lines meet the sides of the triangle, then triangle *XYZ* is the pedal triangle for *P.* The pedal triangle may extend outside the triangle if *ABC* is an obtuse triangle.

The three altitudes of any triangle meet at a common point *O* called the orthocenter of the triangle. (*See* TRIANGLE.) The pedal triangle with respect to *O*, with vertices the feet of the altitudes of the triangle, is often referred to as *the* pedal triangle of a given triangle. It is

also called the orthic triangle. Geometers have shown that if one constructs a nested sequence of three orthic triangles, then the third orthic triangle is similar to the original triangle.

The pedal circle of a triangle with respect to a point *P* is the circle that passes through all three vertices of the pedal triangle with respect to *P.* The pedal circle with respect to the orthocenter *O* is the inscribed circle of the triangle.

In 1775 Italian mathematician Giovanni Fagnano proposed the following problem:

Of all triangles inscribed in an acute triangle, which has the least perimeter?

Using the techniques of CALCULUS, Fagnano was able to prove that the orthic triangle is the desired inscribed triangle. (This can also be proved by pure geometric means.) This observation leads to an interesting result: if *XYZ* is the inscribed triangle of least perimeter, then the two sides *X* to *Y* and *Y* to *Z* are a solution to the famous path-walking problem in OPTIMIZATION theory. Consequently these two sides intercept the "wall" of the given triangle at equal angles, and so represent the path a ball would follow when thrown at that side of the triangle. As the same argument applies to any pair of sides in the orthic triangle, we have:

The orthic triangle represents the closed path of a ball bouncing inside an acute triangle.

Mathematicians have proved that this is the only closed-circuit path a ball could follow inside an acute triangle. For obtuse triangles there is no inscribed triangle of least perimeter.

pentagram (**pentacle, pentalpha, pentangle**) The star-shaped figure formed by the five diagonals of a regular pentagon (or equivalently, formed by extending all sides of a regular pentagon to meet in pairs) is called a pentagram. The Pythagorean Order, a group of followers of PYTHAGORAS, bestowed deep significance to various numbers and geometric shapes, including the pentagram. The figure is often referred to as the "Pentagram of Pythagoras." Its geometric properties were considered divine.

The GOLDEN RATIO appears a number of times in a pentagram. For instance, point *C* divides the length *AD*

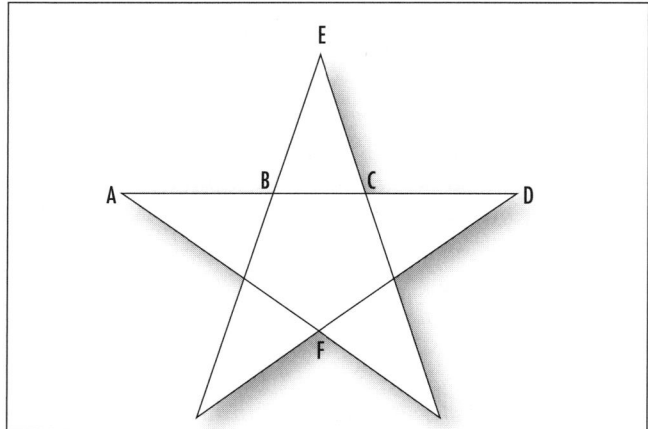

Pentagram

into two parts of ratio equal to the golden mean, and point B divides length AC in the same manner. This, of course, happens on all sides of the star. Also, the lengths AF, DF, AE, and DE are all equal, making AFDE a rhombus.

If the distance from A to D is 1, then:

$$AB = CD = \frac{3-\sqrt{5}}{2}$$
$$BC = \sqrt{5} - 2$$
$$AF = DF = AE = DE = \frac{\sqrt{5}-1}{2}$$

The distance of the center of the pentagram to point C is $\frac{1}{\sqrt{10}}\sqrt{25-11\sqrt{5}}$ and to the point E is $\frac{\sqrt{5}+1}{2\sqrt{5}}\sqrt{5-2\sqrt{5}}$. The height of the pentagram is $\frac{1}{2\sqrt{2}}\sqrt{5+\sqrt{5}}$.

percentage A number that represents a fraction of 100 is called a percentage. For example, 5 percent, written 5%, is the fraction $\frac{5}{100}$, and 125 percent is the fraction $\frac{125}{100}$. The term *percent* comes from the Latin phrase *per centum*, meaning "by the hundred." It is said that the Roman Emperor Augustus (63 B.C.E.–14 C.E.)

levied a tax of one part per 100 on the proceeds of all goods sold at auctions and markets. Since then it has been popular to represent all taxes in terms of percentages.

Any fraction can be converted to a percentage by multiplying by 100. For example, the fraction $\frac{4}{5}$ is interpreted as "four-fifths of 100" and so is written: $\frac{4}{5} \times 100 = \frac{400}{5} = 80$ percent. Decimals can also be represented as percentages by multiplying by 100: 0.327, for example, as a percentage equals $0.327 \times 100 = 32.7$ percent.

To convert a percentage back to a fraction or a decimal, divide by 100 and omit the percentage sign. For example, 85 percent represents "85 parts of 100" and so is the fraction 85/100 or, equivalently, the decimal 0.85.

One hundred percent of a quantity is 100/100 of it, that is, all of it. For example, if 23 students are enrolled in a dance and all 23 appear on a particular day, then 100 percent of the class is present. If only 20 appear, then only the fraction $20/23 \approx 0.87$ of the class is present, that is, 87 percent.

A percentage value larger than 100 percent represents a quantity larger than the original whole. For example, if the Boys Soccer Group sold 1,850 chocolate bars in 2002, and 1,998 chocolate bars in 2003, then the number of bars sold in 2003 compared with the number sold the previous year is $\frac{1998}{1850} = 1.08$ or 108 percent. One would call this an 8 percent increase in chocolate bar sales. In general, if a quantity changes from value a to value b, then the percentage increase of the value is given by the formula: $\frac{(b-a)}{a} \times 100$ percent.

It is said that the symbol % for percentage likely evolved from shorthand notations used by 15th-century Italian clerks for the term *per cento*. Such abbreviations included: per 100, P 00 and Pc°, with the c in the final expression eventually being replaced with a diagonal line. A 1425 manuscript shows the symbol % in use.

The term *permillage* is used for parts per thousand with accompanying symbol: o/oo. For example, a quantity expressed as 543 o/oo represents 543 parts per thousand, or the fraction 543/1,000.

percentage error If the ERROR or uncertainty in a measurement of a quantity is expressed as a PERCENT-

AGE of the total measurement, then the error is called a percentage error. For example, when 1.2 is used as an approximation for the quantity 1.16, then the percentage error is $\frac{0.04}{1.16} \times 100 \approx 3.4\%$. If, in measuring the height of a building as 62 feet, engineers use a device that measures to the nearest half a foot, then the height of the building should be written: 62±0.5 feet. The percentage error of this measurement is: $\frac{0.5}{62} \times 100 \approx 0.81$ percent.

See also RELATIVE ERROR.

percentile (centile, quartile) One of the 99 values that divide a set of data arranged in numerical order into 100 equal parts is called a percentile. For example, the 90th percentile is the value such that 90 percent of the data points are below that value. Often scores in standardized tests are presented in terms of percentiles. For example, if 525 students take an exam and 95 percent of the students receive a score of 73 or lower, then the 95th percentile for that exam is 74.

It is often deemed convenient to divide data sets into four equal parts. The lower (or first) quartile, denoted Q_1, is the 25th percentile. The middle (or second) quartile, Q_2, the median, is the 50th percentile, and upper (or third) quartile, Q_3, is the 75th percentile.

perfect number A whole number that is equal to the sum of its FACTORS—excluding the number itself—is called a perfect number. For example, the numbers 6, 28, 496, and 8128 are each perfect.

$$6 = 1 + 2 + 3 = 2^1(2^2 - 1)$$
$$28 = 1 + 2 + 4 + 7 + 14 = 2^2(2^3 - 1)$$
$$496 = 1 + 2 + 4 + 8 + 16 + 31 + 62 + 124 + 248$$
$$= 2^4(2^5 - 1)$$
$$8,128 = 1 + 2 + 4 + 8 + 32 + 64 + 127 + 254 + 508$$
$$+ 1016 + 2032 + 4064 = 2^6(2^7 - 1)$$

Perfect numbers were deemed to have important numerological properties by ancient scholars, and were extensively studied by the Greeks. In book IX of *THE ELEMENTS*, written around 300 B.C.E., EUCLID proved that if $2^n - 1$ is PRIME, then $2^{n-1}(2^n - 1)$ is a perfect number. At the turn of the 17th century, French philosopher and mathematician MARIN MERSENNE undertook a study to determine which numbers of the form $2^n - 1$ are prime. Any such number is today called a MERSENNE PRIME.

Around 100 B.C.E. Syrian scholar Nicomachos of Gerasa suggested that the converse of Euclid's result is true, that is, that every perfect number *must* be of the form $2^{n-1}(2^n - 1)$ with $2^n - 1$ prime. (In particular, then, there can be no odd perfect numbers.) In 1750 LEONHARD EULER proved that indeed every even perfect number must be of this form, partially proving Nicomachos's claim. Surprisingly, still to this day no one knows whether or not examples of odd perfect numbers exist. This remains a famous unsolved problem. It has been proved that if any odd perfect number exists, it must be larger than a GOOGOL.

Euler's result shows that every even perfect number is a triangular FIGURATE NUMBER. Explicitly, $2^{n-1}(2^n-1) = \frac{2^n(2^n - 1)}{2}$ is the formula for the $(2^n - 1)$-th triangular number.

It is also known that every even perfect number must end with either a 6 or an 8 (this was conjectured by Nicomachos) and that every even perfect number greater than 6 is a sum of consecutive odd cube numbers:

$$28 = 1^3 + 3^3$$
$$496 = 1^3 + 3^3 + 5^3 + 7^3$$
$$8128 = 1^3 + 3^3 + \ldots + 15^3$$

for example, and that every even perfect is one more than a multiple of 9.

The sum of the inverses of all the factors of a perfect number is always 2. For example, $\frac{1}{1} + \frac{1}{2} + \frac{1}{3} + \frac{1}{6} = 2$ and $\frac{1}{1} + \frac{1}{2} + \frac{1}{4} + \frac{1}{7} + \frac{1}{28} = 2$. (This follows from the fact that the sum of all the factors of a perfect number N equals $2N$. Dividing the sum through by N produces the result.)

It is not known if there are infinitely many perfect numbers. Fewer than 50 perfect numbers are known.

A number that is not perfect is classified as either abundant or deficient according to whether the sum of all its factors—excluding the whole number itself—is greater than or less than the number. For example, 18 is abundant, since the sum of its factors is more than 18 (we have $1 + 2 + 3 + 6 + 9 = 21 > 18$), whereas 25 is deficient because its factors sum to less than 25. (We have $1 + 5 = 6 < 25$). All prime numbers are deficient.

See also AMICABLE NUMBERS; GREEK MATHEMATICS.

perigon (round angle) An angle of 360° is called a perigon. It represents one full turn.

perimeter The length of the boundary of a plane figure is called its perimeter. For example, the perimeter of a rectangle is twice its length plus twice its width. The perimeter of a circle is its circumference. Surprisingly, it is possible for a plane figure of finite AREA to have an infinite perimeter.

See also FRACTAL.

period *See* PERIODIC FUNCTION.

period doubling *See* DYNAMICAL SYSTEM.

periodic function A function is periodic if it repeats itself at regular intervals of the variable. For example, the function $\sin(x)$ is periodic because it cycles every 360°, precisely: $\sin(x) = \sin(x + 360°)$ for all values of x. We also have that $\sin(x) = \sin(x + 720°)$ and $\sin(x) = \sin(x + 1080°)$, for example. The smallest positive value p for which a periodic function f satisfies $f(x) = f(x + p)$ for all values x is called the period of the function. The function $\sin(x)$ has period 360°, as does the function $\cos(x)$. (The tangent function $\tan(x)$, however, has period 180°.)

In physics, any phenomenon that repeats itself at regular intervals, such as the swinging of a pendulum, the vibration of a source of a sound, or the rotation of the Earth, is called periodic, and the time it takes for one complete cycle of the phenomenon is called its period. This idea extends to other branches of mathematics as well. For example, the repeating decimal $45.76185185185185185... = 45.76\overline{185}$ has period 3, since three digits are being repeated, and a rotation of 60° about a point in the plane has period 6, since six rotations of this type return points to their original locations.

See also DYNAMICAL SYSTEM; FOURIER SERIES.

permutation (arrangement, order) A specific ordered arrangement of a given collection of objects is called a permutation. For example, a selection of a winner, a first runner-up, and a second runner-up from a group of three finalists in a competition would be a permutation of those three participants. The lists ADBC and BDCA are two permutations of the letters A, B, C, and D.

The MULTIPLICATION PRINCIPLE shows that n distinct objects can be ordered $n!$ different ways. There are thus $4! = 24$ different permutations in all of the letters A, B, C, and D. (*See* FACTORIAL.)

The number of ways to arrange just r objects from a collection of n distinguishable objects ($r \le n$) is denoted P_r^n. These are called "permutations taken r at a time." For example, there are 12 permutations of A,B,C,D taken two at a time: AB, AC, AD, BA, BC, BD, CA, CB, CD, DA, DB, DC. Thus $P_2^4 = 12$. Again, the multiplication principle shows that

$$P_r^n = \frac{n!}{(n-r)!}$$

In particular $P_n^n = n!$.

Permutations of the full set of numbers $\{1,2,...,n\}$ can be classified as either even or odd by counting the number of times a large number appears to the left of a smaller number. For example, in the permutation 25143 of the first five counting numbers, the number 2 appears to the left of 1 ; 5 appears to the left of 1, 4, and 3; and 4 appears to the left of 3. In total, a large number appears to the left of a small number an odd number of times. This permutation is odd.

A transposition is a permutation that interchanges just two objects. For example, 14325 represents a transposition (the 4 and the 2 have switched places). One can check that it is possible to return the entries of any even permutation back to their original order by applying an even number of transpositions, and the entries of an odd permutation by an odd number of transpositions. This is often taken as the definition of what it means to say that a permutation is even or odd.

A permutation in which no element appears in its original location is called a derangement. For example, BCAED is a derangement of A, B, C, D, E. There are precisely

$$n!\left(1 - \frac{1}{1!} + \frac{1}{2!} - \frac{1}{3!} + \cdots + \frac{(-1)^n}{n!}\right)$$

derangements of n objects.

To prove this, let $S_{i_1 i_2 \dots i_k}$ denote the number of permutations that keep the elements $i_1, i_2, \dots i_k$ fixed in place. (Thus the remaining $n - k$ objects may be permuted in any fashion.) We have $S_{i_1 i_2 \dots i_k} = (n - k)!$. There are $\binom{n}{k}$ ways to choose k objects, and by the INCLUSION-EXCLUSION PRINCIPLE there are thus

$$\left(S_1 + S_2 + \dots + S_n\right) - \left(S_{12} + S_{13} + \dots + S_{n-1,n}\right)$$
$$- \left(S_{123} + \dots + S_{n-2,n-1,n}\right) + \dots \pm S_{12\dots n}$$
$$= \binom{n}{1}(n-1)! - \binom{n}{2}(n-2)! + \dots + \binom{n}{n}0!$$

permutations that keep at least one object in its correct location. The formula above follows readily.

The PROBABILITY that a permutation chosen at random is a derangement is this number divided by $n!$, the total number of permutations. If n is large, this is very close in value to the infinite series:

$$1 - \frac{1}{1!} + \frac{1}{2!} - \frac{1}{3!} + \dots$$

which happens to be the TAYLOR SERIES of e^x evaluated at $x = -1$. This solves the famous "hatcheck problem":

> A group of gentlemen check their hats at a country-club cloakroom. The clerk loses records of which hat belongs to whom and starts handing out hats randomly as the men leave. What is the probability that no man receives his own hat?

The answer is approximately $e^{-1} = \dfrac{1}{e} \approx 0.37$.

Permutations are also used to analyze the SLIDE FIFTEEN PUZZLE. The number of permutations of n objects in which each object moves at most one place, either left or right, from its original location is F_{n+1}, the $(n+1)$-th FIBONACCI NUMBER.

See also BINOMIAL COEFFICIENT; COMBINATION; FUNCTION; SUBFACTORIAL.

perpendicular This term is used in any setting to describe two geometric constructs that meet at right angles. For example, two lines are perpendicular if the angle between them is 90°. To "drop a perpendicular" from a point to line is to draw a line segment that starts at the given point and meets the line at right angles. The length of this line segment is the shortest distance between the given point and points on the line. This length is called the perpendicular distance of the point from the line.

The perpendicular bisector of a line segment is the line at right angles to the segment passing through its midpoint. It represents all points in the plane that are equidistant from the two endpoints of the given line segment. Two planes are perpendicular if they meet at right angles.

The term *perpendicular* tends to be used primarily in discussions about lines and planes. Other geometric quantities (such as VECTORs and curved surfaces) might meet at right angles, but mathematicians tend to use the word ORTHOGONAL in these settings. For example, two vectors are orthogonal if the angle between them is 90°. This convention is not steadfast. Often the words *perpendicular* and *orthogonal* are used interchangeably.

See also DOT PRODUCT; NORMAL TO A CURVE; NORMAL TO A PLANE; NORMAL TO A SURFACE; SLOPE.

perspective Two planar figures are said to be in perspective from a point P if, for each point A on one figure, there is a corresponding point B on the other so that the line connecting P to A also meets B. The point P is called the center of perspective.

The notion of perspective played a significant role in the development of artistic techniques in the 15th century. In an attempt to capture a sense of depth in a two-dimensional painted scene, Renaissance artists began drawing objects in the foreground larger than those of the same size in the background. They imagined a distant point at infinity at the back of the scene and drew guidelines emanating from this point across the canvas to aide in creating three-dimensional realism. (Today we call such a construct a central PROJECTION.) Artists and scholars ALBRECHT DÜRER (1471–1528) and GIRARD DESARGUES (1591–1661) were the first to study the mathematics of perspective.

Mathematicians also say that two planar figures are in perspective from a line if corresponding sides of each shape, if extended, meet at points that all lie on a common line. DESARGUES'S THEOREM shows that if two planar figures are in perspective from a point, then they are also in perspective from a line, and vice versa. (Although the theorem is stated only for triangles, any polygonal figure can be subdivided into triangles for which the theorem applies.)

pi The real number, denoted, π, defined as the ratio of the circumference C of a CIRCLE of any size drawn in a plane to its diameter D, is called pi:

$$\pi = \frac{C}{D}$$

Its value is $\pi = 3.141592653589793...$ That the value of this ratio is the same for all planar circles is not immediately obvious. To see why this is the case, note that if any diagram or picture is scaled up or down by a factor k, then all lengths in that picture change by a factor k. In particular, for a picture of a circle, the length of its circumference changes to kC and the length of its diameter to kD. Consequently, the ratio of circumference to diameter remains unchanged: $\frac{kC}{kD} = \frac{C}{D}$. That any two circles in the plane can be thought of as scaled versions of each other thus explains why the value of π is the same for all circles. (It is worth mentioning that the property of scaling used here is a phenomenon of EUCLIDEAN GEOMETRY and that the value of π is not the same for all circles in SPHERICAL GEOMETRY or HYPERBOLIC GEOMETRY. For example, the value of π changes for circles drawn on spheres, ranging in value anywhere between 2 and 3.141592...)

A study of AREA shows that the area of a circle of radius r is given by:

$$A = \pi r^2$$

Thus π can also be defined as the ratio of the area of a circle to its radius squared.

Although it is not standard, one can associate a value π to other shapes as well. For instance, the ratio of the perimeter of a square to its width (short diameter) is the same for all squares. Thus we may say: $\pi_{\text{square}} = 4$. (Using the diagonal of the square as the long diameter produces instead the value $\pi_{\text{square}} = \frac{4}{\sqrt{2}} \approx 2.828$.) The ratio of the perimeter of an equilateral triangle to its diameter (measured as the height of the triangle) is the same for all equilateral triangles. We have: $\pi_{\text{triangle}} = \frac{3}{\frac{\sqrt{3}}{2}} \approx 3.464$. In fact, in precisely this way, by inscribing polygons with increasingly many sides within a circle and calculating the associated values of π for these polygons, ARCHIMEDES OF SYRACUSE (ca. 287–212 B.C.E.) showed that the true value of π for circles lies between $\frac{223}{71} = 3 + \frac{10}{71}$ and $\frac{22}{7} = 3 + \frac{1}{7}$. The approximation $\frac{22}{7}$, correct to two decimal places, is still often used today.

Since the time of antiquity, scholars have attempted to find the exact value of π. The RHIND PAPYRUS from ancient Egypt (dated ca.1650 B.C.E.) describes a procedure for computing the area of a circle that is equivalent to using the approximate value $\frac{256}{81}$ for π. The Babylonians of the same era used the approximate value $\frac{25}{8}$. Around 150 C.E. Greek astronomer CLAUDIUS PTOLEMY established that $\pi \approx \frac{377}{120}$, and 300 years later Chinese scholar ZU CHONGZHI used the improved approximation $\pi \approx \frac{355}{113}$, correct to six decimal places. About 530 C.E., Hindu scholar ĀRYABHATA established the close approximation $\frac{62,832}{20,000} = \frac{3,927}{1,250}$ for π, and around 1150 C.E. BHĀSKARA, after using this estimate multiple times in his work, noted that the $\sqrt{10}$ serves as a sufficiently accurate approximation for most practical purposes.

In 1429 Arab mathematician JAMSHID AL-KASHĪ computed the value of π correct to 16 decimal places, and in 1610, using a polygon with 2^{62} sides to approximate a circle, German mathematician Ludolph van Ceulen accurately computed the first 35 decimal places of π. He devoted most of his life to the task.

In 1767 Swiss mathematician JOHANN HEINRICH LAMBERT proved that π is an IRRATIONAL NUMBER, thereby establishing that the decimal expansion for π will never terminate nor fall into a repeating cycle. Later, in 1873, English scholar WILLIAM SHANKS, after 15 years of work, computed, by hand, the decimal expansion of π to 607 places. Unfortunately he made an error in the 527th place, making the decimal expansion thereafter incorrect, but no one noticed the mistake for almost a century.

In 1882 German mathematician FERDINAND VON LINDEMANN proved that π is a TRANSCENDENTAL NUMBER, establishing once and for all that the problem of SQUARING THE CIRCLE cannot be solved.

With the advent of computing machines in the 20th century, mathematicians could compute more and more digits of π. In 1949 JOHN VON NEUMANN used the U.S. government's ENIAC computer to compute π to the 2,037th decimal place. (It took 70 hr of machine time.) In 1981 Japanese scientists Kazunori Miyoshi and Kazuhiko Nakayama evaluated 2 million decimal places of π, and in 1991, using a homebuilt supercomputer in a New York City apartment, brothers Gregory and David Chudnovsky calculated π to 2,260,321,366 decimal places. Today over 1.24×10^{12} digits of π are known.

The sequence of digits 0123456789 appears in the decimal expansion of π beginning at the 17,387,594,880th decimal place. This is the first, but not the only, appearance of this sequence. The 9876543210 first appears at the 21,981,157,633rd decimal place.

There are many beautiful formulae for π. For instance, VIÈTE'S FORMULA, the GREGORY SERIES, the ZETA FUNCTION, and WALLIS'S PRODUCT show, respectively, that:

$$\frac{2}{\pi} = \sqrt{\frac{1}{2}} \times \sqrt{\frac{1}{2} + \frac{1}{2}\sqrt{\frac{1}{2}}} \times \sqrt{\frac{1}{2} + \frac{1}{2}\sqrt{\frac{1}{2} + \frac{1}{2}\sqrt{\frac{1}{2}}}} \times \ldots$$

$$\frac{\pi}{4} = 1 - \frac{1}{3} + \frac{1}{5} - \frac{1}{7} + \frac{1}{9} - \ldots$$

$$\frac{\pi^2}{6} = 1 + \frac{1}{2^2} + \frac{1}{3^2} + \frac{1}{4^2} + \ldots$$

$$\frac{\pi}{2} = \frac{2}{1} \times \frac{2}{3} \times \frac{4}{3} \times \frac{4}{5} \times \frac{6}{5} \times \frac{6}{7} \times \frac{8}{7} \times \frac{8}{9} \times \ldots$$

The BUFFON NEEDLE PROBLEM also provides another surprising appearance of the π. The Swiss mathematician LEONHARD EULER (1707–83) also showed:

$$\pi = 4\tan^{-1}\left(\frac{1}{2}\right) + 4\tan^{-1}\left(\frac{1}{3}\right)$$

$$\pi = 4\tan^{-1}\left(\frac{1}{7}\right) + 8\tan^{-1}\left(\frac{1}{3}\right)$$

$$\pi = 20\tan^{-1}\left(\frac{1}{7}\right) + 8\tan^{-1}\left(\frac{3}{79}\right)$$

(Similar formulae follow from the general identity:

$$\tan^{-1}\left(\frac{1}{x}\right) = \tan^{-1}\left(\frac{1}{x+y}\right) + \tan^{-1}\left(\frac{y}{x^2+xy+1}\right)$$

for suitable choices of x and y.) Hungarian mathematician PAUL ERDÖS (1913–96) established the following remarkable result:

Beginning with a positive integer n, round it up to the nearest multiple of $n - 1$, and then round the result up to the nearest multiple of $n - 2$, and so on, up until the nearest multiple of 2. Call the result $f(n)$. (We have, for instance, $f(3) = 4$, $f(5) = 10$, and $f(7) = 18$.) Then the LIMIT of the ratio of n^2 to $f(n)$, as n becomes large, is π:

$$\lim_{n\to\infty} \frac{n^2}{f(n)} = \pi$$

A number of basic questions about the interplay between π and Euler's number e remain unanswered. For instance, no one yet knows whether the numbers $\pi + e$, $\frac{\pi}{e}$, or $\log_e(\pi)$ are rational or irrational (nor whether π^π is algebraic or transcendental). It is curious that the quantity $e^\pi - \pi$ has a value extraordinarily close to 20.

Our choice to use the symbol π to denote the ratio of the circumference of a circle to its diameter is due to British mathematician William Jones (1675–1749), who first used it in his 1706 publication *Synopsis palmariorum matheseos*. It is believed that he chose it because π is the initial letter of the Greek word περιφέρεια for "periphery." Euler followed Jones's choice and popularized the use of this symbol in his influential 1736 text *Mechanica*.

The number π has captured the interest of many mathematical enthusiasts. There are clubs across the globe for those who can recite, from memory, the first 100 and even the first 1,000 digits of π. Some people declare March 14 "pi-day," and deem the time 1:59 of that day significant. (This matches the decimal expansion 3.14159...) There is a popular mnemonic for memorizing the first 12 digits of π:

See. I have a rhyme assisting my feeble brain,
its tasks ofttimes resisting.

The letter count of each word in this sentence matches the first 13 digits of the decimal expansion π.

In 1897 amateur mathematician Dr. E. J. Goodwin, of Solitude, Indiana, thought he had discovered a remarkable relationship between the circumference of a circle and its area. Rather than share his discovery with the mathematical community, Goodwin took his "discovery" to his representative in the Indiana General Assembly and persuaded him to introduce the following house bill:

> Be it enacted by the General Assembly of the state of Indiana, that it has been found that a circular area is equal to the square on a line equal to the quadrant of the circumference.

Clearly no politician understood the mathematics of this assertion, for the bill passed and Goodwin was well on his way to persuading the state of Indiana that the area of a circle of circumference C is given by $\left(\dfrac{C}{4}\right)^2$, or, equivalently, that the value of π is exactly 4. Fortunately, due to press attention and the outrage of mathematicians living in the state, the bill was derailed before it reached full senate approval and was made into law. Senator Hubbell, an opponent of the bill who also saw the absurdity of the claim, aptly remarked: "The senate might as well try to legislate water to run uphill."

See also E; EULER'S FORMULA; RADIAN MEASURE; SCALE.

pie chart (**pie graph**)   *See* STATISTICS: DESCRIPTIVE.

pigeonhole principle (**Dirichlet's principle, drawer principle, letterbox principle,** *schubfachprinzip*) Attributed to PETER GUSTAV LEJEUNE DIRICHLET (1805–59), the pigeonhole principle observes that if *n* objects are put in *p* boxes, and *n* > *p*, then at least one box receives two or more objects. The example of housing *n* pigeons in *p* pigeonholes is often cited. This principle can also be stated in the following alternative form:

> If each of *n* objects is assigned one of *p* labels, and *n* > *p*, then at least two objects have the same label.

Thus, for example, if we pull out three socks from a drawer that contains only black socks and white socks (three objects, two labels), at least two of those socks will be the same color.

This straightforward principle has a number of surprising consequences. For instance:

> There exist two (nonbald) people in Sydney, Australia, with exactly the same number of hairs on their heads.

(Label each person by the number of hairs on his or her head. It is known that no human has more than 1 million hairs, and there are more than this many people in Sydney.)

> Any collection of 18 integers contains two integers that differ by a multiple of 17.

(Assign to each integer its remainder when divided by 17. Noting that there are 17 possible remainders—0 through 16—any collection of 18 integers must contain two members with the same remainder. Their difference is a multiple of 17.)

> In any group of six people, there are two who have an identical number of friends in the group.

(Assign to each person the number of friends he or she possesses in the group. The six numbers 0 through 5 are possible, but not both 0 and 5 simultaneously. Thus, among the six in the group, only five labels are possible. Two people must possess the same value.)

The pigeonhole principle can be generalized as follows:

> If each of *nk* + 1 objects is assigned one of *n* labels, then at least *k* + 1 objects are assigned the same label.

(If each label is used at most *k* times, then at most *nk* objects are labeled. As there are more objects than this, some labels must be used more than *k* times.) As consequences we have, for instance:

> Among 2,000 people, at least five were born on the same day of the year.

(Label each person with the day on which he or she was born. As there are 366 possible labels and 2,000 = 366 × 5 + 170 > 366 × 5 + 1, at least five people have the same label.)

If 865 points are scattered on a square sheet of paper 1 ft wide, then some six of those points will be clustered sufficiently close so as to lie in a square 1 in. wide.

(Divide the sheet of paper into 144 squares each of side-length 1 in. and label each point by the square in which it lies. For points that fall on a line between two squares, arbitrarily decide to which square it belongs. As $865 = 144 \times 6 + 1$, one of these small squares contains six points.)

The following principle is sometimes considered a variation of the pigeonhole principle:

If n numbers sum to S, then not all the numbers are larger than $\frac{S}{n}$. Also, not all the numbers are smaller than $\frac{S}{n}$.

As a simple consequence we have:

If wages of five workers summed to $100,000, then at least one worker earned no more than $20,000.

See also COMBINATORICS.

place-value system   *See* BASE OF A NUMBER SYSTEM.

plane   Informally, a plane is a flat surface having no edges and extending infinitely in all directions. More precisely, a plane is a geometric surface, without edges, with the property that any straight line connecting two points in the surface remains in the surface.

Planes are regarded as two-dimensional—within the surface there are only two independent directions of motion: "left and right" and "back and forth." Any other motion (diagonal motion, for example) can be thought of as a combined effect of these two motions.

Sitting in three-dimensional space, a plane is thought of as a geometric object with no thickness. It is possible to write down an equation for such a plane. (*See* VECTOR EQUATION OF A PLANE.)

Plane geometry is the study of relationships between points, lines, and curves lying in the same plane. A plane figure is a figure, such as a square, circle, or a triangle, that lies in a plane. Points that all lie on the same plane are called coplanar.

Two planes sitting in three-dimensional space intersect if they share points in common. If they do, the set of points they share forms a straight line. Two planes that do not intersect are called parallel. Three planes sitting in three-dimensional space can intersect in a point, along a line, or not at all.

In both two- and three-dimensional space, two points P and Q determine a line. In three-dimensional space, three points P, Q, and R determine a plane (under the proviso that P, Q, and R do not themselves lie on a straight line). To see this, consider the line that passes through P and Q. There are infinitely many planes that contain this line. The point R determines which of these planes to choose. (Alternatively, using VECTOR analysis, the POSITION VECTORS $\overrightarrow{PR}$ and $\overrightarrow{QR}$ are two vectors in the desired plane, and so their CROSS PRODUCT $\mathbf{n} = \overrightarrow{PR} \times \overrightarrow{QR}$ is a normal to the plane. This is all we need to write down the VECTOR EQUATION OF A PLANE.) Any three points in three-dimensional space are coplanar.

Four points might or might not all lie on the same plane. This is the reason why four-legged chairs are unlikely to be stable on rough, uneven floors. The tips of the chair legs represent four coplanar points, but the locations at which you wish to place these points on an uneven floor might not be. A three-legged stool, however, will always be stable, as any three points on the uneven ground will be coplanar. For this reason, barn stools used for milking, say, are traditionally made three-legged.

Plato (ca. 428–348 B.C.E.) *Greek Philosophy*   Born in Athens, Greece, philosopher Plato is remembered in mathematics for promoting the notion that mathematical concepts have a real existence independent of human thought. This is consequently linked to the argument that mathematicians *discover* mathematics, rather than *create* it. Plato deemed mathematics as an essential part of a valued education, and he greatly influenced the esteem with which mathematics is regarded in the Western world. He was also the first to describe the five PLATONIC SOLIDS.

Plato was a worldly man very much interested in politics and human affairs. He was a member of the military service during the Peloponnesian War

(431–404 B.C.E.) between Athens and Sparta, and he traveled through Egypt and Italy once democracy was restored to Athens. In Egypt, Plato learned of the water clock and later brought the idea back to Greece. In Italy he encountered the followers of PYTHAGORAS and was greatly influenced by the mathematics they studied. Although Plato made no direct contributions to mathematics, his respect for the subject was profound, and he wrote about the philosophy of the subject at great lengths in his works.

Around 387 B.C.E. Plato returned to Athens to found a school devoted to instruction and research in philosophy and the sciences. He secured a piece of land from the Greek hero Academos, and the school became known as the Academy. Plato presided over the institute for the remainder of his life.

Plato wrote a total of 35 dialogues, fictitious conversations between characters, cleverly structured so as to reveal a philosophical line of thought and investigation. In his dialogue *Phaedo,* Plato described mathematical objects as perfect forms (a line, for instance, really is an object possessing no breadth), and professed that the objects we see and can construct in the physical universe are imperfect imitations of these ideal forms. He also discussed these ideas in his work the *Republic.* It is clear from his dialogues that Plato regarded mathematics as a source of universal truth and that the study of mathematics offered the best tool for fostering intellectual development.

In his work *Timaeus,* Plato described the five Platonic solids as fundamental units of the universe. He suggested, for instance, that each of the four "elements"—earth, fire, air, and water—is appropriately modeled by the shape of a Platonic solid, namely, the CUBE, the TETRAHEDRON, the octahedron, and the icosahedron, respectively. The fifth Platonic solid, the dodecahedron, with its 12 faces matching the 12 signs of the zodiac, represented the universe itself. German astronomer JOHANNES KEPLER (1571–1630) was so smitten with the proposed harmony of the Platonic solids that he later developed a surprisingly accurate model of the solar system based upon them.

Plato's attitude toward, and regard for, the very nature of mathematical thinking had a lasting influence on the entire course of scientific progress in the West. Apart from promoting the value of pure mathematics, Plato also emphasized the need for rigorous proof, thereby laying down a philosophical foundation for others, most notably EUCLID, on which to pursue research. Plato is also credited for limiting the use of geometrical tools to the compass and straightedge alone.

Platonic solid (regular polyhedron) Any convex POLYHEDRON, all of whose faces are congruent regular polygons, is called a Platonic solid. For example, a CUBE, composed of six identical square faces, is a Platonic solid, as is a TETRAHEDRON composed of four identical equilateral triangular faces. Platonic solids have the property that not only are all of its faces identical, but all of its vertices are also alike in the sense that the same number of edges meet at each vertex of the polyhedron.

Regular polygons and regular polyhedra were studied extensively by the classical Greek scholars. In two-dimensional space, regular polygons with any desired number of sides exist. In three-dimensional space, however, there are only a small number of regular polyhedra. As his final proposition in his revered work *THE ELEMENTS,* the geometer EUCLID (ca. 300–260 B.C.E.) showed that there are only five Platonic solids, namely:

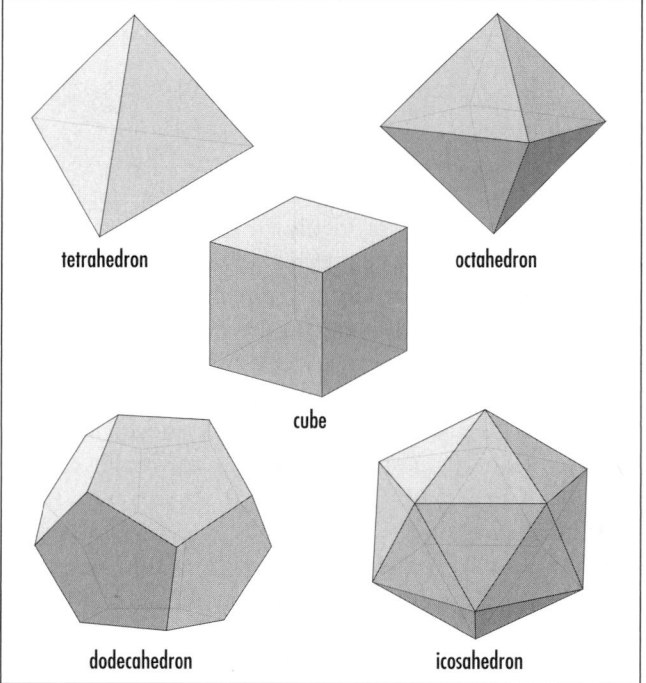

The Platonic solids

1. *Regular tetrahedron,* with four triangular faces and three edges meeting at each vertex
2. *Cube,* with six square faces and three edges meeting at each vertex
3. *Octahedron,* with eight triangular faces and four edges meeting at each vertex
4. *Dodecahedron,* with 12 pentagonal faces and three edges meeting at each vertex
5. *Icosahedron,* with 20 triangular faces and five edges meeting at each vertex

It is not difficult to see why there can be no other Platonic solids. Notice that for any polyhedron, at least three polygonal faces must meet at each vertex of the solid, and that, in order to form a peak at that vertex, the angles of the faces meeting at that vertex must sum to less than 360°. (If the sum of the angles around a vertex is precisely 360°, then the surface would be flat at the vertex. If, on the other hand, the angles sum to more than 360°, then the surface would cave in at that vertex.) Now consider the various regular polygons that might be used to construct a Platonic solid:

Equilateral Triangles: Each angle in an equilateral triangle is 60°. Thus three, four, or five such triangles could surround a vertex of a polyhedron, but not six or more. (The angle sum would no longer be less than 360°.) Each of these possibilities does in fact occur. These are the tetrahedron, the octahedron, and the icosahedron, respectively.

Squares: Each angle in a square equals 90°. Thus only three squares could possibly surround a vertex in a polyhedron. This does in fact occur. It is the cube.

Regular Pentagons: Each angle in a regular pentagon equals 108°, and so only three pentagons could possibly surround a vertex in a polyhedron. This does in fact occur. It is the dodecahedron.

Regular Hexagons and Higher: Each angle in a regular polygon with six or more sides is 120° or greater. No three of these figures could possibly surround a vertex in a polyhedron.

That each Platonic solid can be placed inside a SPHERE—the most perfect three-dimensional figure, according to the ancient Pythagorean sect of 500 B.C.E., in such a way that each vertex of the solid just touches the sphere—was deemed deeply significant in times of antiquity. The Greeks associated each Platonic solid with an "element" of the physical world. The tetrahedron, with its minimal volume per unit surface area, was fire, and the stable cube was solid earth. The less stolid octahedron was air, and the icosahedron, with the maximum volume per surface-area ratio, was water. The Pythagoreans, at first, were not aware of the existence of the dodecahedron, but they attributed the meaning of the entire universe to this figure upon its discovery. (Its 12 faces match the 12 signs of the zodiac.) The word *quintessence,* which today means the best or purest aspect of some nonmaterial thing, derives from the phrase *quinta essential* meaning the fifth element, namely, the dodecahedron representing all. The philosopher PLATO (ca. 428–348 B.C.E.) wrote extensively about the five regular polyhedra and their significance in his work *Timaeus.* This is the reason why we call them Platonic solids today.

German astronomer and mathematician JOHANNES KEPLER (1571–1630) also tried to attribute special meaning to the five Platonic solids. At his time, only five planets were known, and Kepler worked to relate the orbits of these planets in some way to the five special solids. He proposed that the orbit of each planet lay on a sphere and that the distance between successive spheres was precisely such that each of the Platonic solids, in turn, fits snugly between the two spheres, with the inner sphere just touching the faces of the polyhedron, and the vertices of the polyhedron just touching the outer sphere. Kepler later abandoned this theory. (He, in fact, is noted for discovering that the orbit of each planet is not a circle but an ELLIPSE.) It is interesting to note, however, that, allowing for eccentricity, his Platonic solid theory for the first five planets is accurate to within approximately 5 percent.

If we assume that each edge of a Platonic solid is 1 unit in length, then the VOLUME V of each solid is given as follows:

$$V_{\text{tetrahedron}} = \frac{\sqrt{2}}{12}$$

$$V_{\text{cube}} = 1$$

$$V_{\text{octahedron}} = \frac{\sqrt{2}}{3}$$

$$V_{\text{dodecahedron}} = \frac{15 + 7\sqrt{5}}{4}$$

$$V_{\text{icosahedron}} = \frac{5\left(3 + \sqrt{5}\right)}{12}$$

It is worth mentioning that, like the Platonic solids, there are only a finite number of semiregular polyhedra, that is, polyhedra with two types of faces, each a regular polygon with each vertex alike. For example, the classical pattern of 12 pentagons and 20 hexagons on a soccer ball represents a semiregular polyhedron. ARCHIMEDES OF SYRACUSE (287–212 B.C.E.) proved that there are only 13 semiregular polyhedra. For this reason, they are today called the Archimedean solids.

See also CONGRUENT FIGURES; POLYGON.

Playfair's axiom   *See* PARALLEL; PARALLEL POSTULATE.

plot

The act of specifying the location of a point on a set of coordinate axes is called "plotting" the point. If a collection of points are specified by an equation of the form $y = f(x)$, then a graph of the equation is sometimes called a plot.

See also CARTESIAN COORDINATES; COORDINATES; GRAPH OF A FUNCTION; SCATTER DIAGRAM.

plus

The symbol used to denote ADDITION is the plus sign, +. For example, the number 4 increased by the addition of two more units is written 4+2. In general, any operation that is to be interpreted as analogous to addition is denoted with a plus sign. For example, the GROUP operation in an abstract Abelian group is usually denoted +. As an adjective, the + symbol is used to describe a quantity of positive value. For example, +4 is a quantity 4 units greater than zero.

The symbol is believed to be derived as an abbreviation of the Latin word *et* for "and," which was often used to describe addition: 4 *and* 2 make 6, for instance. A 1489 Latin manuscript on arithmetic written by Johannes Widman contains the first known printed use of the symbol. It also contains the symbol "–" for SUBTRACTION, which is believed to have already been in common use in Germany for several decades.

The plus/minus symbol, ±, is used to denote a quantity that which should be both added and subtracted. For example, the two solutions to the QUADRATIC equation $x^2 - x - 1 = 0$ can be written $\frac{1 \pm \sqrt{5}}{2}$. If the choice of operation is important, mathematicians will introduce the symbol ∓. For example, the expression $a \pm b \mp c$ indicates that the second operation is to be different from the first. (Thus the expression can be interpreted as either $a + b - c$ or $a - b + c$, but not $a + b + c$ or $a - b - c$.) Matters are confusing, however. If the symbol ∓ is not used, then all choices of operations are deemed permissible. The expression $a \pm b \pm c$, for instance, could be any one of the four possibilities. There is no special symbol to indicate that both operations must be the same.

The plus/minus symbol was first used by French mathematician Albert Girard in his 1621 text *Tables*.

Poincaré, Jules-Henri (1854–1912) French *Analysis, Topology, Mechanics*

Born on April 29, 1854, in Nancy, France, mathematician Jules-Henri Poincaré is remembered as one of the great geniuses of all time who was active in almost every area of mathematics and physics. Poincaré founded the field of algebraic TOPOLOGY (the application of algebraic techniques to solve problems about space, shape, and form) and discovered an important class of functions called automorphic functions. (These functions, defined in the field of COMPLEX NUMBERS, are ratios of linear functions and have the property of being invariant under various symmetries of the complex plane.) In applied mathematics, he is remembered for his substantial work in the theory of celestial mechanics and for his work in optics, electromagnetism, thermodynamics, quantum theory, and the development of the theory of special relativity. (He is considered a codiscoverer of the special theory with ALBERT EINSTEIN and Hendrik Lorentz.) Poincaré published over 500 memoirs during his lifetime, as well as a number of popular books on the philosophy of mathematics and science.

After working as a mining engineer at Vesoul for several years, Poincaré received a doctoral degree in mathematics from the University of Paris in 1879. His thesis examined applications of DIFFERENTIAL EQUATIONS and served as a springboard for his later work in topology, automorphic functions, and in physics. Poincaré had a talent for recognizing connections between disparate topics of study, allowing him to solve problems from many different perspectives.

Poincaré published his first piece on the topic of topology in 1895 with his text *Analysis situs* (Analysis of position). At the same time he also outlined his method of homotopy theory based, essentially, on the simple idea that, although any loop drawn on the surface of a SPHERE can be continuously shrunk to a point (while remaining on the surface of the sphere), the same is not true for loops drawn on the surfaces of other three-dimensional objects. For example, not all loops drawn on the surface of a TORUS are homotopically equivalent to a point. Thus the study of loops on surfaces leads to a method for distinguishing surfaces. (This is important, for instance, if one is trying to determine the shape of mathematical objects that cannot be visualized.) Poincaré conjectured that any surface that has the same homotopy properties of a sphere is topologically the same as a sphere. This result, known as Poincaré's conjecture, has long been known to hold for one- and two-dimensional surfaces, and, since 1982, for four- and higher-dimensional surfaces. Work on proving the conjecture true for three-dimensional surfaces continues today.

From 1892 to 1899 Poincaré published his three-volume work *Les méthodes nouvelles de la méchanique céleste* (New methods in celestial mechanics), a landmark piece that aimed to characterize all possible motions of mechanical systems.

He was appointed chair of the faculty of science at the University of Paris in 1881, chair of mathematical physics and probability at the University of Sorbonne in 1886, and, in the same year, chair at the prestigious École Polytechnique. He held these chairs until his death in 1912. In 1887 Poincaré was elected, separately, to all five sections of the Académie des Sciences (mechanics, physics, geography, navigation, and geometry), and in 1908 he was also elected to the Académie Française. He also received an extraordinarily large number of awards and honors from other learned institutions around the globe, including the ROYAL SOCIETY SYLVESTER MEDAL in 1901 and the Bruce Medal of the Astronomical Society of the Pacific in 1911.

Poincaré died in Paris, France, on July 17, 1912.

point A location in space (or on a curve or a surface) is called a point. A point is usually specified by its COORDINATES in a coordinate system. For instance, a point in a three-dimensional system of CARTESIAN COORDINATES is given by a triple of numbers (x,y,z). In elementary GEOMETRY, a point is an undefined term but is loosely thought of as a geometric entity having no dimensions. The geometer EUCLID (ca. 300 B.C.E.) gave the vague description of a point as "that which has no breadth."

A collection of lines is said to be CONCURRENT if the lines meet at a common point. A TRIANGLE possesses many interesting points of concurrency.

In arithmetic, the decimal point is the symbol used to separate the integer part of a number from its fractional part. For example, the number 2.53 is read as "two point five three."

See also DECIMAL REPRESENTATION; NINE-POINT CIRCLE; PLOT.

point of contact (tangency point) The single point at which two curves, or two curved surfaces, touch, but do not cross, is called a point of contact. For example, the point at which a ball sitting on a table touches the surface of the table is the point of contact of a SPHERE and a PLANE.

See also TANGENT.

Poisson, Siméon-Denis (1781–1840) French *Probability theory* Born on June 21, 1781, in Pithiviers, France, scholar Siméon-Denis Poisson is remembered for his fundamental work on the theory of PROBABILITY, for the discovery of the distribution named after him, and also for his formulation of the LAW OF LARGE NUMBERS, all completed in the latter part of his mathematical career. Before then Poisson had made significant contributions to the topic of celestial mechanics and to the theory of electricity and magnetism.

Poisson entered the École Polytechnique in Paris in 1798 to study mathematics under the guidance of JOSEPH-LOUIS LAGRANGE (1736–1813) and PIERRE-SIMON LAPLACE (1749–1827), both members of the faculty at the time. After graduating just two years later, Poisson was granted a deputy professorship at the institution and later a full professorship in 1806.

During this early period of his career Poisson studied DIFFERENTIAL EQUATIONS as well as POWER SERIES and their applications to mechanics and physics. Starting in 1808 he produced fundamental results extending the work of Laplace and Lagrange on the motion of the planets, developing new series techniques to approximate solutions to perturbations in their orbits. He published an influential two-volume treatise on the topic of mechanics in 1811 and solved important problems on the theory of gravitation in the years that followed. He found applications of this work to the theory of electromagnetism in 1813 and to the theory of heat transfer in 1815.

Late in his career, Poisson's interests turned toward probability. Fascinated by studies of societal behavior, he began to analyze the probability of a random event occurring within a given interval of time given that its likelihood of happening is very small, but the number of opportunities for it to occur is large. This work led to his discovery of the POISSON DISTRIBUTION, as it is called today, details of which he published in 1837.

Poisson was a prolific writer and published over 300 mathematical works during his career. Unfortunately, scholars at the time generally did not regard his latter work important, and Poisson never enjoyed the full respect of the mathematics community. It was only after his death, on April 25, 1840, that the significance of his many innovative ideas became apparent.

Poisson distribution   *See* BINOMIAL DISTRIBUTION.

polar coordinates   Invented by SIR ISAAC NEWTON (1642–1727) polar coordinates identify the location of a point P in the plane by its distance r from a fixed point O in the plane, called the origin or the pole, and the angle θ the line segment OP makes with a fixed ray placed at O, called the polar axis. The pole is usually taken as the origin of a standard system of CARTESIAN COORDINATES, with the polar axis being the positive x-axis, and the angle θ measured in the counterclockwise sense. The pair of numbers (r,θ) is called the polar coordinates of P.

As an example, the point 1 unit along the x-axis, and 1 unit along the y-axis, written $(1,1)$ in Cartesian

coordinates, is a distance $\sqrt{2}$ from the origin and makes an angle of 45° with the x-axis, and so has polar coordinates $(\sqrt{2},45°)$. By adding multiples of 360° to the angle, one can identify the same point as $(\sqrt{2},405°)$, $(\sqrt{2},765°)$ or even $(\sqrt{2}, -315°)$, for example. This shows that the polar coordinate representation of a point is not unique. The polar coordinates of the origin are not well defined, and this point is usually referred to simply as the point with $r = 0$.

It is possible to convert between polar and Cartesian coordinates. If a point P has Cartesian coordinates (x,y) and polar coordinates (r,θ) then P lies at the apex of a right triangle with a horizontal leg of length x and vertical leg of length y. The length r is the hypotenuse of the triangle and, from TRIGONOMETRY, θ is an angle whose sine is y/r, cosine is x/r and tangent is y/x. Thus, with the aid of PYTHAGORAS'S THEOREM, we have the equations:

$$x = r\cos\theta \qquad r = \sqrt{x^2 + y^2}$$

$$y = r\sin\theta \qquad \theta = \tan^{-1}\left(\frac{y}{x}\right)$$

As a check we see that the point $P = (1,1)$ given in Cartesian coordinates does indeed have polar coordinates given by $r = \sqrt{1^2 + 1^2} = \sqrt{2}$ and $\theta = \tan^{-1}\left(\frac{1}{1}\right) = 45°$. The point Q with polar coordinates $(2,30°)$ has Cartesian coordinates given by: $x = 2\cos(30) = \sqrt{3}$ and $y = 2\sin(30) = 1$. Thus $Q = (\sqrt{3},1)$.

Polar coordinates are useful in describing equations in mathematics that have central symmetry about the origin. For example, in Cartesian coordinates, the equation of a circle of radius five reads: $x^2 + y^2 = 25$. In polar coordinates, this equation reduces to $(r\cos\theta)^2 + (r\sin\theta)^2 = 25$, that is, $r^2(\cos^2\theta + \sin^2\theta) = r^2 = 25$, or simply $r = 5$. The equation of an ARCHIMEDEAN SPIRAL is also elementary in polar coordinates: $r = a\theta$ for some constant a.

A DOUBLE INTEGRAL $\iint_R f(x,y)dA$ is defined to be the volume under the graph $z = f(x,y)$ above a region R in the xy-plane. It can be computed as an iterated integral $\iint f(x,y)dxdy$ with the appropriate limits of integration inserted. To convert this integral to one expressed in polar coordinates, one performs double integration given by $\iint f(r\cos\theta, r\sin\theta)r\,dr\,d\theta$. The appearance of the r in the integrand is explained as follows:

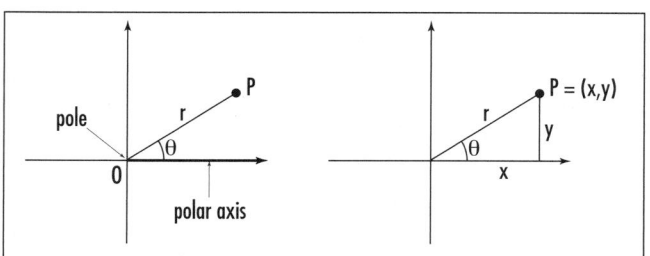

Polar coordinates

Subdivide the region R into small pieces appropriate for examination with polar coordinates. The kth section is thus a region given by values r and θ bounded between fixed values: $r_{k1} < r < r_{k2}$ and $\theta_{k1} < \theta < \theta_{k2}$. It appears as the region between a sector of angle $\theta_{k2} - \theta_{k1}$ of radius r_{k2} and the sector of the same angle but at a smaller radius r_{k1}, and so has area:

$$dA_k = \frac{\theta_{k2} - \theta_{k1}}{2\pi} \cdot \pi r_{k2}{}^2 - \frac{\theta_{k2} - \theta_{k1}}{2\pi} \cdot \pi r_{k1}{}^2$$

$$= \frac{1}{2}(\theta_{k2} - \theta_{k1})(r_{k2}{}^2 - r_{k1}{}^2)$$

$$= \frac{r_{k1} + r_{k2}}{2} \cdot (\theta_{k2} - \theta_{k1})(r_{k2} - r_{k1})$$

$$= \frac{r_{k1} + r_{k2}}{2} \cdot d\theta_k \cdot dr_k$$

where $d\theta_k = \theta_k 2 - \theta_k 1$ and $dr_k = r_k 2 - r_k 1$. The double integral is thus given as a limit

$$\iint_R f(x,y)\,dA = \lim \sum_{k=1}^{n} f(r_k \cos\theta_k, r_k \sin\theta_k)\,dA_k$$

$$= \lim \sum_{k=1}^{n} f(r_k \cos\theta_k, r_k \sin\theta_k) \cdot \frac{r_{k1} + r_{k2}}{2} \cdot d\theta_k \cdot dr_k$$

taken over finer and finer subdivisions of the region R. But in this limit, the values r_{k1} and r_{k2} both approach a common value r, and so the term $\frac{r + r}{2} = r$ appears in the integrand. (Similar calculations explain why one must insert an r when converting an integral to CYLINDRICAL COORDINATES, and a term $\rho^2 \sin\varphi$ when converting to SPHERICAL COORDINATES.)

See also CARL GUSTAV JACOB JACOBI; ROSE.

Pólya, George (1887–1985)

Pólya, George (1887–1985) Hungarian *Education, Geometry, Number theory* Born on December 13, 1887, in Budapest, Hungary, mathematician George Pólya is remembered not only for his influential work in the fields of PROBABILITY theory, mathematical physics, GEOMETRY, COMBINATORICS, and NUMBER THEORY, but also for his landmark work in mathematical education and problem solving. His famous work *How to Solve It* provided heuristic tools for breaking complicated problems into simpler components and developing strategies of attack, and furnished practical insights to finding solutions and understanding the creative process of doing mathematics. The work was immensely influential and was translated into 17 languages, selling over 1 million copies. Pólya is also noted for coining the name CENTRAL-LIMIT THEOREM for the famous fundamental result used in STATISTICS.

Pólya received a doctorate in mathematics from the University of Budapest in 1912 after completing a thesis on the topic of geometric probability. He was interested in not only understanding mathematics, but also the process by which it is discovered. In 1913 he began work on a textbook in ANALYSIS that organized topics and problems not by their subject, but rather by their method of solution. This thinking soon led him to the work on the art of problem solving and the writing of his famous text *How to Solve It*. In brief summary, Pólya developed the following basic four-stage plan to solving a problem:

1. Understand the problem. (Identify the knowns and the unknowns. Draw a diagram or a table. Use the diagram to identify relationships between the variables or do the same with algebraic formulae.)
2. Devise a plan. (Consider if the problem is similar to a previously solved example. Look for a pattern. Consider known formulae that may be applicable. Make guesses and use trial and error. Break the problem into smaller parts or cases. Work backwards from the stated solution. Restate the problem.)
3. Carry out the plan. (Check each step. Prove or verify each step.)
4. Look back, check work, and reflect on the problem. (Does the solution obtained make sense? Can the solution be found via a different, and perhaps shorter, method? Ask "what if" questions. What if the question is changed in a certain way? What if some conditions or statements in the problem are omitted?)

By identifying the thinking process behind solving problems for the first time, Pólya revolutionized the art of mathematics education. He submitted that the greatest good in teaching mathematics comes from providing students the opportunity to discover results for themselves and that the development of problem-solving skills provides the appropriate means for this to occur.

Pólya received many honors throughout his life, including election to the London Mathematical Society, the National Academy of Sciences of the United States, the American Academy of Arts and Sciences, and the Swiss Mathematical Academy. In 1933 he was awarded a Rockefeller Fellowship to support a visit to Princeton University, and in the mid-1940s, Pólya accepted a position at Stanford University, where he stayed for the remainder of his career. He died in Palo Alto, California, on September 7, 1985. His classic text is still used today as the standard model for high school- and college-level courses on problem solving.

polygon A planar geometric figure bounded by a number of straight lines is called a polygon. The name is derived from the Greek language with *poly* meaning "many" and *gonia* meaning "angle." It is usually assumed that no two lines forming the figure intersect (other than at the corners or vertices of the polygon). The polygon is called convex if the interior of the figure lies entirely on one side of each line used to form it. It is called concave otherwise. (For example, a square is a convex polygon, and the shape of a star is a concave polygon.)

The following table gives the names for polygons with different numbers of sides.

Sides	Name
3	Triangle
4	Quadrilateral
5	Pentagon
6	Hexagon
7	Heptagon
8	Octagon
9	Nonagon or Enneagon
10	Decagon
11	Undecagon or Hendecagon
12	Dodecagon
13	Tridecagon or Triskaidecagon
14	Tetradecagon or tetrakaidecagon
15	Pentadecagon or Pentakaidecagon
16	Hexadecagon or Hexakaidecagon
17	Heptadecagon or Heptakaidecagon
18	Octadecagon or Octakaidecagon
19	Enneadecagon or Enneakaidecagon
20	Icosagon
30	Triacontagon

Sides	Name
40	Tetracontagon
50	Pentacontagon
60	Hexacontagon
70	Heptacontagon
80	Octacontagon
90	Enneacontagon
100	Hectogon

A polygon with n sides (an n-gon) also has n interior angles, each of which is less than $180°$ if the polygon is convex. (A concave polygon has at least one interior angle greater than $180°$.)

A DIAGONAL of a polygon is a straight line connecting any two nonadjacent vertices. Drawing *all* the diagonals from a given vertex shows that any convex n-sided polygon can be subdivided into $n-2$ triangles. (The same is true for concave polygons, but one may need to use a different collection of diagonals.) As the sum of the interior angles of a TRIANGLE is $180°$, this shows that the interior angles of any n-sided polygon sum to $(n - 2) \times 180°$. Thus, for example, the interior angles of any quadrilateral sum to $2 \times 180° = 360°$, of any pentagon to $3 \times 180° = 540°$, and so on.

Traversing the boundary of a polygon completes one full turn. This shows that the sum of the exterior angles of any polygon sum to $360°$. (For a concave polygon, one must count left turns as positive and right turns as negative.)

A polygon is called equilateral if all side lengths are equal, and equiangular if all interior angles are equal. A polygon need not be equilateral if it is equiangular (consider a RECTANGLE for example) nor equiangular if it is equilateral (consider the special case of a PARALLELOGRAM called a rhombus). Polygons that are simultaneously equilateral and equiangular are called regular polygons. A SQUARE, for example, is a regular polygon.

The interior angles of a regular n-gon each have value $\dfrac{(n-2)}{n} \times 180°$—the sum of the interior angles divided by the number of angles. Thus, for example, an equilateral triangle has interior angles of $\dfrac{1}{3} \times 180 = 60°$, square angles of $\dfrac{2}{4} \times 180 = 60°$, regular pentagon angles of $\dfrac{3}{5} \times 180 = 108°$, and so on.

The PERIMETER of a regular n-gon, with sides of length s, is ns, and its AREA is $\frac{1}{2} \times perimeter \times r = \frac{1}{2} nsr$. Here r is the APOTHEM of the polygon.

It is possible to construct an equilateral triangle using a straightedge (a ruler without markings) and compass alone. Draw a straight line. Set the compass at a fixed angle and draw a circle with center anywhere along this line. Choose one point of intersection of this circle with the line and draw a second circle of the same radius with this chosen point as its center. The centers of each circle and their point of intersection now form the three corners of an equilateral triangle.

One can also draw a square, regular 5-gon, and a regular 6-gon, but not a regular 7-gon, using compass and straightedge alone. The question of precisely which regular n-gons are CONSTRUCTIBLE this way is an old one.

Given a rectangular piece of paper, it is also possible to create the shape of an equilateral triangle by creasing the sheet. Fold the paper in half along the direction of its longest length, marking this midline as a crease. Bring the bottom left corner of the paper to this midline to form a diagonal crease that passes through the bottom right corner. The location of this left corner on the midline is the apex of an equilateral triangle with the bottom edge of the paper as base.

Only three regular polygons tile the plane: the equilateral triangle, the square, and the regular hexagon. A study of TESSELATION shows that no other regular polygon can do this.

All regular polygons have the following remarkable property:

For any point inside a regular polygon, the sum of the distances of that point from each of the sides is always the same.

Suppose, for example, we choose an arbitrary point P inside a regular hexagon of side-length s lying at distances h_1, h_2,..., h_6 from the sides of the hexagon. Lines connecting P to the vertices of the hexagon divide the figure into six triangles. The area of the polygon is consequently the sum of areas of these triangles, $\frac{1}{2} sh_1 + \frac{1}{2} sh_2 + \frac{1}{2} sh_3 + \frac{1}{2} sh_4 + \frac{1}{2} sh_5 + \frac{1}{2} sh_6$, and so $h_1 + h_2 + h_3 + h_4 + h_5 + h_6$ has value $\frac{2 \times area}{s}$, no matter

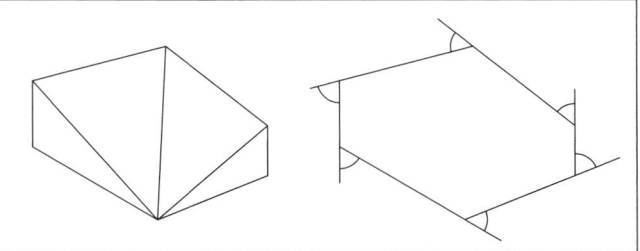

Summing the angles of a polygon

which point is chosen. (In general, the quantity $\frac{2 \times area}{s}$ equals the number of sides of the regular polygon times its apothem.)

The three-dimensional generalization of a polygon is a POLYHEDRON. The generalization into four dimensions is called a polychoron, and into an arbitrary number of dimensions, a polytope.

See also CONCAVE/CONVEX; CYCLIC POLYGON; LONG RADIUS.

polyhedron (plural, polyhedra) A three-dimensional solid figure with a surface composed of plane polygonal surfaces is called a polyhedron. For example, a CUBE, with six square faces, is a polyhedron, as is a TETRAHEDRON with four triangular faces, and any PYRAMID or PRISM, for instance. Each polygonal surface is called a FACE of the polyhedron, and any line along which two faces intersect is called an edge. Any point at which three or more faces meet is called a vertex or a corner of the polyhedron.

POLYGONs and polyhedra were first studied in detail by the ancient Greeks, who also gave them their name: *poly* means "many" and *hédra* means "seat." Thus a polyhedron was considered capable of being seated on any of its faces. In this context, it is usually assumed then that a polyhedron is convex, that is, no plane containing a face of the figure also passes through the interior of the figure. (Consequently a convex polygon can be "seated" on any of its faces on a tabletop.) A polyhedron that is not concave is called convex.

Specific polyhedra are named according to the number of faces they possess. For example, a tetrahedron is any solid figure with four polygonal faces, a pentahedron is one with five faces, and a hexahedron is any polyhedron with six faces. There is no polyhedron with fewer than four faces.

A convex polyhedron is called regular if all of its faces are congruent regular polygons. The geometer EUCLID (ca. 300–260 B.C.E.) proved, in his final volume of *THE ELEMENTS*, that there are only five regular polyhedra: the regular tetrahedron (with four triangular faces), the cube (with six square faces), the octahedron (with eight triangular faces), the dodecahedron (with 12 pentagonal faces), and the icosahedron (with 20 triangular faces). These five regular solids played an important role in the Greek study of geometry, and much mystic significance was ascribed to the figures. Philosopher PLATO (ca. 428–348 B.C.E.) studied and wrote extensively about the five regular polyhedra, and for this reason they are today called the PLATONIC SOLIDS.

A polyhedron is called semiregular if it is composed of two different types of polygonal faces combined together in the same way at each vertex of the polyhedron. For example, the classical pattern drawn on a soccer ball describes a semiregular polyhedron composed of 12 pentagons and 20 hexagons arranged so that each vertex of the figure is surrounded by one pentagon and two hexagons. ARCHIMEDES OF SYRACUSE (ca. 287–212 B.C.E.) proved that there exist only 13 semiregular polyhedra, today called the Archimedean solids.

A polyhedron is said to be stellated if it is built from a regular polyhedron by attaching pyramids to its faces, or faceted if it is formed with these pyramids turned inward.

Swiss mathematician LEONHARD EULER (1707–83) discovered a remarkable formula, EULER'S THEOREM, relating the number of vertices v, edges e, and faces f, of any polyhedron free from holes akin to the hole of a TORUS. We have:

$$v - e + f = 2$$

This formula proves, for instance, that any polyhedron composed of pentagonal and hexagonal faces, not necessarily regular, with three edges meeting at each vertex, must contain precisely 12 five-sided faces. (To see this, let v be the number of vertices of the polyhedron, e the number of edges, p the number of pentagonal faces, and h the number of hexagonal faces. Then $v - e + p + h = 2$. Also, since three edges meet at each vertex, each pentagonal face has five edges, and each hexagonal face six edges, we have $3v = 2e = 5p + 6h$. These equations force p to have value 12, no matter what value h may adopt.)

See also ALTITUDE; BASE OF A POLYGON/POLYHEDRON; CONCAVE/CONVEX; CONE; CYLINDER; FACE; FRUSTUM; HYPERCUBE; NET; PARALLELEPIPED; PRISM; PYRAMID.

polynomial A sum of multiples of positive integer powers of a variable is called a polynomial. The general form of a polynomial is an expression of the form:

$$a_n x^n + a_{n-1} x^{n-1} + \ldots + a_1 x + a_0$$

where $a_n, a_{n-1}, \ldots, a_1, a_0$ are numbers, called the COEFFICIENTS of the polynomial, with leading coefficient a_n assumed to be different from zero. The number a_0, which may be zero, is called the constant term of the polynomial. The highest power n that appears in the expression with nonzero coefficient is called the DEGREE OF THE POLYNOMIAL. For example, $\sqrt{2}x^3 - 5x + 8$ and 5 are polynomials of degrees three and zero, respectively. A polynomial of degree two is called a QUADRATIC, of degree three a cubic, of degree four a quartic, and of degree five a quintic.

One can add two polynomials to produce a new polynomial by collecting like terms, and one can multiply two polynomials using the process of EXPANDING BRACKETS. For example:

$$(x^3 + 2x^2 - 3x + 4) + (x^2 + 4x - 2) = x^3 + 3x^2 + x + 2$$

and

$$(x^2 + 2x + 1)(3x + 5) = 3x^3 + 6x^2 + 3x + 5x^2 + 10x + 5$$
$$= 3x^3 + 11x^2 + 13x + 5$$

If one thinks of the variable x as an unspecified BASE OF A NUMBER SYSTEM, then one can perform the same arithmetical operations on polynomials (such as ELIZABETHAN MULTIPLICATION and LONG DIVISION) as for ordinary numbers. The ratio of two polynomials produces a RATIONAL FUNCTION.

A polynomial function is a function whose outputs are given by a polynomial expression. For example, $f(x) = 5x^3 - 7x - 3$ is a cubic function. It is not possible for a nonzero polynomial function to produce zero as an output for all inputs x. We have:

If $a_n x^n + a_{n-1} x^{n-1} + \ldots + a_1 x + a_0 = 0$ for all values x, then it must be the case that $a_n = a_{n-1} = \ldots = a_1 = a_0 = 0$.

(To see this, write $a_n x^n + a_{n-1} x^{n-1} + \ldots + a_1 x + a_0 = a_n x^n \left(1 + \dfrac{a_{n-1}}{a_n} \cdot \dfrac{1}{x} + \cdots + \dfrac{a_1}{a_n} \cdot \dfrac{1}{x^{n-1}} + \dfrac{a_0}{a_n} \cdot \dfrac{1}{x^n}\right)$. For a large value of x, the term $a_n x^n$ will adopt a very large value, and the expression inside the parentheses will adopt a value close to $1 + 0 + \ldots + 0 + 0 = 1$. Thus the value of the entire polynomial expression will not be zero, unless $a_n = 0$. Repeating this argument shows that all coefficients would then have to be zero.) As a consequence we have:

> If two polynomials produce the same output values for each input value x, then the two polynomials must be identical, that is, their coefficients match.

(The difference of these two polynomials would be a new polynomial that always produces the zero output.) This second statement explains why the process of EQUATING COEFFICIENTS is valid.

The FUNDAMENTAL THEOREM OF ALGEBRA assures us that any degree-n polynomial equation of the form $a_n x^n + a_{n-1} x^{n-1} + \ldots + a_1 x + a_0 = 0$ has precisely n roots (when counted with multiplicity). Some roots may be COMPLEX NUMBERS. There exist arithmetic formulae for finding the roots of any quadratic equation, any CUBIC EQUATION, and any QUARTIC EQUATION. In 1824 algebraist NIELS HENRIK ABEL proved that there are no analogous formulae for solving fifth- and higher-degree polynomial equations.

Evaluating a degree-n polynomial typically requires the computation of $n + (n-1) + \ldots + 2 + 1 + 0 = \dfrac{n(n+1)}{2}$ multiplications. For instance, in the expression

$$2x^3 + 3x^2 + 4x + 5 = 2 \times x \times x \times x + 3 \times x \times x + 4 \times x + 5$$

the multiplication sign "$\times$" appears $3 + 2 + 1 + 0 = 6$ times. This is the formula for the nth TRIANGULAR NUMBER. The process of performing a NESTED MULTIPLICATION reduces the number of products needed to just n. For example, rewriting $2x^3 + 3x^2 + 4x + 5$ as $x(x(2x + 3) + 4) + 5$ reduces the number of multiplications present from six to three. The process of SYNTHETIC DIVISION is intimately connected with nested multiplications.

LAGRANGE'S FORMULA shows that given any $n + 1$ points, drawn in the plane, each with a distinct x coordinates, there exists a polynomial function of degree n whose graph passes through each of those points. Thus it is always possible to "fit" a polynomial function to any finite set of data points. This is useful for the purposes of INTERPOLATION and EXTRAPOLATION.

A polynomial may have more than one variable. For example, $5x^2 y + 7xy^2 - y^3$ is a degree-three bivariate polynomial.

See also BINOMIAL; COMPLETING THE SQUARE; DESCARTES'S RULE OF SIGNS; DIFFERENCE OF TWO CUBES; DIFFERENCE OF TWO SQUARES; DISCRIMINANT; FACTOR THEOREM; FACTORIZATION; HISTORY OF EQUATIONS AND ALGEBRA (essay); MONOMIAL; REMAINDER THEOREM; ROOT; SOLUTION BY RADICALS; TAYLOR SERIES; TRINOMIAL.

polynomial time A computation is said to run in polynomial time if the number of elementary steps required to complete the computation can be expressed as a POLYNOMIAL function of the size of the input. For example, if a basic step is to add or multiply two single-digit numbers, then the computation of multiplying two n-digit numbers via ordinary long multiplication requires at most $4n^2$ steps. (Multiplying each of the pair of digits requires n^2 steps, and summing all results, with carrying, is at most $3n^2$ steps.) Thus long multiplication runs in polynomial time.

Computations that do not run in polynomial time are said to run in exponential time. For example, the task of listing all possible arrangements of n objects grows as n FACTORIAL. As the factorial function will exceed any polynomial function of n for sufficiently large values of n, the operation of listing all possible orders is exponential. Even with the most powerful computers, exponential time computations generally require an infeasible amount of time to complete. Polynomial time algorithms, however, are more practical.

See also NP COMPLETE; TRAVELING-SALESMAN PROBLEM.

polyomino Generalizing the concept of a domino, a polyomino is a shape made by adjoining 1×1 squares along entire edge lengths in such a way that no corner of one square lies at an interior point of another square's edge. A polyomino composed of n squares is called an n-polyomino or simply an n-omino.

Two polyominoes are considered equivalent if one can be picked up, rotated, and possibly flipped, to match the other. Using this notion of equivalence, there is then just one 1-omino (called a "monomino"), one 2-omino (the domino), two 3-ominoes (each called a tromino), and five 4-ominoes (each called a tetromino).

Let $P(n)$ denote the number of distinct n-ominoes. The following table gives the value of $P(n)$ for n from 1 to 12.

n	1	2	3	4	5	6	7	8	9	10	11	12
$P(n)$	1	1	2	5	12	35	108	369	1,285	4,655	17,073	63,600

The exact values for $P(n)$ up to $n = 24$ are known, but finding a general formula for $P(n)$ remains an open problem. In 1966 mathematician David Klarner proved that there is a number K (today called Klarner's constant) such that $\lim_{n \to \infty} \sqrt[n]{P(n)} = K$. (This shows that $P(n)$ has approximate value K^n if n is large, and so the function grows exponentially.) The exact value of K is not known, but mathematicians have established that it lies between 3.9 and 4.649551. (They suspect its value lies close to 4.2.)

The order of a polyomino is the smallest number of identical copies of that polyomino that can be assembled to form a rectangle. If the creation of a rectangle is impossible, then that polyomino is said to have infinite order. The straight tromino has order one (it is itself a rectangle) and the bent tromino has order two. (Two copies of this tromino can interlock to produce a 2×3 rectangle.) The four tetrominoes illustrated

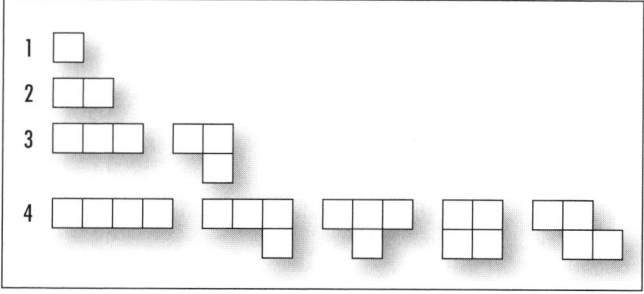

Polyominoes

above have orders 1, 2, 4, 1, and ∞, respectively. (To see that the final tetromino cannot tile any rectangle, consider the placement of the tetromino in the top left corner of the rectangle. Its orientation forces the placement of the tetrominoes below it or to the right of it.) There are no polyominoes of order three. Counting the number of different ways to tile a $2 \times n$ rectangle with dominoes yields the FIBONACCI NUMBERS.

Generalizations of polyominoes to shapes composed of fundamental units other than squares (such as equilateral triangles and regular hexagons) are called polyforms.

See also TESSELLATION.

population   *See* POPULATION AND SAMPLE.

population and sample   STATISTICS is the science of collecting, tabulating, and summarizing DATA obtained from particular systems of study, and making inferences or predictions based on that data. The word *population* is used for the group of all the individuals (or objects or events) that are the subject of the study. A sample is a representative subgroup or subset of the population. For example, in a medical study on the growth rates of 8-year-old children in the United States, the population would be *all* 8-year-old American children. As it is not feasible to examine every child of a particular age, a sample of just 1,000 children might be used for the study.

A sample in which every individual in the population has equal chance of being chosen for the sample is called a random sample. If, in a sample, some portion of the population is represented more heavily than it actually occurs, then the sample is called biased. Biased sampling is to be avoided.

A famous historical example of an erroneous prediction based on biased sampling occurred during the 1936 U.S. presidential elections. The popular publication *Literary Digest,* as part of the sensationalism leading up to the election, conducted a poll to predict the outcome of the race. After interviewing a sample of eligible voters, chosen by drawing names at random from telephone books from across the nation, the editors of the publication concluded that the election was a foregone conclusion—Alfred Landon was to win with a comfortable lead—and they subsequently published much editorial commentary to this effect. It turned out, however, that Landon's opponent, Franklin Roosevelt, won the election by a landslide. Members of the *Digest* did not realize that they had worked with a biased sample—only affluent Americans could afford telephones at the time of the Great Depression and thus be listed in telephone books. This was a class of voter that was more likely to vote Republican. Consequently, the *Digest's* prediction was erroneous. The publication folded in 1937 due to both the sampling fiasco and the difficult times of the depression.

Today, a number of sampling methods are commonly used to help ensure that no bias occurs. These methods include:

Random Sampling

Each subject of the population is assigned a number, and numbers are generated randomly with the aid of a computer to select members.

Systematic Sampling

Each subject of the population is assigned a number, and, starting at a random number, every *k*th member from then on is selected. For example, one might select every 23rd person, starting with the 533rd member.

Stratified Sampling

When a population is naturally divided into groups (such as male/female, or age by decade), selecting a random sample from within each group produces what is called a "stratified sample." Samples produced this way are used to ensure that representatives of each subgroup are present in the study. For example, in a study involving college freshmen and sophomores, one might select 25 students at random from each group—freshman males, freshman females, sophomore males, and sophomore females—to make a sample of 100 students.

Cluster Sampling

If an intact subgroup of a population is used as a representative sample of the entire population, then the sample is called a cluster sample. For example, the set of all freshman females might be used to represent the population of all college students for the purposes of one study, or the 12 eggs in one carton of eggs as representative of all the eggs handled by a particular supermarket.

See also BIAS; STATISTICS: DESCRIPTIVE.

population models   In biology the term *population* means the number of individuals or organisms living in a certain area. For example, the population of Australia is the number of individuals currently living on that continent, and the population of a laboratory yeast culture is the number of organisms present in a particular petri dish. A population model is a mathematical theory used to describe, or predict, how a population size changes over time.

Interest in how populations grow was stimulated in the late 18th century when Thomas Malthus (1766–1834) published *An Essay on the Principle of Population as it Affects the Future Improvement of Society*. Malthus developed a simple model that yielded the troubling conclusion that eventually the human population would reach a size that cannot be sustained with the food resources available on this planet. Although his model oversimplifies matters and has proved to be incorrect for making long-term predictions, the Malthusian model is still useful for understanding short-term growth. His model is developed as follows:

Let $P(t)$ be the population size at time t and assume that over one unit of time (a minute, or a day, or a year) that a certain percentage, say b percent, of the population gives birth to offspring, and another percentage, say d percent of the population, dies. (The number b is called the birth rate and d is called the death rate.) Thus after one unit of time, the population increases by the amount $bP(t) - dP(t)$. We have:

$$P(t + 1) - P(t) = (b - d)P(t)$$

This says that the rate of change of population size is given by a constant $(b - d)$ times the population size. Assuming that the population function $P(t)$ can, for the sake of convenience, be regarded as continuously changing with time, this final statement can be interpreted, in CALCULUS, as a formula:

$$\frac{dP}{dt} = kP$$

where $k = b - d$. (The constant k is called the growth rate.) Thus $P(t)$ is a function whose derivative is a constant times itself. Only exponential functions have this property and so:

$$P(t) = P(0)e^{kt}$$

where $P(0)$ is the population size at time zero. This model predicts that a population size will grow exponentially if the birth rate exceeds the death rate.

Biologists realize that many ecosystems cannot support arbitrarily large population sizes (the Earth can only supply a finite quantity of food resources per year, for example), and so this simple model is not considered realistic. Generally, as a population increases and gets closer to a maximum capacity M, say, the growth rate k decreases. One can model this by no longer assuming that k is constant, but varies as follows:

$$k = M - P(t)$$

for example. (Here, when the population size $P(t)$ is close to M, the growth rate k is indeed small.) This leads to the model:

$$\frac{dP}{dt} = kP = (M - P)P$$

This is called the logistic growth model and was introduced by Dutch biologist Pierre-François Verhulst (1804–49). It is possible to solve this differential equation and obtain an explicit formula for the population function $P(t)$. However, one can quickly describe some features of the population growth without any work. For example,

1. If a population size starts at value M, then $\frac{dP}{dt} = (M - M) \cdot M = 0$. This means there is no change in the population size and the function $P(t)$ forever remains at the value M.
2. If a population size starts at a value greater than M, then $\frac{dP}{dt} = (M - P) \cdot P < 0$. This means that the population size will decrease. (There are not enough food resources to support a large population.)
3. If a population size starts at a value less than M, then $\frac{dP}{dt} = (M - P) \cdot P > 0$. This means that the population size will increase. The rate of increase decreases as the population size approaches M. The graph of the function thus looks like an increasing S-shaped curve trapped above the x-axis and below the constant line $P = M$.

The logistic model works well to describe population changes for simple biological systems (such as a yeast culture), and, surprisingly, also worked well to describe the U.S. population growth between the years 1920 and 1950. However, this is likely coincidental. Human population growth is very difficult to model, given unpredictable factors such as advances in medical technology, wars, and, in modeling a specific country's population, immigration.

position vector   Given a point P in the plane, or in three-dimensional space, the VECTOR represented by the directed line segment $\overrightarrow{OP}$ connecting the origin O to P is called the position vector of P. This vector has the coordinates of P as its components. For example, if P is the point $(2,5)$, then its position vector is the vector $<2,5> = 2\mathbf{i} + 5\mathbf{j}$.

In physics, the position vector of a particle is often denoted by $\mathbf{r}$. It is a function of time t and is usually expressed in the form $\mathbf{r}(t) = x(t)\mathbf{i} + y(t)\mathbf{j} + z(t)\mathbf{k}$, where x, y and z are functions of time. It represents the physical location of the particle at any time t. The DERIVATIVE of the position vector is the VELOCITY vector $\mathbf{v}(t) = x'(t)\mathbf{i} + y'(t)\mathbf{j} + z'(t)\mathbf{k}$, and its double derivative is the acceleration vector $\mathbf{a}(t) = x''(t)\mathbf{i} + y''(t)\mathbf{j} + z''(t)\mathbf{k}$.

See also PARAMETRIC EQUATIONS.

positive   A REAL NUMBER x is said to be positive if it is greater than zero, that is, if $x > 0$. A real number less than zero is called negative.

The product of two positive numbers is again positive. It is surprising that the rules of arithmetic dictate that the product of two NEGATIVE NUMBERS must also be positive. It follows then that for any nonzero (real) number x we must have $x^2 > 0$.

No COMPLEX NUMBER can be deemed positive or negative. For example, if the number i is positive then, by the previous statement, we must have $i^2 > 0$, yielding the absurdity $-1 > 0$. The same conclusion would follow if we were to deem i negative.

An unspecified real number that is positive or possibly zero is called nonnegative. One that is negative or possibly zero is called nonpositive.

See also ORDER PROPERTIES.

postulate Another name for AXIOM. It is customary to call the axioms of EUCLIDEAN GEOMETRY and of ARITHMETIC postulates rather than axioms.

See also PEANO'S POSTULATES.

power series Any SERIES of the form $\sum_{n=0}^{\infty} a_n x^n = a_0 + a_1 x + a_2 x^2 + a_3 x^3 + \dots$ where the coefficients a_n are constant and x is a variable is called a power series. For example, the TAYLOR SERIES of any function is a power series, as is the GEOMETRIC SERIES $1 + x + x^2 + x^3 + \dots$. Whether or not a power series converges depends on which value is chosen for the variable x. For example, the geometric series converges if $-1 < x < 1$ and diverges otherwise. At the very least, every power series converges for $x = 0$.

Mathematicians have proved the following result:

Given a power series $\sum_{n=0}^{\infty} a_n x^n$, precisely one of the following is true:
1. The series converges only for $x = 0$.
2. The series converges (absolutely) for all values of x.
3. There is a positive real number R such that the series converges (absolutely) for all values $-R < x < R$ and diverges for $|x| < R$. (The series might, or might not, converge for each of the values $x = R$ and $x = -R$.)

The value R is called the radius of convergence of the power series. (If the series converges only for $x = 0$, one usually says that the radius of convergence is zero. If the series converges for all values of x, then the radius of convergence is infinite.)

One often makes use of the ratio test from the study of CONVERGENT SERIES to find the radius of convergence of a power series. For example, the ratio test shows that the series $\sum_{n=0}^{\infty} \frac{x^n}{n}$ converges if $\lim_{n\to\infty} \frac{|\frac{x^{n+1}}{n+1}|}{|\frac{x^n}{n}|}$

$= \lim_{n\to\infty} \frac{n}{n+1} |x| = |x|$ is less than 1, and diverges if greater than 1. Thus the radius of convergence for this power series is $R = 1$. (Going further, notice that, for $x = 1$ the series is the HARMONIC

SERIES $\sum_{n=0}^{\infty} \frac{1^n}{n} = 1 + \frac{1}{2} + \frac{1}{3} + \dots$, which diverges, and for it $x = -1$ it is the alternating harmonic series $\sum_{n=0}^{\infty} \frac{(-1)^n}{n} = 1 - \frac{1}{2} + \frac{1}{3} - \dots$, which converges. The power series in question thus converges for $-1 \leq x < 1$.)

See also ABSOLUTE CONVERGENCE.

precision The total number of digits recorded while taking a measurement is called the precision of the measurement. For example, a recorded length of 3.650 meters is precise to four SIGNIFICANT FIGURES. (The final digit zero indicates that the measurement was indeed made to the nearest millimeter.) The number of digits to the right of the decimal point is called the accuracy of the measurement. Thus, for example, the figure 3.650 is accurate to three decimal places.

See also ERROR.

premise A statement that is known or is assumed to be true and on which a logical ARGUMENT is based is called the premise of the argument. A premise could be an AXIOM of a particular mathematical theory, or merely an assumption taken to be true for the purposes of discovering its consequences. For example, the premise that "parallel lines *do* meet at a point" led mathematicians to discover the realm of PROJECTIVE GEOMETRY, a valid and fruitful new approach to understanding ordinary geometry.

prime A whole number possessing just two positive factors is called a prime number, or simply a prime. For example, 7 has only two positive factors, namely 1 and 7, and so is prime. The number 24 has eight positive factors and so is not prime, and the number 1 has only one factor and is not prime. The term *composite* is used to describe numbers greater than 1 that are not prime. (Medieval mathematician FIBONACCI (1170–1250) called prime numbers "incomposite.") It is vital that the number 1 be considered neither prime nor composite for the FUNDAMENTAL THEOREM OF ARITHMETIC to hold true.

The first 25 prime numbers are: 2, 3, 5, 7, 11, 13, 17, 19, 23, 29, 31, 37, 41, 43, 47, 53, 59, 61, 67, 71, 73, 79, 83, 89, 97. These are all the primes smaller than

100. The Greek mathematician EUCLID (ca. 300 B.C.E.) was the first to prove that the list of primes goes on forever. Today we call his particular argument establishing this EUCLID'S PROOF OF THE INFINITUDE OF PRIMES.

Every whole number can be written as a product of prime numbers. (If the number is already prime, it is considered a product with just one term.) For example, the number 100 can be factored as $2 \times 2 \times 5 \times 5$. For this reason, primes are considered the multiplicative building blocks of the natural numbers, and have therefore been the object of much intensive study throughout the centuries. Surprisingly, many basic questions about them remain unsolved. For example, no one knows a simple formula that will generate the prime numbers, or a simple, and computationally feasible, procedure for factoring large numbers into primes. It is not known whether infinitely many TWIN PRIMES exist, or whether, as CHRISTIAN GOLDBACH (1690–1764) conjectured, every even number greater than 2 is indeed a sum of two primes. (This is known as GOLDBACH'S CONJECTURE.) Also, no one knows if there are infinitely many prime numbers of the form $n^2 + 1$, that is, 1 more than a square number, or whether between any two square numbers there must be a prime. (It is known, however, that for any number n greater than 1, there is a prime between n and $2n$.)

In 1791 CARL FRIEDRICH GAUSS (1777–1855) conjectured that the nth prime number has value approximately $n \cdot \ln(n)$ where $\ln(n)$ is the natural LOGARITHM of n. This claim was later proved to be correct and is today called the PRIME-NUMBER THEOREM.

Despite the infinitude of primes, these numbers, in some sense, are very scarce. It is easy to produce an arbitrarily long string of consecutive integers, none of which are prime. For example, making use of the FACTORIAL function, the numbers $1001! + 2$, $1001! + 3$, up to $1001! + 1001$ form a string of 1,000 consecutive composite numbers. (These numbers are, respectively, divisible by $2, 3, \ldots, 1001$.) The SIEVE OF ERATOSTHENES can be used to "sift out" the prime numbers up to any given integer.

Prime numbers, especially large prime numbers, play a key role in CRYPTOGRAPHY. The discovery of new, large prime numbers is of interest to financial institutions and security services needing to develop effective encryption codes. Large prime numbers are discovered with the aid of a computer, and are typically MERSENNE PRIMES. The largest prime number known as of the year 2003 is $2^{20,996,011} - 1$. It is over 6 million digits long.

French mathematician and lawyer PIERRE DE FERMAT (1601–65) proved that if p is a prime number, then $2^{p-1} - 1$ will be divisible by p. If not, it means the number p was not prime to begin with. This is often used as a test to determine whether a number p is a good candidate for being prime. Unfortunately, some numbers can still pass the test without being prime. For example, $2^{340} - 1$ turns out to be divisible by 341, even though the number 341 is not prime ($341 = 11 \times 31$). Composite numbers that pass Fermat's test are called pseudoprimes. The first three pseudoprimes are 341, 561, and 645.

See also COMPOSITE NUMBER; FACTOR.

prime-number theorem   Mathematicians define $\pi(n)$ to be the number of PRIMES less than or equal to n. For example, there are four prime numbers less than 10, and so $\pi(10) = 4$. The quantity $\frac{\pi(n)}{n}$ thus measures the proportion of numbers up to n that are prime. Study of this function is important to understanding the distribution of prime numbers among the natural numbers.

n	$\pi(n)$	$\frac{\pi(n)}{n}$
10	4	0.4
100	25	0.25
1,000	168	0.168
10,000	1,229	0.1229
100,000	9,592	0.09592
1,000,000	78,498	0.078498
10,000,000	664,579	0.0664579
100,000,000	5,761,455	0.05761455
1,000,000,000	45,505,251	0.045505251

At the age of 14, mathematician CARL FRIEDRICH GAUSS (1777–1855) developed a passion for tabulating data about prime numbers. He observed, among other things, that the inverted function $r(n) = \frac{n}{\pi(n)}$ seems to increase in value by constant amount 2.31 with each 10-fold increase of n, at least for large values of n. That is,

$$r(10n) \approx r(n) + 2.31$$

For example, $r(100,000,000)$ equals $\frac{1}{0.05761455} \approx$

17.357 and $r(1,000,000,000)$ equals $\dfrac{1}{0.045505251} \approx$ 19.667. As 2.31 is approximately the natural LOGARITHM of 10 this suggested to Gauss that $r(n)$ is behaving as the natural logarithm function. He conjectured:

$$\frac{\pi(n)}{n} \approx \frac{1}{\ln(n)}$$

for large values of n, but was never able to prove this claim.

One hundred years later JACQUES HADAMARD (1865–1963) and CHARLES-JEAN DE LA VALLÉE-POUSSIN (1866–1962) simultaneously, but independently, proved Gauss's conjecture using sophisticated techniques from analytic NUMBER THEORY. The result is known today as the prime number theorem. It tells us that the nth prime number has value approximately $n \cdot \ln(n)$.

principal *See* INTEREST.

principal axes The most general equation in three variables of second degree is:

$$Ax^2 + By^2 + Cz^2 + Dxy + Exz + Fyz + Gx + Hy + Iz + J = 0$$

with constants A through F not all zero. In three-dimensional space, the graph of such an equation is a surface called a quadric surface. Mathematicians have shown that, by rotating and translating the coordinate axes, it is possible to rewrite the equation with respect to a new set of axes to simplify the form of the equation, reducing it to one of 13 different types. The new coordinate axes are called the principal axes for the quadric.

Six of the possible forms of the equation lead to nondegenerate quadrics:

1. *Ellipsoid:* $\dfrac{x^2}{a^2} + \dfrac{y^2}{b^2} + \dfrac{z^2}{c^2} = 1$

2. *Hyperboloid of one sheet:* $\dfrac{x^2}{a^2} + \dfrac{y^2}{b^2} - \dfrac{z^2}{c^2} = 1$

3. *Hyperboloid of two sheets:* $\dfrac{x^2}{a^2} + \dfrac{y^2}{b^2} - \dfrac{z^2}{c^2} = -1$

4. *Elliptic cone:* $\dfrac{x^2}{a^2} + \dfrac{y^2}{b^2} = \dfrac{z^2}{c^2}$

5. *Elliptic paraboloid:* $\dfrac{x^2}{a^2} + \dfrac{y^2}{b^2} = \dfrac{z}{c}$

6. *Hyperbolic paraboloid:* $\dfrac{x^2}{a^2} - \dfrac{y^2}{b^2} = \dfrac{z}{c}$

The remaining seven possible equations are degenerate quadrics:

1. *Elliptic cylinder:* $\dfrac{x^2}{a^2} + \dfrac{y^2}{b^2} = 1$

2. *Hyperbolic cylinder:* $\dfrac{x^2}{a^2} - \dfrac{y^2}{b^2} = 1$

3. *Parabolic cylinder:* $\dfrac{x^2}{a^2} = \dfrac{y}{b}$

4. *Pair of planes:* $\dfrac{x^2}{a^2} = \dfrac{y^2}{b^2}$ or $\dfrac{x^2}{a^2} = 1$

5. *Single plane:* $\dfrac{x^2}{a^2} = 0$

6. *Line:* $\dfrac{x^2}{a^2} + \dfrac{y^2}{b^2} = 0$

7. *Point:* $\dfrac{x^2}{a^2} + \dfrac{y^2}{b^2} + \dfrac{z^2}{c^2} = 0$

In two-dimensional space, with suitable rotation and translation of the coordinate axes, any nondegenerate quadratic equation $Ax^2 + By^2 + Cxy + D = 0$ can be rewritten as the equation of either an ellipse, a hyperbola, or a parabola, or, in the degenerate cases, a pair of lines, a single line, or a point. The axes in the new coordinate system are again called principal axes.

See also CONIC SECTIONS.

prism Any POLYHEDRON with two faces (the bases) that are congruent polygons lying in parallel planes and such that the remaining faces (the lateral faces) are parallelograms is called a prism. Specifically, a prism is a CYLINDER with a polygonal base.

The lines joining the corresponding vertices of the base polygons of a prism are called lateral edges. If the lateral edges of a prism meet its bases at right angles, then the prism is called a right prism. A prism that is not right is called oblique.

Prisms are named according to their bases. A triangular prism has two triangular bases (and three lateral faces); a quadrangular prism has two quadrilateral bases and four lateral faces. A CUBE is an example of a right quadrangular prism.

The height of a prism is defined to be the distance between the two parallel planes that contain the bases of the figures. The VOLUME V of a prism is given by the area of its base multiplied by the figure's height:

$$V = \text{area of base} \times \text{height}$$

This follows from CAVALIERI'S PRINCIPLE.

A prismatoid is a polyhedron whose vertices lie in one or the other of two parallel planes. The two bases of the figure are not required to be congruent, nor even have the same number of vertices. All lateral faces in a prismatoid are either triangular or quadrilateral. If the number of vertices of each base polygon is the same and each lateral face is a quadrilateral, then the prismatoid is called a prismoid.

prisoner's dilemma In GAME THEORY, any two-person variable-sum game of partial conflict that mimics the following classic scenario is called a prisoner's dilemma:

Two prisoners, held in separate rooms incommunicado, must choose to either confess or deny involvement in a team crime. If both confess, then each will be sentenced to two years of hard labor. If both deny involvement, then each will be sentenced to four years of hard labor. However, if one prisoner denies and the other confesses, then the denial carries just one year of hard labor, and the confession six.

The following tables show the expected payoffs for each prisoner X and Y for each of the four possible outcomes of the game:

Outcomes for Prisoner X

		Prisoner Y C	Prisoner Y D
Prisoner X	C	2 years	6 years
	D	1 year	4 years

Outcomes for Prisoner Y

		Prisoner Y C	Prisoner Y D
Prisoner X	C	2 years	1 year
	D	6 years	4 years

Each prisoner can argue as follows:

I have no indication as to what my partner will do. If he is to choose option C, to confess, then it is to my advantage to choose option D, to deny. If he is to choose option D, then, again, it is to my advantage to choose option D. Either way, I should choose option D.

The choice of D is thus a dominant strategy for each prisoner, and it is likely that both prisoners will deny involvement in the crime. Moreover, it is worth noting that neither player is tempted to deviate from this choice in an attempt to trick his opponent in the game: the risk of being the only confessor inhibits this. Thus the outcome (D,D) is a stable outcome for the game, and both prisoners will likely each be sentenced to four years of hard labor. (Any outcome to a game, such as (D,D) for the prisoner's dilemma, is called a "Nash equilibrium" for the game if no player can benefit by departing unilaterally from it.)

The prisoner's dilemma provides a PARADOX:

In the game of prisoner's dilemma, each player has a dominant strategy that, when used, yields an outcome to the game that is less beneficial than if both were to deviate from the dominant strategy.

This phenomenon is also seen in the predicament of an arms race between two nations: mutual disarmament is of benefit to both nations, but the fear of an opposing nation choosing to defect from such an agreement inhibits cooperation.

Elements of the prisoner's dilemma can be extended to games involving more than two players. For example, a teacher asks each of his students to write on a piece of paper his or her name and either the word *cooperate* or the word *defect*. The students know that candy pieces will be distributed among the class according to the following rules:

If each student chooses to cooperate, then each will receive 10 pieces.

If two or more students defect, then all will be punished. Those that cooperate will receive only five pieces of candy, and the defectors shall receive none.

If, on the other hand, there is a single bold student willing to be the lone defector, then that defector will receive 80 candy pieces, and all other students none.

The Game of Chicken

An interesting variation of the prisoner's dilemma game is the game of chicken. It is modeled on the dangerous driving game in which two drivers head directly toward one another on a single-lane road. Each driver must decide at the last minute whether to swerve to the right, or to not swerve. To bring mathematics into the game, it is convenient to introduce numerical values for the outcomes of the game.

Suppose 10 points are assigned to the winner of the game: she who chooses not to swerve given that the other driver did. As this is embarrassing to the second driver, 0 points are awarded to her. If both drivers decide to swerve, then, as this is only mildly embarrassing to each, assign to each player 5 points. On the other hand, if both players decide not to swerve, then the outcome is disastrous and assign −10 points to each driver in this instance. The outcomes of the game of chicken can be summarized in the following two payoff tables:

Outcomes for Driver 1

		Driver 2 Swerve	Not Swerve
Driver 1	Swerve	5	0
	Not Swerve	10	−10

Outcomes for Driver 2

		Driver 2 Swerve	Not Swerve
Driver 1	Swerve	5	0
	Not Swerve	10	−10

Notice that neither player has a dominant strategy in the game of chicken: his or her best action does depend on what choice the other driver is going to make. Moreover, the game of chicken has two distinct Nash equilibria: (Swerve, Not Swerve) and (Not Swerve, Swerve). (It is to neither player's advantage to deviate from one of these options.) This suggests that a compromise (Swerve, Swerve) is not easy to achieve.

Many conflicts in real life, such as labor-management disputes and international trade conflicts, have the flavor of a game of chicken. As we know, the outcomes of such disputes do indeed vary: one party may decide to "cave in" so as to avoid a disastrous outcome, while many times neither party surrenders, and strikes and wars result.

See also JOHN NASH.

probability   The principles of probability theory were first identified by 16th-century Italian mathematician and physician GIROLAMO CARDANO (1501–76), and later by French mathematicians BLAISE PASCAL (1623–62) and PIERRE DE FERMAT (1601–65). The key idea is that if a situation can be described in terms of possible outcomes, each equally likely, then the probability of any particular outcome is defined to be the number 1 divided by the total number of outcomes. For example, if a die is cast, six outcomes are possible: {1,2,3,4,5,6}. We usually believe that each outcome is equally likely, and so we say that the probability of rolling any one particular outcome, such as a 5 for example, is 1/6.

More generally, if more than one possible outcome is deemed acceptable, then we define the probability of obtaining one of these outcomes to be ratio of the number of desired outcomes to the total number of outcomes. For example, there are three ways to roll an even number when casting a die. Thus we say that the probability of casting a multiple of 2 is 3/6 = 1/2. The chances of casting any of the numbers 1, 4, 5, or 6 are 4/6 = 2/3.

Mathematicians call the set of all possible outcomes of an experiment the SAMPLE SPACE, and any particular set of outcomes (or just a single outcome) an EVENT. For example, in casting a die, the sample space is the set {1,2,3,4,5,6}, and an event could be the subset {2} (rolling a two), for example, or {2,4,6} (rolling an even number). An event is always a subset of the sample space.

If A represents an event, that is, a set of desirable outcomes, then the notation $P(A)$ is used to denote the probability of that event occurring, that is, the probability that the outcome from one run of the experiment will belong to the set A. A value $P(A)$ is always between zero and 1, with value $P(A) = 0$ indicating that event A will never occur and value $P(A) = 1$ indicating that event A will always occur. In casting a die, for example, we have $P(even) = \frac{1}{2}$, $P(\{1,4,5,6\}) = \frac{2}{3}$, $P(a\ multiple\ of\ 7) = 0$, and $P(a\ whole\ number) = 1$.

Assigning probabilities to events requires careful counting. Often the greatest difficulty is identifying

History of Probability and Statistics

Questions in betting and gaming provided much of the early impetus for the development of PROBABILITY theory. In 1654 Chevalier de Méré, a French nobleman with a taste for gambling, wrote a letter to mathematician BLAISE PASCAL (1623–62) seeking advice about divvying up stakes from interrupted games.

For example, suppose, in a friendly game of tennis, two players each lay down a stake of $100 in a gamble to win "best out of five" games, but rain interrupts play after just four games, with one player having won three games, the second only one. What then would be the fair way to divide the $200 pot? Of course the division of money should somehow reflect each player's likelihood of winning the gamble if the series of games were to be finished.

Pascal communicated the concern of analyzing situations like these to his colleague PIERRE DE FERMAT (1601–65), and their subsequent correspondences on the issue represented the birth of the new field of probability theory. Both mathematicians solved de Méré's "problem of points" (using two entirely different approaches, incidentally) and then later worked together to generalize the problem and extend their analyses to other types of games of chance. Their discoveries aroused the interest of other European scholars. In 1656 Dutch physicist-astronomer-mathematician Christiaan Huygens (1629–95) published *De ratiociniis in ludo aleae* (On reasoning in games of chance) summarizing and extending the ideas developed by Pascal and Fermat. He phrased their work in terms of a new notion, that of EXPECTED VALUE. It proved to be very fruitful.

The key principle behind probability theory is the idea that if a situation can be described in terms of possible outcomes that are equally likely, then the probability of any particular outcome occurring is 1 divided by the total number of outcomes. This principle was actually first recognized and discussed more than a century earlier by Italian mathematician and physician GIROLAMO CARDANO (1501–76) in his work *Liber de ludo aleae* (Book on games of chance). This text, however, was not published until 1663, 9 years after Pascal and Fermat had solved de Méré's problem. It is likely that Cardano would be known as "the father of probability theory" had the work been published during his lifetime. Cardano also recognized the LAW OF LARGE NUMBERS.

The Swiss mathematician Jacob Bernoulli (1654–1705) of the famous BERNOULLI FAMILY recognized the wide-ranging applicability of probability in fields outside of gambling. His book *Ars conjectandi* (The art of conjecture), published posthumously in 1713, demonstrated the use of the theory in medicine and meteorology. It was also the first comprehensive text dealing with issues of STATISTICS.

In some sense, probability and statistics represent two sides of the same fundamental situation. Probability explores what can be said about an unknown sample of a known collection. (For example, we know all possible numerical combinations from a pair of dice. What then is the most likely outcome from tossing a pair of dice?) Statistics explores what can be said about an unknown collection given a small sample. (If 37 of these 100 people brush their teeth twice a day, what can be said about teeth-brushing habits of the entire population?) The two fields remained closely intertwined during much of the 18th century and the early part of the next century.

In 1733 ABRAHAM DE MOIVRE (1667–1754) recognized the repeated appearance of the NORMAL DISTRIBUTION in scientific studies and wrote down a mathematical equation for it. It first became apparent from the "randomness" of errors in astronomical observations and in scientific experiments.

The latter half of the 19th century saw significant progress in developing and understanding the theoretical foundations of probability theory. This was chiefly due to the work of French mathematicians-astronomers-physicists JOSEPH-LOUIS LAGRANGE (1736–1813) and PIERRE-SIMON LAPLACE (1749–1827), German genius CARL FRIEDRICH GAUSS (1777–1855), and French mathematician SIMÉON-DENIS POIS-

the fundamental outcomes that are regarded as equally likely. For example, there are 11 possible outcomes for throwing a pair of dice, namely, getting a sum of 2, 3, ..., 11, or 12, but these events are not equally probable. The fundamental outcome here is not the sum of the two numbers on the dice, but rather the *pair* of numbers that the two dice yield—one number (1 through 6) on the first die and a second number (again 1 through 6) on the second. There are 36 equally likely outcomes in all. As six of these pairs have a sum of 7, we can say that the probability of throwing 7 with a pair of dice is $\frac{6}{36} = \frac{1}{6}$. The probability of throwing 10 is $\frac{3}{36} = \frac{1}{12}$, and the probability of throwing 2 is only $\frac{1}{36}$.

SON (1781–1840) who, among other things, mathematically proved the law of large numbers. The most important publication in this era on the theory of probability was Laplace's 1812 text *Théorie analytique des probabilités* (Analytical theory of probability.). In it, Laplace collected and extended everything known on the subject at that time. Russian mathematicians PAFNUTY CHEBYSHEV (1821–1894), Andrei Markov (1856–1922), and Alexandr Lyapunov (1857–1918) further developed the mathematical underpinnings of the subject in the late 19th century.

Basic statistical thought can be deemed as having developed considerably earlier. The ancient Egyptians compiled DATA concerning population and wealth as early as 3050 B.C.E., developing simple techniques to collate and record the numerical information gathered. The ancient Chinese undertook similar studies around 2300 B.C.E. A census was taken in 594 B.C.E. by the Greeks for the purpose of levying taxes, and Athens undertook a population census in 309 B.C.E. The Romans also kept census records, as well as records of births and deaths, and gathered significant quantities of numerical information from geographic surveys taken across the entire empire. Very few statistical records were kept during the period of the Middle Ages, however.

In 1662 John Graunt analyzed birth and death records and produced the first LIFE TABLE. The purpose of the table was to make general observations and predictions about life expectancy for classes of members of a particular population. This work represented a significant step toward analyzing data for the purposes of INFERENCE.

In 1790 the United States took its first decennial census, heralding the return of census taking. Several European nations followed suit soon afterward. Belgian scholar LAMBERT ADOLPHE QUÉTELET (1796–1874) analyzed the nation's records and made important observations about the influence of age, gender, occupation, and economic condition on mortality. In 1835 he attempted to apply probabilistic methods to the study of human characteristics, both physical and behavioral. He used them to give what he hoped was a complete description of the "average man." Although Quételet's work was generally highly respected, his attempt to apply it to the field of behavioral science was met with criticism. In the 1860s, English scholar FRANCIS GALTON (1822–1911) attempted to apply statistics methods to the study of human heredity. His work was influential and helped define statistics as a mathematics discipline in its own right.

At the turn of the 20th century, the corporate world began to recognize the relevance and usefulness of statistics, especially in issues of quality control, economics, insurance, and telecommunications. Many large companies began hiring statisticians.

While working for an English brewing company, industrial scientist WILLIAM SEALY GOSSET (1876–1937) developed the STUDENT'S T-TEST, allowing for the ability to derive reliable information from small samples. (Company policy forbade its employees to publish. Gosset did so in any case, writing under the pseudonym "Student.") English mathematician KARL PEARSON (1857–1936) developed the CHI-SQUARED TEST and is considered the founder of modern hypothesis testing.

RONALD AYLMER FISHER (1890–1962) is considered the most important statistician of the 20th century. His 1925 text *Statistical Methods for Research Workers* transformed statistics into a powerful scientific tool. He clarified many of the mathematical principles on which the discipline is based. Fisher also developed methods of multivariate analysis to properly analyze problems involving more than one variable.

In 1926, pure and applied mathematician JOHN VON NEUMANN (1903–57) founded GAME THEORY—a mathematical framework for analyzing games of chance, such as poker, that involve strategy and choice on the parts of the players. Von Neumann recognized the applications of the theory to economics and social sciences. The work of Nobel Laureate JOHN FORBES NASH, JR., (born 1928) took its applications to economics to a profound level.

See also STATISTICS: DESCRIPTIVE; STATISTICS: INFERENTIAL.

Probability gives a measure of likelihood or frequency of occurrence. If we throw a pair of dice 1,000 times, then we would expect, on average, close to one-sixth of the rolls (around 167 of them) to have a sum of 7; close to 1/12th of them (around 83 rolls) to have sum 10; and close to 1/36th (around 28 of them) to have a sum of 2. This principle is clarified in the LAW OF LARGE NUMBERS.

Computing Probabilities

Two models are often used to help compute probabilities in moderately complicated situations. They can also be used to illustrate two rules of computation.

1. Probability Trees and the Addition Rule

A probability tree is a diagram displaying all the outcomes of a sequence of actions. It is assumed that the

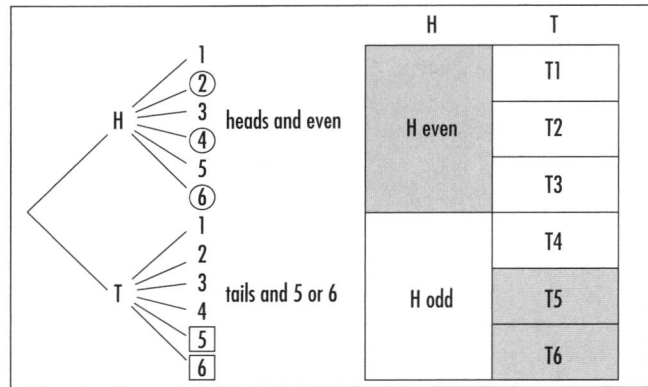

Probability models

outcomes of one action do not affect the outcomes of the actions that follow.

To illustrate, imagine tossing a coin and then casting a die. Each outcome of the coin toss—heads or tails—could be accompanied by any one of the six possible outcomes from casting the die: the numbers 1 through 6. The probability-tree diagram (shown above) is used to make this explicit. In particular, it shows that 12 different results are possible from performing these two actions. Computing the probability of any desired set of outcomes is now straightforward. For example, the chances of tossing a head together with an even number are $\frac{3}{12} = \frac{1}{4}$. Similarly, the probability of tossing a tail together with a 5 or a 6 is $\frac{2}{12} = \frac{1}{6}$.

This model illustrates the additive property in probability theory. In our example:

P(heads and an even number **or** tails and a 5 or a 6)
$$= \frac{3+2}{12}$$
$$= \frac{3}{12} + \frac{2}{12}$$
$= P$(heads and an even number) $+ P$(tails and a 5 or a 6)

In general, if A represents one set of desired outcomes and B another set of outcomes having none in common with A, then

$$P(A \textbf{ or } B) = P(A) + P(B)$$

In particular, the two events "A" and "not A" are disjoint, that is, have no outcomes in common. As we are

certain that A either will or will not occur, $P(A$ **or** not $A) = 1$, we have, by this rule:

$$P(\text{not } A) = 1 - P(A)$$

For example, the probability of *not* rolling a two when casting a die is $P(not\ 2) = 1 - p(rolling\ 2) = 1 - \frac{1}{6} = \frac{5}{6}$.

If two events A and B do have outcomes in common, then the above rule is modified to read:

$$P(A \textbf{ or } B) = P(A) + P(B) - P(A \textbf{ and } B).$$

The term $P(A$ **and** $B)$ is subtracted to counter the double count of outcomes common to A and B.

2. The Square Model and the Multiplication Rule

The square model for probability theory uses a square to represent the set of results in performing an experiment a large number of times. For example, in tossing a coin, we would expect, on average, half the outcomes to be heads and half to be tails. We represent this by dividing the square into two portions of equal area. The left portion now represents a set of experiments in its own right. In next casting a die, we would expect, on average, half of these outcomes to yield an even number and half to yield an odd number. This divides the heads region into two equal subportions. The right portion of the square also represents a set of experiments in its own right. In casting a die we would expect one-sixth of the outcomes to be a 1, one-sixth to be 2, and so on, all the way through to one-sixth of the outcomes being 6. This divides the tails region into six portions of equal area.

Now it is easy to read off probabilities regarding combinations of outcomes. For example, the outcome of "heads followed by an even number" is represented by one half of half the square, that is:

$$P(\text{heads } \textbf{and} \text{ an even number}) = \frac{1}{2} \times \frac{1}{2} = \frac{1}{4}$$

The outcome "tails and a 5 or a 6" is represented by one sixth of half the square plus another sixth of half the square. Thus:

$$P(\text{tails } \textbf{and} \text{ 5 or 6}) = \frac{1}{6} \times \frac{1}{2} + \frac{1}{6} \times \frac{1}{2} = \frac{1}{6}$$

This model illustrates the multiplicative property in probability theory. In our example:

P (heads **and** an even number)

    = half of half the square

    $= \dfrac{1}{2} \times \dfrac{1}{2}$

    $= P(\text{even}) \times P(\text{heads})$

$P(tails$ **and** $\{5, 6\})$

    = one-third of half the square

    $= P(\{5,6\}) \times P(tails)$

In general, if A and B represent two sets of desired outcomes from two different sets of experiments, then the probability of obtaining both A and B when performing the experiments in succession is:

$$P(A \text{ and } B) = P(A) \times P(B)$$

We are assuming here that the outcomes of one experiment have no effect on the outcomes of the second, that is, that the two experiments are INDEPENDENT EVENTS.

Philosophical Difficulties

Probability, as defined thus far, relies on our ability to count outcomes. If the set of outcomes is infinitely large, then the issue of counting is meaningless. Nonetheless we may still have an intuitive understanding of "likelihood" in these situations. For example, in spinning a compass point, we still feel certain that the probability of the pointer landing between north and east is 1/4, even though there are infinitely many places for the pointer to stop in the one desired quarter of the rim of the compass. If an integer is chosen at random, we suspect that the likelihood of it being even is $^1/_2$—after all, there are only two possibilities, even or odd, each representing half the possible outcomes. Thus determining "probability" relies on an ability to assign a relative measure of size to sets of points, or outcomes, even if those sets might be infinite. This is a difficult issue, and one that caused much confusion during the 19th and early 20th century. (See AREA and BERTRAND'S PARADOX.)

A second difficulty lies in the fact that Cardano's definition of probability is circular: the *probability* of any outcome is determined by knowing beforehand which outcomes are equally *probable*.

How Probability Is Understood Today

In tossing a coin, for example, we are generally willing to say that just two outcomes are possible—heads or tails—and we believe that it is appropriate to assign a probability of 1/2 for each occurring. Folks of a contrary disposition may argue, however, that more than two outcomes could occur (the coin might land on its side, for example) and that the values of probability should be assigned differently. Certainly the issues arising in Bertrand's paradox, for instance, show that the notion of randomness is subject to personal understanding. For a meaningful mathematical discussion to take place, it must therefore be agreed upon *beforehand* which outcomes are deemed within the range of possibility, and what the probability of *all* sets of outcomes will be. (This takes Cardano's approach further—not only must the equally likely outcomes be specified, but also all probabilities must be declared at the outset.)

Thus, in any discussion of probability theory, a mathematician today will state at the outset:

1. The sample space S: the set of all outcomes considered possible
2. A probability measure P: a rule that assigns to any event $A \subseteq S$ a number $P(A)$ called the probability of A.

The probability measure is to satisfy these three rules:

i. For any event A, $P(A)$ is a number between 0 and 1.
ii. $P(S) = 1$. (That is, in any run of the experiment, an outcome will occur.)
iii. If two events A and B have no outcomes in common, then $P(A \text{ or } B) = P(A) + P(B)$.

For simple finite models, such as the act of casting a die, this model encodes the approach developed by Cardano, Pascal, and Fermat, but it also extends this thinking to more complex systems. For example, in throwing a dart at a dartboard, probabilities can be defined as ratios of areas. The probability of throwing a bull's-eye, for instance, is the ratio of the area of the bull's-eye to the area of the entire board. This definition satisfies axioms i, ii, and iii above.

The key then to analyzing any random phenomenon is to appropriately define a probability measure. Different probability measures can lead to different results. Physicists and scientists are challenged then to find the measure that best reflects one's intuitive understanding of the phenomenon being discussed.

See also CONDITIONAL PROBABILITY; EVENT; HISTORY OF PROBABILITY AND STATISTICS (essay); KRUSKAL'S

COUNT; MONTE CARLO METHOD; MONTY HALL PROBLEM; ODDS; RANDOM WALK; STATISTICS; TWO-CARD PUZZLE.

probability density function   *See* DISTRIBUTION.

Proclus   *See* EUCLID'S POSTULATES.

product rule   The DERIVATIVE of the product of two functions *f* and *g* is given by the product rule:

$$\frac{d}{dx}\left(f\cdot g\right)=\frac{df}{dx}\cdot g+f\cdot\frac{dg}{dx}$$

For example, we have $\frac{d}{dx}(x\sin x)=1-\sin x+x\cdot\cos x=$ sin*x* + *x*cos*x*. The rule can be proved using the limit definition of the derivative as follows:

$$\frac{d}{dx}\left(f(x)\cdot g(x)\right)=\lim_{h\to0}\frac{f(x+h)\cdot g(x+h)-f(x)\cdot g(x)}{h}$$

$$=\lim_{h\to0}\frac{f(x+h)\cdot g(x+h)-f(x)\cdot g(x+h)+f(x)\cdot g(x+h)-f(x)\cdot g(x)}{h}$$

$$=\lim_{h\to0}\left(\frac{f(x+h)-f(x)}{h}\cdot g(x+h)+f(x)\cdot\frac{g(x+h)-g(x)}{h}\right)$$

$$=f'(x)\cdot g(x)+f(x)\cdot g'(x)$$

Alternatively, one can recognize that the quantity $f(x+h)\cdot g(x+h)$ is the formula for the area of a rectangle, one that contains the smaller rectangle of area $f(x)\cdot g(x)$. Writing a formula for the area of the L-shaped region between the two rectangles, dividing by *h*, and taking the limit as *h* becomes small leads to the same formula for the product rule.

The product rule can be generalized to apply to any finite product of functions. For example, for the product of three functions we have:

$$(f(x)\cdot g(x)\cdot h(x))'=f'(x)\cdot g(x)\cdot h(x)+f(x)\cdot g'(x)\cdot h(x)$$
$$+f(x)\cdot g(x)\cdot h'(x)$$

If one of the functions in a product is a constant *k*, then the product rule shows:

$$(kf(x))'=0\cdot f(x)+k\cdot f'(x)=kf(x)$$

Two applications of the product rule give the second derivative of a product of two functions:

$$(f(x)\cdot g(x))''=f(x)\cdot g''(x)+2f'(x)\cdot g'(x)+f''(x)\cdot g(x)$$

In general, the *n*th derivative of a product of two functions is a sum of products containing BINOMIAL COEFFICIENTS:

$$\left(f(x)\cdot g(x)\right)^{(n)}=\sum_{k=0}^{n}\binom{n}{k}f^{(k)}(x)\cdot g^{(n-k)}(x)$$

This result is called Leibniz's theorem.

See also CHAIN RULE; HIGHER DERIVATIVE; QUOTIENT RULE.

projection   Any mapping of a geometric figure onto a PLANE to produce a two-dimensional image of that figure is called a projection. For instance, the daytime shadow cast by an outdoor object is an example of a projection onto the ground. Since the Sun is a great distance from the Earth, rays of sunlight are essentially PARALLEL, and shadows cast by it are parallel projections. Shadows cast by a single point of light, however, such as the flame of a candle, have different shapes than those cast by the sun. Such projections are called central projections.

French mathematician and engineer GIRARD DESARGUES (1591–1661) observed that the central projection of any CONIC SECTION is another conic section. (For instance, the shape cast on the ground by the circular rim of a flashlight is an ELLIPSE.) This led him to study those properties of geometric figures that remain unchanged under central projections, thereby founding the field of PROJECTIVE GEOMETRY.

In VECTOR analysis, the projection of a vector **a** onto a vector **b** is a vector parallel to **b** whose length is the length of the "shadow" cast by **a** if the two vectors are placed at the same location in space and the "rays of light" casting the shadow are parallel and PERPENDICULAR to **b**. Thus the projection of **a** onto **b** is a vector of the form *x***b**, for some value *x*, with the property that the vector connecting the tip of *x***b** to the tip of **a** is perpendicular to **b**. Using the DOT PRODUCT of vectors, this yields the equation:

$$\mathbf{b} \cdot (\mathbf{a} - x\mathbf{b}) = 0$$

Solving for x gives the formula for the projected vector:

$$\text{Projection of } \mathbf{a} \text{ onto } \mathbf{b} = \frac{\mathbf{a} \cdot \mathbf{b}}{|\mathbf{a}|^2} \mathbf{b}$$

See also MERCATOR'S PROJECTION; STEREOGRAPHIC PROJECTION.

projective geometry The branch of GEOMETRY that examines those properties of geometric figures that remain unchanged by a central PROJECTION is called projective geometry. Informally, the shadows cast by a point source of light, such as the flame of a candle or the small bulb of a flashlight, are central projections, and so projective geometry examines those properties that are the same for the shadow of a geometric object and the object itself. For instance, the "shadow" of a straight line is another straight line, and so the notion of "straight" is a valid concept in projective geometry.

French mathematician and engineer GIRARD DESARGUES (1591–1661), through his studies of PERSPECTIVE in art, was struck by the fact that the image of a CONIC SECTION under projection is another conic section. (For instance, the outline of the circular rim of a flashlight, a conic section, aimed at the ground at an appropriate angle produces the image of either an ELLIPSE, a PARABOLA, or a branch of a HYPERBOLA.) His study of this phenomenon founded the field of projective geometry. Unfortunately, his work was misunderstood at the time and was neglected. The subject was later revived in 1822 by French mathematician Jean Victor Poncelet (1788–1867) with his publication of *Traité des proprietés des figures* (Treatise on the projective property of figures).

DESARGUES'S THEOREM is a fundamental result about perspective. In order to unify special cases of the theorem and to allow for the possibility of two lines being parallel in its statement, Desargues developed the notion of "points at infinity" so that one could appropriately speak of the point of any intersection of *any* two lines in the plane. (Parallel lines are said to intersect at these points at infinity.) In this framework, any two points in the plane determine a line, and any two lines determine a point. Thus Desargues had created a

notion of geometry in which the words *point* and *line* play dual roles, and can be interchanged in any true statement about this geometry to obtain new (true) statements in this geometry. This notion of duality was first properly outlined by Poncelet.

Today, any system of objects, usually called points and lines, satisfying the following five axioms is called a projective geometry:

1. For any two distinct points, there is exactly one line containing them both.
2. For any two distinct lines, there is exactly one point common to both.
3. There is at least one line.
4. Each line contains at least three points.
5. Not all points lie on the same line.

(The last three axioms ensure that the geometry is not trivial.)

A finite projective geometry is any arrangement of a finite number of points and lines that obey these five axioms. For instance, the diagram below, called the Fano plane, is a projective geometry with seven points and seven lines with exactly three points on each line and three lines through each point. (The central triangle is considered a single line.) This is called the symmetric configuration of type 7_3.

There is only one symmetric configuration of type 8_3, three of type 9_3, 10 of type 10_3, and 31 of type 11_3. All the symmetric configurations up to type 18_3 have been determined.

See also HISTORY OF GEOMETRY (essay).

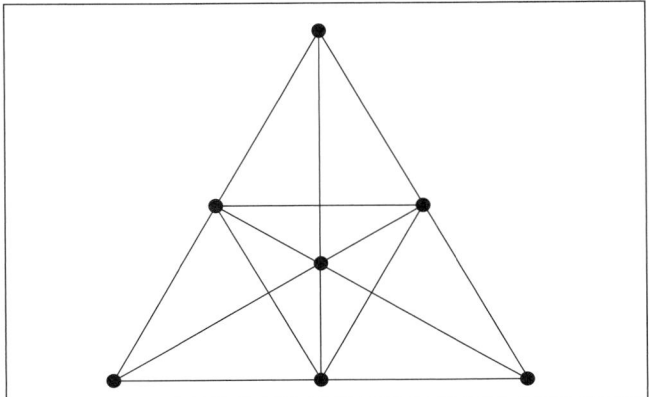

The symmetric configuration of type 7₃

proof A demonstration that one statement follows from another via a process of pure logical reasoning is called a proof. All proofs are based on a set of premises, that is, statements that have been previously established as valid, or are taken as AXIOMS, and follow a sequence of deductions that establish that the final statement presented—the conclusion—is true. Alternatively, one can think of a proof as a written piece demonstrating logical reasoning that convinces all readers of the absolute validity of the result—given the assumptions made as premises.

As an example, the following argument is a proof. It assumes that the basic principles of algebra are understood, and that two basic facts about natural numbers are known, namely, that every natural number is either even or odd, and that every even number is a multiple of 2.

Claim: If n is a natural number such that n^2 is odd, then n itself is odd.

Proof: It cannot be the case that n is even, for then n would be of the form $2k$, making $n^2 = (2k)^2 = 4k^2 = 2(2k^2)$ even, not odd. Thus n must also be odd.

The use of the term *proof* in mathematics differs from the use of the word in the scientific method. Generally, a scientist will put forward a hypothesis to explain a physical phenomenon, and then perform experiments to test that hypothesis. If the results of the investigation consistently conform to the hypothesis (and moreover, if the hypothesis leads to predictions about other physical phenomena that are later confirmed), a scientist may say that a physical principle has been proved. For example, that matter is composed of fundamental elements as listed in a periodic table is a proven scientific theory. It is, however, a result established solely on experience and observation.

Observation, alone, however, does not constitute proof in mathematics, and can lead to erroneous conclusions. For example, one may be tempted to argue that the equation $\sqrt{a^2 + b^2} = a + b$ is a valid algebraic identity since, after all, it works for $a = 1$ and $b = 0$, for instance. After testing all the numbers from 1 to 1,000, for instance, one might conclude that all numbers are smaller than 1 million. (One thousand instances that support a statement surely make it convincing.) Both claims are, of course, absurd. For example, $\sqrt{a^2 + b^2} =$ $a + b$ does not hold for $a = b = 1$, and 1,000,001 is not smaller than 1 million. These examples illustrate why a mathematician will only accept justifications based on logical reasoning and not on observation alone.

There are several different methods of proof. A DIRECT PROOF proceeds linearly from premises to conclusion, whereas an INDIRECT PROOF (also called a PROOF BY CONTRADICTION, or *reductio ad absurdum*) assumes the falsehood of the desired conclusion and shows that to be impossible. Many claims in NUMBER THEORY and GRAPH THEORY are also established by the method of mathematical INDUCTION.

In 1976 two mathematicians Kenneth Appel and Wolfgang Haken presented a computer-generated "proof" of the famous FOUR-COLOR THEOREM. Relying on 1,200 hours of computer computation to verify nearly 2,000 special clauses. No individual, unfortunately, has the means to check the details of the work and be personally convinced of the validity of the argument. The question of whether or not the work of Appel and Harken constituted a "proof" caused great controversy in the mathematics community. For ease of mind, mathematicians today still seek a purely mathematical proof of the result, avoiding computer help altogether.

See also BRUTE FORCE; DEDUCTIVE/INDUCTIVE REASONING; THEOREM.

proof by contradiction (reductio ad absurdum) *See* INDIRECT PROOF.

proof by contraposition *See* CONTRAPOSITIVE.

proportional Two quantities x and y are said to be directly proportional if their RATIO is always constant. This means $\frac{y}{x} = k$, or $y = kx$, for some nonzero constant k called the constant of proportionality. For example, Hooke's law in physics asserts that the force F exerted by a spring extended a small length x is directly proportional to that length: $F = kx$. Here k is called the spring constant.

If y is directly proportional to x, we write $y \propto x$ and say that y varies directly as x. Of course it also follows that $x \propto y$. When the values of y are plotted against the corresponding values of x, the graph obtained is a straight line through the origin.

If a quantity y is proportional to the inverse of a variable x, that is, $y \propto \frac{1}{x}$, then we say that y and x are inversely proportional or that y varies inversely as x. For example, the gas laws in physics assert that gas pressure is inversely proportional to volume.

In algebra, terms in a statement about the ratios of quantities are sometimes called proportionals. For example, in the expression:

$$a : b = c : d$$

the quantity c is the third proportional.

proposition   In mathematics, any statement for which a proof is required or has been provided is called a proposition. In FORMAL LOGIC, any statement that has a truth-value of either true or false (but not both) is called a proposition.

See also THEOREM.

Ptolemy, Claudius (ca. 85–165 C.E.) *Greek Geometry, Trigonometry, Astronomy*   Born in Egypt, around 85 C.E., scholar Claudius Ptolemy is remembered for his epic piece *Syntaxis mathematica* (Mathematical collection), also known as the *Almagest* (The greatest), often deemed as the most significant work in TRIGONOMETRY of ancient times. The work contains accurate tables of "chords" (equivalent to a modern table of sine values), as well as a clear description as to how that table was constructed. (He used a result today known as PTOLEMY'S THEOREM, the geometric equivalent of the trigonometric ADDITION formulae we use today.) It is known that Ptolemy also attempted to prove Eulcid's famous PARALLEL POSTULATE.

The exact location of Ptolemy's birth is unknown. Although born in Egypt, he is referred to as a classical Greek scholar because he followed the scholarly traditions of the Greeks and wrote in that language. His last name, Ptolemy, is Greek, but his first name, Claudius, is Roman. This suggests that he was also considered a Roman citizen.

Accurate translations of all of Ptolemy's works survive today. His most noted work, *Syntaxis mathematica,* comes in 13 volumes, and much of the mathematics developed in it is motivated by concerns

Claudius Ptolemy, an eminent scholar of the second century, wrote the *Almagest,* the most influential work on the topic of trigonometry of ancient times. *(Photo courtesy of the Science Museum, London/Topham-HIP/The Image Works)*

of astronomy. In particular, Ptolemy worked to give detailed descriptions of the motions of the Sun, Moon, and the known planets at the time. He believed that the Earth lay at the center of the solar system.

In Books 1 and 2 of *Syntaxis mathematica,* Ptolemy develops a mathematical theory of compound circular motions (motion in EPICYCLES) as a means to explain the observed motion of the planets. This motivated his need for his accurate table of chords. It is also worth mentioning that in this work Ptolemy used a 360-sided polygon inscribed in a circle to find the following approximation for π correct to three decimal places:

$$\pi \approx \frac{377}{120} \approx 3.14166$$

Book 3 of *Syntaxis mathematica* is concerned with the motion of the Sun. By carefully analyzing the lengths of the seasons and the timing of the solstices and equinoxes, Ptolemy concluded that, although, in his belief, the Sun orbits the Earth, the Earth does not lie at the center of that orbit. He develops a mathematical theory to calculate the distance from that center that the Earth supposedly lies.

Books 4 and 5 examine the motion of the Moon, and Book 6 provides a theory of eclipses. Much of Books 7 and 8 represent a catalogue of over 1,000 stars, and the remaining five books explore his epicyclic theory of planetary motion.

Ptolemy also wrote important scientific works in other fields. His book *Analemma* discusses novel mathematical methods for constructing sundials; his work *Optics* examines properties of color, reflection, and refraction; and his major work *Geography* attempts to map the entire world known at his time, giving measurements as accurate as possible for the latitude and longitude of major cities.

Ptolemy's theorem Second-century Greek astronomer and mathematician CLAUDIUS PTOLEMY proved the following result, now known as Ptolemy's theorem:

If a, b, c, and d are the side-lengths of a QUADRILATERAL inscribed in a circle, and if p and q are the lengths of its diagonals, then $ac + bd = pq$.

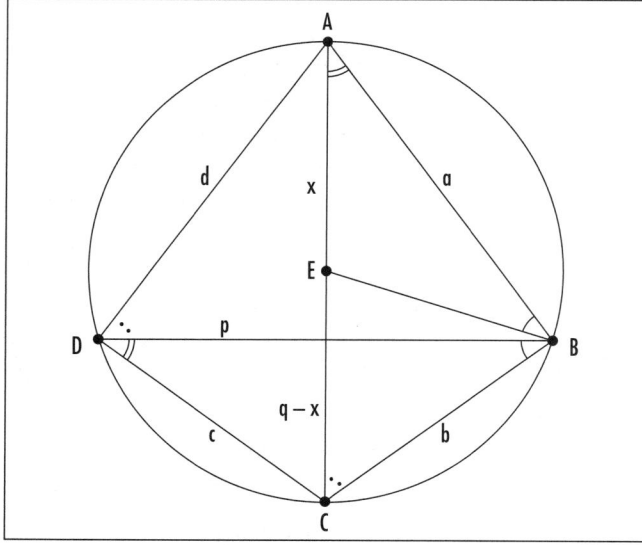

Ptolemy's theorem

It can be proved as follows: according to the standard CIRCLE THEOREMS, angles CAB and CDB shown are equal. Construct line BE so that triangles ABE and CBD are similar. Then $\frac{a}{x} = \frac{p}{c}$. One then checks that triangles ABD and EBC are also similar (angles ADB and ACB are equal), and so $\frac{b}{q-x} = \frac{p}{d}$. Consequently $ac + bd = px + p(q - x) = pq$.

See also BRAHMAGUPTA'S FORMULA; SIMILAR FIGURES.

pure mathematics The study of abstract mathematical systems and structures, without necessarily having practical applications in mind, is called pure mathematics. It has various branches, including ABSTRACT ALGEBRA, GEOMETRY, NUMBER THEORY, CALCULUS, TOPOLOGY, and the topics derived from them, but the distinction from APPLIED MATHEMATICS might not be sharp. For example, EUCLIDEAN GEOMETRY could be analyzed as an abstract study of the relationships between lines, points, and geometric shapes based on the foundations of EUCLID'S POSTULATES, or could, at the same time, be viewed as a study of results that could potentially (and, in fact, has proved to be) useful to architects, surveyors, engineers, and scientists.

Although much of the mathematics developed in the time of antiquity was clearly motivated by practical concerns, the development of mathematics for its own sake was nonetheless of interest to early scholars. For instance, Babylonian tablets from ca. 1600 B.C.E. list large PYTHAGOREAN TRIPLES that could have no practical use. Greek mathematicians of around 400 B.C.E. began to seek rigor, proof, and justification in their mathematical thinking, and ca. 300 B.C.E. EUCLID produced his logically rigorous treatise *THE ELEMENTS*, summarizing all mathematical knowledge known at his time. The unique organization of ideas presented in his work became the key feature of the piece. That, in itself, was seen as an analysis of logical thinking, one that became the paradigm of all mathematical and scientific thinking for the two millennia that followed.

During the 19th century, mathematicians began to search for unifying ideas between distinct branches of algebra and geometry. The general study of structures and operations on them led to the development of abstract algebra, for instance. The development of PARADOXes in SET THEORY and in the foundations of

CALCULUS forced scholars to seek greater levels of rigor and abstraction. Even the nature of logical reasoning itself was examined as an attempt to understand and resolve fundamental paradoxes. The need for abstract ANALYSIS and synthesis was recognized as important, and dichotomy between applied and pure mathematics became more apparent.

Today many research universities possess two departments of mathematics, one considered pure and the other applied. Students can obtain advanced degrees in either field.

See also BABYLONIAN MATHEMATICS; GREEK MATHEMATICS.

pyramid A CONE with a polygon for the base is called a pyramid. Specifically, a pyramid is the solid figure formed by a polygon and a number of triangles, one attached to each side of the polygon, all meeting at a common point called the APEX of the pyramid. The pyramid is classified as a right pyramid if the line connecting the apex of the pyramid to the CENTER OF GRAVITY of the base meets the base at a right angle. (Otherwise it is called an oblique pyramid.) A square pyramid is a right pyramid with a square base.

The study of volumes of cones shows that the volume of a pyramid is just one-third the area of the base of the figure times its height. A TETRAHEDRON is a right pyramid with four equilateral triangles as faces.

See also FRUSTUM.

Pythagoras (ca. 569–475 B.C.E.) *Greek Geometry, Number theory* Born in Samos, Ionia, Greek mathematician and mystic Pythagoras is remembered as founder of the Pythagorean School, which claimed to have found universal truth through the study of number. As such, the Pythagoreans are credited as the first to have taken mathematics seriously as a study in its own right without regard to possible application and practical need. What precisely Pythagoras contributed himself is no longer clear, but the school is acknowledged to have discovered the role of ratios of whole numbers in the musical scale, the properties of FIGURATE NUMBERS, AMICABLE NUMBERS, PERFECT NUMBERS, and the existence of IRRATIONAL NUMBERS. Although PYTHAGORAS'S THEOREM was known to the Babylonians over 1,000 years earlier, the Pythagore-

Pythagoras, leader of a sixth-century B.C.E. sect, claimed to have found universal truth through the study of whole numbers and their ratios. *(Photo courtesy of Topham/The Image Works)*

ans are the first to have provided a general proof of the result.

Very little is known of Pythagoras's life. The sect he founded was half scientific and half religious and followed a code of secrecy that certainly promoted great mystery about the man himself. Many historical writings attribute godlike qualities to Pythagoras and are generally not regarded as accurate portrayals. It is understood that Pythagoras visited with THALES OF MILETUS (ca. 625–547 B.C.E.) as a young man who, most likely, contributed to Pythagoras's interest in mathematics. Around 535 B.C.E., Pythagoras traveled to Egypt and was certainly influenced by the secret sects of the Egyptian priests. (Many of the practices the Pythagoreans followed were the same as those practiced in Egypt—refusal to eat beans or to wear cloth made from animal skins, for instance.)

In 525 B.C.E., Egypt was invaded by Persian forces, and Pythagoras was captured and taken to Babylon.

Eventually, after being released, Pythagoras settled in southern Italy, where he founded his famous school around 518 B.C.E. Both men and women were welcomed as members.

Pythagoras believed that all physical (and metaphysical) phenomena could be understood through numbers. This belief is said to have stemmed from his observation that two vibrating strings produce a harmonious combination of tones only when the ratios of their lengths can be expressed in terms of whole numbers. (Pythagoras was an accomplished musician and made significant contributions to the theory of music.)

Pythagoras, and his followers, studied whole numbers and their ratios, and even went as far as to assign mystical properties to numbers. For instance, they believed that the first natural number, 1, acted as the divine source of all numbers and so was of different stature than an ordinary number. (The number 2 was deemed the first number.) All even numbers were assumed feminine and all odd numbers masculine, and all odd numbers (except 13) represented good luck. The number 4 stood for "justice," being the first perfect square, and 5 marriage, as the union (sum) of the first even number (2) and the first odd number (3). The number 6 was "perfect" since it equals the sum of its factors different from itself.

In GEOMETRY, the Pythagoreans knew that the interior angles of a TRIANGLE always sum to 180° (and, more generally, that the interior angle of an n-sided figure sum to $(n - 2) \times 180°$), and knew how to construct three of the five PLATONIC SOLIDS. They could solve equations of the form $x^2 = a(a - x)$ using geometrical methods, and they developed a number of techniques for constructing figures of a given area. They attributed great mystical significance to the PENTAGRAM because of the geometric ratios it contains. In astronomy, the Pythagoreans were aware that the Earth is round, but believed it lay at the center of the universe. They were aware that the orbit of the Moon is inclined to the equator of the Earth and were the first to realize that the evening star and the morning star were the same heavenly object (namely, the planet Venus).

The exact date, location, and circumstance of Pythagoras's death are not known. Despite his passing, the factions of the original Pythagorean order endured for more than 200 years. Members of the Pythagorean brotherhood so revered their founder that it was considered impious for any individual to claim a discovery for his own glory without referring back to Pythagoras himself. It is said, for instance, that member Hippasus of Metapontum (ca. 470 B.C.E.) was put to death by drowning for announcing his own discovery of the regular dodecahedroan, the fifth Platonic solid. (Other versions of this popular story submit that he was executed for claiming to have discovered that the square root of 2 is an irrational number.)

Pythagoras's theorem (Pythagorean theorem, right-triangle principle) Hailed as the most important result in all of geometry, Pythagoras's theorem states that, for a right-angled triangle, the square of the length of the HYPOTENUSE is equal to the sum of the squares of the other two sides. Geometrically, if one constructs squares on the three sides of a right triangle, then the theorem asserts that the area of the large square equals the sum of the areas of the other two. Algebraically, if the three side-lengths of the triangle are a, b, and c, with length c the hypotenuse, then the theorem asserts that the following relationship holds:

$$a^2 + b^2 = c^2$$

The theorem can easily be proved by arranging four copies of the given right triangle in a big square. As long as the triangles do not overlap, the amount of space around them is always the same, no matter how

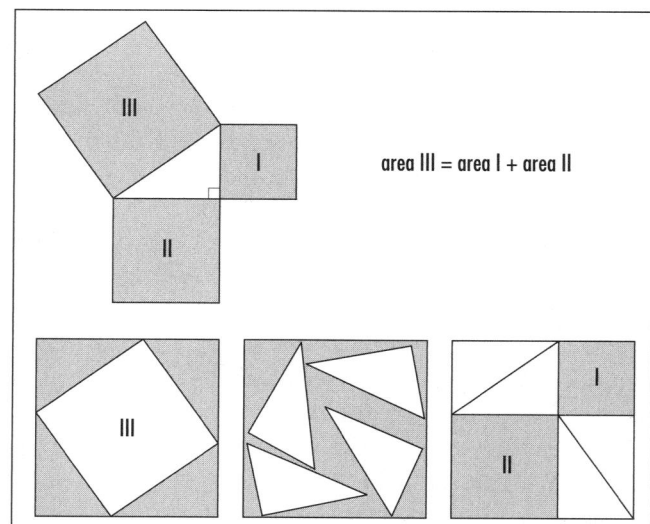

Pythagoras's theorem

the triangles are arranged. In particular, the left and right arrangements show the theorem to be true.

The theorem is named after the Greek philosopher, mathematician, and mystic PYTHAGORAS (ca. 500 B.C.E.). Babylonian tablets dating before 1600 B.C.E. listed tables of numbers a, b and c satisfying the relationship $a^2 + b^2 = c^2$, suggesting that scholars of the time were aware of the result. (It is believed that the Babylonians used 3-4-5 triangles to measure right angles when planning constructions.) The theorem is stated explicitly in the ancient Hindu text *Sulbasutram* (ca. 1100 B.C.E.) on temple buildings, and the diagram of four triangles arranged in a square appears in the ancient Chinese text *Chou Pei Suan Ching* of about 500 B.C.E Although scholars from other cultures may have been aware of the result, Pythagoras is credited as the first to give full and proper explanation as to why the result is true.

Today many different proofs of Pythagoras's theorem are known. The ancient Greeks (thinking solely in terms of geometric constructs) showed how to explicitly divide the largest square into four pieces that could be rearranged to form the two smaller squares. The method of proof presented above is often called the Chinese proof. United States President James Garfield published his own proof of Pythagoras's theorem in 1876 as part of his test to become a high-ranking Mason. Early in the 20th century Professor Elisha Scott Loomis collated and published 367 different demonstrations of the result in his book *The Pythagorean Proposition*. Tiling a floor with squares of two different sizes provides a surprising visual proof of the theorem. (*See* SQUARE.) New proofs of this famous result are still being discovered today.

Consequences of the Theorem

If a right triangle has side-lengths a, b, and c, with the side of length c opposite the right angle, then Pythagoras's theorem asserts $c^2 = a^2 + b^2$, from which it follows that c is larger than both a and b. Thus:

> In a right triangle, the side opposite the right angle is indeed the hypotenuse of the triangle. It is longer than either of the remaining two sides.

Although this observation seems trivial, it has some important consequences:

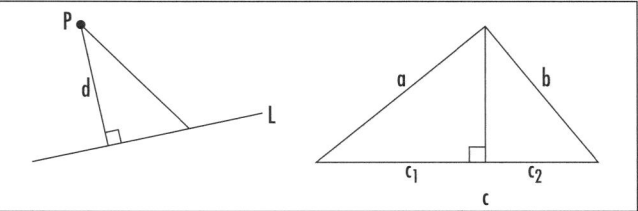

Consequences of Pythagoras's theorem

1. The shortest distance d of a point P from a line L is given by the length of the perpendicular from P to L.

    In the diagram above, any other line segment connecting P to L is longer than d.

    This observation allows one to prove all of the standard CIRCLE THEOREMS.

2. In any triangle, the sum of the lengths of any two sides is larger than the length of the third.

    This result, called the TRIANGULAR INEQUALITY, follows by drawing a perpendicular line from the apex of the triangle. In the diagram above right, we have that $a + b$ is greater than $c_1 + c_2 = c$.

3. The shortest distance between two points in a plane is given by a straight line.

    The triangular inequality shows that the line directly connecting two vertices of a triangle is shorter than the sum of the two remaining side-lengths. Thus, if a path connecting points A and B is composed of straight-line segments, each pair of segments can be replaced with a shorter single straight-line segment. Repeated application of this procedure eventually replaces the path with the straight-line path connecting A to B. If, on the other hand, a path connecting A to B is curved, then one can approximate the curved path by one composed solely of small straight-line segments. As we have just seen, the direct straight-line path connecting A to B is shorter than any such approximation. One can argue that if the approximation is made with some degree of precision, the straight-line path connecting A to B is shorter than the curved path too.

Generalized Pythagorean Theorem

The shapes constructed on the sides of a right triangle need not be squares for Pythagoras's theorem to hold true. For example, if one were to construct equilateral triangles on each of the three sides of a right triangle, then the area of the large triangle would equal the sum of the areas of the two small triangles. The same is true if one were to construct pentagons, semicircles, or any figure on each side of the triangle, as long as the three figures are similar. The reason is straightforward:

> Let F be a figure of area A with one side of length 1, and consider a right triangle with sides of lengths a, b, and c with the side of length c the hypotenuse. Draw scaled versions of the figure F on each of the three sides of the triangle. Then, as area scales as length squared, the area of the large figure is a^2A, and the areas of the two smaller copies are b^2A and c^2A. By the ordinary version of Pythagoras's theorem we have $c^2A = a^2A + b^2A$.

The Converse of the Theorem

The LAW OF COSINES states that, for any triangle, not necessarily a right triangle, with side-lengths a, b, and c, with the angle opposite side c labeled C, the following relation holds:

$$c^2 = a^2 + b^2 - 2ab\,\cos C$$

Thus if an arbitrary triangle were to satisfy Pythagoras's relation $c^2 = a^2 + b^2$, then it must be the case that $\cos C$ equals zero, meaning that C is a 90° angle. This gives the converse to Pythagoras's theorem:

> If an arbitrary triangle with side lengths a, b, and c satisfies $a^2 + b^2 = c^2$, then that triangle is a right triangle (with the right angle opposite the side of length c).

Thus, for example, a triangle with side lengths 3, 4, and 5, is indeed a right triangle precisely because $3^2 + 4^2 = 5^2$. (Many elementary texts in mathematics state without explanation that a 3-4-5 triangle is right.) A study of the properties of acute, obtuse, and right angles in a TRIANGLE provides a very elementary alternative proof of the Pythagorean converse.

Any set of integers a, b, and c, satisfying the relation $a^2 + b^2 = c^2$ is called a PYTHAGOREAN TRIPLE. The triples 5, 12, 13 and 20, 21, 29 are Pythagorean triples and do indeed form the side-lengths of two right triangles.

See also DISTANCE FORMULA; SCALE; SIMILAR FIGURES.

Pythagorean triples   A set of positive integers (a,b,c) satisfying the equation $a^2 + b^2 = c^2$ is called a Pythagorean triple. For example, $(3,4,5)$ and $(48,55,73)$ are two sets of Pythagorean triples. The converse of PYTHAGORAS'S THEOREM shows that any Pythagorean triple (a,b,c) corresponds to the side-lengths of a right triangle, with the side of length c as hypotenuse. Egyptian architects, for example, used knotted ropes to create 3-4-5 triangles and thereby accurately measure 90° angles.

The problem of finding Pythagorean triples is an ancient one. The oldest record known to exist on the topic of number theory, a clay tablet from the Babylonian era (ca. 1600 B.C.E.), contains a table of right triangles with integer sides. That the triple $(4961,6480,8161)$ is listed suggests that the Babylonians had a general method for generating Pythagorean triples and did not rely on trial and error alone to find them. (It also suggests that scholars of the time were also interested in pursuing mathematics simply for the enjoyment of the subject.)

Multiples of any given Pythagorean triple give new triples. For example, from the triple $(3,4,5)$ we obtain new triples $(6,8,10)$, $(9,12,15)$, $(12,16,20)$ and the like. In some sense, these new Pythagorean triples are uninteresting and scholars tend to focus on those triples (a,b,c) for which the numbers a, b, and c share no common factors (other than the number 1). Such Pythagorean triples are called primitive. For example, $(5,12,13)$ is a primitive Pythagorean triple, but $(60,63,87)$ is not.

The Greek mathematician EUCLID (ca. 300 B.C.E.) in book 10 of his text *THE ELEMENTS* completely classified the primitive Pythagorean triples. He showed that any primitive Pythagorean triple must be of the form:

$$(p^2 - q^2,\ 2pq,\ p^2 + q^2)$$

for some positive integers p and q, one even, one odd, with $p > q$ and sharing no common factors. For example, the primitive triple $(3,4,5)$ is obtained by setting $p = 2$ and $q = 1$. His proof of this used only basic principles of arithmetic but was somewhat complicated. Using the theory of COMPLEX NUMBERS, however, the Swiss mathematician LEONHARD EULER discovered in the 1700s an elementary derivation of this result:

> Suppose a, b, and c are three integers satisfying the equation $c^2 = a^2 + b^2$. Write $a^2 + b^2 = (a + ib)(a - ib)$. Since every complex number

has a square root, it must be the case that $a + ib$ equals $(p + iq)^2$ for some complex number $p + iq$. We have $a + ib = (p + iq)^2 = p^2 - q^2 + 2pqi$, from which it follows that $a = p^2 - q^2$ and $b = 2pq$. Also, $c^2 = a^2 + b^2 = (p^2 - q^2)^2 + (2pq)^2 = (p^2 + q^2)^2$ yielding $c = p^2 + q^2$. One now checks that the numbers p and q must be integers with the stated properties.

(To complete the final step, write the integers $(p^2 + q^2) + (p^2 - q^2) = 2p^2$ and $(p^2 + q^2) - (p^2 - q^2) = 2q^2$ each as a product of prime numbers. Use this to show that if p and q are not each themselves integers, then the triple obtained is not primitive.)

Every Pythagorean triple (a,b,c) is a solution to the equation $\left(\dfrac{a}{c}\right)^2 + \left(\dfrac{b}{c}\right)^2 = 1$. Thus, for example, $\left(\dfrac{3}{5}, \dfrac{4}{5}\right)$ is a point on the UNIT CIRCLE $x^2 + y^2 = 1$ with rational coordinates, as are $\left(\dfrac{5}{13}, \dfrac{12}{13}\right)$ and $\left(\dfrac{4961}{8161}, \dfrac{6480}{8161}\right)$. That there are infinitely many Pythagorean triples shows that the unit circle passes though infinitely many points in the plane with rational x- and y-coordinates.

A study of CONTINUED FRACTIONS gives an alternative method for generating Pythagorean triples. If (a,b,c) is a (primitive) triple with "legs" a and b differing by k, then $(2a + b + 2c, a + 2b + 2c, 2a + 2b + 3c)$ is a new (primitive) triple with legs that also differ by k. For example, from the triple $(3,4,5)$, whose first two terms differ by one, we obtain the new triples

$$(3,4,5) \rightarrow (21,20,29) \rightarrow (119,120,169) \rightarrow \ldots$$

From $(5,12,13)$, whose first two terms differ by seven, we obtain the new triples

$$(5,12,13) \rightarrow (55,48,73) \rightarrow (297,304,425) \rightarrow \ldots$$

This procedure will produce all the (primitive) triples with legs that differ by a given amount k.

Although there are infinitely many integer solutions to the equation $a^2 + b^2 = c^2$, Fermat's last theorem shows that there are no nonzero solutions to the companion equations $a^n + b^n = c^n$ for $n > 2$. The planar curves given by $x^n + y^n = 1$ thus never pass through points with rational coordinates.

QED The abbreviation QED is short for *quod erat demonstrandum,* Latin for "which was to be demonstrated." EUCLID of the third century B.C.E. wrote the Greek equivalent of this phrase at the end of each of his proofs to indicate that its conclusion had been reached. Many mathematicians still follow this tradition.

The initials QEF, for *quod erat faciendum* (which was to be done), are sometimes added after the completion of a geometrical construction, and QEI, for *quod erat inveniendum* (which was to be found), after the completion of a calculation.

quadrant In a system of CARTESIAN COORDINATES, the two coordinate axes divide the plane into four regions called quadrants. By convention, the quadrants are numbered, with the first quadrant being the one above the x-axis and to the right of the y-axis (this is the region in which both x and y are positive), and the second, third, and fourth quadrants arranged counterclockwise about the origin.

Many popular science fiction texts describe three-dimensional space as being divided into four quadrants. This is incorrect usage of the term.

One-quarter of the circumference of a circle is sometimes called a quadrant, as is one quarter of the interior of a circle (bounded by two perpendicular radii).

See also OCTANT.

quadratic (quadric) Any expression, function, or equation containing variables raised to the second power, but no higher power, is described as quadratic. Thus a quadratic polynomial, for instance, is a POLYNOMIAL of second degree, and a quadratic equation is an equation formed by setting a quadratic polynomial equal to zero. Such an equation has the form $ax^2 + bx + c = 0$ with a, b, and c either REAL or COMPLEX NUMBERS and a nonzero.

Quadratic equations often arise when computing the dimensions of rectangles. (This explains the use of the prefix "quad" in their name, derived from the Latin name *quattuor* for "four.") For instance, told that a quadrangle of area 55 ft^2 has length 6 units longer than its width leads to the equation (given in terms of the unknown width x) $x(x + 6) = 55$ which is the quadratic equation $x^2 + 6x - 55 = 0$. Ancient Egyptian and Babylonian scholars of ca. 2000 B.C.E. were the first to contend with problems of this type.

The technique of COMPLETING THE SQUARE provides a simple method for solving all quadratic equations. It also leads to the famous quadratic formula:

A quadratic equation of the form $ax^2 + bx + c = 0$ with $a \neq 0$ has solutions given by:

$$x = \frac{-b \pm \sqrt{b^2 - 4ac}}{2a}$$

For example, the quadratic formula shows that $x^2 + 6x - 55 = 0$ has solutions $x = 5$ and $x = -11$. (The positive solution provides an acceptable answer to our quadrangle problem.)

The quantity $b^2 - 4ac$ is called the DISCRIMINANT of the quadratic equation $ax^2 + bx + c = 0$. The quadratic formula shows that if the discriminant is a positive real number, then the equation has two distinct real solutions. If it is zero, then the equation has just one (double) root, and if the discriminant is a negative real number, then there are no real solutions to the equation. (There are, however, two distinct complex solutions.) Given two numbers r_1 and r_2, then $x^2 - (r_1 + r_2)x + r_1 r_2 = 0$ is a quadratic equation with those two numbers as solutions. In general, the quadratic formula shows that the two roots of a quadratic equation $ax^2 + bx + c = 0$ have sum $-\dfrac{b}{a}$ and product $\dfrac{c}{a}$.

A quadratic function is a function of the form $f(x) = ax^2 + bx + c$ with $a \neq 0$. If the constants a, b, and c are real numbers, then the graph of a quadratic function is a PARABOLA, CONCAVE UP if a is positive and concave down if a is negative. Rewriting the equation as:

$$f(x) = a\left(x^2 + \frac{b}{a}x\right) + c$$

$$= a\left(x^2 + \frac{b}{a}x + \frac{b^2}{4a^2}\right) + c - \frac{b^2}{4a}$$

$$= a\left(x + \frac{b}{2a}\right)^2 + c - \frac{b^2}{4a}$$

shows that, if a is positive, the quadratic function has minimal value when $x = -\dfrac{b}{2a}$ (If a is negative, then this produces a maximal value for the function.) The point $(-\dfrac{b}{2a}, c - \dfrac{b^2}{4a})$ represents the vertex of the parabola. The graph crosses the x-axis twice if the quadratic equation $f(x) = ax^2 + bx + c = 0$ has two real solutions, touches the x-axis at a turning point if the two solutions are equal, and does not cross the x-axis at all if there are no real solutions.

Students in high school are often taught to solve quadratic equations by a process of factoring. One notes that the product of two linear terms $Ax + B$ and $Cx + D$ yields a quadratic

$$(Ax + B)(Cx + D) = ACx^2 + (AD + BC)x + BD$$

with coefficients $a = AC$, $b = AD + BC$, and $c = BD$. Here b is a sum of two factors of the product ac. This suggests the following procedure:

To solve a quadratic equation $ax^2 + bx + c = 0$, select two factors p and q of the product ac that sum to b. (This gives $pq = ac$ and $p + q = b$.) Rewrite the quadratic equation as

$$ax^2 + px + qx + c = 0$$

and perform algebra to factor the equation. Begin by selecting a common factor of the first two terms, ax^2 and px, and a common factor of the final two terms, qx and c.

For example, to solve $8x^2 + 2x - 3 = 0$, we seek two factors of -24 that sum to two. This suggests the factors -4 and 6. Rewriting, we obtain:

$$8x^2 + 2x - 3 = 8x^2 - 4x + 6x - 3$$
$$= 4x(2x - 1) + 3(2x - 1)$$
$$= (4x + 3)(2x - 1)$$

It is now clear that the quadratic equation $8x^2 + 2x - 3 = (4x + 3)(2x - 1) = 0$ has solutions $x = -\dfrac{3}{4}$ and $x = \dfrac{1}{2}$.

It is surprising that if a, b, and c are whole numbers, this method will never produce fractional coefficients. One can explain why this is the case using techniques from NUMBER THEORY:

Given a quadratic expression $ax^2 + bx + c$ with integral coefficients, and integers p and q with $pq = ac$ and $p + q = b$, let d be the GREATEST COMMON DIVISOR of a and p, and e the greatest common divisor of q and c. Then the numbers $\dfrac{a}{d}$ and $\dfrac{p}{d}$ are integers sharing no common factors. The same is true for $\dfrac{q}{e}$ and $\dfrac{c}{e}$. Since $ac = pq$ we have $\dfrac{a/d}{p/d} = \dfrac{q/e}{c/e}$. The fraction $\dfrac{a/d}{p/d}$ is presented in reduced form, as is the fraction $\dfrac{q/e}{c/e}$. Since they are the same reduced fraction, their numerators and denominators match. Consequently, $\dfrac{a}{d} = \dfrac{q}{e}$ and $\dfrac{p}{d} = \dfrac{c}{e}$. Factoring the quadratic expression now yields:

$$ax^2 + bx + c = ax^2 + px + qx + c$$

$$= dx\left(\frac{a}{d}x + \frac{p}{d}\right) + e\left(\frac{q}{e}x + \frac{c}{e}\right)$$

$$= dx\left(\frac{a}{d}x+\frac{c}{e}\right)+e\left(\frac{a}{d}x+\frac{c}{e}\right)$$

$$= (dx+e)\left(\frac{a}{d}x+\frac{c}{e}\right)$$

This is a factorization with integral coefficients.

For the quadratic expression $8x^2 + 2x - 3$, we selected $p = -4$ and $q = 6$. Then $d = \gcd(8,-4) = 4$ and $e = \gcd(-3,6) = 3$, again yielding the factorization $(4x+3)(\frac{8}{4}x+\frac{-3}{3}) = (4x+3)(2x-1)$.

A quadratic form is a homogeneous polynomial of degree two. For example, $ax^2 + bxy + cy^2$ is a quadratic form in two variables. A quadratic curve is a curve given by an algebraic equation of second degree. For example, a CIRCLE, given by an equation of the form $(x - a)^2 + (y - b)^2 = r^2$, is a quadratic curve.

See also HISTORY OF EQUATIONS AND ALGEBRA (essay).

quadrilateral (quadrangle, tetragon) Any POLYGON with four sides is called a quadrilateral. For example, any SQUARE, TRAPEZIUM, or PARALLELOGRAM is a quadrilateral. Kites and deltoids are quadrilaterals whose adjacent sides are equal in pairs.

If p and q are the lengths of the diagonals of a convex quadrilateral, and θ is the angle between them, then the AREA of the quadrilateral is given as the sum of the areas of the four triangles they create:

$$\text{area} = \frac{1}{2}xy\sin(\theta) + \frac{1}{2}(p-x)(q-y)\sin(\theta)$$
$$+ \frac{1}{2}y(p-x)\sin(180° - \theta)$$
$$+ \frac{1}{2}x(q-y)\sin(180° - \theta)$$
$$= \frac{1}{2}pq\sin(\theta)$$

Applying the LAW OF COSINES to each the four triangles and summing yields the second equation:

$$b^2 + d^2 - a^2 - c^2 = 2pq\cos(\theta).$$

(Here a, b, c and d are the side-lengths of the quadrilateral as shown in the diagram.) Solving for $\sin(\theta)$ in

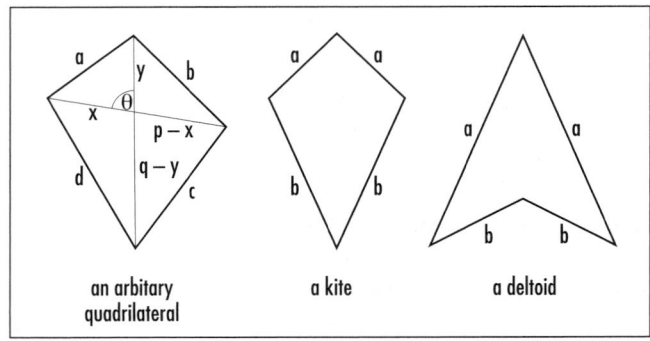

Quadrilaterals

the first equation and for $\cos(\theta)$ in the second and substituting into the equation $\cos^2(\theta)+\sin^2(\theta) = 1$, a standard identity in TRIGONOMETRY, yields one version of BRETSCHNEIDER'S FORMULA for the area of a quadrilateral:

$$\text{area} = \frac{1}{4}\sqrt{4p^2q^2 - \left(b^2 + d^2 - a^2 - c^2\right)^2}$$

Analogous arguments show that these two formulae for area hold for concave quadrilaterals also.

A quadrilateral is called a cyclic quadrilateral if it is a CYCLIC POLYGON, that is, if its four vertices lie on a circle. A consequence of the CIRCLE THEOREMS shows that opposite interior angles of a cyclic quadrilateral sum to 180°. The area of a cyclic quadrilateral is given by BRAHMAGUPTA'S FORMULA.

Every quadrilateral, no matter its shape, provides a TESSELLATION of the plane, that is, copies of any single quadrilateral tile can be used to cover the entire plane without overlap.

See also CONCAVE/CONVEX; PTOLEMY'S THEOREM.

quantifier In English, we frequently encounter statements containing the words *all, some,* and *no* (or *none*). These words are called quantifiers. For example, "All math books are boring," "Some lakes contain fresh water," and "No poet plays the viola" are quantified statements.

Mathematicians reduce all quantified statements to two standard forms by use of either the universal quantifier: "for all," denoted ∀, or the existential quantifier: "there exists," denoted ∃. For example, the statement that all natural numbers of the form $2n + 1$ are odd is written:

$$\forall n,\ 2n + 1 \text{ is odd}$$

that there exists a natural number n satisfying $n \times n = n + n$ can be written:

$$\exists n : n \times n = n + n$$

(a colon is usually read as "such that"), and that there is no natural number n satisfying $n \times n \times n = n + n + n$ as:

$$\neg(\exists n : n \times n \times n = n + n + n)$$

See also ARGUMENT; NEGATION.

quartic equation (biquadratic equation) Any degree-four POLYNOMIAL equation of the form $ax^4 + bx^3 + cx^2 + dx + e = 0$ with $a \neq 0$ is called a quartic equation.

Italian scholar LUDOVICO FERRARI (1522–65), assistant to GIROLAMO CARDANO (1501–76), was the first to find a general arithmetic formula that would solve for x in any quartic equation. His method was published by Cardano in the 1545 epic work *Ars magna* (The great art). French mathematician RENÉ DESCARTES (1596–1650) also found a method of solution, which we briefly outline here.

By dividing through by the leading coefficient a, we can assume that we are working with a quartic of the form:

$$x^4 + Bx^3 + Cx^2 + Dx + E = 0$$

for numbers $B = \dfrac{b}{a}$, $C = \dfrac{c}{a}$, $D = \dfrac{d}{a}$, and $E = \dfrac{e}{a}$.

Substituting $x = y - \dfrac{B}{4}$ simplifies the equation further to one without a cubic term:

$$y^4 + py^2 + qy + r = 0$$

This form of the quartic is called the reduced quartic, and any solution y to this equation corresponds to a solution $x = y - \dfrac{b}{4a}$ of the original equation.

Assume that the reduced quartic can be factored as follows, for some appropriate choice of number λ, m, and n:

$$y^4 + py^2 + qy + r = (y^2 + \lambda y + m)(y^2 - \lambda y + n)$$

EXPANDING BRACKETS and EQUATING COEFFICIENTS consequently yields the equations:

$$n + m = p + \lambda^2$$
$$n - m = \frac{q}{\lambda}$$
$$nm = r$$

Summing the first two equations gives $n = \dfrac{p + \lambda^2 + \frac{q}{\lambda}}{2}$; subtracting them yields $m = \dfrac{p + \lambda^2 - \frac{q}{\lambda}}{2}$; and substituting into the third equation yields, after some algebraic work, a cubic equation solely in terms of λ^2:

$$(\lambda^2)^3 + 2p(\lambda^2)^2 + (p^2 - 4r)(\lambda^2) - q^2 = 0$$

CARDANO'S FORMULA can now be used to solve for λ^2, and hence for λ, m, and n. Thus solutions to the quartic equation:

$$y^4 + py^2 + qy + r = (y^2 + \lambda y + m)(y^2 - \lambda y + n) = 0$$

can now be found by solving $y^2 + \lambda y + m = 0$ and $y^2 - \lambda y + n = 0$ using the QUADRATIC formula.

This method, in principle, is straightforward, but very difficult to carry out in practice. It does show, however, that, if one has the patience, one can indeed write down a formula for the solution of a quartic equation $ax^4 + bx^3 + cx^2 + dx + e = 0$ using only the numbers a, b, c, d, and e, and their roots.

During the 1700s there was great eagerness to find a similar formula for the solution to the quintic (degree-five equation). LEONHARD EULER (1707–83) attempted to find such a formula, but failed. He suspected that the task might be impossible.

In a series of papers published between the years 1803 and 1813, Italian mathematician Paolo Ruffini (1765–1822) developed a number of algebraic results that strongly suggested that there can be no procedure for solving a general fifth- or higher-degree equation in a finite number of algebraic steps. This claim was indeed proved correct a few years later by Norwegian mathematician NIELS HENRIK ABEL (1802–29). Thus—although there is a quadratic formula for solving degree-two equations, Cardano's formula for solving degree-three equations, and Ferrari's and Descarte's methods for solving degree-four equations—there will never be a general formula for solving all equations of degree five or higher.

Of course some degree-five equations can be solved algebraically. (Equations of the form $x^5 - a = 0$, for instance, have solutions $x = \sqrt[5]{a}$.) In 1831 French mathematician ÉVARISTE GALOIS (1811–32) completely classified those equations that can be so solved. The work he conducted gave rise to a whole new branch of mathematics, today called GROUP THEORY.

quartile See PERCENTILE.

quaternions The product of two nonzero REAL NUMBERS is never zero, nor is the product of two nonzero COMPLEX NUMBERS. Both these sets of numbers are said to be division algebras. Since each complex number $a+ib$ corresponds to a point (a,b) in the plane R^2, complex-number multiplication provides a way for defining a multiplication in R^2 that makes R^2 a division algebra:

$$(a,b) \times (c,d) = (ac - bd, ad + bc)$$

(Notice that the "obvious" multiplication given by $(a,b) \times (c,d) = (ac,bd)$ is not satisfactory: $(1,0) \times (0,4)$, for example, gives the zero answer.)

In the early 19th century, WILLIAM ROWAN HAMILTON (1805–65) wondered whether there was a suitable multiplication for R^3 making it, too, a division algebra. Despite his best efforts, he was never able to propose a suitable definition.

Along the way, however, Hamilton was able to provide a suitable multiplication rule for R^4 via the invention of a new number system called the quaternions. These consist of the real numbers together with three new symbols i, j, k, which, like the square root of -1, satisfying the relations:

$$i^2 = j^2 = k^2 = -1$$

along with:

$$i \times j = k \qquad j \times i = -k$$
$$j \times k = i \qquad k \times j = -i$$

$$k \times i = j \qquad i \times k = -j$$

A typical element of the quaternions appears as:

$$a + bi + cj + dk$$

where $a,b,c,$ and d are real numbers. Multiplication of two quaternions is defined by the DISTRIBUTIVE PROPERTY and the relations outlined above. For example,

$$(2 + i) \times (3 + 3j + k) = 6 + 6j + 2k + 3i + 3k - j$$
$$= 6 + 3i + 5j + 5k$$

Hamilton was able to show that the quaternions are a division algebra. Notice that they do not satisfy the COMMUTATIVE PROPERTY: the order of multiplication does affect the answer. For example, $i \times j$ is different from $j \times i$.

In the 1950s mathematicians proved that R^n is a division algebra only for n equal to 1, 2, 4, or 8. These are the real numbers, the complex numbers, the quaternions, and another system called the Cayley numbers (also known as the octonions). These extended number systems are sometimes called hypercomplex numbers.

Quételet, Lambert Adolphe Jacques (1796–1874) Flemish *Statistics, Astronomy* Born on February 22, 1796, in Ghent, Belgium, Adolphe Quételet is often referred to by mathematical historians as the father of modern statistics. Although trained as a mathematician and astronomer, Quételet is remembered for his pioneering work in collecting statistical data and using it to test traditional views on issues of medicine and criminology. His new field of "social mechanics" profoundly influenced European thinking in the social sciences.

Raised in Ghent, Belgium, Quételet received a doctorate in mathematics from the University of Ghent in 1819, having written a dissertation on the theory of CONIC SECTIONS. After teaching mathematics for four years, he moved to Paris in 1823 to begin a study of astronomy. During the course of this work he was introduced to the discipline of PROBABILITY theory and the particular statistical methods astronomers were using to gain accurate measurements of physical phenomena. He began to wonder whether the same techniques apply to human affairs. To test this idea, he undertook a study of data from government records to analyze the numerical consistency of crimes. This work garnered much attention and stimulated general discussions on the nature of free will versus social determinism. He continued his work in analyzing crime and then extended it to analyze mortality, human physique, and social issues. This work produced much controversy among social scientists of the time.

Quételet came to perceive the average value of a human physical or mental quality, or of a general societal characteristic, as a tangible and real "ideal" that nature is trying to create, and he argued that the fact that specific individuals rarely follow these norms is to be expected as usual statistical deviation. His analyses of social phenomena were certainly mathematically sophisticated and gained him much respect as a researcher. Nonetheless, many of his ideas were controversial. In 1835 he published a detailed account of his new social science in *Sur l'homme et le developpement de ses facultés, essai d'une physique sociale* (A treatise on man, and the development of his faculties), which, among other things, detailed the "average man."

The Quételet index (QI) used today, also known as the body mass index (BMI), is derived from Quételet's work. This index of obesity is defined by the formula:

$$QI = \frac{\text{weight in kilograms}}{\left(\text{height in meters}\right)^2} = \frac{704 \times \text{weight in pounds}}{\left(\text{height in inches}\right)^2}$$

An individual with QI-value greater than 30 is considered obese.

Quételet died in Brussels, Belgium, on February 17, 1874. He had a profound effect on the development of working statistics and the shaping of the study of social science. Florence Nightingale considered his work of the stature of a "new bible."

See also HISTORY OF PROBABILITY AND STATISTICS (essay).

quotient In NUMBER THEORY and ARITHMETIC, the result of dividing one integer by another is called the quotient. There might or might not be a remainder. For example, 13 divided by 4 gives a quotient of 3 and a remainder of 1. In general, the quotient of an integer a divided by b is an integer q such that $a = bq + r$, where r is a whole number with $0 \le r < q$. The quotient can be computed with the aid of the floor function: $q = \left\lfloor \dfrac{a}{b} \right\rfloor$

Outside of the arithmetic of integers, a quotient is taken to mean simply the result of dividing a by b (and so the issue of remainders is moot). For example, the quotient of $\sqrt{48}$ and 3 is $\sqrt{48} \div 3 = \dfrac{\sqrt{48}}{3} \approx 2.309$.

The quotient polynomial of two polynomials $p(x)$ and $q(x)$ is a polynomial $Q(x)$ such that $p(x) = Q(x)q(x) + R(x)$ for some polynomial $R(x)$ of degree smaller than that of $q(x)$. For example, dividing $p(x) = x^4 + 2x^3 + 3x^2 + 5x + 5$ by $q(x) = x^2 + 2x + 1$ gives a quotient of $Q(x) = x^2 + 1$ and a remainder of $R(x) = x + 3$:

$$x^4 + 2x^3 + 3x^2 + 5x + 5 = (x^2 + 1)(x^2 + 2x + 1) + (x + 3)$$

Such a computation can be accomplished by the process of LONG DIVISION.

In calculus, the QUOTIENT RULE provides a means for differentiating the ratio of two functions.

See also EUCLIDEAN ALGORITHM; FLOOR/CEILING/FRACTIONAL PART FUNCTIONS; MODULAR ARITHMETIC; REMAINDER THEOREM; SYNTHETIC DIVISION.

quotient rule The DERIVATIVE of the quotient of two functions is given by the quotient rule:

$$\frac{d}{dx}\left(\frac{f(x)}{g(x)}\right) = \frac{f'(x) \cdot g(x) - f(x) \cdot g'(x)}{\left(g(x)\right)^2}$$

For example, we have $\dfrac{d}{dx}\left(\dfrac{\sin x}{x}\right) = \dfrac{\cos x \cdot x - \sin x \cdot 1}{x^2} = \dfrac{\cos x}{x} - \dfrac{\sin x}{x^2}$. The rule can be proved by making use of both the PRODUCT RULE and the CHAIN RULE: simply think of the quantity $\dfrac{f(x)}{g(x)}$ as the product: $f(x) \cdot (g(x))^{-1}$. Consequently:

$$\frac{d}{dx}\left(\frac{f(x)}{g(x)}\right) = \left(f(x) \cdot \left(g(x)\right)^{-1}\right)'$$

$$= f'(x) \cdot \left(g(x)\right)^{-1} + f(x) \cdot \left(\left(g(x)\right)^{-1}\right)'$$

$$= \frac{f'(x)}{g(x)} + f(x) \cdot \left(-\big(g(x) \big)^{-2} \cdot g'(x) \right)$$

$$= \frac{f'(x)}{g(x)} - \frac{f(x) \cdot g'(x)}{\big(g(x) \big)^{2}}$$

which is the quotient-rule formula. It is much easier just to use the product and chain rules for any problem at hand rather than memorize the quotient-rule formula.

radian measure *See* ANGLE.

radius of convergence *See* POWER SERIES.

Ramanujan, Srinivasa Aiyangar (1887–1920) Indian *Number theory* Born on December 22, 1887, in Erode, Tamil Nadu state, India, Srinivasa Ramanujan is considered one of India's greatest mathematical geniuses. With an inexplicable talent for handling SERIES and CONTINUED FRACTIONS, Ramanujan made significant contributions to the field of NUMBER THEORY and offered a whole host of new and fundamentally important formulae that have since found applications in many different branches of science. He is also remembered for his famous collaboration with leading British mathematician GODFREY HAROLD HARDY (1877–1947).

Ramanujan excelled in the topic of mathematics during his early school years. As a teenager he came across a small mathematics text by G. C. Carr called *Synopsis of Elementary Results in Pure Mathematics.* Written in a terse style, and being not much more than a list of 5,000 mathematical formulae, equations, and results, with no proofs or explanations, Ramanujan took it upon himself to work through each and every result and provide his own explanation of it. From that moment on Ramanujan cared only for mathematics. By the age of 17 he had conducted his own investigations on the properties of the HARMONIC SERIES and had calculated EULER'S CONSTANT to 15 decimal places.

In 1904 Ramanujan won a scholarship to attend the Government College in Kumbakonam, but it was revoked the following year because he devoted all his time to mathematics and neglected his other courses. He attended another college in Madras, but never graduated, again for failing to attend to his nonmathematical courses.

As a self-taught scholar, Ramanujan began publishing results in the *Journal of the Indian Mathematical Society.* His brilliant 1911 paper, "Some Properties of Bernoulli's Numbers," on the BERNOULLI NUMBERS garnered him national attention. During this time Ramanujan was supporting himself as a clerk in an accounting office.

Ramanujan soon came to realize that he was working at a mathematical level that was beyond the expertise of anyone he knew in India. Encouraged by friends, he began writing to mathematicians in England for guidance. The renowned number theorist Godfrey Hardy of Trinity College, Cambridge, was the recipient of one of Ramanujan's letters and was suitably impressed by the mathematical content it contained. In 1914 Hardy brought Ramanujan to England, and thus began their extraordinary collaboration.

Hardy was impressed by the intuitive nature of Ramanujan's work. With no formal mathematical education, Ramanujan had only a vague idea of what constituted a mathematical proof, but nonetheless had a profound internal sense of mathematical truth and amazing insight into the workings of numbers. One amusing story, for instance, claims that Hardy once mentioned to Ramanujan that the number of the taxicab he had ridden that day was 1,729, a number that he thought was "rather dull." Ramanujan immediately responded, to the contrary, that 1,729 is not at all dull, being the first of all the positive integers that can be written as the sum of two cubes in two different ways. In an instant, Ramanujan somehow realized that 1,729 equals both $12^3 + 1^3$ and $10^3 + 9^3$, and that no other number smaller than 1,729 has this dual property.

Over the 5 years of their time together in England, Hardy and Ramanujan produced a number of fundamentally important results in number theory. Their most notable work was on the theory of PARTITIONs. During this time Ramanujan was awarded a bachelor of science by research degree (the equivalent of a Ph.D.) from Cambridge, was elected a fellow of the ROYAL SOCIETY of London—England's most prestigious scientific body—and also a fellow of Trinity College.

Unfortunately, Ramanujan suffered from extremely poor health. After developing tuberculosis, Ramanujan returned to India in 1919 but died 1 year later on April 26, 1920.

Ramanujan left behind him a number of unpublished notebooks filled with theorems and claims that mathematicians have continued to study. Mathematics professor George Neville Watson (1886–1965) studied the notebooks and published 14 papers, all under the title of *Theorems Stated by Ramanujan*. He also published an additional 30 papers, all inspired by Ramanujan's work.

random numbers   A sequence of numbers with the property that no next number in the sequence can be predicted from the preceding elements is said to be random. Such a sequence does not follow any regular or repetitive pattern. If the numbers listed come from a finite pool of possible candidates (say, single-digit numbers 0 through 9), then the LAW OF LARGE NUMBERS asserts that, in the long run, each number from that pool should occur equally often.

The traditional method for generating random numbers is to draw numbered balls from a container. It is not possible to generate truly random quantities with a computer—any program is a predetermined set of instructions—but it is possible to create a list of numbers that appear to be random. Numbers generated in this way are called pseudorandom. The following are two popular methods for generating pseudorandom numbers.

Middle-Square Method

Developed in 1946 by JOHN VON NEUMANN (1903–57) and his colleagues while working on the Manhattan Project at Los Alamos Laboratories, the middle-square method works as follows:

1. Select a four-digit number to be the first number in the sequence. (This number is called the seed of the algorithm.)
2. Square this number to produce an eight-digit number. (Add a leading zero if necessary.)
3. Use the middle four digits of this eight-digit number as the next number in the sequence. Repeat.

The result that appears is a seemingly random list of numbers between 0 and 9999. For example, beginning with the seed 7,254 we obtain the sequence 6205, 5020, 2004, 1601, 6320, 9424, 8117, ... Unfortunately, the middle-square method can produce sequences of integers that tend toward zero. For example, beginning with the seed 1049 we obtain the sequence 1049, 1004, 80, 64, 40, 16, 2, 0, 0, 0,...

Linear-Congruence Method

Developed by D. H. Lehmer in 1951, the linear-congruence method uses MODULAR ARITHMETIC to generate a list of pseudorandom numbers. It works as follows:

1. Select three fixed numbers a, b, and m, and an initial value x_0 (the seed).
2. Given a number x_n in the sequence, the next number x_{n+1} in the list is obtained by multiplying x_n by a, adding b, and taking the remainder upon division by m:

$$x_{n+1} = ax_n + b \pmod{m}$$

For example, taking $a = 2$, $b = 3$, and $m = 10$, with seed $x_0, = 1$, produces the sequence 1, 5, 3, 9, 1, 5, 3, 9, 1, ..., which, unfortunately cycles and is far from pseudorandom. Although it is not possible to avoid cycling with this method, one can choose an extraordinarily large value for m so that the length of the cycle produced is extremely long, and the repetition of numbers will not be encountered in a lifetime. (This creates a second problem, however, in that in a truly random sequence, numbers do indeed appear more than once, which is avoided here.) Many computers today use this method with the choice $m = 2^{31}$.

Computer-generated pseudo-random numbers are used in STATISTICS to help facilitate unbiased sampling of a POPULATION.

See also SAMPLE.

random variable   *See* DISTRIBUTION.

random walk   In the theory of statistical mechanics, a random walk is the path traced out by a particle performing a sequence of unit steps in which the direction of each step is chosen randomly. For example, gas molecules randomly jolting back and forth in space are effectively performing three-dimensional random walks. (Their resultant motion is called Brownian motion, which can be used to explain diffusion.)

A child standing on a sidewalk can perform a one-dimensional random walk by taking steps backward and forward according to the results of tossing a coin (say, "heads" means step forward one pace, "tails" step backward one pace). Another child standing in an infinite grid of squares performs a two-dimensional random walk by stepping one place north, south, east, or west according to the rolls of a four-sided die (or of an ordinary die if rolls of 5 and 6 are ignored).

Gamblers betting $1 at a time at a casino are effectively performing random walks—the contents of their wallets increase and decrease in a random fashion, one unit at a time.

We can ask, what are the chances that a gambler will eventually lose all of his or her money? (Or equivalently, what is the probability that a particle performing a one-dimensional random walk will eventually move one place behind "start"?)

This probability can be computed as follows. Suppose the gambler has N dollars in hand and, at each round of play, has a 50 percent chance of going up $1 and a 50 percent chance of losing $1. Let p be the probability that the gambler will *eventually* only have $N - 1$ dollars in hand.

There are two ways for a gambler to lose this dollar, either right away, or to win $1 and then eventually lose $2 later on. This shows:

$$p = P(\text{eventually lose \$1})$$
$$= P(\text{lose \$1 right away, OR, win \$1}$$
$$\text{AND later lose \$2})$$
$$= P(\text{lose \$1}) + P(\text{win \$1})$$
$$\times P(\text{eventually lose \$2})$$
$$= \frac{1}{2} + \frac{1}{2} \times P(\text{eventually lose \$2})$$

To lose $2, a gambler must first lose $1 and then another. Thus:

$$P(\text{eventually lose \$2})$$
$$= P(\text{eventually lose \$1}$$
$$\text{AND eventually lose another \$1})$$
$$= p \times p$$

We thus have the equation: $p = \frac{1}{2} + \frac{1}{2}p^2$, that is, $(p - 1)^2 = 0$, showing that $p = 1$. That is, with absolute certainty, a gambler will eventually go down $1, and then another, and then another, all the way down to ruin. This phenomenon is called GAMBLER'S RUIN.

The above calculation shows that all one-dimensional random walks (with equal chances of stepping backward or forward) eventually return to their starting positions. Mathematicians have shown that two-dimensional random walks (again with equal chances of stepping in any one of four directions) also eventually return to their starting positions with absolute certainty. Surprisingly, there is only a 34 percent chance that a three-dimensional random walk will have the particle return to its starting position.

See also HARMONIC FUNCTION.

rank   In STATISTICS, an arrangement of a set of objects or DATA values in an order dictated by the magnitude or importance of some characteristic is called a ranking of the objects. For example, the arrangement of 20 men in order of height is a ranking, as is the arrangement of 10 numbers from lowest to highest. The position of an object in a list given by a ranking is called the rank of the object. A ranking could be purely subjective, such as the ranking of pies in a taste-testing competition. SPEARMAN'S METHOD and KENDALL'S METHOD can be used to test the degree of association between two different rankings of the same set of objects.

In MATRIX theory, the row rank of a matrix is the maximum number of linearly independent rows the matrix possesses, and its column rank is the maximum number of linearly independent columns it has. The process of GAUSSIAN ELIMINATION shows that these two values always agree, and the common value of these two ranks is called *the* rank of the matrix. An $m \times n$ matrix is said to be of full rank if its rank equals the smaller of m and n. A square matrix of full rank is invertible.

See also INVERSE MATRIX; RANK CORRELATION.

rank correlation Two methods are commonly used to determine whether or not two ranking schemes are well-matched. For example, in a pie-baking contest, two judges might rank five pies as follows:

	Apple	Blueberry	Cherry	Dewberry	Elderberry
Judge 1	5	2	3	1	4
Judge 2	4	1	2	3	5

If the judges followed purely objective criteria, and were free of personal preference in their choices, then one would expect two identical ranking choices. If, on the other hand, the judges followed entirely independent criteria, or no criteria at all (randomly assigning ranks), then one would expect very little correlation between the two lists. The results of this competition seem to lie somewhere between these two extremes.

Kendall's Coefficient of Rank Correlation

In 1938 M. G. Kendall developed one measure of rank association given by a single numerical value τ, adopting values between –1 and 1. A value of 1 indicates a perfect matching in rank values; a value of 0 indicates no consistency in the rank assignments; and a value of –1 perfect disagreement (that is, one judge's top choice is the other judge's least favored choice, and so on.) The numerical value τ is computed by completing the following steps:

1. Rearrange the order of the entrants so that the ranks given by the first judge are in order.
2. Looking now only at the second row, compute the score of each entrant. This is the number of entrants to its right higher in rank minus the number of entrants to its right of lower rank. (The rightmost entrant is assigned a score of zero.)
3. Sum all the scores. Call this sum S.
4. If the number of entrants is n, then the maximum possible sum is $(n - 1) + (n - 2) + \ldots + 2 + 1 + 0 = \frac{1}{2} n(n - 1)$. Kendall's coefficient of rank correlation is the ratio of S to this maximal sum:

$$\tau = \frac{S}{\left(\dfrac{n(n-1)}{2}\right)}$$

In our example, step 1 of the procedure yields the reordered table:

	Dewberry	Blueberry	Cherry	Elderberry	Apple
Judge 1	1	2	3	4	5
Judge 2	3	1	2	5	4

There are two entrants to the right of dewberry with rank higher than 3 (elderberry and apple), and two lower than 3 (blueberry and cherry), thus dewberry has score 0. Similarly, the scores of the remaining pies are: blueberry = 3 – 0 = 3, cherry = 2 – 0 = 2, elderberry = 0 – 1 = –1, and apple = 0. The total sum of scores is $S = 0 + 3 + 2 + (-1) + 0 = 4$. The maximum possible sum is $4 + 3 + 2 + 1 + 0 = 10$. Thus Kendall's rank correlation coefficient here is:

$$\tau = \frac{4}{10} = 0.4$$

This value indicates that the rank assignments are possibly inconsistent.

Spearman's Coefficient of Rank Correlation

An alternative rank coefficient was developed by Charles Spearman in 1904. It is denoted ρ, and is computed by summing the differences squared in the rank of each entrant and then comparing the result with the maximum sum possible.

To illustrate: in the above example, apple received ranks of 5 and 4. The rank difference squared is $(5-4)^2 = 1^2 = 1$. Similarly, the rank difference squared for blueberry is $(2-1)^2 = 1$, and for the remaining pies, in alphabetical order, $(3-2)^2 = 1$, $(1-3)^2 = 4$, and $(4-5)^2 = 1$. Summing all five rank differences squared yields the total score: $D^2 = 1+1+1+4+1 = 8$.

With n entrants, it can be shown that the maximal sum possible is $\dfrac{n(n^2-1)}{3}$. It occurs when the ranks assigned by one judge are in reverse order to the ranks assigned by the second. (To see this, observe how the quantity D^2 changes when the ranks of two entrants are switched. This will show that the quantity D^2 is largest when, for any pair of entrants, their assigned ranks are in reverse order.) For $n = 5$, the maximum sum possible is 40.

Spearman's rank correlation coefficient is given by:

$$\rho = 1 - 2 \cdot \frac{D^2}{\text{maximum score possible}}$$

Notice that a value $\rho = 1$ occurs when $D^2 = 0$, that is, there is perfect agreement. A value $\rho = -1$ occurs when D^2 has the maximum value, i.e., perfect disagreement. In our example, we have $\rho = 1 - 2 \cdot \dfrac{8}{40} = 0.6$. This value again indicates some disagreement and an inconsistency in the two ranking schemes.

rate of change   *See* DIFFERENTIAL CALCULUS.

ratio   The relationship of one numerical quantity to another expressed as a QUOTIENT so as to indicate their relative sizes is called a ratio. We write $a{:}b$ for the ratio of a quantity a to a quantity b, or, alternatively, express the ratio as a FRACTION a/b. The first term of a ratio $a{:}b$ is called the antecedent and the second term the consequent. For example, the ratio of 10 to 4 is written 10:4 or 10/4, which can be simplified to 5/2. (Thus the number 10 is 2 1/2 times as large as 4.) Here, 10 is the antecedent, and 4 is the consequent.

The value of a ratio is unchanged if both terms in the ratio are multiplied or divided by the same quantity. For example, 10:4, 30:12, and 2.5:1, are equiva-

lent ratios. A ratio for which one of its terms is equal to 1 is called a unitary ratio. Thus, 2.5:1, for instance, is a unitary ratio.

If the ratio of two variables y and x is constant, then the two variables are said to be PROPORTIONAL. For example, if y/x has constant value k, then it must be the case that $y = kx$.

In PROBABILITY theory, the ODDS of a game are expressed as a ratio. For example, if in a game of chance a of the possible outcomes are deemed favorable and the remaining b possible outcomes unfavorable, then the odds in favor for the game is the ratio $a{:}b$. Thus for a game with favorable odds of 3:2, for instance, three of the five outcomes are deemed favorable, and there is 3/5, or 60 percent, chance of winning the game.

A number that can be expressed as a ratio of two integers is called a RATIONAL NUMBER. Not all numbers are rational. The FUNDAMENTAL THEOREM OF ARITHMETIC establishes, for example, that $\sqrt{2}$ is not rational. The RATIO TEST is often used to determine the convergence of an infinite SERIES.

In the theory of VECTORs, the ratio theorem (also known as the section formula) states the following:

If a point P divides a line segment AB in the ratio $a{:}b$, then the position vector $\mathbf{p}$ of P is given in terms of the position vectors $\mathbf{a}$ and $\mathbf{b}$ of A and B, respectively, by the formula:

$$\mathbf{p} = \frac{a\mathbf{a} + b\mathbf{b}}{a+b}$$

The ratio notation for numbers can be extended to indicate the relative sizes of more than two quantities. For instance, the ratio $a{:}b{:}c$ states that the ratio of the first quantity to the second is $a{:}b$, that the ratio of the second quantity to the third is $b{:}c$, and that the ratio of the first quantity to the third is $a{:}c$. For example, the numbers 40, 80, and 200 are in the ratio 1:2:5.

See also GOLDEN RATIO.

rational function (rational expression)   Any function $f(x) = \dfrac{p(x)}{q(x)}$ given as a RATIO of two polynomials $p(x)$ and $q(x)$ is called a rational function. Such a function is defined at all values x for which $q(x) \neq 0$. For example, $f(x) = \dfrac{x^2 - 2x + 3}{x^2 - 4}$ is a rational function with a domain of all real values except $x = 2$ and $x = -2$. Any single polynomial $p(x)$ can be regarded as the rational function $\frac{p(x)}{1}$.

If, for a rational function $f(x) = \frac{p(x)}{q(x)}$, the denominator $q(x)$ is zero at $x = a$, then the function either has a vertical ASYMPTOTE at $x = a$ or a REMOVABLE DISCONTINUITY at $x = a$. In the latter case, it must be that the numerator $p(x)$ is also zero at $x = a$. For example, the rational function $\frac{x^2 - 5x + 6}{x - 3} = \frac{(x-3)(x-2)}{x - 3}$, although not defined at $x = 3$, equals the continuous function $x - 2$ at all values different from 3, and so has a removable discontinuity at that point.

If a rational function $f(x) = \frac{p(x)}{q(x)}$ tends to a limit c as $x \to \infty$ or as $x \to -\infty$, then the line $y = c$ is a horizontal asymptote for f. For example, the line $y = 2$ is a horizontal asymptote of $\frac{6x + 5}{3x - 2}$. (Write $\frac{6x + 5}{3x - 2} = \frac{x(6 + 5/x)}{x(3 - 2/x)} = \frac{6 + 5/x}{3 - 2/x}$. This approaches the value $\frac{6 + 0}{3 - 0} = 2$ as $x \to \pm\infty$.)

Every rational function can be rewritten in terms of PARTIAL FRACTIONS.

See also CONTINUOUS FUNCTION.

rationalizing the denominator   The process of eliminating any square-root terms in the denominator of a rational expression, without changing the value of that expression, is called rationalizing the denominator. For example, the quantity $\frac{1}{\sqrt{2}}$ can be rationalized by multiplying both the numerator and denominator by $\sqrt{2}$ to obtain: $\frac{1}{\sqrt{2}} = \frac{1 \times \sqrt{2}}{\sqrt{2} \times \sqrt{2}} = \frac{\sqrt{2}}{2}$.

In general, an expression of the form

$$\frac{1}{\sqrt{a} - \sqrt{b}}$$

is rationalized by multiplying through by the conjugate, $\sqrt{a} + \sqrt{b}$, to produce:

$$\frac{\sqrt{a} + \sqrt{b}}{a - b}$$

During the 1950s and 1960s, before the advent of pocket calculators, it was deemed necessary to always rationalize the denominator of a numerical expression. It has the advantage of making computations relatively straightforward. For instance, although one can read from a book of tables that $\sqrt{2}$ is approximately 1.414, computing the approximate value of $\frac{1}{\sqrt{2}}$ via the process of LONG DIVISION is tedious. However, rewriting $\frac{1}{\sqrt{2}}$ as $\frac{\sqrt{2}}{2}$ allows us to see immediately that this quantity is simply half of $\sqrt{2}$, and so has approximate value 0.707.

Today, the quantity $\frac{1}{\sqrt{2}}$ is regarded as a valid expression of a numerical quantity, and there is absolutely no need to rationalize the denominator. Unfortunately, many high school mathematics programs still insist on rewriting such expressions, even though the reasons for doing so are now obsolete.

rational numbers (rationals)   Any number that can be expressed as a RATIO, a/b, of two integers a and b, with $b \neq 0$, is called a rational number. For example, 2/5 and –6/2 (which is equivalent to –3) are rational numbers. Every rational number is a fraction, and as such, the rules for adding, subtracting, multiplying, and dividing fractions apply to the rationals. The set of all rational numbers is denoted **Q** (for quotient), and the set of all rationals constitutes a mathematical FIELD.

Any real number whose decimal representation eventually repeats in a cycle or terminates (and so can be regarded as containing a repeated cycle of zeros) is a rational. For example, multiplying $x = 0.34$ by 100 yields $100x = 34$, and so $x = \frac{34}{100} = \frac{17}{50}$. Multiplying $y = 2.3181818...$ by 10 and by 1,000 and then subtracting yields:

$$1000y = 2318.181818...$$
$$10y = 23.181818...$$
$$990y = 2295$$

establishing that $y = 2295/990 = 51/22$. Conversely, the process of LONG DIVISION shows that every rational number has a decimal expansion that repeats. (For instance, in dividing 3 by 7, each remainder that appears in the process of long division can only be one of seven numbers—0, 1, 2, 3, 4, 5, 6—and so, eventually, some remainder must appear twice. As soon as this occurs, the process of long division cycles.) This establishes:

> The rational numbers are precisely those numbers whose decimal expansions repeat.

Thus any number with a nonrepeating decimal expansion must be an IRRATIONAL NUMBER. In a similar manner, one can show that the rationals are precisely those numbers with terminating CONTINUED FRACTION expansions.

Between any two rational numbers $\frac{a}{b}$ and $\frac{c}{b}$ lies another rational—their midpoint $\frac{\frac{a}{b} + \frac{c}{d}}{2} = \frac{ad + bc}{2bd}$, for instance. Repeating this reasoning establishes:

> Infinitely many rational numbers lie between any two given rationals.

We say that the set of rationals represents a "dense subset" of the real NUMBER LINE. Moreover, between any two real numbers x and y on the number line, with $x < y$, there exists a rational q that lies between them. (Write x and y as decimal expansions and choose a terminating decimal value that lies between them.)

These density arguments show that there is no smallest positive rational number, that is, there is no "first" rational number on the positive number line. Nonetheless, German mathematician GEORG CANTOR (1845–1918) showed with his famous DIAGONAL ARGUMENT that it is possible to arrange all the positive (and negative) rationals in a list: $q_1, q_2, q_3, \ldots$ This shows:

> The set of rational numbers is COUNTABLE.

With the aid of the GEOMETRIC SERIES $1 + x + x^2 + x^3 + \ldots = \frac{1}{1-x}$ we can now show that the amount of space occupied by the rationals on the number line is zero. (This is surprising given that the rationals form a dense subset of the line.)

Choose $x = 1/10$ and cover the point q_1 on the number line with an interval of length x, the point q_2 with an interval of length $x^2 = 0.01$, the point q_3 with an interval of length $x^3 = 0.001$, and so forth. These intervals likely overlap, but they occupy no more than:

$$x + x^2 + x^3 + \cdots = x \cdot \frac{1}{1-x} = \frac{1}{9}$$

amount of space in total. Thus the set of rational points occupies at most one-ninth of the real number line.

In the same way, by working with $x = 1/100$, we deduce that the rationals occupy at most 1/99 of the space on the number line, or at most 1/999 of the space if we work with $x = 1/1000$. Continuing in this manner, we are forced to conclude that the total amount of space occupied by the rational numbers is zero.

We say:

> The rationals form a set with a measure of zero on the real-number line.

Given that the number line is of infinite length, we must conclude then that "most" numbers are not rational.

See also ALGEBRAIC NUMBER; NUMBER; REAL NUMBERS; TRANSCENDENTAL NUMBER.

ratio test *See* CONVERGENT SERIES.

ray (half-line) The portion of a straight line starting at one point and going on forever in one direction is called a ray. Two rays in a plane starting at the same point create a figure called an ANGLE.

real numbers It is extraordinarily difficult to define precisely what is meant by a real number. Many standard texts in mathematics define a real number to be any RATIONAL NUMBER or any IRRATIONAL NUMBER. Unfortunately, since an irrational number is declared to be any number that is not rational, it is not clear from which pool of numbers the irrational numbers come. Alternatively, one can define a real number to be any number that can be expressed as a decimal expansion. (The rational numbers are then those that have repeating expansions and the irrationals are those that do not.) The difficulty with this approach is that it is not clear how to perform arithmetic operations on these entities. For example, when adding two large integers, 769 + 845, say, one begins with the rightmost digits, adds them, carries a digit if necessary, and proceeds to the next column digits one place to the left to repeat this process again. It is not clear, then, how one should proceed when asked to add two infinite decimal expansions. (What is the rightmost digit with which to begin?)

In the mid-1800s, as mathematicians attempted to establish a sound theoretical footing for CALCULUS, scholars came to realize the necessity of a rigorous understanding of what is meant by a "real number." In particular, in order to validate the three key theorems on which calculus rests (namely, the EXTREME-VALUE THEOREM, the INTERMEDIATE-VALUE THEOREM, and the MEAN-VALUE THEOREM), it was essential that the real-number system be proved "complete," in the sense that no numbers are "missing" from the system. (For example, the square root of 2 is missing from the set of rational numbers).

French mathematician AUGUSTIN-LOUIS CAUCHY (1789–1857) proposed a definition of a real number using Cauchy sequences. (This approach is based on the idea that every irrational number is the LIMIT of a sequence of rational numbers.) Although it is easy to perform arithmetic operations on Cauchy sequences, the completeness of the real-number system created is not immediately clear. In 1872, German scholar JULIUS WILHELM RICHARD DEDEKIND (1831–1916) proposed the notion of a DEDEKIND CUT as a means of defining the real numbers. This approach has the advantage of making the completeness of the real-number system readily apparent, but the disadvantage of making the arithmetic manipulations of the real numbers less natural. Nonetheless, both approaches are valid and are used today as the means for defining the real numbers. The work of GEORG CANTOR (1845–1918), with his famous DIAGONAL ARGUMENT, shows that the CARDINALITY of the reals is greater than that of the rationals. It is not surprising then that a more sophisticated approach to understand the real-number system is needed than that for understanding the rationals.

Today the set of real numbers is denoted **R** and is depicted geometrically as the set of points on a NUMBER LINE. The real numbers constitute a mathematical FIELD.

See also NUMBER; ORDER PROPERTIES.

reciprocal (multiplicative inverse) The number 1 divided by a quantity is called the reciprocal of that quantity. For example, the reciprocal of 2 is 1/2, and the reciprocal of $x^2 + 1$ is $\frac{1}{x^2 + 1}$. The product of a quantity and its reciprocal is always 1.

The reciprocal of any FRACTION $\frac{a}{b}$ is the corresponding fraction with numerator and denominator interchanged: $\frac{1}{\frac{a}{b}} = \frac{b}{a}$. The reciprocal of a COMPLEX NUMBER $x + iy$ is simplified by multiplying the numerator and denominator each by the conjugate of the complex number:

$$\frac{1}{x + iy} = \frac{1}{x + iy} \cdot \frac{x - iy}{x - iy} = \frac{x}{x^2 + y^2} - i\frac{y}{x^2 + y^2}$$

This is the same process as RATIONALIZING THE DENOMINATOR of a fractional quantity.

An equation that is unchanged (that is, has the same set of solutions) if the variables in that equation are replaced by their reciprocals is called a reciprocal equation. For example, $x^2 - 3x + 1 = 0$ is a reciprocal equation, since replacing x by $\frac{1}{x}$ yields $\left(\frac{1}{x}\right)^2 - 3\left(\frac{1}{x}\right) + 1 = 0$ which simplifies to $1 - 3x + x^2 = 0$.

The reciprocal series of a given SERIES is the series whose terms are the reciprocals of the terms of the given series. For example, any HARMONIC SERIES is the reciprocal series of an ARITHMETIC SERIES.

See also INVERSE ELEMENT.

Recorde, Robert (ca. 1510–1558) Welsh *Algebra* Born in Tenby, Wales, ca. 1510 (his exact birth date is not known), mathematician Robert Recorde is remembered for his instrumental work in establishing general mathematics education in England as well as for introducing ALGEBRA to that country. Recorde wrote a number of influential elementary textbooks, all written in English and all using clear and simple terminology. He is also remembered for inventing the symbol "=" to denote equals.

Recorde received a medical degree from Cambridge University in 1545 and practiced medicine in London for several years before being appointed as the general surveyor of mines and monies in Ireland by King Edward VI. It is worth noting that Recorde produced the first coin in England to have the date written with

HINDU-ARABIC NUMERALS (the numbers we use today) rather than ROMAN NUMERALS.

Maintaining an interest in mathematics throughout his life, Recorde wrote a series of instructional texts in mathematics. His first piece, *The Grounde of Artes,* published in 1543, discussed the advantages of the Hindu-Arabic numerals and general arithmetic techniques useful for commerce. He later published an extended version of this work with the same name in 1552, which also explored the theory of WHOLE NUMBERS and RATIONAL NUMBERS. Around this time Recorde also published *Pathwaie to Knowledge,* an abridged version of Euclid's famous work, THE ELEMENTS.

Recorde is perhaps best remembered for his 1557 piece *The Whetstone of Witte,* considered the first significant text on the topic of algebra written in English. Its title is a pun not understood today. Early algebraists used the Latin word *cosa* for "thing," meaning the unknown variable in an equation. This word closely resembles the Latin word *cos* for "whetstone," a stone for sharpening tools. Thus Recorde's title is referring to the art of algebra as a device for sharpening one's wit.

In this famous text, Recorde explains the theory of equations and the arithmetic of square roots. He solves QUADRATIC equations (but dismisses negative solutions). He notes, in particular, that for an equation of the form $x^2 = ax - b,$ the term a equals the sum of the two roots of the equation, and the term b their product. (We see this today by noting that $(x - r_1)(x - r_2) = x^2 - (r_1 + r_2)x + r_1r_2.$) It is in this piece that Recorde introduces the symbol "=" for equality, noting that the use of two parallel line segments is apt "bicause noe 2 thynges can be moare equalle."

Recorde died in London, England, in 1558. The exact date of death is not known.

rectangle A QUADRILATERAL with all four angles being right angles is called a rectangle. A SQUARE is considered a rectangle.

The AREA of a rectangle is the product of its length and its breadth, and its PERIMETER is double the sum of these two quantities:

$$\text{area} = \text{length} \times \text{breadth}$$
$$\text{perimeter} = 2 \times \text{length} + 2 \times \text{breadth}$$

There are only two rectangles with integer side-lengths whose areas and perimeters have the same numerical value: the 3×6 rectangle of area/perimeter 18, and 4×4 square of area/perimeter 16.

The two DIAGONALS of a rectangle are equal in length and bisect each other, and the rectangle is the only quadrilateral with this property. Carpenters measure diagonal lengths to check that their work is rectangular. One can also use this observation to draw surprisingly accurate rectangles on the ground using only two pieces of rope equal in length and a piece of sidewalk chalk: tie the two ropes together at their midpoints and pull them taut to form the two bisecting diagonals of a rectangle. Mark the endpoints of the ropes, these are the corners of the rectangle, and draw its sides using a taut rope as a guide.

See also GOLDEN RECTANGLE.

recurrence relation (**recursive relation, reduction formula**) An equation that allows one to calculate the successive values of a FUNCTION or a SEQUENCE once an initial set of values is given is called a recurrence relation. For example, the FIBONACCI NUMBERS F_n are completely specified by the recurrence relation $F_n = F_{n-1} + F_{n-2}$ for $n > 2$ once we are told that $F_1 = 1$ and $F_2 = 1$. The number of moves required to solve the TOWER OF HANOI PUZZLE with N discs, $T(N)$, is given by the recurrence relation $T(N + 1) = 2T(N) + 1$ with $T(1) = 1$. The recurrence relation $a_n+1 = a_n + 3$ with $a_1 = 7$ defines the arithmetic sequence 7, 10, 13, 16,...

In INTEGRAL CALCULUS, the method of INTEGRATION BY PARTS is often used to establish important reduction formulae. For example, setting $u = x^n$ and $v' = e^x$ in the integral below establishes the relation $\int x^n e^x dx = x^n e^x - n\int x^{n-1} e^x dx$. Repeated application of this formula allows one to eventually evaluate the given integral.

Other reduction formulae include:

$$\int x^n \sin x \, dx = -x^n \cos x + n \int x^{n-1} \cos x \, dx$$
$$\int x^n \cos x \, dx = x^n \sin x - n \int x^{n-1} \sin x \, dx$$
$$\int \sin^n x \, dx = -\frac{1}{n}\cos x \sin^{n-1} x + \frac{n-1}{n}\int \sin^{n-2} x \, dx$$
$$\int \cos^n x \, dx = \frac{1}{n}\sin x \cos^{n-1} x + \frac{n-1}{n}\int \cos^{n-2} x \, dx$$
$$\int (\ln x)^n dx = x(\ln x)^n - n\int (\ln x)^{n-1} dx$$

See also DYNAMICAL SYSTEM; LOGISTIC GROWTH; RECURSIVE DEFINITION.

recursive definition (inductive definition, recursion)
A SEQUENCE a_n is said to be defined recursively if:

1. The first term a_0 is given.
2. An algorithm for computing any term from its predecessor is presented.

For instance, the sequence of powers $1, x, x^2, x^3, \ldots$ can be defined recursively by:

$$a_0 = 1$$
$$a_{n+1} = xa_n$$

and the FACTORIAL function $n!$ can be defined as:

$$0! = 1$$
$$(n + 1)! = (n + 1) \times n!$$

See also DYNAMICAL SYSTEM; RECURRENCE RELATION.

reduced form (lowest terms)   A FRACTION is said to be in reduced form if its numerator and denominator share no common factor (other than one). For example, the fraction 12/25 is in reduced form, whereas 14/21 is not. (The numerator and denominator share 7 as a factor.) Canceling all factors common to the numerator and denominator of a fraction reduces the fraction to one of reduced form. For instance, 14/21 is equivalent to the reduced fraction 2/3.

Mathematicians have proved that if the numerator and denominator of a fraction are chosen at random, then the PROBABILITY that the resultant fraction is in reduced form is precisely $\frac{6}{\pi^2}$ (about 61 percent).

See also CANCELLATION.

reflection   *See* GEOMETRIC TRANSFORMATION; LINEAR TRANSFORMATION.

Regiomontanus (1436–1476) German *Trigonometry, Astronomy*   Born on June 6, 1436, in Königsberg, Prussia (now Germany), scholar Regiomontanus is remembered as author of *De triangulis omnimodis* (On all classes of triangles), published posthumously in 1533, which was the first modern account of TRIGONOMETRY as a discipline independent of astronomy. This work was extremely influential in the revival of the subject in the West.

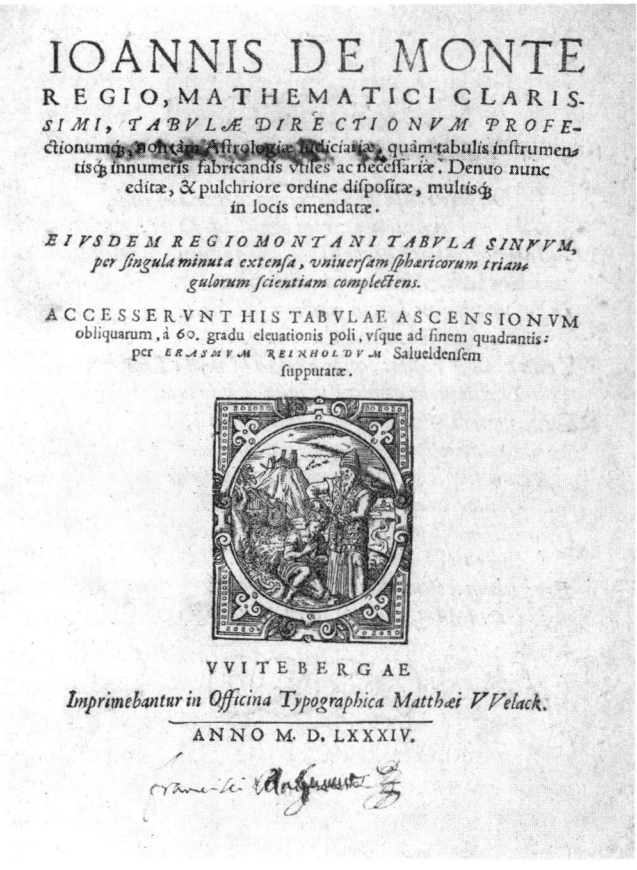

Regiomontanus, a mathematician and astronomer of the 15th century, published *Tabulae*, a text of trigonometric tables important to scholars at that time. *(Photo courtesy of the Science Museum, London/Topham-HIP/The Image Works)*

Although born Johann Müller, Regiomontanus took the name of his birthplace. (Königsberg means the king's mountain," which translates into Latin as "Regiomontanus.") Trained as an astronomer, he was appointed a professorship in the field at the University of Vienna in 1641, and, seven years later, was made astronomer to King Matthias Corvinus of Hungary.

Up until this time, trigonometry was considered only a part of astronomy. Although aware of the Arabic use of the tangent function, Regiomontanus discussed only the sine function in his famous piece. Unlike the ancient Indian mathematicians, he did not think of the sine as a ratio, but instead as a length of a particular line segment drawn for a circle of fixed radius. Using a circle of radius 60,000 units, Regiomontanus presented a large table of sine values, and discussed the means to convert these values to ones appropriate for a circle of a different radius. His work also outlined all the basic theory of trigonometry, along with applications to both plane and spherical geometry. Regiomontanus's work was extremely influential.

Regiomontanus was an able astronomer. In 1472 he made detailed observations of a comet sufficiently accurate enough for scholars to later identify it as Halley's comet. (Halley was the first to compute the length of the orbit of the comet 210 years later.) Regiomontanus also devised a mathematical method for determining longitudinal position on the EARTH by making use of the location of the Moon in the sky. He died in Rome, Italy, on July 8, 1476.

regression If a SCATTER DIAGRAM indicates a relationship between two variables in a scientific study, then any attempt to quantify that relationship is called regression. For example, if an experiment yields data points (x_1, y_1), (x_2, y_2), ..., (x_n, y_n), one can attempt to find a formula $y = f(x)$ that "best fits" the data. The LEAST SQUARES METHOD seeks to minimize the sum of squares:

$$\sum_{i=1}^{n}\big(y_i - f(x_i)\big)^2$$

using techniques of CALCULUS. This approach works particularly well if one suspects a linear relationship $f(x) = ax + b$, but it can be used too for possible quadratic relationships $f(x) = ax^2 + bx + c$, and suspected relationships following other simple formulae.

See also CORRELATION COEFFICIENT.

relation (relationship) Any pairing between elements of one set with some elements of another, or elements of one set with other elements of the same set, is called a relation. For example, consider the set of all the people of the world. Then the "mothered by" correspondence M that associates to each person his or her biological mother is a relation. For the set of real numbers, the "greater than" relation G declares a to be related to b if a is greater than b, and the "circle" relation C sets x as related to y if the two numbers satisfy the equation $x^2 + y^2 = 1$.

If two members a and b of a set are related by a relation R, then we write aRb. For instance, if Brian's mother is Joan, then "Brian M Joan" (that is, Brian is "mothered by" Joan). We also have that $3G2$, since 3 is greater than 2, and $1C\,0$, since the point $(1,0)$ lies on the unit circle. (Note: the "greater than" relation G is usually denoted ">".)

If the relation being discussed is understood, it is also convenient to write two elements that are related simply as an ordered pair. For instance, in the mothering relation M we have (Brian, Joan), but not (Joan, Brian). Mathematicians generally prefer the ordered-pair presentation of a relation and in fact will *define* a relation to simply be any collection of ordered pairs. This has the advantage that if the elements under discussion are numbers, then one can graph the points that satisfy a given relation. For example, under the "greater than" relation G, a pair (a,b) satisfying this relation corresponds to a point in the plane above the diagonal line $y = x$. The set of all pairs (x,y) that belongs to the circle relation C forms a circle in the xy-plane.

A relation is called a FUNCTION if no element in the first set (first coordinate) of the relation is matched with more than one element of the second set (second coordinate). For example, as Brian, as with every human being, can be matched with only one biological mother, the "mothered by" relationship is a function. Since 3 > 2 and 3 > 1, for example, the "greater than" relation is not a function. The "circle" relation C also fails to be a function since, for instance, both $(0,1)$ and $(0, -1)$ belong to the relation.(We have $0C1$ and $0C-1$.) If we consider, instead, the "parabola" relation P defined as the set of all points (x, y) that satisfy $y = x^2$, then this relation is a function: if xPy_1 and xPy_2, then it must be the case that $y_1 = y_2$. (We have $y_1 = y_2 = x^2$.)

relative error If the ERROR or uncertainty in a measurement of a quantity is expressed as a ratio to the total measurement, then the error is called a relative error. For example, when 1.2 is used as an approximation for the quantity 1.16, then the relative error is $0.04/1.16 \approx 0.034$. If, in measuring the height of a building as 62 ft, engineers used a device that measures to the nearest half a foot, then the height of the building should be written 62 ± 0.5 ft, and the relative error of this measurement is $0.5/62 \approx 0.008$.

See also PERCENTAGE ERROR.

relatively prime (coprime) Two integers a and b are relatively prime if their GREATEST COMMON FACTOR is 1. Consequently, the only factor the two numbers have in common is 1. For example, 15 and 28 are relatively prime. Any two consecutive numbers are relatively prime.

Relatively prime numbers play a key role in the famous postage-stamp problem:

> Which postage values can be obtained using 5-cent and 7-cent stamps only?

For example, one can compose a postage value of 22 cents with three 5-cent stamps, and one 7-cent stamp, but the quantity 23 cents cannot be so obtained. One can check that each quantity 24, 25, 26, 27, and 28 is possible. Adding multiples of 5 to each of these numbers then shows that *any* quantity greater than 24 is obtainable. (Below this, one can check that only the values 5, 7, 10, 12, 14, 15, 17, 19, 20, 21, and 22 can be composed.)

In general, given two stamp values a and b, with values a and b relatively prime, one can prove that there is always a number N so that every quantity N and above can be composed as a combination of stamps of each type. This result is not true if a and b are not relatively prime.

Mathematicians have shown that the PROBABILITY that two integers chosen at random are relatively prime is $\frac{6}{\pi^2} \approx 0.61$.

See also COMMON FACTOR; FACTOR; JUG-FILLING PROBLEM.

remainder theorem The process of LONG DIVISION shows that if a POLYNOMIAL $p(x)$ of degree n is divided by a second polynomial $q(x)$ of degree m, one obtains, apart from a QUOTIENT term $Q(x)$, a remainder polynomial $R(x)$ of degree strictly smaller than m:

$$p(x) = Q(x)q(x) + R(x)$$

For instance, dividing $p(x) = x^4 - x^3 + 2x^2 + x + 4$ by the degree-2 polynomial $q(x) = x^2 + 1$ yields a remainder $R(x) = 2x + 1$ of degree 1:

$$x^4 - x^3 + 2x^2 + x + 4 = \left(x^2 + x + 1\right)\left(x^2 + 1\right) + \left(2x + 3\right)$$

In particular, if a polynomial $p(x)$ is divided by a linear term $x - a$ for some constant a, the result remainder must be a polynomial of degree zero, that is, a constant:

$$p(x) = Q(x)(x - a) + b$$

Setting $x = a$ into this equation shows that $p(a) = 0 + b$. Thus the remainder term b is simply the value of the polynomial at $x = a$. This observation is called the remainder theorem:

> If a polynomial $p(x)$ is divided by the term $x - a$, for some constant a, then the remainder term is $p(a)$:
>
> $$p(x) = (x - a)Q(x) + p(a)$$

For example, if $p(x) = x^3 - 7x^2 + 2x + 4$ is divided by $x + 1 = x - (-1)$, the remainder will be $p(-1) = -1 - 7 - 2 + 4 = -6$.

The remainder theorem is useful for finding the factors of a polynomial. For instance, for the polynomial $p(x) = x^3 - 7x^2 + 2x + 4$, we have $p(1) = 1 - 7 + 2 + 4 = 0$, indicating that $(x - 1)$ is a factor: $x^3 - 7x^2 + 2x + 4 = (x - 1)Q(x) + p(1) = (x - 1)Q(x)$.

See also FACTOR THEOREM; FUNDAMENTAL THEOREM OF ALGEBRA.

removable discontinuity *See* CONTINUOUS FUNCTION.

Reuleaux's triangle *See* CONSTANT WIDTH.

Rhind papyrus (Ahmes papyrus) This famous document is the oldest written mathematical text known to exist. Currently housed in the British Museum, this 18-ft long roll of papyrus, 13 in. wide, dates back to ca. 1650 B.C.E. The document was discovered buried in the Egyptian desert sands, near the Valley of the Kings, in the mid-1800s by an unknown Egyptian citizen. Visiting Scotsman Alexander Henry Rhind (1833–63), who had an interest in antiquities, bought the papyrus and transported it to Britain.

The papyrus is a copy of a work that dates back at least 200 years earlier. Although the text consists mainly of numerical problems and solutions, with an emphasis on practical application, it is clear that the

Egyptians of the time held mathematics in high regard. The copyist, whose name is rendered as Ahmes or Ahmose, copied also the grandiose title of the papyrus: *Accurate Rendering: The Entrance into the Knowledge of All Existing Things and All Obscure Secrets*. The purpose of the text seems to be that of a textbook—a series of examples and exercises organized to illustrate certain mathematical skills and techniques. Some of the latter problems discussed, however, have no practical application and clearly indicate a delight in studying mathematics for its own sake.

See also EGYPTIAN FRACTIONS; EGYPTIAN MATHEMATICS; EGYPTIAN MULTIPLICATION.

Richter scale   *See* LOGARITHMIC SCALE.

Riemann, Georg Friedrich Bernhard (1826–1866)
German *Geometry, Analysis, Calculus, Number Theory*
Born on September 17, 1826, in Breselenz, Hanover (now Germany), Bernhard Riemann is remembered as one of the greatest mathematicians of the 19th century. His work in NON-EUCLIDEAN GEOMETRY revolutionized modern thinking on the study of shape and space and provided the tools ALBERT EINSTEIN needed for his description of the universe in the general theory of relativity. Among his many significant contributions to ANALYSIS, Riemann is remembered for his construction of the Riemann integral, which, for the first time, extended the theory of CALCULUS beyond the realm of continuous functions only. Riemann also developed a theory of complex functions (that is, functions whose inputs and outputs are COMPLEX NUMBERS) and studied applications of calculus in this setting. In 1859, while searching for a better approximation for the number of primes as given by the famous PRIME-NUMBER THEOREM, Riemann discovered a generalized version of LEONHARD EULER's important ZETA FUNCTION. He made a deep and insightful conjecture about the properties of this function, the Riemann hypothesis, which, if true, would provide invaluable information about the distribution of prime numbers. Proving (or disproving) the Riemann hypothesis is considered the most important unsolved mathematical problem of today.

Riemann exhibited a keen interest in mathematics at an early age and was encouraged by the director of his high school to pursue this line of study. In 1846

Riemann enrolled at the University of Göttingen to study under the mathematician CARL FRIEDRICH GAUSS (1777–1855). Surprisingly, even with Gauss as a faculty member, Göttingen did not offer a strong program in mathematics, and so, a year later, Riemann transferred to Berlin University, an institution recognized for its strength in mathematics at the time. While there, Riemann developed his general theory of complex variables and functions that later formed the basis of much of his important work.

After completing his undergraduate degree at Berlin, Riemann returned to Göttingen in 1849 to complete a doctoral thesis under the supervision of Gauss. He was awarded a Ph.D. 2 years later in mathematics.

Gauss described Riemann's thesis work as having "a gloriously fertile originality." By viewing the range of outputs of a complex function as a surface in three- or even four-dimensional space, Riemann managed to tie the theory of calculus and analysis to the study of shape, space, and geometry. This innovative approach was immediately recognized as a significant turning point in the advancement of mathematics. On Gauss's recommendation, Riemann was appointed a position at Göttingen, and he commenced work on his habilitation degree. During this time Riemann turned his attention to the study of FOURIER SERIES and questions of integrability. It was during this period that Riemann formulated his famous definition of an integral.

As the final step to completing his degree Riemann was required to give a public lecture. Gauss asked him to speak on the theory of geometry, and on June 10, 1854, Riemann presented to the outside world his groundbreaking work on shape and space.

Riemann devised a general notion of distance in *n*-dimensional space and developed a theory of TENSORs that allowed him to properly define geometric shape. In particular, Riemann developed the notion of curvature tensor to measure, or define, the warping of space. The mathematics Riemann presented on that day proved to be the precise framework Einstein needed for structuring his theory of general relativity. Riemann continued to present and publish groundbreaking work in the years that followed.

On July 30, 1859, Riemann was appointed chair of mathematics at Göttingen, and a few days later he was honored with election to the Berlin Academy of Sciences. By this time Riemann had developed an interest in NUMBER THEORY and the distribution of prime numbers. In a written report to the academy, Riemann presented his work on the famous zeta function and his method of extending the scope of the function to include complex numbers as inputs—another masterpiece. This work completely changed the direction of mathematical research in number theory for the century that followed. Riemann had managed to connect the notions of geometry and space to complex functions and, now, to the study of numbers. This significant achievement provided mathematicians the means to translate insights and advances in one disparate field into results and discoveries in another.

Riemann suffered from ill health most of his life and died of tuberculosis at the age of 39 in Selasca, Italy, on July 20, 1866.

right angle   An ANGLE equal to one-quarter of a complete revolution is called a right angle. Such an angle has measure 90° or $\frac{\pi}{2}$ radians. The corner of a square is a right angle.

One could say that a right angle is the "correct" angle to use in architecture and the construction of buildings. Egyptian architects of 1500 B.C.E. were aware that a 3–4–5 triangle contains a right angle (the converse of PYTHAGORAS'S THEOREM shows this) and used knotted ropes 3 + 4 + 5 = 12 units long to quickly construct these triangles at a building site. A right angle is also the angle made if one were to make a perfect right turn.

See also DEGREE MEASURE; EGYPTIAN MATHEMATICS; RADIAN MEASURE.

right-handed/left-handed system   In three-dimensional space one identifies the location of points by making reference to a set of three mutually perpendicular number lines, usually called the x-, y-, and z-axes, intersecting at a point called the origin. There are two possible ways to orient the axes.

An xyz-coordinate system is called right-handed if, taking the right hand, the positive x-axis points in the direction of the thumb, the positive y-axis in the direction of the index finger, and the positive z-axis in the direction of the (bent) middle finger. An xyz-coordinate system is called left-handed if it follows the directions of these fingers of the left hand instead.

Reversing the direction of any one of the axes, or switching the labels of any two axes, changes the orientation of the coordinate system. Mathematicians have settled on the convention of preferring right-handed systems over left-handed ones.

More generally, three vectors **a, b,** and **c,** in that order, in three-dimensional space form a right-handed system if pointing the thumb of the right hand in the direction of **a,** and the index finger in the direction of **b** has vector **c** lying on the side of the palm of the hand (this is the direction the middle finger would need to curl to point in direction **c**). Alternatively, the three vectors form a right-handed system if the TRIPLE VECTOR PRODUCT **a·** (**b** × **c**) is a positive number.

In two-dimensional space, coordinate axes can again be oriented one of two ways. A set of xy-coordinate axes is said to be positively oriented if a counterclockwise rotation is required to turn the positive x-axis onto the positive y-axis (through the smallest angle possible) and negatively oriented if instead clockwise motion is needed.

See also CARTESIAN COORDINATES; CROSS PRODUCT.

ring   Motivated by the question of what makes arithmetic work the way it does, mathematicians have identified seven key principles satisfied by the operations of addition and multiplication. Today, mathematicians call *any* mathematical system satisfying these basic axioms a ring.

Precisely, a ring is a set R together with two methods for combining elements of R to produce new elements of R, usually called addition "+" and multiplication "*," satisfying the following rules:

1. *Commutative Law for Addition:* For all elements a and b of R, we have: $a + b = b + a$.
2. *Associative Law of Addition:* For all elements a, b, and c of R, we have: $a + (b + c) = (a + b) + c$.
3. *Existence of a Zero:* The set R has an element 0 with the property that $a + 0 = 0 + a = a$ for all elements a in R.
4. *Existence of Additive Inverses:* For each element a in R there is an element, denoted $-a$, such that $a + (-a) = (-a) + a = 0$.
5. *Associative Law of Multiplication:* For all elements a, b, and c of R we have: $a*(b*c) = (a*b)*c$.
6. *Existence of a 1:* There is an element 1 of R with the property that $a*1 = 1*a = a$ for all a in R.
7. *Distributive Laws:* For all elements a, b, and c of R we have: $a*(b + c) = a*b + a*a$ and $(b + c)*a = b*a + c*a$.

Furthermore, we say the ring is commutative if an eighth axiom holds:

8. *Commutative Law for Multiplication:* For all elements *a* and *b* of *R*, we have: *a**b = *b**a.

For example, the set of integers under ordinary addition and multiplication is a commutative ring. Thus any result that is known to follow abstractly from the eight principles outlined above translates to a result about numbers. The set of all functions *f* from the set of real numbers to itself also satisfies the definition of a commutative ring, and so these results also translate to interesting facts in function theory. As a simple example, it is straightforward to prove that the zero element in a ring is unique. (If 0 and 0′ are both zeros, then, by axiom three, 0 = 0 + 0′ = 0′.) Consequently, there is also only one function that can behave as the zero function.

One can impose further conditions on a system. For example, a commutative ring is called an integral domain if a ninth axiom holds:

9. *No Divisors of Zero:* It is never the case that two nonzero elements *a* and *b* of *R* give *a**b = 0.

The set of integers is an integral domain, but the set of functions is not. For example, if *f* is the function that gives the value zero for negative inputs, and the value 1 otherwise, and *g* is the function that gives the value zero for positive inputs, and the value 1 otherwise, then neither *f* nor *g* is the zero function, but their product *f**g is. Any system in MODULAR ARITHMETIC forms a ring but not necessarily an integral domain. For example, $2 \times 3 = 0$ modulo 6.

A commutative ring is called a field if, further, a 10th axiom is satisfied:

10. *Existence of Multiplicative Inverses:* For each nonzero element *a* of *R*, there is an element *b* such that *a**b = *b**a = 1.

The set of integers is not a field. (The number 2, for example, has no multiplicative inverse, since 1/2 is not an integer.) The set of all real numbers under addition and multiplication, however, is a field, as is the set of all the rational numbers and the set of all complex numbers.

It is usually assumed that the elements 0 and 1 in a commutative ring or a field are different. If these two elements are the same, that is, if 0 = 1, then one can prove that the ring *R* contains only this element: *R* = {0}. This ring is called the trivial ring.

See also ABSTRACT ALGEBRA; ASSOCIATIVE; COMMUTATIVE PROPERTY; DISTRIBUTIVE PROPERTY; GROUP; GROUP THEORY; QUATERNIONS; ZERO.

Rolle, Michel (1652–1719) French *Calculus* Born on April 21, 1652, in Ambert, France, Michel Rolle is best remembered for the theorem in CALCULUS that bears his name. Although attracting little attention at the time of its publication, Rolle's theorem is today considered one of the fundamental principles of the subject.

Rolle began his career as a scribe and as an assistant to an attorney. He had little formal education but pursued a personal interest in mathematics all his life. He moved to Paris in 1675 and soon developed a reputation as an expert in arithmetic work. In 1682 he achieved national fame for solving a recreational mathematics problem publicly posed by French mathematician Jacques Ozanam. In honor of this achievement, Rolle was awarded a pension by France's controller of general finance and admission to the Académie Royal des Sciences in 1685.

In 1690 he published the work *Traité d'algèbre* (Treatise on algebra), a text on the theory of equations, in which, among other things, he invented and used the notation $\sqrt[n]{x}$ for the *n*th root of *x*. Rolle also studied GEOMETRY, ALGEBRA, and DIOPHANTINE EQUATIONS. His famous theorem was published one year later in an obscure and little noticed text *Démonstration d'une méthode pour resoudre les égalitez de tous les degrez* (Proof of a method for solving equations of all degrees).

It is ironic that Rolle is today considered a principal figure in the development of calculus. Having studied the emerging subject, Rolle is said to have found the theory unconvincing. He even went so far as to say that calculus is nothing more than a "collection of ingenious fallacies." He died in Paris, France, on November 8, 1719.

See also ROLLE'S THEOREM.

Rolle's theorem In 1691 French mathematician MICHEL ROLLE (1652–1719) established that if a curve intersects the *x*-axis at two locations *a* and *b*, is continuous, and has a TANGENT at every point between *a* and *b*, then there is at least one point in this interval at which the tangent to the curve is horizontal to the *x*-axis. In more stringent mathematical language, his theorem reads:

> For any function *f*(*x*) continuous in a closed interval [*a,b*], differentiable in the open interval (*a,b*), and satisfying *f*(*a*) = *f*(*b*) = 0, there exists at least one value *c* between *a* and *b* such that *f'*(*c*) = 0. That is, the zeros of a differentiable function are always separated by zeros of the derivative.

This result follows readily from the EXTREME-VALUE THEOREM, which asserts that any such function attains a maximum value at some location *c* in the interval (*a*, *b*). The tangent to the curve must be horizontal at any such apex. (A careful study of MAXIMUM/MINIMUM values establishes this.)

Although Rolle's theorem is a special case of the MEAN-VALUE THEOREM, it is usual to prove Rolle's theorem independently and use it as a first step toward proving the more general result.

Roman numerals   Based on a simple tally system similar to the one used by the ancient Egyptians, merchants of the Roman empire of about 500 B.C.E. used letter symbols for powers of 10 and for the intermediate values of 5, and simply grouped symbols together to represent all other quantities. The symbols used were:

I	=	1
V	=	5
X	=	10
L	=	50
C	=	100
D	=	500
M	=	1000

The expression CLXXIII, for instance, represented the number 100 + 50 + 10 + 10 + 1 + 1 + 1 = 173. Although the order of the symbols was not important, it became the convention to list symbols from largest to smallest, left to right.

Initially the symbols D and M were not part of the Roman system. The number 1,000 was written (I), and further applications of round brackets allowed for the expression of even greater quantities. For instance, ((I)) represented 10,000, and (((I))) represented 100,000. Stonemasons introduced the symbols *D* and *M* to simplify their work.

The Romans also introduced other ornamentations to increase the value of a numerical symbol. For instance, vertical bars were used to represent a 100-fold increase, and a bar placed above the symbol represented a 1,000-fold increase. For instance,

$$|X| = 100 \times 10 = 1,000$$
$$\overline{X} = 1,000 \times 10 = 10,000$$
$$|\overline{X}| = 100 \times 1,000 \times 10 = 1,000,000$$

There was no symbol for ZERO in the Roman system.

To avoid the four-fold repetition of symbols (as in the expression CCCCXXXXIIII for 444), a subtractive principle was introduced in the 13th century:

> The placement of a small value immediately to the left of a higher value indicates that that small value is to be subtracted from the higher value.

Thus 4 could be written as IV, 90 as XC, and 444 as CDXLIV. The subtractive principle was subject to two rules:

1. The symbols V, L, and D cannot be used as the numbers to be subtracted.
2. Only one symbol I, X, or C can be placed before a higher number symbol.

Thus, for example, it was not permissible to write IIX for eight. Although not a proper PLACE-VALUE SYSTEM, with the subtractive principle in use, the order of the symbols used was now important.

Performing operations of basic arithmetic with Roman numerals is very awkward. For example, it is not immediate what the solution to the following addition problem should be:

$$\begin{aligned} &\quad \text{XLIV} \\ &+ \text{XVII} \\ &+ \text{XXIX} \end{aligned}$$

That European merchants were comfortable working with the Roman numeral system for well over a millennium suggests that scholars did not use the numeral system to perform calculations, only to record the results. (Arithmetic was performed on a counting board such as an ABACUS.)

The system of Roman numerals remained popular in Western Europe until the 17th century. Although the system was eventually replaced by the HINDU-ARABIC NUMERAL system we use today, it still remains a tradition to use Roman numerals for clock faces, for the inscription of dates on buildings, and for copyright data on films and books, for instance.

See also BASE OF A NUMBER SYSTEM; DECIMAL REPRESENTATION; EGYPTIAN MATHEMATICS; NUMBER.

root (zero) Any value of a variable in an equation that satisfies the equation is called a root of the equation. For example, the equation $2x + 1 = 7$ has root $x = 3$, and the equation $a^3 = 8$ has root $a = 2$. As any equation with a variable x can be written in the form $f(x) = 0$ for some FUNCTION f, the roots of an equation are sometimes called the "zeros" of the associated function f.

If $f(x)$ is a polynomial, then the equation $f(x) = 0$ is called a polynomial equation. The QUADRATIC formula shows that a quadratic equation $ax^2 + bx = c = 0$ has two roots given by:

$$x = \frac{-b \pm \sqrt{b^2 - 4ac}}{2a}$$

The FACTOR THEOREM shows that if $x = r$ is a root of a polynomial equation $f(x) = 0$, then $(x - r)$ is a factor of $f(x)$. The root r is called a simple root if $(x - r)$ is a factor of $f(x)$, but $(x - r)^2$ is not; a double root if $(x - r)^2$ is a factor of $f(x)$, but $(x - r)^3$ is not; and, in general, an nth-order root if $(x - r)^n$ divides $f(x)$, but $(x - r)^{n+1}$ does not. For instance, $x = 2$ is a double root of the equation $2x^3 - 9x^2 + 12x - 4 = 0$, and $x = 1/2$ is a simple root. (We have $2x^3 - 9x^2 + 12x - 4 = (x - 2)^2(2x - 1)$.) The FUNDAMENTAL THEOREM OF ALGEBRA asserts that a polynomial equation of degree n has precisely n (possibly complex) roots if the roots are counted according to their "multiplicity." For instance, the equation $2x^3 - 9x^2 + 12x - 4 = 0$ has three roots, if the solution $x = 2$ is counted twice, as is appropriate.

Numerical methods such as the BISECTION METHOD and NEWTON'S METHOD can be used to find roots of equations to any desired degree of accuracy.

The nth root of a number a, denoted $\sqrt[n]{a}$, is any value x that satisfies the polynomial equation $x^n = a$.

For example, the number 2 is a 10th root of 1,024 (since $2^{10} = 1,024$), and -1 is a sixth root of 1, since $(-1)^6 = 1$. If $n = 2$, then an nth root is called a SQUARE ROOT. If $n = 3$, then it is called a CUBE ROOT.

root test *See* CONVERGENT SERIES.

rose A planar curve shaped like a collection of petals with a common origin is called a rose. In polar coordinates, such a curve has an equation of the form $r = a\sin(n\theta)$ or $r = a\cos(n\theta)$ for some constant a and positive integer n. The angle θ takes values between zero and 360°, and may consequently yield a negative value for r. In this case the corresponding point on the curve is drawn a distance $|r|$ from the origin in a direction opposite to that indicated by angle θ.

If n is odd, the number of petals that appear around the origin is n. If n is even, then $2n$ loops appear. In 1728 Italian mathematician Guido Grandi called these curves "rhodonea."

rotation *See* GEOMETRIC TRANSFORMATION; LINEAR TRANSFORMATION.

rounding If a number has more digits than can be conveniently handled or stored, then it is often convenient to replace the figure by the number closest to it with the desired number of digits. This process is called rounding. For example, the closest two-digit number to 68.7 is 69. Thus 68.7 "rounded to two-digits" is 69. Rounding 5.237 to one decimal place yields 5.2. Rounding 453 to the nearest 10 yields 450, whereas rounding it the nearest 100 yields 500.

It is conventional to "round up" if the number under consideration is equally distant from two approximations. For instance, 9 is deemed to be the closest integer to 8.5.

Errors may be produced if a number is rounded more than once. For example, rounding 78.347 to two decimal places yields 78.35. If we later decide to round to one decimal place, working with the 78.35 yields 78.4, whereas the original number rounded to one decimal place is 78.3. It is essential, then, that all rounding processes be executed in just one step.

The process of simply dropping extra digits is called truncation. For example, when truncated to one decimal place, both 1.834 and 1.8999978 become 1.8. This process is also called "rounding down."

See also ERROR; FLOOR/CEILING/FRACTIONAL PART FUNCTIONS; ROUND-OFF ERROR.

round-off error (**rounding error**) The ERROR produced in ROUNDING a value to a prescribed number of digits is called round-off error. For example, rounding the quantity 6.42 to the nearest decimal place yields 6.4, introducing an error of 0.02.

Round-off errors typically amplify as one performs calculations. For example, rounded to four decimal places, cos 1° has value 0.9998. This suggests that $\frac{1}{1-\cos(1)}$ equals 5,000. This is false. The correct value of this quantity is close to 6,565.8. As calculators round all quantities to eight or 10 decimal places, they too are susceptible to round-off errors and can give erroneous results.

Royal Society (**Royal Society of London**) Founded by royal charter granted by Charles II in 1660, the Royal Society of London is the world's oldest scientific institution still in existence. Its mission today, as it was at the time of its formation, is to disseminate current knowledge about the natural world and to support and promote continued scientific investigation.

Prior to the year 1600, scientists and scholars communicated results through the mail, as well as through the publication of written texts following the invention of the printing press in the 15th century. Such set and rigid formats did not readily allow for the testing of ideas and loose discussion of thought. In an attempt to rectify this, in 1620 Francis Bacon (1561–1626) established an organization called *Novum Organum* (The new instrument) that provided a public forum for scientists, and others, to meet, to discuss scientific ideas and findings, and to learn of the general state of current scientific thinking. Forty years later, the value of such an organization was formally acknowledged, and the Royal Society was formed. The eminent scientists of the day were awarded membership to the society. SIR ISAAC NEWTON was president of the society from 1703 to 1727.

To be elected as a fellow of the Royal Society today is deemed a truly prestigious honor. There are currently 65 Nobel laureates among the Society's 1,300 fellows and foreign members. Each member is given the right to use the letters FRS (fellow Royal Society) after signing his or her name. The society maintains a Web site (www.royalsoc.ac.uk).

The equally prestigious French Royal Academy of Sciences was founded 6 years after the formation of the Royal Society.

Russell, Bertrand Arthur William (1872–1970) British *Foundations of mathematics, Philosophy* Born on May 18, 1872, in Ravenscroft, Wales, mathematical logician and philosopher Bertrand Russell is remembered for his text, cowritten with ALFRED NORTH WHITEHEAD (1861–1947), *Principia Mathematica* (1910,

Bertrand Russell, an eminent mathematician of the 20th century, was the first to discover a fundamental flaw in the elementary theory of sets. *(Photo courtesy of Topham/The Image Works)*

1912, 1913), a monumental work of three volumes that attempted to derive the whole of mathematics from purely logical assumptions. Although not fully successful, the work was highly influential. Russell also discovered a fundamental paradox at the heart of basic SET THEORY, today called RUSSELL'S PARADOX.

Russell studied mathematics and ethics at Trinity College, Cambridge, and developed an interest in the logical foundations of mathematics early in his career. In 1903 he published *Principles of Mathematics,* a text exploring the premise that all of mathematics could be reduced to statements of logic, and that all mathematical proofs can be recast as proofs in the theory of logic. It was during the writing of this work that Russell discovered his famous paradox in set theory, a discovery that made it clear that more work was needed to find a logical foundation to all of mathematics.

His next work, the famous *Principia Mathematica,* was written as an attempt to extend the methods of his *Principles* to a more general ramified theory that could cope with all the disturbing self-referential paradoxes that arose in set theory. Although recognized as a brilliant advancement in understanding the logical underpinnings of all of mathematics, the work still received some criticism from the general mathematics community as being either too *ad hoc,* too stringent, or too weak, and that, at the very least, even more work was needed to achieve the lofty goal Russell had set. In 1930 Austrian mathematician KURT GÖDEL (1906–78) shocked the mathematical world by proving, in his incompleteness theorems, that there can be no logical base to all of mathematics of the type Russell sought and that any attempt to find a basic set of axioms on which to base all of mathematics is *a priori* doomed to failure.

Russell's influence in the development of mathematical logic was profound. He was elected to the ROYAL SOCIETY in 1908 and wrote a number of stunning articles in the field throughout his career. Russell was a political activist and was arrested several times throughout his career for his antiwar activities. (His 1919 article "Introduction to Mathematical Philosophy" was written during a 6-month stint in prison.) Russell taught at City College, New York, during the 1930s and was also a prominent figure in antiwar and antinuclear protests in the United States.

Russell also wrote on a broad range of humanitarian topics, including political science, moral science, and religious studies. He was awarded the Order of Merit in 1949 and won the Nobel Prize for literature in 1950 for his collective writings championing "humanitarian ideals and freedom of thought."

Russell died in Plas Penrhyn, Wales, on February 2, 1970.

Russell's paradox (Russell's antinomy) In 1902 mathematician and philosopher BERTRAND ARTHUR WILLIAM RUSSELL (1872–1970) found a fundamental flaw at the heart of basic SET THEORY. Usually sets are not members of themselves. For instance, if *D* represents the set of all dogs, then *D,* being a set, is not itself a dog. Let us call such sets "normal." Russell noted, however, that not all sets are normal. For example, if *S* represents the set of all sets, then *S* is itself a set and so belongs to *S.* The set of all things that are not dogs is also nonnormal.

Let *N* represent the set of all normal sets. That is, *N* is the set of all sets that are not members of themselves. Now ask: Is *N* normal? If the answer is yes, then *N* is normal, making it an element of *N,* which is precisely what it means to be nonnormal. If the answer is no, then *N* is nonnormal, meaning that *N* is an element of *N,* the set of all normal sets. We have:

$$N \text{ normal} \Rightarrow N \text{ nonnormal}$$
$$N \text{ nonnormal} \Rightarrow N \text{ normal}$$

The set *N* can neither be normal nor nonnormal.

This contradiction is called Russell's paradox, and its discovery foiled mathematicians' attempts to use set theory as the foundational basis of all of mathematics. It also suggested that all of mathematics might be flawed if even the simplest of mathematical theories—set theory—is fundamentally self-contradictory.

Russell also presented several alternative versions of his paradox, including the famous "barber paradox" that he presented in 1919:

> A barber in a certain town has a sign that reads, "I shave all those men, and only those men, who do not shave themselves." Who shaves the barber?

In trying to answer this question, one is mired in the same type of logical impasse as the original paradox.

In 1910, Russell, with the help of ALFRED NORTH WHITEHEAD (1861–1947), published the first volume of a mammoth three-volume treatise that attempted to circumvent the issues raised by the paradox and to find a logical base for all of mathematics. He introduced a notion of "type," demanding that sets and sentences in mathematics belong to a hierarchical structure. At the lowest level there are statements about individual elements. At the next level there are sentences about sets of individual elements, and next, sentences about sets of sets of individual elements, and so forth. Moreover, no sentence is permitted to be at the same level as its subject. Hence any statement that refers to itself (such as the definition of the set N) is simply not allowed in this hierarchical structure. This approach certainly circumvents self-referential paradoxes, but many members of the mathematics community felt, however, that this construct was not particularly satisfying.

Observe that the barber's paradox is easily resolved by noting that the barber could be outside the set of men, namely, the barber could be a woman. Russell's theory of type attempted to mimic this solution by bringing sets "outside of themselves" through the notion of a hierarchy. Mathematicians today are attempting instead to develop a mathematical theory of "context" where the setting in which statements are made is considered important.

Russian multiplication   Also known as peasant multiplication, the following multiplication method is believed to have originated in Russia.

1. Head two columns with the numbers you wish to multiply.
2. Progressively halve the numbers in the left column (ignoring remainders) while doubling the figures in the right column. Reduce the left column to one.
3. Delete all rows with an even number in the left column.
4. Add all the surviving numbers in the right column. This sum is the desired product.

The key to understanding why this method works is to note that every number can be written as a sum of

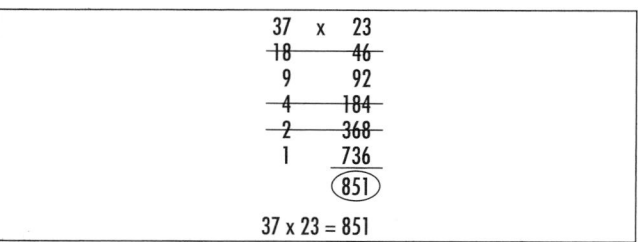

Russian multiplication

the powers of 2: 1,2,4,8,16,32,... For example, 37 = 32 + 4 + 1. That 37 is odd shows that the number 1 is present in this expression. Halving the number twice (ignoring remainders) produces:

$$37 = 32 + 4 + 1 \rightarrow 16 + 2 \rightarrow 8 + 1 = 9$$

That the answer is odd indicates that 4 is present in the expression for 37 as a sum of powers of two. That halving three more times produces the next odd answer (corresponding to 37 halved five times in all) indicates that 32 is also present. Thus, the appearance of odd numbers in the left column indicates which powers of two are used to build the number 37. For the multiplication, note that:

$$37 \times 23 = (32 + 4 + 1) \times 23$$
$$= 32 \times 23 + 4 \times 23 + 1 \times 23$$
$$= 2^5 \times 23 + 2^2 \times 23 + 2^0 \times 23$$

Thus the desired product is obtained by summing the effect of doubling the number 23 zero, two, and five times. This is accomplished in step 4 of the procedure.

This method of course works for any pair of whole numbers you wish to multiply. It is an efficient method of multiplication, and many computers are programmed to compute products this way.

See also EGYPTIAN MULTIPLICATION; ELIZABETHAN MULTIPLICATION; FINGER MULTIPLICATION; MULTIPLICATION; NAPIER'S BONES.

Saccheri, Girolamo (1667–1733) Italian *Geometry*
Born on September 5, 1667, in San Remo, Genoa,
Girolamo Saccheri is remembered for his 1733 publication *Euclides ab omni naevo vindicatus* (Euclid cleared
from every stain), in which he attempted to prove
Euclid's PARALLEL POSTULATE as a consequence of the
remainder of EUCLID'S POSTULATES. Although Saccheri
failed in this attempt, he came narrowly close to being
the first to discover a NON-EUCLIDEAN GEOMETRY.

Saccheri was ordained as a Jesuit priest in 1694
and taught philosophy and theology at various Jesuit
colleges throughout Italy. He pursued mathematics as
an outside interest for a number of years before being
appointed chair of mathematics at Pavia in 1699.

As many scholars throughout the millennia had
also believed, Saccheri was convinced that Euclid's fifth
postulate, the parallel postulate, could be deduced as a
logical consequence of the remaining four axioms.
Although he was aware that no one had succeeded in
showing that this was indeed the case, Saccheri realized
that another approach, an indirect one, could be
adopted. He reasoned as follows:

> Imagine a geometry in which the first four of
> Euclid's axioms are true, but the fifth axiom is
> false. If the fifth postulate can indeed be
> proved from the first four, then we would have
> a geometry with an inherent CONTRADICTION:
> the fifth axiom is both true and false. Thus
> one should study a theory of geometry in
> which the parallel postulate is false and look
> for a contradiction.

Saccheri followed this plan and wrote about it in his
famous 1733 text, but never found the contradiction he
sought. Just over a century later Hungarian mathematician JÁNOS BOLYAI (1802–60) and Russian scholar
NIKOLAI IVANOVICH LOBACHEVSKY (1792–1856), with
their discoveries of HYPERBOLIC GEOMETRY, independently proved that no contradiction will ever be discovered. Although Saccheri had developed the mathematical
theory of hyperbolic geometry to a significant extent, he
cannot be credited for its discovery: he never realized
that it is a consistent mathematical theory in its own
right. Nonetheless the work of Saccheri is appropriately
honored in the field, with many results and theorems
properly attributed to him. He died in Milan, Italy, on
October 25, 1733.

sample   *See* POPULATION AND SAMPLE.

sample space   The set of all possible outcomes from
an experiment or an action is called a sample space.
For example, in casting a die, six scores are possible,
yielding the sample space the set {1, 2, 3, 4, 5, 6}. The
sample space for tossing a coin twice is the set
{head|heads, heads|tails, tails|heads, tails|tails}.

A sample space can be infinitely large. For example, the action of recording heights of individuals has a
continuous array of values for its sample space—values
between 12 in. and 100 in., say—with all fractional
values in between possible.

See also EVENT; PROBABILITY.

scalar In VECTOR analysis, any quantity that is a real number and not a vector is called a scalar. For example, the number 2 is the scalar coefficient of the vector **v** in the vectorial expression 2**v**. The DOT PRODUCT of two vectors (also called their scalar product) is a rule for multiplying two vectors to yield a scalar result.

In physics, any number or measurement that does not involve the concept of direction is called a scalar. For example, length, mass, energy, and temperature are scalar quantities. The notion of speed is also a scalar quantity, but that of VELOCITY is not.

In MATRIX theory, the entries of a matrix are sometimes called scalars. A square matrix with all entries off the main diagonal equal to zero and all entries on the main diagonal equal in value is called a scalar matrix. Such a matrix is a scalar multiple of the IDENTITY MATRIX I.

scale Two geometric figures are said to be scaled versions of each other if it is possible to match the points of one figure with points in the second in such a way that, if P and Q are any two points of the first figure and P' and Q' are the corresponding points of the second, then the ratio of lengths $|PQ| : |P'Q'|$ is always the same fixed positive value k. (The number k is called the scale factor.) SIMILAR FIGURES are scaled figures. Any enlarged or reduced figure produced by a modern photocopier is a scaled version of the original figure.

If a geometric figure is enlarged by a scale factor k, then all lengths in that picture increase (or decrease if $0 < k < 1$) by a factor of k. All ANGLEs in that figure remain unchanged. Consequently, if an $a \times b$ rectangle is enlarged by a scale factor k, then the resultant figure remains a rectangle and has side-lengths ka and kb. Consequently, the AREA of the figure has increased by a factor k^2. As the notion of area of a rectangle defines our notion of area for all geometric figures (such as a triangle, a polygon, or a circle) we have that the areas of all figures increase, upon scaling, by the factor k^2. Similarly, since a rectangular prism of dimensions $a \times b \times c$ scales to become rectangular prism of dimensions $ka \times kb \times kc$, volumes of all geometric solids, upon scaling, increase by a factor of k^3.

These observations have some interesting consequences in the natural world. For instance, the amount of heat loss a mammal experiences is proportional to the amount of surface it has in contact with the air, whereas the amount of food an animal must eat in order to generate body warmth depends on the volume of muscle it possesses. As surface area grows by a factor of k^2 and volume by a factor of k^3, larger mammals possess a smaller surface area per volume ratio than smaller mammals. Thus large mammals (such as polar bears) are better suited to arctic conditions than smaller mammals (such as squirrels)—they lose less heat per unit of body mass and require less food (in relation to their mass) to maintain enough heat production. King penguins in Antarctica are significantly larger than their counterparts in other regions of the world. As another example, the speed a fish can move is governed by the volume of muscle it possesses. Large fish have the advantage of having a smaller surface area to volume ratio than smaller fish. They not only have more muscle power, but also experience less surface-area friction with the surrounding water (per unit of muscle) than their smaller counterparts.

In mathematics, the study of scale plays an important role in defining DIMENSION and is used to analyze FRACTALS. In another context, the markings on the axes of a GRAPH OF A FUNCTION constitute the "scale" of the diagram. Changing the scale of the axes has the equivalent effect of enlarging or reducing the graph itself. The markings on a ruler are also called a scale, and measuring the same object with two rulers of different scales is equivalent to the act of measuring two scaled versions of the figure with a single ruler.

scatter diagram (**scatter plot, Galton graph**) If a scientific study records numerical information about two features of the individuals or events under examination (such as height and shoe size of participating adults, or temperature and pressure at which certain climatic events occur), then that DATA can be represented in a scatter diagram. Specifically, if the pair (x_i, y_i) represents the two numerical facts recorded about the ith individual, then point (x_i, y_i) is plotted on CARTESIAN COORDINATES. A scatter diagram is the resultant graph when all points are displayed. The points are not joined by lines.

A scatter diagram will indicate any relationship between the x and y variables. If the points seem to lie near a straight line, they are said to be linearly correlated. One can then perform a mathematical test to determine the degree of correlation by calculating the value of a CORRELATION COEFFICIENT. If the points seem to lie on another type of curve, then they are non-linearly correlated. Variables that are not correlated will likely produce a diagram with points randomly scattered across the graph. If all but one point seems to follow a well-defined curve, then that exceptional point is called an OUTLIER.

See also LEAST SQUARES METHOD.

scientific notation (exponential notation, standard form) A number x is said to be written in scientific notation if it is expressed as a number between 1 and 10 multiplied by a power of 10:

$$x = p \times 10^n$$

with $1 \le p < 10$ and n an integer, positive or negative. For example, the number 547 is written 5.47×10^2 in scientific notation. The number 0.00106 is written 1.06×10^{-3}.

Scientific notation is useful for writing, comparing and manipulating very large and very small numbers. For example, Sir Henry Cavendish computed the mass of the Earth in 1798 as 6,600,000,000,000,000,000,000 tons. This number is compactly written 6.6×10^{21}. The diameter of an atomic nucleus is about 0.000000000000075 or 7.5×10^{-14} m.

The power of 10 used in scientific notation indicates the number of times that 10 is used as a factor if the exponent is positive, or the number of times 1/10 is used as a factor if the exponent is negative. For example, 3.2×10^3 is $3.2 \times 10 \times 10 \times 10$, which equals 3,200, and 0.0402 is the number 4.02 divided by 10 two times, and so $0.0402 = 4.02 \times \frac{1}{10} \times \frac{1}{10} = 4.02 \times 10^{-2}$.

When the value of a quantity is expressed in scientific notation, then the power of 10 used is called the order of magnitude of the quantity. For example, the speed of light is 3.00×10^8 m/sec, and the speed of sound in air is 3.32×10^2 m/sec. We can readily see that light is about 10^6, or 1 million, times faster than sound. The diameter of the Milky Way galaxy is on the order of 10^{20} m.

In writing the measurement of a quantity in scientific notation, it is assumed that all digits recorded are significant. This makes the ERROR of the measurement apparent. Calculators use the symbol E to display a quantity in scientific notation. For example, on a cal-culator, the expression $1.2345E37$ is to be interpreted as 1.2345×10^{37}. The letter E is an abbreviation for "exponent."

secant A line that intersects a curve, usually at more than one point, is called a secant of the curve. If the secant cuts the curve at two points, the segment of the line between the two points of intersection is called a CHORD of the curve.

In planar geometry, the secant theorem refers to the property that the lengths of two secants to a circle satisfy $a(a + b) = c(c + d)$.

This is proved by identifying similar triangles via the CIRCLE THEOREMS and establishing the correspondence $\frac{c}{a + b} = \frac{a}{c + d}$.

In trigonometry the reciprocal of the cosine function is called the secant function. It is usually written sec x. In a right triangle it equals the ratio of the length of the hypotenuse to that of the side adjacent to the angle x.

secant method *See* NEWTON'S METHOD.

selection *See* COMBINATION.

self-reference A sentence that refers to itself is called a self-referential statement. For example, the statement "This sentence has five words" is self-referential.

Self-referential statements can be problematic because it is often difficult to determine their truth or

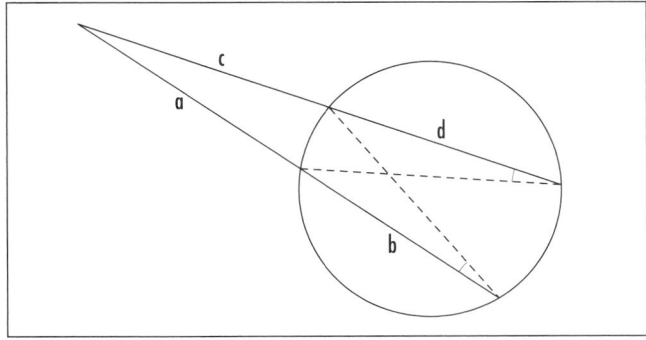

Two-secant theorem

falsehood. Although the statement given above is clearly true, the truth-value of "This sentence is false," for example, is unclear: it cannot be true, for then the statement says it is false, nor can it be false, for then the statement is true! (This example forms the basis for the LIAR'S PARADOX.) The statement "This sentence is true" is equally problematic: it can be both true and false. Self-referential statements are forbidden in formal systems of logic.

Austrian mathematician KURT GÖDEL (1906–78) made clever use of a self-referential statement to prove his famous incompleteness theorems.

See also FORMAL LOGIC; GÖDEL'S INCOMPLETENESS THEOREM; HALTING PROBLEM; JOURDAIN'S PARADOX.

semi-magic square   A square array of numbers in which the sum of the numbers in any row or column is the same is called a semi-magic square. (The array is dubbed "fully magic" if, in addition, the numbers in each of the two diagonals also add to this same sum.) The common sum of the rows and columns of the array is called the magic sum of the array. For example, the array

$$\begin{pmatrix} 2 & 3 & 5 \\ 3 & 7 & 0 \\ 5 & 0 & 5 \end{pmatrix}$$

is a semi-magic square with magic sum 10. (This array, however, is not a MAGIC SQUARE).

Semi-magic squares have a number of remarkable properties in the theory of MATRIX algebra. For instance, suppose A and B are semi-magic squares of the same size with magic sums a and b, respectively. Regard each as a matrix. Then:

1. The matrix sum $A + B$ is again a semi-magic square, with magic sum $a + b$.
2. The matrix product AB is again a semi-magic square, with magic sum ab.
3. The inverse A^{-1}, if it exists, is still semi-magic, with magic sum $1/a$.

These claims can be proved by setting J to be the square matrix with all entries equal to one and noting that a square matrix A is semi-magic with magic sum a if, and only if, $AJ = aJ = JA$.

There is no such thing as a "semi-magic rectangle" if one insists that all entries be positive numbers. Suppose, for instance that an array has n rows and m columns and the entries in each row and column sum to a. Then, by adding together each of the n rows, the sum of the entries in the entire array must be na. By the same token, adding together each of the m columns shows that the sum of the entries in the entire array must also be ma. Consequently, we must have $n = m$. If one permits zero or negative entries, then semi-magic rectangles are possible, but the magic sum must necessarily be zero.

sequence (progression)   A set of numbers arranged in a list, with each number in the list unambiguously specified, is called a sequence. For example, the sequence 3, 5, 7, 9, 13 lists five specific numbers, and the sequence 1, 4, 7, 10, 13, 16, … indicates an infinite list of numbers, with each number being 3 greater than its predecessor. The numbers in a sequence are called its terms or elements.

Sometimes the terms of a sequence are specified as a formula. For example, the sequence 2, 5, 10, …, $n^2 + 1$, … indicates that the nth term of the sequence is given as one more than the number squared. A sequence with finitely many terms is called a finite sequence; one with infinitely many terms is an infinite sequence. A sequence with the nth term given by a_n is denoted $\{a_n\}$, or sometimes (a_n). For example, $\{\frac{1}{n}\}$ represents the sequence $1, \frac{1}{2}, \frac{1}{3}, \frac{1}{4}, \ldots$; $\{(-1)^n\}$ the sequence $-1, 1, -1, 1, \ldots$; and $\{\frac{x^n}{n!}\}$ the sequence $x, \frac{x^2}{2!}, \frac{x^2}{3!}, \frac{x^2}{4!}, \ldots$.

A sequence might also be specified via a RECURRENCE RELATION. For example, the sequence of FIBONACCI NUMBERS $\{F_N\}$ is given by $F_{n+2} = F_{n+1} + F_n$ with $F_1 = 1$ and $F_2 = 1$.

A sequence $a_1, a_2, a_3, \ldots$ is said to converge to a number L if the numbers a_n get closer and closer to L as n becomes large. Such a sequence is called a CONVERGENT SEQUENCE, and the quantity L to which it converges is called the LIMIT of the sequence. For instance, the sequence $\frac{1}{2}, \frac{2}{3}, \frac{3}{4}, \frac{4}{5}, \frac{5}{6}, \ldots$ converges to the value 1.

Convergent sequences play an important role in mathematics. For example, an IRRATIONAL NUMBER can be thought of as a limit of a sequence of rational numbers. (For instance, $\sqrt{2}$ is the limit of the sequence 1, 1.4, 1.41, 1.414, 1.4142,...) As such, it is possible to define a quantity raised to an irrational power as a limit of rational powers, which are well defined. (For example, $3^{\sqrt{2}}$ is the limit of the sequence $3^1, 3^{\frac{14}{10}} = \sqrt[10]{3^{14}}, 3^{1.41} = \sqrt[100]{3^{141}},....$) Any infinite sum that is a SERIES can be thought of as a LIMIT of a sequence of PARTIAL SUMS, and any INFINITE PRODUCT can be considered to be the limit of a sequence of partial products.

See also ARITHMETIC SEQUENCE; BOUND; GEOMETRIC SEQUENCE; HARMONIC SEQUENCE.

series A sum of numbers is called a series. The sum could be finite, such as 2 + 4 + 6 + 8 + 10, for example, or it could be an infinite sum, as for 2 + 5 + 8 + 11 + 14 + ... for instance. Each number in the sum is called a term of the series.

In 1755 LEONHARD EULER introduced use of the Greek letter sigma Σ (S for "sum") to abbreviate the writing of a series. For example, the sum of the first five even numbers can be written $\sum_{n=1}^{5} 2n$ and the sum of all numbers 1 less than a multiple of three as $\sum_{n=1}^{\infty} (3n-1)$. (*See* SUMMATION.)

There are no conceptual difficulties with the notion of summing just a finite collection of numbers. However, understanding what we mean by an infinite sum is a delicate matter—such sums can exhibit very different characters. Some series clearly do not add to a finite value (as for 2 + 5 + 8 + 11 + 14, ..., for example), whereas one could argue that other infinite series do. (The act of walking from one side of the room to the other, for instance, suggests that the series $\frac{1}{2} + \frac{1}{4} + \frac{1}{8} + \frac{1}{16} + \frac{1}{32} +...$ "adds" to the value 1: first walk halfway across the room, and then half the remaining distance, and then the half the distance that remains, and so on.)

In some cases the situation is not at all clear. For example, the series 1 − 1 + 1 − 1 + 1 − 1 + ... of positive and negative terms seems to oscillate: an even number of terms add to zero, an odd number to 1, and it does not seem possible to assign a single sum to this series. (Some mathematicians in the past argued that the sum of this series should be $^1/_2$, a value between zero and 1.)

These different behaviors of infinite series caused scholars much confusion over the centuries. Matters were not properly resolved until AUGUSTIN-LOUIS CAUCHY (1789–1857) introduced the notion of a LIMIT, and was able to apply it to the study of infinite sums. Today we say that an infinite series $\sum_{n=1}^{\infty} a_n = a_1 + a_2 + a_3 +...$ converges to a finite value L if the SEQUENCE of PARTIAL SUMS:

$$S_1 = a_1$$
$$S_2 = a_1 + a_2$$
$$S_3 = a_1 + a_2 + a_3$$
$$\vdots$$

tends to L in the limit as $n \to \infty$. For example, the infinite series

$$\frac{1}{2} + \frac{1}{4} + \frac{1}{8} + \frac{1}{16} + \frac{1}{32} +...$$

has partial sums:

$$S_1 = \frac{1}{2}$$
$$S_2 = \frac{1}{2} + \frac{1}{4} = \frac{3}{4}$$
$$S_3 = \frac{1}{2} + \frac{1}{4} + \frac{1}{8} = \frac{7}{8}$$
$$S_4 = \frac{1}{2} + \frac{1}{4} + \frac{1}{8} + \frac{1}{16} = \frac{15}{16}$$
$$\vdots$$

that approach the value 1. In this sense we say that the sum $\frac{1}{2} + \frac{1}{4} + \frac{1}{8} + \frac{1}{16} + \frac{1}{32} +...$ *equals* 1. (In a practical sense, however, adding the terms of the series only brings us closer and closer to the value 1: we can never attain it. We are physically limited by the fact that we cannot keep adding terms indefinitely.)

An infinite series that converges is called a convergent series. Any series that does not converge to a finite value (either because the value of the sum seems to be infinite, or because the partial sums oscillate) is called divergent. There are a number of tests to determine whether or not a series converges or diverges. (*See* CONVERGENT SERIES.)

Every infinite decimal expansion can be thought of as an infinite series. For example, $0.3333\cdots = \dfrac{3}{10} + \dfrac{3}{100} + \dfrac{3}{1000} + \dfrac{3}{10000} + \ldots$, and the series converges to the value 1/3.

See also ALTERNATING SERIES; ARITHMETIC SERIES; EXPONENTIAL SERIES; GEOMETRIC SERIES; GREGORY SERIES; HARMONIC SERIES; MACLAURIN SERIES; POWER SERIES; SUMS OF POWERS; TAYLOR SERIES; ZENO'S PARADOXES; ZETA FUNCTION.

set theory Loosely speaking, a set is any collection of objects or numbers specified in a well-defined manner. Each item in the set is called an element, or a member, of the set. For example, "dog" is an element of the set of mammals. If an entity a is an element of a set S, we write $a \in S$. If a does not belong to S, we write $a \notin S$.

Sets are typically specified either by listing the elements of the set between a set of braces "{ }", or listing a few elements of the set to indicate a pattern. For example $\{a,e,i,o,u\}$ is the set consisting of the five vowels of the alphabet, and $\{3,6,9,12,\ldots\}$ is the set of all multiples of 3. It may also be possible to define a set as consisting of elements from some universal collection that satisfy a certain property. For example, $\{x \in \mathbf{R} \,|\, x > 5\}$ denotes the set of all real numbers that are greater than 5. (Some mathematicians prefer to use a colon ":" instead of a vertical bar in this notation.)

The order in which the elements of a set are listed is immaterial. For example, $\{A,6,*\}$ and $\{*,6,A\}$ are the same set. Also, elements of a set are listed without repetition. For instance, the set $\{a, a, a, a, a\}$ is the set with a single element a. The EMPTY SET is the set that contains no elements.

Two sets are deemed equal if they possess precisely the same elements. For example, the sets $\{2,4,6,8,\ldots\}$ and $\{n \,|\, n$ is a counting number divisible by $2\}$ are equal sets. A set A is said to be a subset of a set B if every element of A is also a member of B. We write $A \subset B$ if we are certain that the two sets are not equal, and $A \subseteq B$ if equality of the sets is possible. For example, the set of all multiples of 4 is a subset of the set of all multiples of 2.

Although the intuitive notion of a set as a collection of objects is as ancient as the human race, the idea of a set as a formal mathematical concept was not proposed until the 19th century. In his development of BOOLEAN ALGEBRA, British mathematician GEORGE BOOLE (1815–1864) introduced the notion of set as a fundamental tool for the study of FORMAL LOGIC. German mathematician GEORG CANTOR (1845–1918), in his attempts to understand the foundation of all of mathematics, came to regard sets as even more basic and fundamental than the notion of "number." Cantor properly formalized a theory of set manipulations and introduced the striking notion of CARDINALITY. His work led him to profound insights into the nature of finite and infinite sets, leading him to extend the concept of number to include more than one type of INFINITY.

Set Operations

There are a number of basic set manipulations, each of which can be depicted with a VENN DIAGRAM.

Set Intersection: The intersection of two sets A and B, denoted $A \cap B$, is the set of elements common to both A and B. For example, if $A = \{0,2,4,6,8,10,12\}$ and $B = \{0,3,6,9,12,15\}$, then $A \cap B = \{0, 6, 12\}$. Two sets with no elements in common are called disjoint. The intersection of two disjoint sets is the empty set.

Set Union: The union of two sets A and B, denoted $A \cup B$, is the set of all elements that appear either in A or in B, or in both. For instance, in the previous example we have $A \cup B = \{0,2,3,4,6,8,10,12,15\}$.

If two sets A and B each contain a finite number of elements, then the number of elements in $A \cup B$ equals the number of elements in A plus the number of elements in B minus the number of elements in $A \cap B$. (This subtraction counteracts the double counting of the elements that belong to both sets.) This formula is one instance of the general INCLUSION-EXCLUSION PRINCIPLE.

The notions of set intersection and set union can be extended to considerations including more than two sets.

Set Complement: If a set A is a subset of a set B, then the complement of A in B, also called the set difference and denoted $B - A$, is the set of all elements of B that do not belong to A. For example, if $A = \{1,3,5\}$ and $B = \{1,2,3,4,5,6\}$, then $B - A = \{2,4,6\}$. DE MORGAN'S LAWS explain how set complement interacts with intersections and unions of sets.

Philosophical Difficulty

In 1902 British mathematician and philosopher BERTRAND ARTHUR WILLIAM RUSSELL (1872–1970) stunned the mathematical community with his construction of a simple paradox, today called RUSSELL'S PARADOX, that shows that our naïve understanding of the notion of set is fundamentally flawed. Although Cantor believed that set theory is the foundation on which all of mathematics is built, it became clear to mathematicians that the concept of a set and what it means to be an "element of" must remain as undefined terms. In the decades that followed, mathematicians such as ERNST FRIEDRICH FERDINAND ZERMELO (1871–1953) attempted to develop an axiomatic theory of sets (based on undefined terms) that successfully avoids Russell's paradox. To this day, not all mathematicians agree that this goal has yet been achieved.

See also ARGUMENT; AXIOM OF CHOICE; CARDINALITY; CARTESIAN PRODUCT; DIFFERENCE; FINITE.

seven bridges of Königsberg problem *See* GRAPH THEORY.

sexagesimal *See* BASE OF A NUMBER SYSTEM.

Shanks, William (1812–1882) British *Computation* Born on January 25, 1812, in Northumberland, England, William Shanks is remembered for extraordinarily accurate computations of mathematical constants, all done by hand. Shanks evaluated the natural LOGARITHMs of the numbers 2, 3, 5, and 10, each to 137 decimal places, published a table of all the prime numbers up to 60,000, and evaluated all the powers of two of the form 2^{12n+1} for n from 1 to 60. He also evaluated e and EULER'S CONSTANT γ to a large number of decimal places, but he is best remembered for his computation of π to many hundreds of decimal places. He published this result in 1873.

Little is known of Shanks's life. His methods of computation were essentially nothing more than a matter of patience and BRUTE FORCE. For example, Shanks used the following identity to compute 707 digits of π:

$$\frac{\pi}{4} = 4\tan^{-1}\left(\frac{1}{5}\right) - \tan^{-1}\left(\frac{1}{239}\right)$$

using GREGORY SERIES $\tan^{-1}(x) = x - \frac{x^3}{3} + \frac{x^5}{3} - \frac{x^7}{3}$ +.... By substituting in the values $\frac{1}{5}$ and $\frac{1}{239}$, one can begin computing the digits of π.

In 1946 English mathematician D. F. Ferguson used the alternate identity to also compute π:

$$\frac{\pi}{4} = 3\tan^{-1}\left(\frac{1}{4}\right) + \tan^{-1}\left(\frac{1}{20}\right) + \tan^{-1}\left(\frac{1}{1985}\right)$$

He noted that Shanks had made an error in the 528th place and went on to correctly compute π to 808 decimal places.

Shanks died in Houghton-le-Spring, England, in 1882. The exact date of death is not known.

See also E; PI.

Shannon, Claude Elwood (1916–2001) American *Information theory* Born on April 30, 1916, in Michigan, Claude Elwood Shannon is remembered as the founder of the field of INFORMATION THEORY. His seminal article, "The Mathematical Theory of Communication," published with Warren Weaver in 1949, laid down the principles of communication science and introduced the revolutionary idea that information (pictures, words, sounds) could successfully be transmitted through a wire as a sequence of 0s and 1s. (Up to then it was thought that it would be necessary to transmit electromagnetic waves through wires to accomplish this feat.) This paper introduced the term *bit* for the first time. Shannon also developed a mathematical means for measuring the information content of a message.

Shannon was awarded an undergraduate degree in mathematics and electrical engineering from the University of Michigan in 1936. He completed doctoral work at the Massachusetts Institute of Technology in 1940 and worked on the construction of an early type of mechanical computer. He demonstrated, for the first time, that it is possible to combine BOOLEAN ALGEBRA with electrical relays and switching circuits to create a machine that can "do" mathematical logic.

In 1941 Shannon joined AT&T Bell Telephones in New Jersey as a research mathematician and published his famous work 7 years later. This piece also included the mathematical basis of error detection: by adding extra bits to a message, Shannon showed that it is possible to detect and correct errors that occur during the transmission of messages due to noise.

Shannon accepted a faculty position at the Massachusetts Institute of Technology in 1957, but remained a consultant with Bell Telephones. He continued his work on Boolean algebra, applying it to the new field of artificial intelligence. Shannon produced the first effective chess-playing programs.

Shannon received many honors for his work, including the National Medal of Science in 1966 and the Audio Engineering Society Gold Medal in 1985. He died in Medford, Massachusetts, on February 24, 2001.

Sicherman dice   A pair of dice with faces renumbered 1, 2, 2, 3, 3, 4 and 1, 3, 4, 5, 6, 8 are called Sicherman dice, named after their discoverer Col. George Sicherman of Buffalo, New York. These dice have a remarkable property: the PROBABILITY of throwing any particular sum (from 2 to 12) with Sicherman dice matches the odds of throwing that same sum with a

pair of standard dice. Thus Sicherman dice can be used in dice games without affecting the odds of the game.

The tables below show all possible outcomes of rolling a pair of standard dice (top) and a pair of Sicherman dice (bottom), 36 combinations in all for each pair. All sums appear equally often in each table, and so indeed all sum probabilities are identical. (Both tables possess three 10s, for instance. The odds of throwing a 10 are thus the same for each pair: $\frac{3}{36}$ or $\frac{1}{12}$).

The Sicherman dice are unique: there is no other way to renumber the faces of two cubes with *positive* integers yielding two dice with the same sum probabilities as standard dice.

Tetrahedral dice (with each die the shape of a TETRAHEDRON) having faces labeled 1, 2, 2, 3 and 1, 3, 3, 5 produce the same sum probabilities as ordinary tetrahedral dice labeled 1, 2, 3, 4 and 1, 2, 3, 4.

	•	•˙	•˙˙	••	••••	•••
•	2	3	4	5	6	7
•˙	3	4	5	6	7	8
•˙˙	4	5	6	7	8	9
••	5	6	7	8	9	10
•••	6	7	8	9	10	11
•••	7	8	9	10	11	12

	•	•˙	•˙	•˙˙	•˙˙	••
•	2	3	3	4	4	5
•˙˙	4	5	5	6	6	7
••	5	6	6	7	7	8
•˙˙	6	7	7	8	8	9
•••	7	8	8	9	9	10
•••	9	10	10	11	11	12

Comparing ordinary dice with Sicherman dice

sieve of Eratosthenes (Eratosthenes' sieve)   Although there is no known formula for generating PRIME numbers, there is a simple method for "sifting out" the primes between 1 and any given number N. The procedure is called the sieve of Eratosthenes and is attributed to 3rd-century scholar ERATOSTHENES OF CYRENE, a Greek contemporary of EUCLID (ca. 300–260 B.C.E.). The method is performed as follows:

1. List all the positive integers from 2 up to N.
2. Leave the number 2, but cross out every second number after it. This deletes all the multiples of 2 greater than 2.
3. Leave the next remaining number, 3, but cross out every third number after it (if not already deleted). This will delete all the multiples of 3 greater than 3.
4. Leave the next remaining number, 5, and delete all of its multiples.
5. Continue this process by always going to the next remaining number and crossing out the multiples of it that occur further along in the list.
6. The integers not deleted when this process ends are the prime numbers between 2 and N.

Any number k in the list that is not prime can be factored, $k = a \times b$, and so will be deleted when considering the multiples of a, say. Only those numbers in the list that do not factor, that is, the prime numbers, will survive.

Any COMPOSITE NUMBER smaller than N has at least one prime factor smaller than the square root of N. (A number k smaller than N cannot factor as k = a × b with both a and b larger than √N.) This shows that when performing this procedure, one need only delete multiples of prime numbers smaller than the square root of N. For example, to find all the prime numbers from 2 to 100, delete only the multiples of 2, 3, 5 and 7 from the list. This observation simplifies the procedure considerably.

In the 1800s, Polish astronomer Yakov Kulik used this method to find all the prime numbers between 1 and 100,000,000. It took him over 20 years to complete the task. Unfortunately, the library to which Kulik gave his manuscript lost the pages that listed the primes he discovered between 12,642,000 and 22,852,800.

significant figures *See* ERROR.

similar figures Two geometric figures are similar if they are the same shape but not necessarily the same size. (Mirror images are allowed.) More precisely, two POLYGONS are similar if, under some correspondence between their sides and vertices, corresponding interior angles are equal and corresponding sides differ in ratio by a constant factor. The constant of proportionality is called the scale factor.

As an example, any two squares are similar. If the side-length of one square is double the side-length of the other, say, then the two figures have scale factor 2. Any figure enlarged or reduced in size, with the aid of a photocopier say, produces a new figure similar to the first.

Two TRIANGLES are similar if:

1. The three interior angles of one triangle match the interior angles of the other.

This follows, since the LAW OF SINES shows that corresponding sides will also have the same ratio. This is often called the AAA rule.

2. Three sides of one triangle are proportional, by the same scale factor, to the three sides of the other.

This time the LAW OF SINES shows that the three corresponding angles are equal.

3. An angle of one triangle is equal to an angle of the other, and the sides forming the angle in one are proportional to the same sides in the other.

The LAW OF COSINES shows that the third sides of the triangles are in the same proportion.

Identifying similar triangles is often the key step in proving geometric results. The CIRCLE THEOREMS, PTOLEMY'S THEOREM, and the SECANT theorem, for example, demonstrate this.

Any two circles are similar. This observation explains why the value π is the same for all circles.

Two figures that are similar with a scale factor of 1 are called CONGRUENT FIGURES. A figure that is composed of parts similar to the entire figure is a FRACTAL.

See also AAA/AAS/ASA/SAS/SSS; EUCLIDEAN GEOMETRY.

simple interest *See* INTEREST.

Simpson, Thomas (1710–1761) British *Calculus* Born on August 20, 1710, in Leicestershire, England, Thomas Simpson is remembered in mathematics solely for the rule that wrongly bears his name.

With limited education, Simpson began his career as a professional astrologer and confidence man. After an unfortunate incident in 1733 from which he was obliged to leave his home county of Leicestershire, Simpson accepted an evening teaching position in Derby. His interest in mathematics then developed.

Simpson published mathematical articles in the *Ladies' Diary* and soon developed a reputation as a capable scholar in the field of calculus. In 1737 he wrote *A New Treatise of Fluxions* and followed this work with the release of four more influential texts over the following 6 years. His writing garnered him considerable fame in England at the time. In 1743 he was appointed second mathematical master at the Royal Military Academy and in 1735 was elected as a fellow of the ROYAL SOCIETY.

Later in his career, while still working as a mathematics teacher, Simpson also published three best-selling elementary textbooks: *Algebra* (1745), with 10 English editions; *Geometry* (1747), with six English editions, five French editions, and one Dutch edition; and *Trigonometry* (1748), with five English editions.

SIMPSON'S RULE, as it is called today, appeared in his 1743 text *Mathematical Dissertations*. That his name became attached to a technique that he clearly did not invent, but merely described, serves as a testimony to the influence and popularity of his writing. Simpson died in Market Bosworth, England, on May 14, 1761.

Simpson's rule   *See* NUMERICAL INTEGRATION.

simultaneous linear equations   An equation is called linear if each term of the equation containing a variable contains just one variable raised to the first power. For example, $3x + 4 = 6$ and $2x + 7y - z = 11$ are linear equations (but $2xy + z = 5$ and $3x^2 + 4 = 6$, for instance, are not). A collection of linear equations in several variables required to simultaneously hold true for some value of those unknowns is called a system of simultaneous linear equations. For example, the equations

$$5x + 2y = 16$$
$$2x - y = 1$$

form a pair of simultaneous linear equations. This pair has the solution $x = 2$ and $y = 3$.

There are a number of methods for solving pairs of simultaneous linear equations.

Elimination Method: Multiplying each equation by a suitable quantity so that the coefficients of one of the variables match. The solution to the system is then readily apparent if one subtracts the two equations. For example, to solve

$$x + 3y = 7$$
$$3x - y = 11$$

multiply the first equation through by 3 to obtain the equivalent pair:

$$3x + 9y = 21$$
$$3x - y = 11$$

Subtracting yields the equation: $(3x + 9y) - (3x - y) = 21 - 11$, that is, $10y = 10$, indicating that y equals 1. Since $x + 3y = 7$ we now see that x must be 4.

Substitution Method: Solve for one of the variables in one equation and insert the result into the second equation. For example, for the pair

$$x + 3y = 7$$
$$3x - y = 11$$

the first equation gives $x = 7 - 3y$. Substituting into the second equation yields $3(7 - 3y) - y = 11$, that is, $21 - 10y = 11$, or, again, that $10y = 10$. The solution to the system follows as shown in the previous method.

Method of Equating: Solve both equations for one variable and equate the results. In our example, the first equation yields $x = 7 - 3y$ and the second, $x = \dfrac{11 + y}{3}$. Equating yields $7 - 3y = \dfrac{11 + y}{3}$ or $21 - 9y = 11 + y$. Solving for y gives $y = 1$, from which it follows that $x = 4$.

One can also seek a GRAPHICAL SOLUTION. Such an approach shows that any pair of simultaneous linear equations either has a unique solution (as for the pair above), no solution, or infinitely many solutions. It is impossible for a pair of linear equations to have exactly two distinct solutions, for instance.

The process of GAUSSIAN ELIMINATION provides the means to solve systems of linear equations with more than two unknowns. This approach yields key results about the theory of matrices and the study of LINEAR ALGEBRA.

See also HISTORY OF EQUATIONS AND ALGEBRA (essay); MATRIX; SYSTEM OF EQUATIONS.

singular point (singularity)   Any point on a curve at which there is not a single well-defined TANGENT to the curve is called a singular point. It may be that the curve crosses itself at that point, in which case there are two tangents to the curve, or that the point is a CUSP or an ISOLATED POINT, for example.

See also DOUBLE POINT.

SI units   Adopted in 1960 by international agreement, the *Système International d'Unités*, (SI units) is a coherent system of units used for scientific purposes. It is based on seven fundamental base units of which the meter (m), the kilogram (kg), and the second (s) for measuring length, mass, and time, respectively, are the most common in mathematics. There are two supplementary units: the radian used for ANGLE measure and the steradian for measuring a SOLID ANGLE. By multiplying and/or dividing these base and supplementary units, other derived units are obtained. For instance, the unit of area is "square meter" (meter times meter), and the unit of velocity is "meter per second" (meter divided by second). There are 18 specific derived units that have their own names, such as a Newton (N) for a measure of force (given as kilograms times meter per second squared). The following table lists the basic and supplementary units and their symbols.

Quantity	Name	Symbol
length	meter	m
mass	kilogram	kg
time	second	s
temperature	Kelvin	K
amount of a substance	mole	mol
electric current	ampere	A
light intensity	candela	cd
angle (supplementary)	radian	rad
solid angle (supplementary)	steradian	sr

Multiples and fractions of units are defined in multiples of 1,000 and are denoted by special prefixes or symbols. These are shown in the table below. For example, 1 mega-amp equals 10^6 ampere, and 1 millimeter equals 10^{-3} meters.

10^{24}	yotta-	(Y)	Greek: *okákis:* eight times ($24 = 8 \times 3$)
10^{21}	zetta-	(Z)	Greek: *heptákis:* seven times ($21 = 7 \times 3$)
10^{18}	exa-	(E)	Greek: *hexákis:* six times ($18 = 6 \times 3$)
10^{15}	peta-	(P)	Greek: *pentákis:* five times ($15 = 5 \times 3$)
10^{12}	tera-	(T)	Greek: *téras:* monster
10^9	giga-	(G)	Greek: *gígās:* giant
10^6	mega-	(M)	Greek: *mégas:* big
10^3	kilo-	(k)	Greek: *chílioi:* thousand
10^{-3}	milli-	(m)	Latin: *mille:* thousand
10^{-6}	micro-	(μ)	Greek: *mīkrós:* small
10^{-9}	nano-	(n)	Greek: *nānos:* dwarf
10^{-12}	pico-	(p)	Italian: *piccolo:* small
10^{-15}	femto-	(f)	Danish: *femten:* 15
10^{-18}	atto-	(a)	Danish: *atten:* 18
10^{-21}	zepto-	(z)	Greek: *heptákis:* seven times
10^{-24}	yocto-	(y)	Greek: *oktákis:* eight times

The following additional prefixes are also often used:

10^2	hecto-	(h)	Greek: *hekatón:* hundred
10	deka-	(da)	Greek: *déka:* ten
10^{-1}	deci-	(d)	Latin: *decim:* ten
10^{-2}	centi-	(c)	Latin: *centum:* hundred

The metric system, a system of quantifying units of measure in terms of powers of 10, was first established in 1790 by a special Committee of Weights and Measures of the French Academy of Sciences. Surprisingly, bases other than 10 were seriously considered at the time. (There were arguments for using the duodecimal system of base 12, as well as consideration of a base 11 system, hoping to take advantage of PRIME base.) As scientific progress was made over the centuries, the metric system was extended to include measurements of electric current, light intensity, temperature, and molecular quantities. This expanded system of SI units described was adopted at the 11th International General Conference on Weights and Measures in 1960. Minor adjustments to the system were made in subsequent meetings.

skew lines   Two nonparallel straight lines in three-dimensional space are skew if they do not intersect. No pair of skew lines lie in a common plane.

skewness   A measure of the degree of asymmetry of a DISTRIBUTION is called its skewness. If a distribution has a short tail to the left and a long one to the right, then it is called positively skewed, and negatively skewed if the situation is reversed.

See also STATISTICS: DESCRIPTIVE.

slide 15 puzzle   (Boss puzzle, fifteen puzzle)   In 1878 famous American puzzlist Sam Loyd (1841–1911) introduced his "Boss puzzle," today called the slide 15 puzzle. The game consists of 15 tiles, numbered one through 15, held within a 4 × 4 frame with one space left free. Tiles may slide horizontally or vertically into the blank cell, but all moves are confined to the plane of the frame. The challenge is to convert the given arrangement of tiles shown top left to the one shown

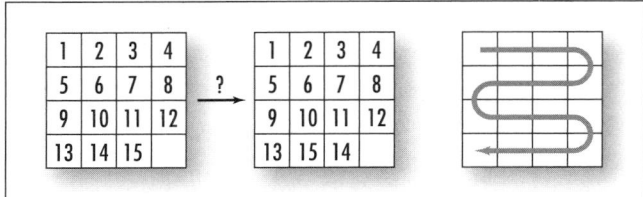

Slide fifteen puzzle

top right, with tiles "14" and "15" switched. Loyd offered a $1,000 prize to the first person submitting a correct solution.

It turns out, as Loyd well knew, that this puzzle is impossible to solve. The theory of even and odd PER-MUTATIONs explains why.

Following the snaked path indicated above, and ignoring the location of the blank cell, one can record any arrangement of tiles as a list of the numbers 1 through 15. For instance, the initial arrangement of tiles appears as the list:

1 2 3 4 8 7 6 5 9 10 11 12 15 14 13

Since a larger number appears to the left of a smaller number nine times, this list represents an odd permutation of the numbers 1 through 15. Notice that if one slides a tile horizontally dur-ing the play of the game, the corresponding list representing the arrangement of tiles does not change. If, on the other hand, one slides a tile vertically, the number of that tile shifts an even number of places up or down the list that repre-sents the arrangement of tiles. Shifting a number one place to the left or right can be effected by a single transposition, two places to the left or right can be effected by two transpositions, and so forth. Thus the result of shifting a number an even number of places to the left or right is the result of applying an even number of transposi-tions, thereby preserving the evenness or oddness of the arrangement of tiles. In playing the game, then—given that the initial arrangement of tiles corresponds to an odd permutation—the arrangement of tiles will forever remain an odd permutation. It is impossible, then, to obtain the arrangement called for by the goal of the puzzle, namely, 1 2 3 4 8 7 6 5 9 10 11 12 14 15 13, an even permutation.

Mathematicians have shown that all arrangements of tiles that correspond to odd permutations of the numbers 1 through 15 can in fact be achieved through the play of the game.

slide rule   One can perform simple additions with the aid of two ordinary rulers. To compute 2.7 + 3.5, for example, place the end (position "0") of one ruler at the location 2.7 along the second ruler. Then read 3.5 units along the first ruler. The corresponding label on the second ruler is then the desired sum 6.2.

Scottish mathematician JOHN NAPIER (1550–1617) discovered LOGARITHMS near the turn of the 17th cen-tury. These functions have the remarkable property of converting computations of multiplication into simpler computations of addition: $\log(N \times M) = \log N + \log M$. In 1622, English mathematician WILLIAM OUGHTRED (1574–1660) realized that two sliding rulers, with labels placed in LOGARITHMIC SCALE, will physically perform the addition of logarithms, and thus allow one to simply "read off" the result of any desired multipli-cation. (As a very simple example, imagine we wished to compute $100 \times 1,000$ with a pair of base-10 loga-rithmic rulers. Note then that the mark with label 100 is placed 2 in. along the ruler, and the mark labeled 1,000 3 in. along the ruler. The sum 2 + 3 is, of course, 5 in. along the ruler. But this fifth position is labeled 10,000, which is indeed the product $100 \times 1,000$.)

Oughtred's mechanical device of two sliding rulers is called a slide rule. The device was inspired by the work of English scholar Edmund Gunter (1581–1626), who had used a single ruler and a pair of pointers to accomplish the same feat.

Slide rules were popular up until the 1970s before the advent of the pocket calculator.

See also NAPIER'S BONES.

slope (grade, gradient)   The slope of a line is a mea-sure of its steepness. This can be determined a number of ways:

1. *Numerically:* Two quantities are said to be linearly related if a unit change in one quantity produces a constant change in the other. The value of that con-stant change is the slope of the relationship. For example, the following table shows the profit made for a company that sells widgets.

No. widgets sold	0	1	2	3	4	5
Profit (in dollars)	0	15	30	45	60	75

The relationship is certainly linear, as each additional widget sold produces the same increase of $15 in profit made. This relationship has slope 15. Notice that in selling two additional widgets, profit is increased by $30, and in selling three additional widgets, profit increases by $45, and so on. In particular, the *ratio* of profit increase to production increase is always the same: $\frac{45}{3} = \frac{30}{2} = 15$. In general, if two quantities x and y are linearly related, the slope of the relationship is computed as:

$$\text{slope} = \frac{\text{change in y values}}{\text{change in x values}}$$

In particular, if the value x_1 produces the output value y_1 and x_2 the output y_2, then:

$$\text{slope} = \frac{y_2 - y_1}{x_2 - x_1}$$

2. *Graphically:* Plotting the points of a linear relationship produces a straight-line graph. The slope of the line is determined by placing the tip of a pencil at any location on the line, moving it horizontally one unit to the right, and measuring the vertical distance the pencil tip must move in order to return to the line. An upward vertical motion is considered positive and a downward vertical motion negative. Graphing the points of the widget/profit example again shows that this linear relationship has slope 15.

It does not matter which point on the line one initially chooses: all locations yield the same value for the slope via this method. (This is only true for straight-line graphs.) One can also make a horizontal run of a length different than 1 to produce a different value for the rise one must take to return to the graph. The ratio "rise over run," however, will always adopt the same value. School students are often asked to memorize this catch phrase as the definition of slope:

$$\text{slope} = \frac{\text{change in y values}}{\text{change in x values}} = \frac{\text{"rise"}}{\text{"run"}}$$

If (x_1, y_1) and $(x_2, y_2$ are *any* two points on the line, then we can interpret the quantity $x_2 - x_1$ as a run with corresponding rise $y_2 - y_1$ and so, again, we see that slope is given by:

$$\text{slope} = \frac{y_2 - y_1}{x_2 - x_1}$$

A line has positive slope if, in moving from left to right, the line rises in value, negative slope if it decreases in value (thus a unit step horizontally to the right requires a negative rise in order to return to the line), and zero slope if it is horizontal.

Traffic highway signs often use a version of this interpretation of slope to warn drivers of the steepness of the road. For example, a sign that reads "15% grade next 7 miles" warns drivers that for each mile traveled for the next 7 miles, the altitude of the road drops by 0.15 miles. (The slope of the road is thus −0.15.)

3. *Algebraically:* The EQUATION OF A LINE is usually written in the form $y = mx + b$, where m and b are numbers. We see that an increase of the x-value by 1 unit causes an increase in the y-value by m units:

If $x \to x + 1$, then $y = mx + b \to m(x + 1) + b$
$= (mx + b) + m = y + m.$

Thus the slope of the line is m. We have:

The slope of a line is simply the coefficient of the x-variable in the equation $y = mx + b$.

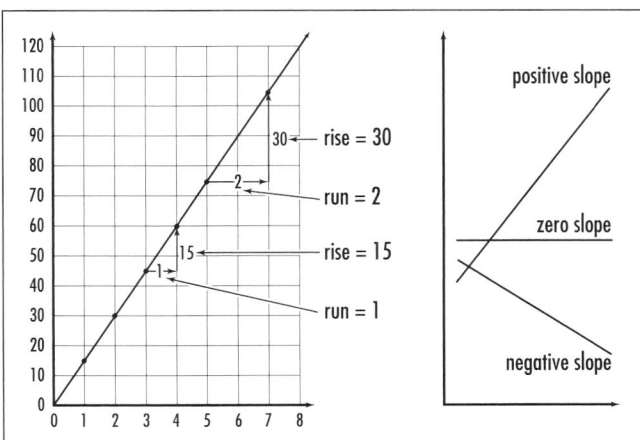

Slope

Two different lines with the same slope are PARAL-LEL. For example, the lines $y = 2x + 3$ and $y = 2x - 5$ are parallel. If two PERPENDICULAR lines have slopes m_1 and m_2 respectively, then the following relationship holds:

$$m_1 \cdot m_2 = -1$$

One can see this by drawing the horizontal and vertical line segments for the run and the rise for the first line (and so $m_1 = \dfrac{\text{rise}}{\text{run}}$), and then rotating that picture counterclockwise 90°. In this new picture, what was the run is now a rise for a perpendicular line, and what was a rise is now a run pointing in the opposite direction. Thus $m_2 = -\dfrac{\text{rise}}{\text{run}} = -\dfrac{1}{m_1}$.

DIFFERENTIAL CALCULUS deals with the issue of defining slope for curves that are not straight lines.

See also COLLINEAR; GRADIENT; TANGENT.

Snell's law   This principle, first described in 1621 by Dutch mathematician Willebrord van Roijen Snell, is best explained through the study of an OPTIMIZATION problem. We phrase it here in a modern setting.

A lifesaver sets off to rescue a swimmer in distress. The drowning swimmer, however, is not directly in front of the lifesaver but off at an angle along the shore. To reach the swimmer, the lifesaver must run across the sand and then dive into the water and swim to the rescue. A question arises: as the lifesaver can run much faster than she can swim, toward which point along the shore should she run so that the total amount of time it takes to reach the swimmer is at a minimum?

Surprisingly, the straight-line path from the lifesaver station to the swimmer does not provide the quickest route.

To analyze this problem, let a, b, α, and β be the distances and angles shown. For ease, assume that the horizontal displacement between the lifesaver and the swimmer is 1 unit. Suppose also that the speed of the lifesaver on shore is r meters per second, and in the water, s meters per second. We wish to find the length x along the shore that produces a path requiring the least amount of time to follow.

According to PYTHAGORAS'S THEOREM, the total distance the lifesaver runs on shore is $\sqrt{a^2 + x^2}$ meters.

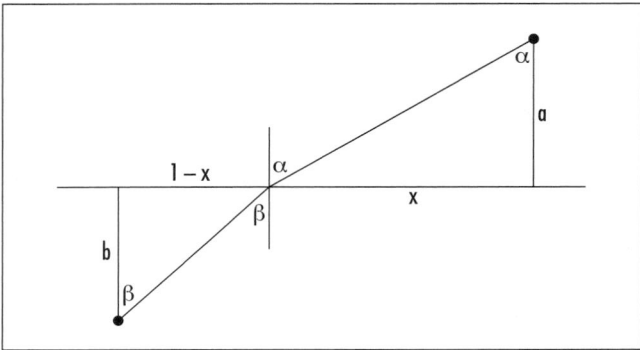

Snell's law

As speed equals distance over time, this takes a total of $\dfrac{\sqrt{a^2 + x^2}}{r}$ seconds to complete. In the same way, we see that the lifesaver will be in the water $\dfrac{\sqrt{b^2 + (1-x)^2}}{s}$ before reaching the swimmer. Thus, for this position x along the shore, the total time $T(x)$ taken to reach the swimmer is given as:

$$T(x) = \frac{\sqrt{a^2 + x^2}}{r} + \frac{\sqrt{b^2 + (1-x)^2}}{s}$$

seconds. We now use the techniques of CALCULUS to find a local minimum for this function. (*See* MAXIMUM/MINIMUM.)

A local minimum of this function can only occur if the first derivative of this function is zero. This yields:

$$T'(x) = \frac{x}{r\sqrt{a^2 + x^2}} - \frac{1-x}{s\sqrt{b^2 + (1-x)^2}} = 0$$

The second-derivative test shows that this does indeed correspond to a minimum. Thus the optimal solution to this problem occurs when the quantity $\dfrac{x}{r\sqrt{a^2 + x^2}}$ equals $\dfrac{1-x}{s\sqrt{b^2 + (1-x)^2}}$. Notice in the diagram that $\sin\alpha = \dfrac{x}{\sqrt{a^2 + x^2}}$ and $\sin\beta = \dfrac{1-x}{\sqrt{b^2 + (1-x)^2}}$ Thus, the lifesaver should run to the position on shore such that the angles α and β satisfy the fundamental relationship

$$\frac{\sin\alpha}{r} = \frac{\sin\beta}{s}$$

In general we have:

If the speed of travel through one medium is *s* units per second, and through a second medium is *r* units per second, then the optimal route of travel across a boundary separating the two media occurs when the angle of incidence α and the angle of refraction β satisfy the relationship:

$$\frac{\sin\alpha}{r} = \frac{\sin\beta}{s}$$

This principle is today called Snell's law of refraction. Snell observed that a beam of light moving from one medium to another, say, from air to water, obeys this law. Light of one color, that is, of one particular wavelength, will travel through the same medium at a speed different from that of a beam of a different color, and so will have its own appropriate angle of refraction. For this reason, white light—the composition of all colors—separates into a rainbow of colors at the interchange of a new medium. Water droplets suspended in the air after a rainstorm separate the colors of sunlight in this way to produce a large rainbow visible from the ground.

soap bubbles The film of a soap solution is elastic and has the property that it always "pulls in on itself" to create a surface of least possible surface area. Noting that a soap bubble, containing a fixed volume of air, always adopts the shape of a SPHERE provides a physical demonstration of the following three-dimensional variation of the ISOPERIMETRIC PROBLEM:

Of all figures with a fixed volume, the sphere has the least surface area.

Polish mathematician Hermann Schwarz (1843–1921) provided a mathematical proof of this observation in 1882.

Many basic questions about arrangements of bubbles separating more than one chamber of air are unresolved. In 2000 mathematicians Frank Morgan, Michael Hutchings, Manuel Ritoré, and Antonio Ros managed to prove that the familiar double bubble seen in nature (two portions of a sphere attached to a curved boundary separating the two volumes) is indeed the shape of minimal total surface area for a configuration enclosing two chambers of air. The corresponding triple-bubble problem remains open.

solid angle Just as an ANGLE in two-dimensional space is the measure of the amount of length of a UNIT CIRCLE "cut off" by two rays describing the angle, a solid angle in three-dimensional space is the amount of surface area of a unit SPHERE intercepted by a bundle of rays emanating from the center of the sphere. A solid angle need not have any particular shape; it is defined by the nappe of a CONE, which can be of any shape.

The unit of measure of a solid angle is called a steradian. Because the surface area of a sphere is $4\pi(1)^2$, there are a total of 4π steradians about a point in space. The measure of a hemisphere is $\frac{1}{2} \cdot 4\pi = 2\pi$ steradians.

Solid angles are used in physics to measure the fraction of the total emission of radiation from a source in space. For example, the Earth, of radius approximately 6370 km, exposes a total surface area of $\frac{1}{2} \cdot 4\pi(6370)^2$ km^2 toward the Sun at any time. As the distance of the Earth from the Sun is approximately 150 million km, we have that the Earth occupies the fraction $\frac{\frac{1}{2} \cdot 4\pi(6370)^2}{4\pi(150\times10^6)^2} \approx \frac{1}{1,110,000,000}$ of the surface area of sphere about the Sun of radius 150 million km. The solid angle of the Earth with respect to the Sun is this fraction of the full 4π steradians of space about the Sun. Notice that this fraction is exceedingly small. Life on Earth is sustained by less than 1 billionth of the total energy emitted by the Sun.

The polyhedral angle of a POLYHEDRON is the solid angle of the region formed by projecting one face of the polyhedron onto a unit sphere.

solid of revolution If the region under a curve $y = f(x)$ between $x = a$ and $x = b$ is rotated through one revolution about the *x*-axis, it generates a three-dimensional figure called a solid of revolution. More generally, any figure in the plane rotated about a line in the plane that does not cut the figure produces a solid of revolution.

For example, rotating a semicircle about its diameter produces a SPHERE. Rotating a circle about an axis that does not cut the circle produces a TORUS.

INTEGRAL CALCULUS is used to calculate the volume of a solid of revolution. For the case of a curve $y = f(x)$ above an interval $[a,b]$ rotated about the x-axis, subdivide $[a,b]$ into small segments given by points $a,x_1,x_2,\ldots, x_n,b$. The region under the curve is approximated as a collection of rectangles of heights $f(x_i)$ and widths $(x_{i+1} - x_i)$. When revolved, each such rectangle produces a disc of radius $f(x_i)$ and width $(x_{i+1} - x_i)$, and consequently volume $\pi(f(x_i))^2 . (x_{i+1} - x_i)$. The volume of the solid in question is then well approximated by the sum of these individual volumes. Using finer and finer approximations shows that, in the limit, the true volume V of the solid of revolution is given by the integral:

$$V = \int_a^b \pi(f(x))^2 \, dx$$

One can use this integration method, called the disc method, to show, for example, that the volume of a sphere of radius r is given by $\frac{4}{3}\pi r^3$ (Use the function $y = \sqrt{r^2 - x^2}$ over the interval $[-r,r]$.) For solids that arise from rotating curves about the y-axis, an analogous integration technique, called the cylindrical shell method, is employed to compute volumes.

By approximating a section of the curve $y = f(x)$ as a series of straight line segments, revolving each segment about the x-axis, and computing the surface area of each small frustum of a cone that results, one can show that surface area S of a solid of revolution is given by the integral:

$$S = \int_a^b 2\pi f(x) \, \sqrt{1 + (f'(x))^2} \, dx$$

(This assumes that the derivative $f'(x)$ is a continuous function.) For example, one can use this formula to show that the surface area of a sphere of radius r is $4\pi r^2$.

In the mid-fourth century C.E., without the aid of formal integral calculus, Greek mathematician PAPPUS OF ALEXANDRIA made two beautiful geometric discoveries.

If a figure in the plane is revolved around an axis that does not cut through the figure, then the volume of the solid of revolution produced equals the product of the area of the region and the distance traveled around the axis by the figure's centroid.

If a segment of a curve in a plane is revolved around an axis that does not cut through the segment, then the surface area of the solid of revolution produced equals the product of the ARC LENGTH of the segment and the distance traveled around the axis by the segment's centroid.

The centroid of a plane figure or of a segment of a curve is its CENTER OF GRAVITY or balance point if we imagine the plane figure as made of uniformly dense material or the curve segment of uniformly dense wire. The centroid of a circle, regarded as a plane figure, for example, is its center. The centroid of just its circumference is also its center.

If a circle of radius r with center R units from the x-axis is rotated about the x-axis, the distance traveled by the centroid is $2\pi R$. Consequently, by Pappus's theorem, the volume V and the surface area S of the torus produced are:

$$V = (\pi r^2) \times (2\pi R) = 2\pi^2 r^2 R$$

and

$$S = (2\pi r) \times (2\pi R) = 4\pi^2 r R$$

See also HYPERBOLOID; PAPPUS'S THEOREMS; PARABOLOID.

solution by radicals   If it is possible to express a solution to a POLYNOMIAL equation in terms of the COEFFICIENTs that appear in the equation under the application of a finite number of additions, subtractions, multiplications, divisions, and root extractions, then we say we have a solution by radicals. For example, the two solutions of a QUADRATIC equation $ax^2 + bx + c = 0$ are given by $x = \dfrac{-b \pm \sqrt{b^2 - 4ac}}{2a}$. These are solutions by radicals. CARDANO'S FORMULA shows that any solution to a CUBIC EQUATION is a solution by radicals, and the work of LUDOVICO FERRARI (1522–65) showed that the same is true of any QUARTIC EQUATION. Certainly some fifth-degree equations have solutions by radicals (the solution of $x^5 - a = 0$, namely, $x = \sqrt[5]{a}$, is a solution by radicals), but the general question of whether or not the solutions of all fifth (or higher) degree equations are solutions by radicals remained an important unsolved question for several centuries. This question is equivalent to asking whether or not there exists a general formula akin to the quadratic formula for quadratic equations that solves all fifth (or higher) degree equations.

The work of ÉVARISTE GALOIS (1811–32) provided the first important steps toward understanding the general properties of solutions to polynomial equations, and Norwegian mathematician NIELS HENRIK ABEL (1802–29) provided the first proof that there can be no general formula for the solution of fifth-degree equations via radicals. This is surprising given that the FUNDAMENTAL THEOREM OF ALGEBRA assures that every fifth-degree equation has precisely five solutions. Abel's result shows then that, in general, not every solution can be expressed in terms of a finite application of simple arithmetic operations on the coefficients of the equation.

As mentioned earlier, some special quintics do have solutions by radicals. For example, ABRAHAM DE MOIVRE (1667–1754) showed that the solutions of a fifth-degree equation of the form:

$$x^5 + ax^3 + \frac{1}{5}a^2x + b = 0$$

(this polynomial is today known as De Moivre's quintic) has five solutions given by:

$$x = \omega^k \times \sqrt[5]{-\frac{b}{2} + \sqrt{\left(\frac{b}{2}\right)^2 - \left(\frac{a}{5}\right)^5}}$$

$$+ \omega^{-k} \times \sqrt[5]{-\frac{b}{2} - \sqrt{\left(\frac{b}{2}\right)^2 - \left(\frac{a}{5}\right)^5}}$$

where $\omega = e^{\frac{2\pi i}{5}}$ is a fifth root of unity and k runs through 0, 1, 2, 3, and 4.

See also GIROLAMO CARDANO; CUBE ROOT/NTH ROOT; SQUARE ROOT; NICCOLÒ TARTAGLIA.

Spearman's method *See* RANK CORRELATION.

sphere The closed surface in three-dimensional space consisting of all those points that are a fixed distance r from a given point O is called a sphere. The length r is called the radius of the sphere, and the point O the center of the sphere. The DISTANCE FORMULA shows that if O has CARTESIAN COORDINATES $O = (a,b,c)$, then the equation of a sphere is given by:

$$(x - a)^2 + (y - b)^2 + (z - c)^2 = r^2$$

It is often convenient to regard a sphere as centered about the origin, in which case its equation reduces simply to $x^2 + y^2 + z^2 = r^2$. This is the three-dimensional analog to the equation of a CIRCLE of radius r in two-dimensions: $x^2 + y^2 = r^2$. (The one-dimensional analog, $x^2 = r^2$, gives two points $x = r$ and $x = -r$ on the number line, and the four-dimensional analog given by $x^2 + y^2 + z^2 + w^2 = r^2$ is a hypersphere in the fourth DIMENSION.)

Any plane through the center of a sphere divides the sphere into two hemispheres. The curve of intersection of such a plane with the sphere is a circle on the surface of the sphere called a great circle. (The equator of the EARTH, for instance, is a great circle.) Any other circle drawn on the surface of the sphere is called a small circle.

Any line segment connecting two points on the surface of the sphere and passing through the center of the sphere is called a DIAMETER, and the two endpoints of any diameter are called ANTIPODAL POINTS. The famous Borsuk-Ulam theorem asserts the following:

If f and g are any two continuous functions on the surface of a sphere (such as the air temperature and air pressure at the surface of the Earth), then there must exist a pair of antipodal points P and Q so that $f(P) = f(Q)$ and $g(P) = g(Q)$.

The sphere can be regarded as a SOLID OF REVOLUTION, obtained by rotating a circle about its diameter one full revolution. As such, PAPPUS'S THEOREMS show that the VOLUME V and the surface area A of a sphere of radius r are given by the formulae:

$$A = \frac{4}{3}\pi r^2$$
$$A = 4\pi r^2$$

These formulae can also be obtained by the techniques of INTEGRAL CALCULUS. (Calculus also shows that the volume of a four-dimensional hypersphere of radius r is $V = \frac{1}{2}\pi^2 r^4$.) One can also use calculus to establish the following surprising result:

If a spherical loaf of bread is sliced into n slices of equal thickness, then the surface area of crust on each slice is the same. (Assume that the slices are made parallel, each perpendicular to a fixed diameter of the loaf.)

The region of a sphere bounded between two parallel planes is called a spherical segment or a zone. If the radius of the sphere is r and the distance between the two planes is h, then the surface area of the zone is $2\pi rh$. (Notice that this formula is indeed independent of where the two planes slice the sphere.) The solid obtained by rotating a sector of a circle about the diameter of the circle is called a spherical sector, and a spherical wedge is any segment of a sphere bounded between two great circles.

A SPHERICAL TRIANGLE is any region on the surface of a sphere bounded by arcs of three great circles, and spherical trigonometry is the study of spherical triangles and the angles formed by intersecting great circles. The study of SPHERICAL GEOMETRY shows that the three angles in any spherical triangle sum to more than 180°. A spherical polygon is the figure formed on the surface of a sphere by three or more arcs of great circles. Mathematicians have shown that the angles in an n-sided spherical polygon sum to value greater than $180(n - 2)°$, but no more than $180n°$.

The famous hairy ball theorem asserts that it is impossible to construct a continuous, nowhere zero, VECTOR FIELD on the surface of a sphere.

A single sphere divides three-dimensional space into two regions: an inside and an outside. Two intersecting spheres divide space into four regions (the region interior to both spheres, two regions each interior to one sphere but not the other, and the outer external region), three intersecting spheres divide space into eight regions, four spheres divide space into 16 regions, and five spheres into 30 regions. In general, N mutually intersecting spheres divide space into

$$\frac{N}{3}(N^2 - 3N + 8)$$

distinct regions. (Thus, for instance, 10 spheres can be arranged in space to produce 260 separate regions.)

In n-dimensional space, the maximal number of nonintersecting spheres (or hyperspheres) of a fixed radius r that can be arranged about a central sphere with each touching or "kissing" the central sphere is called the n-dimensional kissing number. (It is also called the contact number, coordination number, ligancy, or the Newton number.) As it is possible to arrange two line segments at either end of a given line segment, the one-dimensional kissing number is 2. One can arrange six (but not seven) circles about a given circle, and so the two-dimensional kissing number is 6. SIR ISAAC NEWTON (1642–1727) correctly believed that the kissing number in three dimensions is 12 (although this fact was not properly proved until late in the 19th century).

In three-dimensional space there are just three periodic (that is, self-repeating) patterns for stacking an infinite collection of identical spheres. Each arrangement is called a packing. The "face-centered cubic lattice" arranges spheres in layers, with each sphere surrounded by just four spheres within that layer. The "cubic and hexagonal lattices" arrange spheres in a manner such that each sphere in a layer is surrounded by six other spheres within that layer. (The two packings differ in how a third layer of spheres is placed in relation to a first layer.) The density of a packing is the fraction of space occupied by the spheres. The cubic and hexagonal packings each have a density of $\frac{\pi}{3\sqrt{2}} \approx 74.05\%$.

In 1611 German mathematician and astronomer JOHANNES KEPLER (1571–1630) conjectured that the cubic and hexagonal packings are the densest possible. It was not until 1998 that this was finally proved to be the case.

The famous FOUR-COLOR THEOREM asserts that just four colors are needed to paint any map on the surface of a sphere. EULER'S THEOREM shows that if v is the number of vertices in any such map, e the number of edges between countries, and r is the number of regions defined by the map, then $v - e + r = 2$.

As soap film is elastic and wants to "pull in on itself," any soap bubble containing a fixed volume of air adopts a shape that minimizes surface area. That soap bubbles are spherical provides a physical demonstration of the following three-dimensional variation of the ISOPERIMETRIC PROBLEM:

> Of all figures in three-dimensional space of a fixed volume, the sphere has the least surface area.

See also MERCATOR'S PROJECTION; OBLATE/PROLATE; SOAP BUBBLES; SOLID ANGLE; SPHERICAL COORDINATES; TORUS.

spherical coordinates (spherical polar coordinates) In three-dimensional space, the location of a point P can

described by three coordinates ρ, θ, and φ, called the spherical coordinates of P, where ρ is the distance of P from the origin O of a CARTESIAN COORDINATE system, θ is the angle between the x-axis and the projection of the line segment connecting O to P onto the xy-plane (measured in a counterclockwise sense from the x-axis), and φ is the angle from the z-axis to the line segment connecting O to P. The angle θ, called the longitude, takes a value between zero and $360°$, and the angle φ, called the colatitude, takes a value between zero and $180°$. The angles can also be presented in RADIAN MEASURE.

Spherical coordinates are useful for describing surfaces with spherical symmetry about the origin. For example, the equation of a sphere of radius 5 is simply $\rho = 5$. (As the angles θ and φ vary in their allowed ranges, the points of a sphere are described.) The surface defined by the equation $\theta = c$, for some constant c (allowing ρ and φ to vary), is a vertical HALF-PLANE with one side along the z-axis, and the surface $\varphi = c$ is one nappe of a right circular CONE.

A point P with spherical coordinates (ρ, θ, φ) has corresponding Cartesian coordinates (x, y, z) given by:

$$x = \rho \sin \varphi \cos \theta$$
$$y = \rho \sin \varphi \sin \theta$$
$$z = \rho \cos \varphi$$

To see this, project the line segment of length ρ connecting O to P onto the xy-plane. Call the length of this line segment r. Then, according to the POLAR COORDINATE conversion formulae, we have $x = r \cos\theta$ and $y = r \sin\theta$. (Examine the right triangle in the xy-plane with the segment of length r as hypotenuse.) According to the right triangle in a vertical plane, with this line segment of length r as base, the line segment connecting O to P on length ρ as hypotenuse, and a vertical leg of length z, we have $\cos\varphi = \frac{r}{\rho}$ and $\cos\varphi = \frac{z}{\rho}$. Thus $z = \rho \cos\varphi$ and $r = \rho \sin\varphi$, and the conversion formulae now follow.

To convert from Cartesian coordinates to spherical coordinates use:

$$\rho = \sqrt{x^2 + y^2 + z^2}$$
$$\theta = \tan^{-1}\left(\frac{y}{x}\right)$$
$$\varphi = \cos^{-1}\left(\frac{z}{\rho}\right) = \cos^{-1}\left(\frac{z}{\sqrt{x^2 + y^2 + z^2}}\right).$$

(The first equation follows from the standard DISTANCE FORMULA.)

A triple integral of the form $\iiint_V f(x, y, z) dx\, dy\, dz$ over a volume V described in Cartesian coordinates converts to the corresponding integral:

$$\iiint_V f(\rho\sin\varphi \cos\theta, \rho \sin\varphi \sin\theta, \rho \cos\varphi)\, \rho^2 \sin\varphi\, d\rho\, d\theta\, d\varphi$$

in spherical coordinates. The appearance of the term $\rho^2 \sin\varphi$ in the integrand follows for reasons analogous to the appearance of r in the conversion of a DOUBLE INTEGRAL from planar Cartesian coordinates to polar coordinates.

As an example, using radian measure, the volume V of a sphere of radius R can be computed as:

$$V = \iiint_S 1\, dx\, dy\, dz$$
$$= \int_0^R \int_0^\pi \int_0^{2\pi} \rho^2 \sin\varphi\, d\theta\, d\varphi\, d\rho$$
$$= \int_0^R \int_0^\pi 2\pi\rho^2 \sin\varphi\, d\varphi\, d\rho$$
$$= \int_0^R 4\pi\rho^2\, d\rho$$
$$= \frac{4}{3}\pi R^3.$$

See also CYLINDRICAL COORDINATES.

spherical geometry (elliptic geometry, Riemannian geometry) Discovered in 1856 by German mathematician GEORG FRIEDRICH BERNHARD RIEMANN (1826–66), and later slightly modified by FELIX KLEIN (1849–1925), spherical geometry is a NON-EUCLIDEAN GEOMETRY in which the famous PARALLEL POSTULATE fails in the following manner:

> Through a given point not on a given line, there are no lines parallel to that given line.

Riemann used the surface of a SPHERE as a model of this geometry by interpreting the word *line* to mean a great circle on the sphere. Given that in a theory of geometry two lines are meant to intersect at just one point (yet any two great circles intersect at two ANTIPODAL POINTS), it is appropriate then to interpret the word *point* in spherical geometry as an antipodal pair of points on the surface. In this setting, it is now also true that any two distinct points determine a unique line.

Riemann and Klein proved that all but the fifth of EUCLID'S POSTULATES hold in this model and, moreover, that this model is consistent (that is, free of CONTRADICTIONS). This establishes that the parallel postulate cannot be logically deduced as a consequence of the remaining axioms proposed by Euclid.

In spherical geometry all angles in triangles sum to more than 180°, and the ratio of the circumference of any circle to its diameter is greater than π (and this value varies from circle to circle).

See also EUCLIDEAN GEOMETRY; HYPERBOLIC GEOMETRY; PLAYFAIR'S AXIOM.

spherical triangle   A three-sided figure on the surface of a SPHERE bounded by the arcs of three great circles is called a spherical triangle. Unlike plane triangles, the angles in a spherical triangle do not sum to the constant value of 180°. In fact, the sum of angles is always greater than 180° and can be as great as 540°. (This occurs when all three sides of the spherical triangle lie on the same great circle.)

A right spherical triangle has at least one right angle. A birectangular triangle contains two right angles, and a trirectangular triangle has three right angles. (For instance, any triangle that connects the North Pole to two points on the equator of the Earth is birectangular. Such a triangle can also be trirectangular if the points on the equator are positioned appropriately.) A spherical triangle with no right angles is called oblique.

If the angles of a spherical triangle are A, B, and C, measured in degrees, then the surface area of the triangle is given by the formula:

$$\text{area} = \frac{\pi R^2}{180}\left(A + B + C - 180\right)$$

Here R is the radius of the sphere. The quantity $A + B + C - 180$, usually denoted E, is called the spherical excess of the triangle. If the three side-lengths of the triangle are given by a, b, and c, then the spherical excess can also be computed via the following remarkable formula discovered by Swiss mathematician Simon Lhuilier (1750–1840):

$$\tan\left(\frac{E}{4}\right) = \sqrt{\tan(\frac{a+b+c}{4})\tan(\frac{-a+b+c}{4})\tan(\frac{a-b+c}{4})\tan(\frac{a+b-c}{4})}$$

spiral of Archimedes (**Archimedean spiral**)   Studied by the great ARCHIMEDES OF SYRACUSE (ca. 287–212 B.C.E.) in his work *On Spirals*, an Archimedean spiral is the curve traced out by a point rotating about a fixed point at a constant angular speed while moving away from that point at a constant speed. (For example, it is the path traced by a fly walking radially outward from the center of a uniform rotating disc.) The equation of such a spiral has a simple form in POLAR COORDINATES:

$$r = a\theta$$

where a is a positive constant.

A curve given by an equation of the form $r = \frac{a}{\theta}$ is called a hyperbolic spiral. These curves were studied by Johann Bernoulli of the famous BERNOULLI FAMILY. Curves of the form $r = a\theta^{\frac{1}{n}}$ appear as tightly wrapped spirals. A curve given by $\log r = a\theta$ is called a logarithmic spiral.

See also GOLDEN RATIO.

square   In geometry a square is a planar figure with four equal straight sides and four interior right angles. It is simultaneously a RECTANGLE, a PARALLELOGRAM, and a rhombus.

A square with side-length x has AREA $x \times x = x^2$, which explains the term *squaring* for the operation of raising a number or a variable to the second power.

Squares are solutions to ISOPERIMETRIC PROBLEMS of the following classic type:

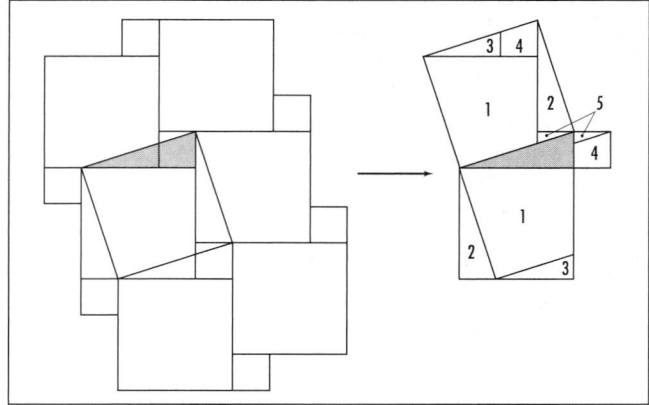

Pythagoras's theorem

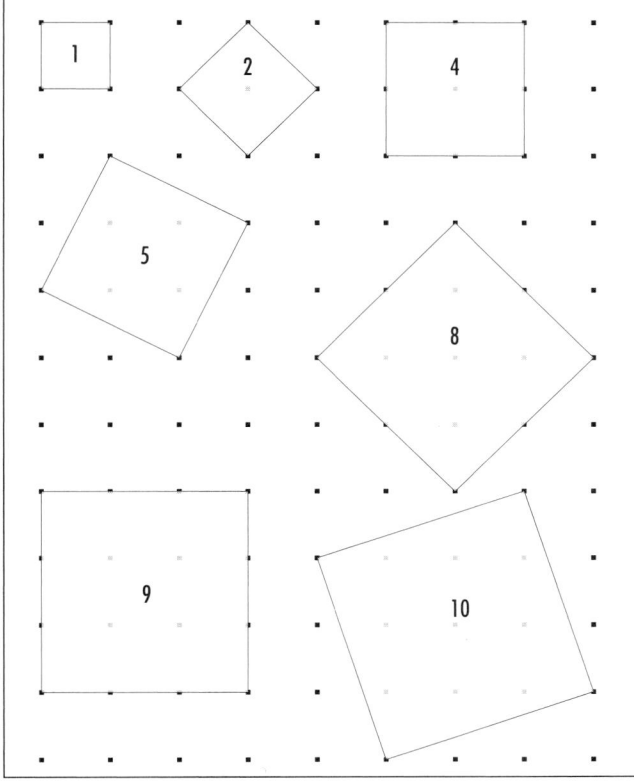

Squares on a lattice

A farmer wishes to build a rectangular pen with 400 ft of fencing. What shape rectangle gives the largest area for this given length of fence?

The answer is a 100×100 square pen. Any other rectangle will have one side shorter than 100 ft, say $100 - x$ ft, a longer width, $100 + x$ ft, and consequently area of $(100 - x)(100 + x) = 10,000 - x^2$ ft², which is less than the area of the square pen.

A square is one of the three regular POLYGONs that provides a TESSELLATION of the plane. Two squares of different sizes also tile the plane. Such a tessellation holds within it a purely visual proof of PYTHAGORAS'S THEOREM.

Using a sheet of graph paper, one can draw squares of areas 1, 2, 4, 5, 8, 9, and 10 units, all with vertices on the lattice points of the paper, but not squares of areas 3, 6, or 7 units.

Pythagoras's theorem shows that one can draw a square of area N if, and only if, N is of the form $N = a^2 + b^2$ for some integers a and b. For example, $9 = 3^2 +$

0^2 and $5 = 2^2 + 1^2$. (The side-length of a square of area 5 spans 2 units across and 1 unit over.) Surprisingly, this set of numbers satisfies the CLOSURE PROPERTY under multiplication.

See also FIGURATE NUMBERS; THEOREM.

square numbers   *See* FIGURATE NUMBERS.

square root   A number x, that, when multiplied by itself, produces a given number a is called the square root of the number a. If $x^2 = a$, then we write $x = \sqrt{a}$. For example, 3 is a square root of 9, since $3 \times 3 = 9$, and zero is a square root of zero, since $0 \times 0 = 0$. Geometrically, the square root of a positive quantity a is the side-length of a square whose area is a.

The invention the symbol $\sqrt{\ }$, called a radical sign, for the extraction of square roots is attributed to German mathematician Christoff Rudolff (1499–1545). The name of the symbol comes from the Latin word *radix* for "root." Any quantity that appears under a radical sign is called a radicand. For example, in the equation $\sqrt{169} = 13$, 169 is the radicand and 13 is the square root.

Any equation of the form $x^2 = a$ with $a \neq 0$ has two distinct solutions. Thus every number different from zero has two distinct square roots. For instance, 3 and −3 are both square roots of 9. By convention, if a is a positive quantity, then $\sqrt{a}$ is used to denote the positive root and $-\sqrt{a}$ the negative square root. For instance, we write $\sqrt{9} = 3$ even though −3 is also a valid square root of 9.

A study of EXPONENTs shows that it is appropriate to define a number raised to the half power to mean the square root of that number. Whether that root is positive or negative is left undefined. Thus, for instance, $9^{\frac{1}{2}} = \pm 3$.

Attempts to define the square roots of negative numbers leads to the invention of the COMPLEX NUMBERS. In the realm of complex numbers, every number (except zero) has exactly two square roots.

The square root of any CONSTRUCTIBLE number is again constructible. Thus, given a line segment of length a drawn on a page, it is possible to construct from it a second line segment of length $\sqrt{a}$ using only the simple tools of a straightedge and a compass.

Greek mathematician THEODORUS OF CYRENE (ca. 425 B.C.E.) used simple geometric arguments to prove that the numbers from $\sqrt{2}$ through to $\sqrt{17}$ (skipping $\sqrt{4}$ = 2, $\sqrt{9}$ = 3, and $\sqrt{16}$ = 4) are each irrational. More generally, one can use the FUNDAMENTAL THEOREM OF ARITHMETIC to show that the square root of a positive integer a is a RATIONAL NUMBER if, and only if, a is a perfect square (that is, $a = b^2$ for some whole number b.) HERON'S METHOD provides a method for computing the square root of any positive real to any prescribed degree of accuracy.

The square-root inequality states that for any positive whole number n we have:

$$2\left(\sqrt{n+1} - \sqrt{n}\right) < \frac{1}{\sqrt{n}} < 2\left(\sqrt{n} - \sqrt{n-1}\right)$$

An exercise in algebra establishes the validity of this statement.

Although the statements:

$$\sqrt{ab} = \sqrt{a}\,\sqrt{b}$$

and

$$\sqrt{\frac{a}{b}} = \frac{\sqrt{a}}{\sqrt{b}}$$

are valid for all numbers a and b (with $b \neq 0$ in the second equation), the equation $\sqrt{a+b} = \sqrt{a} + \sqrt{b}$, in general, is not correct. (Substitute in the values $a = 9$ and $b = 16$, for instance.)

See also CUBE ROOT/NTH ROOT; IDENTITY MATRIX; SOLUTION BY RADICALS; SURD.

squaring the circle (circle squaring, the quadrature of the circle)   One of the problems of antiquity (like DUPLICATING THE CUBE and TRISECTING AN ANGLE) of considerable interest to the classical Greek scholars is the task of constructing a square of the same area as a given circle in the plane. The only tools permitted in the construction are a compass and a straightedge (that is, a ruler with no markings).

The problem has its origins back in the beginnings of mathematics. Ancient Egyptian scholars of 1650 B.C.E. describe in the RHIND PAPYRUS a construction of a square of area nearly equal to that of a circle. (The construction is "exact" if one works with the approximate value 256/81 ≈ 3.1605 for π.) Scholar HIPPOCRATES OF CHIOS (ca. 440 B.C.E.) was interested in the problem and developed several methods for "squaring" a LUNE, a shape made from arcs of circles, but did not succeed in squaring the circle itself. ARCHIMEDES OF SYRACUSE (ca. 287–212 B.C.E.) used his curve the SPIRAL OF ARCHIMEDES to square the circle, providing a solution to the problem using tools more advanced than straightedge and compass alone. APOLLONIUS OF PERGA (ca. 262–190 B.C.E.) also solved the problem with the introduction of special curves and advanced techniques. The problem of constructing a square of the desired area using just a straightedge and compass alone, however, remained unsolved.

The problem garnered considerable notoriety over the centuries that followed. Indian, Chinese, and Arab scholars also attempted to solve the challenge, and succeeded in finding solutions to the problem if one used various approximate values for π. The problem was studied by European scholars of the Renaissance, and it became a popular problem of study for amateur mathematicians during the 18th and 19th centuries. Novices working on the challenge became known as "circle squarers" and often sought fame by submitting supposed solutions to the problem to the prestigious ROYAL SOCIETY of London. In the late 1700s, tired of being inundated with large numbers of tedious and incorrect theses on the subject, the society banned all consideration of alleged "proofs" of squaring the circle. The French Academy of Science followed suit soon afterward.

If we assume that the radius of the given circle is r, then one is asked to produce a square of side-length a so that $a^2 = \pi r^2$. This essentially reduces the problem to one of constructing a length $\sqrt{\pi}$ units long using a straightedge and a compass.

The theory of CONSTRUCTIBLE numbers shows that any quantity of rational length can be constructed, and that if two lengths l_1 and l_2 can be produced, then so too can their sum, difference, product, and quotient, along with the square root of each quantity. If one uses a rational value as an approximation for π, it is not surprising then that one can produce solutions to the problem using this approximate value. Moreover, if π itself is a rational number, then one would expect it possible to solve the problem. As no solution had ever been found, scholars began to suspect that π is irrational.

In the mid-1600s Scottish mathematician JAMES GREGORY (1638–75), in his studies of infinite series, attempted to show that π is not only irrational, but, moreover, that it is a TRANSCENDENTAL NUMBER. This would establish, once and for all, that $\sqrt{\pi}$ cannot possibly be constructed. Unfortunately, Gregory did not succeed in this goal. It was not until a century later that German mathematician JOHANN HEINRICH LAMBERT (1728–77) succeeded in proving that π is irrational, and another century after that that CARL LOUIS FERDINAND VON LINDEMANN (1852–1939) finally proved in 1880 that π is indeed transcendental. As a consequence, Lindemann had proved that the problem of squaring the circle is unsolvable.

It is interesting to note that in 1914 SRINIVASA AIYANGAR RAMANUJAN (1887–1920) used a ruler-and-compass construction akin to squaring a circle to find the following remarkable approximation for π correct to the ninth decimal place:

$$\pi \approx \left(9^2 + \frac{19^2}{22}\right)^{\frac{1}{4}}$$

squeeze rule (sandwich result) This rule asserts that if a function $f(x)$ is sandwiched between two other functions $g(x)$ and $h(x)$, at least for values x close to a number a (that is, we have $g(x) \le f(x) \le h(x)$), and if $\lim_{x \to a} g(x) = L$ and $\lim_{x \to a} h(x) = L$, then it must be the case that $\lim_{x \to a} f(x)$ exists and equals L as well. For example, the inequalities $1 - x \le f(x) \le 1 + x^2$ imply that $\lim_{x \to 0} f(x) = 1$.

Perhaps the most important application of the squeeze rule is the calculation of the limit $\frac{\sin x}{x}$, which arises in computing the DERIVATIVE of the sine function. In the diagram below we notice that the sector is sandwiched between two right triangles. Here the angle x is given in RADIAN MEASURE. We have:

$$\text{area of the small right triangle} = \frac{1}{2} \cdot \cos x \cdot \sin x$$

$$\text{area of the sector} = \frac{x}{2\pi} \cdot \pi 1^2 = \frac{\pi}{2}$$

(This is the fraction $\frac{x}{2\pi}$ of the area of a full circle of radius 1), and

$$\text{area of the large right triangle} \frac{1}{2} \cdot 1 \cdot \tan x.$$

From the picture, we see that $\frac{1}{2} \cdot \cos x \cdot \sin x \le \frac{x}{2} \le \frac{1}{2} \cdot \tan x$, which can be rewritten: $\cos x \le \frac{x}{\sin x} \le \frac{1}{\cos x}$.

Since $\lim_{x \to 0} \cos x = 1$ and $\lim_{x \to 0} \frac{1}{\cos x} = 1$, it follows by the squeeze rule that $\lim_{x \to 0} \frac{x}{\sin x}$ exists and equals 1. This proves:

If x is measured in radians, then $\lim_{x \to 0} \frac{\sin x}{x} = 1$.

(If x is measured in degrees, then the area of the sector is given by $\frac{x}{360} \cdot \pi$. Chasing through the argument above shows that, for x measured in degrees, $\lim_{x \to 0} \frac{\sin x}{x} = \frac{\pi}{180}$)

As an aside, the limit $\lim_{x \to 0} \frac{\cos x - 1}{x}$ follows:

$$\lim_{x \to 0} \frac{\cos x - 1}{x} = \lim_{x \to 0} \frac{(\cos x - 1)(\cos x + 1)}{x(\cos x + 1)}$$

$$= \lim_{x \to 0} \frac{\cos^2 - 1}{x(\cos x + 1)}$$

$$= \lim_{x \to 0} \frac{-\sin^2 x}{x(\cos x + 1)}$$

$$= \lim_{x \to 0} -\frac{\sin x}{x} \cdot \frac{\sin x}{\cos x + 1}$$

$$= -1 \cdot \frac{0}{1 + 1}$$

$$= 0$$

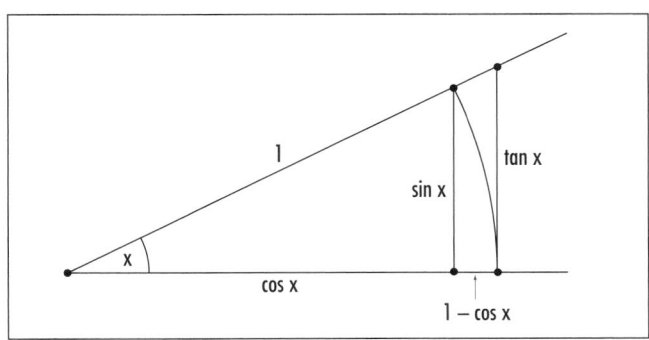

Comparing areas

This key result, needed for computing the derivative of the cosine function, can also be established geometrically by making use of the squeeze rule and the diagram above. (For x measured in degrees we also have $\lim_{x \to 0} \frac{\cos x - 1}{x} = 0$.)

Finally, we note that $\lim_{x \to 0} \frac{\tan x}{x} = \lim_{x \to 0} \frac{\sin x}{x} \cdot \frac{1}{\cos x} = 1 \cdot \frac{1}{1} = 1$ (For x measured in degrees we have $\lim_{x \to 0} \frac{\tan x}{x} = \frac{\pi}{180}$.)

See also APOTHEM; *E*; LIMIT.

standard deviation   *See* STATISTICS: DESCRIPTIVE.

statistics   Statistics is the branch of mathematics concerned with the methods of collecting, tabulating, and summarizing numerical facts (this is called descriptive statistics), and for making inferences and predictions based on these facts (inferential statistics). The numerical information gathered is called DATA, and an individual numerical fact about the data is called a statistic.

For example, a medical study might record the heights of 100 children, all age 8. The average height of the children would be an example of a statistic. Another statistic would be the tallest height recorded or the range of heights observed. Making a judgment based on the data that another child outside of the study is of abnormal height would be an example of using data for inferential purposes.

The word *statistik* was coined by the German political scientist Gottfried Achenwall (1719–72) to mean "a summary of how things stand." It is based on the Latin verb *stare* meaning "to stand."

Statistics is an indispensable tool used in practically every aspect of life today. Weather predictions are based on methods of statistical inference, for example, as are the assessed effectiveness of new drugs, new medical procedures, and other health practices. Statistics is used extensively in government, business, and commerce to analyze opinion polls, campaign and advertising strategies, business operations, pollution control, and other environmental concerns, for example, and as well as in scientific research and economic, political, and sociological studies. Insurance companies analyze LIFE TABLES to make inferences and to set insurance rates. At the turn of each decade, every household in the United States is required to complete a short census questionnaire. Government decisions on the apportionment of representation and of funds are based on the census results. In addition, a small percentage of households must complete a longer questionnaire, from which further statistical inferences about the entire population are made. In leisure, many sports fans follow statistical analyses to assess team and player performance.

Because statistics pervades so many areas of life, study of the subject is now a standard part of many high-school curricula.

See also HISTORY OF PROBABILITY AND STATISTICS (essay); POPULATION AND SAMPLE; STATISTICS: DESCRIPTIVE; STATISTICS: INFERENTIAL.

statistics: descriptive   The science of collecting, tabulating, and summarizing numerical information obtained from observational or experimental studies is called descriptive statistics. For example, a medical study might record the blood types of 100 army inductees and present the information obtained as lists or tables, or perhaps visually via charts, graphs, or frequency diagrams as described below. General features of the DATA, such as the most common blood group observed, or the shape of the frequency distribution observed, can be used to describe and summarize the information. Providing general descriptions of data allows one to draw conclusions about a particular population as a whole. As a simple example, one might extrapolate and deduce that a certain percentage of the entire world's population has a particular blood type.

In the example above, the measurements taken are descriptive and fall into precise categories: type A, type B, type AB, and type O. Such a study is said to be categorical. A numerical study, however, collects numerical information about participants (such as height, age, or weight), and placing the data into categories is a matter of choice. For example, one might wish to arrange the ages of in-patients at a busy metropolitan hospital into categories of decade (ages 0–10 years, 11–20 years, and so on) or some other convenient division. Measurements of height, for example, can adopt a continuous array of values, including fractional values. A study on human growth rates might have its data organized into categories of height ranges, for example, 60.1–65.0 in., 65.1–70.0 in., and so on.

Once categories have been established, there are a number of standard methods for presenting and summarizing data.

Ways of Presenting Data

A frequency distribution is a table or visual chart that shows the frequency (number) of individuals in each possible category considered. For example, of the 100 army inductees tested above, suppose 20 have blood type A, 27 blood type B, 16 blood type AB, and 37 blood type O. This information can be summarized in a table.

This frequency distribution can also be presented visually. A pie chart (also called a pie graph) is a circle divided into sections, one for each category, and each with area in proportion to the frequency of that category. A bar chart is a graph with vertical bars representing the frequency of occurrence for each category. If the data are categorical, then the bars are drawn spaced apart to emphasize the separate nature of the descriptive categories. If the data presented are numerical, the bars are drawn without spaces to indicate that the data come from a range of numerical values, and also to indicate how the data were grouped. In this case, the bar chart is called a histogram.

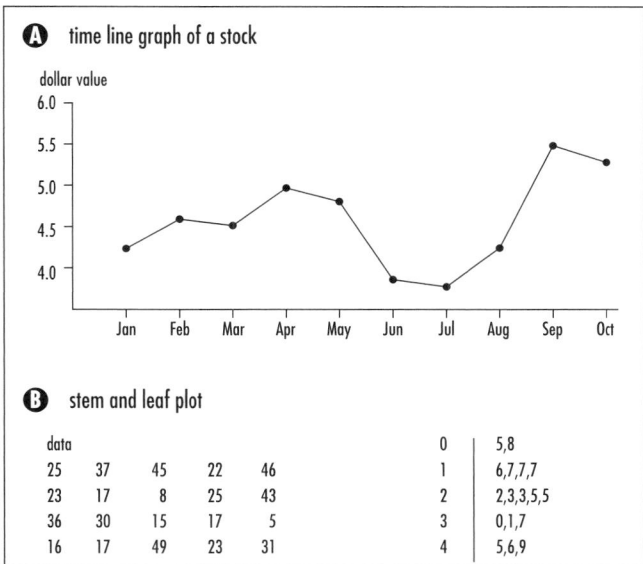

Plotting data

A frequency polygon is similar to a histogram except that a broken line is drawn to connect the midpoint of each class considered at the height of the vertical bar for that class. As a matter of convention, the line is drawn touching the horizontal axis at both sides of the distribution.

A time-series graph depicts the nature of a measurement taken over a period of time. The values of stocks, for example, are often depicted via time-line graphs. Such a diagram does not represent a frequency distribution, however.

Tables of whole-number values are sometimes summarized via stem-and-leaf plots. Each number is divided into two parts: the units digit (the "leaf") and the set of digits to its left (the "stem"). In one column all the stems are listed, and the corresponding leaves are arranged in a second column to the right.

Ways of Summarizing Data

There are three general features statisticians observe to summarize numerical data.

1. *The General Shape of a Histogram:* The following table shows some common frequency distribution shapes. Distributions rarely conform to exact shapes, but statisticians still find it useful to describe the general nature of the distribution.

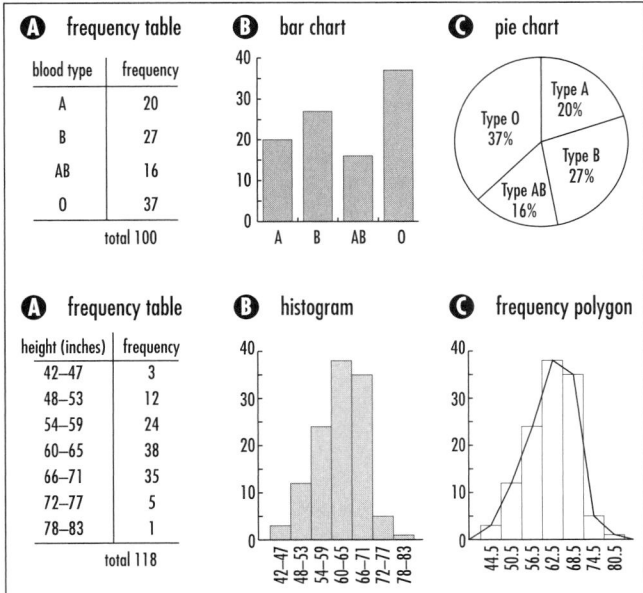

Representing data

A bell-shaped distribution has a single peak and is approximately symmetrical about both sides of the peak. A uniform distribution exhibits an equal number of measures in each category, and J-shaped and reverse J-shaped distributions exhibit increasing and decreasing trends. If a distribution has a single peak, but is not symmetrical, then it is called either positively or negatively skewed. If a distribution has two distinct peaks then it is called bimodal. (For example, incidence of broken limbs relative to age is bimodal, since accidents tend to occur more frequently among children and the elderly than they do the rest of the population.)

2. *Measures of Central Tendency:* A measure of "central tendency" is a single measurement that, in some sense, is typical of the entire data set. It represents the approximate "center" of the frequency distribution. There are four measures commonly in use.

a. The mean or average, usually denoted by the Greek letter μ, is found by summing together all the data values and dividing by the total number of measurements. It is equivalent to the ARITHMETIC MEAN of the data values.

For example, the mean of the four data values 4, 5, 8, 8 is: $\mu = \dfrac{4+5+8+8}{4} = 6.25$. If in another study the value 6 occurs 37 times and the value 9 occurs 20 times, the mean is: $\mu = \dfrac{37 \times 6 + 20 \times 9}{57}$ ≈ 7.05. In general, if a data value x_1 appears f_1 times, the data value x_2 a total of f_2 times, and so on down to the data value x_n appearing a total of f_n times, then the mean is given by:

$$\mu = \frac{f_1 \times x_1 + f_2 \times x_2 + \cdots + f_n \times x_n}{f_1 + f_2 + \cdots + f_n}$$

The sum of differences of each data value from the mean is always zero. For example, for the first data set presented above, we have $(4 - 6.25) + (5 - 6.25) + (8 - 6.25) + (8 - 6.25) = 0$.

The mean is the most commonly used measure of central tendency. (*See also* EXPECTED VALUE.)

b. The mode is the value in the data set that occurs most often. For example, from the 10 data values 3, 6, 5, 3, 1, 6, 5, 3, 8, 3 the mode is 3, and for 4, 5, 8, 8 the mode is 8. A distribution might have more than one mode if two or more scores occur an equal number of times.

The mode is used when the most typical case of a study is desired. For nonnumerical data, the mode is the only measure of central tendency available.

c. The median is the middle value of a sequence of data values, once they are arranged in order from smallest to largest. For example, the median of the data set 3, 3, 5, 6, 7, 16, 16, 19, 37 is 7. If the data set contains an even number of entries, then the average of the middle two values is taken as the median. For example, the median of 4, 5, 8, 8 is $\dfrac{5+8}{2} = 6.5$.

The median is useful for finding the value at the center of the distribution. It divides the data set into two equally sized groups.

d. The midrange of a data set is found by taking the average values of the smallest and largest data values that occur. For example, the midrange of the data set 4, 5, 8, 8 is $\dfrac{5+8}{2} = 6.5$.

The midrange provides a quick estimate to a central value. It is easy to compute, but is highly affected by extremely low or high values in the data set.

3. *Measures of Dispersion:* In order to interpret how well a measure of central tendency is likely to represent an entire group of data values, statisticians must also compute a measure of dispersion. Data values clustered around a central value can be well

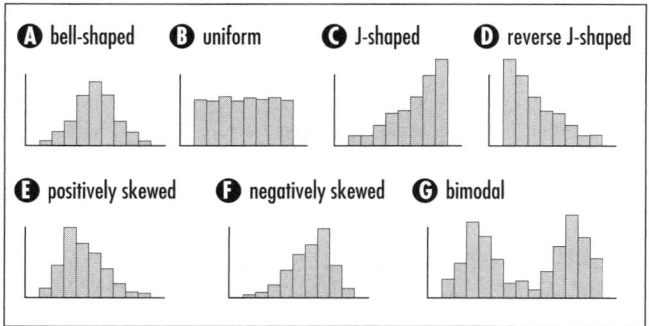

Classifying distributions

summarized by their mean or median, but a set of data values scattered about a large range of values is not well represented by a single measure of central tendency. For example, the two data sets 4.1, 4.1, 4.2, 4.3, 4.4 and 0.1, 0.3, 1.3, 7.8, 11.6 both have mean 4.22. This provides a good representative value for the first data set, but not the second.

The usual statistical measures of scatter are given as follows:

a. The range of a data set is the difference between the lowest and highest values in the set. For example, the range of the data set 4, 5, 8, 8 is 8 − 4 = 4.

   The range is a very simplistic measure of dispersion and does not reveal any information about how the data values are distributed. It is also highly affected by extremely low or high values in the data set. However, the range is often a useful measurement in practical daily issues. For example, weather forecasts usually give the range of temperatures to expect for the day.

b. The deviation of a single data value is the difference between that value and the mean of the data set, written as a positive quantity. For example, in the data set 4, 5, 8, 8 with mean 6.25, the deviation of the data value 5 is $|5 − 6.25| = 1.25$, and the data value 8 has deviation $|8 − 6.25| = 1.75$.

   The average deviation of all data values gives a good measure of overall scatter. For example, the data set 4, 5, 8, 8 has average deviation:

$$\frac{|4 − 6.25| + |5 − 6.25| + |8 − 6.25| + |8 − 6.25|}{4}$$
$$= \frac{2.25 + 1.25 + 1.75 + 1.75}{4} = 1.75$$

A subtle point should be noted. Given n data values $x_1, x_2, \ldots, x_n$, one first computes the mean μ, and then the n deviations: $|x_1 − \mu|, |x_2 − \mu|, \ldots, |x_n − \mu|$. Once the first $n − 1$ of these quantities are computed (and these could turn out to be of any value), the value of the nth quantity, however, is forced—the data set must conform to a mean μ. Thus there are only $n − 1$ "independent" computations to be made. For this reason mathematicians choose to divide the sum of deviations by

$n − 1$ rather than n. Thus a measure of scatter for the data set 4, 5, 8, 8, for example, is computed

$$\frac{|4 − 6.25| + |5 − 6.25| + |8 − 6.25| + |8 − 6.25|}{3}$$

≈ 2.33. If the number of data values is large, dividing by $n − 1$ rather than n will have little effect.

c. When measuring scatter, absolute values are mathematically difficult to work with (especially concerning the theoretical manipulations required in CALCULUS). For this reason, mathematicians prefer to work with deviations squared (again yielding positive quantities), and later applying the square root if desired.

   The variance of a data set is the sum of all deviations squared, divided by one less than the number of data values. For example, the variance of the four data values 4, 5, 8, 8 with mean 6.25 is:

$$\frac{(4 − 6.25)^2 + (5 − 6.25)^2 + (8 − 6.25)^2 + (8 − 6.25)^2}{3}$$
$$= 4.25$$

   Variance is usually denoted by the symbol σ^2 (read "sigma squared"). For n data values $x_1, x_2, \ldots, x_n$, it is given by the formula:

$$\sigma^2 = \frac{(x_1 − \mu)^2 + (x_2 − \mu)^2 + \ldots + (x_n − \mu)^2}{n − 1}$$

Because the deviations are squared, the further a data value is from the mean, the more pronounced its effect on the size of the variance.

d. Denoted by the Greek letter σ, the standard deviation of a data set is the square root of its variance. For example, the standard deviation of the four data values 4, 5, 8, 8 is $\sigma = \sqrt{4.25} \approx 2.06$. In general, for n data values $x_1, x_2, \ldots, x_n$, the standard deviation is given by the formula:

$$\sigma = \sqrt{\frac{(x_1 − \mu)^2 + (x_2 − \mu)^2 + \ldots + (x_n − \mu)^2}{n − 1}}$$

Standard deviation is much more commonly used than variance as a measure of dispersion. It is expressed in the same units as the data values (inches, if the measurements are heights for example, or grams if they are weights) rather than units squared, as is required for variance.

In a NORMAL DISTRIBUTION approximately 68 percent of the data values lie within one standard deviation of the mean (either side), 95 percent within two standard deviations, and 99.7 percent within three. This known as the 68–95–99.7 rule. It shows that standard deviation does indeed give a good indication of how widely the data values in a distribution are scattered. A small standard deviation, for example, indicates that 68 percent of the data values are closely clustered about the mean.

See also DISTRIBUTION; PERCENTILE; SCATTER DIAGRAM; STATISTICS: INFERENTIAL.

statistics: inferential   The science of drawing general conclusions about a population based solely on numerical information gathered from a sample of that population is called inferential statistics. (*See* POPULATION AND SAMPLE.) For example, a medical study might observe that 16 of 100 army inductees have blood type AB. One might then infer that approximately one-sixth of the world's entire population is of this blood type.

An assertion or conjecture about a numerical feature of a population is called a hypothesis. For example, the assertion "one-sixth of the world's population is of blood type AB" is a hypothesis that seems to be supported by the medical study described above. It is unclear, however, whether another study of 600 college seniors, 79 of whom were of blood type AB, also supports the claim.

A statistical test is a mathematical procedure that allows one to determine, to some specified degree of confidence, whether or not the results of a particular study support a hypothesis. The claim being tested for acceptance or rejection is called the null hypothesis. The alternative hypothesis is the assertion to be accepted if the null hypothesis is deemed false.

The principles of statistical testing are well summarized within the following example:

Suppose one is given a coin. We wish to determine whether or not the coin is fair, that is, whether tossing a head is just as likely as tossing a tail. We take as the null hypothesis the statement, "The coin is fair," and alternative hypothesis, "The coin is biased." To test the hypothesis let us say we toss the coin 10 times. Suppose we obtained 10 heads in a row. What should we conclude?

As the chances of tossing 10 heads in a row with a fair coin are very small—the probability of this occurring is $\left(\frac{1}{2}\right)^{10} = \frac{1}{1024} \approx 0.1\%$—we would perhaps conclude that the coin is biased. But there is a small chance that a fair coin could have nonetheless produced this result. To reflect this degree of uncertainty, we can say that we come to the conclusion that the coin is biased with a "99.9 percent level of confidence." The statistical test performed here was a probability calculation. We came to accept the null hypothesis with a 99.9 percent level of confidence.

The rejection of a null hypothesis when in fact it was true is known as a type I error (maybe the coin was fair). If the null hypothesis is accepted despite being false, a type II error is committed. A level of confidence (or significance level) of a statistical test is the probability of committing a type I error. A 95 percent level of confidence is generally deemed acceptable. (As a side note, the chances of tossing nine heads among a series of 10 tosses are $10 \times \left(\frac{1}{2}\right)^{10} = \frac{10}{1024} \approx 1\%$. If, in our experiment, this is what we observed, then we would conclude again that the coin is biased, with a 99 percent level of confidence. Observing eight heads among a series of 10 tosses leads to the same conclusion with a 95.6 percent level of confidence.)

The CENTRAL-LIMIT THEOREM provides the means for performing statistical tests useful for analyzing public surveys and polls. We present here two examples illustrating two common applications.

1. *Estimating Population Proportion:*

In a recent Gallup poll 1,500 people were surveyed, and 45 percent of them agreed that taxes on gasoline should be raised. To what extent does this figure represent the true proportion of all people in this country who hold this view?

Let p represent the true (but unknown) percentage of Americans who believe that gasoline taxes should be raised. Our task is to find the value of p. All we have to work with is the observation that one sample of 1,500 people produced a proportion $\hat{p}$ of 45 percent holding this view.

The central-limit theorem states that, for many different samples of 1,500 people, the statistic $\hat{p}$ will vary in value according to a normal distribution with mean p and standard deviation $\sqrt{\dfrac{p(100-p)}{N}}$ where, in this case, $N = 1,500$. As the value of N is large, the standard deviation is small, meaning that all values of $\hat{p}$ will be closely clustered about the mean value p. In particular, this establishes, as we would expect, that $\hat{p} = 45$ percent is a good estimate for p. The key is to now ask, How good?

From the study of the NORMAL DISTRIBUTION, the 68–95–99.7 rule states that 95 percent of the measurements for $\hat{p}$ fall within a distance of two standard deviations from the mean p. That is, there is a 95 percent chance that our measurement of $\hat{p} = 45$ percent lies within the range of values $p-2\sqrt{\dfrac{p(100-p)}{1500}}$ to $p+2\sqrt{\dfrac{p(100-p)}{1500}}$ (This range of values also contains p, of course, at its center.)

As an approximation, we substitute into these formulae the value $\hat{p} = 45$ percent for p:

$$p-2\sqrt{\frac{p(100-p)}{1500}} \approx \hat{p}-2\sqrt{\frac{\hat{p}(100-\hat{p})}{1500}}$$

$$= 45-2\sqrt{\frac{45\times 55}{1500}} \approx 42.4$$

$$p+2\sqrt{\frac{p(100-p)}{1500}} \approx \hat{p}+2\sqrt{\frac{\hat{p}(100-\hat{p})}{1500}}$$

$$= 45+2\sqrt{\frac{45\times 55}{1500}} \approx 47.6$$

This yields a range of values [42.4, 47.6] that, in this approximation, and with approximately a 95 percent level of confidence, contains the true proportion value p. We call this range of values a 95% confidence interval. If these calculations were performed on a large number of survey results (all involving 1,500 people) then we would be sure that close to 95 percent of the intervals produced contain the true population proportion p.

2. *Estimating a Population Mean:*

In a media study 680 young adults, ages 21 to 25 years, were given a test on current events.

Scores on the test ranged from 0 to 500, indicating a range of knowledge on the topic. The mean score was $m = 170$ (and the standard deviation was 80). On the basis of this sample, what can be said about the mean knowledge level (score) μ of the population of all 19 million young adults?

The central-limit theorem states that, for many different samples of 680 young adults taking the test, the mean score m will vary in value according to a normal distribution with mean μ and standard deviation $\dfrac{\sigma}{\sqrt{N}}$. Here σ is the standard deviation for the entire population (unknown) and N is the sample size ($N = 680$). Since the value of N is large, the standard deviation $\dfrac{\sigma}{\sqrt{N}}$ will be small. This means two things: that the mean $m = 170$ is likely to be close to the true mean value μ, and that using the standard deviation of 80 observed in the sample as an approximation for the true value σ will not seriously alter our calculations. With this said, the 68-95-99.7 rule states that there is a 95 percent chance that our observed value $m = 170$ falls within two standard deviations of the true mean value μ. As an approximation, then, we evaluate:

$$\mu-2\frac{\sigma}{\sqrt{N}} \approx 170-2\frac{80}{\sqrt{680}} \approx 163.9$$

$$\mu+2\frac{\sigma}{\sqrt{N}} \approx 170+2\frac{80}{\sqrt{680}} \approx 176.1$$

yielding a 95 percent confidence interval of [163.9, 176.1] for what would be the mean score if the entire population were to take the test.

In 1908 WILLIAM SEALY GOSSET (1876–1937), publishing under the pseudonym "Student," made a more precise analysis of the distribution of mean values from normal distributions. If $\hat{\sigma}$ is the standard deviation of a sample of size N, Gosset calculated the distribution of values for the sample mean m one would expect using $\hat{\sigma}$ as an approximation for the true standard deviation σ of the population. In particular, he described the distribution of the Z-SCORE of the mean m:

$$\frac{m-\mu}{\left(\dfrac{\hat{\sigma}}{\sqrt{N}}\right)}$$

This is known today as Student's t-distribution (with $N - 1$ degrees of freedom). Student's t-test is used to test whether any difference in the means of two different samples is statistically meaningful. For example, a study might indicate that the mean height of 100 randomly selected basketball players is 8 in. higher than the mean height of 100 baseball players. Student's t-test would test the hypothesis, "Both samples were drawn from the same normal population." If the value 8 in. is deemed too high, the hypothesis would be rejected and the difference in the means would be considered significant.

The F-test tests whether or not two samples come from the same population by focusing instead on the VARIANCE of each sample. If two samples of sizes N_1 and N_2 come from the same normal population, then the ratio of their variances $\dfrac{\hat{\sigma}_1^{\,2}}{\hat{\sigma}_2^{\,2}}$ should be approximately equal to 1. The F-distribution tabulates values of these ratios, and the F-test determines whether the observed ratio for two particular samples has an acceptable value.

See also CHI-SQUARE TEST; CORRELATION COEFFICIENT.

Steiner, Jakob (1796–1863) Swiss *Geometry* Born on March 18, 1796, in Utzenstorf, Switzerland, mathematician Jakob Steiner is remembered for his fundamental contributions to the study of PROJECTIVE GEOMETRY and for his work on the solution of the famous ISOPERIMETRIC PROBLEM. A collection of geometric points are today named in his honor, STEINER POINTS, to acknowledge his work in geometric OPTIMIZATION.

Steiner had no early formal education. It is said that he did not begin to read and write until he was 14, and did not attend any kind of school until age 18. During the latter part of his teen years, however, Steiner demonstrated a talent for mathematics, which earned him admission to the Johann Heinrich Pestalozzi school in Yverdon, Switzerland. Just 2 years after entering the school as a student, he was hired as a teacher of mathematics at the school.

Steiner entered the University of Heidelberg in 1818 and transferred to the University of Berlin 3 years later to pursue a research career in mathematics. His innovative work in geometry was duly noted by the mathematicians of the time. Steiner was awarded an honorary doctoral degree from the University of

Königsberg and the position as chair of mathematics at the University of Berlin in 1834. He held that post for the remainder of his life. He died in Berlin on April 1, 1863.

A number of finite configurations in projective geometry are named in his honor, as well as a geometric surface, the Steiner surface, which has the property that each of its tangent planes slices the surface in a pair of CONIC SECTIONS. His influence on the development of geometric optimization was profound.

Steiner point   Given three points A, B, and C in the plane, a Steiner point for that system is a point P whose sum of distances $AP + BP + CP$ is at a minimum. That such a point always exists was first established by Swiss mathematician JAKOB STEINER (1796–1863). He proved that this point occurs at the location where the angle between each of the line segments AP, BP, and CP is 120°. To see this, first suppose that the point P is placed at a fixed distance from C so that it lies on a circle with C as its center. Then a study of OPTIMIZATION shows that the location on the circle that minimizes the sum $AP + BP + CP$ occurs at the point P where the lines AP and BP make equal angles to the line tangent to the circle at P. Thus the solution to this restricted version of the problem occurs when the angles between line segments AP, BP, and CP are equal. The same occurs for a solution with P a fixed distance from A or P a fixed distance from B. As the solution to the general problem simultaneously solves each restricted version of the problem, all three angles must be equal. Since the three angles sum to 360°, each must therefore equal 120°.)

The Steiner point solves the following road-building problem:

> What design of a road system connects three towns using the minimal total length of road?

The solution is a design that connects each town with a straight segment of road directly to the Steiner point of the three towns.

Steiner also analyzed road-building problems that involve more than three towns. He showed that given N towns, $N \geq 3$, it is necessary to introduce $N - 1$ special points between the towns and draw straight-line segments between these points and the towns in such a way that:

1. Each town is connected to one Steiner point
2. Each Steiner point has three roads emanating from it, equally spaced 120° apart.

See also ISOPERIMETRIC PROBLEM.

stem-and-leaf plot (stem plot) *See* STATISTICS: DESCRIPTIVE.

stereographic projection Consider a PLANE tangent to a SPHERE touching the sphere at its south pole *S*. Then one can map points on the surface of the sphere to points on the plane by drawing straight lines from the north pole *N* of the sphere through points on the surface, and continuing them until they intercept the plane. Every point, except the north pole itself, is thus mapped to a point in the plane. This geometrical transformation of a sphere onto a plane is called the stereographic projection.

Every point on the plane corresponds to a unique point on the sphere, with points farther and farther away from *S* on the plane matching points closer and closer to *N* on the sphere. In this sense, the sphere can be regarded as topologically equivalent to a plane with a single well-defined additional point of infinity attached to it. If the plane is taken to be a representation of the plane of COMPLEX NUMBERS, then the sphere in this construct is usually called a Riemann sphere to honor the work of German mathematician GEORG FRIEDRICH BERNHARD RIEMANN (1826–66) in this field.

A stereographic projection is a CONFORMAL MAPPING. This means that it preserves angles between intersecting curves on the surface of the sphere. Great circles on the sphere, not through *N*, are mapped to circles in the plane, and great circles through *N* are mapped to straight lines. For this reason, geometers often deem it appropriate to regard straight lines as special types of circles.

A gnomonic projection maps points on the southern hemisphere of a sphere onto a plane tangent to *S* by drawing straight lines from the center of the sphere through points on the surface, and continuing them until they intercept the plane. In this model, each point on the equator of the sphere represents a different "point of infinity" attached to the plane.

See also MERCATOR'S PROJECTION; PROJECTION.

Stevin, Simon (1548–1620) Flemish *Arithmetic, Engineering* Born in Flanders, now Belgium, in 1548 (his exact birth date is not known), Simon Stevin is best remembered for *The Tenth*, his influential 1585 text that advocates and explains the use of decimals in all of mathematics and accounting. Although he did not invent the decimal system (it had been used by the Arabs two centuries before), his expository piece on the subject convinced scholars of the time of its merit as an approach to manipulating fractions and real numbers.

Stevin started his career as a bookkeeper and a tax office clerk before entering the University of Leiden at the age of 35. His work on mechanics and engineering garnered him note as an expert in hydrostatics and its related mathematics. Stevin wrote 11 texts in all throughout his academic career, covering topics in arithmetic, algebra, trigonometry, geography, and navigation. In his text *Principles of the Art of Weighing*, Stevin analyzed the geometric addition of forces, developing an approach that simplified the mathematics of mechanics. This approach also formed the basis for the theory of VECTOR analysis developed 200 years later—Stevin had essentially correctly defined vector addition. Stevin also recognized that the distance an object falls in a fixed amount of time is independent of the object's weight. This discovery is normally attributed to the scholar GALILEO GALILEI (1564–1642), but Stevin had reached the same conclusion 3 years before Galileo reported his findings.

By the 15th century, mathematicians in Persia, China, and India were using the decimal system to represent fractions. Stevin recognized the advantages of the system and argued, in his piece *The Tenth*, that adding decimal fractions is just as easy as adding whole numbers. (Summing 0.73 and 0.25, for instance, is no more difficult than adding 73 to 25). He did not, however, use a decimal point in his work, choosing to write 34.875, for instance, as 340817253, circling the digits 0, 1, 2, and 3 to indicate the digits to the left are multiplied by those powers of one-10th. (Scottish mathematician JOHN NAPIER (1550–1617) is responsible for popularizing the use of the decimal point.)

Writing all numbers as decimal fractions had the profound psychological effect of placing all numbers on an equal footing, as it were. The number π, for example, written as 3.141 (at least as an approximation) appears no more important or special than any other real number, 6.234 or 84.668, for instance. One scholar at the time called all numbers consequently "equally boring."

Stevin is also remembered for introducing and popularizing the symbols "+," "−," and "√ ‾" for addition, subtraction, and the square-root operation. He died in The Hague, Netherlands, ca. March 1620.

Stirling's formula   The surprising formula:

$$\lim_{n \to \infty} \frac{n!}{\sqrt{2\pi n}\left(\frac{n}{e}\right)^n} = 1$$

shows that for large values of n, the function $\sqrt{2\pi n}\left(\frac{n}{e}\right)^n$ is a very good approximation for the FACTORIAL $n!$. This limit equation is called Stirling's formula. Although named after the Scottish mathematician James Stirling (1692–1770), the formula was discovered by French mathematician ABRAHAM DE MOIVRE (1667–1754) while attempting to write a formula for the NORMAL DISTRIBUTION curve. Stirling wrote an influential paper on infinite series that helped De Moivre discover this result.

The formula can be derived from WALLIS'S PRODUCT. We present here a simple argument that gives a sense of how the result could be true. Noting that ln $n!$ can be written as a sum of logarithms, we have:

$$\ln n! = \ln 1 + \ln 2 + \cdots + \ln n = \sum_{k=1}^{n} \ln k \approx \int_1^n \ln x \, dx$$

Evaluating the integral, using INTEGRATION BY PARTS, yields:

$$\int_1^n \ln x \, dx = x \ln x - x \Big|_1^n = n \ln n - n + 1 \approx n \ln n - n$$

Thus ln $n! \approx \ln(n^n) - n$, giving $n! \approx n^n e^{-n} = \left(\frac{n}{e}\right)^n$. A more refined argument yields Stirling's result.

Student's t-test   *See* STATISTICS: INFERENTIAL.

subfactorial   In his 1878 study of PERMUTATIONS, mathematician W. Allen Whitworth introduced the notion of a subfactorial. Now denoted n_i, the nth subfactorial of an integer n is given by

$$n!\left(1 - \frac{1}{1!} + \frac{1}{2!} - \frac{1}{3!} + \frac{1}{4!} - \cdots \pm \frac{1}{n!}\right)$$

with 0_i set equal to 1. (*See* FACTORIAL.) The value n_i is obtained by multiplying the previous subfactorial number by n and adding $(-1)^n$. Thus:

$$1_i = 1 \times 0_i - 1 = 1 \times 1 - 1 = 0$$
$$2_i = 2 \times 1_i + 1 = 2 \times 0 + 1 = 1$$
$$3_i = 3 \times 2_i - 1 = 3 \times 1 - 1 = 2$$
$$4_i = 4 \times 3_i + 1 = 4 \times 2 + 1 = 9$$

for example. A study of the number e shows that the ratio $\frac{n_i}{n!}$ approaches the value $\frac{1}{e}$ as n becomes large.

Subfactorials arise in the study of certain permutations called derangements.

See also E.

substitution   The act of replacing all occurrences of a variable by another variable, expression, or numerical value is called substitution. For example, to evaluate the expression $x^2 - 2x + 3$ for $x = 5$, simply replace all occurrences of x with the value 5. This yields the corresponding value of 18 for the expression. (We say that we have "substituted the value 5 for x.") One can also substitute other quantities for x. For instance, replacing each occurrence of x with the quantity $y - 1$ yields the expression $(y - 1)^2 + 2(y - 1) + 3$, which simplifies to $y^2 + 2$. (This particular substitution proves to be fruitful, for it makes clear now that the expression $x^2 - 2x + 3$ will never adopt a value smaller than 2, and that the expression equals this minimal value only when $y = 0$, that is, when $x = 1$.) Substitution techniques are used to solve SIMULTANEOUS LINEAR EQUATIONS.

The method of substitution can be applied simultaneously to more than one variable in an equation. For instance, the AREA A of a triangle with side-lengths a, b, and c, with angle θ between sides of lengths a and b, is given by $A = \frac{1}{2}ab\sin\theta$. The LAW OF COSINES asserts $c^2 = a^2 + b^2 - 2ab \cos \theta$. Solving for sin θ and cos θ in each expression and substituting the results into the equation $\sin^2\theta + \cos^2\theta = 1$ yields HERON'S FORMULA.

The square root of 2 satisfies the equation:

$$\sqrt{2} = 1 + \cfrac{1}{1 + \sqrt{2}}$$

Substituting this formula into itself multiple times yields the following CONTINUED FRACTION expansion for $\sqrt{2}$:

$$\sqrt{2} = 1 + \cfrac{1}{1 + 1 + \cfrac{1}{1 + \sqrt{2}}} = 1 + \cfrac{1}{2 + \cfrac{1}{2 + \cfrac{1}{1 + \sqrt{2}}}}$$

$$= \dots = 1 + \cfrac{1}{2 + \cfrac{1}{2 + \cfrac{1}{2 + \cfrac{1}{2 + \dots}}}}$$

See also INTEGRATION BY SUBSTITUTION; TRANSFORMATION OF COORDINATES.

substitution rule for integration See INTEGRATION BY SUBSTITUTION.

subtraction The process of finding the DIFFERENCE of two numbers is called subtraction. In the elementary ARITHMETIC of WHOLE NUMBERS, subtraction can be thought of as the process of removing a subset from a set. For example, if three apples are removed from a set of eight apples, five apples remain. We write $8 - 3 = 5$.

In a more general context, subtraction is best defined as the inverse operation of ADDITION: the difference $a - b$ is defined to be a quantity that, when added to b, gives the answer a. Here a is called the minuend, b the subtrahend, and the result $a - b$ the difference. (Thus the subtrahend plus the difference gives the minuend.) In some contexts it is convenient to regard subtraction simply as the addition of NEGATIVE NUMBERS. For instance, the difference $8 - 3$ may be viewed as a sum $8 + (-3)$.

The process of subtraction does not satisfy the COMMUTATIVE PROPERTY. For example, $8 - 3$ does not yield the same result as $3 - 8$. (This is clear if one rewrites these differences in terms of sums of negative quantities: $8 + (-3)$ and $3 + (-8)$ are sums of two different pairs of numbers.)

The PLACE-VALUE SYSTEM we use today for writing numbers simplifies the process of subtracting large integers. For instance, subtracting 216 from 589 yields 5

$- 2 = 3$ units of 100, $8 - 1 = 7$ units of 10, and $9 - 6 = 3$ units of 1. Thus, $589 - 216 = 373$. This process is still valid even if one encounters negative quantities of units. For example, $463 - 198$ may be computed as 3 |–3| –5, where vertical bars are used to separate powers of 10. "Borrowing" 1 unit of 100 from the first column, which is equivalent to 10 units of 10 in the second column, allows us to rewrite this as 2 |7| –5. Borrowing 1 unit of 10 from the second column, which is equivalent to 10 single units in the third column, now permits us to rewrite this as 2 |6| 5. Thus we have: $463 - 198 = 265$. Students in schools are usually taught an algorithm that has one borrow digits early in the process of completing a subtraction problem rather than leave this work as the final step. Either method is valid.

The process of subtraction can be extended to FRACTIONS (completed with the aid of computing COMMON DENOMINATORS), REAL NUMBERS, COMPLEX NUMBERS, VECTORS, and MATRICES.

The difference of two real-valued functions f and g is the function $f - g$, whose value at any input x is the difference of the outputs of f and g at that input: $(f - g)(x) = f(x) - g(x)$. For example, if $f(x) = x^2 + 2x$ and $g(x) = 5x + 7$, then $(f - g)(x) = x^2 + 2x - 5x - 7 = x^2 - 3x - 7$.

The subtraction formulae in TRIGONOMETRY assert:

$$\sin(x - y) = \sin x \cos y - \cos x \sin y$$

$$\cos(x - y) = \cos x \cos y + \sin x \sin y$$

$$\tan(x - y) = \frac{\tan x - \tan y}{1 + \tan x \tan y}$$

The mathematician FIBONACCI (ca. 1175–1250) introduced the term *minus* for subtraction in his 1202 text *Liber abaci* (The book of the abacus), which scholars later abbreviated to $\bar{m}$ in their own work. It has been suggested that perhaps the letter "m" was later dropped to leave the bar "–" as the symbol of choice for subtraction. This symbol first appeared in print in Johannes Widman's 1489 book *Behennde und hüpsche Rechnung auf fallen Kauffmannschaften* (Neat and handy calculations for all tradesman).

See also ARITHMETIC.

sufficient condition See CONDITION—NECESSARY AND SUFFICIENT.

summation The process of finding the sum of a collection of numbers is called summation. Mathematicians are often interested in problems involving the sum of a large collection of quantities, and a compact notation for writing such sums is helpful. The Greek letter Σ (representing "S" for "sum"), the so-called sigma notation invented by Swiss mathematician LEONHARD EULER in 1755, can be used to express the

sum of the numbers $a_1, a_2, \ldots, a_n$ as $\sum_{k=1}^{n} a_k$. One reads this as "the sum from $k = 1$ to n of a_k," and computes it via the following procedure:

Write the terms a_k, first with k replaced by 1, and then with k replaced by 2, stopping with k replaced by n. Add together all the terms written: $\sum_{k=1}^{n} a_k = a_1 + a_2 + \ldots + a_n$

Thus, for example, the expression $\sum_{k=1}^{4} k^2$ represents the sum $1^2 + 2^2 + 3^2 + 4^2$, which equals 30.

The letter k used as a subscript in sigma notation is called the index of summation and is a DUMMY VARIABLE: any other letter could be used in its stead. For example, the expressions $\sum_{r=1}^{4} r^2$ and $\sum_{n=1}^{4} n^2$ also represent the same sum $1^2 + 2^2 + 3^2 + 4^2$.

A summation need not begin with index $k = 1$. For example, the expression $\sum_{k=10}^{12} (k^3 - k^2)$ represents the sum $(10^3 - 10^2) + (11^3 - 11^2) + (12^3 - 12^2)$. A summation could contain just a single term, as for $\sum_{k=5}^{5} k^2 = 5^2$, for instance. Any summation of the form $\sum_{k=m}^{n} a_k$ with $n < m$ is considered empty and to have value zero. Finally, sums alternating in sign can be represented by introducing a factor of the form $(-1)^{k-1}$. For example, $\sum_{k=1}^{5} (-1)^{k-1} \frac{1}{k}$ equals $1 - \frac{1}{2} + \frac{1}{3} - \frac{1}{4} + \frac{1}{5}$.

Summations satisfy the following two basic properties:

$$\sum_{k=m}^{n} (a_k \pm b_k) = \left(\sum_{k=m}^{n} a_k \right) \pm \left(\sum_{k=m}^{n} b_k \right)$$

$$\sum_{k=m}^{n} (c \cdot a_k) = c \cdot \left(\sum_{k=m}^{n} a_k \right)$$

where c is a constant. (These statements are patently true when the summations are written out in full.)

The sigma notation is also used to denote SERIES, that is, infinite sums:

$$\sum_{k=1}^{\infty} a_k = a_1 + a_2 + a_3 + \cdots$$

For example, the GEOMETRIC SERIES can be written: $\sum_{k=0}^{\infty} x^k = 1 + x + x^2 + x^3 + \ldots$. If $-1 < x < 1$, then it has value $\frac{1}{1-x}$, in which case we can write: $\sum_{k=0}^{\infty} x^k = \frac{1}{1-x}$.

GOTTFRIED WILHELM LEIBNIZ (1646–1716), coinventor of CALCULUS, used an elongated "S," ∫, both for the sum of a sequence of integers and as a sign of integration (which he thought of as continuous summation).

See also ARITHMETIC SERIES; CONVERGENT SERIES; INFINITE PRODUCT; SUMS OF POWERS.

sums of powers   The following formulae, called the sums of powers formulae, give simple equations for the sums of the first few counting, square, and cube numbers. They can readily be proved by the method of INDUCTION:

$$1 + 2 + 3 + \cdots + n = \frac{n(n+1)}{2}$$

$$1^2 + 2^2 + 3^2 + \cdots + n^2 = \frac{n(n+1)(2n+1)}{6}$$

$$1^3 + 2^3 + 3^3 + \cdots + n^3 = \frac{n^2(n+1)^2}{4}$$

It is also possible to establish these formulae geometrically.

The validity of the first formula can be seen by dividing a $n \times (n + 1)$ grid of squares into two "staircases," each representing the same sum $1 + 2 + 3 + \ldots + n$. This gives:

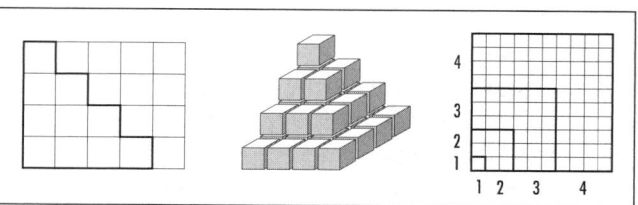

Computing sums of powers

$$2 \times (1 + 2 + 3 + \ldots + n) = n \times (n + 1)$$

In a similar vein, one can show that six copies of a three-dimensional "staircase" stack together to form an $n \times (n + 1) \times (2n + 1)$ rectangular box, thereby establishing the validity of the sum-of-squares formula.

Finally, each L-shaped region in a subdivided $(1 + 2 + \ldots + n) \times (1 + 2 + \ldots + n)$ square has area given by a cube number. Summing together these areas establishes the sum-of-cubes formula.

ARCHIMEDES OF SYRACUSE (ca. 287–212 B.C.E.) knew the formula for the sum of the first n counting numbers. Arab scholars, who translated and preserved the works of Archimedes, were thus also aware of the result. CARL FRIEDRICH GAUSS (1777–1855) is said to have discovered the same formula as a school student, employing a clever trick of writing the same sum forward and backward. (This is now considered a standard technique in the study of ARITHMETIC SEQUENCES.)

In 1713 Swiss mathematician Jacques Bernoulli of the famous BERNOULLI FAMILY searched for a general formula for the sum of the first few kth powers: $1^k + 2^k + 3^k + \ldots + n^k$. He noticed that this sum appears as the coefficient of x^k in the TAYLOR SERIES of

$$k!(1 + e^x + e^{2x} + \ldots + e^{nx})$$

(To see this, write out each term e^{mx} as a series $1 + mx + \dfrac{m^2 x^2}{2!} + \ldots$ Then add and collect all terms containing the expression x^k.) Using the formula for the sum of a GEOMETRIC SEQUENCE, this expression can be rewritten as:

$$k! \cdot \frac{\left(e^x\right)^{n+1} - 1}{e^x - 1} = k! \cdot \frac{e^{(n+1)x} - 1}{x} \cdot \frac{x}{e^x - 1}$$

The first factor here is $k!$, and the Taylor series of the second factor is straightforward:

$$\frac{e^{(n+1)x} - 1}{x} = (n+1) + (n+1)^2 \frac{x}{2!} + (n+1)^3 \frac{x^2}{3!} + \cdots$$

Unfortunately, the Taylor series of the third factor is difficult to compute. Bernoulli decided to simply write:

$$\frac{x}{e^x - 1} = B_0 + B_1 x + B_2 \frac{x^2}{2!} + B_3 \frac{x^3}{3!} + \cdots$$

for some numbers B_0, B_1, B_2, ... yet to be determined. Given this, it now follows that $1^k + 2^k + 3^k + \ldots + n^k$ is the coefficient of x^k in the product:

$$k! \cdot \left((n+1) + (n+1)^2 \frac{x}{2!} + (n+1)^3 \frac{x^2}{3!} + \cdots \right)$$
$$\cdot \left(B_0 + B_1 x + B_2 \frac{x^2}{2!} + B_3 \frac{x^3}{3!} + \cdots \right)$$

Expanding brackets and identifying the coefficient of x^k in this product eventually leads to the remarkable formula:

$$1^k + 2^k + 3^k + \cdots + n^k = \frac{(n+1+B)^{k+1} - B^{k+1}}{k+1}$$

where a quantity B^r is to be interpreted as the value B_r.

The numbers B_0, B_1, B_2, ... are today known as the Bernoulli numbers, and their values are given by:

$$B_0 = 1, \; B_1 = -\frac{1}{2}, \; B_2 = \frac{1}{6}, \; B_4 = -\frac{1}{30}, \; B_6 = \frac{1}{42}, \ldots$$

with $B_k = 0$ for all odd values of k greater than 1. Thus, as an example, we again have:

$$1^2 + 2^2 + 3^2 + \cdots + n^2 = \frac{(n+1+B)^3 - B^3}{3}$$
$$= \frac{(n+1)^3 B_0 + 3(n+1)^2 B_1 + 3(n+1)B_2 + B_3 - B_3}{3}$$
$$= \frac{(n+1)^3 - \frac{3}{2}(n+1)^2 + \frac{1}{2}(n+1)}{3}$$
$$= \frac{n(n+1)(2n+1)}{6}$$

The Bernoulli numbers have since been studied extensively, and many of their properties are now well understood. These numbers make an appearance in evaluating the ZETA FUNCTION at even values. Specifically, mathematicians have shown that:

$$\zeta(2m) = (-1)^{m+1} B_{2m} \frac{(2\pi)^{2m}}{2 \cdot (2m)!}$$

This gives, for example, $\zeta(2) = \frac{\pi^2}{6}$ and $\zeta(4) = \frac{\pi^4}{90}$.

surd A numerical expression containing irrational numbers that arise solely from the operation of taking square or higher roots is called a surd. For example, $\sqrt{5}$, $2 - \sqrt[3]{7}$, and $4^{\frac{2}{5}}$ are all surds. (However, the irrational number $e + \ln 2$, for instance, contains no roots and so is not a surd.) A pure surd contains only irrational root terms (such as $\sqrt{11} + \sqrt[4]{13}$, for instance) and a mixed surd contains both rational and irrational terms (such as $2 - \sqrt[3]{7}$, for instance).

The conjugate of a surd that is the sum of two terms is the difference of those same two terms. For example, the conjugate of $2\sqrt{3} + 4\sqrt{5}$ is $2\sqrt{3} - 4\sqrt{5}$. If only square roots are involved, then the product of a surd and its conjugate is always a rational number. This observation is utilized in the process of RATIO-NALIZING THE DENOMINATOR of a complicated rational expression.

Surds of the form $\frac{a \pm \sqrt{b}}{c}$, where a, b, and c are integers, with b not a perfect square, are sometimes called quadratic surds. The term *surd* is rarely used today and is generally considered obsolete.

syllogism *See* ARGUMENT.

symmetry A figure or an expression is said to be symmetrical if parts of it can be interchanged without changing the figure or the expression as a whole. For example, a geometric figure such as the letter "W" can be divided into two parts, the left half and the right half, which can be interchanged via a REFLECTION about a vertical line to leave the overall shape of the figure unchanged. We say that this figure has reflection symmetry about a vertical line. In the same way, the letter "C" has reflection symmetry about a horizontal line and the letter "S" has ROTATION symmetry about its center point. The figure of a CIRCLE has reflection symmetry about any line that passes through its center and rotational symmetry through any degree of turning about its center. A CUBE has reflection symmetry about any plane through its center parallel to one of its faces.

A geometric figure is said to have n-fold rotational symmetry about an axis of rotation if it is symmetrical about a rotation of $\frac{360°}{n}$ about that axis. For example, a regular pentagon has five-fold rotational symmetry about a line through its center PERPENDICULAR to the page on which the figure is drawn. A TETRAHEDRON has three-fold rotational symmetry about an axis that passes through one of its vertices, and two-fold rotational symmetry about an axis that passes through the midpoints of two opposite edges.

The equation $x^2 + xy + y^2$ is symmetrical in the variables x and y, meaning that interchanging the two variables yields a new equation equivalent to the original. In general, a function f of several variables is said to be totally symmetric if interchanging any two variables does not change the function: $f(x_1, \ldots, x_i, \ldots, x_j, \ldots, x_n) = f(x_1, \ldots, x_j, \ldots, x_i, \ldots, x_n)$ for all values i and j.

The set of all PERMUTATIONS of n letters $x_1, x_2, \ldots, x_n$ forms a GROUP called the nth order symmetric group. Any POLYNOMIAL equation $p(x) = 0$ is symmetrical with respect to permutations of the roots of the equation. If the polynomial is completely factored (as is possible according to the FUNDAMENTAL THEOREM OF ALGEBRA), then any rearrangement of the roots of the equation $(x - \alpha_1)(x - \alpha_2) \ldots (x - \alpha_n) = 0$ does not alter the equation.

A FUNCTION $y = f(x)$ whose graph is symmetrical about the y-axis via a reflection is said to be even. One that is symmetrical about a rotation of $180°$ about the origin is said to be odd. Even functions satisfy $f(-x) = f(x)$ for all values x, and odd functions $f(-x) = -f(x)$.

A square MATRIX A is said to be symmetrical if, for each appropriate value i and j, the entry in the ith row and jth column matches the entry in the jth row and ith column: $A_{ij} = A_{ji}$.

A RELATION is said to be SYMMETRICAL if x being related to y means that y is also related to x. For example, "is a sibling of" is a symmetrical relation among people. (If Lashana is Terell's sibling, then Terell is also Lashana's sibling.) The relation, "is a sister of," however, is not symmetrical.

A figure or expression that is not symmetrical is called asymmetrical.

See also EVEN AND ODD FUNCTIONS; FRIEZE PATTERN.

synthetic division   *See* NESTED MULTIPLICATION.

system of equations   A set of equations in several unknowns—required to be true for particular values of those unknowns—is called a system of equations. For instance, the three equations

$$x^2 - 3xy + z^4 = 11$$
$$3x(y + z) = z^2(x + y)$$
$$x^3 + y^3 + z^3 = 17$$

has the solution $x = 1$, $y = 2$, and $z = 2$. Geometrically, each equation represents a curve or surface in space, and a solution to the system of equations is a common point of intersection of those surfaces. It is possible for a system of equations to have no solutions. (For instance, there are clearly no values of x and y for which $x^2 + y^2 = 1$ and $y^2 + x^2 = 2$.)

Solving an arbitrary system of equations, if at all possible, is usually a very difficult task. Often a mathematician will resort to graphical methods and search for a GRAPHICAL SOLUTION. In special cases, a method of SUBSTITUTION might prove useful. For instance, consider the equations:

$$x^2 + y^2 = 25$$
$$x - y = 1$$

Solving for y in the second equation yields $y = x - 1$. Inserting this value into the first equation gives a QUADRATIC equation in x, namely, $x^2 + (x - 1)^2 = 25$. This can be rewritten $2x^2 - 2x + 1 = 25$ or as $2x^2 - 2x - 24 = 0$. Solving for x yields two solutions, with corresponding values for y. We have $x = 4$ and $y = 3$ as one solution, and $x = -3$ and $y = -4$ as another. (Graphically, we have found the location of the two points at which the straight line $x - y = 1$ intersects the CIRCLE $x^2 + y^2 = 25$.)

The study of LINEAR ALGEBRA and the process of GAUSSIAN ELIMINATION provide effective, straightforward means to solve systems of SIMULTANEOUS LINEAR EQUATIONS.

See also CRAMER'S RULE; HISTORY OF EQUATIONS AND ALGEBRA (essay).

T

tangent A straight line that touches a curve without cutting through it at that location is called a tangent line to the curve, and the location at which the line and curve touch is called the point of contact. With this said, however, this should not be taken as the precise definition of a tangent line. In general, a tangent line is permitted to have more than one point of contact, may cut through the curve at points other than the point of contact, and may even cut through the curve at the point of contact if this point is also an INFLECTION POINT. To include these possibilities, mathematicians define the tangent line to a curve in terms of a LIMIT:

> The tangent line to a curve at a point P is the limiting position of a line through P and through a second point Q on the curve as that point Q slides toward P.

But even this definition has some difficulties, for it may be the case that the limiting position is not well-defined. There is no unique tangent line to the corner of a square, for example.

A curve is called smooth if every point on it has a well-defined tangent line at that point. The theory of DIFFERENTIAL CALCULUS is used to find the SLOPE of a tangent line to a curve.

If a curve crosses itself, then special names are given to the point of intersection according to the behavior of the tangent lines at that point. A point of self-intersection is called a node (or a crunode) if there are two distinct tangent lines at that point, and a tac-point if the two tangent lines coincide.

If two different branches of a curve touch at a point and have a common tangent line at that point, then that point is called a cusp. A single cusp has two branches meeting and terminating at a point, and a double cusp has two branches that each continue through the point of contact. A cusp is classified as being of the "first kind" if the branches of the curve belong to opposite sides of the tangent line, or of the "second kind" if they lie on the same side (at least near the point of contact). If one branch of a double cusp has a point of inflection at the point of contact, then the double cusp is called an osculinflection.

A tangent plane to a surface is a plane that touches the surface without cutting through it. This informal definition can be adjusted and made more precise by considering the limit position of planes passing through three points on the surface. If a solid SPHERE rests on a tabletop, for example, then the surface of the table is a tangent plane for the solid. At the point of contact, all points near that point lie on the same side of the tangent plane. A point on a surface is called a saddle point if points near it lie on either side of the tangent plane at that point. (The shape of the surface at such a point resembles a horse's saddle.)

In geometry, any two figures are said to be tangent at a point P if they touch, but do not cut through each other, at P. In TRIGONOMETRY, the tangent of an angle A, denoted $\tan(A)$, is the ratio $\dfrac{\sin(A)}{\cos(A)}$.

See also NORMAL TO A SURFACE; PARABOLOID.

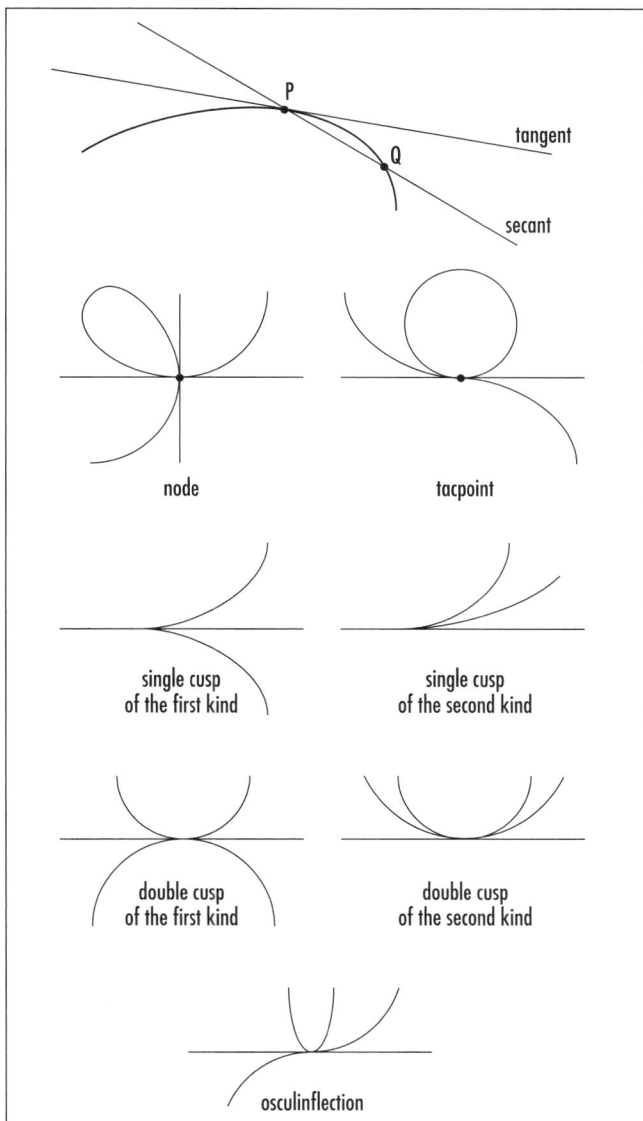

Tangents, secants, and points of contact

scholar to translate and publish EUCLID'S famous treatise *THE ELEMENTS*.

Tartaglia, born Niccolò Fontana, was only a boy when the French invaded his hometown in 1512. During the massacre Tartaglia received severe sword wounds to his jaw and palate and was left for dead in a cathedral. He survived, but thereafter could only speak with difficulty, hence his nickname, Tartaglia, meaning "stammerer."

During the Renaissance, scholars were mostly supported by rich patrons and had to prove their talent by defeating other scholars in public competitions. For this reason, mathematicians would keep their methods secret. In 1535 Tartaglia discovered a means for solving cubic equations of the form $x^3 + ax^2 = b$ and began publicizing his achievement. Earlier, SCIPIONE DEL FERRO (1465–1526) had revealed to his student Antonio Maria Fiore (ca. 1506–60) a method of solving

Niccolò Tartaglia, an eminent mathematician of the 16th century, found a general method of solution to cubic equations. (Photo courtesy of the Science Museum, London/Topham-HIP/The Image Works)

Tartaglia, Niccolò (1499–1557) Italian *Algebra* Born in Brescia, Italy, in 1499 (his exact birth date is not known), scholar Niccolò Tartaglia is remembered as one of the first mathematicians to discover the long-sought solution to the general CUBIC EQUATION. His method was published, without permission, by GIROLAMO CARDANO (1501–76), to whom he had revealed the details in confidence. Tartaglia also published the influential three-volume text *General trattato di numeri et misure* (Treatise on numbers and measures), and was the first Italian cubics of the type $x^3 + cx = d$. Seeking fame, Fiore challenged Tartaglia to a debate and presented him with 30 questions of a type that he was sure Tartaglia would be unable to answer. In the early hours of February 13, 1535, the day of the debate, inspiration came to Tartaglia, and he discovered a general method that solved both types of equations. Tartaglia won the contest with ease.

News of Tartaglia's victory reached Cardano, who, with his assistant LUDOVICO FERRARI (1522–65), was working on the solution to the QUARTIC EQUATION. This work required knowledge of the cubic solution, and so Cardano sought from Tartaglia details of his methods. After much beseeching, Tartaglia eventually revealed his solution to Cardano, under the promise that the methods remain secret. After discovering that another scholar, Scipione del Ferro, had devised identical methods decades earlier, Cardano broke that promise and published the solution, along with the solution to the quartic, in his text 1545 text *Ars magna* (The great art). A bitter dispute between Tartaglia, Cardano, and Ferrari ensued. Today, to honor the achievement of both men, the formula for the solution to the cubic is called the Cardano-Tartaglia formula.

Tartaglia died in Venice, Italy, on December 13, 1557.

tautochrone *See* CYCLOID.

tautology In FORMAL LOGIC, a compound statement that cannot possibly be false by virtue of its structure is called a tautology. For example, "If all the planets are made of cheese, then Mars is made of cheese" is true regardless of the validity of the component statements that "all planets are made of cheese" and "Mars is made of cheese." Tautologies are true purely because of the laws of logic and not because of any known facts about the world. They are therefore statements that contain only definitional information.

See also LAWS OF THOUGHT; TRUTH TABLE.

Taylor, Brook (1685–1731) British *Calculus* Born on August 18, 1685, in Edmonton, England, mathematician Brook Taylor is remembered today for his important contributions to the development of CALCU-

LUS. In his 1715 text *Methodus incrementorum directa et inversa* (Direct and indirect methods of incrementation), Taylor formulated methods for expanding functions as infinite series. These are today known as TAYLOR SERIES. Taylor also invented the technique of INTEGRATION BY PARTS.

Taylor graduated from St. John's College, Cambridge, in 1709 with an advanced degree in mathematics, having already written his first mathematical paper 1 year earlier. Despite his young age Taylor quickly developed a reputation as an expert in the field of calculus. He was elected to the prestigious ROYAL SOCIETY in 1712 and was immediately appointed to a special committee to adjudicate on the issue of whether it was SIR ISAAC NEWTON (1642–1727) or GOTTFRIED WILHELM LEIBNIZ (1646–1716) who had discovered the FUNDAMENTAL THEOREM OF CALCULUS that unites differentiation with integration.

Despite the attachment of his name to the technique, Taylor was not the first to develop a theory of infinite function expansions. Several decades earlier JAMES GREGORY (1638–75), Johann Bernoulli (1667–1748) of the famous BERNOULLI FAMILY, ABRAHAM DE MOIVRE (1667–1754), and others had independently discovered variants of Taylor's expansion theorem. (Taylor was unaware of this body of work when he wrote his famous 1715 text.) The significance of infinite series expansions, however, was not properly recognized until 40 years after Taylor's death when, in 1772, influential French mathematician JOSEPH-LOUIS LAGRANGE (1736–1813) proclaimed it an important basic principle of differential calculus.

As a broad-based scholar, Taylor also wrote on topics in experimental and theoretical physics. He published articles on magnetism, thermometers, vibrating strings, capillary action, and the mathematical laws dictating the motion of the planets. He invented alternative methods for computing logarithms and approximating solutions to algebraic equations. In 1715 he also published *Linear Perspective*, an influential text outlining the mathematical foundations of PROJECTIVE GEOMETRY and the role of the vanishing point in art. He died in London, England, on December 29, 1731.

Taylor series Many functions, such as trigonometric functions and exponential and logarithmic functions, are difficult to manipulate, whereas adding, subtracting, and multiplying POLYNOMIAL functions is relatively straightforward. In 1715 English mathematician BROOK TAYLOR worked to approximate complicated functions with simple polynomials.

Suppose it is indeed possible to approximate a complicated function $f(x)$ as a polynomial:

$$f(x) \approx a_0 + a_1 x + a_2 x^2 + \ldots + a_n x^n$$

One analyzes the situation by first noting that placing $x = 0$ into this formula yields:

$$f(0) = a_0 + 0 + 0 + \ldots + 0$$

This shows that the approximation can be made exact, at least at $x = 0$, by setting $a_0 = f(0)$. To determine the coefficient a_1, differentiate once and then set $x = 0$:

$$f'(x) = 0 + a_1 + 2a_2 x + 3a_3 x^2 + \ldots + na_n x^{n-1}$$
$$f'(0) = a_1$$

This shows that $a_1 = f'(0)$ is a good choice. That is, by setting a_1 to be this value, not only do the values of the function and polynomial match at $x = 0$, but the slopes of the two graphs also match at $x = 0$.

Differentiating another time and setting $x = 0$ (that is, matching second derivatives) yields:

$$f''(x) = 2a_2 + 2 \cdot 3a_3 x + \ldots + n(n - 1)a_n x^{n-2}$$
$$f''(0) = 2a_2$$

and so $a_2 = \dfrac{f''(0)}{2}$. Continuing this way we obtain: $a_3 = \dfrac{f'''(0)}{3!}, \ldots, a_n = \dfrac{f^{(n)}(0)}{n!}$.

Thus a good approximation to the function $f(x)$, at least around the value $x = 0$, would be the polynomial:

$$f(x) \approx f(0) + f'(0)x + \frac{f''(0)}{2!}x^2 + \frac{f'''(0)}{3!}x^3 + \cdots + \frac{f^{(n)}(0)}{n!}x^n$$

The higher the degree the polynomial one uses, the better the approximation would be. Thus the best polynomial of all would be a polynomial of infinite degree, that is, a POWER SERIES:

$$f(x) = f(0) + f'(0)x + \frac{f''(0)}{2!}x^2 + \ldots$$

Mathematicians have proved that if f can indeed be differentiated infinitely many times, then this "approximation" is exact for the range of values the series converges (called its RADIUS OF CONVERGENCE), that is, the function really does equal the infinite sum expressed on the right-hand side of the formula. This formula is called a Taylor series.

As an example, consider the function $f(x) = e^x$. Differentiating and substituting in $x = 0$ yields:

$$f(x) = e^x \qquad f(0) = 1$$
$$f'(x) = e^x \qquad f'(0) = 1$$
$$f''(x) = e^x \qquad f''(0) = 1$$
$$\vdots$$

and so $e^x = 1 + x + \dfrac{x^2}{2!} + \dfrac{x^3}{3!} + \cdots$. A study of power series shows that this series has infinite radius of convergence, and so this equation is valid for all values of x.

The Taylor series of $f(x) = \sin x$ is given by:

$$f(x) = \sin x \qquad f(0) = 0$$
$$f'(x) = \cos x \qquad f'(0) = 1$$
$$f''(x) = -\sin x \qquad f''(0) = 0$$
$$f'''(x) = -\cos x \qquad f'''(0) = -1$$
$$f^{(4)}(x) = \sin x \qquad f^{(4)}(0) = 0$$
$$\vdots$$

$$\sin x = x - \frac{x^3}{3!} + \frac{x^5}{5!} - \frac{x^7}{7!} + \cdots$$

Similarly,

$$\cos x = 1 - \frac{x^2}{2!} + \frac{x^4}{4!} - \frac{x^6}{6!} + \cdots$$

$$\frac{1}{1-x} = 1 + x + x^2 + x^3 + \cdots \quad \text{(valid only for } -1 < x < 1)$$

and

$$\ln(1+x) = x - \frac{x^2}{2} + \frac{x^3}{3} - \frac{x^4}{4} + \ldots \text{(valid only for } -1 < x \leq 1)$$

Since handheld calculators are programmed only to add, subtract, multiply, and divide, Taylor series make it possible to compute values of complicated functions.

(Noting that calculators only display eight or 10 decimal places, one need only use the first few terms of the Taylor series to obtain adequately accurate answers.)

To approximate a function near a value $x = a$ different from zero, one uses a polynomial of the form: $a_0 + a_1(x - a) + a_2(x - a)^2 + a_3(x - a)^3 + ...$ Using the same technique of differentiating and substituting in $x = a$ yields:

$$f(x) = f(a) + f'(a)(x - a) + \frac{f''(a)}{2!}(x - a)^2$$
$$+ \frac{f'''(a)}{3!}(x - a)^3 + \cdots$$

This is called a Taylor series "centered at $x = a$." Taylor series, as discussed above, centered at $x = 0$ are sometimes called Maclaurin series. In 1742 Scottish mathematician COLIN MACLAURIN (1698–1746) wrote an influential text in which he described Taylor's methods. Although Maclaurin made no pretense of having discovered these series centered at zero (he himself acknowledged that they are nothing more than a special case of Taylor's general results), scholars honor his work nonetheless by associating his name with these special series.

Talyor conducted his work in the early 1700s, but it is known that other scholars, such as JAMES GREGORY (1638–75), used power series in the same way decades earlier.

See also DIFFERENTIAL EQUATION; MERCATOR'S EXPANSION.

tensor Just as a VECTOR is a mathematical quantity that describes translations in two- or three-dimensional space, a tensor is a mathematical quantity used to describe general transformations in *n*-dimensional space. Precisely, if the locations of points in *n*-dimensional space are given in one coordinate system by $(x^1, x^2, ..., x^n)$ and in a transformed coordinate system by $(y^1, y^2, ..., y^n)$ (it is convenient to use superscripts rather than subscripts), then a "rank 1 contravariant tensor" is a quantity T, with single components, that transforms according to the rule:

$$T^i_{new} = \sum_{r=1}^{n} \frac{\partial y^i}{\partial x^r} T^r$$

(The coefficients here are PARTIAL DERIVATIVES.) Thus, for example, if the change of coordinates is a translation, $y_i = x_i + a_i$ for some numbers a_i, it follows that $T^i_{new} = T^i$. This means that the quantity T is a quantity, with n components, unchanged by translations. That is, T is indeed a vector.

More complicated transformation rules are permitted for quantities with components given by two or more superscripts (or even subscripts). ALBERT EINSTEIN (1879–1955) used tensor analysis in his general theory of relativity.

tessellation (covering, tiling) A covering of the PLANE with geometric shapes is called a tessellation or a tiling. Usually the shapes, called tiles, are POLYGONs, and the pattern produced is, in some sense, repetitive. Every point in the plane is to be covered by a tile, and two tiles may intersect only along their edges. A location where three or more edges meet at a point is called a vertex of the tessellation. (It is usually assumed that neighboring tiles meet along the full length of their common edge.)

A regular tessellation uses congruent regular polygons, all of one type, as tiles. For example, square tile can be used to cover a plane, as shown below. Four edges meet at each vertex of the tessellation. (Note that four angles of 90° add to a total of 360° around that vertex.) An equilateral triangle also tiles the plane (six triangles, containing angles of 60°, fit around one vertex), as does the regular hexagon. (Three hexagons, containing angles of 120°, fit about a vertex.) As no other regular polygon has appropriate angle values to fit about a vertex, these are the only regular tilings of the plane.

A semiregular tessellation uses congruent regular polygons of more than one kind, arranged so that the arrangement of polygons about every vertex of the tessellation is identical. For example, one can tile the plane with regular hexagons and equilateral triangles so that each hexagon is surrounded by six triangles and each triangle by three hexagons to produce a semiregular tessellation. Mathematicians have proved that there are only eight semiregular tilings of the plane. (If one abandons the restriction that the arrangement of polygons about each vertex be identical, then there are infinitely many such tilings.)

A monohedral tessellation is a tessellation that uses congruent copies of only one type of tile (not necessarily

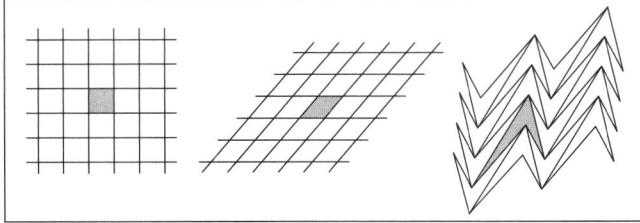

Tessellations

a regular polygon). The diagram above shows that any PARALLELOGRAM tiles the plane. As two copies of the same triangle fit together to form a parallelogram, we have:

> Any triangle provides a monohedral tessellation of the plane.

By distorting the monohedral tessellation with parallelograms, one can show:

> Any quadrilateral, concave or convex, provides a monohedral tessellation of the plane.

Not every pentagon or every hexagon will tile the plane. In his 1918 doctoral thesis, mathematician Karl Reinhardt (1895–1941) classified those hexagons that do tile and found that they fall into three basic types. To this day no one knows how many classes of convex pentagons tessellate the plane. (Fourteen types have currently been identified.) Reinhardt proved that no convex polygon with seven or more sides will tile the plane.

A tessellation of the plane using two SQUAREs of different sizes provides a surprisingly elegant visual proof of PYTHAGORAS'S THEOREM.

tetrahedron (triangular pyramid) Any solid figure (POLYHEDRON) with four triangular faces is called a tetrahedron. If the four faces are congruent equilateral triangles, then the figure is called regular.

The height h and volume V of a regular tetrahedron with edge length a are given by:

$$h = \frac{\sqrt{2}}{\sqrt{3}}a$$

$$V = \frac{\sqrt{2}}{12}a^3$$

and such a regular tetrahedron fits snugly inside a cube of side-length $\frac{a}{\sqrt{2}}$. (Choose a vertex of the cube and draw diagonals on the three faces surrounding that vertex. These coincide with the three edges of an entrapped tetrahedron.)

Although it is possible to stack a finite number of small cubes together to form a larger cube, it is impossible to complete a similar feat for regular tetrahedra: no regular tetrahedron is a union of smaller regular tetrahedra.

The sum of the first n TRIANGULAR NUMBERS is called the nth tetrahedral number. For example, the fifth tetrahedral number is $1 + 3 + 6 + 10 + 15 = 35$. If we place triangular arrays of 1, 3, 6, 10, and 15 dots above one another making use of the third dimension in space, the array of dots produced is a tetrahedron, explaining the name of these FIGURATE NUMBERS. The general formula for the nth tetrahedral number is $\frac{1}{6}n(n + 1)(n + 2)$. The tetrahedral numbers appear as one of the diagonals of PASCAL'S TRIANGLE.

See also NET; PLATONIC SOLID.

Thales of Miletus (ca. 625–547 B.C.E.) Greek *Geometry, Philosophy* Born in the region of Miletus, Asia Minor (now Turkey), Thales is considered the first scientist and philosopher of Western history, at least in the sense of being the first scholar to whom particular scientific and mathematical discoveries have been attributed. Rather than relying on mythology and religion to explain the natural world, Thales searched for rational principles in science. This work led him to also look for unifying principles in geometry and, as such, he was the first scholar to attempt to derive geometric facts by processes of deduction and logical reasoning. He established, for example, fundamental geometrical propositions such as "the base angles of any isosceles triangle are equal" and "the angle in a semicircle is a right angle."

Little is known of Thales's life, and all his written texts have been lost. Nonetheless, scholars that followed Thales made numerous references to his achievements and to his approach to the study of the world. The Greek philosopher PROCLUS (ca. 450 C.E.) claimed that Thales acquired his mathematical knowledge from Egyptian scholars, and centuries later, Hieronymus, a student of Aristotle, wrote that Thales measured the height of the pyramids by observing the length of shadows. (Historians today take this as evidence that Thales was familiar with the principles of similar triangles.) Greek historian Herodotus (ca. 485 B.C.E.) wrote that Thales correctly predicted the eclipse of the sun in 585 B.C.E. (though some scholars today suggest that this may have only been a lucky guess).

Aristotle wrote about Thales's worldview in his text *Metaphysics*. Apparently Thales believed that the Earth was a flat disc floating on an infinite ocean of water. He used this theory to give a rational explanation of why earthquakes occur.

Theodorus of Cyrene (ca. 465–398 B.C.E.) Greek *Geometry, Number theory* Born in Cyrene (now Shahhat, Libya), Theodorus is remembered for his work on IRRATIONAL NUMBERS, proving that not only is $\sqrt{2}$ irrational, but so too are $\sqrt{3}$ and $\sqrt{5}$, as well as the roots of all other nonsquare quantities up to 17.

Theodorus is noted as having tutored the great philosopher PLATO (ca. 428–348 B.C.E.) in the subject of mathematics. Plato later described much of the work of Theodorus in his text *Theaetetus*, and it is chiefly through this document that we know anything of Theodorus's life and work. We learn there, for instance, that Theodorus also studied astronomy, music, and arithmetic.

Unfortunately, Plato did not describe the method by which Theodorus proved his quantities to be irrational, and historians today are puzzled as to why Theodorus stopped his work with the number $\sqrt{17}$. It is likely that Theodorus developed a series of arguments that were highly dependent on the particular number being studied and failed to develop a general approach that dealt with all numbers in one fell swoop. This is all the more curious, given that the standard arithmetic proof of the irrationality of $\sqrt{2}$ was well known to scholars of his day, a proof that can very easily be generalized to numbers other than 2. (If $\sqrt{2} = \frac{a}{b}$, then $a^2 = 2b^2$, creating a contradiction: the PRIME number 2 appears an even number of times in the prime factorization of a^2 and an odd number of times in the factorization of $2b^2$.) One also can establish the irrationality of $\sqrt{2}$ geometrically as follows:

> The EUCLIDEAN ALGORITHM shows that two geometric lengths are COMMENSURABLE (of rational ratio) if repeated subtraction, in turn, of the smaller length from the longer eventually produces two segments of the same length. Thus, two lengths for which this process continues indefinitely must be of an irrational ratio. Noting this, consider line segments of length 1 and length $\sqrt{2}$ forming an isosceles right triangle. With a circular arc, construct a line segment of length $\sqrt{2} - 1$, the difference of these two lengths. Now consider the two segments of lengths 1 and $\sqrt{2} - 1$. Draw a second right triangle with side-length $\sqrt{2} - 1$ as shown. The hypotenuse of his triangle has length $2 - \sqrt{2} = 1 - (\sqrt{2} - 1)$, the difference of the second pair of lengths. As this process can be repeated indefinitely, always subtracting the smaller length from the longer, it must be the case that the two original quantities, 1 and $\sqrt{2}$, are of irrational ratio.

Many historians believe that Theodorus may have developed specific geometric arguments of this type, using triangles, pentagons, and 17-gons, for instance, to show that the numbers he considered are irrational.

theorem (proposition) A statement in mathematics that has been proved true is called a theorem. The name originates from the Greek word *theórema* meaning "a subject for contemplation."

Often theorems are classified in terms of their importance. A lemma is an ancillary theorem, that is, a result proved true for the purposes of later establishing a more important result. (In Greek, *lemma* means "a thing taken.") A corollary is a theorem of immediate consequence, that is, a result that follows from a

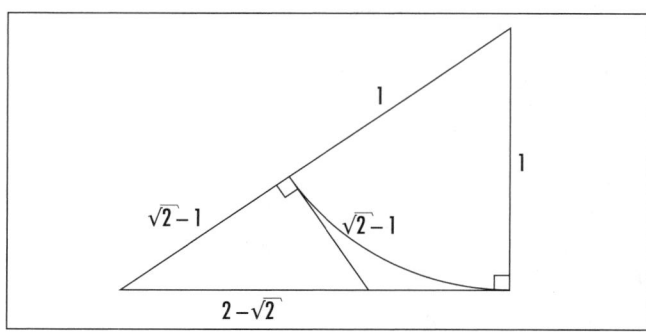

Proving that the square root of two is irrational

previously established theorem requiring little or no further explanation. A conjecture, on the other hand, is a statement yet to be proved or disproved. The following sequence of results illustrates these ideas:

> *Lemma:* A natural number squared is either divisible by four, or leaves a remainder of 1 when divided by 4.

Proof: Let a be a natural number. If a is even, then it can be written in the form $a = 2m$ for some number m. Consequently, $a^2 = (2m)^2 = 4m^2$ is divisible by 4. If, on the other hand, a is odd, then $a = 2m + 1$ for some number m, and so $a^2 = (2m + 1)^2 = 4m^2 + 4m + 1$ is 1 more than a multiple of 4.

> *Theorem:* If a number N equals the sum of two squares, $N = a^2 + b^2$, then N leaves either a remainder of 0, 1, or 2, but not 3, when divided by 4.

Proof: By the lemma, a^2 leaves a remainder of either 0 or 1 when divided by 4, as does the number b^2. It is impossible for the two remainders to sum to a total remainder of 3.

> *Corollary:* It is impossible to write 2,867,039 as a sum of two square numbers.

Proof: This number leaves a remainder of 3 when divided by 4.

> *Conjecture:* It is possible to write 3,457,417,105 as a sum of two square numbers.

This is yet to be proved or disproved.
See also SQUARE.

three-utilities problem A classic puzzle challenges the reader to connect three houses to each of three utility companies—water, electricity, and gas—one pipe between each house and each company, in such a way that no lines or mains cross.

It turns out that this puzzle cannot be solved if the buildings are situated on a plane. A simple heuristic argument gives a sense of why this is the case:

> Any solution to the problem would be a diagram of six houses and nine pipes, dividing the

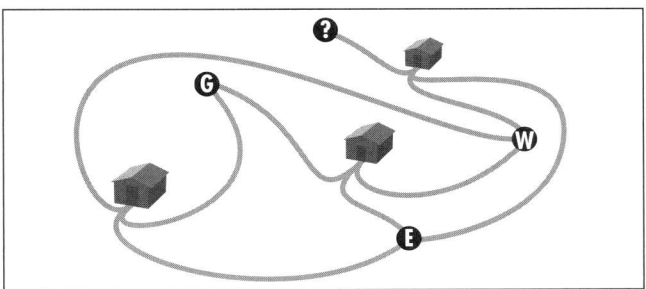

The three-utilities problem

plane into regions. It seems that each region is to be bounded by four pipes. If there are R regions, then $4R$ represents the total number of pipes in the picture, counted twice (each pipe borders two regions). Thus $4R = 18$, suggesting that there must be a nonintegral number of regions. This absurdity demonstrates the impossibility of the task.

Surprisingly the puzzle can be solved if one draws the three houses and the three companies instead on the surface of a TORUS. (One can verify this by using a marker to draw the figures on a bagel, for example.)
See also GRAPH.

time series graph   *See* STATISTICS: DESCRIPTIVE.

topology (analysis situs, rubber-sheet geometry) The study of those properties of geometric shapes and surfaces that remain unchanged when the shapes of those objects are distorted by a continuous DEFORMATION (such as stretching, shrinking, or twisting) is called topology. Unlike a classical geometer, a topologist is not interested in questions of distances and angles, but is only concerned with the relative positions of points. All that is required of a topological transformation is that points that begin close together remain relatively close together. In this viewpoint, a circle and a square (of any size) are topologically equivalent, since either can be continuously deformed into the other, and a number of geometrical properties apply equally well to either object. For example, the statement, "Removing a point from a circle produces a line segment" is valid for the entire class of objects topologically equivalent to a circle.

Any transformation that requires puncturing or tearing a surface, or joining together two disjoint portions of a figure, is not considered a valid topological action. (Tearing a surface, for instance, separates points that were close together.) As it is impossible to deform a SPHERE into a TORUS without puncturing the surface, these two shapes are not topologically equivalent. As other examples, the capital letters C, M, and Z are topologically equivalent, as are the letters D, O, P, and R, but no letter from the first group is topologically equivalent to a letter in the second. No letter in the alphabet is topologically equivalent to the letter B (other than capital B itself).

The study of topology began with LEONHARD EULER (1707–83) and his analysis of the famous SEVEN BRIDGES OF KÖNIGSBERG PROBLEM. The mathematician CARL FRIEDRICH GAUSS (1777–1855) examined the distortion of knots and invariant properties that arise in the study of PROJECTIVE GEOMETRY. In 1895 French mathematician HENRI POINCARÉ (1854–1912) examined these works and published five papers, laying a theoretical framework for a formal study of topology.

The discipline of point-set topology attempts to capture the notion of "closeness" without making mention of distance. In this theoretical approach, one is given a set X and a specified collection of subsets called open sets, which, in some loose sense, are meant to represent those points that are deemed "close" to each other. These subsets are required to satisfy three basic properties:

1. The empty set $\emptyset$ and the entire set X are deemed open.
2. If A and B are two open sets, then their intersection $A \cap B$ is also an open set.
3. The union of any collection of open sets is another open set.

Any set X with a collection of subsets classified as open satisfying the three stated properties is called a topological space. For example, the set of real numbers with unions of open INTERVALS as the open sets is a topological space. So too is the set $X = \{a,b,c,d\}$ with subsets $\emptyset$, $\{a\}$, $\{b,c,d\}$, and X as the open sets. Two topological spaces are said to be equivalent if there is a one-to-one correspondence between the points in one space and the points in the other so that open sets in one space correspond to open sets in the other.

See also KNOT THEORY; MÖBIUS BAND.

torus (anchor ring, donut) A closed curved surface possessing a single hole akin to the surface of a donut or an inner tube is called a torus (plural: tori). A torus can be regarded as a SOLID OF REVOLUTION obtained by rotating a circle in a plane one full revolution about a line that does not intersect the circle. (This circle is called the generating circle.) As such, PAPPUS'S THEOREMS show that the VOLUME V and the surface AREA A of a torus are given by the formulae:

$$V = 2\pi^2 a^2 b$$
$$A = 4\pi^2 ab$$

where a is the radius of the circle rotated about a line, and b is the distance of the center of the the circle from that line. These formulae can also be obtained by the techniques of INTEGRAL CALCULUS

In a CARTESIAN COORDINATE system, if the circle of radius a is positioned in the yz-plane with center at position b along the y-axis and is rotated about the z-axis, then the equation of the resulting torus produced is given by:

$$\left(\sqrt{x^2 + y^2} - b\right)^2 + z^2 = a^2$$

This follows from PYTHAGORAS'S THEOREM applied to a right triangle within the generating circle when it is rotated to a position so that the x-, y-, and z-coordinates of a point on the torus are (x,y,z). (Examine the right triangle of height z with radius a as its hypotenuse.)

If θ denotes the angle of a point on the generating circle, and φ the angle through which that generating circle is rotated about the z-axis, then one obtains the following PARAMETRIC EQUATIONS for the torus:

$$x = (b + a\cos\theta)\cos\varphi$$
$$y = (b + a\cos\theta)\sin\varphi$$
$$z = a\sin\theta$$

Each point on the generating circle of a torus, when rotated about the z-axis, traces out a circle. Thus a

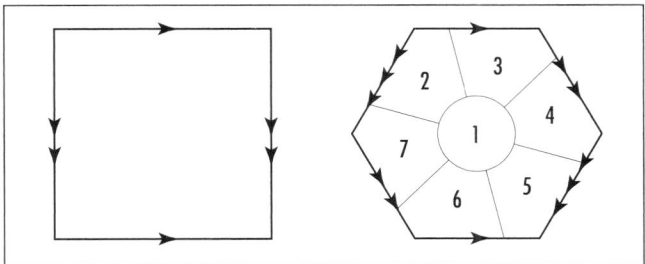

Constructing a torus

torus can be considered the CARTESIAN PRODUCT of two circles: each point on a torus corresponds to a pair of points, one on one circle (the generating circle) and a second on a second circle (the circle traced out by that point on the generating circle). The set of all possible positions of an hour hand and minute hand on a clock, each corresponding to a full circle of values, describes a set of outcomes topologically equivalent to a torus. (Physicists would say that the "phase space" of the two hands of a clock is a torus.)

One can physically construct a torus by sewing together the opposite sides of a square rubber sheet as shown. (One must be sure to align the directions of the arrows indicated.) Sewing together the opposite sides of a hexagon also produces a torus.

This latter construct shows that it is possible to produce a design of seven regions on a torus so that each regions shares a portion of boundary with each of the six remaining regions. Thus some maps on a torus require seven colors to paint if one wishes to ensure that neighboring regions are always of distinct colors. Mathematicians have proved that the situation is never worse than this: seven colors will always suffice to color a toroidal map, no matter how complicated that map may be.

Given a punctured torus made of rubber, such as an inner tube with a small hole, it is physically possible to stretch open the hole and turn the surface inside out through that hole. Surprisingly, the resulting surface is another torus. A double torus is a closed surface with two holes. Everting a double torus through a puncture in its surface again yields the same surface back again, another double torus, and the same result is true of triple tori and multiple tori with any number of holes.

See also FOUR-COLOR THEOREM; SPHERE.

tournament (round-robin tournament, complete digraph)   A competition among a collection of teams is called a tournament if each possible pair of teams plays exactly one match, and each game played results in a win for one team and a loss for the other. (That is, no ties are permitted.) Mathematically, a tournament among n players is a GRAPH with n vertices, with edges drawn between every possible pair of points, directed with arrows pointing from one vertex to another. Here each vertex represents a team, and an edge pointing from vertex A to vertex B indicates a win for team A over team B. The diagram below, for instance, represents a tournament among four teams, A, B, C, and D. We see that team A lost to team D but beat teams B and C.

A graph on n vertices with an edge between every possible pair of vertices directed by an arrow is called a complete directed graph or simply a complete digraph on n vertices.

Structurally, there is essentially only one possible complete digraph on two vertices (although, in terms of team-playing, there are two possible interpretations: either a team A beat a team B, or team B beat team A). For n equal to 3, 4, 5, 6, and 7 there are, respectively, 2, 4, 12, 56, and 456 different complete digraphs on n vertices.

Any team that beat every other team in a tournament is called a source or a transmitter, and any team that lost to all other teams is called a sink or a receiver. (These names reflect the alternative interpretation of a

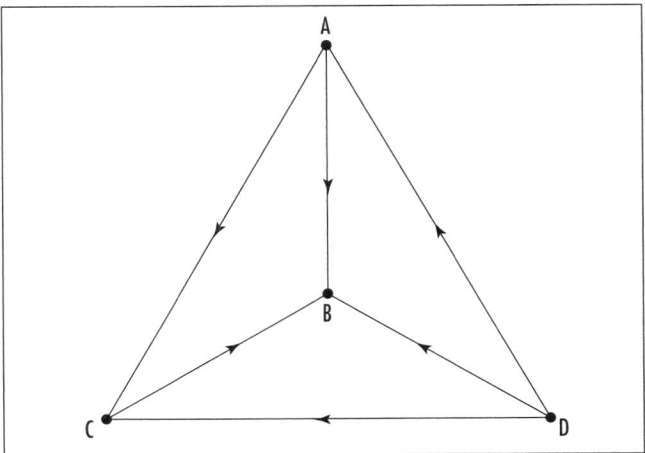

Representing a tournament

digraph as a diagram of information flow.) In the example above, team *D* is a source, and team *B* is a sink. It is impossible for a tournament to possess two different sources or two different sinks. A tournament need not possess either a source or a sink.

Every tournament possesses a HAMILTONIAN PATH. This means that it is possible to find a path through the diagram of a complete digraph that starts at one vertex and follows edges in directions that respect the arrows to visit each and every other vertex exactly once. For example, in the diagram above, the journey $D \rightarrow A \rightarrow C \rightarrow B$ represents a Hamiltonian path. Alternatively, a Hamiltonian path can be interpreted simply as a list of all the teams that played arranged in an order so that the first team on the list beat the second, the second team on the list beat the third, and so forth. The following example illustrates a general method for constructing such a list (or, equivalently, a Hamiltonian path) for any tournament.

Consider the following table outlining the results of a tournament with five teams, *A*, *B*, *C*, *D*, and *E*:

A beat B	C beat B	C beat D	D beat E
A beat C	B beat D	E beat C	
D beat A	B beat E		
A beat E			

and consider the teams one at a time in order to create an appropriate list. First write A. Now if team *B* beat *A,* then it is permissible to write B to the left of *A*. This is not the case, however, and so *B* should be written to the right of *A*. Our list, so far, appears: *A B*.

Now consider team *C* and ask, "Did *C* beat *A*?" As the answer is no, we are not permitted to write *C* to the immediate left of *A*. Did *C* beat *B*? Yes. Thus we are permitted to write *C* to the immediate left of *B*. Our list thus far reads: *A C B*. (We have that *A* beat *C*, and *C* beat *B*.)

Now consider team *D*, and look for the first team on the partially constructed list that *D* beat. This is team *A*. This leads to the new list: *D A C B*.

Finally consider team *E*. The first team on this partially constructed list that *E* beat is *C*. This allows us to place E between A and C to pro-

duce the complete list: *D A E C B*. Notice that it does indeed have the property that each team on the list beat the one succeeding it. This list represents a Hamiltonian circuit in the appropriate digraph.

The total number of games played in a tournament with *N* players is given by the formula:

$$\frac{N(N-1)}{2}$$

(Each of the *N* teams plays *N* − 1 games, but this double counts the total number of games played.) This is the (*N* − 1)th triangular FIGURATE NUMBER.

If *N* is odd, it is always possible to construct an example of a tournament among *N* teams in which each team has the same number of wins. This feat cannot be accomplished if *N* is even. (If each team has *w* wins, then $N \times w$ represents the total number of games played. Consequently, we must have $w = \frac{N-1}{2}$. This is an invalid formula if *N* is even.)

tower of Hanoi (towers of Hanoi, tower of Brahma) Consider three poles and a collection of differently sized discs all placed initially on one pole in order of size, largest at the bottom to the smallest at the top. An ancient challenge demands that all the discs be transferred to a different pole in such a way that:

1. Only one disc is ever moved at a time.
2. No disc ever sits upon another disc of a smaller size.

An eight-disc version of this puzzle was patented and sold as a toy by Edouard Lucas in 1883. He called the puzzle the Tower of Hanoi and wrote that the game was based on a "tower of Brahma" in which priests were given the challenge of moving 64 discs from one pole to another under the same restrictions, moving just one disc a day. The legend claimed that the world would end on the day the task is completed.

A puzzle containing just one or two discs is easily solved. If one knows a method for solving the puzzle with *N* discs, then the solution to the (*N* + 1)-disc version follows readily: transfer the top *N* discs to one pole via the known method, move the large (*N* + 1)th disc to the empty pole, and transfer the *N* discs to the pole containing the large disc. If *T*(*N*) denotes the number of moves required to solve the N-disc version of the puzzle, then it follows from the above procedure that:

$$T(N + 1) = T(N) + 1 + T(N)$$
$$= 2T(N) + 1$$

with $T(1)$ and $T(2) = 3$. This RECURRENCE RELATION has solution:

$$T(N) = 2^N - 1$$

This shows that the 64-disc version of the puzzle requires $2^{64} - 1 = 18,446,744,073,709,551,615$ moves. According to the legend, the end of the world will thus come in approximately 5.05×10^{16} years.

If one wishes to transfer N discs to a specific pole, then one begins solving the puzzle by moving the smallest disc to the desired pole if N is odd but to the third pole if N is even.

If we number the N discs 1, 2,…, N from smallest to largest, then the following table shows the order in which discs are moved to solve the first five small versions of the puzzle:

N	Solution
1	1
2	1, 2, 1
3	1, 2, 1, 3, 1, 2, 1
4	1, 2, 1, 3, 1, 2, 1, 4, 1, 2, 1, 3, 1, 2, 1
5	1, 2, 1, 3, 1, 2, 1, 4, 1, 2, 1, 3, 1, 2, 1, 5, 1, 2, 1, 3, 1, 2, 1, 4, 1, 2, 1, 3, 1, 2, 1

Each line of the table is generated by repeating two versions of the previous solution and inserting the number N in the center. Surprisingly, the rth number in any sequence representing a solution to the puzzle equals 1 more than the exponent of the largest power of 2 that divides the number r. For example, the largest power of 2 that divides 12 is 2^2, and $2 + 1 = 3$ is the 12th number in any sequence representing a solution to the puzzle. Since $16 = 2^4$ is the largest power of 2 that divides the number 10,000, this observation also shows that Brahman priests will move disc $4 + 1 = 5$ on day 10,000.

tractrix (equitangential curve, tractory)   If one drags the front end of a bicycle along a straight path, then the path traced out by the back wheel (assuming the wheels of the bicycle were initially offset) forms a curve called the tractrix. This curve has the property that the endpoint of each line segment of fixed length (the length of the bicycle) TANGENT to the curve meets the straight-line path.

The curve was first studied by Dutch physicist Christiaan Huygens (1629–95), who described the path in terms of an object being dragged by a fixed length of string. He coined the name "tractrix" for the path, deriving it from the Latin word *trahere* meaning "to drag." The curve is also sometimes called the "hundkurve" (German for "hound curve"), inspired by the image of a reluctant dog being dragged on a leash by an insistent owner.

trajectory   The path of a moving object or particle is called its trajectory. The trajectory of a particle is often described via PARAMETRIC EQUATIONS. For example, the equations $x(t) = \cos(t)$ and $y(t) = \sin(t)$ give, at each time t, the x- and y-coordinates of a particle whose trajectory is a UNIT CIRCLE. The trajectory of a ball thrown in the air is an inverted PARABOLA, assuming the effects of air resistance can be ignored.

The term *trajectory* is also used in the theory of DYNAMICAL SYSTEMS as an alternative name for the ORBIT of a point.

transcendental number   *See* ALGEBRAIC NUMBER.

transformation   Any FUNCTION or mapping that changes one quantity into another is called a transformation. Although the term is essentially synonymous with the term *function*, it is usually used only in the context of geometry, where one speaks of a GEOMETRIC TRANSFORMATION and a TRANSFORMATION OF COORDINATES.

Matrix notation is often used to describe geometric transformations of the plane. A point in the plane is represented by a column VECTOR $\begin{pmatrix} x \\ y \end{pmatrix}$ and a geometric transformation as a 2×2 matrix A. The effect of the geometric transformation is given by matrix multiplication. For example, application of the following 2×2 matrices

$$\begin{pmatrix} 1 & \\ & -1 \end{pmatrix}, \begin{pmatrix} \cos\theta & -\sin\theta \\ \sin\theta & \cos\theta \end{pmatrix}, \begin{pmatrix} 1 & 0 \\ 0 & 0 \end{pmatrix}$$

has the effect of, respectively, reflecting points across the *x*-axis, rotating points about the origin through an angle θ, and projecting points onto the x-axis.

In algebra, a change in the form of a mathematical expression, without changing its validity, is called a transformation. For example, the action of EXPANDING BRACKETS transforms the expression $y = (x + 1)^2$ into the equivalent expression $y = x^2 + 2x + 1$.

See also AFFINE TRANSFORMATION; LINEAR TRANSFORMATION; PROJECTION.

transformation of coordinates (transformation of axes)

In coordinate geometry, it is often convenient to change the position of a set of coordinate axes, either by a ROTATION or a TRANSLATION, to simplify the expression of a given curve under study. For example, consider the curve given by the equation $x^2 - 6x + y^2 + 4y = 3$. By COMPLETING THE SQUARE, this equation can be rewritten:

$$(x - 3)^2 + (y + 2)^2 = 16$$

Setting $X = x - 3$ and $Y = y + 2$ this reads:

$$X^2 + Y^2 = 16$$

identifying the curve as a CIRCLE of radius four. In this process we have, in effect, introduced two new coordinate axes, *X* and *Y*, each a translation of one of the original *x*- and *y*-axes. The origin of the new coordinate system lies at the location where $X = 0$ and $Y = 0$, that is, at the point (3,–2).

Changing from the use of CARTESIAN COORDINATES to POLAR COORDINATES, or, in three-dimensional geometry, to SPHERICAL COORDINATES or CYLINDRICAL COORDINATES, is also deemed a transformation of coordinates.

See also PRINCIPAL AXES; TRANSFORMATION.

translation   *See* GEOMETRIC TRANSFORMATION.

transversal (traverse)

A line cutting two or more other lines is called a transversal. When a transversal cuts just two other lines, eight angles are formed. The four angles lying between the two lines are called interior angles, and the four lying outside are called exterior angles. Special names are given to pairs of angles, as shown in the following diagram:

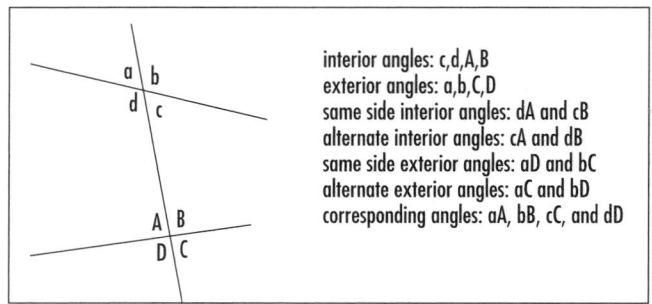

interior angles: c,d,A,B
exterior angles: a,b,C,D
same side interior angles: dA and cB
alternate interior angles: cA and dB
same side exterior angles: aD and bC
alternate exterior angles: aC and bD
corresponding angles: aA, bB, cC, and dD

The angles associated with a transversal

According to the PARALLEL POSTULATE, two lines cut by a transversal are parallel if, and only if, two alternate interior angles are equal (or, equivalently, if any two corresponding angles are equal).

trapezoid/trapezium

A QUADRILATERAL with two sides parallel is called a trapezoid in the United States and a trapezium in the United Kingdom. Matters are confusing, for a four-sided figure with *no* two sides parallel is called a trapezium in the United States and a trapezoid in the United Kingdom. We follow U.S. usage of the terms here.

The area of a trapezoid is given by the formula

$$\frac{1}{2}(b_1 + b_2)h$$

where b_1 and b_2 are the lengths of the two parallel edges, and *h* is the distance between them. (*See* AREA.) The midline or median of a trapezoid is the straight-line segment that joins the MIDPOINTs of the nonparallel sides. An exercise in geometry shows it has length $m = \frac{1}{2}(b_1 + b_2)$. (Use similar triangles.) Thus the area of a trapezoid is the same as that of a RECTANGLE of length *m* and width *h*.

If the two nonparallel sides of a trapezoid are equal in length (and not themselves parallel), we then call the figure an isosceles trapezoid.

See also SIMILAR FIGURES.

trapezoidal rule   *See* NUMERICAL INTEGRATION.

traveling-salesman problem A mathematical problem based on a real-life situation, called the traveling-salesman problem, can be stated as follows:

> There are a certain number of towns connected in pairs by highways. The distances of all the highway routes between towns are specified. Plan a journey that visits all the towns and returns to the starting point while minimizing the total distance traveled.

Alternatively, one can try to minimize the travel costs in moving from town to town, or the time taken to complete the entire journey, for example. The desire to find the efficient routes such as these is an important practical problem. It can lead to significant financial savings for transportation companies and other types of industry.

The traveling-salesman problem is equivalent to the challenge of finding a least-distance Hamiltonian circuit in a GRAPH—a challenge in the field of GRAPH THEORY.

One method for solving the traveling-salesman problem is to simply list all the possible routes and see which one is the most efficient. We can count the number of routes to be checked as follows:

> Suppose there are n towns in all. Starting in one town there are $n - 1$ choices for which town to visit next. Having reached that town, there are $n - 2$ choices for which town to visit second, and so on, all the way down to one final choice of one town to visit before returning to start. Noting that, for the purposes of this problem, traveling a route in the forward direction is equivalent to traveling that route in the reverse direction, we thus have a total of $\frac{(n-1)!}{2}$ possible routes in all to consider.

(This argument makes use of the MULTIPLICATION PRINCIPLE.) For the very smallest values of n, this count is small and it is quite feasible to program a computer to check all the possible routes, but for any reasonably sized practical problem, this formula yields an extraordinarily large count. For instance, the problem for just 20 towns already yields over 60 quadrillion different routes for consideration. Even the fastest computers of today are unable to assist us in finding the most efficient route in any reasonable amount of time.

What is sought is a manageable ALGORITHM that works in a reasonable amount of time. One possible approach is to use the nearest-neighbor algorithm:

> Starting in one town, visit the nearest town first, then visit the nearest town not already visited, and so forth. Return to the start town when no other choice is available.

A computer can easily be programmed to perform this algorithm quickly. Unfortunately, although easy to perform, it is known that this procedure might not necessarily yield the most efficient route overall.

It remains an unsolved problem as to whether there is a simple procedure that is guaranteed to produce the optimal route every time it is tried. Most mathematicians believe that such an algorithm will never be found.

See also NP COMPLETE; OPERATIONS RESEARCH.

tree *See* GRAPH.

triangle (trigon) A POLYGON with three sides is called a triangle. Precisely, a triangle is the closed geometric figure formed by three line segments (the sides of the triangle) joining three noncollinear points in the plane (the vertices of the triangle). Each triangle contains three interior ANGLES.

The PARALLEL POSTULATE shows that the three interior angles of a triangle sum to 180°. This can also be demonstrated physically with a pencil-turning trick:

> Draw a large triangle on a piece of paper and position a pencil in one corner of the triangle against one side. Take note of the direction the tip of the pencil is pointing.
> Rotate the pencil through the angle given by the corner in which it lies. Next slide the pencil along the side of the triangle to the second angle of the triangle. (Notice that this action of sliding does not alter the direction in which the pencil is currently pointing.) Rotate the pencil through this second angle, being sure to remain in the interior of the triangle. Now slide the pencil along the second side of the triangle to position it at the third angle of the triangle. Rotate the pencil through this third angle and slide the pencil back to its original position. Notice that the pencil is now pointing in the opposite direction to its start.

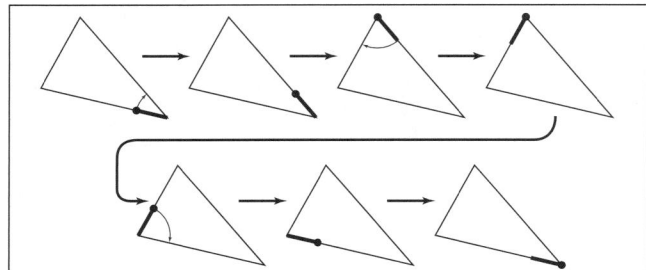

Turning a pencil inside a triangle

This demonstration shows that the net effect of rotating a pencil through each of the three angles of a triangle is to rotate the pencil through half a turn. Thus the three angles of the triangle do indeed sum to 180°. Moreover, it is clear that this argument would work for any triangle one could draw. A similar argument shows that the sum of angles in any quadrilateral is 360° in any five-sided polygon, 540°, and, in general, the sum of angles in any N-sided shape is $(N-2) \times 180°$.

Triangles are classified according to the relative lengths of their sides: A triangle is scalene if each side is of a different length, isosceles if at least two sides are equal in length, and equilateral if all three sides are the same length. A study of the SAS principle shows that the angles OPPOSITE the two sides equal in length in an isosceles triangle have equal measure. As an equilateral triangle is also isosceles, it follows that an equilateral triangle is EQUIANGULAR, that is, all three angles have the same measure (of 60°).

Triangles can alternatively be classified according to their angles. An acute triangle has all three angles less than 90° in measure, a right triangle contains precisely one angle equal to 90°, and an obtuse triangle contains one angle of measure greater than 90°. (It is not possible for a triangle to possess two obtuse angles.)

If a picture of a triangle is oriented so that one side of the triangle is horizontal, then that side is called the base of the triangle. The height of the triangle is the vertical distance between the base of the figure and the vertex of the triangle that does not lie on the base. A study of AREA shows that, if oriented appropriately, the interior of a triangle occupies half the interior of the rectangle that encloses the triangle, with the base as one side of the rectangle. Thus the area of a triangle is given by the formula:

$$\text{area} = \frac{1}{2} \times \text{base} \times \text{height}$$

If the three sides of the triangle are of lengths a, b, and c, then HERON'S FORMULA also shows that the area of the triangle is given by:

$$\text{area} = \sqrt{s(s-a)(s-b)(s-c)}$$

where $s = \dfrac{a+b+c}{2}$ is the semiperimeter of the triangle.

PYTHAGORAS'S THEOREM shows that if a right triangle has side-lengths a, b, and c, with the sides a and b surrounding the right angle, then $a^2 + b^2 = c^2$. The TRIANGLE INEQUALITY follows as a consequence.

For an arbitrary triangle with side-lengths a, b, and c, if the angle between sides a and b is acute (that is, less than 90°), then $a^2 + b^2 > c^2$. If, on the other hand, the angle between sides a and b is obtuse (greater than 90°), then $a^2 + b^2 < c^2$. This follows from the LAW OF COSINES. It can also be seen algebraically as follows:

For the diagram below left, the angle between sides a and b is acute. We clearly have $a > x$ and $b > y$, and so $a^2 + b^2 > x^2 + y^2 = c^2$. For the diagram below right, the angle between sides a and b is obtuse. We clearly have $x > a$. Consequently:

$$\begin{aligned} c^2 &= x^2 + y^2 \\ &= x^2 + (b^2 - (x-a)^2) \\ &= b^2 + 2ax - a^2 \\ &< b^2 + 2a \cdot a - a^2 = b^2 + a^2 \end{aligned}$$

These observations provide a proof of the CONVERSE of Pythagoras's theorem:

If, for a triangle with side-lengths a, b, and c, we have $a^2 + b^2 = c^2$, then the triangle is a right triangle with right angle between the sides of length a and b.

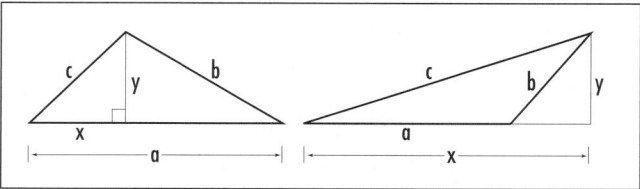

Establishing the converse of Pythagoras's theorem

(It cannot be the case that the angle between sides a and b is acute, for then $a^2 + b^2 > c^2$, which is not true. Nor can the angle between these two sides be obtuse. The only possibility that remains is that these two sides surround a right angle.)

For a right triangle or an obtuse triangle, these observations show that the longest side of such a triangle lies opposite the largest angle of the triangle. The LAW OF SINES can be used to prove that this is true even for acute triangles:

> In any triangle the longest side of the triangle lies opposite the largest angle in the triangle.

Special Properties of Triangles

There are a number of surprising properties of triangles worthy of note.

1. Lines and Concurrency

A PERPENDICULAR line that passes through the MIDPOINT of a given side of a triangle is called a perpendicular bisector of the triangle, and a perpendicular line that passes through the vertex opposite the given side is called an altitude of the triangle. A study of equidistance (*See* EQUIDISTANT) shows that the three perpendicular bisectors of any triangle pass through a common point (called the circumcenter of the triangle), as do the three altitudes of any triangle (through a point called the orthocenter of the triangle). That is, each set of three lines is CONCURRENT.

A MEDIAN OF A TRIANGLE is a line connecting one vertex of a triangle to the midpoint of the side opposite that triangle. It is possible to prove that the three medians of any triangle are also concurrent. (The common point of intersection is called the centroid of the triangle.) The Swiss mathematician LEONHARD EULER (1707–83) showed that the centroid, orthocenter, and circumcenter of any triangle lie in a straight line, today called the EULER LINE.

A study of equidistance also shows that the three lines bisecting the angles of a triangle are concurrent.

2. Equilateral Triangles

Choose any point P inside an equilateral triangle and compute the sum of the distances of this point from each of the three sides of the triangle. Then this sum equals the value of the height of the triangle, no matter where the point P is placed. (To see this, suppose the equilateral triangle has side-length b. Label the vertices of the triangle A, B, and C. Draw line segments connecting P to each vertex of the triangle. Then the area A of the triangle, $A = \frac{1}{2}bh$ where h is the height of the equilateral triangle, is the sum of the areas of the three triangles ABP, ACP, and BCP. Let h_1, h_2, and h_3 be the heights of these three triangles, respectively. We have $\frac{1}{2}bh = \frac{1}{2}bh_1 + \frac{1}{2}bh_2 + \frac{1}{2}bh_3$ yielding $h_1 + h_2 + h_3 = h$.)

3. Tilings

Any triangle can be used for a TESSELLATION of the plane.

4. Napoléon's Theorem

French emperor Napoléon Bonaparte (1769–1821) is alleged to have discovered the following surprising result:

> Given an arbitrary triangle, draw an equilateral triangle on the outside of each side of the triangle. Then the centers of these three equilateral triangles form another equilateral triangle, called the external Napoléon triangle of the original triangle. Moreover, if we flip each equilateral triangle so that it now lies inside the original triangle, then the centers of those triangles again form an equilateral triangle, called the internal Napoléon triangle. The difference of the areas of the external and internal Napoléon triangles equals the area of the original triangle.

5. Integer Triangles

A triangle is called an integer triangle if each side-length of the triangle is a whole number. Thus any triangle that is formed by laying out matchsticks, say, without breaking the matchsticks, is an integer triangle. It is interesting to count the number of integer triangles one can create with a fixed number of matchsticks. For example, there are four different integer triangles with perimeter 11: a 5-5-1 triangle, a 5-4-2 triangle, a 5-3-3 triangle, and a 4-4-3 triangle. Surprisingly, the count *decreases* if you add an extra matchstick to the collection: one can only make three different triangles with perimeter 12 (namely, 5-5-2, 5-4-3, and 4-4-4 triangles).

Mathematicians have proved that the number of different integer triangles one can make with n matchsticks is given by the formula $\left\langle \dfrac{n^2}{48} \right\rangle$ if n is even, and $\left\langle \dfrac{(n+3)^2}{48} \right\rangle$ if n is odd, where the brackets are used to indicate rounding to the nearest integer.

The integer triangle 5-12-13 (a right triangle) has the property that the numerical value of its area is the same as its perimeter: $A = 30$, $P = 30$. The 6-8-10 triangle is the only other right integer triangle with this property. If one relaxes the condition of being right, then there exist many nonright triangles with this property. For instance, a 7-15-20 triangle has area and perimeter each with the numerical value 42. (Use Heron's formula to compute its area.)

One can alternatively search for a pair of integer triangles sharing the same value for perimeter and same value for area. Again there are many such pairs. For example, the 14-18-29 and 8-25-28 triangles each have perimeter 61 and area $210\sqrt{22}$. The 45-94-94 and 49-84-100 also share the same perimeter and the same area.

See also AAA/AAS/ASA/SAS/SSS; BASE OF A POLYGON/POLYHEDRON; BERTRAND'S PARADOX; CIRCUMCIRCLE; EQUILATERAL; FIGURATE NUMBERS; HYPOTENUSE; PEDAL TRIANGLE; TRIANGULAR NUMBERS.

triangle inequality   The proposition that the sum of the lengths of any two sides of a TRIANGLE is greater than the length of the third side is called the triangular inequality. Thus, if a, b, and c are the three side-lengths of a triangle, then each of the following relations hold:

$$a < b + c$$
$$b < a + c$$
$$c < a + b$$

This result follows as a consequence of PYTHAGORAS'S THEOREM. The converse proposition is also true: If three numbers a, b, and c satisfy the three relations above, then it is possible to draw a triangle with side-lengths a, b, and c. (To see this, draw a line segment of length a, draw a circle of radius b with center one endpoint of the line segment, and a circle of radius c with the second endpoint as center. The three relations ensure that these circles intersect at a point P. Then P is the apex of an a-b-c triangle with side-length a as the base.) Thus the construction of a 7-9-12 triangle, for example, is possible, but the construction of a 16-23-40 triangle is not. (The number 40 is not greater than $16 + 23$.)

If any of the relations above is replaced by an equality, $a = b + c$, for instance, then the corresponding triangle is degenerate, meaning that its three vertices lie in a straight line. This observation can be used as follows:

> If the distance: from Adelaide to Darwin is 3,200 km, from Adelaide to Brisbane is 1,200 km, from Brisbane to Canberra is 600 km, and from Canberra to Darwin is 1,400 km, then one can only deduce that all four cities lie on the same straight line.

The triangular inequality can be rephrased as follows:

> The length of any one side of a triangle is less than half the perimeter of the triangle.

(Adding a to both sides of the first inequality, for instance, gives $2a < a + b + c$. The right quantity is the perimeter of the triangle.) This characterization allows one to quickly identify all triangles with integer sides having a given perimeter (as made with matchsticks, for instance). For example, with 11 matchsticks one can make four triangles given by the triples 5-5-1, 5-4-2, 5-3-3, and 4-4-3. (Each number is less than half of 11.) Surprisingly, the count goes down if one adds another matchstick to the collection—there are only three integer triangles with perimeter 12: 5-5-2, 5-4-3, and 4-4-4. Mathematicians have shown that the number of triangles one can produce with n matchsticks is $\left\langle \dfrac{n^2}{48} \right\rangle$ if n is even and $\left\langle \dfrac{(n+3)^2}{48} \right\rangle$ if n is odd, where the angled brackets indicate rounding to the nearest integer.

triangular numbers   *See* FIGURATE NUMBERS.

trigonometry   Contrary to its name, the theory of trigonometry is best motivated as a theory about CIRCLEs, not triangles. (This, in fact, matches the historical development of the subject.) Beginning with the simplest circle imaginable, namely, a circle of radius 1 centered about the origin, one defines two functions, sine and cosine, simply as the x- and y-coordinates of a

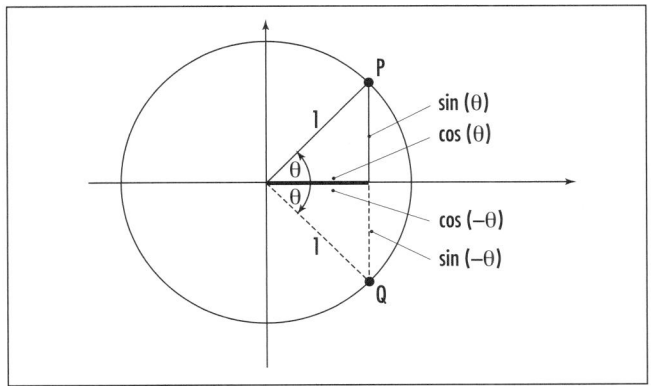

The sine and cosine of a positive and a negative angle

It is possible to compute the sine and cosine of some simple angles. For instance, a point located at angle θ = 0 degrees lies at position (1,0), and so cos(0) = 1 and sin(0) = 0. Similarly, the point at angle θ = 90° has coordinates (0,1) yielding cos(90) = 0 and sin(90) = 1. A point at angle θ = 45° lies at one vertex of an isosceles right triangle, showing that sine and cosine values for 45° are equal. Pythagoras's theorem tells us that $\cos(45) = \sin(45) = \frac{1}{\sqrt{2}}$. Similarly, a point at angle θ = 60 lies at the apex of half an equilateral triangle of side-length 1. We obtain $\cos(60) = \frac{1}{2}$ and, by Pythagoras's theorem, $\sin(60) = \frac{\sqrt{3}}{2}$. These values are reversed for an angle of 30°.

point on that circle located at a particular angle θ. It has become the convention to measure angles in relation to the positive *x*-axis, with a positive angle interpreted as a counterclockwise turn and a negative angle as a clockwise turn. The *y*-coordinate of a point on the circle is the sine (abbreviated "sin") of the angle at which that point lies, and the *x*-coordinate is the companion value cosine (abbreviated "cos"). Thus a point *P* located at angle θ has coordinates *P* = (cos(θ), sin(θ)). A point *Q* located at angle –θ has coordinates *Q* = (cos(–θ), sin(–θ)).

We see from the diagram that the sine and cosine functions satisfy the relations:

$$\cos(-\theta) = \cos(\theta)$$
$$\sin(-\theta) = -\sin(\theta)$$

Also, PYTHAGORAS'S THEOREM gives $(\sin(\theta))^2 + (\cos(\theta))^2 = 1$. Following the convention of writing $(\sin(\theta))^n$ as $\sin^n\theta$, this reads:

$$\sin^2\theta + \cos^2\theta = 1$$

Note that adding or subtracting a multiple of 360° (or, if using radian measure, any multiple of 2π) to an angle does not change the location of the point on the UNIT CIRCLE it represents. We have:

$$\cos(\theta + 360k) = \cos\theta$$
$$\sin(\theta + 360k) = \sin\theta$$

for any whole number *k*.

Connection to Larger Circles and to Triangles

If one enlarges the diagram of the unit circle by a SCALE factor *r*, then all lengths in that diagram increase by that factor *r*. Consequently, the coordinates of a point *P′* located at an angle θ on a circle of radius *r* (still centered about the origin) is given by the values cos θ and sinθ multiplied by *r*. We have *P′* = (*r* cos θ, *r* sin θ).

Within the diagram lies a right triangle with one angleθ and hypotenuse equal to *r*. The side opposite angle θ has length *r* sinθ, and the remaining side adjacent to the angle has length *r* cos θ. Notice that the ratio "opposite over hypotenuse" has value sin θ:

$$\frac{\text{opp}}{\text{hyp}} = \frac{r\sin\theta}{r} = \sin\theta$$

Similarly, the ratio "adjacent over hypotenuse" yields cosθ:

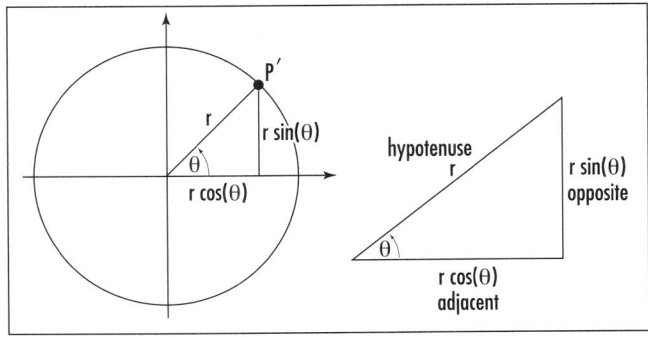

Relating trigonometry to triangles

History of Trigonometry

From very early times, surveyors, architects, navigators, and astronomers have made use of TRIANGLES to determine distances that could not be measured directly. This gave birth to the subject we today know as trigonometry. There are problems in the ancient Egyptian text, the RHIND PAPYRUS of around 1650 B.C.E., that call for the determination of the slope angles of pyramid faces using the equivalent of the cotangent function we use today, and a Babylonian clay tablet from around 1700 B.C.E. contains a table of secant values for the 15 angles between 30° and 45°. The Greek philosopher THALES OF MILETUS (ca. 600 B.C.E.) is said to have made use of similar triangles to determine the height of the Cheops pyramid by comparing the length of its shadow with the length of the shadow of a rod inserted in the ground. ERATOSTHENES OF CYRENE (ca. 250 B.C.E.) computed the circumference of the EARTH using lengths of shadows and a simple geometric argument on angles.

Greek astronomers of ancient times believed that the stars and planets of the night sky moved along circular arcs of a giant celestial sphere, and they worked to develop models that would accurately predict the motion of these objects on the sphere. Rather than phrase matters in terms of angles, which proved to be difficult, Greek astronomers chose to work with measures of straight lengths closely related to angles, namely, the lengths of CHORDS of CIRCLES.

Hipparchus of Rhodes (ca. 200 B.C.E.) constructed a table of such chord lengths for a circle of circumference 21,600 = 360×60 units (which corresponds to 1 unit of circumference for each minute of arc).

In the second century C.E., the mathematician CLAUDIUS PTOLEMY wrote the first extensive treatise on the theory of chords and their use in obtaining information about "spherical triangles," that is, triangles made by great circular arcs on the surface of a sphere. In addition to working out theorems, Ptolemy explained how to construct tables of chord values, and presented his own list of chord values for all angles between zero and 180° in half-degree increments.

The next important step in the development of trigonometry occurred in India. Scholars of the fifth century C.E. had by this time discovered that working with half-chords for half-angles greatly simplified the theory of chords and its applications to astronomy. As shown on the right figure, this approach is almost the same construct as the sine function of today. Whereas we think of sine as a ratio of lengths (the length of the half-chord to the radius), Indian scholars interpreted sine as the actual length of the half-chord. They called this length *jyā-ardha* or simply *jyā,*. Of course, the *jyā,* value of an angle differed for circles of different sizes. The scholars ARYABHATA, BHĀSKHARA II, and others developed astonishingly sophisticated techniques for computing half-chord values.

The Arab scholars of the 10th century took a great interest in the work from India. Mathematician Abu al-Wafa (ca. 950) of Baghdad systemized theorems and proofs of Indian trigonometry and prepared his own comprehensive table of half-chord values. He is also believed to have invented the tangent function, which he called the "shadow," and possibly the secant and the cosecant functions. (Still, all were thought of as specific lengths, not as ratios of lengths.) Arabic scholars did not know how to

$$\frac{\text{adj}}{\text{hyp}} = \frac{r\cos\theta}{r} = \cos\theta$$

As any right triangle can be viewed as coming from a circle with radius equal to the length of the hypotenuse, one can use these ratios as the *definitions* of the sine and cosine of a given angle θ. This is the approach usually taken in introductory texts on the subject. Scholars have given names to all six ratios that appear in a right triangle containing an angle θ:

$$\sin\theta = \frac{\text{opp}}{\text{hyp}}$$

$$\cos\theta = \frac{\text{adj}}{\text{hyp}}$$

$$\tan\theta = \frac{\text{opp}}{\text{adj}} = \frac{r\sin\theta}{r\cos\theta} = \frac{\sin\theta}{\cos\theta}$$

$$\sec\theta = \frac{\text{hyp}}{\text{adj}} = \frac{r}{r\cos\theta} = \frac{1}{\cos\theta}$$

$$\csc\theta = \frac{\text{hyp}}{\text{opp}} = \frac{r}{r\sin\theta} = \frac{1}{\sin\theta}$$

$$\cot\theta = \frac{\text{adj}}{\text{opp}} = \frac{r\cos\theta}{r\sin\theta} = \frac{\cos\theta}{\sin\theta}$$

Here tan stands for tangent, sec for secant, csc for cosecant, and cot for cotangent. These names come from the following observation:

Let *P* be a point on the unit circle located at an angle θ. Draw a vertical tangent line to the translate the Sanskrit word *jyā*, into their texts and simply wrote *jiba* as a close approximation.

In the 12th century, European scholars began translating the Arabic works and soon became acquainted with the extensive theory of trigonometry. Misinterpreting *jiba* as the Arabic word *jaib* for "cove" or "bay," translators wrote the Latin equivalent *sinus* as the name of the half-chord quantity. From this we have the name "sine."

The famous scholar FIBONACCI (ca 1170–1250) also traveled extensively in the Arab countries and wrote of their trigonometry in his famous work *Practica geometriae.* Around 1464, the German astronomer and mathematician REGIOMONTANUS (1436–76) compiled *De triangulis omnimodis,* a compendium of trigonometry of that time. This work was enormously influential, and over the following decades other texts on the topic appeared. German mathematician Georg Joachim Rhaeticus (1514–74) published, in 1551, *Canon doctrinae triangulorum,* which defined, for the first time, all six basic trigonometric functions, and explained how to relate them to right triangles without making any reference to circles. Danish physician Thomas Fincke (1561–1656), in his work *Geometriae rotundi,* coined the terms *tangent* and *secant,* and developed further fundamental trigonometric formulae. The word *trigonometry* itself was invented by German mathematician Bartholomaeus Pitiscus (1561–1613) in his treatise *Trigonometria: sive de solutione triangulorum tractatus brevis et perspicuus.* By this time, trigonometry was certainly regarded as a worthwhile topic of mathematical pursuit, independent of applications to astronomy.

The subject also proved to be useful in the study of algebra. French mathematician FRANÇOIS VIÈTE (1540–1603)

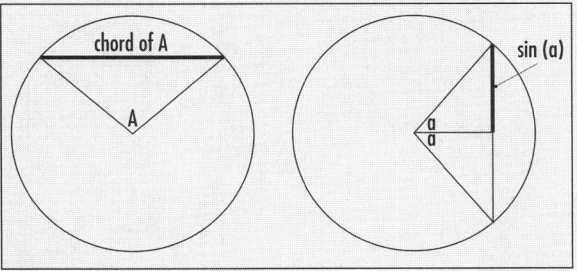

Chords and half-chords

showed, for example, that one could solve certain cubic equations by making trigonometric substitutions. His famous formula for π can be derived with repeated use of trigonometric functions.

Up until this point, sine values, as well as the other trigonometric values, were still regarded as actual line lengths and not ratios of lengths. After the invention of CALCULUS, LEONHARD EULER (1707–83) came to realize that it is appropriate to think of sine not as a physical length, but rather as a function of angle independent of length. He suggested that this could be accomplished by scaling all circles under consideration to unit circles, an operation that is equivalent to dividing all quantities by the radius of the circle. Thus, for the first time, all trigonometric quantities came to be thought of as ratios, In 1748 Euler wrote *Introductio in analysin infinitorum,* which became the dominating textbook on the topic of trigonometry for the century that followed. It essentially outlines the principles of trigonometry as we regard them today.

circle at the point $M = (1,0)$. Extend the radius OP further until it meets the tangent line at Q. Then $\tan \theta$ is the length of the tangent line-segment MQ and $\sec \theta$ is the length of the secant to the circle OQ.

Dividing the Pythagorean identity $\sin^2\theta + \cos^2\theta = 1$ through by $\cos^2\theta$ and by $\sin^2\theta$ yields the identities:

$$1 + \tan^2\theta = \sec^2\theta$$
$$1 + \cot^2\theta = \csc^2\theta$$

The Addition and Subtraction Formulae

The diagram on the following page shows that two congruent copies of a right triangle containing an angle A and two copies of a second right triangle containing an angle B can be arranged in the same large rectangle in two different ways.

The AREA of the region surrounding the four triangles, which must be the same for each diagram, can be computed as the sum of areas of two rectangles or as the area of a PARALLELOGRAM of base of length 1, and height of length $\cos(A - B)$. This yields the difference formula for cosines:

$$\cos(A - B) = \cos A \cos B + \sin A \sin B$$

Substituting in $-B$ for B, yields the addition formula for cosines:

$$\cos(A + B) = \cos A \cos(-B) + \sin A \sin(-B)$$
$$= \cos A \cos B - \sin A \sin B$$

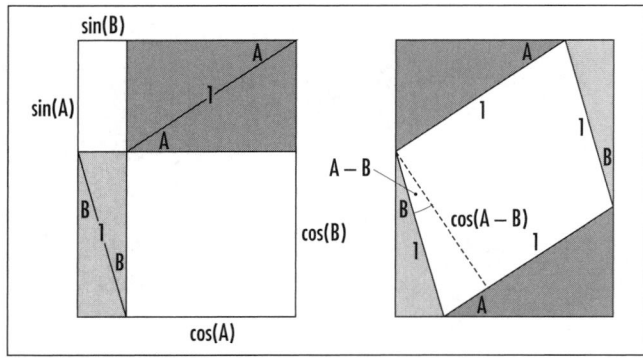

Proving the difference formula

Alternatively, place the angle A at the other corner of each of the triangles containing this angle, and use the same diagrams of four triangles arranged in a rectangle. This yields the sum and difference formulae for sines:

$$\sin(A + B) = \sin A \cos B + \cos A \sin B$$
$$\sin(A - B) = \sin A \cos B - \cos A \sin B$$

These formulae also follow readily from EULER'S FORMULA. We also have the following addition and subtraction formulae for the tangent function:

$$\tan(A + B) = \frac{\tan A + \tan B}{1 - \tan A \tan B}$$

$$\tan(A - B) = \frac{\tan A - \tan B}{1 + \tan A \tan B}$$

These are obtained by writing $\tan(A + B) = \sin(A + B)/\cos(A + B)$, for instance.

See also DE MOIVRE'S FORMULA; DERIVATIVE; HISTORY OF TRIGONOMETRY (essay); HYPERBOLIC FUNCTIONS; INVERSE TRIGONOMETRIC FUNCTIONS; TAYLOR SERIES.

trinomial Any algebraic expression consisting of three terms, such as $a + b + c$ or the quadratic expression $ax^2 + bx + c$, is called a trinomial.

See also MONOMIAL; BINOMIAL; POLYNOMIAL.

triple vector product There are two basic ways to combine the DOT PRODUCT and CROSS PRODUCT operations on VECTORs to form a product of three three-dimensional vectors **a**, **b**, and **c**.

The combination $\mathbf{a} \cdot (\mathbf{b} \times \mathbf{c})$ is called the scalar triple product of the three vectors. The result of this

operation is a real number (scalar) that is positive if **a**, **b**, and **c** form a right-handed system and is negative if, instead, they form a left-handed system. Changing the order of the vectors in this triple product can change the orientation of the system involved. We have:

$$\mathbf{a} \cdot (\mathbf{b} \times \mathbf{c}) = \mathbf{b} \cdot (\mathbf{c} \times \mathbf{a}) = \mathbf{c} \cdot (\mathbf{a} \times \mathbf{b}) = -\mathbf{b} \cdot (\mathbf{a} \times \mathbf{c})$$
$$= -\mathbf{a} \cdot (\mathbf{c} \times \mathbf{b}) = -\mathbf{c} \cdot (\mathbf{b} \times \mathbf{a})$$

The scalar triple product has a nice geometric interpretation:

The absolute value of the scalar triple product equals the volume of the PARALLELEPIPED formed by the vectors **a**, **b**, and **c**.

To see this, note that $\mathbf{b} \times \mathbf{c}$ is a vector perpendicular to the base of the parallelepiped formed by the two vectors **b** and **c**. If θ is the angle between this vector and **a**, then the height of the parallelepiped is the absolute value of $|\mathbf{a}| \cos(\theta)$. Since, according to the cross product, the area of the base is given by $|\mathbf{b} \times \mathbf{c}|$ we have that the volume of the parallelepiped is: base × height = $|\mathbf{a}||\mathbf{b} \times \mathbf{c}||\cos(\theta)|$, which is precisely the formula for the absolute value of the dot product: $\mathbf{a} \cdot (\mathbf{b} \times \mathbf{c})$.

An exercise in algebra shows that the scalar triple product of three vectors can be computed by taking the DETERMINANT of the 3×3 MATRIX whose rows are the entries of the vectors **a**, **b**, and **c**, respectively.

The vector triple product of three vectors is defined to be the combination: $\mathbf{a} \times (\mathbf{b} \times \mathbf{c})$. The result is a vector that lies perpendicular to $\mathbf{b} \times \mathbf{c}$, as well as to **a**. As $\mathbf{b} \times \mathbf{c}$ is perpendicular to the plane formed by **b** and **c**, it follows that $\mathbf{a} \times (\mathbf{b} \times \mathbf{c})$ lies in the **bc**-plane.

See also RIGHT-HANDED/LEFT-HANDED SYSTEM.

trisecting an angle One of the problems of antiquity (like DUPLICATING THE CUBE and SQUARING THE CIRCLE) of considerable interest to the classical Greek scholars

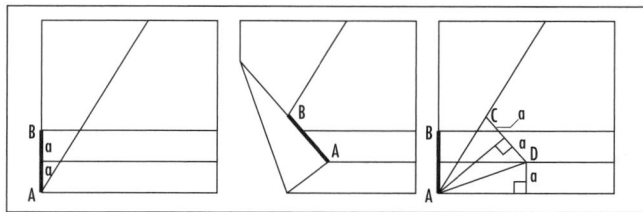

Trisecting an angle

was the task of developing a general procedure that would divide any given angle into three equal parts. The origin of this problem is unknown, though it seems a natural question to ask given that the procedure of bisecting any angle is relatively straightforward. (Given an angle defined by two rays, draw the arc of a circle centered at the vertex of the angle to find two points on the rays at an equal distance from this vertex. Now draw circles of fixed radius with centers at each of these two points. The line connecting the vertex of the angle to the point of intersection of the two circles is an angle bisector.)

Using a compass and a ruler, that is, a straightedge with specific lengths marked along it, HIPPOCRATES OF CHIOS (ca. 440 B.C.E.) developed a straightforward general solution to the problem, as did ARCHIMEDES OF SYRACUSE (ca. 287–212 B.C.E.) some 200 years later. Around this time, for the sake of increased intellectual challenge, scholars decided to add the restriction that only the primitive tools of a straightedge with no markings and a compass could be used in the solution. The difficulty of the problem increased significantly. APOLLONIUS OF PERGA (ca. 262–190 B.C.E.) found a solution to the problem using only the basic tools under the assumption that a HYPERBOLA could be drawn. Unfortunately, this was not deemed an admissible assumption. The problem remained one of the greatest unsolved challenges for two millennia.

In the early 1800s mathematicians focused on the specific problem of trisecting an angle of 60°, or, equivalently, the problem of constructing an angle of 20° using straightedge and compass alone. In particular, if such an angle can be produced, then it is possible to construct a line segment of length $x = \cos20$ (as one of the legs of a right triangle with angle 20°). The trigonometric identity $\cos(3\theta) = 4\cos^3(\theta) - 3\cos(\theta)$, noting that $\cos(60) = 1/2$, shows that the length x must satisfy the equation: $8x^3 - 6x - 1 = 0$.

The theory of CONSTRUCTIBLE numbers shows that any quantity of rational length can be constructed with straightedge and compass, and that if two lengths l_1 and l_2 can be produced, then so too can their sum, difference, product, and quotient, along with the square root of each quantity. Mathematicians showed that any solution x to the equation above cannot be rational and, moreover, in 1837, French mathematician Pierre Laurent Wantzel (1814–48) proved any such number x is not constructible. Thus the general problem of trisecting an angle is unsolvable.

It is interesting to note that in the field of origami, using only the tool of paper folding, it is possible to trisect any given acute angle. Assume the angle is placed in the bottom left corner A of a square piece of paper, with one ray defining the angle being the bottom edge of the sheet, and the other ray, call it L, a crease in the paper. Fold a crease of some arbitrary height parallel to the bottom edge and fold a second parallel crease half this height. Call the distance between these two creases a, and the point on the left edge at height $2a$ point B. Take the bottom left corner of the sheet and fold the paper so that point B lies on the crease L (call the location of this point C) and so that the point A lies on the crease at height a, to define a point D. The line connecting A to D is precisely one-third of the original angle.

(To see why this works, notice that PYTHAGORAS'S THEOREM establishes that line segments AD and BD have the same length. By the symmetry of the folding, length AC is the same as length AD, establishing that triangle CAD is isosceles. Draw the bisector of this triangle from A. This produces a diagram with three congruent right triangles, showing that the angle at A is indeed divided into three equal parts.)

Tristram Shandy paradox This paradox about the infinite is derived from Lawrence Sterne's 1760 novel *Tristram Shandy,* which purports to be part of the protagonist's autobiography. In it, the hero Shandy observes that it has taken him two years to describe his first two days, and so concludes that it will be impossible for him to ever complete the autobiography in full. Philosopher and mathematician BERTRAND ARTHUR WILLIAM RUSSELL (1872–1970) pointed out that if the author were immortal, however, he would be able to complete his goal, still writing at the same rate of progress. (Of course, it would take an infinite amount of time to do so.) Russell argued that a life of infinite length contains just as many years as it does days.

See also CARDINALITY; HILBERT'S INFINITE HOTEL; INFINITY; PARADOX.

trivial solution Any solution to a problem that is obvious, or of no interest in the given context, is called a trivial solution. For example, the famous equation of FERMAT'S LAST THEOREM $x^n + y^n = z^n$ has trivial solutions $x = 0$, $y = 0$, $z = 0$ and $x = 1$, $y = 0$, $z = 1$, and the differential equation $y\dfrac{d^2y}{dx^2} - \dfrac{dy}{dx}\sin(xy) + 2\sqrt{y} = 0$ has trivial solution $y = 0$. Any solution to a problem that may be of interest is called nontrivial. For example, the equation $x^3 - 2x = 0$ has nontrivial solution $x = \sqrt{2}$.

The word *trivial* comes from the Latin word *trivium*, the medieval name for the three least-complicated of the seven subjects of study offered in medieval universities: grammar, rhetoric, and logic. (The remaining four subjects—arithmetic, astronomy, geometry and music—were known as the *quadrivium*.)

Mathematicians often use the word *trivial* to describe a result that requires little or no effort to prove. For example, that any multiple of 4 is divisible by 2 would be deemed a trivial result.

truth table In the field of FORMAL LOGIC, propositional calculus is the name given to the analysis of truth-values of complicated statements built up from simpler statements linked together via the connectives *and; or; if … then …;* and *if, and only if* (called, respectively, the CONJUNCTION, the DISJUNCTION, the CONDITIONAL, and the BICONDITIONAL). One can also consider the NEGATION of a statement.

For example, the statement "The moon is round and is made of green cheese" is a compound statement made of two simpler statements ("The moon is round" and "The moon is made of green cheese") linked together via the connective *and*. The truth-value of the statement as a whole depends on the truth-value of the two individual statements of which it is composed.

In symbols, statements are usually represented as lowercase letters (p,q,r, for example), and the connectives combining them are denoted:

p and q: $p \wedge q$
p or q: $p \vee q$
If p then q: $p \rightarrow q$
p if, and only if, q: $p \leftrightarrow q$
not p: $\neg p$

A compound statement is a statement built up from simpler statements via connectives. For example,

$p \wedge (q \rightarrow r)$ and $p \vee (\neg p)$

are compound statements. (Parentheses are used to indicate the order in which the connectives are to be applied.) For example, if p is the statement "The moon is round," q the statement "The moon is made of green cheese," and r the statement "The moon is edible," then $p \wedge (q \rightarrow r)$ can be interpreted as, "The moon is round AND, IF the moon is made of green cheese, THEN it is edible." The statement $p \vee (\neg p)$ can be interpreted as "The moon either is or is not round."

A truth table is a table showing the truth-value of a statement (typically a compound one) given the possible truth-values of the simple statements of which it is composed. The truth-values of the basic connectives are given as follows:

p	q	$p \wedge q$	$p \vee q$	$p \rightarrow q$	$p \leftrightarrow q$
T	T	T	T	T	T
T	F	F	T	F	F
F	T	F	T	T	F
F	F	F	F	T	T

p	$\neg p$
T	F
F	T

(The truth-values presented here are motivated by intuition. *See* CONJUNCTION, DISJUNCTION, CONDITIONAL, BICONDITIONAL, and NEGATION for details.) The truth-value of *any* compound statement is now completely determined. The procedure is mechanical. For example, the compound statement $p \wedge (q \rightarrow r)$ has the following truth table, given in bold:

p	q	r	p	$\wedge$	$(q$	$\rightarrow$	$r)$
T	T	T	T	T	T	T	T
T	T	F	T	F	T	F	F
T	F	T	T	T	F	T	T
T	F	F	T	T	F	T	F
F	T	T	F	F	T	T	T
F	T	F	F	F	T	F	F
F	F	T	F	F	F	T	T
F	F	F	F	F	F	T	F
			(4)	(5)	(1)	(3)	(2)

(The numbers at the bottom of the table indicated the order in which the columns were computed. Column 5 displays the possible truth-values of the statement $p \wedge (q \rightarrow r)$ as a whole.) Thus if we deem p and r to be true and q false, for example, then we must conclude that $p \wedge (q \rightarrow r)$ as a as a whole is true. (Row three of the table.)

The truth table for $p \vee (\neg p)$ is:

p	p	$\vee$	$(\neg p)$
T	T	T	F
F	F	T	T
	(1)	(3)	(2)

A compound statement that is true irrespective of the truth-values of its component statements (that is, one that has T for every row of its truth table) is called a TAUTOLOGY. Thus $p \vee (\neg p)$, for example, is a tautology. One can also check that $(p \wedge q) \rightarrow (r \vee p)$ is another example of a tautology.

A compound statement that is always false is called a contradiction. For example, $p \wedge (\neg p)$ and $[(p \rightarrow q) \wedge p] \rightarrow (\neg q)$ are contradictions.

Two compound statements are logically equivalent if they have the same truth-values in the corresponding rows of their truth tables. For example, one can check that $p \rightarrow q$ and $(\neg p) \vee q$ have identical truth tables and are hence logically equivalent. The notion of such equivalence is linked to the biconditional.

See also ARGUMENT; LAWS OF THOUGHT.

turning point A point on a differentiable graph $y = f(x)$ at which $f'(x) = 0$ and $f'(x)$ changes sign is called a turning point. Such a point is either a local maximum or minimum for the graph.

See also LOCAL MAXIMUM/LOCAL MINIMUM.

twin primes Two consecutive odd numbers, both of which are PRIME, are called twin primes. For example, the pairs 3 and 5, 17 and 19, and 10,006,427 and 10,006,429 are twin primes. Mathematicians suspect that there are infinitely many pairs of twin primes, but are unable to prove this for a fact. This remains a famous unsolved challenge.

The largest twin primes known, as of the year 2002, are $665551035 \times 2^{80025} - 1$ and $665551035 \times 2^{80025} + 1$.

Two primes that differ by 4, such as 19 and 23 for instance, are called cousin primes. There are other informal names for pairs of primes that differ by six, eight, 10, and 12. The name "twin primes" was coined by the German mathematician Paul Stäckel (1892–1919).

The triple "3, 5, 7" is the only set of three consecutive odd numbers that are each prime. (Any triple of consecutive odd numbers must contain one number divisible by 3.) If the first odd number in the triple is one more than a multiple of 3, then the second odd number of the triple is a multiple of 3. If the first number in the triple is two more than a multiple of three, then the third odd number is divisible by 3.

two-card puzzle The following classic puzzle is often used to demonstrate the need for care when performing PROBABILITY calculations:

Two cards—one red on both sides, the other red on one side and black on the reverse—are placed into a bag. The bag is shaken, and one card is removed. You see that one face of this card is red. What is the probability that the reverse side of that card is black?

One is tempted to answer one-half—after all, the chances of selecting the red/black card is 1 in 2. However, this line of reasoning overlooks a subtlety: the puzzle is really one of CONDITIONAL PROBABILITY, and we are being asked to determine the probability of selecting the double-colored card *given* that the face we see is red. As there are three red faces, only one of which contains black on its reverse side, we conclude that there is actually only a one-third chance of this being the double-colored card. This surprising conclusion can be verified by performing the experiment multiple times.

See also BAYES'S THEOREM; KRUSKAL'S COUNT; MONTY HALL PROBLEM.

U

unary operation An operation or a rule that applies to single elements of a set to produce other (but not necessarily different) elements of the set is called a unary operation. For example, the rule that takes the absolute value of a real number can be thought of as a unary operation on the set of real numbers. The operation that reflects geometric shapes about the x-axis is a unary operation on the set of all geometric figures in the plane. However, the rule that associates each geometric figure with its area, for example, is *not* a unary operation on this set, since the results are numbers, not other geometric figures. A unary operation can be thought of as a FUNCTION from a set to the same set.

See also BINARY OPERATION; OPERATION.

unique solution An equation is said to have a unique solution if there is only one possible value of the variable(s) described in the equation that make the equation true. For example, the equation $x + 4 = 7$ has the unique solution $x = 3$, and the equation $a^2 + b^2 = 0$ has the unique solution, in the realm of the real numbers, $(a,b) = (0,0)$. On the other hand, the equation $x^2 - 9 = 0$ has more than one solution (x could either be 3 or -3), and the equation $x^{44} + 2y^6 = -7$ has no real solutions.

Geometric arguments can sometimes be used to determine whether or not a system of equations has a unique solution. For example, the pair of equations

$$2x + 4y = 5$$
$$x + 3y = 8$$

represents two straight lines of different slopes. As two nonparallel lines must intersect at a unique point, there is a unique solution to the system. A system of three equations representing three planes in space has a unique solution only if those three planes meet at a single point.

Sometimes a solution is called essentially unique if the term *uniqueness* refers only to the underlying structure of the problem and variations of the solution do not affect mathematical content. For example, the solution to the DIFFERENTIAL EQUATION $\frac{dy}{dx} = x$ is essentially unique. One solution is $\frac{1}{2}x^2$, and any other solution to this problem differs from this one only by the addition of a constant. In MODULAR ARITHMETIC, $x = 43$ is, essentially, the unique solution to:

$$x \equiv 3 \pmod 5$$
$$x \equiv 7 \pmod{12}$$

Any other solution to the problem differs from 43 by a multiple of 60.

unit circle The circle of radius 1 centered at the origin is called the unit circle. In CARTESIAN COORDINATES it has equation $x^2 + y^2 = 1$.

The unit circle plays a fundamental role in the development of TRIGONOMETRY: the cosine and sine of an angle θ are defined to be the x- and y-coordinates, respectively, of a point on a unit circle located at an angle θ from the *x*-axis. Consequently the PARAMETRIC EQUATIONS of the unit circle are $x = \cos\theta$ and $y = \sin\theta$, with θ as the parameter.

unknown (indeterminate, variable)   A quantity whose value is to be determined is called an unknown. In mathematics, finding the value of an unknown usually requires solving an equation. For the equation $2x + 5 = 13$, for instance, the variable *x* is the unknown. Algebra shows that this unknown has value 4.

In ancient times, scholars described unknowns and the equations relating them in words. Ancient Egyptian scholars used the word *heap* for an unknown quantity, and many other cultures simply used the word *thing*. Arab scholar MUHAMMAD IBN MŪSĀ AL-KHWĀRIZMĪ (ca. 780–850), from whose book title *Hisab al-jabr w'al muqābala* comes the word *algebra*, used the word *shai* to mean an unknown quantity. When his famous book was translated into Latin, the words *res* and *causa* (depending on the translator) were used for *shai*. German scholars of the Renaissance translated *causa* as *coss*, which the English adopted, calling the study of equations and unknown quantities "the cossic art." This name for algebra was popular for many decades. It literally means "the art of things."

V

Vallée-Poussin, Charles-Jean de la (1866–1962) Belgian *Analysis, Number theory* Born on August 14, 1866, in Louvain, Belgium, Charles-Jean de la Vallée-Poussin is remembered as one of the two mathematicians who independently proved the famous PRIME NUMBER THEOREM first conjectured by CARL FRIEDRICH GAUSS (1777–1855). The second mathematician to prove the result was French mathematician JACQUES HADAMARD (1865–1963). Even though both scholars applied complex function theory to the ZETA FUNCTION to establish the result, the two proofs utilized very different logical arguments.

Vallée-Poussin studied mathematics at the University of Louvain, at the University of Berlin, and at the University of Paris before accepting a faculty position at Louvain, where he stayed for the remainder of his career. He specialized in the field of ANALYSIS and was awarded a prize of a *couronne* from the Belgium Royal Academy for his work in DIFFERENTIAL EQUATIONS in 1892. At this time, Vallée-Poussin was studying newly developed techniques in complex analysis and came to the realization, in 1896, that all the pieces were in place to develop a proof of Gauss's prime-number conjecture. (Hadamard came to the same realization the same year.) Vallée-Poussin also took matters further to make significant contributions to the general study of the Riemann ZETA FUNCTION and the Riemann hypothesis as well as other concerns in analysis. In 1903 he published a very popular text *Cours d'analyse* (A course on analysis) on these topics, written for both the beginner and the expert in the field. (Vallée-Poussin wrote the introductory material in large font and interspersed discussions on advanced technical points in small font.) In subsequent years he wrote four revised editions of the text, as well as the book *Intégrales de Lebesgue* (Lebesgue integration), which is seen as an extension of the piece.

Vallée-Poussin received many honors throughout his career, including election to the Belgian Academy, the American Academy of Arts and Sciences, the Institute of France, and the American National Academy of Sciences. In 1928, after holding the position of chair of mathematics at the University of Louvain for 35 years, the king of Belgium conferred upon him the title of baron. Vallée-Poussin was also a commander of the Legion of Honor and honorary president of the International Mathematical Union. He died in Louvain on March 2, 1962.

See also GEORG FRIEDRICH BERNHARD RIEMANN.

variable A symbol, such as x, y, z, or n, that is used in algebraic equations to represent a quantity whose value is not known or specified is called a variable. For example, in the expression $x^2 + x = 6$, the symbol x represents an unknown number whose sum with its square equals 6. (The methods of ALGEBRA can be applied to show that x must either be 2 or –3.)

Variables are used to codify numeric operations. For example, the operation of squaring the first 10 whole numbers can be written "$y = n^2$ for $n = 1,2,3,...,10$." In this expression, the value of the variable n is not specified, but we are informed that it must be a whole number between 1 and 10 inclusive. The value of the variable y also is not specified, but we are told that its value depends on the chosen value for n. For this reason, y is called a dependent variable: once a value for n has been declared, a value for y then follows. The variable n is called an independent variable because its value is not hinged to the values of any other variables.

Two or more variables are also said to be independent if the value chosen of one variable has no effect on the possible value of the second. For example, if x represents the price of a pound of sugar on any given day, and y the number of students taking math at college, then x and y are independent variables. If each math student were to buy a pound of sugar, then the total money spent, $z = xy$, is a new variable. Here, z is not independent of x or y.

In algebra, symbols are also often used to denote quantities whose values are assumed known. For example, in the expression $ax + b = 0$, it is assumed that a and b are specified numbers, and that x is the value to be determined. (We have $x = -\dfrac{b}{a}$.) Here we are following a convention, first established by RENÉ DESCARTES (1596–1650), stating that lowercase letters at the end of the alphabet (such as x, y, and z) are to represent unknown quantities, and lowercase letters from the beginning of the alphabet (such as a, b, and c) known quantities. (The letters from the middle of the alphabet must be interpreted via context.)

See also CONSTANT; FUNCTION; HISTORY OF EQUATIONS AND ALGEBRA (essay).

variance   *See* STATISTICS: DESCRIPTIVE.

vector   In physics and engineering it is often appropriate to describe physical quantities, such as velocity or force, in terms of both magnitude and direction. For example, wind *speed* is completely specified by a single number that specifies its magnitude (wind strength), but wind *velocity* refers to both the speed of the wind and the direction in which the air is moving. Gravitational force acts on objects on the Earth's surface with a magnitude (strength) dependent on the mass of the object and in a direction pulling the object toward the center of the Earth. Any quantity that is specified by both a magnitude and a direction is called a vector.

Vectors are usually depicted as line segments with arrows indicating the direction of the vector. The length of the line segment indicates the magnitude of the vector. For example, a planar vector of magnitude 2 in the direction east is drawn in the plane as a horizontal line segment, 2 units long, with an arrow pointing to the right. The location at which this line segment starts is not important. All line segments 2 units long pointing to the right represent the same vector. Thus one is free to translate a vector to any starting position in the plane, as long as the length of the vector and its direction remain unchanged.

A vector is usually denoted in textbooks by boldface letters. For example, the symbol **a** might represent one vector and **b** another. However, in some texts, and often in handwritten notes, it is customary to underline letters or use overbars or arrows to indicate that the quantities represented are vectorial: $\underline{a}$, $\bar{a}$, or $\vec{a}$. The magnitude or length of a vector **a** is denoted $|a|$ (or sometimes as $\|a\|$).

A vector can be specified by an ordered pair of numbers. For example, the vector of magnitude 2 in the direction east can be represented as the pair: **a** = <2,0>. This vector "carries particles" 2 units to the right, and no units up or down. The vector **b** = <1,1> "carries particles" 1 unit right and 1 unit up. This represents the vector of magnitude $\sqrt{2}$ (the length of the diagonal of the unit square) in the direction northeast. The vector **c** = <−1,−1> is the same vector but pointing in the opposite direction. The vector **0** = <0,0> is called the zero vector and represents a quantity with no magnitude or direction.

The use of angle brackets to denote vectors is common in physics and engineering, although some texts use parentheses to represent vectors. This can be confusing, for parentheses are also used to denote points in the plane.

If $P = (p_1, p_2)$ and $Q = (q_1, q_2)$ represent two points in the plane, then a vector that "carries a particle" from position P to Q is given by: $<q_1 - p_1, q_2 - p_2>$. This vector is sometimes denoted $\overrightarrow{PQ}$. The point P is called the initial point of the vector, and Q is the terminal point. (However, keep in mind that a vector is not fixed at any specific location. *Any* vector of the same length and direction as $\overrightarrow{PQ}$ represents the same vector, even though the initial and terminal points might be different.)

If **a** = < a_1, a_2 > is a vector, then the numbers a_1 and a_2 are called the components of the vector. The length of the vector is given by PYTHAGORAS'S THEOREM: $|\mathbf{a}| = \sqrt{a_1^2 + a_2^2}$, which is just the DISTANCE FORMULA in this new setting.

Vectors drawn in the plane are called two-dimensional. Three-dimensional vectors, given by triples of numbers, represent quantities with magnitude and direction in three-dimensional space. SIR WILLIAM ROWAN HAMILTON (1805–65) coined the term *vector* for these quantities in his book *Lectures on Quaternions*. The name is based on the Latin word *vectus,* the perfect participle of *vehere,* which means "to carry."

Operations on Vectors

Vector addition combines two (or more) vectors to produce a new vector, called the resultant vector. To add together two vectors **a** and **b,** position vector **b** (without changing its magnitude or direction) so that its initial point lies at the terminal point of **a.** The resultant **a** + **b** is the vector that "carries particles" from the initial point of **a** to the terminal point of **b.** Geometrically, this is the vector that corresponds to the hypotenuse of the triangle formed by **a** and **b.**

Algebraically, if **a** = $<a_1, a_2>$ and **b** = $<b_1, b_2>$, then **a** + **b** = $<a_1 + b_1, a_2 + b_2>$.(If **a** "shifts particles" a_1 units to the right and **b** shifts particles b_1 units to the right, then the combined effect **a**+**b** shifts particles $a_1 + b_1$ rightward. Similarly for the combined vertical motion.) From either viewpoint we see that vector addition is COMMUTATIVE:

$$\mathbf{a} + \mathbf{b} = \mathbf{b} + \mathbf{a}$$

One can also check that vector addition is ASSOCIATIVE:

$$\mathbf{a} + (\mathbf{b} + \mathbf{c}) = (\mathbf{a} + \mathbf{b}) + \mathbf{c}$$

and that the zero vector acts as an identity element: **a** + **0** = **a** = **0** + **a.**

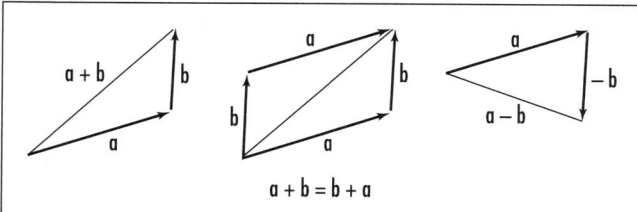

Vector operations

Scalar multiplication multiplies vectors by numbers to produce new vectors. In vector analysis it is customary to refer to numbers as scalars. If **a** is a vector and r a scalar, then $r\mathbf{a}$ is the vector with:

- The same direction as **a** if $r > 0$
- The opposite direction as **a** if $r < 0$
- Length $|r\mathbf{a}|$ equal to $|r|\,|\mathbf{a}|$

Algebraically, if **a** = $<a_1, a_2>$, then $r\mathbf{a} = <ra_1\ ra_2>$. Thus, for example, if **a** = $<6, -8>$, then $\frac{1}{2}\mathbf{a} = <3, -4>$ is the vector half as long, pointing in the same direction; $-\mathbf{a} = <-6, 8>$ is the vector of the same length but pointing in the reverse direction; and $0\mathbf{a} = <0, 0> = \mathbf{0}$ is the zero vector.

Vector subtraction is performed by adding one vector to the negative of the other. For example, if **a** and **b** are vectors, then the vector **a** – **b** is the result of adding –**b** to **a.** Geometrically, this corresponds to reversing the direction of the vector **b** and placing this reversed vector with initial point at the terminal point of **a,** as illustrated below. We have: **a** + (–**a**) = **0.**

The operations of vector addition and scalar multiplication satisfy a number of basic rules, yielding a mathematical system called a VECTOR SPACE.

There are two standard types of multiplication on vectors. These are the DOT PRODUCT and the CROSS PRODUCT.

A vector of length 1 is called a unit vector. In two-dimensional space there are two standard unit vectors **i** = $<1, 0>$ and **j** = $<0, 1>$. Any two-dimensional vector **a** = $< a_1, a_2 >$ can be written as a (linear) combination of these two unit vectors:

$$\mathbf{a} = < a_1, a_2 > = a_1 <1, 0> + a_2 < 0, 1 > = a_1\mathbf{i} + a_2\mathbf{j}$$

Similarly, in three-dimensional space, there are three standard unit vectors: **i** = $<1, 0, 0>$, **j** = $<0, 1, 0>$, and **k** = $<0, 0, 1>$. These too form a BASIS for the space of three-dimensional vectors. Physicists often prefer to represent vectors as combinations of the standard unit vectors.

See also ORTHOGONAL; PARALLELOGRAM LAW; POSITION VECTOR; PROJECTION; TENSOR; TRIPLE VECTOR PRODUCT; VECTOR EQUATION OF A LINE; VECTOR EQUATION OF A PLANE; VECTOR FIELD; VECTOR SPACE.

vector equation of a line The equation of a LINE in two-dimensional space can be written in the form:

$y = mx + b$. Unfortunately, there is no analogous equation for a line sitting in three-dimensional space. Observe, however, that if $P = (p_1, p_2, p_3)$ and $Q = (q_1, q_2, q_3)$ are two points on the line, then $\mathbf{u} = \overrightarrow{PQ} = <q_1 - p_1, q_2 - p_2, q_3 - p_3>$ is a VECTOR with direction along the line, and any other point on the line can be found by placing a scalar multiple of $\mathbf{u}$ at position P. That is, the equation:

$$P + t\mathbf{u} = (p_1 + t(q_1 - p_1), p_2 + t(q_2 - p_2), p_3 + t(q_3 - p_3))$$
$$= (1 - t)P + tQ$$

with t varying over all real numbers, describes all the points on the line. This is called the vector equation of the line. (Technically, one should work with the position vector $\mathbf{p} = <p_1, p_2, p_3>$ rather than the point $P = (p_1, p_2, p_3)$, so that the sum on the left of the above equation is the addition of two like quantities.)

The formula $y = mx + b$, describing the equation of a line in the plane, can also be thought of as a vector equation. Here, points (x, y) on the line are given by:

$$(0, b) + x(1, m)$$

where $(0, b)$ is one point of the line (the y-intercept) and $\mathbf{u} = <1, m>$ is the direction the line takes (1 step over and m steps up).

vector equation of a plane   The mathematical equation of a plane is a formula of the form:

$$ax + by + cz = d$$

where a, b, c, and d are numbers. Every point (x, y, z) satisfying this equation is a point on a plane with VECTOR $\mathbf{n} = <a, b, c>$ as NORMAL TO THE PLANE.

To derive this formula, let $P = (p_1, p_2, p_3)$ be a point on a given plane and $\mathbf{n} = <a, b, c>$ a vector perpendicular to the plane. To find the equation for any other point $Q = (x, y, z)$ on this plane, note that the POSITION VECTOR $\overrightarrow{PQ} = <x - p_1, y - p_2, z - p_3>$ lies in the plane and is consequently perpendicular to $\mathbf{n}$. The DOT PRODUCT $\mathbf{n} \cdot \overrightarrow{PQ}$ is thus zero. This gives the equation:

$$< a, b, c > \cdot < x - p_1, y - p_2, z - p_3 >$$
$$= a(x - p_1) + b(y - p_2) + c(z - p_3) = 0$$

which can be rewritten:

$$ax + by + cz = d$$

where d is just a number.

Notice that the components of $\mathbf{n}$ appear as the coefficients of the variables in the equation. This allows one to quickly "read off" normal vectors to planes. For example, the plane $2x - 3y + 4z = 10$ has normal vector $\mathbf{n} = < 2, -3, 4 >$, as does the plane $2x - 3y + 4z = 6$. The plane $y = 0$ has normal vector $\mathbf{n} = < 0, 1, 0 >$.

To find the equation of the plane with normal vector $\mathbf{n} = < 2, 2, 1 >$, say, and passing through the point $P = (5, -1, 3)$, begin by writing the partial equation: $2x + 2y + z = d$. To find d, substitute the coordinates of the point P. In this example we have: $d = 2 \cdot 5 + 2 \cdot (-1) + 3 = 11$, and so this plane has equation $2x + 2y + z = 11$.

vector field   A function that assigns to every point in space a VECTOR is called a vector field. For example, wind currents on the Earth's surface form a vector field on the surface of the Earth: at every location where there is a vector that describes the wind speed and direction. In three-dimensional space there is a gravitational vector field, given by the strength and direction of gravitational force, that a unit mass would experience when placed at each location in space. The strength and direction of the force varies from point to point.

The hairy ball theorem asserts that any smoothly varying vector field across the surface of a sphere, with vectors lying tangent to the sphere, must have at least one vector that is the zero vector. It shows, for example, that at any instant, there must be some location on the Earth's surface at which the horizontal wind speed is zero. It also shows that it is impossible to comb all the hairs of a tennis ball flat against the surface of the ball in a smooth uniform fashion without ever producing a cowlick. This second interpretation explains the name of the theorem.

See also CURL; DIV.

vector space   The set of two-dimensional or three-dimensional VECTORs come equipped with two fundamental operations: vector addition and scalar multiplication (that is, a multiplication by real numbers). These two operations obey the following rules:

1. Vector addition is COMMUTATIVE:

$$\mathbf{a} + \mathbf{b} = \mathbf{b} + \mathbf{a}$$

2. Vector addition is ASSOCIATIVE:

$$\mathbf{a} + (\mathbf{b} + \mathbf{c}) = (\mathbf{a} + \mathbf{b}) + \mathbf{c}$$

3. There is a zero vector **0** such that:

$$\mathbf{a} + \mathbf{0} = \mathbf{a}$$

4. Every vector **a** has a negative –**a** such that:

$$\mathbf{a} + (-\mathbf{a}) = \mathbf{0}$$

5. For every vector **a** we have:

$$1.\mathbf{a} = \mathbf{a}$$

6. If r is a scalar and **a** and **b** are vectors, then:

$$r(\mathbf{a} + \mathbf{b}) = r\mathbf{a} + r\mathbf{b}$$

7. If r and s are scalars and **a** is a vector, then:

$$(r + s)\mathbf{a} = r\mathbf{a} + s\mathbf{a}$$

8. If r and s are scalars and **a** is a vector, then:

$$(rs)\mathbf{a} = r(s\mathbf{a})$$

A vector space is *any* set V for which it is possible to define a notion of addition (that is, a rule that combines two elements of the set V to produce a new element of V) and scalar multiplication (that is, a means to multiply elements of V by numbers) so that the above eight rules hold. Certainly the set of all two-dimensional vectors forms a vector space, as does the set of all three-dimensional vectors, but examples of vector spaces need not be of this type. For example, let V be the set of all functions from the set of all real numbers to the set of real numbers. One can add any two functions f and *g:*

$$(f + g)(x) = f(x) + g(x)$$

and multiply functions by numbers:

$$(rf)(x) = rf(x)$$

One checks that all eight rules above hold (here **0** is the function that takes the constant value zero), making V a vector space. Also, the set of all 3×3 matrices is a vector space (one can add two matrices and multiply matrices by scalars), as is the set of all COMPLEX NUMBERS. (One can add two complex numbers and one can multiply complex numbers by real numbers.)

Properties of vectors and their algebraic manipulations have been studied extensively by scholars for centuries. That mathematicians have isolated the eight key properties that make vectors work the way they do allows one to immediately apply all that is known about vectors to *any* system, no matter how abstract it may be, that satisfies these eight basic rules. For example, mathematicians have proved that every vector space must have a BASIS. Consequently, there must be a basis for the set of all functions and for the set of all 3×3 matrices. Identifying one possible basis for the set of all functions leads to a study of FOURIER SERIES.

See also GROUP; LINEARLY DEPENDENT AND INDEPENDENT.

velocity   The study of motion examines three fundamental notions: distance, velocity, and acceleration.

The distance traveled by a moving object (also called its displacement) is the total length of the path it moved along. If the object travels along a straight-line path, then its motion is said to be rectilinear. (Motion that is not rectilinear is called curvilinear.) If, in addition to being rectilinear, an object travels equal distances D in equal periods of time T, then its motion is said to be uniform. This is the easiest type of motion to analyze. In this setting, the ratio $\frac{D}{T}$ is constant and is called the (uniform) velocity v of the object, $v = \frac{D}{T}$, and the picture of a velocity-vs.-time graph is a horizontal line at height v. The formula $D = v \times T$, coincidentally, is the equation for the area of the rectangle of width T and height v. This shows, in this simple scenario at least, that displacement equals the area under the velocity graph.

If, in rectilinear motion, the speed of an object changes with time, as when a car accelerates from rest to highway speed, then the analysis of velocity and displacement is more complicated. If $f(t)$ denotes the position of the object at time t, then the quantity $f(t+h)-f(t)$ represents the change in position of the object from time t to a later time $t+h$. The average velocity of the object over this time period is given by:

$$\frac{\text{change in position}}{\text{time taken}} = \frac{f(t+h)-f(t)}{h}$$

Computing the average velocity over smaller and smaller time intervals, that is, in taking smaller and smaller values of h in the formula above, gives the "actual" velocity of the object at time t as read by a speedometer, say. We have:

$$\text{instantaneous velocity} = \lim_{h\to 0}\frac{f(t+h)-f(t)}{h}$$

This, of course, is the formula for the DERIVATIVE of the distance function $f(t)$.

> For an object moving in a straight-line path, the instantaneous velocity of the object is the derivative of the distance formula. That is, instantaneous velocity is the instantaneous rate of change of position.

Now approximate the area under a velocity-vs.-time graph as a collection of narrow rectangles. Since, as we have seen, the area of each rectangle represents the distance traveled over a small period of time, the sum of these areas gives an approximation of total distance traveled by the object. Using narrower and narrower rectangles gives better and better approximations. In the LIMIT we have:

> For an object moving in a straight-line path, the INTEGRAL of the velocity function gives the total distance traveled. That is, the distance traveled is the area under the velocity-vs.-time graph.

If an object in motion follows a curved path, then one usually assigns to velocity not just a magnitude, but also a direction of motion. That is, velocity is considered a VECTOR. The term *speed* is used to denote the distance traveled per unit time, and *velocity* is this number along with an indication as to which direction this motion occurs.

The rate of change of velocity is called acceleration. Its magnitude is given by the first derivative of velocity (and, consequently, the second derivative of displacement). It too is considered a vector and is assigned a direction.

Acceleration a is the first derivative of velocity and the second derivative of distance:

$$a = \frac{d}{dt}\left(\text{velocity}\right) = \frac{d^2}{dt^2}\left(\text{distance}\right)$$

Physicists often follow SIR ISAAC NEWTON's notation of denoting differentiation with respect to time with a dot and of denoting displacement with the letter s. We have: $a = \dot{v} = \ddot{s}$.

If the position of a moving object is given by a set of PARAMETRIC EQUATIONS $x = x(t)$ and $y = y(t)$, then the position vector of the object is the vector $<x(t),y(t)>$, its velocity is the vector $<\dot{x}(t),\dot{y}(t)>$, and its acceleration is the vector $<\ddot{x}(t),\ddot{y}(t)>$.

Scientists at NASA sometimes use the term *jerk* to denote the rate of change of acceleration. Astronauts accelerating at a uniform rate will be pressed back into their seats with a constant force, leading to a smooth ride. Any change in the value of acceleration leads to changes in force pressures.

Acceleration due to gravity, denoted g, is the acceleration with which an object falls freely to earth unimpeded by air resistance. For many centuries it was believed that more massive objects would fall faster than lighter objects, but in 1638, GALILEO GALILEI (1564–1642) demonstrated, by theory and by experiment, that this is not the case: all falling objects accelerate at the same rate irrespective of their mass. (If acceleration were dependent on mass, at what rate would two falling bodies of different mass tied together fall?) The accepted value for g is 9.80665 m/sec/sec, but this magnitude varies at different locations of the Earth due to the fact that the Earth is not a perfect sphere. (At the poles its value is 9.8321 m/sec^2 and at the equator, 9.7799 m/sec^2.)

Venn, John (1834–1923) British *Logic, Probability theory* Born on August 4, 1834, in Hull, England, logician John Venn is remembered for introducing and popularizing the use of diagrams of overlapping circles as a means to represent relations between sets. Although such diagrams were used decades earlier by both GOTTFRIED WILHELM LEIBNIZ (1646–1716) and LEONHARD

EULER (1707–83) as a device to analyze ARGUMENTs, it was not until the publication of Venn's 1881 book *Symbolic Logic* that the practice of using such diagrams became common. Today these diagrams are named in his honor.

Venn graduated from Gonville and Caius College, Cambridge, in 1857 after studying theology and the liberal arts. After working as an ordained priest for several years, Venn returned to the same institution in 1862 to accept a position as a lecturer in moral science. During this time he pursued interests in logic and PROBABILITY theory, and developed a "frequency theory" of probability, which he published in 1866 in his *Logic of Chance*. This work greatly influenced the development of the theory of statistics. Fifteen years later Venn published *Symbolic Logic,* which he followed with *The Principles of Empirical Logic* in 1889. Venn was elected a fellow of the ROYAL SOCIETY in 1883.

After the publication of his work on logic, Venn changed interests and took to researching and writing a comprehensive account of the history of Gonville and Caius College. This was an all-consuming task, and only one volume of the work was published before his death, April 4, 1923.

Venn diagrams have had a profound effect on modern mathematics education. They are often used as a device for encouraging logical thinking at the early stage of a child's intellectual development and are a standard feature in an elementary-school curriculum.

See also VENN DIAGRAM.

Venn diagram A diagram in which mathematical sets are represented by overlapping circles within a boundary representing the universal set is called a Venn diagram. Such diagrams provide convenient pictorial representations of relations between sets. For example, in the diagram above right, a universal set *U* is represented by the interior of a rectangle, and two subsets *A* and *B* of *U* as the interiors of two overlapping circles within the rectangle. The intersection $A \cap B$, the union $A \cup B$, the complement of *B* within *A*, denoted $A - B$, and the universal complement of *B*, denoted B', for instance, are readily apparent.

For example, if *U* is the set of all insects, *A* is the subset of all butterflies, and *B* is the subset of all blue insects, then the shaded region $A \cap B$ represents all blue butterflies, $A \cup B$ all insects that are either blue or

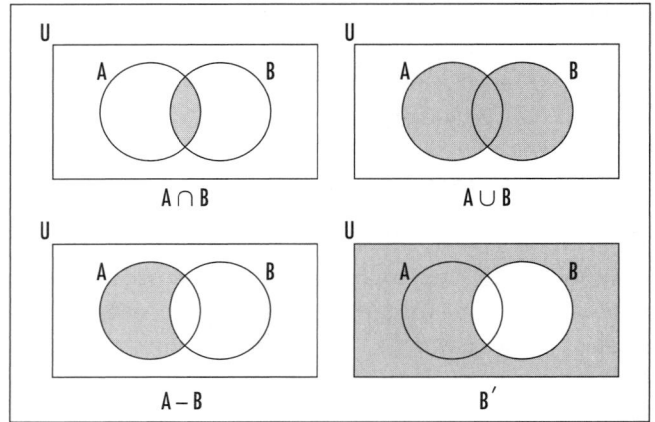

Set operations

are a butterfly, $A - B$ all butterflies that are not blue, and B' all insects that are not blue.

It is possible to demonstrate the validity of an ARGUMENT with the aid of a Venn diagram. Consider, for example, the following line of reasoning:

All logicians are mathematicians. Some philosophers are logicians. Therefore, some philosophers are mathematicians.

In the universal set of all people, let *L* be the set of logicians, *M* the set of mathematicians, and *P* the set of philosophers. The first premise of the argument asserts that *L* is a subset of *M*, and the second that the sets *P* and *L* have a nonempty intersection. This leads to a Venn diagram of the form below.

It is now apparent that *P* and *L* can intersect only inside of M thereby establishing the validity of the conclusion. The German mathematician GOTTFRIED

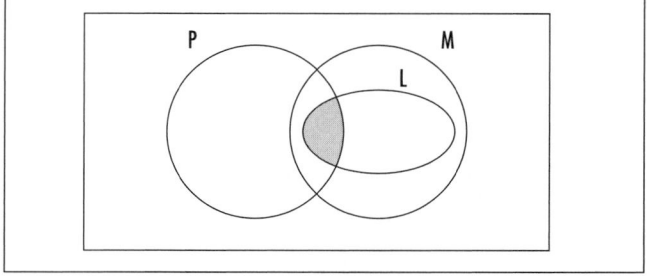

A Venn diagram

WILHELM LEIBNIZ (1646–1716) and the great LEONHARD EULER (1707–83) each used diagrams similar to these to analyze arguments. In the context of FORMAL LOGIC Venn diagrams are often called EULER DIAGRAMS.

A single set *A* divides the universal set into two disjoint pieces, *A* and its complement *A′*. Two subsets *A* and *B* generally divide the universal set into four disjoint pieces: $A \cap B$, $A' \cap B$, $A \cap B'$, and $A' \cap B'$. (This is not the case, however, if *A* is a subset of *B*, *B* is a subset of *A,* or if *A* and *B* do not intersect.) Three subsets *A, B,* and *C,* in their most general arrangement, divide the universal set into eight disjoint pieces. (The diagram produced can be used to illustrate DE MORGAN'S LAWS.)

It is not possible to draw a diagram of four overlapping circles to represent four general subsets *A, B, C,* and *D* in such a way as to make the 16 disjoint pieces of the universal set apparent. For this reason, Venn diagrams are usually only used to illustrate relations between just two or three sets.

See also SET THEORY; JOHN VENN.

Viète, François (Franciscus Vieta) (1540–1603) French
Algebra   Born in Fontenay-le-Comte, France, in 1540 (his exact birth date is not known), scholar François Viète is often referred to as "the father of algebra." His influential 1591 work *In artem analyticam isagoge* (Introduction to the analytic arts) is noted as one of the earliest Western texts on the topic. His practice of using letters as symbols for unknowns represented a first step toward the development of modern algebraic notation and allowed him to make significant advances in the subject. Viète also made important contributions to the field of TRIGONOMETRY.

Viète was never employed as a professional mathematician. After graduating from the University of Poitiers in 1560, he began a career in legal practice, but soon decided to change occupations by accepting a position as a private tutor in 1564. Within this role he pursued an interest in mathematics and soon developed a reputation as a capable scholar. Viète occasionally gave mathematical lectures at institutes of higher learning, and in 1589 was employed by King Henry IV of France to decode secret messages being sent to the enemy of the state, Philip II of Spain.

His first published work, *Canon mathematicus seu ad triangula* (The mathematical canon applied to triangles), appeared in 1579 and was intended as an introduction to mathematical astronomy. In it Viète used the six main trigonometric functions to solve problems relating to plane and spherical triangles, listed tables of values for these functions, and explained the mathematics behind the construction of those tables. Later, in 1593, Viète wrote two other texts on the topic of trigonometry, *Zeteticorum libri quinque* (Algebra volume five), mimicking the fifth book in DIOPHANTUS's *Arithmetica*, and *Variorum de rebus mathematicis responsorum* (Various mathematical problems), which contains his famous formula for π as an infinite product of radicals. He also lectured on the classical problems of DUPLICATING THE CUBE and TRISECTING AN ANGLE.

Today Viète is best remembered for his advances in algebra as presented in his famous 1591 text *Isagoge*, which clearly demonstrated the value of manipulating letters as symbols for both known and unknown quantities as a means to solve algebraic problems. Viète also introduced improved notation for squares, cubes, and other powers, as well as coined the term *coefficient*. With his new approach to algebra, Viète successfully tackled a number of problems that classical scholars had been unable to solve. Viète died in Paris, France, on February 23, 1603.

See also COEFFICIENT; VIÈTE'S FORMULA.

Viète's formula (Vieta's formula)   In 1593 French mathematician FRANÇOIS VIÈTE discovered the following remarkable formula relating π to an infinite product of radicals:

$$\frac{2}{\pi} = \sqrt{\frac{1}{2}} \times \sqrt{\frac{1}{2} + \frac{1}{2}\sqrt{\frac{1}{2}}} \times \sqrt{\frac{1}{2} + \frac{1}{2}\sqrt{\frac{1}{2} + \frac{1}{2}\sqrt{\frac{1}{2}}}} \times \cdots$$

His formula is established by making repeated use of the double angle formulae from TRIGONOMETRY:

$$\sin(2\theta) = 2\sin(\theta)\cos(\theta)$$
$$\cos(2\theta) = 2\cos^2(\theta) - 1$$

Begin by writing:

$$\sin(\theta) = 2\cos\left(\frac{\theta}{2}\right)\sin\left(\frac{\theta}{2}\right)$$

$$= 2^2 \cos\left(\frac{\theta}{2}\right)\cos\left(\frac{\theta}{2^2}\right)\sin\left(\frac{\theta}{2^2}\right)$$

$$= \ldots$$

$$= 2^n \cos\left(\frac{\theta}{2}\right)\cos\left(\frac{\theta}{2^2}\right)\cos\left(\frac{\theta}{2^3}\right)\cdots\cos\left(\frac{\theta}{2^n}\right)\sin\left(\frac{\theta}{2^n}\right)$$

Then:

$$\frac{\sin(\theta)}{\theta} = \cos\left(\frac{\theta}{2}\right)\cos\left(\frac{\theta}{2^2}\right)\cos\left(\frac{\theta}{2^3}\right)\cdots\cos\left(\frac{\theta}{2^n}\right)\times\frac{\sin\left(\frac{\theta}{2^n}\right)}{\left(\frac{\theta}{2^n}\right)}.$$

Using the double-angle formula for cosine, this can be rewritten:

$$\frac{\sin(\theta)}{\theta} = \sqrt{\frac{1}{2}+\frac{1}{2}\cos(\theta)}\times\sqrt{\frac{1}{2}+\frac{1}{2}\sqrt{\frac{1}{2}+\frac{1}{2}\cos(\theta)}}$$

$$\times\sqrt{\frac{1}{2}+\frac{1}{2}\sqrt{\frac{1}{2}+\frac{1}{2}\sqrt{\frac{1}{2}+\frac{1}{2}\cos(\theta)}}}\times\cdots\times\frac{\sin\left(\frac{\theta}{2^n}\right)}{\left(\frac{\theta}{2^n}\right)}$$

still with n radical terms. Put $\theta = \frac{\pi}{2}$. Since $\sin\left(\frac{\pi}{2}\right)=1$ and $\cos\left(\frac{\pi}{2}\right)=0$, this gives:

$$\frac{2}{\pi} = \sqrt{\frac{1}{2}}\times\sqrt{\frac{1}{2}+\frac{1}{2}\sqrt{\frac{1}{2}}}\times\sqrt{\frac{1}{2}+\frac{1}{2}\sqrt{\frac{1}{2}+\frac{1}{2}\sqrt{\frac{1}{2}}}}$$

$$\times\cdots\times\frac{\sin\left(\frac{\pi}{2^{n+1}}\right)}{\left(\frac{\pi}{2^{n+1}}\right)}$$

and since, according to the SQUEEZE RULE, $\frac{\sin(x)}{x}\to 1$ as $x\to 0$, Viète's formula follows.

See also PI, WALLIS'S PRODUCT.

vinculum In the 15th century it was popular to place a horizontal line above, or sometimes below, a group of terms in an expression to indicate that those terms were to be treated as a unit in the evaluation of that expression. The horizontal line was called a vinculum. Today we use parentheses to indicate the order of operations. For example,

$$x - \overline{y + z} = x - (y + z)$$

Use of the vinculum first appeared in the 1484 manuscript *Le triparty en la science des nombres* (Triparty in the science of numbers) written by French physician Nicola Chuquet (1445–88). As printing presses were developed, parentheses were adopted so as to ease typesetting.

Today a vinculum is primarily used to indicate repeating decimals (for example, $\frac{1}{7} = 0.\overline{142857}$ with the quantity 142857 regarded as a repeating unit) and to denote the complex conjugate of a COMPLEX NUMBER.

The horizontal bar used in writing fractions is also sometimes called a vinculum. It too can be thought of as a device for indicating which terms are to be treated as a unit. For example, in the expression $\frac{x+y}{3}$ the entire quantity "$x + y$" is to be divided by three, not x alone nor y alone. Thus $\frac{x+y}{3} = \frac{x}{3} + \frac{y}{3}$.

volume Loosely speaking, the volume of a three-dimensional object or region is "the amount of space it occupies." Such a definition appeals to intuitive understanding. In general, however, it is very difficult to explain precisely just what it is we mean by "space" and the "amount" of it occupied. As with the notion of AREA, this is a serious issue, as demonstrated by the Banach-Tarski paradox that arises in the careful study of that topic.

In any case, given some kind of understanding of what we mean by "area," it is natural to then define the "volume" of a simple object with vertical walls of height h and a base of area A to simply be the product $A \times h$:

volume = area of base × height

For example, a rectangular box a units wide, b units deep, and c units high has a base of area $a \times b$ and thus a volume given by $a \times b \times c$. A cylinder of height h with circular base of radius r has base area πr^2 and thus a volume $\pi r^2 h$.

This point of view imagines volumes as well approximated by stacks of thin layers of "volume," each a horizontal cross-section of solid that is the same size and shape as the base. Like a deck of 52 cards stacked to produce a rectangular box, the volume of a stack of sheets does not change if the pile is skewed: the shape of the deck may change, but its volume does not. This observation was first noted by 17th century Italian mathematician BONAVENTURA CAVALIERI (1598–1647). It leads to a general principle, today called CAVALIERI'S PRINCIPLE:

> Solids of equal height have equal volumes if sections made by planes parallel to the base at the same distance from the base have equal areas.

Techniques of INTEGRAL CALCULUS are used to compute volumes. The methods here are really no different than approximating solids again as stacks of thin cards or, perhaps, as collections of small boxes (whose volumes are known) and taking the LIMIT as better and better approximations are made.

For example, suppose a solid has height h and that the area of a cross section made by a plane parallel to the base at height x is $A(x)$. Imagine we slice the solid into thin "cards" at positions $0 = x_0 < x_1 < \ldots x_{n-1} < x_n = h$. Then the volume V of the solid can be well approximated as the sum of volumes of cards of base area $A(x_i)$ and width $x_{i+1} - x_i$, that is, $V \approx \sum_{i=0}^{n-1} A(x_i)(x_{i+1} - x_i)$. Taking the limit as more and more values between 0 and h are chosen (that is, as thinner and thinner cards are used) gives the true volume of the solid as an integral:

$$V = \int_0^b A(x)dx$$

For instance, the cross section of a sphere of radius r at a distance x from the plane running through the equator is a circle of radius $\sqrt{r^2 + x^2}$ and area $\pi(r^2 - x^2)$. Thus the volume of a sphere of radius r is:

$$\frac{\pi}{2} = \frac{2}{1} \cdot \frac{2}{3} \cdot \frac{4}{3} \cdot \frac{4}{5} \cdot \frac{6}{5} \cdot \frac{6}{7} \cdots$$

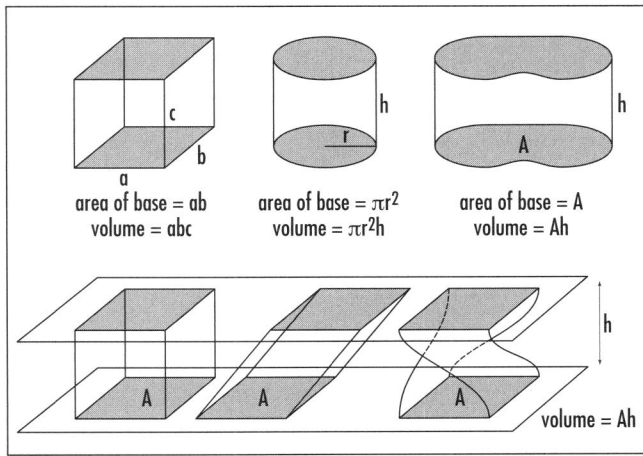

area of base = ab   area of base = πr^2 area of base = A
volume = abc   volume = $\pi r^2 h$ volume = Ah

volume = Ah

The volume of figures

Any cross section of a CONE with base of an arbitrary shape is a scaled version of the same planar shape. At half the distance from the apex of the cone, the cross section has area one-quarter the area of the base. At one-third the distance from the apex, the cross section has area one-ninth the area of the base. In general, if the area of the base is A and the height of the cone is h, then the area of the cross section at a distance x from the apex has area $A(x) = A \times \left(\frac{x}{h}\right)^2$. The volume of any cone with base of area A and height h is thus:

$$V = \int_0^b A\left(\frac{x}{h}\right)^2 dx = \frac{1}{3} Ah$$

The volume of a FRUSTUM can be calculated the same way.

The word *volume* comes from the Latin *volumen*, meaning "scroll," where the size or the bulk of a book eventually led to the use of the word for the size or bulk of any object.

See also DOUBLE INTEGRAL; SCALE; SOLID OF REVOLUTION.

Wallis, John (1616–1703) British *Calculus* Born on November 23, 1616, in Kent, England, mathematician John Wallis is remembered for his substantial contributions to the early development of CALCULUS, which laid the groundwork for SIR ISAAC NEWTON (1642–1727) to later fully develop the subject. His contributions to mechanics also helped Newton formulate the laws of motion. He was considered the leading English mathematician of his time.

Wallis had little exposure to mathematics during the early part of his life. He received a bachelor of arts degree in 1637 and a master's degree in 1640 in theology and was ordained as a minister in 1640.

By happenstance, while working as chaplain during the time of civil unrest in England, Wallis discovered that he could easily decipher the coded messages parishioners shared with him. This demonstrated a talent for mathematical thinking, and Wallis was soon hired by the Parliamentarians to decode Royalist messages.

During this time Wallis read the works of JOHANNES KEPLER and BONAVENTURA CAVALIERI and others, and developed an interest to expand on the ideas of infinitesimal calculus as developed by these authors. In 1656 Wallis wrote his influential text *Arithmetica infinitorum* (Infinite series), which contains, among many things, his famous product formula for $\frac{\pi}{2}$.

Wallis also studied and wrote on the topics of CONIC SECTIONS and ALGEBRA. He was the first to explain the meaning of zero, negative, and fractional powers

John Wallis, an eminent mathematician of the 17th century, used infinitesimals to develop important techniques of computation. Sir Isaac Newton built much of his development of calculus on Wallis's ideas. *(Photo courtesy of the Science Museum, London/Topham-HIP/The Image Works)*

$x^0, x^{-n}, x^{\frac{n}{m}}$) and was one of the first scholars to permit negative and complex solutions to equations. He argued that the number i, representing the square root of -1, does have a place in real-world applications of mathematics. Wallis also introduced the symbol ∞ for infinity.

In addition to working as a research mathematician, Wallis also made significant contributions as a mathematical historian. He restored several ancient Greek texts and presented a comprehensive survey of the history of algebra in his 1685 work *Treatise on Algebra*. He died in Oxford, England, on November 8, 1703.

Wallis helped shape the entire course of mathematical work in England for the latter part of the 17th century and paved the way for Newton to develop his ideas. As a result, Britain became the center of mathematical research in the late 1600s and remained so until the influence of LEONHARD EULER (1707–83) and members of the BERNOULLI FAMILY moved the focus back to continental Europe in the mid 1700s.

See also WALLIS'S PRODUCT.

Wallis's product In 1656 English mathematician JOHN WALLIS discovered the following remarkable expression for the number $\frac{\pi}{2}$ as an INFINITE PRODUCT:

$$\frac{\pi}{2} = \frac{2}{1} \cdot \frac{2}{3} \cdot \frac{4}{3} \cdot \frac{4}{5} \cdot \frac{6}{5} \cdot \frac{6}{7} \cdots$$

He discovered this result while attempting to compute INTEGRALS of the form

$$\int_0^1 (1-x^2)^n \, dx$$

with n not necessarily an integer. (For example, when $n = \frac{1}{2}$, four times this integral equals the area of a circle.) The work of LEONHARD EULER (1707–83) on the ZETA FUNCTION also leads to a proof of Wallis's product, one that is relatively straightforward to follow.

Wallis's formula led English colleague LORD WILLIAM BROUNCKER (1620–84) to discover the following astonishing CONTINUED FRACTION formula for π:

$$\frac{4}{\pi} = 1 + \cfrac{1^2}{2 + \cfrac{3^2}{2 + \cfrac{5^2}{2 + \cfrac{7^2}{2 + \cdots}}}}$$

See also PI.

Weierstrass, Karl Theodor Wilhelm (1815–1897) German *Analysis* Born on October 31, 1815, German scholar Karl Weierstrass is remembered as a leading figure in the field of mathematical ANALYSIS. Throughout his career he emphasized the need for absolute rigor, and, following the efforts of AUGUSTIN-LOUIS CAUCHY (1789–1857), worked to introduce very precise definitions of fundamental notions in the study of CALCULUS. He developed the famous $\varepsilon - \delta$ definition of a LIMIT that we use today, as well as precise clarification of the meaning of continuity, convergence, and of DIFFERENTIALS. To illustrate that intuitive understanding alone never suffices, Weierstrass presented examples of pathological functions that are continuous but have no well-defined tangent lines. Weierstrass also solved the famous ISOPERIMETRIC PROBLEM with his "calculus of variations."

Weierstrass entered the University of Bonn in 1834 to pursue a degree in law and finance, but he never completed the program, choosing to follow instead a study of mathematics at the Theological and Philosophical Academy of Münster in 1839. He began his academic career as a provincial mathematics schoolteacher.

While teaching, Weierstrass published a number of papers on the study of real and complex functions. The bulk of his early work went unnoticed by the mathematics community, and it was not until the publication of his 1854 piece *Zur Theorie der Abelschen Functionen* (On the theory of Abelian functions) that Weierstrass's genius as a mathematician was recognized. He was immediately awarded an honorary doctoral degree from the University of Königsberg and was granted a year's leave from the school to pursue advanced mathematical study. He never returned to school teaching, however. In 1856, at the age of 40, Weierstrass accepted a professorship at the University of Berlin, which he retained for the remainder of his life.

Weierstrass made significant contributions to the study of POWER SERIES and their convergence, and to FOURIER SERIES and their applications to problems in geometry and mechanics. In providing a precise definition of a limit, Weierstrass completely revolutionized mathematical understanding of the founding principles of CALCULUS and modern analysis. In a very real sense, Weierstrass set the standards of rigor that all mathematicians still follow today.

In the study of CONVERGENT SERIES, the comparison test shows that the function:

$$f(x) = \sum_{n=1}^{\infty} \frac{\sin(\pi n^2 x)}{\pi n^2}$$

converges for every value of x. (Compare with series $\sum_{n=1}^{\infty} \frac{1}{n^2}$) Weierstrass was able to show, moreover, that this function is continuous, but that it has no derivative at any irrational value of x. This was perhaps his most famous example of a pathological function that is continuous but possesses no meaningful tangent line at almost all points. That such functions exist, as Weierstrass pointed out, indeed demonstrates the need for uncompromising care in the development of details in all theoretical work. Jules Henri Poincaré (1854–1912) admired his "unity of thought" on this matter.

Weierstrass died in Berlin, Germany, on February 19, 1897.

well-ordered set An ORDERED SET A is said to be well ordered if every nonempty subset of A has a smallest member. For example, the set of natural numbers is well ordered: for any set of natural numbers one cares to describe, there will be a smallest member of that set. On the other hand, the set of all integers is not well ordered. For example, the subset of negative integers has no smallest member.

In 1904 German mathematician ERNST FRIEDRICH FERDINAND ZERMELO (1871–1953) proved that any ordered set can be made well ordered if one is willing to change the ordering on that set. For example, the set of all integers can be made well ordered if one arranges them as follows:

$$0, 1, -1, 2, -2, 3, -3, 4,\ldots$$

(The "smallest" element in the set of all negative integers, for instance, is now –1.) Zermelo needed to invoke a controversial AXIOM OF CHOICE to prove this claim.

Whitehead, Alfred North (1861–1947) British *Logic, Theoretical physics, Philosophy* Born on February 15, 1861, in Kent, England, Alfred Whitehead is best remembered in mathematics for his three-volume work *Principia Mathematica*, written in collaboration with English logician BERTRAND ARTHUR WILLIAM RUSSELL (1872–1970). Whitehead and Russell were hoping to derive the whole of mathematics from purely logical principles.

Whitehead entered Trinity College, Cambridge, in 1880 to study applied mathematics. He worked and taught in this field for 12 years before his interests eventually turned toward pure mathematics and the foundational principles of the subject. In 1891 he commenced work on *Treatise on Universal Algebra*, a project that took him seven years to complete. At the same time, Bertrand Russell entered the college as an undergraduate, and Whitehead immediately recognized him as a brilliant philosopher and capable mathematical scholar. Their collaboration on *Principia Mathematica* began in 1900 and was initially intended to be a one-volume work. Russell's 1901 discovery of his famous set-theory paradoxes, however, forced the two men to reevaluate the project and extend the work to a three-volume treatise. These volumes were published from 1910 to 1913.

During this time Whitehead also published other material, including texts on projective and descriptive geometry and a general popular overview of mathematics, *An Introduction to Mathematics*. Whitehead also wrote pieces on the philosophy of science, and offered an alternative viewpoint to ALBERT EINSTEIN's theory of relativity in his 1922 book *The Principle of Relativity*. The work was not well understood and has been largely ignored.

Whitehead worked as a lecturer at Cambridge for over 30 years before accepting a position as chair of philosophy at Harvard University in 1924. He remained there until his retirement in 1937. He died in Cambridge, Massachusetts, 10 years later on December 30, 1947.

whole number A NATURAL NUMBER is sometimes referred to as a whole number. Matters are confusing because many scholars in mathematics also like to refer to zero as a whole number, and some also like to regard any negative integer as a whole number. There is no standard convention in place in this regard. It is certain, however, that no mathematician would regard a number that is not an integer a whole number.

See also NUMBER; NUMBER THEORY.

Wiles, Andrew John (1953–) British *Algebraic number theory* Born on April 11, 1953, in Cambridge, England, mathematician Andrew Wiles garnered international fame in 1994 as the first person to solve one of the most elusive and difficult mathematical problems of all time, FERMAT'S LAST THEOREM. This notorious problem, posed by French number theorist PIERRE DE FERMAT (1601–65), states that the equation $x^n + y^n = z^n$ has no positive integer solutions if n is an integer greater than 2. Finding a proof of this apparently simple claim has proved to be an extraordinarily difficult challenge, one that has frustrated professional and amateur mathematicians for well over 300 years.

Ever since first reading of Fermat's last theorem at the age of 10, Wiles dreamed of solving it. As a youngster he first tried approaches that he thought Fermat might have followed in thinking about the theorem himself. This proved to be useless. At college, Wiles studied the work of the great 18th- and 19th-century scholars who had worked on the problem, hoping to glean any insights as to how one might approach the challenge. Pursuing mathematics further, Wiles entered Clare College, Cambridge, and in 1980, was awarded a Ph.D. in mathematics. In 1982 Wiles traveled to the United States to take a professorship at Princeton University, New Jersey.

Although Wiles's thesis and early research work was not directly connected to solving Fermat's last theorem, Wiles later learned of some important developments that connected the possible solution of the problem with some new approaches in elliptic curve theory, the topic of his thesis. Upon this news, Wiles abandoned all unrelated research interests to focus exclusively on solving the theorem. Working for 7 years straight, essentially in seclusion, Wiles modified and adapted newly developed advances in many disparate fields to forge a path that would, hopefully, lead to a solution to the problem. In 1993, amidst a flurry of great media excitement, Wiles announced to the mathematical community that he had succeeded. Subsequent careful review of his work, however, revealed a subtle, but damaging, error, and all was thought to be lost.

After another 18 months of concerted effort, with the assistance of colleague Richard Taylor of Cambridge University, Wiles managed to find a way to circumvent the error and produce, at long last, an unflawed proof of the famous result. The proof of the theorem appears in a 1995 volume of the *Annals of Mathematics*. It represents one of the greatest intellectual achievements of the 20th century.

Wiles won many awards for his achievement, including the Wolf Prize in 1995, the Wolfskehl Prize in 1997, the American Mathematical Society Cole Prize in 1997, and the King Faisal Prize in 1998. Because he was over the age limit of 40, Wiles did not receive a FIELDS MEDAL for his work, but he was honored with a silver plaque during the 1998 Fields Medal ceremony. Wiles currently resides in Princeton, New Jersey.

Z

Zeno of Elea (ca. 490–425 B.C.E.) Greek *Philosophy*
Born in Elea, Luciana (now southern Italy), Greek
philosopher Zeno is remembered for his invention of a
number of PARADOXes that significantly influenced, and
challenged, the Greek perception of magnitude,
motion, and continuity. The question of whether or not
matter and time are composed of fundamental indivisible parts ("atoms") was of great interest to philosophers at the time. Zeno managed to devise a variety of
convincing arguments that seem to prove that any view
one wishes to adopt cannot be correct. Four of his
paradoxes in particular garnered certain notoriety.
They remained unresolved for over two millennia.

Essentially nothing is known of Zeno's life, except
for what can be gleaned from the writings of PLATO in
his dialogue *Parmenides*. There we learn that Zeno
studied at the Eleatic School of Philosophy under the
guidance of the founder Parmenides. This sect analyzed
the concept of monism, the idea that "all is one" and
that change and motion are simply illusions and are
not part of an eternal reality. It is believed that Zeno
wrote just one text, his collection of 40 paradoxes on
the nature of time, space, and motion. The text, unfortunately, has not survived, and we learn of its content
through the writings of others. ARISTOTLE describes the
four famous paradoxes in his work *Physics*.

See also ZENO'S PARADOXES.

Zeno's paradoxes In his studies, Greek philosopher
ZENO OF ELEA proposed 40 PARADOXes that challenge
our understanding of time, space, and motion. Four of
his paradoxes have garnered considerable attention for
being particularly troublesome.

With regard to time and space, there are two possibilities: either such magnitudes can be divided into
smaller and smaller parts an unlimited number of
times (that is, space and time each form a continuum),
or there is some fundamental indecomposable unit of
each that can no longer be divided (akin to the notion
that matter is composed of indivisible particles). The
first two of Zeno's famous paradoxes argue that
motion is impossible if the first point of view is
adopted, while the last two argue that motion is again
impossible if the latter perspective is taken. His paradoxes are the following:

1. *Dichotomy:* Assume that time and space
   each are infinitely divisible.

In order to walk across the room, one must first reach
the midway point. But to do that, one must reach the
point one-quarter of the way along. But to get this far,
one must pass through the point one-eighth of the way
across, and before that, the point 1/16th the way along.
As this division can be done indefinitely, it seems then
we can never start our walk across the room—we can
never reach a first point of our journey. Motion is
therefore impossible.

2. *Achilles and the Tortoise:* Assume that time
   and space each are infinitely divisible.

Achilles and a tortoise take part in a race in which the tortoise is given a head start. Consider Achilles' circumstance. To overtake the tortoise, he must first reach the tortoise's starting position. By then, the tortoise will, of course, have moved on to a new position. To overtake the tortoise, Achilles must reach this second position, and, again, by the time he does so the tortoise will have moved farther along. Continuing this way, we see that it is impossible for Achilles to ever overtake the tortoise, no matter how swiftly he moves.

3. *The Arrow Paradox:* Assume that time and space each are composed of discrete fundamental units.

Consider an arrow in flight. At any one fundamental unit of time, the arrow occupies a particular instantaneous position in space. At this moment, it is indistinguishable from a motionless arrow occupying the same location in space during that same unit of time. How then is it that motion is perceived?

4. *The Stadium Paradox:* Assume that time and space each are composed of discrete fundamental units.

Imagine two runners in a stadium moving in opposite directions from the same starting position. Suppose the runners each move at a speed equivalent to one fundamental distance of space for each fundamental unit of time. Ask now, "What does a runner see when he looks behind his shoulder?" The answer: his opponent moving 2 units of space in 1 unit of time. We can only conclude that there must be a unit smaller than the supposedly indivisible unit of time that corresponds to just 1 unit of this motion.

Some comments are in order. It is worth noting that many texts have attempted to resolve Zeno's dichotomy paradox by arguing that the fact that we *can* walk across a room proves that the infinite sum (SERIES) $\frac{1}{2} + \frac{1}{4} + \frac{1}{8} + \ldots$ converges to the finite value 1. This, however, is not relevant to the argument proposed in this first paradox. Zeno is suggesting instead that it is impossible to build such an infinite sum by starting at the "wrong end," as it were. (The construction of an infinite sum, however, is appropriate to analyze the second paradox.)

Some scholars suggest that Zeno's third paradox is an important first step in the study of INFINITESIMALs and issues of DIFFERENTIAL CALCULUS. The standard formula for velocity is $v = \frac{d}{t}$, where d is the distance traveled and t the time taken to do so. At a particular instant, this leads to the meaningless equation $v = \frac{0}{0}$. Here, it is said, Zeno is pointing out the mathematical difficulty of infinitesimals, the same problem that was to haunt the inventors of calculus two millennia later.

Zermelo, Ernst Friedrich Ferdinand (1871–1953)

German *Set theory, Physics* Born on July 27, 1871, in Berlin, Germany, mathematician Ernst Zermelo is remembered for founding axiomatic SET THEORY. In 1908 he published a set of seven axioms designed to overcome the paradoxes posed by BERTRAND ARTHUR WILLIAM RUSSELL (1872–1970) in set theory, and to provide, for the first time, a rigorous and logical foundation for the developing subject. Zermelo's axioms, although later modified by other mathematicians, have remained the starting point on which much continued work on this topic has been based.

Zermelo received a doctorate in mathematics from the University of Berlin in 1894 after completing a thesis exploring issues in calculus, the calculus of variations, and their applications to physics. He continued work in this area at the university for an additional 3 years before moving to Göttingen to study hydrodynamics. He was awarded a lectureship at Göttingen for his outstanding research in this field.

At the time elsewhere in Europe, mathematicians were attempting to make sense of the controversial work put forward by German mathematician GEORG CANTOR (1845–1918). The German mathematician DAVID HILBERT (1862–1943), in his famous Paris lecture of 1900, cited the resolution of Cantor's CONTINUUM HYPOTHESIS as the most challenging and important question of the 20th century. Zermelo was captivated by the problem when he learned of it, and began devoting all his time and efforts to solving it. He soon realized that a first step toward solving the general conjecture would be to establish a second of Cantor's claims, namely, that any set can be a WELL-ORDERED SET.

Zermelo published his first collection of results on this topic in 1901 with the paper "Über die Addition transfiniter Cardinalzahlen" (On the addition of transfinite cardinals), and 2 years later succeeded in proving the well-ordering conjecture in his 1904 article "Beweis, dass jede Menge wohlgeordnet werden kann" (Proof that every set can be well ordered). His success garnered him international fame and recognition as a brilliant mathematician. The University of Göttingen immediately awarded him a full professorship.

In 1905, leaving further work on the continuum hypothesis aside, Zermelo began the task of axiomatizing set theory. By identifying a small set of basic assumptions about the topic, he hoped to resolve some of the paradoxes about infinite sets that were arising in the subject. Unfortunately, he found it necessary to include an "axiom of infinity" asserting, essentially, that infinite sets exist, along with the extraordinarily controversial AXIOM OF CHOICE, which asserts that it is always possible to select, in one fell swoop, one object from each set in any infinite collection of sets. Mathematicians were uncomfortable with each of these assumptions at the time. Moreover, despite his efforts, Zermelo was unable to prove that the axioms he put forward were free from contradictions. Other mathematicians then began to work on this task, and some 15 years later, in 1922, mathematicians Adolf Fraenkel and Thoralf Skolem independently developed a refined system of 10 axioms that were deemed of sufficiently good stead.

Zermelo left Göttingen in 1910 to take an appointment as the chair of mathematics at Zurich University, where he stayed for 6 years. Zermelo spent the remainder of his life in Freiburg. He died there on May 21, 1953.

See also RUSSELL'S PARADOX.

zero (naught) The counting number, denoted 0, used to signify that no objects are present is called zero. This number has the arithmetical property that its addition to any other number does not change that number: $a + 0 = a = 0 + a$ for all values a. Thus zero serves as the additive IDENTITY ELEMENT in arithmetic. Zero is the only real number that is neither POSITIVE nor negative.

It was only recently in the history of mathematics that zero was recognized as a valid and important mathematical entity. In times of antiquity numbers were thought only to represent counts or magnitudes of quantities. Thus if there were no objects to be counted, there was no number to consider (and so speaking of zero as a number consequently had no meaning). The ancient

Egyptian and Greek scholars never used zero in their computations. The ancient Babylonians were the first to utilize a notion of zero by using it, in some sense, as a placeholder when expressing large numbers (much as we use zeros today to distinguish between 102 and 12.) The Hindu mathematician BRAHMAGUPTA (ca. 598–665) is credited today as being the first scholar to acknowledge zero as a valid number and use it in arithmetical computations. His idea was later expanded upon by MUHAMMAD IBN MŪSĀ AL-KHWĀRIZMĪ (ca. 780–850) in his famous development of a theory of algebra. Italian mathematician FIBONACCI (ca. 1170–1250) popularized much of al-Khwārizmī's work in Western Europe, but it was not until the Renaissance that the notion of zero was finally deemed a valid concept.

Arithmetic involving zero can be troublesome. Although any number can be multiplied by zero (to produce the answer zero), no number can be divided by zero. For instance, there can be no answer to the division problem $1 \div 0$. (If $1/0 = x$, then multiplying through by the denominator gives the absurd result: $1 = x \times 0 = 0$.) The quantity $0/0$ is also deemed undefined. (One could argue that the answer to this division problem is 17, since $0 = 17 \times 0$, or, by the same token, that the answer is 2 since $0 = 2 \times 0$. There is no single well-defined solution to $0 \div 0$.)

A counting number is considered even if it can be written as a sum of two equal whole numbers. Since zero equals $0 + 0$, it follows that zero is an even number.

The study of EXPONENTs shows that it is appropriate to define x^0 to be 1 for any real number x other than zero. (For example, $2^0 = 1$ and $173^0 = 1$.) Although it is true that $\lim_{x \to 0^+} x^x = 1$, it is not valid to assume that 0^0 also equals 1. (This LIMIT does not exist if one permits COMPLEX NUMBERS in our considerations, and so the notion of raising zero to the zeroth power is problematic in this setting.) A study of the FACTORIAL shows that it is appropriate to define $0! = 1$.

A ROOT of a function is sometimes called a zero of the function. A zero function is a function whose output is always zero. For example, the function $f(x) = \dfrac{2 - x^{\frac{1}{\log_2 x}}}{x^2}$ defined on positive values of x is always zero. In GAME THEORY, a zero-sum game is a game in which a win for some player or players is always balanced by losses for the remaining players. A zero matrix is a MATRIX all of whose entries are zero.

The number 2^{86} is the largest known power of 2 whose decimal expansion contains no digit equal to zero. (Notice that $2^{10} = 1024$, for instance, has second digit equal to zero.) All the powers of two up to $2^{46,000,000}$ have been checked.

See also BABYLONIAN MATHEMATICS; EVEN AND ODD NUMBERS; GREEK MATHEMATICS; INDIAN MATHEMATICS.

zeta function In 1740 the Swiss mathematician LEONHARD EULER (1707–83) studied infinite series of the form:

$$\zeta(s) = 1 + \frac{1}{2^s} + \frac{1}{3^s} + \frac{1}{4^s} + \cdots = \sum_{n=1}^{\infty} \frac{1}{n^s}$$

The INTEGRAL TEST shows that this series converges if s is a real number greater than 1. Euler called this function, defined for values $s > 1$, the zeta function.

This function is intimately connected to the distribution of PRIME numbers. To see this, recall that the FUNDAMENTAL THEOREM OF ARITHMETIC asserts that every number is a product of a unique set of primes. Thus, in EXPANDING BRACKETS for the following infinite product, selecting one term from each set of parentheses, every integer appears once, and only once, in the infinite sum that results:

$$(1 + 2 + 2^2 + \cdots)(1 + 3 + 3^2 + \cdots)(1 + 5 + 5^2 + \cdots)$$
$$(1 + 7 + 7^2 + \cdots)(1 + 11 + 11^2 + \cdots)\cdots$$
$$= 1 + 2 + 3 + 2^2 + 5 + 2\cdot3 + 7 + 2^3 + 3^2 + 2\cdot5 + 11 + \cdots$$

Although this argument is not mathematically precise, mathematicians have shown that it is valid to perform this procedure on the reciprocals of all the numbers involved, even when raised to the sth power. Consequently:

$$\left(1 + \frac{1}{2^s} + \frac{1}{2^{2s}} + \frac{1}{2^{3s}} + \cdots\right)\left(1 + \frac{1}{3^s} + \frac{1}{3^{2s}} + \frac{1}{3^{3s}} + \cdots\right)$$
$$\left(1 + \frac{1}{5^s} + \frac{1}{5^{2s}} + \frac{1}{5^{3s}} + \cdots\right)\cdots = 1 + \frac{1}{2^s} + \frac{1}{3^s} + \frac{1}{4^s} + \cdots$$

The formula for a GEOMETRIC SERIES shows that this equation can be rewritten:

$$\frac{1}{\left(1 - \frac{1}{2^s}\right)} \times \frac{1}{\left(1 - \frac{1}{3^s}\right)} \times \frac{1}{\left(1 - \frac{1}{5^s}\right)} \times \cdots$$
$$= 1 + \frac{1}{2^s} + \frac{1}{3^s} + \frac{1}{4^s} + \cdots$$

thereby yielding an alternative formula for the zeta function as an infinite product over all prime numbers p:

$$\zeta(s) = \prod_{\text{all primes } p} \left(1 - \frac{1}{p^s}\right)^{-1}$$

Euler managed to compute the value of the zeta function for certain values of s. He showed, for example, that $\zeta(2) = \frac{\pi^2}{6}$, $\zeta(4) = \frac{\pi^4}{90}$, $\zeta(6) = \frac{\pi^6}{945}$, $\zeta(8) = \frac{\pi^8}{9450}$, and $\zeta(10) = \frac{\pi^{10}}{93555}$. (He continued this list up to $\zeta(26)$.) Below we show how Euler computed $\zeta(2)$. To this day, extremely little is known about the values of the zeta function on odd whole numbers.

In 1859 German mathematician GEORG FRIEDRICH BERNHARD RIEMANN (1826–66) showed that the zeta function is well defined even if the argument s is a COMPLEX NUMBER. He showed the series $\sum_{n=1}^{\infty} \frac{1}{n^s}$ converges if the real part of s is greater than 1, and it is possible to extend the definition of the function to incorporate *all* complex values of s. For this reason, the zeta function is often also called the Riemann zeta function.

Riemann was particularly interested in locating the zeros of the zeta function, that is, finding the values of s that yield $\zeta(s) = 0$. He showed that the function has no zeros if $\text{Re}(s) \geq 1$, that its only zeros in $\text{Re}(s) \leq 0$ are at $s = -2, -4, -6, \ldots$, and that it has infinitely many zeros in $0 < \text{Re}(s) < 1$. He called these the "nontrivial zeros." Riemann remarked that it is reasonable to believe that *all* the nontrivial zeros lie on the line $\text{Re}(z) = 1/2$, but offered no proof. This casual comment has become one of the most famous unsolved conjectures of all time. Mathematicians call it the Riemann hypothesis, and proving its truth or falsehood would have profound implications on the study of NUMBER THEORY. (For example, a crucial part of proving the PRIME-NUMBER THEOREM relies on showing that $\zeta(s) \neq 0$ for $\text{Re}(s) = 1$.)

Euler's Computation of ζ(2)

Here we show how Euler established the formula $1 + \frac{1}{2^2} + \frac{1}{3^2} + \frac{1}{4^2} + \cdots = \frac{\pi^2}{6}$. Although the argument is not mathematically rigorous, mathematicians later proved that the issues raised in this approach can be made mathematically sound.

Consider the function $\sin x$ with its infinite number of zeros at locations $0, \pm\pi, \pm 2\pi, \pm 3\pi, \ldots$ The TAYLOR SERIES of the sine function is:

$$x - \frac{x^3}{3!} + \frac{x^5}{5!} - \frac{x^7}{7!} + \cdots$$

which we will regard as a polynomial of infinite degree with the same infinite collection of roots.

The FUNDAMENTAL THEOREM OF ALGEBRA asserts that any polynomial factors into linear terms, one term for each root of the equation. Assuming that the theorem remains valid for infinite polynomials we must have then that:

$$\sin x = x - \frac{x^3}{3!} + \frac{x^5}{5!} - \frac{x^7}{7!} + \cdots$$
$$= x\left(1 - \frac{x}{\pi}\right)\left(1 + \frac{x}{\pi}\right)\left(1 - \frac{x}{2\pi}\right)\left(1 + \frac{x}{2\pi}\right)\cdots$$

or, dividing through by x:

$$1 - \frac{x^2}{3!} + \frac{x^4}{5!} - \frac{x^6}{7!} + \cdots$$
$$= \left(1 - \frac{x}{\pi}\right)\left(1 + \frac{x}{\pi}\right)\left(1 - \frac{x}{2\pi}\right)\left(1 + \frac{x}{2\pi}\right)\cdots$$

(Here we have written the factor corresponding to the root $x = \pi$, for instance, as $1 - \frac{x}{\pi}$ rather than $x - \pi$. This is done to ensure that, in expanding brackets for the right-hand side of the second expression, the result yields a constant term equal to 1, as required.) Combining pairs of terms we have:

$$1 - \frac{x^2}{3!} + \frac{x^4}{5!} - \frac{x^6}{7!} + \cdots = \left(1 - \frac{x^2}{\pi^2}\right)\left(1 - \frac{x^2}{2^2\pi^2}\right)$$
$$\left(1 - \frac{x^2}{3^2\pi^2}\right)\left(1 - \frac{x^2}{4^2\pi^2}\right)\left(1 - \frac{x^2}{5^2\pi^2}\right)\cdots$$

Now consider expanding the brackets on the right-hand side of this expression to obtain terms that yield the quantity x^2. This can only occur by selecting one x^2 term from one set of parentheses, and the term 1 from all remaining parentheses. This gives:

$$-\frac{x^2}{\pi^2} - \frac{x^2}{2^2\pi^2} - \frac{x^2}{3^2\pi^2} - \frac{x^2}{4^2\pi^2} - \frac{x^2}{5^2\pi^2} - \cdots$$
$$= -\frac{1}{\pi^2}\left(1 + \frac{1}{2^2} + \frac{1}{3^2} + \frac{1}{4^2} + \frac{1}{5^2} + \cdots\right)x^2$$

According to the left-hand side of this expression, this quantity must equal $-\frac{x^2}{3!} = -\frac{x^2}{6}$, thereby leading to Euler's formula:

$$1 + \frac{1}{2^2} + \frac{1}{3^2} + \frac{1}{4^2} + \cdots = \frac{\pi^2}{6}$$

As a bonus, consider again the equation:

$$\frac{\sin x}{x} = \left(1 - \frac{x^2}{\pi^2}\right)\left(1 - \frac{x^2}{2^2\pi^2}\right)\left(1 - \frac{x^2}{3^2\pi^2}\right)$$
$$\left(1 - \frac{x^2}{4^2\pi^2}\right)\left(1 - \frac{x^2}{5^2\pi^2}\right)\cdots$$

and put in $x = \frac{\pi}{2}$. This gives:

$$\frac{2}{\pi} = \left(1 - \frac{1}{2^2}\right)\left(1 - \frac{1}{4^2}\right)\left(1 - \frac{1}{6^2}\right)\cdots$$
$$= \left(\frac{1}{2}\cdot\frac{3}{2}\right)\left(\frac{3}{4}\cdot\frac{5}{4}\right)\left(\frac{5}{6}\cdot\frac{7}{6}\right)\cdots$$

which establishes WALLIS'S PRODUCT:

$$\frac{\pi}{2} = \frac{2}{1}\cdot\frac{2}{3}\cdot\frac{4}{3}\cdot\frac{4}{5}\cdot\frac{6}{5}\cdot\frac{6}{7}\cdots$$

See also ANALYTIC NUMBER THEORY; BERNOULLI NUMBERS; FOURIER SERIES; SUMS OF POWERS.

Zhu Shijie   *See* CHU SHIH-CHIEH.

z-score (z-value) If a set of DATA has mean μ and standard deviation σ, then the z-score of a particular data value x is given by:

$$z = \frac{x - \mu}{\sigma}$$

This transformation converts the data values of one set into another set of values with mean 0 and standard deviation 1. It allows one to effectively compare two or more independent sets of data. As an example: a group of freshman college music majors are given exams in performance, theory, and composition. One student, John, wishes to judge how well he fared in each of the categories. The following table summarizes the results:

Category	Mean	Standard Deviation	John's Score
Performance	730	40	670
Theory	380	35	450
Composition	640	25	660

John's z-scores are:

Performance: $z = \frac{670 - 730}{40} = -\frac{3}{2}$

Theory: $z = \frac{450 - 380}{35} = 2$

Composition: $z = \frac{660 - 640}{25} = \frac{4}{5}$

Although John obtained his highest score in performance, this score is $1\frac{1}{2}$ standard deviations below the mean grade in this category. Despite the low number, John scored best in theory, gaining a score 2 standard deviations above his classmates.

If a set of data is believed to be normally distributed with mean μ and standard deviation σ, converting to z-values allows one to compare data with the standard normal distribution. For example, to compute the probability that a measurement taken at random falls within a range of values [a,b], one computes the area under the standard normal curve above the interval $[\frac{a-\mu}{\sigma}, \frac{b-\mu}{\sigma}]$.

This can be found by looking at a table of cumulative distribution values φ(z) and calculating the difference

$$\phi\left(\frac{b-\mu}{\sigma}\right) - \phi\left(\frac{a-\mu}{\sigma}\right).$$

See also NORMAL DISTRIBUTION; STATISTICS: DESCRIPTIVE.

Zu Chongzhi (Tsu Chung Chi) (ca. 430–500) Chinese *Computation, Astronomy* Born in Fan-yang (now Hopeh), China, scholar Zu Chongzhi is remembered for his invention of the Daming calendar, his highly accurate calculation of a value for π, and his derivation of the formula for the volume of a sphere using a method equivalent to that discovered by Italian mathematician BONEVENTURA CAVALIERI (1598–1647) a full millennium later.

Zu Chongzhi developed an interest in astronomy at an early age. As a young scholar he noticed a discrepancy in the position of a sundial's shadow and the calendar in use in China at the time, motivating him to work for many years on an improved calendar system. This work was completed in 462, but it was not until 510, through the efforts of his son, that the Daming calendar was officially adopted.

Turning his attention to mathematics, Zu Chongzhi worked to compute a precise value for the ratio of a circumference of a circle to its diameter, π. Noting, as a start, that the circumference of a hexagon is three times its diameter, and that by cutting off its corners to create a dodecagon yields a ratio slightly larger than 3, Zu Chongzhi continued to shave corners to eventually produce a polygon with 192 sides, which, as he measured, yielded a circumference-to-diameter ratio of 355/113. This represents an approximate value of π accurate to six decimal places. Such precision was not surpassed for another 1,000 years. It is worth noting that Zu Chongzhi made his impressive calculations without the aid of an ABACUS (not used in China until the 1100s) or any other kind of calculating device.

Through the writings of later scholars, historians have determined that Zu Chongzhi wrote at least 51 works. Sadly none survive today. Zu Chongzhi is honored today as one of the few Chinese mathematicians to have a lunar feature named after him—a 28-kilometer-wide crater located 20° north of the lunar equator.

Appendix I

CHRONOLOGY

ca. 50,000 B.C.E.

Animal bones etched with tally marks provide evidence that the Paleolithic people of central Europe were able to count.

ca. 15,000 B.C.E.

Notched animal bones discovered in the Middle East provide further evidence of early counting.

ca. 3000 B.C.E.

The ancient Egyptians improve upon the tally system and devise a numeration system that permits the expression of large numbers with only a few symbols.

ca. 2600 B.C.E.

The Great Pyramid at Giza is built by the Egyptians.

ca. 1800 B.C.E.

Babylonian scholars develop a base-60 PLACE-VALUE SYSTEM of numeration. With it, they perform complex arithmetic work, solve QUADRATIC equations, and compute PYTHAGOREAN TRIPLES.

ca. 1650 B.C.E.

Egyptian scribe Ahmes makes copy of an early mathematical handbook onto a papyrus scroll. British antiquarian Alexander Henry Rhind (1833–63) purchased the scroll in an Egyptian marketplace in 1858, and the scroll is today known as the RHIND PAPYRUS. The text details the method for computing the area of a circle and other basic geometric shapes, the method of "false position" for solving basic linear equations, and a tech-

nique for finding the steepness of a pyramid. The value $256/81 \approx 3.1605$ is used for π.

ca. 600 B.C.E.

Indian scholars write the *Sulba sultras,* a set of religious instructional texts providing detailed geometric methods for the construction of altarpieces. The mathematics described in the texts includes formulae for the areas of basic geometric shapes as well as for the volumes of prisms and cylinders.

ca. 585 B.C.E.

Greek mathematician THALES OF MILETUS founds the earliest known school of philosophy and mathematics. Thales develops seven important geometric propositions, heralding the demand for rigor and proof in Greek mathematical thought.

ca. 569 B.C.E.

PYTHAGORAS of Samos is born. Around 510 B.C.E., Pythagoras founds a secret philosophical and mathematical society that includes both men and women. The Pythagoreans are noted as the first to provide a proof of what is today known as PYTHAGORAS'S THEOREM and are credited with the discovery of IRRATIONAL NUMBERS.

ca. 450 B.C.E.

ZENO OF ELEA proposes a series of paradoxes that challenge the notions of space, time, and motion.

ca. 387 B.C.E.

Greek philosopher PLATO (ca. 428–348 B.C.E.) founds the Academy in Athens. His insistence that mathematics be an integral part of a formal education elevates the status of mathematics within the Western world.

ca. 386 B.C.E.

The oracle of Delos tells the people of Athens that in order to end a plague they must double the size of the cube-shaped altar to the god Apollo. This establishes the problem of DUPLICATING THE CUBE.

ca. 370 B.C.E.

EUDOXUS OF CNIDUS (ca. 400–350 B.C.E.) develops a "method of exhaustion" to determine the areas of simple curved figures. The method is a precursor to the notion of LIMIT developed in 18th-century CALCULUS.

ca. 350 B.C.E.

ARISTOTLE (ca. 384–322 B.C.E.) analyzes the structure of ARGUMENTs and logical reasoning to develop ideas seminal to the field of FORMAL LOGIC.

ca. 310 B.C.E.

Ruler Ptolemy Soter founds the Library of Alexandria. It remains the center of intellectual learning for more than 700 years.

ca. 300 B.C.E.

Greek geometer EUCLID summarizes all mathematical knowledge known at his time in *THE ELEMENTS*. The method of logical deduction and rigor he provides remains the paradigm of mathematical thinking today.

ca. 240 B.C.E.

ARCHIMEDES OF SYRACUSE (ca. 287–212 B.C.E.) uses the method of exhaustion to compute the area under a section of a PARABOLA. He also makes fundamental contributions to the fields of geometry, engineering, astronomy, and hydrostatics. He discovers a method of computing the value π to any degree of accuracy and shows, in particular, that its values lies between 3 10/71 and 3 1/7.

ca. 230 B.C.E.

ERATOSTHENES OF CYRENE (ca. 275–195 B.C.E.) calculates the circumference of the EARTH to be 28,500 miles. He develops the SIEVE OF ERATOSTHENES for computing PRIME numbers.

ca. 220 B.C.E.

APOLLONIUS OF PERGA (ca. 262–190 B.C.E.) develops the study of CONIC SECTIONS. He also develops a theory of EPICYCLES to model planetary motion.

ca. 214 B.C.E.

Construction begins on the Great Wall of China.

ca. 150 B.C.E.

Hipparchus of Nicaea (ca. 190–126 B.C.E.) develops beginning ideas in the theory of TRIGONOMETRY. He uses geometry to calculate the distances of the Sun and the Moon from the Earth.

ca. 100 B.C.E.

Chinese scholars write *JIUZHANG SUANSHU* (Nine chapters on the mathematical arts). The text includes solutions to linear and QUADRATIC equations, the computation of areas and volumes, a statement of PYTHAGORAS'S THEOREM, and the use of NEGATIVE NUMBERS.

ca. 140

Greek astronomer CLAUDIUS PTOLEMY writes the *Almagest*, the most influential work in mathematical astronomy until the 16th century. It includes a table of chord values equivalent to a modern-day table of sines. He uses the value 377/120 for π.

ca. 250

Greek mathematician DIOPHANTUS OF ALEXANDRIA writes *Arithmetica*. He is the first to use symbols to represent unknown quantities.

ca. 320

PAPPUS OF ALEXANDRIA attempts to revive interest in the classical Greek pursuit of mathematics. He writes *Synagoge* (Collections) as a guide to the great Greek works.

ca. 370

HYPATIA (ca. 370–415), the first woman named in the history of mathematics, is born. She becomes head of the Platonist school in Alexandria.

ca. 475

Indian astronomer and mathematician ARYABHATA (ca. 476–550) is born. He develops methods for extracting square roots, summing arithmetic series, and computing chord values. He uses the value 3.1416 for π.

ca. 480

Chinese scholar ZU CHONGZHI (430–500) uses the value 355/113 for π.

ca. 600

Hindu scholars invent DECIMAL REPRESENTATION, the method of numeration we use today to represent numbers.

ca. 640

Hindu mathematician and astronomer BRAHMAGUPTA (ca. 598–665) introduces NEGATIVE NUMBERS and the concept of ZERO as a number into ARITHMETIC.

641

The Library of Alexandria is burned.

ca. 775

Arabic scholars begin translating the great Greek and Indian works into Arabic.

ca. 830

MUHAMMAD IBN MŪSSĀ AL-KHWĀRIZMĪ (ca. 780–850) writes two influential texts founding the field of ALGEBRA. Al-Khwārizmī also promotes the use of the Hindu base-10 system of ARITHMETIC.

850

Indian mathematician Mahavira writes the *Ganita Sera Samgraha* (The compendium of the arithmetic), the first Indian text devoted solely to the topic of mathematics.

1079

Persian scholar OMAR KHAYYAM (ca. 1048–1131) calculates the length of the year to be 365.24219858156 days, correct to the sixth decimal place.

1202

Italian number theorist FIBONACCI (ca. 1170–1250) writes *Liber abaci* (The book of the abacus), introducing HINDU-ARABIC NUMERALS to western Europe. Fibonacci also discovers the sequence 1, 1, 2, 3, 5, 8, 13, 21, 34,...

1360

French mathematician NICOLE ORESME (ca. 1323–82) discovers that the area under a velocity curve corresponds to the distance traveled.

1482

The first printed edition of Euclid's text THE ELEMENTS is produced in Venice, Italy. It becomes the most translated and published textbook of all time.

1498

German scholar Johannes Widmann (1462–98) writes *Mercantile arithmetic* in which the symbols "+" and "−" appear in printed form for the first time.

1533

German scholar REGIOMONTANUS (1436–76) publishes the first comprehensive modern treatise on the topic of TRIGONOMETRY.

ca. 1541

Italian mathematician NICCOLÒ TARTAGLIA (ca. 1499–1557) discovers a general method for solving CUBIC EQUATIONS. He tells his method, in confidence, to GIROLAMO CARDANO (1501–76) and LUDOVICO FERRARI (1522–65), who publish the details without his consent. Ferrari discovers a method for solving QUARTIC EQUATIONS.

1557

Welsh mathematician ROBERT RECORDE (ca. 1510–58) introduces the symbol "=" for equality in the world's first algebra text printed in English.

1585

Dutch mathematician SIMON STEVIN (1548–1620) introduces decimal notation for fractions.

1591

French mathematician FRANÇOIS VIÈTE (1540–1603) writes *In artem analyticam isagoge* (Introduction to the analytical arts), establishing the principles of modern ALGEBRA and modern algebraic notation.

1594

Scottish mathematician JOHN NAPIER (1550–1617) begins his work on arithmetical techniques that eventually led to the discovery of LOGARITHMs. He publishes his work in 1614.

1609

German astronomer and mathematician JOHANNES KEPLER (1571–1630) observes that the planets move in elliptical orbits. Italian scientist GALILEO GALILEI (1564-1642) improves upon the invention of the telescope and begins his own astronomical observations.

1622

British mathematician WILLIAM OUGHTRED (1574–1660) invents the slide rule. He introduces the symbol "×" for multiplication and the abbreviations "sin" and "cos" for sine and cosine, respectively.

1629

French mathematician PIERRE DE FERMAT (1601–65) uses algebra to solve geometric problems but does not publish his results. French mathematician and philosopher RENÉ DESCARTES (1596–1650) later developed similar techniques and is today credited as the founder of this approach.

1635

Italian mathematician BONAVENTURA CAVALIERI (1598–1647) introduces a method of indivisibles for comparing volumes (a precursor to the methods of INTEGRAL CALCULUS) and CAVALIERI'S PRINCIPLE.

ca. 1637

FERMAT introduces modern NUMBER THEORY. He writes a problem in the margin of a text that confounds mathematicians for centuries. FERMAT'S LAST THEOREM was finally solved by ANDREW WILES in 1994.

1639

French mathematician GIRARD DESARGUES (1591–1661) publishes a treatise on his newly discovered PROJECTIVE GEOMETRY. The work is essentially ignored for 200 years.

1654

Mathematician BLAISE PASCAL (1623–62) begins a correspondence with FERMAT about questions of games of chance. Through five consecutive letters, they together create the theory of PROBABILITY.

1662

The ROYAL SOCIETY of London is established. British mathematician LORD WILLIAM BROUNCKER (1620–84) is elected as its first president.

1666

SIR ISAAC NEWTON (1642–1727) develops DIFFERENTIAL and INTEGRAL CALCULUS but does not publish his results until 1711.

1673

German mathematician GOTTFRIED WILHELM LEIBNIZ (1646–1716) develops DIFFERENTIAL and INTEGRAL CALCULUS independently of NEWTON. Leibniz begins publishing his results in 1684, and Newton accuses him of plagiarism. A bitter dispute between the two men ensues, lasting four decades.

1687

Under the urging of astronomer Edmund Halley, Newton publishes *Principia,* today considered one of the greatest scientific works of all time. Newton outlines his laws of motion and the INVERSE SQUARE LAW for gravitation.

1693

Halley compiles the first set of LIFE TABLES and makes use of STATISTICS to analyze birth and death rates.

1696

French scholar MARQUIS DE GUILLAUME FRANÇOIS ANTOINE L'HÔPITAL (1661–1704) publishes the first textbook on CALCULUS.

1703

NEWTON is elected president of the ROYAL SOCIETY of London. Eight years later, after an official investigation, the society concludes that Newton, not LEIBNIZ, is the true inventor of CALCULUS. It is later revealed that Newton, as president, wrote the final proclamation. The verdict is not considered valid today.

1718

French mathematician ABRAHAM DE MOIVRE (1667–1754) publishes *Doctrine of Chances,* the most advanced text on the theory of PROBABILITY of its time. De Moivre later develops the result today known as STIRLING'S FORMULA.

1736

Swiss mathematician LEONHARD EULER (1707–83) solves the SEVEN BRIDGES OF KÖNIGSBERG PROBLEM, thereby establishing the fields of TOPOLOGY and GRAPH THEORY. Throughout his life Euler also discovers, among many accomplishments, the number e, his famous formula relating the trigonometric functions to this number, specific values of the ZETA FUNCTION, and, in geometry, the EULER LINE. Euler also introduces the notion of a FUNCTION and popularizes the use of the symbol π for the ratio of the circumference of a circle to its diameter.

1742

CHRISTIAN GOLDBACH (1690–1764) writes to EULER posing the problem that has since become known as GOLDBACH'S CONJECTURE.

1748

MARIA GAËTANA AGNESI (1718–99) publishes her two-volume survey of elementary and advanced mathematics.

1750

Swiss mathematician GABRIEL CRAMER (1704–52) publishes CRAMER'S RULE.

1767

German mathematician JOHANN HEINRICH LAMBERT (1728–77) proves that π is irrational.

1795

France adopts the metric system.

John Playfair (1748–1819) publishes an equivalent form of the famous PARALLEL POSTULATE, today known as PLAYFAIR'S AXIOM.

1797

CARL FRIEDRICH GAUSS (1777–1855) proves the FUNDAMENTAL THEOREM OF ALGEBRA. Throughout his life, Gauss, among many accomplishments, derives the LEAST SQUARES METHOD, classifies the CONSTRUCTIBLE regular polygons, and makes fundamental contributions to NUMBER THEORY, GEOMETRY, STATISTICS, mathematical physics, and astronomy.

1799

Norwegian surveyor Casper Wessel (1745–1818) publishes the equivalent of an ARGAND DIAGRAM as a means for representing COMPLEX NUMBERS. French bookkeeper and mathematician JEAN ROBERT ARGAND (1768–1822) develops the same method in 1806.

1812

British mathematician and inventor CHARLES BABBAGE (1791–1871) constructs the first mechanical calculator. In 1823 Babbage obtains funds to build the "difference engine," the first digital computer, but the project is never completed.

ca. 1820

Norwegian algebraist NIELS HENRIK ABEL (1802–29) proves that there can be no general formula akin to the famous quadratic formula that solves all fifth-degree polynomial equations. Soon afterward, French mathematician ÉVARISTE GALOIS (1811–32) begins work to classify which fifth- and higher-degree equations can be so solved and consequently founds the field of GROUP THEORY.

1821

French mathematician AUGUSTIN-LOUIS CAUCHY (1789–1857) develops the notion of a LIMIT as an attempt to place CALCULUS on sound mathematical footing. This idea is later refined by German scholar KARL THEODOR WILHELM WEIERSTRASS (1815–97).

1822

French mathematician JEAN BAPTISTE JOSEPH FOURIER (1768–1830) publishes a treatise on the theory of heat and develops the notion of a FOURIER SERIES.

1829

Russian mathematician NIKOLAI IVANOVICH LOBACHEVSKY (1792–1856) and Hungarian mathematician JÁNOS BOLYAI (1802–60) independently discover NON-EUCLIDEAN GEOMETRY.

1843

SIR WILLIAM ROWAN HAMILTON (1805–65) discovers QUATERNIONS. Two years later ARTHUR CAYLEY (1821–95) discovers octonians.

1844

JOSEPH LIOUVILLE (1809–82) discovers the first example of a TRANSCENDENTAL NUMBER.

1854

British scholar GEORGE BOOLE (1815–64) establishes the field of symbolic logic with his development of BOOLEAN ALGEBRA.

German mathematician GEORG FRIEDRICH BERNHARD RIEMANN (1826–66) offers a universal approach to geometry. He discovers SPHERICAL GEOMETRY. He later makes significant advances in the theory of numbers and the study of PRIME numbers.

1858

AUGUST FERDINAND MÖBIUS (1790–1868) and Johann Benedict Listing independently discover the MÖBIUS BAND.

German mathematician JULIUS WILHELM RICHARD DEDEKIND (1831–1916) suggests the notion of a DEDEKIND CUT as a means to properly define the REAL NUMBERS.

CAYLEY introduces the notion of a MATRIX to the study of algebra.

1872

FELIX CHRISTIAN KLEIN (1849–1925) unifies the fields of geometry with his "Erlanger program." He also discovers the KLEIN BOTTLE.

1873

WILLIAM SHANKS (1812–82) computes, by hand, the first 607 decimals of π. He is correct up to the 527th place.

1874

GEORG CANTOR (1845–1918) develops SET THEORY and his theory of CARDINALITY.

1882

FERDINAND VON LINDEMANN (1852–1939) proves that π is transcendental and hence that the challenge of SQUARING THE CIRCLE is impossible.

1883

Françoise-Edouard-Anatole Lucas (1842–91) invents the TOWER OF HANOI puzzle.

1896

JACQUES HADAMARD (1865–1963) and CHARLES DE LA VALLÉE-POUSSIN (1866–1962) independently prove the PRIME NUMBER THEOREM first conjectured by GAUSS in 1792.

1899

German mathematician DAVID HILBERT (1862–1943) provides a complete axiomatic treatment of EUCLIDEAN GEOMETRY.

1900

At the International Congress of Mathematicians in Paris, HILBERT presents his famous list of 23 problems to challenge scholars of the 20th century.

1901

HENRI LÉON LEBESGUE (1875–1941) introduces a revolutionary new approach to INTEGRAL CALCULUS.

1903

Swedish mathematician Nils Fabian Helge von Koch (1870–1924) introduces the KOCH CURVE, the first example of an object later to be classified as a FRACTAL.

1904

JULES-HENRI POINCARÉ (1854–1912) conjectures that any three-dimensional object sharing the same topological characteristics as a SPHERE must indeed be a sphere.

1905

ALBERT EINSTEIN (1879–1955) writes five groundbreaking papers in the field of mathematical physics. The final two papers develop his famous special theory of relativity. Einstein publishes his general theory of relativity in 1916.

1910

BERTRAND ARTHUR WILLIAM RUSSELL (1872–1970) and Alfred North Whitehead (1861–1947) begin publication of their three-volume *Principia Mathematica,* an ambitious attempt to derive all mathematics by logical principles from a small set of beginning AXIOMS.

1913

Indian mathematician SRINIVASA AIYANGAR RAMANUJAN (1887–1920) begins a five-year collaboration with British mathematician GODFREY HAROLD HARDY (1877–1947).

1921

German mathematician AMALIE NOETHER (1882–1935) publishes her theory of RINGS, directing research in algebra away from the study of calculation toward the study of abstract structures.

1925

SIR RONALD AYLMER FISHER (1890–1962) publishes *Statistical Methods for Research Workers,* an influential work that provides the basis for modern experimental design.

1928

James Alexander (1888–1971) develops the Alexander polynomial, the first invariant in KNOT THEORY. Subsequently, John Conway defined the Conway polynomial in 1968, and Vaughn Jones the Jones polynomial in 1985.

1931

KURT GÖDEL (1906–78) stuns the mathematical community by establishing a pair of "incompleteness theorems." Gödel proves that within any logically rigorous system there will necessarily be statements that can neither be proved nor disproved. The goal pursued by RUSSELL and Whitehead is proved unattainable.

1938

CLAUDE ELWOOD SHANNON (1916–2001) establishes that BOOLEAN ALGEBRA can be successfully applied to computer design. He founds the field of INFORMATION THEORY in 1949.

1944

Hungarian-American mathematician JOHN VON NEUMANN (1903–57) and American economist Oskar Morgenstern (1902–77) found GAME THEORY.

1963

Edward Lorenz develops CHAOS theory.

1976

Using 1,200 hours of computer time, Kenneth Appel and Wolfgang Haken prove the FOUR-COLOR THEOREM.

1978

Ronald Rivest, Adi Shamir, and Leonard Adleman develop the RSA public-key encryption system.

1979

Benoit Mandelbrot discovers the MANDELBROT SET. He later founds the field of FRACTAL geometry.

Under the instigation of Pope John Paul II, the Roman Catholic Church opens its files on the Galilean trials. The Church reverses its 17th-century condemnation of the scholar in 1992.

1994

ANDREW WILES, with the assistance of Richard Taylor, proves FERMAT'S LAST THEOREM.

1998

Thomas Hales uses computer methods to establish JOHANNES KEPLER's 1611 conjecture that the cubic close packing and the hexagonal close packing of spheres are the densest packings of spheres. Mathematicians are unable to verify the proof without the aid of a computer.

1999

Hales proves the "honeycomb conjecture," which states that any partition of the plane into regions of equal area has perimeter equal to the design of a honeycomb.

Yasumasa Kanada of the University of Tokyo computes π to 206,158,430,000 decimal places.

2000

Michael Hutchings, Frank Morgan, Manuel Ritoré, and Antonio Ros prove the outstanding "double bubble" conjecture in the theory of SOAP BUBBLES. They establish that the design of minimal surface area that encloses two fixed volumes is indeed the "double bubble" configuration one observes in nature.

2003

Tomás Oliveira e Silva verifies that GOLDBACH'S CONJECTURE holds true for all even numbers between four and 6×10^{16}.

Michael Shafer discovers that the 6,320,430-digit number $2^{20,996,011} - 1$ is PRIME. It is the largest prime and the 40th MERSENNE PRIME known to this date.

2004

Martin Dunwoody announces to the mathematical community that he may have proved the Poincaré conjecture. Mathematicians await the details of his proof.

APPENDIX II

BIBLIOGRAPHY AND WEB RESOURCES

Print

Abbot, E. A. *Flatland: A Romance of Many Dimensions.* Princeton, N. J.: Princeton University Press, 1991.

Abbott, J. C., ed. *The Chauvenet Papers: A Collection of Prize-Winning Expository Papers in Mathematics.* Washington, D. C.: Mathematical Association of America, 1978.

Ahlfors, Lars V. *Complex Analysis.* 3rd ed. New York: McGraw-Hill, 1979.

Aigner, Martin, and Günter M. Ziegler. *Proofs from the Book.* New York: Springer-Verlag, 1999.

Almgren, F. J., Jr., and J. E. Taylor. "Geometry of Soap Films." *Scientific American* 235, no. 1 (1976): 82–93.

Anderson, I. *A First Course in Combinatorial Mathematics.* Oxford, U.K.: Clarendon Press, 1974.

Angel, Allen R., and Stuart R. Porter. *A Survey of Mathematics with Applications.* 6th ed. Boston: Addison-Wesley, 2001.

Anne, C. "Egyptian Fractions and the Inheritance Problem." *The College Mathematics Journal* 29, no. 4 (1998): 296–300.

Anton, Howard, Irl Bivens et al. *Calculus: Brief Edition.* 7th ed. New York: John Wiley & Sons, 2002.

Apostol, Tom M., and Hubert E. Chrestenson et al., eds. *A Century of Calculus, Part 1, 1894–1968,* and *A Century of Calculus, Part 2, 1969–1991.* Washington, D.C.: Mathematical Association of America, 1992.

Appel, Kenneth, and Wolfgang Haken. "The Solution of the Four-Color Map Problem." *Scientific American* 237, no. 4 (1977) 108–121.

Arnold, B. H. *Intuitive Concepts in Elementary Topology.* Englewood, N.J.: Prentice-Hall, 1962.

Ash, J. Marshall, and Solomon W. Golomb. "Tiling Deficient Rectangles with Trominoes." *Mathematics Magazine* 77, no. 1 (2004): 46–55.

Atiyah, M. F., and I. G. MacDonald. *Introduction to Commutative Algebra.* Reading, Mass.: Addison-Wesley, 1969.

Averbach, Bonnie, and Orin Chein. *Problem Solving through Recreational Mathematics.* New York: Dover, 1980.

Ball, W. W. Rouse, and H. S. M. Coxeter. *Mathematical Recreations and Essays.* 12th ed. Toronto: University of Toronto Press, 1974.

Barnsley, M. F. *Fractals Everywhere.* San Diego: Academic Press, 1988.

Barr, Stephen. *Experiments in Topology.* New York: Dover, 1964.

Beckman, Petr. *A History of π.* 3rd ed. New York: St. Martin's Press, 1971.

Belcastro, Sarah-Marie, and Thomas C. Hull. "Classifying Frieze Patterns without Using Groups." *The College Mathematics Journal* 33, no. 2 (2002): 93–98.

Benjamin, Arthur T., and Jennifer J. Quinn. *Proofs that Really Count.* Washington, D.C.: Mathematical Association of America, 2003.

Berlinghoff, William P., and Fernando Q. Gouvêa. *Math through the Ages: A Gentle History for Teachers and Others.* Farmington, Maine: Oxton House Publishers, 2002.

Binmore, K. *Fun and Games: A Text on Game Theory.* Lexington, Mass.: Heath, 1992.

Blitzer, Robert. *Thinking Mathematically.* 2nd ed. Upper Saddle River, N. J.: Prentice-Hall, 2003.

Bondi, C., ed. *New Applications of Mathematics.* London: Penguin Books, 1991.

Borowski, E. J., and J. M. Borwein. *The Harper Collins Dictionary of Mathematics*. New York: Harper Collins, 1991.

Brams, S. J., and A. D. Taylor. *Fair Division: From Cake-Cutting to Dispute Resolution*. Cambridge, U.K.: Cambridge University Press, 1996.

Bruno, Leonard C. *Math and Mathematicians: The History of Math Discoveries around the World*. Vol. 1, A–H, and Vol. 2, I–Z. Farmington Hills, Mich.: Gale Group, 1999.

Bunch, Bryan. *Mathematical Fallacies and Paradoxes*. New York: Dover, 1982.

Cajori, Florian. *A History of Mathematical Notations*. New York: Dover, 1993.

do Carmo, Manfredo P. *Differential Geometry of Curves and Surfaces*. Englewood Cliffs, N.J.: Prentice-Hall, 1976.

Chang, Gengzhe, and Thomas W. Sederberg. *Over and Over Again*. Washington, D.C.: Mathematical Association of America, 1997.

Chartrand, Gary. *Introductory Graph Theory*. New York: Dover, 1977.

Clapham, Christopher. *The Concise Oxford Dictionary of Mathematics*. 2nd ed. Oxford, U.K.: Oxford University Press, 1996.

Cohen, I. Bernard, and George E. Smith, eds. *The Cambridge Companion to Newton*. Cambridge, U.K.: Cambridge University Press, 2002.

Conway, John H. *The Sensual (Quadratic) Form*. Assisted by Francis Y. C. Fung. Washington, D.C.: Mathematical Association of America, 1997.

Conway, John H., and Richard Guy. *The Book of Numbers*. New York: Springer-Verlag, 1996.

Courant, Richard, and Herbert Robbins. *What is Mathematics?* 2nd ed. Revised by Ian Stewart. Oxford, U.K.: Oxford University Press, 1996.

Crystal, David, ed. *The Cambridge Encyclopedia*. Cambridge, U.K.: Cambridge University Press, 1990.

Davenport, H. *The Higher Arithmetic: An Introduction to the Theory of Numbers*. New York: Dover, 1983.

Davis, Philip J., and Reuben Hersh. *The Mathematical Experience*. Harmondsworth, U.K.: Penguin Books, 1981.

Denes, J. *Latin Squares and Their Applications*. New York: Academic Press, 1974.

Devaney, Robert L. *A First Course in Chaotic Dynamical Systems, Theory and Experiment*. Reading, Mass.: Addison-Wesley, 1992.

———. *Chaos, Fractals, and Dynamics: Computer Experiments in Mathematics*. Menlo Park, Calif.: Addison-Wesley, 1990.

Devaney, Robert L., and L. Keene, eds. *Chaos and Fractals: The Mathematics behind the Computer Graphics*. Providence, R.I.: American Mathematical Society, 1989.

Devlin, Keith. *The Joy of Sets*. 2nd ed. New York: Springer-Verlag, 1993.

Doyle, Peter G., and J. Laurie Snell. *Random Walks and Electric Networks*. Washington, D.C.: Mathematical Association of America, 1984.

Dunham, William. *Journey through Genius*. New York: John Wiley & Sons, 1990.

———. *The Mathematical Universe: An Alphabetical Journey through Great Proofs, Problems, and Personalities*. New York: John Wiley & Sons, 1994.

Eves, Howard W. *Mathematical Circles*. Vol. 1. Washington, D.C.: Mathematical Association of America, 2003.

Farris, F. A., and N. K. Rossing. "Woven Rope Friezes." *Mathematics Magazine* 72, no. 1 (1999): 32–38.

Firby, P. A., and C. F. Gardiner. *Surface Topology*. 2nd ed. New York: Ellis Horwood, 1991.

Fisher, Stephen D. *Complex Variables*. 2nd ed. New York: Dover, 1990.

Fomin, Dmitri, and Sergey Genkin et al. *Mathematical Circles (Russian Experience)*. Providence, R. I.: American Mathematical Society, 1996.

Freund, John E., and Gary A. Simon. *Statistics: A First Course*. 5th ed. Englewood Cliffs, N.J.: Prentice Hall, 1991.

Gardner, Martin. *Codes, Ciphers and Secret Writing*. New York: Dover, 1972.

———. *Fractal Music, Hypercards, and More*. New York: W. H. Freeman and Co. 1992.

———. *Mathematics, Magic and Mystery*. New York: Dover, 1956.

———. *Penrose Tiles to Trapdoor Ciphers ... and the Return of Dr. Matrix*. New York: W. H. Freeman, 1989.

Garfunkel, Solomon. *For All Practical Purposes*. 5th ed. New York: W. H. Freeman, 2000.

Gibson, Carol, ed. *The Facts On File Dictionary of Mathematics*. New York: Facts On File, 1985.

Gillespie, Charles Coulston, ed. *Dictionary of Scientific Bibliography*. New York: Charles Scribner's Sons, 1972.

Giordano, Frank R., and Maurice D. Weir et al. *A First Course in Mathematical Modeling*. 2nd ed. Pacific Grove, Calif.: Brooks/Cole, 1997.

Goldman, Jay R. *The Queen of Mathematics: A Historically Motivated Guide to Number Theory*. Wellesley, Mass.: A. K. Peters, 1998.

Golomb, Solomon W. *Polyominoes: Puzzles, Patterns, Problems and Packings*. 2nd ed. Princeton, N.J.: Princeton University Press, 1994.

Gorini, Catherine A. *The Facts On File Geometry Handbook*. New York: Facts On File, 2003.

Graham, Ronald, and Donald Knuth et al. *Concrete Mathematics*. 2nd ed. Boston: Addison-Wesley, 1994.

Grinstead, C. M. and L. J. Snell. *Introduction to Probability*. 2nd ed. Providence, R.I.: American Mathematical Society, 1997.

Gullberg, Jan. *Mathematics from the Birth of Numbers.* New York: W. W. Norton & Co. 1997.

Guy, Richard, and Robert E. Woodrow, eds. *The Lighter Side of Mathematics.* Washington, D.C.: Mathematical Association of America, 1994.

Hadlock, Charles Robert. *Field Theory and its Classical Problems.* Washington, D.C.: Mathematical Association of America, 2000.

Hall, Nina, ed. *Exploring Chaos: A Guide to the New Science of Disorder.* New York: W. W. Norton & Co. 1991.

Hardy, G. H., and E. M. Wright. *An Introduction to the Theory of Numbers.* 5th ed. Oxford, U.K.: Oxford University Press, 1979.

Hass, J. "General Double Bubble Conjecture in R³ Solved." *Focus,* no. 5 (2000): 4–5.

Herstein, I. N. *Topics in Algebra.* 2nd ed. New York: John Wiley & Sons, 1975.

Hildebrandt, S., and A. Tromba. *Mathematics and Optimal Form.* New York: Scientific American Books, 1985.

Hill, Victor E., IV. "President Garfield and the Pythagorean Theorem." *Math Horizons,* (Feb. 2002): 7–9.

Hoffman, Paul. *The Man Who Loved Only Numbers.* New York: Hyperion, 1998.

Honsberger, Ross. *Ingenuity in Mathematics.* Washington, D.C.: Mathematical Association of America, 1970.

———. *Mathematical Diamonds.* Washington, D.C.: Mathematical Association of America, 2003.

———. *Mathematical Gems.* Washington, D.C.: Mathematical Association of America, 1973.

———. *More Mathematical Morsels.* Washington, D.C.: Mathematical Association of America, 1991.

Honsberger, Ross, ed. *Mathematical Plums.* Washington, D.C.: Mathematical Association of America.

Hull, Thomas. "A Note on 'Impossible' Paper Folding." *American Mathematical Monthly* 103, no. 3 (1996): 242–243.

Jacobs, Konrad. *Invitation to Mathematics.* Princeton, N.J.: Princeton University Press, 1992.

James, Ioan. *Remarkable Mathematicians: From Euler to von Neumann.* Cambridge, U.K.: Cambridge University Press, 2002.

Kac, Mark, and Stanislaw M. Ulam. *Mathematics and Logic.* New York: Dover, 1968.

Kaplan, Robert, and Ellen Kaplan. *The Art of the Infinite: The Pleasures of Mathematics.* Oxford, U.K.: Oxford University Press, 2003.

Kay, David C. *College Geometry: A Discovery Approach.* 2nd ed. Boston: Addison-Wesley, 2001.

Kim, Scott. "Hyperspace: Up, Out, and Away." *Discover* 23, no. 10 (Oct. 2002), 82.

Kolmogorov, A. N., and S. V. Fomin. *Introductory Real Analysis.* Translated and Edited by Richard A. Silverman. New York: Dover, 1970.

Konhauser, Joseph D. E., and Dan Velleman et al. *Which Way Did the Bicycle Go? ... And Other Intriguing Mathematical Mysteries.* Washington, D.C.: Mathematical Association of America, 1996.

Maor, Eli. *The Facts On File Calculus Handbook.* New York: Facts On File, 2003.

———. *To Infinity and Beyond: A Cultural History of the Infinite.* Princeton, N.J.: Princeton University Press, 1987.

Massey, William S. *A Basic Course in Algebraic Topology.* New York: Springer-Verlag, 1991.

McDaniel, Michael. "Knots to You." *Math Horizons,* (Nov. 2003): 9–12.

Meisters, G. H. "Lewis Carroll's Day-of-the-Week Algorithm." *Math Horizons,* (Nov. 2002): 24–25, 34.

Mesterton-Gibbons, Michael. *A Concrete Approach to Mathematical Modelling.* New York: John Wiley & Sons, 1995.

Morgan, Frank. "Double Bubble No More Trouble." *Math Horizons,* (Nov. 2000): 2, 30–31.

Mnatsakanian, M. "Annular Rings of Equal Areas." *Math Horizons,* (Nov. 1997): 5–8.

Neilson, William Allan, ed. *Webster's Biographical Dictionary.* Springfield, Mass.: G. & C. Merriam Company, 1976.

Nelson, David, ed. *The Penguin Dictionary of Mathematics.* 2nd ed. London: Penguin Books, 1998.

Nelson, Roger B. "Paintings, Tilings, and Proofs." *Math Horizons,* (Nov. 2003): 5–8.

Newman, James R. *The World of Mathematics: A Small Library of the Literature of Mathematics from A'h-mosé the Scribe to Albert Einstein.* Redmond, Wash.: Tempus Books, 1956.

Olivastro, Dominic. *Ancient Puzzles: Classic Brainteasers and Other Timeless Mathematical Games of the Last 10 Centuries.* New York: Bantum Books, 1993.

Ore, Oystein. *Number Theory and Its History.* New York: Dover, 1976.

Pasles, Paul C. "Some Magic Squares of Distinction." *Math Horizons,* (Feb. 2004): 10–12.

Paulos, John Allen. *Beyond Numeracy.* New York: Knopf, 1991.

———. *Innumeracy: Mathematical Illiteracy and its Consequences.* New York: Hill & Wang, 1988.

Peterson, I. *The Mathematical Tourist: Snapshots of Modern Mathematics.* New York: W. H. Freeman, 1988.

———. *Islands of Truth: A Mathematical Mystery Cruise.* New York: W. H. Freeman, 1990.

———. "Toil and Trouble over Double Bubbles." *Science News* 148 (1995): 101.

Philips, Anthony. "Turning a Surface Inside Out." *Scientific American* 214 (1964): 112–120.

Porkess, Roger. *The HarperCollins Dictionary of Statistics.* New York: HarperCollins, 1991.

Powers, David L. *Boundary Value Problems.* 3rd ed. Fort Worth, Tex.: Harcourt Brace College Publishers, 1987.

Rademacher, Hans, and Otto Toeplitz. *The Enjoyment of Math.* Princeton, N.J.: Princeton University Press, 1957.

Rucker, R. *The Fourth Dimension: Toward a Geometry of Higher Reality.* Boston: Houghton Mifflin, 1984.

Rudin, Walter. *Real and Complex Analysis.* 3rd ed. New York: McGraw-Hill, 1987.

Saari, D. G., and F. Valognes. "Geometry, Voting and Paradoxes." *Mathematics Magazine* 71, no. 4 (1998): 243–259.

Saaty, T. L., and P. C. Kainen. *The Four-Color Problem: Assaults and Conquests.* New York: Dover, 1986.

Sharpes, Donald K. *Advanced Educational Foundations for Teachers.* New York: RoutledgeFalmer, 2002.

Shashkin, Yu A. *Fixed Points.* Providence, R.I.: American Mathematical Society, 1991.

Shifrin, Theodore. *Abstract Algebra: A Geometric Approach.* Englewood Cliffs, N.J.: Prentice Hall, 1996.

Simmons, George F. *Calculus with Analytic Geometry.* 2nd ed. New York: McGraw-Hill, 1996.

Singh, Simon. *Fermat's Enigma.* New York: Random House, 1997.

Smart, James R. *Modern Geometries.* 5th ed. Pacific Grove, Calif.: Brooks/Cole, 1998.

Sondheimer, Ernst, and Alan Rogerson. *Numbers and Infinity: A Historical Account of Mathematical Concepts.* Cambridge, U.K.: Cambridge University Press, 1981.

Stein, Sherman K., and Anthony Barcellos. *Calculus and Analytic Geometry.* 5th ed. New York: McGraw-Hill, 1992.

Stewart, Ian. *Concepts of Modern Mathematics.* New York: Dover, 1995.

Stewart, James. *Single Variable Calculus.* 3rd rd. Pacific Grove, Calif.: Brooks/Cole, 1995.

Strafflin, P. *Game Theory and Strategy.* Washington, D.C.: Mathematical Association of America, 1993.

Tanton, James. "A Dozen Questions about a Donut." *Math Horizons,* (Nov. 1998): 26–31.

———. "A Dozen Questions about the Isoperimetric Problem." *Math Horizons,* (Feb. 2003): 23–26.

———. "A Dozen Questions about the Powers of Two." *Math Horizons,* (Sept. 2001): 5–10.

———. "A Dozen Questions about Squares and Cubes." *Math Horizons,* (Sept. 2000): 29–34.

———. "A Dozen Reasons Why 1 = 2." *Math Horizons,* (Feb. 1999): 21–25.

———. *Solve This: Mathematical Activities for Students and Clubs.* Washington, D.C.: Mathematical Association of America, 2001.

———. "Young Students Approach Integer Triangles." *Focus* 22, no. 5 (2002): 4–6.

Todd, Deborah. *The Facts On File Algebra Handbook.* New York: Facts On File, 2003.

Vakil, Ravi. *A Mathematical Mosaic: Patterns and Problem Solving.* Burlington, Ont.: Brendan Kelly Publishing, 1996.

Wagon, Stan. *The Banach-Tarski Paradox.* Cambridge, U.K.: Cambridge University Press, 1985.

Weeks, J. R. *The Shape of Space.* New York: Marcel-Decker, 1985.

West, Beverly Henderson, and Ellen Norma Griesbach et al. *The Prentice-Hall Encyclopedia of Mathematics.* Englewood Cliffs, N.J.: Prentice-Hall, 1982.

Web Sites

Bogomolny, Alexander. "Cut the Knot!" Available online. URL: http:// www.cut-the-knot.org. Accessed on June 15, 2004. A comprehensive resource of interactive mathematics puzzles and games.

Caldwell, Chris. "The Largest Known Prime by Year: A Brief History." Available online. URL: http://www.utm.edu/research/primes/notes/by_year.html. Accessed on May 1, 2004. Chris Caldwell gives a detailed history of the search for large prime numbers and keeps readers up to date on current finds.

Department of Economics of the New School for Social Research. "The History of Economic Thought Website." Available online. URL: http://cepa.newschool.edu/het/home.htm. Accessed on April 20, 2004. This site provides a comprehensive review of the development of economic thought and practices.

Drexel University. "Math Forum." Available online. URL: http://mathforum.org. Accessed on April 20, 2004. This student-friendly site provides an interactive question-and-answer service for mathematics students and teachers, as well as a host of mathematical resources, essays, and discussions listed by topic.

Frazier, Kendrick. "A Mind at Play: An Interview with Martin Gardner." Committee for the Scientific Investigation of Claims of the Paranormal. Available online. URL: www.csicop.org/si/9803/gardner.html. Accessed on March 1998. An in-depth interview with recreational mathematician Martin Gardner.

Grissinger, Arthur. "Polya's Problem Solving Strategy." Available online. URL: www.lhup.edu/agrissin/polya.htm. Accessed on April 23, 2004. This site provides a brief summary of the problem-solving strategies proposed by George Polya.

Joyce, David. "Mathematics in China." Available online. URL: http://aleph0.clarku.edu/~djoyce/mathhist/china.html. Accessed on September 17, 1995. David Joyce provides a detailed overview of the development of mathematics in China.

Kadon Enterprises, Inc.. "Game Inventor: Martin Gardner." Available online. URL:http:// www.gamepuzzles.com/ martin.htm. Accessed on April 20, 2004. This site presents a short article about the life and work of Martin Gardner.

Kimberling, Clark. "Biographical Studies." Available online. URL: http://faculty.evansville.edu/ck6/bstud. Accessed on April 20, 2004. Clark Kimberling provides a number of in-depth biographical essays on famous mathematicians.

Loy, Jim. "Trisection of an Angle." Available online. URL: http://www.jimloy.com/geometry/trisect.htm. Accessed on April 23, 2004. This site presents, with detailed explanation, the varied techniques that have been proposed throughout the centuries for trisecting arbitrary angles.

Math Academy Online. "Platonic Realms Interactive Mathematics Encyclopedia." Available online. URL: http://www.mathacademy.com. Accessed on April 23, 2004. This site offers mathematics quotes, brief historical notes, and some discussion on mathematical concepts.

Miller, Jeff. "Earliest Known Uses of Some of the Words of Mathematics." Available online. URL: http://members.aol.com/jeff570/mathword.html. Accessed on April 23, 2004. This extensive site attempts to pinpoint the first use of current mathematical terms.

———. "Earliest Known Uses of Various Mathematical Symbols." Available online. URL: http://members.aol.com/jeff570/mathsym.html. Accessed on April 23, 2004. This site attempts to identify the first use of a given mathematical symbol and provide the date and name of the document in which it first appeared.

O'Connor, John, and Edmund Robertson eds. "MacTutor History of Mathematics Archive." Available online. URL: http://www-groups.dcs.st-andrews.ac.uk/~history. Accessed on April 20, 2004. This award-winning site pro-

vides the most comprehensive collection of historical and biographical essays currently available.

Sandifer, C. Edward. "Ed Sandifer's Home Page." Available online. URL: http://vax.wcsu.edu/~sandifer/homepage.html. Accessed January 16, 2004. Ed Sandifer is an expert on the life and work of Leonhard Euler and provides a wealth of invaluable information about this famous scholar on this Web site.

Verrill, H. "Origami Trisection of an Angle." Available online. URL: http://hverrill.net/origami/. Accessed on June 24, 2002. This site provides an accessible overview of the use of mathematics in origami.

Wales, Jimmy, and Larry Sanger, fndrs. "Wikipedia: The Free Encyclopedia: Mathematics." Available online. URL: http://en.wikipedia.org/wiki/Mathematics. Accessed on June 15, 2004. This site serves as a user-friendly resource on mathematical topics.

Weisstein, Eric. "Eric Weisstein's World of Scientific Biography." Available online. URL: http://scienceworld.wolfram.com/biography. Accessed on April 20, 2004. This site provides biographical information on an extensive list of famous mathematicians.

———. "Mathworld." Available online. URL: http://mathworld.wolfram.com. Accessed on April 22, 2004. Perhaps the most comprehensive mathematical resource currently available on the Web, this online encyclopedia provides detailed information on most every aspect of current mathematical research.

Wilkins, David R. "The History of Mathematics." Available online. URL: http://www.maths.tcd.ie/pub/HistMath. Accessed on April 23, 2004. This site provides a directory of Web sites from around the world that relate to the history of mathematics, as well as presents biographies of some 17th- and 18th-century mathematicians.

Appendix III

ASSOCIATIONS

The following organizations provide information about mathematics, mathematics research, and mathematics education of interest to students and teachers.

American Mathematical Society, 201 Charles Street, Providence, R.I. 02904-2294. Telephone: (800) 321 4267. Web site: www.ams.org.

Association for Women in Mathematics, 4114 Computer and Space Sciences Building, College Park, Md. 20742-2461. Telephone: (301) 405 7892. Web site: www.awm-math.org.

Canadian Mathematical Society, 577 King Edward, Suite 109, Ottawa, Ont., Canada K1N 6N5. Telephone: (613) 562 5702. Web site: www.cms.math.ca.

Clay Mathematics Institute, One Bow Street, Cambridge, Mass. 02138. Telephone: (617) 995 2600. Web site: www.claymath.org.

European Mathematical Society, Department of Mathematics, P.O. Box 4 (Yliopistonkatu 5), 00014 University of Helsinki, Finland. Telephone: 3589 1912 2883. Web site: www.emis.de.

Institute for Mathematics and Its Applications, 400 Lind Hall, 207 Church Street, S.E., Minneapolis, Minn. 55455-0436. Telephone: (612) 624 6066. Web site: www.ima.umn.edu.

Institute of Mathematical Statistics, P.O. Box 22718, Beachwood, Ohio 44122. Telephone: (216) 295 2340. Web site: www.imstat.org.

International Mathematical Union, Institute of Advanced Study, Einstein Drive, Princeton, N.J. 08540. Telephone: (609) 683 7605. Web site: www.mathunion.org.

International Society of the Arts, Mathematics, and Architecture, Department of Mathematics, University at Albany, 1400 Washington Avenue, Albany, N.Y. 12222. Telephone: (518) 442 3300. Web site: www.isama.org.

International Statistical Institute, P.O. Box 950, 2270 AZ Voorburg, The Netherlands. Telephone: 31 70 3375737. Web site: www.cbs.nl/isi.

London Mathematical Society, De Morgan House, 57–58 Russell Square, London WC1B 4HS. Telephone: 020 7637 3686. Web site: www.lms.ac.uk.

Math Circle, P.O. Box 313, Jamaica Plane, Mass. 02130. Telephone: (617) 519 6397. Web site: www.themathcircle.org.

Mathematical Association of America, 1529 Eighteenth Street N.W., Washington, D.C. 20036-1385. Telephone: (800) 741 9415. Web site: www.maa.org.

Mathematical Sciences Research Institute, 17 Gauss Way, Berkeley, Calif. 94720-5070. Telephone: (510) 642 0143. Website: www.msri.org.

Mathematics Foundation of America, 129 Hancock Street, Cambridge, Mass. 02139. Telephone: (510) 525 7931. Web site: www.mfoa.org.

Math Forum, 3210 Cherry Street, Philadelphia, Penn. 19104. Telephone: (800) 756 7823. Web site: www.mathforum.org.

National Council of Teachers of Mathematics, 1906 Association Drive, Reston, Va. 20191-1502. Telephone: (703) 620 9840. Web site: www.nctm.org.

Society for Industrial and Applied Mathematics, 3600 University City Science Center, Philadelphia, Penn. 19104-2688. Telephone: (215) 382 9800. Web site: www.siam.org.

St. Mark's Institute of Mathematics, 25 Marlborough Street, Southborough, Mass. 01772. Telephone: (508) 786 6126. Web site: www.stmarksschool.org.

INDEX

Boldface page numbers indicate main entries. Page numbers in *italics* indicate illustrations or diagrams.

A

AAA (angle-angle-angle) rule **1**, 463
AAS (angle-angle-side) rule **1**
abacus *2*, **2–3**, 56, 74, 260
Abel, Niels Henrik **3**, 10, 84, 215, 241, 405, 431–432
  and solutions to polynomial equations 471
Abelian groups 3, 5, 83, 241
Abel Prize 3
abscissa 62
absolute convergence **3–4**, 248, 337
  test for 3–4, 104
absolute error 167
absolute frequency 206
absolute maximum/minimum 330, *331*
absolute value (modulus) **4–5**, 131, 133, 142, 310, 328–329
  of complex number 25, 86
abstract algebra **5**, 10, 314, 354–355
abstract group 68
abundant number 389
Académie Française 8
acceleration 523
Accurate Rendering (Ahmes) 447
Achilles and the tortoise (paradox) 532–533
acnode. *See* isolated point
Acosta, José de 260
actuarial science 5, 312
acute angle **5–6**, 14

acute triangle 506
addend 6
addition **5**, **6**, 205
addition formulae (trigonometry) 6, 41–42
addition law 344, 367
addition-multiplication magic square 327, *328*
addition rule, in probability theory 415–416, *416*
additive function 6
additive identity element (zero) 6, 257
additive inverse, and rings 448
Adleman, Leonard 111
affine geometry **6–7**, 339
affine transformation 7
Agnesi, Maria Gaëtana **7–8**, 58
Ahmes (Ahmose) 155, 447
Ahmes papyrus. *See* Rhind papyrus
Alembert, Jean Le Rond d' *8*, **8–9**, 209
  and Fourier series 202
Alembert's theorem, d'. *See* fundamental theorem of algebra
$\aleph_0$ (aleph null) **60–61**
Alexander, James Waddell 295–296
Alexander polynomial 295–296
algebra **9–11**
  and geometry 124

L'Algebra (Bombelli) 47–48
Algebra (Simpson) 463
algebraic curve 114
algebraic function 11
algebraic notation 51
algebraic number **11**, 60, 97
  and analytic number theory 360
algebraic structure 11
algebra of logic, Boole's 48
Algebra with Arithmetic and Mensuration (Colebrook) 42
algorithm **11**, 294
Al-jam' w'al-tafriq ib hisab al-hind (al-Khwārizmī) 293
Almagest (Ptolemy) 421
alphamagic square 328, *328*
alternate interior/exterior angles *504*
alternating harmonic series 248, 409
alternating knot 295
alternating series **11–12**
alternating-series test 12, 104–105
altitude **12–13**, 105
  of triangle
    concurrency of *12*, 12–13, 165
    and Euler line 174–175, *507*
amicable numbers **13**, 18, 423
Analemma (Ptolemy) 422
analog vs. digital 135–136

Analyse des infiniment petits (L'Hôpital) 40, 310
analysis **13**, 423
Analysis per quantitatum series (Newton) 353
analysis situs. *See* topology
Analysis situs (Poincaré) 398
"The Analyst" (Berkeley) 58
analytic engine, Babbage's 32
analytic geometry 13, 347
analytic number theory **13**, 139–140, 360, 411
analytics 27
anchor ring. *See* torus
ancillary theorem 498–499
angle **13–15**, *14*, 276, 441
  45° (octant) 363
  of elevation/depression 14
  of inclination 261
angle bisector 46, 260
angle brackets 50, 519
angle trisection. *See* trisecting an angle
Annales de Mathématiques (journal) 25
Annals of Eugenics (journal) 387
Annals of Mathematics 531
annulus **15**, 89, 94, 177
anomalous cancellation 56
antiderivative 94–95, 211, 360
antidifferentation **15**, 133, 272
antinomy. *See* Russell's paradox

antipodal points (antipodes) **15**, 471, 473
antisymmetry 365
apex (apices) **15**, 91, 423
Apollonius of Perga **15–16**, 76, 78, 93, 105, 115, 136–137, 189, 239, 252, 253, 476
    and duplicating the cube 149
    and trisecting an angle 513
Apollonius's circle **16**
Apollonius's theorem **16**
apothem (short radius) **16**, 403
Appel, Kenneth 200–201, 420
applied mathematics **17**
appropriately nice functions 315
approximation **17**
Arabic mathematics **17–18**
Arbogast, Louis 249
arc **18**, 76
Archimedean solids 398
Archimedean spiral 239, 400, **474**
Archimedes 18, **18–21**, _19_, _20_, 91, 93, 105, 115, 173, 180, 213, 238–239, 270, 272, 476
    and Archimedean solids 398
    on Fields medal 194
    and his spiral 474
    and semiregular polyhedra 404
    and sum of first _n_ numbers 489
    and trisecting an angle 513
Archimedes' water screw 20, _21_
Archytas **21–22**, 172
arc length **22**, 470
area **22–25**, _23_, _24_
    of annulus 15
    of basic shapes _23_, 23–24, 156, 248, 293
    of circle 76, 222, 392
    of curved figures 172
    of ellipsoid 161–162
    as ill-defined concept 22–23, 24, 526
    of polygon 403
    of quadrilateral, Bretschneider's formula for 52–53, 430
    of quadrilateral inscribed in circle, Brahmagupta's formula for 51–52
    of rectangle 443
    of square 474
    of trapezoid 504
    of triangle 486
area hyperbolic functions **281**

Argand, Jean Robert **25**, 86
Argand diagram 25, 86, 359
argument **25–26**, 86, 199, 524–525
Aristotle **26–27**, 238, 270, 309, 532
    and formal logic 25–26, 199
Aristotle's wheel paradox 375
arithmetic **27**
Arithmetica (Diophantus) 137–138, 189, 190, 239, 253, 525
Arithmetica infinitorum (Wallis) 528
Arithmetica universalis (Newton) 353
Arithmetices principia, nova methodo exposita (Peano) 385
arithmetic–geometric-mean inequality 219, 333
arithmetic mean 333
The Arithmetic of Logarithms (Briggs) 53–54
arithmetic sequence **28**, 139–140
arithmetic series 28
arrangement. _See_ permutation
array **28**
arrow paradox, Zeno's 533
Ars conjectandi (Jacob Bernoulli) 39–40, 414
Ars magna (Cardano) 10, 47, 59–60, 112, 190–191, 494
Artin, Emil 52
Āryabhata (Indian mathematician) 17, **28–29**, 264, 392
Āryabhatiya (Āryabhata) 28–29
ASA (angle-side-angle) rule **1**
"As I Was Going to St. Ives" 156
associative property **29**, 83
    of addition 5
    in Boolean algebra 49
    and groups 241
    of matrices 223
    and order of operation 366
    and rings 448
    of vector addition 520, 522
astroid 115
Astronomiae physicae et geometricae elementa (D. Gregory) 240
Astronomia nova (Kepler) 292
asymmetrical 490
asymptote **29–30**, 234, 440
asymptotic series 39
atto- (10^{-18}) 465

attractor 150
augend 6
automaton _30_, **30–31**, 37, 350–351
automorphic functions 398
average. _See_ mean (average)
axiom (postulate) **31**, 170, 314
axiom of choice **31**, 530, 534

B

Babbage, Charles **32**, **32–33**, 322
Babbage's difference engine _33_
Babylonian mathematics 9, **33–35**, 249, 425, 426, 428
Bacon, Francis 452
Bakhshali manuscript 264
balance point. _See_ center of gravity
La balancitta (Galileo) 213
Banach, Stefan 24, **35**
Banach-Tarski paradox 24–25, 35, 374, 526
band ornament. _See_ frieze pattern
barber paradox 374–375, 453–454
Barbier, Joseph 95
bar chart 479, _479_
"The Bargaining Problem" (Nash) 347
Barrow, Isaac **35–36**, _36_, 54
Der barycentrische Calcül (Möbius) 339
base of exponent 180
base of number system **36–38**, _37_
base of polygon/polyhedron **38**, 506
base-10 logarithms 53
Bayes, Rev. Thomas **38–39**
Bayes's theorem 39, 302
bearing **39**
Behennde unnd hüpsche Rechnung auf fallen Kauffmannschaften (Widman) 6, 487
bel 118
Bell, Alexander Graham 118
bell-shaped distribution 356, 480, _480_
Berkeley, George 58
Bernays, Paul 250
Bernoulli family **39–40**, 50, 284, 529
Bernoulli, Daniel 40, 107, 202
Bernoulli, Jacob 39–40, 58, 414
Bernoulli, Jacob (II) 40
Bernoulli, Jacques 65–66, 489
Bernoulli, Johann 40, 58, 107, 114, 272, 310, 474, 494

Bernoulli, Johann (II) 40
Bernoulli, Johann (III) 40
Bernoulli, Nicolaus (I) 40
Bernoulli, Nicolaus (II) 40
Bernoulli numbers 40, 435, 489
Bernoulli's inequality 267–268
Berry's paradox 374–375
Bertrand, Louis François 40, 71
Bertrand's conjecture 167
Bertrand's paradox _40_, **40–41**, 75, 374, 417
"Beweis, dass jede Menge wohlgeordnet werden kann" (Zermelo) 533–534
Bhāskara II (Bhaskaracharya) 10, 17, **41–42**, 264, 392
    and negative numbers 348
bias **42**
biconditional ("if, and only if" statement) **42**, 91, 514
bifurcation point 151
Bijaganita (Bhāskara) 41–42
bijection 209
bimodal distribution 480, _480_
binary line search. _See_ bisection method
binary numbers 30, **42–43**, 88, 150
binary operation **43**, 257, 279
    cancellation as 56–59
    closure as 79
    commutative property as 83
Binet's formula 193
binomial **43**
binomial coefficient 45, 64–65, 80, 418
binomial distribution **43–44**, _44_, 143, 180
binomial expansion **44–45**
binomial theorem 18, **44–45**, 80–81, 219, 309, 364
Biometrics (journal) 387
biquadratic equation. _See_ quartic equation
birectangular triangle 474
bisection method (dichotomous line search, binary line search) **45–46**, 277, 451
bisector **46**
bit 461
Blatzer, Richard 47
body mass index (BMI). _See_ Quételet index
Bohlen numbers 195
Bolyai, János **46–47**, 171, 219, 255, 318, 355, 455
Bolzano, Bernard Placidus **47**, 276
Bolzano's theorem. _See_ intermediate-value theorem

Bombelli, Rafael **47–48**, 50, 113

Boole, George 10, *48*, **48–49**, 199, 309, 460

Boolean algebra 49, 460, 461–462

border square 328, *328*

Borel, Félix Edouard Émile 216, 307

Borromean rings 49, *49*

Borsuk, Karol 15

Borsuk-Ulam theorem 15, 471

Bosse, Abraham 124

Boss puzzle. *See* slide 15 puzzle

bound **49–50**

bounded above/below, definition of 50

Bourbaki, Nicolas **50**, 168

braces 50, 460

brachistochrone problem 39, 114, 310

brackets **50–51**, 366

Brahe, Tycho 292

Brahmagupta 9, **51**, 114, 264, 348, 534

Brahmagupta's formula **51–52**, 53, 249, 264, 430

Brahmasphutasiddhanta (Brahmagupta) 51, 264

braid 52, *52*

braid group 52

breadth 309

Bretschneider's formula **52–53**

Briggs, Henry **53–54**, 320, 346

Briggsian logarithms 53

Brinkley, John 245

British Association for the Advancement of Science 68

Bromhead, Sir Edward 239–240

Brouillon project d'une atteinte aux evenemens des recontres du cone avec un plan (Desargues) 123–124

Brouncker, Lord William **54**, 529

Brouwer, Luitzen Egbertus Jan 198

Brouwer fixed-point theorem 198

Brownian motion 158

brute force **54–55**, 461

Buffon, Georges 55

Buffon-Laplace problem. *See* Buffon needle problem

Buffon needle problem 55, *55*, 302, 393

Bürgi, Jobst 320

Buridan, Jean 368

butterfly effect 151

C

cake cutting. *See* fair division

Calandrini, Giovanni Ludovico 107

calculator, mechanical 382, *383*

calculus 7, 8–9, **56**, 100, 320
 history of **57–58**, 67, 126, 325, 368
 and optimization 365
 and Snell's law of refraction 468–469

calculus of variations, Weierstrass's 284, 529

calendar, Chinese 537

cancellation **56–59**

cancellation law, for fractions 204–205

Canon doctrinae triangulorum (Rhaeticus) 511

Canon mathematicus seu ad triangula (Viète) 525

Cantor, Georg 11, **59**, 203, 208, 270, 297, 533
 on cardinality 60–62, 129–130, 271, 460–461
 and continuum hypothesis 100
 on infinity 47
 and number theory 63, 359, 367–368, 441, 442

Cantor set 203

Cardano, Girolamo (Jerome Cardan) 10, 47, **59–60**, 112, 190–191, 431, 494
 and imaginary numbers 87
 and negative numbers 348–349
 and probability theory 413, 414, 417

Cardano's formula 112–113, 431, 470, 494

cardinality **60–62**, 100, 106, 129–130, 209, 270, 359
 and number theory 123, 359, 368, 442

cardioid 62, 115

Carr, G. C. 435

"carrying digits" 2, 6

Cartesian coordinates (orthogonal coordinates) **62–63**, 105, 114, 124, 226, 275, 504
 of cardioid 62
 vs. cylindrical coordinates 115
 and direction cosines 138
 vs. polar coordinates 400
 quadrants in 428
 of torus 500

Cartesian product (cross product, external direct product, product set, set direct product) 63

Cartesian space. *See* Euclidean space

Cassini, Giovanni Domenico 370

Castillon, Johann 62

casting out nines **63–64**

Catalan, Eugène Charles **64**

Catalan conjecture **64**

Catalan numbers **64–65**

catenary **65–66**, 255

Cauchy, Augustin-Louis 47, 66, **66–67**, 103, 287, 529
 and definition of limit 269, 312, 459
 and mean-value theorem 335
 and number theory 442

Cauchy-Riemann equations 67

Cauchy-Schwarz inequality 268

Cauchy sequence 442

Cavalieri, Bonaventura Francesco 57, **67**, 74, 293, 527, 528

Cavalieri's principle 57, **67–68**, 115, 412, 527, 537

Cayley, Arthur **68**, 200

Cayley-Hamilton theorem **68–69**

Cayley numbers 359, 432

ceiling/floor brackets 51

ceiling function (least-integer function) **198**

center-limit theorem **69–70**

center of gravity 18–19, **69**, 372, 470

center of mass vs. center of gravity 69

centile. *See* percentile

central angle 77, *77*

central-angle/peripheral-angle theorem 77, *77*

central-limit theorem 143, 401, 482–483
 and normal distribution 356

central projection 418

central tendency, measures of 480

centroid 335, 372, 470, 507

Ceva, Giovanni 70

Ceva's theorem 70

Ceyuan Haijing (Li Ye) 318

chain 365–366

chain rule **70–71**, 88, 133, 138, 258, 274, 281, 282, 378, 433–434

change of variable. *See* integration: by substitution

chaos **71**, 151, 203

chaos game 204

characteristica universalis 309

characteristic polynomial 68–69, 156

characteristic vector. *See* eigenvector

Chebyshev (Tchebyshev), Pafnuty Lvovich **71**, 72, 167, 415

Chebyshev polynomials 71

Chebyshev's theorem 71, **72**, 304

Chen, Jing-Run 230

Chêng Ta-wei 325

chessboard puzzle 224

chicken (game) 218, 413, *413*

Ch'in Chiu-shao **72**, 73

Chinese mathematics **72–74**

Chinese proof 425

Chinese remainder theorem 73

chi-squared test **74–75**, 98, 386, 415

choice set 31

chord 75, 76, 457, 510, *511*

Chords in a Circle (Menelaus) 336

chord theorems 78

chord values, Hipparchus's 239

Chou pei suan ching (anonymous) 73, 425

Chudnovsky, Gregory and David 393

Chuquet, Nicola 526

Chu Shih-Chieh (Zhu Shijie) 73–74, **75**, 383

circle **75–76**, 92, 136, 141–142, 166, 188
 area of 24, *24*
 diameter of 130
 eccentricity of 155

Circles of Proportion and the Horizontal Instrument (Oughtred) 369

circle squaring. *See* squaring the circle

circle theorems 75, **76–78**, 114, 425, 430

circular inversion 225

circumcenter 78, 338, 507

circumcircle 76, **78**, 165, 303–304, 321, 354

circumference
 of circle 75–76
 of earth 29

circumscribe/inscribe **78**

"C¹ Isometric Imbeddings" (Nash) 347

cissoid 136–137

clarity 66–67, 529–530

Clavis mathematicae (Oughtred) 344, 370

clock math 92, 340–341

closed curve 114

closed half-plane 243–244

closed half-space 243–244

closed interval 184, 276, 278

closure property **79**
 of addition 5
 and groups 241
 of matrices 222

cluster sampling 407

coefficient **79**, 404, 525

Cogitata physico-mathematica (Mersenne) 338

Cohen, Paul 101

Colebrook, H. J. 42

Collatz's conjecture 79

The Collected Mathematical Papers of Arthur Cayley 68
Collection (Pappus) 13, 253
collinear **79–80**, 233
Colson, John 7–8
combination (selection, unordered arrangement) **80–82**
combinatorial coefficient 45, 384
combinatorics **82**, 243
commensurable **82**, 172, 238
common denominator **82–83**
common difference 28
common divisor **83**
common factor **83**
common fraction 205
common multiple 83, 305
common notions 170
common ratio 223
commutative group 83
commutative property **83**
  of addition 5
  in Boolean algebra 49
  of dot products 148
  and groups 241
  of multiplication 10, 344
  and quaternions 432
  and rings 448–449
  and subtraction 487
  of vector addition 520, 522
commutative ring 449
comparison test 4, 103
compass and straightedge 95
complement (set operation) 460
complementary angles 14
complete digraph. *See* tournament
complete directed graph 501
complete graph 233
complete induction. *See* induction
completeness law 367
completeness property, of real numbers 47
completing the square *84*, **84**, 428
complex conjugate of complex number 526
complex Euclidean space 170
complex fraction 206
complex numbers (C) **85–87**, 219, 358–359, 408, 522
  completing the square and 84
  and Euler's formula 176
  history of 25, 47–48, 245, 398
  and intersecting circles 76
  multiplication of *86*
  order properties and 367
  polar coordinates of *86*
  and Pythagorean triples 426
  reciprocal of 442

in trigonometry 121–122
  and zeta function 535–536
components of a vector 519
composite 87, 186, 409
composite number **87**, 463
composition (of functions) 69–70, **87–88**, 274, 280
composition (of matrices) 330
compound interest 153–154, 275–276
compound statement 514–515
Comptes Rendu (journal) 66
computer **88**. *See also* calculator, mechanical
  and Boolean algebra 49
  and critical path 109
  history of 32–33, 167, 321, 351
concave **88**
concave polygon 402
concave up/concave down *89*, **89**, 429
concentric 89
"A Concise Outline of the Foundations of Geometry" (Lobachevsky) 318
conclusion (of syllogism) 27
concurrent 89, 261, 335, 338, 399, 507
conditional ("if . . . then" statement) **89–90**, 514
conditional convergence 4, 104
conditional equation 163
conditionally convergent series 4
conditional probability 39, **90–91**, 263
condition—necessary and sufficient **91**
cone **91**
conformal mapping (equiangular transformation, isogonal transformation) **91**, 337, 485
congruence **92**
congruence transformation. *See* isometry
congruent angles 14
congruent figures **92**, 463
congruent triangles 1–2
conical frustum 207
conical helix 248
The Conics (Apollonius of Perga) 15–16, 93, 253
conic sections 18, 76, **92–93**, *93*, 123–124, 149, 238, 419
conjecture 31, 499
conjugate of complex number 86–87
conjugate of sum/difference 131
conjugate of surd 490
conjunction ("and" statement) **93–94**, 514
conjunction circuit *94*
connected **94**

connected graph 233
consistent **94**
constant **94**
constant of integration **94–95**, 272, 273
constant width 75, **95**, 130
constructible **95–97**, *97*, 149, 218, 238, 403, 475, 476, 513
constructivism 297
contact number 472
contingency table *98*, **98**
continued fraction **98–100**, 487
  calculation of π using 54
  and definition of real number 441
  golden ratio as 231
  and Pythagorean triples 427
continuity, defined by Cauchy 66
"Continuity and Irrational Numbers" (Dedekind) 119
continuous function 88, **100**, 184, 276, 386
continuously compounded interest 276
continuous random variable 143
continuum hypothesis 59, 62, **100–101**, 250, 533
contour integral (curvilinear integral, line integral) **101**, 239
contour line **101**
contradiction 94, **101**, 170, 171, 228, 229, 473
contrapositive **101–102**, 265
contrapositive reasoning 26
convergence 66, 324
convergent sequence **102**, 312, 458
convergent series 3, 12, **102–105**, 182, 240, 260, 409, 459, 530
  d'Alembert's ratio test 8, 223
converse **105**, 506
convex 88
convex polygon 402
Conway, John Horton 296
cooperate vs. defect, in prisoner's dilemma 412
coordinate geometry 63, 368
coordinates **105**, 399
coordination number 472
Copernican theory 214
coprime. *See* relatively prime
corner 403
corollary 498–499
correlation, and chi-squared test 74–75
correlation coefficient 79, **106**, 107, 306–307, 386, 456–457
corresponding angles *504*

cosecant function 510
cosine function 139, *509*, 510
cosine rule. *See* law of cosines
"the cossic art" 10, 517
cotangent function 510
countable 11, **106**, 348, 441
countable number. *See* natural number; whole number
counterexample **107**, 120, 229
counterharmonic mean 333
counting number. *See* natural number
Cours d'analyse (Cauchy) 66
Cours d'analyse (Vallée-Poussin) 518
A Course of Pure Mathematics (Hardy) 246
cousin primes 515
covariance 106, **107**, 306
covering. *See* tessellation
Cramer, Gabriel **107–108**
Cramer's rule 107, **108–109**, 223, 282
Crelle, August Leopold 3
Crelle's Journal 3
critical path *109*, **109**
critical point 331
cross product (vector product) 63, **110**, 520
cryptography **111**, 410, 525
cube (hexahedron) **111–112**, 115, 396, 397
cube numbers 195
cube root **112**, 451
cubic equation 34, 84, **112–114**
  discriminant of 141
  history of 59–60, 190–191, 493–494
cubic lattice 472
cubit 309
cumulative distribution function 143
cuneiform tablets 33, *34*
curl 144
curve **114**, 314
curves, Cramer's classification of 108
curvilinear integral. *See* contour integral
cusp 492, *493*
cycle (graph theory) 233
cyclic group 295
cyclic polygon **114**
cyclic quadrilateral 78, 264, 430
cycloid **114–115**, 310
cylinder **115**, 411
cylindrical coordinates **115**, 504
cylindrical helix 248

D

data **116**, 478
Data (Euclid) 169

days-of-the-week formula **116–118**

decibel **118**

decimal representation 36–38, 72–73, 260, 264, 370, 485

declination *261*, **261**

decomposition **118–119**

De configurationibus qualitatum et motuum (Oresme) 368

decrement 262

Decker, Ezechiel de 54

Dedekind, Julius Wilhelm Richard 58, **119**, 120, 172–173, 270, 367, 442

Dedekind cut **119–120**, 172–173, 262, 277, 367
  and bounded real numbers 50
  and definition of real number 442
  and extreme-value theorem 184

De determinantibus functionalibus (Jacobi) 287

deductive logic 420
  Aristotle and 26

deductive reasoning **120–121**

defect vs. cooperate, in prisoner's dilemma 412

deferent, of cycloid 115

deficient number 389

definite integral 148–149, 275

deformation **121**

Degen, Ferdinand 3, 5

degenerate hyperbola 373

degenerate quadrics 411

degree (measure of angle) 14

degree of a polynomial **121**, 404

degree of a vertex (valence) **121**, 233

degrees of freedom **121**

Dehn, Max 165

Delian altar problem. *See* duplicating the cube

deltoid 430, *430*

De methodis serierum et fluxionum (Newton) 352

De Moivre, Abraham 85, 107, **121–122**, 471, 486, 494
  and normal distribution 356, 414, 486

De Moivre's formula (De Moivre's identity) 85, **122**

De Moivre's quintic 471

Démonstration d'une méthode pour resoudre les égalitez de tous les degrez (Rolle) 449

"Démonstration d'un théorème sur les fractions continues périodiques" (Galois) 215

De Morgan, Augustus **122–123**, 200, 265, 270, 349

De Morgan's laws **123**, 460, 525

denominator 204

dense subset 441

denumerable (enumerable, numerable) **123**

denumerable sets 59, 60–62, 100, 106

dependent events 262–263

dependent variable 519

derangement 390–391, 486

De ratiociniis in ludo aleae (Huygens) 414

derivative 132, *132*, 198, 262, 271–272, 418
  and calculation of velocity and acceleration 523
  first and second, and graphing 234

Desargues, Girard **123–124**, 270, 391, 418, 419

Desargues's theorem **124**, 270, 391, 419

Descartes, René 10, 57, **124–125**, *125*, 180, 189, 431–432, 519
  and amicable numbers 13
  and analytic geometry 13
  and coordinate geometry 62–63, 226, 235
  and Euler-Descartes formula 177
  and imaginary numbers 87

Descartes's rule of signs 124, **125–126**

descriptive statistics. *See* statistics: descriptive

determinant **126–128**, 223, 281–282, 287, 309

determinant function 108

De triangulis omnimodis (Regiomontanus) 444, 511

deviation 481

diagonal **128–129**, 376, 402, 443

diagonal argument 59, 60, 100, 106, 123, *129*, **129–130**, 359, 441–442

diagonal Latin square 302

Dialogo (Galileo) 214

diameter 75, 76, **130**, 471

diamond 376, 474

dice, tetrahedral 462

dichotomous line search. *See* bisection method

dichotomy paradox, Zeno's 533

Dido's problem **130**, 284

Dieudonné, Jean 288

"Die von der molekularkinetischen Theorie der Wärme gefurdete Bewegung von in ruhenden Flüssigkeiten suspendierten Teilchen" (Einstein) 158

difference **130–131**, 487

difference engine, Babbage's 32–33, *33*

difference formula 511–512, *512*

difference machine. *See* difference engine

difference of two cubes 131

difference of two squares 9, **131**, 257

differential **131–132**, *132*, 310

Differential and Integral Calculus (Babbage) 32

differential calculus 56, **132–134**, 226, 308–309
  vs. integral calculus 211–212

differential equations **134–135**, 234, 252, 299, 309, 385, 399
  and partial fractions 380

differential geometry 226

differentiation 133, 309

digit **135–136**

digital vs. analog 135–136

digit extraction 183

dihedral **136**

dihedral angle 136

dilation 225, 278, 317

dimension **136**

dimension of a matrix. *See* order of a matrix

Diocles **136–137**

Diophantine equations 54, 99, **137**, 239, 250, 289–290, 360

Diophantus 9, **137–138**, 180, 189–191, 239, 252–253

directed number. *See* integer

directional derivative **138–139**, 233

direction cosines 138

direction numbers 138

direction ratios 138

directly congruent solids 92

direct proof **139**, 265, 420

direct reasoning 26

directrix 155
  of cone 91
  of cylinder 115
  of parabola 373

Dirichlet, Peter Gustav Lejeune **139–140**, 172, 208, 337, 394

Dirichlet's principle. *See* pigeonhole principle

disc method 470

discontinuous functions 100

Discorsi e dimostrazione matematiche intorno a due nuove scienze (Galileo) 213

discrete **140**

discrete mathematics 167

discrete transformation 140

discriminant 84, **140–141**, 429

discriminant of cubic 113

disjoint events. *See* mutually exclusive events

disjunction ("or" statement) **141**, 514

disjunction circuit *141*

disjunctive reasoning 26

dispersion, measures of 480–482

displacement **141**

Disquisitiones arithmeticae (Gauss) 218–219

distance
  of point from line 142
  of point from plane 142
  and velocity 522–523

distance formula 22, 75, **141–142**, 310, 520

distribution 72, **142–143**, *143*, 197, 218
  chi-squared **74–75**

distributive property **143–144**
  in Boolean algebra 49
  of dot products 148
  and Elizabethan multiplication 160
  Pythagoreans and 9
  and quaternions 432
  and rings 448

div (divergence operator) **144**

divergent **144**

divergent sequence 102

divergent series 240, 459

divine proportion. *See* golden ratio

divisibility rules 92, **144–146**

division 146, 205

divisor of zero 147

Doctrine of Change (De Moivre) 122

dodecahedron 396, 397

donut. *See* torus

dot product (inner product, scalar product) 110, 136, 138, **147–148**, 520

double angle formula 525

double cusp 492, *493*

double integral 148, 239, 400, 473

double point 148

double root 140, 451

double torus 501

doubling the cube 149

drawer principle. *See* pigeonhole principle

duality in projective geometry 124

Dudeney, Henry Ernest 165

dummy variable **148–149**, 488

duplicating the cube 18, 21, 97, 136–137, **149**, 166, 172, 239, 525

Dürer, Albrecht **149–150**, 231, 232, 325, 391

dyadic **150**

dyadic fraction 43

dynamical system 71, 88, **150–151**, 203, 296

dynamics 299

E

e (eccentricity of conic) 155

e (Euler's number) 11, **152–154**, 154, 173, 283, 300, 301, 314

earth **154–155**, 166

eccentric **89**

eccentricity **155**, 161, 254, 373

echelon form 220

edge
on graph 233
of polyhedron 403

Egyptian fractions (unit fractions) **155**, 156, 192, 206

Egyptian mathematics **155–156**, 423, 428, 497

Egyptian multiplication 155, **156**

eigenvalue **156–157**, 282

eigenvector (e-vector, latent vector, characteristic vector, proper vector) **156–157**, 282

"Eine neue Bestimmung der Moleküldimensionen" (Einstein) 158

Einstein, Albert *157*, **157–159**, 227, 255, 398, 447, 496, 530

elasticity, mathematics of 227–228

"Élémens d'arithmétique universelle" (Kramp) 187

elementary number theory 359–360

elementary operations 364

elementary row operation 220–222

Elemente der Mathematik (Blatzer) 47

The Elements (Euclid) 1, 78, **159**, 168–169, 171, 172, 183, 226, 238–239, 252, 253, 270
compass and straightedge in 95
and foundations of mathematics 200
perfect numbers in 389
Platonic solids in 396–397

Élements de géométrie (Legendre) 308

Élements de mathématiques (Bourbaki) 50

Elements of Arithmetic (De Morgan) 122

Elements of Geometry (Menelaus) 336

Elements of Geometry (Playfair) 377

elimination method. *See* Gaussian elimination

Elizabethan multiplication **159–160**, *160*, 404

ell 309

ellipse *92–93, 93*, 137, *155*, 160, **160–161**, 239
reflection property of 365

ellipsoid **161–162**, 411

elliptic cone 411

elliptic cylinder 411

elliptic geometry. *See* spherical geometry

elliptic paraboloid 374, 411

empty set (null set, Ø) **162**, 257, 267, 460

Encyclopédie ou dictionnaire raisonné des sciences, des arts, et des métiers (d'Alembert) 8

endpoints, of curve 114

enumerable. *See* denumerable

Enumeratio linearum tertii ordinis (Newton) 353

envy-free fair division 188

epicycle 115

epicycloid 115

Epimenides 311

equality **162**

equal sign (=) 442

equating coefficient **162**, 380, 405

equating real and imaginary parts **162–163**

equation **163**

equation of line **163–164**

equiangular **164–165**, 166, 506

equiangular polygon 402

equiangular transformation. *See* conformal mapping

equidecomposable *165*, **165**

equidistant 76, 78, **165–166**, 260

equilateral 164, **166**, 285, 506

equilateral polygon 402

"Equilibrium Points in N-person Games" (Nash) 347

equipotent (equipollent, equinumerable) sets 60

equitangential curve. *See* tractrix

equivalence 371

equivalent knots 295–296

equivalent sets 60

Eratosthenes 154, **166–167**, 462, 510

Eratosthenes' sieve **462–463**

Erdös, Paul **167**, 393

Erdös number 167

error **167–168**, 197, 451, 457

error detection 461–462

error sum of squares 306

Essai sur une manière de représenter les quantités imaginaires dans les constructions géometriques (Argand) 25

Essay on Conic Sections (Pascal) 382

"Essay on the Application of Mathematical Analysis to the Theory of Electricity and Magnetism" (Green) 239

An Essay on the Principle of Population (Malthus) 407

"An Essay Towards Solving a Problem in the Doctrine of Chances" (Bayes) 38

Euclid 1, 78, **168–169**, 172, 183, 224, 226, 238–239, 252, 253, 265, 270, 314, 324, 410
and compass and straightedge 95
definition of point 399
and formal logic 199, 250
and foundations of mathematics 200
and fundamental theorem of arithmetic 210
and perfect numbers 389
and Platonic solids 396–397
and pure mathematics 422
and Pythagorean triples 426

Euclidean algorithm 11, 82, 159, **169–170**, 210, 237

Euclidean geometry 170, 409
axioms in 31
Hilbert on 250
vs. non-Euclidean geometry 294
and SAS rule 2

Euclidean space (Cartesian space, *n*-space) **170**

Euclides ab omni naevo vindicatus (Saccheri) 455

Euclid of Megara (philosopher) 168

Euclid's postulates 6–7, 159, **170–171**, 226, 255, 455, 474
See also parallel postulate

Euclid's proof of the infinitude of primes 159, **171–172**, 238, 265, 410

Eudoxus 26, 57, **172–173**

Euler, Leonhard 10, 107, 120, 140, 152, 161–163, *173*, **173–174**, 175, 180, 182, 185, 208, 284, 300, 431, 523–525, 529
and continued fractions 99
and continuously compounded interest 276
and Fermat's last theorem 190
and foundations of mathematics 200
and geometry 6, 64, 80
and Goldbach's conjecture 229–230
and graph theory 235–236
and number theory 13, 18, 187, 202, 247, 283, 349, 389, 426
and partition function 381
and pi (π) 393

and topology 500

and unit circle in trigonometry 511

and zeta function 105, 447, **535–536**

Euler circuit 233

Euler-Descartes formula. *See* Euler's theorem

Euler diagrams 26, 362

Eulerian circuit 235–236

Eulerian path 235

Euler line 80, *174*, **174–175**, 354, 507

Euler's brick 174

Euler's constant 174, *175*, **175**, 247, 283, 461

Euler's formula 85–86, 122, 152, 163, 174, **175–176**, 182, 185, 209, 236, 314
and logarithms 320

Euler's polygon division problem, Catalan numbers and 65

Euler's polynomial 174

Euler squares 303

Euler's theorem (Euler-Descartes formula) 80, 174, **176–177**, *177*, 233, 278–279, 358, 404, 472

Euler totient function 174

e-vector. *See* eigenvector

even functions **177**, 490

even numbers **177–178**

event **178**, 413, 417

exa- (10^{18}) 465

excircle 261, 354

excluded middle, law of 304

Exercitatio geometrica (D. Gregory) 240

Exercitationes geometricae sex (Cavalieri) 67

exhaustive subsets 371

existential quantifier ("there exists") 430–431

expanding brackets 9, 131, 162, *179*, **179**, 404, 431
and binomial theorem 45
and distributive property 143
and Elizabethan multiplication 160

expected value (mean, expectation, μ) **179–180**, 334

exponent **180–181**, 263, 366, 475

exponential function 152–153, *181*, **181–182**

exponential inequalities 268

exponential notation. *See* scientific notation

exponential series **182**

exponential time 405

expression **182**

exterior angle *182*, **182–183**, 504, *504*

exterior-angle theorem 182, **183**, *183*, 377

external direct product. *See* Cartesian product
extraction **183–184**
extrapolation **184**, 278, 405
extreme and mean ratio. *See* golden ratio
extreme-value theorem 49, 119, **184**, 330, 442
extremum 184

F

face **185**, 403
face angle 185
face-centered cubic lattice 472
face of graph 185
faceted polyhedron 404
factor **185–186**
factorial 80, 174, **186–187**, 444, 486
factorization 118–119, 187
factor theorem **187**
factor tree 210
Fagnano, Giovanni 387
fair division (cake cutting) **187–188**
fallacy of the converse 26
fallacy of the inverse 26
false position, method of 9, 156
Faltings, Gerd 190
Fano plane 419, *419*
Farey, John 188
Farey sequence (series) **188–189**
al-Farisi 18
F-distribution 484
feedback 150
femto- (10^{-15}) 465
Ferguson, D. F. 461
Fermat, Pierre de 138, 139–140, **189**, 272, 410, 531
   and amicable numbers 13
   and coordinate geometry 63, 226, 234–235
   and figurate numbers 66, 287
   and history of calculus 57
   little theorem of 342
   and probability theory 413, 414, 417
Fermat's last theorem 137, 138, 139–140, **189–190**, 227, 531
Fermat's little theorem 342
Ferrari, Ludovico 10, 60, **190–191**, 431–432, 470, 494
Ferro, Scipione del (Ferreo, dal Ferro) 10, 59–60, 112, **191**, 493–494
Feuerbach, Karl Wilhelm 354
Fibonacci 10, 155, 156, **191–192**, 409, 511
   and Hindu-Arabic numerals 251
   and minus sign 487
   and zero 534

Fibonacci numbers **192–193**, *193*, 197, 384, 443, 458
   and golden ratio 231
   and polyominoes 406
field 5, 85, 442
Fields, John Charles 194
Fields medals **193–194**, 531
fifteen puzzle. *See* slide 15 puzzle
figurate numbers 66, 99, *194*, **194–195**, 287, 381, 389, 423
Fincke, Thomas 511
finger multiplication **195–196**
finite **196**, 270
finite differences 33, 75, **196–197**
finite induction. *See* induction
finite projective geometry 419
finite sets 106
Fiore, Antonio Maria 493–494
first-derivative test 331–332
Fisher, Sir Ronald Aylmer **197–198**, 387, 415
five-fold rotational symmetry 490
five stone problem 43
fixed point **198**
floor/ceiling brackets 51
floor function (greatest-integer function) **198**
fluid mechanics 8, 144, 248, 299
flux 144
fluxion 57–58, **198**, 309, 352
focal chord **198**
focal radius **198**
focus (foci) 155, **199**
   of an ellipse 161
   of hyperbola *253*, 253–254
   of parabola 373
   of paraboloid 374
foot (unit of length) 309
force field 101
Ford, Lester R. 188
formal logic (symbolic logic) 25, 89–90, 94, 101, **199**, 238, 304–305
   Aristotle and 26
   expressions in 182
   founded by Boole 48
   Gödel and 228–229
   set theory and 460
formula **199–200**
Formulario mathematico (Peano) 385
foundations of mathematics 47, **200**
four-color theorem 55, **200–201**, 420, 472
   and Möbius band 340
   and torus 501
four elements (earth, air, fire, water) 397
Fourier, Jean-Baptiste Joseph 139, **201**, 202, 208
Fourier series 139–140, 177, **201–203**, *202*, 307, 522, 530

fourth dimension 136
fractal 136, **203–204**, *204*, 456, 463
fractal shapes
   and Banach-Tarski paradox 24
   and chaos 71
fraction 146, **204–206**, 341, 358, 439, 444
   in binary notation 43
   reciprocal of 442
fractional dimensions 136
fractional exponent 181, 368
fractional part brackets 51
fractional part function 98, 198
Fraenkel, Adolf 534
Français, Jacques, and complex numbers 25
Franklin, Benjamin 325
freedom equations. *See* parametric equations
Frege, Gottlob 199
frequency **206**
frequency distribution 479
frequency polygon 479, *479*
frequency table 479, *479*
frequency theory of probability, Venn's 524
friendly numbers. *See* amicable numbers
frieze pattern *206*, **206–207**
frustum **207**
F-test 484
full linear group **222–223**
full turn 14, 362, **390**
function 139, 173, **207–209**, 208, 445
functional analysis 35
function of a function. *See* composition
function of a function rule. *See* chain rule
fundamental principle of counting. *See* multiplication principle
fundamental property of least time 189
fundamental theorem of algebra 85, 112, 186–187, **209–210**, 359, 405, 451
   and polynomial equations 471
   proofs of 10, 25, 218
   and zeta function 536
fundamental theorem of arithmetic 83, 118–119, 186–187, **210–211**, 283, 409, 439, 476
   proved by Gauss 218
   and zeta function 535
fundamental theorem of calculus 133, 153, **211–212**, 240, 272, 494
   formulated by Leibniz 308
   in history of calculus 57–58

fundamental theorem of isometries **212**, 284
fuzzy logic **212**, 305
fuzzy-set theory 212

G

Galilei, Galileo 65, 93, 114, 125, **213–215**, *214*, 270, 485, 523
Gallai, Tibor 80
galley method 159
Galois, Évariste 10, **215**, 242, 288, 432, 471
Galton, Sir Francis **215–216**, 415
Galton graph. *See* scatter diagram
gambler's ruin **216**, 437
game theory **216–218**, 346–348, 350–351, 415
gamma function 174, 187
Gardner, Martin **218**
Garfield, James 425
Garnier, René 288
Gauss, Carl Friedrich 119, 140, **218–219**, 226, 268
   and fundamental theorem of algebra 209
   and geometry 47, 97, 355–356
   and group theory 242
   and knot theory 295
   and number theory 10, 25, 28, 87, 195, 287, 359, 489
   and prime numbers 243, 410–411, 518
   and probability 356, 414
   and topology 500
Gaussian elimination (pivoting) 119, **219–222**, 282, 288, 438, 491
Gelfond, Aleksandr 11
general form of an equation **222**
general linear group **222–223**, 314
general maximum/minimum *331*
general theory of relativity 158
General trattato di numeri et misure (Tartaglia) 493
generating circle 500
generator 91, 115
geodesic 219, **223**
geodesic dome 223
Geography (Ptolemy) 422
Geometriae pars universalis (J. Gregory) 240
Geometriae rotundi (Fincke) 511
Geometrica organica (Maclaurin) 324
geometric distribution 44
geometric mean 333

geometric progression. *See* geometric sequence

geometric sequence **223–224**, 489

geometric series 103, 163, 409, 441, 488

geometric transformation **224–225**, 258, 316, 503

La géométrie (Descartes) 63, 124–125, 226

"Geometrische Untersuchungen" (Lobachevsky) 318

geometry 124, 125, **225–228**, 238–239

Geometry (Simpson) 463

Germain, Marie-Sophie 190, **226–228**

giga- (10^9) 465

Girard, Albert 209, 398

glide reflection 206, 225, **228**

global maximum/minimum 330

gnomon **228**

gnomonic projection 485

Gödel, Kurt 31, 101, 199, 200, **228–229**, 453, 458

Gödel's incompleteness theorems 101, 199, **229**, 250, 458

Goldbach, Christian **229**, 230, 410

Goldbach's conjecture **229–230**, 250, 359, 410

golden ratio 193, **230–231**, 388–389

golden rectangle **231–232**, *232*

golden triangle 232

Goodwin, E. J. 394

googol/googolplex 232, 389

Gosset, William Sealy **232**, 387, 415, 483

grad 233

grade. *See* slope

gradian 14

gradient 139, **233**, **466–468**, *467*

Graeco-Latin square 302–303

Grandi, Guido 451

graph (network) 94, *233*, **233**, 501

graphical solution 163, **234**, 491

graph of function 114, 207, **234–235**, 275, 368

graph theory 82, 174, **235–236**, 245, 505

Graunt, John 312, 415

gravitation, theory of 399–400

great circle 223, 314

greatest common divisor (greatest common factor) 82, 169–170, **236–237**, 290, 305, 446

greatest-integer function. *See* floor function

Greek alphabet 237, **237**

Greek mathematics 9, **237–239**

Green, George **239–240**

Gregory, David **240**

Gregory, James 57, **240**, 241, 476–477, 494, 496

Gregory series 11–12, **241**, 325, 393, 461

Grelling's paradox 374–375

The Grounde of Artes (Recorde) 443

group 59, **241–242**, 252, 279, 398, 490

group theory 5, 10, 215, **242**, 288, 299, 350, 432

Grundbegriffe der Wahrscheinlichkeitsrechnung (Kolmogorov) 296

Grundlagen der Geometrie (Hilbert) 171, 250

Grundlagen der Mathematik (Hilbert and Bernays) 250

gry 310

Gunter, Edmund 369, 466

H

haberdasher's puzzle 165

Hadamard, Jacques **243**, 411, 518

hairy ball theorem 472, 521

Haken, Wolfgang 200–201, 420

half-chord *511*

half-cone 91–93, 473

half-line. *See* ray

half-open interval 278

half-plane 94, 235, **243–244**, 267, 473

half-space 243–244

Hall, Monty 343

Hall, Philip 244

Halley, Edmund 5, 312, 353, 445

Halley's comet 445

Hall's matching (marriage) theorem **244**

halting problem 88, **244**

Hamilton, Sir William Rowan 10, 87, 236, **244–245**, 359, 432, 520

Hamiltonian circuit 236, 505

Hamiltonian path 236, 502

ham-sandwich theorem **245**

hand (unit of length) 309

handshake lemma 121, 236, **245**

handshakes across a table problem 65

Hardy, Godfrey Harold **246**, 381, 435–436

harmonic function **246**

harmonic mean 21, 333

harmonic sequence (progression) **246–247**

harmonic series 4, 104–105, 174, **247–248**, 368, 409

Haros, C. 188

hatcheck problem 391

al-Haytham 18

Heawood, Percy 200–201

height
  of cone 91
  of cylinder 115
  of pentagram 388
  of prism 412
  of tetrahedron 497
  of triangle 506

Heisenberg, Werner 68

helix 114, 248

hemisphere 471

Hermite, Charles 11, 154, 301, 314

Herodotus 498

Heron **248–249**

Heron's formula 52–53, 78, 248, 249, 336, 486, 506, 508

Heron's method 34–35, 183, 248, 249, 285, 353, 476

hertz (Hz) 206

Hertz, Heinrich 158

Heumann, Casper 312

hexadecimal notation 135

hexagonal lattice 472

hexahedron. *See* cube

hierarchy 454

Hieronymus 497–498

higher arithmetic. *See* number theory

higher derivative **249**

highest common factor. *See* greatest common divisor

Hilbert, David 137, 170, 171, 199, 229, **249–250**, 355, 533

Hilbert's axioms 171

Hilbert's infinite hotel paradox **250–251**, 269

Hilbert space 250

Hindu-Arabic numerals 191–192, 251, 263, 370, 451

Hipparchus 510

Hippasus 172, 283

Hippocrates **251–252**, 322–323, 476, 513

Hippocrates lune *251*

Hisab al-jabr w'al muqābala (al-Khwārizmī) 10, 17, 251, 293, 517

histogram 143, 479, *479*

Historia natural y moral de las Indias (Acosta) 260

HOMFLYPT polynomial 296

homogeneous 252

homomorphism **252**

homotopy 398–399

honeycomb 292, 372

Horner, William George 72

How to Solve It (Pólya) 401

hundkurve 503

Hutchings, Michael 469

Huygens, Christiaan 114–115, 180, 414, 503

Hydrodynamica (Daniel Bernoulli) 40

hydrodynamics 140

hydrostatics 18–19, 20

Hypatia 239, **252–253**

hyperbola 30, 92–93, *93*, 137, 155, 189, 239, *253*, **253–254**, 513

hyperbolic cosine 65–66, 254–255

hyperbolic cylinder 411

hyperbolic functions *254*, **254–255**, 369

hyperbolic geometry 219, 227, **255**, 318, 355–356, 455

hyperbolic paraboloid 374, 411

hyperbolic sine 254

hyperbolic spiral 474

hyperboloid **255–256**, *256*

hyperboloid of one/two sheets 411

hypercomplex numbers 432

hypercube 112, **256**

hypersphere 136, 471

hypocycloid 115

hypotenuse 77, **256**, *509*

Hypothesis physica nova (Leibniz) 309

hypothesis testing 143

I

i (square root of –1) 85, 173, 358–359, 529

i (unit vector) 520

Ibrahim ibn Sinan 18

icosahedron 232, 396, 397

"Idealtheorie in Ringbereichen" (Noether) 355

identity 163, **257**, 304

identity element **257**, 258, 279, 295, 366
  additive 534
  multiplicative 344

identity matrix (unit matrix) 128, 257, **257–258**, 281

identity property 222, 241

image (range) 207, **258**, 481

imaginary numbers 87. *See also* complex numbers (C)

implicit differentiation **258–259**

implicit function **259**

improper factor 186

improper fraction 205

improper integral (unrestricted integral, infinite integral) 104, 144, **259–260**

In artem analyticam isagoge (Viète) 525

Incan mathematics **260**

incenter 260

inch 309

incircle **260–261**, 354

inclination *261*, **261**

inclined plane 261
inclusion-exclusion principle 48–49, **261**, 391, 460
incomposite numbers 409
increasing/decreasing 89, **261–262**
increment 262
indefinite integral 95, 272, 275
independent. *See* pairwise disjoint
independent axiom 171, **262**
independent events 91, 178, **262–263**, 417
independent variable 519
indeterminate. *See* unknown
indeterminate equation 263
index (indices) 263. *See also* exponent
index of summation 488
Indian mathematics **263–265**
indirect proof (proof by contradiction, reductio ad absurdum) 139, 171, **265**, 336, 420
induction 122, 126, 176–177, 261, **265–267**, 270, 420
inductive definition. *See* recursive definition
inductive reasoning **120–121**
Indus inch 263
inequality **267–268**
inference **268**, 415
inferential statistics. *See* statistics: inferential
infinite 196
infinite integral. *See* improper integral
infinite order 406
infinite product 144, **268–269**, 312, 459
infinite series 102, 324, 351, 368, 439
infinitesimal 47, 57–58, 67, **269**, 271, 533
infinity (∞) **269–270**, 529
inflection (inflexion) point 89, 89, 270, 492
information theory **271**
injective function 209
inner product. *See* dot product
inradius 260
inscribe/circumscribe 78
inscribed-angle theorems 77, 77–78
inscribed circle 387
instantaneous value **271**
instantaneous velocity 523
integer (directed number, signed number) **271**, 358
integer triangle 507–508
integral 66, 180, 380, 447, 523
integral calculus 22, 24, 56, 133, **271–273**, 272, 308–309
and Archimedes' method of exhaustion 18

vs. differential calculus 211–212
and volume calculation 470, 527
integral domain 449
"Intégrale, longeur, aire" (Lebesgue) 307
Intégrales de Lebesgue (Vallée-Poussin) 518
integral test, for convergent series 104
integrand 148–149, **273**, 274, 360
integration **15**, 133, 272
  by parts 95, **273–274**, 443, 486
  by substitution (change of variable, substitution rule for integration) **274–275**, 287
intercept 275
intercept form 164, 275
interest **275–276**
interior angle 276, 504, *504*
intermediate-value theorem (Bolzano's theorem) 100, 184, **276–277**, 367, 442
  bisection method and 46
  Bolzano's proof of 47
  and Dedekind cut 119
  and ham-sandwich theorem 245
  in history of calculus 58
  and logarithmic function 320
interpolation 184, 405. *See also* Lagrange's formula
intersection 144, 460
interval **278**
Introduction à l'analyse des lignes courbes algébriques (Cramer) 107–108
"Introduction to Mathematical Philosophy" (Russell) 453
An Introduction to Mathematics (Whitehead) 530
"Introduction to the Doctrine of Fluxions" (Bayes) 38–39
invariant **278–279**
inverse element **279–280**
inverse function (inverse mapping, reverse function) **280–281**
inverse hyperbolic functions **281**
inverse mapping. *See* inverse function
inverse matrix 222, 280, **281–282**
inverse property 5, 49, 223, 241
inverse square law 282
inverse trigonometric functions 280–281, **282–283**
inversion in a circle 225

An Investigation of the Laws of Thought (Boole) 48
irrational numbers 59, 238, **283**, 358, 423
  and definition of real number 441–442
  Kronecker's doubts about 297
  Theodorus's work on 498
irreducible polynomial 186
Isagoge ad locos planos et solidos (Fermat) 189, 226
isogonal transformation. *See* conformal mapping
isolated point (acnode) 283
isometry (congruence transformation) 140, **283–284**
isomorphism 284
isoperimetric inequality 268
isoperimetric problems 130, **284–285**, 469, 472, 474–475, 484–485, 529
isosceles trapezoid (trapezium) 285, 504
isosceles triangle 238, *285*, 285, 506
"Ist die Trägheit eines Körpes von seinem Energieinhalt abhängig?" (Einstein) 158
Istituzioni analitiche (Agnesi) 7
iterated integral 148
iteration 150–151, **285–286**
Iverson, Kenneth 198

J

j (engineers' *i*) 85
j (unit vector) 520
Jacobi, Carl Gustav Jacob **287–288**
Jacobian determinants 287
Jacquard, Joseph-Marie 33
jerk 523
Jiuzhang suanshu (anonymous) 73, 226
Jones, Vaughn 295–296
Jones, William 393
Jones polynomial 296
Jordan, Marie Ennemond Camille 288, 307
Jordan, Wilhelm 288
Jordan canonical form 288
Jordan curve theorem 76, **288**
Josephus problem **288–289**
Jourdain's paradox 289
Journal de Mathématiques Pures et Appliqués (Liouville's journal) 317
Journal for Pure and Applied Mathematics (Crelle's journal) 3
J-shaped distribution 480, *480*
jug-filling problem 137, 170, **289–290**

Julia, Gaston Maurice 203–204
Julia set 203–204
Jungius, Joachim 65

K

k (unit vector in three-dimensional space) 520
al-Kashi, Jamshid Mas'ud 18, **291**, 392
Kasner, Edward 232
Kempe, Alfred Bray 200
Kendall, M. G. 438
Kendall's coefficient 438
Kendall's method 438
Kepler, Johannes 93, **291–293**, *292*, 396, 397, 472, 528
Kepler's laws, Newton and 353
The Key to Arithmetic (al-Kashi) 291
Khandakhadyaka (Brahmagupta) 51
Khayyám, Omar. *See* Omar Khayyám
al-Khwārizmī, Muhammad ibn Mūsā 9, 10–11, 17, 251, **293–294**, 517, 534
kilo- (10³) 465
kinematics 27, 213
kite 430, *430*
Klarner, David 406
Klarner's constant 406
Klein, Felix Christian **294**, 314, 355, 473–474
Klein bottle 200–201, **294–295**, 339–340
Klein-four group (viergruppe) **295**
knot theory 295, **295–296**
Koch, Nils Fabian Helge von 203
Koch curve 203
Kolmogorov, Andrey Nikolaevich **296**
Kramp, Christian 187
Kremer, Gerhard (Mercator) 337
Kronecker, Leopold **296–297**
Kruskal's count **297–298**
Kulik, Yakov 463

L

Laczovich, Miklov 165
laddered exponents 65
Lafforgue, Laurent 194
Lagrange, Joseph-Louis 99, 201, 249, 278, **299–300**, 379, 399, 414
  and group theory 242
  and history of calculus 58
  and mean-value theorem 334
  and square numbers 195
  and Taylor series 494

Lagrange's formula 120, 278, **300**, 405
Lagrange's theorem 299
Lagrangian (unit) 300
Lagrangian description 300
Lagrangian multipliers 300
Lambert, Johann Heinrich 283, **300–301**, 392, 477
Landa, Diego de 332
Langlands, Robert 194
Laplace, Pierre-Simon, marquis de 69, 245, **301–302**, 356, 399, 414
Laplace operator 302
Laplace transform 302
latent vector. *See* eigenvector
lateral surface 115
Latin square 174, **302–303**
latitude 105, 154, 337
lattice 366
lattice method 159
lattice point 164, 166
lattice polygon 164–165, 166
law of averages **303**, 304
law of conservation of energy 309
law of continuity 309
law of cosines 5–6, 16, 78, *303*, **303–304**, 506
law of excluded middle 304
law of falling bodies 213
law of identity 304
law of large numbers 71–72, 303, **304**, 399, 414–415, 436
law of noncontradiction 304
law of sines 1–2, 78, *303*, **303–304**, 463, 507
law of the lever 18, 69, **304**
laws of gravity 353
laws of planetary motion 291–293
laws of thought 26, **304–305**
lawyer paradox 374–375
leading coefficient **305**
least common denominator 83
least common multiple 83, **305**
least-integer function. *See* ceiling function
least squares method 107, 218, 302, **305–307**, 308, 445
Lebesgue, Henri-Léon 58, **307**
LeBlanc, Louis 227
Leçons sur les séries trigonométriques (Lebesgue) 307
Lectiones geometricae (Barrow) 36
Lectures on Quaternions (Hamilton) 520
left derivative **307–308**
left-handed system 369, **448**
left identity 257
left/right, limit from the 313
Legendre, Adrien-Marie 187, **308**

Legendre polynomials 308
Legendre symbol 308
Lehmer, D. H. 436
Leibniz, Gottfried Wilhelm 122, 208, *308*, **308–309**, 523–525
    and calculus notation 249, 488
    and differential 132
    and discovery of calculus 57–58, 211, 227, 240, 272, 352, 494
    and double integrals 148
    and formal logic 199, 228
Leibniz's series. *See* Gregory series
Leibniz's theorem 418
lemma 498–499
length **309–310**
lens 76
Leonardo of Pisa. *See* Fibonacci
Les méthodes nouvelles de la méchanique céleste (Poincaré) 399
Let's Make a Deal! (game show) 343
letterbox principle. *See* pigeonhole principle
"A Letter on Asymptotic Series" (Bayes) 39
Letters to a German Princess (Euler) 174
levers and pulleys 19–20
L'Hôpital, Guillaume François, marquis de 40, 58, **310**
L'Hôpital's rule 40, **310–311**
Lho shu square 325, *326*
Lhuilier, Simon 474
liar's paradox 289, **311**, 374, 458
Liber abaci (Fibonacci) 156, 191–192
Liber de ludo aleae (Cardano) 414
Liber quadratorum (Fibonacci) 192
life tables (mortality tables) **311–312**, 415
ligancy 472
likelihood 197
Lilavati (Bhāskara) 41–42
limit **312–313**
    d'Alembert and 8–9
    and analytic number theory 13
    Bolzano's work on 47
    and chain rule 70–71
    ε - δ definition of 529
    and definition of tangent 492
    derivative as 132
    devised by Cauchy 66
    in history of calculus 58
Lindemann, Carl Louis Ferdinand von 11, 297, **313–314**, 393, 477
line **314**, 315

linear algebra **314**, 491
linear-congruence method 436–437
linear equation 34, 156, 163, **314–315**
linear interpolation 277–278
linearly dependent/independent **315**
Linear Perspective (Taylor) 494
linear programming **316**
linear transformation 7, **316–317**
line integral. *See* contour integral
Liouville, Joseph 11, 59, 215, **317–318**
Liouville's constant (Liouville's number) 11, 317
Listing, Johann 295, 339, 340
Littlewood, John 246
Liu Hui 73
Li Ye (Li Chi, Li Zhi) 73, **318**
Lobachevskian geometry. *See* hyperbolic geometry
Lobachevsky, Nikolai Ivanovich 46–47, 171, 219, 227, *255*, **318**, 355, 377, 455
local maximum/minimum 330–331, *331*, 468
locus (loci) **319**
logarithm 247, 252, 284, **319–320**, 342, 411, 461, 466
    history of 53–54, *55*, 291–293, 345–346
    of matrix 157
logarithmic differentiation 258–259
logarithmic function 152–153, *153*, **320–321**, 337
logarithmic graph 235
logarithmic scale 118, **321**, 369, 466
logarithmic spiral 39, 474
Logarithmorum chilias prima (Briggs) 54
logically equivalent 515
logic gates 88
Logic of Chance (Venn) 524
long division *37*, 37–38, *38*, 146, 349, 404, 433
longitude 105, 154–155, 337, 445
long radius 321
Loomis, Elisha Scott 425
loop (graph theory) 233
Lorentz, Hendrik 398
Lovelace, Augusta Ada 33, **321–322**, *322*
lowest terms. *See* reduced form
Loyd, Sam 465–466
lozenge 376, 474
Lucas, Edouard 502
lune 251–252, *322*, **322–323**
Lyapunov, Aleksandr Mikhailovich 69, 356, 415

M
Maclaurin, Colin 107–108, **324**, 496
Maclaurin series. *See* Taylor series
Madhava of Sangamagramma 264, **324–325**
magic constant 325
magic division square 327, *327*
magic multiplication square 327, *327*
magic rectangle 328, *328*
magic square *149*, 149–150, *325*, **325–328**, *326*, *327*
magnitude **328–329**
al-Mahani 18
major arc 18
major premise 27
Malthus, Thomas 407
Mandelbrot, Benoit 204
Mandelbrot set 204
Manière universelle de Mr. Desargues (Bosse) 124
mapping. *See* function
Markov, Andrei 415
Markov chains 296
The Mathematical Analysis of Logic (Boole) 48
"Mathematical Contribution to the Theory of Evolution" (Pearson) 387
Mathematical Dissertations (Simpson) 464
mathematical induction. *See* induction
"The Mathematical Theory of Communication" (Shannon and Weaver) 461
A Mathematician's Apology (Hardy) 246
Mathematische Grundlagen der Quantenmechanik (von Neumann) 351
matrix (matrices) 109, 119, 126–128, 221–222, 314, **329**
matrix addition 329–330
matrix algebra 68, 458
matrix inverse. *See* inverse matrix
matrix multiplication 241, 330
matrix operations **329–330**
Matyasevic, Yuri 137
maximin strategy 217
maximum/minimum 89, 133, 310, **330–332**, *331*
Mayan mathematics **332–333**
McManus, Chris 118
mean (average) 44, **333–334**, 480, 537
mean. *See* expected value
mean value **334**
mean-value theorem 22, 94–95, 119, 184, 211, 262, 311, **334–335**, 442, 450

Measurement of a Circle
(Archimedes) 18
measures of central tendency
480
measures of dispersion
480–482
measure theory 58, 307
Mécanique analytique
(Lagrange) 299
Mécanique céleste (Laplace)
67, 245
Mechanica (Euler) 173, 393
mechanics 8
median 174–175, 480
median of triangle *335,*
335–336, 338, 507
Meditationes algebraicae
(Waring) 229
mega- (10^6) 465
Melancholia (Dürer) 149, 325
"Mémoire sur les équations
algébriques" (Abel) 5
"Mémoire sur quelques
propriétés remarquables des
quantités transcendantes
circulaires et logarithmiques"
(Lambert) 300
"A Memoir on the Geometric
Representation of Imaginary
Numbers" (Français) 25
*Memoir on the Theory of
Matrices* (Cayley) 68
Menabrea, Luigi 322
Menelaus 17, **336**
Menelaus's theorem 70, *336,*
336
Mercator (cartographer). *See*
Kremer, Gerhard
Mercator, Nicolaus
(mathematician) 337
Mercator's expansion 4, **337**
Mercator's projection 91, **337**
Méré, Chevalier de 414
meridian 154–155
Mersenne, Marin 189,
337–338, 389
Mersenne prime **338,** 389,
410
Metaphysics (Aristotle) 498
meter 465
method. *See* algorithm
method of exhaustion 18, 20,
57, 67, 172
method of indivisibles 67
*Methodus incrementorum
directa et inversa* (Taylor)
494
Metrica (Heron) 248, 249
micro- (10^{-6}) 465
middle-square method 436
midpoint 76, **338**
midrange 480
Mihailescu, Preda 64
mile 309–310
milli- (10^{-3}) 465
minimax strategy 217
minimax theorem 218

minimum. *See*
maximum/minimum
minor arc 18
minor premise 27
minuend 487
minus sign 131, 398, 486,
487
minute (measure of angle) 14
*Mirifici logarithmorum
canonis descriptio* (Napier)
346
Miscellanea analytica (De
Moivre) 122
mixed strategy (game theory)
217–218
mixed surd 490
Miyoshi, Kazunori 393
Möbius, August Ferdinand
338–339, 340
Möbius band (strip) 294–295,
339–340, *340,* 369
Möbius function 339
Möbius inversion formula
339
mode 480
modular arithmetic 59,
71–72, 92, 147, 174, 218,
340–342, 436
modulus. *See* absolute value
monic 305
monohedral tessellation
496–497
monomial **342**
monomino 406
Monte Carlo method
342–343
Monty Hall problem **343**
Mordell conjecture 190
Morgan, Frank 469
Morgenstern, Oskar 216,
350–351
Morley, Frank 343
Morley's theorem **343**
Morrison, Nathan 296
mortality tables. *See* life tables
multi-choosing 81–82
multiplicand 344
multiplication 5, 86, *86,* 146,
205, **343–344,** 370
multiplication law 367
multiplication principle
(fundamental principle of
counting) **344**
and permutations 391
multiplication rule
for exponents 180–181
in probability theory *416,*
416–417
multiplicative identity 257
multiplicative inverse 449
multiplier 344
multivariate analysis 197
mutually exclusive. *See*
pairwise disjoint
mutually exclusive events
(disjoint events) **344**
mutually orthogonal 302

Mysterium cosmographicum
(Kepler) 292

N

Nakayama, Kazuhiko 393
nano- (10^{-9}) 465
Napier, John 74, 293,
319–320, *345,* **345–346,** 466
Napier's bones (rods) 74, *346,*
346, 347
Napier's formulae 346
Napier's inequality 268
Napoléon's theorem 507
nappe 91–93, 473
Nash, John **346–348,** 415
Nash equilibrium 413
*Natural and Political
Observations Made upon the
Bills of Mortality* (Graunt)
312
natural number 106, 129,
348, 358, 385
See also whole number
naught. *See* zero
Nave, Hannibal 191
n-dimensional kissing number
472
n-dimensional space 68, 472
nearest-neighbor algorithm
505
negation ('not' statement)
348, 514
negative coordinates 63
negative infinity 121
negatively oriented 369, 448
negatively skewed 465, 480,
480
negative numbers 48, 51,
348–349, 358, 408
negative slope 467
neo-Pythagorean mean 333
nested multiplication *349,*
349–350, *350,* 405
net (accounting) 350
net (geometry) *350, 350*
net (of hypercubes) 256
net weight 350
network. *See* graph
Neumann, John von **350–351,**
393
neutral element. *See* identity
element
*New Experiments concerning
Vacuums* (Pascal) 382
Newton, Sir Isaac 93, 122,
125, 132, 169, 180, 282,
292, 324, **351–353,** *352,*
452, 523, 528
and binomial theorem 45
and discovery of calculus
57–58, 211, 227, 240,
272, 308–309, 352, 494
and fluxions 198
and polar coordinates 63,
400
and three-dimensional
kissing number 472

Newton number 472
Newton quotient 132, 360
Newton's method 46, 249,
286, **353–354,** 451
Newton's third law of motion
9
A New Treatise of Fluxions
(Simpson) 464
Neyman, Jerzy **354**
n-fold rotational symmetry
490
n-gon 97, 402
Nicomachos of Gerasa 389
nine-point circle **354**
node 148, 233, 493
Noether, Amalie (Emmy) 10,
354–355
n-omino 406
noncontradiction, law of 304
"Non-Cooperative Games"
(Nash) 347
nondegenerate quadrics 411
nondeterministic polynomial
time 357
nonempty set 162
non-Euclidean geometry 159,
171, 227, 255, 314,
355–356, 377, 473–474
discovery of 46, 318, 455
vs. Euclidean geometry
294
nonmeasurable sets 24
nonnegative 408
nonpositive 408
nonrepeating decimals 441
nontrivial solution 514
nonzero-sum game theory 347
normal distribution 69–70,
72, 122, 143, *356,* **356–357,**
414, 482, 483, 486
normal to a curve **357**
normal to a plane **357**
normal to a surface **357**
"Note on the Application of
Machinery to the
Computation of
Astronomical and
Mathematical Tables"
(Babbage) 32–33
"Note sur une équation aux
differences finie" (Catalan)
64
*Notions sur la machine
analytique de Charles
Babbage* (Menabrea) 322
*Nova methodus pro maximis
et minimis* (Leibniz) 309
Nova stereometria doliorum
(Kepler) 292–293
Novum Organum (Bacon) 452
NP complete 88, **357–358**
n-polyomino 406
n-space. *See* Euclidean space
nth dihedral group 242
nth-order general linear group
(GL_n) 222–223
nth-order root 451

*n*th-order symmetric group 490
*n*th root **112**
*n*th root of unity **358**
*n*th-term test 103
n-tuple **358**
null angle 14
null set. *See* empty set
number **358–359**
number-base machine 30
number line **359**, 441, 442
number-naming puzzle 54–55
number systems **358–359**
number theory 82, 139–140, 159, 167, 191, 337–338, **359–360**, 429–430
numerable. *See* denumerable
numerator 204
numerical differentation **360**
numerical integration **360–361**

O

obelus (%) 146
oblate spheroid 154, **362**
oblique **362**
oblique angle **362**
oblique cone **362**
oblique cylinder 115
oblique pyramid 423
obtuse angle 14, **362**
obtuse triangle **362**, 506
obverse **362**
octahedron **396**, 397
octant **363**
octonions 359, 432
odd functions **177**, 490
odd numbers **177–178**
odds (probability) **363**, 439
Oeuvres completes d'Augustin Cauchy 67
Oeuvres de Camille Jordan 288
officer problem 302–303
Omar Khayyám (Umar al-Khāyyamī) 18, 227, **363–364**
"On a Curious Property of Vulgar Fractions" (Farey) 188
"On a New Method in Elementary Number Theory" (Erdös) 167
On Burning Mirrors (Diocles) 136
On Conoids and Spheroids (Archimedes) 18
On Divisions (Euclid) 169
"one-to-one" function 209
On Floating Bodies (Archimedes) 18–19
On Reasoning in a Dice Game (Huygens) 180
On Spirals (Archimedes) 474
On the Economy of Machinery and Manufactures (Babbage) 33

"On the Problem of the Most Efficient Tests of Statistical Hypotheses" (Neyman) 354
"On the Propagation of Heat in Solid Bodies" (Fourier) 201
On the Sphere and Cylinder (Archimedes) 18, 173
"On the Theory of Groups Depending on the Symbolic Equation $\theta^n = 1$" (Cayley) 68
"onto" function 208
open half-plane 243–244
open half-space 243–244
open interval 278
operation **364**
operational precedence. *See* order of operation
operations research (OR) 33, **364**
opposite **364**
opposite angles 364
oppositely congruent solids 92
Optica promota (J. Gregory) 240
Optics (Euclid) 169
Optics (Ptolemy) 422
optics, Newton and 353
Optiks (Newton) 353
optimization 133, 161, 285, 316, 364, **364–365**
  and history of calculus 57
  and inscribed triangles 387
  and Snell's law of refraction 468
  Steiner and 484–485
  von Neumann and 350
OR. *See* operations research
orbit (sequence of iterates) 150
order. *See* permutation
order, of polyomino 406
ordered partition 380–381
ordered set **365–366**, 367, 530
order of a matrix (dimension of a matrix) 258, **366**
order of group 366
order of magnitude 329
order of operation (operational precedence) **366–367**
order properties 267, **367**
ordinal numbers **367–368**
ordinate 62
Oresme, Nicole 63, 208, 234, **368**
orientation **369**
origami 513
origin 105, 359
orthant 363
orthic triangle 387
orthocenter 12, 354, 387, 507
orthogonal **369**, 391
orthogonal. *See* Cartesian coordinates
Osborne's rule **369**

osculation (tacnode) 148
osculinflection 492, *493*
Oughtred, William 344, **369–370**, *370*, 466
outlier 370
oval **370**
ovals of Cassini 370
Ozanam, Jacques 449

P

packing spheres 472
paddle wheel 174
Paganini, Nicolò 13
pairwise disjoint (independent, mutually exclusive) **371**
Pappus 13, 57, 239, 253, **371–372**, 470
Pappus's problem 372
Pappus's theorems **372–373**, 470, 471
  and torus 500
parabola 92–93, *93*, 137, 155, 189, 239, *373*, **373**, 429
  as trajectory 213, 503
parabolic cylinder 411
paraboloid 101, 235, **373–374**
paradox **374–375**, 412, 532
  and pure mathematics 422–423
"Paradoxien des Unendlichen" (Bolzano) 47
parallel **375–376**, *376*, 377
parallelepiped (parallelopiped) **376**
  and scalar triple product of three vectors 512
parallelogram **376**, 430, 474
  area of 23, *23*
  cross (vector) product as 110
parallelogram law (parallelogram rule) 376, **376–377**
parallelotope 376
parallel postulate (Euclid's fifth postulate) 159, 227, 239, 255, 375, *377*, **377–378**, *378*, 473
  and altitudes of triangle 12–13
  and angles of triangle 505
  converse of exterior-angle theorem 183
  and Euclidean vs. non-Euclidean geometry 170–171, 318, 355–356
  Omar Kayyám and 18, 364
parallel projection 418
parametric equations (freedom equations) 164, **378**, 386
  and arc length of curve 22
  of cardioid 62
  of circle 75, 95

of helixes 95
of Möbius band 340
of torus 500
of unit circle 517
parentheses 50–51, 519
parity **378–379**
Parmenides 532
Parmenides (Plato) 532
partial derivative 138, 233, 239, 259, 379
partial differential equations, d'Alembert and 8–9
partial fractions 162, **379–380**
partially ordered set 365–366
partial sum 102, **380**, 459
  and limit 312
partition 358, **380–381**, 436
partition function 381
Pascal, Blaise 74, 189, **381–382**, *382*, 383
  on cycloid 310
  and history of calculus 57
  and Pappus's theorem 372–373
  and probability theory 413, 414, 417
Pascal's distribution 44
Pascal's triangle 45, 65, 195, 196, 382–385, *384*, 497
  and combination 80–81
pathological functions 529
Pathwaie to Knowledge (Recorde) 443
payoff matrix 217
Peano, Giuseppe 265, 385
Peano's curve *385*, **385–386**
Peano's postulates 31, 265, 271, 385, 386
Pearson, E. S. 354
Pearson, Karl 106, 107, 215–216, *386*, **386–387**, 415
peasant multiplication. *See* Russian multiplication
pedal circle, of triangle 387
pedal triangle **387**
Pell, John 54
Pell's equation 54
pencil-turning trick 505–506, *506*
pendulums 114–115, 213
Penny Cyclopedia 122
pentagon 285
pentagram (pentacle, pentalpha, pentangle) 230, **387–388**, *388*, 424
percentage 388
percentage error **388–389**
percentile (centile, quartile) **389**
perfect number 18, 174, 338, **389**, 423–424
perigon (round angle) 14, 362, **390**
perimeter **390**, 403, 443
period doubling 151

periodic function **390**
peripatetics 27
peripheral angle 77, *77*
permillage 388
permutation 82, 127, 186, 209, **390–391**, 490
permutation group 68
perpendicular 369, **391**, 468
perpendicular bisector 46, 174–175, 338, 391, 507
perspective 149–150, **391–392**, 419
peta- (10^{15}) 465
Phaedo (Plato) 396
phyllotaxis 193
Physics (Aristotle) 532
physics, Aristotelian 27
pi (π) 55, **392–394**
  approximations of 20, 28–29, 291, 537
  irrational 283, 300, 477
  not constructible 97
  transcendental 11, 301, 313–314, 393, 477
Piazzi, Giuseppe 219
pico- (10^{-12}) 465
pie chart (graph) 479, *479*
pigeonhole principle **394–395**
Pitiscus, Bartholomaeus 511
pivoting. *See* Gaussian elimination
place-value system 6, 51, 155, 192, 251, 359, 487
planar curve 114
planar graph 233
Planck, Max 158
plane 80, 315, **395**
plane geometry 170
Plane loci (Apollonius) 189
Plato 166, 238, **395–396**, 397, 498, 532
Platonic solids 159, 232, 238, 291–292, 395, *396*, **396–398**, 404, 424
Platonicus (Eratosthenes) 166
Playfair, John 171, 227, 355, 376, 377
Playfair's axiom 46, 171, 227, 355, 376, 377
plot 398
Plouffe, Simon 183
plus 398
plus/minus symbol 398
plus sign 398, 486
Pneumatica (Heron) 248–249
Poincaré, Jules Henri 255, 314, **398–399**, 500, 530
Poincaré disk 255
Poincaré's conjecture 399
point **399**
point of contact (tangency point) **399**
point of inflection. *See* inflection point
point-slope form, of equation of line 164

Poisson, Siméon-Denis 44, 71, **399–400**, 414–415
Poisson distribution 44, 399–400
polar-coordinate graph 235
polar coordinates 22, 39, 63, *400*, **400–401**, 504
  of cardioid 62
  of complex numbers 86, *86*
  of logarithmic spiral 474
  of rose 451
polar form 176
Polignac, Alphonse de 107
Pólya, George **401–402**
polychoron 403
polygon 119, 149, **402–403**, *403*
  area of *23*, 24
  circumcircle of 78
  convexity of 88
  exterior angles of 182
  and tessellation 496–497
polygon division problem 64
polyhedral angle 469
polyhedron 88, 128–129, 176–177, 185, 236, **403–404**
polynomial **404–405**
polynomial equations 163, 215
polynomial time **405**
polyomino **405–406**, *406*
polytope 403
Poncelet, Jean Victor 419
population and sample **406–407**
population mean 483
population model 182, **407–408**
position vector 395, **408**
positive **408**
positively oriented 369, 448
positively skewed 465, 480, *480*
positive slope 467
postage-stamp problem 446
postulate (axiom) 31, 170, **409**
potential 239
power series 121, 324, 399, **409**, 495, 530
P problems vs. NP problems 357–358
Practica geometriae (Fibonacci) 511
precision 66–67, **409**, 529–530
"Preliminary and Elementary Essay on Algebra as the Science of Pure Time" (Hamilton) 245
premise **409**
primary data 116
prime **409–410**, 461
prime factorization 187

prime numbers 186, 389
  and Bertrand's conjecture 71, 167
  vs. composite numbers 87
  and factorization 186–187
  and fundamental theorem of arithmetic 83, 186–187
  infinitude of 171–172
  Mersenne's work on 338
prime-number theorem 71, 219, 410, **410–411**, 447, 535
  proofs of 167, 243, 518
principal axes **411**
Principia (Newton) 293, 351–353, *352*
Principia Mathematica (Russell and Whitehead) 453, 530
Principia philosophiae (Descartes) 125
The Principle of Relativity (Whitehead) 530
The Principles of Empirical Logic (Venn) 524
Principles of Mathematics (Russell) 453
Principles of the Art of Weighing (Stevin) 485
Prior and Posterior Analytics (Aristotle) 27
prism 403, **411–412**
prismatoid 412
prismoid 412
prisoner's dilemma 218, *412*, **412–413**
probability 122, 143, 189, 243, 262–263, **413–418**
  and harmonic functions 246
  Kolmogorov's work on 296
  Monte Carlo method 342–343
  Monty Hall problem 343
  and mutually exclusive events 344
  odds and 363
  Pascal and 382
  and permutations 391
  Poisson and 399–400
  Quételet and 432–433
  and random walks 437
  and set theory 178
  and two-card puzzle 515
probability density function 143
probability models *416*
probability tree 415–416, *416*
procedure. *See* algorithm
Proclus 171, 239, 497
product of matrix 330
product rule 133, 273, 309, **418**, 433–434
product set. *See* Cartesian product

progression. *See* harmonic sequence
projected vector 418–419
projection 337, 391–392, **418–419**
projective geometry 123–124, 270, 339, 372, 382, 409, 418, **419**
prolate spheroid **362**
proof **420**
proof by contradiction. *See* indirect proof
proofs, defective
  that 0 = 1 29
  that 1 = 2 4, 248, 337
  that 1 is the largest integer 284
  that all horses are the same color 266–267
  that the moon is made of cheese 90
proper factor 186
proper fraction 205
properly divergent 144
proper vector. *See* eigenvector
proportion 149, 230–232, 370
proportional **420–421**, 439
proposition **421**
Protagoras 374
p-series test 104
pseudoprime numbers 410
pseudosphere 300
Ptolemy 239, 253, 392, *421*, **421–422**, 510
Ptolemy's theorem 52, 422, **422**
public-key cryptography 111
pure mathematics 159, 168, **422–423**
pure surd 490
pyramid 403, **423**
pyramidal frustum 207
Pythagoras 9, 57, 172, 230, 238, 387, 396, *423*, **423–424**, 425
Pythagoras's theorem 159, 165–166, 208, 226, 238, 256, 314, *424*, **424–426**, *425*, 506
  in Babylonian mathematics 34–35
  in Chinese mathematics 73
  converse of *506*, 506–507
  and distance formula 141
  and law of cosines 303
  and polar coordinates 400
  and Pythagorean triples 426
  tessellation and *474*, *475*, 497
  and vectors 520
The Pythagorean Proposition (Loomis) 425
Pythagorean triples 99, 137, 162–163, 190, 226, 360, 422, **426–427**

Q

QED 159, **428**
QEF/QEI 428
QI. *See* Quételet index
Qin Jiushao. *See* Ch'in Chiu-shao
quadrangle. *See* quadrilateral
quadrant 363, **428**
quadratic **428–430**
quadratic curve 430
quadratic equations 163, 373, 443
quadratic form 430
quadratic formula 84, 200, 451
quadrature of the circle. *See* squaring the circle
Quadrature of the Parabola (Archimedes) 18
quadrilateral 88, 376, *430*, **430**
quadruple (4-tuple) 358
quantifier 199, **430–431**
quantum (quanta) 158
quantum mechanics 159
quartic equation 59–60, 84, 190–191, **431–432**
quartile. *See* percentile
quaternions 241, 244–245, 359, **432**
Quételet, Lambert Adolphe Jacques 415, **432–433**
Quételet index (QI) 433
quintessence 397
quintic equation 3
quipu 260
quotient 85, 146, 204, **433**
quotient rule 133, **433–434**

R

radian 14, 115, 465, 477
radian measure 473, 509
radical sign 475
radius 76
radius-chord theorem 78, *78*
radius of convergence 495
radix 36–38
Rahn, Johann Heinrich 146
Ramanujan, Srinivasa 246, 381, **435–436**, 477
randomness 41
random numbers **436–437**
random sampling 407
random walk 246, **437**
range. *See* image
rank **437–438**
rank correlation **438–439**
ranunculoid 115
ratio 146, 420, **439**
rational function (expression) **439–440**
rationalizing the denominator 85, **440**, 442, 490

rational numbers (Q) 106, 129, 358, **440–441**, 443
vs. algebraic numbers 11
definable as ratio 439
and definition of real number 441–442
denumerability of 123
discreteness of 140
ratio test 103, 182, 439
ratio theorem 439
raw data 116
ray (half-line) **441**
"Real Algebraic Manifolds" (Nash) 347
real line. *See* number line
real numbers (R) 106, 129–130, 358, **441–442**
and Dedekind cut 119–120
and Euclid 172
nondenumerability of 123
receiver 501–502
"Recherches sur diverses applications de l'analyse infinitésimale à la théorie des nombres" (Dirichlet) 140
reciprocal **442**
reciprocal equation 442
Recorde, Robert 162, **442–443**
rectangle 185, 376, **443**, 474
area of 23, *23*
vs. trapezoid 504
rectangular (Cartesian) coordinates 62–63, 105
recurrence relation (recursive relation, reduction formula) **443**, 458
recursive definition (inductive definition, recursion) **444**
reduced form (lowest terms) **444**
reductio ad absurdum. *See* indirect proof
reduction formula. *See* recurrence relation
re-entrant angle 276
reflection 198, 206, 283–284, 490
in a circle 225
and isometry 212
in a line 224–225
in a point 225
reflection property
of ellipse 365
of parabola 373, *373*
reflex angle 14, 362
"Réflexions sur la cause générale des vents" (d'Alembert) 8
reflexivity 365
Regiomontanus **444–445**, 511
regression 107, **445**
regression line 306
regular polygon 164, 166, 402

regular polyhedron. *See* Platonic solid
regular tessellation 496
"Rein Analytischer Beweis" (Bolzano) 47
Reinhardt, Karl 497
Relación de las cosas de Yucatán (Landa) 332
relation (relationship) **445**
relative complement 131
relative error **445**
relative frequency 206
relatively prime 139, 172, 174, **446**
relative maximum/minimum 330–331, *331*
remainder 433
remainder theorem 187, **446**
removable discontinuity 440
repeating decimals 440–441, 526
retrograde motion 173
Reuleaux triangle 95
reverse function. *See* inverse function
reverse J-shaped distribution 480, *480*
Rhaeticus, Georg Joachim 511
Rhind, Alexander Henry 446
Rhind papyrus 9, 155–156, 392, **446–447**, 476, 510
rhomboid 376
rhombus (rhomb) 376, 474
Richter, Charles 321
Richter scale 321
Riemann, Georg Friedrich 58, 227, 270, 272, 355, **447–448**, 473–474, 485, 535
Riemann hypothesis 243, 246, 250, 447
Riemannian geometry. *See* spherical geometry
Riemann integral 58, 447
Riemann sphere 485
Riemann's zeta function. *See* zeta function
right angle 14, 226, **448**
right circular cone 473
right cylinder 115
right derivative **307–308**
right-handed system 369, **448**
right-hand rule 110
right identity 257
right/left, limit from the 313
right pyramid 423
right quandrangular prism 411
right spherical triangle 474
right triangle 506
right-triangle principle. *See* Pythagoras's theorem
ring 5, 144, 252, 271, 354–355, **448–449**
Ringel, Gerhard 200
rise over run (slope) 467

Ritoré, Manuel 469
Rivest, Ron 111
Roberval, Gilles Personne de 57
Rolle, Michel **449**
Rolle's theorem 184, 335, **449–450**
Roman numerals 155, 192, 251, **450–451**
root (zero) **451**, 534
root test 103–104
Ros, Antonio 469
rose **451**
rotation 206, 225, 278, 283–284, 317, 490
round angle. *See* perigon
rounding **451–452**
round-off (rounding) error 452
round-robin tournament. *See* tournament
Royal Society of London 452
RSA encryption method 111
rubber-band problem 247–248
rubber-sheet geometry. *See* topology
Rudolff, Christoff 475
Ruffini, Paolo 10, 72, 431
rule, 68-95-99.7 356
Russell, Bertrand Arthur William 199, 229, *452*, **452–453**, 460–461, 530, 533
Russell's paradox (antinomy) 200, 374, **453–454**, 460–461
Russian multiplication (peasant multiplication) 43, 156, **454**, *454*

S

Saccheri, Girolamo 171, 227, **455**
saddle point 217, 492
salient 276
same side interior/exterior angle *504*
sample 143
sample mean 333
sample space 178, 413, 417, **455**
The Sand Reckoning (Archimedes) 19
sandwich result. *See* squeeze rule
SAS (side-angle-side) rule **1–2**, 463, 506
Scalar **456**
scalar multiplication 329, 520
scalar product. *See* dot product
scalar triple product 512
scale 36–38, **456**
scale factor 463
scalene 506
scatter 481

scatter diagram (scatter plot, Galton graph) 106, **456–457**
Schickard, Wilhelm 382
Schnirelmann, L. 230
schubfachprinzip. *See* pigeonhole principle
Schwarz, Hermann 469
scientific notation 125, 168, **457**
screw propeller 174
secant **457**, *493*, 510
secant theorem 1, 457, *457*
second (measure of angle) 14
secondary data 116
second-derivative test 332
segment bisector 46
Segner, J. A. 64
selection. *See* combination
self-reference 244, **457–458**
semi-magic square 327, **458**
semimajor axis 161
semiminor axis 161
semiregular polyhedron 404
semiregular tessellation 496
sequence (progression) **458–459**
sequence (totally ordered set, chain) 365–366
series **459–460**, 488
set complement 460
set direct product. *See* Cartesian product
set intersection 460
set operations *524*
set theory 144, 422–423, **460–461**
    and axiom of choice 31
    as Boolean algebra 49
    and De Morgan's laws 123
    difference in 131
    and probability 178
    and Russell's paradox 453–454
    von Neumann and 350
    Zermelo's attempt to axiomatize 534
set-theory paradoxes 530
set union 460
seven bridges of Königsberg problem 235, *235*, 500
sexagesimal numbers 33–34
shadows 149
Shamir, Adi 111
Shanks, William 392, **461**
Shannon, Claude Elwood 271, **461–462**
short radius. *See* apothem
Shushu jiuzhang (Ch'in) 72
Siamese method 327
Sicherman, Col. George 462
Sicherman dice 462, **462**
Sidereus nuncius (Galileo) 213
Sierpinski, Vaclav 203
Sierpinski's triangle 203–204, *204*
sieve of Eratosthenes 410, **462–463**

sigma notation 459, 488
sigma squared 481
signed number. *See* integer
significant figures 168, 409
similar figures 456, **463**
similar triangles 1–2
simple fraction 205
simple interest 275
simple root 451
simplification, by cancellation 56
Simpson, Thomas **463–464**
Simpson's rule 361, 464
simultaneous linear equations 219–220, 234, 282, 315, **464**, 486, 491
    and determinant of matrix 126–128
    solved by Cramer's rule 108–109, 128
sine function 291, *509*, 510–511, *511*
    history of 28–29, 51, 444–445
sine rule. *See* law of sines
single cusp 492, *493*
singularity 464
singular point 464
sink 501–502
Sirotta, Milton 232
SI units **464–465**
68-95-99.7 rule 356
Siyuan Yujian (Chu Shih-Chieh) 383
skew curve 114
skew lines 465
skewness 465
Skolem, Thoralf 534
slide 15 puzzle (Boss puzzle, fifteen puzzle) 391, **465–466**, *466*
slide rule 369–370, **466**
slope (grade, gradient) 132, **466–468**, *467*
slope-intercept form 164
smooth vs. non-smooth surface 357
Snell, Willebrord van Roijen 468–469
Snell's law of refraction 365, *468*, **468–469**
soap bubbles **469**, 472
Socrates 374
solar system, mathematical stability of 296
solid angle 15, **469**
solid geometry 170
solid of revolution 57, 372, 382, **469–470**, 500
solution by radicals 3, 215, **470–471**
Soma cube 218
"Some Properties of Bernoulli's Numbers" (Ramanujan) 435
source 501–502
space-filling curve 385

Spearman, Charles 438
Spearman's coefficient 438–439
Spearman's method 438
special theory of relativity 158
specific gravity 18
speed vs. velocity 519
Sphaerica (Menelaus) 336
sphere 76, 136, 166, **471–472**
    connectedness of 94
    diameter of 130
    four-color map on 200–201
    great circle as "straight" 314
    volume of, vs. cylinder 20, 91, 115
sphere packing 292, 472
spherical coordinates **472–473**, 504
spherical excess 474
spherical geometry 225–226, 227, 270, 323, 355–356, 472, **473–474**
spherical helix 248
spherical segment 472
spherical triangle 472, **474**
spheroid 161
spiral of Archimedes 372, **474**, 476
spirals in nature 193
splitting game 279, *279*
square 430, **474–475**
square brackets 50–51
square matrix 329
square model *416*, 416–417
square numbers *194*, 194–195
square pyramid 423
square root 157, 249, 451, **475–476**, 486
squaring the circle (circle squaring, quadrature of the circle) 97, 238–239, 251–252, 313–314, 393, **476–477**
squeeze rule (sandwich result) 16, 321, *477*, **477–478**, 526
SSS (side-side-side) rule **1–2**, 463
Stäckel, Paul 515
stadium paradox 533
stair climbing problem 65
standard deviation 44, 386, 481–482
    vs. mean and z-score 537
standard form. *See* scientific notation
Statistical Methods for Research Workers (Fisher) 197, 415
statistics 414, 437, **478**
    descriptive 116, **478–482**
    inferential 143, 268, 354, 478, **482–484**
        Bayes and 38–39
        and central-limit theorem 69–70

Steiner, Jakob 284, **484**
Steiner point **484–485**
Steiner surface 484
stellated polyhedron 404
stem-and-leaf plot 479, *479*
steradian 15, 465, 469
stereographic projection **485**
Stevin, Simon **485–486**
Stifel, Michael 180, 348–349
Stirling's formula 122, **486**
straight angle 14, 362
straightedge and compass 95
stratified sampling 407
strictly increasing/decreasing 261–262
Student's t-distribution 484
Student's t-test 415, 484
subfactorial **486**
substitution **486–487**, 491
substitution rule for integration. *See* integration: by substitution
subtraction 146, 205, **487**
subtrahend 487
successive doubling 156
Sulbasutram 425
summation (Σ) 173, **487–488**
summation, index of 148
summation problem 65
sums of powers 51, *488*, **488–490**
supplementary angles 14
surd **490**
surjective function 208
"Sur la décomposition des ensembles de points en partiens respectivement congruent" (Banach, Tarski) 35
Sur l'homme et le developpement de ses facultés, essai d'une physique sociale (Quételet) 432–433
"Sur l'intégration des functions discontinues" (Lebesgue) 307
Su-yuan yu-chien (Chu Shih-Chieh) 73–74, 75
syllogism 27
symbolic logic. *See* formal logic
Symbolic Logic (Venn) 524
symmetric difference 131
symmetry 206–207, **490**
Synagoge (Pappus) 239, 371–372
Synopsis of Elementary Results in Pure Mathematics (Carr) 435
Synopsis palmariorum matheseos (Jones) 393
Syntaxis mathematic (Ptolemy) 421–422
systematic sampling 407
system of equations 121, **491**
"Systems of Right Lines in a Plane" (Hamilton) 245

T

Tabulae (Regiomontanus) 444
tacnode (osculation) 148
tacpoint *493*
Tait, Peter 295
tangency point. *See* point of
contact
tangent (geometric) **492**, *493*
tangent (trigonometric) 132,
492, 510
tangent plane 492
tangent theorems 76–77, *77*
Tarski, Alfred 24, 35
Tartaglia, Niccolò 10, 60,
112, 190–191, *493*, **493–494**
task diagram *109*
tautochrone property of
pendulums 114–115, 299
tautology **494**, 515
Taylor, Brook 240, 274, 324,
325, **494**, 495
Taylor, Richard 190, 531
Taylor series 152, 182,
240–241, 268, 271, 315,
324–325, 337, 409, **494–496**
and approximation 16
convergence of 45
Chebyshev's work on
71
and Euler's formula 175
and general binomial
theorem 45
and history of calculus 57
and permutations 391
of sine function 536
Tchebyshev. *See* Chebyshev
technique. *See* algorithm
tensor 447, **496**
The Tenth (Stevin) 485
tera- (10^{12}) 465
tessellation (covering, tiling)
403, 430, 475, **496–497**,
497, 507
tesseract 256
"The Testing of Statistical
Hypotheses in Relation to
Probabilities A Priori"
(Neyman) 354
tetragon. *See* quadrilateral
tetrahedral dice 462
tetrahedral numbers 195, 497
tetrahedron 165, 396, 397,
403, 423, 497
tetromino 406
Teutsche algebra (Rahn) 146
Thabit ibn Qurra 18
Thales 238, 423, **497–498**,
510
Theaetetus (Plato) 498
The Grammar of Science
(Pearson) 387
Theodorus 82, 283, 475–476,
498
Theon of Alexandria 252–253
Theon of Smyrna 99
theorem (proposition)
498–499

theorem of Thales 77
*Theorems Stated by
Ramanujan* (Watson) 436
*Theoria motus corporum
coelestium* (Gauss) 219
*Théorie analytique de la
chaleur* (Fourier) 201
*Théorie analytique des
probabilités* (Laplace) 301,
415
Theorie der Parallellinien
(Lambert) 300
"Théorie des opérations
linéaires" (Banach) 35
*The Theory of Games and
Economic Behavior* (von
Neumann and Morgenstern)
216, 350–351
The Theory of Numbers
(Hardy and Wright) 246
"Theory of Systems of Rays"
(Hamilton) 245
three-body problem 299
three-dimensional coordinates
63, 363
three-dimensional vectors 520
three-fold rotational symmetry
490
three-utilities problem 233,
499, **499**
tiling. *See* tessellation
Timaeus (Plato) 396, 397
time, as fourth dimension
136, 256
time-series graph 479, *479*
topological space 500
topology 121, 307, 339,
499–500
Torricelli, Evangelista 114
torus 94, 399, 470, **500–501**,
501
Euler's formula for 236
Euler theorem and 177
and Jordan curve theorem
76, 288
and three-utilities problem
499
totally ordered set 365–366
tournament (round-robin
tournament, complete
digraph) *501*, **501–502**
towers of Hanoi (Brahma)
443, **502–503**
tractrix (equitangential curve,
tractory) **503**
*Traité analytique des sections
coniques* (L'Hôpital) 310
Traité d'algèbre (Rolle) 449
Traité de dynamique
(d'Alembert) 8
*Traité de la résolution des
équations numériques de
tous les degrés* (Lagrange)
299
*Traité de l'équilibre et du
mouvement des fluides*
(d'Alembert) 8

Traité de mécanique céleste
(Laplace) 301
*Traité des proprietés des
figures* (Poncelet) 419
*Traité des substitutions et des
equations algebraique*
(Jordan) 288
trajectory **503**
transcendental curve 114
transcendental numbers 11,
59, 297, 301, 317–318, 477
transfinite numbers 368
transformation **503–504**
transitive law 367
transitive reasoning 26
translation 206, 225,
283–284
transmitter 501–502
transpose of matrix 330
transposition 390
transversal 336, 375, 377,
504, **504**
trapezoid (trapezium) *23*, 24,
430, **504**
trapezoidal rule for integration
360–361
traveling-salesman problem
55, **505**
traverse **504**, *504*
Treatise of Fluxions
(Maclaurin) 324
A Treatise on Algebra
(Maclaurin) 324
Treatise on Algebra (Wallis)
529
*Treatise on Demonstration of
Problems of Algebra* (Omar
Khayyám) 363
*Treatise on Human
Proportions* (Dürer) 231
*The Treatise on the Chord and
Sine* (al-Kashi) 291
*Treatise on the Equilibrium of
Liquids* (Pascal) 382
Treatise on Universal Algebra
(Whitehead) 530
tree (graph theory) 233
triangle (trigon) 114, **505–508**
area of *23*, *23*, 249
diameter of 130
Euler line in 174–175
exterior angles of 182–183
triangle inequality 267, 425,
506, **508**
triangular numbers *194*,
194–195, 349, 405, 497
trichotomy law 31, 365, 367
trigon. *See* triangle
Trigonometria (Pitiscus) 511
trigonometric functions *254*,
369
trigonometry 16, 28, 239,
444–445, **508–512**, 516–517
Trigonometry (Simpson) 463
*Trigonometry and Double
Algebra* (De Morgan) 122
trinomial **512**

*Le triparty en la science des
nombres* (Chuquet) 526
triple (3-tuple) 358
triple integral 148
triple torus 501
triple vector product 376,
448, **512**
trirectangular triangle 474
trisecting an angle 97, 239,
291, 372, *512*, **512–513**, 525
Tristram Shandy paradox
269, **513**
trivial 183, 514
trivial ring 449
trivial solution **513–514**
tromino 406
truncation 452
truth table 25, *514*, **514–515**,
515
for biconditionals *42*
for conjunction *94*
for contrapositives
101–102
for disjunction *141*
for hypothetical *90*, *90*
for negation *348*
Tsu Chung Chi. *See* Zu
Chongzhi
Turing, Alan 244
turning point **515**
twin primes 410, **515**
twisted curve 114
two-card puzzle 91, **515**
two-chord theorem 78, *78*
two-dimensional vectors 520
two-pancake theorem 245,
277
two-point form 164
two-secant theorem 457, *457*
type, in set theory 454

U

"Über die Addition transfiniter
Cardinalzahlen" (Zermelo)
533–534
"Über die Hypothesen welche
der Geometrie zu Grunde
liegen" (Riemann) 227
"Über die Stabilität des
Gleichgewichts" (Dirichlet)
140
Über die Zahl (Lindemann)
314
"Über einen die Erzeugung
und Verwandlung des Lichtes
betreffenden heuristischen
Gesichtspunkt" (Einstein)
158
Ulam, Stanislaw 15
unary operation 364, **516**
unbounded interval 278
uncertainty, and error
167–168
undefined fraction 205–206
unexpected quiz (hanging)
paradox 374–375, *375*

uniform distribution 480, *480*
uniform motion 522
union (set operation) 144,
    460
unique factorization theorem.
    See fundamental theorem of
    arithmetic
unique solution **516**
unitary ratio 439
unit circle 427, **516–517**
unit denominator rule 204
unit fraction. *See* Egyptian
    fractions
unit matrix. *See* identity
    matrix
unit sphere 469
universal calculus 199
universal language 199
universal mathematics 199
universal quantifier ("for all")
    430–431
unknown (indeterminate,
    variable) **517**
unordered arrangement. *See*
    combination
unordered partition 381
unrestricted integral. *See*
    improper integral
"Untersuchungen über ein
    Problem der Hydrodynamik"
    (Dirichlet) 140

V

valence. *See* degree of a vertex
Vallée-Poussin, Charles-Jean
    de la 411, **518**
Varahamihira 264
variable 517, **518–519**
variance 106, 306, 481, 484
*Variorum de rebus
    mathematicis responsorum*
    (Viète) 525
Veblen, Oswald 288
vector 14, 142, 147, 164,
    310, **519–520**
    and collinearity 80
    decomposition of 119
vector addition 376, 520
vector equation
    of line **520–521**
    of plane **521**

vector field 101, **521**
vector multiplication 329, 520
vector operations *520*
vector product. *See* cross
    product
vector space 5, 71, 201, 243,
    315, 520, **521–522**
vector subtraction 520
vector triple product of three
    vectors 512
velocity 132, 408, 519,
    522–523
Venn, John **523–524**
Venn diagram 123, *524,
    524–525*
Verhulst, Pierre-François 408
vertex 91, 233, 403, 505
vertical angles 364
vibrating strings 8, 299
viergruppe. *See* Klein-four
    group
Viète, François (Franciscus
    Vieta) 10, 113–114,
    348–349, 511, **525**
Viète's formula (Vieta's
    formula) 231, 393, 511,
    525–526
vinculum 81, **526**
Vinogradoff, Ivan 230
Vlacq, Adriaan 54, 320
Voevodsky, Vladimir 194
volume **526–527**, *527*
    of cone 91, 173, 293, 527
    of cube 111
    of cylinder 115, 156
    of frustum 207, 527
    as ill-defined concept 526
    of parallelepiped 376
    of Platonic solids 397
    of prism 412
    of pyramid 173, 293, 423
    of solids of equal height
        67–68
    of sphere 293
    of sphere vs. cylinder 20,
        91, 115
    of tetrahedron 497
von Neumann, John 216,
    218, 415, 436
*Vorlesungen über
    Zahlentheorie* (Dirichlet)
    140

vortex theory, Descartes's 125
vulgar fraction 205

W

Waerden, Bartel Leendert van
    der 355
al-Wafa, Abu 510
Wallace, William 165
wallet paradox 375
Wallis, John 63, 180, 269,
    370, *528*, **528–529**
    and complex numbers 25
Wallis's product 269, 393,
    486, 528, **529**, 536
wallpaper pattern 207
Wantzel, Pierre Laurent 149,
    513
Waring, Edward 229
warping of space 447
Watson, George Neville 436
weak transitivity 365
Weaver, Warren 461
Weierstrass, Karl Theodor
    Wilhelm 5, 58, 284, 297,
    312, **529–530**
Weierstrass's product
    inequality 268
well-ordered set **530**, 533
Wessel, Casper 25, 86
The Whetstone of Witte
    (Recorde) 443
Whitehead, Alfred North 199,
    229, 452–454, **530**
Whitworth, W. Allen 486
whole number 443, **530–531**.
    See also natural number
Widman, Johannes 6, 131,
    398, 487
Wiles, Andrew John 190,
    531
Wilson's theorem 299
word (braids) 52
Wright, E. M. 246

Y

Yang Hui 325
year, length of 333, 363
Yi, Tien-Lien 71
yocto- (10^{-24}) 465
Yorke, James 71

yotta- (10^{24}) 465
Youngs, J. W. T. 200

Z

Zadeh, Lofti 212
Zahlbericht (Hilbert) 250
Zeno 57, 238, 270, **532**
Zeno's paradoxes 269,
    532–533
zepto- (10^{-21}) 465
Zermelo, Ernst Friedrich
    Ferdinand 31, 350, 461,
    530, **533–534**
zero (naught) 5, 6, 41, 51,
    332, **534–535**
    in Boolean algebra 49
    as exponent 534
    factorial of 534
    on number line 359
    as number vs. placeholder
        51, 293
    and rings 448
zero fraction 205
zero matrix 147, 534
zero of function. *See* root
zero-sum game 217, 534
zero-th term, of series 223
zeta function 105, 174, 243,
    246, 250, 393, 447–448,
    489, **535–536**
    and proofs of prime-
        number theorem 518
Zeteticorum libri quinque
    (Viète) 525
zetta- (10^{21}) 465
Zhu Shijie. *See* Chu Shih-
    Chieh
zone 472
z-score (z-value) 357, 483,
    537
Zu Chongzhi (Tsu Chung Chi)
    74, 392, **537**
"Zur Elektrodynamik
    bewegter Körper" (Einstein)
    158
*Zur Theorie der Abelschen
    Functionen* (Weierstrass)
    529